ENCYCLOPEDIA OF FOOD SCIENCE AND TECHNOLOGY

ENCYCLOPEDIA OF FOOD SCIENCE AND TECHNOLOGY

VOLUME 1

Y. H. Hui

Editor-in-Chief

A Wiley-Interscience Publication

John Wiley & Sons, Inc.

New York / Chichester / Brisbane / Toronto / Singapore

Copyright © 1992 by John Wiley & Sons, Inc.

Library of Congress Cataloging in Publication Data:
Encyclopedia of food science and technology / [edited by] Y. H. Hui.
    p.   cm.
  "A Wiley-Interscience publication."
  Includes bibliographical references.

  1. Food industry and trad—Encyclopedias. I. Hui, Y. H. (Yiu H.)
TP368.2.E62 1991
664′.003—dc20   91-22434
                                    CIP

ISBN 0-471-50541-2 (set)

Printed in the United States of America

10 9 8 7 6 5 4 3 2 1

EDITORIAL ADVISORY BOARD

EDITORIAL STAFF

Editor-in-Chief: Y. H. Hui
Editorial Manager: Michalina Bickford
Production Manager: Jenet McIver
Associate Managing Editor: Rose Ann Campise
Designer: Geraldine Spellissy

Assistant Editorial Manager: Lindy J. Humphreys
Assistant Editorial Manager: Cathleen A. Treacy
Production Assistant: John Thomas Steen
Index Editor: Margaret Kennedy
Indexer: Charles Carmony

CONTRIBUTORS

Khalid Abdelrahim, *McGill University, Quebec, Canada,* Thermal processing and computer modeling

S. Abe, *Ajinomoto Company, Kawasaki-Shi, Japan,* Nucleotides

G. Abraham, *Southern Regional Research Center, New Orleans, Louisiana,* Interesterification; Oilseeds and their oils

Nancy J. Alexander, *NCAUR/USDA, Peoria, Illinois,* Genetic engineering, part V: yeasts and ethanol production

George H. Allen, *Humboldt State University, Arcata, California,* Aquaculture

Robert E. Altomare, *General Foods USA, Tarrytown, New York,* Snack food technology

M. A. Amerine, *Consultant, St. Helena, California,* Wine

Burton A. Amernick, *Pollack, Vande Sande and Priddy, Washington, D.C.,* Patents

Cathy Y. W. Ang, *USDA/ARS, Athens, Georgia,* Poultry flavors

APV Crepaco Corporation, *Lake Mills, Wisconsin,* Aseptic Processing: Ohmic Heating; Corrosion and Food Processing; Distillation, Technology and Engineering; Dryers: Technology and Engineering; Evaporators: Technology and Engineering; Food Plant Design and Construction; Heat Exchangers

B. H. Ashby, *U.S. Department of Agriculture, Washington, D.C.,* Refrigerated foods: transportation

Association of Dressing and Sauces, *Atlanta, Georgia,* Salad Dressing

Bonnie Atwood, *U.S. Army Natick, Natick, Massachusetts,* Military food

Jorg Augustin, *University of Idaho, Moscow, Idaho,* Vitamins, Part III: Assay, Procedures

Jerry K. Babbitt, *USDC/NMFS, Kodiak, Alaska,* Crabs and crab processing

Robert J. Baer, *South Dakota State University, Brookings, South Dakota,* Milk and milk products

M. O. Balaban, *University of Florida, Gainesville, Florida,* Supercritical fluid extraction: applications for the food industry

S. M. Balaban, *Monsanto Chemical Company, St. Louis, Missouri,* Sorbic acid

James E. Balthorp, *National Marine Fisheries Service, Charleston, South Carolina,* Marine toxins

Pamela Bankes, *Campden Food and Drink Research Association, Gloucestershire United Kingdom,* Food microbiology and rapid methods

Shai Barbut, *University of Guelph, Guelph, Ontario, Canada,* Poultry processing and product technology

Steven A. Barker, *Louisiana State University, Baton Rouge, Louisiana,* Antibiotics in food of animal origin

Harold J. Barnett, *U.S. Dept. of Agriculture, Seattle, Washington,* Seafoods, flavors and quality

Roger G. Bates, *University of Florida, Gainesville, Florida,* Hydrogen ion activity (pH)

Stephen R. Behr, *Ross Laboratories, Columbus, Ohio,* Enteral feeding systems

James N. BeMiller, *Purdue University, West Lafayette, Indiana,* Carbohydrates; Fiber, diet and food processing; Gums; Lactose; Starches; Sucrose, Sweetners, nutritive

G. R. Bentley, *Meat Industry Research Institute of New Zealand, Hamilton, New Zealand,* Meat science; Meat slaughtering and processing equipment

R. C. Berberet, *Oklahoma State University, Stillwater, Oklahoma,* Integrated pest management

R. G. Berger, *Lebensmittel Chemie der Universität Hannover, Hannover, Germany,* Genetic engineering, part II: food flavors

Daniel Berkowitz, *U.S. Army Natick, Natick, Massachusetts,* Military food

Elizabeth M. Besozzi, *Ross Laboratories, Columbus, Ohio,* Enteral feeding systems

Dilip Bhatia, *Hoffmann-La Roche Corporation, Nutley, New Jersey,* Vitamins

D. Bhatnagar, *SRRC/ARS/USDA, New Orleans, Louisiana,* Aflatoxin: eliminations through biotechnology

Mrinal Bhattacharya, *University of Minnesota, St. Paul, Minnesota,* Extrusion processing: texture and rheology

C. G. Biliaderis, *University of Manitoba, Winnipeg, Manitoba, Canada,* Canola

Darlena K. Blucher, *Humboldt State University, Arcata, California,* History of foods; Kosher foods and food processing

Michael M. Blumenthal, *Libra Laboratories, Piscataway, New Jersey,* Frying technology

B. Borenstein, *Hoffmann-La Roche, Nutley, New Jersey,* Vitamins

J. R. Botta, *Canada Department of Fisheries and Oceans, Newfoundland, Canada,* Lobster: biology and technology

Malcolm C. Bourne, *Cornell University, Geneva, New York,* Water activity: food texture

Robert F. Boutin, *Knechtel Laboratories, Skokie, Illinois,* Confections

Robert E. Brackett, *University of Georgia, Griffin, Georgia,* Food spoilage

J. W. Brady, *Cornell University, Ithaca, New York,* Molecular mechanics: computer modeling of functional properties

John E. Brekke, *Consultant, Escondido, California,* Tropical fruits

James R. Brooker, *USDC/NOAA/NMFS, Silver Spring, Maryland,* United States Department of Commerce

Herbert Brookes, *International Food Information Service, Reading, United Kingdom,* International Development: United Kingdom

John Bumbalough, *Land O'Lakes Corporation, Minneapolis, Minnesota,* Margarine

Sheila M. Campbell, *Ross Laboratories, Columbus, Ohio,* Enteral feeding systems

Charles J. Cante, *General Foods USA, Tarrytown, New York,* Breakfast cereals; Edible films and coatings; Fruit preserves and jellies; Snack food technology

Armand Cardello, *U.S. Army Natick, Natick, Massachusetts,* Military food

M. Caric, *University of Novi Sad, Novi Sad, Yugoslavia,* Processed cheese

Frank G. Carpenter, *Consultant, New Orleans, Louisiana,* Cane sugar

S. Cenkowski, *University of Manitoba, Winnipeg, Manitoba, Canada,* Cleaning-in-place (CIP); Energy usage in food processing plants

Tuu-Jyi Chai, *University of Maryland, Cambridge, Maryland,* Fish and shellfish microbiology

Ping-Yang Chang, *Food Industry Research and Development Institute, Hsinchu, Taiwan,* Sulfites and food

Young-Meng Chang, *National Taiwan University, Taipei, Taiwan,* Seaweed aquaculture: Taiwan

Y. S. Chang, *Council of Agriculture, Taipei, Taiwan,* International Development: Taiwan

Chee-Jen Chen, *Food Industry Research and Development Institute, Hsinchu, Taiwan,* Mushrooms: cultivation

C. S. Chen, *University of Florida, Gainesville, Florida,* Supercritical fluid extractions: applications for the food industry

T. C. Chen, *Mississippi State University, Mississippi State, Mississippi,* Poultry meat microbiology

Wen-Lian Chen, *Food Industry Research and Development Institute, Hsinchu, Taiwan,* Soyfoods, fermented

Kenneth K. Chen, *University of Washington, Seattle, Washington,* Molluscan aquaculture

Cherry Marketing Institute, *Okemos, Michigan,* Cherries and cherry processing

H. Chiba, *Kobe Woman's University, Kobe, Japan,* Peptides

Yoon Hin Chong, *Palm Oil Research, Institute of Malaysia, Selangor, Malaysia,* Palm oil

B. B. Chrystal, *Meat Industry Research Institute of New Zealand, Hamilton, New Zealand,* Meat and electrical stimulation

Fun S. Chu, *University of Wisconsin, Madison, Wisconsin,* Mycotoxin analysis

Victor Chu, *Bio-Rad Laboratories, Richmond, California,* Chromatography, high performance liquid

Warren S. Clark, Jr., *American Dairy Products Institute, Chicago, Illinois,* Evaporated milk; Dry milk; Whey processing: history and development

Thomas E. Cleveland, *SRRC/ARS/USDA, New Orleans, Louisiana,* Aflatoxin: Elimination through biotechnology

William P. Clinton, *Consultant, Monsey, New York,* Coffee

Gus Coccodrilli, *General Foods USA, White Plains, New York,* Breakfast cereals

David L. Collins-Thompson, *University of Guelph, Guelph, Ontario, Canada,* Genetic engineering, part IV: food microbiology

Chris Combs, *Dow Corning/Dow Chemical Corporation, Midland, Michigan,* Foams and silicone in food processing

R. N. Cooper, *Meat Industry Research, Institute of New Zealand, Hamilton, New Zealand,* Meat industry processing wastes: characteristics and treatment

Laurie Creamer, *New Zealand Dairy Research Institute, Palmerston North, New Zealand,* Casein and caseinates

Thomas G. Crosby, *Bunge Edible Oil Corporation, Kankatee, Illinois,* Fats and oils: properties, processing technology, and commercial shortenings

John J. Cunningham, *University of Massachusetts, Amherst, Massachusetts,* Energy

G. W. Cuperus, *Oklahoma State University, Stillwater, Oklahoma,* Integrated Pest Management

Ashim K. Datta, *Cornell University, Ithaca, New York,* Sensors and food processing operations; Thermal sterilization of liquid food

Robert V. Decareau, *Microwave Consulting Services, Amherst, New Hampshire,* Microwave science and technology

Marnie L. De Gregorio, *General Foods USA, White Plains, New York,* Fruit preserves and jellies

J. M. De Man, *University of Guelph, Guelph, Ontario, Canada,* Fats and oils: chemistry, physics, and applications

C. E. Devine, *Meat Industry Research Institute of New Zealand, Hamilton, New Zealand,* Meat electrical stimulation

Frank Di Cosmo, *University of Toronto, Toronto, Ontario, Canada,* Plant products in food, natural

Sila Dikiu-Cansever, *Häagen-Dazs Corporation, Fairfield, New Jersey,* Ice cream and frozen dessert

D. E. Dixon-Holland, *Neogen Corporation, Lansing, Michigan,* Immunological methodology

Richard J. Dole, *USDA/ARS/National Peanut Research Laboratory, Dawson, Georgia,* Peanuts

Joe W. Dorner, *USDA/ARS/National Peanut Research Laboratory, Dawson, Georgia,* Peanuts

Stephen R. Drake, *USDA/ARS, Wenatchee, Washington,* Apples and apple products

William N. Drohan, *American Red Cross, Rockville, Maryland,* Transgenic animals

Grand E. Du Bois, *The NutraSweet Company, Mount Prospect, Illinois,* Sweeteners, Nonnutritive

C. Patrick Dunne, *U.S. Army Natick, Natick, Massachusetts,* Military food

Joe Dunne, *Quest (Biocon), Sarasota, Florida,* Meat tenderizers

Richard A. Durst, *National Institute of Standards and Technology, Gaithensburg, Maryland,* Hydrogen ion activity (pH)

D. J. Ecobichon, *McGill University, Quebec, Canada,* Toxicology and risk assessment

A. F. Egan, *CSIRO, Brisbane, Australia,* Meat starter cultures and meat product manufacturing

EG & G ORTEC, *Oak Ridge, Tennessee,* Radioactivity: measurement

Larry Eils, *National automatic merchandising association, Chicago, Illinois,* Vending

P. Eng, *University of Alberta, Edmonton, Alberta, Canada,* Whey: Processing and manufacture

Phyllis Entis, *QA Laboratories, Toronto, Ontario, Canada,* Membrane filtration systems

Felix E. Escher, *Swiss Federal Institute of Technology, Zurich, Switzerland,* International Development: Switzerland

N. A. M. Eskin, *University of Manitoba, Winnipeg, Manitoba, Canada,*

Peter J. Facchini, *University of Toronto, Toronto, Ontario, Canada,* Plant products in food, natural

Mary E. Feeley, *Häagen-Dazs Corporation, Fairfield, New Jersey,* Ice cream and frozen dessert

P. J. Fellers, *State of Florida Department of Citrus, Lake Alfred, Florida,* Citrus industry; Limonin: grapefruit juice and flavor

P. Fellows, *Intermediate Technology Development Group, Rugby, United Kingdom,* Dehydration; Freeze drying and freeze concentration

Robin M. Fenwick, *New Zealand Dairy Board, Wellington, New Zealand,* Dairy ingredients for food

Matias J. Fernandez-Diez, *Instituto de la Grasa y sus Derivados, Sevilla, Spain,* Olives

Marshall L. Fishman, *USDA/ARS/ERRC, Philadelphia, Pennsylvania,* Pectic substances

Dick Fizzell, *General Foods USA, White Plains, New York,* Breakfast cereals

J. C. Flake, *American Dairy Products Institute, Chicago, Illinois,* Evaporated milk

John D. Floros, *Purdue University, West Lafayette, Indiana,* Mass transfer and diffusion in foods; Optimization methods in food processing and engineering

R. T. Fraley, *Monsanto Chemical Company, St. Louis, Missouri,* Plant biotechnology: crop improvement

Daniel Y. C. Fung, *Kansas State University, Manhattan, Kansas,* Foodborne diseases; Food fermentation; Food microbiology; Microbiology

S. Gaby, *Hoffmann-La Roche, Nutley, New Jersey,* Vitamins

Carol A. Gallagher, *Saint Joseph's University, Philadelphia, Pennsylvania,* Food marketing

Robin Ganse, *The American Lamb Council, Englewood, Colorado,* Lamb

David R. Gard, *Monsanto Chemical Company, St. Louis, Missouri,* Phosphates and food processing

C. S. Gasser, *University of California, Davis, California,* Plant biotechnology: crop improvement

Richard J. George, *Saint Joseph's University, Philadelphia, Pennsylvania,* Food marketing

Donna M. Gibson, *USDA/ARS/SRRC, New Orleans, Louisiana,* Phytases

Marianne H. Gilette, *McCormick and Company, Hunt Valley, Maryland,* Vanilla extract

C. O. Gill, *Agriculture Canada Research Station, Lacoybe, Alberta, Canada,* Meat and modified atmosphere packaging

H. Douglas Goff, *University of Guelph, Guelph, Ontario, Canada,* Food preservation

Juan Gomez-Basuri, *Cornell University, Ithaca, New York,* Fish, minced

H. Gordon, *Hoffmann-La Roche Corporation, Nutley, New Jersey,* Vitamins

J. Richard Gorham, *Food and Drug Administration, Washington, D.C.,* Filth and extraneous matter in food

Harold N. Graham, *Consultant, Englewood, New Jersey,* Caffeine; Tea; Tannins

Bob Graves, *Balchem Corporation, Slate Hill, New York,* Encapsulation techniques

Bruce M. Greenberg, *University of Waterloo, Waterloo, Ontario, Canada,* Photosynthesis

Elizabeth A. Gullett, *University of Guelph, Guelph, Ontario, Canada,* Color and food

James A. Guzinski, *Kalsec, Kalamazoo, Michigan,* Spices and herbs: natural extraction

Norman F. Haard, *University of California, Davis, California,* Marine enzymes

William C. Haines, *Michigan State University, East Lansing, Michigan,* Food industry economic development

Yong D. Hang, *Cornell University, Geneva, New York,* Waste management and food processing

J. J. Harris, *The American Lamb Council, Englewood, Colorado,* Lamb

P. V. Harris, *CSIRO, Cannon Hill, Queensland, Australia,* Meat: texture and rheology

Ronald D. Harris, *Kraft Corporation, Glenview, Illinois,* Cheese

Jim Heard, *Bio-Rad Laboratories, Richmand, California,* Chromatography, high performance liquid

Ronald E. Hebeda, *Enzyme Bio-Systems, Arlington Heights, Illinois,* Syrups

Jim Heber, *Helsingborg, Sweden,* Freezing systems for the food industry

Barbara B. Heidolph, *Monsanto Chemical Corporation, St. Louis, Missouri,* Phosphates and food processing

C. P. Herman, *University of Toronto, Toronto, Ontario, Canada,* Appetite

David H. Hettinga, *Land O'Lakes Corporation, Minneapolis, Minnesota,* Butter and butter products; Margarine

S. Hill, *International Food Information Service (IFIS Publishing) Shinfield, United Kingdom,* International food information service

Edward Hirsch, *U.S. Army Natick, Natick, Massachusetts,* Military food

Allan S. Hodgson, *Bunge Edible Oils Corporation, Kankakee, Illinois,* Hydrogenation

P. G. Hoffman, *McCormick Corporation, Hunt Valley, Maryland,* Vanilla extract

David N. Holcomb, *Kraft Corporation, Glenview, Illinois,* Entropy; Food Microstructure; Rheology

S. Donald Holdsworth, *Campden Food and Drink Research Association, Chipping, Campden, Gloucestershire, United Kingdom,* Energy use in the canning industry

Stanley Holgate, *U.S. Army Natick, Research Development and Engineering Center; Natick, Massachusetts,* U.S. Armed Forces food research and development

Krysten L. Holmes, *Alaska Fisheries Development Foundation Corporation, Anchorage, Alaska,* Surimi: history, economics, and product development

V. H. Holsinger, *USDA, Wyndmoor, Pennsylvania,* Malted milk; Milk, imitation

D. E. Hood, *IUF.ST, Dublin, Ireland,* International Development: Ireland; International Union of Food Science and Technology

Liam Horgan, *Quest (Biocon), Cork, Ireland,* Meat tenderizers

R. J. Hron, *USDA, New Orleans,* Oilseeds and their oils

Fu-Hung, *University of Missouri, Columbia, Missouri,* Extrusion and extrusion and extrusion cooking; Ultrafiltration

Yin Liang Hsieh, *Cornell University, Ithaca, New York,* Fish, minced

Yu-Tsye Li Hsieh, *Cornell University, Ithaca, New York,* Fish, minced

Jack J. Hua, *Food Industry Research and Development Institute, Hsinchu, Taiwan,* Vegetables, pickling and fermenting

Chin-Cheng Huang, *Food Industry Research and Development Institute, Hsinchu, Taiwan*

Y. H. Hui, *Editor-in-Chief, Cutten, California,* Alcohol, polyhydric or sugar; Alkaloids; Canning: regulatory and safety considerations; Cholesterol and heart disease; Cultural nutrition; Dietetics; Enzymes as Food additives; Food utilization; Food processing: Standard industrial classification; Food processing law and regulations, part I: agricultural products; Food processing law and regulations, part II: food and beverages; Faddism; Food additives; Food allergy; HACCP principles for food production; Meat products; Minerals; Nutrition; Packaging, part II: Labeling; Pasta science and technology; Soybeans and soybean processing; Shellfish; Toxicants, natural; U.S. Department of Agriculture; Sanitation and safety

Yen-Con Hung, *University of Georgia, Griffin, Georgia,* Food freezing

W. Jeffrey Hurst, *Hershey Foods Technical Center, Hershey, Pennsylvania,* Laboratory robotics and automation

J. W. Hutchinson, *Proctor and Schwartz Corporation, Horsham, Pennsylvania,* Preproduction process: computer modeling

Keum Taek Hwang, *Cornell University, Ithaca, New York,* Fish, minced

John T. Hynes, *Kraft Corporation, Glenview, Illinois,* Cheese

A. Iannarone, *Hoffmann-La Roche Corporation, Nutley, New Jersey,* Vitamins

Institute of Food Technologists, *Chicago, Illinois,* IFT Awards; The Institute of Food Technologies

Georg Katsushi Iwama, *University of British Columbia, Vancouver, British Columbia, Canada,* Fishes: anatomy and physiology; Fishes: species of economic importance

Robert L. Jackman, *University of Guelph, Guelph, Ontario, Canada,* Proteins: denaturation and food processing

James M. Jay, *Wayne State University, Detroit, Michigan,* Food Borne Microorganisms: detection and identification in foods; Indicator organisms: detection and enumeration in foods

D. S. Jayas, *University of Manitoba, Winnipeg, Manitoba, Canada,* Cleaning-in-place; Energy usage in food processing plants

P. Jelen, *University of Alberta, Edmonton, Alberta, Canada,* Quarg; Whey: processing and manufacture

Peter B. Johnsen, *USDA/ARS/SRRC, New Orleans, Louisiana,* Off-flavors in foods

L. Johnson, *Hoffmann-La Roche Corporation, Nutley, New Jersey,* Vitamins

Thomas A. Johnson, *Häagen-Dazs Corporation, Fairfield, New Jersey,* Ice cream and frozen dessert

R. L. Joseph, *IUF.ST, Dublin, Ireland,* International development: Ireland

Miloslav Kalab, *Food Research Center, Agriculture, Canada, Ottawa, Ontario, Canada*

Mary Kamm, *Shirtsleeve Seminars, Watsonville, California,* Onion and garlic; Statistical process control

I. Kawakita, *Ajinomoto Corporation, Murodomi, Saga, Japan,* Monosodium glutamate

T. Kawakita, *Ajinomoto Corporation, Saga-Ken, Japan,* Nucleotides; Protein: Amino Acids

J. F. Kefford, *Consultant, New South Wales, Australia,* International development: Australia

D. Mark Kettunen, *General Foods USA, Tarrytown, New York,* Snack food technology; Agglomeration and agglomerator systems

G. G. Khachatourians, *University of Saskatchewan, Saskatoon, Canada,* Genetic engineering, part I: principles and applications

J. W. Kiceniuk, *Canada Department of Fisheries and Oceans, Newfoundland, Canada,* Lobster: biology and technology

K. J. Kirkpatrick, *New Zealand Dairy Board, Wellington, New Zealand,* Dairy ingredients for food

William H. Knightly, *Emulsion Technology, Wilmington, Delaware,* Emulsifier technology in foods

Hank Kocol, *California Department of Health Services, Sacramento, California,* Radiation: concepts and risks

Marlene M. Krayl, *University of Toronto, Ontario, Canada,* Plant products in food, natural

C. C. Lai, *Pacteco USA,* Packaging, part III: materials

Rauno A. Lampe, *U.S. Army Natick, Natick, Massachusetts,* Retort pouch

Peter LaMontagne, *Sharples Corporation, Warminster, Pennsylvania,* Centrifuges: principles and applications

Alan G. Lang, *CSIRO, North Ryde NSW, Australia,* Plant cell and tissue culture; food related products

D. R. Lane, *Proctor and Schwartz Corporation, Horsham Pennsylvania,* Preproduction process: computer modeling

Elizabeth Larmond, *Research Branch Agriculture, Canada, Ottawa, Ontario, Canada,* International Development: Canada

A. K. Lau, *The University of British Columbia, Vancouver, British Columbia, Canada,* Evaporation

Harry T. Lawless, *Cornell University, Ithaca, New York,* Taste and odor

Byong H. Lee, *McGill University, Ste. Anne de Bellevue, Quebec, Canada,* Genetric engineering, part II: enzyme cloning

Chang Y. Lee, *Cornell University, Geneva, New York,* Browning reaction, enzymatic; Phenolic compounds

Chong M. Lee, *University of Rhode Island, Kingston, Rhode Island,* Surimi: science and technology

Arnold F. Leestma, *Galloway West, Fond du Lac, Wisconsin,* Sweetened condensed milk

D. Paul Leitch, *U.S. Army Natick Research Development and Engineering Center, Natick, Massachusetts,* U.S. Armed Forces food research and development

Lawrence M. Lenovich, *Hershey Foods Corporation, Hershey, Pennsylvania,* Water activity: microbiology

Hanhua Liang, *Purdue University, West Lafayette, Indiana,* Mass transfer and diffusion in foods

I. Chiu Liao, *Taiwan Fisheries Research Institute, Keelung, Taiwan,* Eels

Eunice C. Y. Li Chan, *University of British Columbia, Vancouver, British Columbia, Canada,* Hydrophobicity in food protein systems

Jerome T. Liebrand, *Pfizer Corporation, Groton, Connecticut,* Acidulants

Michael B. Liewen, *General Mills Corporation, Minneapolis, Minnesota,* Antimicrobials

Rong C. Lin, *Food and Drug Administration, Washington, D.C.,* Water activity: good manufacturing practice

Rudy R. Lin, *Swift-Eckrich Corporation, Oak Brook, Illinois,* Elastin and meat ligaments; Muscle as food

Ted R. Lindstrom, *General Foods USA, White Plains, New York,* Edible films and coatings

Frank F. W. Liu, *Cornell University, Ithaca, New York,* World economic fruits production and utilization

Ming Sai Liu, *Food Industry Research and Development Institute, Hsinchu Taiwan,* Bamboo shoot

Concepcion Llaguno, *Programa Nacional de Technología de Alimentos, Madrid, Spain,* International development: Spain

Göran Löndahl, *Helsingborg, Sweden,* Freezing systems for the food industry

Austin R. Long, *Louisiana State University, Baton Rouge, Louisiana,* Antibiotics in foods of animal origin

C. R. Longdill, *Meat Industry Research Institute of New Zealand, Hamilton, New Zealand,* Meat slaughtering and processing equipment

B. S. Luh, *University of California, Davis,* Vegetable processing

Ted M. Lupina, *Kalsec, Kalamazoo, Michigan,* Spices and herbs: natural extraction

B. Machairidis, *Commercial Counselor, Embassy of Greece, Washington, D.C.,* International development: Greece

L. Machlin, *Hoffmann-La Roche, Nutley, New Jersey,* Vitamins

Joseph M. Madden, *PHS/Food and Drug Administration, Washington, D.C.,* Microbial virulence assessment

Raymond R. Mahoney, *University of Massachusetts, Amherst, Massachusetts,* Enzymology; imbolized enzymes

Yrjo Malkki, *Technical Research Center, Espoo, Finland,* International Development: Finland

Charles H. Manley, *Takasago Corporation USA, Teterboro, New Jersey,* Dairy flavors

Marioni Packing Company, *San Jose, California,* Apricots and Apricot Processing

W. D. Marshall, *McGill University, Ste. Anne de Bellvue, Quebec, Canada,* Pesticide residues in foods

Robert A. Martin, *Hershey Foods Technical Center, Hershey, Pennsylvania,* Chocolate and cocoa products

Robert W. Martin, Jr., *Kraft Technology Center, Glenview, Illinois,* Food microstructure

Samuel A. Matz, *Pan-Tech International Corporation, McAllen, Texas,* Biscuit and cracker technology

Terry L. McAninch, *Birko Corporation, Westminster, Colorado,* Detergents

P. McClean, *North Dakota State University, Fargo, North Dakota,* Plant biotechnology: principles and applications

John McGregor, *Louisiana State University, Baton Rouge, Louisiana,* Cultured Milk Products; Yogurt

W. Mergens, *Hoffmann-La Roche, Nutley, New Jersey,* Vitamins

Alice Meyer, *U.S. Army Natick, Natick, Massachusetts,* Military food

A. J. Miller, *U.S. Department of Agriculture, Philadelphia, Pennsylvania,* Sausages

Stephen S. Miller, *USDA/ARS, Kearneysville, West Virginia,* Apples and apple products

David B. Min, *Ohio State University, Columbus, Ohio,* Fats and oils: flavors

Vikram V. Mistry, *South Dakota State University, Brookings, South Dakota,* Milk and milk products

Keisuke Morimoto, *General Foods USA, White Plains, New York,* Edible films and coatings

D. Darwin Morrell, *USDA/ARS/NRRC, Peoria, Illinois,* Parasitic organisms

Howard R. Moskowitz, *Moskowitz/Jacobs Corporation, Valhalla, New York,* Sensory science: principles and applications

A. S. Mossman, *Humboldt State University, Arcata, California,* Wild games

Tilak Noagodawithana, *Universal Foods Corp., Milwaukee, Wisconsin,* Yeast

S. Nakai, *University of British Columbia, Vancouver, British Columbia, Canada,* Proteins: structure

Ananth Narayan, *U.S. Army Natick, Natick, Massachusetts,* Military food

R. T. Naude, *Animal and Dairy Science Research Institute, Irene, South Africa,* International development: South Africa

K. Rajinder Nuath, *Kraft Corporation, Glenview, Illinois,* Cheese

Maria Nazarowec, *WHIL Research Branch Agriculture Canada, Ottawa, Canada,* International development: Canada

William E. Newman, *Häagen-Dazs Corporation, Fairfield, New Jersey,* Ice cream and frozen dessert

I. L. Nonnecke, *Nonnecke and Associates, Guelph, Ontario, Canada,* Vegetable production

John Norback, *University of Wisconsin, Madison, Wisconsin,* Food processing: technology, engineering, and management

Robert L. Olsen, *Schreiber Foods Corporation, Green Bay, Wisconsin,* Artificial intelligence; Computer applications in food industry; Expert system: cheese defect

Paul H. Orr, *Red River Valley, Potato Research Laboratory, East Grand Forks, Minnesota,* Potatoes and potato processing

Joseph L. Owodes, *Consultant, Sonoma, California,* Alcoholic beverages and human response; Beer; Distilled beverage spirits

Mahesh Padmana B Han, *University of Minnesota, St. Paul, Minnesota,* Extrusion processing: texture and rheology

Raymond L. Page, *Virginia Polytechnic Institute and State University, Blacksburg, Virginia,* Transgenic animals

J. S. Paik, *University of Delaware, Newark, Delaware,* Packaging: part I: general considerations; Packaging, part II: Labeling Chemical

J. S. Palermo, *Monsanto Chemical Company, St. Louis, Missouri,* Sorbic acid

William D. Pandolfe, *APV Gaulin Corporation, Wilmington, Massachusetts,* Colloid mills; Homogenizers

John G. Parsons, *South Dakota State University, Brookings, South Dakota,* Milk and milk products

Attila E. Pavlath, *USDA/ARS/WRRC, Albany, California,* Oxidation

James Payne Smith, Jr., *Humboldt State University, Arcata, California,* Plants, poisonous

Per Oskar Persson, *Frigoscandia Contracting AB, Helsingborg, Sweden,* Freezing systems for the food industry

G. Frank Phillips, *Dr. Pepper Corporation, Dallas, Texas,* Carbonated beverages; Quality control: carbonated beverages

Bernadette Piacel-Llanes, *Häagen-Dazs Corporation, Fairfield, New Jersey,* Ice cream and frozen dessert

Ana Pickering, *Department of Scientific and Industrial Research, Wellington, New Zealand,* International development: New Zealand

Raul H. Piedrahita, *University of California, Davis, California,* Aquaculture: engineering and construction

Mark J. Pietka, *Cadbury Beverages, Stamford, Connecticut,* Quality control; carbonated beverages

George M. Pigott, *University of Washington, Seattle, Washington,* Extraction; Fish and shellfish products; Food engineering; Heat; Heat exchangers; Heat transfer; Thermodynamics

J. Terry Pitts, *Gustafson Corporation, Plano, Texas,* Grains and protectants

Y. Pomeranz, *Washington State University, Pullman, Washington,* Cereals science and technology; Food analysis; Malts and malting; Wheat science and technology

Dorothy Pond-Smith, *Washington State University, Pullman Washington,* Foodservice systems

Margaret A. Poole, *Häagen-Dazs Corporation, Fairfield, New Jersey,* Ice cream and frozen dessert

William Porter, *U.S. Army Natick, Natick, Massachusetts,* Military food

John J. Powers, *University of Georgia, Athens, Georgia,* Food Science and technology; Institute of Food Technologists: history and perspectives; Sensory science: standardization and instrumentation

J. H. Prentice, *Consultant, Axminster, Devon, United Kingdom,* Cheese rheology

R. L. Preston, *Texas Technology University, Lubbock, Texas,* Livestock feeds

Ernst J. Pyler, *Raleigh, North Carolina,* Baking science and technology

Hosahalli Ramaswammy, *McGill University, Ste. Anne de Bellevue, Quebec, Canada,* Packaging, part IV: modified atmosphere packaging, principles and applications; Thermal processing and computer modeling

Swamy Ramaswammy, *McGill University, Ste. Anne de Bellevue, Quebec, Canada,* Thermal processing and food quality

Gerald Reed, *Consultant, Milwaukee, Wisconsin,* Yeasts

A. Renz-Schauen, *Justus Leibig University, Giessen, Germany,* Quarg

M. Ratliff, *Michigan State University, East Lansing, Michigan,* Food industry economic development

Richard A. Roeder, *University of Idaho, Moscow, Idaho,* Animal science and livestock production

Sally A. Rose, *Campden Food and Drink Research Association, Gloucestershire, United Kingdom,* Chilled foods

Mary V. Rosholt, *University of Massachusetts, Amherst, Massachusetts,* Energy

John H. Rupnow, *University of Nebraska, Lincoln, Nebraska,* Proteins: biochemistry and applications

J. M. Russell, *Meat Industry Research Institute of New Zealand, Hamilton, New Zealand,* Meat industry processing wastes: characteristics and treatment

Leif Rynnel, *Helsingborg, Sweden,* Freezing systems for the food industry

H. Sakai, *Ajinomoto Corporation, Mie-ken, Japan,* Proteins: amino acids

C. Sano, *Ajinomoto Corporation, Morodomi, Saga, Japan,* Monosodium glutamate

J. W. Savell, *The American Lamb Council, Englewood, Colorado,* Lamb

Fred W. Schenk, *Corn Products, CPC International, Argo, Illinois,* Corn and corn products

H. B. Schiefer, *Toxicology Research Centre, University of Saskatchewan, Saskatoon, Canada,* Mycotoxins

J. Scheiner, *Hoffmann-La Roche, Nutley, New Jersey,* Vitamins

Gerald T. Schultz, *U.S. Army Natick Research, Development, and Engineering Center, Natick, Massachusetts,* Military food; U.S. Armed Forces food research and development

J. Scott, *Hoffmann-La Roche Corporation, Nutley, New Jersey,* Vitamins

Jeremy D. Selman, *Campden Food and Drink Research Association, Chipping Campden, Gloucestershire, United Kingdom,* Blanching

P. Shatadal, *University of Manitoba, Winnipeg, Manitoba, Canada,* Cleaning-in-place (CIP)

B. J. Shay, *CSIRO, Brisbane, Australia,* Meat starter cultures and meat product manufacturing

Robert L. Shewfelt, *University of Georgia, Griffin, Georgia,* Food crops: sensory evaluation; Food crops: Postharvest deterioration; Food Crops: varietal differences, maturation, ripening, and senescence

Walter S. Sheppard, *USDA/ARS, Beltsville, Maryland,* Beekeeping

Hachiro A. Shimonuki, *USDA/ARS, Beltsville, Maryland,* Beekeeping

W. R. Shorthose, *CSIRO, Cannon Hill, Queensland, Australia,* Meat: texture and rheology

Thomas C. Siewicki, *National Marine Fisheries Service, Charleston, South Carolina,* Marine toxins

Gerald Silvermann, *U.S. Army Natick, Natick, Massachusetts,* Military food

A. P. Simopoulos, *The Center for Genetics, Nutrition and Health, American Association for World Health, Washington, D.C.,* Omega-3 fatty acids

Benjamin K. Simpson, *McGill University, Quebec, Canada,* Marine enzymes; Packaging, part IV: modified atmosphere packaging, principles and applications

Kenneth L. Simpson, *University of Rhode Island, Kingston, Rhode Island,* Carotenoids

Peter J. Slade, *University of Guelph, Guelph, Ontario, Canada,* Genetic engineering, part IV: food microbiology

James Smith, *McGill University, Ste. Anne de Bellevue, Quebec, Canada,* Thermal processing and computer modeling; Marine Enzymes; Packaging, part IV: modified atmosphere packaging, principles and applications

Jeff L. Smith, *University of Guelph, Guelph, Ontario, Canada,* Proteins: denaturation and food processing

P. W. Smith, *Eastern Regional Research Center, U.S. Department of Agriculture, Wyndmoor, Pennsylvania,* Milk imitation

Thomas Smouse, *Archer Daniels Midland Company, Decatur, Illinois,* Salad oils

J. M. Smucker Company, *Salinas, California,* Apricots and apricot processing

J. Frank Smullen, *Hershey Foods Corporation, Hershey, Pennsylvania,* Chocolate and cocoa products; Licorice confectionery

Laszlo P. Somogyi, *Etel Corporation, Berkeley, California,* Fruit dehydration; Vegetable dehydration

S. M. Southwick, *University of California, Davis,* Apricots and apricot processing

John J. Specchio, *Montclair State College, Upper Montclair, New Jersey,* Antioxidants

W. E. L. Speiss, *Federal Research Center for Nutrition, Karlsruhe, Germany,* International Development: Germany

Jeffrey S. Spencer, *A. M. Todd Company, Kalamazoo, Michigan,* Essential oils: citrus and mint

Peter Sporns, *University of Alberta, Edmonton, Alberta, Canada,* Food chemistry and biochemistry; Honey analysis

William J. Stadelman, *Purdue University, West Lafayette, Indiana,* Eggs and egg products

John J. Stanton, *Saint Joseph's University, Philadelphia, Pennsylvania,* Food marketing

Clyde E. Stauffer, *Technical Foods Consultant, Cincinnati, Ohio,* Bakery leavening agents; Bakery specialty products; Emulsifiers, stabilizers, and thickeners; Enzyme assays for food scientists

Kenneth Stevenson, *National Food Processors Association, Dublin, California,* Thermal processing: food canning

K. Stevenson, *National Food Processors Institute, Washington, D.C.,* Aseptic processing and packaging systems

Richard F. Stier, *Libra Laboratories, Piscataway, New Jersey,* Quality assurance

Michael F. Stiles, *University of Alberta, Edmonton, Canada;* Disinfectants; Meat microbiology

Gene Stokes, *California Apricot Advisory, Board, Walnut Creek, California,* Apricot and apricot processing

Ramsey Saithward, *New Zealand Dairy Research Institute, Palmerston North, New Zealand,* Casein and caseinates

Elizabeth D. Strange, *U.S. Department of Agriculture, Philadelphia, Pennsylvania,* Collagens

Jeremiah B. Sullivan, *Degesch America Corporation, Weyers Cave, Virginia,* Fumigants

Sun Garden Packing Company, *San Jose, California,* Apricot and apricot Processing

Taneko Suzuki, *Nihon University, Kanagawa, Japan,* Fish Cakes (Kamboko); Krill

Marsha Swartz, *New Zealand Milk Products Corporation, Santa Rosa, California,* Casein and caseinates

J. E. Swan, *Meat Industry Research, Institute of New Zealand Corporation, Hamilton, New Zealand,* Animal by-product processing

H. Symons, *American Frozen Food Institute, McLean Virginia,* Refrigerated foods; food freezing and processing; Refrigerated foods: food freezing and world food supply; Refrigerated foods: handling and inventory

Bernard F. Szuhaj, *Central Soya Corporation, Fort Wayne, Indiana,* Lecithin

Donna R. Tainter, *Saratoga Specialties, Elmhurst, Illinois,* Spices and seasonings

Irwin Taub, *U.S. Army Natick, Natick, Massachusetts,* Military food

Marilyn Taub, *Association of Official Analytical Chemists, Arlington, Virginia,* Association of Official Analytical Chemists

Harry Teicher, *Monsanto Chemical Company, St. Louis, Missouri,* Phosphates and food processing

Cindy B. S. Tong, *USDA/ARS, Beltsville, Maryland,* Food crops: storage

M. A. Tung, *Technical University of Nova Scotia, Halifax, Canada,* Rheology

A. R. Tricker, *Institute for Toxicology and Chemotherapy, Heidelberg, Germany,* Nitrosamines

George S. Torrey, *Consultant, Brookings, South Dakota,* Starter cultures for cheese and fermented milk

Roger S. Unger, *Kraft Technology Center, Glenview, Illinois,* Food microstructure

Mark L. Unland, *Monsanto Chemical Company, St. Louis, Missouri,* Phosphates and food processing

Walter M. Urbain, *Michigan State University, Sun City, Arizona,* Irradiation of food

F. J. Vaccarino, *University of Toronto, Toronto, Canada,* Appetite

Martien van den Hoven, *Creamy Creations, Rijkevoort, The Netherlands,* Dairy ingredients: applications in meat, poultry, and seafoods

Ben van Valkengoed, *Creamy Creations, Rijkevoort, The Netherlands,* Dairy ingredients: applications in meat, poultry, and seafoods

William H. Velander, *Virginia Polytechnic Institute and State University, Blacksburg, Virginia,* Transgenic animals

John R. Vercellotti, *USDA/ARS, New Orleans, Louisiana,* Off-flavors in foods

Felix Viro, *Kind and Knox Corporation, Sioux City, Iowa,* Gelatin

J. I. Wadsworth, *USDA/ARS/SRRC, New Orleans, Louisiana,* Rice

Neil Walker, *New Zealand Milk Products Corporation, Santa Rosa, California,* Casein and caseinates

Chen Yi Wang, *USDA/ARS, Beltsville, Maryland,* Controlled atmosphere for fresh fruits and vegetables

Pie Yi Wang, *Swift-Eckrich Corporation, Oak Brook, Illinois,* Meat processing: technology and engineering

Shaw S. Wang, *Rutgers University, Piscataway, New Jersey,* Kinetics: Surface (interfacial) tension

Carol K. Waslien, *Consultant, Honolulu, Hawaii,* Muslim dietary laws, nutrition, and food processing

Alley E. Watada, *USDA/ARS, Beltsville, Maryland,* Controlled atmosphere for fresh fruits and vegetables; Food crops: nondestructive quality evaluation; Food crops: storage

E. Waysek, *Hoffmann-La Roche Corporation, Nutley, New Jersey,* Vitamins

Cheng I. Wei, *University of Florida, Gainesville, Florida,* Food surface sanitation

Herb Weiss, *Balchem Corporation, Slate Hill, New York,* Encapsulation techniques

John C. Wekell, *USDA/NOAA/NMFS, Seattle, Washington,* Seafoods: flavors and quality

Cheng I. Wel, *University of Florida, Gainesville, Florida,* Food toxicology

R. C. Whiting, *U.S. Department of Agriculture, Philadelphia, Pennsylvania,* Sausages

Hill Williams, *Monsanto Chemical Company, St. Louis, Missouri,* Sorbic acid

Max W. Williams, *USDA/ARS, Wenatchee, Washington,* Apples and apple products

Sharon Wittinger, *Häagen-Dazs Corporation, Fairfield, New Jersey,* Ice cream and frozen dessert

Dominic W. S. Wong, *USDA/ARS/WRRC, Albany, California,* Food chemistry: mechanism and theory; oxidation

Boh-Keng Wu, *Food Industry Research and Development Institute, Hsinchu, Taiwan,* Mushrooms processing

Jhng-Yang Wu, *Food Industry Research and Development Institute, Hsinchu, Taiwan,* Rice products, oriental

Victor Wu, *USDA/ARS/NCAUR, Peoria, Illinois,* Cereals, nutrients, and agricultural practices

Rickey Y. Yada, *University of Guelph, Guelph, Ontario, Canada,* Proteins: denaturation and food processing

K. L. Yam, *Rutgers University, New Brunswick, New Jersey,* Packaging, part III: materials

Chin Tien Yang, *Food Industry Research Institute, Hsinchu, Taiwan,* Quality control: machine vision system

Varoujan A. Yaylayan, *McGill University, Quebec, Canada,* Flavor chemistry

Samuel H. Yong, *Häagen-Dazs Corporation, Fairfield, New Jersey,* Ice cream and frozen dessert

M. Yoshikawa, *Kyoto University, Kyoto, Japan,* Peptides

J. G. Zadow, *CSIRO, Highett, Victoria, Australia,* Milk and membrane processing

Zaiger's Genetics, *Modesto, California,* Apricots and apricot processing

Votery B. Zemelman, *General Foods USA, White Plains, New York,* Agglomeration and agglomerator systems

B. L. Zoumas, *Hershey Foods Corporation, Hershey, Pennsylvania,* Chocolate and cocoa products

FOREWORD

Life revolves around the production, storage, processing, and distribution of food. A major portion of a person's life is spend in studying, purchasing, preparing, consuming, and discussing food. In addition to food scientists and technologists, many professionals and consumers are dedicated to solving the numerous world food aspects, legalities, and consumption. Central to all of these activities is the availability of information pertaining to one's specific interests and activities. Standard reference materials suffer two major drawbacks. Apart from being scattered over many documents and locations, they are often written in such a manner as to make it difficult for those with nonscientific backgrounds to assimilate.

Wiley-Interscience is to be congratulated for publishing this comprehensive *Encyclopedia of Food Science and Technology* which brings usable information to the fingertips of all interested parties. Key to the success of this *Encyclopedia* was the expert Editorial Advisory Board. This Board, consisting of four individual scientists (Roy Morse, Steven Chang, Merle Pierson, and Stanley Sacharow) and representatives (Hugh Symons, Kenneth Stevenson, and Philip Brandler) from three professional organizations were responsible for selecting topics and providing overall direction.

Dr. Y. H. Hui, with the input of the Board and hundreds of technical experts from worldwide industry, academia, and government, has done an outstanding job of preparing, coordinating, and editing this extensive treatise. The magnitude of the project required four years from conception to publication. As the editor for books and journals in food science, food technology, and food engineering, I appreciate the complexity and work involved in a project of this nature.

Many prominent scientists, engineers, and other food professionals have condensed their fields of expertise into timely articles. The consolidation of this massive amount of data of 3,000 pages (approximately 380 articles) into three volumes, from acidulants to yogurt, is the most comprehensive wealth of knowledge on the subject ever assembled in one reference source. It will be of tremendous value to anyone concerned with food, namely, *everyone!*

Although many of the articles discuss the basics of a given subject, they are written in such a manner as to be easily understood by a nontechnical user. In addition, the well-chosen titles of the articles provide an extensive index that facilitates locating specific item with ease. The volumes will be equally used and appreciated by those using public and private secondary school, university, industry, professional, and personal libraries.

The diversity of the *Encyclopedia* especially is appreciated by someone like myself who is intimately involved with both academia and industry. As a professor at the University of Washington for more than two decades, I know that undergraduate and graduate students will be using the three volumes for both reference material and specific topic familiarization. As the President of Sea Resources Engineering, Inc., a consulting engineering firm that often encounters projects where specific information on a subject area must be obtained rapidly, I have always insisted on a company library that can answer this need. This three-volume text will be an invaluable addition.

It is indeed an honor and privilege to have been chosen as one of the participants contributing to the *Encyclopedia.* I am looking forward to its being available to my students and associates.

GEORGE M. PIGOTT, Ph.D.
Professor and Director
Institute for Food Science and Technology
University of Washington
Seattle, Washington

PREFACE

Although this encyclopedia is designed for food scientists and technologists, it also contains information useful to food engineers, chemists, biologists, ingredient suppliers, and other professionals involved in the food chain.

The topics selected are based on the following frame of reference:

A. BASIC AND APPLIED SCIENCES:

1. Biology: botany, bacteriology, microbiology, mycology, eg, photosynthesis, food microbiology, mycotoxins.

2. Chemistry: biochemistry, physical chemistry, analytical chemistry, organic chemistry, radiochemistry, forensic chemistry, eg, food chemistry, food analysis, interesterification, emulsification.

3. Physics: rheology, thermodynamics, cryogenics, radiophysics, ultrasonics, eg, food rheology, thermodynamics, microwave, irradiation.

4. Nutrition: basic, applied, and clinical nutrition, eg, nutrition, protein, energy, dietetics, nutritional quality and food processing.

5. Psychology: sensory behaviors, eg, food and color, sensory sciences, taste and odor.

6. Medicine: metabolisms, toxicology, heart diseases, eg, food utilization, cholesterol, foodborne diseases.

7. Economics: marketing, development, eg, food marketing, food industry economic development.

8. Integrated sciences: history of foods, food structure.

B. PROCESSING TECHNOLOGY, ENGINEERING, AND THE TWENTY-THREE UNIT OPERATIONS

1. Raw material preparation, eg, apricots and apricot processing, milk and milk products.

2. Size reduction, eg, potato and potato processing, crabs and crab processing.

3. Mixing and forming, eg, confectionery, biscuits and cracker technology.

4. Mechanical separations, eg, meat processing technology, cheeses, poultry product and technology.

5. Membrane concentration, eg, milk and membrane technology

6. Biotechnology, eg, genetic engineering

7. Irradiation, eg, irradiation

8. Blanching, eg, blanching, vegetables processing

9. Pasteurization, eg, yogurt

10. Heat sterilization, eg, aseptic processing

11. Evaporation, eg, evaporation, distillation, heat transfer, heat exchangers

12. Extrusion, eg, extrusion and extrusion cooking, snack foods

13. Dehydration, eg, fruit dehydration, vegetable dehydration, drier technology and engineering

14. Baking and roasting, eg, baking science and technology, breakfast cereals

15. Frying, eg, frying technology, fats and oils

16. Microwave and infrared radiation, eg, microwave science and technology

17. Chilling and controlled- (modified-) atmosphere storage, eg, chilled foods, controlled atmospheres for fresh fruits and vegetables, packaging and modified atmospheres.

18. Freezing, eg, food freezing, freezing systems, refrigerated foods

19. Freeze drying and freeze concentration, eg, freeze drying, driers technology and engineering.

20. Coating or enrobing, eg, fish and shellfish products, poultry processing technology,

21. Packaging, eg, packaging materials, packaging and labeling

22. Filling and sealing of containers, eg, aseptic processing and packaging

23. Materials handling and process control, eg, computer and food processing, artificial intelligence, expert systems

C. FOOD LAWS AND REGULATIONS

1. Basic principles, eg, food laws and regulations

2. Food identity, eg, standards and grades

3. Food chemicals, eg, food additives

4. Health and safety, eg, sanitation

5. Food processing plants, eg, food plant design and construction, food processing: standard industrial classifications.

6. Food plant inspection, eg, food laws and regulations.

Apart from members of the editorial advisory board, hundreds of professionals in the academic, industry, and government have personally counseled me on many aspects of the work. Their combined effort is reflected in the articles in this *Encyclopedia:* appropriateness of the selection of topics; diverse expertise and background of each contributor; the excellent treatment of each article. The users are the best judge of these claims.

A small portion of the text is devoted to cover nontechnical data:

1. Descriptions of selected scientific institutions, eg, Institute of Food Technologists.

2. Food science and technology in selected countries in the world, eg, International Development: South Africa.

3. Research and development in major U.S. government agencies, eg, U.S. Department of Agriculture.

When using this *Encyclopedia,* the following premises should be noted:

1. To serve a professional audience, the average length of each article is 6–7 pages. This length serves two objectives: a reasonable treatment of each subject matter and a reduction on overlapping that frequently associates with numerous short articles in an encyclopedia. Providing more details for each subject matter automatically limits the number of articles to under 400.

2. Although the articles in this encyclopedia do not cover *all* topics in food science, technology, and engineering, major subjects of current interest are included. Also a major effort has been made to assure that many subjects not appearing under an individual heading are discussed under those that do. Their locations are easily identified by using the index, which exhaustively includes every major term.

3. All problems associated with multiple authors in a regular text are magnified in an encyclopedia. In most cases, the apparent missing of a major topic can be traced to an author's failure to deliver for whatever reason. Every effort has been made to provide the missing article, even if it is a short one. Unfortunately, we have failed in some circumstances mainly because of the time element.

4. For organizational purpose, all articles about individual countries are grouped under "International Development." Oral and written invitations were sent to the appropriate offices of embassies or consulates of industrial countries located in Washington, D.C. The countries that responded provided us with those articles appearing in this *Encyclopedia.*

5. Although the articles on food laws and regulations apply to the United States, they are the same for foreign governments or food companies that export foods and beverages to this country. Therefore, their usefulness is obvious.

ACKNOWLEDGMENT

Basically, there are three groups of people who make the completion of this large work possible: the advisors, the contributors, and the production personnel.

The advisors include the members of the editorial board, my friends, colleagues, and other experts. They helped me to formulate the subject matters, select contributors and reviewers, and solve scientific, technical, and engineering matters. Although it is not possible to name them all here, they are the foundation of the project. And I am grateful to them.

The contributors are the components and parts and give substance to the encyclopedia. Their contribution is most appreciated.

The production team puts the parts together and you are the best judge of their professionalism. Personally, I am specially grateful to Michalina Bickford, the managing editor for the encyclopedia.

The support of my family will always be appreciated.

Y. H. HUI
Cutten, California

CONVERSION FACTORS, ABBREVIATIONS AND UNIT SYMBOLS

Selected SI Units (Adopted 1960)

Quantity	Unit	Symbol	Acceptable equivalent
BASE UNITS			
length	meter[†]	m	
mass[‡]	kilogram	kg	
time	second	s	
electric current	ampere	A	
thermodynamic temperature[§]	kelvin	K	
DERIVED UNITS AND OTHER ACCEPTABLE UNITS			
* absorbed dose	gray	Gy	J/kg
acceleration	meter per second squared	m/s^2	
* activity (of ionizing radiation source)	becquerel	Bq	l/s
area	square kilometer	km^2	
	square hectometer	hm^2	ha (hectare)
	square meter	m^2	
density, mass density	kilogram per cubic meter	kg/m^3	g/L; mg/cm^3
* electric potential, potential difference, electromotive force	volt	V	W/A
* electric resistance	ohm	Ω	V/A
* energy, work, quantity of heat	megajoule	Mj	
	kilojoule	kJ	
	joule	J	N·m
	electron volt[x]	eV^x	
	kilowatt hour[x]	$kW·h^x$	
* force	kilonewton	kN	
	newton	N	$kg·m/s^2$
* frequency	megahertz	MHz	
	hertz	Hz	l/s
heat capacity, entropy	joule per kelvin	J/K	
heat capacity (specific), specific entropy	joule per kilogram kelvin	J/(kg·K)	
heat transfer coefficient	watt per square meter kelvin	W/(m²·K)	
linear density	kilogram per meter	kg/m	
magnetic field strength	ampere per meter	A/m	
moment of force, torque	newton meter	N·m	
momentum	kilogram meter per second	kg·m/s	

Selected SI Units (Adopted 1960)

Quantity	Unit	Symbol	Acceptable equivalent
* power, heat flow rate,	kilowatt	kW	
radiant flux	watt	W	J/s
power density, heat flux density, irradiance	watt per square meter	W/m²	
* pressure, stress	megapascal	MPa	
	kilopascal	kPa	
	pascal	Pa	
sound level	decibel	dB	
specific energy	joule per kilogram	J/kg	
specific volume	cubic meter per kilogram	m³/kg	
surface tension	newton per meter	N/m	
thermal conductivity	watt per meter kelvin	W/(m·K)	
velocity	meter per second	m/s	
	kilometer per hour	km/h	
viscosity, dynamic	pascal second	Pa·s	
	millipascal second	mPa·s	
volume	cubic meter	m³	
	cubic decimeter	dm³	L(liter)
	cubic centimeter	cm³	mL

* The asterisk denotes those units having special names and symbols.
† The spellings "metre" and "litre" are preferred by ASTM; however "er-" is used in the Encyclopedia.
‡ "Weight" is the commonly used term for "mass."
§ Wide use is made of "Celsius temperature" (t) defined by

$$t = T - T_0$$

where T is the thermodynamic temperature, expressed in kelvins, and $T_0 = 273.15$ by definition. A temperature interval may be expressed in degrees Celsius as well as in kelvins.
ˣ This non-SI unit is recognized by the CIPM as having to be retained because of practical importance or use in specialized fields.

In addition, there are 16 prefixes used to indicate order of magnitude, as follows:

Multiplication factor	Prefix	Symbol	Note
10^{18}	exa	E	
10^{15}	peta	P	
10^{12}	tera	T	
10^9	giga	G	
10^6	mega	M	
10^3	kilo	k	
10^2	hecto	h[a]	
10	deka	da[a]	
10^{-1}	deci	d[a]	
10^{-2}	centi	c[a]	
10^{-3}	milli	m	
10^{-6}	micro	μ	
10^{-9}	nano	n	
10^{-12}	pico	p	
10^{-15}	femto	f	
10^{-18}	atto	a	

[a] Although hecto, deka, deci, and centi are SI prefixes, their use should be avoided except for SI unit-multiples for area and volume and nontechnical use of centimeter, as for body and clothing measurement.

Conversion Factors to SI Units

To convert from	To	Multiply by
acre	square meter (m^2)	4.047×10^3
angstrom	meter (m)	$1.0 \times 10^{-10\dagger}$
atmosphere	pascal (Pa)	1.013×10^5
bar	pascal (Pa)	$1.0 \times 10^{5\dagger}$
barn	square meter (m^2)	$1.0 \times 10^{-28\dagger}$
barrel (42 U.S. liquid gallons)	cubic meter (m^3)	0.1590
Btu (thermochemical)	joule (J)	1.054×10^3
bushel	cubic meter (m^3)	3.524×10^{-2}
calorie (thermochemical)	joule (J)	$4.184^\dagger$
centipoise	pascal second (Pa·s)	$1.0 \times 10^{-3\dagger}$
cfm (cubic foot per minute)	cubic meter per second (m^3/s)	4.72×10^{-4}
cubic inch	cubic meter (m^3)	1.639×10^{-5}
cubic foot	cubic meter (m^3)	2.832×10^{-2}
cubic yard	cubic meter (m)	0.7646
dram (apothecaries')	kilogram (kg)	3.888×10^{-3}
dram (avoirdupois)	kilogram (kg)	1.772×10^{-3}
dram (U.S. fluid)	cubic meter (m^3)	3.697×10^{-6}
dyne	newton (N)	$1.0 \times 10^{-5\dagger}$
dyne/cm	newton per meter (N/m)	$1.0 \times 10^{-3\dagger}$
fluid ounce (U.S.)	cubic meter (m^3)	2.957×10^{-5}
foot	meter (m)	$0.3048^\dagger$
gallon (U.S. dry)	cubic meter (m^3)	4.405×10^{-3}
gallon (U.S. liquid)	cubic meter (m^3)	3.785×10^{-3}
gallon per minute (gpm)	cubic meter per second (m^3/s)	6.308×10^{-5}
	cubic meter per hour (m^3/h)	0.2271
grain	kilogram (kg)	6.480×10^{-5}
horsepower (550 ft·lbf/s)	watt (W)	7.457×10^2
inch	meter (m)	$2.54 \times 10^{-2\dagger}$
inch of mercury (32°F)	pascal (Pa)	3.386×10^3
inch of water (39.2°F)	pascal (Pa)	2.491×10^2
kilogram-force	newton (N)	9.807
kilowatt hour	megajoule (MJ)	$3.6^\dagger$
liter (for fluids only)	cubic meter (m^3)	$1.0 \times 10^{-3\dagger}$
micron	meter (m)	$1.0 \times 10^{-6\dagger}$
mil	meter (m)	$2.54 \times 10^{-5\dagger}$
mile (statute)	meter (m)	1.609×10^3
mile per hour	meter per second (m/s)	0.4470
millimeter of mercury (0°C)	pascal (Pa)	$1.333 \times 10^{2\dagger}$
ounce (avoirdupois)	kilogram (kg)	2.835×10^{-2}
ounce (troy)	kilogram (kg)	3.110×10^{-2}
ounce (U.S. fluid)	cubic meter (m^3)	2.957×10^{-5}
ounce-force	newton (N)	0.2780
peck (U.S.)	cubic meter (m^3)	8.810×10^{-3}
pennyweight	kilogram (kg)	1.555×10^{-3}
pint (U.S. dry)	cubic meter (m^3)	5.506×10^{-4}
pint (U.S. liquid)	cubic meter (m^3)	4.732×10^{-4}
poise (absolute viscosity)	pascal second (Pa·s)	$0.10^\dagger$
pound (avoirdupois)	kilogram (kg)	0.4536
pound (troy)	kilogram (kg)	0.3732
pound-force	newton (N)	4.448
pound-force per square inch (psi)	pascal (Pa)	6.895×10^3
quart (U.S. dry)	cubic meter (m^3)	1.101×10^{-3}
quart (U.S. liquid)	cubic meter (m^3)	9.464×10^{-4}
quintal	kilogram (kg)	$1.0 \times 10^{2\dagger}$
rad	gray (Gy)	$1.0 \times 10^{-2\dagger}$
square inch	square meter (m^2)	6.452×10^{-4}
square foot	square meter (m^2)	9.290×10^{-2}
square mile	square meter (m^2)	2.590×10^6
square yard	square meter (m^2)	0.8361

Conversion Factors to SI Units

To convert from	To	Multiply by
ton (long, 2240 pounds)	kilogram (kg)	1.016×10^3
ton (metric)	kilogram (kg)	$1.0 \times 10^{3\dagger}$
ton (short, 2000 pounds)	kilogram (kg)	9.072×10^2
torr	pascal (Pa)	1.333×10^2
yard	meter (m)	$0.9144^\dagger$

† Exact.

ABBREVIATIONS AND ACRONYMS

A	ampere	CFR	Code of Federal Regulations
a	atto (prefix for 10^{-18})	cgs	centimeter-gram-second
abs	absolute	CI	Color Index
ac	alternating current, *n.*	*cis-*	isomer in which substituted groups are on same side of double bond between C atoms
a-c	alternating current, *adj.*		
AACC	American Association of Cereal Chemists		
ACGIH	American Conference of Governmental Industrial Hygienists	cl	carload
		cm	centimeter
ACS	American Chemical Society	cmpd	compound
AGA	American Gas Association	CNS	central nervous system
Ah	ampere hour	CoA	coenzyme A
AIChE	American Institute of Chemical Engineers	COD	chemical oxygen demand
AIP	American Institute of Physics	coml	commercial(ly)
alc	alcohol(ic)	cp	chemically pure
alk	alkaline (not alkali)	cph	close-packed hexagonal
amt	amount	CPSC	Consumer Product Safety Commission
amu	atomic mass unit	cryst	crystalline
ANSI	American National Standards Institute	cub	cubic
AOAC	Association of Official Analytical Chemists	D	Debye
		D-	denoting configurational relationship
AOCS	American Oil Chemists' Society	**d**	differential operator
APHA	American Public Health Association	*d-*	*dextro-*, dextrorotatory
API	American Petroleum Institute	da	deka (prefix for 10^1)
aq	aqueous	dB	decibel
Ar	aryl	dc	direct current, *n.*
ar-	aromatic	d-c	direct current, *adj.*
as-	asymmetric(al)	dec	decompose
ASME	American Society of Mechanical Engineers	detd	determined
		detn	determination
ASTM	American Society for Testing and Materials	dia	diameter
		dil	dilute
at no.	atomic number	*dl-*; DL-	racemic
at wt	atomic weight	DOD	Department of Defense
av(g)	average	DOE	Department of Energy
bbl	barrel	DOT	Department of Transportation
Bé	Baumé	DP	degree of polymerization
bid	twice daily	dp	dew point
BOD	biochemical (biological) oxygen demand	dstl(d)	distill(ed)
bp	boiling point	dta	differential thermal analysis
Bq	becquerel	ε	dielectric constant (unitless number)
C	coulomb	*e*	electron
°C	degree Celsius	ECU	electrochemical unit
C-	denoting attachment to carbon	ed.	edited, edition, editor
c	centi (prefix for 10^{-2})	ED	effective dose
c	critical	EDTA	ethylenediaminetetraacetic acid
ca	circa (approximately)	emf	electromotive force
cd	candela; current density; circular dichroism	emu	electromagnetic unit
		en	ethylene diamine

eng	engineering
EPA	Environmental Protection Agency
epr	electron paramagnetic resonance
eq.	equation
esca	electron spectroscopy for chemical analysis
esp	especially
esr	electron-spin resonance
est(d)	estimate(d)
estn	estimation
esu	electrostatic unit
exp	experiment, experimental
ext(d)	extract(ed)
F	farad (capacitance)
F	faraday (96,487 C)
f	femto (prefix for 10^{-15})
FAO	Food and Agriculture Organization (United Nations)
fcc	face-centered cubic
FDA	Food and Drug Administration
FEA	Federal Energy Administration
FHSA	Federal Hazardous Substances Act
fob	free on board
fp	freezing point
FPC	Federal Power Commission
FRB	Federal Reserve Board
frz	freezing
G	giga (prefix for 10^9)
G	gravitational constant = 6.67×10^{11} N·m^2/kg^2
g	gram
(g)	gas, only as in H$_2$O(g)
g	gravitational acceleration
gc	gas chromatography
glc	gas–liquid chromatography
g-mol wt; gmw	gram-molecular weight
GNP	gross national product
gpc	gel-permeation chromatography
GRAS	Generally Recognized as Safe
grd	ground
Gy	gray
H	henry
h	hour; hecto (prefix for 10^2)
ha	hectare
HB	Brinell hardness number
Hb	hemoglobin
hcp	hexagonal close-packed
hex	hexagonal
HK	Knoop hardness number
hplc	high pressure liquid chromatography
HRC	Rockwell hardness (C scale)
HV	Vickers hardness number
hyd	hydrated, hydrous
hyg	hygroscopic
Hz	hertz
IC	inhibitory concentration
ICC	Interstate Commerce Commission
ICT	International Critical Table
ID	inside diameter; infective dose
ip	intraperitoneal
ir	infrared
IRLG	Interagency Regulatory Liaison Group
ISO	International Organization for Standardization
ITS-90	International Temperature Scale (NIST)
IU	International Unit
IUPAC	International Union of Pure and Applied Chemistry
IV	iodine value
iv	intravenous
J	joule
K	kelvin
k	kilo (prefix for 10^3)
kg	kilogram
L	denoting configurational relationship
L	liter (for fluids only) (5)
l-	*levo-*, levorotatory
(l)	liquid, only as in NH$_3$(l)
LC$_{50}$	conc lethal to 50% of the animals tests
LCD	liquid crystal display
lcl	less than carload lots
LD$_{50}$	dose lethal to 50% of the animals tested
LED	light-emitting diode
liq	liquid
lm	lumen
ln	logarithm (natural)
LNG	liquefied natural gas
log	logarithm (common)
LPG	liquefied petroleum gas
ltl	less than truckload lots
lx	lux
M	mega (prefix for 10^6); metal (as in MA)
M	molar; actual mass
m	meter; milli (prefix for 10^{-3})
m	molal
m-	meta
max	maximum
MCA	Chemical Manufacturers' Association (was Manufacturing Chemists Association)
MEK	methyl ethyl ketone
meq	milliequivalent
mfd	manufactured
mfg	manufacturing
mfr	manufacturer
MIBC	methyl isobutyl carbinol
MIBK	methyl isobutyl ketone
MIC	minimum inhibiting concentration
min	minute; minimum
mL	milliliter
MLD	minimum lethal dose
MO	molecular orbital
mo	month
mol	mole
mol wt	molecular weight
mp	melting point
MR	molar refraction
ms	mass spectrum
mxt	mixture
μ	micro (prefix for 10^{-6})
N	newton (force)
N	normal (concentration); neutron number
N-	denoting attachment to nitrogen
n (as n_D^{20})	index of refraction (for 20°C and sodium light)

^{n} (as Bu^n), n-	normal (straight-chain structure)
n	neutron
n	nano (prefix for 10^9)
na	not available
NAS	National Academy of Sciences
NASA	National Aeronautics and Space Administration
nat	natural
ndt	nondestructive testing
neg	negative
NF	*National Formulary*
NIH	National Institutes of Health
NIOSH	National Institute of Occupational Safety and Health
NIST	National Institute of Standards and Technology (formerly National Bureau of Standards)
nmr	nuclear magnetic resonance
NND	New and Nonofficial Drugs (AMA)
no.	number
NOI-(BN)	not otherwise indexed (by name)
NOS	not otherwise specified
nqr	nuclear quadruple resonance
NRC	Nuclear Regulatory Commission; National Research Council
NSF	National Science Foundation
NTP	normal temperature and pressure (25°C and 101.3 kPa or 1 atm)
NTSB	National Transportation Safety Board
O-	denoting attachment to oxygen
o-	ortho
OD	outside diameter
OPEC	Organization of Petroleum Exporting Countries
OSHA	Occupational Safety and Health Administration
owf	on weight of fiber
Ω	ohm
P	peta (prefix for 10^{15})
p	pico (prefix for 10^{-12})
p-	para
p	proton
p.	page
Pa	pascal (pressure)
PEL	personal exposure limit based on an 8-h exposure
pd	potential difference
pH	negative logarithm of the effective hydrogen ion concentration
phr	parts per hundred of resin (rubber)
pmr	proton magnetic resonance
p-n	positive-negative
po	per os (oral)
pos	positive
pp.	pages
ppb	parts per billion (10^9)
ppm	parts per million (10^6)
ppmv	parts per million by volume
ppmwt	parts per million by weight
ppt(d)	precipitate(d)
pptn	precipitation

Pr (no.)	foreign prototype (number)
pt	point; part
PVC	poly(vinyl chloride)
pwd	powder
py	pyridine
qv	quod vide (which see)
R	univalent hydrocarbon radical
(R)-	rectus (clockwise configuration)
r	precision of data
rad	radian; radius
RCRA	Resource Conservation and Recovery Act
ref.	reference
rf	radio frequency, *n.*
r-f	radio frequency, *adj.*
rh	relative humidity
rms	root-mean square
rpm	rotations per minute
rps	revolutions per second
RT	room temperature
S	siemens
(S)-	sinister (counterclockwise configuration)
S-	denoting attachment to sulfur
s-	symmetric(al)
s	second
(s)	solid, only as in H_2O(s)
SAN	styrene–acrylonitrile
sat(d)	saturate(d)
satn	saturation
SBS	styrene–butadiene–styrene
sc	subcutaneous
SCF	self-consistent field; standard cubic feet
Sch	Schultz number
SFs	Saybolt Furol seconds
SI	Le Système International d'Unités (International System of Units)
sl sol	slightly soluble
sol	soluble
soln	solution
soly	solubility
sp	specific; species
sp gr	specific gravity
sr	steradian
std	standard
STP	standard temperature and pressure (0°C and 101.3 kPa)
sub	sublime(s)
SUs	Saybolt Universal seconds
syn	synthetic
t	metric ton (tonne)
t	temperature
TCC	Tagliabue closed up
tex	tex (linear density)
T_g	glass-transition temperature
tga	thermogravimetric analysis
tlc	thin layer chromatography
TLV	threshold limit value
trans-	isomer in which substituted groups are on opposite sides of double bond between C atoms
TSCA	Toxic Substances Control Act
TWA	time-weighted average
Twad	Twaddell

UL	Underwriters' Laboratory	v sol	very soluble
USDA	United States Department of Agriculture	W	watt
USP	*United States Pharmacopeia*	Wb	weber
uv	ultraviolet	Wh	watt hour
V	volt (emf)	WHO	World Health Organization (United Nations)
var	variable		
vol	volume (not volatile)	wk	week
vs	versus	yr	year

ENCYCLOPEDIA OF FOOD SCIENCE AND TECHNOLOGY

A

ABSORPTION. See DISTILLATION: TECHNOLOGY AND ENGINEERING; FOOD UTILIZATION.

ACARICIDES. See HONEY ANALYSIS; PESTICIDE RESIDUES IN FOODS.

ACCIDENTAL CONTACT. See FILTH AND EXTRANEOUS MATTER IN FOOD.

ACID CONVERSION. See SYRUPS.

ACIDIC FOODS. See FOOD MICROBIOLOGY; LOW ACID AND ACIDIFIED FOODS.

ACIDIFICATION. See ACIDULANTS.

ACIDOLYSIS. See INTERESTERIFICATION.

ACIDULANTS

Food acidulants are categorized either as general purpose acids or as specialty acids. General purpose acids are those that have a broad range of functions and can be used in most foods where acidity is desired or necessary. Specialty acids are those that are limited in their functionality and/or range of application.

Citric acid and malic acid are the predominant general purpose acidulants with tartaric and fumaric acids. Fumaric acid is included in this category even though its low solubility limits its potential range of application. However, it is used in popular and widely consumed food products.

All other acidulants fall into the specialty acid category. The most commonly used specialty acids are acetic acid (vinegar), cream of tartar (potassium acid tartrate), phosphoric acid, glucono-delta-lactone, acid phosphate salts, lactic acid, and adipic acid.

CITRIC ACID

Citric acid is the premier acid for the food and beverage industry because it offers a unique combination of desirable properties, ready availability in commercial quantities, and competitive pricing. It is estimated that citric acid worldwide accounts for more than 80% of general purpose acidulants used. It is found naturally in almost all living things, both plant and animal. It is the predominant acid, in substantial quantity, in citrus fruits (oranges, lemons, limes, etc), in berries (strawberries, raspberries, currants) and in pineapples. Citric acid is also the predominant acid in many vegetables, such as potatoes, tomatoes, asparagus, turnips, and peas, but in lower concentrations. The citrate ion occurs in all animal tissues and fluids. The total circulating citric acid in the serum of man is approximately 1 mg/kg of body weight. (1)

Citric acid is manufactured by fermentation, a natural process using living organisms. The acid is recovered in pure crystalline form either as the anhydrous or monohydrate crystal depending on the temperature of crystallization. The transition temperature is 36.6°C. Crystallization above this temperature produces an anhydrous product while the monohydrate forms at lower temperatures. With an occasional exception for technological reasons,

the anhydrous form is preferred for its physical stability and is the more widely available commercial form.

The salts of citric acid that are used in the food industry are sodium citrate dihydrate and anhydrous, monosodium citrate, potassium citrate monohydrate, calcium citrate tetrahydrate, and ferric ammonium citrate, in both brown and green powder. These salts are used either for functional purposes such as buffering or emulsification, as a source of cation for technological purposes, such as calcium to aid the gelation of low methoxyl pectin, or as a mineral source for food supplementation.

Citric acid is a hydroxy tribasic acid that is a white granule or powder. It is odorless with no characteristic taste other than tartness. Its physical and chemical properties are listed in Table 1. Citric acid and its salts are used across the broad spectrum of food and beverage products:

- Beverages
- Gelatin desserts
- Baked goods
- Jellies, jams, and preserves
- Candies
- Fruits and vegetables
- Dairy products
- Meats
- Seafood
- Fats and oils.

MALIC ACID

This is the second most popular general purpose food acid although less than one-tenth the quantity of citric acid used. Malic acid is a white, odorless, crystalline powder or granule with a clean tart taste with no characterizing flavor of its own. The properties of malic acid are listed in Table 1.

Malic acid is made through chemical synthesis by the hydration of maleic acid. An inspection of its structure in Table 1 reveals that malic acid has an asymmetric carbon which provides for the existence of isomeric forms. The synthetic procedure for malic acid produces a mixture of the D and L isomers. The item of commerce, racemic D, L-malic acid does not occur in nature (2) although the acid is affirmed as GRAS for use in foods.

L-malic acid is the isomeric form that is found in nature. It is the predominant acid in substantial quantity in apples and cherries and in lesser quantities in prunes, watermelon, squash, quince, plums, and mushrooms. Like citric acid, L-malic acid plays an essential part in carbohydrate metabolism in man and other animals.

There are no salts of malic acid that are items of commerce for the food and beverage industries. The application for malic and citric acid cover the same broad range of food categories.

Table 1. Properties of General Purpose Food Acidulants

	Citric	Malic	Tartaric	Fumaric
Structure	COOH | CH_2 | HO—C—COOH | CH_2 | COOH	COOH | HOCH | CH_2 | COOH	COOH | HCOH | HOCH | COOH	COOH | CH ‖ HC | COOH
Formula	$C_6H_8O_7$	$C_4H_6O_5$	$C_4H_6O_6$	$C_4H_4O_4$
Molecular weight	192.12	134.09	150.09	116.07
Melting point, °C	153	131	169	286
Solubility at 25°C (g/100 mL of water)	181	144	147	0.63
Caloric value, kcal/g	2.47	2.39	1.84	2.76
Hygroscopicity (resistance to moisture pick up)				
at 66% RH	Fair	Fair	Fair	Good
at 86% RH	Poor	Poor	Poor	Good
Ionization constants				
K_1	8.2×10^{-4}	4×10^{-4}	1.04×10^{-3}	1×10^{-3}
K_2	1.77×10^{-5}	9×10^{-6}	5.55×10^{-5}	3×10^{-5}
K_3	3.9×10^{-6}			

TARTARIC ACID

While tartaric acid has the characteristics of a general purpose acid, fluctuating availability and price have caused users to reformulate where possible. Tartaric acid is a dibasic dihydroxy acid. The product is a white odorless granule or powder that has a tart taste and a slight characteristic flavor of its own. The properties of tartaric acid are listed in Table 1.

Tartaric acid has two asymmetric carbons which permit the formation of a dextro (+) rotatory form, a levo (−) rotatory form, and a meso form which is inactive due to internal compensation. A racemic mixture of dextro- and levo-rotatory forms is also a possibility.

Apart from very limited synthetic production in South Africa, tartaric acid is extracted from the residues of the wine industry. Therefore the natural form, L-(+)tartaric acid, is the article of commerce. The structure in Table 1 is L-(+)tartaric acid.

Potassium acid tartrate, also known as cream of tartar, is the major salt used in the food industry. It is occasionally used as the acid portion of a chemical leavening system for baked goods. It is also used as a doctor in candy making to prevent sugar crystallization by inverting a portion of the sucrose.

Tartaric acid can be used in most food categories and functions that require acidification but in practice it is limited to grape flavored products, particularly beverages and candies. It is also used where high tartness is desired in a highly soluble acid.

FUMARIC ACID

Fumaric acid is also a naturally occurring, organic, general purpose food acid. Although not found ubiquitously or in the concentrations of citric acid, fumaric nevertheless is found in all mammals as well as rice, sugar cane, wine, plant leaves, bean spouts, and edible mushrooms.

Fumaric acid is made synthetically by the isomerization of maleic acid. It is a white crystalline powder that has the clean tartness necessary for a food acidulant. Table 1 shows that fumaric acid is the strongest of the food acids and is also the least soluble. The low solubility of fumaric acid limits its usefulness in foods. In processes where 50% stock solutions are mandatory such as carbonated beverages and jams and jellies, fumaric acid cannot be used.

Fumaric acid is extensively used in noncarbonated fruit juice drinks. Its greater acid strength allows lower use levels than citric acid and its solubility and rate of solution are sufficient for this process. Fumaric is also used in consumer packed gelatin desserts because of its low hygroscopicity. The only salt of fumaric acid that has had any application in the food industry is ferrous fumarate for iron fortification.

FUNCTIONS OF FOOD ACIDS

Food acidulants and their salts perform a variety of functions. These functions are as antioxidants, curing and pickling agents, flavor enhancers, flavoring agents and adjuvants, leavening agents, pH control agents, sequestrants, and synergists. The definitions for these functions are contained in the *U.S. Code of Federal Regulations*. (3) Some of the functions overlap and in any given application an acidulant will often perform two or more functions.

Flavor Enhancer and Flavor Adjuvant

These functions are performed by a majority of the acidulants consumed by the food and beverage industries. Acids

Table 2. Tartness Equivalence

General Purpose Acids	Weight Equivalence
Citric	1.0
Fumaric	0.6–0.7
Tartaric	0.8–0.9
Malic	0.85–1.0

provide a tang or tartness that compliments and enhances many flavors but do not impart a characteristic flavor of their own. The acid itself should have a clean taste and be free of off notes that are foreign to foods. Some acids, such as succinic acid, have a distinctive taste which is incompatible with most food products; and hence, these acids have achieved very little use.

The need for tartness is obvious. Citrus and berry flavors would be flat and lifeless without at least a touch of acidity. However, not all fruit flavors require the same degree of tartness. Lemon candies and beverages are traditionally very sour, while orange and cherry are a little less tart. Flavors like strawberry, watermelon, and tropical fruits require only a trace of acidity for flavor enhancement.

In noncola carbonated beverages, beverage mixes, candies and confections, syrups and toppings, and any application where high solubility is required, citric and malic acids are used extensively. Fumaric acid is used in all ready constituted still beverages for economic reasons. Several acids are suitable for some flavors but not for others. For example, phosphoric acid is used in cola beverages but not in fruit flavored ones. Tartaric acid presents still another category. It has traditionally been used in grape flavored products even though it is suitable for other flavors.

The general purpose acids impart different degrees of tartness that are in part a result of their different acid strengths. Table 2 summarizes the tartness equivalence of the general purpose acids. The relationship shown is based only on tartness intensity and not character of flavor. This relationship can change depending on the formulation ingredients and the particular flavor system being studied. Malic acid, for example, has been claimed to be 10–15% more tart than citric acid in juice based, fruit flavored still beverages. In fruit and berry carbonated beverages, both acids have been perceived as being of equal tartness. Tartness is a difficult property to measure precisely and it must be determined by a trained and experienced taste panel.

Acids have also been used for their effects on masking undesired flavors in foods and food ingredients. Both citric and malic acids and citrate salts are known for their ability to mitigate the unpleasant aftertaste of saccharin. Gluconate salts and glucono-delta-lactone (GDL) have been patented for this function. (4, 5) Claims of enhanced benefits for malic acid over citric acid when used with the new intense sweeteners have been made but definitive advantages have not yet been demonstrated.

pH Control

Control of acidity in many food products is important for a variety of reasons. Precise pH control is important in the manufacture of jams, jellies, gelatin desserts, and pectin jellied candies in order to achieve optimum development of gel character and strength. Precise pH control is also important in the direct acidification of dairy products to achieve a smooth texture and proper curd formation. Increasing acidity enhances the activity of antimicrobial food preservatives, decreases the heat energy required for sterilization, inactivates enzymes, aids the development of cure color in processed meats, and aids the peelability of frankfurters.

Gelatin desserts are generally adjusted to an average pH of 3.5 for proper flavor and good gel strength. However, the pH can range from 3.0–4.0. Adipic and fumaric acids are used in gelatin desserts that are packaged for retail sales. Their low hygroscopicity allows use of packaging materials that are less moisture resistance and less expensive.

In jams and jellies, the firmness of pectin gel is dependent on rigid pH control. Slow set pectin attains maximum firmness at pH 3.05–3.15 while rapid set pectin reaches maximum firmness at pH 3.35–3.45. (6) The addition of buffer salts such as sodium citrate and sodium phosphate assist in maintaining the pH within the critical pH range for the pectin type. These salts also delay the onset of gelation by lowering the gelation temperature. The acid should be added as late as possible in the process. Premature acid addition will result in some pectin hydrolysis and weakening of the gel in the finished product. The acid is added as a 50% stock solution and thus soluble acids are required. Citric is generally used in this application but malic and tartaric are also satisfactory.

The United States Federal Standards of Identity (7) provide for the direct acidification of cottage cheese by the addition of phosphoric, lactic, citric, or hydrochloric acid as an alternate procedure to production with lactic acid producing bacteria. Milk is acidified to a pH of 4.5–4.7 without coagulation, and then after mixing, is heated to a maximum of 120°F without agitation to form a curd. Glucono-delta-lactone is also permitted for this application. It is added in such amounts as to reach a final pH value of 4.5–4.8 and is held until it becomes coagulated. GDL is preferred for this application because it must undergo hydrolysis to gluconic acid before it can lower pH. Thus the rate at which the pH is lowered is slowed, avoiding local denaturation.

The activity of antimicrobial agents (benzoic acid, sorbic acid, propionic acid) is due primarily to the undissociated acid molecule. Activity is therefore pH dependent and theoretical activity at any pH can be calculated. Table 3 shows the effect of pH on dissociation. It can be seen why acidification improves preservative performance and why benzoates are not generally recommended above pH 4.5.

Table 3. Effect of pH on Dissociation[a]

pH	Sorbic	Benzoic	Propionic
3	98	94	99
4	86	60	88
5	37	13	42
6	6	1.5	6.7
7	0.6	0.15	0.7
(pK$_a$)	4.67	4.19	4.87

[a] Percentage of undissociated acid.

Table 4. Canned Vegetables in which Use of Acids is Optional

Asparagus	Celery	Potatoes
Bean sprouts	Greens, collard	Rutabagas
Beans, butter	Greens, dandelion	Salsify
Beans, lima	Greens, mustard	Spinach
Beans, shelled	Kale	Sweet peppers, green
Beet greens	Mushrooms	Sweet peppers, red
Beets	Okra	Sweet potatoes
Broccoli	Onions	Swiss chard
Brussels sprouts	Parsnips	Tomatoes
Cabbage	Peas, black-eyed	Truffles
Carrots	Peas, field	Turnip greens
Cauliflower	Pimientos	Turnips

The use of acid to make heat preservation more effective, especially against spore-forming food spoilage organisms, is an established part of food technology. Under U.S. Federal Standards of Identity (8), the addition of a suitable organic acid or vinegar is required in the canning of artichokes (to reduce the pH to 4.5 or below) and is optional in the canning of the vegetables listed in Table 4. Vinegar is not permitted in mushrooms. Citric acid is specifically permitted as an optional ingredient in canned corn and canned field corn.

The advantage of acidification is especially well illustrated in the canning of whole tomatoes. When the pH of these is greater than 4.5, there is increased incidence of spoilage in the cans. When tomatoes of pH 3.9 are processed at 212°F, only 34 min are required to kill a normal or high spore load without decreases in color and flavor and deterioration of structure. In contrast, at pH 4.8 the cooking must be 110 min. (9)

In the processing of fruits and vegetables, whether for canning, freezing, or dehydration, the prevention of discoloration in the fresh cut tissue is a major concern. Reactions in which polyphenolic compounds are changed by oxidation into colored materials play an important part in this discoloration which may be accompanied by undesirable flavors. The ascorbic acid naturally present in many fruits and vegetables offers some protection, but this is of relatively short duration because of destruction of ascorbic acid by natural enzymes and air. Heating, as applied in blanching, destroys the oxidative enzymes which cause discoloration but may alter flavor and texture if continued sufficiently to completely inactivate oxidative enzymes.

Lowering pH by addition of acid substantially decreases the activity of natural color producing enzymes in fruits. Citric acid also sequesters traces of metals which may accelerate oxidation. Even greater protection is obtained by using citric acid in conjunction with a reducing agent such as ascorbic or erythorbic acid. Some processors have found that a combination of sodium erythorbate and citric acid best serves their needs.

Leavening Agent

The basis for the formulation of effervescent beverage powders, effervescent compressed tablet products, and chemically leavened baked goods is the reaction of an acidulant with a carbonate or bicarbonate resulting in the generation of carbon dioxide. The physical state of some food acids as dry solids is a property appropriate for beverage mixes and chemical leavening systems. In the absence of water, there is essentially no interaction between such acids and sodium bicarbonate. Thus these dry mixes can be stored for long periods.

A desired property for an acidulant in a chemical leavening system is that it react smoothly with the sodium bicarbonate to assure desirable volume, texture, and taste. Leavening acids and acid salts vary quantitatively in their neutralizing capacity. This relationship is shown in Table 5.

The various acids differ in their rate of reaction in response to elevation of temperature. This must be taken into consideration in selecting an acidulant for a particular condition. Under some conditions, a mixture of acidulants may be most suitable to achieve desired reaction times. Table 6 compares the reaction times of GDL and cream of tartar.

Glucono-delta-lactone is an inner ester of gluconic acid. When it hydrolyzes, gluconic acid forms and this reacts with sodium bicarbonate. Although GDL is relatively ex-

Table 5. Neutralizing Value of Various Acidulants Used in Chemical Leavening

Acid	Parts Acid to Neutralize One Part Sodium Bicarbonate
Fumaric	0.69
Glucono-delta-lactone	2.12
Cream of tartar	2.24
Sodium acid pyrophosphate	1.39
Anhydrous monocalcium phosphate	1.20
Monocalcium phosphate monohydrate	1.25
Sodium aluminum sulfate	1.00
Sodium aluminum phosphate	1.00

Table 6. Comparison of Carbon Dioxide Evolution in GDL and Cream of Tartar at Room Temperature

Time, min	GDL, %	Cream of Tartar, %
0.5	12.7	53.7
2	19.5	80.0
5	32.3	92.0
20	69.0	99.4
60	92.4	100.0

pensive, there are certain specialized types of products such as pizza dough and cake doughnuts for which it is eminently suited as an acid component of the leavening system. Cream of tartar (potassium acid tartrate) has limited solubility at lower temperatures. There is a limited evolution of gas during the initial stages of mixing in reduced temperature batters. At room temperature and above, the rate of reaction increases. Because of these characteristics, and its pleasant taste, cream of tartar is used in some baking powders and in the leavening systems of a number of baked goods and dry mixes.

Antioxidants, Sequestrants, and Synergists

Oxidation is promoted by the catalytic action of certain metallic ions present in many foods in trace quantities. If not naturally present in a food, minute quantities of these metals, particularly iron and copper, can be picked up from processing equipment. Oxidation is the cause of rancidity, an off flavor development in fat. It is also responsible for off color development that renders a food unappetizing in appearance. Hydroxy-polycarboxylic acids such as citric acid sequester these trace metals and render them unavailable for reaction. In this regard the acids function as antioxidants.

Hydroxy-polycarboxylic acids are often used in combination with antioxidants such as ascorbates or erythorbates to inhibit color and flavor deterioration caused by trace metal catalyzed oxidation. The ascorbates and erythorbates as well as BHA, BHT, and other approved antioxidants and reducing agents are oxygen scavengers and are effective when used alone. The effect of the combination of a sequestrant, such as a hydroxy-polycarboxylic acid, and an antioxidant is synergistically greater than the additive effect of either component used alone.

Citric acid is the most prominent antioxidant synergist although malic and tartaric acid have been used. In meat products, U.S. Department of Agriculture regulations permit citric acid in dry sausage (0.003%), fresh pork sausage (0.01%), and dried meats (0.01%). A short dip in a bath containing 0.25% citric and 0.25% erythorbic acid improves quality retention in frozen fish. This treatment is also applicable to shellfish to sequester iron and copper that catalyze complex blueing and darkening reactions.

Untreated fats and oils, both animal and vegetable, are likely to become rancid in storage. Oxidation is promoted by the catalytic action of certain metallic ions such as iron, nickel, manganese, cobalt, chromium, copper, and tin. Minute quantities of these metals are picked up from processing equipment. Adding citric acid to the oil sequesters these trace ions, thereby assisting antioxidants to

prevent development of off flavors. Although the oil solubility of citric acid is limited, this can be overcome by first dissolving it in propylene glycol. The antioxidant can be dissolved in the same solvent so that the two can be added in combination.

Curing Accelerator

The acids approved by the U.S. Department of Agriculture for this function in meat products (10) must be used only in combination with curing agents. In addition to ascorbates and erythorbates, the approved acids are:

1. Fumaric acid to be used at a maximum of 0.065% (1 oz–100 lb) of the weight of the meat before processing.
2. GDL to be used at 8 oz to each 100 lb of meat in cured, comminuted meat products and at 16 oz per 100 lb of meat in Genoa salami.
3. Sodium acid pyrophosphate not to exceed, alone or in combination with other curing accelerators, 8 oz per 100 lb of meat nor 0.5% in the finished product.
4. Citric acid or sodium citrate to replace up to 50% of the ascorbate or erythorbates used or in 10% solution to spray surfaces of cured cuts.

In conjunction with sodium erythorbate or related reducing compounds, GDL accelerates the rate of development of cure color in frankfurters during smoking. This permits shortening smokehouse time by one half or more and products have less shrinkage and better shelf life.

The special property of GDL upon which these advantages depend is its lactone structure at room temperature. In this form there is no free acid group and the GDL can thus be safely added during the emulsifying stage of sausage making without fear of shorting out the emulsion. Under the influence of heat in the smoking process, the ester hydrolyzes rapidly and is converted in part to gluconic acid. This lowers the pH of the emulsion during smoking, providing conditions under which sodium erythorbate or other reducing compounds (erythorbic acid, ascorbic acid, and sodium ascorbate) react with greater speed to convert the nitrite of the cure mixture into nitric oxide. The nitric oxide, in turn, acts upon the meat pigment to form the desired red nitrosomyoglobin.

CONCLUSION

The many functions and broad range of applications of food acidulants makes the selection of the most suitable acid for a food product a matter of serious concern. The physical and chemical properties of food approved acidulants must be an essential part of the knowledge of those food technologists who make the decision.

BIBLIOGRAPHY

1. E. F. Bouchard and E. G. Merritt in M. Grayson, ed., Kirk-Othmer, *Encyclopedia of Chemical Technology*, Vol. 6, 3rd ed., John Wiley & Sons, Inc., New York, 1979, pp. 150–179.
2. *Code of Federal Regulations*, Title 21 § 184.1069.
3. *Code of Federal Regulations*, Title 21 § 170.3(o).

4. U.S. Pat. 3,285,751 (Nov. 15, 1966) Artificial Sweetener Composition, P. Kracauer (to Cumberland Packing Company).

5. U.S. Pat. 3,684,529 (Aug. 15, 1972) Sweetening Compositions, J. J. Liggett (to William E. Hoerres).

6. M. Glicksman, *Gum Technology in the Food Industry*, Academic Press, Inc., Orlando, Fla., 1969, p. 179.

7. *Code of Federal Regulations*, Title 21 § 133.129.

8. *Code of Federal Regulations*, Title 21 § 155.200.

9. C. T. Townsend, cited in S. Leonard, R. M. Pangborn, and B. S. Luh, *Food Technology*, **13**, 418 (1959).

10. *Code of Federal Regulations*, Title 9 § 318.7(c)(4).

JEROME T. LIEBRAND
Pfizer Inc.
Groton, Connecticut

ACUTE TOXICITY. See TOXICOLOGY AND RISK ANALYSIS.

ADDITIVES. See EMULSIFIERS, STABILIZERS, AND THICKENERS; FOOD ADDITIVES; FOOD LAWS AND REGULATIONS; NUTRITIONAL QUALITY AND FOOD PROCESSING.

ADHESIVES. See FOOD ADDITIVES.

ADULTERATION. See CITRUS INDUSTRY; FILTH AND EXTRANEOUS MATTERS IN FOOD; FOOD LAWS AND REGULATIONS.

ADVERTISING. See FOOD MARKETING.

AERATION. See AQUACULTURE ENGINEERING AND CONSTRUCTION; YEASTS.

AFLATOXIN: ELIMINATION THROUGH BIOTECHNOLOGY

Aflatoxins are members of a diverse family of poisonous fungal metabolites known as mycotoxins (myco = fungus). The current knowledge about aflatoxins has been extensively reviewed (1). The aflatoxins, produced by two common molds, *Aspergillus flavus* and *A. parasiticus*, have received increased attention from the food industry and the general public for two main reasons. First, certain members of the aflatoxin family (aflatoxin B1) not only are extremely toxic to animals and humans, but also are the most carcinogenic of all known natural compounds. In fact, aflatoxin B1 is second in carcinogenicity only to the most carcinogenic family of chemicals known, namely, the synthetically derived polychlorinated biphenyls (PCBs). Second, aflatoxin contamination has received significant publicity, since the incidence of these compounds in food and feed is ubiquitous and has occurred throughout the United States as well as the rest of the world (2).

The metabolic process leading to aflatoxin synthesis has no obvious physiological role in primary growth and metabolism of the organism and therefore is considered to be secondary metabolism (3). As yet, there is no confirmed biological role of aflatoxins in the survival of the fungal organism. However, since aflatoxins have been shown to be toxic to certain potential competitor microbes (1) in the ecosystem, a survival benefit to the fungi producing aflatoxins is implied. Aflatoxins are also toxic to insects and animals. Theories have been proposed about the possible biological role of aflatoxins or related compounds as deterrents to insect feeding activity on fungal overwintering bodies (sclerotia) (4).

The aflatoxins causing the most severe economic losses in pre- and postharvest contamination of foods and feeds

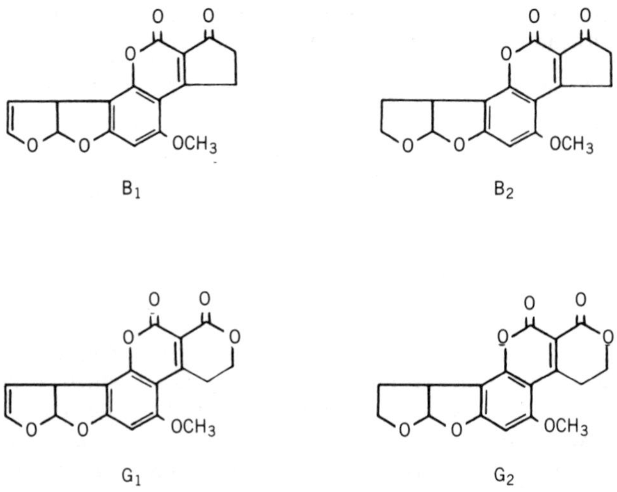

Figure 1. Structures of aflatoxins that commonly contaminate foods and feeds before harvest.

are aflatoxins B1, B2, G1, and G2 (Fig. 1). Special emphasis in this article will be on aflatoxin B1 (Figs. 1 and 2), since this compound is the most toxic and carcinogenic of the four and is also the most studied. However, it should be mentioned that there are a variety of chemical derivatives of the four basic aflatoxins (1) that can be produced synthetically using physicochemical methods or are produced after contact with enzymes in living systems, for example, the detoxifying enzymes (cytochrome P450) present in human and animal liver tissue. Also, a class of naturally occurring toxic compounds known as the sterigmatocystins (sterigmatocystin and *O*-methylsterigmatocystin, Fig. 2) are produced by several Aspergilli (5). These compounds, considered to be in the aflatoxin family, are aflatoxin precursors and are also known to contaminate foods and feeds. The aflatoxins and the sterigmatocystins, being food toxicants, are regulated in certain regions of the world, particularly in the United States and Europe. In general, procedures for eliminating aflatoxin B1 would be applicable to controlling B2, G1, and G2 and other members of the aflatoxin family that contaminate food and feed.

OCCURRENCE OF AFLATOXIN CONTAMINATION OF FOODS AND FEEDS

Aflatoxins can be produced on practically any food or feed (2) when stored improperly, for example, in grain silos. Special care by growers is necessary to achieve (before storage) and then maintain low moisture levels in the stored product, since the aflatoxin-producing molds can thrive at moisture levels lower than those required by many fungal species growing on plant material (1). However, with some effort, the grower can achieve control of aflatoxin contamination of postharvest stored food or feed by properly drying the grain or seed before storage and maintaining a low water potential in the stored product, a condition less conducive to growth of aflatoxigenic molds on the crop substrate. When aflatoxin contamination does occur, there are procedures for detoxifying these com-

polyketide   norsolorinic acid   averantin

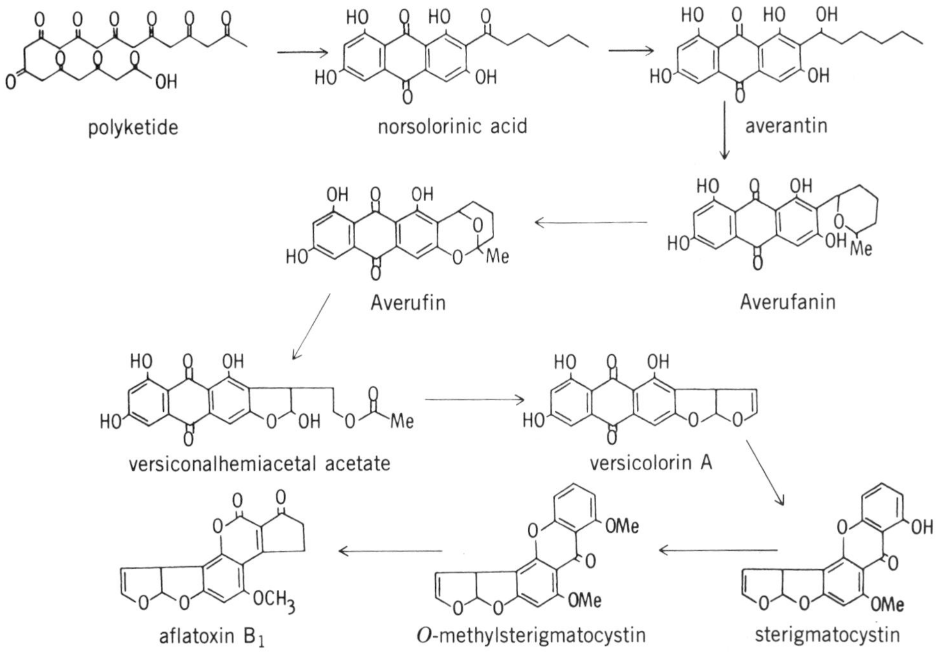

Averufin   Averufanin

versiconalhemiacetal acetate   versicolorin A

aflatoxin B$_1$ O-methylsterigmatocystin   sterigmatocystin

Figure 2. Aflatoxin biosynthetic pathway.

pounds by an ammoniation procedure (6). Ammoniation has proven to be effective in destroying aflatoxins in cottonseed and corn. However, the procedure adds to the production costs of the grower, and, more importantly, the by-products of ammoniation are still under investigation for any potentially harmful effects to health. If the ammoniation procedure proves to be safe for use in detoxification of aflatoxin, the procedure would make a larger proportion of aflatoxin-contaminated crops fit for consumption, although at a somewhat higher cost to the grower or processor.

Aflatoxin contamination often occurs on crops before harvest due to growth of aflatoxin-producing molds in developing crop tissues. Therefore, methods are needed by the grower to prevent infection of crops with aflatoxin-producing fungi. Preharvest control of aflatoxin-producing fungi would not only prevent preharvest contamination of the crops but also should prevent a postharvest bloom of dormant aflatoxigenic fungi during shipment or storage of the product. Periodic droughts can result in preharvest conditions, which are largely unmanageable by the grower, that are conducive to aflatoxigenic mold growth on the crops; the drought of 1988 resulted in extensive losses in the corn crop due to a major aflatoxin outbreak (7). Since preharvest losses from aflatoxin contamination are the least manageable by the grower compared to postharvest losses, the discussion in this article will deal primarily with known (conventional) and hypothetical (novel) methods that can be utilized in controlling preharvest aflatoxin contamination.

REASONS FOR PREHARVEST CONTAMINATION

Hot, dry growing conditions are conducive to aflatoxin contamination of several crops before harvest, including cottonseed, corn, and peanut. Two theories that might explain why drought conditions promote aflatoxin con-

tamination in preharvest crops are: (1) the hot, dry environmental conditions exclude other benign fungal competitors growing epiphytically (on the surface of the plant), and/or (2) drought stress is harmful to the plant and compromises host-plant defenses against fungal attack. If environmental conditions are correct, A. flavus and A. parasiticus residing on living or dead crop material will produce millions of reproductive structures known as conidial spores, which are extremely buoyant and can be carried for miles in air currents. The fungal spores land inside natural openings or sites of insect injury on the plant, germinate to produce growing fungal mycelia, and infect developing seed where aflatoxin is subsequently produced.

CONVENTIONAL METHODS TO CONTROL PREHARVEST AFLATOXIN CONTAMINATION

Conventional methods to manage preharvest aflatoxin contamination have centered around

1. Managing drought conditions through irrigation (when possible).
2. Using pesticides to inhibit growth of aflatoxigenic molds or to reduce populations of insects causing host-plant damage and providing a portal of entry into plant tissues by aflatoxin-producing fungi.
3. The use of varieties demonstrating resistance to attack by these fungi.

The host-plant resistance approach is still in the experimental stages of development. The discovery of varieties of corn that demonstrate a partial resistance to invasion by aflatoxin-producing fungi is encouraging (8).

Conventional control methods, such as cultural practices (eg, irrigation), pesticides, and resistant crop varieties, have reduced but not eliminated preharvest aflatoxin contamination in corn and cottonseed. Growers have used insecticides to reduce insect injury of the crops and entry into injury sites by aflatoxin-producing fungi; this practice has reduced aflatoxin incidence in cottonseed and corn. However, chemical pesticides have to be used in excessive amounts to achieve even partial control. Other conventional strategies proven to be effective in reducing aflatoxin levels in crops, such as irrigation, add to the grower's expense in cultivating the crop and probably will not totally eliminate aflatoxin in any case. Through good cultural practices the grower can reduce infection of the crop by aflatoxin-producing fungi to low levels, but even these low levels of preharvest infection can lead to unacceptable levels of aflatoxins in crops; a guideline of 20 ppb (20 parts of aflatoxin per billion parts of food or feed substrate) is the maximum aflatoxin level allowed by the Food and Drug Administration (FDA) for interstate shipments of foods and feeds. Moreover, with the expected onset of more stringent guidelines throughout the world that restrict aflatoxin levels in foods (3 to 5 ppb in some parts of Europe), the development of novel methods used in conjunction with conventional methods probably will be required to solve this serious problem. Fundamental knowledge about the biochemical and genetic mechanism(s) of aflatoxin biosynthesis may suggest methods to eliminate preharvest aflatoxin contamination through new tools in biotechnology.

THE MOLECULAR BASIS OF AFLATOXIN BIOSYNTHESIS

Intensive research efforts are underway to understand the molecular basis of aflatoxin biosynthesis; the information produced by these studies will be used to devise strategies to interrupt this process. A complete understanding of the aflatoxin pathway intermediates, enzyme proteins, and genes that govern the complex process of aflatoxin biosynthesis is being sought. The regulation of a complex biosynthetic pathway can occur at any of several stages between the DNA or the gene (where the basic genetic information of the cell is stored) and the production of the final product, ie, aflatoxin (Fig. 3). In addition, there probably are several genes and gene products (enzymes), and there certainly are several chemical intermediates involved in the complex aflatoxin biosynthetic pathway (Fig. 2). Each of these genes and gene products is subject to complex gene regulatory mechanisms. Living organisms have regulatory mechanisms for controlling (through induction or derepression) the transcription of genes into mRNA, translation of the mRNA into possible biosynthetic pathway enzymes, or posttranslational modifications of proenzymes to yield the active biosynthetic enzyme. Some enzyme activities might be subject to allosteric regulation resulting from binding of intermediates produced by other steps in the biosynthetic sequence to critical binding sites on the enzyme proteins. The regulation at the gene level for aflatoxin biosynthesis is largely unknown and investigations are underway to gain information on this process.

Before any investigation on regulation of a complex biosynthetic pathway can be accomplished, the pathway intermediates must be extensively characterized and placed in proper order in the pathway. Eight of the biosyn-

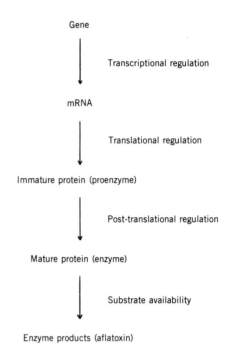

Figure 3. Potential stages of molecular regulation of aflatoxin biosynthetic enzymes.

thetic pathway steps between norsolorinic acid (NOR) and aflatoxin B1 have been determined in *A. parasiticus* (9, 10) (Fig. 2). It is known that acetate is the basic subunit in aflatoxin from experiments that demonstrated incorporation of ^{14}C-labeled acetate into aflatoxin. A multiple-step reaction sequence occurring before NOR is hypothetical and is thought to involve a chain of reactions that assemble acetate units into a polyketide backbone, based on predictions of the most likely organic reaction mechanisms (11–13). The polyketide then undergoes a variety of condensation reactions to yield the anthraquinone intermediates/pigments, norsolorinic acid (NOR), averantin (AVN), averufanin (AVNN), averufin (AVR), versiconal hemiacetal acetate (VHA), and versicolorin A (VER A) in the aflatoxin pathway (Fig. 2). The anthraquinone intermediates in the aflatoxin B1 pathway were elucidated by examining non-aflatoxin-producing mutants of *A. parasiticus* with mutational blocks at various steps in the pathway. Several of these mutants in the early studies (9, 14, 15) were found to accumulate yellow-orange compounds in the fungal mycelium, easily visible when screening colonies growing on agar plates. The pigments were isolated, structurally characterized, radiolabeled, and fed into parent aflatoxin-producing strains. The radiolabel in the pigments was shown to be incorporated into aflatoxins, thus demonstrating that the pigments were intermediates in the aflatoxin pathway. The anthraquinone intermediates were placed in their proper order in the pathway by their relative efficiencies of incorporation into aflatoxin B1. Similarly, the radiolabel from xanthone intermediates near the end of the pathway, namely, sterigmatocystin (ST) and *O*-methylsterigmatocystin (OMST), was incorporated into aflatoxins with greater efficiencies than the anthraquinones, since they are closer to the final product in the pathway and are less subject to losses from shunt pathway metabolism. In this way, the aflatoxin B1 biosynthetic pathway was established and the order of the pathway intermediates (Fig. 2) determined (9, 10, 12).

Knowledge gained on the biosynthetic steps between NOR and aflatoxin B1 has been instrumental in the detection and isolation of aflatoxin pathway enzymes; the identity of enzyme substrates and products is required before the activity of enzymes can be monitored, thus leading to their isolation and purification. The literature on the postulated aflatoxin pathway enzymes has been extensively reviewed (16). One of the pathway enzymes that was detected in an early study was shown to convert VHA to VER A (Fig. 2) (17); this discovery was significant in that the synthesis of VER A involves the formation of the bisfuran moeity known to be involved in the carcinogenicity of chemicals in the aflatoxin family (17, 18). Two aflatoxin pathway enzymes have been isolated (19), a methyltransferase (MT) and an oxidoreductase (OR), catalyzing ST to OMST and OMST to aflatoxin B1 conversions, respectively. The MT has been purified to homogeneity, extensively characterized, and used in the production of an antiserum probe for use in studying the regulation of this aflatoxin pathway enzyme; the OR has been partially purified (10).

The regulatory mechanism that triggers the biosynthetic sequence producing aflatoxin is unknown. In general, it is hypothesized that secondary metabolites are assembled from primary metabolites carried over from primary metabolic processes (3); when primary metabolism slows, these leftover primary metabolites (such as acetate) provide the building blocks for assembly of the complex heterocyclic intermediates (in the case of the aflatoxin biosynthetic pathway). It has not been well established whether the enzymes involved in the formation of aflatoxin and other secondary metabolites are carried over from primary metabolic processes or they are synthesized when needed specifically for secondary metabolic functions. Also, the mechanism of molecular regulation of these pathway enzymes at key stages between the DNA and aflatoxin synthesis (Fig. 3) is unknown. Regulatory studies of the aflatoxin pathway have been initiated (10) using previously discovered biosynthetic intermediates and the recently isolated biosynthetic enzymes as markers (19) to monitor the process of aflatoxin formation in developing A. parasiticus cultures.

Aflatoxin B1 levels increase sharply in liquid shake cultures of A. parasiticus about 2 days after inoculation of spores in the growth medium (Fig. 4), at which time mycelial growth rate declines. When the AVN-accumulating mutant blocked in aflatoxin synthesis was observed under the same culture conditions, the yellow-orange AVN pigment could be detected (visually) first in the fungal mycelium slightly earlier than two days after inoculation (Fig. 4). Thus both an aflatoxin precursor (AVN) and the final pathway product, aflatoxin B1, appear at the time when fungal growth has begun to slow down. These are, therefore, secondary metabolites. When MT and OR activities were monitored in these time-course studies, they were not detected in young (1-day-old) mycelia and were observed to appear slightly before aflatoxin synthesis began; their activities peaked after about 4 days in liquid shake

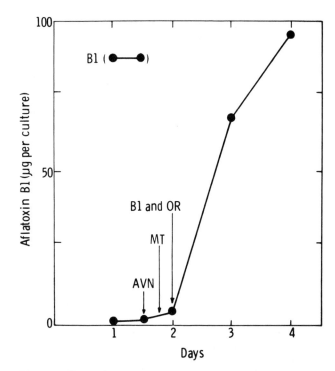

Figure 4. Time of appearance in a liquid culture of A. parasiticus of the aflatoxin precursor, averantin (AVN), aflatoxin B1 (B1) and two enzymes involved in the late stages of aflatoxin B1 synthesis, a methyltransferase (MT) and an oxidoreductase (OR). Time of first appearance of the various metabolites and enzymes is indicated by downward arrows.

culture (Fig. 4) (19). Thus MT and OR fit the predicted pattern of specific secondary metabolic enzymes since they are not detected in young, actively metabolizing fungal mycelia and they peaked well after fungal growth rate slowed. Further investigations of the MT protein have indicated that this enzyme is newly (de novo) synthesized during the late growth phase of the fungus (20). Therefore it seems likely that at least two of the aflatoxin pathway enzymes (MT and OR) are not present during or involved in primary metabolism and are induced by a presently unknown process during the initiation of aflatoxin synthesis. The induction of MT probably is not regulated by a posttranslational activation of a preexisting inactive form of MT (Fig. 3), since MT protein was detected de novo at the same time the enzyme (MT) activity was detected in the fungal mycelia (20). In addition, the appearance of aflatoxin B1 (Fig. 4) is not regulated by the availability of enzyme substrates since in some studies enzymes were supplied with substrates added exogenously to the submerged fungal cultures. Therefore the production of aflatoxin B1 is regulated by the appearance of the enzyme proteins.

It is likely that, in addition to MT and OR, several of the other enzymes in the aflatoxin pathway as well are not required in the primary metabolic processes and are therefore not required for the growth and life of the fungus. A collection of A. parasiticus mutants blocked at various steps in the aflatoxin biosynthetic pathway, ie, at the polyketide → NOR, NOR → AVN, AVNN → AVF, VER A → ST, and OMST → aflatoxin B1 conversion steps, has been studied. These mutants grow quite well in pure culture and on their natural crop substrates. Therefore, it seems unlikely that the enzymes lacking in these mutants could be involved in primary metabolic processes required for the life of the organism; otherwise the mutations outlined above would have been lethal to the organism. Other enzymes catalyzing several of the other steps in the aflatoxin biosynthetic pathway are currently being purified in several laboratories (10, 16, 17).

NOVEL METHODS TO ELIMINATE AFLATOXIN IN FOODS AND FEEDS

The fact that aflatoxin pathway enzymes appear to be disposable and not required for the life of the fungus opens the door to some intriguing possibilities for the control of aflatoxin in crops through novel methods of biotechnology. One possibility, with the extensive knowledge now available about the structures of the pathway intermediates, would be the synthesis of inhibitor analogues based on structures of the pathway metabolites. In theory, aflatoxin pathway enzyme inhibitors could be synthesized by a molecular design that affect only specific enzymes involved in aflatoxin synthesis, and not enzyme systems required for primary metabolism of organisms exposed to the inhibitor chemicals. In this way, it might be possible to design a new class of ecologically safe pesticides that are specifically inhibitory to aflatoxin synthesis and are nontoxic to plants or animals consuming the plants.

There are several plant derived, natural product inhibitors of aflatoxin synthesis (21–23), with unknown mechanisms of action, that could be subject to the same molecular design strategy for development of ecologically safe pesticides. These natural product inhibitors also could serve as markers for enhancement of aflatoxin resistance traits in plants through classical plant breeding or new molecular engineering techniques.

Yet another novel area of future research, which has the potential to eliminate aflatoxin from preharvest or even postharvest crops, involves the use of biocompetitive agents consisting of non-aflatoxin-producing fungal strains. Aggressive strains of A. flavus have been discovered that produced little or no aflatoxin during invasion of cottonseed (24, 25); this discovery implies that the aflatoxin trait is independent of the aggressiveness of the fungal strain during invasion of host-plant tissues. Recently strain aggressiveness was correlated with production of a specific plant cell wall degrading enzyme activity by A. flavus (26). Production of a pectin degrading enzyme (pectinase) was associated with aggressiveness in certain strains of A. flavus; fungal strains exhibiting low aggressiveness were lacking this pectinase, suggesting that ability to produce this enzyme may be an important trait for aggressiveness and invasion of boll tissues (26). In biocompetition studies, the aggressive nontoxigenic strains of A. flavus that produced this enzyme (26) were capable of excluding toxin-producing strains during invasion of cotton bolls (24).

Recent progress in isolation of aflatoxin pathway enzymes will allow eventual cloning of aflatoxin pathway genes through new techniques in fungal molecular biology (9, 27). Cloned aflatoxin genes, subtly modified or inactivated by genetic engineering, could be used as probes to target aflatoxin genes on the fungal chromosome and replace or inactivate them by homologous recombination; new techniques in fungal genetic engineering are now available to replace any gene of interest with homologous DNA probes (27). Thus technology now exists to bioengineer A. flavus or A. parasiticus to produce safe and stable nontoxigenic strains for potential use as biocompetitive agents.

The advantage of the biocompetitive agent strategy is that it could potentially be used on all crops with an aflatoxin problem and at relatively low cost. However, more fundamental knowledge about the molecular genetic basis for toxigenesis and competitiveness and aggressiveness in A. flavus is necessary in order to select or bioengineer fungal strains that are truly nontoxigenic and have the other desirable characteristics of a biocontrol agent. Bioengineered fungal strains of A. flavus and A. parasiticus could provide the fundamental genetic material for development of a unique group of biocontrol agents. The biocontrol strains would be unique in that they are the same species as the aflatoxigenic fungal pests and should therefore occupy the same ecological niche as the pests the grower is trying to control. In addition, the crop pests A. parasiticus and A. flavus are uniquely amenable to development of biocontrol fungi through domestication by genetic engineering, since the undesirable aflatoxin trait apparently is not required for aggressiveness (25) and theoretically could be removed without affecting the competitiveness of the strain. An additional advantage to using members of the A. flavus or A. parasiticus group as biocompetitive agents is that these fungi typically do not colonize plant tissues to a degree that results in significant yield losses. Thus this proposed research is the first suggested utilization of domesticated biocontrol strains of fungi that are the same species as the pest, but which can be engineered to remove undesirable traits (aflatoxin production) without affecting desirable characteristics (competitiveness).

CONCLUSION

Significant knowledge is now available about the molecular basis for aflatoxin biosynthesis that could suggest novel approaches in solving the aflatoxin problem through biotechnology. Several aflatoxin pathway intermediates and enzymes have been identified and some knowledge is available about the regulation of aflatoxigenesis. Therefore the cloning of aflatoxin pathway genes is now feasible using the pathway enzymes as markers, which would provide the means to genetically engineer fungi through precision techniques to remove the undesirable trait of aflatoxin production. In addition, preliminary investigations have identified at least one fungal enzyme involved in aggressiveness during invasion of cotton bolls. Thus genes for aggressiveness could also be cloned, their activity modified by genetic engineering, and then used as tools in bioengineering of fungal strains for use in agriculture as superior biocompetitive agents to significantly reduce or eliminate aflatoxin contamination. The novel strategies outlined here will complement the conventional techniques of the modification of agroecosystems, which have resulted in only partially controlling the aflatoxin contamination process.

BIBLIOGRAPHY

1. R. W. Detroy, E. B. Lillehoj, and A. Ciegler, "Aflatoxin and Related Compounds," in A. Ciegler, S. Kadis, and S. J. Ajl, eds. *Microbial Toxins,* Vol. 6, Academic Press, New York, 1971, pp. 3–178.

2. C. F. Jelinek, A. E. Pohland, and G. E. Wood, "Review of Mycotoxin Contamination. Worldwide Occurrence of Mycotoxins in Foods and Feeds—An Update," *Journal of the Association of Official Analytical Chemists* **72**, 223–230 (1989).

3. V. S. Malik, "Genetics and Biochemistry of Secondary Fungal Metabolites," *Advances in Applied Microbiology* **28**, 27–116 (1982).

4. D. T. Wicklow and O. L. Shotwell, "Intrafungal Distribution of Aflatoxins among Conidia and Sclerotia of *Aspergillus flavus* and *Aspergillus parasiticus*," *Canadian Journal of Microbiology* **29**, 1–5 (1982).

5. R. J. Cole and R. H. Cox, "Sterigmatocystins," in *Handbook of Toxic Fungal Metabolites,* Academic Press, New York, 1981, pp. 68–77.

6. D. L. Park, L. S. Lee, R. L. Price, and A. E. Pohland, "Review of the Decontamination of Aflatoxins by Ammoniation: Current Status and Regulation," *Journal of the Association of Official Analytical Chemists* **71**, 587–596 (1988).

7. S. Kelman, "Spreading Poison: Fungus in Corn Crop, A Potent Carcinogen Invades Food Supply," *Wall Street Journal,* 1A, 10–11 (Feb. 23, 1989).

8. N. W. Widstrom, "Breeding Strategies to Control Aflatoxin Contamination of Maize through Host Plant Resistance," in M. S. Zuber, E. B. Lillehoj, and B. L. Renfro, eds., CIMMYT, Mexico City, 1987, pp. 212–220.

9. J. W. Bennett and S. B. Christensen, "New Perspectives on Aflatoxin Biosynthesis," *Advances in Applied Microbiology* **19**, 53–92 (1983).

10. D. Bhatnagar, T. E. Cleveland, and E. B. Lillehoj, "Enzymes in Late Stages of Aflatoxin B1 Biosynthesis: Strategies for Identifying Pertinent Genes," *Mycopathologia* **107**, 75–83 (1989).

11. J. D. Bu'Lock, "Polyketide Biosynthesis," in E. Haslam, ed., *Comprehensive Organic Chemistry,* Vol. 5, *The Synthesis and Reactions of Organic Compounds,* Pergamon Press, Oxford, 1979, pp. 927–987.

12. K. K. Maggon, S. K. Gupta, and T. A. Venkitasubramanian, "Biosynthesis of Aflatoxins," *Bacteriological Reviews* **41**, 822–855 (1977).

13. C. A. Townsend, "Progress towards a Biosynthetic Rationale of the Aflatoxin Pathway," *Pure and Applied Chemistry* **58**, 227–238 (1986).

14. R. Singh and D. P. H. Hsieh, "Aflatoxin Biosynthetic Pathway: Elucidation by Using Blocked Mutants of *Aspergillus parasiticus*," *Archives of Biochemistry and Biophysics* **178**, 285–292 (1977).

15. J. W. Bennett and L. S. Lee, "Mycotoxins—Their Biosynthesis in Fungi: Aflatoxins and Other Bisfuranoids," *Journal of Food Protection* **42**, 805 (1979).

16. M. F. Dutton, "Enzymes and Aflatoxin Biosynthesis," *Microbiological Reviews* **52**, 274–295 (1988).

17. N. C. Wan and D. P. H. Hsieh, "Enzymatic Formation of the Bisfuran Structure in Aflatoxin Biosynthesis," *Applied Environmental Microbiology* **39**, 109–112 (1980).

18. J. J. Dunn, L. S. Lee, and A. Ciegler, "Mutagenicity and Toxicity of Aflatoxin Precursors," *Environmental Mutagenesis* **4**, 19–26 (1982).

19. T. E. Cleveland, A. R. Lax, L. S. Lee, and D. Bhatnagar, "Appearance of Enzyme Activities Catalyzing Conversion of Sterigmatocystin to Aflatoxin B1 in Late-Growth-Phase *Aspergillus parasiticus* Cultures," *Applied Environmental Microbiology* **53**, 1711–1713 (1987).

20. T. E. Cleveland and D. Bhatnagar, "Evidence for de Novo Synthesis of an Aflatoxin Pathway Methyltransferase near the Cessation of Active Growth and Onset of Aflatoxin Biosynthesis by *Aspergillus parasiticus* Mycelia," *Canadian Journal of Microbiology* **36**, 1–5 (1990).

21. L. L. Zaika and R. L. Buchanan, "Review of Compounds Affecting the Biosynthesis of Bioregulation of Aflatoxins," *Journal of Food Protection* **50**, 691–708 (1987).

22. D. Bhatnagar and S. P. McCormick, "The Inhibitory Effect of Neem (*Azadirachta indica*) Leaf Extracts on Aflatoxin Synthesis in *Aspergillus parasiticus*," *Journal of the American Oil Chemists' Society* **65**, 1166–1168 (1988).

23. H. J. Zeringue, Jr., and S. P. McCormick, "Relationships between Cotton Leaf-Derived Volatiles and Growth of *Aspergillus flavus*," *Journal of the American Oil Chemists' Society* **66**, 581–585 (1989).

24. P. J. Cotty, "Prevention of Aflatoxin Contamination with Strains of *Aspergillus flavus*," *Phytopathology* **79**, 1153 (1989).

25. P. J. Cotty, "Virulence and Cultural Characteristics of two *Aspergillus flavus* Strains Pathogenic on Cotton," *Phypathology* **79**, 808–814 (1989).

26. T. E. Cleveland and P. J. Cotty, "Reduced Pectinase Activity of *Aspergillus flavus* Is Associated with Reduced Virulence on Cotton," *Phytopathology* **79**, 1208 (1989).

27. J. W. Bennett and L. L. Lasure, "Gene Manipulations in Fungi," Academic Press, New York, 1985, pp. 558.

THOMAS E. CLEVELAND
D. BHATNAGAR
SRRC/ARS/USDA
New Orleans, Louisiana

AGED AND PROCESSED MEAT. See MEAT PRODUCTS; MEAT AND ELECTRICAL STIMULATION; MEAT SCIENCE.

AGGLOMERATION AND AGGLOMERATOR SYSTEMS

Agglomeration is a process commonly used to obtain particles of larger size, with shorter dissolution time, better resistance to attrition, limited dusting, and more attractive appearance. It is well recognized that dry food powders are difficult to handle (segregation and flowability problems), to store (caking), and to dissolve in water (wetting and dispersion). In order to overcome these problems, blends of food ingredients may be processed in controlled temperature and/or humidity conditions for a time sufficient to form sticky particle surfaces. If the bonds are strong enough to hold particles together, an agglomerated product is obtained.

PROPERTIES OF UNAGGLOMERATED POWDERS

Numerous food powders, both one-component (coffee, sugar, milk, etc) and multicomponent systems (beverages, desserts, soups, etc), experience significant change in particle size distribution during storage, transportation, or processing. Attrition causes reduction in average particle size while aggregation is associated with formation of larger particles. Fines generated by attrition may either form clusters or coat larger particles (the latter process is called plating). While interparticle adhesion depends on particle size usually, the ratio between adhesion and weight is inversely proportional to the square of particle size (1). As a result, this ratio is two orders of magnitude higher for particles of 10 μm than for particles of 100 μm. Dry food powders with average sizes of 80–100 μm are usually free flowing, while powders having sizes below 20–30 μm become cohesive and form secondary particles (clusters) of larger size with a restricted wettability.

Adhesion without formation of bridges between the adjacent particles occurs as a result of either van der Waals or electrostatic forces; the other type of adhesion, associated with formation of bridges, is relatively stronger.

Both types of food powders, free flowing and cohesive, may undergo segregation during their storage, transportation, and handling. Primarily because of the differences in particle size and also in density, shape, and resilience, fine particles migrate to the bottom while large particles find themselves at the top of the vessel. As a result, some minor components of beverage blends (colors, flavors, vitamins) may become unevenly distributed between packages. Two processes are being used to overcome segregation effectively: wet mixing and drying with or without consequent agglomeration (2).

CAKING

Agglomeration via caking may occur even unintentionally since blends of particles are always exposed for some time to the ambient environmental conditions (temperature and/or humidity). For example, food powders which include lipids (soups, sauces, baking mixes) may undergo caking if the temperature exceeds the melting point of the lipids. As a result, sticky liquid bridges are formed. Once cooled, the lipids recrystallize, liquid bridges between particles become solid, and caking is reinforced.

While starchy and proteinaceous components are relatively insensitive to the environmental conditions, the soluble components of food powders (sugars, salts) absorb moisture and eventually change their state from solid to liquid. This transition, called the glass-transition or melting temperature, initiates caking—an undesirable type of agglomeration.

The ability of sugars to soften depends on their history (conditions under which they were produced and stored). These conditions are responsible for the formation of crystalline or amorphous structure. Amorphous sugars absorb moisture at much lower relative humidity and have lower glass-transition temperatures than crystalline sugars (3).

While a stable, crystalline structure is formed at equilibrium conditions, the amorphous one is created at non-equilibrium conditions. Relatively slow moisture withdrawal during carefully controlled crystallization (nuclei formation and crystal growth) leads to the development of a crystalline structure. Fast moisture withdrawal from a solution of carbohydrate via spray drying, roller drying, or freeze drying helps to produce mainly the amorphous form; even the mechanical impact of milling of sugar crystals produces an amorphous surface capable of recrystallization after absorbing water (4). Upon recrystallization, amorphous sucrose gives up water which facilitates formation of bridges between particles and initiates caking. To prevent caking, the glass-transition temperature may be effectively raised by the addition of high molecular weight components to the blend (5).

Stickiness of the liquified outer surface causes particles to adhere to each other, form bridges, and grow if the viscosity of this surface layer is sufficient to hold particles together. Sticky-point temperature as a function of moisture content has been measured for mixtures of sucrose and fructose and later extended to coffee and a mixture of maltodextrin, sucrose, and fructose (6, 7). Since the magnitudes of the viscosities at the sticking point were relatively constant for each powder, the mechanism of sticking and agglomeration has been interpreted as viscous flow driven by surface energy. In all cases, viscosity of the liquified powder surface is inversely proportional to sticky-point temperature which, in turn, is inversely proportion to moisture content. Addition of maltodextrin to a sucrose–fructose blend increases its glass-transition point (Fig. 1).

Uncontrolled caking may be effectively suppressed with the addition of anticaking agents (tricalcium phosphate, magnesium oxide, calcium silicate, etc) which absorb a portion of moisture from the blend (reduce amount of available moisture) and, as a result, increase its softening point. Although total moisture content of the blend with or without anticaking agent stays virtually unchanged, it is relative humidity (RH) (generated by the blend in a sealed chamber) that reflects the amount of available moisture: blend with added anticaking agent generates lower RH than blend without anticaking agent (Fig. 2). The effectiveness of the anticaking agents depends largely on their water holding capacity, so that with an unlimited source of humidity (open storage), their impact is lessened.

Even packaged food powders may undergo caking in-

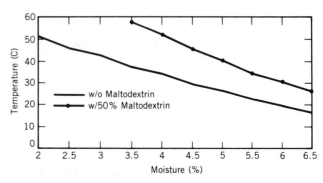

Figure 1. Sticky-point temperature vs moisture content of sucrose/fructose blends at the ratio of 7 : 1.

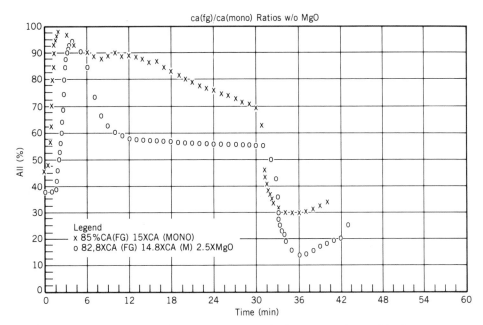

Figure 2. Relative humidity vs time in the headspace of mixer/blender.

fluenced by the environment inside their packages. Being relatively isolated, the headspace inside the package is affected not only by the surface moisture of the particles and temperature in the warehouse, but by the permeability and heat conductivity of the package film. Variations in the temperature and humidity outside of the packaged material often accelerate an exchange in surface moisture between the ingredients and initiate caking.

Humidity and temperature of the air are well correlated with the state of the particle's surface. With a relatively short time of exposure and slow diffusion of moisture from the surface of the particle to its core, the surface may absorb moisture and melt while a major portion of the particle stays virtually dry. As a result, caking and agglomeration may occur even without noticeable changes in total moisture content in the dry blend. In the summer, elevated humidity and temperature of the air correspond to higher moisture content at the surface of stored particles; agglomeration of these particles requires much shorter time and/or lower temperature than for particles processed in the winter.

Measurement of moisture content at the surface of the particles is a very difficult technical task. However, surface moisture can be easily monitored via relative humidity generated by a sample of the powder placed in a sealed vessel. Seasonal changes in the environment inevitably affect surface moisture of the particles and, in turn, make significant changes in RH generated by the powder.

PROPERTIES OF AGGLOMERATED PRODUCT

Agglomeration, in contrast to caking, is a deliberate process of particle enlargement conducted within limited time/temperature/humidity conditions. These controlled conditions may facilitate formation of liquid bridges between adjacent particles adequate to hold them together.

Humid air provides more uniform moisture distribution than the method of spraying water onto the product. After being dried, the liquid bridges solidify and form a product with the desirable size, density, friability, dissolution rate, flowability, and appearance.

Agglomeration resulting in improved wetting (penetration of liquid into the porous system due to capillary action), sinking, dispersion, and dissolution of the particles is called instantizing (8). Agglomerates with instant properties should be dissolved completely within a few seconds; it is acceptable, however, that a small amount of disintegrated particles stays suspended and disguised. Successful attempts are being made to quantify some instant properties of food particles (dried skimmed and whole milk) (9). Two instruments, one measuring light transmissions affected by the dispersed and/or residual portions of a sample and the other measuring dispersibility gravimetrically, have been used to objectively test some instant products.

Agglomerated particles may partially disintegrate when subjected to a more or less violent impact (vibration, shaking) during storage, transportation, and handling. In the case of agglomerated coffee, attrition is the primary cause in separation of fines from the outer surface of agglomerates (10). Formation of so-called secondary particles (shattering) from the disintegrated pieces is also taking place.

Particle size distribution of agglomerated products should be measured using an adequately representative sample and with minimum disruption. (care should be taken to prevent swelling, dissolution, disintegration). Well recognized methods for size analysis are sieving (dry for particles larger than 30–40 μm and wet for smaller particles); light microscopy (may be followed with image analysis) and scanning electron microscopy. Other instrumental methods include electrical impedance of particles (Elzone Counter for particles smaller than 1000 μm) and laser diffraction patterns (Malvern for particles smaller than 1800 μm). The last method is more versatile since it allows the measurement of larger particles with a dry feeder in a relatively nonviolent way, and is also applicable for fines in a dispersing liquid.

The strength or resistance of agglomerated products to attrition may be measured by comparing its size distribution before and after rotation for a limited time inside a cylinder (friabilator). Other properties of agglomerates (bulk or free flow density, flowability, cohesion, angles of spatula and repose) are usually measured with the Hosokawa (MicroPul) tester.

EXAMPLES OF AGGLOMERATED FOOD PRODUCTS

Recent developments of agglomerated (instantized) products include:

Agglomerated maltodextrin (Malta Gran) and dextrose (Unidex)—both may be used as carriers for flavors, colors, and nonnutritive sweeteners in instant beverages and desserts (11, 12).

Agglomerated soy protein isolates for high protein beverage blends and for better dispersibility in meat emulsions.

Agglomerated prejelled starches and gums as soup thickeners.

Agglomerated whey protein concentrates and calcium caseinates for dairy blends, all developed by IFT, Inc. (13).

Powders such as egg proteins, cocoa, or various fibers will have improved dispersibility after agglomeration in the presence of maltodextrin or surfactants (14).

Agglomerated food products are inseparable from some specific processes which were used to obtain these products. For example, an agglomerated instant coffee having a roasted and ground appearance was produced by a process comprising (a) milling of spray-dried instant coffee to produce a powder with an average particle size of 25–75 μm; (b) adjusting the cohesiveness of the powder so that it will flow and will bind together with slight compaction (by adding coffee oil, or colloidal particles, or controlling the electrostatic forces of the powder, increasing the moisture content, and combinations of these methods); (c) forming regularly shaped, loosely bound, structurally intact clusters with sizes of 800–2100 μm from powder; (d) fusing the outer surface of the clusters to a depth of 5–30 μm in a free-fall condition with low velocity steam; and (e) drying and screening the fused clusters to produce an agglomerated instant coffee having density of 0.20–0.28 g/cc, acceptable hardness of less than 8 units, color of 17–24 Lumetron units, and average agglomerate size of 800–1300 μm (15). This method actually employs the formation of a preagglomerated product outside of the agglomerator tower, in contrast to the well known methods which rely on collision of particles facilitated by a high velocity steam.

A method for making agglomerated bits containing aspartame includes preblending aspartame and a bulking

agent (maltodextrin) to form a premix (16). The latter is then mixed with other dry ingredients (flavors, starches, binders, dispersing agents, and vitamins) to form a dry mix. Liquid ingredients (vegetable oil and water) are blended into a dry mix with a ribbon blender or a paddle mixer to form moistened clumps. These granules must be dried in a forced-air convection oven and then screened to obtain a desirable particle size distribution. While starches (tapioca, corn, potato, modified wheat) and gums are used as the binders in forming agglomerates, baking soda and maltodextrin assist in the dispersion of the agglomerated product. These bits are suitable for use in home cooked grain cereals and other foods.

Agglomerated potato granules can be prepared from a mixture of potato granules, egg white solids, and water (17). After the wet premix is formed, a gentle sieving, drying, and crushing are used to obtain agglomerates with the desired size and density.

Agglomerated bread crumbs may be produced from a starch containing raw material (flour, meal) and water in a continuous pellet mixer (18); subsequent baking in a humidified atmosphere (to control the desired gelatinization) and sizing (cutting) of the agglomerates are utilized. A controlled retrogradation (recrystallization of starch) occurs due to the controlled cooling process.

Agglomerated beverage blends having aspartame as a sweetener may be manufactured to prevent clumping of aspartame and to improve its water dispersibility (19). Agglomeration has been conducted in a jacketed blender so that a heating or cooling fluid may be passed through the jacket while the blender is rotated to provide adequate mixing. The blending time and temperature (but not, however, humidity) were controlled to obtain a desired agglomerate size distribution.

Aqueous, sugary syrups (honey, high fructose corn syrup, invert sugar, corn syrup, etc.) were dehydrated using thin film drying in the presence of binders (soy protein and ungelatinized starch which was partially gelatinized in situ) (20). A spray of water was added during tumbling. The resultant agglomerates were dried and then slightly coated with a high melting point fat, apparently to prevent caking.

Agglomerates of garlic, onion, and their mixtures were produced in an upright chamber (21). Wetting of particles was provided by atomized water in the upper part of this chamber, followed by drying of the agglomerated product with air in the lower part of the chamber.

An agglomerated milk product was prepared by spraying a concentrate of milk into a stream of drying gas directed against the surface of a fluidized layer of already spray dried particles (22). Adjustment of temperatures, flow rates of drying air, and residence times allowed better efficiency for skim milk, whole milk, and whey particles.

Agglomerates of meat analogues were prepared by extrusion cooking of soy concentrate, comminution of the extrudate, mixing it with a water slurry of binder, frying the mixture in edible fat or oil to produce an agglomerated mat, and sizing (23). Particles with the desired size distribution were used for meat-type sauces.

A porous and pelletized food product may be formed by premixing two or more ingredients, one of which is capable of forming sticky bonds after being moistened by an aqueous medium (24). This interaction occurs when the mixture of particles is tumbled and rolled on a pelletizing disc; the adhered particles form pellets or wet aggregates. Examples of a dry mix may include sugars, starches, dried milk products, proteinaceous materials, dehydrated juices, and powdered coffee concentrates. Some of these products (sugars, starches) may become self-adherent in contact with water and are used to form agglomerates. In order to provide pellets with controlled porosity, the particulate mixture also includes a chemical leavening system (sodium bicarbonate and leavening acid). Once moist agglomerates are in contact with hot air, two processes take place: drying and formation of gaseous carbon dioxide (due to a reaction between the bicarbonate and leavening acid). The resulting pellets have a porous, cellular structure with a crisp, crunchy, and friable texture.

The vast majority of gelatin dessert mixes require the use of hot water to dissolve the gelatin and an extended time (3–4 h) to prepare the meal. However, if gelatin-containing mix with a limited moisture content of 1–3% is agitated and slowly heated to 190–195°F, it forms agglomerates (25). Subsequent cooling helps to form so-called cold-water-soluble gelatin, which dissolves and disperses in water at 40–50°F. Agglomeration of gelatin mixes (ie, sucrose, gelatin, citric acid) was conducted in a jacketed rotating blender.

All of the agglomeration methods described include several common stages: (a) adjustment of the distance between the particles through milling and/or mixing of the ingredients; (b) formation of the sticky bridges between the adjacent particles via wetting and/or heating of the blend (wet agglomeration); (c) solidification of the bridges through drying and cooling; and (d) control of size and density of agglomerates by sieving and/or crushing.

AGGLOMERATOR SYSTEMS

A well-known example of an agglomerator system is described in reference 26. This system is used to alter and control some physical properties (size, density, color) of spray-dried particles. After grinding (Fig. 3), spray-dried particles are treated in a feed port with steam to make them sticky at the surface and to form clusters. These clusters pass from the feed port into the top section of the agglomerating tower, which is supplied with hot air to facilitate the melting of wet bonds. The agglomerates are dried in the bottom portion of the tower and then cooled and screened. A critical part of the agglomeration process is in the feed port section (Fig. 4). A uniformly distributed curtain of ground powder moves downward where it interacts with jets of steam at right angles to the curtain. Steam wets the particles and fuses them into agglomerates. A typical production rate of the agglomerator system is 500–2000 lb/h of coffee with a size of 8–40 mesh.

There are five major types of equipment used for agglomeration of food and powders.

Spray Drying with Recycle

One widely utilized system of agglomerating powders is to introduce a predried material into the spray dryer along

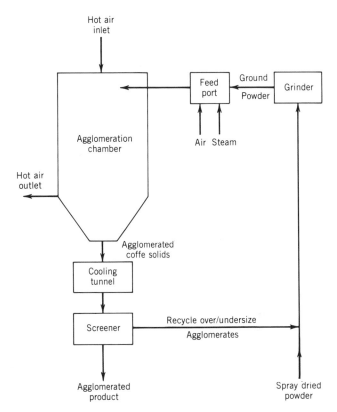

Figure 3. Steam fusion agglomeration process.

with the freshly wetted substance with which it forms an agglomerate. This system is often accompanied with a screening device or air classifier that facilitates the separation of the fine powder for the recycle. Typically the predried material in this process might contain 5–8% moisture.

Filtermat Process

The filtermat process is an adaption of spray drying with recycle in that the product is partially dried in a spray-dry portion of the machine to a moisture content of approxi-

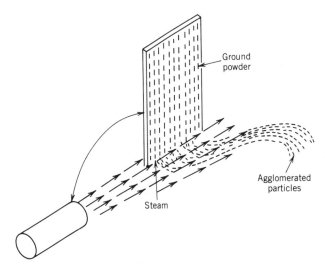

Figure 4. Agglomerator feed port.

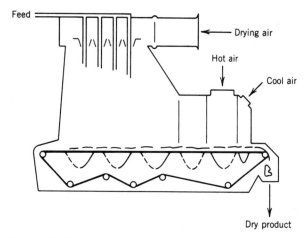

Figure 5. Filtermat dryer/agglomerator (dec international).

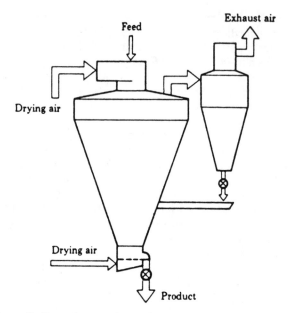

Figure 7. Spray dryer with attached fluidized bed agglomerator. Courtesy of Niro Atomizer.

mately 15–25% whereupon it is deposited by gravity onto a porous moving bed and further dried to approximately 5–10% moisture. There is sufficient moisture present in the intermediate dried product so that agglomerates are formed in the final drying process. These are then classified and/or size reduced to the desired size. A schematic of this device is shown in Figure 5 (27).

Fluidized Bed Agglomerators

Fluidized bed agglomerators such as those provided by APV Anhydro utilize either a moisture product feed or a rewet system so that sufficient moisture is present to ag-

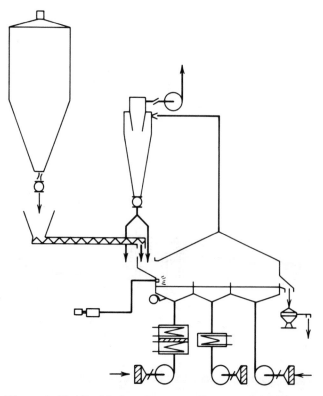

Figure 6. Fluidized bed agglomerator. Courtesy of APV Anhydro.

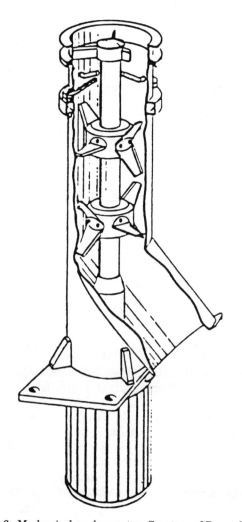

Figure 8. Mechanical agglomerator. Courtesy of Bepex Schugi.

glomerate the product. There are three fluidized zones in a typical unit. These include the entry or wetting zone, a drying zone, and a cooling zone. A schematic drawing of a fluidized bed agglomerator is shown in Figure 6 (28).

Combination Technologies

There are several combination technologies available as well. Examples of these are provided by both APV Anhydro and Niro Atomizer. This type of equipment combines spray drying with fluidized bed agglomerators in ways that are tailored to produce specific finished product properties. Partially dry product is gravity fed from the spray dryer directly to the fluidized bed bottom section where it is agglomerated with a previously produced product and dried to final moisture content in a continuously agitated area. A schematic drawing of this type agglomerator is shown in Figure 7 (28, 29).

Mechanical Agglomerators

There are also several mechanical agglomerators that utilize mechanical mixers to provide the liquid addition, product interaction, and mixing to facilitate the agglomeration process. A notable unit provided by Bepex Schugi is shown in Figure 8. This unit has extremely short residence times (approximately 1 s) and has considerable flexibility in the types and amounts of feed stock. The Schugi system also employs a flexible housing so that product build-up on the interior walls of the agglomerator is minimized, if not eliminated (30).

BIBLIOGRAPHY

1. H. Schubert, "Food Particle Technology. Part 1: Properties of Particles and Particulate Food Systems," *Journal of Food Technology* **6**, 1–32 (1987).

2. G. Barbosa-Canovas and co-workers, "Segregation in Food Powders," *Biotechnology Progress* **1**(2), 140–146 (June 1985).

3. B. Makower and W. B. Dye, "Equilibrium Moisture Content and Crystallization of Amorphous Sucrose and Glucose," *Agricultural and Food Chemistry* **4**, 72–77 (1984).

4. J. M. Flink, "Transition in Dried Carbohydrates," in M. Peleg and E. B. Bagley, eds., *Physical Properties of Foods*, AVI Publishing Co., Westport, Conn., 1983.

5. E. C. To and J. M. Flink, "Collapse," *Journal of Food Technology* **13**, 551–594 (1978).

6. G. E. Downton and co-workers, "Mechanism of Stickiness in Hygroscopic, Amorphous Powders," *Industrial and Engineering Chemistry Fundamentals* **21**, 447–451 (1982).

7. D. A. Wallack and C. J. King, "Sticking and Agglomeration of Hygroscopic, Amorphous Carbohydrate and Food Powders," *Biotechnology Progress* **4**(1), 31–35 (Mar. 1988).

8. H. Schubert, "Processing and Properties of Instant Powdered Foods," in P. Linko and others, eds., *Food Process Engineering*, Elsevier Applied Science Publishers Ltd., London, pp. 657–684, 1980.

9. H. Schubert, "Principles of Agglomeration," *International Chemical Engineering* **21**(3), 363–377 (1981).

10. J. Malave-Lopez and M. Peleg, "Patterns of Size Distribution Changes During the Attrition of Instant Coffees," *Journal of Food Science* **51**(3), 691–694 (1986).

11. "New Ingredients," *Food Technology* **43**(1), 169 (Jan. 1989).

12. "A Sweetener for Specialized Applications," *Food Engineering* **59**(9), 56 (Sept. 1986).

13. "Instantizing Ingredients," *Food Engineering* **57**(3), 58–59 (Mar. 1984).

14. D. D. Duxbury, "Innovative Ingredient Properties Enhanced, Customized by Agglomeration, Coating Process," *Food Processing* **49**(5), 96–100 (May 1988).

15. U.S. Pat. 4,594,256 (June 10, 1986), V. B. Zemelman and co-workers (to General Foods Corp.).

16. U.S. Pat. 4,741,910 (May 3, 1988), J. Karwowski and A. M. Magliacano (to Nabisco Brands, Inc.).

17. U.S. Pat. 4,797,292 (Jan. 10, 1989), J. DeWit (to Nestec S.A.).

18. U.S. Pat. 4,344,975 (Aug. 17, 1982), W. Seiler (to Gebruder Buhler AG).

19. U.S. Pat. 4,554,167 (Nov. 19, 1985), R. M. Sorge and co-workers (to General Foods Corp.).

20. U.S. Pat. 3,941,893 (Mar. 2, 1976), E. F. Glabe and co-workers (to Food Technology, Inc.).

21. U.S. Pat. 4,394,394 (Jul. 19, 1983), L. J. Nava and N. L. Ewing (to Foremost-McKesson, Inc.).

22. U.S. Pat. 4,490,403 (Dec. 25, 1984), J. Pisecky and co-workers (to A/S Niro Atomizer).

23. U.S. Pat. 4,447,461 (May 8, 1984), P. J. Loos and co-workers (to the Procter and Gamble Co.).

24. U.S. Pat. 4,310,560 (Jan. 12, 1982), R. C. Doster and S. L. Nelson (to Carnation Co.).

25. U.S. Pat. 4,571,346 (Feb. 18, 1986), D. M. Lehmann and co-workers (to General Foods Corp.).

26. S. Barnett, "Spray and Freeze Drying of Coffee Extract," *Proceedings of 4th International Symposium on Agglomeration*, Toronto, Canada, June 2–5, 1985, pp. 527–534.

27. S. Mignat, "Spray Drying and Instantization—A Survey," *Dragoco Report, Flavoring Information Service* **4**, 105–116 (1988).

28. B. Bjarekull and co-workers, APV Product Literature, APV Anhydro, Attleboro Falls, Mass.

29. O. Hicks and co-workers, Niro Atomizer Product Literature, Niro Atomizer, Columbia, Md.

30. P. Koenig and co-workers, Bepex Corp. Product Literature, Bepex Corp., Minneapolis, Minn.

Valery B. Zemelman
D. Mark Kettunen
General Foods USA
White Plains, New York

ALCOHOLIC BEVERAGES AND HUMAN RESPONSE

Alcoholic beverages are, in essence, flavored solutions of ethanol. The flavors may come from grains, as in beer; or from grapes and other fruit, as in wine; or from any source of carbohydrates, grains, sugar, or grapes, as in whiskey, rum, and brandy. In addition, consumers may add their own flavors, as lime with some beers or fruits with some wine or carbonated sodas with distilled spirits. The spectrum of flavors is wide indeed. But the purpose of drinking any of these is to supply ethanol in measured doses to the user.

Ethanol is as unique as humanity itself. It is a food but requires no digestion. It acts on many organs in the body but has no cellular receptors as do all other drugs. It is stable in the atmosphere to any chemical change, whereas all other foods will undergo some kind of decomposition. It is the only food produced solely by microbial action. It enters any cell in the body, freely, without any transport mechanism. All other foods (and all other substances except water) require a transport mechanism to enter any cell. It provides energy more rapidly than any other food.

This article inquires more closely into these and other aspects of ethanol. Although alcohol is a generic term for a large group of related substances, so common is ethanol that the term alcohol has been usurped for it and will be used here from now on to mean ethanol. So common is the drinking of alcoholic beverages that the word drink or drinker implies the drinking of alcoholic beverages and not any others.

People have always needed a release from reality. From earliest recorded history this release has come quite effectively from alcohol. It must have been discovered by accident, and probably in more than one place. It is readily produced from any *saccharous* source, is pleasant tasting, and not prone to any pathogenic divergence.

Whether alcohol appeared first from grapes as wine or from grain as beer or from honey as mead is not known. The catalyst that converts any of these into alcohol is ubiquitous. A recipe for beer has been found on a clay tablet from Mesopotamia some 4000 yr old. It was probably known during the new Stone Age, some 6000 yr ago. All but three or four of the many cultures that have survived to modern times knew alcohol. It is absent from polar people and Australian aborigines.

Probably the nature of the alcohol in any culture depended on the prevalence of the source. In cool northern Europe it was likely to be beer or mead. In the Near East it may have been beer or wine. In the Far East it was probably beer. In early cultures the making of alcohol was so cherished that it fell under the domain of the priest and clergy. Vestiges of this still remain in many monasteries in Europe.

GENERAL METABOLISM OF ALCOHOL

The first step in the metabolism of alcohol is a dehydrogenation to acetaldehyde.

$$\begin{array}{ccc} \text{H H} & & \text{H H} \\ \text{HC—C—OH} & \rightleftharpoons & \text{HC—C=O} \\ \text{H H} & & \text{H} \end{array}$$

This is mediated by the enzyme alcohol dehydrogenase, with nicotine adenine dinucleotide (NAD^+) as hydrogen acceptor. The reaction is reversible, and the reverse reaction is the last step in the process by which alcohol is produced by yeast.

This reaction is followed by the oxidation of the aldehyde to acetate, brought about by another enzyme, aldehyde dehydrogenase, again with NAD^+. This reaction has never been reversed. The acetate in turn joins with coenzyme A to form the ever-present acetyl CoA. This can take part in the citric acid cycle and be oxidized to CO_2 and H_2O. This scheme is shown in Figure 1.

Alcohol dehydrogenase (ADH) exists in about 20 forms, each with differing activity toward ethanol and to other alcohols. These isozymes vary in concentration among diverse ethnic groups, no doubt accounting for different sensitivities to alcohol by different peoples. All forms have zinc as the core metallic element. ADH is found in all tissues, including red and white blood cells and the brain. Before 1970, it was thought that ADH existed only in the liver, but that is certainly not the case. That it is present in many isosteric forms is probably rooted in the many functions it performs and the many needs it satisfies in metabolism.

Aldehyde dehydrogenase (ALDH) also exists widely in humans. Cytoplasmic ALDH is the same in all people, whereas the mitochondrial ALDH does differ among people, with that found in Asians being less active than the form found in whites. But it is probable that mitochondrial ALDH is not nearly as important in oxidizing acetaldehyde as is cytoplasmic ALDH. Because very little acetaldehyde is found circulating in the blood even after high alcohol intake, it is assumed that the rate-limiting step in alcohol metabolism is the first step—its dehydrogenation to acetaldehyde.

HOW ALCOHOL IS CONSUMED

Wine

The fermentation of the juice of grapes produces a wine containing about 12% alcohol by volume (10% by weight). The stoichiometry of fermentation,

$$\begin{array}{ccc} \text{C}_6\text{H}_{12}\text{O}_6 \rightarrow & 2\text{C}_2\text{H}_5\text{OH} & + 2\text{ CO}_2 \\ 180 & 92 & 88 \end{array}$$

dictates that a 22°Brix grape juice will give an alcohol solution of somewhat more than 10% by weight. Perhaps by evolutionary coincidence, a 10% alcohol solution is close to the limit that most yeasts can produce.

Because most countries forbid the addition of water to grape juice before fermentation, wines worldwide are very similar in alcohol content. Champagnes, which are fermented twice, may be a little higher, say 14% by volume. Fortified wines, such as port and sherry, are wines to which brandy has been added at some stage. These may contain as much as 20% alcohol by volume.

Other fruits may also be fermented to wine, but only apple juice, as cider, has found even limited acceptance.

Beer

The solution from which beer is made, called wort, is made at the brewery and not in nature. As such, its concentra-

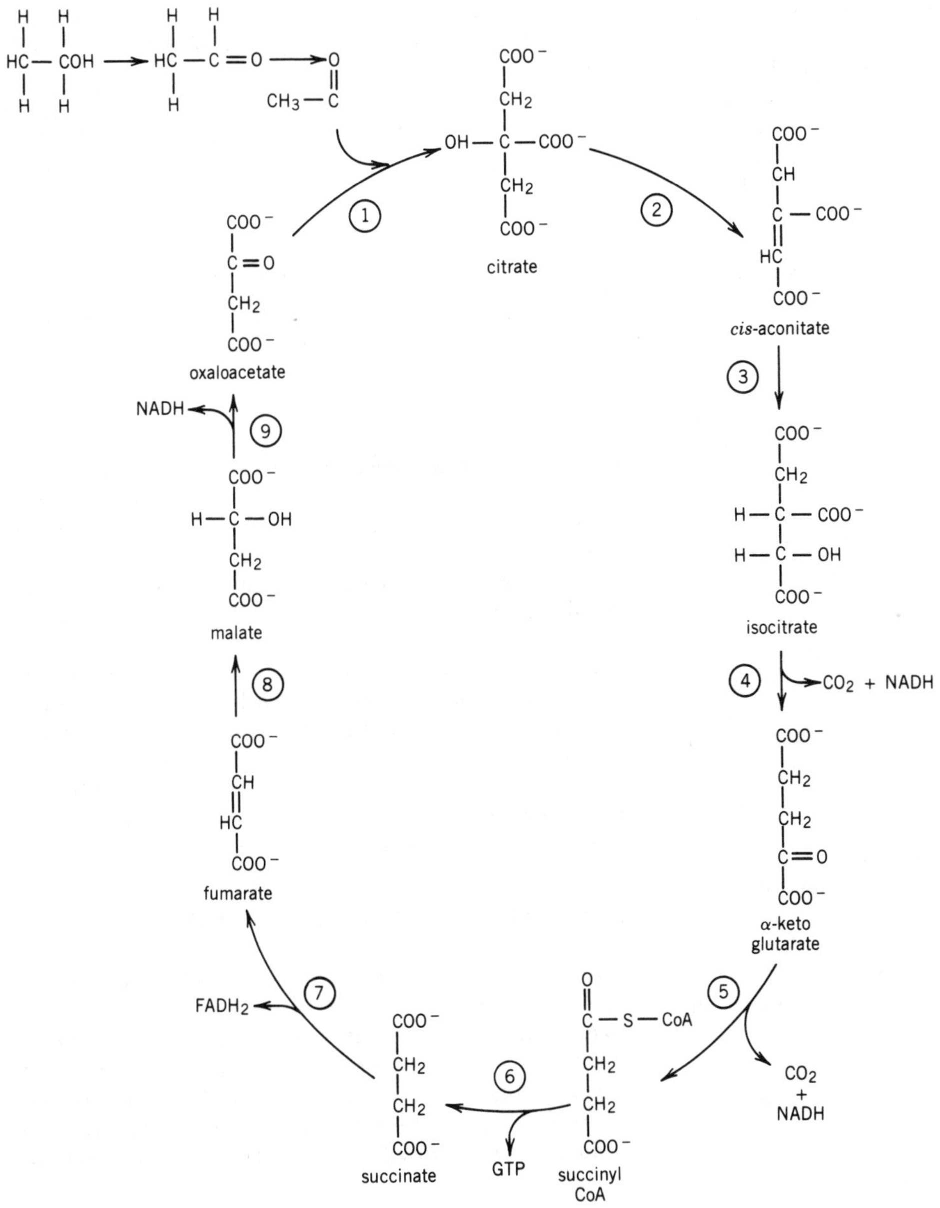

Figure 1. The citric acid cycle.

tion and fermentability (see BEER) are designed for the beer being made and may vary widely. But the concentration of the wort only varies between 12 and 18°Brix, and the fermentability between about 60% and 90%. So beer is never as high in alcohol as wine. Typically, beer contains 5% alcohol by volume (4% by weight). So-called malt liquors contain about 6% alcohol by volume.

Mead

As honey is about 82% solids, it must be diluted to about 20°Brix and then fermented. The high price of honey and the rather weak flavor of its fermented solution has not permitted mead to be more than a historical curiosity.

Distilled Spirits

These spirits are distilled at a rather high proof, be they from grains (whiskey or vodka), molasses (rum), or wine

(brandy), and are diluted to a marketable strength. The normal concentration is about 80 proof or 40% by volume or 32% by weight. Whiskeys are usually consumed after mixing with water, often carbonated, and also flavored. As purchased, wine is 2½ stronger than beer, and whiskey is 3 or more times stronger than wine.

ABSORPTION OF ALCOHOL

Alcohol is absorbed into the bloodstream mainly from the upper small intestine. Although it enters the stomach, there is only a limited exit into the blood from that organ.

Emptying of the stomach's contents is controlled by the pyloric sphincter muscle at the base of the stomach. It opens when the pH of the contents falls below 3. The presence of proteins, which act as buffers, or fats, which delay access of the stomach's enzymes and acid to their substrates, delay the fall in pH and thus also the exit of food or drink from the stomach.

All the blood that nourishes the entire digestive tract, and that in turn carries all the digested foodstuffs from the tract, first enters the liver via the portal vein. Thus alcohol absorbed from the stomach or the small intestine goes first to the liver. This probably accounts for the old and incorrect belief that alcohol is oxidized to acetaldehyde only in the liver.

Alcohol consumed at high concentrations is absorbed more rapidly than when consumed in diluted forms. The presence of carbon dioxide also accelerates the absorption. Fatigue and exercise delay the absorption.

It has been recently found that a considerable amount of alcohol is metabolized in the stomach to acetaldehyde, and that the quantity oxidized is higher in men than in women and lower in alcoholic men and women. The blood alcohol curves for alcohol administered by mouth or intravenously for nonalcoholic and alcoholic men and women are shown in Figure 2. Of course, the alcohol metabolized in the stomach can have no effect elsewhere. This so-called first-pass metabolism accounts for the more ready effect that alcohol has on women than on men.

A comparison among whiskey, wine, and beer showing the maximum blood alcohol concentrations for the three beverages is shown in Table 1. Wine and whiskey reach about the same maximum concentration, but whiskey is faster. Beer achieves a lower maximum, and it takes longer to get there.

At a lower alcohol level, the same relative relations are maintained, but the maximum concentrations attained are lower (Fig. 3).

Table 2 shows the relationship between body weight, amount of beer and whiskey to reach certain blood alcohol levels, and the time at which the maximum concentration

Table 1. Effects of Alcoholic Beverages on the Maximum Blood Alcohol Levels and the Time at Which These Levels Are Reached[a]

	c_{max}, %	t_{max}, min
Whiskey neat	0.108	57
Whiskey and water	0.107	71
Whiskey and soda water	0.104	51
Table wine	0.111	75
Beer	0.086	103

[a] Alcohol administered at a rate of 0.75 g/kg body weight. Equal to an all-at-once consumption of either four 12-oz. bottles of beer (for a 160-lb person) or two-thirds of a 750 mL bottle of 12% alcohol wine, or 3⅔ shots (1½ oz each) of 80-proof whiskey.

is reached. For example, an average person weighing 175 lb (79.5 kg) who consumes four 12-oz bottles of beer will have a maximum blood alcohol level of 0.08% in about 95 min.

Another set of experiments compared the sequential blood alcohol levels in a group of 50 men weighing between 145 and 175 lb who consumed 44 g of alcohol (four bottles of beer or four shots of whiskey) at once. The results in Figure 4 show that while whiskey produces a maximum blood alcohol level of 0.085, beer only gives a maximum of 0.045.

Fig. 5 shows the maximum blood alcohol concentration, in percent of various alcoholic beverages. The effect of dilution on whiskey action is very marked.

PHYSIOLOGICAL ACTIONS OF ALCOHOL

Alcohol is one of only two substances to which human cells have never found a need to bar entrance. The only other freely moving substance is water. Because there is no barrier to entrance, there are no known cellular receptors to alcohol. This free movement indicates that the body does not consider alcohol toxic or in any way foreign. It is this facile movement that permits alcohol to be excreted in the

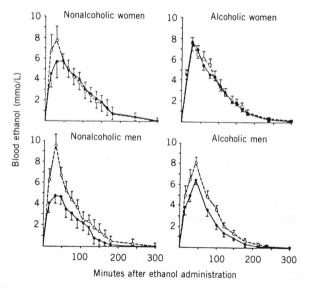

Figure 2. Effects of sex and chronic alcohol abuse on blood ethanol concentrations. Ethanol was administered orally (solid lines) or intravenously (dashed lines) in a dose of 0.3 g/body weight. The shaded area represents the difference between the curves for the two routes of administration (the first-pass metabolism).

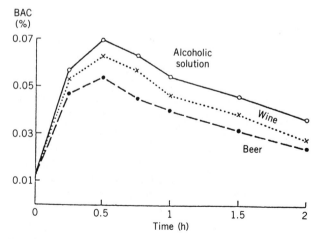

Figure 3. Blood alcohol curve after ingestion of 0.5 g alcohol per kg body weight in form of a diluted alcoholic solution (12.5%), of wine (11%), and of beer (5.5%) (mean value of 13 subjects).

Table 2.

Body Weight, lb (kg)	g Alcohol per kg wt	To Provide Given Amount of Alcohol		Time to Reach max. alc, min		Max. Blood Alcohol Level, %	
		No. of 12-oz Bottles of Beer (5%)	No. of 1-1/2-oz Shots of Whiskey (80 Proof)	Beer	Whiskey	Beer	Whiskey
120 (54.5)	0.25	<1	1	30	15	0.03	0.04
	0.50	2	2	50	35	0.06	0.075
	0.75	2-7/8	3	95	60	0.08	0.11
	1.0	3-3/8	4	150	95	0.10	0.14
	1.5	5-3/4	6	(240)	165	0.11	0.17
150 (68.2)	0.25	1-1/4	1-1/4	30	15	0.03	0.04
	0.50	2-1/2	2-1/2	50	35	0.06	0.075
	0.75	3-1/2	3-1/2	95	60	0.08	0.11
	1.0	4-3/4	5	150	95	0.10	0.14
	1.5	7-1/4	7-1/2	(240)	165	0.12	0.17
175 (79.5)	0.25	1-1/2	1-1/2	30	15	0.03	0.04
	0.50	2-4/5	3	50	35	0.06	0.075
	0.75	4	4-1/2	95	60	0.08	0.11
	1.0	5-1/2	6	150	95	0.10	0.14
	1.5	8-1/4	9	(240)	165	0.12	0.17
200 (90.9)	0.25	1-5/8	1-3/4	30	15	0.03	0.04
	0.50	3-1/4	3-1/2	50	35	0.06	0.75
	0.75	4-3/4	5	95	60	0.08	0.11
	1.0	6-1/2	6-3/4	150	95	0.10	0.14
	1.5	9-1/2	10-1/4	(240)	165	0.12	0.17

urine and in breath and also permits it to be found and measured in the breath.

There is many-faceted evidence that alcohol is a product of normal metabolism in humans; it has even been found in teetotalers. The amount is low, about 0.001%.

Effect of Alcohol on Fluid Balance

The initial effect of alcohol is to produce a slight retention of water, followed by a more marked diuresis of water and sodium. This latter effect is due to an inhibition by alcohol of the pituitary antidiuretic hormone (vasopressin). This diuresis occurs only when the concentration of blood alcohol content is rising. The concerted drinking of wine or whiskey, but not beer, requires the inhibition of other fluids to maintain homeostasis.

Effect of Alcohol on Nutrition

Alcohol is readily metabolized and yields 6.9 kcal/g. The rate of metabolism varies from person to person and ranges from 50 to 100 mg/kg of body weight per hour. Alcohol does not affect the use of dietary protein and is as effective as carbohydrates in protecting the body from loss of protein.

Alcohol does not raise heat production or CO_2 produc-

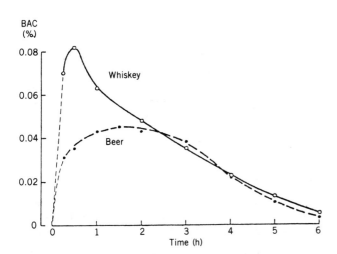

Figure 4. Blood alcohol curve after ingestion of 44 g alcohol in form of whiskey and of beer by 50 subjects, weighing between 65 and 79 kg (1).

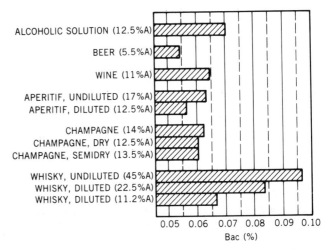

Figure 5. Blood alcohol maximum after ingestion of 0.5 g alcohol per kg body weight in form of several alcoholic beverages with different dilutions (2).

tion, so it has no thermogenic effect. There is no increase in alcohol metabolism with exercise, nor a decrease with rest.

Alcohol has only a minor effect on appetite. Distilled spirits consumed in concentrated form decrease gastric motility, whereas dilute alcohol increases gastric motion. These will appear to affect appetite.

In addition, during a meal, the palate-cleansing action of wine and beer will promote eating and lessen the feeling of satiety. It is probably the tannin in wine and the hops in beer that are responsible for this action. Yet heavy drinkers are far less likely to be obese than nondrinkers. Among a large group of low-income women in their 30s, 41% of nondrinkers were obese, whereas only 22% of heavy drinkers were so rated. Among men, the ratio is 31% obese for nondrinkers and 16% for drinkers.

Effects of Alcohol on the Central Nervous System

The primary effect of alcohol in the body is a sequential depressant action on the central nervous system. From the spinal cord rises the medulla oblongata, the pons, and the midbrain, the control centers for the autonomic nervous system, which regulates reflexes and those functions beyond control of the will. Above the midbrain lies the thalamus and hypothalamus, whose dominion includes functions not willfully regulated, such as body temperature, metabolism of fats, and blood sugar level.

Above all this, and more recent in the evolution of the species, lie the cerebellum and the cerebral cortex. The former has sway over voluntary muscular movements and manual skills, as well as balance and speech. The cerebral cortex, the corolla of the mind, is master of all that distinguishes humans and makes them distinguished.

Alcohol, which has ready access to all cells and crosses the almost impenetrable blood-brain barrier as if it were a sieve, exerts its earliest action on the cerebral cortex. Its action is not as a stimulant but as a tranquilizer or depressant. It lessens inhibitions and increases confidence. It replaces discontent with discovery, bashfulness with bravado, and cowardice with courage. Alcohol negates a no and affirms a yes.

These effects, which people had sought long before the tensions of modern life were imposed on them, occur at blood levels between 0.02 and 0.05%. Whatever pressures and problems were mankind's lot in Biblical days, they were surely different in scope and pervasiveness than those we confront today. But alcohol was used then and was considered of priestly stature. In many cultures, its production was limited to the clergy.

Apparently, the evolution of the cerebral cortex, with its awesome powers of the mind, brings with it a need to lessen its potent sway over our actions. This alcohol does with speed and thoroughness.

At blood alcohol levels between 0.05 and 0.10%, the sphere of its influence expands to the cerebellum. Balance and speech become less than normal. The gait becomes unsteady and speech uncertain. It is at the higher of these levels that ability to operate machinery is impaired.

At 0.2%, balance is more seriously affected and speech nearly incoherent. Neurotransmission in the cerebellum is disturbed. At increasing blood alcohol concentrations, the effects become deeper and parts of the brain developed earlier in evolution are also effected.

At 0.3–0.4%, most of the behavior that we call human is sharply curtailed if not lost. Anesthesia sets in. In fact, alcohol was the earliest anesthetic, before ether was discovered.

At 0.5–0.6% a remarkable shut-off valve operates and wakefulness is not maintained. This normally prevents any further action, because even the autonomic nervous system, controlling basic functions of respiration and heart beat, are denervated at a level above 0.6% and death ensues. But sleep normally prevents further inhibition, so fatal levels are not attained except with rapid drinking of strong whiskey. Fatal levels cannot be achieved with wine or beer.

Effect on the Digestive Tract

Alcohol slightly increases gastric and intestinal mobility. It also has a slight positive but very transient effect on appetite. It has no effect on the absorption of fats or cholesterol from the intestinal tract. Some of the very slight appetite stimulation may not even be due to alcohol, but to other factors present in some alcoholic beverages—juniper in gin, bitters in some aperitifs, hops in beer.

Effect on Cardiovascular System

Alcohol slightly dilates peripheral capillaries, leading to blush and a feeling of warmth. The effects are transitory and meager.

There is much clinical evidence that alcohol prevents or alleviates some attacks of angina pectoris. Increases in coronary blood flow have been reported. The lower likelihood of coronary attacks among moderate drinkers of alcohol has been repeatedly found. The mechanism for these actions is, however, unclear. Moderate drinking is taken as less than 80 g of alcohol a day (six 12-oz bottles of beer, 750 mL of wine, or 250 mL of 80-proof whiskey).

Effect of Alcohol on men vs Women

Alcohol per se has the same effects on either sex. However, a given amount of alcohol has a greater effect on women than on men for several reasons.

First, women are usually smaller than men, so a given volume of an alcoholic beverage will have a greater effect on the smaller person.

Second, women have a lower water content in their bodies than do men. Thus the alcohol, which is only soluble in water and not in fat, will reach a higher concentration in women. This difference in water content, on an equal weight basis, is about 12%.

And last, and probably most important, women have a lower gastric activity of alcohol dehydrogenase than do men. They therefore do not destroy as much alcohol in their stomach as do men, allowing more of it to be absorbed into the blood. This phenomenon, only recently discovered (1990), is known as first-pass metabolism. It may differ by 20–30% between men and women. Thus the bioavailability of alcohol is considerably greater in women than in men.

Alcohol as Medicine

Before the advent of the modern pharmacopoeia, alcohol was prescribed for many human ills. It dulls pain, summons sleep, soothes the spirit, and tames the teeming mind. Each of these functions can now be accomplished by a coterie of drugs, so alcohol is rarely recommended for these ills. But no one substance does all of what alcohol does, and people in all cultures have found alcohol to be the preferred, and possibly less noxious, means to tranquility. Furthermore, as a controlled way to peace and pleasure, alcohol has no peer.

Regulations in the United States prohibit any declaration of therapeutic value for any alcoholic beverage, so each generation must rediscover the value of alcohol. Many countries permit some therapeutic claim for alcohol, such as a bedtime relaxer or tension reducer.

As medicine, alcohol differs from drugs. All drugs are detoxified in the liver, ie, the liver so chemically modifies the drug as to render it inactive and able to be excreted, usually by the kidneys. This detoxification usually involves oxidation, hydrolysis, or sulfation. But alcohol is not detoxified, it is metabolized, just as is any foodstuff. No drug supplies energy; alcohol does. It may be more valid to consider alcohol as a food that has effects on the body rather than as a drug. As was pointed out earlier, alcohol has free entrance to every cell in the body; no drug has. All drugs operate by attaching themselves to a specific cellular receptor; alcohol needs none and has none.

Alcohol has at least one clinically proven and prescribed medicinal function. It is the only antidote to poisoning with ethylene glycol (antifreeze). Quick administration of alcohol, usually intravenously, makes all alcohol dehydrogenase sites act on the alcohol, harmlessly, instead of acting on the glycol, which produces very toxic oxalic acid. The glycol, not oxidized, is slowly excreted by the kidneys. The preferential reaction of alcohol dehydrogenase with alcohol permits this therapeutic action.

There are foods other than alcohol that affect the brain and human behavior. The amino acid tryptophane is a soporific and has been used as such. Meals high in carbohydrates produce high tryptophane blood levels. This they do by producing an influx of insulin into the blood, which in turn mobilizes some amino acids from plasma to muscle, leaving a higher concentration of tryptophane in the plasma. Histidine, threonine, tyrosine, and choline also are neuroactive, but their action is compromised by complex factors.

There are no confirmed allergies to alcohol per se. Its free entrance to all cells and ubiquitous presence mitigate against any allergic reaction.

Alcohol the Day After

The day after a serious session of drinking is often one of mental and physical fatigue and headache. In the United States, these symptoms are lumped together and called a hangover. In other languages the term is more descriptive, a wooden mouth in French, for example. The symptoms relate to the known constituents of alcoholic beverages and the known physiology of alcohol.

Alcohol masks fatigue, so a night of drinking will lead to excessive activity, which will be revealed by fatigue and malaise the next day. Rest always allows return to the normal state. The dry mouth or excessive thirst is the result of the diuretic effect of alcohol (and possibly the very salty nibbles that often accompany drinking). Drinking water or fruit juice, and a little time, alleviates this condition.

The headache may result from the fatigue and the dehydration, but may also result from the congeners of distilled spirits and of wine. The congeners are simple compounds produced by yeast concomitant with alcohol. They range from ethyl acetate to amyl alcohol. Also, particularly in some wines, and in brandy, there may be measurable amounts of methyl alcohol. These levels of methyl alcohol are not toxic per se, but when metabolized to formaldehyde may produce headaches. The drinking of a small amount of some alcoholic beverage the next day, possibly beer or vodka, will induce the liver to return to oxidizing alcohol and allow the methanol to be eliminated by the kidneys.

Alcohol and Alcoholism

Alcohol, although it effects many organs and systems, is singularly responsible for only one disease, alcoholism. Those who have this disease are called alcoholics.

An alcoholic has been defined as a person who has a compulsion to drink and cannot stop drinking once begun, to the point where that person's social and economic life are adversely affected. There is no certain cause of this disease, although a genetic factor may well be involved. An alcoholic is a heavy drinker, but not all heavy drinkers are alcoholics. Drinking because they want to characterizes drinkers, heavy or light, but drinking because they have to is the hallmark of alcoholics. In an alcoholic, drinking sets up a chain reaction, which terminates only with sleep. There is no clear evidence that an alcoholic metabolizes alcohol any differently than others, but there appears to be, in alcoholics, a strong predilection toward alcohol metabolism, rather than carbohydrate or fat metabolism, as the major energy source.

Alcoholism may lead to several other pathological conditions. It is important to understand that alcoholics may derive up to or even above 50% of their daily caloric requirements in the form of alcohol. The human system can only tolerate such univalent load with carbohydrates. Either fat or protein in such excess will lead to problems of a different nature, but none the less severe.

The major problem facing alcoholics is a series of metabolic and structural changes in the liver. The initial change is the accumulation of fats in the liver, resulting in a fatty liver. This is easily reversible. It occurs because, in the presence of alcohol, the liver uses it as a preferential source of energy and the fatty acids, which it would otherwise use, accumulate. This change is followed, after some years, by structural changes in the liver, including scarring or necrosis of the liver tissue, the mark of cirrhosis. About 20% of long-term alcoholics develop cirrhosis.

Alcohol consumption equivalent to about 50% of the daily caloric intake, for from 5 to 20 yr, is necessary to induce cirrhosis in susceptible persons. Cessation of drinking often leads to reversal of the cirrhotic condition.

The male to female ratio of alcoholics in the United States is now 4:1. Although alcohol is a necessary component in alcoholism, it is no more correct to say that alcohol is the cause of alcoholism than it would be to say that marriage is the cause of divorce. The roots of alcoholism are social, metabolic, and genetic. No medical treatment exists. The use of drugs that produce unpleasant symptoms if alcohol is consumed is not widespread because of toxic effects. Societal treatment, in groups of peers such as Alcoholics Anonymous, has proved most beneficial.

Another problem of alcoholism relates to a condition known as fetal alcohol syndrome. Pregnant women who are alcoholics may give birth to a child of subnormal weight and intelligence. This is fairly rare and occurs only if the alcoholic bouts take place at crucial times during gestation. Because any drinking may lead to a session of heavy drinking, it has become prudent to warn pregnant women to avoid alcohol.

The Salubrity of Alcohol

The vast majority of consumers of alcohol—as beer, wine, or distilled spirits—do so because it makes them feel better. It relieves stress and smooths the way from work to winding down. Not for naught has the time in a tavern after work and before the trip home come to be called the happy hour.

In addition to general relief from stress and care, moderate drinkers have been found to show genuine improvement in health. Recall that the ill effects of alcohol are confined to those who consume, regularly and for many years, alcohol in excess of 80 g a day. This amounts to almost two cases (twenty-four 12 oz packages) of beer a week, or more than a 750 mL bottle of wine a day, or six $1\frac{1}{2}$ oz shots of whiskey a day.

The consumption of moderate amounts of alcohol has been found to decrease stepwise the risk of coronary heart disease. The lowest level of consumption produces the least salutary effect, and the highest levels, more than 31 g but less than 80 g a day, reduce the risk to only 20–75% of that in nondrinkers. These results have been found in many studies, in many lands, among men and women, and among various socioeconomic strata. The Framingham Study, the longest and largest research of its kind, has always found this correlation between alcohol consumption and coronary heart disease—moderate alcohol use lowers the risk of coronary heart disease.

The recent (1989) Physicians Health Study, in which 22,000 physicians were studied prospectively, to determine the effect of aspirin on cardiovascular mortality showed incidentally that daily alcohol consumption significantly reduced the overall mortality rate over that produced by aspirin alone.

Physicians using alcohol daily and taking aspirin had a myocardial infarction rate of 1% whereas those taking aspirin alone had a 1.4% rate. This 40% improvement was among the best of any other risk factor studied. The synergistic effect of moderate alcohol consumption was not expected.

Alcohol consumption has been found to increase high density lipoproteins (HDL), which in turn has been shown to decrease atherosclerosis. It may well be that this is the major factor in the benevolent action of alcohol intake. But the relief of tension and stress may also moderate other determinants of coronary disease.

Alcohol and Drugs

Alcohol has a potentiating effect on some drugs, particularly sedatives. This may be drug specific, but also in many cases is due to the fact that the liver, where all drugs are detoxified, may prefer to metabolize alcohol and the drugs may persist in the bloodstream longer than they would normally. Repeating the drug dosage, in, say, 4 or 6 h, may lead to an overdosage if the drug has not been removed by the liver.

Those taking drugs for any chronic ailments should find out if alcohol interferes with the action or metabolism of those drugs.

BIBLIOGRAPHY

1. C. F. Gastineau, W. J. Darby, and T. B. Turner, *Fermented Food Beverages in Nutrition,* Academic Press, New York, 1979.

2. Roueche, B., *The Neutral Spirit: A Portrait of Alcohol,* Little, Brown, Boston, 1960.

General References

C. Leake and M. Silverman, *Alcoholic Beverages in Clinical Medicine,* World Publishing Co., Cleveland, 1966.

National Research Council (U.S.), *Diet and Health,* National Academy of Sciences, National Academy Press, Washington, D.C., 1989.

D. A. Roe, *Alcohol and the Diet,* AVI Publishing Co., Westport, Conn., 1979.

E. Rubin, "Alcohol and the Cell," *Annals of the New York Academy of Sciences,* **492,** p. 1–7; 181–190 (1987).

F. A. Seixas, K. Williams, S. Eggleston, "Medical Consequences of Alcoholism," *Annals of the New York Academy of Sciences,* **252,** p. 11–19; 63–78, 85–100 1975.

F. A. Seixas, S. Eggleston, "Alcohol and the Central Nervous System," *Annals of the New York Academy of Sciences,* **215,** p. 10–36; 325–330 1973.

JOSEPH L. OWADES
Consultant
Sonoma, California

ALCOHOL, POLYHYDRIC OR SUGAR

SUGAR ALCOHOLS

The sugar alcohols bear a close relationship to the simple sugars from which they are formed by reduction and from which their names are often derived. The polyols discussed here contain straight carbon chains, each carbon atom usually bearing a hydroxyl group. Also included are polyols derived from disaccharides. Most of the sugar alcohols have the general formula $HOCH_2(CHOH)_n\text{-}CH_2OH$,

$n = 2$–5. They are classified according to the number of hydroxyl groups as tetritols, pentitols, hexitols, and heptitols. Polyols from aldoses are sometimes called alditols. Each class contains stereoisomers. Counting meso and optically active forms, there are 3 tetritols, 4 pentitols, 10 hexitols, and 16 heptitols, all of which are known either from natural occurrence or through synthesis. Of the straight-chain polyols, sorbitol and mannitol have the greatest industrial significance.

Physical Properties

In general, these polyols are water-soluble, crystalline compounds with small optical rotations in water that have a slightly sweet to very sweet taste. Selected physical properties of many of the sugar alcohols are listed in Table 1.

Polymorphism has been observed for both D-mannitol and sorbitol with three different forms for each hexitol. Bond lengths of crystalline pentitols and hexitols are all similar, with an average distance for C—C of 0.152 nm and C—O of 1.43 nm. Conformations in the crystal structures of sugar alcohols are rationalized by Jeffrey's rule that the carbon chain adopts the extended, planar zigzag form when the configurations at alternate carbon centers are different, and is bent and nonplanar when they are the same. Conformations that avoid parallel C—O bonds on alternate carbon atoms are adopted. Very little, if any, intramolecular hydrogen bonding exists in the crystalline sugar alcohols, but extensive intermolecular hydrogen bonds are found, usually with involvement of each hydroxyl group in two hydrogen bonds, one as a donor and one as an acceptor.

The small optical rotations of the alditols arise from the low energy barrier for rotation about C—C bonds permitting easy interconversion and the existence of mixtures of rotational isomers (rotamers) in solution.

The weakly acidic character of acyclic polyhydric alcohols increases with the number of hydroxyl groups, as indicated by the pK_a values in aqueous solution at 18°C (Table 2). At 60°C, the pK_a value of sorbitol is 13.00.

In aqueous solution, sugar alcohols influence the structure of water, presumably by hydration of the polyol hydroxyl groups through hydrogen bonding, as indicated by effects on solution compressibility, vapor pressure, enthalpies of solution, dielectric constant, and Ag-AgCl electrode potential. Compressibility measurements indicate that mannitol in aqueous solution is hydrated with two molecules of water at 25°C. Osmotic coefficients are related to the number of hydrophilic groups per molecule, those of sorbitol being larger than those of erythritol.

Occurrence

D-Arabinitol (lyxitol) is formed in lichens, in a variety of fungi, in the urediospores of wheat stem rust, in the dried herbage of the Peruvian shrub pichi, along with D-mannitol, dulcitol, and perseitol, and in the avocado. It is formed by fermentation of glucose and in 40% yields using blackstrap molasses. Studies with ^{14}C-labeled glucose show that the yeast converts glucose C-1 to C-1 and C-5 in D-arabinitol and glucose C-2 to C-1, C-2, and C-4 of D-arabinitol. D-Arabinitol is formed by catalytic hydrogenation of D-arabinose in the presence of Raney nickel and from the γ-lactones of D-arabinonic and D-lyxonic acids by reduction with sodium borohydride. L-Arabinitol is synthesized by the reduction of L-arabinose, which is abundant in nature (see chemistry below).

D,L-Arabinitol can be prepared by the action of hydrogen peroxide in the presence of formic acid on divinyl carbinol and, together with ribitol, form D,L-erythro-4-pentyne-1,2,3-triol, $HOCH_2CHOHCHOHC{\equiv}CH$.

Xylitol is found in the primrose and in minor quantity in mushrooms. It can be obtained from glucose in 11.6% overall yield by a sequential fermentation process through D-arabinitol and D-xylulose. Xylitol is synthesized by reduction of D-xylose catalytically, electrolytically, and by sodium amalgam. D-Xylose is obtained by hydrolysis of xylan and other hemicellulosic substances obtained from such sources as wood, corn cobs, almond shells, hazelnuts, and olive waste. Isolation of xylose is not necessary; xylitol results from hydrogenation of the solution obtained by

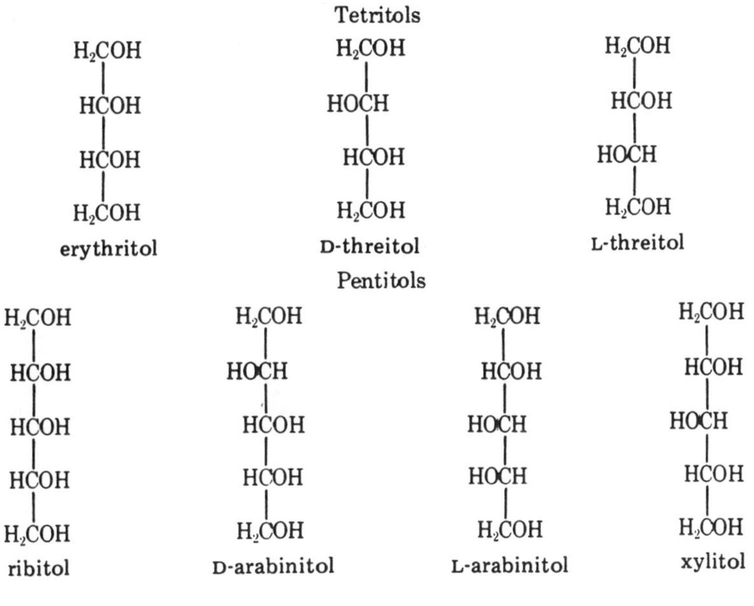

Tetritols

erythritol   D-threitol   L-threitol

Pentitols

ribitol   D-arabinitol   L-arabinitol   xylitol

Table 1.   Physical Properties of the Sugar Alcohols

Sugar alcohol	CAS Registry Number	Melting point, °C	Optical activity in H$_2$O, $[\alpha]_D^{20-25}$	Solubility g/100 g H$_2$O[a]	Heat of solution in water, kJ/mol[b]	Heat of combustion, constant vol, kJ/mol[b]
Tetritols						
erythritol	[149-32-6]	120	meso	61.5	23.3	−2091.6
threitol	[7493-90-5]					
D-threitol	[2418-52-2]	88.5–90	+4.3	very sol		
L-threitol	[2319-57-5]	88.5–90	−4.3			
D,L-threitol	[6968-16-7]	69–70				
Pentitols						
ribitol	[488-81-3]	102	meso	very sol		
arabinitol	[2152-56-9]					
D-arabinitol	[488-82-4]	103	+131[c]	very sol		
L-arabinitol	[7643-75-6]	102–103	−130[c]			−2559.4
D,L-arabinitol	[6018-27-5]	105				
xylitol	[87-99-0]	61–61.5 (metastable) 93–94.5 (stable)	meso	179		−2584.5
Hexitols						
allitol	[488-44-8]	155	meso	very sol		
dulcitol (galactitol)	[608-66-2]	189	meso	3.2 (15°C)	29.7	−3013.7
glucitol	[26566-34-7]					
sorbitol (D-glucitol)	[50-70-4]	93 (metastable) 97.7 (stable)	−1.985	235	20.2	−3025.5
L-glucitol	[6706-59-8]	89–91	+1.7			
D,L-glucitol	[60660-56-2]	135–137				
D-mannitol	[69-65-8]	166	−0.4	22	22.0	−3017.1
L-mannitol	[643-01-6]	162–163				
D,L-mannitol	[133-43-7]	168				
altritol	[5552-13-6]					
D-altritol	[17308-29-1]	88–89	+3.2	very sol		
L-altritol	[60660-58-4]	87–88	−2.9			
D,L-altritol	[60660-57-3]	95–96				
iditol	[24557-79-7]					
D-iditol	[23878-23-3]	73.5–75.0	+3.5			
L-iditol	[488-45-9]	75.7–76.7	−3.5	449		
Disaccharide alcohols						
maltitol (4-*O*-α-D-gluco-pyranosyl-D-glucitol)	[585-88-6]		+90			
lactitol (4-*O*-β-D-galacto-pyranosyl-D-glucitol)	[585-86-4]	146	+14			

[a] At 25°C unless otherwise indicated.
[b] To convert kJ/mol to kcal/mol divide by 4.184.
[c] In aqueous molybdic acid.

Table 2. Acyclic Polyhydric Alcohols: pK$_a$ Values in Aqueous Solutions at 18°C

Alcohol	pK$_a$
Glycerol	14.16
Erythritol	13.90
Xylitol	13.73
Sorbitol	13.57
Mannitol	13.50
Dulcitol	13.46

Source: See statement under bibliography.

acid hydrolysis of cottonseed hulls. Xylitol also is obtained by sodium borohydride reduction of D-xylonic and γ-lactone and from glucose by a series of tranformations through diacetone glucose.

Sorbitol (D-glucitol) was discovered initially in the fresh juice of mountain ash berries in 1872. It is found in the fruits of apples, plums, pears, cherries, dates, peaches, and apricots, in the exudate of flowers of apples, pears, and cherries, and in the leaves and bark of apples, plums, prunes, the genus *Fraxinus,* and the genus *Euonymus.* Small amounts are found in the plane tree, the African snowdrop tree, and in various algae. Because it occurs to a very small extent in grapes, assay of the sorbitol content of wine has been used to detect the adulteration with other fruit wines or apple cider. An anhydride of sorbitol, polygallitol (1,5-sorbitan), is found in the *Polygala* shrub. Sorbitol occurs in the overwintering eggs of the European red mite, affording protection against freezing.

Sorbitol is synthesized commercially by high-pressure hydrogenation of glucose, usually with a nickel catalyst. Catalyst promoters include magnesium salts, nickel phosphate, and iron. Other heterogenous catalysts used for glucose hydrogenation include cobalt, platinum, palladium, and ruthenium. Reduction of glucose to sorbitol also can be effected using ruthenium dichlorotriphenyl phosphine as a homogenous hydrogenation catalyst, preferably in the presence of a strong acid such as HCl. To form sorbitol, glucose is usually hydrogenated in the pH range of 4–8. Under alkaline conditions, glucose isomerizes to fructose and mannose; hydrogenation of the fructose and mannose yields mannitol as well as sorbitol. In addition, under alkaline conditions the Cannizzaro reaction occurs with formation of sorbitol and gluconic acid; gluconic acid formation during hydrogenation can be minimized if anion exchange resins in the basic form are the source of alkalinity.

glucose   sorbitol

Although aqueous solutions are customarily used, the monomethyl ethers of ethylene glycol or diethylene glycol also are satisfactory solvents. Electrolytic reduction of glucose formerly was used for the manufacture of sorbitol. Both the γ- and δ-lactones of D-gluconic acid may be reduced to sorbitol by sodium borohydride. Sorbitol results from simultaneous hydrolysis and hydrogenation of starch, cotton cellulose, or sucrose.

D-Mannitol is widespread in nature. It is found to a major extent in the exudates of trees and shrubs such as the plane tree (80–90%), manna ash (30–50%), and olive tree. The manna ash, *Fraxinus rotundifolis,* formerly was cultivated in Sicily for the mannitol content of its sap. Mannitol occurs in the fruit, leaves, and other parts of various plants. This hexitol is present in pumpkin, hedge parsley, onions, celery, strawberries, the genus *Euonymus,* the genus *Hebe,* the cocoa bean, grasses, lilac, *Digitalis purpurea,* mistletoe, and lichens. Mannitol occurs in marine algae, especially brown seaweed, with a seasonal variation in mannitol content that can reach more than 20% in the summer and autumn. It is found in the mycelia of many fungi and is present in the fresh mushroom to the extent of about 1.0%. There is a direct relation between mushroom yield and mannitol content. Microbial formation of D-mannitol occurs with fungi or bacteria, starting with glucose, fructose, sucrose, or the tubers of Jerusalem artichokes. The precursor of mannitol in its biosynthesis in the mushroom *Agaricus bisporus* is fructose. Mannitol is produced from glucose in 44% yield after 6 days by aerobic fermentation with *Aspergillus candidus* and in 30% yield after 10 days with *Torulopsis mannitofaciens.* It is formed by submerged culture fermentation of fructose with *Penicillium chrysogenum* in 7.3% conversion. Using sodium acetate as the sole carbon source, *Aspergillus niger* forms D-mannitol as well as D-arabinitol, erythritol, and glycerol. Small quantities of mannitol are found in wine.

Reduction of D-mannose with sodium borohydride or electrolysis leads to D-mannitol in good yield. Pure D-mannose is not yet commercially available, but it can be obtained by acid hydrolysis of the mannan of ivory nut meal in 35% yield and from spent sulfite liquor or prehydrolysis extracts from conifers through the sodium bisulfite mannose adduct or methyl α-D-mannoside. Reduction of fructose leads to sorbitol and D-mannitol in equal parts. Sucrose, on reduction under hydrolyzing conditions, yields the same products in the ratio of three parts of sorbitol to one of D-mannitol. Commercially, D-mannitol is obtained by the reduction of invert sugar. In alkaline media, glucose, fructose, and mannose are interconverted; all the mannose formed can be reduced to mannitol. Mannitol can be prepared by hydrogenation of starch hydrolyzates in alkaline media in the presence of Raney nickel. Mannitol, because of its lower solubility, usually is separated from sorbitol by crystallization from aqueous solution. The two hexitols also can be separated chromatographically on a column of calcium poly(styrenesulfonate), which preferentially retains sorbitol. Both the γ- and δ-lactones of D-mannonic acid are reduced catalytically to D-mannitol.

L-Mannitol does not occur naturally but is obtained by

the reduction of L-mannose or L-mannonic acid lactone. It can be synthesized from the relatively abundant L-arabinose through the L-mannose and L-glucose cyanohydrins; conversion to the phenylhydrazines, which are separated; liberation of L-mannose; and reduction with sodium borohydride. Another synthesis is from L-inositol (obtained from its monomethyl ether, quebrachitol) through the diacetonate, periodate oxidation to the blocked dialdehyde, reduction, and removal of the acetone-blocking groups.

D,L-Mannitol has been obtained by sodium amalgam reduction of D,L-mannose. The identical hexitol is formed from the formaldehyde polymer acrose by conversion through its osazone and osone to D,L-fructose (α-acrose) followed by reduction.

Dulcitol (galactitol) is found in red seaweed, in shrubs of the *Euonymus* genus, the genus *Hippocratea,* the genus *Adenanthera,* the physic nut, the parasitic herb *Cuscuta reflexa,* and in the mannas from a wide variety of plants. It is produced by the action of yeasts and is found in beer. D-Galactose is reduced to dulcitol catalytically, electrolytically, and chemically. Reduction of hydrolyzed lactose leads to both dulcitol and sorbitol. Dulcitol, which is relatively insoluble, is isolated by crystallization from the aqueous reduction mixture. Prehydrolysates of larch wood, which contain large quantities of galactose as well as glucose, arabinose, and xylose, can be simultaneously hydrolyzed and hydrogenated to a product containing 80% dulcitol, affording a pure product after repeated crystallization. Dulcitol is formed together with D-galactonic acid by treatment of D-galactose with alkali in the presence of Raney nickel. Reduction of D-galactonic acid γ-lactone with sodium borohydride leads to dulcitol. Photolysis of 1-deoxy-1-S-ethyl-1-thio-D-galactitol and its sulfoxide in methanol forms dulcitol. Oxidation of a mixture of meso and racemic 1,5-hexadiene-3,4-diol with hydrogen peroxide and formic acid (trans addition of hydroxyl) leads predominantly to dulcitol, with some D,L-iditol.

Maltitol (4-O-α-D-glucopyranosyl-D-glucitol), formed by catalytic hydrogenation of maltose, has been obtained as both a noncrystalline powder and a viscous liquid.

Lactitol (4-O-β-D-galactopyranosyl-D-glucitol) is obtained by sodium borohydride reduction or catalytic hydrogenation of lactose. Potentially large quantities of this sugar alcohol are available from lactose obtained from whey.

Disaccharide alcohols

maltitol

lactitol

CHEMICAL PROPERTIES

Anhydrization

The sugar alcohols can lose one or more molecules of water internally, usually under the influence of acids, to form cyclic ethers. Nomenclature is illustrated by the hexitol derivatives; monoanhydro internal ethers are called hexitans, and the dianhydro derivatives are called hexides. The main dehydration involves loss of water from the primary hydroxyl groups, forming tetrahydrofuran derivatives with the configuration of the starting alditol as shown with sorbitol. Small amounts of the 2,5-anhydrides form at the same time with inversion at the 2- or 5-positions (eg, 2,5-anhydro-L-iditol from sorbitol). Hexitols anhydrize faster than pentitols or tetritols; rates of dehydration depend on configuration, being slower where a noninvolved hydroxyl can interact with a leaving group. (see chemistry top of next page)

Loss of a second molecule of water occurs on heating sorbitol, 1,4-sorbitan, or 3,6-sorbitan with concentrated sulfuric or hydrochloric acid, forming 1,4:3,6-dianhydro-D-glucitol (isosorbide). Mannitol and iditol under similar conditions anhydrize, respectively, to isomannide and isoidide. In isomannide the hydroxyl groups are oriented toward each other (*endo*), in isoidide they are oriented away from each other (*exo*), and in isosorbide one hydroxyl is *endo* and the other *exo*. Xylitol loses two moles of water to form 1,3:2,5-dianhydroxylitol.

Esterification

Both partial and complete esters of sugar alcohols are known. The most important method for the preparation of partial fatty esters involves the interaction of polyols and fatty acids at 180–250°C. During direct esterification of the sugar alcohols, anhydrization occurs to varying degrees, depending on the conditions. Thus, esterification of sorbitol with stearic acid leads to a mixture of stearates of sorbitan and isosorbide as well as of sorbitol. Unanhydrized esters may be prepared by reaction with acid anhydrides or acid chlorides or by ester interchange reactions. In general, use of an excess of these reagents leads to esterification of all hydroxyl groups. Sorbitol hexanicotinate is prepared from the action of nicotinic acid chloride and sorbitol in the presence of pyridine. Completely substituted formate esters result from reaction of pentitols and hexitols and concentrated formic acid in the presence of phosphorus pentoxide. Primary hydroxyl groups react with esterifying reagents appreciably more rapidly than secondary hydroxyls. As a consequence, it is possible to prepare ester derivatives involving only the primary hydroxyls, as in the formation of D-mannitol 1-monolaurate and 1,6-dilaurate by reaction of D-mannitol with lauroyl chloride at 100°C. The *endo* hydroxyl of isosorbide, which is involved in intramolecular hydrogen bond formation, is more easily esterified than the *exo* hydroxyl.

Cyclic carbonates result from polyols by transesterification with organic carbonates. Thus sorbitol and diphenylcarbonate in the presence of dibutyl tin oxide at 140–150°C form sorbitol tricarbonate in quantitative yield.

Mannitol hexanitrate is obtained by nitration of mannitol with mixed nitric and sulfuric acids. Similarly, nitration of sorbitol using mixed acid produces the hexanitrate when the reaction is conducted at 0–3°C and at −10 to −75°C; the main product is sorbitol pentanitrate. Xylitol, ribitol, and L-arabinitol are converted to the pentanitrates by fuming nitric acid and acetic anhydride. Phosphate esters of sugar alcohols are obtained by the action of phosphorus oxychloride and by alcoholysis of organic phosphates. The 1,6-dibenzene sulfonate of D-mannitol is obtained by the action of benzene sulfonyl chloride in pyridine at 0°C. To obtain 1,6-dimethanesulfonyl-D-mannitol free from anhydrides and other by-products, after similar sulfonation with methane sulfonyl chloride and pyridine, the remaining hydroxyl groups are acetylated with acetic anhydride, the insoluble acetyl derivative separated, followed by deacetylation with hydrogen chloride in methanol. Alkyl sulfate esters of polyhydric alcohols result from the action of sulfur trioxide-trialkylphosphates as in the reaction of sorbitol at 34–40°C with sulfur trioxide-triethyl phosphate to form sorbitol hexa(ethylsulfate).

Etherification

The reaction of alkyl halides with sugar polyols in the presence of aqueous alkaline reagents generally results in partial etherification. Thus, a tetraallyl ether is formed on reaction of D-mannitol with allyl bromide in the presence of 20% sodium hydroxide at 75°C. Treatment of this partial ether with metallic sodium to form an alcoholate, followed by reaction with additional allyl bromide, leads to hexaallyl D-mannitol. Complete methylation of D-mannitol occurs, however, by the action of dimethyl sulfate and sodium hydroxide. A mixture of tetra- and pentabutyloxymethyl ethers of D-mannitol results from the action of butyl chloromethyl ether. Completely substituted trimethylsilyl derivatives of polyols, distillable in vacuo, are prepared by interaction with trimethylchlorosilane in the presence of pyridine. Hexavinylmannitol is obtained from D-mannitol and acetylene at 25.31 MPa (250 atm) and 160°C.

Reaction of olefin oxides (epoxides) to produce poly(oxyalkylene) ether derivatives is the polyol etherification of greatest commercial importance. Epoxides used include ethylene oxide, propylene oxide, and epichlorohydrin. The products of oxyalkylation have the same number of hydroxyl groups per mole as the starting polyol. Examples include the poly(oxypropylen) ethers of sorbitol and lactitol usually formed in the presence of an alkaline catalyst such as potassium hydroxide. Reaction of epichlorohydrin and isosorbide leads to the bisglycidyl ether. A polysubstituted carboxyethyl ether of mannitol has been obtained by the interaction of mannitol with acrylonitrile followed by hydrolysis of the intermediate cyanoethyl ether, —OH → —OCH₂CH₂COOH.

Acetyl Formation

In common with other glycols, the sugar alcohols react with aldehydes and ketones to yield cyclic acetals and ketals. Five-membered rings are formed from adjacent hydroxyls, and six-membered rings result from 1,3-hydroxyls. From the hexitols, mono-, di-, or triacetals or ketals may be obtained. Acetal formation is extensively used to protect hydroxyl groups during transformations in the polyol and carbohydrate series as conditions for the formation and removal of the cyclic acetal linkages are relatively mild and proceed without inversion of configuration at asymmetric centers.

Oxidation

Sorbitol is oxidized by fermentation with *Acetobacter suboxydans* to L-sorbose, an intermediate in the synthesis of ascorbic acid. The same organism, *Acetobacter xylinium*, and related bacteria convert erythritol to L-erythrulose, D-mannitol to D-fructose, and allitol to L-ribohexulose. These results are generalized in case of the more specific *Acetobacter suyboxydans* by the Hudson-Bertrand rules, which state that in a *cis*-glycol with a D-configuration, a secondary hydroxyl adjacent to a primary hydroxyl is oxidized to a ketone. A similar stereospecific oxidation is that of an L-secondary hydroxyl adjacent to a primary hydroxyl in a polyol by diphosphopyridine nucleotide, which is catalyzed by L-iditol dehydrogenase; sorbitol is oxidized to fructose, L-iditol to sorbose, xylitol to D-xylulose, and ribitol to D-ribulose.

Careful oxidation with aqueous bromine produces mixtures of aldoses and ketoses. Thus, sorbitol is converted to a mixture of D-glucose, D-fructose, L-glucose, and L-sorbose. Aldoses and ketoses also result from ozone oxidation of sorbitol and mannitol. Reducing sugars are formed from the action of hydrogen peroxide in the presence of ferrous salts on erythritol, D-mannitol, dulcitol, or sorbitol. Ribitol, D-arabinitol, and xylitol are oxidized by mercury (II) acetate to mixtures of 2- and 3-pentuloses. Permanganate, manganese dioxide, chromic acid, and nitric acid completely oxidize polyols to carbon dioxide.

Reduction

Sorbitol and mannitol are each converted by the action of concentrated hydriodic acid to secondary iodides (2- and 3-iodohexanes). The results of this reduction were used in early proofs of structure of glucose and fructose. Catalytic hydrogenolysis of the polyols results in breaking of both carbon-to-carbon and carbon-to-oxygen bonds. Thus, hydrogenolysis of sorbitol leads to the formation of ethylene glycol, propylene glycol, glycerol, erythritol, and xylitol. Ethylene glycol, propylene glycol, glycerol, erythritol, and monohydric alcohols are formed by the hydrogenolysis of xylitol.

Metal Complexes

The sugar alcohols form complexes in solution with most metal ions. In aqueous solution, calcium, strontium, and barium are more strongly complexed than are sodium, potassium, and magnesium. Solid magnesium complexes of sorbitol and other polyols are prepared with magnesium ethoxide; sorbitol alcoholate complexes of sodium and lithium precipitate from anhydrous ethanol. Polyol and carbohydrate complexes of alkali and alkaline earth metals have been reviewed. The stability of some polyol metal complexes is such that precipitation of the metal hydroxide from solution is inhibited. Thus, addition of sorbitol or D-mannitol to aqueous solution of sodium ferric tartrate prevents precipitation of ferric oxide. D-mannitol prevents precipitation of titanium (IV) hydroxide; at pH values above 12, precipitation of calcium, strontium, and cupric hydroxides is partially inhibited by sorbitol. Copper, iron, and cobalt all complex with cellobiitol. Tetritols, pentitols, and hexitols complex with copper in aqueous solutions of both cupric acetate and basic cupric acetate. Separation of mannitol from a mixture with dulcitol (resulting from reduction of wood sugar mixtures containing mannose and galactose) is achieved by selective complexation of dulcitol with ferric salts.

Isomerization

Isomerization of sorbitol, D-mannitol, L-iditol, and dulcitol occurs in aqueous solution in the presence of hydrogen under pressure and a nickel–kieselguhr catalyst at 130–190°C. In the case of sorbitol, D-mannitol, and L-iditol, a quasiequilibrium composition is obtained regardless of starting material, with equilibrium concentrations of 41.4% sorbitol, 31.5% D-mannitol, 26.5% L-iditol, and 0.6% dulcitol. With the same catalyst and isohexides establish an equilibrium at 220–240°C and 15.2 MPa (150 atm) of hydrogen pressure, having the composition of 57% isoidide, 36% isosorbide, and 7% isomannide.

ANALYSIS

Analytical separation of the sugar alcohols from each other and from similar materials, such as carbohydrates, is done chromatographically. Sorbitol, eg, is readily separated from glycerol, erythritol, and other polyols by chromatography on paper using butanol–water as developing solution, and on thin layers of silicic acid using butanol-acetic acid-ethyl ether–water as developing solution. Improved separations of sugar alcohols and carbohydrates have been obtained by including boric acid or phenylboronic acid in the solvent, thereby forming esters with the polyols. Paper electrophoresis of polyols and carbohydrates in molybdate, tellurate, germanate, or stannate also enables useful separations. Although polyols and carbohydrates can be separated by column chromatography on silicate absorbents, partition chromatography on ion exchange resins is of particular value for quantitative separations of these classes of materials. Use of cation exchange resins in the lithium form and anion exchange resins in the sulfate or molybdate form permit the separation of pentitols, hexitols, heptitols, and disaccharide alcohols as well as related carbohydrates. Direct gas chromatographic determination of free sugar alcohols in biological media, such as fermentation cultures, has been reported. A mixture of glycerol, erythritol, D-arabinitol, and xylitol can be separated and each polyol determined by this procedure. Erythritol and other sugar alcohols are used as stationary phases for the gas chromatographic separation of volatiles in beer and wine.

Separated polyols are detected by a variety of reagents, including ammoniacal silver nitrate, concentrated sulfuric acid, potassium permanganate, lead tetraacetate, and potassium telluratocuprate. A mixture of sodium metaperiodate and potassium permanganate can be used to detect as little as 5–8 μg of mannitol or erythritol.

Conversion to acetates, trifluoroacetates, butyl boronates, trimethylsilyl derivatives, or cyclic acetals offers a means both for identifying individual compounds and for separating mixtures of polyols chiefly by gas-liquid chromatography (GLC). Thus, sorbitol in bakery products is converted to the hexaacetate, separated, and determined by GLC with a flame ionization detector; aqueous solutions of sorbitol and mannitol are similarly separated and determined. Sorbitol may be identified by formation of its monobenzylidene derivative and mannitol by conversion to its hexaacetate.

The sugar alcohols can be determined by periodate oxidation followed by measurement of the formaldehyde and/or formic acid liberated or titration of the excess periodate. Small quantities of sorbitol in biological fluids have been determined by this method. Measurement of the heat liberated in periodate oxidation of sorbitol in foodstuffs is the basis of a thermometric determination. Sorbitol is determined in wine and vinegar by precipitation with o-chlorobenzaldehyde forming the tris(o-chlorobenzylidene) derivative. Sugar alcohols may be analyzed colorimetrically after reaction with p-hydroxybenzaldehyde or p-dimethylamino-benzaldehyde, thiourea, and concentrated sulfuric acid. After complexation with copper, sorbitol and mannitol are determined by iodometric titration of excess cupric ion. Enzymatic assays have been described for several polyols, including sorbitol, ribitol, and erythitol. Although nonspecific, one of the most valuable procedures for the quantitative analysis of polyols is the determination of hydroxyl number. This method involves reaction with acetic anhydride, followed by measurement of the acetic acid liberated.

Polarimetric analysis of sorbitol and mannitol in the presence of each other and of sugars is possible because of their enhanced optical rotation when molybdate complexes are formed and the higher rotation of the mannitol molybdate complex under conditions of low acidity. The concentration of a pure solution of sorbitol may be determined by means of the refractometer. Mass spectra of trimethylsilyl ethers of sugar alcohols provide unambiguous identification of tetritols, pentitols, and hexitols and permit determination of molecular weight.

MANUFACTURE OF SORBITOL, MANNITOL, AND XYLITOL

Sorbitol is manufactured by catalytic hydrogenation of glucose using either batch or continuous hydrogenation procedures. Corn sugar is the most important raw material, but other sources of glucose, such as hydrolyzed starch, also may be used. Both supported nickel and Raney nickel are used as catalysts. In the continuous procedure, a 50% solution of dextrose in water is prepared and transferred to a mixing tank. The catalyst, nickel on diatomaceous earth, is added to the glucose solution and the resulting slurry is pumped to the reactor after being heated to about 140°C. Hydrogen is introduced into the reactor at a pressure of approximately 12.7 MPa (125 atm) concurrently with the sugar solution. Spent catalyst is collected on a filter and is separately regenerated for reuse. The sorbitol solution is purified in two steps: (1) by passage through an ion-exchange resin bed to remove gluconate as well as other ions and (2) by treatment with activated carbon to remove trace organic impurities. The solution of pure sorbitol is concentrated in a continuous evaporator to a solution containing 70% solids, meeting the latest United States Pharmacopeia (USP) standards. Crystalline sorbitol is obtained by further concentration and crystallization. It is sold in both pelleted and powder forms.

When invert sugar is used as a starting material, sorbitol and mannitol are produced simultaneously. Mannitol crystallizes from solution after the hydrogenation step owing to its lower solubility in water. It is sold as a white, crystalline powder or free-flowing granules meeting the latest USP standards.

Extraction of mannitol from seaweed is a method of lesser importance commercially. In one method starting with this source, whole seaweed is steeped at 20°C in water that has been acidified to pH 2 with sulfuric acid. After filtration, the extract is neutralized and made alkaline with lime, precipitating calcium, and magnesium sulfates together with some colloids. The filtrate or centrifugate from this operation is dialyzed; mannitol and mineral salts pass through the semipermeable membrane. The dialyzate is concentrated, and mannitol crystallizes from solution on cooling. If the weight ratio of mannitol-to-alkali-chlorides in the dialyzate is less than 1, this ratio is attained by addition of pure mannitol before fractional crystallization is begun.

Xylose is obtained from sulfite liquors, particularly from hardwoods, such as birch, by methanol extraction of concentrates of dried sulfite lyes, ultrafiltration, and reverse osmosis, ion exchange, ion exclusion, or combinations of these treatments. Hydrogenation of xylose to xylitol is carried out in aqueous solution, usually at basic pH, with Raney nickel catalyst at a loading of 2%, at 125°C and 3.5 MPa (515 psi).

BIOLOGICAL PROPERTIES

Mannitol, sorbitol, and xylitol are accepted for use in foods by the FDA in accordance with the regulations published for each polyol.

Acute oral toxicity values in mice for mannitol and sorbitol are given in Table 3. The acute oral LD_{50} value for xylitol in mice is 25.7 g/kg. To some toxicologists, these toxicity values are classified as relatively harmless. Ingestion of 10 g/day for 1 month by a normal human subject of either mannitol or sorbitol resulted in no untoward effects. Xylitol given to healthy humans for 21 days in increasing doses up to 75 g/day produced no adverse effects. Mannitol and sorbitol have mild laxative effects. Where reasonably foreseeable consumption may result in a daily ingestion of 20 g of mannitol or 50 g of sorbitol, foods must bear the statement Excess consumption may have a laxative effect. The limiting dose of xylitol for production of diarrhea in humans is 20–30 g, but tolerance usually develops on continued administration.

Absorption of mannitol, sorbitol, and xylitol from the intestinal tract is relatively slow compared to glucose. In humans, approximately 65% of orally administered mannitol is absorbed in the dose range of 40–100 g. About one-third of the absorbed mannitol is excreted in the urine, and the remainder is oxidized to carbon dioxide. After an oral dose of 35 g of sorbitol, normal or mildly diabetic human subjects excreted 1.5–2.7% in the urine and oxidized 80–87% to carbon dioxide. Human subjects, after oral ingestion of up to 220 g of xylitol per day, excreted less than 1% of the dose in the urine, indicating efficient metabolism similar to the results of animal experiments.

Table 3. Acute Oral Toxicity of Mannitol and Sorbitol in Mice, LD_{50}, g/kg

Mice	Mannitol	Sorbitol
Male	22.2	23.2
Female	22.0	25.7

The first metabolic product from mannitol and sorbitol is fructose and from xylitol it is D-xylulose. Fructose, sorbitol, and xylitol are principally metabolized in the liver independent of insulin. Although fructose, sorbitol, and xylitol are used as glucose precursors by the liver (and the subsequent metabolism of glucose requires insulin), blood glucose concentration is increased only slightly following their oral or intravenous administration.

USES

The hexitols and their derivatives are used in many diverse fields, including foods, pharmaceuticals, cosmetics, textiles, and polymers.

Aqueous sorbitol solutions are hydroscopic and are used as humectants, softeners, and plasticizers in many different types of formulations. The hydroscopicity of sorbitol solutions is less than that of glycerol but greater than that of sugar solutions. In crystalline form, sorbitol does not absorb moisture greatly below the level of about 70% relative humidity. Above this level, sorbitol is deliquescent and will dissolve in absorbed water. Mannitol is considerably less hygroscopic in its crystalline form. Many applications of mannitol take advantage of its low hygroscopicity and its resistance to occlusion of water.

Sweetness is often an important characteristic of sugar alcohols in food and pharmaceutical applications. The property of sweetness is measured in a variety of ways with corresponding variability in ratings. Based on one or more test methods, erythritol and xylitol are similar to or sweeter than sucrose. Sorbitol is about 60% as sweet as sucrose, and mannitol, D-arabinitol, ribitol, maltitol, and lactitol are generally comparable to sorbitol.

The partial fatty acid esters of the hexitols, usually anhydrized, find extensive use in surface-active applications such as emulsification, wetting, detergency, and solubilization. Anhydrized sorbitol or mannitol moieties are versatile building blocks that confer hydrophilic properties on surfactants containing them. The less-expensive sorbitol derivatives are used more extensively than the analogous mannitol compounds. Fatty acid esters of hexitans tend to be oil soluble and to form water-in-oil (w/o) emulsions. The hydrophilic character of sorbitan fatty esters is enhanced by attachment of poly(oxyethylene) chains. As the poly(oxyethylene) chain length is increased, the tendency for water solubility increases and oil solubility decreases. Hexitan fatty esters with long poly(oxyethylene) chains tend to form o/w emulsions.

FOODS

Sorbitol imparts body and texture as well as some sweetness to frozen desserts. Sorbitol is used in frozen desserts for diabetics where its slow rate of absorption, followed by conversion to fructose in the liver, results in a prolonged, slow supply of fructose, considered to be of advantage to the diabetic. For frozen dairy desserts and mixes, FDA regulations permit sorbitol to be used as a nutritive sweetener at levels not to exceed good manufacturing practice, defined as 17%. In the manufacture of sugarless chewing gum, both mannitol and crystalline sorbitol provide the water-soluble solids. In some instances, a 70% aqueous solution of sorbitol is used in this application to provide the proper plasticity. Sorbitol resists fermentation to acids by microorganisms in the mouth and, therefore, is believed not to contribute to the incidence of dental caries. Neither sorbitol nor mannitol in 5 or 10% solution supported caries activity in rats. In chewing-gum experiments with humans lasting a year, the incidence of caries in a xylitol group was 0.33 compared to 3.76 in a sucrose group.

In artificially sweetened canned fruit, addition of sorbitol syrup provides body. Sorbitol has the property of reducing the undesirable aftertaste of saccharin. It sequesters metal ions in canned soft drinks and sequesters iron and copper ions in wines, thereby preventing cloudiness from compounds of these metals. Spray drying a solution of mannitol and acetaldehyde gives a nonhygroscopic powder useful as a flavor enhancer for fruit-flavored gelatins or beverages. D-Arabinitol has been formulated in jams and other foods as a sweetening agent of low caloric value. Maltitol and lactitol increase viscosity and confer sweetness in beverages and other foods. Carotenoids and edible fats are stabilized by sorbitol, which also prolongs the storage life of sterilized milk concentrates. Mixtures of sorbitol or mannitol and a fat, applied as an aqueous syrup to snacks and cereals, confer a rich mouthfeel and keep them crisp when immersed in milk. Both mannitol and sorbitol, at 1% concentration, inhibit the growth of *Aspergillus niger* in an intermediate moisture (15–40%) food system containing raisins, peanuts, chicken, and nonfat dry milk. Freezer burn of rapidly frozen livers is prevented or greatly reduced by dipping in 25% sorbitol solution. Incorporation of sorbitol into frankfurter meat improves color, taste, and shelf life and facilitates removal of the casings. Sorbitol, sometimes in combination with propylene glycol, texturizes pet foods by its humectant properties.

In candy manufacture, sorbitol is used in conjunction with sugars to increase shelf life. It is used in making fudge, candy cream centers, soft and grained marshmallows, and in other types of candy where softness depends on the type of crystalline structure. The function of sorbitol in this application is in retarding the solidification of sugar often associated with staleness in such candy. In butter creams, an additional benefit is involved in flavor improvement by its sequestering action on trace metals. Sorbitol may be used in diabetic chocolates. Crystalline sorbitol or crystallized blends of sorbitol and mannitol constitute a sugarless confection, as does sorbitol containing up to 5% citric acid.

Sorbitol is used as a humectant and softener in shredded coconut, having a decided advantage over the invert sugar often used, as darkening of the product does not occur. A small quantity of sorbitol, as the 70% aqueous solution, added to peanut butter has been shown to reduce dryness and crumbliness and improve spreadability. Nuts coated with a blend of mannitol and sorbitol, from an aqueous solution or slurry of the hexitols, are roasted at 177–205°C. The hexitol coating on the nuts immobilizes salt applied during cooling and will not flake off. No oxidative degradation of the hexitols occurs on prolonged heating, and there is no significant decrease in hexitol content; only traces of mannitans and sorbitans are formed.

Sorbitan fatty esters and their poly(oxyethylene) derivatives are used as shortening emulsifiers. In cakes and cake mixes, emulsification of the shortening is improved, resulting in better cake volume, texture, grain, and eating qualities. These emulsifiers are used in icings and icing bases as well as in pressure-packed synthetic cream-type toppings. In ice cream, they confer improved body and texture and provide dryness and improved aeration. Poly(oxyethylene) sorbitan monooleate and the corresponding tristearate are used as emulsifiers either separately or as blends in ice cream and other frozen desserts. Crystallization of cane sugar is accelerated, the fluidity of the crystallizing mass improved, and the sugar yield increased by addition of poly(oxyethylene) sorbitan monostearate.

PHARMACEUTICALS

Mannitol finds its principal use in pharmaceutical applications. It is used as a base in chewable, multilayer, and press-coated tablets of vitamins, antacids, aspirin, and other pharmaceuticals, sometimes in combination with sucrose or lactose. It provides a sweet taste, disintegrates smoothly, and masks the unpleasant taste of drugs such as aspirin. Tablets containing mannitol retain little moisture because of its low affinity for water.

Sorbitol finds use as a bodying agent in pharmaceutical syrups and elixirs. The use of sorbitol in cough syrups reduces the tendency of bottle caps to stick owing to crystallization of the sugar present. Enhanced stability is conferred by sorbitol in aqueous preparations of medicaments such as vitamin B_{12}, procaine penicillin, and aspirin. Stable suspensions of biologicals, such as smallpox vaccine, are obtained in a medium containing polydimethylsiloxane, mannitol, and sorbitol. Inclusion of sorbitol in aqueous suspensions of magnesium hydroxide prevents flocculation and coagulation even when subjected to freeze–thaw cycles. A gel base in which other ingredients can be incorporated to make w/o creams is produced by combining 70% sorbitol solution with a lipophilic surfactant. Crystalline sorbitol is used as an excipient where it gives a cool and pleasing taste from its endothermal heat of solution. Use of crystalline and powdered sorbitol enable preparation of troches with different degrees of hardness by direct compression. Sorbitol is used in enema solutions.

A major pharmaceutical use of poly(oxyethylene) sorbitan fatty acid esters is in the solubilization of the oil-soluble vitamins A and D. In this way, multivitamin preparations can be made that combine both water- and oil-soluble vitamins in a palatable form.

Sorbitan sesquioleate emulsions of petrolatum and wax are used as ointment vehicles in skin treatment. In topical applications, the inclusion of both sorbitan fatty esters and their poly(oxyethylene) derivatives modifies the rate of release and promotes the absorption of antibiotics, antiseptics, local anesthetics, vasoconstrictors, and other medicaments from suppositories, ointments, and lotions.

Poly(oxyethylene) sorbitan monooleate, also known as Polysorbate 80, has been used to promote absorption of ingested fats from the intestine.

Manufacture of vitamin C starts with the conversion of sorbitol to L-sorbose. Sorbitol and xylitol have been used for parenteral nutrition following severe injury, burns, or surgery. An iron-sorbitol-citric acid complex is an intramuscular hematinic. Mannitol administered intravenously and isosorbide administered orally are osmotic diuretics. Mannitol hexanitrate and isosorbide dinitrate are antianginal drugs.

BIBLIOGRAPHY

This article was adapted from "Alcohols, polyhydric (sugar)," F. R. Benson, in M. Grayson, ed., Kirk-Othmer, *Encyclopedia of Chemical Technology,* Vol. 1, 3rd ed., John Wiley & Sons, Inc., New York, 1979. Note: All specific citations have been removed. A user must refer to the original text to obtain specific references.

General References

R. W. Binkley, *Modern Carbohydrate Chemistry,* Marcel Dekker, New York, 1988.

A. T. Branen, and co-workers, *Food Additives,* Marcel Dekker, New York, 1990.

R. J. Lewis, Sr., *Food Additives Handbook,* Van Nostrand Reinhold, New York, 1989.

B. Middlekauff, and P. Shubik, *Food Regulatory Handbook,* Marcel Dekker, New York, 1990.

F. R. Senti, "Food Additives and Contaminants," in M. E. Shils and V. R. Young, eds., *Modern Nutrition in Health and Disease,* 7th ed, Lea & Febiger, Philadelphia, 1988.

R. C. Weast, Ed., *CRC Handbook of Data on Organic Compounds,* 2nd ed., CRC Press, Boca Raton, Fla., 1989.

N. R. Williams, *Carbohydrate Chemistry,* Vol. 20, Part 1, CRC Press, Boca Raton, Fla., 1988

Y. H. HUI
EDITOR-IN-CHIEF

ALIMENTARY SYSTEM. See FOOD UTILIZATION.
ALLIIN. See ONION AND GARLIC.

ALKALOIDS

CLASSIFICATION AND PROPERTIES

Alkaloids are naturally occurring substances with a particularly wide range of structures and pharmacologic activities. They may be conveniently divided into three main categories: the true alkaloids, the protoalkaloids, and the pseudoalkaloids.

The true alkaloids have the following characteristics: they show a wide range of physiological activity, are usually basic, normally contain nitrogen in a heterocyclic ring, are biosynthesized from amino acids, are of limited taxonomic distribution, and occur in the plant as the salt of an organic acid. Exceptions are colchicine, aristolichic acid, and the quaternary alkaloids. The protoalkaloids, also known as the biological amines, include mescaline and N,N-dimethyltryptamine. They are simple amines synthesized from amino acids in which the nitrogen is not in a heterocyclic ring. The pseudoalkaloids, those not derived from amino acids, include two major series of compounds: the steroidal and terpenoid alkaloids (eg, conessine) and the purines (eg, caffeine). Most alkaloids occur in the Angiosperms, the flowering plants, but they are also found in animals, insects, marine organisms, microorganisms, and the lower plants. The only common characteristic of alkaloid names is that they end in "ine" (except camptothecin).

Alkaloids are preferably grouped by their biosynthetic origin from amino acids (eg, ornithine, lysine, phenylalanine, tryptophan, histidine, and anthranilic acid) rather than their heterocyclic nucleus. The pseudoalkaloids, which are not derived from amino acids and clearly cannot be classified this way, are best organized in terms of their parent terpenoid class, eg, diterpenoid and steroidal alkaloids, or as purines or ansamacrolides.

Physical Properties

Most alkaloids are colorless crystalline solids with a defined melting point or decomposition range (eg, vindoline and morphine). Some alkaloids are amorphous gums and some are liquids (eg, nicotine and conine) and some are colored (eg, berberine is yellow and betanidine is red).

The free base of the alkaloid is normally soluble in an organic solvent; however, the quaternary bases are only water soluble, and some of the pseudo- and protoalkaloids show substantial solubility in water. The salts of most alkaloids are soluble in water.

The solubility of alkaloids and their salts is of considerable significance in the pharmaceutical industry, both in the extraction of the alkaloid from plant or fungus and in the formulation of the final pharmaceutical preparation. Solubility is also of considerable significance in the clinical distribution of an alkaloidal drug.

Chemical Properties

Most alkaloids are basic, which makes them extremely susceptible to decomposition, particularly by heat and light.

Ornithine-Derived Alkaloids

Ornithine-derived alkaloids include the tropanes (atropine, l-hyoscyamine, l-scopolamine, and cocaine), the Senecio alkaloids, and nicotine.

Tropane. Tropane alkaloids are derived from plants in the Solanaceae, Erythroxylaceae, and Convolvulaceae families. These alkaloids comprise two parts: an organic acid and an alcohol (normally a tropan-3α-ol). The pharmacologically active members of this group include atropine, the optically inactive form of l-hyoscyamine, which is isolated from deadly nightshade (*Atropa belladonna*); l-hyoscyamine and l-scopolamine, which are found in the leaves of *Duboisia metel* L., *D. meteloides* L., and *D. fastuosa* var. *alba*.

The tropane alkaloids are parasympathetic inhibitors.

For example, atropine acts through antagonism of muscarinic receptors, the receptors responsible for the slowing of the heart, constriction of the eye pupil, vasodilation, and stimulation of secretions. Atropine prevents secretions (eg, sweat, saliva, tears, and pancreas) and dilates the pupil. Atropine is used to reduce pain of renal and intestinal cholic and other gastrointestinal tract disorders, to prolong mydriasis when necessary, and as an antidote to poisoning by cholinesterase inhibitors. Small doses produce respiratory and myocardial stimulation and decrease nasal secretion, and the drug has little local anesthetic action. Hyoscyamine and scopolamine have mydriatic effects. They are also used in combination as sedatives, in anti-motion-sickness drugs, and in antiperspirant preparations.

Cocaine. Cocaine is a potent central nervous system (CNS) stimulant and adrenergic blocking agent. It is extracted from South American coca leaves or prepared by converting ester alkaloids to exgonine, followed by methylation and benzoylation. It is too toxic to be used as an anesthetic by injection, but the hydrochloride is used as a topical anesthetic. It has served as a model for a tremendous synthetic effort to produce an anesthetic of increased stability and reduced toxicity.

cocaine

Senecio Alkaloids. Senecio alkaloids possess a pyrrolizidine nucleus and occur in the genera Senecio (Compositae), Heliotropium and Trichodesma, and Crotalaria. They are biosynthetically derived from two units of ornithine in a manner similar to some of the lupin alkaloids. Certain of the alkaloids having an unsaturated nucleus are potent hepatotoxins.

Nicotine. Nicotine is toxic, soluble in water, and a constituent of tobacco. The lethal human dose is ca 40 mg/kg. Pharmacologically, there is an initial stimulation followed by depression and paralysis of the autonomic ganglia. The biosynthesis of nicotine is well established.

nicotine

Lysine-Derived Alkaloids

Lysine-derived alkaloids contain the pyridine nucleus or its reduced form, piperidine. They include alkaloids derived from the Areca or Betel nut, Lobelia alkaloids, and those derived from pomegranate or club mosses.

Arecoline, a colorless liquid alkaloid that has a pronounced stimulant action (in large doses, paralysis may occur) is found in the Areca or Betal nut. As the hydrobromide it is used as a diaphoretic and anthelmintic.

Lobelia inflata, known as Indian tobacco, contains lobeline, which is similar to, but less potent than, nicotine in pharmacologic action and is used as an emetic; the sulfate salt is used in antismoking tablets.

The root of *Punica granatum* contains alkaloids such as pelletierine and pseudopelletierine, which are formed from lysine and acetate. Pelletierine is toxic to tapeworms and is used as an anthelmintic.

The club mosses, *Lycopodium* spp., produce polycyclic alkaloids such as lycopodine, whereas *Hydrangea* spp. yield febrifugine, an active antimalarial agent. Anabasine is found in *Haloxylon persicum* Bunge; this alkaloid has antismoking and respiratory muscle stimulation action similar to lobeline and is also used as a metal anticorrosive.

A host of complex alkaloids such as lupinine, sparteine, cytisine, and matrine are found in the lupins, a large plant family of the Leguminosae. Sparteine paralyzes motor nerve endings and sympathetic ganglia. The sulfate is used as an oxytoxic and the adenylate derivative is used to treat cardiac insufficiencies. Cytisine, found in the seeds of the highly toxic plant *Cytisus laburnum* L., is a strongly basic alkaloid that produces convulsions and death by respiratory failure.

pelletierine   **lupinine**

Anthranilic Acid-Derived Alkaloids

Anthranilic acid-derived alkaloids exhibit great structural diversity. This group includes dictamnine, platydesmine, vasicine, cusparine, and rutecarpine. Vasicine has oxytocic activity.

Phenylalanine- and Tyrosine-Derived Alkaloids

Phenylalanine- and tyrosine-derived alkaloids are by far the most numerous group of alkaloids, ranging from simple phenethylamines to the very complex dimeric benzylisoquinolines and the highly rearranged Cephalotaxus alkaloids.

Ephedrine. Ephedrine, from the Chinese drug Ma Huang, is soluble in water, alcohol, chloroform, and diethyl ether. It melts over the range of 33–42°C, depending on the water content. Little of the ephedrine of United States commerce is obtained from natural sources. Ephedrine is produced commercially through a biosynthetic process. Ephedrine has mydriatric effects and demonstrates adrenaline-like activity. It causes a rise in arterial blood pressure, increased secretions, and dilated pupils; it is a monoamine oxidase inhibitor and is used in nasal decongestants.

Peyote. Peyote, the small cactus of the Indians of north central Mexico, contain over 60 constituents. It is used as a hallucinogen in religious ceremonies and for medicinal purposes. A principal constituent is mescaline, a simple trimethoxyphenethylamine.

mescaline

Ipecac. Ipecac, derived from *Cephaelis ipecacuanha* (native to Brazil), contains emetine and cephaeline, both of commercial significance. Emetine exhibits profound pharmacologic effects including clinical antiviral activity and is used in the treatment of amebic dysentery. Its side effects are cardiotoxicity, muscle weakness, and gastrointestinal problems including diarrhea, nausea, and vomiting. Two synthetic routes to emetine are of commercial importance: the Roche synthesis, which produces dehydroemetine, and the Burroughs-Wellcome synthesis. In handling emetine and its products, exposure should be limited as it can cause severe conjunctivitis, epidermal inflammation, and asthma attacks in susceptible individuals.

Isoquinoline and Related Alkaloids

Isoquinoline and related alkaloids are the largest group of alkaloids derived from phenylalanine (or its hydroxylated derivatives) and corresponding β-phenylacetaldehydes. The alkaloids are prevalent in the plant families Fumariaceae, Papaveraceae, Ranunculaceae, Rutaceae, and Berberidaceae. There are many structural types, including the simple tetrahydroisoquinolines; the benzylisoquinolines; the bisbenzylisoquinolines, such as the *dl-* and *d*-isomers of tetrandrine, which exhibit anticancer activity; the proaporphines, such as glaziovine, which is an antidepressant; the aporphines, such as glaucine; the aporphine–benzylisoquinoline dimers, such as thalicarpine, which shows cytotoxic and antitumor activity; the oxoaporphines; the protoberberines, a group of over 70 alkaloids such as xylopinine, berberine, canadine, and corydaline, known for such diverse pharmaceutical uses as tranquilizers, CNS depressants, antibacterial and antiprotozoal agents, anticancer agents, and alpha adrenergic blockers; the benzophenanthridines, a group of 30 alkaloids such as fagaronine and nitidine, which exhibit antitumor properties; the protopines; the phthalideisoquinolines, a group that includes narcotine, which possesses antitussive activity, depresses smooth muscles, and is not narcotic, and hydrastine, which is used as an astringent in mucous membrane inflammation; and the homoaporphines.

The Opium Alkaloids

The opium alkaloids number over 25, some of which are of commercial importance and major significance. Opium is the air-dried milky exudate from incised, unripe capsules of *Papaver somniferum* L. or *P. albumen* Mill (Papaveraceae). Notable opium-derived alkaloids include morphine, codeine, thebaine, noscapine, and papaverine.

Morphine is the most important alkaloid. It is isolated from opium. Along with its salts, it is classified as a narcotic analgesic and is strongly hypnotic. Side effects include constipation, nausea, and vomiting in addition to habituation, reduced power of concentration, and reduction in fear and anxiety. Respiration is also deepened.

Codeine is the methyl ether of morphine, and thebaine is one of the methyl enol ethers of codeinone. Codeine pharmacologically resembles morphine but is weaker, less toxic, and exhibits less depressant action (it does not depress respiration in normal therapy). It is used in the treatment of minor pain and as an antitussive. Codeine is available as a free base or the sulfate or phosphate salt. Thebaine is a convulsant poison rather than a narcotic.

morphine R = H
codeine R = CH$_3$

Papaverine, a smooth-muscle relaxant that is neither narcotic nor additive, occurs in *P. somniferum* to the extent of 0.8–1.0%, commonly accompanied by noiscopine (narcotine). Most papaverine used is synthetically produced. The glyoxylate salt is used in treating arterial and venous disorders.

Amaryllidaceae

Amaryllidaceae alkaloids include galanthamine, margetine, and narciprimine. Galanthamine is a water-insoluble crystalline alkaloid that exhibits powerful cholinergic activity and analgesic activity comparable to morphine. It has been used to treat diseases of the nervous system. Its derivatives show anticholinesterase, antibacterial, and CNS depressant activity. Narciclasine, margetine, and narciprimine exhibit anticancer activity.

Colchicine

Colchicine, also known as hermodactyl, surinjan, and ephemeron, has some of the most unusual solubility characteristics of any alkaloid: it is soluble in water, alcohol, and chloroform, but only slightly soluble in ether or petroleum ether. Colchicine-type alkaloids are present in ten other genera of the Liliaceae and 19 species of Colchicum. Reviews of the chemistry of colchicine and related compounds and their history and pharmacology are available.

Colchicine has the ability to artificially induce polyploidy or multiple chromosome groups. It is also used to suppress gout.

Cephalotaxus

Cephalotaxus alkaloids are found in the Japanese plum yews, *Cephalotoxus* spp. The esters, such as harringtonine and homoharringtonine, are potent antileukemic agents. The absolute configuration of the ester moiety has been determined. The α-hydroxy ester is essential for *in vivo* antileukemic activity.

cephalotaxine R = H

harringtonine R = — $\overset{\text{OH}}{\underset{\text{CH}_2\text{CO}_2\text{CH}_3}{\text{COC(CH}_2)_2\overset{\text{OH}}{\text{C}}(\text{CH}_3)_3}}$

deoxyharringtonine R = — $\overset{\text{OH}}{\underset{\text{CH}_2\text{CO}_2\text{CH}_3}{\text{COC(CH}_2)_2\text{CH(CH}_3)_3}}$

Securinine

Securinine, isolated from *Securinega suffruticosa* Rehd, is similar to strychnine in action, but exhibits lower toxicity, stimulates respiration, raises blood pressure, and increases cardiac output. The chemistry, pharmacology, and biosynthesis of securinine and related compounds have been reviewed.

Tryptophan-Derived Alkaloids

Tryptophan-derived alkaloids, which occur in the families Apocynaceae, Rubiaceae, and Loganiaceae, have recently become of great interest, particularly those derived from tryptamine and a monoterpene unit. There are many diverse structural types in this group. Attempts to determine the important details of the biosynthetic interconversions of these compounds have been reported.

The simplest indole alkaloids are derived from tryptamine itself. These include indole-3-acetic acid, a potent plant-growth stimulator; serotonin (5-hydroxytryptamine), a vital mammalian product that inhibits or stimulates smooth muscles and nerves; N-acetyl-5-methoxytryptamine (melotonin), a constituent of the pineal gland with melanophase-stimulating properties; 5-methoxy-N,N-dimethyltryptamine, a constituent of the hallucinogenic Virola snuffs; psilocybine, a hallucinogenic found in the mushroom *Psilocybe mexicana* Heim.

The harmala alkaloids, such as harmine and harmaline, are powerful monoamine oxidase inhibitors, previously used in the treatment of Parkinsonism. Harmine is the active ingredient of the narcotic drug yage.

Ellipticine and derivatives show anticancer activity.

Several alkaloids of *Calycanthus* spp. are the products of dimerization or trimerization of simple tryptamine residues, such as folicanthine.

The main indole alkaloid skeletons are derived from tryptamine and a C_{10} unit. Examples include corynanthine; yohimbine, which has hypotensive and cardiostimulant properties and is used to treat rheumatic disease; ajmaline, which has coronary dilating and antiarrhythmic properties; decarbomethoxydihydrovobasine, which shows vasodilating and hypotensive activities whereas related compounds exhibit antiviral activity; akuammicine; tabersonine; catharanthine; rhynchophylline; vindoline; dihydrovobasine; 10-methoxyibogamine, whose acyl derivatives exhibit analgesic and anti-inflammatory activity; and strychnine. This structural diversity has been the source of intense biosynthetic interest.

Physostigmine

Physostigmine, found in the perennial West African woody climber, *Physostigma venenosum* Balfour, is pharmacologically similar to pilocarpine and is a reversible cholinesterase inhibitor used to treat glaucoma.

The Ergot Alkaloids

The ergot alkaloids are obtained from ergot, the dried sclerotium of the fungus *Claviceps purpurea* (Fries) Tulsane (Hypocreaceae). Ergot alkaloids are produced by isolation from the crude drug grown in the field, by extraction from saprophytic cultures, and by partial and total synthesis.

The ergot alkaloids act pharmacologically to produce peripheral, neurohormonal, and adrenergic blockage, and to produce smooth-muscle contraction as well. The two medicinally important ergot alkaloids are ergotamine and ergonovine.

lysergic acid

Three main groups of ergot alkaloids exist:

1. The clavine type, a group of over 20 alkaloids, which is water insoluble and does not give lysergic acid on hydrolysis. This group includes elymoclavine, agroclavine (a potent uterine stimulant), and chanoclavine-I.

2. The aqueous lysergic acid derivatives such as ergonovine, which in its maleate salt (or as methyl ergonovine maleate) is the drug of choice to treat postpartum hemorrhage.

3. The peptide ergot alkaloids, a group of water insoluble lysergic acid derivatives. This group includes ergotamine, ergocornine, and ergocryptine. Ergotoxine, a mixture of three peptide ergot alkaloids, possesses strong sympatholytic action and is used as a peripheral vasodilator and antihypertensive. Dihydroergotoxine is used for vascular disorders in the aged.

Some ergot alkaloids (eg, 2-bromo-α-ergocryptine) stimulate prolactin release and are being evaluated for treatment of breast cancer.

Catharanthus and Vinca Alkaloids

Catharanthus and Vinca alkaloids, usually discussed together, are quite distinct. The most important alkaloids of the Catharanthus genus are vincaleukoblastine, leurocristine, and leurosine, all antileukemic agents. Vincaleukoblastine and leurocristine are used clinically. The most important alkaloid of Vinca is vincamine, used to treat hypertension, angina, and migraine headaches. Alkaloids of this type produce marked hypotensive effects and curare-like action. The ethers of vincaminol are potent muscle relaxants.

Rauwolfia

Rauwolfia alkaloids include reserpine, the first tranquilizer, rescinnamine, and deserpidine. Reserpine is a sedative and tranquilizer useful in treating hypertension. It is also used as a rodenticide.

Strychnine

Strychnine, from the seeds of many Strychnos species, is a widely known poison (although, in fact, it is only moderately toxic). Pharmacologically, strychnine excites all portions of the CNS; it is a powerful convulsant and death results from asphyxia. It has no therapeutic uses in Western medicine, although its nitrate is used in treating chronic aplastic anemia.

Cinchona Alkaloids

Cinchona alkaloids, derived from the dried stem or root bark of various Cinchona species, include quinine and quinidine. These alkaloids are bitter tasting white crystalline solids, sparingly soluble in water. Quinine is toxic to many bacteria and other unicellular organisms and was the only specific antimalarial remedy until the Second World War. It is a local anesthetic of considerable duration. Quinine is commonly used as the sulfate and dihydrochloride. Quinidine, produced by the isomerization of quinine or found in Cuprea bark, is more effective on cardiac muscle than quinine and is used to prevent or abolish certain cardiac arrhythmias.

Camptothecine

Camptothecine, isolated from the Chinese tree *Camptotheca acuminata* Decsne, is used to treat cancer in the People's Republic of China.

Histidine-Derived Alkaloids

Histidine-derived alkaloids include pilocarpine and saxitoxin. Pilocarpine stimulates parasympathetic nerve endings and is used to treat glaucoma. The main commercial source of pilocarpine is *Pilocarpus microphyllus* Stapf., known as Marnham jaborandi. Saxitoxin is an extremely toxic neuromuscular blocking agent found in the so-called coastal red tides of North America.

Monoterpenoid Alkaloids

Monoterpenoid alkaloids include chaksine, a guanidine alkaloid from *Cassia lispikula* Vahl, which induces respiratory paralysis in mice; β-skytanthine, which is tremorigenic; cantleyine, derived from a monoterpene before loganin; and those derived from secologanin, such as gentianine, which exhibits hypotensive, anti-inflammatory, and muscle-relaxant actions, gentioflavine, gen-

---

Table 1. Effects of Natural Food Alkaloids on Humans and Animals[a]

Alkaloid(s); Identified or Potential Food Sources
    Clinical Effects
        Toxic Symptoms

Amatoxins (α-, β-, γ-amanitins, amanin, amanullin); toxic mushrooms [Death Cup (*Amanita phalloides, A. verna, A. virosa*)]
    Liver and kidney damage (man)
        Delayed vomiting, abdominal pain and diarrhea, coma, death; LD[b] (man, oral) = 0.11 mg/kg

Arecoline; Areca nut or Betal nut (*Areca catechu*)
    Parasympathomimetic agent, cathartic (horses); cholinergic (man)
        Flushing, perspiration, bronchial spasms, contraction of pupils, diarrhea, dyspnea, and collapse; LD[b] (mouse, subcutaneous) = 102 mg/kg

Caffeine; coffee, tea, cola nuts, guarana
    Cardiac, respiratory, and psychic stimulation, diuresis; vascular cephalalgesia (man)
        Nausea, restlessness, vomiting, insomnia, tremors, tachycardia, cardiac arrhythmia (man); LD_{50}[c] (rat, oral) = 19.6 mg/kg

Capsaicine; red peppers
    Skin irritant
        Sweating, salivation (mammals)

Carapine; papaya (*Carica papaya*)
    CNS depression (mammals)
        Bradycardia (mammals)

Caulerpicin; marine algae (*Caulerpa sertulariodes, C. racemosa* var. *clavifera*)
    Anesthesia (mammals)
        Numbness of lips and tongue (man)

Chavicine; black and white pepper (*Piper nigrum*)
    Sharp peppery taste (man)
        No information

Choline; brain, egg yolk, bivalve mollusks (*Callista brevisiphonata*), areca nut
    Lipotrophic agent, antihypercholesterolemic agent
        Nausea, vomiting, diarrhea (man); LD[b] (rabbit, subcutaneous) = 450 mg/kg

Convicine; fava bean (*Vicia sativa* or *V. fava*)
    Inhibition of glucose-6-phosphate dehydrogenase, decreased or reduced glutathione levels in red blood cells
        Hemolysis, growth retardation (rat), hemoglobinemia (dog)

Curry alkaloids (murrayanine, mukoeic acid, girinimbine murrayacine, koenimbine, koenine, koeniginine, koenigine, mahanimbine, cyclomahanimbine, curryanine); curry plant (*Murraya koenigii* Spreng)
    For skin eruptions
        No information

Cystisine; milk [contaminated from *Cystisus laburnum, Sophora secundiflora* (mountain laurel)]
    No information
        Excitement, sweating, incoordination, convulsion, death (man, pig, cattle, horse); LD[b] (as nitrate, dog, subcutaneous) = 4.3 mg/kg

3,4-Dihydroxy-L-phenylalanine (DOPA); fava bean (*Vicia fava*)
    Decrease in reduced glutathione in red blood cells (man)
        No information

Dopamine; banana, avocado, cephalopods
    Hypertensive agent, increased cardiac output
        No information

Dioscorine; wild yam (*Dioscorea hirsutus, D. hispida*)
    CNS stimulation: analeptic, diuretic, expectorant
        Sialorrhea, nausea, vomiting, diarrhea, confusion, cold sweat, pallor, clonic convulsion, paralysis, asphyxia (man), emetic hemolytic agent; LD[b] (mouse, intraperitoneal) = 130 mg/kg

Eptatretin; Pacific hagfish (*Polistotrena stouti*)
    Cardiac stimulant (frog, dog)
        None, due to rapid detoxication

Table 1 (*continued*)

Alkaloid(s); Identified or Potential Food Sources
    Clinical Effects
        Toxic Symptoms

Ergot alkaloids (Eergonovine, ergotamine, ergosine, ergocristine, ergocyptine, ergocornine, ergosinine, ergocristinine, etc); Barley, rye (contaminated from ergot produced by the fungi *Claviceps paspali* and *C. purpurea*)
    Uterine stimulation, analgesic for migraine headache
        Tachycardia, hypertension, vomiting, diarrhea, mental confusion, hallucinations, convulsions, gangrene, gastrointestinal (GI) disturbance; LD[b] (ergotamine, rat, intravenous) = 60 mg/kg
Gelsemine; Honey [contaminated from yellow jasmine (*Gelsemium sempervirens*) nectar]
    Uterine stimulation, CNS stimulant
        Dizziness, dimness of vision, mydriasis, nausea, muscular debility, unusual prostration, weak pulse, dyspnea (man); minimum LD (rabbit, subcutaneous) = 0.15 mg/kg
Glucobrassicin; broccoli, brussels sprouts, cabbage, cauliflower, kohlrabi, radish, rutabaga
    Goitrogenesis (rabbit, rat)
        Hyperplasia of the thyroid; kidney and liver enlargement (rabbit, rat)
Goitrin; cabbage, rape, rutabaga
    Goitrogenesis (rat)
        Hyperplasia of thyroid, liver and kidney enlargement (rat)
Gramine; barley (*Hordeum vulgare*)
    Vasoactivity, stimulation of intestines and uterus (rabbit)
        Hypertension, clonic convulsion, and excitation of respiratory center
Histamine; derived from beer, chocolate, fish, sauerkraut, wine, and yeasts
    Vasoactivity (mammals)
        Headache (man), hypotension (man, mammals); LD_{50}[c] (mouse, intraperitoneal) = 12.7 g/kg, LD[b] (monkey, intravenous) = 52 mg/kg
Hordenine; germinated barley (*Hordeum vulgare*), sorghum (*Sorghum vulgare*), millet (*Panicum miliaceum*)
    Sympathomimetic intestinal stimulation, respiratory stimulation (cat, dog); uterine stimulation (guinea pig)
        Hypertension (man, dog), psychostimulation, convulsion, respiratory inhibition (man), LD[b] (as sulfate, dog, oral) = 1.9 g/kg
Islanditoxin; rice (contaminated from *Penicillium islandicum*)
    Carcinogenic hepatotoxin (mouse)
        Liver damage, liver cancer (mouse, rat); LD_{50}[c] (mouse, oral) = 5.0 mg/10 g
Laminine; marine algae (*Laminaria* spp.) and others
    Hypotension (rabbit); depression of contraction of smooth muscles (mouse, guinea pig)
        Hypotension (rabbit)
Methyl pyrazine (various derivatives); green peas (*Pisum sativum*)
    No information
        No information
Methyl pyrroline; black pepper (*Piper nigrum*)
    No information
        No information
Mimosine; *Leucaena glauca*
    No information
        Alopecia (mammals)
Murexine (urocanyl choline); shellfish (*Muricidae* spp.)
    Excitation of respiratory center, neuromuscular blocking, muscle relaxation (vertebrates and invertebrates)
        Muscular and respiratory paralysis (vertebrates and invertebrates); LD_{100} (as oxalate, mouse, subcutaneous) = 310 mg/kg
Muscaridine; mushroom fly agaric (*Amanita muscaria*)
    Uterine stimulation (rabbit, guinea pig)
        Hashish or alcohol-like intoxication (man)
Muscarine; mushroom fly agaric (*Amanita muscaria* and other mushrooms)
    Uterine stimulation (rabbit, guinea pig)
        Sialorrhea, lacrimation, diaphoresis, nausea, vomiting, bradycardia, convulsions, coma, death; LD_{100} (mouse) = 16 μg/day
Nicotine; tomato
    Respiratory stimulation (man); uterine stimulation (cat, rabbit, pig); cerebral and visceral ganglial stimulation (man)
        Hypotension (dog); respiratory depression and paralysis (cat); hyperglycemia, convulsion, dizziness, nausea (man), etc; LD (man, oral) = 40 mg/70 kg; LD_{50}[c] (rat, oral) = 60 mg/kg
Norepinephrine; banana (particularly the pulp), orange (trace), potato
    Vasoactivity sympathomimetic, adrenergic
        Hypertension
Pahutoxin; fishes [Hawaiian boxfish (Pahu), Ostraciontidae family]
    Neuromuscular effects (mammals)
        Hemolysis (mammals); ataxia, respiratory distress, coma (mouse); minimum LD[b] (mouse, intraperitoneal) = 0.25 mg/g
Phallotoxins (phalloidin, phalloin, phallisin, phallicidin); toxic mushrooms [Death Cup (*Amanita phalloides, A. verna, A. virosa*)]
    Liver and kidney damage (man)
        Vomiting, abdominal pain, diarrhea, coma, and death (man); LD (man, oral) = 3 mg/kg

Table 1 (*continued*)

Alkaloid(s); Identified or Potential Food Sources
    Clinical Effects
        Toxic Symptoms

β-Phenylethylamine; Mushrooms, bitter almonds
    Respiratory stimulation, intestinal relaxation, symphatomimetic (man)
        Hyperventilation, hypertension, hypotension (cat); LD_{50} (mouse, intraperitoneal) = 350 mg/kg
Piperine; black pepper (*Piper nigrum*)
    No information
        No information
Pyrrolizidine alkaloids; wheat and other cereals, legume (contaminated from *Crotalaria laburnifolia, C. striata*)
    No information
        Liver damage, carcinogenesis, venoocclusive disease (mammals); glaucoma (rat, mouse); $LD_{50}{}^c$ (mouse) = 20.0 mg/kg
Sanguinarine; mustard oil and cereal grains (contaminated from *Argemone mexicana*)
    Expectorant; for chronic eczema and skin cancers
        LD^b (as sulfate, intraperitoneal, rat) = 450 mg/kg
Saxitoxin; shellfish, bivalve mollusks, and other pelecypods (contaminated from the dinoflagellates *Gonyaulax catenella* and *G. tamarensis*)
    Neural stimulation (mammals); hypotension (cat and rabbit); myocardial and respiratory depression (mammals)
        Peripheral paralysis, tingling and numbness of lips, respiratory failure (man and other mammals); $LD_{50}{}^c$ (man, oral) = 10–20 μg/kg
Serotonine; avocado, banana, pineapple, octopus, papaya, plantain, passion fruit, red plum tomato
    Respiratory stimulation (dog); respiratory inhibition (cat)
        Hypertension (man); CNS depression (most animals); inhibition of ovulation (rabbit); teratogenesis (mouse)
Solanine (solanidine); Irish potato (*Solanum tuberosum*), tomato (*Lycopersicon esculentum*)
    Acetylcholine esterase inhibition (mammals)
        Drowsiness, hyperesthesia, dyspnea, vomiting, diarrhea (man); LD^b (rabbit, intravenous) = 25 mg/kg
Tetrodotoxin; Pufferfish (Tetradontidae and Diodontidae families)
    Neuromuscular blocking (mammals); hypotension (cat); respiratory inhibition, hypothermia (dog)
        Weakness, dizziness, pallor and paresthesia, nausea, vomiting, sweating, salivation, muscular paralysis, cyanosis, respiratory paralysis, death (man); minimum LD (cat, subcutaneous) = 11 mg/kg; $LD_{50}{}^c$ (mouse, oral) = 325 mg/kg
Tetramine; shellfish (Buccinidae and Cymatidae families)
    Curare-like effects, hypotension, bradycardia (mammals, frog)
        Salivation, lacrimation, miosis, peristalsis (mouse); motor paralysis (mammals)
Threobromine; cacao bean; cola nuts; tea
    Diuresis, myocardial stimulation, vasodilation, respiratory stimulation (cat)
        GI distress; $LD_{50}{}^c$ (cat, oral) = 220 mg/kg
Theophylline; tea
    Diuresis, hypertensive cephaloanalgesia (man)
        Nausea and vomiting, vertigo, insomnia, flushing, convulsions, death; $LD_{50}{}^c$ (mouse, oral) = 550 mg/kg
Tomatine; tomato juice
    No information
        LD (rat, oral) = 1 g/kg
Toxoflavin; Indonesian bongkrek (contaminated from *Pseudomonas cocovenenans*)
    No information
        $LD_{50}{}^c$ (mouse, oral) = 800 mg/kg $LD_{50}{}^c$ (mouse, intravenous) = 2.0 mg/kg
Trigonelline; green peas (*Pisum sativum*), coffee, soybeans, potatoes
    No information
        LD^b (rat, subcutaneous) = 4.9 g/kg
Tryptamine; plum (red and blue), orange, tomato
    Vasoactivity, musculotropic (rabbit); intestinal and uterine contraction (rabbit, guinea pig)
        Hypertension, headache (man)
Tyramine; avocado, banana, octopus, orange, spinach, potato, tomato
    Respiratory stimulation (rat, cat); vasoconstriction (mammals)
        Paroxysmal hypertension, intracerebral hemorrhage (man); mydriasis, bradycardia (dog); hypothermia (cat); LD (rabbit, intravenous) = 250 mg/kg
Vicine; fava bean (*Vicia sativa* or *V. fava*)
    Inhibition of glucose-6-phosphate of dehydrogenase; decrease in reduced glutathione
        Hemolysis, growth retardation (rat); hemoglobinemia (dog)

---

[a] Ref. 1.
[b] LD = lethal dose.
[c] LD_{50} = median lethal dose.
Source: See reference under S. W. Pettetier.

tiatibetine, pedicularine, and actinidine, a potent feline attractant.

Diterpene Alkaloids

Diterpene alkaloids are not of commercial or therapeutic significance, but some have potent pharmacological activity, eg, aconitine and Erythrophleum alkaloids.

Steroidal and Triterpene Alkaloids

Steroidal and triterpene alkaloids are found in the plant families Solanaceae, Liliaceae, Apocynaceae, and Buxaceae. There are four main groups based on the botanical source: the Veratrum, Solanum, Holarrhena and Funtumia, and Buxus alkaloids.

The Veratrum alkaloids include jervine, protoveratrine A, and protoveratrine B; the latter two produce pronounced bradycardia and a fall of blood pressure by stimulation of vagal afferents. The Solanum alkaloids are of interest as potential sources of steroids. Examples of these alkaloids are tomatidine and solanidine. Some Solanum alkaloids exhibit fungistatic activity. Biosynthetically, the alkaloids are derived from acetate and mevalonate.

tomatidine

solanidine

Purine Alkaloids

Purine alkaloids are derivatives of the xanthine nucleus and include caffeine, theophylline, and theobromine, the principal constituents of plants used throughout the world as stimulating beverages.

Caffeine has the structure 1,3,7-trimethylxanthine. It is derived from cola, coffee (qv), tea (qv), guarana, and maté. Theophylline, 1,3-dimethylxanthine, is found in tea. Theobromine, 3,7-dimethylxanthine, is found in cocoa and tea. The xanthine derivatives have pharmacological properties in common: central nervous system (CNS) and respiratory stimulation; skeletal-muscle stimulation; diuresis; cardiac stimulation; and smooth-muscle relaxation. Caffeine is used to increase CNS activity; it acts on the cortex to produce clear thought and to reduce drowsiness and fatigue. Theophylline is used in smooth-muscle relaxants. Theophylline, as the ethylenediamine salt, is used in preference to caffeine in cardiac edema and in angina pectoris.

Miscellaneous Alkaloids

Coniine is an extremely toxic alkaloid that induces paralysis of the motor nerve endings and is the primary toxic constituent of poison hemlock. It was the first alkaloid to be synthesized. Carpaine, a crystalline macrocyclic alkaloid that induces bradycardia, depresses the CNS and is a potent amoebicide. Alkaloids are found in the poisonous Amanita species of mushrooms, such as α- and β-amanita toxins, ibotenic acid, muscimol, and muscazone. Maytansine and related ansamacrolides are potent antileukemic agents. Surugatoxin, found in the carnivorous gastropod *Babylonia japonica*, produces a pronounced mydriatic effect, sometimes resulting in death.

INGESTION AND HUMAN HEALTH

The effects of natural alkaloids on humans and animals vary. This article is concerned mainly with food alkaloids and their clinical effects when ingested intentionally or accidentally; Table 1 summarizes the information.

BIBLIOGRAPHY

The section on classification and properties has been adapted from reference 1. Check original article for specific reference citations.

1. G. A. Cordell, "Alkaloids," in M. Grayson, ed. *Kirk-Othmer Concise Encyclopedia of Chemical Technology,* Wiley-Interscience, New York, 1985.

General References

J. N. Hathcock, ed., *Nutritional Toxicology,* Vols. 1, 2, Academic Press, Orlando, Fla., 1987.

P. M. Newberne, "Naturally Occurring Food-Borne Toxicants," in M. E. Shils and V. R. Young, eds., *Modern Nutrition in Health and Disease,* 7th ed, Lea & Febiger, Philadelphia, Pa., 1988.

S. W. Pelletier, ed., *Alkaloids: Chemical and Biological Perspectives,* 6 Vols., Wiley-Interscience, New York, 1983–1988.

S. L. Taylor and R. A. Scanlon, *Food Toxicity,* Marcel Dekker, New York, 1990.

Y. H. Hui
EDITOR-IN-CHIEF

AMINO ACIDS. See CANOLA; CASEIN AND CASEINATES; CONTROLLED ATMOSPHERES FOR FRESH FRUITS AND VEGETABLES; HYDROPHOBICITY IN FOOD PROTEIN SYSTEMS; MICROBIOLOGY; NUTRITIONAL QUALITY AND FOOD PROCESSING; TEA; ENTRIES UNDER GENETIC ENGINEERING; PROTEINS.

AMMONIATION. See AFLATOXIN: ELIMINATION THROUGH BIOTECHNOLOGY.

ANIMAL BY-PRODUCT PROCESSING

Domesticated animals are grown and slaughtered for meat for human consumption. Normally, only 30–40% of the animal is utilized for human food (meat cuts, edible offals, processed products). Although this represents by far the most important product of meat processing plants in monetary terms, economics of operating the meat slaughter plant, together with the cost of pollution abatement and/or disposal of inedible material from the slaughter operation, demand that the inedible material be utilized profitably.

Many products can be made from the nonmeat parts of the animal (Fig. 1), and various by-products often contribute significantly to the meat plant's profitability. Their commercial value is often higher than the sum of the running expenses and the margin required for the meat plant to operate profitably. Also, because there is a worldwide shortage of animal protein, it is essential to maximize the use of these raw materials. This applies both to products used directly for human consumption as well as to protein-containing material that can be processed and fed back to animals.

In most countries everything produced by or from the animal except the dressed carcass is considered a by-product. (In some countries the terms offal and by-products are interchangeable.) Hides and pelts are an obvious by-product and are discussed elsewhere. A partial, but not all-encompassing, list of animal by-products has been given (1–4).

By-products are usually classified as edible or inedible. Edible by-products, such as heart, liver, tongue, oxtail, kidney, brain, sweetbreads, and tripe, that are segregated, chilled, and processed under sanitary conditions are called variety meats (or in some countries, offals). Chitterlings and natural casing (intestines) and fries (lamb or calf testicles) can also be eaten. In some countries the blood and/or blood fractions from healthy animals, processed hygenically under conditions specified by the appropriate regulatory authority, can be an edible by-product. Some by-products must be processed or refined, eg, stomachs for tripe, bones and skin pieces for gelatin manufacture, fatty tissue for edible fat, before they can be eaten. There is a sizable international trade in edible by-products because they are a very economical source of high-quality protein.

Most noncarcass material, if cleaned, handled, and processed in an appropriate manner, could be edible. Custom, religion, palatability, and reputation of these products usually limit their use for human consumption. How a meat processor classifies a specific product depends on both the possible utilization of that product and the availability of a potential market. For example, many potentially edible by-products are downgraded to an inedible use because a profitable market does not exist.

Animal by-products are often underused. Their effective utilization depends on the following: a practical commercial process to convert the animal by-product into a usable commodity; actual or potential markets for the commodity produced; a large enough volume of economically priced raw material in one location for processing; a method for storing perishable material before and after processing; and, often, the availability of highly technically trained workers.

By-product yields (as a percentage of the animal's slaughter weight) depend on the species and category of animals slaughtered, the degree of animal processing, and the end form of the by-product. Edible offals, material for pet food, and blood for edible use can represent up to 20% of carcass weight. When by-products are further processed (eg, rendered into meal and processed fat), one-third to two-thirds of the original weight of raw material is lost as water.

BLOOD

Whole blood is a liquid with a dry matter content of 18–20%, of which 90–95% is protein. Animal type and age may slightly influence blood composition. Blood undergoes complex biochemical reactions (clotting, coagulation, syneresis) once it has been released. Uncoagulated blood can be fractionated into plasma (a pale yellow fluid) and red blood cells (a fraction containing hemoglobin). Serum can be obtained when clotted blood synereses.

Blood is an obvious source of high-quality protein. Whole blood and the processed fractions (especially plasma) can be used as food ingredients. The dark red color of whole blood and red blood cells usually limits the use of these products in foods. Several processes exist for manufacturing decolorized blood fractions. The edible use of blood proteins has been reviewed (5–7).

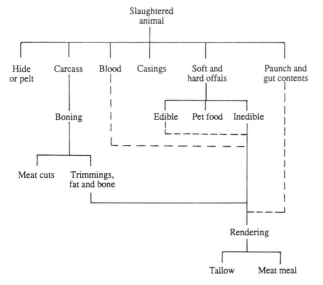

Figure 1. Flow diagram of material from a slaughtered animal.

The high costs and the logistics of collecting blood hygenically usually preclude extensive use of blood for edible products. Blood represents 3–4% of the animal's live weight, so large quantities are produced daily at slaughter facilities, even small ones. Whole blood has a high biological oxygen demand (250,000 mg/L). It is usually processed into an inedible blood meal, partly to produce some income but mainly to prevent a major pollution problem. Because blood is usually not collected hygienically, it may contain urine and ingesta, and it usually has been diluted with wash water. Thus the dry matter content can be as low as 10%, especially if there is poor water management.

In commercial practice, proteins in the blood are coagulated, separated, and dried to produce blood meal, a cheap source of animal protein high in lysine with a moisture content of 8–10%. The three most common methods of processing inedible blood are:

1. Apply indirect heat to the whole blood to boil off most of the water. This method is very energy inefficient and produces a denatured product with very low solubility. It is often used at smaller, older plants, or at plants that have not updated their equipment.

2. Inject live steam to bring the temperature to about 90°C to coagulate the blood proteins; then remove most of the water in a decanter and dry the solids (eg, in a ring dryer with hot circulating gases, a rotating drum cooker, or a batch dryer). This method uses less energy than method (1) because about half of the water is removed mechanically. A denatured protein powder is produced, the solubility of which depends on the type of dryer used; powders produced in ring dryers usually are more soluble. The high temperatures and/or long times used during processing lower the nutritional value of the blood meal.

3. Concentrate whole blood by ultrafiltration, then dry the concentrate in a spouted-bed dryer (8). A very soluble powder is produced.

The yields of dried blood depend on the processing parameters (9). Aging the blood (leaving for 12 or more hours before processing) can increase yields. Any water added to the raw blood reduces yields. If the steam coagulator is not working efficiently, the losses of protein in the blood water from the centrifuge can increase.

Blood and its components (serum, albumin, red blood cells, hemoglobin) can be used in food, animal feeds, laboratory reagents, medical preparations, industrial uses, and as fertilizer. Blood albumin can be a substitute for egg albumen. Dried blood meal is used as a protein supplement in livestock feed. It is deficient in the amino acids tryptophan and isoleucine, but is high in lysine, although lysine availability is affected by the drying method. Laboratory uses for blood are as a nutrient for tissue culture media and as a necessary ingredient in some agars for bacteriological use. Many blood components are isolated from whole blood and used in chemical and medical analyses or as nutritive supplements. Industrial uses of blood are as an adhesive and for its film-forming properties in the paper, lithographic, plywood, veneering, fiber, plastics, and glue industries. As a fertilizer, blood meal contributes nitrogen, aids humus formation, and improves soil structure.

Whole blood can be stabilized chemically, eg with urea, ammonia, or metabisulfite. The preserved blood can be held without refrigeration and fed to pigs. Blood can also be preserved by adding sulfuric acid.

Blood char or blood charcoal is the carbon component of whole blood or blood meal. Blood char is produced by treating 20% of the weight of whole blood or 50% of the weight of blood meal with activating agents and heating in air-tight containers to 650 to 750°C for 6 to 8 h. Blood char contains 80% carbon and is used for absorption of gases, as an industrial decolorant, and as an antidote for chemical poisoning.

RENDERING

Animal tissues and organs are composed essentially of water, fat, and protein, with some minerals. The term rendering refers to a variety of processes that are used to separate the water, fat, and protein components, as far as is practicable, into commercial products.

Fatty tissue can be located anywhere in connective tissue and is made up of cells containing fat that has been deposited as an animal matures. The fat cells are surrounded by reticular fibers. The fat cannot be released from animal tissue until the supporting structure has been broken.

Until the 1850s, the highly perishable by-products from meat slaughtering were considered waste and were buried. The rendering industry developed to convert these materials into farm fertilizer and realize a profit from the operation. Proteinaceous by-products yielded nitrogen fertilizer, whereas bone produced phosphate fertilizer. Today the rendering industry produces hundreds of useful products that can be broadly classified as edible and inedible fats, fine chemicals, meat meals, and bone meals.

Animal by-products used by renderers consist of excess fat, bones, hoofs, and soft offals (viscera). Further processing of carcasses yields bones and fat trimmings, increasing the quantity of material for rendering. The material for rendering can represent 30 to 60% of the weight of a slaughtered animal. In developed countries, with centralized slaughter of large numbers of animals, a large volume of material is available for renderers. This has led to the development of sophisticated equipment and processes.

Basic Principles

The basic purpose of rendering is to produce stable products of commercial value, free from disease-bearing organisms, from raw material that is often unsuitable or unfit for human consumption. The two basic processes in rendering are separation of the fat and drying of the residue. The most common method used to rupture fat cells in the tissue is heat, although enzymic and solvent-extraction rendering processes are also used.

A large proportion of the raw material is viscera (soft offal). Because paunch and gut contents contain chemicals that can adversely affect fat quality, for most rendering processes the viscera must first be washed. The viscera are

Table 1. Best-Known and Most-Used Rendering Processes

			Temperature Range, °C
Dry	Batch	Cookers	105–130
	Continuous	Keith	105–140
		Stord Bartz with centrifuges	110–140
		Duke	120–140
	Continuous with evaporator	Carver-Greenfield	90–105
		TM-1	120–140
		Retrofit	120–140
		Stord Bartz with presses	120–140
Wet	Batch	Digesters	90–110
	Semicontinuous	Centrimeal	125–140
		Instant meal	125
	Continuous low temperature systems	Atlas	
		Balfour	
		Centribone	60–95
		MLTR	
		Pfaudler	
		Stord Bartz wet pressing	

then cut to reduce their size; the smaller the particle size, the quicker the mass and heat transfer. However, in some rendering processes it may be difficult to get even heat transfer in the processing vessel if the particles are too small. Size reduction is usually done in devices with rotating knives or anvils (termed hogors or pre-breakers when producing large particles, grinders or mincers when producing fine-particle material) or with rotating hammer devices (hammer mills). Hoofs and bones (hard offals) need to be reduced in size but usually do not require washing. The raw material is then processed to separate the fat from the nonfat phase.

Processes

A rendering process can generally be classified as wet or dry, depending on whether the fat is removed from the raw material before or after the drying operation. Processes can operate in a batch, semicontinuous, or continuous mode, and some of the newer rendering systems are classified as low-temperature rendering (LTR) systems because of their milder heat treatment. Table 1 lists the best-known and most-used of the many rendering systems, divided according to process type. Bracketed systems produce products of similar quality when processing similar raw material. There is no one best process for all applications; the most appropriate method will often depend upon the application. Profitability of any rendering system can be maximized by ensuring maximum yields and by obtaining the best possible product quality. Regular and planned maintenance of equipment will minimize repairs and maintenance costs. Rendering added water (from excessive washing and hosing) is wasteful, because more energy and a larger rendering capacity is required for a given throughput of solid material.

Wet Rendering. Wet rendering is an old processing method. In older systems, the pre-ground raw material is cooked in a closed vertical tank (termed a digester, autoclave, or cooker) under pressure by direct steam injection, usually to 380–500 kPa for 3–6 h (Fig. 2). Operators use past experience to gauge the end point. The pressure is then slowly released and the liquid and solid phases are allowed to settle. The fat that has floated to the top is drawn off, and can be polished in disk centrifuges to remove residual water and particles (fines). The water phase (liquor) in the cooker is drained off, then the solid material (greaves) is removed and can be pressed or centrifuged to remove additional liquid (stickwater) before being dried. The liquor and stickwater may be further processed to remove fines, which can be recycled to the greaves, and residual fat. Wet rendering produces good quality fat (but only if the viscera are cut and washed). However, it requires long cook times, is labor intensive, and has significant losses (up to 25% of the solids may be lost in the stickwater). It is energy intensive, but heat can be recovered from the vent steam.

A more modern wet rendering system is semicontinuous and involves cooking raw material in a conventional dry rendering cooker under pressure (to ensure steriliza-

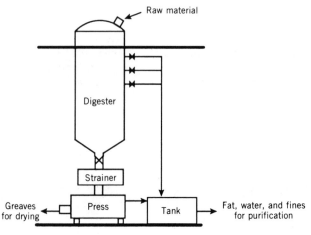

Figure 2. Wet rendering system.

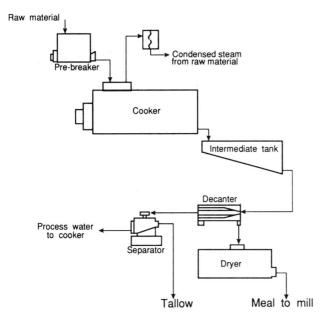

Figure 3. Semicontinuous wet rendering system.

tion) for a short time, then processing the cooked material in decanters to separate the liquor from the wet solids (Fig. 3). The meal is dried in continuous dryers and the fat is separated from the liquor in disk centrifuges. Process water is evaporated in multiple-effect evaporators and the concentrate is added to the wet solids. The system produces high-quality fat and low-fat meal, and less energy is used than in conventional wet or dry rendering systems. However, capital costs, repair costs, and maintenance costs are high.

Dry Rendering. Both batch and continuous processes exist (Fig. 4). The material is heated in a horizontal, steam-jacketed vessel until most of the water has evaporated. The vessel has an agitator, which also may be steam heated. The evaporated water is usually condensed to recover heat and reduce atmospheric pollution. In batch systems the raw material can be subjected to 200 to 500 kPa for some specified time to sterilize the material and/

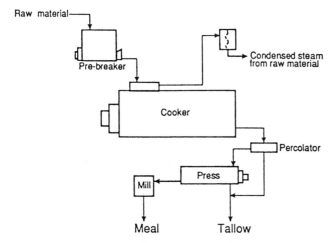

Figure 4. Dry rendering system.

or hydrolyze wool and hair. It may take up to 3 h to produce a dry material, and the end-point temperature is often 120 to 140°C. The stage at which pressure is applied can influence the ease of further processing. Cooking times in continuous systems depend on cooker volume, its heat-transfer capability, and the characteristics and feed rate of the raw material.

Cooker contents are discharged into a percolator to remove free-draining fat; then the solid material is pressed (a continuous operation) or centrifuged (a batch operation) to remove additional fat. The material is then ground into meat meal. The fat is polished in disk centrifuges to remove fines and moisture.

Dry batch rendering has the following advantages:

Little material is lost.

Cooking, pressurizing, and sterilizing can be carried out in the same vessel.

Different cookers can be assigned for processing different raw material, and hence for producing different grades of fat.

Heat can be recovered from the vent steam.

The disadvantages are:

The fat is usually of poorer quality than that from wet or LTR systems.

The high temperatures used produce fines, which can pass into the fat, degrading its quality, and can be lost in the effluent from the polishing centrifuges.

The meal has a higher fat content than meal from wet and LTR systems.

Raw material must be washed (producing effluent) to produce good-quality fat.

Indiscriminate hosing and inadequate draining of washed material add extra water, increasing evaporation loadings and hence energy requirements.

It is difficult to process high-protein, low-fat material such as slaughter wastes from calves and young lambs.

It is difficult to control the end point of the cooks, therefore fat quality can be variable.

It is difficult to keep the processing area clean and tidy.

The process is not completely enclosed, therefore cooked products can be recontaminated.

Energy usage is high, especially if vent steam is not recovered as hot water.

Dry rendering cookers are not efficient dryers.

The process is labor intensive.

Continuous dry rendering has most of the advantages of batch dry rendering, uses less labor, and requires less space. It has the disadvantages that the system cannot be pressurized; therefore sterilizing and hydrolyzing cannot be done. Tallow quality tends to be lower than with batch operations because higher end-point temperatures are usually used.

Low-Temperature Rendering (LTR). In the wet and dry rendering systems discussed so far, the raw material is

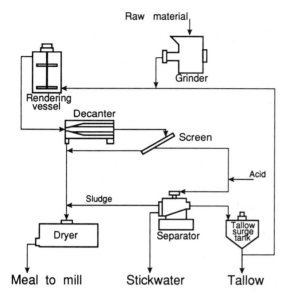

Figure 5. MIRINZ low-temperature rendering (MLTR) system.

subjected to high temperatures for long times. Near the end of the cooking process in dry rendering, when most of the free water has evaporated, the solids (cracklings) are essentially frying in the fat. With such high temperatures, the raw material must be washed to ensure that paunch contents and other dirt does not downgrade the color of the fat. Operators often overdry the cracklings to make pressing easier, so dry-rendered meal can be high in fat and low in moisture.

Low-temperature rendering (LTR) systems were developed in the late 1970s to overcome some of the disadvantages of dry rendering systems. Heat treatment is minimized and phase separation is carried out at low temperatures (70 to 100°C). Systems are usually continuous, so material flow through the size-reduction equipment, heating units, phase separating units, evaporators, and dryers must be balanced using surge bins and/or variable speed drives. Raw material, which can be unwashed, is minced, then heated to 70 to 95°C in a unit called a

coagulator, preheater, melting section, or rendering vessel. The phases are then mechanically separated in decanters or presses.

In the MIRINZ low-temperature rendering (MLTR) system (Fig. 5), processing 5 tonnes of raw material per hour, the rendering vessel has a volume of 1 m³, yet the residence time for the raw material, along with some recycled fat, to reach 80 to 95°C using indirect heat (in coils in the vessel) is just 6 to 8 min (10). The resulting liquor and solids are separated in a decanter, and the wet solids are then dried, often in a direct-fired rotary dryer, to give a low-fat meal. The liquor is separated into high-quality fat and an aqueous phase (stickwater). Product loss in the stickwater should be low.

In other low-temperature systems (eg, the Atlas or Stord Bartz wet pressing systems), raw material is heated indirectly to about 90 to 95°C over a 30 to 60 min period in a preheater, then pressed in a twin-screw press (Fig. 6). Drainings from the preheater and press are further heated to coagulate any soluble protein, then centrifuged to remove fines. The fines and pressed solids are dried, usually in contact dryers.

The stickwater from LTR systems can be concentrated by ultrafiltration or in evaporators, then dried. The steam side of the evaporator is often supplied with waste vapors from the cooker/dryer to save energy. Heat can be recovered from the stickwater and from the vapors from the dryer and used to preheat incoming raw material or to provide hot water.

The advantages of LTR systems are:

Energy requirements are usually about half those of dry rendering systems.

The raw material may not need to be washed.

A low-heat treatment is used, so high-quality fat is produced.

The meal has a low fat content and a high nutritional value because heat treatment is minimal.

Labor requirements are low.

The system can be easily automated.

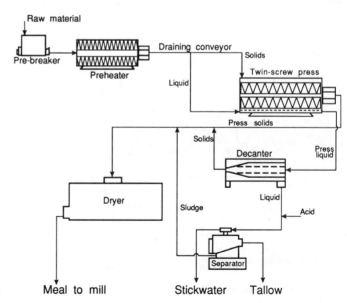

Figure 6. Wet pressing low-temperature rendering system.

However, the systems have high capital costs, they may have high repair and maintenance costs, and they usually require highly trained technical operators.

Uses of Fat

The two important fats of (land) animals used commercially are lard (fat from pigs) and tallow (fat from sheep and beef animals). Lard is made from specified, clean, sound tissues of healthy pigs, whereas rendered pork fat can be made from any fatty tissue. Edible tallow (dripping) is manufactured from specified edible fatty tissue. Oleo stock (premier jus) is a high-grade tallow prepared by low-temperature wet rendering of the fresh internal fat from beef. Inedible tallows (usually defined as fat with a melting point >40°C) and greases (melting point <40°C) are produced in many grades from inedible fatty tissue and dead stock.

Animal fats are composed of triglycerides—three fatty acids esterified to glycerol. Fat quality is measured by titer; free fatty acid (FFA); FAC color (standard set by the Fat Analysis Committee of the American Oil Chemists' Society); bleach color; and moisture, insoluble impurities, and unsaponifiable matter (MIU). Other tests that can be specified include saponification number, iodine value, peroxide value, and smoke point.

The fatty acid chain length and degree of saturation of the carbon bonds affect the fat's hardness or melting point (titer); the longer the chain length and the more saturated the fatty acid, the higher the titer. Fats of different species of animals and from different sites in the body have different titers. Type of feed can affect the titer, but rendering method does not.

When fat molecules break down, free fatty acids are released, so FFA content, which is usually expressed as percentage of free oleic acid, indicates the degree of spoilage that has occurred. To minimize FFA values, poor-quality raw material should be segregated from good-quality material, and material should be processed promptly. If material cannot be processed promptly, the raw material should be kept whole (unbroken) as long as practicable, to minimize microbial and enzymic activity, and preserved by cooling or by adding acid. Processing equipment and storage tanks should be kept clean.

The factors affecting color of fats include animal breed, feed, age, and condition; source of raw material; and presence of contaminants (feces, gut contents, etc). Fats can be almost white, yellow, or green (from contact with chlorophyll in digested plant material), or red or brown (from overheating or contact with blood). Processing parameters during rendering (temperature, time) influence fat color.

Some color components can be removed by bleaching with activated clay, and the color of the bleached sample read in a Lovibond tintometer. Fat bleachability indicates the temperature and handling conditions that the tallow has been subjected to: the cleaner the raw material and the lower the temperatures used, the lighter the bleach color.

The MIU value indicates fat purity. Moisture content should be as low as possible, because microbial and enzymic activity can hydrolyze fat at fat–water interfaces. Insoluble impurities such as protein fines, ground bone, and manure that were not completely removed during processing may form colloidal fines that are not removed by settling or centrifuging. Trace amounts of copper, tin, and zinc can cause fat oxidation, and any polyethylene that has melted during rendering can adversely affect industrial processes that utilize fats. Unsaponifiable matter is fatty components such as cholesterol and gums that cannot be converted into soap by the use of alkali. Such matter reduces soap yields and affects catalyst efficiency. Unsaponifiable matter can also impart objectionable odors as well as downgrade a tallow.

The saponification number indicates the average length of fatty acid chains, and the iodine number indicates the degree of unsaturation. These values can be used to identify types of fats and oils. The peroxide value indicates the degree of rancidity. Unrancid tallow with good oxidative stability will have a low peroxide value. The smoke point is inversely related to FFA content, and is the temperature to which the fat may be heated before it begins to smoke.

Many different grades of tallow can be produced, and most countries have their own trading standards for tallow and grease. Tallow price reflects the prices obtained for other oils and fats, especially soy and palm oil. Vegetable oils are usually used for edible purposes and tend to be higher priced than tallow, which is used mainly for inedible processes such as soap manufacture, oleochemical production, and as an energy source in animal feed. Fats are a useful component of animal feeds, because they have about twice the energy content of protein and carbohydrates. They also reduce dust, improve color and texture, enhance palatability, increase pelleting efficiency, and reduce machinery wear during feed production. Tallow can be used as an oleochemical feedstock, for example, for fatty acid production, and then used to make abrasives, candles, cement additives, cleaners, cosmetics, paints, polishes, perfumes, detergents, plastics, synthetic rubber, and water-repellent compounds. Tallow and its methyl esters can also be used as a fuel oil. Considerable amounts of tallow are still used for soap making. Glycerine, a co-product of soap or fatty acid production, can be purified and used as a chemical in its own right.

Edible tallow and lard can be used in margarine, shortenings, and cooking fats. Tallow tends to give a better flavor to fried foods and is more stable during the cooking process than vegetable oils. However, because hard fats such as tallow can be associated with heart disease, the overall consumption of edible tallow has declined.

Uses of Meals

The material remaining after water and fat have been removed can have one of many names, depending on the rendering method and/or raw material used. These names include tankage; meat-meal tankage; digester tankage; wet-rendered tankage or feeding tankage (finely ground, dried residue from wet-rendered material low in hair, hoof, horn, manure, and paunch contents); digester tankage with bone; meat and bone meal digester tankage; meat and bone meal tankage or feeding tankage with bone (of higher phosphorus content than feeding tankage); meat meal (usually from dry rendering processes); meat and bone meal or meat and bone scrap (of higher phosphorus content than meat meal). Feather meal is made from finely ground, wet-rendered feathers. It is very digestible, but not nutritionally well balanced. Steamed bone meal (from wet rendering) or bone meal (from dry rendering) is defatted, dried, and ground bones suitable for animal feeding. Poultry meal is similar to meat meal in composition, appearance, and nutritional value, but made from poultry by-products. Fish meal, a high-quality meal, is similar to meat meal and will vary in composition depending on the type of fish processed.

The type of raw material rendered and the rendering process used influence the composition of the meal produced. Meals made from proteinaceous material have a high nitrogen content (the crude protein content is usually >50%) and are used in animal feeds. They also contain calcium, phosphorus, and fat. Meals made from material with a high bone content, such as bone meal and low-quality meat meal, have a relatively small market. One main use is as a fertilizer, but these meals are now being replaced by mineral fertilizers.

Most meat meal is sold as meat and bone meal with a typical composition of 50% minimum crude protein, 4–10% moisture, 8–16% fat, and 20–30% ash. The price received for a meal usually depends on its crude protein content, although some buyers may specify a minimum digestibility and availability of amino acids.

FINE CHEMICALS AND PHARMACEUTICALS

Animals have ductless (endocrine) glands that secrete hormones, and ducted glands and organs that release enzymes and other biologically active molecules. Many biological chemicals can be recovered. A description of the animal glands and the medicinal and pharmaceutical uses of by-products has been reviewed (11, 12).

The glands and tissues used for producing fine chemicals represent a small portion of the animal's live weight. Age, sex, and species of the animal determine the content of the active material. Most glands are very perishable and must be processed quickly to limit autolysis and microbial activity. The logistics of collecting sufficient raw material to operate a fine chemical plant or process usually limits operations to either larger meat plants or to centralized processing of material from many small plants.

SAUSAGE CASINGS

Intestines can be processed into natural casings used in sausage manufacture (13). Casing quality is affected by handling and cleaning procedures. Factors such as the age and species of the animal, breed, fodder consumed, and the conditions in which the animals are raised also affect casing quality and value. Animal casings naturally come in a wide variety of different shapes and sizes. The preference for a particular type of casing varies from country to country.

Reconstituted casings are manufactured from hide pieces (14), and artificial casings are often made from cellulose. Animal intestines are also used for the manufacture of surgical sutures and strings for musical instruments and tennis rackets. Intestinal tract not used for these purposes is converted into pet food or is rendered to yield meat meal and tallow.

PET FOOD

Inedible animal tissue and rendered material can be used in both wet and dry pet foods. Pet foods require high-quality ingredients, which means that renderers may have to segregate raw material. Generally, pet-food manufacturers are using less fatty tissue and tallow, and more protein-rich tissue and/or meat meal in their formulations.

BIBLIOGRAPHY

1. American Meat Institute Committee on Textbooks, *By-products of the Meat Packing Industry,* Institute of Meat Packing, University of Chicago, Chicago, 1958.
2. A. Levie, *The Meat Handbook,* 2nd ed., AVI Publishing Co., Inc., Westport, Conn., 1967.
3. F. Gerrard, "What Is Offal?" *Meat Trades Journal* (4398) 8 (Sept. 14, 1972).
4. P. Filstrup, *Handbook for the Meat By-products Industry.* Alfa-Laval Slaughterhouse By-products Department, Denmark, 1976.
5. C. W. Dill and W. A. Landmann, "Food Grade Proteins from Edible Blood," in A. M. Pearson and T. R. Dutson, eds., *Edible Meat By-products* (Advances in Meat Research, Vol. 5), Elsevier Applied Science, New York, 1988, pp. 127–145.
6. C. L. Knipe, "Production and Use of Animal Blood and Blood Proteins for Human Food," in A. M. Pearson and T. R. Dutson, eds., *Edible Meat By-products* (Advances in Meat Research, Vol. 5), Elsevier Applied Science, New York, 1988, pp. 147–165.
7. V. M. Gorbatov, "Collection and Utilization of Blood and Blood Proteins for Edible Purposes in the USSR," in A. M. Pearson and T. R. Dutson, eds., *Edible Meat By-products* (Advances in Meat Research, Vol. 5), Elsevier Applied Science, New York, 1988, pp. 231–274.
8. Q. T. Pham, "Behaviour of a Conical Spouted-bed Dryer for Animal Blood." *Canadian Journal of Chemical Engineering* **61,** 426–434 (1983).
9. J. E. Swan, *Maximising Yields from Conventional Blood Processing,* Meat Industry Research Institute of New Zealand Bulletin No. 10, Hamilton, New Zealand, 1985.
10. T. Fernando, "The MIRINZ Low-temperature Rendering System," *Proceedings of the 22nd Meat Industry Research Conference,* Meat Industry Research Institute of New Zealand Publ. No. 816, Hamilton, New Zealand, 1982, pp. 79–84.
11. H. W. Ockerman and C. L. Hansen, *Animal By-product Processing,* Ellis Horwood, Chichester, England, 1988.
12. R. J. Banis, "Pharmaceutical and Diagnostic By-products," in J. F. Price and B. S. Schweigert, eds., *The Science of Meat and Meat Products,* 3rd ed., Food and Nutrition Press, Inc., Westport, Conn., 1987.
13. R. E. Rust, "Production of Edible Casings," in A. M. Pearson and T. R. Dutson, eds., *Edible Meat By-products* (Advances in Meat Research, Vol. 5), Elsevier Applied Science, New York, 1988, pp. 261–274.
14. L. L. Hood, "Collagen in Sausage Casings," in A. M. Pearson, T. R. Dutson, and A. J. Bailey, eds., *Collagen as a Food* (Advances in Meat Research, Vol. 4), AVI Publishing Co., New York, 1987, pp. 109–129.

General References

Reference 11 gives a good overview.

A. M. Pearson and T. R. Dutson, eds., *Edible Meat By-products* (Advances in Meat Research, Vol. 5), Elsevier Applied Science, New York, 1988.

D. Swern, ed., *Bailey's Industrial Oil and Fat Products,* Vols. 1–3, 4th ed., John Wiley and Sons, Inc., New York, 1982.

"Meat Animal By-products and their Utilization," in J. F. Price and B. S. Schweigert, eds., *The Science of Meats and Meat Products,* 3rd ed., Food and Nutrition Press, Inc., Westport, Conn., 1987.

"Red Meat and Meat Products," in I. A. Wolff, ed., *CRC Handbook of Processing and Utilization in Agriculture,* Vol. 1, *Animal Products,* CRC Press, Inc., Boca Raton, Fla., 1982.

J. E. SWAN
The Meat Industry Research
Institute of New Zealand (Inc.)
Hamilton, New Zealand

ANIMAL FEEDS. See ANIMAL SCIENCE AND LIVESTOCK PRODUCTION; ANIMAL BY-PRODUCT PROCESSING; FOODBORNE DISEASES; FOOD PROCESSING: STANDARD INDUSTRIAL CLASSIFICATION; LIVESTOCK FEEDS.

ANIMAL SCIENCE AND LIVESTOCK PRODUCTION

The discipline of animal science is concerned with the application of the biological and physical sciences to the efficient production of livestock. Food and fiber are the primary products of livestock production in the United States. Examples of products derived from animal agriculture include meat, milk, eggs, wool, mohair, leathers, pharmaceuticals, and products used in research laboratories throughout the world.

BUSINESS ASPECTS OF LIVESTOCK PRODUCTION

Livestock are raised on a wide variety of terrains, climates, and management systems. However, concentrations of livestock raising often border on regions of grain raising (cattle and pigs are heavily concentrated in the Midwestern states of Iowa, Kansas, Nebraska, and Texas) or located near population centers (dairy cattle are concentrated in the Northeastern states, Wisconsin, and California). Fifty percent of the land area of the United States is classified as grazing or range area that cannot be used to cultivate crops (1). Ruminant livestock (cattle and sheep) are efficient utilizers of the millions of acres of land in the United States too rough or dry to grow crops.

Agriculture is the largest industry in the United States, accounting for 15% of the gross national product. About 2 million people are directly involved in production agriculture with ca 1.37 million agricultural operations involving beef or dairy cattle (1).

Beef Cattle

The beef cattle industry is the largest segment of American agriculture, accounting for almost 25% of all farm marketings (2). Cattle are raised in every state, including Hawaii and Alaska. Commercial cattle operations tend to be large and diverse enterprises that involve many types and breeds of cattle. The beef cattle industry in the United States is distinctly segmented into three production schemes: cow-calf (both commercial and purebred), stocker (intensive grazing), and feedlot. The term feeders is sometimes used to describe cattle, pigs, and lambs going into intensive feeding programs. Feeder cattle production is the goal of commercial (nonpurebred) cow-calf and stocker operations. Feeder cattle may be either steers or heifers and are the animals of highest value in terms of beef production. The singular goal of purebred cattle production is the selective breeding and selection of superior offspring to attain distinguishable characteristics of the breed, such as carcass merit, production efficiency, or reproductive traits.

Stocker cattle are weaned calves grazed on grass for inexpensive growth. This period is characterized by growth of lean tissues and frame size. In the Midwest, stocker cattle may be grazed on small grain pastures, grain stubble, or legume pastures before being placed in feedlots for fattening. Stocker cattle are usually owned for less than 1 year.

Fattening steers and heifers in feedlots involves feeding nearly market-sized cattle moderately high to high energy rations (grain) until they have reached a sufficient finish (fattening) to produce a carcass that will grade USDA choice. Large feedlots predominate in the Western states and account for a large percentage of fed-beef marketings. Midwestern feedlots tend to be more plentiful, but are generally smaller and produce a smaller percentage of fed-beef marketings. Most fat steers and heifers are slaughtered at 15–24 months of age.

Bulls are sometimes fed for slaughter and are usually referred to as bullocks. Bulls generally gain weight more rapidly and are more feed efficient than steers or heifers. Bulls also tend to be leaner. Older cows culled from beef or dairy cow herds are also sometimes fed grain for short periods of time before slaughter to increase the palatability (juiciness and tenderness) characteristics of the meat.

Dairy Cattle

Dairy farms are highly specialized agricultural enterprises, with considerable financial and labor input and returns. Dairying as a secondary enterprise on farms and ranches is uncommon. The number of dairy farms and dairy cattle in the United States has steadily declined during the past 20 years. The number of farms with dairy cattle decreased about 90% from 2 million in 1950 to 205,000 in 1989 (3, 4). However, this decrease has been offset by an approximately 50% increase in milk production (Fig. 1). Modern dairy cattle produce on average 15,000 lb of milk per lactation.

The price dairy farmers receive for whole (fluid) milk has been, historically, on hundred weight basis according to the percentage of butterfat. However, as the value of butterfat has declined as the most valuable component of whole milk, the price differential has increased less than the price per pound. The decrease in demand for butterfat is directly attributable to the introduction of margarine (containing plant oils), increased consumption of cheeses (made from whole milk), and processed food containing

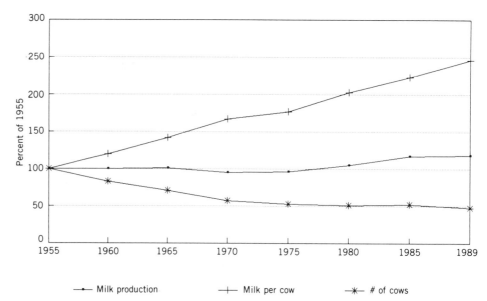

Figure 1. Index of milk production, milk per cow, and cow numbers in the United States (3, 4).

—•— Milk production   —+— Milk per cow   —*— # of cows

nonfat dry milk. Also, there has been a strong trend for health-conscious people to eat and drink only low fat and non-fat dairy products such as low fat cottage cheese, low fat yogurt and skim milk.

Locations of dairy farms have traditionally been close to major markets. This is because of the high cost of transporting whole milk, which contains a high proportion of water, and because milk is relatively perishable. However, in recent years location of dairies near consumers has been a less important consideration than the high cost of real estate near major markets. Also, conflicts between large dairies and home owners over water quality, dust, flies, and odors have forced dairy farms to relocate to less populated areas. Improved processing techniques and more rapid transportation networks also have contributed to decreased costs in moving milk to markets.

Swine

Swine production units are located in every state of the United States, but are generally concentrated in or near feed-grain-growing regions of the central United States. The corn belt states of Iowa, Illinois, Minnesota, Indiana, and Nebraska are leaders in market hog production (5). Swine farrowing was traditionally a spring and fall operation, but with the development of improved housing and nutrition, production of swine is a year-round business. This has tended to equalize the monthly marketings of pigs and reduced the seasonal fluctuations in market price.

The trend in pork production, as in other agricultural operations, has been toward larger and more specialized operations. There is also a trend toward more confinement units in the production of pigs. Swine production systems usually can be described by one of three basic systems: farrowing to finishing, feeder pig production, and finishing feeder pigs.

Finished pigs are usually sold directly to the processor by the producer. Selecting the market and timing the presentation of pigs for slaughter are two of the most important decisions made by the swine producer. Most pigs are

sold on a live-weight basis, but some are sold based on carcass yield and grade. Computerized hog-marketing systems are also being used in some parts in the United States.

Sheep

The products of sheep production businesses are primarily lambs and mutton for meat and wool. Most of the sheep in the United States are raised in the Western states. Texas, California, Colorado, South Dakota, Wyoming, and Utah account for more than 50% of the sheep produced (6). Sheep production systems are generally classified as farm flocks or range flocks. Predators, such as coyotes and wild dogs, and disease are some of the common problems associated with raising sheep. Lamb mortality rates as high as 20% are not uncommon. Most feeder lambs are sold to feedlots weighing between 65–80 lb. Ewes culled from flocks are also marketed for slaughter. The seasonality of breeding for sheep unfortunately causes a shortage of fresh lamb during certain times of the year.

THE DISCIPLINE OF ANIMAL SCIENCE

The multiple and interdisciplinary needs of farm animal production that focus on growth, reproduction, and lactation are at the center of animal science programs in research, teaching, and extension. Animal science is primarily an applied science that incorporates advances in science and technology from all disciplines. Nearly all land-grant universities, as well as several state and private colleges with agricultural programs, offer a bachelor of science degree in animal science. Most animal science departments in land-grant universities also offer graduate programs leading to master of science (M.S.) and doctor of philosophy degrees (Ph.D.). The Ph.D. degree is a research-based science degree.

Curriculum at the undergraduate level is designed to prepare students for professional careers in animal agriculture. The student majoring in animal science usually takes specialized courses, which include nutrition, breeding and genetics, biochemistry, animal reproduction, agricultural economics, animal health, meat science, dairy science, and animal production and products courses.

The Journal of Animal Science, The Journal of Dairy Science, and, *Meat Science* serve as the repository for advances in the science of livestock production as well as their products.

PRINCIPLES OF ANIMAL REPRODUCTION AND BREEDING

Productivity of livestock depends on reproductive efficiency and is usually measured by the number of offspring produced by an animal or herd per unit of time. Therefore, management of reproductive cycles is critical for obtaining maximum efficiency of reproducing animals.

Animal Reproduction

Dairy and beef cattle are derived from common ancestors and thus have common reproductive characteristics. The period of time from one ovulation to the next ovulation is called the estrous cycle. The estrous cycle in cattle is on average 21 days in length and is characterized as being one in which the cow does not permit mating to occur except at the time of ovulation. This period of time in which mating is permitted is relatively short, being in the order of 16–18 h. The estrous cycle in cattle is continuous (polyestrous) throughout the year and is not seasonal as is the case in some other ruminant-type animals such as sheep, deer, antelope, and elk. Cattle ovulate 10–14 h after the end of estrus, and normally one follicle ovulates per estrous cycle. The gestation period is 283 days, and twins occur only 1% of the time.

Sheep are generally seasonally polyestrous, with recurring estrous cycles during the fall of the year followed by a prolonged quiescent period. Some breeds of sheep that originated in equatorial regions and are subject to less variations in temperature and photoperiod have longer breeding seasons. The period of sexual receptivity in sheep lasts for 24–36 h, but may vary widely. The length of the estrous cycle is 14–19 days. Seasonal breeding in animals is associated with an increased frequency and size of episodic releases of luteinizing hormone (7). Ovulation normally occurs near the end of estrus. Ovulation rate is influenced by breed and nutrition, but twinning is extremely common in sheep. Gestation is 150 days in length.

The pig, like the cow, is polyestrous. The period of sexual receptivity in pigs lasts about 40–60 h and may be influenced by season, breed, and endocrine dysfunction (8). The length of the estrous cycle is 21 days. Ovulation rates are strongly influenced by weight of the gilt at breeding and the amount of inbreeding. Gestation in the pig is 118 days and average litter size is in the order of 9–12 pigs per litter.

The estrous cycle in all livestock is characterized by profound changes in behavioral patterns and blood hormone profiles. Cyclic changes during the estrous period reflect the secretory functions and interdependence of the ovary, uterus, hypothalamus, and pituitary gland. The estrous cycle is controlled by ovarian hormone secretions (estrogens and progesterone) and may be subdivided into a follicular phase, an ovulatory phase, and a luteal phase (Fig. 2). The ovary functions to provide fertilizable ova and a balanced ratio of steroid hormones to facilitate development of the reproductive tract for migration of the early embryo and successful implantation in the uterus.

The male reproductive system in livestock is comprised of two testes (producing both sperm and the male sex hormone testosterone), excretory ducts (epididymis, vas deferens, and ejaculatory duct), and accessory structures (prostate, seminal vesicles, bulbourethral glands, and penis). The scrotum, containing two testes, provides for efficient regulation of testicular temperature. Descent of both testes from the abdomen to their proper scrotal location is necessary for maximal fertility. Failure of one or both testes to descend is a common reproductive organ defect in livestock, especially in swine, where the condition is a hereditary defect transmitted by the male and is referred to as cryptorchidism (9).

Reproductive management of the male primarily involves maintenance of health and nutrition to optimize sperm production and libido. Methods for assessing individual male fertility, such as breeding soundness evaluations, are subjective and poorly correlated with pregnancy rates.

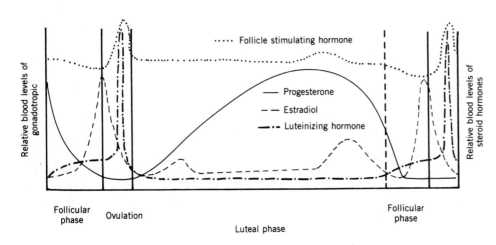

Figure 2. Cyclic changes in ovarian steroid and gonadotropic hormones in the ewe.

Breeds of Livestock

Decisions about breeding are some of the most important decisions a livestock producer must make. The livestock breeder must consider the heritability of a characteristic, such as prolificacy, feed conversion, milk yield, or carcass merit, for example. The livestock breeder also must consider the merits of inbreeding, outbreeding, crossbreeding, and the relative merits of different breeds of animals.

The development of modern breeds of livestock began in the 1700s. The origin of cattle and sheep breeds can be traced to Europe and the British Isles. For classification purposes there are four basic categories of modern beef and dairy cattle. The classification system for cattle is not consistent because two of the categories reflect geographical origin of the cattle and the other two reflect the purpose of the cattle.

1. British and continental breeds (beef): Includes Angus, Hereford, Shorthorn, Charolais, Limousin, Chianina, Gelbvieh.
2. North American breeds (beef): Brahman, Brangus, Beefmaster, and Santa Gertrudis.
3. Dual-purpose breeds (beef and dairy): Ayrshire Milking Shorthorn, Red Poll, and Brown Swiss.
4. Dairy breeds: Guernsey, Holstein, and Jersey.

The foundation of modern-day pigs can be traced to European and Asiatic strains of pigs. Modern swine are raised to produce lean high-quality pork. However, not long ago, breeds of swine were classified as lard-type and bacon-type. Although the individual breeds of swine continue today, all swine producers strive for the same high-quality meat-type pig.

As in the case of cattle, the classification system for breeds of sheep is inconsistent. The Rambouillet and Merino breeds were developed for the production of the fine wool characteristic, but much lamb and mutton comes from these breeds. Likewise, income derived from wool represents a significant portion of the value of the meat-type breeds.

Animal Breeding

Crossbreeding is one of the quickest and most economical methods for increasing total beef or pork production. Heterosis, also sometimes commonly referred to as hybrid vigor, is the result of crossbreeding. Increased performance and vigor is not limited to animal breeding, but also results in hybrid plant production. Hybrid increases are seen in the first generation (F1) crossbred animals. F1 animals are the result of crossbreeding two purebred lines of livestock. Increased performance benefits may be as high as 25% for F1 calves. Hybrid chickens and crossbred pigs have contributed to the increased economics of poultry and pork production.

Three general crossbreeding systems are used to produce market animals: (1) Mating females of one breed with males of another to produce a F1 market animal; (2) three-breed terminal crossbreeding, where F1 females produced according to (1) are mated to bulls of a third breed to produce a three-breed market animal with a max-

imal level of heterosis (alternatively, the F1 female could be mated to one of the original parent breed to produce a backcross market animal); (3) rotational crossbreeding, where the breed of the sire is rotated among the three original breeds. Rotational breedings usually only use three breeds because the heterosis gained by the addition of a fourth breed is small.

PRINCIPLES OF ANIMAL NUTRITION AND FEEDING

The proper feeding of livestock is a matter of supplying them with the correct amount of nutrients essential for reproduction, growth, or lactation. Nutrients are chemical elements and compounds required by the cells of the animal's body in support of the three basic functions: (1) structural matter for building and maintaining the body; (2) a source of energy for work, thermogenesis (heat production), and fat deposition; and (3) regulating body processes or the synthesis of body regulators. An auxiliary function of nutrients would be their use in milk production.

The Nutrients

Nutrients required for livestock, just like for humans, can be categorized into six functional or chemical classes: carbohydrates, proteins, fats, minerals, vitamins, and water. All carbohydrates or saccharides are related structurally and chemically and contain the same amount of gross energy. Carbohydrates are comprised primarily of hexose ($C_6H_{12}O_6$) and pentose ($C_5H_{10}O_5$) molecules. Tetrose and triose molecules are present in small quantities but are generally not important in animal nutrition. Carbohydrates are usually categorized as monosaccharides, disaccharides, and polysaccharides based on how many hexose and pentose molecules are linked together. Common monosaccharides, also called simple sugars, consist of glucose, fructose, and galactose. Disaccharides consist of two monosaccharides linked together with hydroxyl groups of each sugar unit. The common monosaccharides include sucrose (table sugar), maltose, and lactose. Polysaccharides have the empirical formula $(C_6H_{10}O_5)_n$ and contain large polymers of monosaccharides. Polysaccharides important in livestock nutrition include starch, glycogen, and fiber (hemicellulose, cellulose, and lignin).

Carbohydrates are important energy sources in livestock feeds, comprising 65–80% of plant dry weight. Carbohydrate as a class of nutrients is usually divided into two groups: nitrogen free extract (NFE) and fiber. Fiber is what remains after a feed has been boiled in dilute acid or alkali and roughly approximates the amount of carbohydrate poorly digested in the animal's intestinal tract. NFE represents the soluble, readily digestible carbohydrate portion of a feedstuff. Corn grain, eg, contains only 2.2% fiber and 70% NFE. Good-quality alfalfa hay contains 26% fiber and 46% NFE. Cellulose is not an efficient energy source for nonruminant livestock (swine and poultry) but can be readily digested by the bacteria of the rumen. Because of this characteristic, ruminant livestock occupy an important niche in utilizing a potentially wasted feed resource.

Lipids are water-insoluble organic molecules that can be extracted from plant and animal tissues with nonpolar solvents such as benzene and ether. Lipids contain 2.25 times more energy per unit weight than either carbohydrate or protein. The term lipid is used in a general sense and used interchangeably with fat and oils. Livestock rations are generally low in fat, with most grains and roughages containing less than 5% lipid. Fats may sometimes be added to a ration, especially for swine and poultry, and high producing dairy cattle when higher energy rations are desired.

Lipids are composed of carbon, hydrogen, and oxygen, as are carbohydrates, but contain less oxygen. Lipids have several important biological functions, including storage and transport forms of energy and components of cell surface membranes. The fatty acid composition of body fat may be altered by the composition of dietary fats. This is especially true for swine and poultry.

Protein constitutes the most expensive portion of livestock feed. Protein primarily consists of 20 α amino acids linked together by peptide bonds. Both plant and animal tissues contain a diversity of proteins with variable amounts of amino acids. Ten amino acids cannot be synthesized by animal tissues or can be synthesized only to a limited extent. These 10 amino acids are referred to as dietary essential amino acids (Table 1). The primary function of dietary protein in livestock rations is to supply amino acids, the building blocks of proteinaceous body tissues. Ruminant livestock do not have dietary amino acid requirements as such, but depend on ruminal bacterial protein synthesis to supply these nutrients. Because of this, ruminant animals are able to utilize nonprotein nitrogen, such as urea, to replace part of the natural protein in the ration. The use of rumen bypass protein supplements, such as corn gluten meal and blood meal, has been shown to increase growth in beef cattle and milk production in dairy cattle (10, 11). Swine ration's are usually balanced for the first two or three most-limiting amino acids as well as for balancing the ration for crude protein.

Sixteen or more minerals are required by livestock. Of these, seven are classified as macrominerals (Ca, P, K, Na, S, Mg, Fe), with the remaining constituting microminerals (Co, Cu, F, I, Fe, Mn, Mo, Se, Zn). A description of the minerals function and deficiency characteristics in livestock nutrition has been reviewed in Ref. 12.

Vitamins are dietary-essential organic molecules that are required in minute quantities by livestock. Vitamins are usually classified into two groups: water soluble and fat soluble (Table 2). Approximately 15 vitamins are known to function in animal metabolism, but only several of these are needed in the ration of livestock because synthesis of certain vitamins occurs in the animal. Animal

Table 1. Dietary Essential Amino Acids Required by Swine and Poultry

Arginine	Lysine	Tryptophan
Histidine	Methionine	Valine
Isoleucine	Phenylalanine	Glycine (poultry)
Leucine	Threonine	Proline (poultry)

Table 2. Fat Soluble and Water Soluble Vitamins Required by Livestock

Fat Soluble	Water Soluble
Vitamin A	Thiamine
Vitamin D	Riboflavin
Vitamin E	Niacin
Vitamin K	Pyridoxine
	Biotin
	Pantothenic acid
	Vitamin B_{12}
	Choline
	Folic acid
	Inositol
	Para-aminobenzoic acid

nutritionists formulating livestock rations are especially concerned about vitamins A and D for ruminants and certain B vitamins for nonruminants. Ruminants do not generally require B vitamin supplementation because the bacteria of the rumen synthesize sufficient quantities. Vitamin E is normally present in ample quantities in natural feeds, but may be supplemented to ruminants and nonruminant livestock receiving processed grain diets. Vitamin E supplementation of cattle diets also has been shown to stabilize the red color of fresh beef and help prevent oxidative rancidity (13). Livestock normally do not need ascorbic acid supplementation where adequate quantities are synthesized in the tissues. Ref. 12 outlines a review of vitamins in livestock nutrition.

Digestion

Digestion involves the physical and chemical preparation of feed in the gastrointestinal tract to absorption-ready nutrients. The digestive systems of livestock are anatomically and functionally similar, with the very important distinction of the ruminant having a large, four-compartment stomach (Fig. 3). The abomasum or the true stomach is functionally similar to the stomach of monogastrics. The rumen, the largest of the four compartments, with a capacity approaching 50 gal, functions as a fermentation vat and contains billions of bacteria and protozoa. The presence of the bacteria and the enzymes that they secrete enables the ruminant to efficiently use cellulose from fibrous plants and feedstuffs. The microorganisms also synthesize amino acids to make bacterial protein and supply the ruminant animal with essential amino acids.

RED MEAT PRODUCTION

The slaughtering procedure for beef and pork usually involves immobilization, bleeding, removing hair (pigs) or skin (cattle and sheep), removing viscera—trachea, esophagus, lungs, heart, stomach, intestines, and reproductive organs. Immobilization of meat animals, except those killed for kosher meat, is accomplished according to humane, accepted methods (captive bolt or electric shock, eg) regulated by state and federal laws. Most commercial slaughtering plants use a movable rail where immobilized animals are individually hung by their shanks and moved

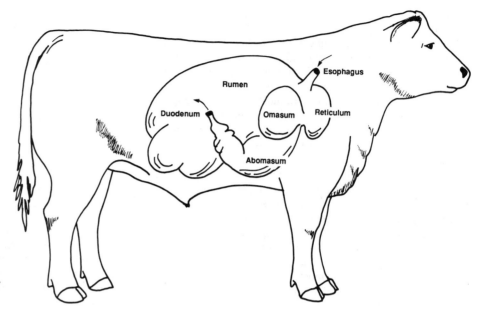

Figure 3. The digestive system of the polygastric ruminant.

along a deassembly line to dress the slaughtered animal. Electrical stimulation of fresh beef, pork, veal, and lamb carcasses is sometimes employed to improve tenderness (14). To facilitate rapid chilling of the carcass, beef carcasses are entirely split along the backbone, while pig carcasses are usually split to the neck.

The term carcass is what remains after dressing and usually refers to both sides of the same slaughtered animal. A side is one of the two halves of a split carcass. Beef quarters are the portions remaining after splitting a side of beef between the 12th and 13th rib. Wholesale cuts are divisions of the quarters including the round, full loin or short loin and sirloin, flank, rib, short plate, chuck, brisket, fore shank. Primal cuts are any of the wholesale cuts except for short plates, briskets, flanks, and fore shanks.

The principal cuts of beef, pork, and lamb are shown in Figures 4, 5, and 6. Most beef is sold as carcass sides or quarter or wholesale cuts to retailers, wholesalers, chainstore breaking plants, and restaurants. However, an increasing portion of beef is sold as 60–80 lb saw or knife-ready subprimal cuts shipped in boxes. Processors nearly always reduce pork carcasses to wholesale and retail cuts at the plant because hams and bellies, and sometimes other cuts, are smoked and cured. Most lamb is sold as intact carcasses, but some carcasses are reduced to wholesale cuts.

Significant quantities of beef are ground, tested for lean content, and blended to meet federal and state quality requirements for fat content. Low-quality carcasses, such as cutter and canner grades of beef, pale and soft pork carcasses, and cull grades of lamb and mutton, are not sold as wholesale or retail cuts but are instead boned-out at the plant. The edible meat from these carcasses may be sold as hamburger, or used in the preparation of a variety of meat products, such as canned meat products or sausage.

The USDA through the Food Safety and Inspection Service (FSIS) is responsible for ensuring the wholesomeness of the meat supply. FSIS assurance includes ante-

mortem and postmortem inspection and compliance with sanitation and temperature, ingredients, and label requirements. Meat bearing the U.S. inspected and passed stamp informs the buyer the meat was processed in a federally approved plant and is wholesome and safe for human consumption. Additionally, the carcass may be graded for meat yield and quality. The yield grade is an estimate of the portion of the carcass that is edible meat. Quality grades are intended to identify meat on the basis of palatability and cooking attributes. Factors used to establish quality grades include kind or species of animal, class or physical attributes, maturity, marbling, firmness of flank, color and structure of lean, and conformation, fleshing, and finish of the carcass. USDA quality grades for various kinds of meat are given in Table 3.

The Food and Drug Administration (FDA) is responsible for ensuring the safety of the drugs used in livestock production and food additives. Both the FDA and the Federal Trade Commission have been involved with monitoring and regulating claims made during advertising of food products.

MILK PRODUCTION

Dairy cattle are normally milked two or three times per 24 h day. Following freshening (parturition), maximum milk production is normally reached after 45 days. A slow, but steady decline in milk produced by the cow occurs after peak yield until she is dried. Normal lactation periods for commercial dairy farms is 305 days lactation and 60 days dry (Fig. 7). The dry period allows for the cow to replenish her body reserves and rest her mammary tissue. Because a calf must be produced every 12.5 to 13.5 months in order to meet this idealized production schedule, cows are bred at approximately 80–90 days of lactation.

Feeding dairy cattle represents approximately 50% of the cost of milk production. The percentage of protein and fat may be influenced by feeding and management prac-

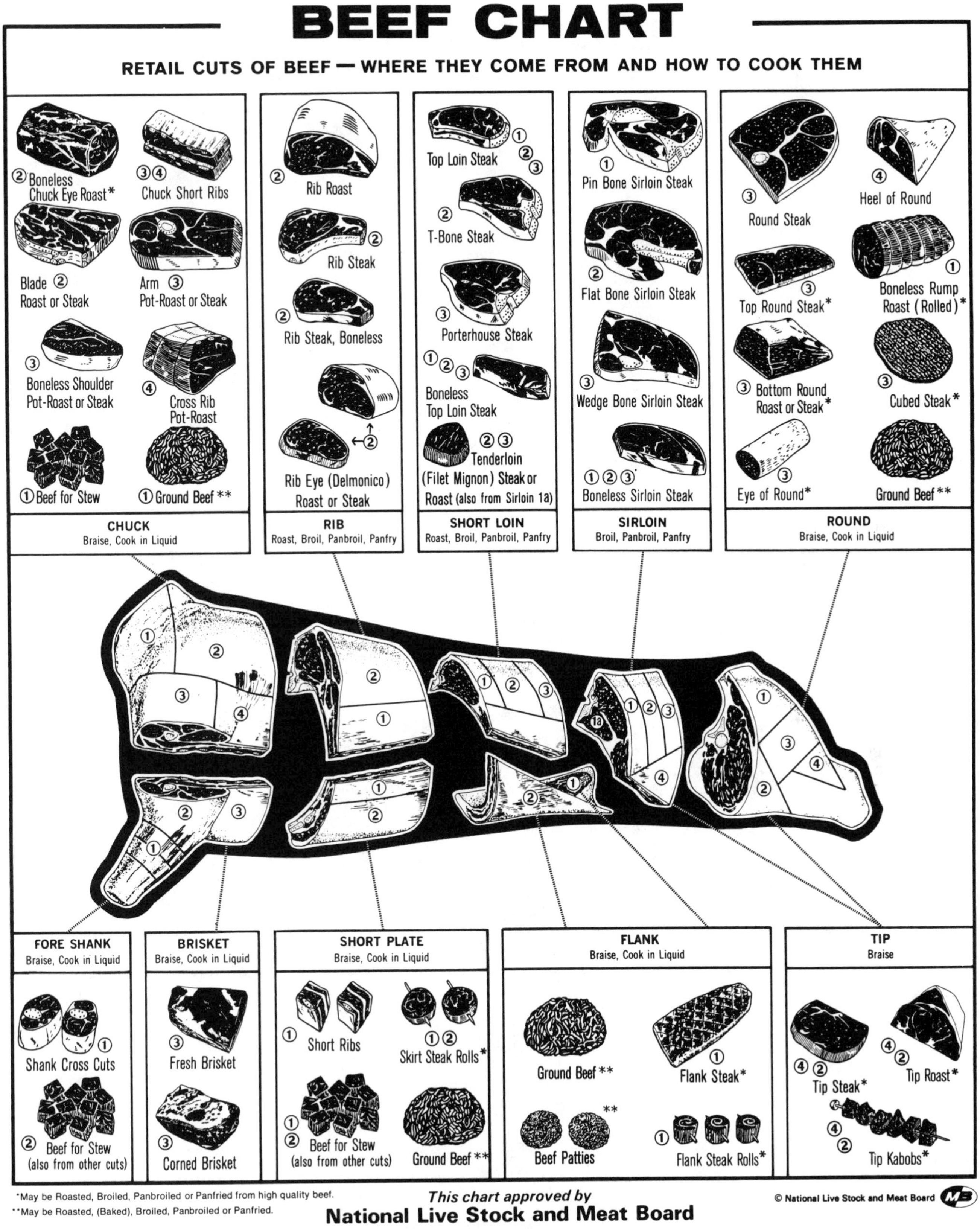

BEEF CHART

RETAIL CUTS OF BEEF — WHERE THEY COME FROM AND HOW TO COOK THEM

CHUCK
Braise, Cook in Liquid

- ② Boneless Chuck Eye Roast*
- ③④ Chuck Short Ribs
- Blade ② Roast or Steak
- Arm ③ Pot-Roast or Steak
- ③ Boneless Shoulder Pot-Roast or Steak
- ④ Cross Rib Pot-Roast
- ① Beef for Stew
- ① Ground Beef**

RIB
Roast, Broil, Panbroil, Panfry

- ② Rib Roast
- ② Rib Steak
- ② Rib Steak, Boneless
- ② Rib Eye (Delmonico) Roast or Steak

SHORT LOIN
Roast, Broil, Panbroil, Panfry

- ①②③ Top Loin Steak
- ② T-Bone Steak
- ③ Porterhouse Steak
- ①②③ Boneless Top Loin Steak
- ②③ Tenderloin (Filet Mignon) Steak or Roast (also from Sirloin 1a)

SIRLOIN
Broil, Panbroil, Panfry

- ① Pin Bone Sirloin Steak
- ② Flat Bone Sirloin Steak
- ③ Wedge Bone Sirloin Steak
- ①②③ Boneless Sirloin Steak

ROUND
Braise, Cook in Liquid

- ③ Round Steak
- ④ Heel of Round
- ③ Top Round Steak*
- ① Boneless Rump Roast (Rolled)*
- ③ Bottom Round Roast or Steak*
- ③ Cubed Steak*
- ③ Eye of Round*
- ③ Ground Beef**

FORE SHANK
Braise, Cook in Liquid

- ① Shank Cross Cuts
- ② Beef for Stew (also from other cuts)

BRISKET
Braise, Cook in Liquid

- ③ Fresh Brisket
- ③ Corned Brisket

SHORT PLATE
Braise, Cook in Liquid

- ① Short Ribs
- ①② Skirt Steak Rolls*
- ①② Beef for Stew (also from other cuts)
- ②③ Ground Beef**

FLANK
Braise, Cook in Liquid

- Ground Beef**
- ① Flank Steak*
- ① Beef Patties**
- ① Flank Steak Rolls*

TIP
Braise

- ④② Tip Steak*
- ④② Tip Roast*
- ④② Tip Kabobs*

*May be Roasted, Broiled, Panbroiled or Panfried from high quality beef.
**May be Roasted, (Baked), Broiled, Panbroiled or Panfried.

This chart approved by
National Live Stock and Meat Board

© National Live Stock and Meat Board **MB**

Figure 4. Wholesale and retail cuts of beef.

RETAIL CUTS OF PORK

WHERE THEY COME FROM AND HOW TO COOK THEM

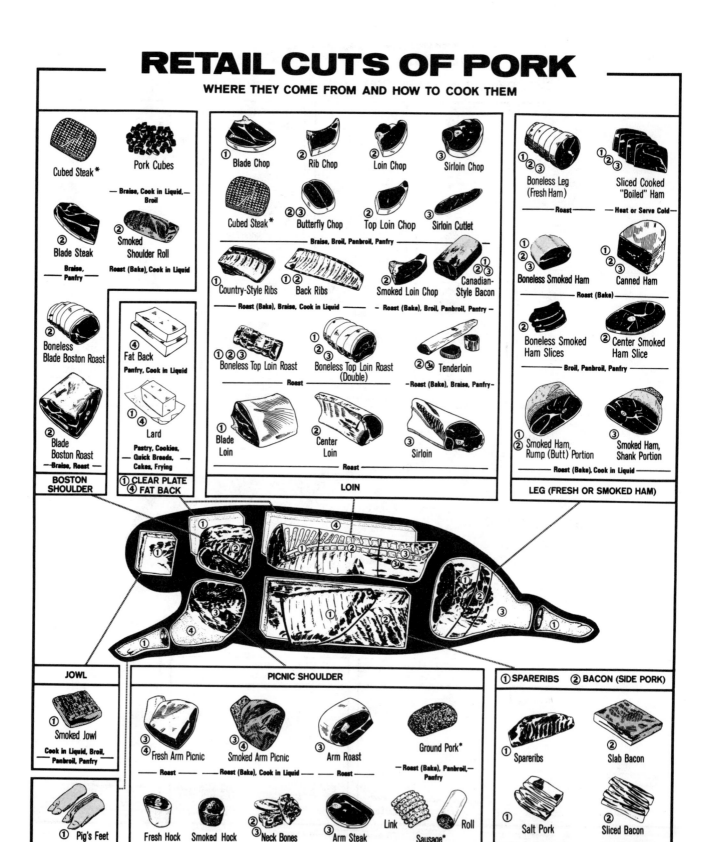

BOSTON SHOULDER

Cubed Steak*

Pork Cubes

— Braise, Cook in Liquid,— Broil

② Blade Steak

Braise, Panfry

② Smoked Shoulder Roll

Roast (Bake), Cook in Liquid

② Boneless Blade Boston Roast

② Blade Boston Roast

— Braise, Roast —

① CLEAR PLATE
④ FAT BACK

④ Fat Back

Panfry, Cook in Liquid

①④ Lard

Pastry, Cookies, Quick Breads, Cakes, Frying

LOIN

① Blade Chop

② Rib Chop

② Loin Chop

③ Sirloin Chop

②③ Cubed Steak*

②③ Butterfly Chop

② Top Loin Chop

③ Sirloin Cutlet

— Braise, Broil, Panbroil, Panfry —

① Country-Style Ribs

①② Back Ribs

Smoked Loin Chop

②③ Canadian-Style Bacon

— Roast (Bake), Braise, Cook in Liquid — — Roast (Bake), Broil, Panbroil, Pantry —

①②③ Boneless Top Loin Roast

①②③ Boneless Top Loin Roast (Double)

②③ⓐ Tenderloin

— Roast — — Roast (Bake), Braise, Panfry —

① Blade Loin

① Center Loin

③ Sirloin

— Roast —

LEG (FRESH OR SMOKED HAM)

①②③ Boneless Leg (Fresh Ham)

①②③ Sliced Cooked "Boiled" Ham

— Roast — — Heat or Serve Cold —

①②③ Boneless Smoked Ham

①②③ Canned Ham

— Roast (Bake) —

① Boneless Smoked Ham Slices

② Center Smoked Ham Slice

— Broil, Panbroil, Panfry —

①② Smoked Ham, Rump (Butt) Portion

③ Smoked Ham, Shank Portion

— Roast (Bake), Cook in Liquid —

JOWL

① Smoked Jowl

Cook in Liquid, Broil, Panbroil, Panfry

① Pig's Feet

— Cook in Liquid, Braise —

PICNIC SHOULDER

④ Fresh Arm Picnic

③④ Smoked Arm Picnic

③ Arm Roast

Ground Pork*

— Roast — — Roast (Bake), Cook in Liquid — — Roast — — Roast (Bake), Panbroil,— Panfry

Fresh Hock

Smoked Hock

② Neck Bones

③ Arm Steak

Link

Roll

Sausage*

— Braise, Cook in Liquid — — Cook in Liquid — — Braise, Panfry — — Panfry, Braise, Bake —

① SPARERIBS   ② BACON (SIDE PORK)

① Spareribs

② Slab Bacon

① Salt Pork

② Sliced Bacon

Bake, Broil, Panbroil, Panfry, Cook in Liquid

— Bake, Broil, Panbroil,— Panfry

*May be made from Boston Shoulder, Picnic Shoulder, Loin or Leg.

This chart approved by
National Live Stock and Meat Board

Figure 5. Wholesale and retail cuts of pork.

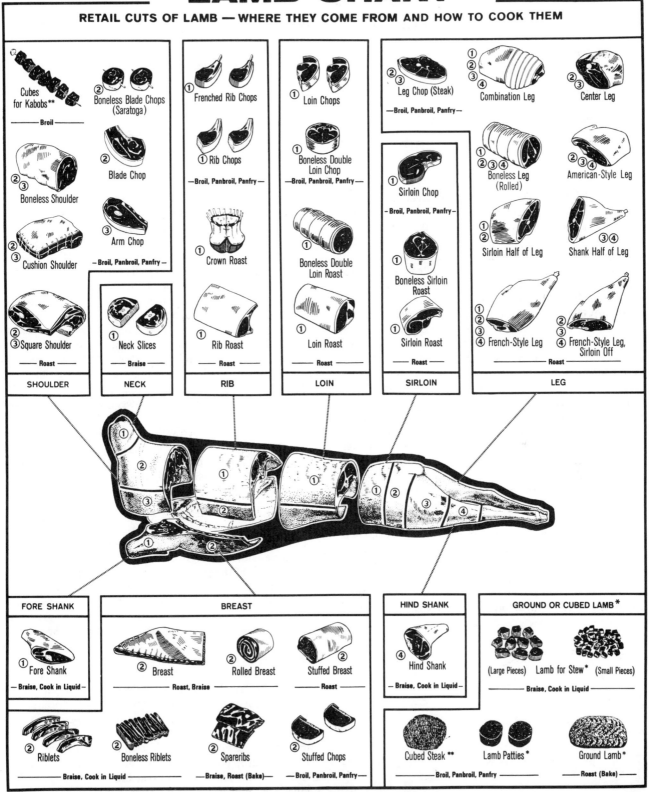

LAMB CHART

RETAIL CUTS OF LAMB — WHERE THEY COME FROM AND HOW TO COOK THEM

SHOULDER

Cubes for Kabobs**
— Broil —

② Boneless Blade Chops (Saratoga)

② Blade Chop

③ Arm Chop
— Broil, Panbroil, Panfry —

②③ Boneless Shoulder

②③ Cushion Shoulder

②③ Square Shoulder
— Roast —

NECK

① Neck Slices
— Braise —

RIB

② Frenched Rib Chops

① Rib Chops
— Broil, Panbroil, Panfry —

① Crown Roast

① Rib Roast
— Roast —

LOIN

① Loin Chops

① Boneless Double Loin Chop
— Broil, Panbroil, Panfry —

① Boneless Double Loin Roast

① Loin Roast
— Roast —

SIRLOIN

②③ Leg Chop (Steak)
— Broil, Panbroil, Panfry —

① Sirloin Chop
— Broil, Panbroil, Panfry —

① Boneless Sirloin Roast

① Sirloin Roast
— Roast —

LEG

①②③④ Combination Leg

②③ Center Leg

①②③④ Boneless Leg (Rolled)

②③④ American-Style Leg

①② Sirloin Half of Leg

③④ Shank Half of Leg

①②③④ French-Style Leg

②③④ French-Style Leg, Sirloin Off
— Roast —

FORE SHANK

① Fore Shank
— Braise, Cook in Liquid —

② Riblets
— Braise, Cook in Liquid —

BREAST

② Breast

② Rolled Breast

② Stuffed Breast
— Roast, Braise — — Roast —

② Boneless Riblets

② Spareribs
— Braise, Roast (Bake) —

② Stuffed Chops
— Broil, Panbroil, Panfry —

HIND SHANK

④ Hind Shank
— Braise, Cook in Liquid —

GROUND OR CUBED LAMB*

(Large Pieces) Lamb for Stew* (Small Pieces)
— Braise, Cook in Liquid —

Cubed Steak **

Lamb Patties *

Ground Lamb *
— Broil, Panbroil, Panfry — — Roast (Bake) —

* Lamb for stew or grinding may be made from any cut.

**Kabobs or cube steaks may be made from any thick solid piece of boneless Lamb.

This chart approved by
National Live Stock and Meat Board

© National Live Stock and Meat Board

Figure 6. Wholesale and retail cuts of lamb.

57

Table 3. USDA Grades for Meat[a]

Kind	Class	Grade Names
Beef quality grades	Steer, heifer, cow	Prime, choice, good, standard, commercial, utility, cutter, canner
	Bullock	Prime, choice, good, standard, utility
Beef yield grades	All classes	1, 2, 3, 4, 5
Calf quality grades		Prime, choice, good, standard, utility, cull
Veal quality grades		Same as calf
Lamb and mutton quality grades	Lamb, yearling lamb	Prime, choice, good, utility
Lamb yield grades	All classes	1, 2, 3, 4, 5
Pork carcasses	Barrow, gilt	U.S. No. 1, U.S. No. 2, U.S. No. 3, U.S. No. 4, U.S. utility
	Sows	U.S. No. 1, U.S. No. 2, U.S. No. 3, U.S. No. 4, U.S. utility, cull

[a] Ref. 15.

tices on dairy farms. Dairies may select feeding programs to produce fat and produce percentages that best suit milk pricing schemes. However, because milk flavor is contained in the lipid component of milk, feed flavors are the most common causes of off-flavor in milk. Color of milk may also vary among breeds. Guernsey cattle milk, eg, has a more yellow color than milk from Holstein cattle.

The term for the facilities used for the milking of cows is the milking parlor. Modern milking parlors must meet rigid sanitation requirements, keep cows comfortable, and be labor efficient. In comparison to the Federal Meat Inspection Service, state and local laws regulate cow health, sanitation, and milk-handling equipment. Dairy cows are no longer milked by hand but instead by modern machines that rapidly move milk from the milking parlor to refrigerated storage tanks. Milking machines function to remove milk from the cow's udder by applying partial vacuum to the teat end. This vacuum is alternately applied by pulsation consisting of milk and rest cycle 45–60 times per minute.

Milk moves from the cow through a series of stainless-steel pipes to a bulk cooling–storage tank. Milk must be cooled rapidly to approximately 40°F to prevent bacterial growth. Some states allow for in-line rapid cooling coils. However, thorough cleaning of the coils must be ensured.

Most dairies sell their milk directly to processors or through cooperative milk marketing associations. Fluid milk is usually marketed and priced under federal milk marketing orders, in accordance with the Agricultural Marketing Act of 1937. The objective of the federal order is to promote and maintain orderly milk marketing conditions for dairy farmers and assure consumers of an adequate supply of milk.

Milk production is somewhat seasonal, with more milk being produced in the spring. The price of milk paid to dairy farmers is usually on the hundred weight basis adjusted for percentage of components. Milk protein and fat are usually considered. The pricing scheme will consider local market factors and final use of the milk. For example, cheese plants usually will consider milk protein in establishing prices paid to producers. Milk purchased for the manufacture of cheese may be priced on a cheese yield formula that estimates the quantity of final product.

Almost 50% of the milk produced by farms is processed for consumption as whole, low-fat, or skim milk. All milk is pasteurized to destroy disease-producing microorganisms that may be present in unprocessed milk. The pasteurization process involves heating the milk to 161°F for 15 s in a continuous flow process.

FUTURE DEVELOPMENTS

The productivity of modern livestock production was made possible through basic and applied research. The application of the principles of genetics, discovery of vitamins and other nutrients necessary for health and productivity of livestock, and the control of reproduction have made large contributions to the economy of producing food. However, for livestock production and its products to remain economically competitive the emergence of new technologies must continue.

Several new technologies and discoveries with applications in livestock production hold promise for enhancing the competitiveness of United States livestock production. Recombinant DNA technology is largely responsible for the production of commercially abundant supplies of somatotropin, also commonly known as growth hormone, to enhance milk and meat production. Porcine somatotropin (PST) administration in pigs increases rate of gain, feed efficiency, and carcass muscling, while significantly reducing carcass fat deposition (16). This technology, along with other methods to regulate lean tissue growth, such as β agonists, will help the pork and beef industry produce meat that is leaner and thus help to reduce fat consumption in the American diet.

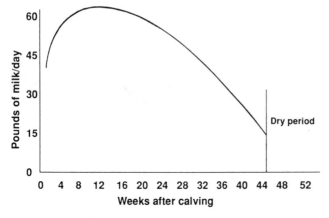

Figure 7. Typical lactation curve.

Bovine somatotropin (BST) administration to dairy cattle to enhance milk production has been shown to be efficacious. Depending on dosage of somatotropin administered, milk production will be enhanced by 10–40% without significant changes in the composition of the milk (17). However, at this time, concern about the impact of this technology on the small dairy farmer is confounding the introduction of somatotropin for use in dairy cattle in the United States. There is little doubt that BST will be used in Eastern Europe, Africa, and Central and South America to enhance milk production.

Artificial insemination and embryo transplantation of superior lines of swine, sheep, and dairy and beef cattle will continue to increase in popularity among livestock breeders. The ability of scientists to develop improved embryo freezing techniques will greatly enhance the efficacy of this technology. The successful introduction of transgenic livestock with improved rates of growth, carcass characteristics, lactation, and disease resistance will influence the methods and efficiency of livestock production (18).

BIBLIOGRAPHY

1. *Factsheet, "The Beef Industry,"* National Cattlemen's Association, Englewood, Colo., 1990.

2. C. Lambert, *Beef Industry Facts,* National Cattlemen's Association, Englewood, Colo., 1990.

3. *Dairy Producer Highlights,* National Milk Producers Federation, Arlington, Va. 1982.

4. *Agricultural Statistics,* U.S. Department of Agriculture, U.S. Government Printing Office, Washington, D.C., 1990.

5. *Fact Book of Agriculture,* U.S. Department of Agriculture, Miscellaneous Publication No. 1063, 1989.

6. *Agricultural Statistics,* U.S. Department of Agriculture, U.S. Government Printing Office, Washington, D.C., 1989.

7. P. Yuthasastrakosal and co-workers, "Release of LH in Anoestrous and Cyclic Ewes," *Journal of Reproduction and Fertility* **50,** 319–321 (1977).

8. L. Anderson, "Pigs," in E. S. E. Hafez, ed., *Reproduction in Farm Animals,* 4th ed., Lea & Febiger, Philadelphia, 1980.

9. R. Ashdown and J. Hancock, "Functional Anatomy of Male Reproduction," in Ref. 8.

10. *Ruminant Nitrogen Usage,* National Research Council, National Academy Press, Washington, D.C., 1985.

11. R. Preston and S. Bartel, "Quantification of Escape Protein and Amino Acid Needs for New Feedlot Cattle," *Journal of Animal Science* **68** (Suppl. 1 6th Ed.), 531 (1990).

12. M. Jurgens, *Animal Feeding and Nutrition,* Kendall/Hunt Publishing Company, Dubuque, Iowa, 1982.

13. C. Faustman and co-workers, "Improvement of Pigment and Lipid Stability in Holstein Steer Beef by Dietary Supplementation with Vitamin E," *Journal of Food Science* **54,** 858–863 (1989).

14. Riley and co-workers, "Palatability of Beef from Steer and Bull Carcasses as Influenced by Electrical Stimulation, Subcutaneous Fat, Thickness and Marbling," *Journal of Animal Science* **56,** 592–597 (1983).

15. M. Judge, E. Aberle, J. Forrest, H. Hedrick, R. Merkel, *Principles of Meat Science,* 2nd ed., Kendall/Hunt Publishing Company, Dubuque, Iowa, 1989.

16. E. Kanis and co-workers, "Effect of Recombinant Porcine Somatotropin on Growth and Carcass Quality on Growing Pigs: Interaction with Genotype, Gender and Slaughter Weight," *Journal of Animal Science* **68,** 1193–1200 (1990).

17. P. Eppard and co-workers, "Effect of Dose of Bovine Growth Hormone on Lactation of Dairy Cows," *Journal of Dairy Science* **68,** 1109–1115 (1985).

18. G. Seidel, "Characteristics of Future Agricultural Animals," in J. W. Evans and A. Hollaender, eds., *Genetic Engineering of Animals: An Agricultural Perspective,* Plenum Press, New York, 1986.

General References

Reference 15 is also a good general reference.

D. Acker, *Animal Science and Industry,* 3rd ed. Prentice-Hall, Inc., Englewood Cliffs, N.J., 1983.

M. Jurgens, *Animal Feeding and Nutrition,* 6th ed., Kendall/Hunt Publishing Company, Dubuque, Iowa, 1988.

National Research Council, *Designing Foods: Animal Product Options in the Market Place,* National Academy Press, Washington, D.C., 1988.

J. Blakely, and Bade, D. *The Science of Animal Husbandry,* 4th ed., Reston Publishing Company, Inc., Reston, Va, 1985.

RICHARD A. ROEDER
University of Idaho
Moscow, Idaho

ANTHROPOLOGY. See HISTORY OF FOODS; CULTURAL NUTRITION.
ANTIBIOTIC RESISTANCE. See ENTRIES UNDER GENETIC ENGINEERING.
ANTIBIOTIC SENSITIVITY. See FOOD MICROBIOLOGY.
ANTIBIOTICS. See MICROBIOLOGY; ANTIBIOTICS IN FOOD OF ANIMAL ORIGIN.

ANTIBIOTICS IN FOODS OF ANIMAL ORIGIN

In 1877 Pasteur and Joubert discovered that growth of *Bacillus anthracis* could be inhibited by the presence of other microorganisms. This discovery led to the isolation of pyacyanase, the first antibiotic. In 1928 penicillin was discovered by Fleming when he observed that the outgrowth of an agar culture of *Staphylococcus aureus* was inhibited by mold. In 1932 Domagk discovered that the dye prontosil red, the first sulfonamide, demonstrated antimicrobial properties. Since these early discoveries the number of antimicrobials either isolated from natural sources or synthesized in the laboratory has grown tremendously. Antibiotics, in the narrowest sense, are products produced by living organisms that are not toxic to the producing organism but are capable of inhibiting the growth of or terminating other organisms. However, there are many synthetically manufactured compounds that inhibit or kill organisms as effectively as natural antibiotics. The chemistry and mode of action of these synthetic antimicrobial compounds are well documented. While there are virtually thousands of known antimicrobials, only a few are marketed for use in present-day food animal production.

Treatment of food animals with therapeutic and subtherapeutic dosage forms of antibiotic–antimicrobial drugs has increased in the last decade. This has been due to the advent of modern mass production operations that involve the maintenance of thousands of animals simultaneously and requires the utilization of carefully formulated and medicated diets that serve to maximize growth, minimize production costs, and provide an acceptable consumer product that is wholesome and affordable. Animal feeds that contain antibiotics and other antimicrobials as a prophylactic are now used routinely in beef, swine, chicken, and turkey production. In this way the antibiotics–antimicrobials help to maintain the optimal health of the animals so treated. Antibiotics may be coadministered with antimicrobial drugs resulting in a net increase in drug effectiveness compared to individual drug administrations. Strictly speaking, sulfonamides are not antibiotics. However, for purposes of this article sulfonamide antimicrobials will be collectively termed antibiotics.

The aminoglycoside, beta-lactam, ionophore, macrolide, sulfonamide, tetracycline, and other antibiotics are an integral part of food animal production. Regulations governing the use, dosage, and withdrawal times for many of the members of these antibiotic classes in animal production have been established in the United States by federal law, as outlined in the *Code of Federal Regulations,* Title 21, for each compound (1). The U.S. Food and Drug Administration (FDA) regulates the use of antibiotics in animal production while the U.S. Department of Agriculture (USDA) monitors residue levels by testing animal-derived products for antibiotics. Specific information regarding manufacturers, new animal drug application (NADA) codes, approved dosage forms and directions for proper use can be found in the *Code of Federal Regulations,* Title 21 and other published sources (2).

Antibiotics may manifest themselves as residues in animal-derived human foods if improperly used or if withdrawal times have not been observed for treated animals. Animal-derived human foods that contain violative antibiotic residues may pose a potential human health hazard. These potential health hazards can be broken down into three broad categories (3): toxicological, microbiological, and immunopathological. Toxicological concerns relate to the direct toxic effect of the compound on the consumer, resulting in physiological abnormalities, an example being sulfamethazine, which has recently been shown to produce cancer in laboratory animals (4). Microbiological concerns relate to the transmittance of antibiotic resistance. Antibiotics have the potential to act as a selective force that favors the emergence of resistant pathogenic bacteria in the natural flora of meat consumers. An example is a reported outbreak of salmonella poisoning in humans believed to be associated with a resistant strain in hamburger obtained from culled cattle that had been treated routinely with chloramphenicol (5). Finally, there are immunopathological mechanisms, whereby the drug serves as an antigen, demonstrates allergenic properties, and may result in hypersensitivity reactions to the drug that has sensitized some individuals. Penicillin is a prime example. Antibiotics are the principal compounds of concern in the federal residue control strategy (6). Such routine monitoring for antibiotic residues in animal-derived foods is intended to minimize the occurrence of violative antibiotic levels in the food supply. Such a capability is dependent on the analytical methods utilized.

Methods for the determination of antibiotics can include but are not limited to bioassay (ba), thin layer chromatography (tlc), gas chromatography (gc), liquid chromatography (lc), and various immunoassay (ia) techniques. Each of these methods has found utility in antibiotic determinations and many have been shown to be precise, specific, or sensitive, depending on which analytical technique, detector, visualization method, and compound are used. In this regard, a method utilized to detect an antibiotic residue at low nanogram antibiotic per gram of sample (ng/g) or even picogram antibiotic per gram of sample (pg/g) levels may be confounded by the presence of interferences found in the sample matrix or in the sample extract. Thus analytical capability is governed ultimately by the sample preparation and extraction steps. The isolation of such residues from a complex biological matrix poses unique problems to the analyst and because of the number of antibiotics being utilized in animal production the need for multiresidue determinations exists. An emerging trend is the development of rapid multiresidue–multidrug class isolation techniques that result in clean, interference-free extracts and that minimize cost, time, and expendable material requirements, enabling the analyst to test for multiple drugs isolated from one sample. In conjunction with improved isolation methods the need to screen samples rapidly and accurately for antibiotic residues exists. Such antibiotic screening protocols and analytical capability can be enhanced by rapid, reproducible, efficient, and cost-effective residue isolation techniques.

SAMPLE PREPARATION AND RESIDUE ISOLATION

Sample preparation requirements are dependent on the particular analysis that is to be performed. For example, ba or ia procedures for antibiotics may require little or no sample preparation, whereas sample preparation for tlc, lc, and gc procedures can be a major limiting factor of the analyses.

Liquid samples such as milk, blood, urine, saliva, or other body fluids can be assayed directly for antibiotics by ba or ia techniques, or the samples may require only minor cleanup steps (centrifugation, pH adjustment, or a protein precipitation) prior to the assay. In some cases the antibiotic residues that may be present in the sample at low concentrations may require an analyte-concentration step prior to the analysis.

Residue enrichment methods that concentrate the antibiotic residue prior to an analysis may involve solvent extractions, column chromatography, or solid-phase extraction (spe) techniques. The utility of solvent–solvent extractions is limited because of the polar characteristics of many antibiotics such as beta-lactams, aminoglycosides, macrolides, polyether ionophores, and tetracyclines. Ion-exchange column chromatography techniques utilized to isolate polar ionizable antibiotics are only marginally effective. Solid-phase extraction appears to hold more promise as a routine residue enrichment approach. Methods that attempt to circumvent cleanup steps by assaying samples for antibiotics directly are only rarely efficient, except for ba or ia techniques of some liquid samples such as milk.

Sample-preparation steps may not be necessary in some cases for antibiotic determinations but their use can often facilitate more accurate analyses. Therefore, it is incumbent on the analyst to obtain the cleanest sample extract possible. Clean extracts from some liquids and most nonliquid matrices such as muscle or organ tissues are more difficult to obtain and generally require an extraction step and multiple manipulations to isolate the antibiotic residue from the sample with high percentage recoveries. Unfortunately, sample extracts obtained in this manner may contain naturally occurring inhibitors or interferences that could affect the analysis. Thus further cleanup of extracts is usually required prior to performing the analysis. Because of the polar nature of many antibiotics, buffered aqueous solutions are routinely employed for antibiotic extractions and it is difficult, if not impossible, to isolate antibiotic residues from these aqueous extracts by partitioning with organic solvents. Thus drying of the aqueous extract, a time-consuming process, or the use of other preparative steps such as column chromatography or spe techniques may be required before an analytical determination can be made. A direct assay of the aqueous extract may be possible if the concentration of the antibiotic in the extract is sufficiently high or if the analytical method is particularly sensitive. As the complexity of the sample and the number of sample cleanup steps increases, the utility of techniques such as ba, ia, and tlc for quick antibiotic screening purposes decreases. In addition, extracts suitable for ba or ia procedures may not be sufficiently clean for more sophisticated tlc, lc, or gc determinations, thereby requiring additional and different residue isolation procedures for confirmatory techniques.

Sophisticated techniques based on lc, gc, lc–mass spectrometry (ms), and gc–ms usually require more rigorous sample preparation to isolate the antibiotic free from interferences found in the sample extract. These classic isolation techniques can be laborious and time-consuming. Classic isolation techniques for aminoglycoside (7), beta-lactam (8–10), chloramphenicol (11), ionophore (12), macrolide (10, 13), sulfonamide (14–17), and tetracycline (18) antibiotics generally involve their extraction from biological matrices with large volumes of extracting solvents, chemical manipulations such as pH adjustments and protein precipitations, centrifugations, back washing, and the evaporation of large volumes of organic solvents. This approach limits the usefulness of many classic isolation techniques for multiresidue determinations by lc and gc.

Analytical capability is, therefore, limited to a large extent by interferences present in the sample extract. While state-of-the-art analytical techniques can detect picogram levels of pure antibiotic standard compounds, this same level of detection may not be achievable for antibiotics obtained from the extraction of biological samples. Specifically, coextracted interferences may hinder antibiotic determinations by tlc, lc, or gc because they may have similar detector and chromatographic characteristics or they may exist in such a large quantity that they overwhelm the detection method used. Interferences in ba or ia techniques can contribute to cross-reactions that may lead to false-positive or false-negative determinations. Thus a major limiting factor associated with antibiotic determinations is not the available analytical capability but the sample-preparation steps required to extract the antibiotics. The sample-preparation steps should ideally result in clean biological extracts that contain the antibiotic residue with high percentage recoveries and that have minimal interferences that might limit the choice of analysis.

The need to test for more antibiotic residues in more foods requires rapid, rugged, and multiresidue isolation techniques allowing the analyst to test for multiple drugs in the same sample. Classic residue isolation techniques have not been able to meet this challenge. Matrix solid-phase dispersion (mspd) techniques (19–25), recently developed for the isolation of drug residues from animal matrices, have the potential to greatly enhance many antibiotic residue isolation protocols. In mspd the sample (0.5 g) is dispersed onto octadecasilyl polymeric–derived silica beads [C-18 reversed-phase packing material (2 g), 1,000 m^2 surface area, theoretical]. The dispersion mechanism, utilizing a mortar and pestle, involves the disruption, unfolding, and rearrangement of matrix constituents by mechanical and hydrophobic forces onto the C-18 beads. Lipid and lipophilic materials associate with the lipophilic C-18, allowing the more hydrophilic components and protein regions to extend outward away from the nonpolar, inner C-18–lipid region. Water and more polar constituents preferentially associate with the hydrophilic ends. A column fashioned from the C-18–sample matrix blend can be eluted sequentially with solvents (8 mL) of different polarities to effectively remove interferences in one solvent and elute the target residue in a different solvent. The process can be envisioned as an exhaustive extraction process whereby a large volume of solvent is passed over a thin layer of sample. Mspd has been utilized for the isolation of beta-lactams from beef tissue (20); sulfonamides from milk (21), infant formula (22), and pork muscle tissue (23); as well as for chloramphenicol (24) and tetracycline (25) isolations from milk. The theoretical aspects describing the disruption, unfolding, and rearrangement of matrix constituents onto the C-18 have been published (19–25).

Advances in multiclass–multiresidue isolation procedures, which are rapid, rugged, generic in nature, and free from interferences and which facilitate the isolation of multiple antibiotic residues from one sample, will greatly enhance analytical determinations of antibiotics.

ANALYTICAL METHODOLOGY

Bioassays

Bioassays are used routinely to test for violative levels of antibiotics in milk, animal tissues, and feeds. Bioassays involve the inhibition of growth of specific bacterial spores or viable bacteria in the presence of a sample or sample extract that contains antibiotic residues. Bioassays may also utilize the measurement of labeled antibiotic analyte bound to receptors, a ligand assay, on vegetative bacterial cells. Bioassays have been utilized to detect aminoglycosides (26–29), beta-lactams (27, 29–31), chloramphenicol (29, 32–34), ionophore (12), sulfonamide (26, 27), and tetracycline (26, 27, 29, 30, 35) antibiotics in foods.

For example, the swab test on premises (stop) procedure (27) is used by federal meat inspectors to test for chloramphenicol, tetracycline, aminoglycoside, penicillin G, and sulfonamide antibiotics. The stop method involves taking a sterile cotton swab and macerating the target organ or tissue with the noncotton end. The size of the macerated zone should be slightly larger in diameter and deeper than the cotton end of the swab. The cotton end is then inserted into the disrupted area and allowed to absorb fluids (30 min). The fluid-soaked cotton swab is then placed onto a nutrient agar plate that has been inoculated with a lawn of specific bacterial spores and the plates are then incubated (16–22 h). Standards of known concentration are run in parallel. A zone of microbial growth inhibition on the sample plate is an indication of the presence of an antibiotic and the size of the zone of inhibition can be semicorrelated with the size of the zone of inhibition for a given concentration of pure antibiotic standard. The sample is positive if the size of the zone of inhibition is similar to that for a known pure standard. Unfortunately, the stop procedure detects only the presence of inhibitors, not their specific identity. Other antibiotics or naturally occurring inhibitors may contribute to the size of the zone of growth inhibition observed, indicating a positive sample although the amount of the individual antibiotic compounds present may be less than the violative level.

The classic disk assay procedure (31) is similar to the stop procedure for the same antibiotics except that a filter paper disk is placed on the inoculated nutrient agar, the liquid sample or suitable extract is added to the disk and then the plate is incubated for a minimum of 2.5 h or until a zone of inhibition can be observed. The zone of inhibition may be enhanced by dyeing techniques (36, 37), which aid in its detection. Semiquantitative determinations by the disk assay can be accomplished, provided incubation times are increased (16–22 h). The disk assay procedure is the official method described in the Pasteurized Milk Ordinance (PMO) (38) for antibiotic testing in milk. The disk assay procedure suffers limitations similar to those of the stop procedure. However, the PMO allows for the use of any method that gives results equivalent to the disk assay method; therefore, many states utilize alternative techniques such as the color reaction test (crt) to test for antibiotics in milk. The Delvo test (36) is an example of a crt technique used to determine antibiotics, specifically beta-lactams, in milk. The test involves placing the milk sample onto agar containing a viable strain of *Bacillus,* nutrients, and pH indicators. If the color of the agar changes from purple (basic) to yellow (acid) after incubation (1.5 h) then no penicillin is present to prevent the outgrowth of the acid-producing bacteria. Color reaction tests can be rapid and simple to perform.

The microbial receptor assay (mra) (Charm test) technique (35) can be used to detect beta-lactam, macrolide, and aminoglycoside antibiotic classes. The mra method involves the use of C-14 or tritium isotopically labeled analyte to displace nonlabeled analyte from the bacterial receptors located on vegetative cells under a standard set of conditions. An equilibrium condition between unlabeled and isotopically labeled analyte results, allowing quantitative measurement of the amount of antibiotic present by an appropriate radiometric method. However, the identity of the compound is not known and must be determined by other methodology.

The thin layer chromatography–bioautography of tlb bioassay technique (12, 29, 37) uses traditional thin layer chromatography to separate sample constituents on silica or microcrystalline cellulose chromatography (mcc) plates. Different antibiotics will migrate on the plates according to their chemical characteristics and the developing solvent utilized. The developed plates are covered with a spore-inoculated nutrient agar and then incubated (16–24 h). Zones of growth inhibition observed for different locations on the plate after incubation indicate the presence of antibiotics. Aminoglycoside (29), beta-lactam (29), chloramphenicol (29), ionophore (12, 29), macrolide (29, 37), and tetracycline (29) antibiotics have been assayed by tlb. The tlb method is more specific and sensitive (33) than either the stop or classical disk assay procedures because it takes advantage of the ability of tlc to separate 14 different antibiotics from each other as well as from other impurities that may be present in the sample. Therefore, zones of inhibition can be more closely correlated to different antibiotics. The tlc-developing solutions are designed to optimize separations between similar antibiotics. However, the separation of up to 14 different antibiotic residues can only be accomplished by utilizing three separate tlc plates. Furthermore, extensive sample-preparation and extraction steps are employed in the tlb procedure that are similar to those employed for more sophisticated analytical determinations. Because of the incubation time required the tlb procedure is not as rapid as the crt or the disk assay methods for screening purposes. However, the tlb procedure has an advantage over the stop or disk assay procedures for screening purposes because it can provide for more precise determinations between different antibiotics. Absolute confirmation of the antibiotics is not possible by tlb and time requirements are in excess of many sophisticated, more definitive, liquid and gas chromatographic techniques.

The stop, disk assay, crt, and tlb procedures are valuable screening tools. The first three techniques are relatively cheap, easy to perform, and require a minimal amount of equipment and technician training. However, they lack specificity and sensitivity and/or they may be subject to interpretive errors in the presence of naturally occurring inhibitors. Furthermore, each of these four methods requires confirmational testing of positives. The mra method using isotopically labeled antibiotics can be expensive and requires special handling and equipment; but it can provide for quantitative determinations. These bioassays, if used judiciously, can minimize the number of samples screened by more costly analytical methods, but they cannot replace such methods for absolute confirmations.

Immunoassay

Immunoassay techniques are based on classic antibody–antigen reactions whereby the antibody will bind with its corresponding antigen (antibiotic) and result in visible turbidity if reacted in solution or form visible immunoprecipitation in a gel at the location where the antigen and antibody meet. Visible end point immunoprecipitation that occurs at low antibiotic concentrations may be difficult to see and may require detection by more sensitive nonvisual means.

Immunoassay techniques such as enzyme immunoassay (eia), radio immunoassay (ria), enzyme-linked immunosorbent assay (elisa), fluorescence polarization immunoassay (fpia), particle-concentration immunoassay (pcia), particle-concentration fluorescence immunoassay (pcfia), quenching fluoroimmunoassay (fia), and latex-agglutination inhibition immunoassay (laia) require the measurement of by-products produced by linked enzyme systems or the measurement of radioactive, fluorescence, metal, or latex labels that have been attached to one of the reactants (39). The displacement between the unlabeled and labeled antigen or antibody allows for true measurement of the immunoprecipitate concentration, which corresponds to the concentration and type of analyte present in the sample extract. Ria determinations of chloramphenicol (40, 41) as a residue in eggs, milk, and meat were comparable to values obtained by gc. The eia determination of monensin (42), a polyether antibiotic in urine, serum, and fecal extracts, may be applicable to food extracts as well. A review of ia techniques such as ria, fia, fpia, and laia for aminoglycoside determinations in body fluids has been published (43). The recent use of a monoclonal-antibody–based agglutination test (spot test) for beta-lactams in milk (44) and elisa for chloramphenicol determinations in swine tissue (45), milk (46), and sulfamethazine in swine tissue (47) may prove to be adaptable to other antibiotic determinations as well, provided that specific antibodies can be developed for such antibiotics.

Immunoassays can be sensitive, class specific, accurate, and provide a means for rapidly screening samples for antibiotics; the future development of monoclonal antibodies will allow for more specificity in immunoassay determinations. Radioimmunoassay techniques that require radioisotopes may be subject to specific regulations, require expensive counting equipment, and pose disposal problems. Reagent kits can be relatively expensive and have a limited shelf life. In this regard, nonisotopic immunoassays such as elisa, fpia, pcia, or pcfia, and monoclonal-based ias will in all likelihood play an increasingly important role in antibiotic screening immunoassay determinations.

At present, ia techniques for antibiotics approved for use in animal production are limited and have minimal utility for aminoglycoside, beta-lactam, chloramphenicol, and sulfonamide determinations, but they have the potential for use as screens for all the major antibiotic classes. Immunoassay techniques have been limited to liquid samples such as milk because of the difficulties associated with the isolation of antibiotic residues from matrices such as muscle or organ tissues. Increased use of ia techniques to screen for antibiotics in tissues is directly dependent on the development of rapid residue isolation techniques that isolate the antibiotic from the tissue and provide for an extract that is free from cross-reacting components and nonspecific binding (NSB) factors. The development of such isolation techniques will further the use of ia techniques for the screening of antibiotics in tissues.

Immunoassay techniques are more specific than micro-biological based bioassays and would allow for the rapid screening of large numbers of samples. With the need to screen more samples for more drugs, the ia techniques will become more important in residue control protocols. Immunoassay techniques can greatly reduce the number of samples that are presently screened by more sophisticated analytical techniques and have the potential to replace many bioassay and tlc methods, provided rapid and efficient antibiotic extraction procedures can be developed. As a regulatory tool, ia techniques for antibiotic screening coupled to highly specific and accurate techniques, such as lc, gc, lc–ms, or gc–ms for confirmations of positives, mark the future with respect to a rapid and reliable residue control strategy.

Thin Layer Chromatography

Thin layer chromatography techniques have been used in chemical separations for decades. Thin layer chromatography is rapid and inexpensive, can be highly sensitive depending on the compound examined and the visualization technique employed, and is easy to use and versatile. It can be adapted to separations of all classes of antibiotics by utilizing different sorbents and solvent-developing systems. Aminoglycoside (28, 29), beta-lactam (29, 45, 48), chloramphenicol (29), ionophore (12, 29), macrolide (13, 29, 37), sulfonamide (49, 50) and tetracycline (35, 51, 52) antibiotics have been successfully assayed by tlc. However, the utility of tlc for the detection of picogram levels of antibiotics is limited and thus tlc techniques have come to play only a minor role in antibiotic analyses.

The sulfa on site (sos) tlc procedure for sulfonamide determinations in swine urine was developed by the USDA–FSIS and is presently in use in 100 of the largest swine-slaughtering facilities in the United States (53). Thin layer chromatography techniques can complement other antibiotic assay techniques such as tlc–bioautography bioassays (12, 29, 37) to provide for qualitative and semiquantitative determinations, but is limited to a large degree by the cleanliness of the sample that is to be analyzed. Impurities in the sample can interfere with the chromatography of the antibiotic by altering the migration of the antibiotic on the tlc plates compared to pure standards, and these impurities can negate visualization techniques. Thin layer chromatography determinations also require confirmations of suspect residues isolated from the tlc plates to determine if the residue is indeed an antibiotic. In this regard bioassay and immunoassay techniques provide for a more precise determination of the presence of antibiotics in a sample extract. Thin layer chromatography will perhaps play a decreasing role in antibiotic residue control strategies where low-level detection is required for some antibiotics, but should not be eliminated from the analyst's tools utilized for the isolation and purification of compounds from animal-derived matrices.

Liquid Chromatography

The convenience and versatility of liquid chromatography has led to its adoption as the analytical method of choice for the determination of many drugs, especially antibiotics. Liquid chromatography, using selective detectors, can give reproducible results that are specific, sensitive, and precise. The polar characteristics of antibiotics make them well suited to lc procedures, in which mobile phase solvent systems and columns can be varied to facilitate specific antibiotic determinations.

Recent publications describing lc methods for the analysis of aminoglycoside (7, 28, 54), beta-lactam (8–10, 13, 20, 55, 56), chloramphenicol (11, 24, 57, 58), ionophore (12, 13), macrolide (10, 13), sulfonamide (14, 16, 21–23, 59–61), and tetracycline (18, 33, 51, 56, 62–64) antibiotics underscore the utility of lc determinations for the analysis of these compounds as residues in foods and other biological matrices.

Food extracts containing residues of sulfonamide (12, 16, 21–23), beta-lactam (8, 13, 20, 55), chloramphenicol (11, 24), tylosine (13) and spiramycin (10) macrolides, and tetracycline (18, 33, 51, 56, 64) antibiotics may be analyzed directly by ultraviolet (uv) detection at picogram–nanogram levels because they have characteristic uv absorbances and large extinction coefficients. Photodiode array uv detection of antibiotics can provide the analyst with uv spectra of suspect peaks (65) and thus serve as a preconfirmational screening tool. However, some beta-lactams and macrolides, which have a maximum absorbance in the low uv (210–240 nm) range or relatively small extinction coefficients, can be more difficult to analyze by lc–uv because of coextracted interferences that absorb readily in this range. Optimizing lc chromatographic conditions to separate a particular antibiotic residue from interferences that may be present in the extract severely limits lc techniques in terms of multiresidue antibiotic determinations.

Aminoglycoside and most macrolide antibiotics that have low or nonexistent uv-absorbing properties may require the formation of derivatives to aid in their detection or the use of alternative detection methods. Benzene sulfonyl chloride (66) and 1-fluoro-2,4-dinitrobenzene (67) have been used to prepare uv derivatives of aminoglycosides for analyses by lc. Detection methods for antibiotics that do not require making derivatives are ideal because reaction conditions for many derivatizations can be difficult to optimize. However, until improved or new detection methods are developed for antibiotics, derivatives will continue to be used to facilitate sensitive and selective detections, especially for aminoglycosides.

The analyses of aminoglycoside (7), beta-lactam (9), macrolide (13), and tetracycline (35) antibiotics can also be facilitated by the formation of fluorescent derivatives. The ionophore lasalocid has native fluorescence due to its salicyclic acid-type aromatic moiety and, therefore, can be analyzed directly in food extracts at nanogram levels by fluorescence detection (12). Fluorescent derivatives of tetracyclines (35) can be made by simply complexing the tetracycline with different metal ions. The tetracycline–metal ion complexes have unique excitation and emission wavelengths that can be measured to quantitatively determine tetracycline concentrations in the food extract. In the case of aminoglycoside antibiotics, which do not have native fluorescence, fluorescent derivatives must be made to facilitate their detection. The preparation of fluorescent derivatives can require exacting reaction conditions, additional equipment in the form of reactors and delivery pumps, and in many cases the removal of reaction reagents before an analysis can be made. However, this cannot be avoided in the case of aminoglycosides because no other suitable chromatographic method presently exists for their detection at nanogram levels. Fluorescence detection of o-phthalaldehyde (OPA) or fluorescamine derivatives has been shown to give accurate and reliable values for aminoglycoside residues in animal tissue (7, 28). The derivatization of the aminoglycoside amino group with a fluorogenic agent can be accomplished precolumn or postcolumn. The added cost of the postcolumn reactor and solvent pump needed to deliver the derivatization solution may be disadvantageous. Alternatively, precolumn derivatization reaction mixtures may require the removal of derivatizing reagents prior to the analyses. Specific applications may require either approach and are dependent on the type of food that is to be extracted and the inherent interferences present.

Fluorescent derivatives (9) of penicilloaldehyde products obtained by enzymatic hydrolysis of beta-lactam rings and reaction with dansyl hydrazine have been reported for eight neutral beta-lactams. This is a novel approach for beta-lactam determinations and might be applicable to other antibiotics that are inherently unstable and difficult to analyze. Techniques that serve to characterize unique enzymatic or degradative products of antibiotics may be a useful tool for many antibiotic determinations.

Liquid chromatography of antibiotics will continue to be the method of choice for most antibiotic determinations. Coupling of photodiode array uv, fluorescence, electrochemical and other yet to be developed detectors in tandem can provide information in terms of retention times, structures, and characteristic uv and fluorescent spectra. Information obtained by this approach may be sufficient to confirm the presence of specific antibiotics if coupled with positive results obtained by immunoassay techniques for the specific antibiotic in question. Liquid chromatography–mass spectrometry is not presently suitable for low-level detection of the major antibiotic classes, but advances in this area, the use of tandem detection methods, and specific monoclonal-based immunoassay screening techniques would contribute significantly to overall antibiotic residue control strategies needed to insure a safe and wholesome food supply.

Gas Chromatography

Gas chromatography coupled to specific detectors can provide valuable information about the retention time, structure, chemical characteristics, and identity of compounds. Gas chromatography has been utilized for antibiotic determinations but is limited to some extent by the molecular weight, high polarity and the relative lack of thermal stability of many antibiotics. Chemical derivatives can serve to impart greater stability and volatility to antibiotics. However, controlling reaction conditions to insure the formation of consistent derivative products with high yields can be difficult to accomplish. These difficulties have limited the usefulness of gc for routine antibiotic determinations.

Presently there are no suitable gc methods for the routine determination of aminoglycoside, beta-lactam, and most ionophore, macrolide, and tetracycline antibiotics. Gas chromatography of trimethylsilyl (TMS) derivatives of aminoglycoside (68), lasalocid ionophore (11), and tetracycline (69) antibiotics have been reported. However, the usefulness of this technique for low-level determinations of these antibiotics in food extracts is limited because of the need to control silylating reaction conditions carefully, the formation of multiple TMS antibiotic derivative products, the abundance of TMS interferences contributed by the sample, and the lack of stability of TMS derivatives.

Gas chromatography methods for chloramphenicol have been reviewed (11). Gas chromatography utilizing electron capture (ECD), flame ionization (FID) and thermionic (TID) detection of TMS and heptafluorobutyl derivatives of chloramphenicol isolated from muscle, liver, kidney, and milk resulted in picogram–nanogram detection giving results comparable to radioimmunoassay determinations (40). Gas chromatography of chloramphenicol residues in food extracts can provide for confirmations of suspect residues and complement lc determinations.

Gas chromatography methods for sulfonamides have also been reviewed (14). Sulfonamide determinations in animal tissues (17, 70) by gc have been accomplished by analyzing volatile methylated or acylated derivatives. Methylation of sulfonamides at the N-1 position followed by acylation at the N-4 position can provide for stable volatile derivatives suitable for gc analysis. However, it can be difficult to control reaction conditions necessary to optimize this two-step derivative reaction. Low yields and the formation of multiple derivative products may result in nonrepresentative and misleading data. Although methods for the quantitative determination of sulfonamides (15, 17, 61) have been reported, it has been noted (15) that some sulfonamides determined in this manner resulted in low and variable recoveries. Because of this variability, it is necessary to be careful when quantitatively evaluating data obtained by this technique.

Advances in supercritical fluid chromatography (sfc) may eventually lead to applications involving antibiotic determinations. At present, sfc is limited as a routine analytical tool to nonpolar substances (71). Analysis of polar antibiotics by sfc will require advances in sfc hardware and columns. In addition, the solubility characteristics of different antibiotics in different supercritical media will require extensive laboratory investigation to define optimal conditions in terms of pressures and polar mobile phase modifiers that may be required:

Thus gas chromatography methods have minimal utility for antibiotic determinations. Gas chromatography will remain a supplemental analytical tool for antibiotic residue determinations until new derivatization schemes, reagents, and refined reaction conditions are developed. The development of such innovations can, however, advance present marginal gc antibiotic techniques to full-fledged quantitative antibiotic confirmatory methods, which will contribute significantly to an integrated antibiotic residue control strategy.

Lc–ms and Gc–ms

Mass detectors coupled to chromatographic techniques such as lc and gc can provide for the unequivocal identification of compounds. Unfortunately, sensitivity limitations associated with lc–ms have precluded its use in antibiotic residue analyses where low-level determinations are required, and gc–ms techniques for antibiotics have been applied predominantly to sulfonamide determinations. Gas chromatography–mass spectrometry of volatile sulfonamide derivatives (14, 17, 61, 70) may provide for the confirmation of suspect residues but may be inadequate as a quantitative determinative regulatory tool. Ultimately, however, any data gleaned from lc–ms or gc–ms determinations of antibiotic residues may be useful in antibiotic determinations. The refinement of lc–ms and gc–ms techniques, advances in ms sensitivity, more efficient lc–ms interfaces, and future developments may allow ms to become a routine confirmatory tool that will greatly enhance antibiotic residue determinations.

FUTURE TRENDS

Microbiological and bioassays have historically served to screen foods for the presence of antibiotics. While these assays have proven to be quite sensitive and inexpensive they are far too nonspecific and cause too lengthy a delay in obtaining data. This delay may allow antibiotic contaminated animal-derived foods to be marketed and consumed with drug detection being accomplished after the fact. This policy is inherently inadequate but has arisen from the limitations of the available analytical technologies. In this sense, bioassays and other methodologies that require lengthy waiting periods and prolonged and complex analytical procedures cannot meet the increasing need to perform rapid drug screening and confirmation on more samples for more drugs.

The advent of more specific, sensitive, and simple to perform immunoassays should overcome many of the problems of performing rapid screening and early detection of antibiotic contamination in animal-derived foods. In this regard, simple elisa, card, or test-strip assays that are suitable for use at production sites, slaughterhouses, and packing plants could dramatically reduce the occurrence of antibiotic violations and preclude the problems that can arise from antibiotic contamination of the food supply.

These and other more complex immunoassay methods could be performed on milk, urine, blood, or other body fluids. However, new methods for the rapid isolation of drugs from tissues may also make such immunoassays more directly applicable to the screening of this complex sample material. Matrix solid-phase dispersion (mspd) techniques have shown promise in providing a simple, rapid, and generic method for performing drug isolations from tissues and milk. Thus the combination of ia and mspd may prove to be a useful approach to the more rapid screening and analysis of tissues and animal derived products for antibiotics as well as other drugs. A further advantage of mspd is the capability to isolate a class of drugs or several classes of drugs from a single sample. In this regard the advances in immunoassay screening technology must be matched by advances in extraction procedures. Extraction methodology that is too narrow in extraction capability leads to a need to develop a specific method for each antibiotic drug. Extraction methods that are multidrug and multidrug class specific will have the greatest utility for performing screening analyses in the future, whether such screens are conducted by immunoassay, tlc, hplc, or gc.

Of these analytical techniques ia and tlc will be best utilized for rapid screening, whereas hplc and gc may be the best methods for subsequent confirmation and quantitation of antibiotic residues. However, hplc methods that are capable of separating and detecting several drug residues obtained, perhaps, from a single multidrug class extraction may also prove to be useful screening techniques. The usefulness of such screens is directly related to the efficiency of recovery of a multidrug extraction method, the elimination of background interferences, and the ability of various detectors to indicate the presence or absence of a drug with adequate sensitivity.

For the antibiotics, gc will remain of limited value but should always be considered and applied where feasible. It is expected that sfc will eventually play a major role in the extraction and analysis of compounds such as the antibiotics. However, current results are less than promising and the awaited applications may be long in development.

BIBLIOGRAPHY

1. *Code of Federal Regulations,* 21 CFR "Food and Drugs," U.S. Government Printing Office, Washington, D.C., 1989.
2. S. F. Sundlof, J. E. Riviere, and A. L. Craigmill. *Food Animal Residue Avoidance Databank. Trade Name File. A Comprehensive Compendium of Food Animal Drugs,* 3rd ed, Institute of Food and Agricultural Sciences, University of Florida, Gainesville, Fla., 1989.
3. Food and Agriculture Organization of the United Nations. *Residues of Veterinary Drugs in Foods,* FAO Food and Nutrition Paper **32.** Food and Agriculture Organization of the United Nations, Rome, Italy, 1985.
4. National Center for Toxicological Research, *Chronic Toxicity and Carcinogenicity of Sulfamethazine in B6C3F1 Mice,* technical report, National Center for Toxicological Research, Jefferson, Ark., 1988.
5. J. S. Spika, S. H. Waterman, G. W. S. Hoo, M. E. St. Louis, R. E. Pacer, S. M. James, M. L. Bisset, L. W. Mayer, S. Y. Chiu, B. Hall, K. Freene, M. E. Potter, M. L. Cohen, and P. A. Blake, "Chloramphenicol-Resistant *Salmonella newport* Traced through Hamburgers to Dairy Farms," *New England Journal of Medicine* **316**(10), 565–570 (1987).
6. J. Brown, ed., U.S. Department of Agriculture/Food Safety and Inspection Service. *Compound Evaluation and Analytical National Residue Program Plan,* USDA, FSIS Science and Technology Program, Washington, D.C., 1988.
7. V. K. Agarwal, "High Performance Liquid Chromatographic Determinations of Gentamicin in Animal Tissue," *Journal of Liquid Chromatography* **12**(4), 613–628 (1989).
8. J. O. Miners, "The Analysis of Penicillins in Biological Fluids and Pharmaceutical Preparations by High Performance Liquid Chromatography: A Review," *Journal of Liquid Chromatography* **8**(15), 2827–2843 (1985).
9. R. K. Munns, W. Shimoda, S. E. Roybal, and C. Viera, "Multiresidue Method for Determination of Eight Neutral β-Lactam Penicillins in Milk by Fluorescence-Liquid Chromatography," *Journal of the Association of Official Analytical Chemists* **68**(5), 968–971 (1985).
10. T. Nagata and M. Saeki, "Determination of Ampicillin Residues in Fish Tissues by Liquid Chromatography," *Journal of the Association of Official Analytical Chemists* **69**(3), 448–450 (1986).
11. E. H. Allen, "Review of Chromatographic Methods for Chloramphenicol Residues in Milk, Eggs and Tissues from Food Producing Animals," *Journal of the Association of Official Analytical Chemists* **68**(5), 990–999 (1985).
12. G. Weiss and A. MacDonald, "Methods for Determination of Ionophore Type Antibiotic Residues in Animal Tissues," *Journal of the Association of Official Analytical Chemists* **68**(5), 971–980 (1985).
13. W. A. Moats, "Chromatographic Methods for Determination of Macrolide Antibiotic Residues in Tissues and Milk of Food-Producing Animals, *Journal of the Association of Official Analytical Chemists* **68**(5), 980–984 (1985).
14. W. Horwitz, "Analytical Methods for Sulfonamides in Foods and Feeds. II. Performance Characteristics of Sulfonamide Methods," *Journal of the Association of Official Analytical Chemists* **64**(4), 814–824 (1981).
15. A. J. Manuel and W. A. Steller, "Gas-Liquid Chromatographic Determination of Sulfamethazine in Swine and Cattle Tissues," *Journal of the Association of Official Analytical Chemists* **64**(4), 794–799 (1981).
16. B. L. Cox and A. Krzeminski, "High Pressure Liquid Chromatographic Determination of Sulfamethazine in Pork Tissue," *Journal of the Association of Official Analytical Chemists* **65**(6), 1311–1315 (1982).
17. R. M. Simpson, F. B. Suhre, and J. W. Shafer, "Quantitative Gas Chromatographic Mass Spectrometric Assay of Five Sulfonamide Residues in Animal Tissue," *Journal of the Association of Official Analytical Chemists* **68**(1), 23–26 (1985).
18. R. B. Ashworth, "Liquid Chromatographic Assay of Tetracyclines in Tissues of Food Producing Animals," *Journal of the Association of Official Analytical Chemists* **68**(5), 1013–1018 (1985).
19. S. A. Barker, A. R. Long, and C. R. Short, "A New Approach to the Isolation of Drug Residues From Animal Tissues," in W. Huber, ed., *Proceedings of the Sixth Symposium on Veterinary Pharmacology and Therapeutics,* Blacksburg, Va., 1988, pp. 55–56.
20. S. A. Barker, A. R. Long, and C. R. Short, "Isolation of Drug Residues from Tissues by Solid Phase Dispersion," *Journal of Chromatography* **475,** 353–361 (1989).
21. A. R. Long, C. R. Short, and S. A. Barker, "Multiresidue Method for the Isolation and Liquid Chromatographic Determination of Eight Sulfonamides in Milk," *Journal of Chromatography* **502**(1), 87–94 (1990).
22. A. R. Long, L. C. Hsieh, M. S. Malbrough, C. R. Short, and S. A. Barker, "A Multiresidue Method for the Isolation and Liquid Chromatographic Determination of Seven Sulfonamides in Infant Formula," *Journal of Liquid Chromatography* **12**(9), 1601–1612 (1989).
23. A. R. Long, L. C. Hsieh, M. S. Malbrough, C. R. Short, and S. A. Barker, "Multiresidue Method for the Determination of

Sulfonamides in Pork Tissue," *Journal of Agricultural and Food Chemistry* **38**(2), 423–426 (1990).

24. A. R. Long, L. C. Hsieh, A. C. Bello, M. S. Malbrough, C. R. Short, and S. A. Barker, "Method for the Isolation and Liquid Chromatographic Determination of Chloramphenicol in Milk," *Journal of Agricultural and Food Chemistry* **38**(2), 427–429 (1990).

25. A. R. Long, L. C. Hsieh, M. S. Malbrough, C. R. Short, and S. A. Barker, "Matrix Solid Phase Dispersion (MSPD) Isolation and Liquid Chromatographic Determination of Oxytetracycline, Tetracycline and Chlortetracycline in Milk," *Journal of the Association of Official Analytical Chemists* **73**(3), 379–384 (1990).

26. R. B. Read, Jr., J. G. Bradshaw, A. A. Swartzentruber, and A. R. Brazis, "Detection of Sulfa Drugs and Antibiotics in Milk," *Applied Microbiology* **21**(5), 806–808 (1971).

27. R. W. Johnston, R. H. Reamer, E. W. Harris, H. G. Fugate, and B. Schwab, "A New Screening Method for the Detection of Antibiotic Residues in Meat and Poultry Tissues," *Journal of Food Protection* **44**(11), 828–831 (1981).

28. B. Shaikh and E. H. Allen, "Overview of Physical Chemical Methods for Determining Aminoglycoside Antibiotics in Tissues and Fluids of Food Producing Animals," *Journal of the Association of Official Analytical Chemists* **68**(5), 1007–1013 (1985).

29. E. Neidert, P. W. Saschenbrecker, and F. Tittiger, "Thin Layer Chromatographic/Bioautographic Method for Identification of Antibiotic Residues in Animal Tissues," *Journal of the Association of Official Analytical Chemists* **70**(2), 197–200 (1987).

30. J. R. Bishop, A. B. Bodine, G. D. O'Dell, and J. J. Jansen, "Quantitative Assay for Antibiotics Used Commonly in Treatment of Bovine Infections," *Journal of Dairy Science* **68**(11), 3031–3036 (1985).

31. American Public Health Association, *Standard Methods for the Examination of Dairy Products,* Port City Press Inc., Baltimore, Md., 1985.

32. C. J. Singer and S. E. Katz, "Microbiological Assay for Chloramphenicol Residues," *Journal of the Association of Official Analytical Chemists* **68**(5), 1037–1041 (1985).

33. G. O. Korsrud and J. D. MacNeil, "A Comparison of Three Bioassay Techniques and High Performance Liquid Chromatography for the Detection of Chlortetracycline Residues in Swine Tissues," *Food Additives and Contaminants* **5**(2), 149–153 (1987).

34. C. D. C. Salisbury, J. R. Patterson, J. D. MacNeil, T. E. Feltmate, F. F. Tittiger, J. Asselin, and W. D. Black, "Survey of Chloramphenicol Residues in Diseased Swine," *Canadian Journal of Veterinary Research* **52**, 15–17 (1988).

35. M. Riaz, "The Quantitative Analysis of Tetracyclines," *Journal of the Chemical Society of Pakistan* **8**(4), 571–583 (1986).

36. S. Williams, ed., *Official Methods of Analysis,* 14th ed., Association of Official Analytical Chemists, Inc., Arlington, Va., 1984.

37. M. Petz, R. Solly, M. Lymburn, and M. H. Clear, "Thin-Layer Chromatographic Determination of Erythromycin and Other Macrolide Antibiotics in Livestock Products," *Journal of the Association of Official Analytical Chemists* **70**(4), 691–697 (1987).

38. Food and Drug Administration, *PMO, Grade A Pasteurized Milk Ordinance,* Public Health Service Publication No. **229,** Washington, D.C., 1985.

39. B. A. Morris, M. N. Clifford, and R. Jackman, eds., *Immunoassays for Veterinary and Food Analysis—Part 1,* Elsevier Science Publishing Co., Inc., New York, 1988.

40. D. Arnold, D. Vom Berg, A. K. Boertz, V. Mallick, and A. Somogyi, "Radioimmunological Determination of Chloramphenicol Residues in Musculature, Milk and Eggs," *Archiv für Lebensmittelhygiene* **35,** 131–139 (1984).

41. D. Arnold and A. Somogyi, "Trace Analysis of Chloramphenicol Residues in Eggs, Milk and Meat: Comparison of Gas Chromatography and Radioimmunoassay," *Journal of the Association of Official Analytical Chemists* **68,** 984–990 (1985).

42. M. E. Mount, D. L. Failla, and S. Wie, "Enzyme Immunoassay for the Feed Additive Monensin," In Ref. 38.

43. M. C. Rouan, "Antibiotic Monitoring in Body Fluids," *Journal of Chromatography* **340,** 361–400 (1985).

44. J. J. Ryan, E. E. Wildman, A. H. Duthie, and H. V. Atherton, "Detection of Penicillin, Cephapirin and Cloxacillin in Commingled Raw Milk by the Spot Test," *Journal of Dairy Science* **69,** 1510–1517 (1986).

45. C. Van de Water, N. Haagsma, P. J. S. Van Kooten, and W. van Eden, "An Enzyme Linked Immunosorbent Assay for the Determination of Chloramphenicol Using Monoclonal Antibody," *Zeitschrift für Lebensmittel-Untersuchang und-Forschung* **185,** 202–207 (1987).

46. J. F. M. Nouws, J. Laurensen, and M. M. L. Aerts, "Monitoring Milk for Chloramphenicol Residues by an Immunoassay (Quick-Card)," *Veterinary Quarterly* **10**(4), 270–272 (1988).

47. D. E. Dixon-Holland and S. E. Katz, "Competitive Direct Enzyme-Linked Immunosorbent Assay for Detection of Sulfamethazine Residues in Swine Urine and Muscle Tissues," *Journal of the Association of Official Analytical Chemists* **71**(6), 1137–1140 (1988).

48. T. Saesmaa, "Identification and Purity Determination of Benzathine and Embonate Salts of Some Beta-Lactam Antibiotics by Thin Layer Chromatography, *Journal of Chromatography* **463**(2), 469–473 (1989).

49. J. Sherma, W. Bretschneider, M. Dittamo, N. DiBiase, D. Huh, and D. P. Schwartz, "Spectrometric and Thin-Layer Chromatographic Quantification of Sulfathiazole Residues in Honey," *Journal of Chromatography* **463**(1), 229–233 (1989).

50. B. Wyhowski de Bukanski, J. M. Degroodt, and H. Beernaert, "A Two-Dimensional High Performance Thin-Layer Chromatographic Screening Method for Sulphonamides in Animal Tissues," *Zeitschrift für Lebensmittel-Untersuchung und-Forschung* **187**(3), 242–245 (1988).

51. H. Oka, Y. Ikai, N. Kawamura, K. Uno, and M. Yamada, "Improvement of Chemical Analysis of Antibiotics X. Determination of Eight Tetracyclines Using Thin Layer and High Performance Liquid Chromatography," *Journal of Chromatography* **393,** 285–296 (1987).

52. Y. Ikai, H. Oka, N. Kawamura, M. Yamada, K. Harada, and M. Suzuki, "Improvement of Chemical Analysis of Antibiotics. XIII. Systematic Simultaneous Analysis of Residual Tetracyclines in Animal Tissues Using Thin-Layer and High Performance Liquid Chromatography," *Journal of Chromatography* **411,** 313–323 (1987).

53. R. L. Ellis, "Less Expensive and Faster Than Laboratory Tests," *Association of Official Analytical Chemists Newsletter, The Referee* **13**(2), 9 (1989).

54. C. Stubbs and I. Kanfer, "High-Performance Liquid Chromatography of Erythromycin Propionyl Ester and Erythromycin Base in Biological Fluids," *Journal of Chromatography* **427**(1), 93–101 (1988).

55. M. Margosis, "Quantitative Liquid Chromatography of Ampicillin: Collaborative Study," *Journal of the Association of Official Analytical Chemists* **70**(2), 206–212 (1987).

56. K. Tyczkowska and A. L. Aronson, "Ion-Pair Liquid Chromatographic Determination of Some Penicillins in Canine and Equine Sera," *Journal of the Association of Official Analytical Chemists* **71**(4), 773–775 (1988).

57. G. Knupp, G. Bugl-Kreickmann, and C. Commichaw, "A Method for the Verification of Chloramphenicol Residues in Animal Tissues and Eggs by High Performance Liquid Chromatography with Radioimmunological Detection (HPLC-RIA)," *Zeitschrift für Lebensmittel-Untersuchung und-Forschung* **184**, 390–391 (1987).

58. M. F. Pochard, G. Burger, M. Chevalier, and E. Gleizes, "Determination of Chloramphenicol Residues with Reverse-Phase High Pressure Liquid Chromatography. Use of a Pharmacokinetic Study in Rainbow Trout with Confirmation by Mass Spectrometry," *Journal of Chromatography* **490**, 315–323 (1987).

59. M. M. Aerts, W. M. Beek, and U. A. Brinkman, "Monitoring of Veterinary Drug Residues by a Combination of Continuous Flow Techniques and Column-Switching High Performance Liquid Chromatography. I. Sulfonamides in Egg, Meat and Milk Using Post-Column Derivatization with Dimethyl-aminobenzaldehyde," *Journal of Chromatography* **435**(1), 97–112 (1988).

60. R. Malisch, "Multi-Method for the Determination of Residues of Chemotherapeutics, Antiparasitics and Growth Promoters in Foodstuffs of Animal Origin. Part I. General Procedure and Determination of Sulfonamides," *Zeitschrift für Lebensmittel-Untersuchung und-Forschung* **182**, 385–399 (1986).

61. G. D. Paulson, A. D. Mitchell, and R. G. Zaylskie, "Identification and Quantitation of Sulfamethazine Metabolites by Liquid Chromatography and Gas Chromatography-Mass Spectrometry," *Journal of the Association of Official Analytical Chemists* **68**(5), 1000–1006 (1985).

62. W. A. Moats, "Determination of Tetracycline Antibiotics in Tissues and Blood Serum of Cattle and Swine by High Performance Liquid Chromatography," *Journal of Chromatography* **358**, 253–259 (1986).

63. F. Kondo, S. Morikawa, and S. Tateyama, "Simultaneous Determination of Six Tetracyclines in Bovine Tissue, Plasma and Urine by Reverse-Phase High Performance Liquid Chromatography," *Journal of Food Protection* **52**(1), 41–44 (1988).

64. I. Norlander, H. Johnsson, and B. Österdahl, "Oxytetracycline Residues in Rainbow Trout Analyses by a Rapid HPLC Method," *Food Additives and Contaminants* **4**(3), 291–295 (1987).

65. D. W. Hill and K. J. Langner, "HPLC Photodiode Array UV Detection for Toxicological Drug Analysis," *Journal of Liquid Chromatography* **10**(2, 3), 377–409 (1987).

66. N. E. Larsen, K. Marinelli, and A. M. Heilesen, "Determination of Gentamicin in Serum Using Liquid Column Chromatography," *Journal of Chromatography* **221**, 182–187 (1980).

67. K. Tsuji, J. F. Goetz, W. Van Meter, and K. A. Gusciora, "Normal-Phase High Performance Liquid Chromatographic Determination of Neomycin Sulfate Derivatized with 1-fluoro-2,4-dinitrobenzene," *Journal of Chromatography* **175**, 141–152 (1979).

68. K. Tsuji and J. H. Robertson, "Gas-Liquid Chromatographic Determination of Neomycins B and C," *Analytical Chemistry* **41**, 1332–1335 (1969).

69. K. Tsuji and J. H. Robertson, "Formation of Trimethylsilyl Derivatives of Tetracyclines for Separation and Quantitation by Gas-Liquid Chromatography," *Analytical Chemistry* **45**(12), 2136–2140 (1973).

70. S. J. Stout, W. A. Stellar, A. J. Manuel, M. O. Poeppel, and A. R. DaCumha, "Confirmatory Method for Sulfamethazine Residues in Cattle and Swine Tissues Using Gas Chromatography-Chemical Ionization Mass Spectrometry," *Journal of the Association of Official Analytical Chemists* **67**(1), 142–144 (1984).

71. M. V. Novotny, "Recent Developments in Analytical Chromatography," *Science* **246**, 51–57 (1989).

Austin R. Long
Steven A. Barker
Louisiana State University
Baton Rouge, Louisiana

ANTIFOAMS. See Foams and silicone in food processing.

ANTIMICROBIALS

Antimicrobial food additives are an efficient, cost-effective, and often the only effective way to control fungal growth in foods. Antimicrobial food additives are basically chemicals that prevent or interfere with microbial growth. These chemicals may be found naturally occurring in certain foods, such as some organic acids and essential oils, or may be added to foods during processing (1). Antimicrobial food additives should never be used as a substitute for good manufacturing practices or proper sanitation procedures. Their proper use is as a processing aid to complement the above practices. Obviously, antimicrobial food additives must be safe for human consumption and their use is limited by law in most countries to relatively low levels and to specific foods. The various antimicrobial food additives are briefly reviewed in this article.

ORGANIC ACIDS

Organic acids have been used for years to control food spoilage. They find wide use because of solubility, taste, and low toxicity. The mode of action of organic acids is attributed to depression of intracellular pH by ionization of the undissociated acid molecule or disruption of substrate transport by alteration of cell membrane permeability (2, 3). In addition to inhibiting substrate transport, organic acids may inhibit NADH oxidation, thus eliminating supplies of reducing agents to electron transport systems (4).

Because the undissociated portion of the acid molecule is primarily responsible for antimicrobial activity, effectiveness depends on the dissociation constant of the acid and the pH of the food to be preserved. Because the dissociation constant of most organic acids is between pH 3 and 5, organic acids are generally most effective at low pH values (1). This, along with solubility properties, determines the foods in which organic acids may be effectively used.

A few fungal species possess mechanisms of resistance to organic acid preservatives. *Saccharomyces bailii* is resistant to high concentrations of sorbic and benzoic acids (3–5). Some molds in the genus *Penicillium* are able to grow in the presence of high concentrations of sorbic acid and decarboxylate sorbic acid to 1,3-pentadiene, a volatile compound with an extremely strong kerosenelike odor (2, 6, 7). In cases where resistance to, or metabolism of, an organic acid is a problem, other preservative systems must be used.

Sorbic Acid

Sorbic acid and its potassium salt are the most widely used forms of the compound and are collectively known as sorbates. The salt forms are highly soluble in water, as is the case with all organic acids. Their most common use is preservation of food, animal feed, and cosmetic and pharmaceutical products as well as technical preparations that come in contact with the human body. Methods of application include direct addition into the product; dipping, spraying, or dusting of the product; and incorporation into the wrapper (2, 8).

Typical use levels in foods range from 0.02% in wine and dried fruits to 0.3% in some cheeses. Foods in which sorbate has commercially useful antimicrobial activity include baked goods (cakes and cake mixes, pies and pie fillings, doughnuts, baking mixes, fudges, and icings), dairy products (natural and processed cheeses, cottage cheese, and sour cream), fruit and berry products (artificially sweetened confections, dried fruits, fruit drinks, jams, jellies, and wine), vegetable products (olives, pickles, relishes, and salads), and other miscellaneous food products (certain fish and meat products, mayonnaise, margarine, and salad dressings).

Environmental factors such as pH, water activity, temperature, atmosphere, type of microbial flora, initial microbial load, and certain food components, singly or in combination, can influence the activity of sorbate. Together with preservatives such as sorbic acid they often act to broaden antimicrobial action or increase it synergistically. Use of other preservatives in combination with sorbate can also broaden or intensify antimicrobial action. The length and temperature of storage are other important considerations. If growth of spoilage or pathogenic organisms is inhibited but the microorganism is not killed, growth will eventually resume under proper conditions. The length of inhibition will vary with storage temperature as well as with any of the other factors discussed.

Sorbic acid is a broad-spectrum antimicrobial that is effective against yeasts and molds and some bacteria. The antimicrobial effect of sorbate is greater at pH 5.0 or lower than at higher pH values. Sorbic acid has little antifungal activity at pH values higher than 5.5–6.0. Above this pH range, little of the acid is in the antimicrobially active undissociated form. However, sorbic acid has a relatively high dissociation constant compared to benzoic or propionic acids and is, therefore, usually the most effective of the organic acids at pH levels of 5.0 and higher.

Benzoic Acid

Benzoic acid also has widespread use in the food industry. It occurs naturally in raspberries, cranberries, plums, prunes, cinnamon, and cloves (1). As an antifungal food additive, the water-soluble sodium and potassium salts and the fat-soluble acid form are suitable for foods and beverages with pH below 4.5. Benzoates have little effect at neutral pH values. They are not as effective as sorbates at pH 5.0, but effectiveness increases at lower pH values.

Benzoic acid is active against yeasts and molds, including aflatoxin-forming microorganisms (9). The acid form is often added to the fat phase or the sodium salt to the water phase of products such as salad dressings, mayonnaise, pickled vegetables, fruit products, and fruit drinks. Because benzoate can impart a fairly strong bitter off-flavor, it is often used in combination with sorbate. This mixture is usually more effective in inhibiting yeasts and molds than the comparable level of either preservative alone. In addition, the mixture is less offensive organoleptically than benzoate alone.

Propionic Acid

This organic acid inhibits molds but not yeasts. It occurs in some foods as, result of natural processing. It is present in Swiss cheese at concentrations up to 1%, where it is produced by the bacterium *Propionibacterium shermanii* (1). Because yeasts are typically unaffected, the acid can be added to bread dough without interfering with leavening.

In the food industry, propionic acid is often used as a sodium or calcium salt (10). Propionates are used primarily to inhibit molds in bakery goods. In addition to their antimycotic properties, propionates will inhibit *Bacillus mesentericus,* the rope-causing bacterium. Propionates are also used to a limited extent to inhibit mold growth in process cheeses.

The antimicrobial activity of propionic acid is weak compared to the other organic acids; therefore, they must be used in relatively high concentrations to be effective. As with other organic acids, the pH value of the food to be preserved affects the antimicrobial activity. Because of its low dissociation constant, propionic acid is active in a pH range similar to that of sorbic acid.

MEDIUM-CHAIN FATTY ACIDS

Generally, fatty acids are most effective as inhibitors of gram-positive bacteria and yeasts, although some fatty acids exhibit antimycotic activity (1). It has been observed that fatty acid derivatives reduced growth and aflatoxin production by *Aspergillus* sp. (11).

Polyhydric alcohol fatty acid esters have great potential for use as emulsifiers in food formulations. They also possess antifungal properties and, therefore, may exert a preservative effect in foods. Strong fungistatic activity of glycerol monocaprate and glycerol monolaurate toward *Aspergillus niger, Penicillium citrinum, Candida utilis,* and *Saccharomyces cerevisiae* has been demonstrated (12). Sucrose monocaprate and sucrose monolaurate were found to be slightly inhibitory to a spoilage film-forming yeast inoculated into a soy sauce substrate (13). Six sucrose esters substituted to different degrees with a mixture of palmitic and stearic acids were examined for antifungal properties (14). Growth of *Aspergillus, Penicillium, Cladosporium,* and *Alternaria* sp. was inhibited in media containing 1% of sucrose esters.

ANTIBIOTICS

Natamycin

Natamycin, formerly called pimaricin, is an antibiotic that possesses strong antifungal properties, yet is not active against bacteria. Its use is currently allowed in several countries. Commercially it is used primarily in the preservation of cheese as a surface coating.

Researchers have demonstrated that natamycin is active at low concentrations against many of the fungi known to cause food spoilage. Levels of 1–100 ppm added to laboratory media have been reported to be effective for inhibiting mold growth (15). Similar concentrations have been reported to be effective in cottage and Cheddar cheeses while solutions of 1,000–2,000 ppm are effective as dips for cheese. Natamycin was found to inhibit aflatoxin formation by molds only when growth was completely inhibited (9).

Nisin

Nisin is active against gram-positive bacteria exclusively according to most reports and is used primarily to inhibit growth of *Clostridium botulinum* (9). However, it has been reported that nisin at levels of 5 and 125 ppm delayed growth of *Aspergillus parasiticus* in culture media, but after long periods of incubation, the mold grew faster and produced more aflatoxin in nisin-containing cultures than in cultures without it (16).

METABOLITES FROM LACTIC ACID BACTERIA

A wide variety of raw foods are preserved by lactic acid fermentation, including milk, meat, fruits, and vegetables. Reduction of pH and removal of large amounts of carbohydrates by fermentation are the primary preserving actions that lactic acid bacteria (LAB) provide to a fermented food. These actions are largely ineffective in preventing growth of fungi in foods. However, it has also been recognized that LAB are capable of producing inhibitory substances other than organic acids (lactate and acetate) that are antagonistic toward other microorganisms (17). The antibacterial properties of LAB are well documented (17). Several LAB, typically of the genera *Streptococcus* and *Lactobacillus,* are known to produce antibacterial substances. These substances include hydrogen peroxide, diacetyl, bacteriocins, and other undefined inhibitory substances. Antifungal properties of LAB have received little attention; however, several metabolites of LAB have been reported to have antifungal activity (17, 18).

Diacetyl

Diacetyl is a metabolic end product produced by some species of LAB. It is best known for its buttery aroma that it imparts to cultured dairy products (17). Its antimicrobial action has been investigated (19, 20). It was reported that a concentration of 200 ppm was inhibitory to yeasts and gram-negative bacteria, and 300 ppm was inhibitory to molds and gram-positive bacteria. Acidity of the growth medium was shown to have a direct effect on the antimicrobial activity of diacetyl. The compound was clearly more effective as an antimicrobial agent below pH 7.0 than above this value. The reasons for pH-associated antifungal activity is not clear. Because effective concentrations of diacetyl impart a sharp odor of butter, potential for use in foods as an antimicrobial agent is limited. However, its use as a utensil sanitizer and in wash or rinse water for certain products is more feasible (20).

Microgard

Microgard (Wesman Foods, Inc., Beaverton, Oreg.) is grade A skim milk that has been fermented by *Propionibacterium shermanii* and then pasteurized (21). The product has been shown to prolong the shelf life of cottage cheese by inhibiting psychotropic spoilage bacteria. The product is also antagonistic toward some yeasts and molds (17). Microgard consists of propionic acid, diacetyl, acetic acid, and lactic acid (22).

Bacteriocins

Bacteriocins are protein-containing macromolecules that exert a bactericidal mode of action on susceptible bacteria (17). Bacteriocins produced by LAB are well documented in recent literature. The bacteriocin-producing ability of the bacteria in the genera *Lactobacillus, Leuconostoc,* and *Pediococcus* has been the focus of much study. Characteristics common to these bacteriocins are that their genetic determinants appear to be plasmid-borne and they are active against a broad spectrum of gram-positive bacteria. The ability of these bacteriocins to inhibit many foodborne pathogens, including *Listeria monocytogenes,* make them attractive as potential food-preservation agents.

HERBS AND SPICES

Herbs and spices are widely used to impart flavor to foods. It is generally accepted that certain herbs and spices have antimicrobial activity and may influence the keeping quality of foods to which they have been added. However, they are not currently used with the primary purpose of providing a preservative effect.

One of the earliest reports to describe the preservative action of cinnamon, cloves, mustard, allspice, nutmeg, ginger, black pepper, and cayenne pepper found that cinnamon, cloves, and mustard were most effective and ginger, black pepper, and cayenne pepper were least effective (23). The effect of spices and their essential oils on growth of several test organisms, including *Aspergillus* and *Penicillium* species was studied (24). It was concluded that spices used in amounts as employed normally for ordinary foods were insufficient as preservatives. However, when used in larger amounts, cinnamon, cloves, and allspice retarded mold growth. It was reported that cinnamon in concentrations as low as 0.02% inhibited mold growth and aflatoxin production in culture media and cinnamon bread (25).

Cloves, cinnamon, mustard, allspice, garlic, and oregano at the 2% level in potato dextrose agar have been shown to completely inhibit mycotoxigenic molds for up to 21 days (26). Combinations of different levels of potassium sorbate in combination with cloves showed an enhanced or possibly synergistic inhibitory effect on the growth of molds, indicating the possibility of using spices and commercial antifungal agents together in small amounts to obtain antifungal activity.

In most cases, herbs and spices are not effective antifungal agents when used alone in amounts normally added to foods. However, when used in combination with other preservative systems, they can be valuable contributors to an antifungal system consisting of interacting physical and chemical preservatives.

ESSENTIAL OILS

The antimicrobial activities of extracts from several types of plants and plant parts used as flavoring agents in foods and beverages have been recognized for many years. Certain of these essential oils have antifungal properties. The effects of garlic and onion against yeasts have been documented (27), and other studies have shown these extracts to be inhibitory to molds (28–31). When the effects of lemon and orange oils on *Aspergillus flavus* were examined, it was found that citrus oils added to grapefruit juice or glucose–yeast extract medium at a concentration of 3,000–3,500 ppm suppressed growth and aflatoxin production (32). When orange oil was added to either medium at concentrations up to 7,000 ppm, growth and aflatoxin production were greatly reduced although still evident. It was shown that essential oil of allspice, cinnamon, clove, oregano, savory, and thyme were inhibitory to food and industrial yeasts (27).

PHENOLIC ANTIOXIDANTS

Phenolic antioxidants have been shown by several researchers to possess antifungal activity. It was demonstrated that in a glucose salts medium, 1,000 ppm butylated hydroxyanisole (BHA) inhibited growth and aflatoxin production of *Aspergillus parasiticus* spores, and >250 ppm inhibited growth and aflatoxin of *A. parasiticus* mycelia (33). However, it was also found that at 10 ppm BHA, total aflatoxin production was more than twice that of the control, with virtually no effect on mycelial weight. These results indicate that at high levels, BHA may serve as an effective antifungal agent; however, at low levels BHA may actually stimulate aflatoxin production. The effects of BHA and butylated hydroxytoluene (BHT) on several strains of *A. flavus* were tested (34). On solid agar, BHT was not inhibitory at any level tested. BHA inhibited growth and aflatoxin production.

Because the primary use of these compounds in foods is as antioxidants, their effectiveness as antifungal agents in food systems has not been adequately studied. While results of experiments in growth media indicate that these compounds exert antifungal effects, extrapolation of these results to food systems should be done with caution.

Interaction of these compounds with food components will undoubtably affect their antifungal properties.

GASES AND MODIFIED ATMOSPHERES

Elimination of oxygen is often used as a control measure for inhibiting growth of molds and aerobic bacteria. Exclusion of oxygen will not prevent growth of yeasts. Studies on bakery products have demonstrated that atmospheric O_2 levels must be reduced to 0.1–1.0% to effectively inhibit growth of molds. In studies on toasted bread, it was demonstrated that visible mold would occur in 3 days in air; 5 days in 99% N_2 and 1% O_2; and >100 days in 99.9% N_2 and 0.1% O_2, 99% CO_2 and 1.0% O_2, 99.8% CO_2 and 0.2% O_2, and 100% CO_2 (35). This study demonstrates that although molds are considered to be aerobic organisms, certain species have the ability to grow at very low O_2 concentrations. Effectively controlling molds by simple gas flushing can be difficult in practice. Chemical oxygen scavengers can be used in place of or to supplement gas flushing (36). Oxygen scavengers will also give protection against package leaks and infiltration of O_2 through the package.

Carbon dioxide exerts antimicrobial action that is in addition to simple exclusion of O_2 and is more effective than inert gases such as nitrogen. The gas probably exerts antimicrobial activity by altering intracellular pH levels (37). Recent research has shown carbon monoxide to have potential use with foods. CO has been shown to inhibit yeasts and molds that cause postharvest decay in fruits and vegetables (38). The potential toxicity of this compound to workers requires special handling procedures.

Sulfur dioxide is broadly effective against yeasts, molds, and bacteria (39). It is used extensively for controlling growth of undesirable microorganisms in fruits, fruit drinks, wines, sausages, fresh shrimp, and pickles. The antimicrobial activity of SO_2 is associated with the unionized form of the molecule. Therefore, it is most effective at pH values <4.0, where this form is predominate (38).

Ethylene oxide has been widely used to reduce microbial contamination and kill insects in various dried foods. The gas has been used to treat gums, spices, dried fruits, corn, wheat, barley, dried egg, and gelatin (40). Concern over toxicity of residues have limited the use of this gas in recent years.

Propylene oxide has been less studied than ethylene oxide. However, it appears that its antifungal effects are similar (38). Yeasts and molds are more sensitive to the gas than bacteria. Propylene oxide has been used as a fumigant for control of microorganisms and insects in bulk quantities of goods such as cocoa, gums, processed spices, starch, and processed nut meats (40).

INTERACTION OF FACTORS

Many of the antifungal agents reviewed above must be used at extreme concentrations or levels in a food to be effective when used alone. However, a variety of factors can prevent growth of fungi. While the fungi tend to be more tolerant of adverse environmental conditions than bacteria, combining inhibitory factors such as temperature, water activity, or pH with antifungal agents can result in considerable improvement of the microbial stability of foods. Suitable combinations of growth-limiting factors at subinhibitory levels can be devised so that certain microorganisms can no longer proliferate.

Sorbic acid at a concentration of 1,000 ppm at pH 7.0 will not inhibit mold growth. However, if the pH is lowered to 5.0, growth of most molds will be inhibited (6). Antioxidants such as BHA and BHT have been shown to potentiate the action of sorbic acid (41). In general, antifungal food additives become more effective as environmental conditions move away from the optimum for a particular organism.

The level of a single growth-limiting factor that will inhibit a microorganism is usually determined under conditions in which all other factors are optimum. In preserving foods, more than one factor is usually relied on to control microbial growth. The addition of a substance that in itself does not give full inhibition can effectively preserve products in the presence of other subinhibitory factors. The effect of superimposing limiting factors is known as the hurdles concept (41).

Little information is currently available on combining subinhibitory factors to preserve foods. It is time consuming and expensive to design preservative systems using the hurdles concept by random design. Predictive modeling can be used to test the consequences of a number of factors changing at the same time. With proper design and interpretation, preservative systems can be designed rapidly and efficiently (42). With greater emphasis on development and marketing of refrigerated foods by the food industry, these methods will become more widely used and accepted. However, product challenge studies should be conducted to verify the effectiveness of a combination of subinhibitory factors.

BIBLIOGRAPHY

1. L. R. Beuchat and D. A. Golden, "Antimicrobial Occurring Naturally in Foods," *Food Technology* **43**, 134–142 (1989).

2. M. B. Liewen and E. H. Marth, "Growth and Inactivation of Microorganisms in the Presence of Sorbic Acid: A Review," *J. Food Prot.* **48**, 364–375 (1985).

3. A. D. Warth, "Mechanism of Resistance of *Saccharomyces balii* to Benzoic, Sorbic and Other Weak Acids Used as Food Preservatives," *Journal of Applied Bacteriology* **43**, 215–230 (1977).

4. E. Fresse, C. W. Sheu, and E. Galliers, "Function of Lipophilic Acids as Antimicrobial Food Additives," *Nature* **241**, 321–325 (1973).

5. J. I. Pitt and K. C. Richardson, "Spoilage by Preservative-Resistant Yeasts," *CSIRO Food Res. Q.* **33**, 80–85 (1973).

6. M. B. Liewen and E. H. Marth, "Inhibition of Penicillia and Aspergilli by Potassium Sorbate," *J. Food Prot.* **47**, 554–556 (1984).

7. W. J. Tsai, M. B. Liewen, and L. B. Bullerman, "Toxicity and Sorbate Sensitivity of Molds Isolated from Surplus Commodity Cheeses," *J. Food Prot.* **51**, 457–462 (1988).

8. G. Cirilli, "Preservation of Bread with Sorbic Acid," *Industrie Alimentari.* **15**, 67–68 (1976).

9. L. L. Ray and L. B. Bullerman, "Preventing Growth of Potentially Toxic Molds Using Antifungal Agents," *J. Food Prot.* **45**, 953–963 (1982).

10. F. Sauer, "Control of Yeasts and Molds with Preservatives," *Food Technology* **31**(2), 66–67 (1977).

11. J. R. Chipley, L. D. Story, and J. J. Kabara, "Inhibition of *Aspergillus* Growth and Extracellular Aflatoxin Accumulation by Sorbic Acid and Derivatives of Fatty Acids," *Journal of Food Safety* **2**, 109–115 (1981).

12. N. Kato and I. Shibasaki, "Comparison of Antimicrobial Activities of Fatty Acids and Their Esters," *Journal of Ferment. Technol.* **53**, 793–795 (1975).

13. N. Kato, "Antimicrobial Activity of Fatty Acids and Their Esters Against a Film-Forming Yeast in Soy Sauce," *Journal of Food Safety* **3**, 321–325 (1981).

14. D. L. Marshall and L. B. Bullerman, "Antimicrobial Activity of Sucrose Fatty Acid Ester Emulsifiers," *Journal of Food Science* **51**, 468–470 (1986).

15. H. Gourama and L. B. Bullerman, "Effects of Potassium Sorbate and Natamycin on Growth and Penicillic Acid Production by *Aspergillus ochraceus*," *J. Food Prot.* **51**, 139–144, 155 (1988).

16. A. E. Yousef, S. E. El-Gendy, and E. H. Marth, "Growth and Biosynthesis of Aflatoxin by *Aspergillus parasiticus* in Cultures Containing Nisin," *Zeitschrift fuer Lebensmittel-Untersuchung und-Forschung* **171**, 341–343 (1980).

17. M. A. Daeschel, "Antimicrobial Substances from Lactic Acid Bacteria for Use as Food Preservatives," *Food Technology* **43**(1), 165–166 (1989).

18. V. K. Batish, S. Grover, and R. Lal, "Screening Lactic Starter Cultures for Antifungal Activity," *Cultured Dairy Prod. J.* **24**(2), 21–25 (1989).

19. J. M. Jay, "Antimicrobial Properties of Diacetyl," *Applied and Environmental Microbiology* **44**, 525–532 (1982).

20. J. M. Jay, "Effect of Diacetyl on Foodborne Microorganisms," *Journal of Food Science* **47**, 1829–1831 (1982).

21. G. H. Weber and W. A. Broich, "Shelf Life Extension of Cultured Dairy Foods," *Cultured Dairy Prod. J.* **21**(4), 19–23 (1986).

22. N. Al-Zoreky, *Microbiological Control of Food Spoilage and Pathogenic Microorganisms in Refrigerated Foods*, M.S. thesis, Oregon State University, Corvallis.

23. C. Hoffman and A. C. Evans, "Use of Spices as Preservatives," *J. Ind. Eng. Chem.* **3**, 835 (1911).

24. F. M. Bachman, "The Inhibitory Action of Certain Spices on the Growth of Microorganisms," *J. Ind. Eng. Chem.* **8**, 620 (1916).

25. L. B. Bullerman, "Inhibition of Aflatoxin Production by Cinnamon," *Journal of Food Science* **39**, 1163–1165 (1974).

26. M. A. Azzouz and L. B. Bullerman, "Comparative Antimycotic Effects of Selected Herbs, Spices, Plant Components and Commercial Antifungal Agents," *J. Food Prot.* **45**, 1298–1301 (1982).

27. D. E. Conner and L. R. Beuchat, "Effects of Essential Oils from Plants on Growth of Spoilage Yeasts," *Journal of Food Science* **49**, 429–434 (1984).

28. M. A. Collins and H. P. Charles, "Antimicrobial Activity of Carnosol and Ursolic Acid: Two Antioxidant Constituents of *Rosmarinus-officinalis*," *Food Microbiology* **4**, 311–316 (1987).

29. V. Moleyar and P. Narasimham, "Fungitoxicity of Binary Mixtures of Citral, Cinnamic Aldehyde, Menthol and Lemon Grass Oil Against *Aspergillus niger* and *Rhizopus stolonifer*,"

Lebensmittel-Wisenshaft und- Technologie **21**, 100–102 (1988).

30. A. Sharma, G. M. Tewari, A. J. Shrikhande, S. R. Padwal-Desai, and C. Bandyopadhyay, "Inhibition of Aflatoxin-Producing Fungi by Onion Extracts," *Journal of Food Science* **44**, 1545–1547 (1979).

31. M. R. Tansey and J. A. Appleton, "Inhibition of Fungal Growth by Garlic Extract," *Mycologia*. **67**, 409–415 (1975).

32. G. G. Alderman and E. H. Marth, "Inhibition of Growth and Aflatoxin Production of *Aspergillus parasiticus* by Citrus Oils," *Z. Lebensm. Unters-Forsch.* **160**, 353–358 (1976).

33. H. C. Chang and A. L. Branen, "Antimicrobial Effect of Butylated Hydroxyianisole (BHA) and Butylated Hydrotoluene (BHT)," *Journal of Food Science* **40**, 349–351 (1975).

34. D. Y. C. Fung, S. Taylor, and J. Kahan, "Effects of Butylated Hydroxianisole (BHA) and Butylated Hydrotoluene (BHT) on Growth and Aflatoxin Production of *Aspergillus flavus*," *Journal of Food Safety* **1**, 39–51 (1977).

35. G. Cerny, "Retardation of Toast Bread by Gassing," *Chem., Mikrobiol., Technol. Lebensm.* **6**(1), 8–10 (1979).

36. J. P. Smith, B. Orraikv, W. J. Koersen, E. D. Jackson, and R. A. Lawrence, "Novel Approach to Oxygen Control in Modified Atmosphere Packaging of Bakery Products," *Food Microbiol.* **3**, 315–320 (1986).

37. L. G. Corral, L. S. Post, and T. J. Montville, "Antimicrobial Activity of Sodium Bicarbonate," *Journal of Food Science* **53**:981–982.

38. M. K. Wagner and L. J. Moberg, "Present and Future Use of Traditional Antimicrobial," *Food Technology* **43**(1), 134–142 (1989).

39. J. D. Dziezak, "Preservatives: Antimicrobial Agents," *Food Technology* **40**(9), 104–107 (1986).

40. International Commission on Microbiological Specifications for Foods, *Microbial Ecology of Foods., Vol. 1. Factors Affecting Life and Death of Microorganisms*, Academic Press, Inc., Orlando, Fla., 1980.

41. V. N. Scott, "Interaction of Factors to Control Microbial Spoilage of Refrigerated Foods," *J. Food Prot.* **52**, 431–435 (1989).

42. T. A. Roberts, "Combinations of Antimicrobial and Processing Methods," *Food Technology* **43**(1), 156–163 (1989).

MICHAEL B. LIEWEN
General Mills Inc.
Minneapolis, Minnesota

ANTIOXIDANTS

U.S. citizens expect their food system to be the finest in the world. To provide the large, urban population with food in its best condition, food scientists employ many tools, including food additives. Recently, the necessity of food additives has been generally recognized (eg, increase in shelf life, retention of nutrients, and prevention of bacterial spoilage). However, controversy exists over the safety of many of these additives. The most misunderstood and controversial food additives are antioxidants.

Antioxidants are defined by the U.S. Food and Drug Administration (FDA) as preservatives that specifically retard deterioration, rancidity, or discoloration due to oxidation. After examining the causes of oxidation and the adverse affects resulting from it, the role of antioxidants and the types used in the food supply will be discussed. Oxidation of food occurs when oxygen is added to unsaturated sites of molecules. Oxygen, light, heat, heavy metals, pigments, alkaline conditions, and degree of unsaturation are catalysts in this process.

The mode of action of most antioxidants is not widely understood; therefore, a description of how antioxidants work follows. The oxidation of lipids, for example, occurs at the unsaturated sites of the fatty acids on a triglyceride. Off-flavors and odors are the result of the formation of aldehydes and ketones that are produced by an oxidation reaction. This rancidity of the fat not only results in food waste but also in the formation of potentially toxic by-products. Other deterioration reactions include the discoloration of pigments, loss of product flavor and odor, changes in texture, and reduction in nutritional value due to loss of vitamin activity (1).

The use of antioxidants prevents or minimizes the phenomenon of oxidation in foods. One way this is accomplished is through the donation of electrons or H+ atoms to terminate a free-radical chain. The formation of a complex between the antioxidant and the fat chain contributes to the antioxidant's inhibitory effect. To be effective, antioxidants must be added as early as possible in the manufacturing process or to the finished fat.

CLASSES OF ANTIOXIDANTS

Two classes of antioxidants are free-radical scavengers and metal sequestrants. Free-radical scavengers include butylated hydroxyanisol (BHA), butylated hydroxytoluene (BHT), *tert*-butylhydroquinone (TBHQ), propyl gallate (PG), tocopherols, and ascorbic acid. These compounds react with free radicals to create less-reactive elements. Metal sequestrants, otherwise known as chelaters, precipitate a metal or suppress its reactivity by occupying all coordination sites (2). Phosphates, ethylenediaminetetraacetic acid (EDTA), and citric acid are common metal sequestrants. EDTA's mode of action includes formation of highly stable complexes by its sequestering action on iron, copper, and calcium. Maximum chelating efficiency occurs when higher pH values are indicated and the carboxyl groups are dissociated (1).

To extend the shelf life of foods containing a high percent of both unsaturated and saturated fats, antioxidants are used as processing aids and preservatives. BHA and BHT are fat-soluble antioxidants. BHA is used in controlling oxidation of fatty acids and is found widely in cereals, lard, and confectionary products (3). When BHA and BHT are used together, they exert a synergistic effect (1).

Another antioxidant, TBHQ, protects frying oils against oxidation. It is also used in conjunction with hydrogenation, or the conversion of a liquid to a solid fat. The inclusion of a chelating agent such as citric acid in the TBHQ antioxidant system can also further enhance the stabilizing properties. This combination is used primarily for the protection of vegetable oils and shortenings. It should be noted that TBHQ is ineffective in baking applications due to heat denaturation.

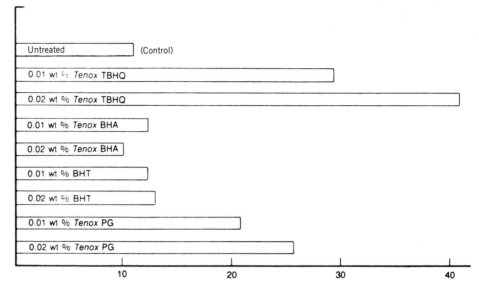

Figure 1. Effectiveness of TBHQ antioxidant in refined soybean oil. Peroxide value is defined as milliequivalents (meq) of peroxide per kilogram of fat; AOM stability—as hours to develop peroxide value of 70 meq. Courtesy of Eastman Chemicals Division, Kingsport, Tenn (5).

PG is a synthetic antioxidant that is slightly water soluble. It functions in the stabilizing of animal fats and vegetable oils. However, PG loses its effectiveness under heat conditions. Therefore, it is unsuitable for frying applications that involve temperatures in excess of 190°C. Thus its applicability in food processing is severely limited.

A side benefit of phenolic antioxidants are their antimicrobial properties. Phenolic antioxidants have been shown to have antimicrobial activity against bacteria, molds, viruses, and protozoa. Some strains of *Salmonella typhimurium* were completely inhibited by 150 ppm BHT and 400 ppm BHA. In addition, *Staphococcus aureus* growth was inhibited 93–100% by 150 ppm BHA. Moreover, concentrations of 100 ppm BHA or greater inhibited the formation of *S. aureus* enterotoxin. Some strains of *Clostridium perfringens* were completely inhibited by 150 ppm BHA. BHA also inhibits spoilage organisms such as *Pseudomonas*. BHA also has antifungal properties and in-

hibits the production of aflatoxins in some strains of *Aspergillus* (4).

The data presented here were generated to evaluate the effectiveness of TBHQ, BHA, BHT, and PG at Eastman Laboratories (Kingsport, Tenn.) (5). Stability was determined by the active oxygen method (aom), a test developed in which oil or fat is heated and held at 97.8°C. A constant airflow is sparged through the liquid and peroxides are measured during exposure (5).

Figure 1 compares the effectiveness of various antioxidants in refined soybean oil. These data indicate the greatest antioxidative effect was achieved with 0.02 wt % TBHQ followed by PG, BHT, and BHA, respectively, at 0.02%. Similar results were reported in tests using refined cottonseed oil (Fig. 2) and crude safflower seed oil (Fig. 3).

Research (5) also produced data on antioxidant effectiveness in edible animal fats and butterfat (Figs. 4 and 5). All antioxidants increased the aom stability time as compared to the control, the most effective was TBHQ. In

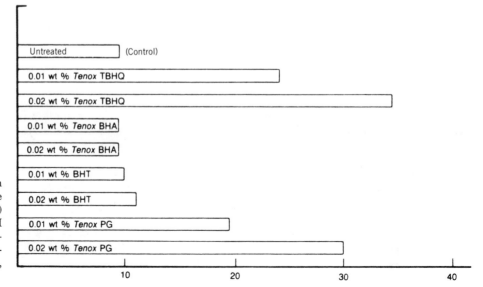

Figure 2. Effectiveness of TBHQ in refined cottonseed oil. Peroxide value is defined as milliequivalents (meq) of peroxide per kilogram of fat; AOM stability—as hours to develop peroxide value of 70 meq. Courtesy of Eastman Chemicals Division, Kingsport, Tenn (5).

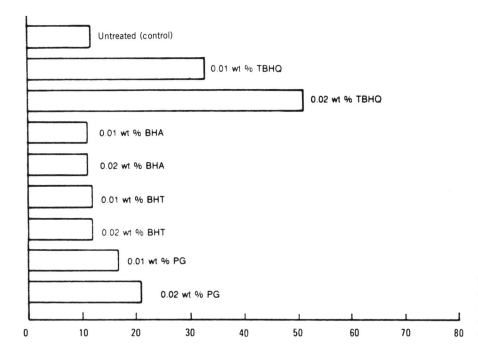

Figure 3. Effectiveness of TBHQ in crude safflower-seed oil; AOM stability—as hours to develop peroxide value of 70 meq. Courtesy of Eastman Chemicals Division, Kingsport, Tenn (5).

addition, good carry through has been reported (5) in most potato chips and baked products (Table 1, Fig. 6). The beneficial effects of the various antioxidants are reported in pastry, cracker, and anhydrous butterfat. Antioxidant potency in lemon, peppermint, and orange oils has been reported as well (5). Figure 7 shows that all of the antioxidants exhibited better potency than the control; TBHQ was the most effective followed by BHA, BHT, and PG, respectively (5).

The regulations governing the use of antioxidants and amounts of synthetic antioxidants are presented in Table 2. These regulations are issued by the FDA and permit the use of TBHQ alone or BHA or BHT together at concentrations not to exceed 0.02 wt % of the lipid content of the food (Table 2).

HEALTH RISKS

Safety evaluation of direct and indirect antioxidants is indeed an important consideration, because of the chemical changes that occur during food processing and storage. As a result of the extent that chemical alterations of the additives and reactions with food components occur in processing, the exposure to humans of these anthropogenic compounds of unknown biological activity is inevitable (6). To document more information about the reactions of BHT in food, several research projects involving the chemical breakdown of BHT, BHA, and TBHQ were published (6). Results given in Figure 7 demonstrate that because of the inherent thermal instability of the peroxide linkage of BHT, there were measurable levels of peroxy compounds throughout the experiment in the frying process.

In another study (6), BHT was converted to HBHT (Hydro butylated hydroxytoluene) via oxidation when heated above 160°C in the presence of oxygen (Fig. 8). Further-

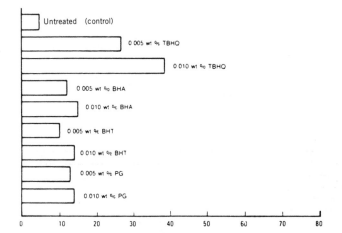

Figure 4. Effectiveness of TBHQ in poultry fat; AOM stability—as hours to develop peroxide value of 20 meq. Courtesy of Eastman Chemicals Division, Kingsport, Tenn (5).

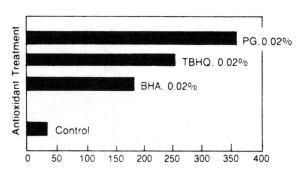

Figure 5. Comparison of antioxidant performance in anhydrous butterfat; AOM stability—as hours to reach a peroxide value of 20 meq. Courtesy of Eastman Chemicals Division, Kingsport, Tenn (5).

Table 1. Carry Through into Baked Products (Pastry and Crackers)[a]

Antioxidant Treatment of Lard, wt %		Oven Storage Life of Baked Goods, As Days to Develop Rancid Odor			
		Pastry		Crackers	
		37.8°C (100°F)	62.8°C (145°F)	37.8°C (100°F)	62.8°C (145°F)
Untreated (control)		10	2	9	3
Tenox TBHQ	0.005	14	2	27	7
	0.010	25	3	—	10
Tenox BHA	0.005	35	8	125	12
	0.010	96	21	218	22
BHT	0.005	40	5	90	10
	0.010	—	10	132	14
Tenox PG	0.005	23	2	—	3
	0.010	28	5	33	6
Tenox TBHQ + *Tenox* BHA	0.010 0.010	236	38	236	23
Tenox TBHQ + BHT	0.010 0.010	167	16	255	25

[a] Courtesy of Eastman Chemicals Division, Kingsport, Tenn.

more, mass spectrometry analysis yielded identification of BHT, 3,5-di-*tert*-butyl-5-hydroxybenzaldehyde (DBHB), and 4-methyl-4-hydroxy-2,6-di-*tert*-butyl-cyclohexa-2,5-dione (MHCD) as thermolysis breakdown products of HBHT (Fig. 9). The oxidation of BHA and TBHQ to form TBBQ was reported in fat used to prepare french fried potatoes (Fig. 10) (6). These studies generated information that showed that phenolic antioxidants are subject to complex chemical changes and reactions in foods: organic chemistry plays an obvious role in the formation of these intermediates and products; however, the safety levels in foods remain questionable.

NATURAL ANTIOXIDANTS

Lately, food scientists have identified many types of natural antioxidant that are alternatives to synthetic ones. Tocopherols (vitamin E) are natural antioxidants found in plants and show their greatest potency as antioxidants in animal fats, carotenoids, and vitamin A. Unsaturated vegetable oils with their inherent tocopherol content do not benefit much from the addition of tocopherols. Tocopherols are often lost during processing due to their instability. Vitamin C, ascorbic acid, is a free-radical scavenger that when added to food systems acts as an effective antioxidant. Ascorbic acid is then oxidized to form dihydroascorbic acid thereby asserting its antioxidant action. Ascorbic acid alone has minimal antioxidant activity, however, when used in combination with other antioxidants, it functions as a synergist by promoting their antioxidant effects (1). Two naturally occurring antioxidants are found in *B*-carotene and ginger rhizome. *B*-Carotene has been used in prevention of lipid oxidation and ginger rhizome has been applied to meat products. Both are beginning to be widely used in the food industry.

New experiments *in vitro* show that *B*-carotene belongs to a previously unknown class of biological antioxidants.

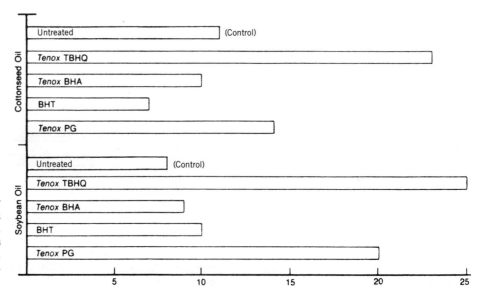

Figure 6. Carry through of stability into potato chips (crisps) deep fried in oil containing 0.02% antioxidant. Oven storage life of potato chips as days to develop rancid odor at 63°C. Courtesy of Eastman Chemicals Division, Kingsport, Tenn (5).

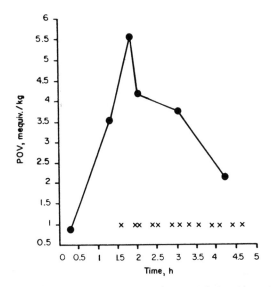

Figure 7. Peroxy value (milliequivalents per kg) vs time at frying temperature (X indicates batch of potatoes fried). Courtesy of the Association of Official Analytical Chemists (6).

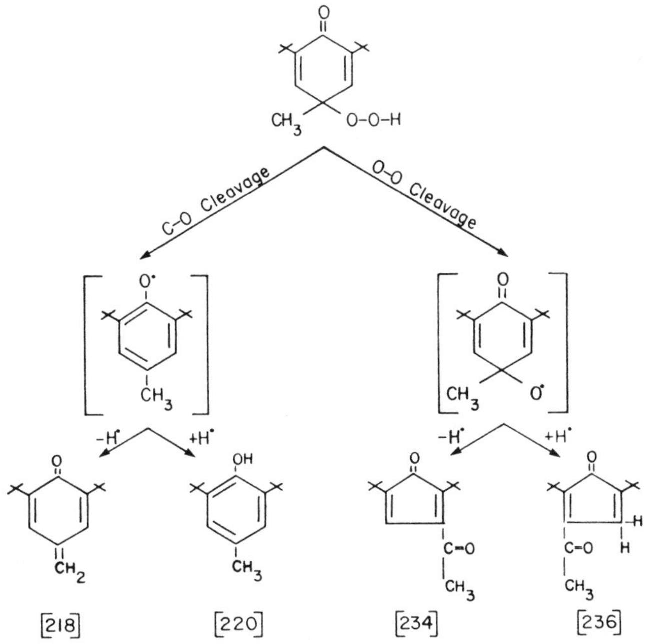

Figure 8. Routes of decomposition of HBHT (mol wt in brackets). Courtesy of the Association of Official Analytical Chemists (6).

The ability of B-carotene to quench singlet oxygen molecules and free radicals demonstrates its usefulness in prevention of lipid oxidation (7).

Another natural antioxidant includes citric acid, a metal sequestrant, that is used in a variety of products such as seafood. When used with ascorbic acid, it inactivates enzymes that lead to deterioration. Browning is also prevented when enzymatic copper is chelated with citric acid. Antioxidation and inactivation of enzymes are achieved in fruits and vegetables where citric acid is used with antioxidants to inhibit color and flavor deterioration (1).

Combined with other antioxidants, citric acid prevents oxidative rancidity in meats. In fats and oils it chelates metal ions.

Phosphates are derivatives of phosphoric acid and short-chain phosphates have the best sequestering ability. Phosphoric acid is synergistic with other antioxidants to guard against oxidative rancidity in vegetable oils.

Many spices also have antioxidant properties including clove, oregano, sage, vanilla, and rosemary (8). In fact, a study comparing rosemary extract in turkey sausage reported that the extract exhibited greater antioxidant activity than BHA and was equal to BHT. The addition of rosemary extract to the meat was shown to increase protection from oxidation during cooking as indicated by the thiobarbituric acid test and volatile aroma analysis (9). In

Table 2. Antioxidants Listed in Food Additives Status List on Regulations[a]

Chemical or Common Name	Status	Tolerance	Specified Use or Restrictions
Anoxomer	REG	5,000 ppm	—
BHA	GRAS	0.02% of fat or oil	Essential oils of foods including oleomargarine and margarine
BHT	GRAS	0.02%	Potato granules same as BHA
	FS	33 ppm in rice	Enriched or parboiled
	FS	0.02%	Soda H_2O
	FS	<0.02% in oleomargarine	Margarine, oleomargarine
Dilauryl thiodipropinate	GRAS	0.02% of fat or oil	—
Ethoxyquin	REG	3.0 ppm	Poultry fat, uncooked liver
		0.5 ppm	Uncooked muscle
		0.5 ppm	Eggs
		100 ppm	Chili powder, paprika
TBHQ	REG	<0.02% of lipids	Essential oils of foods including margarine
THBP	REG	<0.02% of lipids	
THBP	REG	<0.02% of lipids	

[a] Table promulgates under Section 401 (Food Standards), 409 (Food Additives), and 512 (Animal Drugs for Feel Use) of the Food, Drug and Cosmetic Act. Courtesy of Eastman Chemicals Division, Kingsport, Tenn.

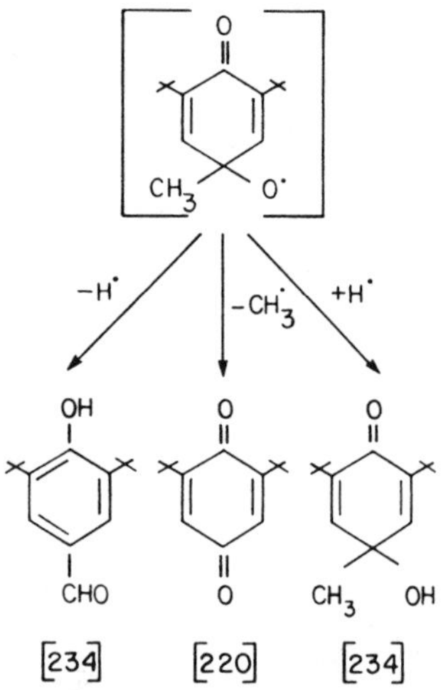

Figure 9. Reaction of alkoxy radical to HBHT to form three products found in the thermolysis product mixture (mol wt in brackets). Courtesy of the Association of Official Analytical Chemists (6).

another study, the effectiveness of onion peel as an antioxidant was investigated. It was found that a 2% yellow onion peel–hot water extract was an effective means of reducing rancidity in both shark and mackerel (10).

Antioxidants are also produced in some methods of food processing and should be called natural. For example, sulfur-containing amino acids such as cystein exhibit antioxidant properties when heated (11). Although natural antioxidants may be helpful, they are of limited value because they are not sufficiently effective or widespread.

A principal cause of muscle food deterioration is oxidative rancidity due to the oxidation of unsaturated fatty acids catalyzed by hemoproteins as well as nonheme iron (12). Ginger rhizome is one of the most popular spices in Oriental cuisine. Investigations in Japan report that ginger or ginger extract added to lard or other foods demonstrates reasonably strong antioxidant properties. The antioxidants effectiveness of ginger depends on the kind of preparation, pH, and concentration of fat in the muscle. The shelf life of products determined by 2-thiobarbituric

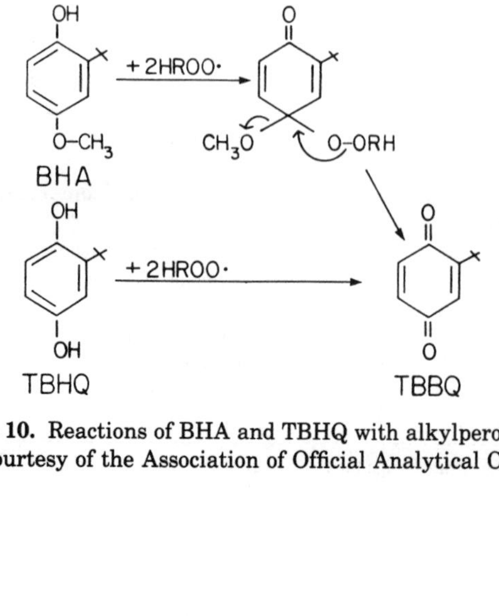

Figure 10. Reactions of BHA and TBHQ with alkylperoxy radicals. Courtesy of the Association of Official Analytical Chemists (6).

acid (TBA) value demonstrated improvement with the inclusion of ginger extract (12).

REGULATORY ISSUES

Various regulations exist for the use of antioxidants in the U.S. food system. The FDA requires that ingredient labels carry and list the antioxidants as well as their carriers. This must be followed by an explanation of their purpose (1). Usage levels of antioxidants are generally limited to 0.02%. This amount is calculated on the basis of the fat content of the recipient food. The preservatives discussed have been instrumental in preventing the effects of lipid oxidation in foods.

Sulfites will not be discussed in detail because they have been thoroughly investigated over the last several years. Sulfites do act as antioxidants on many fresh vegetables, beverages, and dried-food products. The FDA has since banned added sulfites in fresh produce and vegetables. In addition, it is difficult in many cases to distinguish between added sulfites and naturally occurring sulfites in food.

CONSUMPTION ISSUES

As important as antioxidants sound in the overall picture of food preservation there are some valid health questions that have been raised. The average U.S. citizen consumes somewhere between 5 and 10 lb of food additives each year. Some of these additives can cause allergic reactions. Although these points are well taken, consumers should not be concerned. Most scientists agree that food additives, at the current usage level, are not causing any direct health problems for the general public. Food processors are required to use the 100-fold margin of safety rule. This rule states that the level of a food additive cannot exceed 1/100 of the maximum amount determined in toxicological studies not to cause cancer.

Allergic responses have been linked to antioxidants such as BHA and BHT. However, they appear infrequently and occur less often than allergies caused by foods. Antioxidants have a valid place in the food supply. It has only been through the regulated, judicious use of food additives such as antioxidants, that the United States food supply has become the safest, most nutritious, most wholesome, and most abundant in the world.

BIBLIOGRAPHY

1. J. D. Dziezak, "Antioxidants: The Ultimate Answer to Oxidation," *Food Technology* **40,** 94–100 (1986).
2. J. R. Mahoney and E. Braf, "Role of Alpha-Tocopherol, Ascorbic Acid, Citric Acid, and EDTA as Oxidants in Model Systems," *Journal of Food Science* **51,** 1291 (1986).
3. H. Graham, *Safety of Foods*, AVI Publishing Co., Inc., Westport, Conn., 1980, pp. 672–673.
4. A. L. Branen, Antimicrobial Properties of Phenolic Antioxidants and Lipids," *Food Technology*, 1986.
5. Eastman Laboratories, Kingsport, Tenn., 1987.

6. T. Fazio and J. Sherma, *Food Additives Analytical Manual,* Vol. 2, Association of Official Analytical Chemists, Arlington, Va., 1987.

7. G. W. Burton and K. U. Ingod, "B-Carotene: An Unusual Type of Lipid Antioxidant," *Science* **224**, 569 (1984).

8. Pearson 1982.

9. B. Shai and co-workers, "Antioxidant Properties of *Rosemary olcoresin* in Turkey Sausage," *Journal of Food Science* (50), 5,1356 (1985).

10. M. Younathan and co-workers, "Control of Heat Induced Oxidative Rancidity in Refrigerated Shark and Mackerel," *Journal of Food Science* (48), 176–178 (1983).

11. V. Packard, *Processed Foods and the Consumer,* University of Minnesota Press, Minneapolis, 1976.

12. Y. B. Lee, Y. S. Kim, and C. R. Ashmore, "Antioxidant Property in Ginger Rhizome and Its Application to Meat Products," *Journal of Food Science* **51**, 20–23 (1986).

JOHN J. SPECCHIO
Montclair State College
Upper Montclair, New Jersey

APPETITE

The term appetite does not have a single widely accepted meaning. Colloquially, it refers to a motivational state or desire, ordinarily for food but more generally for almost any reinforcer. Often the desire is not for food per se but for a specific food or nutrient (eg, salt). Because much of the relevant literature stems from research on nonhuman animals, appetite often cannot be determined from verbal reports, but must be inferred from so-called appetitive behavior, in which the animal actually consumes available food. In humans, complexities arise when we consider that people often desire food but do not consume it even when it is available (as happens with dieters) and, obversely, people often consume food for which they have little or no reported desire (as occasionally happens under conditions of social pressure).

Some investigators have distinguished between hunger (ie, a depletion of energy or a specific nutrient) and appetite (ie, a desire not directly responsive to physiological depletion but rather to psychological factors). To avoid such epistemological quandaries, the focus will be on what factors affect eating. This approach begs the question of motivational state by connecting aspects of the internal and external environment to overt consummatory behavior.

The factors that affect eating have not been systematically cataloged, let alone integrated into a complete systems approach to eating. Some physiological researchers accept as an axiom that all eating reflects hunger and that the termination of eating necessarily reflects satiety (or at least the absence of hunger, which may or may not be the same thing). Accordingly, they prefer to assume that the final common pathway leading to eating passes through the motivational way station of hunger. A more agnostic approach will be taken here, an approach in which eating may occur in the absence of hunger (and vice versa); indeed, it has been suggested that research subjects (as well as researchers) often infer that hunger caused them to eat; this inference, however, may be based on plausible lay theories of eating (derived in turn from prevailing popular research) rather than from any direct experience of hunger. That is, people may assume that they ate because they were hungry, rather than accept the less obvious or plausible explanation that they ate because of social influence or simply because the food was there. As is evident, both scientific and lay meanings of the terms appetite and hunger are problematic. This explains (if not justifies) the decision to bypass appetite and hunger and focus on eating.

The analysis is divided into internal and external influences on eating, corresponding roughly to physiological versus environmental factors. Some factors, such as the palatability of the available food, depend jointly on the properties of food, the history of the organism, and the acute physiological state of the organism; such factors, obviously, do not fit neatly into the internal–external categorization. The internal and external distinction is also threatened by findings that demonstrate that through conditioning, external signals can stimulate hunger, or at least feeding (1). Likewise, the perception of attractive food triggers (cephalic) anticipatory digestive reflexes that in turn may affect feeding (2). By the same token, deprivation or other physiological manipulations can alter receptivity to external influences. Any comprehensive analysis of eating must contend with these mutual, interactive influences, but current research only hints at some of these complexities. In the absence of such a fully integrated view of eating, this article will be confined to a less satisfactory and somewhat arbitrary listing of influences.

PHYSIOLOGICAL INFLUENCES

As indicated, most laypeople and many researchers subscribe to an essentially physiological view of feeding. The feeding system is activated by various physiological events, most notably the progressive depletion over time of energy or more specific nutrients. This system is usually conceptualized as involving a central regulation of signals from the periphery, with eating as a tool in the service of homeostasis.

METABOLIC AND HORMONAL INFLUENCES

Metabolic factors governing the control of eating are generally separated into those associated with turning on eating (hunger) and those associated with turning off eating (satiety). Hunger (usually accompanied by eating) develops during the fasting phase of metabolism, after the ingestion and absorption of nutrients has been completed. Satiety (usually accompanied by cessation of eating) develops during the absorptive phase of metabolism, when a meal is being ingested and while nutrients are being absorbed from the intestine.

The principal sources of energy for all tissues are glucose and fatty acids, which may be drawn directly from the bloodstream. Energy reserves are stored in three different forms: protein, fat, and glycogen, which must be converted to glucose and fatty acids to be used. During the absorptive phase, metabolism is directed at the accumulation of these reserves; glucose and amino acids are converted to glycogen and fat, and amino acids are also stored as protein.

In the fasting phase, metabolic processes are aimed at utilizing energy. After a meal, blood glucose is used for energy needs, but as the supply of available glucose diminishes, energy reserves must be converted to usable forms. Glycogen is converted back to glucose and fat to fatty acid. These processes are associated with hunger. In the event of long-term deprivation or starvation, there is an increasing demand for energy derived from the conversion of amino acids to glucose, and fatty acids to ketones. These conversion processes provide the substrate for internal influences on eating.

Initiating Eating

Low levels of circulating glucose (and a corresponding higher rate of glycogen to glucose conversion) are important in the initiation of eating. Experimental manipulations that decrease glucose availability are a powerful means of stimulating eating (3). It is important to note, however, that low levels of circulating glucose is not an entirely accurate characterization of the metabolic trigger for eating. Consider untreated diabetics, who have high levels of circulating glucose but are chronically hungry, because the glucose is not metabolically available owing to low levels of insulin, the hormone necessary for glucose uptake into cells. Thus it is the availability of glucose, not its mere presence, that is the critical signal. Accordingly, because of its importance in determining glucose availability, insulin is considered an important hormone involved in the initiation of eating.

Recent research attempting to characterize more precisely the connection between glucose levels and feeding indicates that meal onsets are preceded by small transient declines in blood glucose (4). If this decline in blood glucose is prevented, no meal occurs; feeding will not begin until a transient decline in blood glucose is experienced, reinforcing the notion that a drop in glucose availability is causally related to eating. Of special interest is the finding that the glucose-drop signal that initiates eating is of surprisingly short duration and small magnitude.

Whereas glucose appears to be important for regulating eating in the short term (ie, meal to meal), lipids appear to play a more important role in regulating food intake in the long term (5). During fasting, the body's fat reserves are converted to fatty acids for energy; this process probably represents an additional signal for turning on eating.

Terminating Eating

Physiological factors responsible for the cessation of eating can be divided into stomach and hormonal factors. Each of these contributes to what is typically referred to as satiety.

Stomach factors. The stomach produces two types of satiety signal: one associated with the nutrient content of the meal and the other with the volume of the meal (6). It seems likely that these two meal-related signals are detected by separate mechanisms in the gut, because rats can regulate caloric intake independently from meal volume and meal volume can produce satiety cues that do not depend on caloric content.

Nutrient receptors in the stomach respond to the caloric density of a meal. A large but low-calorie meal would activate a large number of nutrient receptors but at a low rate, whereas a small but calorically dense meal would activate a small number of nutrient receptors at a high rate. In addition to this indirect effect on nutrient receptor activation rate, meal volume also induces satiety directly by activating stretch receptors in the stomach.

Current research indicates that stomach distension cues (via activation of stretch receptors) and caloric cues (via activation of nutrient receptors) are transmitted through separate channels. Stomach distension cues are transmitted to the brain by way of the vagus nerve, while stomach nutrient cues do not require the integrity of the vagus nerve and may, therefore, involve hormonal messengers (6).

Hormonal Factors. Although a number of different hormones have been implicated in satiety, the present discussion will briefly describe the influence of glucagon and cholecystokinin (CCK), the two hormones that have received the most attention as likely satiety agents. Glucagon is produced by islet cells of the pancreas as well as by cells in the gut. Animals treated with glucagon eat smaller meals and animals treated with antibodies to glucagon eat larger than normal meals, suggesting that endogenous glucagon may contribute to meal termination. Because glucagon stimulates the conversion of glycogen (the stored form of glucose which cannot be directly utilized for energy) to glucose, the resultant increased levels of glucose released into the hepatic vein from the liver may stimulate glucose receptors and, therefore, decrease hunger signals.

Choleystokinin has also received considerable attention as a signal associated with postprandial satiety (7). CCK is released from the duodenum (upper portion of the small intestine) when food reaches the stomach. Animal and human studies indicate that systemic administration of CCK inhibits food intake through the normal behavioral sequence that characterizes satiety, without affecting water intake, implying that the behavioral effects of CCK are specific to feeding. Physiological studies indicate that CCK exerts its inhibitory effects on feeding through activation of vagus nerve input to the brain.

Brain Systems Controlling Eating

The Hypothalamus. Current notions regarding the organization of the brain's feeding system derive from early animal studies aimed at identifying brain regions responsible for turning on or turning off feeding. One of the original findings in this regard was that destruction of the medial hypothalamus, in particular the ventromedial hypothalamic nucleus (VMH), produced a behavioral syndrome characterized by severe overeating and ultimately obesity (8). Humans with VMH tumors displayed a similar behavioral profile, confirming the importance of the VMH for eating across species.

Initially, the VMH was viewed as the brain's satiety center, but it is now recognized that both the psychological and neural characteristics of satiety are far too complex to be accounted for solely by the VMH. However, it is generally agreed that the VMH and its connections are important for relaying satiety signals from the periphery to neural systems involved in turning off feeding.

Recent evidence indicates that VMH lesions also destroy fibers projecting to the pituitary stalk and brain stem, both of which are important in the control of various hormones. Of relevance here are findings showing that VMH lesions increase insulin secretion and decrease glucagon secretion, a combination associated with overeating. Thus the overeating associated with VMH lesions may be secondary to the peripheral hormonal and metabolic effects of the lesions rather than a direct result of the missing VMH.

In contrast to the inhibitory role ascribed to the VMH, the lateral hypothalamus (LH) has become associated with stimulation of eating. Animal experiments reveal that lesions of the LH produce aphagia and adipsia whereas stimulation of the LH produces eating responses (7). Although LH lesions produce aphagia and adipsia, artificially fed animals can eventually recover and maintain close to normal eating habits; this indicates that the central nervous system is sufficiently plastic to enable other brain areas to take over, to some extent, the feeding functions of the LH. It should be noted, however, that although eating deficits can eventually recover, LH-lesioned rats seldom return to their prelesion weight. This finding suggests that the LH contributes to the establishment and maintenance of a set point for body weight which may be altered permanently by the lesion.

Other Hypothalamic Structures Involved in Eating. Another important structure implicated in eating behavior is the paraventricular nucleus (PVN) of the hypothalamus. In fact, many of the effects associated with VMH manipulations have more recently been interpreted as PVN effects, based on the discovery that VMH lesions destroy fibers originating from the PVN, some of which are associated with control of hormonal functions described above and others of which project to brain-stem mechanisms involved in the control of eating. Moreover, it has been demonstrated that PVN lesions have effects on eating similar to those of VMH lesions. For these reasons, as well as for reasons discussed below, researchers have begun to view the PVN as a critical structure involved in feeding.

Neurotransmitters and Eating. Any description of the neural basis of eating would be incomplete without a discussion of the neurotransmitter systems implicated in eating. Neurotransmitters are responsible for interneuronal communication. Thus the activity of cells within particular brain regions is determined by the neurotransmitters that stimulate or inhibit them.

Norepinephrine, endorphins, and neuropeptide Y are transmitters that act in the PVN to facilitate food intake (9). Of special interest here is the finding that the stimulatory eating effects of norepinephrine in the PVN appear to be specific to carbohydrates (9). Elsewhere, growth hormone–releasing factor in the medial preoptic area of the hypothalamus stimulates eating (10). Neurotransmitter signals involved in inhibiting eating include dopamine and norepinephrine in the perifornical region of the hypothalamus as well as serotonin in the PVN (9). Interestingly, serotonin's inhibitory effect, like norepinephrine's stimulatory effect, is carbohydrate selective (9).

Although exactly how these transmitter signals are integrated during normal eating is still unclear, it is assumed that these chemical signals reflect food-relevant information derived from the periphery. For example, consider the regulation of carbohydrate intake. A high carbohydrate meal increases tryptophan uptake in the brain and ultimately serotonin synthesis, for which tryptophan is the precursor. Increased serotonin activity, as discussed above, suppresses food intake, particularly carbohydrates. Thus increased carbohydrate intake eventually feeds back negatively on further carbohydrate intake (probably through serotonin's actions in the PVN), providing a mechanism for the short-term regulation of carbohydrate intake.

Extrahypothalamic Influences on Feeding. Nonhypothalamic brain systems have also been implicated in the control of eating. In particular, activation of the nigrostriatal and mesolimbic dopamine pathways, which use dopamine as a neurotransmitter, is associated with stimulatory effects on eating. These pathways originate in the midbrain and project rostrally to striatal and limbic structures in the forebrain, with some of the axons involved coursing through the lateral hypothalamus en route. This anatomical fact has been taken to suggest that some of the feeding decrements associated with LH lesions may in fact be due to destruction of ascending dopamine neurons.

Although the role of the nigrostriatal and mesolimbic dopamine pathways in feeding is not entirely understood, it is of interest to note that these pathways are crucial for reward and reinforcement functions in general (11), suggesting that activation of these systems may reinforce the sensory–motor associations involved in eating and also increase the rewarding properties of food or eating (12, 13). Thus contrary to the conventional homeostatic view of eating regulation, these particular signals may be associated with nonregulatory influences on eating such as taste hedonics or learned eating patterns.

CONDITIONED APPETITES

Ordinarily, the physiological substrate of appetite is thought to respond to energy considerations as outlined above. However, these reactions are also subject to conditioning, such that particular environmental stimuli may, through repeated pairing or reinforcement, come to elicit the physiological states associated with the onset or offset of eating. For instance, it has been demonstrated that previously neutral stimuli repeatedly paired with the presentation of food can induce stated animals to continue eating (1). Visual, olfactory, or social cues may induce feeding in people who are not otherwise hungry; such cues may likewise terminate eating in individuals who are not otherwise sated. It has been suggested that such external, conditioned cues do not control behavior directly, but rather operate on the physiological substrate of hunger and satiety; for example, the mere sight of attractive food cues may induce physiological reactions (eg, insulin release) associated with hunger (14). Even temporal cues may stimulate appetite; people become hungry as mealtime approaches, but if for some reason the meal is skipped, hunger may subside. This area of research promises to become increasingly important in the analysis of appetite.

SPECIFIC APPETITES AND NUTRIENT DEPLETIONS

The organism must concern itself not only with overall energy balance, but also with its inventory of the nutrients crucial to its well-being. Many well-known diseases (eg, scurvy and rickets) may appear in otherwise well-fed individuals who lack a particular vitamin or mineral. Specific appetites refer to hardwired cravings for certain substances the absence of which is detected and rectified wherever the food supply permits. It should be noted that whereas some deficits (eg, salt) produce clear cravings, not all deficiencies are detectable by the suffering organism.

Recently, speculation has arisen that the organism may regulate nutrients (especially macronutrients) not so much for their medicinal value as for their psychoactive effects. Most notably, a carbohydrate-craving hypothesis has been championed (15). It is argued that a deficiency of carbohydrates affects neurotransmitter levels to create a negative affective state; accordingly, individuals may learn to (over)eat carbohydrates in an attempt to bolster their mood, especially when they are feeling upset or depressed. This hypothesis has been offered to explain the overeating characteristic of bulimia nervosa and some forms of obesity.

SENSORY AND PERCEPTUAL FACTORS

Flavor

The term flavor is intended to encompass both gustatory (taste) and olfactory (smell) properties of food. Palatability is a matter of both aspects of flavor; when people say that a food tastes good, they tend to underestimate the contribution of smell. The palatability of food, which corresponds to how positively its flavor is rated and also includes certain secondary properties of the food such as its texture, is a major determinant of intake. Organisms are basically finicky, avoiding indiscriminate consumption in favor of selective consumption (where possible) of preferred foods. The finding that nonhuman animals consume more highly palatable food is compromised by the fact that palatability must often be inferred from intake itself; but humans also show a strong tendency to consume more of foods that they independently rate as more pleasant.

Palatability has a direct effect on intake, but whether palatability overrides other potentially competing considerations in controlling intake remains an active research question. It has been observed that the duration of the meal is largely a matter of the palatability of the available food but the likelihood of a rat initiating feeding is more closely tied to caloric deprivation (16). In short, palatability may affect some meal parameters more than others. In humans extremes of food deprivation or satiety mute the effects of flavor; starving people tend to ignore the flavor of food and eat a great deal of unpalatable food if that is all that is available, whereas fully sated people eat very little of even quite palatable food. Interestingly, research indicates that at least under certain qualifying conditions hungry organisms do not display reduced finickiness; indeed, finickiness may be exaggerated both with respect to acceptance of good-tasting food and rejection of bad-tasting food (17). An apparent paradox has been identified in which "the animal eats . . . for taste when he needs calories [ie, when deprived]" (18). This paradox may be more apparent than real, in that the flavor of food is ordinarily confounded with its caloric density (ie, good-tasting foods are calorically denser, leading one humorist to describe the calorie as a term that scientists use to measure how good something tastes). Accordingly, the hungry organism may become more finicky as a way of maximizing energy intake in an energy deficiency crisis.

The potency of palatability as a determinant of eating may vary in different types of individual. Obese people, for instance, have long been regarded as prone to overeat highly palatable foods (rather than food per se); this view was crystallized in the externality theory of obesity, which argues that obese people's eating is controlled virtually entirely by environmental cues, including taste, whereas normal-weight people's eating is more responsive to physiological cues of hunger and satiety (19). The observation that obese people are especially responsive to taste (19) may be meshed with the observation that hungry organisms are especially responsive to taste (18) if it is postulated that most obese people are chronically hungry as a result of their partially successful attempts to reduce their weight (20).

Other eating problems have likewise been associated with extreme partiality toward highly palatable foods. Binge eating in normal-weight individuals has been interpreted as targeted specifically on highly palatable carbohydrates, which elevate serotonin levels and ultimately, mood (15). Conceivably, this view might be extended in the direction of proposing that some flavors are deemed palatable precisely because of their central hedonic consequences.

Sensory Factors

If eating offset were associated solely with the accumulation of particular nutrient stores, then it would be difficult to account for the "dessert effect." That is, despite a satiating meal (fully repleting nutrient stores), people can still find room for dessert. This effect can probably best be explained in terms of sensory specific satiety, which posits that organisms can become satiated to a particular flavor without showing satiation to other normally preferred flavors (21). Thus the individual may become satiated to the sensory characteristics of the main course while not being satiated to the sensory characteristics of the dessert.

Collateral support for the role of nongut sensory factors in satiety is found in studies of sham-feeding (esophagotomized) animals. In these animals, food is chewed and swallowed but does not reach the gut, and the nutrients are therefore not absorbed. Of special interest here is the fact that although these animals eat larger than normal meals (providing evidence for the involvement of postingestional factors in regulating eating offset), they do eventually stop eating. This has been interpreted to indicate that sensory factors (taste, olfaction, and texture) of the food are sufficient to induce satiety-like effects on eating. A full explication of satiety will eventually have to incorporate and reconcile sensory and postingestional factors.

Perceptual Salience

Does the presence of food act merely to allow individuals to satisfy their preexisting hunger, or might the sight of food actually stimulate the appetitive drive, which would not have existed in the absence of food? One experiment found that normal-weight individuals ate the same amount (just under two sandwiches) regardless of whether they were initially presented with one sandwich (and had to obtain additional sandwiches from a refrigerator) or three sandwiches (22). This study suggests that the visual salience of food has a minimal impact on consumption. Obese subjects in this study, however, ate considerably more when presented with three sandwiches than when presented with only one, suggesting that visual cues have a strong effect on them. This finding was interpreted as support for the externality theory of obesity, with the visual salience of food representing an external cue *par excellence*. (One important implication of this study is that in the absence of excitatory external cues, obese people should eat less than normal-weight people.) Another experiment found that normal-weight subjects ate the same amount of cashews regardless of whether the cashews were brightly illuminated or difficult to see (and regardless of whether they were instructed to think about the cashews or about something else), whereas obese subjects ate above-normal amounts of the visually salient cashews and below-normal amounts of the dimly illuminated cashews (23). Obese subjects also ate more when instructed to think about the cashews, reinforcing the point that it is awareness of food that matters and not simply optical effects. It should be noted that some researchers interpret these findings to mean that obese subjects eat in direct response to external cues and not because such cues stimulate hunger, which in turn drives eating. However, it has been suggested that exposure to food cues may trigger anticipatory (cephalic phase) reflexes at the physiological level, such that the sight or smell of palatable food may literally make the fat person hungry (14). It seems unlikely that the perceptual salience of food has no effect on normal-weight individuals, but research evidence for such an effect has not yet appeared.

Appetizer Effects

Does intake itself sometimes stimulate more intake? The term appetizer would suggest so, and common experience would seem to confirm that appetite can be stimulated by small delicacies. No compelling research has been conducted on this topic in humans, but if strong appetizer effects were confirmed, then the presumptive physiological mediation of such effects (ie, the appetizer triggers an anticipatory digestive reflex, which in turn demands satisfaction in the form of a complete meal) should be distinguished from alternative explanations (eg, the appetizer initiates a behavioral response sequence that is not hunger driven but that perseverates until satiety occurs). The role of behavioral momentum in feeding and other repetitive action patterns has not been adequately explored, although it has been argued that obese subjects overeat because of an inability to terminate repetitive behavior sequences (especially eating) once they have begun (24). This would suggest that appetizer effects might be stronger in the obese.

Preload studies, in which subjects are initially given a fixed amount of food and then permitted to continue *ad lib*, show a complex pattern. When subjects are classified as dieters (obese or normal weight) or nondieters, opposite reactions are observed, with nondieters eating *ad lib* in inverse proportion to preload size and dieters eating more following large preloads than following small preloads. The nondieter pattern would seem to contradict expectations based on an appetizer effect, and the data from dieters likewise do not fit in well, because a small preload (appetizer) would be expected to trigger the greatest amount of subsequent eating. Interpretations of these effects in dieters emphasize deliberate cognitive control of intake and its disruption, rather than the sort of analysis ordinarily invoked to explain the hypothetical appetizer effect (25).

EMOTIONAL FACTORS

It is readily agreed that emotional disturbance affects appetite and eating, but the precise nature of this effect is variable. In nonhuman animals, arousal tends to promote feeding. Nonspecific activation may render them more responsive to external, food-related stimuli, which direct behavior toward eating (26). (Such external responsiveness is presumably not specifically directed at food-related stimuli; rather, it focuses on whatever powerful behavior-relevant stimuli may dominate the animal's immediate perceptual environment, and might yield sexual, aggressive, or other stimulus-bound behavior depending on the environmental configuration that obtains when activation occurs.) It has been suggested that any sort of activation or drive induction might accentuate stimulus-bound eating in humans (27). Such stimulus-bound eating would not necessarily increase intake but rather make it more responsive to environmental conditions; if the food were bad-tasting, it might suppress eating. Such a mechanism might explain why hungry organisms sometimes eat less of a bad-tasting food than nondeprived organisms do.

Studies of arousal in humans tend to focus on particular valenced emotions (eg, anxiety, depression, and anger). In general, such negative emotions have developed a reputation for suppressing eating, even when palatable food is available. This effect is usually attributed to the sympathomimetic effects of emotional distress; these effects presumably act to suppress peripheral hunger signals. Clinical depression is indexed by a loss of appetite, among other criteria. Still, some depressed individuals tend to eat more and gain weight (28). Indeed, emotional agitation tends to increase eating among many people, to the point where emotional disturbance is often cited as a prime cause of weight gain and failure to maintain dietary restraint. How can these contradictory effects be reconciled?

It does appear to be the case that strong negative emotions produce physiological effects that oppose appetite. (It seems conceivable that minor agitation or disturbances might stimulate eating in humans much as seems to be the case in other animals, but such weak manipulations have not been explored.) Powerful distress (eg, intense clinical depression) probably suppresses appetite for virtually everyone. Short of overwhelming distress, however, emotional agitation appears to have opposite effects depending on the type of individual involved. Normal eaters, whose behavior is at least partially governed by the sort of hunger and satiety signals discussed above, tend to respond to distress by eating less. Individuals who ordinarily attempt to suppress their eating (ie, dieters, including perhaps most of the obese), however, tend to eat more when they are upset (and possibly when elated, although studies of the effects of positive moods are so few as to defy conclusiveness). It seems plausible that individuals who react to emotional distress by overeating would have good reason to attempt to diet, but this explanation of the association between dieting and overeating under distress does not account for the initial overeating. It has been argued that dieting is the primary cause of the association (29). Dieters, because they must overcome normal physiological controls on eating if they are to succeed in reducing their intake, eventually dissociate their eating from such controls, in favor of cognitive calculations of appropriate intake. Their intake is deliberately inhibited, but such inhibitions are vulnerable to disruption by (among other factors) emotional agitation. The emotionally distressed dieter, then, becomes disinhibited and overeats, unconstrained by the sort of normal satiety considerations that depend on physiological signals and conditioning, both of which are functionally atrophied in the dieter. The effects of distress may be exacerbated or attenuated depending on the type of distress involved; for example, physical threats tend to have a stronger suppressive effect, whereas ego threats tend to have a more facilitative effect (30). The exploration of overeating under conditions of distress remains a prominent concern of researchers studying obesity and especially bulimia nervosa.

SOCIAL FACTORS

Other people influence an individual's eating in a variety of ways. Social facilitation refers to increased intake in the presence of others; this usually occurs when others are eating substantial amounts as well (31). More typically social inhibition of eating is encountered, for example, being observed by noneating others, or eating along with others who eat very little, tends to reduce consumption. More generally, people seem to be highly vulnerable to social influences on eating, from indirect pressure (such as conformity to a model or experimental confederate) to more direct influences (such as requests or demands that a certain amount of food be eaten). These social effects are relatively powerful compared to other experimental manipulations of factors affecting eating and lend support to a nonhomeostatic view of eating, at least in the short term. Eating seems to be acutely responsive to social influences that bear little relation to an individual's physiological state or needs. Homeostatic considerations may correct for such short-term deviations; for example, if a person undereats in a public spotlight, he or she may compensate by eating more later in a private setting. Another consideration is the likelihood that such social effects as copying others or not gorging when being watched serves a broader biological survival purpose. It is possible to benefit from others' experiences, and people may look to social guides when the appropriate amount or type of food to be consumed is ambiguous, as it often is. Furthermore, social norms may demand some sort of equity in the distribution of the food supply. Still, it is clear that such social factors may have a profound effect on eating, and this effect may act in direct opposition to purely physiological needs and signals. Even 24-hour food-deprived experimental subjects, who ought to eat in response to powerful internal signals, are strongly influenced by the eating patterns of a model (32).

Another sort of social influence on eating occurs when people eat in a particular way to convey a certain impression. This impression-management view of eating focuses on eating as a self-presentational strategy (33). For example, women eat less when they are paired with a man than when paired with another woman, presumably because they wish to convey a more feminine impression to the man, and eating lightly is part of the feminine stereotype. Again, this sort of influence on eating seems to have little to do with the exigencies of internal state.

COGNITIVE FACTORS

The impact of both internal (eg, hunger) and external (eg, food salience) factors on eating may be mediated in most cases by cognition; people may decide to eat up on consideration of the allure of such pressures. Because such internal and external factors have seemingly direct effects on eating (that is, because these factors can be interpreted as affecting eating directly, without the mediation of decisions that add little in the way of explanation) most analyses of eating do not bother to include deliberation or conscious choice as important elements. When we find that people eat in a manner opposed to both internal and external pressures, however, some sort of deliberative element seems necessary. If someone were to continue eating an unpalatable food despite being sated, it would probably be necessary to have access to that individual's phenomenology to help provide an explanation. More commonly, people fail to eat despite their evident hunger and the availability of palatable food. Such dietary restraint demands an analysis that adds a set of mental factors to the control of eating in addition to physiological and environmental stimuli to which the person responds more or less reflexively (34). Such cognitive dietary calculations do not arise spontaneously, of course; they themselves are learned and may be reinforced or extinguished. But the prevalence (in humans, at least) of eating patterns that depart radically from the straightforward application of our knowledge of physiology and environmental stimulus control requires that attention be paid to willful opposition to such signals.

INTERACTIONS AMONG RELEVANT FACTORS

A variety of examples have been presented of the nonadditive influence of factors affecting eating. For instance, hunger does not simply add to palatability as a determinant of eating; hungry people may eat more good-tasting food, as expected, but they may eat less bad-tasting food, depending on circumstances. Likewise, social influences may combine with, mask, or oppose hunger and satiety considerations in eating. An attempt has been made to reconcile the various influences on eating in a boundary model (35). The basic premise is that physiological influences will predominate when the person is particularly hungry or sated and that environmental and cognitive considerations will be most influential when the person is indifferent (neither hungry nor sated). This model has not been successful in predicting behavior, but its emphasis on the need to assess the influence of one factor (eg, social influence) in the context of other factors (eg, hunger–satiety) has proven fruitful experimentally. Research on the influence of one particular factor in isolation seems likely to misrepresent the importance of that influence; under certain conditions, its influence will be exaggerated, whereas under others it may be suppressed.

CONCLUSIONS

The scientific study of appetite has yielded a wide range of interesting phenomena. Still, the field is a painfully long way from achieving any sort of satisfying integration, in which the mutual and combined influences of various factors acting simultaneously might be predicted or explained. This conclusion is especially warranted in the case of human eating behavior, where social and cognitive factors are particularly evident. Such factors may apply as well to other animals. It can only be hoped that the countervailing opportunities of human and animal research will eventually be synthesized into a comprehensive analysis of the determinants of feeding and appetite.

BIBLIOGRAPHY

1. H. P. Weingarten, "Conditioned Cues Elicit Feeding in Sated Rats: A Role For Learning in Meal Initiation," *Science* **220**, 431–433 (1983).

2. T. L. Powley, "The Ventromedial Hypothalamic Syndrome, Satiety, and a Cephalic Phase Hypothesis," *Psychological Review* **84**, 89–126 (1977).

3. G. A. Bray and L. A. Campfield, "Metabolic Factors in the Regulation of Feeding and Body Energy Storage," *Metabolism* **24**, 99–117 (1975).

4. L. A. Campfield and F. J. Smith, "Functional Coupling Between Transient Declines in Blood Glucose and Feeding Be-

havior: Temporal Relationships," *Brain Research Bulletin* **17**, 427–433 (1986).

5. N. Mrosovsky, "Body Fat: What Is Regulated?" *Physiology and Behavior* **38**, 407–414 (1986).

6. M. F. Gonzalez and J. A. Deutsch, "Vagotomy Abolishes Cues of Satiety Produced by Gastric Distention," *Science* **212**, 1283–1284 (1981).

7. G. P. Smith, "Gut Hormone Hypothesis of Postprandial Satiety," in A. J. Stunkard and E. Stellar, eds., *Eating and Its Disorders,* Raven, New York, 1984, pp. 67–76.

8. S. P. Grossman, "Contemporary Problems Concerning Our Understanding of Brain Mechanisms That Regulate Food Intake and Body Weight," in Ref. 7, pp. 5–14.

9. S. Liebowitz, "Brain Monoamines and Peptides: Role in the Control of Eating Behavior," *Federation Proceedings* **45**, 1396–1403 (1986).

10. F. J. Vaccarino and M. Hayward, "Microinjections of Growth Hormone-Releasing Factor Into the Medial Preoptic Area/ Suprachiasmatic Nucleus Region of the Hypothalamus Stimulate Food Intake in Rats," *Regulatory Peptides* **21**, 21–28 (1988).

11. F. J. Vaccarino, B. B. Schiff, and S. E. Glickman, "Biological View of Reinforcement," in S. B. Klein and R. R. Mowrer, eds., *Contemporary Learning Theories,* Erlbaum, Hillsdale, N.J., 1989, pp. 111–142.

12. K. R. Evans and F. J. Vaccarino, "Amphetamine and Morphine-Induced Feeding: Evidence for Involvement of Reward Mechanisms," *Neuroscience and Biobehavioral Reviews* **14**, 1–14 (1990).

13. N. M. White, "Control of Sensorimotor Function by Dopaminergic Nigrostriatal Neurons: Influence on Eating and Drinking," *Neuroscience and Biobehavioral Reviews* **10**, 15–36 (1989).

14. J. Rodin, "Has the Distinction Between Internal Versus External Control of Feeding Outlived Its Usefulness?" in G. Bray, ed., *Recent Advances in Obesity Research,* Vol. II, Newman Publishing, London, 1978, 75–85.

15. R. J. Wurtman, "Neurotransmitters, Control of Appetite, and Obesity," in M. Winick, ed., *Control of Appetite,* John Wiley & Sons, Inc., New York, 1988, pp. 27–34.

16. J. Le Magnen, "Advances in Studies of the Physiological Control and Regulation of Food Intake," in E. Stellar and J. M. Sprague, eds., *Progress in Physiological Psychology,* Vol. 4, Academic Press, Orlando, Fla., 1971.

17. P. Pliner, C. P. Herman, and J. Polivy, "Palatability as a Determinant of Eating: Finickiness as a Function of Taste, Hunger and the Prospect of Good Food," in E. D. Capaldi and T. L. Powley, eds., *Taste, Experience, and Feeding,* American Psychological Association, Washington, D.C., in press.

18. H. L. Jacobs and K. N. Sharma, "Taste Versus Calories: Sensory and Metabolic Signals in the Control of Food Intake," *Annals of the New York Academy of Sciences* **157**, 1084–1125 (1969).

19. S. Schachter, "Some Extraordinary Facts about Obese Humans and Rats," *American Psychologist* **26**, 129–145 (1971).

20. R. E. Nisbett, "Hunger, Obesity, and the Ventromedial Hypothalamus," *Psychological Review* **79**, 433–453 (1972).

21. B. J. Rolls, E. T. Rolls, and E. A. Rowe, "The Influence of Variety on Human Food Selection and Intake," in L. M. Barker, ed., *The Psychobiology of Human Food Selection,* AVI Publishing Co., Inc., Westport, Conn., 1982, pp. 101–122.

22. R. E. Nisbett, "Determinants of Food Intake in Obesity," *Science* **159**, 1254–1255 (1968).

23. L. Ross, "Effects of Manipulating Salience of Food upon Consumption by Obese and Normal Eaters," in S. Schachter and J. Rodin, eds., *Obese Humans and Rats*, Lawrence Erlbaum Assoc., Maryland, 1974, 43–52.

24. D. Singh, "Role of Response Habits and Cognitive Factors in Determination of Behavior of Obese Humans," *Journal of Personality and Social Psychology* **27**, 220–238 (1973).

25. C. P. Herman and J. Polivy, "Studies of Eating in Normal Dieters," in B. T. Walsh, ed., *Eating Behavior in Eating Disorders,* American Psychiatric Association Press, Washington, D.C., 1988, pp. 95–111.

26. T. W. Robbins and P. J. Fray, "Stress-Induced Eating: Fact, Fiction or Misunderstanding?" *Appetite* **1**, 103–133 (1980).

27. C. P. Herman, J. Polivy, E. Werry, and S. T. McGree, unpublished data.

28. J. Polivy and C. P. Herman, "Clinical Depression and Weight Change: A Complex Relation," *Journal of Abnormal Psychology* **85**, 338–340 (1976).

29. C. P. Herman, J. Polivy, C. Lank, and T. F. Heatherton, "Anxiety, Hunger, and Eating Behavior," *Journal of Abnormal Psychology* **96**, 264–269 (1987).

30. C. P. Herman, T. F. Heatherton, and J. Polivy, unpublished data.

31. S. L. Berry, W. W. Beatty, and R. C. Klesges, "Sensory and Social Influences on Ice Cream Consumption by Males and Females in a Laboratory Setting," *Appetite* **6**, 41–45 (1985).

32. S. J. Goldman, C. P. Herman, and J. Polivy, unpublished data.

33. D. Mori and P. L. Pliner, "'Eating Lightly' and the Self-Presentation of Femininity," *Journal of Personality and Social Psychology* **53**, 693–702 (1987).

34. C. P. Herman and J. Polivy, "Restrained Eating," in A. J. Stunkard, ed., *Obesity,* Saunders, Philadelphia, 1980, 208–225.

35. C. P. Herman and J. Polivy, "A Boundary Model for the Regulation of Eating," in Ref. 7, pp. 141–156.

C. P. HERMAN
F. J. VACCARINO
University of Toronto
Toronto, Canada

APPLES AND APPLE PRODUCTS

Apples have been grown by man since the dawn of history. They are mentioned in early legends, poems, and religious books. The fruit which the Bible says Adam and Eve ate in the Garden of Eden is believed by many to have been an apple. The ancient Greeks had a legend that a golden apple caused quarreling among the gods and brought about the destruction of Troy. The Greek writer Theophrastus mentions several cultivars grown in Greece in the fourth century B.C. Apple trees were grown and prized for their fruit by the people of ancient Rome.

The apple species, *Malus pumila,* from which modern apple developed had its origin in southwestern Asia in the area from the Caspian to the Black Sea. The stone age lake dwellers of central Europe used apples extensively. Remains found in their habitations show they stored apples fresh and also preserved them by cutting and drying in the sun. The apple was brought to America by early European settlers. Most everyone is familiar with the Johnny Appleseed story of how apples were carried west with early settlers.

THE UNITED STATES AND WORLD APPLE PRODUCTION

The apple is the most widely grown fruit. Apple trees of one cultivar or another grow all over the world except the hottest and coldest regions. Commercial apple production for the United States is 8–10 billion lb/year; the total annual world production is about 60 billion lb. Apple production in the United States is concentrated in the states of Washington, California, Michigan, New York, Pennsylvania, West Virginia, Virginia, and North Carolina, these states produce two-thirds of the total U.S. production; New England, Eastern, Central, and Western states, produce the other one-third. A breakdown of 1986 and 1987 production by states is given in Table 1. Distribution of apple production in North America, South America, Europe, and Oceania is presented in Table 2. Depending on the climate apple production can vary by 15–20% from year to year.

Production (1987–1988) in the United States was about 30% higher than the previous year because of ideal climatic conditions and increased size of young bearing trees. Canada produced about 1.1 billion lb (0.5 million MT) of apples in 1987. Production is centered in the provinces of British Columbia, Ontario, and Quebec. The United States and world production of apples will continue to increase until about 1992 after which it will level off or decline due to over production. Countries not shown in Table 2 such as China, Korea, Russia, Poland, Romania, and others, produce large quantities of apples and some may export them in the future.

Production and utilization of apples in the United States has risen from 5.9 million lb in 1972 to 10.5 million lb in 1987. In 1987, fresh consumption was 60% of the total, and 40% was processed. Of the later 36% was canned, 7% was dried, 8% was frozen, 46% was utilized as juice, and 3% was used in other miscellaneous products such as vinegar, wine, and jam. Except for apple juice and dried fruit there was little change in the quantity of processed apple products even though the production in the last 15 years increased. Nearly one-half of the apples grown in the United States are produced in the Pacific coastal states of California, Oregon, and Washington and only 20% of them are processed. Over 50% of the apples produced in the Eastern states are processed. In the United States some processors have had to import apple products, particularly in the forms of concentrate, to insure an adequate supply of product for their manufacturing facility.

APPLE CULTIVARS

There are hundreds of apple cultivars, many of them are shown with color plates in *The Apples of New York* (3). Approximately 20 cultivars are now grown commercially in the United States. More than 90% of the production is represented by 14 cultivars (Table 3). Five of the 14 culti-

Table 1. United States Production in Pounds of Apple by States for 1986–1987

State	Millions of Pounds	State	Millions of Pounds
Arizona	9.5[a]/—	New Hampshire	50.0/50.0
California	535.0/650.0[b]	New Jersey	100.0/80.0
Colorado	17.6/125.0	New Mexico	6.0/12.6
Connecticut	46.0/45.0	New York	900.0/880.0
Delaware	27.0/26.0	North Carolina	120.0/390.0
Georgia	29.0/50.0	Ohio	90.0/150.0
Idaho	94.0/149.0	Oregon	105.0/210.0
Illinois	90.0/103.0	Pennsylvania	620.0/460.0
Indiana	37.0/72.0	Rhode Island	5.5/5.0
Iowa	5.4/10.0	South Carolina	30.0/45.0
Kansas	2.9/12.0	Tennessee	8.5/15.0
Kentucky	3.9/21.0	Utah	34.0/68.0
Maine	86.0/75.0	Vermont	48.0/44.0
Maryland	85.0/40.0	Virginia	450.0/481.0
Massachusetts	92.0/96.0	Washington	3100.0/4800.0
Michigan	700.0/1050.0	West Virginia	230.0/180.0
Minnesota	18.0/26.0	Wisconsin	53.0/65.0
Missouri	37.0/53.0	Total	7865.3/10,538.6

[a] Ref. 2.
[b] Ref. 1.

vars, Delicious, Golden Delicious, McIntosh, Rome Beauty, and Granny Smith account for most of the world apple production. Newer cultivars not shown in Table 3 are beginning to appear in fruit markets. Some of these are Gala, Fuji, Jonagold, Braeburn, and Lady Williams. Most of the new commercial plantings are selected red strains of the above cultivars. Some cultivars, ie, Gala mature in 100 days or less while others, ie, Lady Williams, grown in Western Australia, require over 200 frost free days to reach maturity. Some cultivars are very winter and frost hardy while others are very tender. Some cultivars, ie, Delicious require long cold winters to break dormancy, others, ie, Anna, a cultivar grown in Israel, can be grown in mild Mediterranean type climates.

Washington State grows the most apples in the United States, over 3 billion lb compared to about 900 million lb in New York (Table 1). The trend in the United States is to plant newer apple cultivars. Consumers are requesting high quality apples with distinctive tastes. A projection of new apple trees planted from new plantings is presented in Table 4. This trend in Washington State in tree planting will diversify apple production away from the present dominance of Delicious, Golden Delicious, and Granny Smith. Similar trends are occurring in other states. California does not produce many Delicious, but acreage of Gala and Fuji are increasing. In the future, planting densities will increase when new plantings are made, therefore, yield will also increase.

Origin of the Current Popular Cultivars

The original Delicious apple was discovered as a chance seedling in 1881 by Jessee Hiatt near Peru, Iowa. Presently over 100 strains of Red Delicious originated as natural selections or mutations and propagated by growers and nurserymen. It is of interest that the world's two most popular apples, Delicious and Golden Delicious are not the result of an organized fruit breeding program. Stark Bros. Nursery, Louisiana, Missouri played a major role in the discovery and commercial development of these apple cultivars. The Golden Delicious apple originated as a chance seedling ca 1903 on a mountainside on the A. H. Mullins farm in Clay County, West Virginia. A comprehensive history of these apple cultivars may be obtained from Stark Bros. Nursery. A detailed description of many of the world's apple cultivars is available (4, 5).

The third most popular apple in the world, Granny Smith, also originated as a chance seedling in Mrs. Smith's back yard in Australia thus the name, Granny Smith. It is a very firm green tart apple ideal for apple pie and juice production.

The newer cultivars, Gala, Jonagold, and Fuji came from breeding programs. Gala originated in New Zealand and is a cross between Golden Delicious and Kidd's Orange. Jonagold originated in New York State and is a cross between Golden Delicious and Jonathan. Fuji came from Japan and is a cross between Delicious and Ralls Janet. Several red mutations of these cultivars have been selected and are now grown and becoming available to the consumer.

PROCESSING APPLES

Cultural Management and Handling

All of the apple cultivars grown commercially are used to some extent in processed products. Some cultivars such as York Imperial are grown almost exclusively for processing. Only sound ripe fruit should be used for processed products. Processing quality can be affected by decay, damage, maturity, firmness, color, soluble solids, acids,

Table 2. Production of Apples in Specified Countries for 1986–1987[a]

Continent & Country	Millions of Pounds
North America:	
Canada	822
Mexico	1,278
United States	7,294
South America:	
Argentina	2,032
Chile	1,137
Total	3,169
European Community:	
Belgium and Luxembourg	493
Denmark	113
France	3,783
Germany Fed. Rep.	3,830
Greece	640
Italy	4,043
Netherlands	759
Spain	1,769
United Kingdom	621
Total	16,051
Other Europe:	
Austria	548
Hungary	2,133
Norway	89
Sweden	79
Switzerland	310
Turkey	3,962
Yugoslavia	1,229
Total	8,350
Total Europe	24,400
Africa:	1,066
Total	1,066
Asia:	
Japan	1,918
Oceania:	
Australia	692
New Zealand	690
Total	1,382
Grand Total	41,330

[a] Ref. 2.

Table 3. Apple Production in the United States for 1986–1987

Cultivar	Millions of Pounds	Main Region
Delicious (all strains)	3010/4660	West
Golden Delicious	900/1702	West
McIntosh	830/693	East
Rome Beauty	430/634	East
Granny Smith	350/443	West
Jonathan	315/402	Central
York	280/285	East
Stayman	245/219	East
Winesap	140/171	West
Cortland	140/131	East
Newton	140/178	West
Rhode Island Greening	133/115	East
Northern Spy	122/129	East
Gravenstein	70/107	West

Table 4. Apple Cultivar Planting Trends in Washington State

Cultivar	1986, %	1988, %	Estimate 1990, %
Delicious	35.0	29.4	30.7
Golden Delicious	5.7	7.9	6.4
Granny Smith	23.3	13.6	3.9
Rome Beauty	4.5	4.1	5.1
Gala	6.3	3.8	23.7
Jonagold	2.3	7.9	2.2
Red Fuji	.3	3.5	8.7
Others	22.6	29.8	19.3

and other chemical compounds, such as tannins, contained in the fruit (6). The cultivar used in processing will be dictated to some degree by the quality of the product to be produced. Many of the apples grown for the fresh market have some imperfections such as skin blemishes or off shapes and are utilized by the processors. These are perfectly good quality apples and are in high demand. Up to 20% of the Delicious and other fresh market apples are processed. Delicious apples that are firm, sweet and juicy yield a large volume of high quality juice. Although sauce can be produced using Delicious apples the product would not be of good quality particularly in relation to texture and color. Golden Delicious on the other hand not only makes a good quality juice but produces a high quality sauce or sliced processed product. Cultivars used in processed products are determined by availability of the raw product, quality of the product produced, and market demand. Cultivars such as Golden Delicious, Rome Beauty, Granny Smith, McIntosh, and others may have more than 20% of the volume diverted to processing.

Apples may be grown specifically for delivery to the processor, a practice common among orchards in the Eastern United States but, most apples sold to the processor are salvaged from fruit grown for the fresh market. Production costs for processing apples have been reported to be lower than costs for fresh market apples (7). This is not true. Because a premium price is paid for large, bruise free apples delivered to the processor, growers must give full attention to the cultural management details similar to those given apples grown for the fresh market. Production practices for apples vary with the climate and soils in which they are grown. Space does not permit a detailed description of these practices. Interested readers are referred to several of the many books (7–9) and extension publications (10–15) that are available. For specific recommendations in your area consult the State Experiment Station Fruit Specialist.

While the cultural management techniques for fresh and processed apples are quite similar, a few important differences should be recognized. Many of the best processing cultivars are prone to biennial bearing. Growers must follow an annual and aggressive spray thinning program to encourage annual production of large fruit. Heavy nitrogen application has been recommended to force trees growing processing apples into earlier and higher production. This technique may not yield the highest quality fruit. For example, Golden Delicious produced at a leaf N level of about 2% are superior in quality to those produced at a higher leaf N level (2.2% or above). Detailed pruning is not common in processing orchards. The most common tree forms are modified leader or a free-form (no training) system. This contrast with orchards producing fresh market apples where trees are trained to well defined shapes most commonly central leader or open center which permits maximum sunlight penetration. Skin color is not of primary concern in growing processing apples. Light is important, however, for strong flower bud development and renewal pruning, the practice of making large cuts to open up an area in the tree's canopy to light and new wood, is often used in pruning these orchards. The principle disease problems facing processing growers are scab, fruit rots, and cork spot, a physiological disorder associated with low fruit calcium.

Apples for processing should be harvested at optimum maturity for fresh market storage and handling. Only in a few instances are apples harvested with the processed product in mind. To date the majority of the apple crop is harvested by hand. Large bins (about $4 \times 4 \times 2.5$ ft high) holding 750–1050 lb of fruit have replaced the traditional 42 lb capacity wooden field crate. Fruit is picked in canvas bags or lined buckets, placed in the palletized bins, loaded with fork lifts on trucks or stacked for transport by special bin carriers to the packing house or processing plant. Mechanical harvesters which shake the tree and catch the falling fruit have not been perfected for apples. It is estimated that less than 10% of the apples in the world are harvested by mechanical methods (6).

Processing of apples, particularly in the Northwest, is a salvage operation. As a result those apples processed are picked and stored in the same manner as fruit destined for the fresh market. Few, if any, processors can utilize all the fruit as it is delivered to the plant. Early in the season some fruit to be processed will be stored directly in the palletized bins in regular atmosphere storage without the benefit of refrigeration. This type of storage is for short term and only limited to the plants processing capacity for short periods. The bulk of the processing apples are culls from the fresh market packing line. The volume available depends on fresh market demand. Early in the season there are large quantities of fruit available from refrigerated storage continuing through January and early February. Refrigerated storage temperatures range from $-1°–4°C$ depending on the cultivar in question.

Delicious and Golden Delicious are very temperature tolerant and store well at the colder temperatures. Many cultivars like McIntosh and Rhode Island Greening do not store well below 2°C and can show symptoms of chilling injury when stored for long periods below their optimum storage temperature.

After January, processing apples are available from controlled atmosphere storage. Controlled atmosphere storage normally consists of a modified atmosphere, 2–3% oxygen and 1–4% carbon dioxide in conjunction with refrigerated temperatures. Again, depending on the cultivar stored, both the atmosphere and storage temperature must be adjusted for the cultivar in question. Processing apples from controlled atmosphere storage are generally in excellent condition. These apples are capable of producing high quality packs of whole, sliced, or sauced product.

The desired quantities of the product are not always available for processing because of the high cost of controlled atmosphere storage.

Apples from both refrigerated and controlled atmosphere storage are capable of producing quality products (16). The product produced and the grade desired must be taken into consideration by the manufacturer when considering apples from not only the different types of storage but directly from the field as well. The processor may choose to hold the fruit at elevated temperature to allow for further maturation development (softening, color change, etc). Some cultivars such as Delicious require additional press aid and filtration as they advance in maturity and become softer. Different grades of sauce can be manufactured from the same cultivar depending on the type of storage and condition when processed.

Apple Juice

By far the largest volume of processed apple products is in the form of juice. Approximately 713,000 tons of apples are processed into juice in the State of Washington alone (17). Apple juice is processed and sold in many forms. Fresh apple juice or sweet cider is considered to be the product of sound ripe fruit that has been bottled or packaged with no form of preservation being used other than refrigeration. This type of fresh juice is normally sold in markets not too distant from the producer. Apple juice is sweet cider that has been treated by some method to prevent spoilage. Shelf-stable or processed apple juice can be natural juice, clarified juice, crushed juice, or frozen apple concentrate. Natural juice is juice as it comes from the press and generally only ascorbic acid is added to preserve color. It is then bottled and heat processed. Some forms of natural apple juice are produced with the use of heat pasteurization only. Apple juice that has been clarified with some form of filtration before bottling and pasteurization is the most popular apple juice product produced in the United States. Crushed apple juice is a product with a high pulp content. This crushed apple juice is produced without the aid of a cider press by passing coarsely ground apples through a pulper and desecrator before pasteurization. Frozen apple juice concentrate can be either natural or clarified juice concentrated to 70 Brix.

The character of the apple juice is directly related to the cultivar and maturity of the apple used to make the product. Juices produced in the East are more acid than those juices produced in the West (6). This flavor difference is directly related to the cultivars that are predominantly grown in these areas. In both eastern and western juice manufacturing facilities the juice product is a blend of the juice from two or more cultivars. This blending procedure allows for a more uniform product throughout the season and from season to season. Regardless of the cultivar used only sound ripe fruit showing no decay should be used in juice products. Immature apples produce a juice lacking in flavor and very astringent. Over mature apples are very difficult to press and clarify or filter. Most commercial cultivars of apples will produce an acceptable juice, particularly when blended.

Apples for juice are dumped either by the bulk truck load or palletized bins into a tank of water with a leaf-eliminator section. The apples should be thoroughly washed to remove dirt, chemicals, and other foreign material, inspected and any damaged or decayed fruit removed. After the initial wash, fruit is rinsed in clean water. Sorting of apples to remove damage or decayed fruit is mandatory. If not removed, damaged or decayed fruit may impart off-flavors to the product and pose a health hazard.

Before processing, whole apples are ground into a mash or pulp for extraction. This mashing process is accomplished with either a hammer or grating mill. The hammer mill is most often used in the United States. These mills crush the apple to proper consistency depending on the maturity of fruit. When using firm fruit for juice small particles are desired. As the season progresses and the apples become softer, pressing becomes more difficult; thus bigger particles of pulp are preferred for pressing.

Equipment used to press or extract juice from fruit are of several types and many variations (18). The pressing process can be batch or continuous depending upon the type of press used. More common types of presses are hydraulic, pneumatic, screw basket, and belt presses. The hydraulic press is a batch type operation that is very labor intensive but requires no press aid and the juice has a low level of solids. Although the hydraulic press is one of the older types of juice presses, it is still in wide spread commercial use today. Other types of presses are modern automated versions of the basic hydraulic press. These newer presses allow for a greater percentage of juice extraction from a given volume of apple pulp but require press aids in the form of added papers or rice hulls. Juice yields from the different types of presses vary greatly (70–90%) and depend on the type of equipment, maturity of the fruit, press aid, press time, and in some instances the addition of pectic enzymes to the apple mash.

After pressing and before the filtration process, apple juice is treated to remove suspended solid material (19). This material, if not removed, will clog filters, reduce production, and can result in a haze in the final juice product. Enzyme treatments are widely used to remove pectinaceous material and clarify the juice. Tannins are removed from apple juice with the use of gelatin treatments which can be used in combination with enzyme treatment, bentonite, or by itself. This gelatin treatment will remove suspended colloidal material from the juice. In some cases tannic acid is added to the juice before gelatin treatment to prevent color and flavor changes. Bentonite, another fining procedure, can be added along with the heat treatment to remove excess protein. After this fining process, the juice is filtered.

Many types of juice filters are available and their capacity can accommodate any scale of production. To obtain the desired product color and clarity most juice manufacturers use a filter aid in the filtration process. This filter aid helps prevent blocking of the filters. As the fruit matures, more filter aid will be required. Several types of filter aid are available, the most commonly used is diatomaceous earth or cellulose type materials. Before filtration, centrifugation may be used to remove a high percentage of the suspended solids. In some juice plants centrifugation is used instead of filtration. This centrifugation process produces a product that is not as clear as filtered juice, this process allows more or less continuous

production. Centrifugation used with filtration reduces the amount of filter aid required.

Pressure, vacuum, and membrane filters are available and all can produce an acceptable product. The type of filter used must match the capacity to maintain plant production. The filtration process is critical not only from production consideration, but the quality of the end product. Both pressure and vacuum filters have been used with success in juice production (18). Membrane (ultrafiltration) filtration is a recent development. Ultrafiltration based on membrane separation has been used with good results to separate, clarify, and concentrate various food products. Ultrafiltration of apple juice cannot only clarify the product, but depending on the size of the membrane can remove some microorganisms.

Preservations of apple juice can be refrigeration, pasteurization, or chemical treatment or membrane filtration. By far the most common is heat pasteurization based on temperature and time of exposure. The juice is heated and poured hot into containers (cans or bottles) and hermetically sealed. When containers are closed hot, a vacuum develops that also aids in the prevention of microbial growth. After pasteurization the juice product can also be stored in bulk containers, but aseptic conditions must be maintained to prevent microbial spoilage. Chemicals such as benzoic or sorbic acid and sulfur dioxide are generally used only to increase the shelf-life of unpasteurized juice used either for bulk storage, or as an aid in helping to preserve refrigerated products.

Canned Apple Products: Applesauce and Slices

Canned applesauce and apple slices rank second in importance among processed apple products. Approximately 35% of the total raw processed crop in the United States is utilized for applesauce and canned slices. In 1988 this represented about 535,481 MT or 15% of the total United States apple crop (1). Seventy-six percent is used for sauce, 12% for slices, and 12% for other canned products such as apple pie filling, whole baked apples, apple rings, and glazed apples. The major areas processing canned apple products in the United States are the Appalachian area (North Carolina, Virginia, West Virginia, and Pennsylvania) and New York. This region produces about 66% of the processed canned product; Michigan cans 13% and the remainder is processed on the west coast (California, Oregon, and Washington) (20). Per capita consumption of sauce and slices for the period 1978–1983 averaged 1.4 kg fresh apple weight, a decline from previous years. This compares with 8.1 kg per capita consumption for fresh raw apples during the same period (1). Imported applesauce and slices is not a significant factor in the United States canned apple market, representing less than 1% of the United States production of canned apple products (21).

Almost all apple cultivars can be used in processing for sauce but only a select few are considered ideal. Quality attributes in raw apples that produce a high quality finished product are described by LaBelle (22). Desirable characteristics in apples for sauce include high sugar and acid solids, aromatic, bright golden or white flesh, variable grain or texture, and sufficient water-holding capacity. In the Appalachian region the most important sauce-type apples are York Imperial, Golden Delicious, Jonathan, Stayman, Rome, and Winesap. New York uses primarily Rhode Island Greening, Northern Spy, Twenty Ounce, Cortland, and to a lesser extent Mutsu and Monroe. In the west, particularly California, Gravenstein and Yellow Newton are used along with Granny Smith and Golden Delicious. McIntosh, though not considered an ideal sauce apple, is used in the northeast, because it is so plentiful. It is generally blended with three or four other cultivars, a common technique used by processors to maintain a uniform product in taste and texture. A typical blend might be primarily York (more than 50%) with Golden Delicious and Rome each contributing a lesser percentage. Some 100% McIntosh sauce is made for the New England market. For many years Northern Spy was a favorite of processors in Michigan and the northeast but production of this cultivar has decreased significantly in recent years because of poor productivity. Jonathan is commonly used for processing sauce in Michigan. Rome is popular in a number of areas because the tree crop is heavy and because the apple's shape is well suited to machine peeling, but for sauce Rome is less desirable than most cultivars because of poor flesh color (23). Sauce made with a high percentage of Rome will have an off-color and weak, runny texture. Processors in the Appalachian region consider York the ideal processing apple. York has a very firm creamy yellow flesh that produces a high quality sauce with grainy texture and good color. The fruit resists bruising and stores exceptionally well, characters favored by processors. York apples have a small core and thus yield a high percentage of processed product when peeled, cored, and trimmed (24). Golden Delicious, a popular fresh market apple, is also processed in large quantities. Its high soluble solids and resistance to oxidative browning in the flesh make it attractive for sauce and slicing.

Apples for canned slices must be firm, maintain integrity of the flesh when sliced, and have good color. York, Stayman, Golden Delicious, Northern Spy, Rhode Island Greening, Yellow Newton, and Jonathan are preferred for making slices. Sweetness is less important in making slices than in sauce. Regardless of whether apples are for sauce or slices the most important factors are fruit quality and maturity. Sound, mature eastern Delicious can produce a high quality processed apple slice (7).

The ideal processing apple has been characterized as a perfect sphere, 3 in. in diameter with a small core, thin, light-colored skin, firm flesh (pressure test of 89 N on October 1; 67 N on June 1 from common storage), high soluble solids (13 Brix +), mildly acidic (0.2–0.25%), of pleasing taste (such as Northern Spy), with a long supple stem that is strongly attached to the fruiting spur until the day of harvest at which time it could be easily detached (25). If such an apple existed, processors would probably emphasize freedom from defect, size, and shape in their grade and pricing structure.

Apples for sauce, slices, and other canned products are received and handled by the processor in a similar manner. When a load of fruit is received a representative sample is taken for grading and testing. A group of simple tests are used to segregate fruit for immediate processing, common storage, or controlled atmosphere storage. The standard tests include flesh firmness, soluble solids (Brix), and internal ethylene level. Processing apples are graded after peeling into four categories based on the percentage of trim waste and presence of major defects: U.S. No. 1— less than 5% trim waste; No. 2—5 to 12% waste; Cider— more than 12% trim waste; and Culls—decay, worms, or other major defects. Prices paid to the grower are based on grade and size; large fruit command a premium price. Some processors also downgrade for bruises. Fresh bruises are generally not considered serious since they don't interfere with the finished product however, if the fruit is stored, bruised tissue becomes corky and may appear as a defect in the finished product. Tests to predict the quality of finished product from raw-product indices have not been too successful (26,27).

Apples for sauce or slices are cleaned, washed, graded (includes sized and blended), peeled, cored, reel washed, trimmed, and inspected before delivery to the designated processing line. Automatic peeling and coring machines have replaced the once common hand-fed peelers. Automatic peelers require firm fruit (above 53 N) since soft fruit tends to spin off these peelers. Some processors use chemical peelers (NaOH or KOH) which produce a low trim waste. A third method now used by some processors relies on high-pressure steam peeling.

Labor shortages and higher production costs have encouraged apple processing plants to become highly automated. Electronics have enabled a number of hand labor tasks to be automated (28). The potential for using robotics in several facets of apple processing has been described (29). Apples selected and prepared for sauce are chopped and fed into a cooker. Ascorbic acid, salt, granulated sugar, corn starch, or other desired ingredients are added just before cooking or immediately after the raw product exists. Live steam is used to cook the raw product. Cooking softens the fruit tissue and inactivates polyphenoloxidase which is responsible for enzymatic browning. Time, temperature, and raw product must be controlled to produce sauce of good texture, color, and consistency. After cooking, applesauce is passed through a finishing screen which removes debris and defines texture (smooth or grainy); large screens produce a more grainy sauce, ie, baby food sauce is finished with fine screens (0.16 cm). Sauce is usually packed in glass or metal cans immediately after finishing. Some processors now pack an individual molded plastic single-serving size container. Minimum fill temperature is 93°C; containers filled below 88°C are heat processed to destroy bacteria, molds, and yeast. In addition to regular applesauce many processors produce specialty products such as natural sauce (no sugar added), chunky sauce, and a mixture of sauce and other fruit such as apricot, peach, or cherry.

Steps for processing apple slices are similar to those used for sauce production with a few notable exceptions. Slice packs generally consist of a single cultivar thus eliminating the need for blending. Texture is very important, therefore, apples with firm flesh and high quality are desired for making slices. Processors desire a uniform slice size within a pack. Consistency of slice size can be controlled by using fruit within a preselected size range. The slicing operation is usually an integral part of the peeling and coring process. Apple slices contain about 25% occluded gases which are removed by vacuum treatment.

The space left behind is filled with water, salt, antioxidants and/or sugar. Apple slices are blanched with steam pressure just before container filling. Several automated systems to vacuumize and blanch apple slices have been described (30,31). Cans (No. 303 or No. 10) are closed with a steam-vacuum process after adding hot water or syrup. Containers are heat processed in boiling water for 20–35 min depending on their size. Immediately after heat processing the finished canned product is cooled in water and cased.

Less than 1% of the processed apples are canned as whole baked or glazed apples or specialty-products such as spiced rings. Large (7.0–7.6 cm), firm, symmetrical fruit such as Rhode Island Greening, Rome, or Stayman are best suited for baking. These apples are cored, partially peeled and baked either in the can (short method) or before canning (long method) (20). A 40–50 Brix syrup is used as a cooking and/or filling media. Baked or glazed apple products and rings require more hand labor than applesauce or slice processing.

Refrigerated, frozen, or dehydrofrozen apple slices are prepared much like canned apple slices except they are not heat processed. To prevent enzymatic browning sliced raw product is subjected to one of several available anti-browning treatments. Fresh and refrigerated slices may be treated with SO_2 (0.2–0.4%) alone or in combination with $CaCl_2$ (0.1–0.2%). Ascorbic acid has been substituted for SO_2 with good results (32). Calcium-treated apples appear to resist enzymatic browning and microbial spoilage better than non-Ca treated slices (33). Fresh slices, if blanched, will resist browning up to 48 h, however, blanching does result in loss of sugar, acid, and flavor which can produce a blander product.

Apple slices to be frozen or dehydrofrozen are usually treated in an acid sulfite bath. Vacuum processing is also used by some processors. Evacuated air is replaced with salt, ascorbic, and citric acid and sometimes a sugar syrup solution. Dehydrofrozen apple slices are dehydrated and frozen to less than 50% of their original weight and volume. Processed slices are packed in cardboard containers or large metal cans with polyethylene liners and rapidly frozen in forced-air freezers before storing. Frozen slices are thawed and soaked in a combined solution of sugar, $CaCl_2$, and ascorbic acid or SO_2. The advantages in processed dehydrofrozen slices over regular frozen slices have been noted (33).

Apple Butter

Apple butter is processed much like sauce except a slower batch cooking is used to produce a more stable product. Small apples, tailings (peel waste) and lower quality fruit may be used in making apple butter. More sugar is used in this process than in sauce.

Quality Control

Quality control is maintained throughout processing beginning with information on growers' spray programs, maturity indices of fruit fed into processing lines, blending as it relates to finished product consistency, on-line measurements such as trim and coring efficiency, processing temperature, and container condition including closing, headspace, and labeling. Finished products are examined and tested to insure United States standards for grades of sauce or canned apples are met.

Dried Apple Products

Drying has been used for centuries to preserve food products. Dried apples are convenient to handle, store, and use (34). Under proper storage conditions they are almost immune to spoilage. Two types of dried apple products are recognized under United States standards—dried apples and dehydrated apples. Dried apples are irregular segments of dried or evaporated apple fruit, the moisture content of which is not more than 24% by weight. Dried apples may be treated with sulfur to retard discoloration. Dehydrated apples are apple tissue in which most of the moisture has been removed. Grade A dehydrated apples cannot contain more than 3.0% moisture by weight and Grade B not more than 3.5%. A Substandard dehydrated apple contains more than 3.5% moisture. Maximum allowable SO_2 level in dehydrated apples is 1,000 ppm.

Dried apple products are cut into pie or sauce pieces, flakes, rings, or wedges. Most good processing cultivars are acceptable for drying. A desirable characteristic in apples for drying is a high sugar/water ratio (19). Sulfur dioxide is the primary agent used to control enzyme activity and preserve the color of dried apple tissue. A number of factors affect drying including size and geometry of pieces, temperature, humidity, air velocity and pressure within the drier, and wet-bulb depression (35). A unique method for producing dehydrated apples, is explosion puffing developed by the USDA, Agricultural Research Service. In this process, partially dehydrated apple pieces are heated in a closed rotating cylindrical container called a gun until the internal pressure has reached a predetermined value. At this point the gun is discharged instantly to atmospheric pressure producing a highly porous piece of apple tissue (36,37).

NUTRITIONAL VALUE OF APPLES

Fresh apples are about 84% water, fiber (0.7–0.8%), carbohydrates (15%), potassium (115 mg/100 g), and malic acid (38). Fructose, sucrose, and glucose are the most abundant sugars. Fresh apples are considered moderate in energy value and low in protein, lipid, and vitamin content. The nutritive value of most processed apple products is similar to the fresh raw product. Dried or dehydrated apples have a higher energy value per gram tissue due to the concentration of sugars.

Acknowledgements: The authors wish to acknowledge the technical assistance of Karen Burkhart, Raw Fruit Coordinator, Knouse Foods Cooperative, Inc., Peach Glenn, Pa, for suggestions and constructive criticism.

BIBLIOGRAPHY

1. International Apple Institute, "1988 Apple Marketing Clinic Production and Utilization Analysis," *Apple News* **19**(3), 1A–27A (1988).

2. *U.S. Department of Agriculture Statistics 1987,* U.S. Government Printing Office, Washington, D.C. p. 187.

3. S. A. Beach, N. O. Booth, and O. M. Taylor, *The Apples of New York,* Vols. I, II, New York Agricultural Experiment Station, J. B. Lyon Co., Albany, N.Y., 1905.

4. W. H. Upshall, ed., *North American Apples: Varieties, Rootstocks, Outlook,* Michigan State University Press, East Lansing, Mich., 1970.

5. J. Bultitude, *Apples, A Guide to the Identification of International Varieties,* University of Washington Press, Seattle, Wash., 1983.

6. D. L. Downing, ed., *Processed Apple Products,* Van Nostrand Reinhold, Co., Inc., New York, 1989.

7. N. F. Childers, *Modern Fruit Science,* Horticultural Publications, Gainesville, Fla., 1983, pp. 141–145.

8. H. B. Tukey, *Dwarfed Fruit Trees,* MacMillan Publishing Company, New York, 1964.

9. M. N. Westwood, *Temperate-Zone Pomology,* W. H. Freeman and Company, San Francisco, Calif., 1978.

10. Anon., "Establishing and Managing Young Apple Orchards," *United States Department of Agriculture Farmers' Bulletin,* No. 1897, 1972.

11. D. R. Heinicke, "High-Density Apple Orchards—Planning, Training, and Pruning," *United States Department of Agriculture Handbook,* No. 458, 1975.

12. W. J. Lord, and J. Costante, "Establishing and Management of Compact Apple Trees," *University of Massachusetts Cooperative Extension Service Publication,* C–102, 1977.

13. C. G. Forshey, D. C. Elfving, and R. T. Lawrence, "The Planting and Early Care of the Apple Orchard," *New York Agricultural Experiment Station Information Bulletin,* No. 65, 1981.

14. C. G. Forshey, "Cultural Practices in the Bearing Apple Orchard," *New York Agricultural Experiment Station Information Bulletin,* No. 160, 1980.

15. J. E. Swales, "Commercial Apple-Growing in British Columbia," *Minister of Agriculture,* Victoria, B.C., 1971.

16. S. R. Drake, J. W. Nelson and J. R. Rowers, "The Influence of Controlled Atmosphere Storage and Processing Conditions on the Quality of Applesauce from Golden Delicious Apples," *Journal of the American Society of Horticultural Science* **104,** 68–70 (1979).

17. P. E. Nelson, and D. K. Tresler, *Fruit and Vegetable Juice Processing Technology,* 3rd ed., AVI Publishing., Westport, Conn., 1980.

18. WASS, *Washington Fruit Survey,* Washington Agricultural Statistics Service, Olympia, Wash., 1986.

19. R. M. Smock and A. M. Newbert, *Apples and Apple Products,* Interscience Publishers, New York, 1950.

20. R. C. Wiley and C. R. Binkley, "Applesauce and Other Canned Apple Products," in D. L. Downing, ed., *Processed Apple Products,* Van Nostrand Reinhold Co., Inc., New York, 1989, pp. 215–238.

21. R. R. Miller, "Competitive Fruit Situation," *Apple News* **19**(3), 5–8 (1988).

22. R. L. Labelle, "Apple Quality Characteristics as Related to Various Processed Products, in R. Teranishi and H. Borrena Benitez, eds., *Quality of Selected Fruits and Vegetables of North America,* American Chemical Society, Washington, D.C., 1981, pp. 61–76.

23. R. D. Way and M. R. McLellam, "Apple Cultivars for Processing," in D. L. Downing, ed., *Processed Apple Products,* Van Nostrand Reinhold Co., Inc., New York, 1989, pp. 1–29.

24. H. A. Rollins Jr., "The York Imperial Apple," *Fruit Varieties Journal* **43**(1), 2–3 (1989).

25. W. G. Huehn, *Processed Apple Industry Overview,* National Fruit Products Co., Inc., Winchester, Va., 1987.

26. R. C. Wiley and A. H. Thompson, "Influence of Variety, Storage and Maturity on the Quality of Canned Apple Slices," *Proceedings of the American Society for Horticultural Science* **75,** 61–84 (1960).

27. R. C. Wiley and V. Toldby, "Factors Affecting the Quality of Canned Applesauce," *Proceedings of the American Society for Horticultural Science* **76,** 112–123 (1960).

28. U.S. Pat. 3,950,522 (April 13, 1976) J. R. Cogley.

29. J. Yang and T. O'Connor, "Possible Application of Robotics in Apple Processing," in *Proc. Processed Apples Institute Research Seminar,* University of Maryland, College Park, Md., 1984, p. 12.

30. A. S. Ellett, "Process and Apparatus for Continuous Deaeration of Fruits," *Food Process Review* **21,** 93–99 (1968).

31. W. L. Keifer, "Process for the Blanching of Apples," *Food Process Review* **21,** 77–79 (1963).

32. J. D. Ponting, R. Jackson, and G. Watters, "Refrigerated Apple Slices, *Journal of Food Science* **37,** 434–436 (1972).

33. G. C. Hall, "Refrigerated, Frozen, and Dehydrofrozen Apples," in D. L. Downing, ed., *Processed Apple Products,* Van Nostrand Reinhold, Co., Inc., New York, 1989, pp. 239–256.

34. L. P. Somogyi and B. S. Luh, "Dehydration of Fruits," in J. G. Woodruff and B. S. Luh eds., *Commercial Fruit Processing,* 2nd ed., AVI Publishing Co., Westport, Conn., 1986, pp. 353–405.

35. G. C. Hall, "Dried Apple Products," in D. L. Downing, ed., *Processed Apple Products,* Van Nostrand Reinhold, Co., Inc., New York, 1989, pp. 257–278.

36. N. H. Eisenhardt, R. K. Eskew, and J. Cording, Jr.," Explosion Puffing Applied to Apples and Blueberries" *Food Engineering* **36**(6), 53–55 (1964).

37. J. F. Sullivan, J. C. Craig, Jr., R. P. Konstance, E. M. Egoville, and N. C. Aceto, "Continuous Explosion Puffing of Apples," *Journal of Food Science* **45,** 1550–1555, 1558 (1980).

38. C. Y. Lee and L. R. Mattick, "Composition and Nutritive Value of Apple Products," in D. L. Downing, ed., *Processed Apple Products,* Van Nostrand Reinhold, Co., Inc., New York, 1989, pp. 303–322.

Max W. Williams
Stephen R. Drake
USDA/ARS
Wenatchee, Washington

Stephen S. Miller
USDA/ARS
Kearneysville, West Virginia

APRICOTS AND APRICOT PROCESSING

China is given credit for being the ancestral home of the apricot (*Prunus armeniaca*) where it grew wild in several parts of that vast land. Chinese records note its cultivation there as early as 2200 BC. Apricot trees spread through the Persian Empire thence to the Mediterranean in Persian trading ships. The Roman Empire imported and then grew apricots by the time of Christ. Some scholars suggest that the famous Golden Apples of antiquity were really apricots, a theory supported, in part, by the climatic and land conditions of the region. The apricot thrives in a Mediterranean-type climate.

By the end of the sixteenth century the English were enjoying this colorful fruit; Captain John Smith reported that apricots were growing in Virginia in 1629. It was in California, however, that they really took to the land and climate when Spanish missionaries introduced apricot trees into their mission gardens. In 1792 the first commercial orchard was planted in Santa Clara. Apricots are produced in most of the principal fruit-growing areas of the world (Table 1).

BOTANY AND BIOLOGY

The apricot belongs to the Rosaceae family and most cultivated apricots belong to the species *Prunus armeniaca* L. Closely related species include *P. mume* Sidb. and Zucc., the Japanese apricot; *P. dasycarpa* Ehrh., the black apricot; *P. brigantiaca* Vill., the Briancon apricot from the French alps; *P. ansu* Komar; *P. sibirica* L.; and *P. mand-*

Table 1. Worldwide Apricot Production[a]

Area	Production (1,000 MT)	
	1986	1988
Continent		
Africa	187	236
North America	52	101
Latin America	32	26
Asia	826	753
Europe	728	958
Oceania	35	38
World total	*1,860*	*2,112*
Country		
Turkey	350	345
Italy	186	189
USSR	178[b]	200[b]
Spain	150	164
France	115	97
Greece	77	146
Morocco	70	71
Syria	62	94
Romania	60	45
Iran	56	50
Pakistan	54[b]	63[b]
United States	50	95
Hungary	40[b]	30[b]
Algeria	40[b]	40[b]
Afghanistan	37[b]	37[b]
Iraq	33[b]	32[b]
South Africa	27[c]	67[b]
Lebanon	26[b]	27[b]
Bulgaria	26	35
Australia	26	29
Egypt	22[b]	26[b]
India	21[b]	17[b]
Tunisia	18	21
Argentina	12	13
Israel	10	10

[a] Ref. 1.
[b] FAO estimate.
[c] Unofficial figure.

shurica (Maxim.). The apricot is diploid ($2n = 16$, $x = 8$). Flowers are borne singly or doubly at a node on short stems (peduncles). Flowers have about 30 stamens with one pistil. The solitary flowers are white or pinkish. Most commercial cultivars in the United States are self-fertile, however, Perfection and Riland are examples of self-incompatible cultivars. The bearing habit of the apricot is intermediate between a plum and a peach. Trees generally produce vigorous upright growth, but not as upright as plum. Floral initiation occurs in late spring-summer. Apricot cultivars currently available require approximately 300–1,200 h of chilling (temperatures below 7.2°C) depending on the cultivar. In years with insufficient winter (December and January) chilling, bud drop can occur. Most flowers and fruit are borne laterally on short shoots called spurs, however, some smaller percentage of the flowers are borne on one-year-old shoots. Spurs are usually productive for approximately three to five years. Apricots are usually the first deciduous fruit trees to produce flowers in the spring after almond and are, therefore, subject to frost damage. Trees bloom over an approximately one- to two-week period, depending on weather conditions, and flowers are followed by the appearance of leaves, which are simple, alternate, and serrated; round–ovate to ovate; and sharp pointed. Apricots produce more flowers than are needed to insure the production of an adequate crop and often require the removal of small fruit (thinning) to insure adequate fruit size.

The apricot is a stone fruit (drupe). Other stone fruits in the genus *Prunus* include almond, cherry, nectarine, peach, and plum. Other drupe fruits include olive, coconut, and mango. Botanically, a drupe fruit is a fleshy, one-seeded fruit that does not split open of itself, with the seed enclosed in a stony endocarp, which is called a pit. The fruit of apricot consists of a stony endocarp, a fleshy mesocarp, and an outer exocarp (skin). Fruit are generally yellow to orange, often with a reddish blush and have a fresh weight range from 30 to 80 g. The fruit, once commercially dried, weigh approximately one-fifth to one-sixth of their initial fresh weight. The apricot is one of the finest of the stone fruits, but requires careful handling when picked fresh to insure that its quality characteristics can be appreciated by consumers.

HORTICULTURE

Production

California produces ca 97% of U.S. grown apricots. Washington State is the second largest producer followed by Utah, Colorado, Arizona, Michigan, and New Jersey. The production of fruit is adapted to special environmental conditions where fungal disease is minimal and winter chilling requirements insure consistent crop production. Major producing areas in California include Stanislaus, San Joaquin, Merced, Contra Costa, Yolo, Solano, San Benito, Santa Clara, Fresno, Kern, Kings, and Tulare counties.

California apricot acreage reached its peaks in 1928 with 83,000 bearing and 13,000 nonbearing acres. However, improved cultural practices were responsible for the state's largest crop in 1934, when only 66,000 acres produced 324,000 short tons. Since then, acreage in California has declined as has fruit volume. Today the state produces between 90,000 and 120,000 tons on about 17,000 acres and averages 6.5 tons to the acre. Yield per acre has improved, but absolute fruit production has dwindled, as has per capita consumption. The U.S. Department of Agriculture (USDA) recorded consumption of 1.33 lb of apricots per person in 1970 and currently is reporting less than half of that.

About 15% of the crop is sold for fresh market consumption, 22% of the tonnage is in dried apricots, 10% is processed for frozen fruit, and about 3% is used for baby food. The remaining 50% goes into some form of canning: retail canned; institutional–food service canned; concentrate and juice for apricot nectar, ingredients for bakery goods, toppings, fillings, etc. Apricot preserves are made mostly from frozen apricots.

Varieties and Rootstocks

The varietal situation in regard to apricots is less complex than that of peach or plum. Apricot breeding in the United States has not been intensive, and as a result fewer varieties are commercially available. Descriptions of some commercial California apricot varieties are listed in Tables 2 and 3.

Apricots are grown commercially on apricot seedling, peach (*P. persica*), and plum (*P. cerasifera*, myrobalan seedling and myrobalan 29C; *P. cerasifera* × *P. munsoniana?*, marianna 2624) roots. The main plum rootstocks include myrobalan seedling, myrobalan 29C, and marianna 2624. Plum rootstocks are characteristically resistant to wet and heavy (clay) soil conditions. The union between apricot and plum root reportedly is not as sound as apricot on peach or apricot root and occasionally some apricot varieties break off at the union during heavy wind storms. Apricots growing on plum roots may not be as productive as those grown on apricot or peach roots. Peach root for apricot is quite popular. The stocks include those propagated from Lovell seedling and Nemaguard seedlings. Apricots grown on peach are productive, but peach cannot tolerate wet and heavy soil conditions as well as plum. Peach roots are very susceptible to *Phytophthora* root and crown roots. Apricots grown on apricot seedling roots are not as common as those on peach roots. Apricot roots are not as resistant to wet soil or *Phytophthora* as plum and also do not have resistance to nematodes as does Nemaguard peach root. However, apricot roots are thought to occasionally impart better fruit quality and productivity to apricots.

The Citation rootstock, developed by Zaiger at Modesto, Calif., is a peach × plum hybrid and has shown promising results in initial testing. The rootstock may impart some tree-size control to selected apricot varieties.

Culture and Harvesting

Apricot trees grow vigorously on various soils and when mature may require more space than peaches. Tree planting distances range from 24 by 24 ft (7.3 by 7.3 m) to 8 by 16 ft (2.4 by 4.9 m) in some experimental commercial orchards. There are no commercially suitable dwarfing rootstocks for apricot production and current planting distance is often dictated by soil type and the grower's management expertise. Trees planted at higher densities per acre generally can yield commercial crops of higher economic value at an earlier stage in their development, but as the trees age and fill their allotted space, branch crowding may occur, limiting light interception in certain canopy regions. Light-limiting conditions in canopy regions may lead to small fruit size and reduce yield. In addition, higher costs may be incurred by pruning more trees per acre with dense canopies.

Trees are generally planted in California from January to March in moist soils. Trees are headed back to 24–32 in and little pruning is done until the first dormant season. Trees are normally trained to an open center tree form, which requires heading and thinning cuts on limbs to develop the desired tree form in the first three years of growth. Nitrogen applications in young orchards are variable, depending on grower experience and soil fertility. Cropping on some varieties can begin in the third year, but can be influenced by pruning practice. More commonly, crops are borne in the fourth year. Apricot fruit is harvested primarily by hand, but fruit from mature bearing trees used for certain types of processing can be harvested by machine. Pruning after cropping has commenced is directed toward maintaining adequate light penetration into the canopy, crop regulation, and renewal of fruiting wood for future crop production. Summer pruning is practiced to avoid damage from the Eutypa fungal dieback disorder. Dormant pruning is more common. The vigorous growth on treetops is often removed by machine, making an orchard appear flat-topped. This practice lowers tree height and is relatively inexpensive. Apricot trees are thought to require similar amounts of water to those of other stone fruit. Preharvest irrigation is practiced to insure that fruit size is not reduced from water deficit. In addition, severe water stress during the postharvest period has been shown to reduce the production of flowers, fruit, and yield in the following season. In general, 3–5 acre-ft of water are applied to apricot trees per season in California.

There are many pests and diseases affecting apricot trees growing in California. Insect pests include the peach twig borer (*Anarsia lineatella*), peach tree borer (*Sanninoidea exitiosa*), Pacific flat-headed borer (*Chrysobothris mali*), branch and twig borer (*Polycaon confertus*), shot hole borer (*Scolytus rugulosus*), red humped catepillar (*Schizura concinna*), scale insects, mites, aphids, fruittree leaf roller (*Archips argyrospila*), green fruit worms, orange tortrix (*Argyrotaenia citrana*), European earwig (*Forficula auricularia*), codling moth (*Laspeyresia pomenella*), and Western tussock moth (*Orgyia vetusta*). Regular dormant sprays containing oils plus parathoin or diazinon control peach twig borer and scale insects and also keep many of the above-listed pests below levels at which economic damage may be caused. Insects feeding on foliage and fruit can generally be controlled by insecticides such as diazinon, parathion, or carbaryl. Integrated pest management control techniques that depend on natural predators and monitoring population dynamics have resulted in reduced use of pesticides for apricot production.

There are three fungal diseases that are most damag-

Table 2. Commercial Apricot Varieties in California

Variety	Characteristics
Castlebrite	Released by the USDA in Fresno, Calif., as a very early ripening fresh shipping variety. The fruit is yellow-orange, attractive with a red blush. It is firm, medium in size, but does not possess fine eating quality.
Derby, Derby Royal	Principally grown in the Winters district of California and ripens about one week before Royal. It originated from a chance seedling and was first planted in 1895 (1). It is suitable for fresh shipping, but not for drying because the stone can cling to the flesh. The fruit may ripen unevenly. At least two distinct strains of the Derby have been recognized around Winters with differing fruit characteristics.
Flaming Gold	Origin in Modesto, Calif., by Zaiger. The variety was introduced in 1967 and was a seedling of Perfection. The fruit ripens early and has been used for fresh shipping primarily.
Flavor and Spring Giant	Large firm, fresh-shipping fruit, yellow-orange, developed by Zaiger. Early maturing about the third and fourth weeks of May.
Katy	Large- to medium-size fruit developed by Zaiger. The fruit is yellow, firm, and an early-ripening fruit used for fresh shipping.
Modesto	Origin in Le Grand, Calif., by Anderson. The variety was introduced in 1964 and was an F_2 open-pollinated seedling of Perfection. The fruit is firm fleshed, medium to large size, often with a red blush, and orange fleshed. The tree bears regularly and the fruit ripens about the time of Royal and several days ahead of Patterson. Modesto is resistant to pit burning.
Patterson	Origin in Le Grand, Calif., by Anderson. The variety was introduced in 1968 and was an F_2 seedling of Perfection × unknown. The variety is firm, has orange flesh and skin, and sometimes is found with a red blush. The fruit is medium to large, and produces consistently good crops, it appears to be less affected by mild winters than the Tilton variety, and the fruit ripens evenly. Fruit ripens three to four days later than Tilton. Patterson is resistant to pit burning. Patterson is a good variety for processing and makes a good dried product and is fair for fresh eating fruit.
Perfection	Origin in Waterville, Wash., by the Goldbecks. The variety was developed from a chance seedling planted in 1911 and the variety was introduced in 1937. The parentage was unknown. The fruit is large, oval, light yellow-orange skin, the flesh is yellow-orange, with no blush. The fruit is firm, resists pit burning, is fair in quality, and is self-unfruitful. The fruit matures about a week to ten days before Royal.
Pinkerton	Origin in Le Grand, Calif., by Anderson. The variety was introduced in 1967 and was an F_2 seedling of Perfection. The tree is a regular productive bearer. The fruit is medium in size, the flesh is yellow and firm. The fruit is freestone and it ripens about three weeks before Royal. Pinkerton has been used as an early fresh shipping apricot.
Royal and Blenheim	The Royal and Blenheim varieties are considered together because they have lost their separate identities. However, the Blenheim was thought to have originated with a woman named Shipley in Blenheim, UK (Shipley's Blenheim), while the Royal, which closely resembles the Blenheim, is an old French variety described first in *Bon Jardinier* in 1826. It originated in the Royal Garden of the Luxembourg. The Royal–Blenheim is one of the leading varieties in California and is used fresh, dried, canned, and frozen. The variety tends to alternate bear and unless properly thinned may have small fruit. It has excellent flavor when fully matured. It is rather soft and difficult to ship. The flesh is orange, and the skin is yellow-orange, it often may have a red blush. The fruit tends to pit burn and is not well adapted for the interior valleys of California. It can develop a condition known as fog spot in the coastal regions of California.
Tilton	The variety originated as a seedling in Kings County, California, by Tilton in 1885. Tilton is one of the leading apricots for processing in California but, is currently being replaced by Patterson. Tilton is adapted for drying, shipping, and canning, however, the fresh flavor is only fair. It ripens a week to 10 days later than Royal. The tree tends to alternate bear and is affected by mild winter temperatures. When properly thinned it may bear medium- to large-size fruit. The fruit is orange to yellowish and often has a red blush.
Tracy	Origin in Le Grand, Calif., by Anderson. The variety was introduced in 1971 and was an F_2 open-pollinated seedling of Perfection. It has medium- to large-size fruit. It ripens several days later than Patterson, has firm flesh, and resists pit burning.
Westley	Origin in Le Grand, Calif., by Anderson. The variety was introduced in 1973 and was an F_2 of Perfection × Tilton. The fruit is medium to large, flesh is orange, medium firm, good flavor, used for some fresh and drying. The tree tends to alternate bear, and is later ripening than Tilton or Patterson.

ing to apricots. Brown rot (*Monilinia laxa* and *M. fructicola*) affects blossoms, twigs, and fruit. The blossom and twig blight phase begins with death of blossoms and then spurs and leaves. Infection moves from flowers into twigs; small cankers may form with gum exuding from the base of flowers. The disease reduces the number of flowers and kills fruiting spurs, resulting in crop loss if left unchecked. Brown rot can affect ripening fruit, resulting in

reductions in marketable yield. Several fungicides are available for control and they are applied during bloom or two to three weeks before harvest to protect ripening fruit. Apricot varieties are being selected that may have resistance to brown rot.

Another principal disease of apricot is shot hole (*Stigmina carpophila*). Shot hole disease causes spots on fruit and holes in leaves, which are noticeable in the

Table 3. California Acreage of Selected Apricot Varieties[a]

Variety	Acres
Castlebrite	986
Derby	221
Flaming gold	157
Improved flaming gold	118
Katy	319
Modesto	578
Patterson	3,404
Perfection	55
Royal–Blenheim	7,305
Royal Derby	76
Tilton	5,922
Westley	174
Other	584

[a] Refs. 3, 4.

spring. Severe infection can cause leaf drop and reddish spots appearing on ripening fruit. Control can be achieved by using dormant copper sprays and fungicides during bloom.

Eutypa (Cytosporina, Gummosis) dieback, named after the perfect (sexual) fruiting stage of the fungus Eutypa armeniacae, causes death of limbs. Infection occurs through pruning wounds, and, in general, two to five years later fruit production is reduced and the infected limb is girdled by the growing canker and dies. Fruiting bodies of the Eutypa fungus discharges spores during rain, with the greatest numbers occurring in October and April. Viable spores can travel in the air for great distances and infect fresh pruning wounds. Pruning wounds are susceptible to infection for two to four weeks after pruning. The summer in California is virtually rain free and there is almost no risk of infection at this time. Summer pruning completed by early September allows for

pruning wounds to heal and reduces the risk of infection by the Eutypa fungus. Other diseases include jacket rot (Botrytis cineria, Sclerotinia sclerotiorum), oak root fungus (Armillaria mellea), crown gall (Agrobacterium tumefaciens), verticillium (Verticillium sp.), and bacterial canker (Pseudomonas syringae).

NUTRITION, IMPORTS, AND PROCESSING

Apricots are among the most nutritious and health-promoting fruits produced by nature. Tables 4–7 detail the nutritional content of the apricot as well as its economic character. Dried apricots contain the second highest level of fiber, next to dried figs.

The United States is the seventh largest producer of apricots. Turkey, Italy, USSR, Spain, Greece, and France surpass the United States in commercial production at last count (1987, Table 1). Exports of apricots to the United States from various countries exceeded 45,000 fresh short tons, or almost half of California's total production in 1988. Of these fresh tons, 38,756 were exported as dried apricots, which, at a fresh to dry fruit weight ratio of 6 to 1, equaled 6,626 dry tons landing in U.S. ports. Turkey shipped 86% of this tonnage.

Canned fruit imports in 1988 totaled 3,965 short tons, with 1,415 of these tons coming from Spain and 827 from Israel. Israel continues to increase exports of apricots to the United States.

Fresh apricot shipments to the United States equaled 1,434 short tons, with 1,190 tons coming from Chile. Argentina, Brazil, and Peru are expected to become serious off-season exporters in the future.

Approximately 85% of all California apricots are processed. California represents nearly 100% of commercially processed fruit in the United States. Growers produce fruit that is processed in five different ways: canned, nec-

Table 4. Nutritional Information Per 1-Cup Serving of Apricots[a,b]

Nutrient	Heavy Syrup	Light Syrup	Juice Pack	Water Pack	Uncooked Dried	Cooked Dried
Calories	210	160	120	65	310	211
Protein, g	1	1	1	1	5	3
Carbohydrates, g	55	41	31	16	80	55
Fat, g	0	0	0	0	1	0
Potassium, mg	361	349	409	465	1,791	1,222
Sodium, mg	10	10	9	7	13	9

Percent of U.S. Recommended Daily Allowance (US RDA)

Nutrient	Heavy Syrup	Light Syrup	Juice Pack	Water Pack	Uncooked Dried	Cooked Dried
Protein	*[c]	*[c]	*[c]	*[c]	10	6
Vitamin A	60	70	80	60	188	118
Vitamin C	10	10	20	15	6	6
Thiamine	2	2	2	2	10	4
Riboflavin	2	2	4	4	18	10
Niacin	4	4	4	4	*[c]	*[c]
Calcium	2	2	2	2	4	4
Iron	4	4	4	4	30	20
Phosphorus	2	2	4	2	15	10
Magnesium	4	4	6	4	15	10

[a] Ref. 5.
[b] Contains no cholesterol.
[c] Less than 2% USRDA.

Table 5. Nutritional Information Per 100-g Serving of Apricots[a,b]

Nutrient	Heavy Syrup	Light Syrup	Juice Pack	Water Pack	Dried Uncooked	Dried Cooked	Fresh	Frozen Sweetened	Nectar[c]
Calories	83	63	48	27	238	85	48	98	56
Protein, g	0.53	0.53	0.63	0.71	3.65	1.30	1.40	0.70	0.37
Carbohydrates, g	21.47	16.49	12.34	6.39	61.75	21.90	11.12	25.10	14.39
Fat, g	0.08	0.05	0.04	0.16	0.46	0.16	0.39	0.10	0.09
Vitamin A, IU	1,230	1,322	1,691	1,293	7,240	2,363	2,612	1,680	1,316
Potassium, mg	140	138	165	192	1,378	489	296	229	114
Sodium, mg	4	4	4	3	10	3	1	4	3
Iron, mg	30	39	30	32	4.70	1.67	0.54	0.90	0.38
Ascorbic Acid, mg	3.10	2.70	4.90	3.40	2.40	1.60	10.00	9.00	0.60
Fiber, g	0.40	0.41	0.39	0.42	2.95	1.05	0.60	0.60	0.19

[a] Ref. 5.
[b] Contains no cholesterol.
[c] Without ascorbic acid added.

tar, dried, frozen, and baby food. For the purpose of describing these methods only canning, drying, and freezing will be included, because the production of nectar and baby food are similar to the process for producing apricot concentrate or purée by canners.

CANNED APRICOTS

The majority of apricots for canning are grown in Stanislaus and San Joaquin counties. The major apricot-processing plants are located in this general area. Tilton and Patterson varieties are most widely used for canning and the Patterson will soon replace Tilton as the principal variety.

Prices paid to growers by canners are negotiated by the Apricot Producers of California, located in Modesto, Calif. Current prices are based on a price per ton for U.S. No. 1s with a downward sliding scale of five grades below this level. Grading is carried out by a third-party inspection program approved by both parties.

Fruits are picked by hand or mechanically shaken and dumped into 600-lb bins. These bins are delivered to a weighing station where they are graded, weighed, and then delivered to the canning plant. Next, the apricots are mechanically dumped into a water bath containing chlorinated water. They travel down an inspection belt for removal of any nonapricot material (leaves, sticks, and stones) and immature, green, overripe, or rotted fruits. Fruits are then size graded with the small fruits, unsuitable for canning, collected and directed toward eventual use in concentrate and nectar-type products. These inspected and sized fruits are delivered to mechanical apri-

cot cutters, which align the fruits so that they are cut along their sutures. Cut fruits are then opened and the loose pits drop through a perforated stainless-steel plate. Cut and pitted fruits are then delivered to the sizer and grader to be mechanically sorted, resulting in uniformly sized fruits. Customers expect a specific number of units of cut apricots of uniform size in each can. It is therefore necessary to mix various fruit sizes to result in a specified number of units per can. Once the cans have been filled, a topping medium—for example, sugar water or fruit juice—is added, and the cans are immediately seamed. The product is then ready to be cooked, rendering it commercially sterile. Later, when the product reaches the warehouse, it will be labeled and cased according to the customer's request. Canned apricots are packed in heavy syrup, light syrup, juice pack (apricot or pear), and water pack. Can sizes include 8-oz buffet size; 14–16-oz #300s; 16–17-oz #303s; 29-oz #2½s, and 104–117-oz, #10s. The latter size is packed for institutional–food service customers.

Fruits unsuitable for cutting are delivered to a thermal screw where they are heated to 99°C and delivered to a pulping unit, which removes the pit. The heated pulp is then pumped through a series of finishers that will remove some of the fibrous material, such as the skin of the

Table 6. Comparative Nutritional Levels of Canned, Juice—Packed with Skin, 1-Cup Serving

	Vitamin A, R.E.	Vitamin C, mg	Potassium, mg
Apricots	419	12	409
Pears	1	4	238
Peaches	95	8	317
Fruit cocktail	76	7	235

Table 7. Cost per Serving of Canned Apricots and Peaches, in Dollars[a]

Price per Case (6 #10 Cans)	Apricots (4 Halves)	Peaches (2 Halves)
19.00	0.117	0.158
20.00	0.123	0.166
21.00	0.129	0.175
22.00	0.135	0.183
23.00	0.141	0.191
24.00	0.148	0.200
25.00	0.154	0.208
26.00	0.160	0.216
27.00	0.166	0.225
28.00	0.172	0.233
29.00	0.179	0.241
30.00	0.185	0.250

[a] Ref. 6.

apricot. Finished juice is then ready for the evaporation process or delivered to the nectar room where it will be mixed with sugar, water, and citric acid. The nectar cans are filled and sealed similarly to the cut fruit canning process and delivered to the warehouse. Fruit juice destined for the evaporator will be condensed over twofold, resulting in 32% fruit concentrate. The product is then canned in 110-oz cans or aseptically filled in 55-gal drums. The concentrated product is ideally suited for nectar reconstitution or as an ingredient in various sauces.

Critical parameters for quality canned fruit are cooking time and temperature, (soft or overripe appearance), uniformity of size, and unsightly blemishes. The concentrate product must be heated to deactivate enzymes, yet not be overcooked, resulting in a dark color.

There are a few marketers of apricot nectar that purchase off-grade apricots directly from the grower and process their own concentrate or purée according to their own particular specifications. The concentrates are then diluted to an acceptable level as a drinkable apricot juice called nectar.

DRIED APRICOTS

General Information

Fruit drying is one of the oldest food-preservation techniques known to man and is now yielding some of the newest forms of ingredients for products. Its essential feature is that the moisture content of the fruit is reduced to a level below that at which enzymatic or microbial damage occurs.

Although one of the principal reasons for sun drying fruit is preservation, other factors play an important part in the selection of that process. Significant reductions of weight and bulk are important, but the most important reasons for sun drying are the unique flavor and texture of the finished dried apricot. The special flavor and texture are winning over a new generation of consumers who use dried apricots as snacks because of the fruit's flavor and nutritional value.

Dried apricots are produced from plump, ripe, fresh fruit. The fruit is usually sun dried between June and August. The Blenheim and Tilton apricot is the principal variety used for drying. Other varieties, such as the Patterson and Tilton, are dried, but most do not have the unique combination of high flavor density and solids-to-acid ratio that creates the luscious sweet–tart flavor of the California dried apricot.

Sun Drying of Fruits

The drying of apricots parallels methods used for processing other dried fruits (figs, peaches, pears, etc). Stage of maturity of the starting fruits is important in selecting fruits for drying. If fruits are picked too early, color and flavor are lacking in the final product. When overripe, the final product loses shape and becomes slablike in appearance. Much of the fine flavor of California dried apricots is due to the fact that the dried fruits can be produced from fully tree-ripened fruit, something not readily possible with canned or frozen fruit.

Preparing fresh fruits for sun drying is simple. The common predrying treatments applied to apricots are (1) selection and sorting of fresh fruit, (2) washing, (3) cutting into halves and removal of pits, (4) spreading of fruit on drying trays with the cut surface upward, (5) sulfuring with burning sulfur or gaseous sulfur dioxide, (6) placing of trays in full sun in the dry yard.

Sulfuring of Apricots

For many years sulfur dioxide (SO_2) has been used to preserve the color of dried fruits. It is the only chemical added to dried apricots. Sulfur dioxide is generally recognized as safe for use in dried fruits by the Food and Drug Administration. Apricots prepared for drying are usually exposed to gaseous SO_2 before being put in the sun for drying, however, treatment with fumes of burning elemental sulfur is also practiced.

In addition to preventing enzymatic browning, SO_2 treatment reduces degradation of carotene and ascorbic acid, which are valuable nutrients in apricots, thus nutritional qualities are preserved.

The SO_2 treatment of apricots for drying to retain their natural color must be closely controlled so that enough SO_2 is present to maintain the physical and nutritional properties of the product throughout its life. Fruits high in carotene content, such as apricots, require high SO_2 levels to retain natural color. The storage life of the dried apricots is directly proportional to the SO_2 level.

Sulfured dried apricots will contain a SO_2 level of 2,500–3,000 ppm. The amount of SO_2 must also be controlled because the SO_2 level permitted in fruit is regulated and varies from country to country. Sulfur dioxide in the fruits begins to dissipate as soon as it is applied and continues to diminish throughout storage, distribution, and retail shelf life of the product.

Sun Drying of Apricots

The art of sun drying of fruit as a method of food preservation has remained unchanged from ancient times. The practice is limited to climates with hot sun and dry atmosphere, and to certain fruits such as grapes, peaches, figs, apricots, and pears.

After sulfuring, the trays are placed in the drying yard for sun drying. Sun drying is complete when the apricots have a moisture content of 15–20%. Drying time can vary depending on the condition of the fruit, air moisture, and constancy of sun exposure. To dry fruits, the cut and sulfured fruits are left in full sun for 5–10 days, followed by further drying, usually in stacked trays, away from direct sunlight for a sufficient time to bring moisture up to the desired level 27%.

After drying, the apricots are transferred to boxes to cure and to bring about equilibrium of moisture content. This requires from two to three weeks or even longer. The dried apricots are then ready for grading and final preparation for packing.

Screening, Inspection, and Washing

Size grading is required to obtain fruit of desired size. It is accomplished by passing dried fruits over a shaking, per-

forated metal screen and collecting the varied sizes separately. Each fruit size fraction passes on to the final inspection operation.

The dried product is inspected to remove low-quality pieces, discolored pieces, or other imperfections such as mashed or torn fruit. Hand sorting of sun-dried fruits is necessary. Sorting is carried out by inspectors picking out the undesirable pieces while the product moves along on a conveyer. In addition to inspectors, magnetic devices are installed over the belts to remove any metal contaminants, and electronic metal detectors are used to reject portions of lots containing ferrous or nonferrous metals not removed by the other means.

Prior to packing, chopping, or grinding, the apricots are rigorously washed to remove dust, leaf particles, etc. Washing consists of a presoak in water followed by mechanical scrubbing and rinsing. After washing, the fruits are spread on trays for a second treatment with SO_2 to insure optimum color and keeping qualities. This postwashing treatment is accomplished with gaseous SO_2 or fumes from burning sulfur. At this point, control is critical to attain the proper final level of SO_2 in the finished product to meet shelf life requirements or requirements specified by the customer for bulk packs.

Sizes of Dried Apricots

The terms used to designate apricot sizes can be confused with those used to designate quality levels in certain other foods (canned vegetables for example). It is important to remember that the name designations shown in Table 8 refer to apricot size only, not overall quality.

Storage of Dried Apricots

Dried apricots are protected from product decomposition due to microbial or enzymatic deterioration. The relatively low moisture level, high natural sugar level, high acid content, sulfur dioxide level, and low pH precludes these types of spoilage.

The most common mode of dried apricot deterioration, which leads to economic loss, is the darkening of the product due to gradual loss of sulfur dioxide. This darkening is not spoilage in the usual sense, but does cause the product to become unappealing and unsalable. Flavor and nutri-

tional value also suffer as the darkening proceeds. The fruits take on a carmelized flavor.

The loss of sulfur dioxide cannot be eliminated entirely, but it can be controlled so that it is of little consequence in normal commercial storage. The single most important factor in determining the storage life in dried specialty fruits is temperature of the storage space. Temperature is so critical that the storage life is cut approximately in half for every 1°C increase. Dried apricots may be frozen with no adverse effects; however, dried frozen fruits should be allowed to defrost in a low humidity area prior to use.

The following conditions and tips will help to obtain maximum storage life for dried fruits:

1. Store dried fruits at 4.4°C and 75% RH for excellent keeping for at least six to nine months (for washed and resulfured apricots).
2. Keep temperature and humidity constant.
3. Be sure that the product is well wrapped and is not exposed to air.
4. Protect dried fruits from high-intensity direct light.

FROZEN APRICOTS

Apricots for freezing include the Patterson, Blenheim, and Tilton varieties, with the Patterson the principal fruit, replacing the Blenheim. There are also some Modestos being purchased.

Frozen apricots are processed for three different product types: sliced halves, slices, and a multiple-scored apricot that is then sliced. The first two types are sold primarily to bakers, ice-cream makers, and frozen dessert makers. The third type is processed for use in preserves (jams and jellies) and is called machine pitted.

Halves and Slices

Receiving. The apricots are received in approximately 800-lb bins. Each truckload is weighed and assigned a lot number. Bins from each lot are visually sampled for quality. The apricots are checked for ripeness, insect penetration, rot, flesh damage, and excess foreign materials such as leaves. Fruits are placed in high-temperature cold storage or left in receiving for ripening. As needed, bins of fruits are brought to the processing plant by truck.

Dumping. Bins are dumped at regular intervals by the manually operated hydraulic bin dumper.

Leaf Roller. The apricots pass over a series of parallel rollers. The rollers are separated by small gaps through which leaves and twigs may fall. A worker is stationed there to remove leaves, foreign material, rot, and green fruits. The green fruits are held for rerunning when properly ripened.

Shaker–Washer. From the leaf roller the apricots fall into the shaker–washer. The shaker–washer agitates the apricots in water and propels them under freshwater sprays.

Table 8. Names, Tolerances, and USDA Designations for Apricot Size Classification

Name Description	Size Tolerance	USDA Size Designation
Jumbo	>1⅜-in. diameter	1
Extra fancy	1¼–1⅜-in. diameter	2
Fancy	1⅛–1¼-in. diameter	3
Extra choice	1–1⅛-in. diameter	4
Choice	1³⁄₁₆–1-in. diameter	5
Standard	<1³⁄₁₆-in. diameter	6
slabs	No size limitation, applies to slab apricots. Slabs generally have flattened, broken, or misshapen appearance caused by drying very ripe fruit.	Not applicable

First Inspection Belt. The shaker–washer discharges the apricots onto the first inspection belt. The fruits are examined for rot, leaves, and green color. From the first inspection belt the fruits drop onto a cleated elevator, which carries the fruits to a cross-belt.

Halver. The cross-belt delivers the apricots to one of two apricot halvers. A shaker–feeder receives the apricots. Rotating scrubbers clean the apricots and arrange them into five single-file lanes. The apricots from the five lanes are fed into pickup pockets. The pickup pockets feed the apricots to V-trough belts in time with orienting fingers. The orienting fingers rotate the apricots so that the suture is aligned with blades that then cut the apricots in half in line with the suture.

Cut-Up Pit Shaker. A cross-belt carries the apricot halves to a conveyor that transports them to a cup-up pit shaker. As the apricot halves move across the shaker pan they pass over holes that permit the pits to fall through as they are shaken loose. The flesh material left on the pits is used to make purée. As the apricot halves are discharged from the cup-up pit shaker they are aligned with the pit cavity facing up.

Pit-Inspection Belt. (This step is eliminated for sliced halves.) The cup-up pit shaker distributes the apricot halves onto the pit-inspection belt. On the belt the apricots are inspected for pits and soft fruits. Soft fruits are placed into sort-out buckets and used to make purée.

Gang Blade Slicer. From the pit-inspection belt the apricots are transported to a gang blade slicer. The slicer has parallel circular bars separated by a $\frac{1}{2}$-in. gap. The slicer cuts the halves into strips.

Final Inspection Belt. The sliced apricots fall onto a final inspection belt. The apricots are inspected for pits, blemishes, and harmless extraneous material.

Fill–Weigh Station. Fruits from the final inspection belt are carried to a fill–weigh station. The fruits are placed in a prelabeled tin and weighed. A depressor is placed on top of the fruits and 60°Brix syrup with ascorbic and citric acid is added. The container is weighed again. A lid is placed on the can and it is passed through a can washer. The cans are coded, palletized, and tagged.

Cold Storage. The completed pallet is promptly moved to cold storage for freezing.

Machine Pitted (for Use in Preserves)

Receiving and Dumping. This is the same as the process for halved and sliced fruits.

Air Cleaner. The apricots pass under an air cleaner. The air cleaner creates a vacuum that removes loose leaves and debris.

Leaf Roller. The apricots pass over a series of parallel rollers. The rollers are separated by small gaps through which leaves and twigs may fall. A worker is stationed there to remove leaves, foreign material, rot, and green fruits. Green fruits are held for rerunning when properly ripened.

Wash Tank. From the leaf roller the apricots fall into the wash tank where they are immersed in water. A rotating paddle propels the apricots through the water. A mesh elevator lifts the apricots out of the tank.

Cross-Belt. The mesh elevator drops the apricots onto a cross-belt. A worker may be stationed there to remove leaves. From the cross-belt the fruits are diverted onto inspection belts.

Inspection. The inspection belt is divided into sections by metal bars. As the fruits move up the outer sections they are examined for stems, blemishes, and leaves. The inspected fruits are thrown into the center section of the belt. Blemished fruits and leaves are dropped down a chute to the inside of the inspection belt where they are carried to a waste dumpster. When the apricots reach the end of the inspection belt they fall onto a cleated elevator, which carries them to an overhead conveyor.

Elliott Pitter. From the overhead conveyor the apricots are diverted into two Elliott Pitters. The Elliott Pitter removes the pit from the apricot. A pounder feeds the apricots into the top of the machine. The apricots are squeezed between a rubber roller and a serrated steel roller. The pits are squeezed out and they fall onto the shaker below the rollers. The pitted apricots adhere to the steel roller until they are scraped off by metal fingers and discharged from the front of the machine. Pits and apricot pulp are discharged from the back of the machine. The pulp is passed through a finisher to make purée.

Purée. The Elliott Pitter discharges pits and apricot pulp onto a cross-belt, which carries them to a screw conveyor. The screw conveyor carries the pits to a 0.25-in. pulper finisher, which separates the fruit pulp from the pits. The pits discharged from the pulper finisher are conveyed to a pit bin.

A solution of citric and ascorbic acids from the acid-holding tank is pumped into the finisher. The flow of acid starts when pits make contact with a high–low probe located in the bottom of the screw conveyor. The volume of acid flow is regulated by an adjustable valve. The amount of acid used is recorded by a meter located in the quality control laboratory.

The pulp from the 0.25-in. finisher is pumped to a 0.02-in. finisher. It consists of several rotating brushes on metal bars that force the pulp between them and a 0.02-in. screen. The purée is forced through the screen. The residual pulp is forced out of the finisher and placed in a water dumpster. From the 0.02-in. finisher the purée is pumped to the purée-holding tank.

Final Inspection. Fruits discharged from each Elliott Pitter fall onto an inspection belt. Workers here remove pits, stems, blemishes, and harmless extraneous material. Fruits from the two inspection belts fall onto a cross-belt.

Fruit Pump and Acid Addition. A cross-belt carries the fruits to a hopper situated above a fruit pump where a solution of citric and ascorbic acids are added. As fruit fills the hopper it contacts a high–low probe. When contact is made, the fruit pump starts and the flow of acid begins.

Slicer. The fruits are pumped to the slicer where they are cut into strips by parallel circular blades.

Fill–Weigh Station. At the fill–weigh station, purée from the holding tank is weighed out into a 55-gal drum. If the apricots are very soft the addition of purée may be omitted. The drum is moved onto a scale under the slicer and fruits are added. Acid solution is sprayed on top of the fruits by means of a hand-held sprayer. The drum is moved onto another scale where it is weighed again. A sugar cap is weighed out and placed on top of the fruit. A lid is placed on the drum and it is labeled and palletized.

Cold Storage. The drums of finished product are moved to cold storage for freezing.

PRODUCTIONS AND CALIFORNIA ECONOMICS

Productions

At the height of California apricot production the principal segment of usage was in sun-dried apricots. This was due primarily to the dried product's ability to remain shelf stable for a relatively long period of time and its versatility as an out-of-hand snack and cooking and baking ingredient, and its sweet–tart flavor density.

After World War II the canning industry began its dynamic growth and dried apricots gave way to the canned version because of its growing acceptance as a convenient and economical fruit product as well as its greater similarity to the fresh fruit. The dried apricot continued to be the largest usage type through the late 1940s but lost its position to canned apricot products (which included baby food, purée, and nectar) in 1950 and never regained its plurality. Frozen apricots were introduced to the American public in 1943, diverting about 4,000 tons out of a total utilized crop of 78,300 tons, which reflected the worst crop failure on record until then. Apricots going into frozen production peaked in 1945 at 26,400 tons out of a total of 159,000 tons and has never regained that level. Today, frozen apricots account for about 10% of the crop.

Fresh apricots, that is, sold as fresh in retail outlets, have never developed into a major usage factor. The high point in sales was reached in the disastrous 1943 crop year with 21% of total normal production. Most years since 1909 have held at less than 10% of the total crop. Since 1984, however, with substantially more marketing effort behind fresh apricots, this product category has been maintaining a 12–17% share of salable tonnage.

In 1988, the last year reporting actual tonnage by use type, the 93,000 tons of apricots reported to the California Apricot Advisory Board were distributed as shown in Table 9.

Table 9. Distribution of Apricots Produced in the United States, 1988[a]

Product	Tonnage	Percent
Canned[b]	42,000	45
Dried	23,000	25
Fresh	15,000	16
Frozen	10,000	11
Baby food	3,000	3
Total	*93,000*	*100*

[a] Ref. 7.
[b] Includes canned apricots, nectar, and purée.

Economics

The continuing decline in the production and consumption of apricots in the United States since 1944 can be expected to continue unless a number of influences are introduced, eliminated, increased, or reduced.

The production decline has, in recent years, been aided and abetted by increasing imports from other apricot producing countries. Both Turkey, with dried apricots, and Spain, with canned apricots, have increased their shipments significantly in the last six years. Turkey now ships more dried apricots to the United States than the United States produces and Turkey ships less than a third of its annual output to the United States. Spain, and now Israel, have carved out significant shares of the American institutional–food service market, supplied mostly in #10 (7 lb) cans.

While this competition, growing primarily from lower priced products, has seriously hurt the California apricot industry, the paradox is that it has also helped the industry. Although U.S. growers and marketers have not been able to slow the decline in the consumption of apricots, foreign competition has. The influx of lower priced product has increased the consumption of apricots at those price levels. This has indirectly benefited the California grower by sampling new users and previous users who dropped out because of increasing retail prices for California fruit. In the 18-yr period from 1970 to 1988 per capita consumption of apricots declined from 1.3 lb/person to 0.85 lb. Although this is a drastic drop in usage, it would have been significantly worse had not foreign imports increased in that same period from 4,753 fresh tons to 45,155 tons—a tenfold growth—which represents one-third of the total tonnage available in the United States in 1988. Over 90% of the imported fruits were in dried apricots, with Turkey responsible for the great majority.

Several other factors have created this decline in the production and consumption of California apricots:

1. Pricing changes that occurred in 1972 reversed a decades-old history of canned apricots being priced below that of canned peaches and pears. These latter two products have had a competitive price advantage ever since.

2. The canned fruits and vegetables decline began in 1974. It was still in effect in 1987, and this, along with the price change turnaround, exacerbated the downturn in apricot consumption and, subse-

quently, production. Until 1984, when Turkey exported 8,500 tons of dried apricots (51,000 fresh tons), the principal downturn of U.S. apricots came from the major product, canned apricots. This decline was worsened by the demise of a number of small California canners whose processing volume was not absorbed by remaining canners. This situation became cataclysmic in the early 1980s with the closing of two large canners, California Canners and Growers and Sacramento Foods. Their capacity closed down about 25–30% of apricot canning production. This volume was not replaced by the surviving factors. Today there are only four major canners in business where 19 canners were operating 20 years ago.

3. Urbanization has created a demand for prospective realty acreage that has increased the value of farmland in certain key apricot growing areas in California. Farmers have sold out, or sold what land they required, to meet their financial needs. This was especially damaging to the dried apricot business with the dynamic urban growth of San Jose and Santa Clara.

4. Undependable supply brought on by some alternate bearing trees and smaller production volume have combined to make consistent supply a problem every few years. This has led to a lack of steady supply for export markets, loss of retail distribution, and a reluctance to commit foodservice menus to apricots.

All of this has fueled a price spiral that reduces the customer base each year.

FUTURE DEVELOPMENT

Short term, this industry needs a longer season to create more exposure for fresh apricots as a sampling device. The fresh market is only six weeks long at enough volume to maintain minimal grocer distribution. Other popular summer stone fruits have three to four times this seasonal market window. The industry is beginning to look at varieties that can be grown earlier and later than the May through June period.

Improved varieties are needed to provide the desirable characteristics necessary to incite growers to plant more trees and processors to increase their processing capacity for apricots. Among the characteristics being sought through genetic experimentation are:

1. A series of varieties that will mature every 10 days from May 1 to mid-September.

2. Apricots that will grow with a wider range of adaptability in regard to climate conditions, soil type, etc.

3. Apricots with firmer flesh texture, storage and shipping ability for mechanization and handling and that retain a good flavor density.

4. Multipurpose apricots with higher soluble solids that can be shipped fresh, canned, or dried and are flavorful.

This last characteristic is exemplified by the Patterson apricot variety. While not a new apricot, until recently it has not been a grower favorite. Pattersons first became a variety of significant volume in 1988. The Patterson is an all-around fruit, acceptable as a fresh, canned, dried, and frozen product. It is more dependable year to year for a consistent yield, and that yield is considerably greater than most of the apricots grown in volume. When trees are pruned for size the Patterson is generally a larger fruit than most commercial varieties, making it more acceptable to the consumer and the grocer.

The California apricot industry must also look to advancements in pomological and cultural experiments and practices overseas. Countries in Europe and elsewhere have been working diligently to improve the quality and quantity of the apricot for a number of years.

With improvements from within the total industry, U.S. consumers and their purchasing agent, the grocer, will together increase demand, and thereby production, for apricots. The apricot itself contains the elements for successful growth—a highly nutritious portable package of deliciousness that still evokes a nostalgic recollection from a large segment of consumers: "Apricots, oh, that's my favorite fruit!"

BIBLIOGRAPHY

1. C. O. Hesse, "Apricot Culture in California," *California Agricultural Experiment Station, Extension Service Circular* **412,** 1952.

2. FAO Yearbook Production Vol. 42, Rome, Italy, 1988.

3. California Agricultural Statistics Service, California Fruit & Nut Acreage, Sacramento, CA, June, 1989.

General References

C. H. Bailey and L. F. Hough, "Apricots," in J. Janick and J. N. Moore, eds., *Advances in Fruit Breeding*, Purdue University Press, West Lafayette, Ind., 1975, pp. 367–383.

D. S. Brown, "Effects of Irrigation on Flower Bud Development and Fruiting in the Apricot," *Proceedings of the American Society for Horticulture Science* **61,** 119–124 (1953).

Marketing California Apricots 1988, California Federal–State Market News Service, Sacramento, Calif., 1989.

W. H. Chandler, M. H. Kimball, G. L. Philp, W. P. Tufts, and G. P. Weldon, "Chilling Requirements for Opening of Buds on Deciduous Orchard Trees and Some other Plants in California," *University of California Agricultural Experiment Station Bulletin* **611,** 1937.

F. J. Chittenden, ed., "Apricot," *The Royal Horticultural Society Dictionary of Gardening,* Vol. 1, Clarendon Press, Oxford, UK, 1956, pp. 155–156.

F. M. Coe, "Apricot Varieties," *Utah Agricultural Experiment Station Bulletin* **251,** 1934.

P. Crossa-Raynaud and J. M. Andergon, "Apricot Rootstocks," in R. C. Rom and R. F. Carlson, eds., *Rootstocks for Fruit Crops,* John Wiley & Sons, Inc., New York, 1987, pp. 295–320.

L. H. Day, "Rootstocks for Stone Fruits," *California Agricultural Experiment Station, Extension Service Bulletin* **736,** 1953.

U. P. Hedrick, *Cyclopedia of Hardy Fruits.* Macmillan, New York, 1922.

O. Lilleland and J. G. Brown, "Fruit Thinning. I. The Relationship of Fruit Size in Unthinned Apricot Trees to Crop and

Season," *Proceedings of the American Society for Horticultural Science* **37,** 165–172 (1939).

S. E. McGregor, *Insect Pollination of Cultivated Crop Plants,* Agricultural Handbook No. **496,** Agricultural Research Service, U.S. Government Printing Office, Washington, D.C., 1976.

J. R. Magness and E. Bostelsmann, "How Fruit Came to America," *The National Geographic Magazine,* **100**(3), 325–377 (1951).

W. P. Tufts, "Pruning Deciduous Fruit Trees," *California Agricultural Extension Service Circular* **112,** 1939.

M. N. Westwood, *Temperate-Zone Pomology,* W. H. Freeman & Co., New York, 1978.

N. I. Vavilov, "The Mountainous Districts as Home of Agriculture," trans. by N. I. Vavilov, "Studies on the Origin of Cultivated Plants" (Russian and English), *Bul. Appl. Bot. Plant Brdg.* **16**(2), 218–220 (1926).

N. I. Vavilov, "Phytogeographic Basis of Plant Breeding," trans. by K. S. Chester, "The Origin, Variation, Immunity and Breeding of Cultivated Plants," *Chronica Botanica* **13**(1/6), 13–54 (1951).

S. Yunwei, H. Puchao, L. Jiarui, X. Jianguo, and X. Haibo, "The Germplasm Resources of Fruit Trees in Northwest China," in *International Symposium on Horticultural Germplasm, Cultivated and Wild, (Abstracts),* International Academic Publishers, Beijing, China, 1988, p. 90.

Gene Stokes
California Apricot Advisory
Board
Walnut Creek, California

S. M. Southwick
University of California
Davis, California

Sun Garden Packing Company
San Jose, California

Mariani Packing Company
San Jose, California

J. M. Smucker Company
Salinas, California

Zaiger's Genetics
Modesto, California

AQUACULTURE

The relative role of aquaculture in meeting human food needs can be appreciated through a comparison of the quantitative and qualitative differences between land and water coverage of the earth's surface. Roughly three-quarters of the earth is covered by saline waters of open oceans, seas, and estuaries. The other one-fourth is land surface, with only a small percentage of the land covered with freshwaters. Of the total estimated water volume of the biosphere (1.4 billion cubic meters), the oceans contribute more than 97% of the total, with water associated with continents the remainder (ground water, 2.08%; freshwater lakes, 0.009%; saline lakes, 0.008%; and rivers, 0.00009%).

Historically, humans have concentrated along the edges between water and land (ecotones). The diversity of plant and animal communities that has evolved in these ecotones and the high organic productivity of such communities all made the edge of aquatic zones a place of easy exploitation for early human societies in meeting food and fiber needs. Although intertidal edges along rocky shores and sandy beaches tend to be the most inhospitable of ocean aquatic environments, adjoining shallow waters and associated wetlands of estuaries are the most productive environments. Nutrients in incoming freshwaters are mixed with ocean waters to provide wide salinity gradients in space and time. The sun's energy is utilized to the maximum in these relatively shallow waters by both microscopic plants (phytoplankton) and large plants (macrophytes) such as sea grasses, mangroves, and seaweeds. Similarly in freshwater habitats, river edges, swamps, marshes, and submerged weed beds along lake edges characterize these highly organically productive areas. Thus it is not surprising that human inhabitants have concentrated along these water edges where human needs for food, water, transportation, and aesthetic values are readily available.

In the most recent millennia, despite such attractions to the water's edge to human populations, the contribution of aquaculture to overall human food consumption is minor compared to land-based agriculture. On a weight basis, aquatic products lag behind agriculture and forest-derived products (rough ratio of 8 : 1.5 : 0.5) in meeting world food needs. Although there is no precise estimate of the contribution of aquaculture to world food production, informed guesses suggest that 10 to 12% of total world aquatic food production is from aquaculture. Such a low contribution today is not unexpected. When technologies such as irrigation and pumping ground waters moved agriculture away from their historic flood-plain locations occupied by early civilizations, there was relatively less competition for the same space from other competing human activities as compared to aquaculture. Historically there was a relatively large amount of dry land available for exploitation. Even if aquaculture technologies had advanced as fast as those of agriculture, there would be more competition by aquaculture for space because of the site-specific nature of aquaculture, as discussed later. Undoubtedly one of the major differences between food production from land and from water is this more highly site-specific nature of aquaculture, arising from the need for water of suitable temperature, of sufficient volume, and of proper chemical composition for cultured organisms. Another important difference between agriculture and aquaculture is that agriculture usually grows stationary objects (plants) on a two-dimensional surface (land), whereas aquaculture usually involves motile organisms (fish, crustaceans) living in a three-dimensional fluid (water). Of course, there are important exceptions to such a broad categorization (livestock and birds in agriculture, and molluscs and seaweeds in aquaculture). Despite an innate limitation in the quantity of food produced, aquatic food production has always had a major role in the quality of the food provided. Fish, crustaceans, and bivalves are a major source of high-quality animal protein which is very important in providing essential amino acids in human nutrition, especially where subsistence husbandries are the major source of human food.

DEFINITION

Both very inclusive and very restricted definitions of aquaculture are now used in the literature. A recent author defines aquaculture as the manipulation of the aquatic environments, either natural or artificial, in order to augment the production of aquatic organisms, such as fish, crustaceans, molluscs, and algae, which are useful to humans (1). Webster's definitions illustrate the older, narrower concepts: (a) the art of cultivating the natural produce of water; (b) the raising or fattening of fish in enclosed ponds; and (c) a form of hydroponics (2). The *Encyclopaedia Britannica* treats aquaculture as fish culture only, with the aquaculture of bivalves and crustaceans relegated to other sections of the encyclopedia (3). Its definition is: "In the broad sense fish culture can be considered a branch of animal husbandry, if the latter term is used to include both the rearing of domestic animals and the harvesting and management of wild game." The maintenance of fish indoors in aquariums was not treated as a form of aquaculture since the activity was considered purely recreational.

The past several decades have experienced an explosive expansion in aquaculture worldwide as a means of increasing human food supplies. Aquaculture has been slowly supplementing traditional capture fisheries, which have been suffering declines in catch from overfishing, natural declines, habitat deterioration, or a combination of all three (4). Recent writers believe that any single comprehensive treatise on aquaculture as broadly defined is now impossible (5) and specialized works on aquaculture are reflecting both broadened and narrowed approaches (6, 7). Most recent publications tend to restrict the meaning of aquaculture only to those culturing activities whose products are to be used directly as human food. Such definitions exclude a wide range of husbandry that produces products which indirectly or directly contribute to human food resources (aquarium industry, public and private fee-fishing recreational lakes and ponds, recreational fishing in artificially constructed private and public water bodies, juvenile fish released into public waters for enhancing both commercial and recreational harvest, and private ocean-ranching) (8). The magnitude of some of these operations is impressive (9). In the United States alone, based on preliminary studies, perhaps 200–300 million dollars is generated by the operators of recreational farm ponds (150,000), fish-out operations (4,000), and commercial fish farms (5,000), with the total aquabusiness associated with the products about one billion dollars as an educated guess.

Governmental and scientific disciplines view aquaculture differently, depending on the commonly accepted definition held by the agency or discipline. Thus the Food and Agriculture Organization of the United Nations has departments of Agriculture and of Fisheries, with aquaculture treated under fisheries. In the United States, commercial aquaculture for food is located in the Department of Agriculture, while recreational fisheries aquaculture is under the Bureau of Sport fisheries and Wildlife (Department of Interior), and the culture of juvenile anadromous Pacific salmon is within the National Marine fisheries Service, a unit under the National Oceanic and Atmospheric Administration (Department of Commerce). Thus aquaculture scientists are now defining their field as narrowly as possible to avoid ambiguities, a laudable effort but one which is difficult since all taxonomies tend to have fuzzy edges. For example, no single simple definition could include the lake fish farming of Poland and of other central European countries. This aquatic food production system involves intensive stocking of juvenile fish into highly managed natural and artificial bodies of water harvested by both recreational and commercial fisheries operating over the entire water body (8). Even the distinctions between agriculture (producing food on land) and aquaculture (producing food from water) overlap where land and water species are integral parts of a single system (rice–fish culture) (10, 11).

HISTORY

The simplest forms of aquatic husbandry extant appear to have come directly from the simplest forms of early fishing, presumably the hand gathering of aquatic products from natural bodies of water. Many varieties of plants and animals are known to colonize heavily or to be attracted to submerged objects such as logs, brush, stones, and rock outcrops or to utilize such objects temporarily during some stage of the life of the plant or animal. Artificially duplicating or manipulating structures known to be used by plants or animals in shallow-water areas, such as intertidal zones of coastal areas, would significantly increase the concentration of aquatic species available for hand capture. A reasonable hypothesis can be made that early humans living along the shallow margins of seas, lakes, and streams engaged in simple manipulation of natural objects to increase the amount of useful products that could be gathered by hand.

Such primitive (simple) techniques continue to be the basis for successful seaweed and mollusc culture systems in which hard objects are artificially placed in intertidal areas (stone culture) to provide attachment sites for desired plants and animals. Many primitive aquacultural technologies still in productive use in many parts of the world must have developed from early human exploitation of such a simple technology. A second pathway may have followed from observing that waters receding from any natural shallow basin, such as riverine pools, tidal pools, oxbow lakes along large streams, and deep areas of wetlands, marshes, and swamps, leave large assemblages of aquatic species stranded after normal events (eg, daily tidal recessions, episodic floods, and recession of stream flows from seasonal hydrological events). Fish concentrated by such natural processes undoubtedly would attract hunter-gatherer peoples, and certainly would be a supplemental source of food for agriculturists living adjacent to streams and lakes. Such "flood fisheries" are still important today (11). A pioneer American limnologist (Stephen Forbes) reported observations on the relative fertility of oxbow lakes along the Illinois River after recession of spring flood water, and an associated abundance of trapped fish. Herodotus, a Greek historian, wrote in 5c BC, about the flood fisheries of the Nile, and the same Nile that irrigated the fields deposited upon them, and its innudation, thousands of fish in shallow pools; even the same net with which the peasant fished during the day was used around his head at night as a double protection against mosquitoes (12).

It would be a short step for early agriculturists to deliberately impound water in natural basins where only small openings allowed trapped waters back into a parent stream, lake, or ocean. This technique of culturing fish can be observed today in Thailand and other southeast Asian countries where systems of aquaculture have originated from similar ancient methods of fish trapping. During monsoon flood seasons, fish from headwater areas follow flood waters down irrigation canals and ultimately into rice fields. Fish that enter the deeper water areas of the rice field grow well and subsequently are caught for food after the rich harvest. An extension of this system was developed by constructing small ponds fitted with narrow entrance channels along dikes and canals. These ponds have bottoms dug lower than the adjacent stream or canal. This configuration attracts fish during floods. The narrow channels are easily blocked and the fish seined at leisure after several months of growth. Obviously it would be a logical step to artificially fertilize the pond or to feed the trapped fish, which, of course, is now the basis for more complicated (modern and intensive) aquaculture enterprises. Ancient royal Hawaiian saline-water fish ponds followed this technique. Excellent descriptions of these ponds and of other systems have been published (13).

A probable evolutionary process of aquaculture is suggested by the history of milkfish (Chanos chanos) culture in Java. Apparently pond culture was an outgrowth of netting adult milkfish in mangrove swamps, followed by replacing the swamps with dikes and control structures and stocking the ponds with fry caught along coastal beaches. The practice is reportedly at least 600 years old, and 180,000 such ponds (tambahs) of 2–2.5-ha average size are currently in operation as family businesses (9). A similar evolutionary process has been documented in the development of the valli or fish reservoirs of the Comacchio region of Italy, located on the northwestern coast of the Adriatric Sea.

Any detailed tracing of the early prehistory of aquaculture is hampered by poor preservation of artifacts for archeological recovery and study. Since aquaculture appears intimately related to floodplains, structures and artifacts are readily swept away or buried under sedimentation from catastrophic floods or other events producing massive floodplain debris. The considerable early association by humans with aquatic habitats as a major source for gathering food can be inferred from the generalized lists of tools tabulated by archaeologists for both Old Stone Age and New Stone Age cultures. Along with axes, scrapers, needles, awls, etc, there are always fish hooks. Shells of mollusks also appear ubiquitously. These objects usually are classified as articles of adornment, but obviously they are also evidence for a massive reliance on shellfish for food (in kitchen middens along coastal Netherlands and in American mound cultures of the lower Mississippi River). Fish culture, like other forms of husbandry, can flourish and expand only under settled conditions. Earliest records, presumably of some sort of formal aquaculture, come from paintings, mosaics, bas-reliefs, and carvings on tombs and temples associated with ancient riverline-based civilizations (Egypt, Babylonia). It is not surprising, therefore, that the earliest extant written records on aquaculture are Chinese. Archeological records dating back ca 15,000–20,000 yr provide evidence that grass carp (Ctenopharyngoden idella) were regularly eaten by Chinese mountain cave dwellers, and that skull bones of the species were used as primitive ornaments. Artifacts dated to the next 5,000–10,000 yr commonly include hooks and spears made of bones and net weights of stone and clay. There are Chinese records estimated to have been written ca 1000 BC on the behavior and growth of fish in ponds. About 400 BC a text on carp culture for food was written by Fan Li documenting the ancient tradition of pond culture in China. There are copies in China and in the British Museum and a translation at the University of Maryland.

Whether aquaculture originated first in China or in other areas of ancient civilization in southeast Asia remains an intriguing question. With the worldwide increase in interest in aquaculture it is possible that examination of the archeological evidence is already underway for clues to such historical questions. The recent discovery and excavation of earth ponds along the shore of a former lake in Poland (estimated to have been constructed around 5000 BC) suggest widespread and independent development of aquaculture practices.

Important human activities are reflected in religious motifs. Thus the human fish-egg sculptures of goddesses of neolithic societies located along the middle Danube River and its tributaries as early as 7000 BC reflect early capture fisheries but also, conceivably, primitive aquaculture (14). The aquacultural histories of 31 countries are now available (8). It may be found that aquaculture rivals agriculture in antiquity.

AQUACULTURE SPECIES

A large variety of aquatic animals and plants are cultured for food throughout the world. A basic source for such information is the yearly statistics on the nominal catches of aquatic species published by the Food and Agricultural Organization (FAO) of the United Nations (15). Estimates lump catches made for all purposes (commercial, industrial, and subsistence) but do not include recreational fisheries. The statistics lump production from mariculture (salt water), aquaculture (freshwater), and other kinds of culture producing food and fiber products. Thus only a very rough idea of the amount and kinds of organisms cultured for food throughout the world can be made by examining FAO statistics. FAO's most inclusive statistical divisions are freshwater fishes, diadromous fishes (which live in freshwater and reproduce in saline waters, or vice versa), marine fishes, crustaceans, molluscs, aquatic mammals, miscellaneous aquatic animals, miscellaneous aquatic animal products, and aquatic plants. Systematically, these aquatic organisms are listed as Piscis (fishes, which include jawless chordates such as the lampreys; sharks, skates, and rays; and the bony fishes), Crustacea (shrimps, prawns, spiny lobsters, and crabs), Mollusca (abalone, clams, mussels, and oysters), Mammalia (whales and dolphins), Amphibia (frogs, turtles, and crocodiles), Protozoa (a heterogeneous grouping of mainly marine invertebrates such as sponges, annelid worms, and sea urchins), and Algae (seaweeds, algae, and other aquatic plants). Production from capture fisheries and aquaculture combined is reported as a total of 840 species items, most of these being fish groups.

The diversity of fishes [at least 21,000 species (16)] potentially and actually used in aquaculture is reflected in 511 categories (species, genera, or other groupings) of items representing 103 different families of fishes found in the FAO statistics (15). The vast majority of these families are marine and brackish water in distribution except the family Cyprinidae (carps, barbels, and minnows), which are freshwater in distribution. Cyprinids, along with other freshwater groups (characids and the cichlids), represent three of the four most abundant families of living fishes (2,050, 850, and 700 estimated identified species) (17). This is a large percentage of the world's known fish species, which is variously estimated at 15,000–40,000 species, with most recent estimates ranging from 21,000 to 25,000 species (16–18). This relatively high diversity of cyprinids potentially available for culture in freshwater is reflected by the 35 species of cyprinids in a list of 69 total species reported cultured in the Indo-Pacific region in 1962 (19). Over half of the more than 500 freshwater fish species of China are used as food, with the carps providing the majority of the 50 types most important economically (20). There is also a wide variety of brackish and marine species of fish either cultured or potentially available for aquaculture throughout the world (18).

The species selected for aquaculture in different parts of the world vary widely because of economic and cultural factors. In the United States, where there is an abundant animal protein supply from beef, poultry, chicken, and hog production, only six categories provide the bulk of production (Table 1) (21). Cyprinids, either native or exotic, are cultured primarily as bait for hook-and-line anglers and for the aquarium trade. Not only do most American food habits consider minnows and suckers too bony, but they are generally viewed as trash fish. Cyprinids such as the carps, which tolerate enriched environments generally unsuitable to native species, reinforce such attitudes. Cyprinids also have bones attached to their ribs which project into the lateral muscles, making them objectionable to eat for some people.

Not every species of fish, even if highly desirable as a food fish, possesses the biological characteristics suitable for its cultivation. A list of desirable characteristics for a good culture species has been compiled (Table 2) (18). One such group is the *Tilapia spp.* (family Cichlidae), which live primarily in freshwater and brackish water habitats and are widely distributed and cultivated throughout the world. Characteristics that make these species suitable to aquaculture are excellent growth rates on low protein diets, whether cropping natural aquatic production or receiving supplementary food; toleration of wide ranges in environmental conditions; strong resistance to disease; amenability to handling and captivity; a short generation time; easy breeding in captivity; wide human acceptance for food because of their high palatability; and a long history of use by humans from capture fisheries (7).

The characteristics that make a fish a desirable species for culture also apply to other aquatic animals such as crustaceans and molluscs. These latter forms, however, generally have more complex life histories than fish. Trout and salmon (salmonids) have relatively simple life histories. Large yolky eggs can be stripped readily from salmonids and easily fertilized artificially. On hatching, young salmonids closely resemble adults and essentially eat a protein diet, although the types of food eaten change with the increasing size of the fish. Thus artificial pellets that need only to be changed in size can be manufactured for feeding all stages of salmonids. In contrast, crustaceans and bivalves have small nonyolky eggs that cannot be stripped from the animals; they must be induced to spawn. The eggs in turn produce only minute-sized young that metamorphose through a number of juvenile stages that have different nutritional requirements. Only in the older stages can the organisms be readily fed on a standard manufactured artificial food. Insuring a steady supply of high quality "seed" (juvenile organisms) for crustacean and mollusc grow-out operations continues to be a major research and development area for promoting crustacean and molluscan aquaculture success.

Purely local food preferences, income levels, availability of water and land, and the need for preserving endangered species have stimulated investigation of species that possess only some of the desirable features required for a commercially usable aquaculture species. For example, sturgeon are cultured to replenish stocks captured for caviar production in the USSR, whereas in the western United States, sturgeon culture is now being studied as a potential commercial food source and to provide juveniles to maintain wild stocks depleted by increased recreational fishing. Recently alligator farming has developed because alligators possess all of the desirable characteristics for an aquaculture species, plus the advantage that virtually full utilization can be made of the skin and carcass. Expansion in industrialized societies of home aquariums could easily translate into backyard aquaculture, as widely practiced

Table 1. Estimated United States Private Aquaculture Production, 1985[a]

Category	Metric Tons
Catfish[b]	123,344
Salmon[c]	38,320
Crawfish	29,545
Trout	23,129
Baitfish	11,276
Oysters	10,215
Clams	722
Mussels	422
Marine shrimp	200
Freshwater prawns	121
Other species[d]	6,364

[a] Ref. 21.

[b] Includes catfish not sold to processors.

[c] Includes returns from government and nonprofit hatcheries.

[d] Includes species such as sturgeon and abalone. Data shown are live weight for consumption, except for oysters, clams, and mussels, which are meat weight.

Table 2. Desirable Characteristics of Successful Aquatic Animals Employed in Aquaculture[a]

Factor	Description and Examples
Reproductive habits	Fry readily available by harvesting wild stocks (milkfish (*Chanos chanos*): mullet (*Mugil* spp.); shrimp).
Eggs and larvae	Desirable species produce fewer and larger eggs generally and therefore have less delicate larvae [salmonids (trout and salmon); catfish]. Species that provide protection to their eggs generally are easier to culture (*Tilapia* spp. hatch eggs in their mouths; shrimp carry eggs on body).
Feeding habits	Species that feed low on the food chain (eg, on bacteria and algae) can be supplied with low-cost feeds, with the product sold profitably to humans existing on low-protein subsistence diets (eg, tilapias and milkfishes).
	Species that feed high on the food chain, requiring a high-quality protein feed of relatively high cost, are economically feasible primarily to supply restaurants and luxury products (trout, salmon, catfish).
	Bivalve molluscs feed low on the food chain and are also high prized by humans for food at both luxury and subsistence levels, making bivalve molluscs extremely desirable as aquaculture species.
	Polyculture (rearing of several species together) generally uses species that feed low on the food chain (carps, tilapias, mullets) with carnivorous (predatory) species added for biological management purposes.
Adaptability to crowding	Crowding of animals in aquatic systems produces problems that become exacerbated in aquaculture systems. Accumulations of waste products can induce stress, inhibit growth, and cause outbreaks of disease.
	Normally territorial channel catfish (*Icatalurus punctatus*) lose territorial behavior under culture but continue to grow rapidly with adequate food. Trout can be grown under extremely high densities if water flows are ample, temperatures cool, and oxygen content of the water high.

[a] Ref. 18.

in Oriental countries (Indonesia, China, Japan, etc). Even the Romans developed and argued over the relative functions of their *dulces* (freshwater ponds kept for fish to be sold for profit) and the *salsae* or *maritimae* (sea water ponds kept by rich nobles to provide recreation).

WORLD PRODUCTION

Aquaculture's contribution to world food supplies can be inferred from the nominal world catches of aquatic species in the environmental categories in which aquaculture makes significant contributions. Such analyses suggest that in 1985 aquaculture production represented over 10% of world food supplies (Table 3). Most of this production occurred on the Asian continent or closely adjoining regions (Table 4). Aquaculture contributions to world food supplies increased in all major categories over the 1975–1985 decade, with crustaceans exhibiting nearly an order of magnitude increase in production (Table 5). Identifying precisely which species are the major contributors worldwide requires a knowledge of the error and bias in published statistics which is beyond the range of this article. Relative importance can be assigned, however, by comparing the world catch of those species that have known aquacultural importance. In 1985 the FAO statistics reported nominal world catches as follows (in millions of metric tons, MMT):

Pacific cupped oyster, 0.57

Tilapias and other cichlids, 0.43

Common carp, 0.41

Pink salmon, 0.30

American cupped oyster, 0.27

Chum salmon, 0.27

Milkfish, 0.24

Bombay duck, 0.13

Varying portions of these production figures are not from aquaculture but from capture fisheries.

Precise figures for comparing aquaculture production by country are unavailable; however, China, the USSR, and Japan are usually considered the three leading aquaculture producers. One of the few studies reporting on worldwide aquaculture production by country showed

Table 3. Reported Total Nominal Catches by Category and Production Areas, 1985[a]

Category	Millions of Metric Tons
World	
catches in inland waters	11
catches in marine fishing areas	17
Total	*86*
Freshwater fishes	8.4
Asian inland fishng areas	6.7
Marine fishing areas, molluscs	5.8
Inland fishing waters	10.1

[a] Ref. 15.

Table 4. Revised Estimates of Aquaculture Production in 1985 by FAO (Thousands of Metric Tons)[a]

FAO Region	Finfishes	Crustaceans	Molluscs	Seaweed	Other	*Total*
Africa	61	0.1	0.4	0.0	0.0	*61*
America, North	198	34	161	0.2	0.0	*393*
America, South	28	33	2	5	0.0	*68*
Asia and Oceania	3794	199	2140	2768	28	*8929*
Europe	341	0.3	495	4	0.0	*841*
USSR	296	0.0	0.0	0.0	0.0	*296*
Totals	*4718*	*266*	*2799*	*2777*	*28*	*10587*

[a] Ref. 21.

marked differences in the nonfish species groups cultured by various countries (Table 6): seaweeds were of great importance in aquaculture production in Asian countries; Japan leads in oyster culture; Spain in mussels; the Republic of Korea in clams; Japan in scallops; and shrimp and prawns are major contributions in India, Indonesia, and Thailand.

The leading freshwater aquatic species produced throughout the world in 1985 was the common carp (*Cyprinus carpio*) (411,000 metric tons, MT), with eight other types of cyprinids contributing from 2,000 to 57,000 MT (15). Over 100,000 MT of production could not be assigned by FAO to any category except Cyprinidae, so all the figures represent minimal production. The USSR produced over half of the reported carp catch (215,000 MT), with European countries contributing between 70,000 and 100,000 MT from six major producers. The milkfish (*Chanos chanos*) and the Bombay duck (*Harpodon nehereus*) are important freshwater species cultured principally in the Philippines, India, and Indonesia. In 1985 *Sarotherodon* and *Tilapia* accounted for over half of the identified Cichlidae reported (15).

Trout (*Salmo* spp.) are an important aquaculture species throughout Europe and North America (Table 7). The historical production of Atlantic salmon (*S. salar*) from open-ocean and coastal fisheries is rapidly being augmented by aquaculture using net-pens and farming of diked and netted fjords and embayments. Similarly, open-ocean and coastal fisheries for Pacific salmon (*Oncorhynchus* spp.) based on native stocks have been enhanced by juveniles (smolts) released by government and private agencies in salmon-ranching operations (Table 8). Pink and chum salmon are the principal species ranched by Japan and the USSR. The importance of these species for salmon ranching is explained by Japanese reports that one unit weight of juvenile fish (smolts) released into the north Pacific ocean yields 80 unit weights of adult chum salmon at harvest.

The United States, although fifth in world production of aquatic food products in 1985 (4.8 MMT), produces only a relatively small volume of aquaculture products, with catfish by far the most important species by weight (Table 1). Salmon and oysters are equally important by value because of the relatively higher price received per unit. A sizeable portion of the salmon and trout production in the United States is released for enhancing recreational fisheries. The catfish aquaculture industry in the United States has shown the greatest growth, expanding by an order of magnitude during the 1975–1985 decade (Table 9).

Other locally more important cultured fish species and their production in thousands of metric tons in 1985 include *Abramis brama*, 57; *Puntius (Barbus)* spp., 47; *Rutilus* spp., 27; *Rastrinrobala hasselti*, 26; and *Crassius auratus*, 19. All are freshwater species of the family Cyprinidae.

Italy has had a major increase in production of trout, while Norway, through a national effort of promoting the rearing of Atlantic salmon in fjords, has become a model for integrated high-technology operations (Table 7). An increase in Italian trout production since 1970 was primarily due to the allocation of large amounts of spring water for aquaculture. The largest farm has one spring of 4,000 L/s (63,360 gal/min), with a total spring flow for all the operations of 111,000 gal/min of water at the optimum temperature for trout (54–55°F). These Italian site-specific conditions duplicated the growth seen in the U.S. trout industry, 90% of which is located in the flood plain of the Snake River of Idaho, where large volumes of spring water (14.4°C, 58°F) emerge from the base of a large basalt formation along the banks of the river. The largest single spring produces 920 L/s (14,500 gal/min) of water.

AQUACULTURE SYSTEMS

Aquaculturists utilize a wide variety of containers for growing fish or other aquatic organisms in addition to a diversity of management practices in reaching culture objectives. Seven general types of culturing techniques (17) are listed below. Those considered primitive are marked with an asterisk:

1. * Transplanting organisms from poor to better growing grounds, the least intensive form of aquaculture.

2. Transplanting organisms either as wild or hatchery stocks to exotic places (rainbow trout and Chinook salmon to New Zealand from North America; coho

Table 5. Increase in Estimated World Aquaculture Production of Finfish and Crustaceans (Thousands of Metric Tons)[a]

Year	Finfishes	Crustaceans	Molluscs
1975	2629	30	1961
1980	3207	75	3300
1985	4718	266	2799

[a] Ref. 21.

Table 6. Estimated World Production of Invertebrates and Seaweed through Aquaculture, 1975

Category	Country	Metric Tons	Category	Country	Metric Tons
Shrimps and prawns	India	4,000	Mussels	Spain	160,000
	Indonesia	4,000		Netherlands	100,000
	Thailand	3,300		Italy	30,000
	Japan	2,800		France	17,000
	Ecuador	900		Germany, FRG	14,000
	Taiwan	500		Korea, ROC	5,600
	Singapore	100		Chile	1,300
	Korea, ROC	30		Yugoslavia	300
Oysters	Japan	230,000		Philippines	200
	United States	129,000		New Zealand	150
	France	71,000		Tunisia	60
	Korea, ROC	56,000	Clams	Korea, ROC	25,000
	Mexico	45,000		Taiwan	14,000
	Thailand	23,000		Philippines	33
	Taiwan	13,000	Scallops	Japan	63,000
	Australia	9,200			
	Canada	5,100	Cockles and	Malaysia	28,000
	UK	3,000	other mulluscs	Taiwan	1,200
	Spain	2,300		Korea, ROC	700
	Netherlands	1,500		Philippines	11
	Chile	900	Seaweeds	Japan	503,000
	Philippines	800		China	300,000
	New Zealand	700		Korea, ROC	245,000
	Senegal	200		Taiwan	7,000

[a] Ref. 22.

and rainbow trout to the Great Lakes of the United States from the Pacific coast, carp from China into central Europe during the Middle Ages; Japanese oysters to many parts of the world).

3. * Inducing fish and invertebrates to enter special enclosures where they are trapped and held until ready for harvest.

4. * Culturing trapped fish as above and fertilizing water and/or installing devices to control water flows in and out of an enclosure. (This is a widespread technique used for culturing milkfish and shrimp in southeast Asian countries.)

5. * Constructing artificial enclosures that exclude wild fish. Fertilization of water to improve the level of natural food production along with controlled stocking. (This is the traditional subsistence type

aquaculture in China which evolved into the sophisticated systems now being widely studied and adapted to all parts of the world.)

6. Heavy supplementary feeding of animals cultured in ponds, in staked or floating cages, or in containers made of concrete, plastic, wood, or waste materials (catfish farming in the United States and Europe; trout farming in the United States, Europe, and Japan; common carp culture worldwide; Atlantic salmon pen culture in Norway; production of juvenile salmon reared to migratory size for release to grow in open ocean areas; salmon ranching in the USSR, Japan, and Alaska).

7. * Suspending objects in water from rafts or attaching them to the bottom (mussels raised on ropes suspended from rafts in the Galician bays of Spain

Table 7. Trout Production by Country Reported to FAO, 1985 (Thousands of Metric Tons)[a]

Scientific Name	Common Name	Nominal World Production	Leading Country	Nominal Production
Salmo salar	Atlantic salmon	48	Norway	31
			Scotland	7
Oncorhynchus mykiss (Salmo gairdneri)	Rainbow trout	86	France	25
			Japan	17
			United States	0.8
			Canada	0.2
Salmo spp.	Trout	80	Italy	33
			Denmark	24
			Spain	17

[a] Ref. 15.

Table 8. Estimated Harvest from Salmon Ranching by Country, 1984[a,b]

Country	Estimated Harvest in Metric Tons
Japan	108,000
USSR	64,000
United States	21,000

[a] Ref. 4.
[b] Data courtesy W. J. McNeil.

and the inland seas of New Zealand, on stakes and poles in France, and recently in the inland seas of New Zealand; seaweed rope culture from rafts in protected bays and estuaries of mainland China, Japan, Taiwan, and the Philippines).

Of the seven systems listed, five are essentially primitive (simple) and indicate that successful aquaculturists heed the laws of thermodynamics even while drawing on the recent large accumulations of scientific data and advanced industrial technologies.

Current technical literature arranges the same culture systems into two broad types of culturing: intensive and extensive culture. Intensive culture involves keeping organisms under high densities and feeding formulated diets, often containing high levels of animal protein, derived from fish meats and meals. Japan, Europe, and the United States practice intensive aquaculture. Extensive culture involves growing of organisms in ponds on natural foods produced either by the inherent fertility of the water or by artificial fertilization. China, Africa, and the Far East practice extensive aquaculture and account for most of the world's aquaculture production. Intensive aquaculture is characteristic of developed countries where there are relatively good supplies of energy and no real shortage of protein, whereas extensive aquaculture is mostly practiced in areas suffering from protein shortage where the aquaculture products represent an important nutritional component in subsistence diets. The categories, however, overlap where pond systems have high degrees of aeration of the water and are coupled with high-quality plant or animal protein feeds, making such systems very intensive.

Pond Culture

The art and science of culturing aquatic species in earthen ponds has been organized in a variety of ways. One approach (11) lists 13 topics: water, soil, pond construction and management, pond management, fertilizers, supplementary feeding, brackish water, running water, rice fields, yields, genetics, disease, and public health. Similar categories are used in China to describe their approaches to improving aquaculture production. Covering all pond-culture topics in a single article is impossible but there are a few factors that generally determine the success and long-term viability of traditional pond culture systems for fish. These are cost of feed, rates of stocking, and the time and size of harvest of fish. Early systematic research on maximizing production in ponds (11) used curves showing the final mass (standing crop and/or carrying capacity) with increasing stocking rates in unfertilized, fertilized, and fertilized ponds with supplemental feeding (Fig. 1). Only with heavier rates of stocking can the maximum carrying capacity be attained. This general approach has been applied to different species, culture systems, and local conditions. Israeli pond culturists (23) at a research station at Dor investigated production techniques by studying short-period fish growth rates (1–2 weeks) according to the size (weight) of fish attained under different rearing regimes (unfertilized ponds, fertilized ponds, fertilized ponds with fish fed sorghum grain, and fertilized ponds with fish fed pellets) (Fig. 2). Each treatment showed an increasing average fish size up to a point where fish growth ceased, necessitating a more intensive treatment to increase fish growth and size. Fertilized ponds with supplementary feeding of protein-rich pellets produced fish growth of 10 g per fish per 1–2 week period, with individual fish averaging 1,000 g. Israel pond researchers using both single species and polyculture systems are reporting increasingly larger rates of production from experimental systems. In ponds stocked with common carp, silver carp, and grass carp, using daily application of liquid cow manure, production levels have reached 8 metric tons per year in experimental studies. A recent publication on Chinese polyculture lists two principal methods of pond management (20): (1) the stock, harvest, and restock system, and (2) a multigrade conveyer culture in which stocking and harvest are continuous processes based on growth rates of the species in the polyculture system. In both systems the frequency of harvest (removal) and rate of stocking is species dependent and modified by the size of fish required by local markets, the general fertility of the pond water, and the quantity and quality of the feed available. In the United States the federal government in assisting aquaculture development

Table 9. Estimated Production of Catfish in the United States[a]

Year	Ref. 21	Ref. 4
1975	9	7.3
1980	21	21
1983	62	62
1985	87 (123)[b]	

[a] Refs. 4, 21.
[b] Reference 4 lists 87 MT (Table 5) and 123 MT (Table 6) as reported from statistics compiled by the Department of Agriculture and Department of Commerce, respectively.

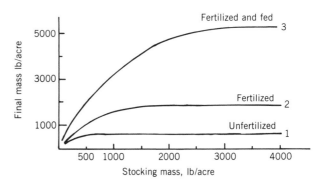

Figure 1. Curves showing relationship between stocking rate per unit area of pond and the final mass per acre under three conditions (11).

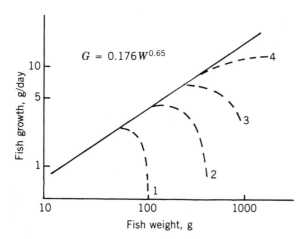

Figure 2. Relationship between the short-period fish growth (1–2 weeks) and fish weight for different treatments at the Fish Culture Research Station, Dor, Israel. Treatments: (1) unfertilized ponds and no supplementary feed; (2) fertilized ponds without supplementary feed; (3) fertilized ponds with feeding of sorghum grains; (4) fertilized ponds with feeding of protein-rich pellets (23).

has organized a computer storage and information retrieval system that succinctly outlines all the steps necessary to minimize the time for development of a viable commercial aquaculture scheme (24) (Fig. 3). Such a systematic approach is necessary for developing success in any food producing operation, whether on land or in water. Extensive systematic research is also carried out by federal, state, and private agencies where aquaculture is a part of managing commercial and recreational fisheries (25–29).

Modern Technologies

The bulk of aquaculture production worldwide comes from both intensive and extensive pond culture (28, 29). In industrially developed countries, however, possessing capital, energy, and high technology capabilities, venture cap-

ital has stimulated research and development of new culture systems designed for production species that bring high prices per unit volume of product (crustaceans such as shrimps and prawns and bivalves such as oysters and mussels) (30). These are species favored by the restaurant trade or species with strong appeal to limited groups of people based on cultural traditions and preferences (fish products sales during Lent or national holidays). Newer culture systems, particularly important in meeting speciality demands, have also exploited certain industrial technologies which provided materials and equipment which could be adapted to aquaculture. Plastics technology has been especially important in developing pumps, valves, pipes, fitting, etc that are light in weight, sturdy, flexible, easily plumbed, noncorrosive, and nontoxic to plant and animal life.

Such materials have been a major factor in the massive expansion of home aquariums for tropical marine species. The aquarium industry in the United States now generates a GNP rivaling the commercial fishing and aquaculture trade, and the industry is equally important in Western Europe and Japan. Plastics technology now allows the culture of aquatic species in self-contained systems using only a minimal amount of water in totally artificially maintained conditions at sites far removed from the species' natural habitats. Such systems have also used technologies developed by biological oceanographers and marine biologists based on artificial seawater solutions required for studies of phytoplankton, seaweed, marine invertebrate embryos, and the early-life-history stages of fish and invertebrates. Not only has there been an infrastructure of technical capabilities available for transfer to aquaculture, but stimulus for funding of new commercial aquaculture enterprises has come from economic and cultural factors. Rising incomes of many groups of people have allowed greater sale of higher priced species. Improved transportation and refrigeration have expanded markets locally and allowed aquabusinesses to transport perishable aquatic products readily worldwide. When market prices to consumers for traditional species rise excessively due to either overfishing or increasing operating

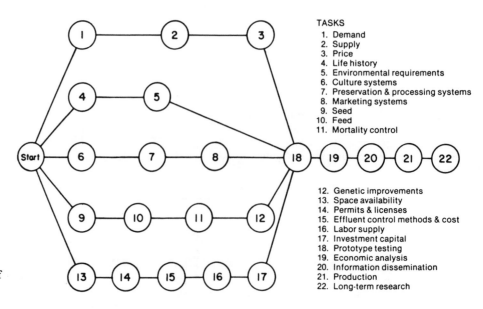

Figure 3. Scheme for development of an aquaculture project.

TASKS
1. Demand
2. Supply
3. Price
4. Life history
5. Environmental requirements
6. Culture systems
7. Preservation & processing systems
8. Marketing systems
9. Seed
10. Feed
11. Mortality control
12. Genetic improvements
13. Space availability
14. Permits & licenses
15. Effluent control methods & cost
16. Labor supply
17. Investment capital
18. Prototype testing
19. Economic analysis
20. Information dissemination
21. Production
22. Long-term research

costs, the potentials for substitute products through aquaculture using high-technology approaches will continue to stimulate research and development.

The literature on aquaculture has exploded in the past decade. Advancing knowledge and the status of the world aquaculture industry can be understood by reference to specialized books, monographs, journals, and government publications and to works listing these sources (5).

Extensive background on virtually all facets of aquaculture are found in the proceedings of two world aquaculture conferences organized by the FAO, Fisheries Division (May 1966, Rome, Italy, and June 1976, Kyoto, Japan) (28, 29). FAO has also sponsored regional aquaculture conferences. A number of recent national conferences are also sources of information and generally can be located through either extension or educational agencies of state or national agricultural departments. International conferences dealing with single species (eg, tilapias, carps) are now beginning to occur with increasing frequency.

LIMITATIONS TO AQUACULTURE

There are many constraints on the development of profitable aquaculture enterprises (Fig. 3) (24). A discussion of all factors limiting aquaculture is beyond the scope of this article, especially as every location and operation will have unique problems. There are, however, several major factors that are important to understanding limitations to aquaculture operations.

Space Availability

Aquaculture is a very site-specific activity. Major restrictions on siting an aquaculture venture are related to the availability of water and space. Water of suitable quantity and quality inherently constrains aquaculture to limited sites. For example, salmon, trout, and other related species suffer fatalities or have their growth reduced at the very low concentrations of molecular ammonia (NH_3) that come from metabolic products produced by the fish. Intensive systems historically have relied on the use of large volumes of water at suitable temperatures to flush away metabolic products and uneaten protein feeds. As noted previously, major trout producing enterprises in the world are associated with limited areas with suitable water (Po Valley in Italy; banks of the Snake River in Idaho). In industrially developed countries, most large spring-water flows have been appropriated, largely for agriculture or domestic water supply. This fact has stimulated intensive research or the initial stages of commercial development of intensive culture systems using high-technology wastewater treatment and recirculating technologies using limited water supplies. Capital costs and operating costs associated with the energy necessary to run pumps, in addition to feed and labor costs, limit such operations to species which bring in high unit value for the products reared.

For marine species, intensive culture systems using modern materials for pens, nets, floats, etc are restricted to protected areas. With interception on the high seas of local Atlantic salmon stocks, as well as competition in nearshore marine fisheries from many countries, Norway adopted as a national policy the production of salmon in their protected inland fjord waters. Similar national support and development of net-pen culture was given to Japanese aquaculturists using protected inland bays and seas to culture a marine scombrid, the yellowtail. The yellowtail has always enjoyed high consumer acceptance, allowing for a profitable operation despite a high unit cost of production.

The requirement that land acquisition costs or land rents not be excessive as part of overall operation costs remains as important to aquaculture as to agriculture. Two interesting examples of site-specific considerations related to this constraint are provided by the catfish industry in the United States and the shrimp industry of Equador. Both are relatively recent intensive aquaculture programs. The major location of the U.S. catfish industry is the flat agriculture land of the lower Mississippi River drainage, where large ponds can be constructed with modern machinery and supplied with water pumped from local aquifers or from abundant surface waters. In addition, catfish has enjoyed wide consumer acceptance in the southern, central, and adjoining regions of the United States. Catfish ponds also have the advantage of being restored to agricultural use as economics might dictate. In Eccuador shrimp aquaculture depends on the flat land of the mangrove swamps. Historically, the mangrove has been the nursery ground for local shrimp stocks. As shrimp have increased in export value, economics has shifted the production of shrimp from a catch fishery to large corporate aquaculture ventures. This requires converting mangrove swamps to ponds. Seed still comes from wild stocks that require the mangrove for rearing. Obviously either sufficient seed from culture will be required or pond expansion will have to cease, or both. Nevertheless, the operations are very site-specific with expansion under present operating conditions problematic.

Site-specific aquaculture requirements are also a serious limiting factor associated with ocean ranching of both andromous Pacific salmon and andromous trout. Pacific salmon ranching occurs ubiquitously around the rim of the northern Pacific Ocean, while Atlantic salmon ranching is distributed in streams entering the northern portions of the Atlantic Ocean in both North America and Europe. For Pacific salmon, the Japanese concentrate on the chum salmon, the Russians on both pink and chum. Japan has a complex mixture of both private and government hatcheries, all directed toward direct commercial utilization. Russian ventures are mainly state operated. In the United States ocean-ranching policies are dictated by the states. Alaska fosters cooperative nonprofit ocean-ranching ventures and Washington prohibits ocean ranching other than by native Americans. Oregon allows ranching under restrictions, whereas California requires a permit from the state legislature, which has restricted ranching to one experimental venture. Canada appears heading toward pen-culture as a social program to provide employment for indigenous citizens.

Most culturing of andromous salmonids in the United States and Canada is by government agencies and is directed toward enhancement of existing recreational and net fisheries or mitigation of losses of native stocks from pollution or blockage to traditional spawning grounds or rearing habitats. Choice of sites for ocean ranching by private ownership, where the activity is permitted, not only requires flat land and water of sufficient flow and quality, but must be located to minimize impacts on native runs from hatchery diseases, competition for food in freshwater and estuary rearing, and the potential risk of genetic deterioration in native stocks. Whether government-operated mitigation and enhancement operations are to perform in the same manner as required for private aquaculture operations involves biopolitics, a topic beyond the scope of this article.

Feeds and Fertilizers

A major aquaculture constraint is availability of inexpensive, readily available, and reliable sources of cheap fertilizers for extensive aquaculture, and nutritious protein-based manufactured feeds for intensive operations. Increased use of natural processes, such as eutrophication from wastewaters, seem to be offering ways to ameliorate food and fertilizer costs. Where agriculture and aquaculture are integrated into a single system, nutrients trapped in the pond muds from fecal matter of cultured organisms can be placed back on adjacent land for growing plants. Plant residues are then fed back to cultured species. The same principle acts on a major scale in bays and estuaries, in what has been described as biological traps. The process accounts for the massive production of mussels from floating rafts located in the Galacian bays of northern Spain, and for the explosive production in the last decade of milkfish in the Laguna de Baja near Manila in the Philippines. Both of these water bodies are fertilized by agricultural runoff and discharge of domestic wastes associated with human populations in the areas. The ability of species of fish such as the carps and tilapias to extract oxygen for their metabolic needs at low oxygen tensions and to feed on plant proteins allows profitable operations in both intensive and extensive aquaculture operations. Similarly, bivalves feeding low on the food chain virtually eliminate feed costs as a major limitation to successful operations using eutrophic systems. Polyculture systems such as those practiced in Israel, China, and other Asian and southeast Asian countries are alternative methods of managing around the problem of costly feeds and fertilizers.

Nontechnical Constraints

A variety of factors not directly related to the art and science of culturing plants and animals can prevent or restrict aquaculture enterprises. Administrative policies, public attitudes, esthetic considerations, and industrial development are examples. In developing countries where the treatment of industrial wastewaters would reduce export profits, aquaculture enterprises both close to and far from an industrial effluent discharge can suffer losses in wild seed stocks from toxic substances released. Producing replacement seed from brood stocks may require extensive research over a long period of time. Loss of local cultured species, along with overfishing of native species important to local communities, often leads to the introduction of exotic species to compensate for local fishery and aquacul-

ture production declines. Carps, salmon, tilapias, oysters, and prawns have been widely introduced throughout the world. Such introductions of exotics having good culture characteristics and high economic value are now coming into conflict with fisheries management agencies charged with maintaining the integrity of native flora and fauna (25–27). In some industrialized countries with adequate animal protein resources, locating fish farms in community scenic areas are factors that can restrain an aquaculture development. Additional constraints can occur in nonindustrialized countries from local food prejudices, monopoly in an economic infrastructure, and/or lack of planning based on the critical role of women in food production systems. The latter factor is now slowly being more adequately considered by world lending institutions in evaluating aquaculture development projects.

CONCLUSION

Use of fresh water for purposes of aquaculture, to provide protein for burgeoning human demands, is not a very promising endeavor because of the relatively low efficiencies of conversion of solar to potential chemical energy. Terrestrial agriculture, in spite of its high freshwater requirements, is in general more efficient. The greatest demands for fresh water, however, are in the area of technology. Much of this water is returned to the environment in a seriously degraded form or removed from a source, such as groundwater, with slow renewal times and lost or reduced from the region of need (31).

This quantitative approach to the overall potential of aquaculture in freshwaters is true, but there are a number of qualitative factors that will continue to make aquaculture extremely important to meeting human nutritional needs. In many low-income countries, and a few industrial ones, fish are the principal source of protein. A few scraps of fish in a bowl of rice is enough to meet minimum daily protein requirements. Recent findings show that lack of essential amino acids in early childhood is not only deleterious to growth but impairs mental capacity. This effect carries over into adult life, an even more serious consequence than reduced growth and impaired health. Most inland areas of the world lack adequate transport and refrigeration systems and thus do not have marine fish products available unless they are dried or frozen. Until refrigeration technology becomes widespread, local catch fisheries, and, increasingly, aquaculture, are the major practical means of supplementing protein in the diets of vast numbers of human beings in the world. In China especially, but also in India, the aquaculturist is also the agriculturist. In these areas integrated aquaculture–agriculture systems have evolved because of limited land, patterns of ownership, and conservative production strategies which are rooted in trial-and-error experiences that have practical value. Pond muds are rich in nutrients, thus serving as an organic fertilizer for land on adjacent pond banks and fields. Aquatic and terrestrial crops are rotated, benefiting production from each crop in turn. Such efforts in nonindustrialized or low-income countries are labor intensive, practices not considered efficient in industrialized countries with large holdings and mechanized operations. There are growing signs, however, that aquaculture is now slowly being integrated even into large agribusiness operations for a number of salient ecological reasons. There are many species of food fish which are tolerant of a wide range of physical and chemical properties in water. This is especially so for total dissolved solids (salinities), which is a major problem for irrigation agriculture. Fish could become part of advanced biological treatment systems for water leading into and draining from irrigation enterprises. The approaches to highly eutrophic lakes and ponds are often occupied by emergent aquatic plants which are proving to be highly adapted to processing degraded wastewaters. Such marsh–fish systems can be operated in low-lying land, often marginal either for agriculture or as industrial sites. Low-lying sites are often subject to floods and buildings often have to meet costly earthquake construction standards. Thus the adaptation of historically proven integrated agriculture–aquaculture systems in many parts of the world to ameliorating industrially induced freshwater degradation will slowly come about when the cost of doing otherwise becomes too much for society to tolerate, and when demonstration projects show financial as well as environmental benefits (31).

In the economy of human survival throughout the world, food production in back-yard fish ponds and home gardens, fuelwood gathering, etc are difficult to estimate but are undoubtedly substantial. There are people in Los Angeles, California, who have filled their swimming pools with soil and planted gardens. Should climate shifts actually occur from the greenhouse effect, as predicted by computer models, what a marvelous opportunity for the urban swimming pool to contribute to local food needs in future water-short areas. The pools could become aesthetic fish ponds with Japanese koi, polycultured with catfish or rainbow trout, in a managed system. Aquaculture may not solve the food problems of the world, but site-specific contributions will be significant enough to make many entrepreneurs happy and give increasing numbers of amateur aquaculturists great satisfaction.

BIBLIOGRAPHY

1. A. Jones, "Historical Background, Present Status, and Future Perspectives on the Aquaculture Industry on a Worldwide Basis," *IFAC Automation and Data Processing in Aquaculture, Trondheim, Norway, 1986.*

2. *Webster's Third New International Dictionary of the English Language Unabridged,* Merriam-Webster, Inc., Springfield, Mass., 1981.

3. W. E. Ricker, "Fish Culture," in W. Benton, ed., *Encyclopaedia Britannica,* Chicago, 1970.

4. L. R. Brown, "Maintaining World Fisheries," in *State of the World 1985,* W. W. Norton and Co., New York and London, 1985.

5. W. McLarney, *The Freshwater Aquaculture Book,* Hartley and Marks, Vancouver, B.C., Canada, 1984.

6. G. H. Allen and R. L. Carpenter, "The Cultivation of Fish with Emphasis on Salmonids in Municipal Wastewater Lagoons as an Available Protein Source for Humans," in F. M. D'Itri, ed., *Wastewater Renovation and Reuse,* Marcel Dekker, Inc., New York, 1977.

7. R. S. V. Pullin and R. H. Lowe-McConnel, "The Biology and Culture of Tilapias," in *ICLARM Conference Proceedings 7,* International Center for Living Aquatic Resource Management, Manila, Philippines, 1982.

8. E. E. Brown, *World Fish Farming: Cultivation and Economics,* 2nd ed., AVI Publishing Co., Inc., Westport, Conn., 1983.

9. E. E. Brown and J. B. Gratzek, *Fish Farming Handbook. Food, Bait, Tropicals and Goldfish.* AVI Publishing Co., Inc., Westport, Conn., 1980.

10. United States National Research Council, *Food, Fuel and Fertilizer from Organic Wastes,* National Academy Press, Washington, D.C., 1981.

11. C. F. Hickling, *Fish Culture,* Faber and Faber, London, 1962.

12. W. Durant, *The Story of Civilization: 1. Our Oriental Heritage,* Simon and Schuster, New York, 1935.

13. E. M. Borgese, *Seafarm. The Story of Aquaculture,* Harry N. Abrams, Inc., New York, 1980.

14. M. Gimbutas, *The Goddesses and Gods of Old Europe,* University of California Press, Berkeley, 1982.

15. *Yearbook of Fishery Statistics,* Vol. 60, Food and Agriculture Organization of the United Nations, Rome, 1987.

16. D. M. Cohen, "How Many Recent Fishers Are There?" *Proc. Calif. Acad. Sci., 4th Ser.* **38:**341–346. (1970).

17. J. S. Nelson, *Fishes of the World,* 2nd ed., John Wiley & Sons, Inc., New York, 1984.

18. J. E. Bardach, J. H. Ryther, and W. O. McLarney, *Aquaculture: The Farming and Husbandry of Freshwater and Marine Organisms,* John Wiley & Sons, Inc., New York, 1972.

19. S. L. Hora and T. V. R. Pillay, *Handbook on Fish Culture in the Indo-Pacific Region* (Fish Biol. Tech. Paper No. 14) Food and Agriculture Organization of the United Nations, Rome, 1962.

20. Z. Lin and co-workers, *Pond Fish Culture in China,* China National Bureau of Aquatic Products, Guangzhou, China, 1980.

21. R. J. Rhodes, "Status of World Aquaculture: 1988," in *Aquaculture Magazine, Buyer's Guide '89 and Industry Directory,* Ashville, North Carolina, 1988.

22. T. V. R. Pillay, "The State of Aquaculture 1974," in T. V. R. Pillay and W. A. Dill, eds., *Advances in Aquaculture,* Fishing News Books Ltd., Farnham, Surrey, England, 1979.

23. B. Hepher, "Supplementary Feeding in Fish Culture," *Proceedings of the 9th International Congress on Nutrition,* Vol. 3, Karger, Basel, Switzerland, 1975.

24. J. B. Glude, ed., *NOAA Aquaculture Plan,* U.S. Government Printing Office, Washington, D.C., 1977.

25. R. D. Smitherman, W. L. Shelton, and J. H. Grover, *Symposium on Culture of Exotic Fishes,* Department of Fisheries and Allied Aquaculture, Auburn University, Auburn, Alabama, 1978.

26. R. H. Stroud, ed., "Fish Culture in Fisheries Management," American Fisheries Society, Bethesda, Md., 1986.

27. W. R. Courtenay, Jr., and J. R. Stauffer, *Distribution, Biology, and Management of Exotic Fishes,* The John Hopkins University Press, Baltimore, 1984.

28. T. V. R. Pillay, *Proceedings of the FAO World Symposium on Warmwater Pond Fish Culture,* Vols. 1–5, Food and Agriculture Organization of the United Nations, Rome, 1968.

29. T. V. R. Pillay and W. A. Dill, eds., *Advances in Aquaculture.* Fishing News Books Ltd., Farnham, Surrey, England, 1979.

30. K. K. Chew, "Review of Recent Molluscan Culture," in M. Bilio, H. Rosenthal, and C. J. Sinderman, eds., *Realism in Aquaculture: Achievements, Constraints, and Perspectives,* European Aquaculture Society, Bredene, Belguim, 1986.

31. R. G. Wetzel, *Limnnology,* Saunders, Philadelphia, 1975.

General References

K. H. Alikunhi, "Synopsis of Biological Data on Common Carp *Cyprinus carpio* (Linnaeus) 1785. Asia and the Far East," *FAO Fisheries Synopsis, No. 31,* Food and Agriculture Organization of the United Nations, Rome, 1966.

F. Allen and co-workers, *Bioeconomics of Aquaculture,* Elsevier Science Publishing Co., Inc., New York, 1984.

J. G. Balchen, ed., *Automation and Data Processing in Aquaculture,* Pergamon Press, Oxford and New York, 1987.

G. W. Bennett, *Management of Artificial Lakes and Ponds,* Reinhold, London, 1962.

M. Bilio, H. Rosenthal, and C. J. Sinderman, eds., *Realism in Aquaculture: Achievements, Constraints, Perspectives, World Conference on Aquaculture, Venice, Italy, 21–25 September 1981,* (review papers), European Aquaculture Society, Bredene, Belgium, 1986.

C. D. Boyd, *Water Quality Management for Pond Fish Culture,* Elsevier, Amsterdam, The Netherlands, 1988, 318 pp.

E. E. Brown, *World Fish Farming: Cultivation and Economics,* AVI Publishing Co., Inc., Westport, Conn., 1983.

K. J. Chen and W. G. Co, *Prawn Culture,* Westpoint Aquaculture Corp., Dagupan City, Philippines, 1988, 323 pp.

I. Cheston, *Business Management in Fisheries and Aquaculture,* Fishing News Books Ltd., Farnham, Surrey, England, 1984.

A. G. Coche, *Simple Methods for Aquaculture: Soil and Freshwater Fish Culture,* FAO, Rome, and Unipub, Ann Arbor, Mich., 1985.

P. Edwards, "Aquaculture: A Component of Low Cost Sanitation Technology," *UNDP World Bank Integrated Resource Recovery Project* (World Bank Technical Paper No. 36), World Bank, Washington, D.C., 1985.

Fan Li, *The Chinese Fish Culture Classic,* T. S. Y. Koo, trans., Contr. No. 489, Chesapeake Biological Laboratory, University Maryland, Solmons, Maryland.

J. B. Forbes, "Directory of Aquaculture Information," *Bibliographies and Literature of Agriculture No. 25,* National Agricultural Library, Beltsville, Md., 1982.

J. B. Forbes and C. N. Bebee, *Literature for United States Aquaculture: 1970–1982,* National Agricultural Library, Beltsville, Md., 1983.

S. A. Forbes, "The Lake as a Microcosm," *Illinois Natural History Survey Bulletin No. 15,* 1925, pp. 537–550, as reprinted in: L. E. Keup, W. M. Ingram and K. M. Mackenthum. Biology of Water Pollution. U.S. Dept. Interior, Federal Water Pollution Control Administration, 2nd Ed., Cincinnati, Ohio, 1968.

J. Halver, ed., *Special Methods in Pond Fish Husbandry,* Halver Corporation, 1984.

J. Halver, ed., *Fish Nutrition* 2nd ed., Academic Press, New York, 1988.

M. Huet, *Textbook of Fish Culture: Breeding and Cultivation of Fish,* 2nd ed., rev. by J. Timmerman, Fishing News Books Ltd., London, 1988.

J. E. Huguenin and J. Colt, *Design and Operating Guide for Aquaculture Seawater Systems,* Elsevier, New York, 1989.

J. V. Huner, ed., *Crustacean and Mollusk Aquaculture in the United States,* AVI Publishing Co., Inc., Westport, Conn., 1985.

V. B. Jhingran and R. S. V. Pullin. *A Hatchery Manual for the Common, Chinese and Indian Major Carps,* International Center for Living Aquatic Resource Management, Asian Development Bank, Bangkok, Thailand, 1988.

R. Kirk, *A History of Marine Fish Culture in Europe and North America,* Fishing News Books Ltd., Farnham, Surrey, England, 1987.

P. Korringa, *Farming Marine Fishes and Shrimps,* Elsevier, Amsterdam, The Netherlands, 1976.

E. Leitritz and R. C. Lewis, *Trout and Salmon Culture (Hatchery Methods),* California Fish and Game Department, Fish. Bull. No. 164, Sacramento, CA. 1976.

C. A. Lembi and J. R. Waaland, *Algae and Human Affairs,* Cambridge University Press, Cambridge, 1988.

T. Leuring, H. A. Hoppe, and O. J. Schmid. *Marine Algae. A Survey of Research and Utilization,* Cram, De Gruyter & Co, Hamburg, 1969.

P. R. Limburg, *Farming the Waters,* Beaufort Books, New York, 1980.

G. Logsdon, *Getting Food from Water: A Guide to Backyard Aquaculture,* Rodale Press, Emmaus, Pa., 1978.

T. Lovell, *Nutrition and Feeding of Fish.* AVI Van Nostrand, New York, 1989.

J. P. McVey, ed., *CRC Handbook of Mariculture. Vol. 1: Crustacean Culture,* 4th ed., Boca Raton, Fla., 1989, 546 pp.

R. G. Michael, ed., *Managed Aquatic Ecosystems,* Elsevier Science Publishing Co., Inc., New York, 1987.

National Academy of Sciences, *Nutrition Requirements of Trout, Salmon and Catfish,* U.S. National Academy of Sciences, Washington, D.C., 1973.

National Academy of Sciences, *Making Aquatic Weeds Useful: Some Perspectives for Developing Countries.* U.S. National Academy of Sciences, Washington, D.C., 1976.

National Academy of Sciences, *Nutrient Requirements of Warmwater Fishes,* U.S. National Academy of Sciences, Washington, D.C., 1977.

National Research Council, *Aquaculture in the United States: Constraints and Opportunities,* National Academy of Sciences, Washington, D.C., 1978.

W. C. Olsen, "Major Aquaculture Associations, Education and Research Resources in the United States," *Bibliographies and Literature of Agriculture No. 26,* National Agricultural Library, Beltsville, Md., 1983.

T. V. R. Pillay, *Planning Aquaculture Development: An Introductory Guide,* Fishing News Books Ltd., Farnharm, Surrey, England, 1974.

R. G. Piper and co-workers, *Fish Hatchery Management.* U.S. Department of the Interior, Fish and Wildlife Service, Washington, D.C., 1982.

G. Post, *Textbook of Fish Health,* TFH Publications, Neptune City, N.J. 1987.

G. I. Pritchard, *Proceedings of the National Aquaculture Conference: Strategies for Aquaculture Development in Canada,* Department of Fisheries and Oceans, Ottawa, 1984.

R. S. V. Pullin and Z. H. Shehadeh, *Integrated Agriculture—Aquaculture Farming Systems,* International Center for Living Aquatic Resources Management, Los Baños, Philippines, 1988.

R. S. V. Pullin and co-workers, *2nd Annual International Symposium on Tilapia in Aquaculture,* International Center for Living Aquatic Resource Management, Manila, Philippines, 1988.

N. K. Pushkar and C. J. Siandermann, *Drugs and Food from the*

Sea: Myth or Reality, University of Oklahoma Press, Norman, 1987.

M. Reisner, *Cadillac Desert. The American West and Its Disappearing Water,* Viking Penguin Inc., New York, 1986.

R. J. Roberts and C. J. Shepherd. *Handbook of Trout and Salmon Diseases,* Fishing News, Farnham, Surrey, England, 1986.

H. Rosenthal, *Bibliography on Transplantation of Aquatic Organisms and Its Consequences on Aquaculture and Ecosystems,* European Mariculture Society, Bredene, Belgium, 1978.

O. H. Rosenthal and O. H. Oren, *Research on Intensive Aquaculture* (Spec. Publ. No. 6), European Mariculture Society, Bredene, Belgium, 1981.

S. Sarsig, "Synopsis of Biological Data on Common Carp *Cyprinus carpio* (Linnaeus) 1958. Near East and Europe," *FAO Fisheries Synopsis, No. 312,* Food and Agriculture Organization of the United Nations, 1966.

G. I. Shpet, "Comparative Efficiency of Fish Culture and Other Agricultural Activities per Unit of Area Used," *Hydrobiological Journal,* 8(3), 46–51 (1972).

J. C. Sindermann and D. V. Lighter, eds., *Disease Diagnosis and Control in North American Marine Aquaculture,* Elsevier, Science Publishing Co. Inc., New York, 1988.

S. H. Spotte, *Fish and Invertebrate Culture. Water Management in Closed Systems,* 2nd ed., Wiley-Interscience, New York, 1979.

R. Stickney, *Culture of Non-salmonid Fishes,* CRC Press, Boca Raton, Florida, 1986.

R. R. Stickney, *Principles of Warmwater Aquaculture,* John Wiley & Sons, Inc., New York, 1979.

T. Tamura, *Marine Aquaculture,* Parts I and II, M. I. Watanabe, trans., National Technical Information Service, Springfield, Va., 1960.

J. G. Thorpe, ed., *Salmon Ranching,* Academic Press, London, 1980.

C. H. Townsend, "The Private Fish Pond—A Neglected Resource," *Transactions of the American Fisheries Society,* 43, (1914).

C. S. Tucker, *Channel Catfish Culture.* Elsevier, Amsterdam, The Netherlands, 1985.

Aquaculture: Shellfish: 1977–September, 1982, U.S. National Technical Information Service, Springfield, Va., 1982.

A. Usui, *Eel Culture,* Fishing News Books Ltd., Farnharm, Surrey, England (reprint) 1974.

A. S. Watson, *Aquaculture and Algae Culture: Processes and Products,* Noyes Data Corp., Park Ridge, N.J., 1979.

F. W. Wheaton, *Aquaculture Engineering,* John Wiley & Sons, Inc., New York, 1977.

George H. Allen
Humboldt State University
Arcata, California

AQUACULTURE: ENGINEERING AND CONSTRUCTION

The origins of aquaculture are ancient but the explicit application of engineering principles to aquaculture is quite recent. Only during the past 15 to 20 years has there been visible research activity in this field. At the present time the number of aquacultural engineering practitioners is still small, as is the number of educational programs that offer aquacultural engineering courses and degree programs. The development of aquacultural engineering has followed a path analogous to that followed by agricultural engineering 50 to 75 years earlier. Typically, engineers trained in related disciplines have acquired enough understanding of the biological components of aquaculture production systems to be able to design, evaluate, and manage these systems. The early activities have led to the training of students, graduate and undergraduate, in the combination of engineering and biological principles that would allow them to function effectively as aquacultural engineers.

Aquaculture encompasses a wide range of activities, culture techniques, and species. The application of engineering principles to the different forms of aquaculture varies widely, from the sophisticated, explicit engineering needed for an intensive system that incorporates water reuse, to the simple, nonformal engineering used for extensive, subsistence type aquacultural operations. Some of the processes described here will, therefore, be applicable only to certain types of production systems.

DEFINITION

Aquacultural engineering is the application of engineering principles to the design, construction, and management of systems for the production of aquatic animals or plants. Specific areas covered include the overall system design and management; the physical system used to hold the animals or plants; the supporting structures; processes, monitoring equipment, and techniques needed to insure adequate environmental quality in the production system; and the facilities and equipment needed to handle fish and other materials such as feed. In the context of this article, fish may be a finfish such as trout or catfish, or a crustacean such as shrimp or lobster, or a mollusc such as oysters or clams. As a research discipline, aquacultural engineering also deals with the evaluation of the effects of water quality and other environmental conditions on fish, as well as the effect that fish have on the water. The effects on the fish that are of concern include growth rates, susceptibility to disease, and reproductive success. In addition, traditional engineering disciplines ranging from environmental to mechanical to electrical engineering are used in aquaculture.

SYSTEM DESIGN

An aquaculture system must be designed to provide a healthy environment for the target fish or plant, and this must be achieved within constraints inherent in the choice of location where the facility is to be constructed. Other conditions that need to be taken into account include marketing, economics, and regulatory restrictions. The procedure for the development of an aquaculture system is summarized in Figure 1 and follows stages of data collection, evaluation, design, and construction (1–3). Background information needed includes data on the biological (or bioengineering as they are sometimes referred to) characteristics of the target species, as well as data on the site, possible markets for the product, and regulations affecting aquaculture at the chosen site. The biological

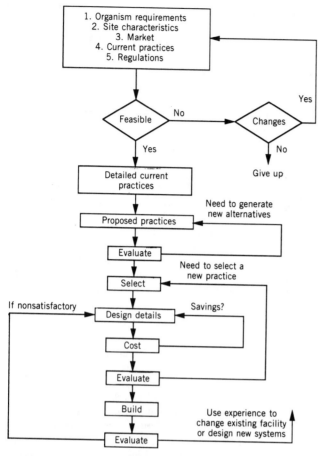

Figure 1. Flow diagram of the design process followed for aquaculture operations.

data needed to quantify the relationship between the target organism and its environment include information on tolerances and optimum levels for various water quality parameters, the effect of target animal activity on water quality, space and water velocity requirements, reproductive characteristics, and feeding behavior. Site data are primarily related to water availability and to the quality of the water, but other factors are important, such as soil characteristics (soil permeability, physical properties affecting possible construction, and fertility), topography, climate, other land uses, transportation, and other infrastructure. Site characteristics may be considered at the level of a specific aquaculture site or at the regional planning level, where attempts have been made to incorporate satellite imaging into the site characterization process (4). Desirable marketing information includes not only the type of species that is saleable, but also the particular requirements imposed by the market on product size, level of processing, seasonal fluctuations in demand, etc. Lastly, regulatory aspects of aquaculture must not be overlooked. Restrictions on land and water use are augmented by considerations of species approvals, and in some cases by the lack of familiarity of regulators with aquaculture and their unwillingness to consider it as a form of agriculture rather than as an industrial enterprise.

A last topic on which background information should be gathered is an overall review of the methods and proce-

dures used for the culture of the species of interest. Wherever possible, this should focus on conditions that are similar to those of the planned facility.

From this point on, the design process follows fairly well established overall guidelines for engineering design. A preliminary evaluation is carried out in which the background information is reviewed and an initial decision is made about the possible viability of the operation or of the need to make changes and collect additional background information.

Once a decision has been made to proceed with the planning of the facility, a detailed review of the practices used by others in the culture of the target species is undertaken. This review should be based on published information as well as on personal contacts with other producers, with extension agents, and with researchers. As a first step, several alternative designs may be generated and evaluated to select the most promising one for further development. After evaluation and selection, the details of the design are completed, including estimated costs and preparation of construction documents. Evaluation of the plans and facilities should be carried out at the various stages of the process to minimize the probability of errors, as well as to insure maximum functionality and operational efficiency for the investment. Evaluation of the completed installation can be used in the preparation of future designs or in the possible expansion of the facility.

HOLDING SYSTEMS

The types of impoundments or holding systems used in aquaculture are determined by the requirements of the particular species and by the production and management practices being employed. Typical systems include ponds, raceways, tanks, and cages.

Ponds

Pond size, shape, and construction technique vary widely (5–7). Ponds tend to be shallow, less than 1.5 m deep, and vary in area between less than 100 m² to over 10 ha (Fig. 2). The pond bottom is sometimes lined with concrete or plastic liners to prevent water losses to infiltration. The ponds may be built by excavation, by construction of dykes, or more commonly by a combination of the two techniques. In all cases, pond design should incorporate structures for water inflow for filling and maintenance flows, and for effluents including flows during pond drainage and emptying.

Raceways

Raceways are relatively narrow, long channels with water constantly entering through one end of the channel and leaving through the opposite end (Fig. 3). Raceways are often built of concrete, and the normal dimensions are in a ratio of 30:3:1 (length:width:depth). Raceways are the most common holding system for the production of trout and are usually approximately 30 m long. Flow in a raceway approaches plug flow characteristics, creating a gradient of water quality along the raceway. Water quality is best near the head of the raceway where the water enters, and decays toward the drain side of the raceway. Raceways are often constructed in series of up to six raceways

Figure 2. Pond being used for the culture of shrimp with supplemental mechanical aeration.

(8) with some form of reaeration between raceways to replenish some of the oxygen used by the fish.

Tanks

Most aquaculture tanks are round (Fig. 4), ranging in diameter from less than 1 m to over 10 m. Water depth is

Figure 3. Trout in a concrete raceway.

normally maintained between 0.5 m for the smaller tanks and 1 to 2 m for the larger tanks. Other tanks may be rectangular with rounded corners, while others have the approximate shape of a racetrack (9). The two most common materials used for tanks are fiber glass and concrete.

The size and shape of tanks used in aquaculture vary greatly, but the common characteristic is that flow approaches the ideal continuously stirred tank reactor (CSTR) model in which water quality within the tank is approximately uniform. To achieve this uniformity, water is introduced along the periphery of the tank through one or more inlets that normally inject the water tangentially, and the water effluent is collected from the center of the tank. By adjusting the type of inlet used and its location the operator may determine the range of water velocities found in the tank without necessarily changing flow rates. An important feature is that, by adjusting the circulation pattern and water velocity to maintain particulate matter in suspension, tanks may be made to be self-cleaning (9). Flow rates in tanks are normally selected to result in residence times ranging from 10 to 15 min to several hours and the value used depends on the level of water treatment (normally aeration) taking place within the tank and on the fish biomass being held.

Cages

Although cages have been used in aquaculture for many years, they have received a great deal of attention recently, especially with the development of the salmon industry in Norway and other countries (Fig. 5). Conventional cage technology is based on a rigid frame that is either floating or fixed, and to which a cage made of some type of netting material is attached. Traditionally, cages had been rectangular and relatively small, with total volumes of less than 10 m³. The recent growth of the industry, the availability of new materials and technologies, and the presence of competing users for coastal resources have resulted in the development of large cage systems suitable for open sea deployment (10). Some of these cages may be as large as 32 m in diameter and 10 m deep, and are designed to hold up to 100,000 kg of fish (11). The engineering requirements for cage construction and an-

Figure 4. Round concrete tanks used for the culture of sturgeon. Effluent from the tanks is directed to the ponds in the background where a second crop (of catfish in this case) is raised.

choring become very specialized as the cage size and exposure to wind, waves, and currents increases. Static loads are easily quantified and are caused by the weight of the fish and the structure itself, but transient loads may be orders of magnitude greater than static loads and are very difficult to determine accurately. Frequent cage maintenance is required to prevent fouling of the net from restricting the flow of water. Exchange of water through the net must be sufficient to maintain adequate water quality for the fish.

QUANTIFYING PROCESSES

As was mentioned in the definition of aquacultural engineering, many engineering disciplines can be applied to the design of aquaculture systems. In this section some of the critical processes that are unique in an aquaculture system and that define design parameters will be de-

scribed. The key processes in the design of aquatic systems have to do with the primary water quality factors that affect the cultured organism and with how the organism affects water quality.

As a starting point in the description of important processes in an aquaculture operation one can consider how a fish relates to its environment. Fish consume oxygen from the water and release waste products from their metabolism back into the water. The main waste products of concern are ammonia, carbon dioxide, and particulate and dissolved organics. Depletion of dissolved oxygen below a certain level, which is species and size dependent, results in increased stress to the fish, and may ultimately result in suffocation. Similarly, the accumulation of waste products may reach toxic levels, as in the case of ammonia and carbon dioxide; or it may alter the pH of the water, again as a result of carbon dioxide accumulation; or it may affect growth and reproductive behavior as a result of the accumulation of dissolved organics such as hormones; or it

Figure 5. Cages used for the culture of Atlantic salmon in Norway.

may affect the overall oxygen balance of the system by the oxygen demand created by the decomposition of the particulate and dissolved organics.

A major difference between aquaculture and land-based food production systems is that in an aquaculture system the animal or plant takes its nutrients and oxygen from and releases its waste products into the same environment: the water. Maintaining water quality in an aquaculture system is based on providing sufficient oxygen to satisfy the needs of the fish and on removing metabolites to prevent their buildup to toxic levels. Design of systems to achieve these goals requires the quantification of processes that take place within the system, especially those related directly to the target species. The relative importance of the various consumption and production processes depends on the type of aquatic production system being considered. In general, the shorter the residence time of the water in the production system, the more important are the fish-related processes with respect to other biological processes taking place. If, on the other hand, the residence time is long (over one day and as long as one year), as is the case in most ponds, biological and chemical processes that are not directly associated with the target product's metabolism become very important in determining water quality. As an example, dissolved oxygen in a tank with a short residence time will be determined by the concentration in the influent water, the fish biomass being held in the tank, the amount of oxygen being added to the system through aeration mechanisms, and the oxygen consumption by the fish. In a pond, on the other hand, many processes can have significant effects on dissolved oxygen concentration. These processes include photosynthetic oxygen production, respiration by all the organisms in the ponds (phytoplankton, zooplankton, fish, bacteria, and sediment decomposition processes), and gas exchange between the water and the atmosphere.

Quantification, design, and management of aquaculture production systems is most predictable in those systems with short residence times, where the majority of water quality changes are associated with fish activity. Pond systems, which often have long residence times, are often difficult to quantify and their management remains very much an art rather than a science.

Design of systems is ultimately based on mass balances of the critical water quality factors mentioned above. The general form of the mass balance equation for a particular substance (eg, dissolved oxygen or ammonia) in a control volume can be written as

$$\text{Accumulation} = \text{inflow} + \text{production} - \text{outflow} - \text{consumption} \qquad (1)$$

For simplicity in calculations and because of incomplete information on the time dependence of production and consumption rates, the mass balance is often simplified to consider steady-state conditions in which accumulation becomes zero, and equation 1 can be written as

$$\text{inflow} + \text{production} = \text{outflow} + \text{consumption} \qquad (2)$$

The inflow and outflow terms may be expressed as the product of the concentration of the substance being observed and the flow rate:

$$\text{inflow} = QC_i$$
$$\text{outflow} = QC_o$$

where Q = flow rate (m^3/d)
C_i = influent concentration (g/m^3)
C_o = effluent concentration (g/m^3)

In a fully mixed rearing impoundment, the effluent concentration is approximately equal to the concentration at any point within the impoundment and is the concentration to which the target species is exposed. These production and consumption functions are dependent on the species size, and condition of the fish, and on the environmental conditions to which they are exposed. In addition, these rates fluctuate through diel cycles in response to activity levels associated primarily with feeding and digestion. In general, there is little quantitative information on how the various factors affect the rates of oxygen consumption and metabolite production, and for purposes of design it is common to relate these rates to feeding levels (12):

$$DO_{\text{cons}} = (\text{feed})DO_{\text{rc}}$$
$$DCO_{2\text{prod}} = (\text{feed})DCO_{2\text{rc}}$$
$$DTAN_{\text{prod}} = (\text{feed})DTAN_{\text{rc}}$$
$$DSOLIDS_{\text{prod}} = (\text{feed})DSOLIDS_{\text{rc}}$$

where DO_{cons} = rate of oxygen consumption by fish within the impoundment (g/d)
feed = amount of feed applied to the impoundment (g/d)
DO_{rc} = oxygen consumption per unit of feed applied (g/g)

and similarly for carbon dioxide (CO_2), total ammonia nitrogen (TAN), and SOLIDS productions. Feeding rate is often expressed as a proportion of the fish biomass applied per day, and tables are available for various species and ages (13). Values normally used for the consumption of oxygen and production of carbon dioxide, ammonia, and solids per unit mass of feed (g/g) are 0.20, 0.28, 0.03, and 0.30, respectively (12). These values should be changed when more specific information is available for the particular species under consideration.

WATER CONDITIONING

Water conditioning operations may be used to treat the water and increase the fish biomass that may be held for a given water flow rate. The rates of oxygen consumption and metabolite production and the tolerances of fish to concentrations of these substances are such that oxygen is the first limiting factor encountered. If aeration is used to replenish oxygen, then the second limiting factor reached may be either carbon dioxide or ammonia accumulation, depending on characteristics of the water supply (in particular, alkalinity and pH of the water) and on the fish tolerances.

Oxygenation

Oxygen sources differ between ponds and other types of holding systems. Possible sources of oxygen in a pond are production by aquatic plants, including phytoplankton, mechanical reaeration, and reaeration across the water surface with minimum contributions by the water exchange. The relative importance of the three depends on the type of production system being used. In raceways and tanks, the oxygen sources are limited to the water supply and some form of mechanical reaeration.

Oxygen Production by Phytoplankton. Phytoplankton produce oxygen through photosynthesis. The rate of oxygen production by phytoplankton is very difficult to quantify but conceptually may be expressed as

$$DO_{phyto} = DO_{equi}(phyto)(max)(nutri)(light)(temp)$$

where DO_{phyto} = oxygen production rate by phytoplankton $(g/m^3\ h)$

DO_{equi} = oxygen production per unit of phytoplankton biomass growth (g/g)

phyto = phytoplankton concentration (g/m^3)

max = maximum phytoplankton growth rate under optimum conditions (h^{-1})

nutri = nutrient limitations on growth

light = light limitation on growth

temp = temperature limitation on growth

A multitude of expressions have been proposed for the various limitation terms (ref. 14), but their use in aquaculture is limited to research computer models.

The overall effect of phytoplankton on oxygen concentration in a pond is further complicated by their respiration at night. This combination of oxygen production by phytoplankton during the day and consumption at night by phytoplankton, fish, and other organisms results in the dissolved oxygen cycle observed in ponds (Fig. 6). Maintaining a balance between the phytoplankton production and consumption terms in ponds becomes the primary goal of water quality management in most pond production systems.

Oxygen Production by Mechanical Aeration. Oxygen transfer into the water takes place as a result of a driving force equal to the difference between the saturation concentration and the actual concentration found in solution, a mass transfer coefficient, and the area of contact between the liquid and gas phases. The mass transfer coefficient and the area of contact are normally combined into an overall gas transfer coefficient, K_{La}, that may be determined experimentally. The equation describing the rate of oxygen transfer can be written as

$$DO_{trans} = (DO_{sat} - DO)\,K_{La}V$$

where DO_{trans} = dissolved oxygen transfer rate $(g/m^3\ h)$

DO_{sat} = saturation concentration of dissolved oxygen (g/m^3)

DO = dissolved oxygen concentration in solution (g/m^3)

K_{La} = overall gas transfer coefficient

V = volume over which the aeration is taking place

Saturation concentration is determined by the composition of the gas phase, by temperature, by atmospheric pressure, and by the presence of dissolved substances in the water. Saturation concentration for fresh water as a function of temperature may be expressed as (16)

$$DO_{sat} = 14.652 - 0.41022T + 0.007991T^2$$
$$- 0.000077774T^3 \tag{3}$$

where T = water temperature (°C)

Equation (3) and tables of dissolved oxygen saturation are normally prepared for an atmosphere of 20.9% oxygen. Enrichment of the atmosphere will result in a corresponding increase in saturation dissolved oxygen concentration. Enrichment may be achieved by aerating with pure oxygen.

An important difference between aeration systems used for aquaculture and those used in water and wastewater treatment is that the driving force in aquaculture tends to be lower, reducing the rate of oxygen transfer for a given aerator. The lower driving force is caused by the requirement of most fish for dissolved oxygen concentrations above 3 to 4 g/m^3, compared with the usual limit of just above zero in wastewater treatment.

Common types of aerators used in aquaculture may be classified as surface, gravity, and diffuse types (9,17). Surface aerators spray water into the air or beat the water, increasing the turbulence and area of contact between the water and air. Gravity aerators rely on the fall of water for aeration. Different types of structures such as plates and baffles may be used in gravity aerators to increase transfer rates. A particularly useful and common gravity aerator is the packed column aerator (PCA), in which water is introduced at the top and air or another gas is introduced at the bottom of a column filled with a medium designed to maximize turbulence and transfer area. Design equations for the use of PCAs in aquaculture have

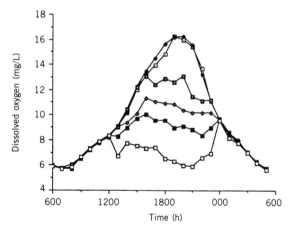

Figure 6. Changes in dissolved oxygen concentration in a fish pond. The lines indicate the concentration DO at depths ranging from 0.05 m to 1.0 m in a 1.1-m-deep pond being used for raising catfish (15). Key: ⊡, 5 cm; ◆, 20 cm; ■, 40 cm; ◆, 60 cm; ■, 80 cm; ▢, 100 cm.

been proposed (18). In diffuse aeration, gas bubbles are introduced into the water and transfer takes place between the bubble and the water. The simplest type of diffuse aerator is the airstone, in which bubbles are released at the bottom of a tank or pond. This type of aerator tends to be inefficient in conventional aquaculture systems because of the shallow water depths used and the resulting short contact times between the bubble and the water. Variations on the basic concept of diffuse aeration have been developed, including some systems, designed to be used as part of a pipeline, that pressurize and trap bubbles, allowing for longer contact times and, in some cases, complete dissolution of the bubbles into the liquid (19).

There has been a great deal of interest recently in the use of pure oxygen in aquaculture systems. These systems rely on either liquid bottled oxygen or the on-site production of oxygen. The use of pure oxygen makes possible the addition of large amounts of oxygen to the water and an increase in fish biomass held in a given water supply.

Oxygen Production by Surface Reaeration. The transfer of oxygen by reaeration is important in ponds where it can result in substantial net gains or losses of oxygen, depending on whether the dissolved oxygen concentration is below or above saturation. Estimates of reaeration rates may be obtained from (20).

$$DO_{reaer} = (DO_{sat} - DO)K_{reaer}/D \qquad (4)$$

where DO_{reaer} = reaeration rate (g/m^3)
$K_{reaer} = 0.03v_w^{0.5} - 0.0132v_w + 0.0015v_w^2$
v_w = wind velocity (m/s)
D = pond depth (m)

An implicit assumption in equation (4) is that conditions in the pond are uniform and there is a minimum of stratification.

Ammonia Removal

Ammonia is toxic to aquatic animals and is also the waste product of their protein metabolism. The sensitivity of aquatic animals to ammonia concentration depends on species, life stage, level of stress, and other environmental conditions. Ammonia exists in water as the equilibrium product of ammonium ion and un-ionized ammonia:

$$NH_4^+ \leftrightarrow NH_3 + H^+, \qquad pK_a = 9.25 \text{ at } 25°C$$

Un-ionized ammonia (NH$_3$) is the toxic form (21). Approximate estimates of the concentration of un-ionized ammonia in fresh water can be obtained from measures of total ammonia (TA, NH$_3$ + NH$_4^+$ concentrations) and pH as (22)

$$[NH_3] = [TA](1/10^{(pK_a - pH)})$$

where $[NH_3]$ = un-ionized ammonia concentration (mol/L)
$[TA]$ = total ammonia concentration (mol/L)

Removal of ammonia from aquaculture systems may be achieved by ion exchange or by biological filtration. Ion exchange is carried out with a natural zeolite (clinoptilo-

lite) that has a high affinity for ammonium ions. Removal capacity of the resin varies widely, but a common design value is around 1 mg NH$_4^+$ per gram of resin (23). Clinoptilolite resin may be regenerated with brackish or salt water, making the use of the resin suitable for fresh water applications only. Ion exchange columns need to be designed to incorporate the downtime involved with the recharging and reconditioning. In addition, unless properly managed, these columns have a tendency to be colonized by bacteria and become biological filters.

The biological removal of ammonia is by the process of nitrification carried out by two groups of bacteria, nitrosomonas which change ammonia to nitrite, and nitrobacter which complete the reaction to nitrate. Nitrate is toxic to fish only at very high concentration, and therefore is a suitable end product for ammonia and nitrite, the toxic forms of nitrogen. The nitrification process is relatively slow and sensitive to temperature, pH, and ammonia concentration. To achieve practical nitrification rates, high bacterial biomass must be maintained and this is normally achieved with some form of attached growth filter (24). There are many types of filters used for ammonia removal, and they include downflow, upflow, rotating disks, submerged, trickling, and fluidized beds (24–26). Removal rates are highly variable and this is an area of active research at the present time.

Solids Removal

Removal of solids produced by aquaculture may be attempted to make the water suitable for reuse or to meet discharge guidelines. Solids account for a high percentage of the biochemical oxygen demand (BOD) in aquaculture effluents; they tend to be highly variable in size and break up easily if subjected to mechanical forces. Solids are formed from uneaten feed and fecal material. Their density tends to be very close to that of water.

Removal is achieved by filtration with particulate filters (eg, sand filters) or screens (eg, stainless steel) or by sedimentation. The high organic content of the particulates makes them highly biodegradable, and filters need to be backwashed frequently to prevent clogging by a mat of biological slime. Screens are commonly used in aquaculture, yielding a predictable particle size range of removal. Commercial systems are available in which the screens are continuously cleaned and backwashed, resulting in low maintenance. Simple settling tanks or ponds with overflow rates as high as 175 m/day can result in solids removal rates as high as 85% (21). Aquaculture settling ponds are normally operated without sludge removal, relying on biological decomposition to prevent excessive accumulation of sludge.

COMPUTER MODELING (WATER QUALITY)

The modeling of water quality in aquaculture systems has been attempted primarily for pond systems (27). These systems have generally been modeled based on the assumption of uniformity of water quality throughout the pond. The models developed have been designed primarily to simulate dissolved oxygen concentration and may be classified in two general categories: empirical and mechanistic. The empirical models have been developed from statistical analysis of data obtained from ponds. Mechanistic models are based on mass balance equations (see equations 1 and 2) where rate coefficients for the various processes taking place are determined empirically or from analysis of published literature.

Recent developments in monitoring and data acquisition equipment have made possible the collection of data from several points in a pond at various times during a diel cycle. These data have been used to quantify the extent of stratification in ponds and to develop models to simulate temperature and dissolved oxygen at the various depths in the pond (15,28). The field measurements and models have shown clearly that stratification can cause differences between surface and bottom water quality in aquaculture pools that can exceed 10°C and 20 g/m³ of dissolved oxygen. These differences are particularly important for the culture of benthic animals, and new management practices are being developed based on these findings.

Computer modeling in aquaculture remains largely a research activity. There has been a great deal of recent interest in developing management systems that incorporate models for ponds and also for intensive tank and raceway systems. These management systems will include automatic monitoring and control systems coupled with computer models. In some cases, the computer models are taking the form of expert systems.

FUTURE DEVELOPMENT

The aquaculture farm of the future will take many forms. The difference from today's farms will be in the level of intensity of water use, in the degree of control over the culture environment exerted by the operator, and in the predictability of growth and overall production rates. The application of engineering will become more and more important as the aquaculture industry continues to grow and diversify. Developments will be in new culture systems; materials; equipment; and water quality monitoring, treatment, and control. The efficient use of natural resources, especially water, will be a primary constraint on new aquaculture operations. To make more efficient use of water, higher fish biomass will be maintained per unit of water volume and per unit of water use. The safe rearing of this large biomass will require continuous monitoring of water quality and the use of alarms and backup systems. In most cases these will be all computer controlled.

Improvements in water treatment are also needed to make possible reductions in water use and the commercial viability of culture systems based on water reuse. New, more efficient, and reliable filtration systems for the removal of particulate and dissolved organics and of ammonia will be developed.

Systems for inventoring and handling fish are needed. Obtaining accurate estimates of number and size of fish in a culture system are practically impossible today as they are based on sampling or on the use of growth functions and initial stocking values. Equipment and techniques that can be used to count animals and estimate their size with a minimum of disruption and stress to the fish will have to be developed.

BIBLIOGRAPHY

1. R. D. Mayo, "A Format for Planning a Commercial Model Aquaculture Facility" (presented at Northeast Fish and Wildlife Conference, February 25–28, 1974, Great Gorge, N.J.), *Technical Reprint No. 30,* Kramer, Chin and Mayo, Inc., Consulting Engineers, Seattle, Wash., 1974.

2. B. G. Shepherd, "The Biological Design Process Used in the Development of Federal Government Facilities during Phase I of the Salmonid Enhancement Program," *Salmonid Enhancement Program. Canadian Technical Report of Fisheries and Aquatic Sciences 1275,* Department of Fisheries and Oceans, Vancouver, B.C., Canada, 1984.

3. C. M. Brown and C. E. Nash, "Planning an Aquaculture Facility—Guidelines for Bioprogramming and Design," *Aquaculture Development and Coordination Programme Report ADCP/REP/87/24,* Food and Agriculture Organization of the United Nations, Rome, 1988.

4. "Applications of Remote Sensing to Aquaculture and Inland Fisheries" (Ninth UN/FAO International Training Course in Co-operation with the Government of Italy), *Remote Sensing Centre Report RSC Series 27,* Food and Agriculture Organization of the United Nations, Rome, 1985.

5. C. R. dela Cruz, "Fishpond Engineering: A Technical Manual for Small- and Medium-Scale Coastal Fish Farms in Southeast Asia," *South China Sea Fisheries Development and Coordinating Programme Manual No. 5,* United Nations Development Programme, Food and Agriculture Organization of the United Nations, Manila, Philippines, 1983.

6. M. Huet, *Textbook of Fish Culture, Breeding and Cultivation of Fish,* 2nd ed. (H. Kahn, trans.), Fishing News Books Ltd., Surrey, England, 1986.

7. W. McLarney, *The Freshwater Aquaculture Book, a Handbook for Small Scale Fish Culture in North America,* Hartley and Marks Publishers Inc., Point Roberts, Wash., 1984.

8. E. Leitriz and R. C. Lewis, "Trout and Salmon Culture (Hatchery Methods)," *California Department of Fish and Game, Fish Bulletin 164,* Sacramento, Calif., 1976.

9. F. W. Wheaton, *Aquacultural Engineering,* John Wiley & Sons, Inc., New York, 1977.

10. M. C. M. Beveridge, *Cage Aquaculture,* Fishing News Books Ltd., Surrey, England, 1987.

11. J. Gunnarsson, "Bridgestone HI-SEAS Fish Cage Design and Use," in *Aquaculture Engineering, Technologies for the Future,* Symposium Series No. 111, The Institution of Chemical Engineers, Rugby, England, 1988, pp. 133–141.

12. J. Colt, "An Introduction to Water Quality Management in Intensive Aquaculture," in *Oxygen Supplementation: A New Technology in Fish Culture,* Bulletin 1, U.S. Department of the Interior, Fish and Wildlife Service, Region 6, Denver, Colorado, 1986.

13. R. G. Piper, I. B. McElwain, L. E. Orme, J. P. McCrare, L. G. Fowler, and J. R. Leonard, *"Fish Hatchery Management,"* United States Department of the Interior, Fish and Wildlife Service, Washington, D.C., 1982.

14. S. Zison, W. B. Mills, D. Deimer, and C. W. Chen, *Rates, Constants, and Kinetics Formulations in Surface Water Quality Modeling,* EPA-6003 78-105, United States Environmental Protection Agency, Environmental Research Laboratory, Athens, Ga., 1978.

15. T. M. Losordo, "Characterization and Modeling of Thermal and Oxygen Stratification in Aquaculture Ponds," Ph.D. Dissertation, University of California, Davis, 1988, 416 pp.

16. H. L. Elmore and T. W. Hayes, "Solubility of Atmospheric Oxygen in Water," *Journal of the Sanitary Engineering Division, American Society of Civil Engineers,* **86**(SA4), 41 (1960).

17. C. E. Boyd and T. Ahmad, "Evaluation of Aerators for Channel Catfish Farming," *Alabama Agricultural Experiment Station Bulletin 584,* Auburn University, Auburn, Ala., 1987.

18. G. Hackney and J. Colt, "The Performance and Design of Packed Column Aeration Systems for Aquaculture," *Aquacultural Engineering* 1, 275–295 (1982).

19. R. E. Speece, M. Madrid, and K. Needham, "Downflow Bubble Contact Aeration," *Journal of the Sanitary Engineering Division, American Society of Civil Engineers,* 97(SA4), 433–441 (1971).

20. R. B. Banks and F. F. Herrera, "Effects of Wind and Rain on Surface Reaeration," *Journal of the Environmental Engineering Division, American Society of Civil Engineers* 103(EE3), 489–504 (1977).

21. J. E. Huguenin and J. Colt, *Design and Operating Guide for Aquaculture Seawater Systems,* Elsevier, Amsterdam, The Netherlands, 1989.

22. W. Stumm and J. J. Morgan, *Aquatic Chemistry, an Introduction Emphasizing Chemical Equilibria in Natural Waters,* 2nd ed., John Wiley & Sons, Inc., New York, 1981.

23. W. Bruin, J. W. Nightingale, and L. Mumaw, "Preliminary Results and Design Criteria from an On-Line Zeolite Ammonia Removal Filter in a Semi-Closed Recirculating System," in L. J. Allen and E. C. Kinney, eds., *Proceedings of the Bio-Engineering Symposium for Fish Culture,* American Fisheries Society, Bethesda, Maryland, 1981, pp. 92–96.

24. G. H. Kaiser and F. W. Wheaton, "Nitrification Filters for Aquatic Culture Systems: State of the Art, *Journal of the World Mariculture Society* 14, 302–324 (1983).

25. G. L. Rogers and S. L. Klemetson, "Ammonia Removal in Selected Aquaculture Water Reuse Biofilters," *Aquacultural Engineering* 4, 135–154 (1985).

26. J. F. Muir, "Recirculated Water Systems in Aquaculture," in J. F. Muir and R. R. Roberts, eds., *Recent Advances in Aquaculture,* Westview Press Inc., Boulder, Colo., 1982, pp. 357–446.

27. R. H. Piedrahita, "Modeling Water Quality in Aquaculture Ecosystems," in D. E. Brune and J. R. Tomasso, eds., *Aquaculture and Water Quality,* World Aquaculture Society, Baton Rouge, La, 1990.

28. T. P. Cathcart and F. W. Wheaton, "Modeling Temperature Distribution in Freshwater Ponds," *Aquacultural Engineering* 6, 237–257 (1987).

Raul H. Piedrahita
University of California
Davis, California

ARTIFICIAL INTELLIGENCE

Artificial intelligence is defined as the application of computer technology to resemble human thought and action. The main areas of artificial intelligence include robotics, machine vision, voice recognition, natural language processing, and expert systems. The term artificial intelligence was first used in 1956 at a conference held at Dartmouth College in Hanover, New Hampshire. The basis of the conference was the conjecture that every feature of intelligence can be so precisely described that a machine can be made to simulate it (1). Since that time the field has grown and developed. Many of the lofty early claims have been dismissed and replaced with more pragmatic and profitable applications. Because of the success and interest in expert-system technology, it will receive a more detailed description.

Robotics along with computer vision have been applied in food-processing operations for sorting and evaluation. Robots are used in manufacturing, performing such functions as welding and painting. Through artificial intelligence techniques robots can function in unstructured situations, performing tasks such as identifying objects and selecting and assembling them into a unit.

Machine vision is the ability of a machine to look at or see objects and to differentiate between those objects intelligently. A vision system can observe defects even at microscopic levels. Because of its success, the implementation of machine vision is extensive. In the food industry machine vision is used for sorting purposes in quality control applications.

Voice recognition systems respond to the human voice instead of a keyboard or other input. In certain industrial settings such as package routing, voice recognition has been successful. However, because of the wide variety in human voice patterns, voice recognition has to be constrained to a narrow range of applications or subjects.

Natural language processing allows computer software to understand normal conversational commands. Natural language is one of the least-implemented artificial intelligence technologies. A major reason for this is a lack of improvements in the technology. Natural language programs are useful in some well-defined applications. In data-base applications, natural language programs translate queries from natural language to some data-base query language.

Expert systems are computer programs designed to solve problems requiring experts. An expert system includes a knowledge base and an inference engine. The knowledge base is the accumulation and representation of knowledge specific to a particular task. It contains all the system's task-specific information. It is often in the form of rules, but methods such as frames and semantic nets are also used. The inference engine is the system's machinery for selecting and applying knowledge from the knowledge base to the specific problem. In conventional programs the knowledge base is part of the program, but with an expert system it is separate. This distinction makes it possible to substitute a new knowledge base for a new task in place of the existing knowledge base. Expert systems can be updated readily as new knowledge is discovered. In the field of artificial intelligence, knowledge-based expert systems have received most of the attention in recent years. Expert systems take advantage of an expert's heuristics or knowledge from experience. This enables the program to more quickly narrow the search for a solution. The combination of heuristics and learned principles helps account for the value of an expert. Usually an expert system is developed through an iterative process where an initial program is prepared and then developed and improved with additional items of knowledge. Expert systems cannot exactly model human problem-solving processes, but rather they attempt to interact with expert thought concerning specific areas of knowledge. Most expert systems allow confidence factors in order to express uncertainty. Uncertainty may arise from either the stored knowledge base data or user responses.

Although an expert system is useful for arriving at a solution from numerous options, its range of focus needs to be narrow to avoid an excessively complex system. The economic advantages of going to the time and expense of developing an expert system need to be established. If the problem is very simple or if there are ample resources already available to solve the problem within the company, then the investment may not be worthwhile. Some problems are too broad or general to be solved by an expert system. In order to be solved by an expert system, the problem would have to be solvable by an expert in the field. Expert systems work well with programs where symbolic logic and rules are required.

Advantages and Disadvantages

Expert systems can benefit problem-solving situations in which humans suffer from cognitive overload or fail to monitor all available information. It can be difficult to simultaneously manipulate all relevant information to obtain the optimal solution. An expert system may be more consistent in certain situations in which there is little time to think. Expert systems can often deliver solutions more quickly than their human counterparts. They are also readily available at any time. Human experts are able to better use their skills on more difficult problems beyond the ability of the expert system. The process of building a knowledge base can help to clarify and organize an expert's logic and thinking process. A disadvantage of an expert system is that it can be costly to develop. It also needs to be maintained and updated to remain current.

Knowledge Representation

Semantic networks are collections of nodes interrupted by arcs or links. Biological evolution charts are examples of semantic networks. Expert systems can be based on these types of knowledge representations. Another approach in developing a knowledge base is to use object-attribute-value triplets. In this case the object is the physical entity or concept, the attribute is the characteristic, and the value is the specification of the characteristic. Rule-based expert systems are commonly used. Rules can be in an object-attribute or an object-attribute-value form. They include a premise and a conclusion. These are often referred to as the antecedent and consequent using IF and THEN statements. Frames are descriptions of an object with slots for all appropriate information. Frames have the capability to structure a knowledge base.

Inference Engine

An inference engine uses the knowledge base along with responses from the user to solve problems. Inference engines use the process of forward chaining, backward chaining, or a combination of the two. Reasoning from facts to conclusions is called forward chaining. The infer-

ence engine takes information and proceeds forward through the knowledge base looking for a valid path. It is often referred to as data driven. Reasoning from conclusions to facts is called backward chaining. The inference engine starts with goals and works backward until a clear path is found. There are usually fewer possible outcomes, and it is referred to as goal driven. As the inference engine examines rules and facts, it adds new facts where possible. It also decides the order in which inferences are made.

Software

Symbolic processors are most suitable for rule-based systems. Symbolic languages process symbols rather than numbers. Languages such as LISP allow symbolic processing on general-purpose computers. PROLOG is another common symbolic language. Expert system shells allow expert systems to be developed using English-language rules. Also, they can be written in algorithmic languages such as PASCAL or C. Many programs are currently on the market with widely varying prices.

PROCESS CONTROL

Early process-control systems were connected to the system through an RS-232 port. Data were transferred from the control system to the expert system. The operator would receive the results of the analysis from the computer. Currently the expert system is the control system without a separate specialized computer.

Self-Tuning Controllers

Proportional-integral-derivative or PID controllers are widely used in process control. The addition of an expert system will advise the controller when to adjust itself. These devices use pattern recognition algorithms, calculate control loop performance on-line, and readjust controller settings automatically. They do not require operator intervention. With pattern recognition, responses are observed and necessary PID adjustments are made. PID controllers can also eliminate unwanted oscillations using frequency filtering. This is achieved by forcing the outputs of two filters to conform to a certain ratio. Self-tuning controllers can also use model-based references where the difference in response between the process and the model is used to tune the controller and also update the model. Expert-system tools can help monitor a process and notify an operator of an out-of-specification situation. They become even more useful when they provide the operator with information on how to correct the problem.

Control-System Configurators

The arrangement of equipment and controls is a complex, time-consuming assignment and many factors need to be considered. With control-system configuration, operating conditions and parameters, objectives, and constraints such as equipment limitations can be entered into a process model. The configuration best satisfying all requirements is then presented to the user; some type of graphic display is commonly used. An expert system can respond to user questions, query the user for additional information, and provide a report of the interaction along with a process and instrumentation diagram.

Diagnostic or Fault Analyzers

Fault analyzers respond to alarms and other process inputs to diagnose abnormal conditions. They can find a previously determined pattern in a set of alarm messages. Rather than receiving a number of alarms, the operator will receive a simplified message stating the recommendation. He can then take the required corrective action.

APPLICATION EXAMPLES

Expert systems have been used in a variety of applications. These include electronic troubleshooting, determination of molecular structure from mass spectrometer data, infectious disease diagnosis and therapy selection, and operations in symbolic mathematics.

A processor of canned soup has developed a diagnostic program dealing with malfunctions of a large 68,000 can retort or pressure cooker (2,3). The objective was to obtain knowledge of an expert employee who was near retirement and make it available to plant maintenance personnel. A rule-based expert system was designed and developed to be applied to all the company's hydrostatic and rotary retorts. It includes startup and shutdown procedures and has been installed in eight of the company's plants. Other systems have been developed to diagnose problems in filling and can-closing equipment.

An expert system was designed to permit rapid analysis of cheese samples for comparison to a standard reference (4). It operates in MS/DOS mode using PROLOG and has mainly been applied to Sicilian cheeses.

An expert system has been developed that can identify foods according to variety and origin (5). The system includes identification tools that associate a confidence degree to the final hypothesis. The program can work with 50 chemical parameters whose data have been previously acquired by the analyst. The system has been verified with 144 oil samples using 50 analytic variables such as acidity, color, fatty acid, sterols, and triglycerides.

An expert system for food labeling has been reported (6). The system consults a source of food regulations and proposes one or several food names for a product under evaluation. It then produces a food label based on the given information. The expert system uses a data base on food composition to code and retrieve nutritional information. Information on factors such as sugar content, food source, type of processing, and other user information can also be used. It is particularly useful as a tool for people who are not familiar with food regulations.

A retail specialty bakery chain uses artificial intelligence in several of the company's operations (7). An integrated artificial program known as the Retail Operations Intelligence (ROI) system was developed. The ROI system includes modules entitled Daily Sales Capture, Sales Reporting and Analysis, Labor Scheduler, Form-Mail, Time Collection, Repair and Maintenance, Interviewing, Skill Testing, Daily Production Planner, Inventory, Tax and Licensing, and Lease Management. Each module or component runs on a PC with 640 K of memory and a 20 MG

hard disc. Nationwide data are collected with a minicomputer. The production planner, as an example, checks special dates that may affect sales, historic records, projects how much of a particular product can be expected to be sold during a particular time, and how much raw material to order. Hourly progress is charted and continual recommendations are given for increasing sales based on hourly progress. Even factors such as the weather are considered.

An example of an expert system that is indirectly related to food processing is one that is used in the chemical industry for a chlorine vaporizer (8). Liquid chlorine is vaporized with steam. The liquid chlorine level is controlled by adjusting the steam flow. An erroneous low-level signal to the controller could reduce the steam flow. The vaporizing container and the vapor line could fill with chlorine, leading to a dangerous situation. With the expert system the available data can be analyzed to detect the erroneous signal and warn the operator with an alarm.

A substantial portion of the computer resources in a food-processing plant are devoted to office information systems. Expert systems are used to diagnose the causes of malfunctioning office hardware and software. One such program enables workers and specialists to locate the source of problems with office computer equipment. Using about 200 rules, the system guides workers through a series of questions. The system then recommends methods of solving the problem. A major objective of the program is to allow highly trained specialists to concentrate on the more difficult problems.

FUTURE DEVELOPMENT

Expert systems will help penetrate organizational layers and become an important part of computer-integrated manufacturing. Frame-based shells will become more available for microcomputers. As software becomes easier to use, the distinction between dependent experts in a particular field and knowledge engineers who construct the knowledge base will diminish. The number of expert-systems vendors will probably decrease. Programs will become more standardized in the features they offer. An increased use of expert systems with robotics and process controllers is likely to occur. As the food industry comes to realize the proper application of expert systems, their use will increase.

BIBLIOGRAPHY

1. E. Charniak and D. McDermott, *Introduction to Artificial Intelligence*, Addison-Wesley, Reading, Mass., 1985.

2. R. A. Herrod, "Industrial Applications of Expert Systems and the Role of the Knowledge Engineer," *Food Technology* **43**(5), 130–132, 134 (1989).

3. R. K. Tyson and R. A. Herrod, "Capturing 32 Years on Computer," *Chilton's Food Engineering* **57**, 69–71 (Dec. 1985).

4. C. Russo, C. M. Lanza, and F. Tomaselli, "Use of Expert Systems in the Quality Control of Typical Sicilian Cheeses," *Industrie Alimentare* **28**(268), 119–122, 130 (1989).

5. R. Aparicio, "The SEXIA Expert System," *Journal of Chemometrics* **3**(suppl. A), 175–192 (1988).

6. C. M. A. Geslain-Laneelle, A. P. Soyeux, and M. H. Feinberg, "Expert System for Food Labeling," *Food Technology* **43**(4), 98, 100, 103 (1989).

7. H. P. Newquist, III, "Lisp Chips to Chocolate Chips," *AI Expert* **4**(11), 55–58 (1989).

8. K. W. Goff, "Artificial Intelligence in Process Control," *Mechanical Engineering* **107**(10), 53–57 (Oct. 1985).

ROBERT L. OLSEN
Schreiber Foods, Inc.
Green Bay, Wisconsin

ARTIFICIAL SWEETENERS. See SWEETENERS, NONNUTRITIVE.

ASEPTIC PROCESSING AND PACKAGING SYSTEMS

In aseptic processing, packages and food product are sterilized in separate systems. The sterile package is then filled with sterile product, closed, and sealed in a sterile chamber. Because aseptic processing is a continuous operation, the behavior of one part of the system can affect the overall performance of the entire system. As a result, there are numerous critical factors associated with aseptic processing and packaging often requiring automated control systems. Process establishments for aseptic systems must consider not only the sterilization of the product, the processing equipment, and downstream piping, but also the sterilization of the packaging material and the packaging equipment and the maintenance of sterile conditions throughout the aseptic system.

DEFINITIONS

To assist in the discussion of aseptic processing and packaging systems, some definitions are presented here. You should be familiar with these terms and their particular significance to aseptic systems.

Aseptic describes a condition in which there is an absence of microorganisms, including viable spores. In the food industry, the terms aseptic, sterile, and commercially sterile are often used interchangeably.

Aseptic system refers to the entire system necessary to produce a commercially sterile product contained in a hermetically sealed container. This term includes the product processing system and the packaging system.

Aseptic processing system refers only to the system that processes the product and delivers it to a packaging system.

Aseptic packaging system refers to any piece of equipment that fills a sterile package or container with sterile product and hermetically seals it under aseptic conditions. These units or systems may also form and sterilize the package.

BASIC ASEPTIC SYSTEM

Figure 1 is a diagram of a simplified aseptic system. Raw or unprocessed product is heated, sterilized by holding at

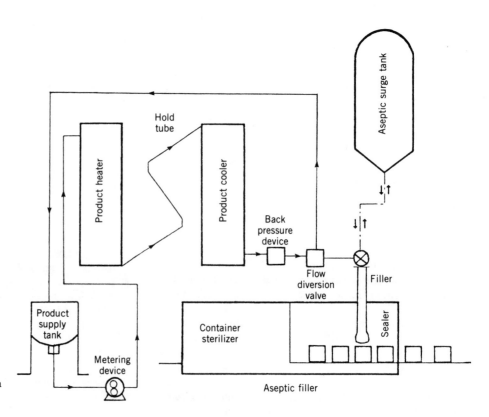

Figure 1. Simplified diagram of an aseptic processing system.

high temperature for a predetermined amount of time, then cooled and delivered to a packaging unit for packaging. Commercial sterility is maintained throughout the system, from the moment of product heating to the discharge of hermetically sealed containers.

Achieving successful aseptic processing of foods requires as a minimum:

1. Equipment that can be brought to a condition of commercial sterility.
2. Commercially sterile product.
3. Commercially sterile packages.
4. A commercially sterile environment within the packaging machine in which to bring sterile product and packages together and hermetically seal the packages.
5. Monitoring and recording of critical factors.
6. Proper handling of finished packages to ensure container integrity.

DESCRIPTION OF THE ASEPTIC PROCESSING SYSTEM

Although the equipment for aseptic processing systems varies, all systems have certain common features:

1. A pumpable product.
2. A means to control and document the flow rate of product through the system.
3. A method of heating the product to sterilizing temperatures.
4. A method of holding product at an elevated temperature for a time sufficient for sterilization.
5. A method of cooling product to filling temperature.
6. A means to sterilize the system before production and to maintain sterility during production.
7. Adequate safeguards to protect sterility and prevent nonsterile product from reaching the packaging equipment.

Pre-Production Sterilization

Producing a commercially sterile product cannot be assured unless the processing system and filler have been adequately sterilized before starting production. It is important that the system be thoroughly cleaned before sterilization; otherwise the process may not be effective.

Some systems, or portions thereof, use saturated steam for sterilization. However, for most systems, equipment sterilization is accomplished by circulation of hot water through the system for a sufficient length of time to render it commercially sterile. When water is used, it is heated in the product heater and then pumped through all downstream piping and equipment up to (and generally past) the filler valve on the packaging unit. All product contact surfaces downstream from the product heater must be maintained at or above a specified temperature by continuously circulating the hot water for a required period of time.

Surge tanks are generally sterilized with saturated steam rather than hot water owing to their large capacity. Although sterilization of surge tanks may occur separately, it is usually conducted simultaneously with hot water sterilization of the other equipment.

To control aseptic system sterilization properly, it is necessary that a thermometer or thermocouple be located at the coldest point(s) in the system to ensure that the proper temperature is maintained throughout. Thus, the temperature-measuring device is generally located at the most distant point from the heat exchangers. Timing of the sterilization cycle begins when the proper temperature is obtained at this remote location. If this temperature should fall below the minimum, the cycle should be restarted after the sterilization temperature is reestablished. Recording devices are recommended to provide a permanent continuous record to show that the equipment is adequately sterilized before each production run.

Flow Control

Sterilization time or residence time, as indicated in the scheduled process, is directly related to the rate of flow of the fastest moving particle through the system. The fastest moving particle is a function of the flow characteristics of the food. Consequently, a process must be designed to ensure that product flows through the system at a uniform and constant rate so that the fastest moving particle of food receives at least the minimum amount of heat for the minimum time specified by the scheduled process. This constant flow rate is generally achieved with a pump, called a timing or metering pump.

Timing pumps may be variable speed or fixed rate. The pumping rate of the fixed rate pump cannot be changed without dismantling the pump. Variable speed pumps are designed to provide flexibility and allow for easy rate changes. When a variable speed pump is used, it must be protected against unauthorized changes in the pump speed that could affect the rate of product flow through the system. A lock or a notice from management posted at or near the speed-adjusting device warning that only authorized persons are permitted to make adjustments is a satisfactory way to prevent unauthorized adjustments.

Product Heating

A product heater brings the product to sterilizing temperature. There are two major categories of product heaters in aseptic food processing: direct and indirect.

Direct heating, as the name implies, involves direct contact between the heating medium (steam) and the product. Direct heating systems can be one of two types: steam injection or steam infusion.

Steam injection introduces steam into the product in an injection chamber as product is pumped through the chamber (Fig. 2). Steam infusion introduces product through a steam-filled infusion chamber (Fig. 3). These systems are currently limited to homogeneous, low viscosity products.

Direct heating has the advantage of very rapid heating, which minimizes organoleptic changes in the product.

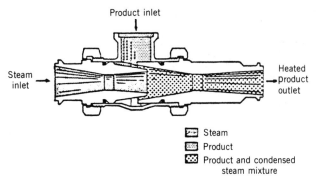

Figure 2. Steam injection.

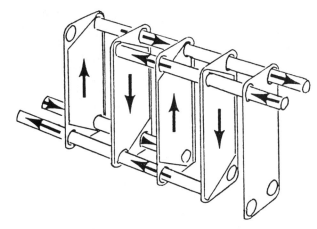

Figure 4. Plate heat exchanger.

The problems of fouling or burn on of product in the system may also be reduced in direct heating systems compared to indirect systems.

There are also some disadvantages. The addition of water (from the condensation of steam in the product) increases product volume. Because this change in volume increases product flow rate through the hold tube, it must be accounted for when establishing the scheduled process. Depending on the product being produced, water that was added as steam may need to be removed. Water removal is discussed under product cooling. Steam used for direct heating must be of culinary quality and must be free of noncondensable gases. Thus, strict controls on boiler feed water additives must be followed.

The other major category of product heaters is indirect heating units. Indirect heating units have a physical separation between the product and the heating medium. There are three major types of indirect heating units: plate, tubular, and swept-surface heat exchangers.

Plate heat exchangers (Fig. 4) are used for homogeneous liquids of relatively low viscosity. The plates serve as both a barrier and a heat transfer surface with produc-

tion on one side and the heating medium on the other. Each plate is gasketed, and a series of plates are held together in a press. The number of plates can be adjusted to meet specific needs.

Tubular heat exchangers (Fig. 5) employ either two or

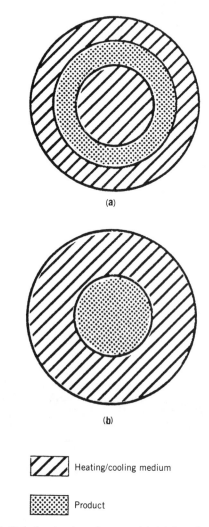

Figure 3. Steam infusion.

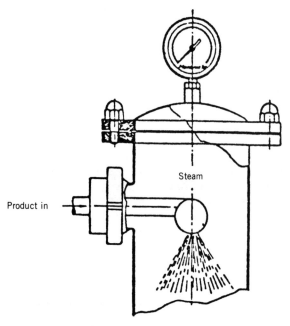

Figure 5. Tubular heat exchanger: (a) triple tube; (b) double tube.

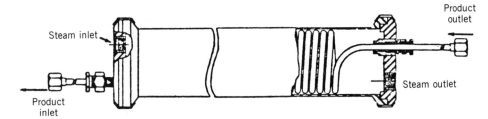

Figure 6. Shell and tube heat exchanger.

three concentric tubes instead of plates as heat-transfer surfaces. Product flows through the inner tube of the double-tube style and through the middle tube of the triple-tube style, with the heating medium in the other tube(s) flowing in the opposite direction to the product. In a shell and tube heat exchanger (considered a type of tubular exchanger), the tube is coiled inside a shell (Fig. 6). Product flows through the tube while the heating medium flows in the opposite direction through the shell. As with plate heat exchangers, tubular heat exchangers are used for homogeneous products of low viscosity.

Scraped-surface heat exchangers are normally used for processing more viscous products (Fig. 7). The scraped-surface heat exchanger consists of a mutator shaft with scraper blades concentrically located within a jacketed, insulated heat exchange tube. The rotating blades continuously scrape the product off the wall. This scraping reduces buildup of product and burn on. The heating medium flowing on the opposite side of the wall is circulating water or steam.

Some systems incorporate the use of product-to-product heat exchangers. These devices are either plate or tubular heat exchangers with product flowing on both sides of the plates or through both sets of tubes. This process allows the heat from the hot, sterile product to be transferred to the cool, incoming, nonsterile product. Energy and cost savings can be significant by recycling the heat from sterile product.

When a product-to-product regenerator is used, the regenerator must be designed, operated, and controlled so that the pressure of the sterilized product in the regenerator is at least 1 psi greater than the pressure of any nonsterilized product in the regenerator. This differential helps ensure that any leakage in the regenerator will be from sterilized product into nonsterilized product. In addition, an accurate differential pressure recorder must be installed on the regenerator. One pressure sensor has to be installed at the hot sterilized product outlet and another at the cooler nonsterilized product inlet. Upon installation, the recorder must be tested against a known accurate standard pressure indicator and must be retested at least once every 3 mo, or as often as necessary to ensure proper functioning. Pressure differential maintenance and recording is also recommended for heat exchangers employing product-to-water regeneration and may be a requirement in certain instances.

Hold Tube

Once the product has been brought to sterilizing temperature in the heater, it flows to a hold tube. The time required for the fastest product particle to flow through the hold tube is referred to as the residence time. The residence time must be equivalent to or greater than the time necessary at a specific temperature to sterilize the product and is specified in the scheduled process. Hold-tube volume, which is determined by hold-tube diameter and length, combined with the flow rate and flow characteristics of the product, determines the actual residence time of the product in the hold tube. Because the hold tube is essential for ensuring that the product is held at sterilization temperatures for the proper time, certain precautions must be followed:

1. The holding tube must have an upward slope in the direction of product flow at least 0.25 in./ft to assist in eliminating air pockets and prevent self-draining.

2. If the holding tube can be taken apart, care should be taken that all parts are replaced and that no parts are removed or interchanged to make the tube shorter or different in diameter. Such accidental alterations could shorten the time the product remains in the tube.

3. If the holding tube can be taken apart, care should be exercised when reassembling to ensure that the gas-

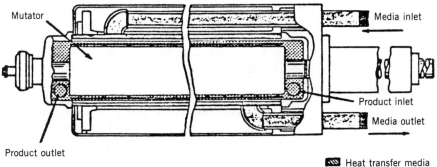

▨ Heat transfer media
▨ Product

Figure 7. Scraped-surface heat exchanger.

kets do not protrude into the inner surface. The tube interior should be smooth and easily cleanable.

4. There must be no condensate drip on the tube, and the tube should not be subjected to drafts or cold air, which could affect the product temperature in the holding tube.

5. Heat must not be applied at any point along the hold tube.

6. The product in the hold tube must be maintained under a pressure sufficiently above the vapor pressure of the product at the process temperature to prevent flashing or boiling because flashing can decrease the product residence time in the hold tube. The prevention of flashing is usually accomplished by use of a back-pressure device.

The temperature of the food in the holding tube must be monitored at the inlet and outlet of the tube. The temperature at the inlet of the tube is monitored with a temperature recorder-controller sensor that must be located at the final heater outlet and must be capable of maintaining process temperature in the hold tube. A mercury-in-glass thermometer or other acceptable temperature-measuring device (such as an accurate thermocouple-recorder) must be installed in the product sterilizer holding tube outlet between the holding tube and the cooler inlet. An automatic recording thermometer sensor must also be located in the product at the holding tube outlet between the holding tube and cooler to indicate the product temperature. The temperature-sensing device chart graduations must not exceed 2°F (1°C) within a range of 10°F (6°C) of the required product sterilization temperature.

Product Cooling

Product flows from the hold tube into a product cooler that reduces product temperature before filling. In systems that use indirect heating, the cooler will be a heat exchanger that may be heating raw product while cooling sterile product. Those systems that use direct heating will typically include a flash chamber or vacuum chamber. The hot product is exposed to a reduced pressure atmosphere within the chamber, resulting in product boiling or flashing. The product temperature is lowered, and a portion of all of the water that was added to the product during heating is removed by evaporation. On discharge from the flash chamber, product may be further cooled in some type of heat exchanger.

Maintaining Sterility

After the product leaves the hold tube, it is sterile and subject to contamination if microorganisms are permitted to enter the system. One of the simplest and best ways to prevent contamination is to keep the sterile product flowing and pressurized. A back-pressure device is used to prevent product from boiling or flashing and maintains the entire product system under elevated pressure.

Effective barriers against microorganisms must be provided at all potential contamination points, such as rotating or reciprocating shafts and the stems of aseptic valves. Steam seals at these locations can provide an effective barrier, but they must be monitored visually to ensure proper functioning. If other types of barriers are used, there must be a means provided to permit the operator to monitor the proper functioning of the barrier.

Aseptic Surge Tanks

Aseptic surge tanks have been used in aseptic systems to allow the packer to hold sterile product before packaging. These vessels, which range in capacity from about 100 gal to several thousand gallons, provide flexibility, especially for systems in which the flow rate of a product sterilization system is not compatible with the filling rate of a given packaging unit. If the valving that connects a surge tank to the rest of the system is designed to allow maximum flexibility, the packaging and processing functions can be carried out independently, with the surge tank acting as a buffer between the two systems. A disadvantage of the surge tank is that all sterile product is held together, and if there is a contamination problem, all product is lost. A sterile air or other sterile gas supply system is needed in order to maintain a protective positive pressure within the tank and to displace the contents. This positive pressure must be monitored and controlled to protect the tank from contamination.

Automatic Flow Design

An automatic flow diversion device may be used in an aseptic processing system to prevent the possibility of potentially unsterile product from reaching the sterile packaging equipment. The flow diversion device must be designed so that it can be sterilized and operated reliably. Past experience has shown that flow diversion valves of the gravity drain type should not be used in aseptic systems owing to the possibility of recontamination of sterile product. Because the design and operation of a flow-diversion system is critical, it should be done in accordance with recommendations of an aseptic processing authority.

The flow diversion valve should divert product automatically if a deviation occurs. A few examples of situations that may cause a diversion are temperature at the hold tube dropping below the scheduled minimum, inadequate pressure differential in the regenerators, or the packaging unit dropping below minimum operating specifications.

ASEPTIC PACKAGING SYSTEMS

Basic Requirements

Aseptic packaging units are designed to combine sterile product with a sterile package, resulting in a hermetically sealed shelf-stable product. As with aseptic processing systems, there are certain features common to all aseptic packaging systems. The packaging units must

1. Create and maintain a sterile environment in which the package and product can be brought together.
2. Sterilize the product contact surface of the package.
3. Aseptically fill the sterile product into the sterilized package.
4. Produce hermetically sealed containers.
5. Monitor and control critical factors.

There are a wide variety of packaging systems that satisfy these requirements in many different ways. The following discussion concentrates on those requirements common to all aseptic packaging systems, using the features mentioned previously as a basis for discussion.

Sterilization Agents

Sterilization agents are used in aseptic packaging units to sterilize the packaging material and the internal equipment surfaces to create a sterile packaging environment. In general, these agents involve heat, chemicals, high-energy radiation, or a combination of these. For aseptic packaging equipment the sterilization agents used must effectively provide the same degree of protection in terms of microbiological safety that traditional sterilization systems provide for canned foods. This requirement applies to both the food contact surface of the packaging material and the internal machine surfaces that constitute the aseptic or sterile zone within the machine. The safety and effectiveness of these agents must be proven and accepted or approved by regulatory agencies for packaging commercially sterile low-acid or acidified foods in hermetically sealed containers. Food processors considering use of an aseptic packaging unit should request written assurances that the equipment has passed such testing and that the equipment and sterilizing agents are acceptable to the regulatory agencies for their intended use.

Heat is the most widely used method of sterilization. Steam or hot water is commonly used and referred to as moist heat. Superheated steam or hot air may also be used in certain situations and is referred to as dry heat. Dry heat is a much less effective sterilization agent than moist heat at the same temperature. Systems that use moist heat operate at elevated pressures compared to dry heat systems, which operate at atmospheric pressures. Other methods may be used to generate heat, such as microwave radiation or infrared light. As new methods are developed, they will have to be evaluated by aseptic processing authorities.

Chemical agents such as hydrogen peroxide are often used in combination with heat as sterilization agents. The Food and Drug Administration (FDA) regulations specify that a maximum concentration of 35% hydrogen peroxide may be used for food contact surfaces. If hydrogen peroxide is used as a sterilant, the packaging equipment must be capable of producing finished packages that also meet FDA requirements for residuals. Not more than 0.5 ppm hydrogen peroxide may be present in tests done with distilled water packaged under production conditions. These regulations apply also to U.S. Department of Agriculture (USDA) regulated products.

Other sterilants, such as high energy radiation (uv light, gamma or electron beam radiation), could be used alone or in combination with existing methods. Completely new alternative sterilants may be developed in the future. Whatever methods are developed, they will have to be proven effective in order to protect the public health and will be compared with existing methods.

Aseptic Zones

The aseptic zone is the area within the aseptic packaging machine that is sterilized and maintained sterile during production. This is the area in which the sterile product is filled and sealed in the sterile container. The aseptic zone begins at the point where the package material is sterilized or where presterilized package material is introduced into the machine. The area ends after the seal is placed on the package and the finished package leaves the sterile area. All areas between these two points are considered as part of the aseptic zone.

Before production, the aseptic zone must be brought to a condition of commercial sterility analogous to that achieved on the packaging material or other sterile product contact surfaces. This area may contain a variety of surfaces, including moving parts composed of different materials. The sterilant(s) must be uniformly effective and their application controllable throughout the entire aseptic zone.

Once the aseptic zone has been sterilized, sterility must be maintained during production. The area should be constructed in a manner that provides sterilizable physical barriers between sterile and nonsterile areas. Mechanisms must be provided to allow sterile packaging materials and hermetically sealed finished packages to enter and leave the aseptic zone without compromising the sterility of the zone.

The sterility of the aseptic zone can be protected from contamination by maintaining the aseptic zone under positive pressure of sterile air or other gas. As finished containers leave the sterile area, sterile air flows outward, preventing contaminants from entering the aseptic area. The sterile air pressure within the aseptic zone must be kept at a level proven to maintain sterility of the zone. Air or gases can be sterilized using various sterilization agents, but the most common methods are incineration (dry heat) and/or ultrafiltration.

Production of Aseptic Packages

A wide variety of aseptic packaging systems are in use today. These are easily categorized by package type:

1. Preformed rigid and semirigid containers, including
   a. Metal cans
   b. Composite cans
   c. Plastic cups
   d. Glass containers
   e. Drums
2. Web-fed paperboard laminates and plastic containers.
3. Partially formed laminated paper containers.
4. Thermoform–fill–seal containers.
5. Preformed bags or pouches.
6. Blow-molded containers.

A number of different packaging systems are represented in these categories. Not all of these systems, however, are being used in the United States for aseptic applications.

Containers in these categories may be sterilized by a variety of means. For example, one system utilizing metal cans uses superheated steam to sterilize the containers. In other systems, preformed plastic cups may be sterilized by hydrogen peroxide and heat or by saturated steam. Systems using containers formed from paperboard laminates also utilize hydrogen peroxide and heat or hydrogen peroxide and ultraviolet irradiation to sterilize packages. Thermoform–fill–seal containers may be sterilized by the heat of extrusion (dry heat) or by hydrogen peroxide and heat. Plastic pouches or bags may be sterilized by gamma irradiation, by the heat of extrusion, or by chemical means such as hydrogen peroxide.

Research is now being conducted to explore alternative sterilization methods for most categories of packaging. Thus, it can be said with some certainty that currently familiar equipment may not be state of the art tomorrow. Nevertheless, whatever equipment or sterilant or packaging material is used, the monitoring and control of critical factors will be vital to successful operation.

INCUBATION

Incubation is defined as the holding of a sample at a specified temperature for a period of time for the purpose of permitting or stimulating the growth of microorganisms that may be present. Routine product incubation programs are recommended as a check on the overall quality and sterility of aseptic products. If the packer is producing an item that is covered by the inspectional authority of the (USDA), routine incubation of samples is mandatory and the exact program should be established in cooperation with the regulatory representative. Incubation tests should be conducted on a representative sample of containers of product from each packaging code. Records of the incubation tests should be kept.

RECORD REQUIREMENTS

Accurate record keeping is essential to successful operation of aseptic systems. Automatic recording and monitoring systems are relied on more heavily than in traditional processing because of the inherent complexity and numerous critical factors associated with aseptic systems.

The kind of recording or monitoring system used is determined by the variability of individual factors during operation. For example, machine packaging rate is often a preset function with relatively little variability. When machine packaging rate is a critical factor, operator monitoring may be sufficient to ensure that the rate stays within critical factor limits. On the other hand, critical factors such as temperature, which could be affected by a variety of other machine functions, can exhibit greater

Container Size _____

Filler Used   _____

Production Process

(Time/Temp.) _____

Chart No. _____   Page ____ of ____

_____   Date _____

_____   Lot. No._____

| Product/Code | Production | | | CPM/ GPH | HOMO PSI | Holding Tube | | | | Visual | |
| | | | | | | Inlet | Outlet | | Pre- Heat | Regen- PHSIG | Steam |
	Start	Stop	Time			°F	T/C °F	Hg °F	°F	Diff.	Seal Check

Sterilization of	°F	Time	°F	Time
Processing System				

COMMENTS: _____

_____

_____

Figure 8. Processing log—product processing system.

variability. Thus, the need for automatic recording or monitoring of temperature would be greater.

Production records for an aseptic system consist of production logs and recording charts from both the aseptic processing system and the aseptic packaging system. Recording charts are a continuous record of the aseptic system's performance. Operators must mark the charts clearly, in ink, to indicate the time that sterilization and production begin and end. If production is stopped for any reason, such as a drop in process temperature, loss of sterile air pressure, or packaging problems, the time of day that production is stopped and restarted should be noted on the chart.

The production log is a written record of the aseptic system's operation. The operator must note in the log the time of day that events occur, along with any problems and related corrective action. In addition, the appropriate regulations must be consulted to ensure that all required information is included in the log.

Examples of production log entries may include the following information (Fig. 8):

1. Temperatures at the hold-tube outlet taken from the recorder chart and from the temperature indicating device.

2. Maintenance of a proper pressure differential if a product-to-product regenerator is used

3. Adequate operation of steam seals.

4. Information relating to additional critical factors.

Although entries for the aseptic packaging system would change according to the type of equipment, the following information could be included (Fig. 9).

1. Temperature in container sterilizer tunnels (metal containers)

2. Container rate through the packaging unit.

3. Temperature and pressure in the air sterilizer.

4. Temperature of heated air or hydrogen peroxide.

5. Hydrogen peroxide concentration at start and end of production.

6. Any additional critical factors.

For aseptic surge tanks, a record must be kept to show that the tank has been sterilized and that the sterile air within the tank is maintained at the proper pressure. If the air is sterilized by heat, a record of the incinerator temperature is necessary. If the air filter is sterilized, a

Chart No. _____   Page ____ of ____

_____   Date _____

Lot No. _____

| Product/Code | Production | | | Sterile Air | | H_2O_2 | System | | | Heating Element °F |
	Start	Stop	Time	Temp. °F	Pressure	Level	Low cons.	Wet Guard	PSM	

Sterilization of Filler	Sterile Air Temp. °F	Heater Draw	Time

H_2O_2 Concentration: _____

Start of Production: _____

End of Production: _____

COMMENTS: _____

_____

_____

Figure 9. Processing log—aseptic filler.

record must show that the filter and sterile air supply system was sterilized at the required temperature and time at the beginning of the operation and that the filter cartridge was changed at proper intervals.

Container Integrity

The integrity of aseptic containers must be ascertained by inspection and testing in order to assure that the containers will maintain a hermetic seal during handling, distribution, and storage. The concept of maintaining container integrity is part of good manufacturing practice and an important function of every good quality control and assurance program.

Because there is such a wide variety of containers, the test methodology may be designed for a particular type of package. Both the FDA and USDA require that container integrity be determined through regular inspection procedures.

SUMMARY

1. Aseptic is a term that describes the absence of microorganisms and may be used interchangeably with commercial sterility.

2. An aseptic processing system consists of a timing pump, a means to heat the product, a hold tube, and a means to cool the product.

3. The processing system from the hold tube past the fillers is brought to a condition of commercial sterility before the introduction of product into the system.

4. The thermal process occurs in the hold tube where flow rate, or residence time, and temperatures are critical factors.

5. In the event of a process deviation, flow diversion devices are used to prevent potentially inadequately processed product from reaching the filler.

6. Aseptic packaging machines create and maintain an aseptic zone (sterile environment) in which sterilized containers are filled and sealed.

7. Sterilization agents such as heat, chemicals, high-energy radiation, or a combination of these, can be used to sterile packages or machine surfaces.

8. Accurate record keeping is essential to successful operation of aseptic processing and packaging operations.

9. For government regulations, consult the appropriate offices in the FDA and USDA.

K. STEVENSON
National Food Processors
Association

ASEPTIC PROCESSING: OHMIC HEATING

The recent development of the ohmic heater represents a major advance in the continuous processing of particulate food products.

The ohmic heater uses resistance heating within the flow of electrically conducting liquid and particulates to provide heat and is capable of handling food products containing particulates of up to 25 mm.

When combined with an aseptic container or bag-in-box filling system this new process offers the food processor the possibility of producing high-quality ready-prepared meals containing large quantities of undamaged chunks of meat and vegetables which can be stored at ambient temperatures for long periods of time. The process has also been successfully applied to the sterilization of diced and sliced fruit at solids concentrations of up to 80% drained weight.

PRINCIPLE OF OHMIC HEATING

The ohmic heating effect occurs when an electric current is passed through a conducting product (Fig. 1). In practice, low-frequency alternating current (50 or 60 Hz) from the public main supply is used to eliminate the possibility of adverse electrochemical reactions and minimize power supply complexity and cost.

In common with microwave heating, electrical energy is transformed into thermal energy. However, unlike microwave heating, the depth of penetration is virtually unlimited and the extent of heating is governed by the spatial uniformity of electrical conductivity throughout the product and its residence time in the heater. For most practical purposes the product does not experience a large temperature gradient within itself as it heats, and liquid and particulate are heated virtually simultaneously. The requirement to overprocess the liquid to ensure sterility at the center of a large particulate as with scraped surface and tubular heat exchangers is therefore reduced. This results in less heat damage to the liquid phase and prevents overcooking of the outside of the particulates.

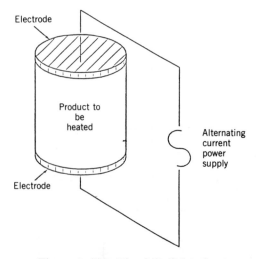

Figure 1. Principle of ohmic heating.

Another major advantage is that there are no heat transfer surfaces, which reduces the possibility of deposit formation and the transfer of burned particles to the finished product. Nor is there a need for mechanical agitation of the product which can cause loss of true particulate identity.

The applicability of ohmic heating is dependent on product electrical conductivity. Most food preparations contain a moderate percentage of free water with dissolved ionic salts and hence conduct sufficiently well for the ohmic effect to be applied. The system will not directly heat fats, oils, alcohols, bone, or crystalline structures such as ice.

DESIGN OF THE OHMIC HEATER

The ohmic heater column typically consists of four or more electrode housings machined from a solid block of polytetra fluoroethylene (PTFE) and encased in stainless steel, each containing a single cantilever electrode (Fig. 2). The electrode housings are connected using stainless steel interconnecting tubes lined with an electrically insulating plastic liner. Suitable lining materials include poly(vinylidene fluoride) (PVDF) and glass. These flanged tube sections are bolted together and sealed with food grade rubber gaskets.

The column is mounted in a vertical or near vertical position with the flow of product in an upward direction. A vent valve positioned at the top of the heater ensures that the column is always full.

The column is configured such that each heating section has the same electrical impedance and hence the interconnecting tubes generally increase in length toward

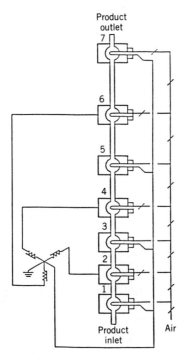

Figure 2. Design of an ohmic heater column.

the outlet. This is so because the electrical conductivity of food products usually increases with increase in temperature; indeed, for aqueous solutions of ionized salts there is a linear relationship between temperature and electrical conductivity.

This phenomenon is attributed to increased ionic mobility with increase in temperature and also applies to most food products. Exceptions could be products in which viscosity increases markedly at higher temperatures, such as those containing ungelled starches.

MEASUREMENT OF ELECTRICAL CONDUCTIVITY OF PARTICULATE FOOD PRODUCTS

Conductivity measurements of products with large particles are often impossible using commercially available cells because of their small internal clearances.

As consistent results can only be obtained using a cell having a cross-sectional area considerably greater than that of the largest particle, APV has developed a special cell to overcome this problem, and can now accurately measure the conductivity of products containing particulates up to 35-mm cubes.

TEMPERATURE CONTROL OF THE OHMIC HEATER

Ohmic heating plants are supplied with a fully automatic feedforward temperature control system. Pure feedback is not satisfactory due to the time constant in the heater column.

Inlet changes which will affect the final product outlet temperature are changes in inlet temperature, mass flow rate, and product specific heat capacity. In the feedforward control system, a microprocessor scans these variables including the last, which is an operator entered or computer generated variable. It continuously computes the electrical power required to heat the product and compares this value with the signal from a power transducer on the output side of the transformer. Feedback monitoring is used to prevent any long-term drift in outlet temperature.

ASEPTIC PROCESSING USING THE OHMIC HEATER

The ohmic heater assembly can be visualized in the context of a complete product sterilization or cooking process. A typical line diagram of such a plant is shown in Fig. 3. In common with other types of continuous systems, careful design of the ancillary process equipment is necessary and, once heated, the product must be cooled by more conventional means such as scraped surface or tubular heat exchangers. The latter are generally preferred for particulate processing in order to maximize the greatest advantage of ohmic heating, that of minimal structural damage to the particulates.

Presterilization of the ohmic heater assembly, holding tubes, and coolers is carried out by recirculation of a solution of sodium sulfate at a concentration which approximates the electrical conductivity of the food material

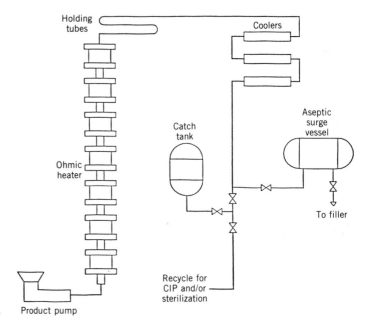

Figure 3. Ohmic plant line diagram.

which will subsequently be processed. Sterilization temperatures are achieved through the passage of electrical current and back pressure is controlled using a back pressure valve. The aseptic storage reservoir, interface catch tank, and connecting pipework to the filler are sterilized by traditional steam methods.

The use of sterilizing solution of similar product electrical conductivity minimizes adjustment of electrical power during subsequent change to product, thus ensuring a smooth and efficient changeover period with little temperature fluctuation.

Once the plant has been sterilized, the recycled solution is cooled, using a heat exchanger in the recirculation line. When steady-state conditions are reached, sterilizing solution is run to drain and product introduced into the hopper of a positive displacement feed pump. Typically, this might be an auger-fed mono or rotary or a Marlen reciprocating piston pump. Product is usually prepared in premix vessels which can incorporate preheating or blanching operations.

Back pressure during the changeover period is controlled by regulation of top pressure in a catch tank using sterile compressed air or nitrogen. This tank serves to collect the sodium sulfate–product interface. Once the interface has been collected, product is diverted to the main aseptic storage vessel where top pressure is similarly used to control back pressure in the system.

Back pressure is maintained at a constant 1 bar when sterilizing high-acid food products at temperatures of 90–95°C. A 4-bar back pressure is used for low-acid food products where sterilization temperatures of 120–140°C are necessary. Safety features are incorporated to ensure that power is automatically switched off should there be any loss of pressure.

The use of pressurized storage vessels has proved to be an extremely effective method of controlling back pressure in pilot-plant heaters where throughputs are typically less than 1,600 lb/h. In larger systems operating in excess of 4,000 lb/h an aseptic positive displacement pump

downstream of the cooler can be used as an alternative, depending on the composition of the food product. This alternative overcomes the requirement for two aseptic storage vessels in order for the system to continuously feed product to the filling machine.

Product is progressively heated to the required sterilization temperature as it rises through the ohmic heater assembly. It then enters an air-insulated holding (and/or cooking) tube before being cooled in a series of tubular heat exchangers.

More rapid cooling can be achieved with some products in which the final particulate level will be less than 40%. This system involves a combination of ohmic and traditional heat treatment and takes advantage of the capability of the ohmic heater to process products containing up to 80% particulates. During preparation, the product is formulated into a high-concentration particulate stream and a separate liquid stream. The liquid stream is conventionally sterilized and cooled in a plate or tubular heat exchanger system before being injected into the particulate stream leaving the holding tube of the ohmic heater. Such an operation has the advantage that it reduces the capital and operating cost for a given throughput. It also allows the electrical conductivity of the carrier fluid in the ohmic heater to be more closely matched to that of the particulate, should they be markedly different, by manipulation of the dissolved ionic contents of the two processing streams. This assists in ensuring that both particulates and carrier fluid are heated at the same rate.

After cooling, product enters the main storage reservoir prior to aseptic filling.

It is not usually necessary to carry out intermediate cleaning and/or resterilization of the plant when several different products are processed. This is due to the almost complete absence of fouling with most food materials. After one product is processed, the plant is flushed with a food-compatible liquid or base sauce before introducing the next product. The catch tank is used to collect product–sauce interfaces.

Higher product quality (than in can sterilization)
Lower energy costs
Lower cost packaging
More attractive packaging
Lower storage costs (than frozen or chilled)

Figure 4. Advantages of aseptic processing.

When food processing is complete, the power is turned off and the plant rinsed with water. This is followed by cleaning using a 2% (w/v) solution of caustic soda recirculated at 60–70°C for 30 min. Heating of the cleaning solution is by standard methods and not by using the ohmic principle. This is because the electrical conductivity of the cleaning solution is too high for the ohmic principle to be applied.

A range of ohmic heaters incorporating different sizes of electrode housings and interconnecting spacer tubes are soon to be available to process food products at throughputs of up to approximately 5,500 lb/h depending on the required temperature rise. A larger unit with 4-in.-ID spacer tubes is currently under development for processing 10,000 lb/h of product.

PRODUCT QUALITY

Included in the list of perceived advantages of continuous aseptic processing (Fig. 4) is that of improved product quality, with the basic standard being set by products sterilized by traditional canning methods.

These quality standards include microbiological safety (ie, process lethality), cooking effects, and nutrient and vitamin retention.

Unlike other processes requiring heat penetration, either from the outside to the center of a can (in-can sterilization) or from the outside of the particulate to the center (tubular or swept surface), ohmic processing has the ability to provide exceptionally fast heating of all of the product virtually simultaneously; this provides a high degree of microbiological security, as there is a very narrow range of lethality from the surface to the center of the particulates.

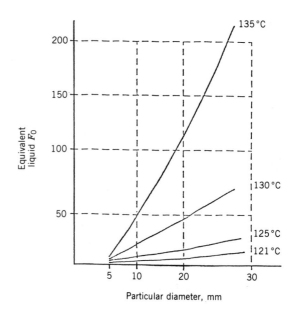

Figure 6. Liquid F_0 necessary to obtain $F_0 = 5$ in center of particulate.

Challenge tests carried out (on now commercially available products) by the Campden Food Research Group (United Kingdom) have involved particulates of meat and vegetables up to 19-mm cubes; an average of 10 million spores of *Bacillus stearothermophilus* per particle embedded in a food–alginate mixture were included in batches of mince and mushrooms in bolognese sauce and in kidney, peas, and gravy in order to determine the lethalities at the center and the surface of a range of particulate sizes.

The results obtained (Fig. 5) demonstrate the high level of microbiological security of the process and the very narrow range of lethality from the surface to the center of the particulates.

The cook value (C_0) of a process is also a function of

	Particle Type	Number of Particles	F_s Range, min	Mean	Standard Deviation
Run 1	Mince (5 mm)	41	23.6–34.3	27.9	2.2
	Mushroom (19 mm)	16	34.9–40.3	37.9	1.4
	Mushroom (3 mm)	6	31.9–41.8	36.0	4.1
Run 2	Pea (5 mm)	28	33.7–46.4	38.5	2.8
	Kidney (19 mm)	30	47.2–62.8	54.8	3.8
	Kidney (3 mm)	20	45.6–56.2	50.2	3.2

Figure 5. Lethality tests of APV ohmic system on *Bacillus stearothermophilus*.

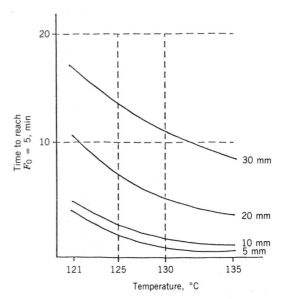

Figure 7. Time taken to reach $F_0 = 5$ in center of particulate.

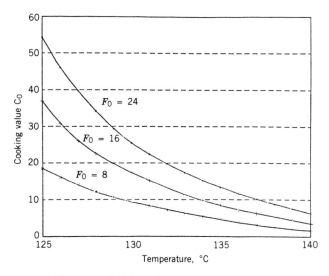

Figure 8. Cooking values versus temperature.

time and temperature (dependent upon the viscosity of the product). For small cans, typical F_0 values for viscous products can range from 5 to 300, resulting in high cook values. Larger cans will show even greater variations in F_0 values and higher degrees of overcook in order to achieve satisfactory F_0 values at the center.

Continuous systems using heat penetration will have wide variations in F_0 values from the center of the particulate to the surface, the variation depending greatly upon the size of the particulate, the ratio of solids to liquid (as the carrier fluid must provide the excess heat required to penetrate the particulate), and the sterilization temperature chosen.

Figures 6 and 7 show the F_0 value given to the carrier fluid to achieve a minimal F_0 value of 5 in the center of the particulate and the processing times required at various temperatures. It can be seen that for particulates larger than 10 mm a considerably wide range of F_0 values results, irrespective of the sterilization temperature chosen. This results in undesirable overcooking of the carrier fluid and the outside surface of the particulates. For example, processing a 19-mm particle to a minimum F_0 value of 5 at a sterilization temperature of 135°C will require the carrier fluid to have an equivalent F_0 treatment of 108, as the heat penetration time will have been on the order of 4 min.

In contrast, ohmic heating can achieve a high F_0 level with a low cook value and low nutrient/vitamin destruction. Figure 8 shows the effect of processing temperature on F_0 and C_0 (cooking effect). The C_0 value is based on a Z value of 33 and is calculated as

$$10^y[T]$$

where $y = \left[\dfrac{°C - 100}{Z}\right]$

T = time, min.

At a processing temperature of 140°C, a virtually uniform F_0 of 24 equates to a likewise uniform C_0 value of less than 5.

This relationship allows such products as diced potato to be subjected to an F_0 value greater than 24 and still not be fully cooked; thus the particle retains its sharp edges

Figure 9. Ohmic plant.

Figure 10. Ohmic processed ham and sweet corn pasta.

Figure 11. Ohmic processed fruit compote (black currant and apple).

Figure 12. Ohmic processed ratatouille.

Figure 13. Ohmic processed drained winter soup.

throughout the process. Indeed, such products processed through the ohmic system therefore require further cooking (as part of the reheat process) by the end user.

The ability to heat particulates uniformly, without mechanical damage, with lower nutrient and vitamin loss, and with no fouling of heat transfer surfaces leaves no doubt that ohmic heating will play a major role in the growing requirements for aseptic food products containing large particulates, as it provides the food processor with an opportunity to produce new high-quality, long-life products that were not possible with alternative sterilization techniques.

Typical of such high-quality added-value products are those already fully tested, which contain meat chunks, shrimps, baked beans, mushrooms, diced or sliced vegetables, pasta, whole strawberries, black currants, blackberries, and sliced kiwi fruit. Figure 9 shows an ohmic plant. Figures 10, 11, 12, and 13 show ohmic processed ham and sweet corn pasta, fruit compote (black currant and apple), ratatouille, and drained winter soup, respectively.

APV Crepaco, Inc.
Lake Mills, Wisconsin

Adapted from "Continuous Aseptic Processing Using the Ohmic Heating Process," by D. F. Dinnage, *dfi News* 11(3):6–13, 1989. Copyrighted APV Crepaco, Inc. Used with permission.

ASPARTAME. See Sweeteners, nonnutritive.

ASSOCIATION OF OFFICIAL ANALYTICAL CHEMISTS

The Association of Official Analytical Chemists (AOAC) is an international organization established to ensure the development, testing, validation, and publication of uniform, precise, and accurate chemical, biological, and physical methods for the analysis of foods, vitamins, food additives, pesticides, drugs, cosmetics, plants, feeds, fertilizers, hazardous substances, air, water, and any other products, substances, or phenomena affecting the public health and safety, the economic protection of the consumer, or the protection of the quality of the environment (1). Other objectives of AOAC are to promote uniformity and reliability in the statement of analytical results; to promote, conduct, and encourage research in the analytical sciences related to agriculture, public health, and regulatory control of commodities in relevant fields; and to afford opportunities for the discussion of matters of interest to scientists engaged in relevant pursuits (2).

These objectives are accomplished by AOAC through its methods validation, publication, meeting, short course, and regional section programs.

HISTORY

The Association of Official Agricultural Chemists was formed in 1884 to adopt uniform methods for regulatory analysis of fertilizers. By 1887, AOAC was also adopting and publishing methods of analysis for feeds and dairy products. After passage of the U.S. Pure Food and Drug Act in 1906, AOAC extended its activities into many areas outside of agriculture. In 1965, the name was changed to the Association of Official Analytical Chemists (3).

MEMBERSHIP

There are four classes of AOAC membership: Individual, Honorary, and Emeritus for individuals, and Sustaining Membership for organizations. As of November 1990, there were 3,750 Individual Members, one Honorary Member, and 221 Sustaining Members.

When AOAC was established, membership was limited to analytical chemists connected with state regulatory agencies exercising control over fertilizer manufacture. Membership was gradually extended, and today, any individual with a degree in science who is or has been engaged, directly or indirectly, in analysis or analytical research in areas of interest to AOAC is eligible for membership (1).

Sustaining Membership was begun in 1980 to encourage governmental agencies, from the local to the national level; colleges and universities; firms; and other organizations with an interest in the work of AOAC to provide financial and scientific support to the association.

PROGRAMS

Methods Validation Program

AOAC's primary program is the coordination of a methods validation process whereby over 800 member-volunteers, working in their official and professional capacities as staff scientists in organizations worldwide, propose and participate in the development, collaborative study, and adoption of methods as official AOAC methods. AOAC methods are used to check compliance with government regulations, and for quality control, product label verification, research, and teaching purposes.

To be approved for adoption by AOAC, methods must undergo an interlaboratory collaborative study and perform with accuracy and precision under usual laboratory conditions. The development and validation of each method is organized and directed by an analyst designated as Associate Referee for the specific subject under investigation.

As of 1990, some 650 Associate Referees appointed by the association are responsible for over 700 topics. Associate Referees are selected for their knowledge, interest, and experience in the subject field. The Associate Referee operates under the scientific guidance of a General Referee, who is responsible for several topics within a subject area.

The Associate Referee reviews the literature and selects a method, modifying it as needed, or develops or adapts a method used in laboratory, testing it thoroughly before designing a collaborative study. The samples to be analyzed in a collaborative study are prepared and distributed to collaborators, who are analysts with experience and interest in the general subject. All collaborators are instructed to follow the method exactly as written. Results are reported to the Associate Referee who compiles and evaluates them, and, if the data are satisfactory, recommends adoption to the General Referee and Committee Statistician. If they recommend adoption, the method is referred to the Chairman of the Official Methods Board, whose approval results in consideration of adoption of the method at First Action status by the Official Methods Board.

"First Action" methods and supporting data are published in the *Journal of the AOAC* and are subject to scrutiny by the analytical community for at least two years before the method is presented to the membership for a vote on "Final Action" adoption. All "First Action" and "Final Action" adopted methods are published in *Official Methods of Analysis of the AOAC.*

Publication Program

The main publication of AOAC is *Official Methods of Analysis of the AOAC,* a compendium of over 1,800 adopted methods. This volume is considered to be the analytical methods bible by thousands of analysts who use and depend upon it. An updated edition is published every five years with annual supplements published in the intervening years.

The association's peer-reviewed bimonthly periodical, the *Journal of the AOAC,* publishes basic and applied research in the analytical sciences related to foods, drugs, agriculture, and the environment. AOAC publishes monographs on subjects of interest to analysts as well as a monthly newsletter.

Meeting Program

AOAC holds its Annual International Meeting and Exposition at a different location in the late summer or early fall of each year. The meeting affords many opportunities for attendees to report research, exchange scientific information, and discuss analytical problems and solutions. Technical programs include symposia, poster presentations, a panel presentation and roundtable discussion on regulatory topics, a forum for general discussion, short courses on subjects of interest to analytical laboratories, and an exhibit of analytical equipment, services, and publications. Workshops and conferences on special topics are held as needed.

Short Course Program

Since 1984, AOAC has offered four to six two-day intensive short courses a year on laboratory quality assurance, field and laboratory sampling, laboratory safety, statistics, and laboratory waste disposal and compliance.

Regional Section Program

The AOAC regional section program, begun in 1981, was established to provide low cost, close-to-home opportunities for analytical scientists to meet and exchange scientific information. There are eleven sections located in the United States, Canada, Europe, and the Mediterranean area as of Fall 1990. Each holds at least one meeting a year.

Awards Program

AOAC provides annual awards for scientific achievement, scientific potential, and service to the association. The $2,500 Harvey W. Wiley Award, honoring Harvey W. Wiley, the father of the U.S. Pure Food and Drug Act and a founder of AOAC, goes to a scientist or group of scientists who have made an outstanding contribution to analytical methodology in an area of interest to AOAC. The Harvey W. Wiley Scholarship Award provides a $1,000 scholarship to an undergraduate student planning to continue to study and work in an area important to AOAC. All other AOAC awards—the Fellow of the AOAC, General Referee of the Year, Collaborative Study of the Year, and Associate Referee Awards—are provided to member-volunteers for service and scientific contributions to the work of the association.

VOLUNTEER/COMMITTEE STRUCTURE

Over 25% of AOAC's members serve in a volunteer capacity. They serve on the Board of Directors, a governing body of officers and directors concerned with administration and policy making; the Official Methods Board, which oversees the methods validation and approval process; the Editorial Board, which oversees publication development and quality; eight methods committees, each of which coordinates an area of AOAC methods validation, such as foods, pesticides, drugs, etc; and 17 standing committees, which advise the Board of Directors on a variety of topics; They also serve as Associate Referees, topic experts, who implement and coordinate actual method development and testing; General Referees, multiple topic experts, who oversee the Associate Referees for their topic areas; liaison officers, who coordinate with other groups concerned with methods development; advocates and representatives, who recruit individual and organization membership and participation; collaborators, who take part in collaborative studies of methods seeking adoption; and, as members of the regional section executive committees, which direct, plan, and implement regional section affairs.

BIBLIOGRAPHY

1. *Bylaws,* Association of Official Analytical Chemists (revised Sept. 13, 1990).
2. *Handbook for AOAC Members,* 6th ed., Association of Official Analytical Chemists, Arlington, Va., 1989.
3. K. Helrich, *The Great Collaboration,* Association of Official Analytical Chemists, Arlington, Va., 1982.

MARILYN TAUB
Association of Official
Analytical Chemists
Arlington, Virginia

ASTRINGENCY. See Taste and odor; Seafood flavors and quality; Entries under Sensory science.

AUSTRALIA. See International development

AUTOMATION. See Baking science and technology; Laboratory robotics and automation; Cleaning-in-place; Expert system; Artificial intelligence; Computer applications in the food

industry; Meat slaughtering and processing equipment; Entries under Quality control.

AUTOOXIDATION. See Antioxidants: food additives: oxidation: entries under Fats and oils.

AWARDS. See IFT awards

AZEOTROPES. See Distillation: technology and engineering.

B

BAKER'S YEAST. See YEASTS.

BAKERY LEAVENING AGENTS

All baked foods are leavened. Even unleavened bread such as matzohs is raised by the steam generated during baking. Normally, leavening is thought of as being due to the products of fermentation (carbon dioxide, alcohol) or to CO_2 generated by chemical reactions. However, the expansion of air incorporated into a dough during mixing and the formation of steam, from water in the dough, also cause the baking piece to expand. If this does not happen, the product is dense and unappetizing. This article discusses leavening obtained by chemical reactions as well as from yeast fermentation.

CHEMICAL LEAVENING

The common form of chemical leavening involves a food acid and soda (sodium bicarbonate). The soda dissolves readily in the aqueous phase of the dough or batter. When the acid dissolves, the hydrogen ion reacts with bicarbonate ion, releasing carbon dioxide (CO_2), which expands the baking piece. Water-vapor pressure increases as the internal temperature of the bakery item increases. Air incorporated during dough mixing also expands. Finally, certain materials, notably ammonium bicarbonate, decompose on heating and generate gases that leaven the product.

Air

During dough mixing air is incorporated as numerous small bubbles (1,2). The actual amount varies, but with an ordinary horizontal mixer a bread dough contains about 15–20% air by volume. If yeast fermentation is occurring, this volume shrinks by 20% owing to oxygen depletion (3), and during baking the nitrogen remaining expands by 17% owing to temperature rise, so the net leavening contribution due to air is nil. In the absence of yeast, thermal expansion is still only 17%, so the overall effect is small.

In cookie doughs and cake batters the amount of air retained is variable. In a standard sugar cookie dough, eg, the dough contains about 10% air by volume (4). Cake batter specific gravity may vary from 0.60 to 1.10 g/mL, depending on the formulation and emulsifier system used. The density of air-free batter is about 1.30 g/mL (5), so the amount of air in the batter may range from 15 to 55% by volume. Even at the highest level, however, the thermal expansion of this air contributes only slightly to the final volume of the cake.

Nevertheless, it is important to get good air incorporation in the dough or batter during mixing, because the air bubbles serve as nuclei for other leavening gases. The internal pressure produced by the surface tension of a gas bubble is inversely related to bubble diameter. If a small bubble and a large bubble are near each other, the gas diffuses from the small one to the large one, and the small bubble disappears. If no gas bubbles are already present in, say, a cake batter when baking is begun, then the CO_2 that is formed by the chemical reactions in the batter has no place to go, because a new bubble cannot form (with a zero radius the surface tension is infinite). If the cake batter contains relatively few air bubbles, the CO_2 diffuses to these, and the resulting cake has a coarse, open grain, often accompanied by tunnels. With many small air nuclei, the grain is fine and close.

Surfactants aid incorporation and subdivision of air into dough (2). Emulsifiers help incorporate air into shortening during the preparation of cookie doughs and cake batters. In all cases the presence of many air bubbles for nucleation is an important part of any leavening action.

Steam

The formation of steam by evaporation of water would seem to be important only when the temperature in the baking piece exceeds 100°C in items such as cookies, crackers, and snacks such as tortilla chips. Also, it has been suggested that the water present in roll-in margarine, used in making puff pastry, evaporates and helps to give the characteristic open structure of this product. Extruded goods, of course, depend almost entirely on steam for their expansion.

The vapor pressure of water increases with increasing temperature, and this increased pressure can expand the gas bubbles in cake batter and in bread. In calculations of the expansion of sponge-cake batter (which contained no soda for chemical leavening), it was found that the theoretical curves (which took into account the thermal expansion of the air) fitted the experimental measurements within a few percent (5). When theoretical expansion for bread dough in the oven (see section on oven spring below) was calculated, it was found that water evaporation accounted for 60% of the total volume increase (6).

Ammonium Bicarbonate

Ammonium bicarbonate is often used as a supplementary leavening agent for cookies and snack crackers. At room temperature, dissolved in the dough water, it is stable, but when the temperature reaches 40°C (104°F) in the early stages of the oven, it decomposes:

$$NH_4HCO_3 \rightarrow NH_3 + CO_2 + H_2O$$

One mole of ammonium bicarbonate (79 g, or 2.8 oz) gives 2 mol of gas (44.8 l, or ca 1.6 ft^3).

This potential for extensive gas formation means that ammonium bicarbonate must be uniformly distributed throughout the dough; the presence of small undissolved pieces would give rather large blowouts in the product. Ammonium bicarbonate is usually dissolved in a few quarts of warm water and then added to the mixer along with the other water.

Because ammonia is water soluble, this leavener is applicable only in low-moisture products. If the finished moisture is above 5% (say, in a soft cookie), the water will retain some of the NH_3 and the product has a characteristic ammonia taste. In contrast to this general rule, ammonium bicarbonate is sometimes added to eclair and popover doughs (*patê de choux*). In this case the combination of high baking temperatures, thin walls, and a large internal cavity allows enough space for ammonia to diffuse out of the baked good.

Sodium Bicarbonate

Baking soda (sodium bicarbonate) has been the workhorse chemical leavener of baked goods for well over a century. In most cases it is combined with a leavening acid to form CO_2, but soda itself undergoes thermal decomposition:

$$2 \text{ NaHCO}_3 + \text{heat} \rightarrow \text{Na}_2\text{CO}_3 + \text{CO}_2 + \text{H}_2\text{O}$$

This reaction usually requires rather high temperatures ($>120°C$) to happen at an appreciable rate, so using soda as a sole leavening agent is restricted mainly to cookies and snack crackers in which the internal temperatures can approach this range.

The more usual reaction is with hydrogen ions from leavening acids:

$$\text{NaHCO}_3 + \text{H}^+ \rightarrow \text{Na}^+ + \text{CO}_2 + \text{H}_2\text{O}$$

Soda is readily soluble in water (saturation solubility of 6.5% at 0°C, 14.7% at 60°C) and dissolves in the dough or batter water during mixing. The rate of reaction is governed by the rate of dissolution of the leavening acid (see below); no H^+ is present until the acid dissolves and ionizes. Of course, the soda can react with acidic ingredients in the formula such as buttermilk and even flour itself. Chlorinated cake flour, which is more acidic than ordinary pastry flour, will neutralize about 0.27 g of soda per 100 g of flour.

Soda is a mild alkali; the pH of an aqueous solution is about 8.2. Increasing the amount of soda in a formula raises the pH of the dough, eg, in raising the dough pH of saltine crackers to 7.4. Excess soda (more than that neutralized by available leavening acid) is added to devil's food cake batter to raise the pH of the finished cake crumb to about 7.8 and generate the dark chocolate color desired (cocoa is lighter colored in mild acid, darker colored in mild alkali).

Sodium bicarbonate is available in various granulations (Table 1); it is important to choose the correct one for the application. If the ingredient is added at the mixer, and the dough or batter will be processed fairly quickly, then rapid dissolution is wanted and No. 3 (fine powdered) or No. 1 (powdered) would be appropriate. At the other extreme is the inclusion of soda in a dry mix (eg, cake, cake doughnut) that is packaged and stored for a period of time before use. In this case thermal decomposition of the dry powder becomes a factor. Grade No. 1 (powdered) decomposes 50% faster than grade No. 5 (coarse granular) in this situation. A loss of 2–4% per week at 50°C (122°F) and 0.5–1% at 30°C has been reported. A common practice is to add 5–10% more than needed when the dry mix is blended, in the hope that when the mix is used (with uncontrolled variables of age and storage temperatures), the amount of soda will be adequate to perform as expected. It would be preferable to use grade No. 4 (granular) or No. 5 (coarse granular) for this application and reduce the amount of overage in the dry mix.

Potassium Bicarbonate

Recently this material has become commercially available as a substitute for sodium bicarbonate; it is functionally equivalent but lowers the sodium content of the finished product. As shown in Table 1 the granulation grade presently offered is intermediate between No. 4 (granular) and No. 5 (coarse granular) sodium bicarbonate. The effect on pH and the reactions to CO_2 are the same. Because the molecular weight is greater (100.11 for $KHCO_3$ vs 84.01 for $NaHCO_3$), 19% more is required to get the same effect; replace 10 lb of sodium bicarbonate with 11.9 lb (11 lb 14$\frac{1}{2}$ oz) of potassium bicarbonate.

Leavening Acids

The traditional acidulants for baking were vinegar (acetic acid), lemon juice (citric acid), cream of tartar (potassium acid tartrate), or sour milk (lactic and acetic acids). In all cases the acid reacted with the soda as soon as they were mixed and the baker relied on batter viscosity to retain the CO_2 until the cake or cookie was baked in the oven. In 1864, monocalcium phosphate hydrate was patented for use in making a commercial baking powder. Later (ca 1885) sodium aluminum sulfate began to be used; this compound has a low solubility in water at room temperature and essentially none in the aqueous phase of a batter, so it does not release CO_2 until late in the baking cycle. The combination of the two leavening acids gave rise to double-acting baking powder. During the first third of the twentieth century a number of phosphate compounds

Table 1. Sodium Bicarbonate Granulation

Grade Number		Cumulative % Retained, Minimum–Maximum							
	USS sieve:	60	70	80	100	140	170	200	325
	Microns:	250	210	177	149	105	88	74	44
1. Powdered					0–2			20–45	60–100
2. Fine granular				0–Tr	0–2			70–100	90–100
3. Fine powdered					0–Tr	0–5		0–20	20–50
4. Granular				0–Tr	0–2			80–100	93–100
5. Coarse granular		0–8	0–35		65–100		95–100		
Potassium bicarbonate		0–5			40–60			80–100	

Table 2. Reactions of Leavening Agents

Leavening Acid	NV[a]	EV[b]	Reaction
Monobasic calcium phosphate (MCP)	80[c]	1.25[d]	$3\ Ca(H_2PO_4)_2 \rightarrow Ca_3(PO_4)_2 +$ $3\ HPO_4^{2-} + H_2PO_4^- + 7\ H^+$
Sodium acid pyrophosphate (SAPP)	72	1.39	$Na_2H_2P_2O_7 \rightarrow 2\ Na^+ + P_2O_7^{2-} + 2\ H^+$
Sodium aluminum phosphate (SAlP)	100	1.00	$NaH_{14}Al_3(PO_4)_8 \cdot 4\ H_2O + 5\ H_2O \rightarrow$ $3\ Al(OH)_3 + Na^+ + 4\ H_2PO_4^- +$ $4\ HPO_4^{2-} + 11\ H^+$
Sodium aluminum sulfate (SAS)	104	0.96	$NaAl(SO_4)_2 + 3\ H_2O \rightarrow Al(OH)_3 +$ $Na^+ + 2\ SO_4^{2-} + 3\ H^+$
Dicalcium phosphate $\cdot$ 2 H_2O (DCP $\cdot$ Di)	35	2.86	$3\ CaHPO_4 \cdot 2\ H_2O \rightarrow Ca_3(PO_4)_2 +$ $HPO_4^{2-} + 6\ H_2O + 2\ H^+$

[a] Neutralizing value, grams of $NaHCO_3$ neutralized by 100 g of the leavening acid.

[b] Equivalence value = 100/NV; grams of leavening acid to neutralize 1 g of $NaHCO_3$.

[c] NV is 83 for anhydrous MCP, 80 for the monohydrate form.

[d] EV is 1.20 for anhydrous MCP, 1.25 for the monohydrate form.

were studied, and the range of leavening acids available today to formulators was developed. These acids have the advantage of controlled and predictable rates of CO_2 release and account for essentially all commercial leavening acids sold today. Several good reviews on the properties and uses of leavening acids have been published (7–10), which give more details about various specific applications.

Table 2 lists the leavening acids and the chemical reactions that occur when they dissolve in water to liberate H^+. Of use to the formulator are the neutralizing value (NV) and the equivalence value (EV). Suppliers characterize their products by NV, which is defined as the pounds of sodium bicarbonate that are neutralized by 100 lb of the acid. When developing product formulas, it is more convenient to use EV; eg, if 2% soda is included in the formula, 2.78% sodium acid pyryphosphate (SAPP) is added to achieve exact equivalence.

Neutralizing value is determined by experimental titration (7,8). The yield of H^+ shown in Table 2 corresponds to NV, although there are some discrepancies. For instance, dicalcium phosphate dihydrate (DCP $\cdot$ Di) should yield 2.15 H^+ to have an NV of 35. One that is even more puzzling is the reaction of SAPP. The pyrophosphate ion has four ionization constants, the first two of which are neutralized during manufacture of disodium pyrophosphate. The additional ionization constants for the dihydrogen dianion formed on solution in water are 5.77 for pK_3 and 8.22 for pK_4. In a cake batter at pH 7 the dianion should yield only 1.1 H^+, equivalent to an NV of 42, but in actual fact NV is 72 as shown in the table. Obviously other, unrecognized, reactions occur that increase the acidifying potential of SAPP. Neutralizing value (and EV) are for practical guidance; the reactions are for interest and edification.

Reaction Rates. Two kinds of reaction rates are of interest to bakers: (1) immediate, at room temperature; and (2) during baking. Both are functions of the solubility of the leavening acid. The first rate, called the dough reaction rate (DRR), is measured by mixing all the dry ingredients (cake flour, nonfat dry milk, shortening, salt, soda, and leavening acid); adding water; stirring briefly; and then measuring the evolution of gas at 27°C for 10 min (11). In the absence of leavening acid there is a blank reaction of about 20% of the soda due to acidity of the flour and dry milk. Monocalcium phosphate monohydrate (MCP $\cdot$ H_2O) reacts about as quickly as cream of tartar, ie, it is a very fast acting acid. Anhydrous MCP is about 80% as reactive. Dicalcium phosphate dihydrate (DCP $\cdot$ Di) is essentially unreactive under these conditions, as are sodium aluminum phosphate (SAlP) and sodium acid pyrophosphate for refrigerated dough (SAPP–RD). The various grades of SAPP (see below) show increasing extents of reaction, with the reaction rate for sodium aluminum sulfate (SAS) being somewhere in the middle of the SAPPs (8).

Baking (temperature-dependent) reaction rates measure the rate of CO_2 release from a dough at various temperatures. The rate is actually governed by the solubility of the acid. A slow acid such as SAlP is nearly insoluble at room temperature, so it cannot ionize to give H^+ for reaction with soda; as the temperature rises, the material dissolves and starts to perform its function. Figure 1 shows the rates for several leavening acids (7,8). The ranking of acids is the same as that given by the DRR test, but Figure 1 is useful in choosing acids for giving lift at various points in the baking cycle. For instance, MCP hydrate generates CO_2 in a cake batter during mixing, whereas SAlP starts to leaven the cake only midway through the baking cycle. Dicalcium phosphate dihydrate is effectively unreactive (insoluble) until the temperature reaches about 80°C, near the end of the bake cycle. However, it has value in triggering a late release of CO_2, which helps prevent dipped centers or fallen layer cakes.

Sodium acid pyrophosphate is provided in various grades ranging from fast to slow baking reaction rates. Although the basic molecule is the same in all cases, Ca^{++} slows the dissolution rate of SAPP granules. Thus the inclusion of milk in a SAPP-leavened formulation (eg, a pancake batter) delays the time of leavening slightly. The various reaction grades are made by intentionally adding a certain amount of Ca^{++} to the disodium salt during manufacture (adjustments in granule size also control dissolution rate). The grade designations used by Monsanto

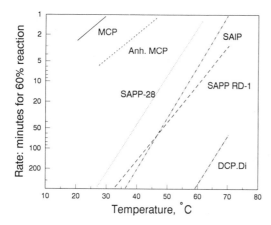

Figure 1. The rate of reaction of various leavening acids as a function of temperature. Rate is expressed as the time to achieve 60% of the total reaction with sodium bicarbonate (on a logarithmic scale). Monocalcium phosphate monohydrate (MCP) is the fastest-acting acid, while dicalcium phosphate dihydrate (DCP·Di) is the slowest leavening acid (8). Adapted with permission.

Chemical Co. range from SAPP-43 (the fastest) through 40, 37, and 28 to SAPP RD-1 (the slowest). The corresponding grades marketed by Stauffer Chemical Co. (now a division of Rhône-Poulenc) are named Perfection, Donut Pyro, Victor Cream, BP Pyro, and SAPP #4. The slowest grade is used in making refrigerated biscuits and doughs, whereas for other uses (cakes, cake doughnuts, biscuits, and pancakes) a combination of SAPP acids is used to get continuous leavening throughout the baking cycle.

The high reaction rate of MCP may be moderated by coating it with a somewhat insoluble material. This coating dissolves slowly with time, allowing the MCP to then dissolve and form H^+. Instead of 70% of the reaction occurring during the first 10 min after mixing, coated MCP typically gives 20% reaction immediately and then the other 50% over the next 30 min.

The slow-reacting SAlP leavening acids are useful in pancake and waffle batter for restaurant use, where little or no evolution of CO_2 is desired during the holding period between uses.

Formula Balance. Some of the ions found in leavening acids may influence other properties of certain baked

goods. The pyrophosphate ion gives a certain taste that is detectable by some people. Aluminum ion plays a role in developing optimum layer cake structure (12), and SAlP and/or SAS is useful as part of the leavening system. Sodium aluminum sulfate is reported to contribute to rancidity in dry cake mixes stored for a long period of time. Calcium ion also helps set protein structure and contributes to fine grain in cakes and cake doughnuts. One leavening acid may be selected over another with an equivalent reaction rate just to take advantage of these ion effects.

Developing a formulation with a properly balanced chemical leavening system is part science and part art—preplanning followed by trial and adjustment. The selection of timing of the leavening acids, the ratio between the various ones that might be used, the total amount of CO_2 required, and the amount of excess of soda desired are all factors that interplay with other ingredients and structural functionalities in the baked goods. Achieving the correct leavening balance in a particular formula requires a certain amount of trial and error.

Baking Powder

In the latter part of the nineteenth century baking powders were developed that combined both parts of the chemical leavening system—baking soda and a leavening acid. The first baking powder contained only one leavening acid, anhydrous MCP, and reacted fairly quickly in cake batters, biscuits, etc. Soon afterwards, sodium aluminum sulfate was incorporated; this gives leavening action during the middle of the baking process, in addition to the early action due to the MCP. This double-acting baking powder was a great success.

In Table 3 some typical compositions of baking powders are shown. In addition to the soda and acid, various fillers and calcium salts are also included. The level of potential leavening gas is nearly constant: it is the amount of CO_2 equivalent to 28% sodium bicarbonate in the single-acting powder, and carbon dioxide equivalent to 30% soda in all the double-acting types. Thus these powders are interchangeable from the standpoint of the amount of leavening achieved during baking.

Household double-acting baking powder is based on MCP·H_2O and SAS, whereas the commercial types contain MCP·H_2O and SAPP as the leavening acids. The use of a combination of different grades of SAPP in place of

Table 3. Baking Powder Compositions[a]

	Single-acting	Double-acting types					
		Household			Commercial		
Sodium bicarbonate, granular	28.0	30.0	30.0	30.0	30.0	30.0	30.0
Monocalcium phosphate, hydrate		8.7	12.0	5.0	5.0		5.0
Monocalcium phosphate, anhydrous	34.0						
Sodium aluminum sulfate		21.0	21.0	26.0			
Sodium acid pyrophosphate					38.0	44.0	38.0
Cornstarch, redried	38.0	26.6	37.0	19.0	24.5	26.0	27.0
Calcium sulfate	13.7						
Calcium carbonate			20.0				
Calcium lactate					2.5		

[a] Composition given as percentage of each ingredient in the baking powder.

SAS allows better control of the timing of CO_2 release in the oven, which is sometimes an advantage in commercial bakery production.

BIOLOGICAL LEAVENING

The use of yeasts in food processing is prehistoric in origin. Baking, brewing, and enology all depend on the ability of yeasts to carry out anaerobic fermentation of sugars, yielding CO_2 and ethanol. In brewing and wine-making, alcohol is the prime product of interest; in baking the leavening effect of CO_2 and ethanol is more important. In this article only the leavening action of yeast in baked goods is discussed.

Oven Spring

The use of yeast in bread production gives a product containing many small gas cells—leavened bread. By the end of the proofing period the aqueous phase of the bread is saturated with CO_2 and the volume has roughly doubled owing to the pressure of CO_2 that has diffused to air cells or bubble nuclei. During the first part of the bake cycle the loaf expands; then at some temperature the matrix sets, expansion stops, and the rest of the baking time is for the purpose of starch gelatinization, crust coloring, flavor development, etc. The increase in loaf volume during baking is called oven spring (13).

The magnitude of oven spring depends on two factors: (1) generation and expansion of gases; and (2) the amount of time available for loaf expansion before the structure sets. The first factor is primarily a function of yeast fermentation; the second one is affected by dough components such as shortening, surfactants, gluten protein, and flour lipids.

The influence of several dough components on finished volume has been shown (14,15). A key observation is shown in Figure 2 (15,16). Dough with or without shortening expands at the same rate initially, but at 55°C the structure of the fat-free dough begins to set, slowing the expansion rate. In the presence of 3% shortening, dough continues to expand up to 80°C; the longer time at the initial expansion rate results in a larger final volume. Similar results were found in the presence of 0.5% monoglyceride or 0.5% sodium stearoyl lactylate (14).

Much of the CO_2 formed during fermentation is lost to the atmosphere (17). If the internal gases in a bread were at equilibrium with the atmosphere, there would be no excess CO_2 pressure inside the gas cells. The fact that there is such a pressure (as shown by increased dough volume) is due to inhibition of diffusion of CO_2 from the gas cells to the surrounding air. The causes of this inhibition are not known, but increased viscosity due to water-soluble pentosans and/or the developed gluten phase has been suggested. At the end of proof ca 35% of the CO_2 formed during fermentation is present in gas cells, ca 5% is dissolved in the aqueous phase, and 60% has been lost to the atmosphere (18). When the solubility of CO_2 in bread dough was measured experimentally, a value of 0.81 mL (equivalent) dissolved per gram of dough was found (19). Others calculated 0.32 mL (equivalent) dissolved per

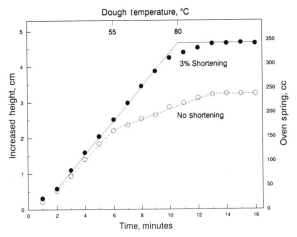

Figure 2. The increase in volume of bread dough in a resistance-heating oven. Doughs contained 3% (flour basis) shortening, or no added shortening. The initial rate of dough expansion is the same for both doughs, but in the absence of shortening some (as yet unknown) reaction takes place at ca 55°C which slows volume increase, followed by a reaction at a temperature above 80°C which sets the bread structure. In the presence of shortening the initial rate of dough expansion is maintained until dough temperature is near 80°C, then the reactions occur which set the bread structure (16). Adapted with permission.

gram of dough, based on the known solubility of CO_2 in water (17).

A mass balance calculation for a pup loaf based on 100 g of flour (173.5 g of dough) found that CO_2 and air occluded during mixing accounted for 370 mL of loaf volume at the end of proofing, and in the oven this gas phase expanded by 62 mL (17). Dissolved CO_2, driven out of the aqueous phase owing to rising temperature, contributed 43 mL to oven expansion, and CO_2 generated by increased yeast activity during early stages of baking amounted to 4 mL. The total volume expansion in the oven from CO_2 and air was 109 mL, but the actual increase in loaf volume was 360 mL. (If the solubility data of Ref. 19 are used, the volume of CO_2 expelled from the aqueous phase would be 109 mL, and the total volume expansion due to CO_2 and air would be 175 mL).

The difference, 251 mL (or 185 mL, if Ref. 19 data are used) must be supplied by the evaporation of ethanol and water. The amount of dough expansion that could be expected owing to evaporation of water, based on the vapor pressure of water as internal dough temperature rises was calculated (6). This is a slight overestimate, as the aqueous phase of the dough is roughly 0.5 M in salt and 0.3 M in sugar, as well as containing water-soluble proteins, flour pentosans, and fermentation acids. These solutes lower the vapor pressure of the water by roughly 2%. The amount of ethanol formed during fermentation and proofing usually amounts to ca 3% by volume based on the aqueous phase, or about 0.65 M. By Raoult's law, the vapor pressure in the gas phase in bread due to ethanol (during the stages before structural setting) would only be ca 1.2% of the vapor pressure of pure ethanol at those temperatures. Thus, although pure ethanol has a higher vapor pressure than pure water at temperatures from 30° to 80°C, the partial pressure of water in bread dough is much larger than that of ethanol, and water vapor is probably much more important than ethanol vapor as a leavening gas in bread.

To summarize, CO_2 contributes ca 25% of the gas pressure responsible for oven spring in the standard white loaf considered; expansion of air accounts for ca 15%; and water vaporization makes up the rest. The actual magnitude of oven spring depends on other dough components that govern the temperature at which rapid expansion ceases; the higher that temperature, the greater the increase in loaf volume during baking.

BIBLIOGRAPHY

1. J. C. Baker and M. D. Mize, "The Origin of the Gas Cell in Bread Dough," *Cereal Chemistry* **18**, 19–34 (1941).

2. R. C. Junge, R. C. Hoseney, and E. Varriano-Marston, "Effect of Surfactants on Air Incorporation in Dough and the Crumb Grain of Bread," *Cereal Chemistry* **58**, 338–342 (1981).

3. N. Chamberlain, "The Use of Ascorbic Acid in Breadmaking," in J. N. Councell and D. H. Hornig, eds., *Vitamin C,* Applied Science, London, 1981.

4. J. L. Vetter, D. Blockcolsky, M. Utt, and H. Bright, "Effect of Shortening Type and Level on Cookie Spread," *Technical Bulletin of the American Institute of Baking* **6**(10), 1–5 (1984).

5. M. Mizukoshi, H. Maeda, and H. Amano, "Model Studies of Cake Baking. II. Expansion and Heat Set of Cake Batter During Baking," *Cereal Chemistry* **57**, 352–355 (1980).

6. A. H. Bloksma, "Rheology of the Breadmaking Process," *Cereal Foods World* **35**, 228–236 (1990).

7. T. P. Kichline and J. F. Conn, "Some Fundamental Aspects of Leavening Agents," *Bakers Digest* **44**(4), 36–40 (1970).

8. J. F. Conn, "Chemical Leavening Systems in Flour Products," *Cereal Foods World* **26**, 119–123 (1981).

9. H. M. Reiman, "Chemical Leavening Systems," *Bakers Digest* **51**(4), 33–34, 36, 42 (1977).

10. D. K. Dubois, "Chemical Leavening," *Technical Bulletin of the American Institute of Baking* **3**(9), 1–6 (1981).

11. J. R. Parkes, A. R. Handleman, J. C. Barnett, and F. H. Wright, "Method for Measuring Reactivity of Chemical Leavening Systems," *Cereal Chemistry* **37**, 503–518 (1960).

12. N. B. Howard, "The Role of Some Essential Ingredients in the Formation of Layer Cake Structures," *Bakers Digest* **46**(5), 28–30, 32, 34, 36–37, 64 (1972).

13. J. C. Baker, "Function of Yeast in Baked Products. Effects of Yeast on Bread Flavor," in C. S. McWilliams and M. S. Peterson, eds., *Yeast—Its Characteristics, Growth, and Function in Baked Products,* National Academy of Science, National Research Council, Washington, D.C., 1957.

14. W. R. Moore and R. C. Hoseney, "Influence of Shortening and Surfactants on Retention of Carbon Dioxide in Bread Dough," *Cereal Chemistry* **63**, 67–70 (1986).

15. W. R. Moore and R. C. Hoseney, "The Effects of Flour Lipids on the Expansion Rate and Volume of Bread Baked in a Resistance Oven," *Cereal Chemistry* **63**, 172–174 (1986).

16. R. C. Junge and R. C. Hoseney, "A Mechanism by Which Shortening and Certain Surfactants Improve Loaf Volume in Bread," *Cereal Chemistry* **58**, 408–412 (1981).

17. W. R. Moore and R. C. Hoseney, "The Leavening of Bread Dough," *Cereal Foods World* **30**, 791–792 (1985).

18. R. C. Hoseney, "Gas Retention in Bread Doughs," *Cereal Foods World* **29**, 305–308 (1984).

19. G. E. Hibbert and N. S. Parker, "Gas Pressure-Volume-Time Relationships in Fermenting Doughs. I. Rate of Production and Solubility of Carbon Dioxide in Dough," *Cereal Chemistry* **53**, 338–346 (1976).

CLYDE E. STAUFFER
Technical Foods Consultant
Cincinnati, Ohio

BAKERY SPECIALTY PRODUCTS

The main items produced by commercial bakeries, in terms of total tonnage, are bread and rolls, cakes, cookies, and crackers, with bread (either pan bread or hearth-baked types) representing the largest part of this production. Over many centuries many other types of food products, using wheat flour as their main ingredient, have been developed. This article discusses several of these products that are available to consumers today.

NATIVE FLAT BREADS

The earliest form of wheat bread is made by grinding cereal grain(s), adding water, mixing to a coherent mass, and allowing this mass to rest and ferment for a period of time. Then small portions are taken, flattened into a sheet, and baked quickly on a hot surface, usually either the floor or the wall of a heated oven. The process is usually a sourdough type; ie, a portion of the fermented dough is retained at the end of the day's baking and mixed in with the next day's dough, thus inoculating it with the yeast and other fermentative microorganisms that produce ethanol, carbon dioxide, and organic acids as they metabolize and grow.

These breads are produced widely throughout the Middle East and the Indian subcontinent. Depending on details of formulation and production, they are called chappati, nan, roti, tannouri, tamees, korsan, shamuf, and several other names. The simplest formula consists of whole wheat meal (frequently stone-ground by hand in the kitchen), salt, water, and sourdough starter. At the other extreme, the dough also contains ground pulses, sesame seeds, shortening (either ghee or sesame seed paste), honey or other sweetener, and spices (anise, cardamom, curry powder, dill, etc). Although wheat is the major cereal grain used, ground rice, barley, maize, or sorghum are also sometimes included at up to 20% of the weight of wheat meal or flour.

Production may be either in the home or in small commercial establishments. Home production is done strictly by hand. Commercial bakeries have a mechanical slow-speed mixer and perhaps a set of motorized sheeting rollers. The oven is usually heated by building a wood fire inside it, raking out the coals when it is hot, and baking on the floor of the oven. In the home, I have seen ovens in the shape of an urn, about 3 ft high with an 18-in. opening in the top and a small opening in the bottom for feeding the fire. When the sides of the oven are hot, the sheeted dough is inserted through the top and plastered (by hand) against the oven wall. Dexterity and practice is required for this, because the oven temperature approaches 500°F. The baked bread is removed with wooden tongs. In the commercial hearth oven, insertion and removal is performed using a wooden peel.

As is common practice in most parts of the world, the bread is produced, purchased, and consumed on a daily basis, so staling and development of mold is not a problem. The meat and vegetable dishes, in the countries where flat breads are common, tend to be stews and purées, and freshly baked flat bread is an ideal and flavorful accompaniment to such a meal.

A similar sort of bread called tortilla, based on maize, is common in Latin American countries and recently in the United States. The maize (corn, in United States terminology) is soaked in dilute lime (calcium hydroxide) for up to 24 h to remove the hull. The endosperm is ground, while wet, to make masa, then shaped into balls, flattened, and baked as described above. These differ from the flat breads described above in the way in which the grain is processed and in having no fermentation step. The leavening of the tortilla, such as it is, is by steam. Tortillas can also be made using wheat flour. Commercial flour tortillas are usually leavened by the addition of a small amount of sodium bicarbonate. Commercial production of tortillas, to accompany Mexican food specialties, incorporates a number of refinements; space does not allow a full discussion of these factors in this article.

PITA BREAD

Another type of Middle Eastern bread, related to the flat breads, is pita (pocket, balady, burr) bread. A wheat flour is used, with an extraction rate of 75–82%, depending on the specific type of bread being made. The strength of this flour is greater than that of the flour (or wholemeal) used in flat breads. Usually the bread is made by a sourdough process, but with a rather large (20%) portion of sourdough, so that fermentation is more rapid than with the usual flat bread. (Pita doughs can be made adding yeast to each dough, as for regular bread production.) The absorption is somewhat greater than for regular bread dough. The dough is allowed to ferment for about 1 h after mixing, then it is divided into pieces weighing about 3 to 4 oz each. These receive a short intermediate proof, are flattened into circles about ⅛ in. thick and 6–9 in. in diameter, and allowed to proof for 30–45 min.

The proofed pieces are baked for a short time (1.5 to 2 min) in an extremely hot oven (700–900°F). The high oven heat causes formation of a top and bottom crust almost instantaneously after the piece is inserted. As the heat penetrates the dough the ethanol and water rapidly evaporate and the vapors, trapped by the crust, cause the bread to balloon. As the bread cools, after removal from the oven, the internal pressure dissipates and it collapses to the familiar round, flat shape, but with the internal "pocket" still present.

PIZZA DOUGH

Pizza is basically an open-faced sandwich made on a leavened flat bread, in which the topping and the bread are baked together. It has become a popular baked food worldwide, and, although the familiar tomato sauce/sausage/cheese topping is the most popular, fish, ground lamb, puréed legumes, and a wide variety of chopped vegetables are also used. The possible variety of pizzas is limited only by the availability of ingredients and the ingenuity of the chef.

Pizza crusts are of two types—thin cracker-type crusts and thick bread-type crusts—and are mixed shortly before use, or are mixed and retarded in a commissary for later use. These variations lead to differences in dough formulas and handling. The flour used for pizza dough should be a good grade of bread flour with 12.0–12.5% protein. The thin-crust dough for immediate use requires even more protein, so rather than carry a 14% protein flour for this dough and a 12% flour for thick-crust doughs, it is more convenient to add 2–3% of vital wheat gluten to the 12% flour. Gluten generally is not used in commissary doughs unless flour of an adequate strength is not available. The gluten protein forms a seal at the crust/topping interface, minimizing formation of a soggy layer during baking.

Dough for immediate use receives little fermentation time beyond a 10–15 min floor time. It should be somewhat undermixed (just beyond the "cleanup" stage) and quite warm (33–38°C, 90–100°F). These doughs often resist sheeting and tend to shrink back in the oven. To counteract this effect, a reductant, either L-cyteine or sodium bisulfite, is used. Cornmeal also reduces dough elasticity. Adding it does away with the need for reductants and improves crust color, flavor, and eating properties. Sodium stearoyl lactylate improves the crust volume of thick-crust pizzas, giving closer grain and a more tender crust.

The variations in water absorption between the two types of crusts are consonant with the descriptions cracker-type crust and bread-type crust. If absorption is too high, the dough is sticky, whereas if it is too low the dough tears when it is sheeted.

Commissary pizza doughs are mixed, divided, rounded, and immediately placed in the retarder at 5–8°C (40–45°F). The dough ball should double its volume in 24 h at this temperature. If the volume increase is greater than this the dough temperature (usually around 25°C, 78°F) should be reduced; if the volume increase is less than double, mix the doughs to a warmer temperature. Slight adjustments in yeast levels may be made to get the amount of fermentation to the desired point, but too much yeast in the dough may result in large blisters on the crust when it is sheeted and baked in the pizza shop.

BAGELS

Bagels supposedly originated in Vienna, during the period (late 17th century) when the city was fighting off the Turks. Jan Sobieski, the king of Poland and a famous horseman, was a major factor in the defeat of the Turks, and a hard roll in the shape of a stirrup (bügel in German), immersed briefly in a boiling-water bath before being baked to preserve the shape, was developed in his honor. In later times the shape was simplified and the name corrupted to bagel, but the same production procedures are still followed.

Bagels traditionally are made from a lean, low-absorption straight dough, using a high-protein (ca 13.5–14%), clear, hard-red spring-wheat flour. In recent years numerous adjustments have been made to the basic formula, in the direction of making the bagel richer (eg, adding whole eggs), softer (by the addition of shortening), or sweeter (by adding honey and/or raisins). The main feature that separates bagels from other bakery items (besides the short boiling-water treatment) is the type of fermentation schedule used.

A combination of proofing (35–40°C, 95–105°F) and retarding (5–10°C, 40–50°F) is used to obtain the desired product volume as well as the internal and external characteristics. If the bagels are overproofed they collapse in the oven; if they are underretarded they tend to ball, ie, expand and close the center hole. Also, tiny blisters on the surface of the finished bagel are considered desirable. These blisters are due to the presence of lactic acid formed during retarding, and if they are absent (but the bagels are otherwise acceptable), 0.1–0.5% lactic acid may be added to the dough.

The sequence and timing of the fermentation steps is open to experimentation. Proofing may be done before retarding, after retarding, or both. Retarding times may be as short as 8 h or as long as 24 h. Specific process parameters must be worked out by the baker, depending on the formula and equipment being used, to make the product that is most acceptable to customers.

Just before baking the proofed bagels are placed in a boiling-water bath; they float on the top of the bath, and are boiled for about 1 min, being flipped halfway through. This treatment gelatinizes the starch on the surface of the bagel and causes the formation of the hard, shiny crust when the bagel is baked. The usual baking procedure is directly in a hearth oven, although they also may be baked on sheet pans. Various toppings (poppy seed, sesame seed) are applied to the boiled (and still moist) bagel; other flavorants such as dried onion are mixed in with the dough.

PRETZELS

Considered to be the world's oldest snack foods, pretzels reputedly originated around the year 600 in a monastery in the Italian Alps. Supposedly, the monk in charge of the bakery formed strips of bread dough into a shape intended to resemble arms folded across the breast in prayer. When baked, these pretiola (little reward in Italian) were given to children who learned their prayers properly. Later, these baked bread snacks became popular north of the Alps, where the name was germanicized to bretzels, a spelling still used by certain pretzel bakers.

Soft pretzels, presumably identical with those made by the Italian monk, are still popular today. They have a moisture content similar to that of bread and a short shelf life. Hard pretzels were supposedly discovered when a bakers' apprentice forgot to remove the last trays of pretzels from the oven at the end of the day, and the overnight residence in the cooling oven dried them, giving them a hard, crisp texture. The rest of this discussion concerns itself with the low-moisture (2–3%) hard pretzel.

Pretzel dough is extremely stiff, made with a low-protein flour (that from soft white winter wheat is preferred), and only 38–42% water. The only other ingredients are yeast (ca 0.25%) and roughly 1% each of salt, shortening, and dry malt. This dough is mixed in a horizontal sigma-arm mixer, allowed to ferment for up to 4 h, then processed.

The dough is divided into small pieces, elongated into a thin strip, and twisted into the typical pretzel shape by a special machine. The formed pretzels receive a short intermediate proof, then are carried through a caustic bath, which contains 1.25 ± 0.25% sodium hydroxide, held at about 190°F. After exiting this bath, salt is sprinkled on the tops and the pretzels enter the oven. Here they are baked at about 450°F for 4–5 min. The caustic bath gelatinizes the starch on the surface of the pretzel, giving the crisp, shiny crust. In the oven all residual caustic is converted to sodium bicarbonate. The baked pretzels are then piled onto slower-moving conveyors that carry them through a drying stage, where they remain for 30–90 min (depending on the size of the piece) at a temperature of approximately 250°F. This treatment slowly reduces the moisture to about 2%, and also allows equilibration so that the finished pretzel does not check (form small cracks on its surface).

Variations on this production scheme are many. A machine is available that extrudes the dough through a die already shaped in the twisted form, so tying is not necessary. By changing the die the machine will also make straight rods, which are a popular form for this snack. Also, the exact dough formulation and extent of fermentation varies among manufacturers, the details being closely held as proprietary information, each company sure that its particular formula makes the best possible pretzel. As with any low-moisture snack, the shelf life is governed by the quality of the packaging; the loss in consumer acceptance is directly related to moisture uptake, with about 3.5% moisture marking the upper limit of acceptability.

ENGLISH MUFFINS

The nomenclature of this item is anomalous, as it was developed in the United States; the term English applied to them is a highly successful marketing ploy. They are a disc-shaped baked product, usually around 3–4 in. in diameter and about three-quarters of an in. high. A good general description is: "good eating muffins are relatively tough, chewy, and honey-combed with medium to large size holes ($\frac{1}{8}$–$\frac{1}{4}$ in. diameter). Flavor is bland and somewhat sour. Side walls are straight and light colored. The edge between the sides and the flat, dark brown top crust is gently rounded, not sharp." They are baked in a covered griddle cup on a highly automated line. The main features of importance are that the dough must be soft so it flows to fill the cup, and the formulation must lead to the formation of the large internal holes.

A typical English muffin dough contains 83–87% water (flour basis equals 100%), 2% sugar, 1.5% salt, 5–8% yeast, and 0.5–0.7% calcium propionate. Vital wheat gluten at 1–2% is sometimes added to increase the chewiness of the finished muffin. At this absorption level the dough is very soft, and it must be cold (20°C, 68°F) when taken from the mixer so that it does not stick when it is divided, rounded, and deposited from the intermediate proofer into the griddle cups. Dusting material, usually a blend of corn flour and cornmeal, is used rather liberally (3–4% of the dough weight) to facilitate the various transfers.

Fermentation of the dough piece after it is divided and rounded is rather short, usually about 30 min. Three factors—short time, cold dough, and high calcium propionate levels—all tend to retard fermentation, so a rather high yeast level is required. Leavening in the griddle cups is due to CO_2, ethanol, and steam; in addition, some bakers add 0.5% baking powder to get an early kick in the cups, so that the mold is quickly filled by the dough.

The finished English muffin has a rather high moisture content (about 45%) and is prone to mold formation. Calcium propionate is used at the elevated level mentioned to help overcome this problem, and it is also added to the dusting flour. Another method is to spray the muffin with a potassium sorbate solution as it leaves the oven. If this is done the propionate may be omitted from the dough, which in turn allows a decrease of about 1% in the amount of yeast.

In some instances the internal grain porosity is still less than desired, and the addition of protease enhances the openness of the crumb. This additive also increases pan flow, so absorption might be reduced by 1–2%, which might or might not be desirable. Sometimes a tangy, acidic taste is wanted (sourdough muffins), and a dry sour base or vinegar (acetic acid) is added to the dough. The increased acidity (lower pH) of the finished product enhances the antimold action of calcium propionate. Another popular variety contains raisins, added in amounts equal to 25–50% of the flour weight.

DOUGHNUTS

Doughnuts are fried in deep fat rather than baked in an oven as are most other bakery products. They are related to another product, fritters, which are made with a soft, rich dough containing pieces of fruit (apples) or vegetables (corn), dropped by spoonfuls into hot fat. An apocryphal story relates that a 19th century Yankee ship captain loved apple fritters, but had troubles holding onto them when at the ship's wheel during high seas. His cook, anxious to please, got the idea of making the fritters with a hole in the center so the captain could stick them on the handspokes of the wheel during busy moments. Although there is a certain implausibility about this story, the fact remains that doughnuts, in one form or another, are extremely popular sweet snacks or desserts.

There are three basic types of doughnuts: cake doughnuts, yeast-raised doughnuts, and French crullers. Cake doughnuts are chemically leavened; yeast-raised doughnuts use a sweet yeasted dough; and French crullers are steam leavened. These will be discussed separately.

Cake Doughnuts

Batters for this product are a leaner, lower-sugar version of layer cake batters. A typical regular cake doughnut mix contains (on 100% flour basis) about 40% sugar and 13% nonemulsified shortening. Other ingredients frequently found in doughnut mixes are defatted soy flour, nonfat dry milk, potato starch, and dried egg yolk. These ingredients give a tender eating quality to the finished product, as well as limiting fat absorption during frying. The protein ingredients decrease fat absorption, while potato starch helps retain moisture in the finished doughnut and gives a longer shelf life. Lecithin is sometimes used to enhance the wetting of the dry mix during batter preparation, and occasionally monoglyceride is added to make a more tender final product. Other emulsifiers are not used, as they increase fat absorption without conferring any quality enhancement.

Most cake doughnuts are produced by extruding (or depositing) the batter from a reservoir directly into the hot fat fryer. Bench-cut doughnuts are made using the same recipe, but with a reduced amount of water. This gives a dough (rather than a batter) that can be rolled out on the workbench and cut into rings using a doughnut cutter. This process takes much more hand labor than depositing and is usually used only in home baking or in small, traditional doughnut shops.

In commercial bakeries and most doughnut shops doughnut batter is deposited through a cutter into the frying fat. Proper batter viscosity is extremely important in order to attain uniformity of flow through the cutter and uniform spread in the fryer. Batter viscosity is governed by the amount of water used in mixing (usually about 70% based on flour, or 38–44% based on total dry mix weight) and the way that water is partitioned between flour, sugar, soy flour, and the other ingredients present. Sometimes high-viscosity gums (guar, locust bean, sodium carboxymethylcellulose) are added to the dry mix at levels of 0.1–0.25%. The gum helps give uniform viscosity from batch to batch and also acts as a water binder that reduces fat absorption and lengthens shelf life.

Leavening in cake doughnuts is provided by 1.5% soda (flour basis) and 2.1% sodium acid pyrophosphate (SAPP). The fast-reacting grades of SAPP are required, from SAPP-37 to SAPP-43 (the fastest grade). When the doughnut is first deposited in the hot (190°C, 375°F) frying fat the ring of dough sinks and is supported on a bottom plate. Leavening begins, and after 3–7 s the doughnut rises to the surface and frying of the first side commences. As heat penetrates the dough piece it slowly expands; after the first side is fried, the doughnut is flipped and the second side is fried. During this time, expansion continues, but the diameter of the doughnut has been fixed because the first side has set (the starch has gelatinized and proteins have denatured). Thus, second-stage expansion tends to be into the center hole and the typical star is formed (or the doughnut balls and fills the center hole if conditions are incorrect).

When the finished doughnut is cut crosswise, three regions are readily seen. The central core is surrounded by a more porous leavened ring, but this porosity tends to be greater on the second side than the first. If the core is inadequately leavened, it will be dense and gummy, and the doughnut will not have a good eating quality. Part of the art of the doughnut formulator is in choosing leavening acids that strike the best compromise to achieve moderate porosity of the two sides along with at least some opening up of the core.

Yeast-Raised Doughnuts

These doughnuts are made using a rather lean, sweet dough. After fully developing the gluten of the dough, it is sheeted, cut into rings, and the doughnuts are proofed for 30–45 min. The raised doughnuts are then introduced into the fryer, where they are fried for about 1 min on each side. (Because they are already leavened, they float on the fat throughout the frying cycle.) They are then removed, the surface fat is drained, and the fried doughnuts are (most often) glazed with a simple sugar/water glaze. If they are to be iced or otherwise coated, they are usually cooled before these operations are carried out.

Yeast-raised doughnuts can be made in many different shapes. In addition to rings, the dough is also frequently cut into strips and twisted into various fanciful shapes before proofing and frying. In the form of rectangles about 2 by 4 in., they are iced and topped with chopped nuts or fruit to make Long Johns. In the form of discs, the fried piece can be filled with jelly or custard. This product, filled with raspberry-flavored jelly, is known as a Berliner in Austria and Germany. (This explains why the Germans roared with laughter when, in 1963, President Kennedy proclaimed, "Ich bin ein Berliner." He was telling them that he was a jelly-filled doughnut.)

French Crullers

These doughnuts are made using eclair dough (see below). The dough is deposited in a ring shape onto a sheet of parchment paper, and after a few moments to allow it to set, the parchment is dipped into the frying fat and the rings of dough are loosened from the paper. The dough can also be deposited using a specially shaped cutter that produces a fluted surface on the ring of dough. The dough piece expands to a much greater extent than does cake doughnut batter, and the finished piece is much less dense than other doughnuts. It is frequently glazed with a butter/rum flavored glaze.

ECLAIRS

The dough for eclairs and French crullers (pâte de choux) is a paste made in a rather unusual procedure. Water (two parts) and shortening (one part) are placed in a kettle and brought to a rolling boil. After the fat is melted, a pinch of salt is added, flour (1.5 parts) is added, and the mixture is stirred briskly. The gelatinized flour paste is cooled to about 150°F, and two parts whole egg is added in several increments, with continuous stirring. The final paste should be soft enough to deposit from a pastry bag, yet firm enough to hold its shape once deposited. If it is too stiff, it may be softened by the addition of a small amount of milk or water.

The finished dough is deposited on sheet pans lined with parchment paper, then baked in a hot (425°F) oven until the eclairs are crispy. Leavening action is due to the formation of steam in the interior of the piece. If inadequate leavening occurs, a small amount of ammonium bicarbonate (no more than 1% of the weight of the water used) may be dissolved in a little water and mixed in at the end of the dough-making procedure. The finished piece should be crisp on the outside, hollow on the inside, and with no soggy layer on the interior wall.

Eclairs are usually filled with a chilled custard, either plain or a French custard (butter-flavored). They are then iced, with chocolate being the most popular flavor, or dusted with confectioners (very finely powdered) sugar. To prevent bacterial growth, custard-filled eclairs should be refrigerated until served.

LAYERED DOUGH PRODUCTS

Numerous bakery products have a unique texture, in that the baked dough is present in many fine layers, rather than one continuous matrix. This effect is obtained by placing a layer of fat between two layers of dough, rolling out this sandwich to a thin sheet, folding the sheet over onto itself into three or four layers, and repeating this rolling and folding process two or three more times. If done properly, the final sheet consists of 50–100 layers of dough interspersed with an equal number of layers of fat. When baked, the dough layers set up more or less independently, giving the flaky texture of the finished product.

Puff Pastry

Puff pastry dough is used for a large number of culinary items, from basic pastries such as fruit turnovers and napoleons, to patty shells for holding a variety of fillings for appetizers, to dough for enclosing fish or meat, such as beef Wellington. It needs no fermentation, but does require some care in the roll-in process to obtain a flaky final product.

The dough for this product is a flour/water dough (no yeast), containing about 6% shortening and 6% whole eggs. After it is fully developed it is sheeted out, fat is spread over two-thirds of the dough, which is then folded to give a sandwich having three layers of dough separated by two layers of fat. This is sheeted out and folded as described above. Usually the dough is refrigerated after the second sheeting and folding, to keep the fat in a plastic state. If the dough becomes too warm the fat melts and gets incorporated into the dough itself, and the layering effect is lost. The fat usually used is a special puff-paste margarine, having a broad plastic range, and containing about 15% water. In the oven the water (which is finely emulsified in the fat) forms steam, which helps give the puff to puff pastry. Thus, puff pastry is a steam-leavened product.

Croissants

Croissants are another legacy of the siege of Vienna. The story is that the Turks were busy digging a tunnel under the city walls, but were overheard by a baker in the early hours of the morning, as they were digging under his shop. He alerted the authorities, and a counter-tunnel foiled the plot. After the siege was lifted, the baker made rolls in the shape of a crescent (the emblem on the Turkish flag), which were sold to celebrate the victory, and which became a Viennese tradition. Marie Antoinette introduced these Kipfeln to the French court, and they became popular under the French name croissant.

A typical croissant dough is basically a straight dough, containing 10% sugar, 1% salt, and 4% margarine (flour basis), with a fairly high (9%) yeast level and slightly below normal absorption. Croissant dough is mixed cold (18–20°C, 64–68°F). This practice, together with the slight underabsorption, retards fermentation. Proofing temperatures are also relatively low, not exceeding the melting point of the fat (ca 37°C, 98°F). For these reasons a rather high level of yeast is used in the dough.

Numerous variations on the basic mixing, sheeting, and roll-in process are used, most of which seem to be based more on tradition than on actually yielding superior product. After mixing, the dough is allowed to ferment for up to 1 hr, then the fat is rolled in, essentially as described for puff pastry. The traditional fat for croissants is unsalted butter, although many commercial companies today use a bakers' margarine that does not get as hard as butter at retarder temperatures, and so spreads more uniformly during the sheeting and folding process. The usual fat-to-dough ratio is ca 25%, ie, 1 lb of butter to 4 lb of dough.

The rolled-in dough is sheeted out, then cut into triangles having a height about twice the width of the base. This triangle is rolled up, starting from the base, the shaped dough piece is placed on a sheet pan, and the ends are pulled down to give the typical crescent shape. After proofing for 1–3 h, it is baked and cooled. Care must be taken that the piece is not underproofed, otherwise the finished product loses its flakiness and is tough, chewy, and unpalatable. Croissants are frequently sliced horizontally and used to make sandwiches, with a wide variety of fillings being offered by various specialty shops.

Danish Pastry

Danish dough is a rich, sweet, yeasted dough made with a higher than usual absorption. The dough is quite soft, mixed cold (62–64°F), and not fully developed in the mixer. Gluten development takes place during the roll-in process. The dough is divided into pieces of a suitable size (eg, 16 lb) and then retarded for several hours. Some fermentation takes place during this time. The cold dough is then sheeted out and fat is rolled-in, as described above. Specialty shops use butter or bakers' margarine for the roll-in fat, but all-purpose shortening works equally well. The dough sheet is usually given the first two sheetings and folds, then returned to the retarder for 6–12 h. It is then given one more sheeting and three-fold, and returned to the retarder for 24–48 h for further fermentation.

Danish pastry is almost invariably made by sheeting out the rolled-in dough, cutting into various shapes, then adding some sort of sweet filling (fruit, nuts, cinnamon, etc) and folding or rolling up the dough to retain this filling during proofing and baking. The variety of shapes and fillings is limited only by the ingenuity of the baker. The proofed pieces are sometimes washed lightly with diluted egg just before baking, to give a shine to the finished piece. Alternatively, the piece may be sprayed with a light sugar syrup as it exits the oven, to give the same effect. If a Danish coffee cake, for example, is to receive an icing, this is usually a rather thin icing that is drizzled over the top of the coffee cake in a random pattern, rather than applied in a solid layer (as on sweet rolls). A good Danish pastry is one of the foundation stones of any small retail bakery today.

PIES

The term pie (like biscuit) means something rather different in the United States than it does on the other side of the Atlantic Ocean. In Europe a pie usually contains meat, potatoes, and other vegetables and may have a crust of puff pastry, of mashed potatoes, or a flour dumpling. It is normally the main dish of a meal. In the United States, on the other hand, a pie is for dessert, or is a sweet snack. It need not be a baked food, eg, the ice cream pies that are now so popular. In at least one instance (Boston cream pie) it is more properly a filled, layered sponge cake. Having said that, this article will discuss pies primarily in terms of the traditional one-crust or two-crust baked American dessert pie, and, briefly, the fried pie, a related, traditional American dessert.

A pie crust is one of the simplest of baked doughs, comprising only flour, shortening, salt, and water. The ability to combine these few ingredients properly to produce a flaky, tender pie crust, however, is rare, as attested by the number of leathery, soggy, nearly inedible pie crusts that are served to diners in United States restaurants every day. The first step is to cut the shortening into the flour, ie, mix the two and, using a fork or similar mixing tool, reducing the particle size of the shortening to small lumps, perhaps one-quarter the size of a pea. Then a measured amount of cold water (even ice water, in hot weather), with the salt dissolved in it, is added, and the dough is gently mixed until the flour is wetted and the dough forms a lumpy, slightly sticky ball. The dough should then be allowed to rest for a time, preferably in a refrigerator, to allow the water to equilibrate with the flour protein.

An amount of this dough just sufficient to form the crust (either top or bottom) is then rolled out into a circle. An absolute minimum of dusting flour is used, and it is best to employ a cloth sleeve on the rolling pin and a flat cloth on the bench top, both of which have been rubbed with flour and the excess flour removed. Once the desired diameter of crust is achieved, the circle of dough is loosely rolled up on the rolling pin and transferred to the pie pan, where it is gently shaped and the edges sealed (if it is the top crust, and the filling is in place).

The main mistake in making a flaky, tender pie crust

lies in excessive working of the dough and consequent development of the gluten in the flour. This can happen if too much water (or water that is warm) is added to the dough, and it is then mixed until the water disappears. In addition, using excessive dusting flour, and rolling the dough more than once, will contribute to gluten formation. In large-scale pie production (either in a commercial bakery, or in an institutional setting) the dough described above may be too tender for the kind of handling it receives, and a certain amount of gluten development will be done on purpose, in the interests of production efficiency. Although this speeds up the operating line, it produces the leathery crusts that are abandoned on dessert plates across the country.

Fillings for pies are usually either some kind of fruit in a starch-stabilized juice matrix or a custard. The formulation of a good pie filling deserves an entire article to itself, and will not be explored here. However, the stabilizing system should be such that during baking the filling will not boil or foam excessively, to the point where it overflows the side of the pie. If the pie crust at the filling/crust interface does not bake rapidly enough, it will soak up moisture from the filling, and the final crust will be soggy. This may be due to using a cold filling in the pie or having too low an oven temperature.

One type of pie avoids this pitfall, by adding the filling after the bottom crust is baked. An example is strawberry pie; a single crust is baked in the pie pan, then the fruit is placed in the cooled crust and a strawberry-flavored starch glaze is poured over the top. The pie is then refrigerated until served.

Fried Pies

These desserts are traditional in the Southeastern part of the United States, although they are now produced commercially and sold throughout the country. A pie crust with a slightly developed gluten is used. A circle is rolled out, fruit filling is placed on one half the circle, the other half of the crust is folded over, and the edges are sealed. This is then fried in a half inch of hot fat, in a heavy skillet. In the commercial version, the shaped and filled piece is fried in a deep fat (doughnut) fryer. A top conveyor belt keeps the pie submerged throughout its journey through the hot fat. The crust absorbs some of the frying fat, tenderizing it so that it is not leathery, as it would be if baked. The commercial fried pie is given a sugar glaze before being cooled, wrapped, and dispatched to the consumer. The traditional fried pie, on the other hand, is usually eaten while still warm, fresh from the kitchen, and any glaze or other topping would be superfluous.

BIBLIOGRAPHY

General References

American Institute of Baking, Technical Bulletins: *Bagels*, **8**(11), 1986; **10**(4), 1988; *Croissants*, **8**(10), 1986; *Donuts*, 1(6,7), 1979; **2**(7), 1980; **4**(9), 1982; *English Muffins*, 1(1), 1979; *Pizza Dough*, **1**(11), 1979; **8**(12), 1986.

S. A. Matz, *Bakery Technology and Engineering*, AVI Publishing Co., Westport, Conn., 1972.

E. J. Pyler, *Baking Science & Technology,* 3rd ed., Sosland Publishing Co., Merriam, Kans, 1988.

W. J. Sultan, *Practical Baking,* 5th ed., Van Nostrand Reinhold, New York, 1990.

CLYDE E. STAUFFER
Technical Foods Consultants
Cincinnati, Ohio

BAKING SCIENCE AND TECHNOLOGY

It is obviously not possible to provide extensive details about the science and technology of baking in 15,000 words. However, at the suggestion of the editorial team, the author has selected parts of his two-volume text and created an article that provides a brief review of this important topic in food science and technology.

This area of study can be conveniently divided into five parts: elements of food science, ingredients of baking, fundamentals of baking technology, aspects of cake baking, and bakery equipment. Information on cookie and cracker manufacturing is presented in another article in this encyclopedia.

A thorough knowledge of the elements of basic food science entails the study of carbohydrates, fats and oils, proteins, enzymes, yeasts, molds, and bacteria, together with physical chemistry and colloidal systems. Details about each of these different subjects are provided in other articles in this encyclopedia under the appropriate headings.

Seven groups of ingredients are of significance in baking: wheat flour, miscellaneous flours, sugars and syrups, bakery shortenings, dairy products and blends, eggs and egg products, and water. Details for these topics can be found in other articles in the encyclopedia. However, brief discussions of the specific effects of shortenings, dairy products, and eggs on baking are provided in the next section.

THE INGREDIENTS OF BAKING

Fats

Fats are used in baking primarily for the tenderness and shortness they impart to bakery products. This function is of particular importance in such products as biscuits, wafers, and cookies, and in many types of cake in which shortening is used at significant levels. The tenderizing effect is due to the ability of fat to lubricate the structure of the baked product by being dispersed in films and globules in the dough or batter during mixing and thereby preventing the starch and protein components from forming a continuous, three-dimensional gluten network.

The shortening value of a fat is generally determined by a device called a shortometer. This instrument measures the breaking strength of standard wafers baked with the shortening. In general, the softer fats, such as lard, are rated superior in shortening value to the harder fats, whereas, on the other hand, liquid oils exhibit very little shortening ability, presumably because their lack of plasticity tends to favor their dispersion in the form of globules rather than of films in the dough.

Aerating Function. In the production of certain cakes, notably yellow and white layer cakes and pound cakes, shortening plays a decisive role in determining the structural characteristics of the products as developed by leavening. Even though chemical leavening agents, which generate carbon dioxide gas and thereby aerate the baking cake batter, are used in such cakes, the entrapment and uniform dispersion of considerable amounts of air during the mixing process is required to produce a cake with proper volume, grain, and texture. The incorporation of air during creaming is solely a function of the shortening, for it is the fat that entraps the air in the form of minute cells and bubbles.

Fat is the only major ingredient in the cake batter that does not undergo drastic changes during baking. Cake batters consist of mixtures of certain dry ingredients, such as flour, sugar, salt, baking powder, and plastic fat and of liquid ingredients, such as egg, milk, and water. On mixing, the sugar, salt, and baking powder will dissolve in the liquid ingredients and the resultant solution will mix freely with the flour. No such intimate absorption occurs with fat, which is dispersed throughout the batter in the form of small, irregularly shaped particles. The plastic character of the shortening assures its dispersion in the form of films or clumps rather than of globules, as is the case with oil, and this creates a much larger fat surface than is possible with liquid oils. In the case of fluid shortenings, which lack the plasticity of solid shortenings, the presence of crystalline triglycerides and balanced emulsifier systems that exhibit relatively high melting points ensures the dispersion of the fat in the form of minute and stable fat globules that, because of their small individual size, appear to create a comparably large fat interface.

The leavening action of the fat-entrapped air in cakes containing no chemical leaveners is far greater than would be expected solely on the basis of the thermal expansion of the air during baking. The large expansion in cake volume on baking can come only from a greatly increased water-vapor pressure within the air cells at high temperatures. Water vapor, however, cannot exert a leavening action without preexisting air bubbles.

Creaming Quality. The ability of fat to absorb air during mixing is called its creaming quality. Because the fat-entrapped air plays a principal role in cake quality, it is evident that a fat's creaming quality governs its suitability as a cake shortening. Although a fat's deficiency in air absorption may be ameliorated sufficiently by the use of baking powder or other chemical leaveners to yield a satisfactory cake volume, the cake's grain will be coarse and marked by large pockets, as the action of the baking powder will be largely uncontrolled in the absence of minute air cells. It is, therefore, of considerable importance for the baker to be able to test the creaming quality of a shortening and evaluate its suitability.

Considerable variations in creaming properties are encountered with different fats. To cream satisfactorily, fats appear to require the presence of certain highly saturated glycerides. Partial hydrogenation effectively improves the creaming qualities of oils and soft fats. Similar improvements may be obtained by the addition of small proportions of hard fat. Thus, whereas lard does not cream well without prior treatment, its creaming quality may be effectively improved by a limited hydrogenation or the addition of hydrogenated fat. By the same token, fluid shortenings require the presence of high-melting emulsifiers, as well as crystalline triglycerides, to produce maximum aeration. Butter will generally be found to be inferior in creaming property compared to high-grade vegetable or compound-type shortenings.

Stabilizing Function. The stabilizing function of fats, or their ability to impart to cake batters sufficient strength to prevent their collapse during baking, is intimately related to their aerating function. Considered from a physical standpoint, the cake batter is an emulsion in which fat forms the internal, or discontinuous, phase and the remaining ingredients, such as sugar, flour, milk, and eggs, form the external, or continuous, phase. In the absence of proper aeration, this emulsion will become very thin and sloppy. This condition is caused by the diluting ingredients, especially the high sugar content, that greatly weaken the structure of the batter. The solid fat portion of the shortenings exhibits a certain degree of inherent stability. When the fat is creamed, the incorporation of air gives rise to innumerable air cells that impart considerable mechanical strength to the batter. This, in turn, counteracts its tendency to collapse of its own weight during baking just before setting by the oven heat when the protein structure coagulates and assumes a rigidity of its own. The finer the cellular structure of the batter, the greater will be its mechanical strength. This is another reason why optimum aeration and fat distribution should be a primary goal of the creaming operation. Given the importance of an optimum dispersion of fat in a batter, it is evident that any factor that tends to maintain this dispersion assumes equal importance. Although it is true that plastic shortening does not coalesce in a batter as would liquid oil, it nevertheless does accumulate into larger agglomerates that greatly reduce the effectiveness of the emulsion. The higher the liquid and sugar contents of a batter, the more difficult it becomes to maintain the emulsion in a highly dispersed state. Thus, with ordinary shortening, the maximum combined milk and sugar content cannot exceed much more than 40% if good cake is to result. With the use of surfactants in shortenings to improve their emulsifying properties, the sugar and milk may make up as much as 50–55% of the total ingredients. The explanation for this difference in the behavior of the two types of shortening lies in the fact that emulsifier-containing shortenings become more finely dispersed in batters of high sugar and liquid contents and thereby impart greater stability to them.

Eating and Keeping Qualities. Eating quality represents an essential attribute of foods in general. The concept of eating quality covers such sensual impressions as smell, taste, and cutaneous sensations. The taste, flavor, tenderness, moistness, and similar properties of cakes and other baked products are thus decisive factors that govern the acceptance or rejection of the product by the consumer. Fats contribute to eating quality principally by imparting shortness and tenderness to the baked goods. They function secondarily by making possible the use of increased proportions of such enriching ingredients as sugar, milk, and eggs that enhance the taste of the products. Finally, if the fats are flavored, as in the case of butter and some lards, they also contribute importantly to the organoleptic characteristics of the product.

The keeping quality of a product is a measure of the degree to which it retains the characteristics of freshness over a period of time. The extent of time over which a product may be reasonably expected to maintain freshness varies with its type. Thus, the rate at which bread, for example, loses freshness differs considerably from that at which crackers become stale. The palatability of bread is seldom expected to extend for more than 4–5 days after baking, whereas crackers are normally consumed weeks and months after baking. The rate of staling in a given product may be influenced by the ingredients and method of production. In the case of cakes, for example, the inclusion of relatively large proportions of fat tends to reduce the rate of staling, or at least the appearance of those changes that are normally associated with true staling, such as loss of moisture, tenderness, and flavor. It is not uncommon to have cakes rich in shortening retain their palatability for a week or more, although there is an inevitable loss of quality with the passage of time. Cakes in which an insufficient amount of shortening is used will stale more rapidly and will, therefore, have a reduced shelf life.

Milk Products

Nonfat dry milk (NFDM), obtained by separating the butterfat from whole fluid milk and drying the defatted portion by either spray drying or roller drying, represents a valuable baking ingredient that improves both the nutritional value and general physical quality of bakery product in which it is used in functionally significant amounts. Formerly used in bread products at levels up to 6%, based on flour, its use by bakers has declined sharply in recent years, primarily because of its drastic rise in cost, but also because functionally equivalent, but less expensive, dairy blends and milk substitutes have become available.

When nonfat dry milk is used at a level of 6%, based on flour weight, certain changes in the physical properties of the dough and in the qualitative properties of the bread become apparent. The recommendation of 6% was originally based on the fact that it will yield bread that is equivalent to a product in which all the added liquid is skim milk.

It has been well established that milk solids influence dough absorption, mixing requirements, fermentation rate, the bromate requirements of flours, baking times and temperatures, and the physical properties of bread.

The recommended procedure for adding dry milk products and dairy blends to the dough in the mixer is to place the milk on top of the flour just before mixing. This prevents the formation of lumps that may occur when the dry milk is put into direct contact with water in the mixer.

There appears to be no advantage to reconstituting NFDM before its addition to the dough, so the dairy product is normally added in its dry form.

Doughs containing 6% NFDM normally require longer mixing than do milk-free doughs, indicating that milk solids exert a strengthening effect. Practical experience also indicates that doughs containing milk should be mixed somewhat slacker as they tend to tighten up during fermentation as a result of progressive hydration of the milk solids. Given good quality NFDM, a general rule is to increase the dough absorption by a percentage equal to that of dairy product used.

Both the fermentation tolerance and the fermentation time of a dough are increased by NFDM. Whereas the fermentation time is readily controlled by the baker, fermentation tolerance, which represents that period of active fermentation during which bread of acceptable quality will result, is a far more critical factor. By extending the fermentation tolerance, milk products act to stabilize the dough and thereby contribute materially toward the day-to-day uniformity of bread.

Normally, NFDM is added at the dough stage in the sponge-and-dough process. However, some or all of it may be added to the sponge if the flour (1) has a low-protein content or lacks strength, (2) exhibits excessive amylolytic activity, (3) has a short fermentation tolerance, or (4) tends to break down easily. With flours that possess the reverse characteristics, NFDM is more properly used in the dough.

Doughs with 6% NFDM should be set at somewhat higher temperatures than is normal with no-milk doughs, eg, 81°F (27°C) in winter, and 78–79°F (25–26°C) in summer. Such doughs generally also require somewhat longer recovery periods in the overhead proofer than do milk-free doughs. Usually, an extension of 2–3 min will prove adequate.

Proper recovery of the dough pieces is essential if uniformly molded loaves are to result. The final pan-proof of milk doughs is also slightly more protracted as compared to water doughs. The proof box temperature should not exceed 100°F (38°C), and the relative humidity should be maintained slightly lower than is the case with milk-free doughs.

There is a tendency for bakers to underbake bread containing NFDM, partly in order to achieve a softer crumb, and partly because a more rapid crust color formation results from the high lactose content present in the milk products. If baking time is adjusted solely on the basis of crust browning, bread with more than minimum amounts of NFDM or whey solids will tend to be underbaked. Doughs with high levels of residual sugar, such as are contributed by dairy products, color more rapidly than do doughs with limited levels of residual sugar. In high-sugar doughs, regardless of the source of sugar, the degree of crust browning cannot, therefore, be taken as an indication of the degree of baking.

Eggs

Eggs and their products constitute important ingredients in a wide range of bakery products because they contribute significantly to the nutritional value of the baked foods and markedly improve the physical and organoleptic properties of the products in which they are used. In cake making, eggs exert a binding action, are capable of leavening five to six times their weight of other ingredients, exhibit a considerable emulsifying effect, and, in the case of the yolk, exert a marked shortening action, and they contribute to the flavor of the baked product.

The pleasing color that eggs impart to baked goods has long been accepted as a mark of superior quality. Eggs improve the cell structure of the product, maintain it during the baking process, and reduce moisture loss from the baked product, thereby extending its shelf life. Egg yolks impart a rich yellow color to cakes and are often used to fortify whole-egg blends to obtain a deeper color or an increase in emulsifying action.

Egg Whites. Egg whites find their major application in the production of angel food cakes and meringues. They also serve as a functionally useful ingredient in layer cakes, pound cakes, and cookies, and in hard rolls and hearth-baked products, to which they impart a crispier crust. Consisting primarily of proteins, egg white excels as a foaming agent, binder, and emulsifier in these products. Dried egg whites may be readily reconstituted to the liquid form by combining with water in a ratio of 1:7 on a weight basis. However, it is generally preferable to dry-blend them with other dry ingredients and to add the water of reconstitution with the other liquids during the mixing stage. In angel food formulations, a principal purpose of the use of cream of tartar or other acid salts is to produce a pure-white color in the crumb that otherwise tends to take on a yellowish cast. Acid salts additionally help create a firmer egg–sugar foam structure, thereby improving cake volume. Optimum foam stability and cake volume are obtained when the pH of the egg white is adjusted to a range of 5.8–6.4. When the foam is made with dried egg white it should be whipped to a stiffer and drier peak than when fresh or frozen egg white is used.

Frozen egg whites are quite alkaline, with a pH of 8.5–9.5. When they are replaced with dried egg white in an angel food cake formula originally designed for frozen egg white, a reduction in the acid salt level by one-third to one-half is required. Certain other ingredient modifications have also been found to be helpful. For example, improved texture quality is obtained when 5–20% of the granulated sugar in the formula is replaced with powdered sugar. Substitution of approximately one-third of the flour with wheat starch will also increase cake volume, moistness, and eating quality.

Egg whites are used in layer cakes primarily for their binding, and secondarily for their emulsifying, properties. Dried whites produce somewhat thinner batters than do frozen whites, leading to a tendency by bakers to reduce the amount of formula liquid so as to obtain a batter consistency comparable to that yielded by frozen whites. However, regardless of batter appearance, it is essential that the full complement of formula liquids be added if cakes with lower volume, greater toughness, and poorer texture are to be avoided.

Whole Eggs and Yolk. The functional properties of standard and stabilized whole egg products are for all practical purposes identical. Moreover, there are no essential differences in the binding and emulsifying properties of dried plain whole egg and yolk and those of fresh liquid eggs when used in bakery production. They all impart the same flavor, color, and nutritional value. The dried egg products, however, lack foaming properties, except under special conditions of use, such as at elevated temperatures.

Dried whole eggs are reconstituted by combining one part by weight with three parts of water. For egg yolk solids, the corresponding ratio is 1:1.25. Reconstituted whole eggs and yolks exhibit a much lower viscosity than do thawed frozen eggs, whose high viscosity is caused primarily by the gelation of the yolk during freezing and frozen storage. Whereas frozen plain whole egg is rather fluid when thawed, thawed plain yolk tends to exhibit a rubbery consistency. Gelation of yolk during the freezing process can be inhibited or totally prevented by such pretreatment as homogenization and the addition of 10% sucrose and sodium chloride. However, either of these added ingredients may restrict the future use of the yolk to specific food products. Alternative additives include corn syrup, glycerin, phosphates, and some other carbohydrates. When yolks are required in a specific formula, frozen sugared yolk is commonly used.

Dried plain yolks, when reconstituted, resemble fresh yolk in their viscosity. In their dry form, they are widely used in doughnut and sweet dough mixes.

Commercial dried yolks affect batter fluidity variably, depending on the degree of heat-induced alterations that occurred during preheating of the liquid yolk, the drying process, or the storage of the solids at moderate-to-high temperatures. In general, commercial dried yolks that yield the least fluid doughnut batters are those that have the lowest protein solubility and highest reconstituted viscosities. Yolk solids that have undergone little or no processing change exhibit maximum protein solubility, produce the most fluid batters, require the lowest water level in the batter for normal operating extrusion pressure, result in the highest fat absorption, and yield the highest quality in the fried doughnuts.

The general preference of bakers for eggs with yolks of dark color has led to experimentation with poultry feeding designed to control yolk color. Increasing the xanthophyll content of the poultry diet markedly darkens the color of the yolk. Baking tests with dark-colored yolks have yielded sponge cakes that were judged to be significantly higher in moistness than were the control cakes. At the same time, however, the cakes made with dark-colored yolk were considered to be less desirable in flavor, less tender, and to have a coarser texture than the control cakes. They also produced a gummy mouthfeel.

In yellow layer cake production, dried whole eggs are generally blended with the other dry ingredients on the low speed of the mixer for about 2 min. This is followed by the addition of shortening and one-half of the total formula water and mixing on medium speed for 3–5 min. Finally, the remaining water is added in two portions and mixing continued on low speed until creaming is accomplished. As in the case of white layer cakes, the batter consistency of yellow layer cakes will be more fluid when using dried whole egg than frozen eggs. Essentially the same procedure is followed for incorporating dried whole egg in pound cake batter.

Blends of eggs with carbohydrates offer even greater versatility than do plain whole egg and yolk products in their application to bakery products, as they can advantageously replace whole egg or yolk in sponge cakes, layer cakes, pound cakes, cookies, sweet doughs, and doughnuts.

Bulk Liquid Eggs. Fresh liquid eggs in bulk, properly chilled and delivered to the bakery in tank cars or smaller containers, also find extensive use. They obviously require a different mode of handling by both the egg processor and the baker than do frozen eggs. The following general guidelines for handling fresh liquid eggs have been proposed by the American Institute of Baking: Once the eggs have been cooled following pasteurization, they must not be allowed to rise above 40°F (4.4°C) during their transit to, and storage in, the bakery. When delivery is made by tank car, the discharge valve on the car must be thoroughly sanitized before connecting it to the bakery's receiving and storage system. Before acceptance of the shipment, samples should be taken for testing for the possible presence of off-colors and off-odors and objectionable microorganisms. Finally, the baker should obtain from the egg processor a guarantee that the eggs being delivered are Salmonella-free and conform to all federal, state, and local food laws and health department regulations pertaining to their chemical and general bacteriological quality.

Recent Developments. Among more recently developed egg products is an instantized dried egg white that contains some sucrose. The product is readily adapted to bakery formulations that call for egg white and, because of the greater availability of its functional proteins, can be used at significantly lower total egg solids levels.

Another recent development in frozen eggs is IQF (instant quick frozen) pellets. Measuring approximately $1'' \times \frac{3}{4}'' \times \frac{5}{16}''$ in size, the pellets are formed between stainless-steel belts in cup-shaped sections and frozen instantly at a very low temperature. The egg pellets maintain their identity when held under frozen conditions.

The introduction of preweighed batch-sized packages of egg products that can be added directly to the mixer in the bakery offers attractive labor savings and improved accuracy. Many bakeries have standardized their formulas based on 30-lb cans of frozen whole eggs and egg whites. Packages of dried egg white solids, equivalent to a 30-lb can of frozen egg white, are now available; their contents can be dry blended immediately and directly into the first stage of an angel food cake operation.

Where continuous supplies of liquid eggs are required, available reconstitution equipment can convert dried egg solids to the liquid product automatically. In one such device, known as a powder horn and available in various sizes to fit individual bakery requirements, the dried egg product is fed into a funnel, to be picked up by the dissolving fluid, reconstituted, and pumped into a holding tank. When more than one ingredient is to be reconstituted, they may be added at the funnel to ensure homogeneity in the final liquid dispersion.

DOUGH FORMULATION AND THE DOUGH-MIXING PROCESS

Bread, in its simplest form, requires but four ingredients: flour, water, yeast, and salt. For the most part, however, conventional white pan bread and most specialty breads include optional ingredients for the purpose of enhancing the product's over-all quality. Thus, such attributes as loaf volume, crumb softness, grain uniformity, silkiness of texture, crust color, flavor and aroma, softness retention, and, most important of all, nutritive value, can all be improved to varying degrees by the judicious addition of appropriate optional ingredients. The materials that are either required or may be optionally included in specified amounts in the production of various standardized bread products are legally defined by the Food and Drug Administration.

A representative formula for white pan bread will include the following percentages of required and optional ingredients, all based on flour as constituting 100%: Flour, 100; water, 64; yeast, 2.75; salt, 2.1; sweetener solids, 7.25; shortening, 2.3; dairy blends, 2; yeast food, 0.5; protease enzyme, 0.25; emulsifier, 0.5; dough conditioner or strengthener, 0.5; and preservative, 0.2.

Mixing may be most simply defined as the act of combining or blending into a homogeneous mass the diverse ingredients that are destined to become part of the processed end product. In the production of bread and other yeast-raised bakery products, it initiates the long series of complex changes and interactions of such diverse components as water, starch, protein, lipids, enzymes, salt, sweeteners, yeast, and oxidizing and reducing agents by bringing them into intimate contact with each other through the agency of physical work to ultimately result in a dough or batter. Two preconditions must be met for the production of a dough with the right properties: an appropriate proportioning of the individual ingredients as established in a well-balanced dough formula and a homogeneous distribution of these ingredients throughout the dough mass.

In its essentials, dough mixing involves the combining and blending of the formula ingredients and then applying sufficient physical work to the mixture to transform it into a cohesive mass with the requisite viscoelastic properties. In large-scale commercial bakery production, such major ingredients as flour, dairy blends, and dry sweeteners are normally weighed by automatic scales that feed directly into the mixers, while water, liquid shortening, and liquid sweeteners are piped into the mixer through meters that can be preset to deliver specified volumes. The small-volume ingredients are usually weighed out individually on small, more-sensitive scales, or, in the case of yeast foods, enrichment ingredients, dough conditioners, enzymes, and other highly reactive materials, they may be added in the form of tablets, wafers, or premeasured small packets made of soluble edible films. Yeast, depending on the form in which it is used, ie, whether compressed, active dry, or instant active dry, may be added either crumbled, rehydrated as a slurry, or in its dry state.

In addition to achieving a thorough dispersion of the ingredients into a homogeneous mixture, the dough-mixing process in breadmaking has the further important objective of physically developing the gluten proteins into a coherent three-dimensional structure that will impart to the dough the desired degree of plasticity, elasticity, and viscosity. Significant physical effects during the initial mixing phase are hydration of the flour particles and the incorporation of atmospheric oxygen. As the dough develops, many complex physical, colloidal, and biochemical changes occur that transform the dough into a complex viscoelastic polymer system.

Among mixing methods being practiced commercially, the most prevalent is the sponge-and-dough process that involves two mixing stages, namely one of the sponge and the other of the dough. The sponge-mixing stage aims at a homogeneous ingredient dispersion and flour hydration and is normally of relatively brief duration, whereas the more critical phase of dough development is reserved for the more extensive mixing of the final dough. In the straight-dough method, as well as in those systems that employ various forms of liquid preferments, there is but one mixing stage in which complete dough development must be achieved. It is the dough-development phase of the dough-mixing process that establishes the major basis for a quality loaf.

DOUGH FERMENTATION

Before dough can yield a light, aerated loaf of bread, it must be fermented for a sufficiently long time to permit the yeast to act on the assimilable carbohydrates and convert them into alcohol and carbon dioxide as the principal end products. In conventional practice, the sponge or dough is discharged from the mixer into a greased dough trough in which it undergoes fermentation in bulk, usually in an atmospherically controlled room. In liquid ferment systems, yeast fermentation is initiated in so-called preferments, brews, or broths, which generally, but not always, contain some of the total formula flour. In chemically developed doughs, on the other hand, fermentation is limited almost entirely to the dough pieces undergoing the final proofing stage. Our attention here will center primarily on the bulk fermentation of plastic sponges and straight doughs.

The three major purposes of fermentation in a sponge or dough are leavening or aerating of the dough mass, maturation or physicochemical modification of some of the dough constituents, and flavor development. Major attention in the past has centered on the yeast's leavening action in the dough, although in recent years the formation of bread flavor precursors has received more intensive investigation. The actual role played by yeast in dough maturation remains the least understood aspect of dough fermentation. Yeast, like any living organism, must be provided with a supportive environment before it can function properly; therefore, such factors as an adequate level of moisture, moderate temperatures, a proper degree of acidity, and an ample supply of fermentable carbohydrates and assimilable nitrogenous substances, as well as certain essential minerals, form the basic requisites for a proper fermentation. Yeast itself brings about changes in the dough environment in the course of fermentation, such as a depletion of fermentable substances; accumulation of waste products in the form of ethanol and other alcohols, carbon dioxide, and acids and esters; a modification of the pH conditions; and a softening or mellowing of the gluten character. The most apparent physical change marking the course of fermentation in a dough is the steady increase in the volume of the dough mass. The sponge in the dough trough expands to four to five times its original volume before it recedes, assuming at the same time a light, spongy character.

These effects, in their totality, bring about a transformation in the physical character of the dough that is designated as dough ripening or maturing. Proper dough maturity represents that point in fermentation at which the dough possesses the optimum spring and elastic properties that will yield the best-quality bread of which a given flour is capable.

There is no significant increase in the number of yeast cells under the fermentative conditions that prevail in sponges, doughs, and liquid ferments, although there is an increase in the budding of cells and the size of the individual yeast cells. The rate at which yeast generates carbon dioxide is influenced largely by the prevailing temperature of the medium, with higher temperatures accelerating the gassing rate, and to a lesser degree by such factors as the nature of the carbohydrate supply, pH, and ethanol concentration. In practice, sponge doughs are generally set to ferment at temperatures of 74–78°F (23–26°C). Full maturation of the sponge will be reached in 3–4.5 h, during which time the temperature will rise by some 10°F (5.6°C). Straight doughs, which are used mainly in retail bakeries and in the production of variety breads, are generally set at slightly higher temperatures to counteract the retarding effect of such dough ingredients as salt, milk solids, and mold inhibitors on yeast activity.

The use of continuous mixing processes, in which liquid preferments are an inherent feature, has decreased sharply because the character of the bread produced by this method has failed to gain widespread consumer acceptance. The process is currently employed on a limited scale for the production of soft hamburger and hot dog buns. On the other hand, the liquid ferment process continues to expand its applications. It represents essentially a modification of the traditional sponge-dough method, the major difference being that the plastic sponge is replaced by a pumpable liquid sponge or ferment that contains either no flour or only a portion of the flour that is normally used in the plastic sponge.

The so-called no-time dough processes either involve high-speed intensive mechanical dough development or rely on chemical dough development in which approved reducing and oxidizing chemical reagents play a primary role. Although used extensively in different parts of the world, these methods have made only limited inroads into commercial baking operations in the United States. Their primary purpose is to circumvent the lengthy bulk fermentation stage by imparting to the dough the desired physical character by means other than time-consuming bulk fermentation.

DOUGH MAKEUP OPERATIONS

The basic function of a bakery's makeup department is the division of the fermented bulk dough into individual dough pieces of proper weight that, when molded, proofed, and baked, will yield the desired finished product. At one time, this operation was done entirely by hand and involved cutting the bulk dough into pieces of appropriate loaf size, rounding the pieces on a flour-dusted dough bench to seal the surface with a gas-retaining skin, and, following a brief rest, degassing and molding them into the desired shape, ready for panning and proofing. All these operations are now performed by a series of specialized machines. The makeup department of a bread bakery typically features a dough divider and rounder that operate in tandem, an intermediate or overhead proofer in which the rounded dough pieces are given a brief rest, and a molder. These machines are interconnected by belt conveyors. More often than not, the molder is operated in conjunction with a panning device that automatically deposits the shaped dough pieces into the baking pans. Interposed between the molder and the oven is the final proofer. It is at this stage that the dough pieces undergo their final intensive fermentation that results in their optimum volume expansion before they enter the oven. In practice, the pans containing the molded dough pieces are placed on portable racks that are then transferred into a relatively large, insulated, and air-conditioned chamber, or final proofer, whose interior is maintained at a temperature within a range of about 100–115°F (37.8–46°C) and relative humidities of 75 to 90%. The final proof is generally carried out over a period of 60–65 min, by the end of which time the dough pieces in the pans will have expanded to a predetermined volume or height.

THE BAKING PROCESS

The final step in breadmaking is the actual baking process in which the raw dough piece, under the influence of heat, is transformed into a light, porous, readily digestible and flavorful product. The various reactions that underlie this transformation are both basic—they irreversibly alter the structural nature of the major dough constituents—and highly complex—they involve a vast series of physical, chemical, and biochemical interactions. The most apparent effects produced by oven heat on the dough piece are an expansion of its volume, the formation of an enveloping crust, the inactivation of yeast and enzymatic activities, and the coagulation of the dough's protein and partial gelatinization of its starch. Simultaneously, there occurs an extensive stabilization of the otherwise rather sensitive colloidal dough system. These basic transformations are accompanied by the formation of new flavor substances, such as caramelized sugars, pyrodextrins, and melanoidins, as well as a broad range of aromatic compounds comprising aldehydes, ketones, various esters, acids, and alcohols.

The production of bread with the requisite quality attributes presumes a carefully controlled baking process. Thus, the rate of heat application and the amount of heat supplied, the humidity level within the baking chamber, and the duration of the bake all exert a vital influence on the final quality of the bread. Although many of the chemical and physical changes that take place during bread baking are only partially understood, our knowledge in this area has been expanding dramatically in recent years.

Baking Stages

Modern ovens are generally designed to convey the baking loaf on either trays or a traveling hearth through a series of zones in which it is exposed for definite time periods to different temperature and humidity conditions. In normal practice, the first stage of baking, at a temperature of about 400°F (204°C), lasts about 6.5 min and comprises one-fourth of a total baking time of 26 min. During this period, the temperature of the outer crumb layers increases at an average rate of 8.5°F (4.7°C) per minute to a level of about 140°F (60°C). The first observable change produced by the oven heat is the almost instantaneous formation of a thin and initially expandable surface skin.

At first, the rise in temperature accelerates enzymatic activity and yeast growth. There is a perceptible increase in loaf volume, the so-called oven rise, that results from the continued aerating process that is produced by the rapid expansion of carbon dioxide gas. At about 122–140°F (50–60°C), most enzymes undergo thermal inactivation and the yeast and other microorganisms are killed. At this temperature, all the carbon dioxide gas has been released from solution and contributes to loaf expansion. The surface skin thickens, loses its elasticity, and begins to acquire the first signs of a brown coloration. There is then a rather sudden expansion of the dough by about one-third of its original volume: the so-called oven spring.

The second and third phases of baking together last some 13 min, or about one-half of the total bake time. During this period, the oven temperature is held constant at about 460°F (238°C). The crumb temperature rises at a rate of 9.7°F (5.4°C) per minute during the second stage until it attains a level of 209–210°F (98.4–98.9°C) by the start of the third stage, and then remains constant. This temperature level coincides with the maximum rates of moisture evaporations, starch gelatinization, and protein coagulation. The dough interior is progressively transformed into a crumb structure from its outer to its inner portions by the penetrating heat. As the crust temperature reaches 300–400°F (150–250°C), the crust begins to assume its typical brown coloration. The final, or finishing, oven zone is maintained at a constant temperature of 430–460°F (221–238°C) and serves to firm up the cell walls and develop the desired crust color. Representing the final one-fourth of the total baking time, this finishing stage is marked by the volatilization of certain organic substances that is designated as the bake-out loss.

These temperatures and durations of the individual baking phases are representative of conventional baking practice; however, considerable deviations are encountered. Such factors as oven design, weight or volume of product, crust character and color, level of residual crumb moisture, all have a bearing on the actual baking temperature and time. Product size in particular is an important determinant of baking time, with the smaller loaves of pan bread achieving the optimum degree of bake quicker than the larger loaves at an appropriately regulated temperature. Thus, if a 16-oz loaf requires 18–20 min to bake adequately, the time for a 24-oz loaf will normally be 20–22 min.

Heat Transmission

In conventional baking processes, heat is transmitted to the dough piece in three different ways, namely, by radiation, convection, and conduction. Although all three modes play significant roles in baking, their relative importance depends largely on the type and design of the oven.

Radiant heat consists principally of invisible infrared rays that emanate from the heated internal surfaces of the oven. Its behavior differs from the other types of heat in two distinct ways: It is blocked in its transmission by any intervening opaque object, and it is highly responsive to the absorptive properties of the product which is exposed to it.

Convected heat is that which is transferred by means of intermixing fluid media, such as air, water vapor, or combustion gases. In ovens, convected heat is distributed through the baking chamber by the turbulence of the atmosphere and is transferred by conduction to the dough piece when the hot air contacts the dough surface.

Conducted heat is that which is transmitted by physical contact from one body to another or from one part to another in the same body. Thus, the side and bottom crusts in pan bread result mainly from the heat that is transmitted by the walls of the pan, and the gradual heating of the interior of the dough piece during baking is also largely the result of heat conduction.

A dough piece that is exposed to oven heat will undergo rapid warming on its surface and progressively slower warming of its interior mass as the distance from the surface increases. One study has shown that the temperature of the dough surface reaches 302°F (150°C) during the first third of the baking cycle at a constant temperature, and that thereafter its rate of climb slows perceptibly until it reaches 356°F (180°C) or higher at the end of baking. The temperature of the crumb never exceeds 210°F (99°C), and this level is reached after variable time intervals that depend on the distance between the point of measurement and the crust. The center of the loaf does not attain the maximum temperature until near the end of the bake. The mechanism that accounts for the temperature of the crumb failing to rise to the boiling point of water is the greater rate of heat loss caused by the evaporation of water as compared with the rate of heat absorption by the dough.

The time required for the crumb interior to reach any given temperature level is measurably influenced by the oven temperature (Fig. 1). Thus, the time required for the crumb interior to pass through the critical temperature range of 131–203°F (55–95°C) is increased by 2.2 min when the baking temperature is lowered from 475°F (246°C) to 355°F (179°C). Oven temperatures within the range of 385 and 445°F (196 and 229°C) are required for acceptable baking results.

Role of Oven Steam

Steam performs several necessary functions during the initial stages of baking. These include preserving the extensibility of the surface skin on the dough piece over the critical periods of continued oven rise and oven spring; imparting a desirable gloss to the crust, particularly of hearth loaves and hard rolls; and promoting heat penetra-

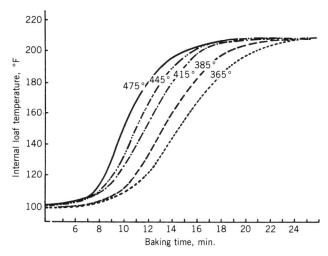

Figure 1. Rates of temperature increase within the loaf at various oven temperatures.

tion into the loaf interior. In a dry oven, the dough piece will almost immediately experience rapid evaporation of water from its exposed surface. This causes the premature formation of a dry, inelastic outer shell that restricts loaf expansion or gives rise to unsightly tears in the finished crust, or both. Because of an inadequacy of moisture, the surface layer of starch undergoes pyrolysis rather than partial gelatinization and the crust will not acquire the desired gloss. Finally, the evaporative process absorbs considerable heat from the surface, thereby slowing the rate of heat penetration into the loaf interior. These effects are minimized by proper steam conditions in the oven. When these exist, there is at first moisture condensation on the dough's surface as it enters the oven. This moisture, given the porous nature of the surface, penetrates into the surface skin and keeps it flexible during the initial baking stage; it further promotes starch gelatinization and the concomitant gloss formation in the crust; and, finally, it facilitates more rapid heat flow into the loaf.

Oven Spring

The rather sudden expansion of the dough loaf by about one-third of its original volume that occurs during the initial stage of baking is referred to as the oven spring. This volume increase, initiated by the penetration of heat into the dough mass, involves the interplay of several elementary physical phenomena. According to Charles's gas law, heat applied to a gas at constant volume increases the pressure of the gas. If the gas is confined in an elastic or expandable vesicle or cell, one visible effect of this increased pressure is an expansion in the volume of the cell. A dough piece contains millions of minute gas cells confined within elastic gluten walls. When heat is applied, the pressure of the gas within the cells increases and the elastic gluten walls permit the individual cells to expand.

A second effect of heat is to reduce the solubility of gases. A major proportion of the carbon dioxide generated by yeast fermentation is dissolved in the dough's aqueous phase. As the temperature of the dough is increased to 120°F (49°C) and beyond, the carbon dioxide held in solution is released and migrates into existing cells, thereby adding to their internal pressure and, concomitantly, to their size.

A third effect of heat is to change the physical state of liquids with low boiling points into a gaseous state by volatilization or distillation. Although ethanol is quantitatively the major low-boiling liquid present in dough, other dough constituents that are volatile at temperatures below 175°F (79°C) include many of the organic acids, esters, aldehydes, and ketones. They are transformed into vapor early in the baking process, thereby contributing to the gas pressure within the cells and to their expansion.

A fourth factor that affects oven spring is the reciprocal relation between the gas pressure in a cell and the cell radius. It has been observed that small gas cells require considerably greater pressures to expand than do large cells. Hence, once the pressure in the minute gas cells of the dough exceeds a certain critical limit, the restrictive forces of the cell walls suddenly give way and the cells undergo an abrupt expansion. Because the crust has begun to form by the time this stage is reached, one readily observable effect of the oven spring is the appearance of the characteristic break and shred on the upper sides of the pan loaf.

Fermentation improves the baking quality of the dough by increasing the cells' gas-retention ability. In baking experiments, bread obtained from fermented dough had a fine-grained, thin-walled cell structure, whereas bread from unfermented dough exhibited a coarse grain and thick cell walls. When, however, suitable oxidants were added to the unfermented dough, its oven spring and cell structure equaled those of the fermented dough.

Baking Reactions

Starch Gelatinization. During baking, the starch granules begin to swell at a temperature of about 104°F (40°C). The viscoelastic properties of dough are largely supplanted by fluidity by the time the temperature attains the range of about 122–150°F (50–65°C). This rheological change is caused principally by enzymatic starch degradation before swelling becomes significant. During swelling and gelatinization, the starch granules avidly absorb the free water and protein-held water of the dough. But even though their granular structure undergoes considerable deformation and becomes quite flexible at temperatures of 140–158°F (60–70°C), a large proportion of the granules remains intact because of the limited supply of water. One study has suggested that during starch swelling and gelatinization a part of the linear fraction of starch dissolves and diffuses out of the granules and into the surrounding aqueous medium where it becomes concentrated as the amount of interstitial water is reduced by continued starch swelling. This dissolved linear fraction sets up into a gel on cooling and appears to play a significant role in bread staling because of its pronounced tendency to retrograde.

Protein Denaturation. The gluten proteins are present in dough in a hydrated state, binding an estimated 31% of the total dough absorption water. They participate in the formation of the dough structure by providing the matrix in which are embedded strings of small starch granules. As the temperature of the crumb reaches about 140–158°F (60–70°C), the proteins begin to undergo thermal denaturation that reduces their water-binding capacity to practically zero. Baking thus causes the transfer of the water from the proteins to the starch in the course of gelatinization. In dough, the gluten films provide the main structural element that sustains the loaf volume. When the dough temperature rises above 165°F (74°C), heat denaturation transforms the gluten films surrounding the individual gas vacuoles into a semirigid structure by interaction with the swollen starch. As the gas cells expand, the flexible starch granules within the cell wall are elongated, thereby enabling the gluten film to become thinner and ultimately to rupture. By this time, however, starch swelling has advanced sufficiently to prevent a collapse of the dough structure.

Enzyme Activity. The amylases begin to hydrolyze starch during baking with the onset of swelling, their rate of action approximately doubling for each 18°F (10°C) rise in temperature. At the same time, the inactivating effect of the higher temperatures begins to exert itself until ultimately amylolysis is stopped. Amylase activity during the early stages of baking produces a twofold effect: It renders the dough more fluid and thereby promotes dough expansion; and it increases the levels of both dextrins and maltose. The latter sugar is partly fermented by the yeast whose activity is also increased at the start of baking. Inadequate amylase activity, or the exclusive reliance on heat-labile fungal amylases, results in a reduced loaf volume. Conversely, excessive amylase activity may produce overexpansion of the loaf volume and even cause the loaf to collapse completely. One study has found that inactivation of malt α-amylase takes place between 149–194°F (65–95°C), with the most rapid inactivation occurring at 154–181°F (68–83°C), a temperature range that is traversed in about 4 min during baking. β-Amylase is inactivated rapidly at 134–161°F (57–72°C), a range that lasts less than 2.5 min. Fungal α-amylase is most active at about 122°F (50°C) and is largely inactivated at 140°F (60°C).

The activity of the proteolytic enzymes present in dough also increases with the rise in temperature until the inactivation level has been reached. Although fungal proteases used in dough conditioners are relatively heat labile, the proteolytic enzymes of cereals, such as of barley and wheat, have been shown to retain most of their activity in the dough at temperatures up to 158°F (70°C), at which level far-reaching protein denaturation occurs.

Water Movement. When baking is carried out in a moisture-saturated oven atmosphere, there occurs a slight water uptake by the dough during the first few minutes as the steam condenses on its surface. However, once the surface temperature exceeds the dewpoint of the baking chamber atmosphere and crust formation begins, the moisture in the outer loaf layers is transformed into steam, part of which escapes through the incipient crust, while some migrates into the loaf interior and condenses.

As baking proceeds, the moisture content of the crust and outer layers up to 3–4 mm is reduced to some 5%, whereas in the interior crumb it remains relatively constant at 43.5–45.1%, similar to that of dough before baking. Immediately on removal of the bread from the oven, there is a rapid moisture migration toward the dry crust layer. Evaporation from the crust continues during loaf cooling until the water content of the bread as a whole is reduced to the legal limit of 38%. According to one study, most of the water in bread is loosely bound and the crumb's water activity (A_w) is a rather high 0.95–0.98.

Cell-Structure Formation. The crumb structure of bread consists of a porous and resilient protein-starch-lipid matrix that encloses, in honeycomb fashion, minute gas cells that make up most of the loaf volume. The character of the cell structure is influenced primarily by processing conditions that prevail before baking. A fast fermentation rate in the intermediate proofer makes it difficult for the molder to degas the dough piece effectively. This may lead to an irregular crumb structure with large holes in the baked loaf. Rapid final proofing tends to produce a cell structure described as young, which is typical of underfermented doughs; whereas slow proofing yields a structure that is designated as old, which typifies overfermented doughs. Both under- and overmixing may cause cell-structure characteristics that are generally associated with young doughs.

The cellular structure of proofed dough originates from the gas bubbles that are created during mixing from either occluded or entrapped air and subsequently reduced in size and dispersed by the molder's sheeting action. The proper setting of the molder to ensure uniform gas bubble dispersion is a significant factor in creating an ideal cell structure. Equally important is the ratio of pan size to the scaled dough weight. The smaller the cubic displacement of the pan in relation to the weight of the dough piece, the easier it is to obtain a fine, uniform cell structure. With pans that are too large for the scaled dough piece, there is a tendency to overproof in order to achieve adequate loaf height. Such overproofing results in an open grain with round cells that are typical of an old dough. Correct baking conditions also have a major influence on the cell structure. For example, if the crust is formed prematurely, it will restrict loaf volume expansion and create thermal stresses within the crumb that disrupt the cells and result in a heavy-walled, coarse, open, and irregular crumb structure. An ideal crumb structure has been defined as one in which the cells are small, fairly thin-walled, slightly elongated, uniform in size, free from large holes, and possessing a smooth, soft, velvety feel when touched lightly with the top of the fingers.

Flavor Development. When a dough is baked by means of dielectric or microwave heating, which produces a crustless loaf, little flavor development occurs and the resultant bread has a rather flat, yeasty flavor that is similar to that of dough. From this it follows that much of the characteristic bread flavor is formed in the loaf's crust region from whence it diffuses into the crumb portion and is retained there by absorption.

Bread flavor can be defined as the psychological reac-tions to the physiological stimuli produced by a multitude of chemical compounds present in both the crumb and crust of bread, supplemented by such cutaneous stimuli as crumb softness and resilience and crispness of the crust.

The mildly acid taste of conventional white bread derives from such water-soluble organic acids as acetic, lactic, propionic, and pyruvic that are formed by yeast and bacterial fermentations. The aroma of fresh bread is a composite of such volatile alcohols as ethyl, amyl, and isoamyl, in addition to such organic acids as caproic, isocaproic, lauric, myristic, and palmitic, and a variety of ethyl esters. The carbonyl compounds, present as aldehydes and ketones, are produced mainly by the Maillard reaction that takes place primarily in the crust region. From the crust, they then migrate into the crumb on cooling.

The flavor of white bread is derived from (1) ingredients, (2) yeast and bacterial fermentation products, (3) mechanical and biochemical degradations, and (4) thermal reaction products. Such major flavor attributes as sweetness, saltiness, wheaty and starchy characteristics originate largely in the formula ingredients. Mixing, hydration, and enzymatic modifications and hydrolysis of proteins and starch alter the original raw starchy and wheaty notes into a smoother and blander flavor. This, in turn, supplements other flavors that are subsequently formed.

When bread is held at room temperature over a period of time, the integrity of its flavor blend weakens progressively, and only some of the flavor components survive. After 96 h the bread has become stale and its flavor has suffered a loss in amplitude, sweet aromatics, and sweet taste; its sour impression has increased; the yeasty character has deteriorated; and its flour-based flavor has changed.

Although the significance of yeast and bacterial fermentation products for the character and intensity of bread flavor has been questioned in recent years, the fact remains that the typical bread aroma cannot be duplicated without fermentation. According to one study, of the approximately 4 g of ethanol produced by fermentation per pound of white bread, some 2–3 g is retained in the freshly baked bread. Such less volatile alcohols as isoamyl alcohol, isobutanol, ispropanol, and propanol, though present in trace amounts only, possess highly potent flavor notes and are thought to exert a major influence on the bread's flavor profile.

The extremely complex nature of bread flavor may be gauged by the following summary of the different categories of flavor compounds that have been detected in white bread by various investigators.

Alcohols: ethanol, isobutanol, *n*-propanol, isoamyl alcohol, and *d*-amyl alcohol.

Acids: acetic, propionic, butyric, isobutyric, valeric, lactic, isovaleric, caproic, heptanoic, octanoic, nonanoic, capric, pyruvic, hydrocinnamic, benzilic, itaconic, and levulinic.

Esters: ethyl acetate, ethyl lactate, ethyl succinate, ethyl itaconate, ethyl pyruvate, ethyl levulinate, and ethyl hydrocinnamate.

Aldehydes: formaldehyde, propionaldehyde, *n*-valeraldehyde, 2-methylbutanal, 2-ethylhexanal, benzaldehyde, furfural, 2-butanal, acetaldehyde, isobutanaldehyde, isovaleral, *n*-hexanal, crotonaldehyde, pyruvaldehyde, and hydroxymethyl furfural.

Ketones: acetone, methyl *n*-butyl, ethyl *n*-butyl, diacetyl, acetoin, and maltol.

THERMAL REACTIONS

The thermal browning reactions that occur in the crust have long been known to be responsible for the development of the crust color and flavor and for the ultimate flavor of the crumb of bread. These reactions are of two distinct types: caramelization and carbonyl-amino reactions that lead to melanoidin formation. Both are nonenzymatic and nonoxidative in nature, but require the presence of heat, with caramelization being by far the more energy intensive.

Caramelization is the transformation of sugars, under the influence of heat, from colorless, generally sweet substances into compounds varying in color from pale yellow to dark brown, and in flavor from mild and pleasant caramel to burnt, bitter, and acrid. The Maillard reaction, whose end products are melanoidins, involves specifically the interaction of free amino groups of amino acids, peptides, or proteins with free reducing sugars.

ENERGY REQUIREMENTS

A frequently cited theoretical value for the amount of heat needed to bake 1 lb of bread is 235 Btu. However, actual heat requirements can differ considerably, as they are influenced by such variable factors as the specific heat of dough, which may range from 0.65 to 0.88; the temperature of the dough entering the oven, which can vary from 90 to 110°F (32–43°C); differences in moisture evaporation rates; and variations in oven temperatures. When all these variables are taken into account, the calculated theoretical value for the required Btu per pound of bread will fall anywhere within the range of 150 to 250.

In actual practice, such additional factors as type of oven, kind of fuel used, method of heat application, amount of steam used, and level of efficiency of oven operation all enter importantly into the calculation of energy requirements. Actual measurements have shown that the heat input required per pound of bread will range from a highly efficient 325–400 Btu in modern ovens to a rather wasteful 1,780 Btu in old coal-fired brick peel ovens.

Baking Conditions

Baking temperatures normally encountered in the production of most breads range from 375 to 450°F (191–230°C). A notable exception is pumpernickel, which is baked in closed pans for up to 25 h at temperatures of 212–375°F (100–191°C) under saturated steam conditions. The humidity levels within the various zones of the baking chamber will vary from a high moisture condition in the front section of the first zone, created by the injection of low-pressure saturated steam, to a relatively low atmospheric moisture content in the subsequent oven zones. The baking time for bread may range anywhere from 18 to 35 min, depending on the oven temperature, loaf weight, and type of product. Thus, the actual baking process, as is true with all the preceding stages of breadmaking, is circumscribed by time, temperature, and humidity as the principal control factors.

More specifically, ordinary white pan bread requires a baking time of about 1 min/oz of dough at a temperature of 425–450°F (218–232°C), with steam injection for the first 0.5–2 min of baking. Thus, a 1-lb loaf of bread, scaled to 18 oz of dough, will take about 18 min to bake. With larger loaf sizes, the unit bake time needs to be extended by a fraction and the oven temperature lowered by a few degrees to account for the relatively constant heat-transmission rate within the loaf.

Pullman bread as a rule is baked slightly longer than open-top pan bread. Rye breads and hard rolls call for both higher sustained temperatures, 440–460°F (227–238°C), and rather copious amounts of steam at the outset of baking, as these conditions promote the formation of a smooth, glossy crust. Doughs containing high levels of sweeteners should be baked at somewhat lower temperatures to avoid premature and excessive caramelization that leads to dark crusts. The same applies to white bread containing 4–5% of dry milk solids or other dairy-type additives. The prevalent practice is to limit steam injection to the very front section of the first baking zone and to rely on moisture evaporation from the baking product to maintain an appropriate atmospheric moisture level in the subsequent oven zones.

Most large bakeries no longer pipe steam to the oven unless breads requiring a hard or glossy crust form part of the product mix. In such cases, experience has shown that best steaming results are obtained by injecting low pressure (2–5 psi) saturated steam at a rate of 200–500 ft/min over the baking product during the first minute of baking. Excessive steam turbulence around the product should be avoided, as this will interfere with moisture condensation on the dough surface and thereby negate the purpose of the steam injection.

Baking temperatures and baking times are influenced to a marked degree by product formulation. In general, relatively lean formulas call for higher baking temperatures and a shorter baking time than do richer formulas. The reason for this lies in the fact that the sugars and dairy ingredients present in greater amounts in rich formulations enter readily into thermal browning reactions. This often results in an excessively dark crust before the crumb has reached its optimum bake. Similar effects can be observed with young or underfermented doughs that retain an excess of residual sugar. On the other hand, in lean doughs, as well as in old doughs, these thermolabile ingredients are either absent or present in only minimal amounts, so that crust coloration must depend to a greater extent on the brown pyrolysis products of starch and dextrins. The higher baking temperatures required in such cases must be compensated for by somewhat shorter baking times to avoid excessive bake-out losses.

Because of these variables, as well as others that relate to oven design and operational procedures, it is impossible to establish optimal baking conditions with a high degree of precision. These can be determined only by a practical study of the actual oven performance in individual plants.

Bread Cooling

Bread requires proper cooling before slicing and packaging in order to avoid (1) difficulties in slicer operation that often result in crippled loaves, and (2) undesirable moisture condensation within the package that is likely to occur when hot bread is wrapped or bagged. Although no unanimity of opinion exists as to what represents optimum bread cooling, the general consensus is that the interior crumb temperature should be reduced to 95–105°F (35–40°C), and that this should be accomplished in as short a time as possible and without excessive evaporative moisture loss.

The atmospheric and temperature conditions that prevail during bread cooling must be so maintained as to ensure that the moisture content of bread reaching the consumer does not exceed the legal limit of 38%. The evaporative moisture loss that has to take place during cooling is, therefore, related to the magnitude of the bake-out loss that occurred in the oven. In general, an additional moisture loss of 2% on cooling may be regarded as average.

Convection Cooling

Convection cooling, which is the simplest method, takes the form of a rod- or slat-type conveyor of the spiral or multitier design, housed either in a box or, more generally, enclosed in an overhead tunnel that is equipped with an air exhaust system. Bread from the depanner is transferred to the conveyor at the top run of the cooler unit and then descends in a series of loops until it reaches the discharge end, ready for slicing and packaging. An exhaust fan located in the uppermost part of the cooler removes the heat radiated by the bread. Fresh air is drawn in at the discharge end of the cooler, moving slowly upward and past the cooling loaves, thereby creating a dual cooling effect by convection and accelerated evaporation. The average cooling time is reduced to 2–2.5 h by this system. This is by far the most prevalent method of bread cooling encountered in practice. Although this system does not provide an accurate control of moisture loss by the cooling loaf, some regulation of the over-all cooling process is possible by appropriate adjustments. Thus, the cooling time may be shortened and the rate of moisture loss accelerated by increasing the velocity of the air currents drawn past the loaves. At the same time, higher air velocities will tend to perceptibly reduce the actual moisture loss of the bread by shortening the cooling time.

Conditioned Air Cooling

In the second system of bread cooling, the product is exposed to conditioned air that is maintained at such dry-bulb and wet-bulb temperatures as to produce effective loaf cooling within about 90 min. Although this system is available in various designs, the most common one consists of a boxlike enclosure through which racks charged with the hot loaves are conveyed in a straight line. Recommended cooling conditions include air temperatures within a range of 72 to 78°F (22–25.5°C), a relative humidity of 85%, and an air velocity sufficient to bring about a temperature rise of 15–20°F (8–11°C) in the air at the exhaust point. Under such conditions, a 1.25-lb loaf of bread will cool to an internal temperature of 90°F (32°C) in approximately 90 min. If higher internal loaf temperatures are acceptable, these will be reached sooner, of course. Thus, a level of 100°F (38°C) is attained in about 65 min, and of 110°F (43°C) in about 52 min. The cooling times actually obtained depend to some degree also on the specific volume of the bread.

Because both the temperature and relative humidity of the cooling medium are held relatively constant in air-conditioned bread cooling, the rate of moisture loss from the cooling loaf is essentially predetermined at the start of the cooling cycle for as long as the relative humidity of the external air remains constant. At high ambient air humidities, the efficiency of the air-conditioned bread coolers may be impaired unless sufficient refrigeration capacity is available to remove the excess moisture from the incoming air. The amount of refrigeration required to counteract extreme humidity conditions that occur sporadically and for only relatively brief periods during summer months may preclude, on economic grounds, the additional safeguard afforded by large refrigeration systems.

Vacuum Cooling

The third method of bread cooling involves the application of a vacuum to the bread. This greatly accelerates the vaporization of its free moisture, which, by absorbing energy in the form of the latent heat of vaporization, rapidly reduces the bread's temperature. This method of cooling, however, finds very limited application at present.

KEEPING PROPERTIES OF BREAD

Most bakery products with a moist, spongy crumb, when they are stored at ambient temperatures, undergo a progressive deterioration of quality that is commonly designated as staling. In general, the higher the practical moisture content of the baked product in its fresh state, the more pronounced are the changes in its properties that occur on staling. Thus, products such as bread, yeast-raised sweet goods, and cakes stale much more markedly than do cookies and crackers that possess much lower initial moisture contents. In fact, the shelf life of the latter-type products appears to depend much more on the stability of the shortening and any possible flavor materials used in their production than on the deteriorative reactions that are usually associated with staling.

As losses resulting from bread staling are of great economic importance, much attention has been focused on this problem. Practical efforts to retard the bread-staling process or minimize its effects have centered mainly on modifications of the bread production process and on the use of antifirming or antistaling agents and moisture-retaining substances in the dough formulation. Most scientific studies have attempted to elucidate the chemical reaction mechanism of staling in an effort to devise measures that would minimize or eliminate its adverse effects.

CAKING MAKING

The major ingredients used in cake baking are for the most part the same as those that find application in the production of bread-type products or differ from them at best only in degree. Materials such as eggs, milk products, sweeteners, shortenings, emulsifiers, salt, and even flour are either identical to those used in yeast-raised products, or vary from them only in functional rather than basic properties. Two categories of products that are more or less restricted to cake baking include the chemical leavening agents and certain flavor, spice, and coloring materials. The composition and functional role of the major ingredients used in baking are reviewed in some detail in other articles in this encyclopedia.

Cake products are difficult to define precisely because of their wide variety and the broad range of their formulations. Essentially, they are products that are leavened mainly by baking powder, but occasionally also by air incorporation, as in the case of foam-type cakes, and by yeast. They usually contain relatively high levels of such enriching ingredients as sugar, shortening, eggs, milk, and flavorings, in addition to soft wheat flour, and are consequently characterized by a sweet taste, a short and tender texture, and pleasing flavors and aromas. They may be classed into two broad categories, ie, shortening-based cakes whose crumb structure is derived from the fat-liquid emulsion that is created during batter processing; and foam-type cakes that depend for their structure and volume primarily on the foaming and aerating properties of eggs. A third category of products that is frequently classed with cakes are the sweet dough products, eg, sweet rolls, coffee cakes, puff pastry, Danish pastry. These are generally yeast-leavened and frequently contain various fruit, jam, and nut fillings and toppings.

The quality of cakes in all instances is governed by three major parameters, namely, the special suitability of the ingredients for the type of cake being made, an appropriate and properly balanced formula, and the adherence to optimal mixing and baking procedures.

In a very general sense, the basic cake ingredients may be grouped into five broad categories according to the functional roles they perform: (1) tougheners, or structure builders; (2) tenderizers; (3) moisteners; (4) driers; and (5) flavorings. These categories do not constitute exclusive classifications, as many of the commonly used ingredients perform several of these roles in cakemaking. Thus, flour, while serving as the principal structure-forming ingredient, also acts a a drying agent because its major components—starch and protein—exert a marked moisture-absorbing property at some stage of cake baking. Similarly, sugar serves principally as a flavoring ingredient, but may also, depending on whether it is used in its crystalline form or a liquid form such as syrup, honey, or molasses, act either as a drying or a moistening agent, and also as a tenderizer by diluting the proteins of flour. Shortening is the primary tenderizer, but also functions as a moisture and flavor retainer. Milk, depending again on whether it is used in the dry or the liquid form, will perform corresponding drying or moistening functions, in addition to contributing to the cake's flavor. Eggs perform a multiplicity of functions in cake baking, depending on the specific nature of the egg product used, as well as on the type of cake. Fresh egg white, eg, contains sufficient moisture to serve as a principal moistener, whereas its albumen content contributes significantly to structure formation. Fresh egg yolks, in contrast, contain much less moisture but consist of about one-third of liquids and one-sixth of protein, and hence their role extends to that of moistener, structure builder, and tenderizer. The effects produced by whole eggs resemble more closely those obtained from egg yolks than from egg whites. Finally, dried egg products act primarily as driers and structure formers.

Cake Formula Balance

The incorporation of appropriate amounts of the individual ingredients is an essential requisite to the production of high-quality cakes. In general, properly balanced cake formulas for the production of shortening-based cakes in commercial baking will have the weight of sugar exceed the weight of the flour, the weight of the eggs exceed the weight of the shortening, and the weight of the combined liquids, eg, milk, eggs, and water, exceed slightly the weight of the sugar. Accordingly, the majority of the different types of yellow and white cakes contain an average of 125% of sugar based on the flour level in the formula, with some containing as much as 140%.

For foam-type cakes, which do not contain shortening and which are represented by such products as angel food cake and sponge cakes, different formula balancing rules apply. In the case of angel food cake, a well-balanced formula will call for the weight of the sugar to equal the weight of the egg whites, and the weight of the flour to represent about one-third of the weight of the sugar. In sponge cakes, in which whole eggs rather than just egg whites are added, formula balancing calls for the amount of sugar either to be equal or to slightly exceed the amount of whole eggs, the weights of all the liquids to exceed the weight of sugar by one-fourth, the weight of either the sugar or the whole eggs to exceed that of the flour, and the combined weights of the eggs and flour to be greater than the combined weights of the sugar and liquids other than whole eggs.

Cake Batter Mixing Methods

The primary purpose of cake batter mixing is to bring about a thorough dispersion and mutual emulsification of the various cake ingredients, accompanied by the entrapment and size reduction of air cells and a minimum development of the gluten from the flour proteins. Depending on the type of cake being produced, the mixing procedure will differ in such aspects as the order of ingredient incorporation, duration and mixing rate of the various stages in multistage methods, temperature of the ingredients, and other factors. In very general terms, batter-type cake mixing procedures may be differentiated into the creaming method, the blending method, and the single stage method, although there exists a goodly number of variant procedures. In the creaming method, the shortening and the sugar are first combined at low or medium mixing speed to form a uniform, aerated mass, followed by the incorporation of the eggs with continued creaming action,

Figure 2. A representative continuous cake batter mixing system consisting of, left to right, a continuous mixer, slurry holding tank, and slurry mixer. Photograph courtesy of Oakes Machine Corp.

the process being completed with the addition of the milk and the flour in alternate small portions.

In the blending or flour-batter method, the shortening and flour are creamed to a fluffy mass in one bowl, with the simultaneous whipping of the eggs and the sugar to a semifirm foam at medium speed in a second bowl. The sugar–egg foam and the creamed flour are then combined, with the gradual addition of the milk.

The single-stage method, as its name implies, calls for blending all the major ingredients into a homogeneous mixture at one time and at low speed, followed by mixing at medium speed, and finishing off again at low speed.

In large-volume operations, cake batters are generally mixed by continuous batter mixers, of which two basic types find the most extensive applications, ie, the compact rotor-stator mixing chamber type, and the tubular swept-surface mixer. Both types require the use of a premixer or slurry mixer in which all the dry and wet ingredients are blended into a homogeneous fluid mixture, which is then pumped into a slurry holding tank and from thence into the mixing chamber of the continuous mixer where it is emulsified at high speed and aerated by the incorporation of purified compressed air to yield a batter of the desired specific gravity (Fig. 2). The specific gravity of cake batters has long been recognized as a major determinant of the tenderness, grain, texture, and volume of the finished cake.

Baking the Cake

The mixed cake batter is deposited into appropriate cake pans and conveyed directly to the oven with a minimum loss of time so as to avoid a potential loss of the aeration that is being generated in the batter by the dissolved leavening agents. Oven temperatures for cake baking normally range from 325 to 400°F (163 to 204°C), with the temperature actually selected depending on such factors as the sugar and milk levels of the formula, fluidity of the batter, pan size, etc. Baking times will range from 50 to 65 min at the lower end of the temperature range for pound cakes to as short as 15 to 20 min for the small cup cakes which are normally baked at the higher temperatures.

BAKING EQUIPMENT

Commercial baking began as a small-scale activity of limited scope in which doughmaking and dough processing were performed manually. With the progressively expanding scale of operation, there arose a need for machinery that would carry out those tasks that had begun to exceed human capabilities, eg, the mixing of large doughs, or that would perform them more rapidly, more efficiently, or more economically. As a result there occurred at first a gradual mechanization of individual processes, to be followed by an increased automation of major bakery operations, and ending finally in the ever-extending use of computerized controls over nearly every phase of production, packaging, and distribution.

Present-day computers are programmable electronic devices that possess far-reaching capacities for accepting, storing, retrieving, and processing data. Because of their unvarying performance, once they have been properly programmed, they have brought the accuracy of functions

Figure 3. Two outdoor flour silos, with the one on the right featuring the manifold of pneumatic lines that serve both silos from a single unloading site. Photograph courtesy of Fred D. Pfening Co.

control to previously unattainable levels and are increasingly replacing electromechanical controls. Programmable controllers may be applied to time, to sequence, and to direct the operations of entire systems, such as bulk handling and in-plant blending of ingredients, pastry dough laminating, conveyorized proofing and baking, and robotized pattern-forming basket or tray loading. They are also used to an even greater extent to monitor the performance of individual equipment units, such as dough mixers, metering and weighing devices, and pan stackers/unstackers.

A comprehensive treatment of the principles of computer technology is beyond the scope of this article. Readers interested in this subject can consult appropriate textbooks or any of the several service organizations that specialize in this area. Occasional references will, however, be made in the discussions that follow to the applicability of programmable controllers to specific types of equipment and systems.

When studying baking equipment, two considerations are important: operational functions and design principles. Major equipment categories include:

1. Ingredient-handling equipment (pneumatic conveyors, storage bins and silos, sifters, liquid-bulk systems, weighing and batching systems).

2. Dough-handling and processing equipment (mixers, dividers, rounders, molders, dough pumps and extruders, dough laminators, proofers, etc).

3. Baking equipment (ovens, oven loaders and unloaders, fryers, etc).

4. Product-handling equipment (depanners, coolers, slicers, packaging machinery, conveyors, etc).

5. Miscellaneous equipment (refrigerators, freezers, pan washers, and dough troughs).

It is beyond the scope of this article to review all the equipment units that make up the production facilities of a modern bakery. Consequently, only the principal equipment categories will be briefly indicated.

Bulk-Handling Equipment

All larger bakeries are at present equipped with internal or external flour storage bins of various designs, normally of sufficient capacity to meet the plant's production requirements for at least 1 wk. Whether the flour in such cases is delivered by flour tank truck or rail tank car is governed by the volume of flour being processed and on whether the bakery has access to a railroad siding. Most flour storage installations consist of several bins or silos. A common method of loading multiple bins is to manifold the pneumatic lines at or near the unloading site. Each bin's conveying line terminates at the manifold and can be connected directly to the flexible hose of the unloading device.

Figure 4. A 340-qt vertical mixer that can perform a wide variety of mixing actions. Photograph courtesy of AMF Inc.

Figure 5. A large, fully automatic dough mixer, shown with a bowl in its tilted discharge position. Photograph courtesy of Peerless Machinery Corp.

Flour storage bins are available in various configurations and sizes. Where outdoor flour storage is dictated by a lack of floor space in the baking plant, the most commonly selected container is the cylindrical silo (Fig. 3). For in-plant storage, square or rectangular bulk containers of suitable size are frequently installed, as they offer the most effective use of the available floor space.

Practically all flour and granular sugar conveying systems are of the pneumatic type in which air is used to move the solid material from one point to another. Material movement is induced by creating a pressure differential between the two points, either by increasing the air pressure at one end, or by decreasing it to produce a partial vacuum at the other. Depending on which procedure is applied, either a pressure or a vacuum system results. In practice, a combination of both is frequently encountered to exploit the inherent advantages of both.

To an ever-increasing extent, bakeries have converted from the use of dry sugars to liquid sweeteners, as the latter are available in a variety of blends with various proportions of sucrose, invert sugar, dextrose, and corn syrup; are more conveniently handled; and are more efficiently weighed or metered. Bulk systems for liquid sweeteners usually consist of several stationary storage tanks of suitable size, complete with permanent piping for

Figure 6. Tweedy high speed batch dough mixer with proportional weighing system and microprocessor controls. Photograph courtesy of Tweedy of Burnley, Ltd.

Figure 7. Heavy duty six-pocket dough divider. Photograph courtesy of APV Baker, Inc.

substituted by appropriate mechanical dough development has led to the introduction of several types of ultra-high-speed batch mixers. These are capable of fully developing doughs in 5 min or less by an energy input of 5 Wh/lb within that time span (Fig. 6). In currently operational systems with hourly capacities of up to 11,000 lb of dough, the mixing bowl or drum is cylindrical in shape and mixing is done by an impact plate or an impeller rotating at speeds of 280–350 rpm, with one particular system attaining speeds of 1,200 rpm.

Makeup Equipment

The dough, following its proper development, next enters that stage of the production process in which it is divided into individual product units of specific weight and shaped into its final form. The standard equipment involved in these operations include, in sequential order, (1) the divider, which scales the bulk dough into units of predetermined weight (Fig. 7 and Fig. 8); (2) the rounder, which imparts a spherical form to the unsymmetrical dough pieces that emerge from the divider and simultaneously seals their raw cut surfaces with a fine skin to prevent excessive loss of the evolving carbon dioxide gas (Fig. 9); (3) the intermediate or overhead proofer in which the rounded dough pieces receive a brief rest to recover from

both receiving the sweeteners from the delivery vehicle and for pumping them to use points.

Except for fats whose plastic character is essential to their proper functioning—such as shortenings used in some cakes, in icings, and pie crusts and as roll-in fat in Danish pastry-type products, as well as such water-fat emulsions as margarine—all shortenings can be fluidized for liquid bulk handling. Liquid shortening or oil deliveries are made either in tank trucks of 4,000–6,000 gal capacity or in rail tank cars of 8,000 to as high as 20,000 gal capacity.

Dough-Mixing Equipment

Mixer types used by commercial bakeries fall into two basic categories: vertical mixers and horizontal mixers. Vertical mixers usually feature variable-speed drives that permit agitator speeds ranging from 40 to 370 rpm (Fig. 4). Such mixers, when equipped with appropriate agitators, are able to perform such varied tasks as dough development, batter homogenization, and foaming and whipping. Horizontal mixers, employed chiefly for doughmaking in bread production, normally are of the two-speed variety. The mixing arms in such mixers, when in the slow speed mode, rotate at half the rated maximum speed, eg, at 35 rpm in a mixer whose maximum speed is 70 rpm (Fig. 5).

Batch capacities of horizontal mixers range from about 100 lb for the smallest units to some 2,500 lb for the largest mixers. Corresponding power requirements range from about 5 hp to as high as 125 hp. Vertical mixer bowl capacities normally range from 20 to 340 qt, with drive motors having power ratings within the 2–10 hp range.

The recognition that bulk dough fermentation may be

Figure 8. Newly developed rotary dough divider showing circular dough extrusion port and rotating knife at left, and electronic control panel at right. Photograph courtesy of AMF Inc.

Figure 9. Umbrella, or inverted cone-type rounder. Photograph courtesy of APV Baker, Inc.

the physical abuse they have sustained during dividing and rounding; and (4) the molder, which sheets and molds the dough pieces into their final loaf form, expelling most of its gas in the process (Fig. 10). The made-up dough pieces, placed into pans or on peel boards or baking sheets, next pass into the final proofer in which they undergo a vigorous final fermentation under near-optimum environmental conditions until they have attained the appropriate height or volume. From the final proofer they are then conveyed directly into the oven for baking.

Oven Equipment

The basic types of oven in current use are the reel oven, single-lap tray ovens, double-lap tray ovens, tunnel-band ovens, and the most recently introduced conveyorized ovens. They may be heated either by direct-firing or by indirect-firing systems.

Reel ovens, of which only relatively few are still in use, consist of a reel structure that revolves vertically around

a horizontal axis within the baking chamber and supports the baking trays in Ferris wheel fashion. In traveling tray ovens, the reel has been replaced with two parallel, endless chains that carry trays through the baking chamber, with a third chain acting to stabilize the trays' horizontal position during the baking cycle. In single-lap ovens, the trays travel back and forth in a single loop through each of the oven's heat zones (Fig. 11). The double-lap tray oven employs essentially the same method for conveying and stabilizing the trays through the baking chamber, except that in this instance the trays travel through four successively superimposed heat zones rather than just two. The tunnel oven features a long, low baking chamber, which may vary in length from 100 to 400 ft, through which a motor-driven conveyor, carrying the baking hearth, passes in a straight line. The conveyorized oven consists of an enclosure within which an endless conveyor is arranged in spirals or tiers about the periphery and to which heat is supplied by means of ribbon burners that are mounted in groups directly under the conveyor grids.

The two principal systems of heating conventional ovens are (1) by direct firing, whereby fuel combustion occurs within the baking chamber, usually by the use of ribbon burners; and (2) by indirect firing, which uses separate combustion chambers. Attempts at improving oven efficiency include the installation of air-circulating systems, special air impingement devices, or of electrode grids that create a corona wind effect by means of an electrostatic field.

CONCLUSION

It may be appropriate at this time to point out that as a result of continuing efforts to upgrade the sanitary and safety aspects of bakery equipment, standards that reflect desirable practices in these areas have evolved over the years. Thus, the Baking Industry Sanitation Standards Committee (BISSC) has periodically issued proposals that facilitate the sanitary maintenance of bakery equipment on the one hand, and minimize or totally prevent the inadvertent contamination and infestation of baking ingredi-

Figure 10. A cross-grain molder/panner that shapes the dough piece into its final form and automatically deposits it in the baking pan. Photograph courtesy of Teledyne Readco.

Figure 11. Single-lap, direct gas fired-tray oven with automatic oven loader. Photograph courtesy of APV Baker, Inc.

ents and materials, on the other. The recommended design features and sanitary practices cover all types of bakery equipment. They are periodically revised and updated to comply with changing federal, state, and local regulations and to reflect the changes in equipment design. Bakers take care to familiarize themselves thoroughly with these standards and take them into account whenever they acquire new machinery or undertake modifications of existing equipment.

Similarly, the American National Standards Institute has been a strong advocate of safety measures and practices as they apply to bakery equipment. Ideally, the proposed safety standards have as their goal the total elimination of the hazards that are inevitably associated with the operation and servicing of bakery machinery. Both common sense and humanitarian considerations require that these standards be carefully observed in the design and operation of all equipment so as to provide a maximum degree of employee safety.

BIBLIOGRAPHY

This article was adapted from and used with permission from E. J. Pyler, *Baking Science and Technology*, 3rd ed., Vols. I and II, Sosland Publishing Co., Merriam, Kans., 1988. Note: all specific citations have been removed. The reader should refer to the original text to obtain specific references.

Current and past issues of the following journals provide valuable information on baking science and technology: *Bakery Production and Marketing, Baking and Snack Systems, Cereal Chemistry,* and *Cereal Foods World.* In addition, the *Technical Bulletins* of the American Institute of Baking, Manhattan, Kans., and the *Proceedings of Annual Meetings* of the American Society of Bakery Engineers, Chicago, Ill., are good sources of information.

General References

V. Bauman, *Baking, The Art and Science,* Baker Tech Inc., Calgary, Alberta, Canada.

BISSC Certified Standards, Baking Industry Sanitation Standards Committee, New York, 1971.

R. C. Hoseney, *Principles of Cereal Science and Technology,* American Association of Cereal Chemists, St. Paul, Minn., 1986.

International Symposium on Advances in Baking Science and Technology, Kansas State University, Manhattan, Kans., 1984.

S. A. Matz, *Bakery Technology and Engineering,* 2nd ed., AVI Publishing Co., Westport, Conn., 1972.

S. A. Matz, *Ingredients for Bakers,* 1987; *Formulas and Processes for Bakers,* 1987; *Equipment for Bakers,* 1988; *Bakery Technology: Packaging, Nutrition, Product Development, QA,* 1989. Pan-Tech International, McAllen, Tex.

B. S. Miller, ed., *Variety Breads in the United States,* American Association of Cereal Chemists, St. Paul, Minn., 1981.

Y. Pomeranz, *Cereal Science and Technology,* VCH Publishers, New York, 1987.

Y. Pomeranz, ed., *Wheat Chemistry and Technology,* 3rd ed., American Association of Cereal Chemists, St. Paul, Minn., 1988.

E. J. Pyler, *Baking Science and Technology,* 3rd ed., Sosland Publishing Co., Kansas City, Mo., 1988.

Safety Requirements for Bakery Equipment. American National Standards Institute, Inc., New York, 1971.

P. R. Whitely, *Biscuit Manufacture,* Elsevier Publishing Co., London, 1971.

ERNST J. PYLER
Raleigh, North Carolina

BAMBOO SHOOT

FAMILY AND RELATED SPECIES

Bamboo belongs to the family Gramineae. The emerging shoots of several species, of the genera *Dendrocalamus, Bambusa, Phyllostachys,* and others are used as a vegetable (1).

PRACTICAL DESCRIPTION

The members of the family Gramineae are tall woody shoots of perennial grasses. The reproductive organs resemble the grasses, particularly the oat (*Avena sativa*). The anthers are borne on long filaments resembling those of corn (*Zea mais*) (2).

ORIGIN, DISTRIBUTION, AND PRODUCTION

The majority of the edible bamboos originated in China, Japan, and Southeast Asia. Some are wild and others are cultivated. The use of bamboo shoots outside of this area is restricted (1). The important edible species include the tropical clump bamboos, *Bambusa oldhami* Nakai and *Dendrocalamus latiflorus* Munro (Fig. 1), and the spreading bamboos, *Phyllostachys edulis* (Fig. 2), *P. pubescens,* and *P. makinoi,* which are not confined to the tropics.

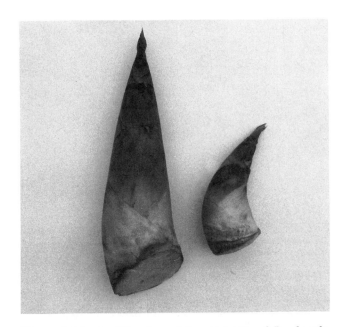

Figure 1. Shoots of *Bambusa oldhami* (right) and *Dendrocalamus latiflorus* (left).

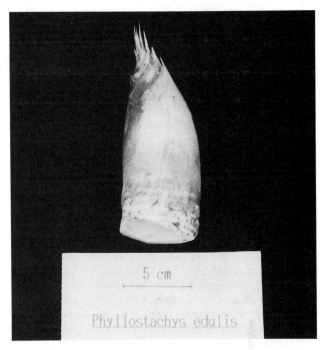

Figure 2. Shoot of *Phyllostachys edulis.*

The planted area of edible bamboos in Taiwan is ca 30,000 ha with an average yield of 12 t/ha. The annual production of bamboo shoots in 1989 was 401,152 t (3).

CULTURE

The cultivation of bamboo has been practiced in the Orient for thousands of years. Bamboo is propagated by the transfer of clumps or rhizomes as shown in Figure 3 (4). Some bamboos can be propagated from culm cuttings. In the life cycle of a clone, the bamboo flowers once, and the top dies. Parts of the same clone, even when transferred to different areas and conditions, will flower at the same

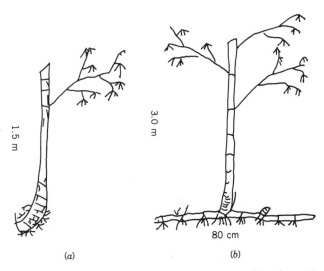

Figure 3. Rhizomes transferred for propagation of bamboos (a) *Bambusa* sp, or *Dendrocalamus* sp; (b) *Phyllostachys* sp.

time (2). Very few bamboo flowers produce viable seeds, but the collection of seeds is important for the breeding of an entirely new seedling, which can start a new life cycle of 50–60 yr (5).

HARVEST

The harvest of bamboo for its edible shoots is an ancient and highly specialized act (6). The grower must know what proportion of the young shoots may be cut and when to cut without endangering the vigor of the parent plant. The usual practice is to cover and mound up the clumps with soil, rice hull, rice straw, bamboo leaves, etc so that the emerging shoots are in the dark. Polyethylene (PE) or poly(vinyl chloride) (PVC) sheets have recently been used on top of the mounds (7). Exposure of shoots to light causes greening and bitterness and these shoots are mainly used for processing. The shoots are ready for cutting when the tips are just emerging from the surface of the soil in the mound, if left much later they will become woody. First the mud is removed by hand and then the shoot is cut off with a sharp knife or spade designed for the purpose (Fig. 4) (5).

In Japan, bamboo shoots are forced by growing under mulch heated with electric cables placed 6–8 cm below the soil. About 2–3 cm of rice straw are placed on the soil surface, an additional 4–5 cm layer of soil is put on top, and finally plastic sheeting is laid on top of all. The soil temperature is kept between 13 and 15°C. By the mulching procedure, bamboo shoots can be harvested almost a month earlier than those not mulched (2).

STORAGE

Harvested bamboo shoots are highly perishable with a respiration rate over 100 mg CO_2/kg·h at 20°C (8). The fiber content increases quickly from the cut end toward the tip. The quality and yield of canned bamboo shoots depend greatly on the holding condition of raw materials before processing (9,10). Rapid precooling and storage at low temperature and high humidity conditions can prevent quality deterioration and extend the shelf life of the bamboo shoot from 1–2 days to 7–10 days (11). In Taiwan, harvested bamboo shoots are washed and packed in bamboo baskets for nearby markets or in PE-lined cartons, which are shipped to the wholesale fresh markets in Taipei.

PRODUCTION SEASON

In Taiwan markets shoots of *B. oldhami* Nakai are available from May to early October. They are called green bamboo shoots and are sweet and tender. Green bamboo shoots are popular in summer when other vegetables are in shortage. *Dendrocalamus latiflorus* Munro has a longer production season from late March to early November. The average weight of each shoot is 2 kg. Green bamboo shoots weigh 0.65 kg. The shoots of *D. latiflorus* are cut when 50–100 cm above the ground (Fig. 5) and are mainly used for processing because they are relatively bitter and tough.

Phyllostachys edulis yields winter shoots, available from November to February, which are smaller, sweeter, and of better quality than spring shoot, available from March to May. Winter shoots of *P. edulis* are considered the best flavored shoots and are precious among winter vegetables due to small production.

FOOD USES

In the preparation of shoots for use, the sheaths and tough basal portions are removed, and the tender shoots are

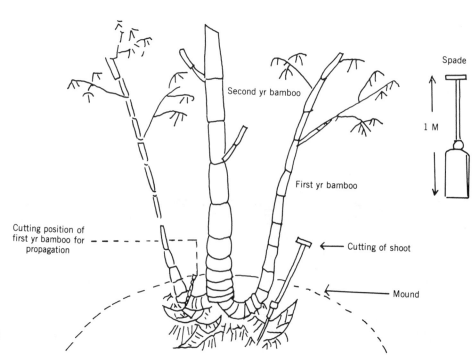

Figure 4. Shooting and harvesting position of *Bambusa oldhami*.

Figure 5. Harvest of *Dendrocalamus latiflorus* shoot.

Table 1. Approximate Composition of *Bambusa oldhami* and *Phyllostachys edulis* in a 100-g Edible Portion

Constituent	Composition	
	B. oldhami	*P. edulis*
Energy (cal/100 g)	60	22
Macroconstituent, 100%		
Water	83	92
Protein	3.9	2.4
Fat	1.1	0.3
Ash	0.7	1.0
Fiber	1.3	0.9
N-free extract	10.3	3.2
Minerals, (mg %)		
Ca	44	43
Mg	21	45
P	39	38
Fe	3.4	2.2
Na	98	26
K	330	435
Vitamins, (mg %)		
A (IU)	30	20
B_1	0.02	0.08
B_2	0.03	0.07
Niacin	2.23	1.58
C	3	3

then sliced or cut lengthwise and boiled in water. The still crispy slices can be used in many dishes. Upon tasting, if the shoot is bitter, it is boiled again to remove the bitterness. Raw bamboo shoot is acrid, and boiling removes the acridity. The bitterness is from cyanogenic glucosides, which, when ingested, is poisonous because cyanide is released. Parboiling in water leaches the compound to render the shoots nontoxic (2). Green bamboo shoots and the winter shoots of *P. edulis* are not bitter because they are harvested before they emerge from the soil surface. Both shoots can be directly cooked with other food without precooking.

In Taiwan, shoots of *D. latiflorus* Munro, which are much larger in size, are used for canning or fermentation, fewer go to fresh market. Canned bamboo shoots are mainly for export. Dried fermented shoots are prepared from fresh bamboo shoots of ca 1 m long, the green portion is removed before processing. After blanching and fermenting for about one month the shoots are sun dried and sliced (Fig. 6). The final product has a crispy, tender texture and good flavor from the lactic acid fermentation (12).

NUTRITIONAL VALUE

Bambusa oldhami and *Phyllostachys edulis* are two varieties sold in Taiwan fresh markets. The approximate nutritional composition in a 100-g edible portion is given in Table 1 (13).

BIBLIOGRAPHY

1. R. M. Ruberté, "Leaf and Miscellaneous Vegetables," in F. W. Martin, ed., *Handbook of Tropical Food Crops*, CRC Press, Inc., Boca Raton, Fl., 1984.

2. M. Yamaguchi, "Bamboo Shoot," *World Vegetables*, AVI Publishing Co., Westport, Conn., 1983.

3. Department of Agriculture and Forestry, "Crop Production," in *Taiwan Agricultural Yearbook*, Taiwan Provincial Government, Taichun, Taiwan, 1990.

4. B. N. Lee, "Bamboo Shoot," *Farmers Guide. Harvest Farm Magazine* Ltd., Taipei, Taiwan, 1980 pp. 885–888.

5. G. M. Leu, "Disappearing of Bamboos" *Harvest* **35**, 16–17 (June 28, 1985).

6. G. A. C. Herklots, *Vegetables in South-East Asia*, George Allen & Unwin Ltd., London, 1972.

7. M. T. Chang, "Cultivation of Bamboo Shoot," *Harvest* **36**, 42–45 (Aug. 1, 1986).

8. M. L. Liao and co-workers, "Respiration Measurement of Some Fruits and Vegetables of Taiwan," *FIRDI Technical Report* **E-66**, Hsinchu, Taiwan, 1982.

9. T. R. Yen and co-workers, "Quality Improvement of Canned Bamboo Shoot," *FIRDI Technical Report* **237**, Hsinchu, Taiwan, 1982.

10. R. Y. Chen and co-workers, "Postharvest Handling and Storage of Bamboo Shoots," *FIRDI Technical Report* **294**, Hsinchu, Taiwan, 1983.

Figure 6. Sun drying of fermented shoots of *D. latiflorus*.

11. R. Y. Chen and co-workers, "Control of Postharvest Quality of Bamboo Shoot," *FIRDI Technical Report* **423,** Hsinchu, Taiwan, 1986.

12. G. T. Huang and co-workers, "Improvement on the Quality of Dried Fermented Bamboo Shoots," *FIRDI Technical Report* **E-121,** Hsinchu, Taiwan, 1985.

13. FIRDI, "Table of Taiwan Food Composition," *FIRDI Technical Report* **24,** Hsinchu, Taiwan, 1971.

MING-SAI LIU
Food Industry Research and
Development Institute
Hsinchu, Taiwan

BATCH DISTILLATION. See DISTILLATION: TECHNOLOGY AND ENGINEERING.
BATCH DRYERS. See DRYERS: TECHNOLOGY AND ENGINEERING.
BATCH PAN. See EVAPORATES: TECHNOLOGY AND ENGINEERING.

BEEKEEPING

HISTORY

Honey bees may represent the quintessential beneficial insect. While producing well-known products of the beehive such as wax and honey, honey bees also provide valuable pollination for many agricultural crops. Based on the diversity of species, most scientists believe the honey bee genus *Apis* originated in Southeast Asia (1). New species of *Apis* are still being described from this region today. The four species that are most widespread and well known are the Western honey bee (*Apis mellifera*), the Eastern honey bee (*A. cerana*), the giant honey bee (*A. dorsata*), and the dwarf honey bee (*A. florea*).

In nature, both *A. mellifera* and *A. cerana* are cavity-nesting species and thus have been collected by humans and successfully transferred and managed in artificial domiciles or hives. The natural range of *A. mellifera* is Europe, Africa, and the Middle East, although it was introduced into the New World (the Americas, Australia, and New Zealand) by early European colonists and is now widely established. The distribution of *A. cerana* is limited to Asia. *Apis mellifera* is usually considered to be a better honey producer and in some regions of Asia is used by beekeepers as a replacement for the native *A. cerana*.

It is likely that the first beekeepers were honey hunters or honey gatherers, as depicted in a paleolithic rock painting found in eastern Spain dating from around 7000 BC (2). These gatherers collected honey by robbing wild bee nests. Honey hunting is still widely practiced in Asia to obtain honey from *A. dorsata* and to a lesser extent from *A. florea* nests. It can be postulated that an early stage in the development of beekeeping may have involved the cutting of a log section enclosing a bee colony and transporting the log to the dwelling of the gatherer. There the log and bees would provide a convenient source of honey and brood for periodic robbing by the new owner. The use of log sections as beehives still persists in some African regions. The first man-made hives were probably made of clay, straw, or tree bark. In the Middle East honey bees are still being kept in clay pottery, mud, and straw hives. Some of these hive materials are known to have been used some 6,000 years ago.

The advent of what might be considered the modern era of beekeeping came about with the discovery of "bee space" and the subsequent development of movable-frame beehives. The person generally credited with publicizing the significance of bee space was Langstroth who, in 1851, discovered that bees would refrain from building wax connections between neighboring combs or combs and the side of the beehive if the space separating them was approximately ⅜ in. (3). This seemingly modest discovery led directly to the evolution of hives that had fully removable frames that enclose and support the beewax combs. These frames could be easily manipulated by the beekeeper and removed for the purpose of examining the brood or extracting honey. The latter part of the nineteenth century saw the development of an amazing array of inventions based on this technology, including machines for removing wax (uncapping) and extracting honey from the new removable frames to, finally, the establishment of a standard-size beehive (Langstroth hive) that remains largely unchanged to the present day.

BIOLOGY OF HONEY BEES

All members of the genus *Apis* are eusocial, which means truly social. To entomologists this term indicates that there is cooperative care of the young by individuals of the same species, an overlap in generations, and production of a sterile or nonreproducing worker caste. This type of social development is uncommon considering the fact that there are over 1 million named insect species. Living as a colony, eusocial species like honey bees provide more food and protection for themselves and their young than can any solitary individual. It is because honey bees store food for future needs that humans have been able to exploit them.

Scientists subdivide the wide-ranging species *A. mellifera* and *A. cerana* into a number of races. These races differ in behavioral and physical characteristics that are the result of adaptations of the species over time, following isolation in geographical regions. In fact, such races are sometimes called geographic races (4). In *A. mellifera* some races can be physically distinguished only through the use of computer-assisted morphometric (body measurement) analysis. To describe honey bee races, specialists use a trinomial system of nomenclature as opposed to the traditional binomial. For example, the Italian honey bee race's trinomial is *A. mellifera ligustica*. Other examples are *A. mellifera mellifera*, a race from Northern Europe and the first honey bee race known to have been introduced into the New World (5), and *A. mellifera carnica*, the Carniolan honey bee from Southern Austria and Yugoslavia.

A colony of honey bees may contain more than 30,000 adult individuals during the summer when the population peaks. For life stages present in the colony are the eggs, larvae, pupae, and adults. These stages can be further subdivided by the two sexes, females and drones. The females may be further divided into two castes, queens and workers.

Workers

Workers, as their name implies, do the work of the colony. The worker caste builds the combs used to house developing eggs, larvae, and pupae (brood) and to store honey and pollen. They forage outside the colony for nectar (carbohydrate) and pollen (protein and fat) to provide food for the colony. Workers also feed the brood, queen, and drones and defend the colony from intruders. When a worker bee stings, the barbed stinger usually pulls out of her abdomen and remains attached to her victim, and the worker dies soon afterward.

Sexually, workers are females with 32 chromosomes and usually lack functional ovaries. However, in certain situations, such as the absence of a queen, it is possible for workers to lay eggs. Laying workers are incapable of laying fertile eggs since they don't mate with drones to obtain sperm. As a consequence they lay eggs that only produce haploid drones. This leads to the eventual death of the colony. In rare situations, laying workers can produce females impaternate (without fathers). This parthenogenetic means of reproduction is found in highest frequency in the race A. mellifera capensis.

During the busy foraging season, the life span of a worker is about six weeks. They literally work themselves to death. However, workers reared late in the fall often live through the winter into the following summer.

Queens

Queens are also females. They have the capacity to mate and lay fertilized eggs. Usually each colony has a single queen. However, under certain conditions colonies may have more than one queen. A few commercial beekeepers intentionally manipulate beehives so they can maintain a two-queen colony that is sometimes more productive than a single-queen colony. Queens can live for as long as five years but most commercial beekeepers replace them every two years. The queen is the egg-laying female of the colony. The maximum number of eggs a queen lays varies throughout the seasons. It is estimated that a good queen can lay 1,500–2,000 eggs a day. A colony will produce new queens if its queen becomes a drone layer (depletes her sperm supply) or is injured or if the colony is about to swarm. New queens have to be raised in specially constructed cells. The workers feed the developing larva in such a way that a queen, rather than a worker, is produced.

Drones

The drones are males and their primary function is to mate with virgin queens. They are raised in specially constructed cells that are larger than the worker cells. Genetically drones are haploid, having a chromosome number of 16, in contrast to the diploid status of the workers and queens, who have a chromosome number of 32. In temperate areas, drones are produced in large numbers, usually beginning in April, and are maintained until the first killing frost. A typical colony in the summer will have 1,000–2,000 drones. Populous colonies sometimes will keep their drones into late November. Some scientists suspect that drones have functions other than mating but no evidence has been forthcoming to support this viewpoint.

Mating

In nature, the mating of drones and queens occurs in flight. The queen mates only at one early period of her life, during the so-called nuptial flight(s). About seven days after she emerges from her cell as an adult, she is considered sexually mature and will leave the hive to mate. A queen mates with about 17 drones. Queens are believed to fly to a location known as a drone congregation area (DCA) to mate. To this day, it is not possible to predict with certainty what geographical conditions or other attributes are needed to establish a DCA. The results of recent work done with radar suggest that DCA's may be busy intersections in a complex network of drone pathways. Drones mortally injure themselves during mating, after which they fall to the ground and die.

Swarming

The process of swarming represents the natural method of colony reproduction. Under certain conditions, often associated with the seasonal growth cycle of the colony, workers begin to raise additional queens in the colony. At some point before these virgin queens emerge from their cells, the old queen will depart from the nest with approximately half of the workers and drones, and attempt to establish a colony at a new homesite. The propensity for swarming and the conditions leading up to this behavior vary within and among races of honey bees. Sometimes one or more of the newly emerged virgin queens will also depart with a group of workers and drones in an afterswarm, also attempting to establish a new colony.

DISEASES AND PESTS

Like all living animals, the honey bee is subject to the ravages of diseases and pests. The diseases range from viral to fungal in origin. One troublesome brood disease is called American foulbrood disease. This disease brought about the establishment of many state bee inspection programs. American foulbrood disease is caused by Bacillus larvae. Only the spore stage of this bacteria is infective and only in honey bee larvae less than three days of age. Another important brood disease is European foulbrood disease, caused by Melissococcus pluton. In both cases, beekeepers sometimes use an antibiotic (oxytetracycline) for disease prevention and treatment. Another bacterial disease of minor importance is powdery scale disease, which is caused by Bacillus pulvifaciens.

All the known fungal diseases that affect the brood are of minor significance except for chalkbrood disease, which is caused by Ascosphaera apis. Until 1968, no chalkbrood disease had been found in the United States. Presently, chalkbrood is reported present in most of the states (6), and many beekeepers report reductions in their honey crop due to this disease. There is no effective chemical control for chalkbrood.

There are a number of viruses that affect honey bees.

Most virus diseases of bees are considered of minor importance and no treatment is necessary. Two brood diseases of viral origin are sacbrood and filamentous virus disease (formerly believed to be of rickettsial origin).

Varroa jacobsoni, a mite parasitic on honey bees, was found recently in the United States for the first time. This mite is believed to be as devastating to *A. mellifera* as all the known bee diseases combined. One reason for this conclusion is that, except for a fungus disease of minor importance, this is the only parasite that affects both the brood and adult stages. *Varroa jacobsoni* was originally restricted to *A. cerana,* where it reportedly causes little damage (7), but in recent years it has successfully begun to attack *A. mellifera* and spread throughout much of that species' range.

Adult Diseases

In 1984, the parasitic mite, *Acarapis woodi* (tracheal mite) was found in this country for the first time. Some scientists believed, based on European experience, that this mite would not be important. However, many colonies in the United States have been killed by this mite. There is some evidence that honey bees from the United States are more susceptible to *A. woodi* than current European honey bee populations (8). Menthol crystals have been recently registered by the Environmental Protection Agency for use in honey bee hives to control these mites.

Nosema disease is caused by the microsporidia, *Nosema apis.* This protozoan shortens the life span of adult bees and consequently reduces the honey production of colonies by 30–40%. The effects of this disease are difficult to diagnose and most beekeepers do not realize that their honey bees are infected. Fumagillin is used by some beekeepers for the treatment of this disease.

Chronic bee paralysis is a virus-induced disease that is occasionally seen in honey bee colonies. This disease mimics the effect of pesticide kill, and many beekeepers may misdiagnose this condition. Fortunately, the disease is not serious. It is generally believed that the effect of this disease can be abated by requeening the colonies.

There are other minor bee diseases that affect the health of the honey bees but are of little economic significance. Noninfectious conditions caused by pesticides and poisonous plants are serious problems in some years. A more detailed discussion of this subject has been published (9).

POLLINATION

The contribution of honey bees to agriculture far exceeds the value of honey production. Whereas honey bees produce an average of 200 million lb of honey a year valued at $150 million, the value of crops pollinated by the honey bees is in excess of $10 billion (10).

PRODUCTS OF THE HIVE

Most people think only of honey when honey bees are mentioned. The number of useful products of the hive on a worldwide basis include been venom, royal jelly, propolis, pollen, and beewax. The larvae and pupae of the honey bee can also be used as a diet item. However, in most cases, bee brood is used as a novelty food or is crushed in the process of extracting honey and consumed with the honey.

Bee Venom

Bee venom is used in the United States primarily for desensitizing individuals allergic to honey bee stings. In some countries bee venom is also used for the treatment of arthritis. There are specific devices used by beekeepers for the collection of bee venom. One device generates a mild electric current to irritate the honey bees enough to sting a plastic surface. The surface material to be stung must be one that honey bees can sting and still withdraw their barbed stinger. After airdrying, the bee venom is then scraped off the underside of the plastic material.

Propolis

Propolis or bee glue is a resinous material collected by the bees from trees. This material is used by honey bees to seal small openings or cracks and for mummification of foreign objects that are too large for the bees to remove (such as a mouse that is killed in the beehive). Because the bees also use propolis to cement together wooden parts of man-made beehives, most beekeepers consider propolis a nuisance and maintain bees that use less of the substance. Propolis has been shown to possess antimicrobial properties and is used in some European countries as the active ingredient in tinctures and throat lozenges. Propolis can be scraped from wooden hive parts and can also be harvested through the use of a perforated plastic mat that is placed between the supers (wooden sections of beehives). After the bees deposit propolis on it, the mat is removed and the propolis harvested from the mat. Periodically, bee journals carry advertising for propolis, but as a rule the demand for this material is limited.

Royal Jelly

Royal jelly (bee milk) is the food given to young honey bee larvae and it is this food that differentiates workers from queens. Only worker larvae selected to become queens are fed royal jelly throughout their developmental period. Royal jelly is used in cosmetic products and food supplements. In some countries it is believed that royal jelly possesses medicinal properties. The collection of royal jelly is labor intensive. In commercial production, larvae are grafted (moved into a queen cell) as if to produce queens, so that nurse bees are induced to feed the larvae liberal amounts of royal jelly. Three days after grafting, the royal jelly can be harvested from the cells.

Pollen

Pollen is consumed primarily as a diet supplement. Its protein, vitamin, and mineral content is variable, depending on the season and plant source. In a survey of a mixed sample of bee-collected pollen from seven states, the average protein content of pollen on a dry weight basis was 24.1% (11). However, when individual plant sources were analyzed, the range was 7.02–29.87% protein (12).

No precise data are available on the amount of pollen needed by one colony over an entire year. This would depend on the population and brood-rearing activities. Nevertheless, a good estimate is 44 lb (13). Because most colonies collect pollen in excess of their immediate needs, beekeepers can supplement their income by collecting pollen.

Pollen is collected by a pollen trap, a device with a screen that literally scrapes the pollen off the hind legs of a worker bee. Pollen traps must be emptied frequently, as pollen is an attractive food source for many insects and mites and it will absorb moisture and decay. Some of the pollen collected by beekeepers is used to feed honey bee colonies during dearth periods. Processing of pollen is generally a simple matter after its removal from the hive. The first step is to dry the pollen, usually by air-drying. Then foreign debris, such as insect parts and the remains of diseased bees, is removed. In some cases pollen pellets are also separated by color.

The market for pollen is not especially lucrative. Some foreign pollen is imported into the United States for sale in health food stores. There are no reliable figures on the size of this trade.

Beeswax

One of the most useful products of the hive is beeswax. Honey bees produce beeswax through internal body chemical processes, using honey as a starting point. It is estimated that it takes approximately 8 lb of honey to produce 1 kg of beeswax. In terms of the costs of both products, it is economically more profitable for a beekeeper to produce honey. Because beekeepers must remove the beeswax cappings from honey combs prior to extracting honey, beeswax is usually considered to be a by-product of honey production. Cappings and old combs can be boiled in water to extract the beeswax, which floats on the surface and hardens when the mixture is allowed to cool. Beeswax ranges in color from almost white through light green or tan to golden yellow. Like honey, the color of beeswax from cappings varies according to the floral source, while beeswax from old combs is usually darkest.

The majority of beeswax produced in the United States is used in candle making and cosmetics. Other uses include furniture polish, lubricants, jewelry making (lost-wax casting), batik, egg coloring, sewing, and honeycomb foundation (used by beekeepers).

Honey

Honey can be defined simply as a plant exudate of floral or extrafloral origin that is modified by honey bees and is stored in honeycombs. Nectars collected by the honey bees are processed into honey. Principally, this ripening procedure involves enzymatic inversion of sucrose into fructose and glucose and reduction in the water content of nectar. Other changes occur in the process but not much is known about them. The color, odor, and taste of honey is strongly influenced by the floral source. Depending on the floral source and location, honey is traditionally harvested whenever the combs are filled and capped with wax. Both the stage of the ripening process achieved before the honey is harvested and the ambient relative humidity affect the moisture content of the final product.

Honey yields are extremely variable and two colonies placed side by side seldom produce identical-size honey

crops. The production of honey is dependent on a number of factors. In addition to differences in nectar production among plant species, honey production is influenced by such factors as ground moisture, air temperature (day and night), wind conditions, and rainfall, because they affect both plants and bee flight. Honey bees exhibit preferences among plant species and may collect nectar from many sources. Consequently, honey from only one plant source is extremely rare. For instance, honey that is advertised as tulip poplar frequently contains honey from a number of floral sources though the flavor is primarily that of tulip poplar. As a result, many beekeepers label their honey as mountain honey, wildflower honey, and so forth.

Honey is available in many forms, although the most popular and well known is extracted honey in the familiar queen-line jars. Extracted honey is harvested from the combs by first removing the beeswax capping and then extracting the cell contents by centrifugal force. Some beekeepers heat extracted honey (63°C for 30 min) to destroy yeasts and to reduce granulation. When honey granulates, the remaining liquid portion of the honey increases in moisture content and the likelihood of fermentation also increases. Other beekeepers only heat their honey enough to aid in the process of filtration (38°C) and still others do not heat their honey at all. The filtration process is somewhat variable, ranging from merely straining the honey to remove dead bees, comb, and other large debris, to filtration that removes bits of beeswax and even pollen.

The USDA has established a system of grading honey based on percentage of solubles, clarity, flavor, aroma, and absence of defects (beeswax, propolis, etc). In addition to grades, honey is also classified by color: water white, extra white, white, extra light amber, light amber, amber, and dark amber. Honey color varies according to floral source. There have been attempts to develop worldwide honey standards (14). Meanwhile, in the United States the USDA Standards for Grades of Extracted Honey and Comb Honey are currently being used (Table 1).

In addition to liquid-extracted honey, many products using honey are available on grocery store shelves. These include honey roasted nuts, cereals, baked goods, ice cream, and candy. Honey is widely used in baked goods

Table 1. Average Composition of 490 Samples of Honey[a]

Characteristics	Average Value
Moisture	17.20%
Levulose	38.19%
Dextrose	31.28%
Sucrose	1.31%
Maltose	7.31%
Higher sugars	1.50%
Undetermined	3.10%
pH	3.91%
Lactone–free acid	0.335%
Ash	0.169%
Nitrogen	0.041%
Free acid	22.03 meq/kg
Lactone	7.11 meq/kg
Total acid	29.12 meq/kg
Diastase value	20.80

[a] Ref. 15.

because of its hygroscopic properties, which tend to prolong freshness in such products. To the purist, however, comb honey is the ultimate form for eating. Within this category of packaging there are several subdivisions, such as whole combs, cut combs, chunk honey, and comb honey (round and square). Honey sold in combs is unheated, unfiltered, and unlikely to be adulterated.

The typical beekeeper removes only the honey that is surplus to the colony's needs. In the United States, the commercial beekeeper's colonies produce an average of 50 kg of honey annually. In good years, beekeepers report honey yields of 200 kg or more in some areas of the United States and Canada.

FUTURE DEVELOPMENT

In recent years the honey bee industry in the United States has faced many difficult problems. Foreign honey imports and lower honey prices coupled with increased costs of production have created a considerable financial challenge. Because of the recent establishment of two parasitic mites (*Acarapis woodi* and *Varroa jacobsoni*) in the United States, Canada and Mexico have banned imports of United States package bees and queens, an action severely damaging to that segment of the industry. The arrival of Africanized honey bees in late 1990 also creates some uncertainty concerning the future of the honey bee industry. The reliance of United States agriculture of honey bees, however, makes it imperative that these problems be solved. The pollination of many crops cannot be effectively completed without the placement of honey bee colonies in the orchards or fields. Fortunately, the demand for one of the direct products of the insect, honey, shows signs of increasing. The formation of the National Honey Board, with its stated goal to expand domestic and foreign markets for honey and to develop and improve markets, bodes well for the future of commercial and hobby beekeeping in the United States.

BIBLIOGRAPHY

1. F. Ruttner, *Biogeography and Taxonomy of Honeybees.* Springer-Verlag, New York, 1988.
2. E. Crane, *The History of Honey,* in E. Crane, ed., *Honey, A Comprehensive Survey,* Heinemann, London, 1976, pp. 439–488.
3. F. Naile, *America's Master of Bee Culture The Life of L. L. Langstroth,* Cornell University Press, Ithaca, N.Y., 1942.
4. F. Ruttner, *Races of Bees,* Dadant and Sons, eds., *The Hive and the Honey Bee,* Dadant and Sons, Hamilton, Ill., 1975.
5. W. S. Sheppard, "A History of the Introduction of Honey Bee Races into the United States," *American Bee Journal* **129,** 617–619, (1989).
6. D. Menapace and W. T. Wilson, "The Spread of Chalkbrood in the North American Honey Bee, *Apis Mellifera,*" *American Bee Journal* **116,** 570–573 (1976).
7. A. Dietz and H. R. Hermann, *Biology, Detection and Control of* Varroa jacobsoni: *A Parasitic Mite of Honey Bees,* Lei-Act Publishers, Commerce, Ga., 1988.
8. B. Adam, *In Search of the Best Strains of Bees.* Dadant and Sons, Hamilton, Ill., 1983

9. R. J. Barker, "Poisoning by Plants," in R. A. Morse, ed., *Honey Bee Pests, Predators and Diseases,* Cornell University Press, Ithaca, N.Y., 1978, pp. 276–296.
10. W. S. Robinson, R. Nowogrodzski, and R. A. Morse, "The Value of Honey Bees as Pollinators of U.S. Crops, Part II," *American Bee Journal* **129,** 477–487 (1989).
11. E. W. Herbert, Jr., and H. Shimanuki, "Chemical Composition and Nutritive Value of Bee-Collected and Bee-Stored Pollen," *Apidologie* **9,** 34–40 (1978).
12. F. E. Todd and O. Bretherick, "The Compositions of Pollens," *Journal of Economic Entomology.* **35,** 312–316 (1942).
13. W. J. Nolan, "Colony Production and Honey Crops," *Gleanings in Bee Culture* **52,** 366–368 (1925).
14. A. Fasler, "Honey Standards Legislation," in Ref. 2.
15. J. W. White, Jr., M. L. Riethof, M. H. Subers, and I. Kushnir, "Composition of American Honeys," *USDA Technical Bulletin* **1261,** 1962.

General References

Refs. 2 and 4 are good general references.

R. A. Morse, ed., *The ABC and XYZ of Bee Culture,* A. I. Root Company, Medina, Ohio, 1989.

R. A. Morse and T. Hooper, *The Illustrated Encyclopedia of Beekeeping,* E. P. Dutton, New York, 1985.

S. E. McGregor, *Insect Pollination of Cultivated Crop Plants,* USDA Handbook No. 496, U.S. Government Printing Office, Washington, D.C., 1976.

R. G. Stanley and H. F. Linskens, *Pollen: Biology, Biochemistry, Management,* Springer-Verlag, New York, 1974.

Hachiro A. Shimanuki
Walter S. Sheppard
USDA/ARS
Beltsville, Maryland

BEER

Beer is traditionally, and in many countries, legally, defined as an alcoholic beverage derived from barley malt, with or without other cereal grains, and flavored with hops.

Barley malt is barley that has been purposely sprouted, allowed to partly germinate, and then dried or kilned. This three-step process is called malting. Because barley is virtually the only cereal that is malted, it is often just referred to as malt. The drying stabilizes the malt and allows beer to be produced year round and in places far removed from where barley is grown. In thus differs from wine, which is only produced seasonally and only near grape-growing areas.

Other cereal grains that are used to make beer are rice, corn, sorghum, and wheat. The choice, depends on the individual brewer and often on the availability of the different cereals.

HISTORY

Beer was likely a casual discovery when a bowl of grain that got wet, and then very wet, accidentally fermented with airborne yeast. It has been made intentionally for more than 6,000 years and is indigenous to most temperate and tropical cultures.

The oldest known written recipe was found on a 4,000-year-old Mesopotamian clay tablet, and it was for beer (Fig. 1). The Babylonians made 16 kinds of beer, using barley, wheat, and honey. The Egyptians described a "beer of truth" and a "beer of eternity".

Beer making was an important trade in ancient Mesopotamia, Egypt, Greece, and Rome. In less-settled cultures in Europe, Africa, China, India, and South America some kind of fermented grain beverages were made on a small scale for home consumption. It may well have been discovered that brewing beer makes any water safe for human consumption.

Large-scale brewing began in medieval Europe as a result of increased travel and trade and the attendant increase in inns and taverns offering food and lodging. Monasteries began brewing beer, as well as wine and liqueurs, and accepted voyagers as guests. This helped spread knowledge about beer.

In North America, beer was introduced with the early English and Dutch settlers. The Pilgrims landed in Plymouth, instead of going farther south, because they had run out of beer. Many of the founders of the United States were active brewers or maltsters.

All the beer of the early settlers was actually ale, fermented by strains of yeast, *Saccharomyces cerevisiae,* that are airborne and are also used in making bread and wine. Ales are produced at ambient temperatures.

In the 1840s German brewers came to the United States and began brewing lager beer, fermented by strains of a yeast *S. carlsbergensis* that exists only in

THIS IS THE OLDEST WRITTEN RECIPE IN THE WORLD. AND IT'S FOR BEER.

Cuneiform tablet from Nippur, Iraq, ca. 1750 BC, excavated by The University Museum, University of Pennsylvania.

Around 2,000 B.C., scribes worked with pointed sticks and slabs of wet clay to record history's first known recipe for beer.

Figure 1. Mesopotamian clay tablet. Courtesy of the Beer Institute.

breweries and is used only for making beer. A characteristic of this yeast is that it can only function at temperatures below 70°F. For this reason, the early lager breweries were built near rivers or lakes that froze in the winter. Ice was cut from these and stored in cellars for refrigeration during the summer. Thus New York, St. Louis, and Philadelphia became centers of lager brewing. By the 1870s mechanical refrigeration was invented and used first in a brewery in Brooklyn.

Lager beer has almost completely replaced ale as the beverage of choice, and today ale is only produced in the United Kingdom and parts of Canada, with only token amounts elsewhere. In the United Kingdom ale is referred to as beer and what the rest of the world calls beer is called lager. In what follows, beer well to be used to refer to both ale and beer.

THE MAKING OF BEER

Beer production involves three distinct but interrelated stages. First is the preparation of an extract of the malt and the grains selected. This extract is called wort (pronounced to rhyme with "hurt"). This step takes place in the brewhouse and is often referred to as brewing. It takes about 4–10 h. A photograph of a typical brewhouse is shown in Figure 2.

The next stage is fermentation, the conversion of this liquid by yeast into beer. This is a temperature-controlled process that takes 3–10 d.

Figure 2. Brewhouse, showing lauter tubs above and kettles below. Courtesy of Anheuser-Busch, Inc.

Fermentation is usually conducted in large stainless steel vessels that hold the volume of an entire brew, or several brews. Their size may be between 7,000 and 375,000 gal, and they may be horizontal rectangular in shape, or vertical cylindrical, with or without a conical bottom. The conical bottom allows simple gravity expulsion of yeast. A set of cylindroconical tanks are shown in Figure 3. The rectangular tanks require manual raking of yeast for reuse in subsequent brews.

The final stage is finishing or the refining of this liquid into salable beer by the brewery. It may take 2–25 d. It takes place in tanks equal in size to the fermentation tanks or multiples of the fermenter size. Aging tanks, preassembled on site, are shown in Figure 4.

Each stage offers the brewery many choices. It is for these reasons that there are not many beers, perhaps none, that are exactly alike. The finished beer may be packed in large containers, holding 2–31 gal, or into small ones, holding 7–40 oz.

RAW MATERIALS OF BREWING

Water

Water is, only quantitatively, the major component of beer. Its requirements for brewing are fairly simple. First, it must meet local or international standards for potable water. The World Health Organization's standards for drinking water are shown in Table 1.

The second requirement is that it not be too alkaline. A maximum alkalinity of 50 ppm (as calcium carbonate) is acceptable.

A third requirement is that it be hard and contain calcium. The level of calcium considered most desirable is 100 ppm (as calcium), but lesser amounts may also be used. The water in Pilsner, Czechoslovakia, had these attributes, neutrality and hardness, and soon became the standard by which all beer is made.

All water supplies are either surface water or groundwater. Surface water is derived from rainfall or snow and may be quite pure. If it is collected in reservoirs near the source of rain or snow, it is usually soft and neutral. If it is collected in rivers, its purity will depend on the distance between the collection and distribution points and the uses it suffers in between. Such water may need purification with sand and charcoal.

Groundwater comes from wells or springs and is usually tasteless but may be quite alkaline or hard, or both.

Table 1. World Health Organization Standards for Drinking Water

	Maximum Permissible Concentration mg/L
Chlorides	60
Sulfates	400
Calcium	200
Magnesium	150
Total dissolved solids	1,500

Figure 3. Cylindroconical tanks. Courtesy of Paul Mueller Company.

Such water may need acidification to remove alkalinity and, less often, softening if it is too hard. Such water rarely needs treatment with activated charcoal.

Barley Malt

Malted barley is the principal ingredient of beer. It furnishes almost all the required elements for making beer. Malt provides starch and sugars, which will, partly or completely, become alcohol during fermentation; and protein and amino acids, which will supply nutrition for the yeast, as well as color, foam, and flavor in the finished beer.

Barley used to make malt is almost always selected from special strains that have been developed and grown specifically for this purpose. There are two major types of barley, distinguished by the number of rows on the stalk. Six-rowed barley is grown mainly in the midwest of the United States and Canada, but now also in Europe and Australia. Two-rowed barely is grown in the far west, and also in Europe and other barley growing regions. The two-rowed barleys are considered by some to give a smoother beer. A drawing of a barley kernel is shown in Figure 5.

The husk, which protects the seed and is composed of the palea and lemma, is, in barley, alone among the cereals, fused to the testa and the pericarp. It is this fusion

Figure 4. Aging tanks assembled on site. Courtesy of Paul Mueller Company.

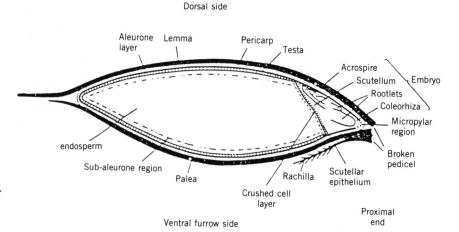

Figure 5. Schematic longitudinal section of a barley grain, taken to one side of the furrow, showing the disposition of the parts.

that permits barley to be malted efficiently and to be the universal basis for making beer. Grains that lose their husks at germination tend to become moldy when malted in bulk.

The malting of barley begins with steeping, in which the barley, in tanks, is soaked in aerated water. The steeped water is changed several times. The steeping is complete when the barley reaches a moisture content of 45%. It is then transferred to long, horizontal compartments, through which moist air is blown. In 4–6 d, the barley germinates and starts to grow as if it were to become a barley plant. During this time the barley produces starch-splitting enzymes, notably α-amylase and β-amylase, and also proteinases and cellulases. The amylases break down some of the starch in the barley; the proteinases degrade some of the proteins to amino acids, and the cellulases soften the cell walls. All these enzymes are secreted by the aleurone layer and migrate into the endosperm. These changes, which collectively are called modification, prepare the barley for its use in brewing.

At the time when the modification is considered complete, the malt is transferred to a kiln and dried. The temperature at which the malt is dried determines the color and, partly, the flavor of the beer from which it will be made. Kilning stabilizes the malt and allows it to be shipped widely and to be stored for a year or more.

The analysis of a typical malt is shown in Table 2.

UNMALTED CEREALS (ADJUNCTS)

Cereals such as rice, corn, and sorghum are often used to complement or attenuate the malt. These grains do not

Table 2. Analysis of Barley Malt, %

Moisture	4.0
Starch and dextrins	52.5
Simple sugars	9.5
Total protein	13.0
Soluble protein	5.4
Cellulose	6.0
Other fiber	10.0
Fats	2.5
Minerals	2.5

contribute to flavor, color, or foam but serve only to supply carbohydrates to the wort. Their use probably arose when malt was in short supply. They are widely used throughout the world. Only in parts of Germany, in Switzerland, and in Greece is their use prohibited by law. This prohibition, which dates to 1516 in Bavaria, as the *Reinheitsgebot*, originally was intended as a means to police the taxation of beer because malt could only be made in a malthouse. The prohibition on the use of other sources of carbohydrates precluded the surreptitious increase in beer production. It remains to this day as a vestigial reminder of days past that has served to prevent the importation into those three countries of beers made with unmalted cereals. However, the European Common Market has overruled this law and such beers may now be imported into the all-malt countries.

The quantity of unmalted cereals may vary from 10 to 50% of the total mash, depending on the beer to be produced. Increased use gives a lighter, less bodied beer. If rice or sorghum is used, it must be ground before mashing. If corn is chosen, then a physically separated fraction of the ground corn kernel that contains mainly starch and protein and very little corn oil is used.

Starch is the major component of both malt and the other cereals. It is a complex polysaccharide made up of about 35% of a straight-chain polymer of glucose known as amylose and about 65% of a branched polymer, amylopectin. The links in all of amylose are 1,4′ bonds between neighboring glucose units. The bonds in amylopectin are

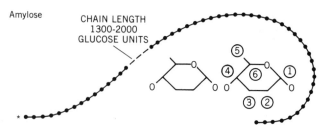

Figure 6. Structure of the linear starch fraction. Each black circle represents a glucose unit, and the aldehydic terminus of the chain is indicated by an asterisk. Interglucose bonding is α-1,4, as shown in the inset, with carbon atoms in the glucose unit numbered in conventional fashion.

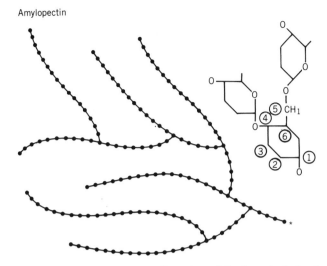

Amylopectin

Figure 7. Structure of a small section of the branched starch fraction. Branching is effected through an α-1,6 linkage, as shown in the inset.

both 1,4′ and 1,6′,—the latter at the branch points. Schematic diagrams of amylose and amylopectin are shown in Figures 6 and 7, respectively.

SPECIAL MALTS

Some malts are made with special properties, mainly differing in color and flavor. A caramel, or crystal, malt is made by heating a wet malt with steam under pressure. The resulting malt is used with regular malt to produce dark beers. A so-called chocolate malt is a variant of crystal malt. These malts provide color and special flavor to the beer. Malt may also be darkened by kilning at higher temperatures (200°F or higher) or for longer times. Such malt only imparts more color to the beer.

Wheat may be malted, with difficulty, but it produces an ordinary beer unless a yeast is used that by itself pro-

duces a spicy flavor. This has found some favor in parts of Germany and Belgium.

HOPS

Hops, *Humulus lupulus*, is a truly remarkable plant. Although it originally grew wild, and still does, it is intensively cultivated in a few countries for use only in beer. The major hop-growing countries are Germany, the United States, the Soviet Union, Czechoslovakia, Yugoslavia, and the United Kingdom. Figure 8 shows hops growing in Germany. Hop picking by hand in Czechoslovakia is shown in Figure 9.

The use of hops in beer only goes back some 500 years. Many plants had been used to flavor beer, until it was discovered that hops not only makes beer pleasantly flavored but also controls the growth of spoilage bacteria. The use of hops is now universal in beer.

Hops contain a group of compounds, humulones, which are very insoluble in water, but which undergo a chemical rearrangement during the brewing process to form an isomeric group of compounds called isohumulones. The isohumulones are soluble in water and impart to beer a palate-cleansing bitterness that provides beer with its unusual property of drinkability. The isohumulones in beer can be readily analyzed, permitting a quantitative measure of their presence and thus of the beer's bitterness. Not many flavor compounds in other foods are so easily measured. A diagram of humulone and its conversion product isohumulone is shown in Figure 10. The analogues of humulone (and isohumulone)—cohumulone, adhumulone, and prehumulone—differ only in the number of carbon atoms in the side chains.

In addition, hops contain a volatile oil containing many odoriferous compounds, some of which survive into the finished beer. Some varieties of hops, such as Clusters and Galena, are used primarily for bitterness, whereas others, such as Hallertau and Cascade, are used for their aroma.

The essential oil of hops is, as in all plant materials, or terpenic origin. The major terpenes are myrcene (**1**), a

Figure 8. Hops growing in Germany. Courtesy of S. S. Steiner.

Figure 9. Hand picking of hops. Courtesy of S. S. Steiner.

diterpene; humulene (**2**), and caryophyllene (**3**) sesquiterpenes (Fig. 11). Oxidation products of these hydrocarbons, such as humaladienone and caryophyllene epoxide have been found and are probably more important to the flavor of beer.

Also present in hop oil are alcohols, such as linalool and geraniol; ketones, such as undecanone-2; and esters, such as geranyl butyrate. These survive into beer and are important components of the hoppy aroma of some beers.

Hops is a perennial vine that grows to more than 20 ft in one season. They are trained to high wires held by long poles. The flowers are mechanically picked and dried in a kiln. Hops are used in several forms. The dried cones, compressed into bales weighing 200 lb, are used by many brewers. In this state, hops must be kept refrigerated. Alternatively, the hop cones can be milled to a powder and then recompressed into pellets. These are easier to use at a brewery, but also need refrigeration. Lastly, the hops may be extracted, with an organic solvent such as hexane or alcohol, or with liquid carbon dioxide, and the extract used. These do not need refrigeration and are popular in tropical countries. A cluster of Fuggle hops is shown in Figure 12. Some varieties of hops grown commercially are listed in Table 3, with the country producing the major quantities indicated.

A typical analysis of hops is shown in Table 4.

YEAST

Yeast is a unicellular microscopic organism that is fairly distinctive in being able to metabolize sugars either to carbon dioxide and water, in the presence of air; or to alcohol and carbon dioxide, in the absence of air. It is this latter trait that is used in all alcoholic fermentations.

One genus of yeast, Saccharomyces is used in brewing. But there are two species. One is *S. Cerevisiae,* used in producing ales, and also for making bread, wine, and whiskey. It is a fairly hardy yeast and survives in the atmosphere.

The other is *S. carlsbergensis* (uvarum) used only for lager beers and found no where else but in breweries. It is temperature-sensitive and does not survive in the atmosphere.

Yeast reproduces asexually and multiplies severalfold during a normal fermentation. Because yeast enters a commercially sterile liquid, wort, it may be reused many times in successive brews without danger of contamination. Beer is the only fermented product that starts with a sterile medium. A photograph through an electron microscope of a dividing yeast cell is shown in Figure 13.

The two species of yeast differ in many biochemical characteristics. Table 5 gives some of these differences.

Most breweries have their own strains of yeast and

Figure 10. Humulone and isohumulone.

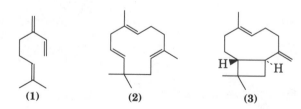

Figure 11. The terpenes (**1**) myrcene, (**2**) humulene, and (**3**) caryophyllene.

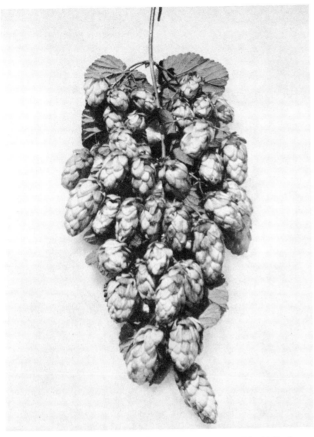

Figure 12. A cluster of Fuggle hops. Courtesy of S. S. Steiner.

Table 3. Hop Varieties

United States	Europe[a]	Australia
Clusters	Northern Brewer (UK & G)	Pride of Ringwood
Galena	Fuggle (UK)	
Willamette	Golding (UK & SU)	
Fuggles	Brewers Gold (UK & G)	
Chinook	Hallertau (G)	
Tettnang	Tettnang (G)	
Hallertau	Hersbruck (G)	
Nugget	Jura (G)	
Cascade	Spalt (G)	
	Styrian (Y & SU)	
	Saaz (Cz)	

[a] Key: G = Germany   Y = Yugoslavia
            SU = Soviet Union   Cz = Czechoslovakia
            UK = United Kingdom

Table 4. Chemical Composition of Hops

Moisture	10%
Total resins	17–20%
Volatile oils	0.3–1.2%
Polyphenols	2–5%
Waxes and lipids	3%
Ash	7%
Cellulose	55%

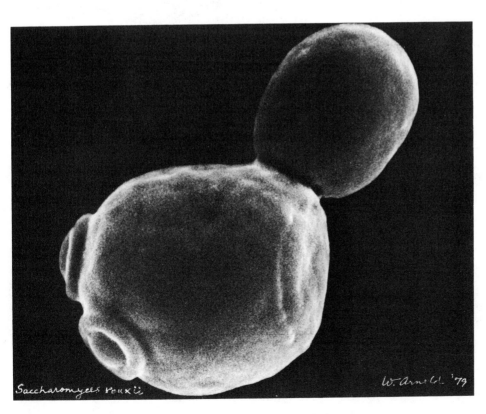

Saccharomyces rouxii

W. Arnold '79

Figure 13. Dividing yeast cell.

Table 5. Biochemical Characteristics of Yeast Used in Beer Making

	S. carlsbergensis	S. cerevisiae
Ferment melibiose	+	–
Ferment raffinose	+++	+
Utilize ethanol aerobically	–	+
Produce hydrogen sulfide	+++	+
Attitude after fermentation	Settles to bottom	Rises to top

continue to use them indefinitely. Very often a pure yeast culture is maintained and propagated in each plant, or, more often, in the parent plant if the brewery has more than one plant.

THE BREWING PROCESS

The Brewhouse

Milling. Grains are normally received in breweries in bulk and transferred to silos. Malt needs to be crushed before it is used. This is done just before use. The aim in milling is to crush the endosperm of the barley, but leave the husk as intact as possible. These husks act as the filtration medium during later processing. A six-roller malt mill is shown in Figure 14.

Corn as received in a brewery does not need any treatment before use. It has already been crushed by a corn miller and separated pneumatically. The lighter germ of the corn is conveyed in a different airstream from the heavier, starchy endosperm. It is this latter fraction that is used by breweries.

Rice needs merely to be crushed before use.

Mashing. This step determines the ultimate structure of the finished beer. A typical mashing cycle requires the use of four vessels, a cooker, mash tub (tun in Europe), lauter tub, and kettle. Examples of some brewhouse control panels are shown in Figure 15.

The cooker is a simple vessel, with an agitator and some means of being heated. The lauter tub is a complex vessel with a specially perforated false bottom through which the mash is strained or filtered. It has variable speed rakes, with adjustable flights that can be raised and lowered. The husks of the malt serve as the filtration medium in the lauter tub. A cut-out diagram of a lauter tub is shown in Figure 16. The kettle is also complex. It is heated internally with steam passing through coils and a percolator, or directly with a flame; or externally in a steam boiler. Steam-heated kettles are shown in Figure 17

There are several systems of mashing, each with certain advantages and each setting the requirements for the brewhouse equipment. The most common system in the United States and most other countries is a double-mash system. The corn or rice is boiled in the cooker, with a small amount of malt to lower the viscosity. The main portion of malt is admitted to the mash tub, and, after a stand at about 50°C for proteolysis, the temperature is raised by adding the contents of the boiling cooker. The

Figure 14. Six-roller malt mill. Courtesy of Wittemann Hasselberg.

combined mash is allowed to rest at the temperature that results from this mixing for a variable time, and then the temperature is raised selectively until the final mashing off temperature, usually 75°C, is reached.

In the United Kingdom and its brewing satellites the most common mashing system is a simple infusion method. In this system, the mashing process takes place at a single temperature. This system developed with mash tubs that could not be heated. It is not nearly as versatile as the double-mash system. Mashing, with or without corn or rice, boiled separately in a small vessel, takes place at about 65°C.

In Germany, and its brewing followers, still a third system, called decoction brewing, is common. Here too the mash tub cannot be heated, but a smaller decoction vessel can. The malt is transferred into the mash tub at a low temperature. A small portion of this mash is pumped to the decoction vessel and boiled. The boiled mash is pumped back into the main mash vessel. This serves to raise its temperature to a level that depends on the relative volumes in both vessels. This transfer is repeated

Figure 15. Brewhouse control panels. Courtesy of Wittemann Hasselberg.

once or twice (double decoction or triple decoction), leading to a series of rising temperatures in the mash tub. This process is rather lengthy and has not found wide use.

The reactions that occur in the mash tub are very complex. At a low temperature, about 50°C, proteolytic and cytolytic enzymes in the malt are operative. The corn or rice are not present at this stage and don't need to be. Proteolysis is necessary to supply amino acids for the growth of yeast and also to break down large, insoluble proteins to simple soluble peptides for foam enhancement. Cytolysis is required to reduce the viscosity of the barley gums (soluble fiber) to accelerate the later filtration. After some 10 to 60 min, the temperature is raised by the addition of the boiled cooker (or decoction vessel). The resulting temperature may be held for 10–60 min, and the tem-

perature raised again, finally ending at 75°C. During these elevated temperatures, two amylolytic enzymes, and α-amylase and β-amylase, cooperate to hydrolyze most of the starch to fermentable sugars. Beta-amylase, which operates best at about 63°C, attacks starch from its nonreducing end and splits off the disaccharide, maltose. It cannot proceed past a 1,6′ linkage, and so depends on the concurrent action of α-amylase. Alpha-amylase, which operates best at 70°C, splits any 1,4′ linkage at random, and so provides a supply of nonreducing ends for the β-amylase to operate on. Cooperatively, the action of these two enzymes can result in the splitting of about 65–75% of the starch into mono-, di-, and trisaccharides.

The mash temperature is raised to 75–77°C when the desired action of the amylolytic enzymes has been

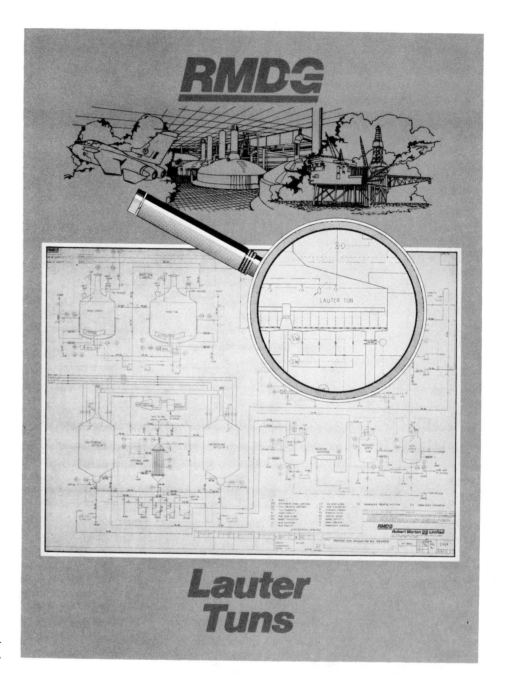

Figure 16. Cut-out diagram of lauter tub. Courtesy of Robert Morton.

achieved. This arrests any further enzymatic action and prepares the mash for the next step, lautering.

Lautering (Straining or Filtration). The converted mash is transferred to a vessel that will permit separation of the liquid (wort) from the insoluble solids (husk, fiber) of the malt and grains. Three different types of vessels accomplish this separation. The one most widely used is a lauter tub, a large cylindrical vessel with a slotted false bottom and a set of movable and adjustable flights. The mash is pumped into this tub, allowed to settle for a few minutes, and then liquid is allowed to flow to the next stage, the kettle. The first runnings are always cloudy and are run back onto the mash. When it runs clear, it is run to the kettle. After the clear liquid, called first wort, stops running, fresh water (sparge water) is sprayed onto the grain bed to elute the wort adhering to the grains.

Sparging continues until the next vessel is full, the concentration of the wort has reached the desired value, or the solids in the effluent are too dilute to be worthwhile. In some breweries, these last runnings are recovered and used for a succeeding brew. At the end of sparging, the rakes are lowered, their angle changed, and the spent grains are pushed into now-open large holes at the bottom of the lauter tub. These grains have a ready market as a valuable animal feed.

Another device used to separate wort from grain is a mash filter, a plate-and-frame press with plastic cloth as the barrier. These are faster than lauter tubs, but lack flexibility to grain depth. They have found some use in the United States, but wider use outside that country.

A third device, patented by Anheuser-Busch, is a Strainmaster, a modified lauter tub with perforated tubes projecting into the grain bed that serve to collect the wort.

Figure 17. A row of brew kettles. Courtesy of Miller Brewing Co.

This also is more rapid than a lauter tub. They are used almost exclusively in the United States.

The Kettle. The clear wort running from any of the grain-separation vessels is collected in a large vessel, the kettle, and boiled. Heat is normally supplied by steam in a set of coils and a center-mounted percolator. Alternatively, the vessel may be heated by direct fire or by an external collandria heated by steam.

During the heating period hops are added, in one or several portions, and at varying times. The varieties, the quantity, and the duration of their boiling time, all affect the flavor of the finished beer.

The boiling of the wort serves many vital functions:

1. It extracts the resin from the hops.
2. It isomerizes the humulones to soluble isohumulones.
3. It volatilizes most, but not all of the volatile oils in the hops.
4. It stops all enzymatic action.
5. It sterilizes the wort.
6. It concentrates the wort.
7. It removes grainy odors from the malt and other cereals.
8. It darkens the wort and produces flavor-inducing melanoidin reactions between the amino acids and simple sugars.

At the end of the boil, the wort is passed through a strainer to remove hop leaves, if whole hops were used, or directly to a tank in which the wort is allowed to settle. Recently, tangential entrance to this tank was found to produce a rapid settling of the precipitate that always forms in the kettle. This precipitate, called trub, contains hop bitter resins, proteins, and tannins and needs to be eliminated from the beer.

After settling some 20–30 min the wort is passed through a heat exchanger and cooled to the temperature desired for fermentation. Immediately after cooling, the wort is aerated to provide oxygen for the yeast.

Yeast, from a preceding brew, is added (pitched) in measured amounts to the aerated wort. The usual quantity is 1 lb of liquid yeast per barrel of wort. This will give a count of about 12 million yeast cells per milliliter. The second stage of the brewing process now begins.

Fermentation

Yeast enters a cool solution containing oxygen, fermentable sugars, and various nutrients. The yeast quickly absorbs the oxygen as well as minor nutrients that it requires, such as phosphate, potassium, magnesium, and zinc. It then begins to metabolize sugars and amino acids. The glycolytic pathway that yeast uses is common to many higher organisms, and only the ability to split pyruvic acid into acetaldehyde and carbon dioxide makes yeast distinctive. The scheme for the fermentation of glucose is shown in Figure 18.

The temperature during fermentation is very rigidly controlled. Because fermentation itself is exothermic, the fermentation vessels in all beer fermentations are cooled by a refrigerated fluid circulating through coils in the tanks or through jackets surrounding the tanks. The process usually takes 5–9 d.

During this time the yeast also metabolizes amino acids and makes three to four times as much yeast as was pitched. The excess yeast, over that used for succeeding brews, is collected and sold for use in dietary supplements and flavors for soups and snack foods.

The metabolism of amino acids is orderly and proceeds in a specific sequence. Certain amino acids are metabolized first, followed by the others, as shown in Table 6.

S. carlsbergensis has a marked ability to flocculate when it completes fermentation. This tendency allows the yeast to settle quickly when fermentable sugars are ex-

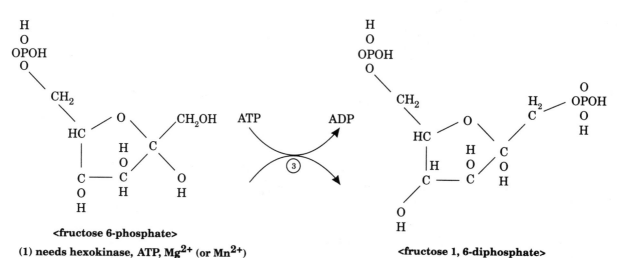

(a) Phosphorylation

(1) needs hexokinase, ATP, Mg^{2+} (or Mn^{2+})
(2) needs glucosephosphate isomerase
(3) needs phosphofructokinase—rate limiting

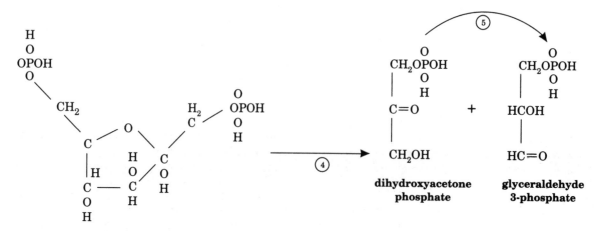

(4) needs aldolase, Zn^{2+}, is a reverse aldol condensation
(5) needs triose isomerase

(b) Splitting of hexose

Figure 18. Chemistry of alcoholic fermentation.

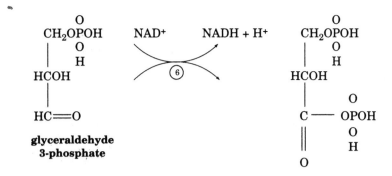

**glyceraldehyde
3-phosphate**

**1,3 - diphosphoglyceric
acid**

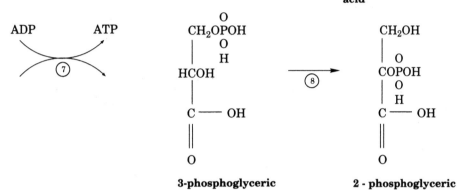

**3-phosphoglyceric
acid**

**2 - phosphoglyceric
acid**

(6) needs glyceraldehyde phosphate dehydrogenase and NAD⁺
(7) needs phosphoglyceric kinase
(8) needs phosphoglyceromutase

(**c**) Oxidation without oxygen

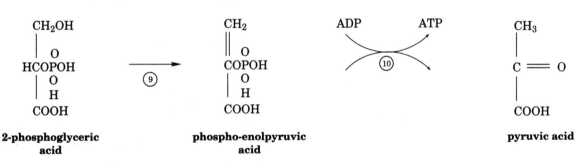

**2-phosphoglyceric
acid**

**phospho-enolpyruvic
acid**

pyruvic acid

(9) needs enolase and Mg²⁺
(10) needs pyruvate kinase

(**d**) Formation of pyruvic acid

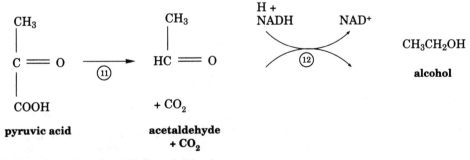

pyruvic acid

**acetaldehyde
+ CO₂**

alcohol

(11) needs carboxylase, Mg²⁺ and thiamin
(12) needs alcohol dehydrogenase and Zn²⁺

(**e**) Production of alcohol and carbon dioxide

Figure 18. (*continued*)

Table 6. Metabolism of Amino Acids

Group	Extent of Absorption	Acid
Group A	Almost completely absorbed within 24 hours	Glutamic acid Glutamine Aspartic acid Asparagine Serine Threonine Lysine Arginine
Group B	Absorbed gradually during fermentation	Valine Methionine Leucine Isoleucine Histidine
Group C	Absorbed only after Group A amino acids are gone	Glycine Phenylalanine Tyrosine Tryptophane Alanine Ammonia
Group D	Absorbed only slightly	Proline

hausted and makes collection and reuse of the yeast fairly simple. Some brewers hasten this settling by using centrifuges to collect their yeast.

S. cerevisiae, on the other hand, first rises to the top of its fermenting tank, where it may be collected by skimming, but then settles to the bottom and may also be collected there. Certain strains of *S. cerevisiae* settle more quickly and have been chosen by many brewers of ales and stouts.

Finishing

The fermented beer may now be finished in one of several ways. The simplest and most widely used is merely to transfer the beer to another tank, chilling it en route. This stage, called ruh (rest), allows much of the still suspended yeast to settle and also removes some harsh, sulfury notes. Furthermore, the yeast removes some undesirable flavor compounds, notably diacetyl, which were produced during the earlier fermentation. Ruh normally takes 7–14 d.

An alternative way of finishing beer is to move it to another tank sometime before it is completely fermented. Again, it is chilled en route, and the so-called secondary fermentation is allowed to proceed at a much lower temperature, 0–3°C. Much the same cleansing action occurs, but the early transfer ensures the presence of sufficient yeast to cleanse the beer more efficiently. It takes 2–4 w. It has not found wide favor, probably because of the need to transfer the beer at times that could be inconvenient.

A third, and most elegant way, is to transfer the fermented beer to another tank, chilling it less, and treating it with a small quantity of beer that just started to ferment a day earlier. This subjects the fermented beer not just to a specific quantity of yeast, but to very active yeast that very effectively cleanses the fermented beer. This process, called krauesening, is used by very fastidious brewers. It normally takes 3–5 w.

After any of these finishing processes, the beer is filtered, always in the cold, to remove insoluble particles

and yeast. This filtration uses diatomaceous earth as the filter medium. The beer may be filtered again, and this filtration may be a sterile filtration to remove all yeast and lactobacilli or another diatomaceous earth filtration. A sterile filtration may involve cotton fibers, a porous plastic sheet, or a ceramic filter as the retaining barrier.

Beer Packaging

Beer is packaged in either cans or bottles or in large containers or kegs for use in restaurants and taverns. The packaging machinery for beer has become very sophisticated and fast. Because oxygen changes the flavor of beer (see later), modern fillers evacuate the bottle before it is filled and replace the air with CO_2. Then the beer enters against a counter pressure of CO_2. Bottles are being filled at more than 20/s, and cans at more than 30/s. A high-speed can filler is shown in Figure 19.

If such packaged beer has not been sterile-filtered, it must be pasteurized, as beer is a fertile medium for many microbes. The pasteurization may be done just before filling, a so-called bulk pasteurization, or after filling in long tunnels with hot-water sprays. Bulk pasteurization takes about a minute, tunnel pasteurization about an hour.

Kegs may be filled with diatomaceous earth-filtered beer, in which case they are kept refrigerated; or they may be filled with sterile filtered or bulk pasteurized beer, in which case they do not need refrigeration.

Problems of Beer

The complexity of beer makes it prone to several problems. First and foremost is that the delicate flavor of beer is adversely affected by oxygen, so great pains are taken in the brewery and in the packaging operation to minimize the pick up of oxygen. When yeast is present, it will scavenge oxygen and keep the beer oxygen-free. But after filtration beer is very prone to pick up oxygen. At every transfer, steps are taken to avoid oxygen pickup, which is monitored closely. The filling operation has been mentioned earlier.

Correctly packaged beer has a flavor shelf life of about 4 m: if kept cold, it will last longer. The change in flavor with oxidation is difficult to describe because it is unique to beer. It has been described as papery, bready, or cardboard. But mainly it loses the fine appeal of fresh beer.

Another problem of beer is that it develops a peculiar, skunky aroma if packaged in clear or green bottles and is exposed to light. The problem has been eliminated with the discovery that brown bottles filter the offending wavelengths of light. The aroma is caused by the action of light on isohumulone, which splits off an unsaturated hydrocarbon, which in turn reacts with a trace of hydrogen sulfide in the beer to give a mercaptan.

A third problem is that beer is a good medium for some nonpathogenic microbes and so must be sterile-filtered or pasteurized. Either of these treatments effectively eliminate the problem.

Another shortcoming that has largely been eliminated is the tendency for beer to become cloudy when it is chilled. This haze is caused by a reversibly insoluble protein–polyphenol complex. It is removed in one of several ways. The protein may be solubilized by the enzyme pa-

Figure 19. High-speed can filler. Courtesy of H&R Inc.

pain, or the polyphenols may be removed by absorption onto silica gel. These treatments are effective and widely used.

Light Beer

The enzymes present in barley malt cannot completely degrade starch to fermentable sugars. So all regular beers contain dextrins, fragments of starch held by 1,6′ linkages and too large to be fermented by yeast. There is an enzyme, however, that can break up all the starch—amyloglucosidase.

A true light beer is one in which all the starch has been made fermentable. This is accomplished by the use of amyloglucosidase, either in the brewhouse to work with the α-amylase and β-amylase or during fermentation to work after the α-amylase and β-amylase.

Some light beers are made by merely extending the action of the α-amylase and β-amylase. This produces a beer reduced in dextrins but not free of them. The legal definition of a light beer in this country permits this.

Analysis of Beer

Typical analysis of some American beers are shown in Table 7.

Types of Beer

Beer may be classified in many ways, some of which are given here.

I. By type of yeast
   A. Lager yeast, *S. carlsbergensis*
      1. By color
         a. Light to amber color—lager beer
         b. Dark color—porter
         c. Dark and strong—bock beer
      2. By process
         a. Ruh beer
         b. Krauesened beer
   B. Ale yeast, *S. cerevisiae*
      1. By color
         a. Light to amber color—ale
         b. Dark colored—stout
II. By properties of the water
   A. Alkaline, hard water—Munich type
   B. Neutral, hard water—Pilsen type
      (All beers made now are Pilsen type)
III. By alcohol content
   1. No-alcohol beer
   2. Low-alcohol beer (shank beer)

Table 7. Analysis of American Beers

	Standard Beer	All Malt Beer	Light Beer
Water	92	91	6
Alcohol, wt%	3.8	3.9	3.3
Carbohydrates, %	3.4	3.8	<1.0
Protein, %	0.3	0.6	0.2
Acids, as lactic, %	0.1	0.1	0.1
Ash, %	0.1	0.2	0.1
Isohumulones, ppm	15	25	15.

Table 8. Vitamins in Beer

Vitamin	Vitamin Content of Beer, mg/L	% Minimum Daily Requirements Supplied by 1 L
Thiamin	0.02	<2
Riboflavin	0.3	20%
Pantotenic acid	1.0	25%
Pyridoxine	0.5	20%
Biotin	0.007	7%
Cyanocobalamin	0.1	3%

    3. Beer, ale
    4. Malt liquor
    5. Barley wine
IV. By market category
    1. Super premium
    2. Premium
    3. Standard, regional
    4. Generic

Dietary Aspects

Beer, being made from grains, has considerable nutritional value. Table 8 shows the vitamin content of a typical beer.

Beer also contains the trace minerals zinc, copper, and manganese. It is also very low in sodium, 12–18 mg sodium per liter, and high in potassium, 110–125 mg potassium per liter. This ratio of sodium to potassium has made for the inclusion of beer in the diets of patients with fluid retention problems. The well-known diuretic effect of beer is also a factor in many diets.

PLANT OPERATIONS

Breweries are operated most efficiently on round-the-clock schedules. The limiting factor in production is almost always capacity of the brewhouse, mainly the lauter tub and kettle. Some breweries will have one malt mill, one cereal cooker, one mash tub, two lauter tubs, and two kettles. With a mash filter or Strainmaster, only one may be required. Large breweries will have more than one brewing line. The maximum volume of beer that can be produced from the largest practical kettle (31,000 gal) is about 3–4 million barrels (90–120,000,000 gal). Any plant making more than this has more than one brewing line.

Round-the-clock brewhouse operation also demands 24-h manning in the fermentation cellars. Although fermenting and aging tanks are always above street level, the rooms they are in are referred to as cellars, from the days when the absence of mechanical refrigeration made underground storage necessary to keep the fermenting beer cool.

The speed of packaging equipment allows the so-called bottleshop section of a brewery to run on less than round-the-clock operation, even if the rest of the brewery is on a 24-h schedule.

The many choices involved in the brewing process, and the actual day-to-day running of the brewery (but not the bottleshop) are under the domain of the brewmaster. All else in a brewery, including the bottleshop, are the domain of the plant manager.

QUALITY CONTROL

Quality are those attributes of a product that makes it desirable to a consumer. In the production of beer there are many opportunities for undesirable properties—in aroma, in taste, and in appearance—to develop. Brewers have learned to be very careful about their products.

Quality control begins with the raw materials used in brewing. The water used is monitored for taste, alkalinity, hardness, and trace metals.

Barley malt is controlled very closely. The variety of barley is always specified, and sometimes the region in which it is grown. The malting process is controlled indirectly, by the analysis of the finished malt. The kilning cycle is specified, and the temperatures on the kiln are controlled.

Analyses exist for malt that reveal the size and vitality

Table 9. Analyses of Two-Row and Six-Row Malts

		Two-Row Malt	Six-Row Malt
Assortment: On $\frac{7}{64}$ in., %		61.4	47.5
	$\frac{6}{64}$	30.7	43.3
	$\frac{5}{64}$	7.0	9.2
Through $\frac{5}{64}$		0.9	1.0
Growth:	0–$\frac{1}{4}$	0	0
	$\frac{1}{4}$–$\frac{1}{2}$	1	2
	$\frac{1}{2}$–$\frac{3}{4}$	5	6
	$\frac{3}{4}$–1	84	83
	Over 1	10	9
Moisture, %		4	4
Extract, fine grind, as is, %		77.3	75.2
Extract, coarse grind, as is, %		76.0	73.8
Diastatic power		133	142
α-Amylase		51	55
Total protein, %		13.7	13.8
Soluble protein, %		5.8	5.6

Figure 20. A view of a taste panel. Courtesy of Miller Brewing Co.

of the barley used, the extent to which the barley is germinated, how well it was modified or malted, and how well it will perform in the brewery. A typical analysis of a two-row and six-row malt is shown in Table 9.

The assortment indicates the size and grade of the barley used to make this malt. Uniformity in size is as important as absolute values. The growth shows how many dead kernels there were in the barley and how extensively and how uniformly the barley germinated.

The extract values, which give the amount of soluble matter in the malt, is an important economic factor. The difference in extract between a coarsely ground and a finely ground malt indicates how well the barley was modified; the smaller the difference the better the modification.

The diastatic power and α-amylase value measure the β-amylase and α-amylase, respectively. The ratio of soluble protein over total protein indicates how well the barley grew during malting and can predict yeast behavior in the brewery.

Hops are specified by variety and country or state in which they are grown. Samples from lots that could be purchased are examined visually by the brewer. Acceptance depends on aroma and appearance. Analysis for humulones accompanies the sample and may be a criterion also.

Corn and rice are judged by extract values and low oil content. In the brewery, control begins with the wort produced in the kettle. This is judged for solids content, color, bitterness, pH, degree of fermentability, and sterility.

The fermentation process is monitored for rate of fermentation (decrease in specific gravity) and temperature and biologically for the presence of lactobacilli. The aging process is checked for gravity and microbiologic state.

After filtration, the oxygen content is monitored, as is the carbon dioxide level. Just before packaging, and possibly again in the package, the beer is analyzed completely to maintain standards and uniformity.

Tasting is an integral part of quality control. In many breweries beer is tasted just before it is transferred from one tank to another. It is again tasted just before packaging, and again, critically, in the package. Taste panels are selected from brewery workers and are trained to detect certain flavors in beer and to describe them. These panels operate daily, or twice daily, with rotating membership. A view of a taste panel is shown in Figure 20.

WORLD BEER PRODUCTION

Beer is made in almost every country in the world. It probably ranks next to bread in the ubiquity of its produc-

Table 10. World Beer Production in 1989 (in 1,000 hL)

	1989
The Americas	
United States	233,619
Brazil	55,000
Mexico	38,677
Canada	22,710
Colombia	18,000
Venezuela	11,000
Argentina	6,100
Peru	5,400
Cuba	3,333
Chile	2,765
Ecuador	1,700
Dominican Republic	1,467

Table 10. (*continued*)

	1989
Bolivia	1,000
Jamaica	1,000
Paraguay	1,000
Panama	980
Guatemala	900
Costa Rica	809
Honduras	738
El Salvador	680
Uruguay	650
Puerto Rico	592
Trinidad	450
Nicaragua	300
Netherlands Antilles	165
Guyana	120
Barbados	120
Surinam	117
Santa Lucia	85
Bahamas	84
Haiti	70
Martinique	63
Belize	51
St. Vincent	38
Guadeloupe	30[a]
Grenada	30
St. Kitts	18
Total	*409,661*
Europe	
West Germany	93,200
UK	60,140
USSR	66,000[a]
Spain	27,200
East Germany	24,800
Czechoslovakia	22,684
France	20,900
Netherlands	18,813
Romania	14,000
Belgium	13,168
Poland	12,380
Yugoslavia	11,107
Italy	10,383
Hungry	9,388
Austria	9,200
Denmark	8,600
Bulgaria	7,000[a]
Portugal	6,810
Ireland	6,401
Sweden	4,586
Switzerland	4,133
Finland	3,914
Greece	3,700[a]
Norway	2,205
Luxembourg	611
Malta	163
Albania	100[a]
Iceland	66
Total	*450,650*
Africa	
South Africa	21,000
Nigeria	7,000
Cameroon	4,736
Kenya	3,900
Zaire	3,173
Zimbabwe	1,800
Ivory Coast	1,360
Zambia	945

Table 10. (*continued*)

	1989
Burundi	916
Gabon	800
Rwanda	710
Ethiopia	706
Congo	700[a]
Angola	700
Ghana	614
Tansania	538
Upper Volta	500[a]
Morocco	500
Namibia	498
Egypt	460
Togo	438
Botswana	415
Tunisia	400
Algeria	366
Central African Republic	325
Lesotho	313
Mozambique	300
Madagascar	250
Swaziland	213
Benin	206
Mauritius	180
Senegal	163
Malawi	150[a]
Liberia	143
Reunion	135
Uganda	130
Chad	113
Niger	100[a]
Mali	80[a]
South Yemen	60
Seychelles	53
Sierra Leone	43
Guinea	20
Gambia	16
Total	*56,160*
Far East	
Japan	61,005
Peoples Republic of China	60,000
Philippines	13,650
South Korea	10,500
Taiwan	4,000[a]
Vietnam	2,000[a]
Thailand	1,900
Hong Kong	1,666
India	1,723[a]
North Korea	1,000[a]
Indonesia	992
Malaysia	709
Singapore	441
Iran	100[b]
Mongolia	100[a]
Sri Lanka	71
Burma	50[a]
Nepal	50
Laos	10
Pakistan	9
Bangladesh	5[a]
Cambodia	5[a]
Total	*160,006*
Near East	
Turkey	2,500
Israel	700
Iraq	400[a]

Table 10. (continued)

	1989
Cyprus	314
Lebanon	130[a]
Syria	90[a]
Jordan	37
Total	4,171
South Pacific	
Australia	18,700
New Zealand	3,890
Papua-New Guinea	494
Figi	170[a]
Tahiti	121
New Caledonia	59
Samoa	54
Solomon Islands	—
Total	23,488

[a] Estimate
[b] Nonalcoholic

tion. Table 10 lists of the production, in 1,000 hectoliters, in these countries.

GLOSSARY

Apparent extract. The density, usually measured with a hydrometer, of a fermenting or fermented liquid. In beer it measures the extent to which the wort has fermented. The value reflects the residual solids in the beer, modified by the alcohol present (which lowers the reading).

Barley malt. A legally required ingredient in beer; it is barley that has been wetted and induced to germinate and then heated to stop the germination. The germination partly digests the starch and protein in the barley and produces some enzymes. The temperature at which the terminal drying is done determines the color of the resulting beer. Barley may be either two-rowed or six-rowed and may be either blue or white. Malt provides carbohydrates, protein, and yeast nutrients to the brewing process.

Bitterness unit or isohumulones. A quantitative measure of the amount of isohumulones—the major bitter compound from hops—in beer, in milligrams per liter.

Corn grits, corn flakes, corn starch, corn syrup, rice, milo, and sorghum. Ingredients used as adjuncts to dilute the barley malt; they are only a source of carbohydrates.

Enzyme. A protein, made by a living cell or organism, that catalyzes or promotes a chemical reaction. Most of the processes of life, and most of those in brewing, are mediated by enzymes.

Hops. The flower of a perennial vine *Humulus lupulus* grown in the United States only in Washington, Oregon, Idaho, and California. Many varieties exist, with differing aromas and bittering qualities. The United States grows the varieties called Clusters, Northern Brewer, Fuggles, Bullion, and Cascades.

Krauesen. A stage after primary fermentation in which about 15% by volume of freshly fermenting beer is added to beer that has completed its fermentation. The term, as a noun, is also used to describe the foamy head that appears on top of a fermenting tank at the height of fermentation.

Original gravity. The concentration of solids in the wort from which beer is made. In the United States and most other countries, the concentration is expressed in percent solids, or °Plato or °Balling (after the workers who laboriously determined these values). Thus a wort with 12% solids would be said to have a gravity of 12°P (or 12°B), and the beer made from it would be said to have an original gravity of 12°P. An analysis of beer permits you to calculate the original gravity of the wort from which it was made, without reference to the wort. In the United Kingdom, the concentration of solids in the wort is expressed as the actual specific gravity. A wort with 12% solids would have a specific gravity of 1.048 and would be said to have a gravity of ten forty-eight.

Real extract. The actual percentage of solids in a beer, determined after removing the alcohol present.

Ruh. The period of rest between fermentation and filtration during which the beer clarifies itself and loses some harsh notes.

Wort. The liquid from which beer is made; a warm-water extract of malt and other grains.

Yeast. The microorganism that converts simple sugars to alcohol and carbon dioxide. Two types of yeast are used in brewing—*Saccharomyces carlsbergensis* for lager beers and *S. cerevisiae* for ales. Most brewers have their own strain of these yeasts. During a normal brewers fermentation, yeast multiplies three- to fourfold, yielding more than enough yeast for successive brews. Yeast also produces many compounds that have a marked influence on the flavor of the beer.

BIBLIOGRAPHY

Briggs, D. E., J. S. Hough, R. Stevens, and T. W. Young, *Malting and Brewing Science,* Vol. 1, *Malt and Sweet Wort;* Vol. 2, *Hopped Wort and Beer,* Chapman and Hall, New York, 1981.

Broderick, H. M., *The Practical Beer,* Master Brewers Association of the Americas, Madison, Wis., 1977.

Broderick, H. M., *Beer Packaging,* Master Brewers Association of the Americas, Madison, Wis., 1982.

Burgess, A. H., *Hops,* Interscience Publishers, New York, 1964.

European Brewery Convention, Fermentation and Storage Symposium, *E.B.C. Monograph V,* European Brewery Convention, Amsterdam, The Netherlands, 1978.

European Brewery Convention, Symposium on the Relationship Between Malt and Beer, Helsinki, *E.B.C. Monograph VI,* European Brewery Convention, Amsterdam, The Netherlands, 1980.

European Brewery Convention, Flavour Symposium, Copenhagen November 1981, *E.B.C. Monograph VII,* European Brewery Convention, Amsterdam, The Netherlands, 1981.

Findlay, W. P. K., *Modern Brewing Technology,* Macmillan, London, 1971.

Pollock, J. R. A., *Brewing Science,* Vol. 1, 1979; Vol. 2, 1981; Vol. 3, 1987, Academic Press, Orlando, Fla.

Preece, I. A., *The Biochemistry of Brewing,* Oliver & Boyd, London. 1954.

Rose, A. H., and J. S. Harrison, *The Yeasts,* Vol. 2, Academic Press, Orlando Fla., 1987.

Strausz, D. A., *The Hop Industry of Eastern Europe and the Soviet Union,* Washington State University Press, Pullman, Wash., 1969.

JOSEPH L. OWADES
Consultant
Sonoma, California

BEVERAGES, CARBONATED. See CARBONATED BEVERAGES.

BIOASSAY. See ANTIBIOTICS IN FOODS OF ANIMAL ORIGINS; entries under VITAMINS.

BIOTECHNOLOGY. See CANOLA; ENZYMOLOGY; MARINE ENZYMES; MOLECULAR MECHANICS: COMPUTER MODELING OF FUNCTIONAL PROPERTIES; PLANT BIOTECHNOLOGY; TRANSGENIC ANIMALS; YEASTS; entries under FOOD SCIENCE AND TECHNOLOGY; GENETIC ENGINEERING.

BISCUIT AND CRACKER TECHNOLOGY

A modern definition of biscuits would have to be considerably different than the one contained in most standard dictionaries if it is to encompass the wide range of products understood by the baking industry and consumers to be included in that category. In the United States the biscuit is a chemically leavened bread roll, is generally circular in outline and flat in profile, and is rather similar in composition to some British scones. This product is sometimes called a baking powder biscuit. In the UK and most of the rest of the English-speaking world biscuits are the two types of product called crackers and cookies in the United States. The former are nonsweet products used like bread while the latter items include a vast array of dessert foods characterized mainly by being baked in small pieces (usually less than an individual serving) and having a texture or consistency that is drier, chewier (or crisper), and denser than most cakes. An important characteristic of biscuits is that they usually have a much longer shelf life than moister baked products such as bread and cake.

Except for their lower moisture content, biscuits are similar in composition to breads and cakes, and the range of ingredient percentages in biscuit formulas will be found to overlap with formulas for breads and cakes. Manufacturing processes often involve procedures and equipment that are not much different from those used for other bakery products, although the forming and depositing devices may be highly specialized.

INGREDIENTS

Flour

The principle structure-forming ingredient used in biscuit doughs is wheat flour. Very few cookies or crackers are made without any wheat flour, and these are usually quite atypical in organoleptic characteristics. Vital wheat gluten can replace wheat flour to a limited extent.

There are many different kinds of wheat flour, and the specifications of this ingredient must be carefully chosen if it is to impart satisfactory machining properties to the dough and desirable appearance and eating quantities to the finished biscuit. Cracker doughs require a relatively high protein content flour made from hard red winter wheat or spring wheat, whereas some cracker sponges and most cookies, will be based on a soft wheat flour or a low protein fraction from hard wheat.

Soft wheat flours suitable for biscuits may vary in protein content (mostly gluten) from 7.0 to 7.5% (for cookies) to 10% or more (for cracker sponges). They may range from 50% extrashort patent cake flour to intermediate and long patents, straights, stuffed straights, and a variety of clears. They may be unbleached, as in cookie or pastry flours, or heavily bleached to a pH range of 4.5–4.8 for certain specialty items such as moist cookies. The cracker sponge flours may contain added malt (wheat or barley).

Specifications for cookie flour will usually include requirements for protein content, moisture content, ash, color, particle size, starch damage, pH, odor, and flavor. Tests will be required to reveal contamination by insects, fungi, pesticides, and other unwanted materials. It is also common to specify certain rheological tests, which are expected to correlate with dough response to processing conditions. Finally, the flour must yield cookies of specified characteristics when it is used in a dough that is prepared and baked under standardized conditions.

Flours other than wheat can be used in crackers and cookies. Rye flours are used primarily in snack crackers to give a typical color and flavor, or, occasionally for economic reasons. Corn flour is a fairly uncommon additive, but may be found in some snack crackers. Sorghum flour, cottonseed meal, soybean meal, triticale flour, barley flour, and other cereal and noncereal powders have occasionally been suggested as additives for enhancing nutritional or other properties of cookies and crackers. These nonwheat ingredients have structure-forming properties that vary from nil to poor and must be supplemented with wheat flour or vital gluten for most purposes. Rice flour has been used to make cookies suitable for persons who are allergic to wheat gluten; it is incapable of forming a typical dough.

Shortenings

Shortenings are essential components of most crackers and cookies. The type and amount of shortenings and emulsifiers in the formula affect both the machining response of the doughs and batters and the quality of the finished products. Coatings (such as chocolate) and fillings (such as sandwich crèmes) are dependent on fats and oils to furnish the structural part of the adjunct.

Natural fats and oils suitable for shortenings include butter, lard, beef fats, and vegetable shortenings. The latter materials are purified and modified soybean, cottonseed, coconut, palm, and corn oils. The processing of shortenings may include some or all of the steps of refining (to remove contaminants), deodorizing, winterizing (to remove high–melting-point fats), bleaching, hydrogenation, fractionation, and blending. The purpose of these operations is to yield a shortening that is bland in flavor, essentially colorless, and free of aromas; has a melting point and solid fraction index within the desired range for the application for which it is destined; and has other characteristics desired by the purchaser. Butter, and to a lesser extent lard and beef fats, have natural flavors and physical qualities that are considered desirable in some applications and that are normally retained through their processing.

The function of shortenings in baked goods is primarily to modify the physical properties or texture of the finished product, making it more tender or flakier and, in some cases, giving it a glossier, more appealing appearance. Shortenings may also affect the rheological characteristics of the doughs.

Sweeteners

All cookies and many crackers contain some form of sweetener. The quantity of sweetener is usually such that it has significant effects on the texture and appearance of the product, as well as on its flavor. Machining properties and response of the dough piece to oven conditions are also related to the type and quantity of sweetener employed. In fermented goods, sugars serve as substrates from which yeast and the other microorganisms elaborate carbon dioxide and the flavoring substances characteristic of these products.

Commercial sucrose, cane or beet sugar, functions not only as a nutritive sweetener, but also as a texturizer, coloring agent, and as a means of controlling spread during baking. In marshmallow, jellies, and other adjuncts of relatively high moisture content, sucrose has the valuable property of delaying microbiological spoilage when it is present in a high enough concentration.

Sucrose can be obtained in various particle sizes or in the form of syrups. When added to the doughs as granulated or powdered sugar, its particle size influences the dimensions of the finished cookie by controlling the extent to which the dough piece spreads as it is heated.

The syrups are easier to handle, being adaptable to fluid transfer systems of pumps, pipes, valves, and meters, but they occasionally cause sanitation problems or undergo spoilage. They can also crystallize in the pipes, causing problems that are difficult to correct. Pure sucrose will not form a stable aqueous solution of greater than about 67% concentration at room temperature, and this is not enough to insure resistance to all forms of microbiological growth. If the sugar is partially inverted, more concentrated solutions can be prepared and they are more resistant to spoilage.

Invert sugar solutions are sweeter than pure sugar solutions of the same concentration and are less likely to crystallize. The solids portion of these syrups consists of about equal quantities of fructose and glucose, except that commercial invert sugar solutions will almost always include a significant percentage of sucrose. If the syrup is prepared by inverting sucrose, rather than from corn syrup, readily flowable liquids of up to 76% solids content can be obtained.

Corn syrups are sweeteners prepared by hydrolyzing corn starch with acids and enzymes. The standard types contain glucose and varying amounts of maltose and larger glucose polymers and other carbohydrates, plus minor amounts of impurities. In preparing the high-fructose varieties, manufacturers isomerize part of the glucose to fructose. Because fructose is sweeter than glucose, a substantial improvement in flavoring power results. High-fructose corn syrups can be substituted for invert sugar syrups in many applications, but they are not identical.

Regular corn syrup is classified into four types on the basis of dextrose equivalent (DE), an analysis for reducing sugar content. Total solids content is another specification of importance. Syrups and cookies that contain a high reducing sugar content have a more improved sweetness than cookies with a similar content of low DE syrup. In addition, the browning of the crust and, for that matter the interior of the cookie, is facilitated by high contents of reducing sugar from any source. Low DE corn syrups can increase the apparent viscosity of doughs and batters and make the finished product chewier. It is helpful in firming marshmallows and meringues.

Molasses is a concentrated sugar syrup containing some of the impurities, or flavoring and coloring materials, resulting from the sugar-refining process. The reason for using molasses in cookies is to obtain the typical color and flavor it provides. There are many cookies for which molasses is the chief characterizing ingredient. Molasses is available in various degrees of darkness, and the browner the color the more bitter, less sweet, and more impure the product is. True blackstrap, the final residue of the refiners' art, is not suitable for human consumption.

Commercial brown sugar is granulated sugar which has been coated with a small percentage of molasses. No commercial brown sugar available in this country is made by removing partially refined granulated sugar from the refining process. There are, however, partially refined sugars available from foreign refiners, for example, Demarara, that occasionally provide some economic advantages over pure corn or beet sugar.

Honey is primarily an invert syrup with various impurities that give typical flavors (different flavors for different types) and some color. Various concentrated fruit juices, principally grape and apple, treated to have low flavor and light color, are being offered as natural sweeteners. They are more expensive than sucrose or corn syrup.

Caramel color is a widely used brown pigment made from acid- or alkali-treated corn syrup (or sugar syrup). It is used in many types of cookie because it is relatively inexpensive, gives a true brown shade, and contributes practically no flavor to the finished product. It is not sweet.

Synthetic sweeteners such as saccharin and aspartame can be used in certain dietetic biscuits, but the problem with biscuits is that the sweetener constitutes a considerable part of the bulk and contributes to the texture of the cookie, so that some kind of bulking agent must be used to replace it, and this is not always practical.

Leavening Agents

A leavening agent is a substance or system that expands or lightens the dough or batter at some stage of its processing. The leavening effect is absolutely essential to the formation of a finished product having the appearance and eating qualities that are required by consumers. A totally unleavened product would be a solid block of material having a hard texture quite unexpected in a bakery product. Leaveners familiar to every baker are baking powder and yeast. Air and water vapor are also leaveners in that they cause expansion of the product during baking, provided the dough or batter has a structure capable of retaining the gas. Ammonium bicarbonate is a chemical additive that has leavening and other effects on cookie doughs.

Yeast acts on certain sugars to form alcohol and carbon dioxide. It is the latter compound that expands and modifies bread doughs during their fermentation stages. It is essential to proper leavening of all types of goods that the dough or batter have physical properties enabling it to entrap the gas and prevent it from escaping to the atmosphere. In the case of bread dough, gluten from the flour absorbs water to form extensible membranes that inflate like microscopic balloons to decrease the density of the dough mass. The meringue, which is the basis of angel food cake, relies on egg albumen to initially entrap the air, which is beaten into the mixture; it then becomes dependent on the strengthening effect of wheat gluten in the later stages of mixing and baking.

Few cookies depend on yeast leavening, but sponges for soda crackers, saltines, and certain snack crackers require a long fermentation period to develop their typical flavor and the acid that will later react with sodium bicarbonate to yield additional leavening gas. Sponges are flour, water, and yeast mixtures (sometimes other ingredients are added) that are fermented for a rather long time before being mixed with the rest of the cracker ingredients.

The great majority of cookies contain some sort of chemical leavening system such as baking powder, which is a mixture of sodium bicarbonate and an acid-reacting compound such as cream of tartar (potassium acid tartrate). In addition, diluents such as starch and other additives will be included in commercial baking powders to standardize their strength and delay storage deterioration. When developing cookie dough formulas, most technologists use sodium bicarbonate and acid-reacting materials in proportions adjusted to meet the pH and acid potential of the basic dough, rather than rely on premixed baking powder. The goal is usually to get a finished product with a pH close to neutrality, although this generality may be modified for special products.

Ammonium bicarbonate is a self-reacting leavener that decomposes as the dough becomes hot, giving off ammonia and carbon dioxide. It is acceptable only for products baked nearly to dryness, because water tends to entrap the ammonia, giving high-moisture cookies an unpleasant aroma. It can be very effective in increasing spread.

Many cookies undergo an increase in volume (decrease in density due to leavening) that is quite small compared to that of bread. For example, shortbread cookies of the traditional type may contain no added soda and do not seem to increase in size during baking. An examination of their interior shows, however, that some expansion has taken place as a result of air mixed into the dough and of the water vapor evolution occurring as dough temperature approaches the boiling point.

Other Ingredients

Ingredient water can have a significant effect on dough properties, but most plants soon develop formula modifications that offset any undesirable effects that might result from use of the water from their particular supply system. The water from any potable source, such as the municipal pipelines or a well, can be regarded as legally suitable for incorporation in a biscuit formula, but it should not be necessary to say that any water that has an undesirable odor or flavor should be subjected to a purification process before it is mixed into a dough. The pH of the ingredient water supply, especially from municipal sources, can in some cases vary widely throughout the week or day, and this can cause the physical properties of the dough to change, adversely affecting the machining of this material.

Milk and eggs are not common ingredients in cookies, partly because they are relatively costly. When present, they are likely to constitute only a small percent of the product. Eggs have a substantial effect on the physical properties of the dough or batter, acting as structure formers, leaveners (due to entrapment of air and water vapor), emulsifiers, and lubricants. Egg whites are better structure formers and leaveners, whereas yolks are more effective as emulsifiers and lubricants. Eggs do affect flavor, not always positively, and yolks add color, usually a desirable result. Nonfat milk has ambivalent results on structure, while whole milk generally has a weakening effect due to the fat. Both milk and eggs add good-quality protein to the biscuit, although in the present atmosphere of cholesterolphobia their nutritive appeal may be, on balance, negative.

Salt is an important ingredient. Not only is it an essential flavoring ingredient in any product intended for general distribution, but it has effects on the physical characteristics of doughs. Some of these results may be favorable, such as the slight toughening effect it has on some doughs, and others are considered unfavorable, such as its retarding effect on yeast-leavening systems. The large granules of topping salt are essential on saltines, snack crackers, and pretzels. Particle size and shape are important parameters in salt specifications.

Crackers and especially cookies have a distressing tendency to turn rancid because of their relatively high fat content, their low moisture content, and their fairly long shelf life. Antioxidants can counteract this tendency to a considerable degree. Common antioxidants are butylated hydroxyanisol (BHA), butylated hydroxytoluene (BHT), and tert-butylhydroquinone (TBHQ). Amounts that can be added are restricted by federal regulations. Citric acid and phosphoric acid are sometimes useful for chelating metal ions that would otherwise accelerate rancidity.

There are so many characterizing ingredients such as dried fruits, nuts, seeds, spices, cocoa, colors, flavors, etc. that it is useless to attempt to discuss these materials here. A full discussion of all the materials useful in cookie and cracker formulations has been published (1).

FORMULAS AND PROCEDURES

In general, it can be said that biscuit dough consist of a structure-forming ingredient such as flour, a sweetener such as sugar, a texture-modifier such as shortening, a leavener to expand the product during baking (thereby improving both texture and appearance), and various characterizing ingredients such as flavors, colors, nuts, and fruits. Saltines and the like will have small amounts of sweeteners and are normally leavened with yeast. Sweet biscuits (cookies) will contain moderate to large amounts of sugar and shortening and are leavened mostly with sodium bicarbonate systems or, less frequently, with ammonium bicarbonate; they may be combined with adjuncts such as icings, fillings, fruits, nut pastes or pieces, and chocolate.

Cookie formulas are generally classified according to the kind of equipment used to form the individual pieces. Stamping machines, rotary cutters, rotary molders, wire-cut machines (and similar extruding devices), and depositors are used for more than 90% of the cookies produced in the United States. The type of equipment being used sets limits on the rheological qualities and, therefore, on the composition of the doughs.

Deposit cookies are the machine-made counterparts of hand-bagged cookies and many published formulas for the latter can be easily adapted to factory requirements. General rules are that deposit cookies should contain (on the basis of flour as 100%), about 35–40% sugar, 65–75% shortening, and 15–25% liquid whole eggs. The flour should be milled from soft wheat, and it should be unbleached, with 8–8.5% protein.

Although wire-cut cookie equipment is rather tolerant as far as dough consistency is concerned, there are still some rheological requirements that must be observed. The doughs should be sufficiently cohesive to hold together as they are extruded through orifices, and yet they must be relatively nonsticky and short enough so that they separate cleanly when cut by the wire. These formulas may contain several times as much sugar as flour, and shortening up to 100% of the flour. Doughs may be almost as soft as some cake batters or too stiff to be easily molded by hand. The softest wire-cut doughs overlap deposit doughs in consistency, while the other extreme is close to the consistency of some rotary molded doughs.

Whatever the mixing procedure, it must be sufficient to cause a uniform distribution of ingredients. Many biscuit doughs are rather tolerant in the amount and type of mixing that will yield satisfactory performance and products, but saltines and the like must be developed (or brought to a stage of maximum strength) by the mixing process. For high-sugar formulas a preliminary creaming stage, in which the sugar and shortening are beaten together for a time before the rest of the ingredients are added, is said to insure uniformity and cause a finer grain in the finished cookie. Most lump-type ingredients (eg, chocolate chips) are best added at a late stage of the mixing.

Brownies are one of the few kinds of cookies that can be baked in continuous sheets on a oven band, then cut into rectangular pieces during or after cooling. Brownies are made with high proportions of invert syrup and other hygroscopic ingredients so that they are soft and chewy when fresh and retain this texture fairly well over a period of several weeks when properly packaged. The dough, which is often quite soft, is extruded directly onto the oven band as a sheet of uniform thickness and no other forming operations are performed.

The manufacturing process for crackers and saltines consists of the following steps: weighing the sponge ingredients, mixing the sponge, fermenting the sponge, mixing the sponge with the remainder of the ingredients, laminating the dough, sheeting the dough to a required thickness, cutting and embossing the dough sheet, removing scrap, sprinkling salt crystals on the dough, baking, breaking the dough sheet into whatever multiples are required for packaging, wrapping, and packaging. The manufacturing steps for yeast-leavened snack crackers will resemble the foregoing sequence to a considerable extent, with variations depending on the need for special shapes or content of exotic ingredients.

Stamping machines (reciprocating cutters) and rotary cutting machines must be fed in continuous sheets of dough. An important requirement of these machines is that the scrap that is generated must be removed in one piece if the operation is to be efficient. Also, the sheet thickness must be held within a narrow range so that the weight of the dough pieces will not vary significantly. The dough must have some elasticity and be cohesive enough to bear its own weight if it is to form a sheet that will retain its continuity and not tear. Excessive elasticity creates problems with shrinkage of the pieces and difficulties in maintaining uniform piece weight. To obtain these characteristics, the dough must contain a substantial amount of wheat flour to provide the gluten that will give the dough strength and elasticity. In processing, the dough must be developed by a mixing technique that orients the gluten molecules or must be repeatedly sheeted and layered. The content of ingredients that weaken the dough, such as sweeteners and shortenings, must be kept relatively low. Moisture content, for all practical purposes the amount of ingredient water added, should be enough to allow full hydration of the gluten without weakening the dough excessively.

Nearly all biscuits are baked. A few recipes for fried cookies and crackers can be found in the literature, but these do not appear to have been commercialized to an appreciable extent. Other types of cooking are not used.

EQUIPMENT

The type of equipment that can be used to process a given biscuit dough is highly dependent on the consistency of the dough. The types of dough that can be processed satisfactorily with one or more of the kinds of available equipment include everything from soft fluffy batters and wet flowable mixtures to very stiff claylike masses and tough but elastic doughs resembling bread doughs in many respects.

Mixers

An extremely wide range of mixing machines are used for preparing biscuit doughs and batters. The elastic cohesive doughs used for soda crackers and the like can be developed in the same type of horizontal mixers commonly employed for bread and roll doughs. Vertical mixers with paddle agitators are used for batch mixing of many kinds of cookie doughs. Dense, claylike doughs, such as used for some rotary molded cookies, may have to be mixed in sigma-bladed horizontal mixtures with high-horsepower motors and sturdy construction.

Laminators

Many of the doughs processed through cutting machines first undergo multiple sheeting and folding steps. The basic principle is that of forming a relatively thin sheet of dough by passing it between two rollers (several pairs of rollers may be present in one sequence), folding this thin sheet so there is a stack of several layers, then sheeting the stack so the various layers are incorporated together. This laminating property gives the dough special properties that enhance its performance in the cutting machine and give the final product special characteristics. If a saltine is examined closely, traces of this layering process will be found. Rather complicated machines have been developed to do this laminating process.

Forming Equipment

Each type of forming equipment has been designed to process doughs or batters falling within a particular range of physical characteristics. The action or method of shaping the dough of necessity sets certain limits on the appearance of the products. It is convenient to separate this discussion into cookie equipment and cracker equipment, because the machines used in forming these two types of product are quite different.

Cookie Equipment

The three broad types of forming or shaping device that are used to manufacture the great majority of commercial cookies are (1) extruders that push the dough or batter through a constricting orifice, exemplified by deposit machines, bar presses, and wire-cut equipment; (2) rotary molders that shape the dough in die cavities cut into the surface of a metal cylinder; and (3) stamping machines or rotating cutters that cut shaped pieces from continuous sheets of dough. The last type will be described below.

A simple rotary molding machine consists of a hopper, a feed roll, a cylindrical die, a knife or scraper, a cloth or woven plastic belt (also called a web or apron), and a rubber-covered compression roller. There will also be a frame, motors, controls, etc. These machines may be permanently affixed to the oven frame or they may be constructed as demountable attachments for cutting machine lines. They may also be mounted on casters so that the equipment can be moved out of the way when some other forming device is being used.

The curved surface of the metal die cylinder is covered with engraved cavities having the shape desired for the dough piece. Alternatively, plastic die cavities may be fastened to the surface. Pressure is exerted on the dough in the hopper by one or more grooved rollers near the bottom of the hopper. This causes the dough to be forced into the die cavities as the die roller rotates beneath a slit at the bottom of the hopper. The cylinder rotates past the knife or scraper (which forms the flat bottom on the dough piece) until it contacts the cloth belt passing beneath it. The belt is forced against the die by a rubber-covered roller, which can be adjusted to vary the pressure. As the cylinder lifts off the belt in its continuing rotation, the dough piece adheres to the belt more strongly than it does to the cavity and is, therefore, drawn out of the die. The dough piece is carried by the belt to the oven band where transfer is completed by various kinds of simple mechanisms. Two critical features are the necessity for completely filling the cavity and for getting more adhesion to the cloth than to the die. There are limits to the adjustments that can be made on the machine to achieve these operational essentials, so the physical characteristics of the doughs can vary only slightly.

Extruders vary widely in complexity, from simple equipment consisting of a hopper fitted with feed rollers that press the dough through adjustable slits to complicated devices that extrude deposit cookie batters through orifices moving in predetermined patterns. The most common type of machine consists of a hopper with one or more feed rollers that force the dough through an array of tubes usually called die cups. These dies may have orifices of different shapes: square, round, oval, scalloped, etc. In wire-cut machines, disks are sliced from the continuously extruded cylinder of dough and allowed to drop onto the oven band or transfer belt.

In deposit machines, the batter is extruded intermittently through shaped nozzles. Separation into cookie-size portions is achieved by lowering the oven band during the time the extrusion is stopped, this causes the batter on the band to be pulled away and separated from the batter still in the extrusion orifice. Then the band is moved up toward the nozzle and extrusion begins again. These machines are relatively simple, but they are also quite demanding in regard to dough characteristics. In more advanced machines the nozzles can be moved to form various patterns such as curves, wavy fingers, swirls, and circles. A second depositor can be synchronized with the first one to put jelly or some other filling on top of the cookie.

A bar press machine (sometimes called rout presses) extrudes continuous strings or strips of dough directly onto the oven band. Separation of these bands into individual cookie pieces can be made either before or after baking by a cutting device. The die plate may be inclined in the direction of the extrusion, so that the ribbon is supported for a longer period of time, an arrangement that reduces breaking or thinning of the dough strand due to gravitational pull. The die orifices are usually slots with a straight lower edge to give a flat bottom to the cookie and a grooved top edge to give a ribbed upper surface. Height of the strip can be varied by moving one of the slot edges.

Wire-cut machines represent an advance in complexity over the bar press and deposit machines in that they include a device that cuts off pieces of the extruded dough as it is emerging from the die orifice. The cutoff device consists of a wire or blade that is quickly drawn through the dough by a reciprocating harp. The harp is simply a frame or support attached to a mechanism that moves it back and forth beneath the die cups. The dough emerges from the cups in approximately cylindrical form if round cookies are being made, but the cross section of the dough strand can be modified by changing the shape of the orifice. Several of the die cups will be held on a bar that fits snugly into a channel at the bottom of the hopper. Wire-cut machines can probably handle more types of dough than any other cookie-forming apparatus, but they cannot operate satisfactorily on elastic, extensible doughs.

Cracker Equipment

Vertically reciprocating cutters, or stamping machines, are used mostly for crackers, although embossed cookies are also made by these devices. Rotary cutters employ metal cylinders with cutting dies attached to the curved surface; two of these may be placed in tandem to perform two different operations on the same piece of dough, such as cutting the outline and docking the piece.

Ovens

It is frequently said that the biscuit factory must be designed around the oven, because the limitations of this device affect nearly every other operation in the bakery. The baking equipment found in nearly all large biscuit factories will be some form of band oven, a type of oven that is ideally suited for rapid and uniform processing of small pieces of dough. Small operations may use reel ovens, deck ovens, rotary ovens, or just about any other kind of baking equipment. This discussion will concentrate on the band oven.

The band oven consists of an endless steel belt passing through a baking chamber that can be heated directly or indirectly. The belt, or band, may be a solid strip of steel or a band of woven steel wire. The earlier mesh bands had a fairly open weave, but most of the recent versions have a closely woven structure. Perforated steel bands and wire mesh bands allow steam to escape from the bottoms of dough pieces and help to prevent gas pockets and other distortions. They also tend to retard spread.

Most steel bands are low-carbon alloy steels, which have been cold rolled. The cold-working process generally increases the yield strength and tensile strength and reduces ductility. The hardened steel is tempered to restore its ductility. There is an obvious need for a straight, flat band with minimum distortion and maximum uniformity of composition. Expansion and contraction of the band caused by temperature changes is compensated for by pneumatic cylinder adjustments of the drums at each end of the oven, or by heavy counterweights on one of the drums.

The band must be of uniform thickness throughout. Solid steel bands probably average about 0.16 cm in thickness. Wire mesh bands will be nearer 0.56 cm in thickness. Relatively thin bands allow use of smaller drive drums and insure rapid response to heat input but, of course, wear out sooner.

There are large steel cylinders or drums at each end of the oven, outside the baking chamber proper, that carry the band and keep it taut. As it passes through the oven, the band rides on brackets or rollers that not only support the belt and keep it level but also nudge the band gradually into its proper track if it develops a tendency to move to one side. One of the bands is turned by a powerful, variable-speed electric motor and the other drum (at the exit end) rotates freely.

The baking chamber consists of a series of modular units, eg, 10 ft long, 7 ft wide, and 5 ft wide. Heating means are supplied above and below the band. Thick insulation covers all surfaces that are not occupied by air input and output ducts, ports, and other fixtures. Heat is often provided by ribbon burners above and below the band in direct-heated ovens. Open flame or surface combustion (ceramic) elements have been used, but the former type is more common. Gas, oil, or even diesel fuel is burned, depending on the limitations of the heating system and the market price of the fuel. Electrically heated ovens function satisfactorily, but are economically impractical in most locations.

Indirect-heated ovens burn the fuel in a chamber separate from the baking tunnel, and the hot air is transferred to the band by fans and natural convection. The length of the oven is divided into zones, which permit different temperatures to be applied to the product during different cooking stages.

Electronic ovens of both the dielectric and microwave varieties have been used to cook biscuits, but are not widely employed at the present time because of design and safety problems as well as economic disadvantages. Microwave heating has had a considerable success in finishing or drying soda crackers after they have been baked in a conventional gas oven. This form of energy application enables the manufacturer to reduce moisture in the cracker without undue browning of the cracker's surface. No doubt further successes will be achieved when plant designers learn how to take full advantage of the special features of electronic heating.

Other Equipment

Sugar wafers consist of top and bottom layers of a characteristically crisp (and rather bland) wafer sheet enclosing a layer of flavored crème consisting mainly of shortening and sugar. Sometimes several alternating layers of crème and wafer are combined in a single cookie, and the cookie may be enrobed with chocolate. The wafer sheets are baked in a specialized oven called a wafer oven. The oven consists essentially of an enclosed chamber through which pairs of plates, which can be brought together and separated as required, are transported. At the beginning of the process, a fluid batter is deposited on one of the plates, then the other plate in the pair is immediately brought near to it to form a closed chamber within which the batter rapidly expands under the influence of heat. The plates are patterned on their contiguous surfaces so the baked wafer emerges with a sort of waffle pattern. In fact, there are many points where the process resembles home baking of waffles. The baked wafers are removed from the griddle, cooled, spread with crème, topped with another wafer, cut into the desired shapes, and packed.

Sandwiching machines assemble two cookie wafers (usually of the molded type, but sometimes wire cut) and a layer of crème to form a sandwich cookie, the prototypical example of which is the Oreo. Many variations in shapes and flavors are possible; in fact, crackers can be combined with cheese or peanut butter filling to make a nonsweet snack. The cookies, called base cakes, are received at the sandwiching machines from cooling conveyors leading from the oven or from chutes that are hand fed from storage boxes. Vibrating conveyors jog the rows of cookies into a stacked-on-edge position. Half of the rows carry cookies with the top (having an embossed design) trailing while the other rows are of cookies with the design leading. Cakes are fed by the conveyor into magazines or chutes in the proper position for sandwiching; they are removed one at a time by means of double pins on double chains. As the bottom cakes travel through the machine with their embossed side downward, they receive deposits of soft crème extruded through a rotating sleeve having shaped orifices. As the bottom base cake with its crème deposit reaches the second set of magazines, the top cakes are dropped onto the crème. Then the sandwich is gently pressed together to insure adherence of the components and to establish uniform thickness of the finished cookies. A much more complete discussion of biscuit equipment has been published (2).

PACKAGING, STORAGE, AND DISTRIBUTION

Most biscuits will have moisture contents of less than 5%, some of the crisper varieties (such as saltines) averaging around 2%. Exceptions are the chewy cookies such as brownies and the cakelike cookies such as hermits, which when fresh will have moisture contents of more than 10%. Within the range of normal ambient relative humidities as experienced in the United States, the drier cookies and crackers will tend to absorb water vapor whereas the chewier cookies will lose moisture, so it is highly desirable to package all biscuits in containers having low rates of moisture vapor transfer if extended shelf lives are to be achieved.

Deterioration of cookies and crackers generally takes the form of loss of flavor or acquisition of stale flavors due to chemical changes (such as oxidation) and loss of volatiles by evaporation, and undesirable texture changes due to loss of moisture or uptake of moisture, depending on whether the biscuit is expected to be chewy or crisp, respectively. Sometimes the appearance also changes because the surface reflects and refracts light differently due to physical changes such as crystallization of sugars, solidification or liquification of fat fractions, or formation of empty vacuoles as moisture evaporates.

Microbiological changes are usually not particularly important causes of storage deterioration of biscuits, because the low water activity of the substrate strongly inhibits growth of yeasts, bacteria, and molds. Occasionally, fermentation by osmophilic yeasts will take place in adjuncts such as marshmallow and jellies, but formula adjustments to decrease the water activity of these materials, plus improvements in sanitation, can often remedy such problems. An extensive discussion of packaging and preservations methods for biscuits and other baked goods has been published (3).

CURRENT AND FUTURE TRENDS

New product introductions in the United States in recent years have emphasized health and nutritional claims (not always the same things) and high-priced gourmet variations. Health claims are faddish in nature and vary according to the enthusiasm of the moment. High fiber content seems to be one of the most popular approaches to health improvement, although increased calcium and decreased sugar are other approaches to health-with-happiness. High fiber content can be achieved with heavy additions of wheat, oat, or corn bran, or more exotic ingredients such as psyllium seed husks, purified cellulose (from wood, sugar beet pulp, etc), and pectinous materials.

Gourmet cookies are achieved by formula variations that give highly enriched cookies, heavy in chocolate, fruit preserves, butter, and the like; by decorative enhancements; and by addition of perishable components such as whipped cream, buttercream fillings, and other adjuncts that are practical only for freshly prepared items or for cookies that are distributed in frozen form. The boutique cookie shops that appeared in shopping malls and other high-traffic locations during the 1970s provide an opportunity for supplying consumers with freshly baked cookies that have characteristics different from anything that can be achieved by factory-baked and shelf-stable products.

It can be expected that future developments will follow new dietary fads as they become entrenched in the popular imagination. The exact direction cannot be predicted, but products of reduced caloric content or useful for some other reason in a dietary regimen intended to reduce body weight will always be popular.

Gourmet products will probe the most advanced frontiers of richness and elaborate design, so that the differences between candy bars, ice cream confections, French pastries, Danish, and cookies will be blurred. Various ethnic specialties, such as modifications of baklava or babas, may be found to have popular appeal.

There will always be a reliable (but limited) market for specialty cookies that look terrible and taste worse but are homemade or are claimed to be healthful, natural, or organic and these market niches will continue to be exploited with varieties that can make combinations of such claims. No doubt claims about biscuits that are environmentally correct or contain recycled ingredients will be forthcoming.

BIBLIOGRAPHY

1. S. A. Matz, *Ingredients for Bakers,* Pan-Tech International, Inc., McAllen, Tex., 1987.

2. S. A. Matz, *Equipment for Bakers,* Pan-Tech International, Inc., McAllen, Tex., 1988.

3. S. A. Matz, *Bakery Technology,* Pan-Tech International, Inc., McAllen, Tex., 1989.

General References

D. J. R. Manley, *Technology of Biscuits, Crackers and Cookies,* Ellis Horwood Ltd., Chichester, UK, 1983.

S. A. Matz and T. D. Matz, *Cookie and Cracker Technology,* 2nd ed., AVI Publishing Co., Inc., Westport, Conn., 1978.

S. A. Matz, *Formulas and Processes for Bakers,* Pan-Tech International, Inc., McAllen, Tex., 1988.

W. H. Smith, *Biscuits, Crackers, and Cookies,* Magazines for Industry, Inc., New York, 1972.

P. R. Whiteley, *Biscuit Manufacture,* Elsevier Publishing Co., London, UK, 1971.

SAMUEL A. MATZ
Pan—Tech International, Inc.
McAllen, Texas

BITTER. See Taste and odor.

BIVALVE MOLLUSKS. See Fish and shellfish products; Shellfish; Entries under aquaculture.

BLANCHING

The blanching process typically utilizes temperatures around 75–95°C for times of about 1–10 min, depending on the product requirements. It is a necessary pretreatment for many vegetables in order to achieve satisfactory quality in dehydrated, canned, and frozen products. A blanching process may be needed if there is likely to be a delay in reaching enzyme inactivation temperatures or, as in freezing preservation, if such temperatures are never achieved. The process should ensure the required reduction of enzyme activity that otherwise might cause undesirable changes in odor, flavor, color, texture, and nutritive value during frozen storage. Another major effect is the removal of intercellular gases. This reduces the potential for oxidative changes in the food and allows the achievement of suitable headspace vacua within cans. As a heat process, blanching may result in some reduction in microbial load, and texture may be improved. Vegetable matter tends to shrink because of loss of turgor, which can aid the achievement of the required fill weight. Undesirable losses of heat sensitive nutrients may be caused, and, in water blanching, soluble constituents may be leached, resulting in large volumes of effluent (1).

BLANCHING METHODS

Water Blanching

Water blanching is the most widely used method. There are four basic designs of water blancher.

(a) The tubular blancher in which particulate vegetable matter is transported by pumped water through a system of tubes for the required residence time. Direct steam injection is used to heat the water to blanch temperature, while tube length and product flow rate govern the blanch time.

(b) The rotary screw blancher in which a central screw rotates, at a speed that moves the product forward to give the required blanch time, within a static drum containing the blanch water.

(c) The rotary blancher, which comprises a drum with a scroll attached to the inner wall, rotating on trunnions.

(d) The thermascrew blancher in which the rotating central screw is hollow and contains the heating medium.

Other designs have been developed for various products, such as the countercurrent blanching tower in which product and hot water are moved in opposite directions to promote rapid heat transfer. Product characteristics vary considerably and types (b) and (c) have been used widely. For rotary water blanchers, by a relatively simple scale-up of the basic design, ie, increasing blancher length, blancher capacity is increased for a range of blanch times. Estimates of product outputs can be predicted from data based on trials, as shown by the example in Figure 1 and Table 1.

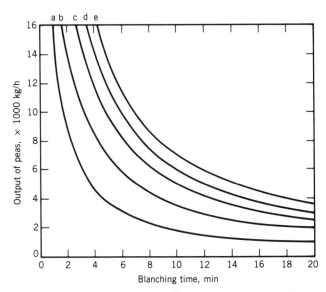

Figure 1. Typical example of relation between product output and blanch time for various products blanched in various sizes of rotary water blancher. For a blancher diameter ≈ 2 m, blancher length is approximately a = 3.1 m, b = 4.5 m, c = 5.9 m, d = 7.4 m, and e = 8.8 m.

Capital and running costs are relatively low for water blanchers, and thermal efficiency has been reported to be as high as 60% (2). Improvements in design have been directed at reducing energy and water consumption to minimize leaching and effluent. Mean heat utilization for a blancher with an average spinach capacity of 58,000 kg/h has been estimated at 39%, assuming a value of 758 kJ/kg for raw spinach leaves free from adhering water (3). Energy consumptions of a tubular water blancher and a water blancher with a screw conveyor have been measured as a basis for suggesting energy conserving modifications. The former required 0.54 MJ/kg and the latter 0.91 MJ/kg, indicating the importance of complete steam condensation. A study of water heating by heat exchange (68% efficient) and by steam injection (17% efficient) confirmed the energy savings potential of heating by using a heat exchanger to minimize the escape of steam rather

Table 1. Multiply Graph Outputs by Appropriate Factor for These Products[a]

Product	Factor
Potatoes, whole	1.15
french fried	1.00
Carrots, whole	1.10
Beans, soaked	1.08
Peas, soaked	1.08
fresh	1.00
Green beans, cross cut	0.70
whole	0.60
Cauliflower, florets	0.50
Spinach, leaf	0.30
Brussels sprouts	0.30
Cabbage, leaf	0.30

[a] See Figure 1.

Table 2. Comparison of Water Blanchers[a]

Type of Blancher	Product	Energy Efficiency, %	Energy, MJ/kg Product	Effluent Volume or Mass/Ton Product
Ordinary water blancher, calculated values	Peas	60		240–384 L
Ordinary water blancher	Peas		4.8	1.080–3.600 kg
Ordinary water blancher			4.5–6.8	400–1.800 kg
Tank water blancher	Spinach	30.7	1.5	
Ordinary screw blancher	Cauliflower	31.3	0.9	
Tubular blancher	Beans	34.7	0.54	
Pilot screw blancher with steam injection	Cauliflower	16.8	2.09	
Pilot screw blancher with heat exchanger	Cauliflower	67.6	0.51	
Screw blancher	Peas		2.09	4.000 L
Cabinplant blancher & cooling section	Peas		0.27	240 L

[a] Ref. 13.

than by steam injection (4). Energy consumption can be reduced by minimizing the flow rate of water to the blancher (5), and a comparison of energy consumptions and effluent volumes of various water blanchers is shown in Table 2.

Loss of solubles may be reduced by maximizing product to water ratio, by recycling the blanch water, and by minimizing fresh water addition. This allows the solute concentration in the blanch water to rise toward that of the vegetable cell sap, the isotonic condition thus minimizing net solute loss (6). Tissue damage may promote more rapid loss of solutes and oxidative changes, for example, in peas (7), and hence mechanically induced product movement may be significant both to nutrient quality and product break-up of the more delicate vegetables such as broccoli. The Cabinplant integrated blancher–cooker (Denmark) utilizes minimal water volumes, heat exchangers, and a design to optimize many of the above points, and is the most sophisticated water blancher available (Fig. 2) (8).

One advantage of water blanching is that processing aids such as sodium chloride and sodium bicarbonate may be added to the blanch water to obtain the preferred quality of, for example, cabbage (9). However, blanch waters can also provide a good microbial growth medium. Total microbial loads may be reduced on heating by 10^4–10^5 per gram of product, but numbers of thermoduric organisms may not be reduced. It is therefore important that blanchers be easily accessible for cleaning, be constructed of stainless steel, and that dead spaces where food splashes may collect be designed out wherever possible.

Steam Blanching

As a means of reducing leaching losses, much attention has been given to the development of steam blanchers. The first major development of steam blanching was that of the Individual Quick Blanch (IQB) system (10). This consisted of a 25-s exposure to steam of a single layer of diced carrot, followed by 50 s in a deep bed to allow equili-

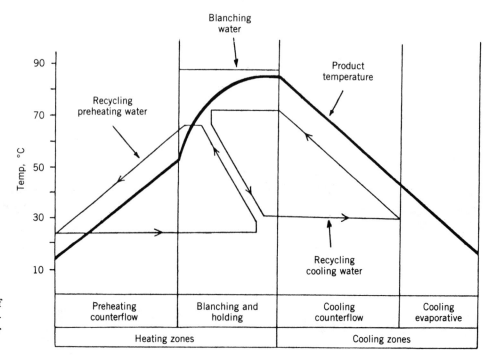

Figure 2. Representation of flow of water and temperatures in a Cabinplant integrated blancher–cooler with heat exchanger.

Table 3. Comparison of Steam Blanchers[a]

Type of Blancher	Product	Energy Efficiency, %	Energy, MJ/kg Product	Effluent Volume or Mass/Ton Product
Ordinary steam blancher, calculated values	Peas	5		191–313 L
Ordinary steam blancher	Peas		6.7–9.0	200 kg
Ordinary steam blancher	Spinach	13	2.12	
Ordinary steam blancher with water cooling	Peas			5.151 L
Steam blancher with end seals	Spinach	27	0.95	
Steam blancher with water curtains	Spinach	19	1.56	
Vibratory spiral blancher	Beans	85		28 L
Individ. quick blancher	Peas			225 L
Steam blancher with end seals and steam recycling	Spinach	31	0.91	
Steam blancher with thermocompression			0.32–0.49	
IQB-prototype K1 & evap. cooling	Peas		0.27–0.32	130–150 kg
IQB-prototype K2 & evap. cooling			0.32–0.56	75–81 kg

[a] Ref. 13.

briation of temperature throughout the carrot pieces. A venturi thermocycle blancher was described, which was claimed to reduce steam consumption by 50%, maintain a uniform temperature throughout the blanching area, and, because of the water steam traps, provide a vapor-free plant environment (11). Further development of the IQB concept involved an integrated blanching–cooling system comprising vertical helical vibratory conveyors and condensate spray cooling, although problems were encountered when handling nonparticulate material such as leafy green vegetables (12).

In spite of the improved performance of such blancher designs regarding energy and leaching, capital costs and other variable costs have been so high that the conventional water blancher has remained the most competitive system. For many large-scale operations this may no longer be true, and Table 3 compares the performance of various steam blanchers (13). The most recently developed energy efficient steam blancher is that manufactured by the Atlantic Bridge Company (ABCO), Canada (14). The principle of operation of the ABCO K2 was developed from the earlier IQB concept.

Hot-Gas Blanching

Hot-gas blanching has been studied because of the potential for reducing leaching and, in particular, for reducing effluent. One report indicated that for spinach, green beans, corn on the cob, and leeks the retention of water-soluble vitamins was much the same as for other commercial blanching operations. Excessive surface drying in some cases and the presence of atmospheric oxygen, which could promote oxidative changes, were disadvantages. Also, it was shown that operational costs could be much higher than water blanching for some vegetables (15).

Microwave and Electroconductive Blanching

A number of studies of microwave blanching have been carried out to explore the potential for reduced leaching and rapid heating, particularly for vegetables of large cross section such as potatoes, corn on the cob, and brussels sprouts (16). On an industrial scale, processing advantages have been identified, but in the past it was concluded that these were outweighed by the high electricity costs (17). Until recently microwave generators were limited to about 30 kW, with the microwave blancher incorporating steam preheating and humidity control for optimum efficiency. However, with the recent development of 60-kW generators, which also permit more uniform heating, the cost effectiveness for blanching operations needs reviewing (18). There has been limited interest to date in electroconductive blanching (19).

OBJECTIVES FOR BLANCHER CONTROL

Close control of blanch time and temperature influences uniformity of product quality as well as energy consumption. For example, where a vegetable does require blanching prior to freezing, it is now being established that the blanch treatment required depends largely on the heat stability of those enzymes directly responsible for the main deteriorative changes in a given product during frozen storage. Hence energy may be wasted if the blanching conditions are loosely controlled to inactivate peroxidase enzymes, whereas the less heat stable lipoxygenases are the relevant enzymes to inactivate (20).

Accepting that there will always be variations in the vegetable raw material in terms of the physical and thermal properties, consistently uniform blanching conditions will result in consistent blanching effects. This may be less easy to achieve where one blancher is to be used to meet the requirements for a range of vegetable products. Consistently controlled blanching also permits the prediction of the degree of enzyme inactivation (21) and of leaching losses of various solutes using suitable models for water blanching (22).

EFFECTS OF BLANCHING

Water blanchers have been shown to be the main operation contributing to the solids content of cannery wastes, and some recent studies have considered the treatment and concentration of waste blanch water. The use of water as the blanch medium also allows the addition of certain chemicals as processing aids, such as citric acid, and this is reviewed elsewhere (1).

Weight

In some instances weight losses can be excessive; for example, when blanching mushrooms, weight losses in excess of 19% and volume losses between 11 and 15% have been recorded (23). It has been shown that weight loss from vegetable tissue during water blanching occurs by two main mechanisms. In the typical temperature range of 50 to 55°C the cytoplasmic membranes that enclose the cell contents become disorganized, and on loss of turgor the cells contract and express some cell solution. Simultaneously, the damaged cell membranes allow free diffusion of solutes out of the cells. Because of continued diffusion of solutes out of the tissue during the blanch time, the net tissue weight loss also increases. The example of whole peas blanched in the laboratory at 85°C is shown in Figure 3. The kinetics of mushroom shrinkage have been described by three apparently first-order reactions (24). Artificially altering the water content of the vegetable prior to blanching will also influence subsequent weight loss, although the effect on the diffusional loss of solutes is small (25).

The blanching of dried vegetables has been studied since it was found that the standard rehydration step of around 18 h in cold water could be avoided by an extended blanch of 10–70 min at 71–85°C, depending on the vegetable. During this time sufficient water was taken up to compensate for lack of soaking, with resulting improved overall quality (26).

Nutrients

Some loss of nutrients occurs in all cases of blanching, but water blanching may promote excessive losses, typically up to 40% for minerals and certain vitamins, 35% for sugars, and 20% for protein (1). The most commonly measured nutrient is ascorbic acid because of its high water solubility, susceptibility to oxidation, and ease of analysis. Loss of vitamin C from spinach has been reported as being greater than 70% after 3–5-min blanching (27). Spinach blanched for a range of times at different temperatures was found to lose vitamin C and thiamin following first-order reaction kinetics (28). The decimal reduction times decreased as temperature increased, and Z values (temperature range over which the decimal reduction time changes tenfold) were about 65°C for both these vitamins.

Losses of vitamin C from peas during water blanching have been studied extensively. Temperatures of 35–97°C for times of 15 s to 25 min were used, and the changes in ascorbic acid and dehydroascorbic acid levels were measured (7,29). It was also found that vitamin C concentration was higher in the seed coat (30). A typical 1-min blanch at 97°C resulted in 28% loss of ascorbic acid from Dark Skinned Perfection peas. A broader study showed that at 94–99°C the K1 blanch–cool system (see also Table 3) had a profound effect on ascorbic acid content of some vegetables but not others. For example, large losses were observed for peas and broccoli but not green beans. An analysis of the beans showed that ascorbic acid was concentrated in the seeds (38 mg/100 g fresh wt), which are protected by the pod (8.4 mg/100 g fresh wt) (31). This highlights the need to study the location and distribution of nutrients within various tissues and to study how the concentrations vary with maturity and variety.

Toxic Constituents

In addition to the leaching of nutrients, toxic constituents naturally present in the vegetable may also be leached. The level of nitrates in foods has caused concern because of potential toxicity, and high levels of nitrites may cause methaemoglobinaemia in infants. Nitrate levels in spinach petioles have been shown to be as much as 5 to 10 times greater than in leaves, but water blanching subsequently reduced levels by more than 50% (32). Any nitrite formed by reduction was eliminated, and it was concluded that a combination of leaf selection and blanching could reduce nitrate levels to 25% of the original content. Nitrate levels in carrots were reported for raw carrot, 339 mg nitrate/kg; water washed at 17°C, 226 mg/kg; blanched at 65°C, 187 mg/kg; blanched at 80°C, 181 mg/kg; and blanched at 95°C, 165 mg/kg (33).

The calcium uptake from blanch water influences the ratio of soluble and insoluble oxalates (32). Although excess calcium chloride in the blanch water reduced soluble oxalate in spinach, it also adversely affected color. The influence of washing and blanching on the cadmium content of several vegetables has been studied. Contents initially were 0.121–0.379 mg % for spinach, 0.029 mg % for green beans, and 0.038 mg % for peas. Although 0.1 ppm cadmium was recommended as the maximum in the fresh weight, some spinach still contained higher levels after blanching (34).

Contaminants

Blanching can significantly reduce the level of contaminating microorganisms, which is especially important

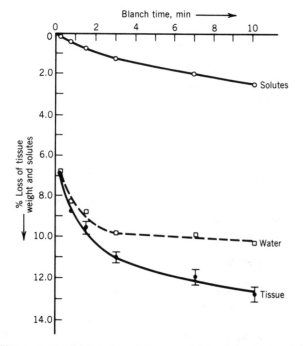

Figure 3. Percentage loss of tissue weight, solutes, and water (by difference) from whole pea samples after the given blanch time at 85°C (means of two duplicated replicates) (25).

prior to freezing or dehydration. A 3-min blanch of soy sprouts in boiling water reduced the total bacteria to ≤14,000/g, coliforms to <10/g, and salmonella to zero (35). However, the numbers of thermoduric organisms may not be reduced, and *Bacillus stearothermophilus* was isolated from blanch water used at 90°C for 5-min treatments prior to canning peas in an investigation of the cause of spoiled packs (36). Because blanching may render the vegetable more easily infected by microorganisms, hygienic handling is important.

The content of pesticide residues in vegetables has been studied recently. Commercial blanching was shown to remove 50% of DDT (dichlorodiphenyltrichloroethane) and 68–73% of carbonyl residue from green beans (37). The fate of di-syston has been studied during the processing of potatoes (38), and a 15-min blanch at 100°C reduced the levels of Aldrin, heptachlor epoxide, and Endrin in Irish and sweet potatoes (39). Blanching for 3 min at 100°C reduced levels of DDT isomers in turnip greens (40). The effects of water cooking on the pesticide residues in spinach have been studied (41).

Laboratory canning operations were highly effective in reducing strontium and cesium concentrations in beans and kale. However, blanching of sweet potatoes appeared to result in a transfer of radioactivity from the peel to the core, suggesting that skins of contaminated potatoes should be removed prior to thermal treatment (42).

Enzymes

Peroxidase has often been used as a blanching efficiency indicator enzyme because it is the most heat resistant enzyme and is easy to measure (43). Recent work has shown that a significant but not well-defined proportion of active peroxidase can be left in many vegetables after blanching and a long storage life can still be achieved (44). Less than 5% residual peroxidase activity did not affect quality during storage of carrot, cauliflower, french bean, onion, leek, and swede stored at −20 or −30°C for 15 months (20). Good quality was retained in carrots after storing at −20°C when palmitoyl-CoA hydrolase was inactivated by blanching, even though catalase and peroxidase activity was present. Although this hydrolase was a better indicator enzyme, there was no cheap, easy method of measurement. The complete inactivation of peroxidase does correlate well with the achievement of best quality in peas, but the best sensory quality of green beans was achieved prior to complete peroxidase inactivation.

A study on brussels sprouts found that the activities of polyphenoloxidase and peroxidase were related to the size of the sprouts, and increased from the center to the outer layers. Reducing activity to about 1% of the initial level required 7 min at 98°C for large sprouts, 5 min for medium, and 4 min for small. Residual activity increased toward the center of the sprout (45). The inactivation of lipid-degrading enzymes has also been shown to be very important to final sensory quality (46).

Color, Flavor, and Odor

The blanching operation promotes the thermal degradation of the blue-green chlorophyll pigments to the yellow-green pheophytins (47,48). This has been demonstrated in peas, and it has been shown that the addition of sodium carbonate or calcium oxide or alkalinization with sodium hydroxide during a 3-min blanch at 90°C increased chlorophyll stability. Addition of calcium chloride had the opposite effect. Iron and tin ions promoted, and copper and tin inhibited, chlorophyll decomposition; chlorophyll *b* was more resistant to heat than chlorophyll *a* (49).

Steam blanching of ground red peppers gave a major improvement in color in the first 10 min. This effect was greatest in material that had not been stored, in which the ripening process was not terminated. Because of changes in the conjugated system of carotenes, the red proportion increased and the yellow decreased (50). The effect of different types of blanch on the color of sliced carrots using hot water, in-can steam, direct steam, and microwaves resulted in similar color changes. These arose from changes in ultrastructure, especially in chromoplasts, from which liberated carotenes dissolved into lipids, and some were lost into the water (51).

The use of sodium acid pyrophosphate has been reported in blanch solutions, especially for potato and cauliflower to prevent discoloration during storage of the processed products. A cause of the discoloration may be the reaction between *ortho*-dihydroxyphenols with ferrous ions, forming a pigment in the ferric form on oxidation. Pyrophosphate is unique among condensed phosphates because of its strong ferric ion binding ability.

When peaches were microwaved a uniform browning of the skin resulted because the high levels of polyphenoloxidase in the outer fruit parts received the least heating effect. To prevent this a combination of microwave heating to blanch the inner parts and lye peeling to scald the outer parts was proposed (52).

Soluble or volatile flavor components may be lost during blanching. However, the resulting enzyme inactivation and oxygen removal may aid greater flavor retention during subsequent frozen storage. Unblanched carrot, cauliflower, and french bean developed off-odor after 9, 3, and 6 months, respectively, at −30°C (53). Unblanched onion, leek, and swede did not develop detectable off-flavor or off-odor, and no changes in total lipid content were found. In onion, no lipoxygenase or peroxidase has been found.

Structure and Texture

Blanching may induce both physical and biochemical changes in the structure and textural properties, depending on blanch time and temperature, and type and state of the vegetable. Blanching green beans at 90°C for 30 to 240 s did not seem to influence the degree of cell damage (54), and no tissue disruption in carrots was observed after blanching for 3 min at 100°C or cooking for 10 min at 100°C. This suggested that physiological and chemical rather than physical changes may have occurred to cause softness of the tissues. Subsequent freezing did cause cell disruption (55).

It has been observed that the effects of moisture and heating during blanching lead to a swelling of the cell walls, for example, of green bean pods (150–200%), and to a beginning of separation of the individual cells. This is attributed to the extraction of protopectin from the middle lamella. It was found that not only short blanching but also boiling to doneness preserved the histological structure well. No evidence of any cell wall rupture was observed during cooking, and it was concluded that earlier suggestions that cell wall rupture did occur probably arose through poor methodology in preparing sections (56).

The effects of boiling, steaming, and pressure cooking on the loss of pectic substances, and also the effect of water hardness on potato structure, have been reported (57). Steaming and pressure cooking gave similar results, but boiling resulted in significantly greater losses followed by a characteristic leveling off with time. Steaming probably caused less cell wall damage than boiling because of the lack of water to wash away degraded pectic substances. Boiled potatoes retained significantly more pectic substances when cooked with hard water. Surface to volume ratio was increased by slicing, thus increasing the rate of calcium diffusion into the tissue and of degraded pectic substance out of the tissue.

French fries have been preblanched in disodium dihydrogen pyrophosphate to sequester ions, such as calcium, and prevent discoloration, and this treatment did not reduce protopectin stability or cause deterioration of French fry texture. Low-specific-gravity potatoes were also studied because these gave a soft texture after frying. After 15 min at 70°C in 0.5% calcium chloride the best firming results were obtained, although still not equal to the firmness of high-specific-gravity potatoes (58). It was shown that blanching increased shear value and springiness but decreased hardness of asparagus. The optimum blanch times at 100°C were 135 s for stalk diameters of 16–25 mm, 120 s for 12–15 mm, 105 s for 9–11 mm, and 90 s for 6–8 mm (59).

The texture and color of beans were improved by a stepwise blanch treatment involving high and low temperatures (60). Others found that low-temperature blanching at 74°C for 20–30 min resulted in firmer carrots. Their studies confirmed the conclusion that the increase in firmness was caused by the effects of pectin methyl esterase, which was activated at low temperature and inactivated at a higher temperature (61). The rate of thermal softening of these vegetables follows first-order kinetics (62). Mung bean shoots were blanched at 75°C for 30 s to activate pectinesterase and then held at 55°C for 30 min to maximize deesterification. Such firming has been applied to tomato, potato, and cauliflower. Subsequent addition of calcium salts to the canning liquor caused insoluble calcium pectate gels in the cell walls, thus firming the cell walls (63). The noncooking effects of salts added after cooking on the texture of canned snap beans has been studied (64). Blanching may also affect the dietary fiber content of vegetables (65).

Intercellular Gases

Blanching induces the expansion and removal of intercellular gases from within vegetable tissues. This reduces the potential for subsequent oxidative changes and permits the attainment of adequate headspace vacua in cans. Air removal from the surfaces of a vegetable may change the apparent color shade. When peas of various sizes were blanched between 66 and 95°C for a few seconds up to 10 min it was found that removal of internal gases had largely occurred within the first minute. Gas volume varied from 0.47 mL/100 g grade 2 peas to 1.21 mL/100 g grade 7 peas (66).

The gas and volume changes in sliced green peppers and beans were measured during blanching in steam at 1.5 atm and in water at 100, 90, and 80°C. Initially the peppers contained 14% gas, which was reduced to 2.8–3.8% by steam or water at 100°C, but only to 3.2–5.8% at the lower temperatures. Beans initially contained 19.3%, which was reduced to 1.1–4.1% by steam or water at 100°C, and to 4.2–6.4% at the lower temperatures. Maximum gas reduction was achieved after 2 min with beans and after 1 min with peppers (67). Further data are available for red cabbage and carrots (68).

Several early blanching studies reported that steam blanching was slower than water blanching. Since the heat transfer coefficient of condensing steam is at least an order of magnitude greater than that of hot water to the same surface, it was postulated that the noncondensable gases mixed with steam accumulated at the surfaces being heated, thus interfering with the heat transfer from the steam. Blancher designs where the noncondensable gases are not swept away continually and removed from the system might be prone to a stagnant layer of gas accumulating at the surfaces, thus reducing heat transfer rate. Studies using pure steam revealed that the most important source of noncondensable gases was from the interior of the vegetables being blanched. Foaming was observed, demonstrating that heat transfer had to occur through an insulating layer of foam from the steam to the vegetable. A surface heat transfer coefficient for carrot in steam at atmospheric pressure was given as 1136 $W/m^2 \cdot K$ (69). The work showed that the rate of heat transfer was increased by vacuum pretreatment and decreased by pressure pretreatment. Results indicated that some means of degassing vegetable particles prior to steam blanching would reduce the heating time much more effectively than increases in steam velocity.

Heat and Mass Transfer

The calculation of heat transfer rates during blanching requires a knowledge of the geometry of the vegetable piece, the blanching conditions, including the initial food temperature and blanch medium temperature, and the thermal properties of the vegetable. For unsteady-state heat conduction, numerical solutions to Fourier's general law are available for the central temperature history of the three elementary shapes: an infinite slab, an infinite cylinder, and a sphere (70). Thermokinetic models for enzyme inactivation have also been studied using unsteady-state heat conduction procedures (71).

The two contributions to the total resistance to mass transfer in the blanching of vegetables are the surface resistance due to convection and the internal resistance due to mass diffusion. These may be represented by Fick's first and second laws, together with a mass balance at the interface. When there is sufficient agitation of the blanching liquid, the surface resistance becomes small, and it can be assumed that the total resistance is due to only the internal resistance. Then only the solution to Fick's second law is required, and this is available for the infinite slab, infinite cylinder, and sphere, with the mean concentration obtained after integration with respect to position as a function of time, again in nondimensionalized form (72).

These solutions were originally developed for drying applications over a wide range of concentrations; however, in the case of blanching vegetables, the concentrations are much smaller and near to zero. The solutions have been recalculated for smaller increments in this range of concentrations using the first ten terms for the three series (6). Where the apparent diffusion coefficient for a solute is known at various temperatures it is possible to predict the changes in concentration that will occur under defined conditions of blanching (73,74).

BIBLIOGRAPHY

1. J. D. Selman, "The Blanching Process," in S. Thorne, ed., *Developments in Food Preservation,* Vol. 4, Elsevier Applied Science, London, 1986.

2. J. L. Bomben, "Effluent Generation, Energy Use and Cost of Blanching," *Journal of Food Process Engineering* **1**, 329–341 (1977).

3. A. Schaller, "Heat Utilisation in Water Blanching of Spinach Leaves," *Confructa* **17**, 264–273 (1973).

4. T. R. Rumsey, E. P. Scott, and P. A. Carroad, "Energy Consumption in Water Blanching," *Journal of Food Science* **47**, 295–298 (1982).

5. M. A. Rao, H. J. Cooley, and A. A. Vitali, "Thermal Energy Consumption for Blanching and Sterilisation of Snap Beans," *Journal of Food Science* **51**, 378–380 (1986).

6. J. D. Selman, P. Rice, and R. K. Abdul-Rezzak, "A Study of the Apparent Diffusion Coefficients for Solute Losses from Carrot Tissue during Blanching in Water," *Journal of Food Technology* **18**, 427–440 (1983).

7. J. D. Selman and E. J. Rolfe, "Effects of Water Blanching on Pea Seeds, II: Changes in Vitamin C Content," *Journal of Food Technology* **17**, 219–234 (1982).

8. M. Togeby, N. Hansen, E. Mosekilde, and K. P. Poulsen, "Modelling Energy Consumption, Loss of Firmness and Enzyme Inactivation in an Industrial Blanching Process," *Journal of Food Engineering* **5**, 251–267 (1986).

9. C. Srisangnam, D. K. Salunkhe, N. R. Reddy, and G. G. Dull, "Quality of Cabbage, I: Effects of Blanching, Freezing and Freeze-Dehydration on the Acceptability and Nutrient Retention of Cabbage," *Journal of Food Quality* **3**, 217–231 (1980).

10. M. E. Lazar, D. B. Lund, and W. C. Dietrich, "A New Concept in Blanching," *Food Technology* **25**, 684 (1971).

11. C. R. Havighorst, "Venturi Tubes Recycle Heat in Blancher," *Food Engineering* **45**(6), 89–90 (1973).

12. J. L. Bomben, "Vibratory Blancher-Cooler Saves Water, Heat and Waste," *Food Engineering International* **1**(7), 37–38 (1976).

13. K. P. Poulsen, "Optimisation of Vegetable Blanching," *Food Technology* **40**, 122–129 (1986).

14. D. Atherton and J. B. Adams, "New Blanching Systems," *Technical Memorandum No. 319,* Campden Food and Drink Research Association, Chipping Campden, England, 1983.

15. J. W. Ralls, H. J. Maagdenberg, N. L. Yacoub, M. E. Zinnecker, J. M. Reiman, H. D. Karnath, D. M. Homnick, and

W. A. Mercer, "In-Plant Hot-Gas Blanching of Vegetables," *Proceedings of the 4th National Symposium of Food Processing Wastes,* Environmental Protection Agency, Syracuse, N.Y., pp. 178–222, 1973.

16. W. C. Dietrich, C. C. Huxsoll, and D. C. Guadagni, "Comparison of Microwave, Conventional and Combination Blanching of Brussels Sprouts for Frozen Storage," *Food Technology* **24**, 613–617 (1970).

17. A. C. Cross and D. Y. C. Fung, "The Effect of Microwaves on Nutrient Value of Foods," *Critical Reviews in Food Science and Nutrition* **16**, 355–381 (1982).

18. W. Stolp, D. J. van Zuilichem, and F. Westbrook, "Microwave Blanching of Fresh Mushrooms," *Vaedingsmiddelentechnologie* **22**(10), 45–47 (1989).

19. R. L. Garrote, E. R. Silva, and R. A. Bertone, "Blanching of Small Whole Potatoes in Boiling Water and by Electro-conductive Heating," *Lebensmittel-Wissenschaft und Tecnologie* **21**(1), 41–45 (1988).

20. P. Baardseth, "Quality Changes in Frozen Vegetables," *Food Chemistry* **3**, 271–282 (1978).

21. R. P. Singh and G. Chen, "Lethality-Fourier Method to Predict Blanching," in P. Linko, Y. Malkki, J. Olkku, and J. Larinkari, eds., *Food Process Engineering,* Vol. 1, Applied Science, London, 1980, pp. 70–74.

22. S. M. Alzamora, G. Hough, and J. Chirife, "Mathematical Prediction of Leaching Losses of Water Soluble Vitamins during Blanching of Peas," *Journal of Food Technology* **20**, 251–262 (1985).

23. T. R. Gormley and C. MacCanna, "Canning Tests on Mushroom Strains," *Irish Journal of Food Science and Technology* **4**(1), 57–64 (1980).

24. M. Konanayakam and S. K. Sastry, "Kinetics of Shrinkage of Mushrooms during Blanching," *Journal of Food Science* **53**, 1406–1411 (1988).

25. J. D. Selman and E. J. Rolfe, "Effects of Water Blanching on Pea Seeds, I: Fresh Weight Changes and Solute Loss," *Journal of Food Technology* **14**, 493–507 (1979).

26. UK Patent Application GB2088695 A (1982), I. J. Farrow, Cleary & Co. Ltd, "Blanching Dried Vegetables."

27. H. Koaze, T. Okamura, K. Ishibashi, K. Hironaka, and S. Kato, "Studies on Frozen Vegetables, IV: Effect of Blanching on Retention of Reduced Ascorbic Acid, Chlorophyll and Peroxidase Activity of Frozen Spinach," *Refrigeration* (Reito) **55**, 771–776 (1980).

28. K. Paulus, R. Duden, A. Fricker, K. Heintze, and H. Zohm, "Influence of Heat Treatment of Spinach at Temperatures up to 100°C, II: Changes of Drained Weight and Contents of Dry Matter, Vitamin C, Vitamin B1 and Oxalic Acid," *Lebensmittel Wissenschaft Technologie* **8**(1), 11–16 (1975).

29. J. D. Selman, "Review—Vitamin C Losses from Peas during Blanching in Water," *Food Chemistry* **3**(3), 189–197 (1978).

30. J. D. Selman and E. J. Rolfe, "Studies on the Vitamin C Content of Developing Pea Seeds," *Journal of Food Technology* **14**, 157–171 (1979).

31. D. B. Cumming, R. Stark, and K. A. Sanford, "The Effect of an Individual Quick Blanching Method on Ascorbic Acid Retention in Selected Vegetables," *Journal of Food Processing and Preservation* **5**(1), 31–37 (1981).

32. B. L. Bengtsson, "Effect of Blanching on Mineral and Oxalate Content of Spinach," *Journal of Food Technology* **4**, 141–145 (1969).

33. T. Mari and I. Binder, "Reduction of the Nitrate Content of Carrots by Various Blanching Methods," *Hütoipar* **25**(1), 7–9 (1978).

34. H. J. Beilig and H. Treptow, "Influence of Processing on Cadmium Content of Spinach, Green Beans and Peas," *Berichte über Landwirtschaft* **55**, 809–816 (1977).

35. G. Marcy and W. Adam, "Soy Sprout Salad as a Health Hazard," *Archiv für Lebensmittelhygiene* **28**(5), 197–198 (1977).

36. A. C. Georgescu and M. Bugulescu, "Thermophilic Organisms in Insufficiently Sterilized Canned Peas," *Industria Alimentara* **20**, 251–253 (1969).

37. E. R. Elkins, F. C. Lamb, R. P. Farrow, R. W. Cook, M. Kawai, and J. R. Kimball, "Removal of DDT, Malathion and Carbaryl from Green Beans by Commercial and Home Preparation Procedures," *Journal of Agricultural and Food Chemistry* **16**, 962–966 (1968).

38. M. G. Kleinschmidt, "Fate of Di-syston in Potatoes during Processing," *Journal of Agricultural and Food Chemistry* **19**, 1196–1197 (1971).

39. J. M. Solar, J. A. Liuzzo, and A. F. Novak, "Removal of Aldrin, Heptachlor Epoxide and Endrin from Potatoes during Processing," *Journal of Agricultural and Food Chemistry* **19**, 1008–1010 (1971).

40. J. G. Fair, J. L. Collins, M. R. Johnston, and D. L. Coffey, "Levels of DDT Isomers in Turnip Greens after Blanching and Thermal Processing," *Journal of Food Science* **38**, 189–191 (1973).

41. G. Melkebeke, M. van Assche, W. Dejonckheere, W. Steurbaut, and R. H. Kips, "Effect of Some Culinary Treatments on Pesticide Residue Contents of Spinach," *Revue de l'Agriculture* **36**, 369–378 (1983).

42. C. M. Weaver and N. D. Harris, "Removal of Radioactive Strontium and Caesium from Vegetables during Laboratory Scale Processing," *Journal of Food Science* **44**, 1491–1493 (1979).

43. D. C. Williams, M. H. Lim, A. O. Chen, R. M. Pangborn, and J. R. Whitaker, "Blanching of Vegetables for Freezing—Which Indicator Enzyme to Choose," *Food Technology* **40**(6), 130–140 (1986).

44. H. Böttcher, "The Enzyme Content and the Quality of Frozen Vegetables, II: Effects on the Quality of Frozen Vegetables," *Nahrung* **19**, 245–253 (1975).

45. E. Pogorzelski, J. Rotsztejn, and J. Berdowski, "Effect of Blanching Conditions on Polyphenoloxidase and Peroxidase Activities of Brussels Sprouts," *Przemysl Fermentauyjny i Owocowo-Warszawny* **25**(5/6), 38–39 (1981).

46. P. Baardseth and E. Naesset, "The Effect of Lipid-Degrading Enzyme Activities on Quality of Blanched and Unblanched Frozen Stored Cauliflower, Estimated by Sensory and Instrumental Analysis," *Food Chemistry* **32**, 39–46 (1989).

47. J. Abbas, M. A. Rouet-Mayer, and J. Philippon, "Comparison of the Kinetics of Two Pathways of Chlorophyll Degradation in Blanched or Unblanched Frozen Green Beans," *Lebensmittel Wissenschaft und Technologie* **22**(2), 68–72 (1989).

48. N. Muftugil, "Effect of Different Types of Blanching on the Color and the Ascorbic Acid and Chlorophyll Contents of Green Beans," *Journal of Food Processing and Preservation* **10**(1), 69–76 (1986).

49. B. Segal, "Chlorophyll Decomposition in Green Peas during Processing," *Industria Alimentara* **21**, 493–497 (1970).

50. K. Klyamov, "Effect of Blanching on Colour of Ground Red Peppers," *Bilgarski Plodove Zelenchutsi i Konservi* **8**, 25–27 (1975).

51. S. Mirza and I. D. Morton, "Blanching of Carrots and Nutrient Loss," *Journal of the Science of Food and Agriculture* **25**, 1043 (1974).

52. C. Avisse and P. Varoquaux, 'Microwave Blanching of Peaches," *Journal of Microwave Power* **12**(1), 73–77 (1977).

53. P. Baardseth, "Meat Products and Vegetables (Thermal Processing), in *Physical, Chemical and Biological Changes in Food Caused by Thermal Processing,* Applied Science Publishers, London, 1977, pp. 280–289.

54. C. Buonocore and G. Crivelli, "Structural Changes in Vegetable Products as a Function of Freezing Time, V: Observations on Green Beans," *Industrie Agrarie* **8**, 225–230 (1970).

55. A. R. Rahman, W. L. Henning, and D. E. Westcott, "Histological and Physical Changes in Carrots as Affected by Blanching, Cooking, Freezing, Freeze-drying and Compression," *Journal of Food Science* **36**, 500–502 (1971).

56. M. Grote and H. G. Fromm, "Fine Structural Analysis of the Morphological Changes Involved in the Blanching, Cooking, Dehydration and Rehydration of Green Bean Pod Tissue," *Zeitschrift für Lebensmittel Untersuchung und Forschung* **166**, 203–207 (1978).

57. D. E. Johnston, D. Kelly, and P. P. Dornan, "Losses of Pectic Substances during Cooking and the Effect of Water Hardness," *Journal of the Science of Food and Agriculture* **34**, 733–736 (1983).

58. A. S. Jaswal, "Effects of Various Chemical Blanchings on the Texture of French Fries," *American Potato Journal* **47**(1), 13–18 (1970).

59. T. Motohiro and N. Inoue, "Manufacture of Canned Asparagus, IV: Changes in Shear Value, Springiness and Hardness of Asparagus on Heat Treatment," *Journal of Food Science and Technology (Japan)* **20**(1), 1–4 (1973).

60. E. Steinbuch, "Technical Note: Improvement of Texture of Frozen Vegetables by Stepwise Blanching Treatments II," *Journal of Food Technology* **12**(4), 435–436 (1977).

61. C. Y. Lee, M. C. Bourne, and J. P. van Buren, "Effect of Blanching Treatments on the Firmness of Carrots," *Journal of Food Science* **44**, 615–616 (1979).

62. M. C. Bourne, "Effect of Blanch Temperature on Kinetics of Thermal Softening of Carrots and Green Beans," *Journal of Food Science* **52**, 667–668, 690 (1987).

63. A. J. Taylor, J. M. Brown, and L. M. Downie, "The Effect of Processing on the Texture of Canned Mung Bean Shoots," *Journal of the Science of Food and Agriculture* **32**, 134–138 (1981).

64. J. P. van Buren, "Effects of Salts Added after Cooking on the Texture of Canned Snap Beans," *Journal of Food Science* **49**, 910–912 (1984).

65. M. Nyman, K. E. Palsson, and N. G. Asp, "Effects of Processing on Dietary Fibre in Vegetables," *Lebensmittel Wissenschaft und Technologie* **20**(1), 29–36 (1987).

66. R. S. Mitchell, D. J. Casimir, and L. J. Lynch, "Blanching of Green Peas, II: Puncturing to Improve Efficiency of Gas Removal," *Food Technology (Champaign)* **23**, 819–822 (1969).

67. S. I. Jankov, "Physical Changes in Sliced Green Peppers, Beans and Peas during Blanching, *Confructa* **15**(2), 88–92 (1970).

68. C. Kluge, K. Melhardt, and L. Linke, "Studies on Gas Escape during the Thermal Treatment of Vegetables," *Lebensmittelindustrie* **33**, 234–237 (1986).

69. C. C. A. Ling, J. L. Bomben, D. F. Farkas, and C. J. King, "Heat Transfer from Condensing Steam to Vegetable Pieces," *Journal of Food Science* **39**, 692–695 (1974).

70. P. J. Schneider, *Conduction Heat Transfer,* 6th ed., Addison-Wesley, Reading, Mass., 1974.

71. J. A. Luna, R. L. Garrote, and J. A. Bressan, "Thermo-kinetic

Modelling of Peroxidase Inactivation during Blanching–Cooling of Corn on the Cob," *Journal of Food Science* **51**, 141–145 (1986).

72. A. B. Newman, "The Drying of Porous Solids: Diffusion Calculations," *Transactions of the AIChE* **27**, 310 (1931).

73. G. Hough, S. M. Alzamora, and J. Chirife, "Effect of Piece Shape and Size on Leaching of Vitamin C during Water Blanching of Potato," *Journal of Food Engineering* **8**, 303–310 (1988).

74. P. Rice, J. D. Selman, and R. K. Abdul-Rezzak, "Nutrient Loss in the Hot Water Blanching of Potatoes," *International Journal of Food Science and Technology* **25**, 61–65 (1990).

JEREMY D. SELMAN
Campden Food & Drink
Research Association, Chipping
Campden, Gloucestershire,
United Kingdom

BLEACHING. See HYDROGENATION; SOYBEANS AND SOYBEAN PROCESSING; PALM OIL; OILSEEDS AND THEIR OILS; OLIVES; ENTRIES UNDER FATS AND OILS.

BLOOD. See ANIMAL BY-PRODUCT PROCESSING; KOSHER FOODS AND FOOD PROCESSING.

BOTULISM. See FOODBONE DISEASES.

BREADINGS. See EXTRUSION AND EXTRUSION COOKING.

BREAKFAST CEREALS

Breakfast cereals may be classified as hot cereals, requiring some preparation such as the addition of hot water, or ready-to-eat (RTE) cereals, usually consumed with cold milk. RTE cereals are so-called because they are precooked using one of a number of processes. Corn, wheat, rice, oats, or other grains may be puffed, flaked, shredded, or extruded; coated with sugars; enriched with micronutrients; or enhanced with fruits to provide a convenient, nutritious, and readily digestible breakfast food.

The history of breakfast cereals parallels the growth of the Kellogg Company and the Postum Cereal Company, the latter becoming the nucleus of General Foods (1,2). Health and nutrition were the driving forces in the development of breakfast cereals. At the turn of the century, people were looking for lighter, more digestible foods as they moved into less physically demanding work environments. The heavy fried and starchy foods of the pioneer days became a major cause of stomach disorders among the sedentary segment of the population (2).

About this time Dr. J. H. Kellogg was superintendent of the Battle Creek sanitarium. Kellogg, with his brother W. K. Kellogg, developed the first flaked cereal as a health food to feed the sanitarium patients. C. W. Post was a patient at the sanitarium and was most intrigued at the commercial possibilities of the health foods being served there. When he regained his health, Post went on his own to develop the cereal products that are still known today as Grape Nuts and Post Toasties.

From humble beginnings, breakfast cereals became a large industry in the United States. In 1987 it was a $5.3 billion business, with 90% coming from the RTE segment and 10% from the hot cereal category. The per capita consumption of breakfast cereals in 1987 was 11.7 lb for RTE and 3.6 lb for hot cereals (3).

READY-TO-EAT BREAKFAST CEREALS

Ready-to-eat breakfast cereals can be described by grouping them into a few basic cereal types: puffed, flaked, shredded, baked, and agglomerated.

Puffed Cereals

Puffed cereals are whole grains, grain parts, or a grain dough that has been formed into a specific shape and then expanded by subjecting it to heat and pressure to produce a very light and airy grain product. The two types of process that are normally used for puffed cereals are gun puffing and extrusion.

Gun Puffing. Gun puffing consists of a chamber (gun) that is charged with a quantity of grain material. The chamber is subjected to high-pressure steam (100–250 psig) for 5–45 s. The pressure is then released very rapidly giving a sound reminiscent of a gun. The shot of grain is propelled into an expansion vessel. The change from the high-pressure steam to atmospheric conditions causes the grain piece to expand to about 3–10 times its original volume. The puff is then dried to a final moisture content of 1–3% (4,5).

Extrusion. Extrusion employs either a single- or twin-screw extruder. A single flour or a mixture of grain flours is fed into the extruder. These flours may have been preconditioned with water or steam before extrusion. Other ingredients may be added to the extruder such as water or sugar syrups. The design of the screw, barrel elements, and die of the extruder impart desired characteristics to the finished product. Heating or cooling is accomplished through mechanical energy from the screw (6) and through jacketing the extruder barrel. Pressures in the extruder range from 500 to 2500 psig and temperatures range from 250 to 400°F. As the material exits from the end of the extruder it expands to release steam from the product. The produce is then dried to 1–4% moisture.

Puffed cereals are often coated with a sugar solution that may be flavored and dried to give a presweetened cereal. This coating is normally done in a rotating drum or screw auger where the puffed cereal is fed at a controlled rate. Sugar syrup is sprayed onto the cereal in the desired amount. The coated cereal is dried to remove excess moisture.

Flaked Cereals

Flaked cereals are cooked whole grains, parts of grains, or flours that have been pelletized or agglomerated, flattened into a flake, and dried or toasted. A normal process involves blending the grain with a flavoring syrup that may contain sugar, water, malt, or salt as well as other ingredients or combinations thereof. The blend is cooked in a steam atmosphere in a batch or continuous cooker. Cooking conditions involve moistures between 25 and 40%, time between 30 min and 3 h, and steam pressures of 15–40 psig. The cooked material is dried (pelletized and sized if required) to the desired flaking moisture, normally between 12 and 22% moisture. The material is tempered for 1–24 h to equilibrate moisture throughout the grain piece. After tempering, the pieces are fed through flaking rolls to get the desired flake thickness, normally between 0.010 and 0.035 in. After flaking, the wet flakes go to toasting ovens where the product moistured is quickly brought down to 2–3%. At times, grain flakes (especially rice) are oven puffed. This requires that the wet flakes be subjected to a very high temperature (up to 600°F) for a short period of time, often 1 min or less (7). This is done effectively in a fluidized bed dryer.

Shredded Products

Shredded products are whole grains or mixtures of grains that have been shaped into thin strands or a crosshatch shape, then layered, cut into biscuits or geometric shapes, and finally dried or toasted. For a whole grain such as wheat, the grain is cooked either in atmospheric conditions with an excess of water or in a pressurized steam vessel. Moisture content at this point is approximately 40–43% (8). After cooking, the grain is allowed to temper for several hours to equilibrate; it is then fed into shredding rolls consisting of one smooth roll and one grooved roll. The smooth roll forces the cooked grain into the grooved roll. A comb scrapes the grain material out of the grooves and deposits the strands onto a conveyer belt. Some 18–20 layers of shreds are deposited one on top of another from a bank of shredding rolls, and then cut into bite-size pieces. The pieces go through an oven to bake and dry (4,6).

Crosshatched types of products (chex) are normally made with rice or corn plus a flavoring syrup. Cooking is done in a batch or extrusion cooker and the shredding rolls are basically the same except for the addition of a horizontal groove on the grooved roll which is slightly less deep than the vertical groove. The layered web is then cut into bite-size pieces and puffed in a fluidized bed dryer to develop the characteristic light texture.

Baked Cereals

Baked cereals are the nugget type or rotary mold type. They are normally made in a typical bakery process. The nugget type involves blending grain flours, yeast, salt, and water together in a dough mixer, allowing the dough to proof, forming the dough into loaves or sheets, baking in an oven, then breaking the baked grain material into chunks, and drying to a final moisture of approximately 2%. The chunks are ground into finished food size (8). The rotary mold type utilizes a cookie baking line where the ingredients—grain, water, sugar, and shortening—are made into a dough. Pieces are formed by a cookie mold with a shape and size appropriate for RTE cereal. These are baked and dried in a typical baking oven.

Agglomerated Cereals

Agglomerated cereals are blends of grain products such as rolled oats, crisp rice, and toasted rolled wheat that have been glued together using a sugar or maltodextrin syrup

as a binding agent. Vegetable oil is added to prevent sugar absorption into the agglomerated piece and to improve crispness retention in milk. The process involves blending the ingredients together in a ribbon blender or continuously in a mixer auger. The product is spread onto the belt of a traveling screen dryer (4) and moisture is reduced to approximately 2–3%. The sheets of agglomerated grain are sized off the dryer through the use of a grinder and then readied for packaging. Ingredient addition and sequencing are important to insure good agglomeration of the product without causing sugar lumps or bricking of the product in the package.

HOT BREAKFAST CEREALS

Oats

Hot breakfast cereals are usually made of oat flakes. The two principal factors that make oats different from other cereals are its retention of its whole grain identity and its nutlike flavor. Products vary from rolled oats (whole oat flakes), which require five or more minutes to prepare, to instant products, which can be prepared in a bowl with hot water.

In making oat flakes, the oat grain is first heated with dry steam, which reduces its moisture content to approximately 6%. Enzymes that produce rancid flavors during storage are deactivated during steaming. A high-speed impact operation separates the dry friable hull from the groat (oat berries). Even a small percentage of whole oats remaining is unacceptable because the hull fraction in oat flakes is unpalatable. Whole rolled oats are produced by flaking the whole groat.

Steel cutting reduces the groats for further processing. Another steaming step softens the cut groats for the rolling step. The size of the flake is determined by the type of rollers used to flatten the groats. Instant cook-in-the-bowl oats are produced from highly polished rollers with a narrow gap.

Oat flakes vary in sizes from 0.02 to 0.03 in. for rolled oats to 0.011 to 0.013 in. for instant oat flakes. Thicker flakes require longer cooking and retain more flake integrity (6,9).

Wheat

Hot wheat breakfast cereal is made with farina, which is a fraction of middlings from hard wheat. Soft wheat is not suitable because it produces a product that loses particle integrity too rapidly during cooking. Particle size is critical for quality. United States standards require that 100% of the product pass through a 20-mesh sieve and not more than 10% pass through a 45-mesh sieve. Farina must be cooked in boiling water for several minutes to insure complete starch gelatinization.

Instant farina cooks in about 1 min with boiling water. Proteolytic enzymes are used to instantize farina by opening avenues for easier penetration of water into particles to facilitate hydration during cooking (6).

Corn

In the southern portion of the United States a breakfast cereal made from corn called grits or hominy grits is popular. Grits are produced by dry milling white corn and are essentially small pieces of endosperm (6).

NUTRITIONAL CONSIDERATIONS

Breakfast Cereals and Dietary Recommendations

Dietary recommendations from the government and major health organizations suggest that the intake of dietary fat and cholesterol should decrease and the caloric intake from complex carbohydrates should increase (10). One way this can be accomplished is to increase the amount of cereals and grains in the diet. A RTE breakfast cereal is right in line with these dietary recommendations.

A recent General Foods study showed that adults who consumed breakfast, especially one that incorporated RTE cereal, had a better overall diet quality than those who skipped breakfast. In addition, RTE cereal users usually have lower daily intake of fat and cholesterol (11).

Another survey showed that 95% of the survey population consumes some kind of food item in the morning. Eighty-nine percent of these people eat breakfast. RTE cereals were the second most frequently consumed meal, behind a food group that included rolls, muffins, and toast (12). A third survey showed that 88% of the survey population consumed breakfast, and approximately 32% consumed cereal, primarily RTE cereal (13).

Breakfast cereals are, therefore, an excellent vehicle for delivery of calories from complex carbohydrates and, at the same time, reduce fat calories. Figure 1 shows the distribution of calories in a complete cereal breakfast (1 oz cereal, 12 fl oz lowfat milk, two slices toast with spread, and 6 fl oz orange juice) compared with a bacon and egg breakfast (one egg, two slices bacon, one slice toast with spread, 6 fl oz orange juice, and 8 fl oz lowfat milk). The cereal breakfast is lower in fat and higher in carbohydrates than the bacon and egg breakfast.

Fortification

Cereal fortification has been practiced in the industry for decades and most RTE cereals today are fortified with vitamins and minerals. In 1974 the Food and Drug Administration (FDA) proposed guidelines on cereal fortification that were later incorporated into a general fortification regulation (14). This regulation, along with other guidelines (15), serves as the cornerstone for cereal fortification. Cereal manufacturers today use the following guidelines for specific nutrient fortification.

1. The intake of the nutrient is below the desirable level in the diets of a significant number of people.
2. The food used to supply the nutrient is likely to be consumed in quantities that will make a significant contribution to the diet of the population in need.
3. The addition of the nutrient is not likely to create an imbalance of essential nutrients.
4. The nutrient added is stable under conditions of storage.
5. The nutrient is physiologically available from the food.
6. There is reasonable assurance against excessive intake to a level of toxicity.

Other factors considered when fortifying include the estimated frequency of consumption of the food, the target consumer group, current health issues, and the dietary role of the product.

Breakfast cereals therefore make a significant contribution to the overall diet quality of an individual through their ingredient and fortification profiles. They provide a logical way to translate dietary recommendations to increase carbohydrate intake and decrease fat consumption; at the same time they provide a means for enhancement of vitamins and minerals in the diet.

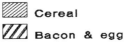

☒ Cereal
☒ Bacon & egg

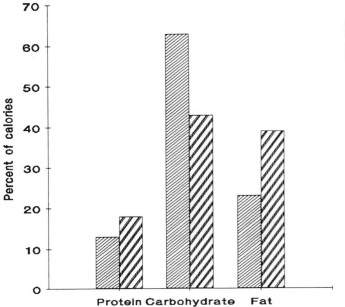

Figure 1. Percent calories in a cereal breakfast vs a bacon and egg breakfast.

STABILITY OF BREAKFAST CEREALS

Breakfast cereals should remain stable for up to one year when stored under reasonably cool and dry conditions. RTE cereals should remain crisp without developing off-flavors in storage. They should stay crisp in milk for at least 3–5 min.

Texture Stability

Loss of crispness (staling) in RTE cereals is associated with moisture pickup. Fresh RTE cereals have a moisture content of 2–3% and are very crisp. A good indicator of sensory crispness is water activity (A_w), which is the ratio of the vapor pressure of water over the food material to the vapor pressure of pure water, both measured at the same temperature. Fresh cereals generally have an A_w of about 0.20. As the moisture content goes up, the A_w increases until it reaches a critical value; for most cereals, this is around 0.45 (16). Beyond the critical A_w, the cereal becomes stale and unacceptable.

In the case of fruited cereals such as raisin bran, the raisin, which contains up to 18% moisture, will transfer its moisture during storage to the cereal, which has about 2–3% moisture. As long as the critical A_w of the cereal is not exceeded, the cereal will remain crisp. But the raisin could become unacceptably hard.

Some remedies to keep fruits soft while retaining cereal crispness include controlled addition of moisture to the cereal to modulate moisture migration from the fruit. In some cases this could make the cereal lose crispness. Infusion of fruits with edible humectants such as glycerol will keep the fruits soft while preventing the cereal from becoming stale (16,17). Coating RTE cereals with hydrophobic ingredients such as fats and oils, or incorporation of ingredients like magnesium stearate in the formulation, can significantly improve the bowl life of the cereal (18,19).

Flavor Stability

Breakfast cereals can develop off-flavors during storage. Unsaturated fatty acids and other compounds containing unsaturated linkages, such as vitamins, may undergo autooxidation, resulting in oxidative rancidity. Other reactions that can cause off-flavors are fat hydrolysis and reversion (8,20). Susceptibility to oxidation depends on the type of cereal grain used. Breakfast cereals made from oats are more susceptible to oxidative rancidity because oats contain higher levels of oil, about 7%. Other breakfast cereals, such as those made from rice, wheat, and corn, contain much less oil and thus are more stable in storage.

Antioxidants are used to retard autooxidation in breakfast cereals. Butylated hydroxyanisole (BHA) and butylated hydroxytoluene (BHT) are the most common phenolic antioxidants used. Federal regulations allow BHT and BHA to be added directly to breakfast cereals up to 50 ppm. The more common practice is to add antioxidants in the wax liner of the cereal box. These phenolic antioxidants, volatile at room temperature, diffuse from the liner to the product, thus providing protection from oxidation (21).

Nutrient Stability

Breakfast cereals are usually fortified with vitamins and minerals up to 25% of the U.S. recommended daily allowance or higher. These micronutrients, especially vitamins A, D, and C, are not stable during cereal processing. Hence, micronutrient fortification is done after processing, just before packaging (8,22). Properly maintained temperature and humidity will protect micronutrients during storage.

BREAKFAST CEREAL PACKAGING

The forerunner of today's RTE cereal packaging was introduced about 1900 by Nabisco in packaging the Uneeda Biscuit. For the consumer, a folding carton with an internal liner became a convenient alternative to cracker-barrel–type bulk packaging. Although today's packaging resembles that of the early 1900s, technological changes have occurred in packaging materials and equipment as a result of more demanding product specifications, consumer needs, distribution, merchandising, and regulatory and production requirements.

The major functions of the RTE cereal package are containment, preservation, identification, and integrity. Most RTE cereal packaging systems comprise three components: primary (liner), secondary (folding carton), and tertiary (corrugated shipping container).

Primary Package

The cereal liner's principal function is preservation. The liner evolved over time from an uncoated, bleached sulfate paper to a more protective waxed glassine paper. Plastic films were introduced in the mid-1970s. These coextruded films provide superior product protection via improved moisture and grease barriers and seal integrity. In 1985 films were made with bidirectional barrier properties to retain product aroma while preventing external contamination (23). Cereal liners can be modified to preserve other product attributes. Phenolic antioxidants can be incorporated into both waxed paper liners and plastic film coextrusions to minimize oxidative rancidity, which may impart off-flavors to the product (21). Titanium dioxide is generally added to plastic films to provide varying degrees of opacity where transparency is not desired.

Secondary Package

The principal functions of the folding carton are liner containment and communication (graphics). Most cereal cartons are produced from a machine clay-coated newboard comprising multiple layers of recycled paper fiber and are printed by either offset lithography or rotogravure. A second method prints clay-coated sulfite paper before lamination to an uncoated, special-bending newsboard using adhesives or waxes. Other less-economical, higher quality constructions using virgin fiberboard include coated, solid, bleached sulfate and coated, solid, unbleached sulfate.

Tertiary Package

Maintenance of the integrity of the retail package through the distribution cycle is the main function of the corrugated shipping case. Shipping cases are generally regular slotted containers able to withstand a burst test ranging from 150–275 lb. Top-to-bottom compression strength of the case is important to prevent crushing during handling and storage. Minimum standards for corrugated containers have been established for rail and truck shipments and are covered under specific uniform classification codes (24–26).

Cereal Packaging Line

The bag-in-the-box (BNB) packaging system is replacing the double package liner packing lines. BNB packaging uses vertical form–fill–seal baggers coupled with horizontal cartoners. Product is deposited in the bag (liner) during bag formation and is sealed prior to carton insertion. In contrast, traditional packing lines form a liner and carton around a mandrel prior to product fill. BNB equipment is more efficient and flexible, operates with wider liner material selection, and produces a higher quality cereal liner.

BIBLIOGRAPHY

1. G. Carson, *The Cornflake Crusade,* Rinehart & Co., New York, 1957, pp. 147–175.

2. J. Berry, *Inside Battle Creek,* rev. ed., Battle Creek Public Schools, Brochure No. **6,** Battle Creek, Michigan, 1989.

3. The Staff, "State of the Food Industry, Breakfast Cereals," *Food Engineering,* **60**(6), 86 (1988).

4. R. Fast, "Breakfast Cereals: Processed Grains for Human Consumption," *Cereal Foods World* **32**(3), 241–244 (1987).

5. R. Daniels, *Breakfast Cereal Technology,* Noyes Data Corp., Park Ridge, N.J., 1974.

6. R. Hosney, *Principles of Cereal Science and Technology,* American Association of Cereal Chemists, St. Paul, Minn., 1986, pp. 293–304.

7. D. Tressler and W. Sultan, *Food Products Formulary,* Vol. 2, AVI Publishing Co., Westport, Conn., 1975.

8. N. Kent, *Technology of Cereals,* Pergamon Press, Elmsford, NY, 1983, pp. 144–153.

9. F. H. Webster, "Oat Utilization: Past, Present, and Future," in F. H. Webster, ed., *Oats: Chemistry and Technology,* American Association of Cereal Chemists, Inc., St. Paul, Minn., 1986, pp. 413–426.

10. Committee on Diet and Health, *Diet & Health: Implications for Reducing Chronic Disease Risk,* National Academy Press, Washington, D.C., 1989.

11. K. Morgan, M. Zabik, and G. Stampley, "The Role of Breakfast in Diet Adequacy of the U.S. Adult Population," *Journal of the American College of Nutrition* **5,** 551–563 (1986).

12. J. Markle, F. Vellucci, and G. Coccodrilli, *Food Selection Patterns of Individuals Who Consume a Food Item in the Morning,* Internal Research Report, General Foods USA, White Plains, N.Y., 1985.

13. J. Heyback and F. Vellucci, *Clustering Patterns of Food Intake at Breakfast,* Internal Research Report, General Foods USA, White Plains, N.Y., 1985.

14. 21 *CFR* Part 104, *Federal Register* **45**(18) 6314–6324 (1980).

15. AMA Council on Foods and Nutrition and the Food and Nutrition Board of the National Academy of Science—National Research Council "Improvement of Nutritive Quality of Foods," *Journal of the American Medical Association* **205**(12), 160 (1968).

16. U.S. Pat. 4,256,772 (Mar. 17, 1981), A. Shanbhag and A. Szczesniak (to General Foods Corp.).

17. U. S. Pat. 4,103,035 (July 25, 1978), C. Fulger and T. Morfee (to Kellogg Co.).

18. U.S. Pat. 3.484,250 (Dec. 16, 1969), W. Vollink and M. Steeg (to General Foods Corp.).

19. U.S. Pat. 4,588,596 (May 13, 1986), D. Bone and co-workers (to the Quaker Oats Co.).

20. R. B. Coulter, "Extending Shelf Life by Using Traditional Phenolic Antioxidants," *Cereal Foods World* **33**(2), 207–210 (1988).

21. R. Sims and J. Fioriti, "Antioxidants as Stabilizers for Fats, Oils, and Lipid Containing Foods," in T. E. Furia, ed., *CRC Handbook of Food Additives,* Vol. 11, 2nd ed., CRC Press, Inc., Boca Raton, Fl., 1980, pp. 13–56.

22. R. H. Anderson and co-workers, "Effects of Processing and Storage on Micronutrients in Breakfast Cereals," *Food Technology* **30**(5), 110–114 (1976).

23. W. Drennan, "Barrier Resins Key New Package Development," *Plastics Packaging,* 17–21 (July–Aug. 1988).

24. *Uniform Freight Classification,* Uniform Freight Classification Committee, Chicago, Ill., annual.

25. *National Motor Freight Classification,* American Trucking Associations, Washington, D.C., annual.

26. *Coordinated Freight Classification,* The New England Motor Rate Bureau, Inc., Burlington, Mass., annual.

General References

M. Bakker, *Encyclopedia of Packaging Technology,* John Wiley & Sons, Inc., New York, 1986.

S. A. Matz, *The Chemistry and Technology of Cereals as Food & Feed,* AVI Publishing Co., Westport, Conn., 1959.

S. A. Matz, *Cereal Technology,* AVI Publishing Co., Westport, Conn., 1970.

Dick Fizzell
G. Coccodrilli
Charles J. Cante
General Foods USA
White Plains, New York

BREVETOXIN. See Marine toxins.
BREWERIES. See Beer.

BROWNING REACTION, ENZYMATIC

The change in color following mechanical or physiological injury of fruits and vegetables is due to oxidative reaction of phenolic compounds by phenolase and the reaction product, o-quinone, to various polymerized oxidation products. In food processing, such color change is commonplace during preparation for canning, dehydration, freezing, or storage. Formation of dark brown color causes the product to become unattractive and is accompanied by undesirable changes in flavor and a reduction in nutritive value of the products. Enzyme-catalyzed oxidative browning has long been recognized since the report in 1895 that the changes in color of fresh apple cider were due to oxidation of tannin by the oxidase present in apple tissue (1).

PHENOLASE

Definition

Phenolase is of particular importance in the processing and marketing of horticultural products because of the brown discolorations of bruised fruits and vegetables resulting from the action of this enzyme. Not only is there undesirable color formation, but browning is accompanied by a loss of nutrient quality and the development of undesirable flavors. On the other hand, browning by phenolase is a desired reaction in certain foods such as tea, coffee, cocoa, prunes, and dates.

Phenolase is a generic term for the group of enzymes that catalyze the oxidation of phenolic compounds to produce brown color on cut surfaces of fruits and vegetables. Phenolase, or commonly polyphenol oxidase, is a copper-containing enzyme that belongs to a group of oxidoreductases. It is one of the most damaging enzymes with respect to quality maintenance of fresh fruits. Depending on the substrate specificity, International Enzyme Nomenclature has designated monophenol monooxygenase or tyrosinase as EC 1.14.18.1, diphenol oxidase or catechol oxidase or diphenol oxygen oxidoreductase as EC 1.10.3.2, and laccase or p-diphenol oxygen oxidoreductase as EC 1.10.3.1. The first two enzymes, tyrosinase and diphenol oxidase, which is often referred to as catechol oxidase, occur in practically all plants. Phenolase from banana, tea leaves, tobacco leaves, and clingstone peaches has been reported to oxidize o-diphenols only, whereas the enzyme(s) from potato, apple, sugar beet leaf, broad bean leaf, and mushrooms have both types of activity. Laccase is less frequently encountered as a cause of browning in fruits and vegetables than is catechol oxidase.

Catechol oxidase catalyzes two distinct reactions: (1) insertion of oxygen in a position ortho to an existing hydroxyl group, usually followed by oxidation of the diphenol to the corresponding quinone,

and (2) oxidation of o-diphenol to quinone,

It was reported in 1944 that the phenolase preparations obtained from mushrooms had the ability to catalyze oxidation of monophenols in addition to catalyzing oxidation of polyphenols (2). However, the ratio of the rate of monophenol oxidation to that of polyphenol oxidation varied considerably among different mushroom sources. Since the 1930s reports have shown a wide range of phenolase activities in many fruits and vegetables in different ratios. Today it is known that enzyme preparations from a number of other sources possess these two activi-

ties in different ratios and that the ratios may change during isolation and purification. Most catechol oxidase preparations from potatoes, apples, and beans possess both activities, while those tea leaf, tobacco, mango, banana, pear, and sweet cherry have been reported not to act on monohydroxy phenols (3).

Laccase oxidzes o- and p-diphenols,

In addition, it oxidizes various other p-diphenols, m-diphenols, and p-phenylenediamine, but laccase does not oxidize tyrosine. Detailed reviews of these enzymes have been published (4,5).

Catechol oxidase was first described in 1895; it was demonstrated that the darkening of mushroom was due to enzymatic oxidation of tyrosine (6). In addition to its general occurrence in plants, catechol oxidase is also found in some bacteria, fungi, algae, bryophytes, and gymnosperms (7–10). Laccase was first observed in the latex of the Japanese lacquer tree (11).

Catechol oxidase appears in almost all tissues of plants; however, significant differences in both the level of catechol oxidase activity and in the concentration of its substrates have been observed among different cultivars of fruits and vegetables. The enzyme has been found in a variety of cell fractions, both in organelles and in the soluble fraction of the cell (12,13). The level of catechol oxidase often changes markedly during the development of the plant (14–16) and may be significantly affected by growing conditions.

Assay

Qualitative and quantitative colorimetric tests and manometric determinations have been used to detect and measure enzyme activity. Because phenolases use molecular oxygen to oxidize phenolic substrates, their activity can be determined by measuring the rate of substrate disappearance or the rate of product formation. When based on the rate of substrate disappearance, oxygen absorption is usually measured either manometrically in a Warburg respirometer, or polarographically with an oxygen electrode. The most popular method of assay is to follow the initial rate of formation of the quinone spectrometrically by measuring the optical density. Care must be taken to restrict measurement to the initial phase of the oxidation, because the reaction soon slows down. Some researchers object to the use of spectrophotometric analysis because it measures the secondary reaction products of phenolase and the secondary reactions are influenced by many factors difficult to control. Thus the presence of ascorbic acid creates a lag phase and lowers the values obtained, while autooxidation products of polyphenols may increase the levels of enzyme activity. It is important to note that the reaction changes with time and temperature and depends on the substrate type and concentration, on the pH of the reaction mixture, and on the buffer used (17,18). A wide variety of substrates can be used with spectrophotometric methods: catechol (19), pyrogallol (20), or chlorogenic acid (21). It has to be taken into account that the reaction products formed from the oxidation of various phenols have absorption maxima at different wavelengths, that the substrate may undergo autooxidation, and that an excess of some substrates causes strong inhibition of the enzyme (21). Other methods for measuring phenolase activity are based on disappearance of ascorbic acid, a reaction that is directly proportional to enzyme activity (22), and a spectrophometric method using Besthorn's hydrazone (23) or 2-nitro-5-thiobenzoic acid (24).

It is important to differentiate between catechol oxidase and laccase on the one hand and peroxidase on the other. Because peroxidative oxidation of phenols is often mistaken for catechol oxidase or laccase reaction, peroxides should be removed from the reaction mixture by addition of catalase and alcohol to prevent any oxidation of phenols (4,25). For the determination of laccase, syringaldehyde (26) and 2,6-dimethoxyphenol (27) have been used as substrates of laccase. The catechol oxidase and laccase activities can be differentiated by the use of cinnamic acid derivatives to inhibit catechol oxidase and cationic detergents to inhibit laccase (28).

Location

The vascular elements of many fruits and vegetables darken rapidly when they are cut and exposed to air. The relative catechol oxidase activity of different parts of apricots, apples, or potato plants varies. Catechol oxidase is considered to be an intracellular enzyme in plants except in a few cases where it is found in the cell wall fraction. The enzyme is located in a variety of cell fractions, both in organelles where it may be tightly bound to membranes and in the soluble fraction (12,13). Histochemical studies using the electron microscope and density-gradient centrifugation support the idea that catechol oxidase is bound within chloroplast lamellae and grana and in mitochondria (4). Conversion of particulate forms of the enzyme to soluble forms occurs in fruits when they are exposed to stress conditions or during ripening and storage. Browning reactions that take place after the disruption of tissues rich in catechol oxidase may cause binding of soluble phenolase to a particulate fraction. The microbial extracellular laccase has been studied extensively but the location of laccase in higher plants is not well known. Although it is believed that the enzyme is found in a cytoplasmic soluble form, more information is needed on its location and distribution in plant tissues.

Preparation

Because of the extremely low concentrations of catechol oxidase and laccase in plants and their instability, preparation and purification of catechol oxidase and laccase are not easy. Oxidation reactions taking place during the isolation of the enzyme, due to natural substrates within the plant, result in changes in enzyme properties as well as in apparent multiplicity. Such reactions can be partially prevented by isolation under nitrogen gas or by using reducing agents such as ascorbic acid or cysteine, or by using phenol-adsorbing agents such as polyvinylpyrrolidinone or polyethylene glycol (29,30). The binding of catechol oxidase to membranes in many tissues further complicates its isolation. Solubilization of acetone powder preparations or extracting with detergents also result in modification of the structure and properties of the enzyme. To minimize these changes, all extractions steps should be carried out at temperature below 0°C, preferably at -20--30°C. Communion and homogenization are often carried out in liquid nitrogen or under a nitrogen atmosphere (31). Acetone precipitation followed by buffer extraction is one of the methods most often applied (3).

For the purification of the enzyme, several methods have been used, which vary according to the enzyme source and the degree of purity to be attained. Most often, precipitation with ammonium sulfate of different saturation, gel chromatography on Sephadex columns, or ion-exchange chromatography with DEAE-cellulose or DEAE-Sephadex are applied. Recently, hydrophobic gel chromatography using Phenyl-Sepharose CL-4B has been used effectively (32,33). The separation of enzymes by electrophoresis is now a standard procedure. Sodium dodecylsulfate techniques and electroblotting of catechol oxidase have also been used successfully (34,35).

Physiochemical Properties

Pure catechol oxidase is colorless and laccase is blue. Numerous reports suggest that catechol oxidase contains one copper atom per polypeptide chain or subunit. Other reports indicate that the functional unit of the enzyme contains a pair of copper ions. However, the single polypeptide chain of purified laccase is reported to contain four atoms of copper. As already mentioned, laccase can be distinguished from catechol oxidase by its substrate specificity and its specificity to inhibitors. As with all enzymes, the activity of phenolases is influenced by concentrations of substrate and enzyme, temperature, pH, and salts.

Molecular Weight. Most reports on the molecular weight of the catechol oxidase are based on estimates on partially purified preparations using gel filtration or acrylamide gel electrophoresis. Therefore, a wide range of values are reported. Often crude or partially purified preparations show a multiplicity of forms that may have resulted from association–dissociation reactions. The molecular weight of mushroom catechol oxidase has been reported to be between 116,000 and 128,000 daltons (29). The enzyme consists of four identical subunits of molecular weight of ca 30,000, each containing one copper atom (36). Two types of polypeptide chain, heavy (mol wt 43,000) and light (mol wt 13,400), were reported in mushroom (37). Apple catechol oxidase shows three forms with different molecular weights: 30,000–40,000; 60,000–70,000; and 120,000–130,000 (38). Avocado, banana, tea leaves, and other sources also showed multiple molecular weights over a wide range. The predominant form of catechol oxidase in grape had a molecular weight of 55,000–59,000 but on storage it dissociated into 31,000–33,000 and 20,000–21,000 subunits (38). Based on recent knowledge regarding the multiplicity of catechol oxidase, it is believed that many previous observations of multiplicity were due to secondary reactions, such as tanning during enzyme preparation.

The molecular weight of laccase was also reported to vary according to the source. *Rhus vernicifera* laccase has been reported to vary between 101,000 and 140,000 depending on the species and variety. The molecular weight of the fungal laccase varies between 56,000 and 390,000; peach laccase was reported to have a molecular weight of 73,500. These data suggest that laccases are a heterogenous group of glycoproteins, having a basic structural unit of molecular weight between 50,000 and 70,000, which can undergo aggregation to give larger units.

pH and Temperature Optima. The optimum pH for the activity of an enzyme preparation from any one source usually varies with different substrates and is characteristic of the substrate as well as the enzyme preparation. The optimum pH of catechol oxidase activity has a relatively wide range, pH 5.0 to 7.0. Some preparations were reported to be inactive below pH 4.0. The type of buffer and the purity of the enzyme affect the pH optimum; also, isoenzymes have different pH optima. Enzyme preparations obtained from the same fruit or vegetable at various stages of maturity have been reported to differ in optimum pH of activity as well. The pH optimum of laccase activity depends on the enzyme source and on the substrate. The stability of the enzyme is reported to be dependent on buffer and other factors in the medium. Optima between pH 4.5 and 7.5 have been reported for enzymes from various fungi and peaches (39).

The optimum temperature for catechol oxidase activity depends essentially on the same factors as the pH optimum. Generally, the activity of catechol oxidase increases as the temperature rises to temperatures of maximum activity (25–35°C); activity declines rapidly at temperatures of 35–45°C, depending on the enzyme source and substrate (3). Like most enzymes, the application of heat (70–90°C) to plant tissues or enzyme solutions for a short period caused rapid inactivation. Exposures to temperatures below zero may also affect activity. Concentrated solutions of the enzyme in dilute phosphate buffer at a pH near neutrality can be held without detectable loss of activity of several months at 1°C or when frozen at −25°C. However, the enzyme slowly losses activity even in the frozen state. Thermotolerance of catechol oxidase also depends on the source of the enzyme. It is difficult to compare heat stability data of purified enzyme preparations with those of the enzyme when in tissues or juice. Different molecular forms of catechol oxidase from the same source may have different thermostabilities. Laccase is usually more sensitive to inactivation by heat than is catechol oxidase.

Activators and Inhibitors. The activation mechanism of catechol oxidase from different sources is not well understood, but it has been suggested that conformational changes of protein and perhaps protein association or dissociation are involved in the process. Anionic detergents, such as sodium dodecylsulfate, were found to reactivate catechol oxidase in crude and partially purified preparations from different avocado cultivars (40). Short exposures to acid pH or urea also resulted in a severalfold reversible activation of grape and bean catechol oxidase.

There are two principal types of catechol oxidase inhibitor: reagents that interact with the copper in the enzyme and compounds that affect the site for the phenolic substrate. The metal-chelating agents such as cyanide, diethyldithiocarbamate, carbon monoxide, dimercaptopropanol, sodium azide, phenylthiourea, and potassium methyl xanthate are all inhibitory. Other reducing agents, sulfate, cysteine, and thio compounds are also effective inhibitors. Benzoic acid and some substituted cinnamic acids were effective competitive inhibitors of the catechol oxidase from several fruits and potatoes (41,42). Soluble polymers such as polyvinylpyrrolidinone act as competitive inhibitors of catechol oxidase. Some other natural inhibitors of catechol oxidase from various tissues have been reported.

The activity of laccase is not inhibited by CO or phenylhydrazine, both of which inhibit catechol oxidase. The enzyme is usually inhibited by the copper chelator, diethyldithiocarbamate, as well as by cyanide, azid, and EDTA. This response to inhibitors permits an easy differentiation between laccase and catechol oxidase.

Substrate Specificity. While catechol oxidases from animal tissues are relatively specific for tyrosine and dopa, those from fungi and higher plants act on a wide range of monophenols and o-diphenols. There are individual differences among phenolases from different sources. The rate of oxidation of o-diphenols by catechol oxidase increases with increasing electron-withdrawing power of substituents in the para position. o-Diphenol substitution ($-CH_3$) at one of the positions adjacent to the −OH groups prevents oxidation. These positions should remain free for oxidation to take place. As mentioned earlier, some catechol oxidase preparations from some plants lack cresolase activity, which may be due to changes in the structure of the protein during preparation. Some arguments regarding the physical relationships between the cresolase and catechol oxidase functions suggested that both functions are catalyzed by a single site while other imply the participation of two sites, either on the same enzyme molecule or on different ones. Although catechol oxidase could oxidize a wide range of phenolics, the individual enzymes tend to prefer a particular substrate or certain type of phenolic substrate. The affinity of plant catechol oxidases of the phenolic substrates is relatively low. The K_m is high, usually around 1 mM, which is higher than that of most fungi and bacteria (0.1 mM). The affinity to oxygen is also relatively low, ranging from 0.1 to 0.5 mM.

Mechanism of Oxidation. In spite of extensive studies on phenolases, its reaction mechanism is still unclear. One of the reasons is that the mechanism in the case of phenolic compounds is complicated. As stated above, the cresolase reaction apparently involves a hydroxylation reaction, which differs from the oxidation of the o-diphenols by the catecholase reaction. For a long time there was a difference of opinion about whether hydroxylation of monophenols and oxidation of o-dihydroxyphenols were catalyzed by the same enzyme. The oxidation mechanism of o-diphenol to o-benzoquinone in equation 1 is the sum of several other intermediate reactions.

Because copper is the prosthetic group, the catalytic activity of the phenolases is based on the cupric–cuprous valency change. On isolation of the enzyme in its natural state, the copper is in the cuprous form, but in the presence of o-dihydroxy phenols the copper would be oxidized to cupric copper. The substrate is oxidized by losing two electrons and two protons. The two electrons are taken up by the copper of the enzyme, which then passes into the cuprous state. The cuprous enzyme rapidly transfers the two electrons to oxygen, which immediately forms water with the two protons that were liberated and the enzyme returns to the cupric state. This two-state reaction can be represented by

$$4Cu^{2+}(\text{enzyme}) + 2\text{ catechol} \rightarrow$$
$$4Cu^{+}(\text{enzyme}) + 2\text{ }o\text{-quinone} + 4H$$

$$4CU^{+}(\text{enzyme}) + 4H^{+} + O_2 \rightarrow 4CU^{2+}(\text{enzyme}) + 2H_2O$$

Recent understanding of the complete amino acid sequence of fungal catechol oxidase and the site of interaction with the phenolic substrate made it possible to create models that include a binuclear center. Histidine residues appear to link at least one of the two active sites of the copper atoms. Monophenols bind to one of the Cu^{2+} atoms, while diphenols bind to both of them (43,44). As stated, catechol oxidase can undergo monomer–dimer–tetramer transitions and therefore it is not clear which of the forms is being investigated. Catechol oxidase can undergo conformational changes induced by the substrate oxygen and by pH. The changes in conformation are accompanied by changes in the K_m of the enzyme for both its substrates, oxygen and diphenol.

A two-step reaction mechanism by laccase has been reported (45). The first step in the oxidation of quinol by laccase is the formation of the semiquinone, with transfer of an electron from the substrate to the copper in the enzyme. The second step is a nonenzymatic disproportionation reaction between two semiquinone molecules to give one molecule of quinone and one of quinol. The function of copper and electron transfer in the reaction mechanism has been studied by using inorganic ions, electron paramagnetic resonance, and spectrophotometric methods (5,46).

The Brown Color: o-Quinones. Although the mechanism of pigment formation from phenolic substrates is not understood completely, the general course of pigmentation is known to involve enzymatic oxidation, nonenzymatic oxidation, nonoxidative transformations, and polymerizations. The o-quinones formed from phenolic compounds by phenolases are the precursors of the brown color. The o-quinones themselves possess little color, but they are among the most reactive intermediates occurring in plants. They take part in a secondary reaction, bringing about the formation of more intensely colored secondary

products. The most important secondary reactions are coupled oxidation of the substrates oxidized, complexing with amino compounds and proteins, and condensation and polymerization. The principal reaction of o-quinones in browning reactions is the one leading to the formation of the unstable hydroxyquinones. The hydroxyquinones polymerize readily and are easily further oxidized nonenzymatically to a dark brown pigment.

Tyrosine is oxidized to dopa quinone by catechol oxidase and then proceeds to melanin as shown in chemistry top of page 228.

Tyrosine is first converted into dihydroxyphenylalanine (dopa) which is then oxidized to the corresponding dopa quinone. The dopa quinone, on intramolecular rearrangement, is converted into 5,6-dihydroxy indole-2-carboxylic acid, which on further oxidation is converted into red, 5,6-quinone indole carboxylic acid; then, finally, the black melanins are formed. Quinones also react readily with simple amines, such as

$$o\text{-benzoquinone} + \text{glycine} \rightarrow 4\text{-}N\text{-glycyl-}o\text{-benzoquinone}$$

This reaction product is the intermediate responsible for the deamination of glycine with the concomitant formation of deeply colored pigments. o-Quinones also react with the proteins and sulfhydryl compounds, such as cysteine, and produce dark colored insoluble products.

Dimerization or polymerization of o-quinones that lead to the colored products is common. The formation of dimers or oligomers of o-quinone by condensation of a hydroxyquinone with a quinone was recently demonstrated. The enzymatically generated caffeoyl tartaric acid o-quinones were shown to oxidize other phenols, such as 2-S-glutathionyl caffeoyl tartaric acid and flavans, by coupled oxidation mechanisms, with reduction of the cafeoyl tartaric acid quinones back to caffeoyl tartaric acid (47). The o-quinones formed by enzymatic or coupled oxidation can also react with a hydroquinone to yield a condensation product (48–50). It is possible to regenerate the original phenolic from an intermediate by reduction provided that oxidation and subsequent transformation have not gone too far. In later stages the browning is no longer reversible.

BROWNING CONTROL IN FOODS AND FOOD PROCESSING

Extensive studies have been carried out to control enzymatic browning ever since it was recognized in 1895 that the changes in color occurring in freshly pressed cider were enzymatic (1). The practical control of enzymatic browning in foods has been carried out by several methods. The method of choice depends on the food product and the intended use. Enzymatic browning in fruits and vegetables can be controlled by selecting cultivars that are least susceptible to discoloration either because of the absence of the specific phenolic substrate or because the substrate or enzyme is present at low concentration. It is also possible to select fruits and vegetables at stages of maturity when discoloration is at a minimum. Other methods include the removal of oxygen from fruit and vegetable

Melanin

tissues as well as from the atmosphere surrounding the food; the addition of acids to reduce the pH and thus reduce phenolase activity, the addition of antioxidants or reducing substances; the addition of or treatment with permissible inhibitors, and heat inactivation of the phenolases. The use of antioxidants during processing and heat inactivation of the enzyme are widely used practices that have been moderately successful.

All fruits susceptible to browning should be processed as quickly as possible. Heating destroys the enzymes responsible for the reaction; thus when fruit is canned or made into jams or jellies, browning stops as soon as the fruit is heated sufficiently to denature the enzyme. The exact temperature necessary varies with enzyme, rate of heating, pH, and other factors such as size of the fruits. Unfortunately, some undesirable effects on quality often result from adequate heat processing. In general, cherry, peach, and apricot take 2–3 min in boiling water, and apple, pear, and plum take 4–5 min to completely inactivate the enzymes responsible for browning. Vegetables such as beans and peas also take 3–5 min. The optimum blanching conditions for strawberry, black currant, sour cherry, and prune range from 1.5 to 3 min at 85°C.

In the preparation of fruit for freezing, sugars and sugar solution have been used to sweeten and to exclude direct contact of the fruit tissue with molecular oxygen. The sugar solutions inhibit discoloration by reducing the concentration of dissolved oxygen and by retarding the rate of diffusion of the oxygen from the air into the fruit

tissues. Concentrated sugar solutions also exert an inhibiting effect on fruit phenolases. The pH also affects the rate of the browning reaction: acid dips are sometimes used to lower the pH and by this method delay or retard browning. The effect on oxidases of many salts and compounds such as sulfur dioxide, hydrogen sulfide, hydrocyanic acid, and thiourea has been studied.

In the home preparation of fruits, pineapple juice and lemon juice have long been used to prevent browning, Pineapple juice has a relatively high concentration of sulfhydryl compounds, which are active antioxidants, whereas lemon juice contains relatively high amounts of both citric and ascorbic acids.

Heat treatments and the application of sulfur dioxide or sulfites are common commercial methods for inactivating phenolases. Phenolase activity can be inhibited by the addition of sufficient amounts of acidulants such as citric, malic, or phosphoric acids to yield a pH of 3 or lower. Oxygen can be eliminated by vacuumization or by immersing the plant tissues in a brine or syrup. Phenolic substrates can be protected from oxidation by reaction with borate salts, but these are not approved for food use.

For products like vegetables, which will eventually be cooked, heat inactivation by steam or hot water blanching is the most practical method of inactivating phenolases. Overblanching should be avoided because of loss of firmness and nutrients. Most commercial blanching of vegetables has been based on the residual activity of the peroxidase because peroxidase is one of the most stable enzymes in plants. It has been generally accepted that if peroxidase is destroyed it is quite unlikely that other enzymes will have survived.

There are, however, problems associated with the use of peroxidase as an indicator of an adequate blanch in that several enzymes other than peroxidase have been shown to be largely responsible for quality deterioration during frozen storage of different vegetables (51,52). Therefore, optimization of blanching of individual products must be established to tailor the process to each type of raw material and product desired. Heat inactivation of phenolases in fruit products has been applied to fruit juices and purées and to fruit intended for such products. Because the enzyme is labile to heat at 85°C and above, the higher the temperature, the shorter the time required for inactivation. To minimize the undesirable changes due to excess heating, optimum temperature–time requirements for phenolases must be established. Rapid cooling after enzyme inactivation is also necessary for best quality retention.

Sulfur dioxide is an effective inhibitor of phenolases and has been used for many years. It readily reacts with compounds such as aldehyde and other carbonyl-containing molecules. These reaction products are ineffective against phenolases and, therefore, a sufficient amount of free SO_2 must be maintained. To be effective in preventing phenolase activity, SO_2 must penetrate throughout the tissues. Sulfurous acid penetrates better than the bisulfite form. Apple slices are subjected to a vacuum after immersion in a solution of SO_2 to enhance penetration in some commercial operations. Penetration of sulfite into slices is also influenced by pH. The use of excess SO_2 will produce both undesirable flavors and an excessively soft product. In recent years the safety of sulfites in foods has been questioned because of their hazard to certain asthmatics. Since August 9, 1986, the Food and Drug Administration has required a label declaration on any food containing more than 10 ppm of sulfiting agent.

Since phenolase is a metalloprotein with copper as its prosthetic group, it can be inhibited by metal-chelating agents such as cyanide, carbon monoxide, sodium diethyldithiocabamate, 2-mercaptobenzthiazol, azid, potassium methyl xanthate, or EDTA, as discussed above. Other chemicals such as benzoic acid, sodium chloride, potassium chloride, potassium bromide, cysteine, glutathione, and adenosine triphosphate are known to prevent browning. Soluble polymers such as polyvinylpyrrolidinone act as competitive inhibitors of phenolases. Recently, an acidic polyphosphate, Sporix, and cyclodextrin have been tested as antibrowning agents. However, many of these are not in commercial use.

Ascorbic acid and its derivatives such as erythorbic acid, ascorbic acid 2-phosphate, ascorbic acid triphosphate, and ascorbic acid 6-fatty acid esters are reported to control browning. Ascorbic acid is an effective reducing agent, as it reduces the o-quinones formed by phenolases to the original o-dihydroxyphenolic compounds. Ascorbic acid alone or in combination with citric acid has been used widely by the food industry. Its prevention of browning lasts as long as any residual ascorbic acid remains and, therefore, stabilized forms of ascorbic acid should be effective sulfite substitutes.

Another promising approach involves enzymes that transform the substrates of phenolases by methylation or oxidative cleavage of the benzene nucleus. o-Methyl transferase methylates the position of 3,4-dihydroxy aromatic compounds in the presence of a methyl donor, converting phenolase substrates into inhibitors of phenolases; for example, caffeic acid to ferulic acid.

The exclusion of oxygen as a means of controlling enzymatic browning is generally used in combination with other methods. For example, retail packs of frozen peach slices maintain their high quality when packaged in ascorbic acid–containing syrup combined with hermetically sealed containers in which the oxygen has been removed from the head space. Osmotic dehydration using sugar or syrup also inhibits enzymatic browning and protects flavors.

BIBLIOGRAPHY

1. M. Lindet, "Sur l'oxydation du tannin de la pomme a cidre," *Comptes Rendus des Seances de l' Academie des Sciences* **120**, 370–372 (1895).

2. J. M. Nelson and C. R. Dawson, "Tyrosinase," *Advances in Enzymology* **4**, 99–152, (1944).

3. L. Vamos-Vigyazo, "Polyphenol Oxidase and Peroxidase in Fruits and Vegetables," *Critical Review in Food Science and Nutrition* **15**, 49–127 (1981).

4. A. M. Mayer and E. Harel, "Polyphenol Oxidases in Plants," *Phytochemistry* **18**, 193–215 (1979).

5. A. M. Mayer, "Polyphenol Oxidases in Plants—Recent Progress," *Phytochemistry* **26**, 11–20, (1987).

6. G. Bertrand, "Sur la recherche et la presence de la laccase dans les vegetaux." *Comptes Rendus Hebdomadaires des Sciences de l'Academie des Sciences, Paris* **121**, 166–168 (1895).

7. K. Prabhakaran, "Properties of Phenoloxidase in *Mycobacterium leprae*," *Nature* (London) **218**, 973–974 (1968).

8. R. D. Tochner and B. J. D. Meeuse, "Enzymes of Marine Algae. 1. Studies on Phenolase in the Green Algae, *Monostroma fuscum*," *Canadian Journal of Botany* **44**, 551–561 (1966).

9. M. von Poucke, "Copper Oxidase in Thalli *Marchantia pollumorpha*," *Physiological Plantarum* **20**, 932–945 (1967).

10. R. C. Cambie and S. M. Bocks, "A p-Diphenol Oxidase from Gymnosperms," *Phytochemistry* **5**, 391–396 (1966).

11. H. Yoshida, "Chemistry of Lacquer (*Urushi*). Part I." *Journal of the Chemical Society* **43**, 472–486 (1883).

12. E. Harel, A. M. Mayer, and Y. Shain, "Catechol Oxidases from Apples, Their Properties, Subcellular Location and Inhibition," *Physiological Plantarum* **17**, 921–930 (1964).

13. A. M. Mayer, "Nature and Location of Phenolase in Germinating Lettuce," *Physiological Plantarum* **14**, 322–331 (1961).

14. E. Harel, A. M. Mayer, and Y. Shain, "Catechol Oxidases, Endogenous Substrates and Browning in Developing Apples, *Journal of the Science of Food and Agriculture* **17**, 389–392 (1966).

15. K. W. Wissemann and C. Y. Lee, "Polyphenoloxidase Activity During Grape Maturation and Wine Production," *American Journal of Enology and Viticulture* **31**, 206–211 (1980).

16. Y. Co Seteng and C. Y. Lee, "Changes in Apple Polyphenoloxidase and Polyphenol Concentration in Relation to Degree of Browning," *Journal of Food Science* **52**, 985–989 (1987).

17. H. S. Mason and C. I. Wright, "The Chemistry of Melanin. V. Oxidation of Dihydroxyphenylalanine by Tyrosinase," *Journal of Biological Chemistry* **180**, 235–247 (1949).

18. W. S. Pierpoint, "The Enzymatic Oxidation of Chlorogenic Acid and Some Reactions of the Quinone Produced," *Biochemical Journal* **98**, 567–580 (1966).

19. V. Khan, "Some Biochemical Properties of Polyphenol Oxidase from Two Avocado Varieties Differing in Their Browning Rates," *Journal of Food Science* **42**, 38–43 (1977).

20. L. Vamos-Vigyazo, K. Vas, and N. Kiss-Kutz, "Studies into the *o*-Diphenol Oxidase Activity of Potatoes. I. Development of an Enzyme Activity Assay Method," *Acta Alimentaria Academiae Scientarum Hungaricae* **2**, 413–429 (1973).

21. K. Mihalyi and L. Vamos-Vigyazo, "A Method for Determining Polyphenol Oxidase Activity in Fruits and Vegetables Applying a Natural Substrate," *Acta Alimentaria Academiae Scientarum Hungaricae* **5**, 69–85 (1976).

22. M. El-Bayoumi and E. Frieden, "A Spectrophotometric Method for the Determination of the Catecholase Activity of Tyrosinase and Some of Its Applications," *Journal of the American Chemical Society* **79**, 4854–4859 (1957).

23. F. Mazzocco and P. G. Pifferi, "An Improvement of the Spectrophotometric Method for the Determination of Tyrosinase Catecholase Activity by Besthorn's Hydrozone," *Analytical Biochemistry* **72**, 643–647 (1976).

24. H. Esterbauer, E. Schwarzl, and M. Hayn, "A Rapid Assay for Catechol Oxidase and Laccase Using 2-Nitro-5-thiobenzoic acid," *Analytical Biochemistry* **77**, 486–494 (1977).

25. A. M. Mayer, E. Harel, and Y. Shain, "2,3-Naphthalenediol, a Specific Competitive Inhibitor of Phenolase," *Phytochemistry* **3**, 447–451 (1964).

26. R. J. Petroski, W. Peczynska-Czoch, and J. P. Rosazza, "Analysis, Production, and Isolation of an Extracellular Laccase from *Polyporus anceps*," *Applied and Environmental Microbiology* **40**, 1003–1006 (1980).

27. V. Kovac, "Effet des moisissures du genre penicillium sur l'activite laccase du botrytis cinerea." *Annales de Technologie Agricole* **28**, 345–355 (1979).

28. J. R. L. Walker and R. F. McCallion, "The Selective Inhibition of *ortho*- and *para*-Diphenol Oxidases," *Phytochemistry* **19**, 373–377 (1980).

29. D. Kertesz and R. Zito, "Mushroom Polyphenol Oxidase. I. Purification and General Properties," *Biochimica et Biophysica Acta* **96**, 447–462 (1965).

30. W. D. Loomis and J. Battaile, "Plant Phenolic Compounds and the Isolation of Plant Enzymes," *Phytochemistry* **5**, 423–438 (1966).

31. N. O. Benjamin and M. W. Montgomery, "Polyphenol Oxidase of Royal Ann Cherries: Purification and Characterization," *Journal of Food Science* **38**, 799–806, (1973).

32. K. W. Wisseman and M. W. Montgomery, "Purification of d'Anjou pear (*Pyrus communis* L.) Polyphenol Oxidase," *Plant Physiology* **78**, 256–262 (1985).

33. K. W. Wisseman and C. Y. Lee, "Purification of Grape Polyphenoloxidase with Hydrophobic Chromatography," *Journal of Chromatography* **192**, 232–235 (1980).

34. E. L. Angleton and W. H. Flurkey, "Activation and Alteration of Plant and Fungal Polyphenoloxidase Isoenzymes in Sodium Dodecylsulfate Electrophoresis," *Phytochemistry* **23**, 2723–2725 (1984).

35. M. Hruskocy and W. H. Flurkey, "Detection of Polyphenoloxidase Isoenzymes by Electroblotting and Photography." *Phytochemistry* **25**, 329–332 (1986).

36. S. Bouchilloux, P. McMahill, and H. S. Mason, "The Multiple Forms of Mushrooms Tyrosinase. Purification and Molecular Properties of the Enzymes," *Journal of Biological Chemistry* **238**, 1699–1707 (1963).

37. K. G. Strothkamp, R. L. Jolley, and H. S. Mason, "Quaternary Structure of Mushroom Tyrosinase," *Biochemical and Biophysical Research Communications* **70**, 519–524 (1976).

38. E. Harel, A. M. Mayer, and E. Lehman, "Multiple Forms of *Vitis vinifera* Catechol Oxidase," *Phytochemistry* **12**, 2649–2654 (1973).

39. A. M. Mayer and E. Harel, "Laccase-Like Enzyme in Peaches," *Phytochemistry* **7**, 1253–1256 (1968).

40. V. Kahn, "Latency Properties of Polyphenol Oxidase in Two Avocado Cultivars Differing in Their Rate of Browning," *Journal of the Science of Food and Agriculture* **28**, 233–239 (1977).

41. J. R. L. Walker and E. L. Wilson, "Studies on the Enzymic Browning of Apples. Inhibition of Apple-*o*-Diphenol Oxidase by Phenolic Acids," *Journal of the Science of Food and Agriculture* **26**, 1825–1831 (1975).

42. J. R. L. Walker, "The Control of Enzymic Browning in Fruit Juices by Cinnamic Acid," *Journal of Food Technology* **11**, 341–345 (1976).

43. M. E. Winkler, K. Lerch, and E. I. Solomon, "Competitive Inhibitor Binding to the Binuclear Copper Active Site in Tyrosinase," *Journal of the American Chemical Society* **103**, 7001–7003 (1981).

44. M. A. Augustin, H. M. Ghazali, and H. Hashim, "Polyphenoloxidase from Guava (*Psidium guajava* L.)," *Journal of the Science of Food and Agriculture* **36**, 1259–1265 (1985).

45. T. Nakamura, "On the Process of Enzymatic Oxidation of Hydroquinone," *Biochemical and Biophysical Research Communications* **2**, 111–113 (1960).

46. R. A. Holwerda, S. Wherland, and H. B. Gray, "Electron Transfer Reactions of Copper Protein," *Annual Review of Biophysics and Bioengineering* **5**, 363–396 (1976).

47. V. Cheynier, C. Osse, and J. Rigaud, "Oxidation of Grape Juice Compounds in Model Solutions," *Journal of Food Science* **53**, 1729–1732 (1988).

48. V. L. Singleton, "Oxygen with Phenols and Related Reactions in Musts, Wines, and Model Systems: Observations and Practical Implications," *American Journal of Enology and Viticulture* **38**, 69–77 (1987).

49. V. Chenyier, N. Basire, and J. Rigaud, "Mechanism of *trans*-Caffeoyltartaric Acid and Catechin Oxidation in Model Solutions Containing Grape Polyphenoloxidase," *Journal of Agricultural and Food Chemistry* **37**, 1069–1071 (1989).

50. J. Oszmianski and C. Y. Lee, "Enzymatic Oxidation Reaction of Catechin and Chlorogenic Acid in a Model System," *Journal of Agricultural and Food Chemistry*, in review.

51. M. H. Lim, P. Velasco, R. A. Pangborn and J. R. Whitaker, "Enzyme Indicators of Adequate Blanching," in D. S. Reid, *Proceedings International Conference on Technical Innovations in Freezing and Refrigeration of Fruits and Vegetables, University of California, Davis,* Food Science and Technology, University of California, Davis, July 9–12, 1989, pp. 67–72.

52. C. Y. Lee, N. L. Smith, and D. E. Hawbecker, "An Improved Blanching Technique for Frozen Sweet Corn on-the-Cob," in *Proceedings International Conference on Technical Innovations in Freezing and Refrigeration of Fruits and Vegetables, University of California, Davis,* Food Science and Technology, University of California, Davis, July 9–12, 1989, pp. 85–90.

CHANG Y. LEE
Cornell University
Geneva, New York

BROWNING REACTION, NONENZYMATIC. See CARBOHYDRATES; BLANCHING; OXIDATION; FOOD ADDITIVES; ANTIOXIDANTS; FOOD PROCESSING; FOOD CHEMISTRY: MECHANISM AND THEORY; ENZYMOLOGY; FOOD SPOILAGE; FOOD PRESERVATION; DEHYDRATION; FOOD FREEZING; FREEZE DRYING AND FREEZE CONCENTRATION; NUTRITIONAL QUALITY AND FOOD PROCESSING; HEAT; BROWNING REACTION, ENZYMATIC; SYRUPS; Entries under THERMAL PROCESSING; FATS AND OILS; PROTEINS.

BUTTER AND BUTTER PRODUCTS

The art of butter making dates back to times immemorial. Reference to the use of butter for sacrificial worship, for medicinal and cosmetic purposes, and as a human food may be found long before the Christian Era. Documents indicate that, at least in the Old World, the taming and domestication of animals constituted the earliest beginnings of human civilization and culture. There is good reason to believe, therefore, that the milking of animals and the origin of butter making aforedate the beginning of organized and permanent recording of human activities.

The evolution of the art of butter making has been intimately associated with the development and use of equipment. With the close of the eighteenth century the construction and use of creaming and butter-making equipment, other than that made of wood, began to receive consideration and the barrel churn made its appearance.

By the middle of the nineteenth century attention was given to improvement in methods of creaming. These efforts gave birth to the deep-setting system. Up to that time, creaming was done by a method called shallow pan. The deep-setting system shortened the time for creaming and produced a better quality cream. An inventive Bavarian brewer in 1864 conceived the idea of adapting the principle of the laboratory centrifuge. In 1877 a German engineer succeeded in designing a machine that, although primitive, was usable as a batch-type apparatus. In 1879 engineers in Sweden, Denmark, and Germany succeeded in the construction of cream separators for fully continuous operation (1).

In 1870, the last year before introduction of factory butter making, butter production in the United States was 514 million pounds, practically all farm made. Authentic records concerning the beginning of factory butter making are meager. It appears that the first butter factory was built in Iowa in 1871. This beginning also introduced the pooling system of creamery operation (1).

Other inventions that assisted the development of the butter industry included the Babcock test (1890), which accurately determines the percentage of fat in milk and cream; the use of pasteurization to maintain milk and cream quality; and the use of pure cultures of lactic acid bacteria and refrigeration to help preserve cream quality.

BUTTER CONSUMPTION

Production and consumption continues a long-declining trend (Tables 1 and 2). A dramatic shift occurred, starting in 1985, to the table spreads category of products (less than 80% fat) from full-fat butter, margarine, and blends

Table 1. Market Shares for Butter, Margarine, Blends, and Spreads for 1980–1990[a]

Year	Butter	Margarine	Blends	Spreads
1980	25	74	1	—
1985	22	73	3	2
1990[b]	17	48	2	33

[a] Refs. 2, 3.
[b] Estimate.

(2,3). The spreads category encompasses all non-Standard-of-Identity table spreads (ie, from 0 to 79.9% fat).

In a 1984 survey the most important barriers to increased butter sales were listed in the following order (2):

1. Price (opinion of an overwhelming majority when butter is compared to margarine).
2. Health (negative consumer attitudes toward cholesterol and saturated fats are increasing).
3. Poor spreadability.
4. Inadequate promotional spending.
5. Product innovation in margarine and spreads.
6. Legislation and regulatory restrictions.

Butter manufacture continues to serve as the safety valve for the dairy industry. It absorbs surplus milk supply above market requirements for other dairy products. Milk not required by the demand for these products overflows into the creamery, is skimmed, and the cream is converted to butter. When the milk supply for other products runs short of their demand, milk normally intended for butter making is diverted into the channels where needed. Even though consumption patterns have dramatically changed over the years, the butter industry thus provides a never-failing balance wheel that takes up the slack in the relationship of supply and demand for all other dairy products.

DEFINITIONS AND GRADING OF BUTTER

Standards for butter in the United States were established by an Act of Congress and are supported by the U.S. Department of Agriculture (USDA) standards for grades of butter. In the revised standards the following definitions apply. Butter refers to the food product usually known as butter, which is made exclusively from milk, cream, or both; with or without common salt; and with or

Table 2. Per Capita Consumption of Butter for the Years 1950–1990[a]

Year	Pounds Per Capita
1950	10.7
1960	7.7
1970	5.4
1980	4.5
1990[b]	4.0

[a] Refs. 3–5.
[b] Estimate.

without additional coloring matter. The milk fat content of butter is not less than 80% by weight, allowing for all tolerances. Cream refers to the cream separated from milk produced by healthy cows. Cream is pasteurized at a temperature of not less than 73.9°C for not less than 30 min, or it can be pasteurized at a temperature of not less than 89°C for not less than 15 s. There are other approved methods of pasteurization that give equivalent results (6).

The cream may be cultured by the addition of harmless lactic acid bacteria to enhance flavor, natural flavors obtained by distilling a fermented milk, or cream may be added to the finished butter. In addition, color, derived from a Food and Drug Administration (FDA) approved source, may be used.

There are three U.S. grades of butter: AA, A, and B. Butter is graded by first classifying its flavor organoleptically. In addition to the overall quality of the butter flavor itself, the standards lists 17 flavor defects and the degree to which they may be present for each grade. If more than one off-flavor is noted, assignment is made on the basis of the flavor resulting in the lowest grade. This grade is then lowered by defects in the workmanship and the degree to which they are apparent. Disratings are characterized by negative body, flavor, or salt attributes, which are fully described in the standards. Butter that does not meet the requirements for U.S. Grade B is not graded. To bear the USDA seal, the finished product must fall within the following microbiological specifications: proteolytic count not more than 100 per gram, yeast and mold not more than 20 per gram, and coliform not more than 10 per gram. Butter should be stored at 4.4°C or lower or at less than −17.8°C, if it is to be held for more than 30 days (7).

Legal requirements for butter vary considerably in different countries. For example, in Europe butter must contain 82% fat and in France it may contain a maximum of 16% moisture (7). In some tropical parts of the world, milk fat is used in nearly anhydrous form because it is less susceptible to bacterial spoilage. This product is known as ghee. In the Middle East and India ghee is prepared from heated cow or buffalo milk.

COMPOSITION

The concentration of milk fat results in a dairy product called butter. The concentration is a two-stage process with the first stage being effected in a mechanical separator, which causes a 10-fold concentration of the fat into an intermediate product, cream. The second stage is the churning process, which results in a further 2-fold concentration of the fat.

In butter manufacture, the preparation of the cream is effected quite independently of the butter-manufacturing stage. Today most butter plants receive their cream from other operations rather than directly from the farm, as was the case in the past. Because of general improvement in sanitation practices, the receipt of fresher milk and cream, and advances in the knowledge and understanding of deleterious handling conditions, the quality of cream is constantly improving.

The composition of milk fat is the most important factor affecting the firmness of butter and, therefore, its spreadability. The composition of milk fat changes primarily according to the feed; therefore, the entire problem is connected to the animal's diet. Today, the fatty acid composition of milk fat produced in various countries is rather accurately known, along with its seasonal variations. In Europe the amount of saturated fatty acids is generally highest in winter and lowest in summer or fall (Table 3) (9). Green fodder decreases the amount of saturated fatty acids and correspondingly increases the amount of unsaturated fatty acids. The differences between the maximum and minimum values can be fairly large. For palmitic and oleic acids, the quantitatively most important fatty acids, a difference of more than 10% between the maximum and minimum values was found in some cases. This makes it understandable that there are also significant differences in the physical characteristics of the butter. The structure of the triglycerides in the milk fat, along with the fatty acid composition, is important in determining the physical characteristics of the fat, because the softening point of fat has been found to rise as the result of interesterification (10).

Textural characteristics of butter are significantly dependent on milk fat composition and method of manufacture. It is possible, therefore, knowing the chemical composition of milk fat, to select the appropriate technological parameters of butter making to improve the texture. To obtain butter with constant rheological characteristics and to control the parameters of the butter-making process it is necessary to take account of the difference in the chemical compositions and the properties of the milk fat in various seasons.

Cream Quality

Raw cream should be processed without delay to minimize deterioration of quality. If the cream is to be held for more than two hours between separation and processing, it should be cooled below 5°C. For holding periods exceeding one day it may be advantageous to heat, treat, or pasteurize the cream.

Table 3. Compositional Characteristics of Summer and Winter Milk Fat, 1970[a]

Samples Investigated (N = 140)	Fatty Acids				
	Volatile	Saturated	Monounsaturated	Polyunsaturated	Iodine Number
Total	10.98	56.50	29.81	2.50	32.2
Summer	9.49	58.82	33.53	3.14	36.8
Winter	12.45	59.15	26.15	1.86	27.7

[a] Ref. 8.

To make high-quality butter, the cream must be of high quality. The flavor of butter is the most important organoleptic property and it depends in no small part on the flavor quality of the cream from which it is made. There is no real alternative to tasting in assessing flavor and this simple but important procedure should be conducted routinely as part of the overall quality-control program.

Fat Content

The percentage of fat in the cream must be known and controlled. It influences fat losses in churning. Knowledge of the fat content assists in yield estimations for operational conditions in continuous manufacture. A number of satisfactory procedures are available, with the Babcock test being the most common.

The chemical composition of the triglycerides, which comprise milk fat, varies throughout the year depending on the stage of lactation and the cow's diet. This causes a cyclic change in the melting properties of the fat. In the control of the butter-making process and the physical properties of the finished butter this factor must be monitored. The term melting property is used rather than softness or hardness because these more correctly refer to an altogether different attribute of solids. A number of procedures have been used to follow the seasonal change in the melting properties. The iodine number, refractive index and differential scanning calorimetry, or pulsed nmr spectroscopy can be used to prepare a melting curve. However, the expense and complexity of these melting curve techniques precludes the approach in most quality-control situations (11). The traditional chemical determinations for fat, saponification value and Polenske value, are of limited value. They are scarcely relevant for quality control and the information they represent can be more usefully quantified by the determination of the fatty acid profile using gas chromatography.

Today the use of stainless steel has essentially eliminated the exposure of the fat to copper and iron. The presence of copper and, to a lesser extent, iron can catalyze oxidative deterioration of butter during storage, particularly in the presence of salt and a low pH.

Many procedures have been used or proposed to assess the microbiological quality of the milk or cream. Generally, microbiological tests are performed to determine the hygiene of production and storage conditions or for safety reasons. Tests include total counts and counts for specific classes of microorganisms such as yeasts and molds, coliforms, psychotrophs, and pathogens such as *Salmonella*. Rapid-screening tests based on dye reduction or direct observation using a microscope or automatic total counters are also in use.

Lipase Activity

Lipolysis in butter after manufacturing caused by thermoresistant lipase enzymes created in milk or cream by psychotrophic bacteria is an increasing problem. Based on a determination of the lipase activity in cream the keeping quality of manufactured butter in regard to lipolysis can be predicted with reasonable accuracy. A similar prediction for sweet cream butter can be based on activity determination in the serum phase (12).

Consistency

Consistency has been defined as "that property of the material by which it resists permanent change of shape and is defined by the complete force flow relation" (9). This implies that the concept of consistency includes many aspects and cannot be expressed by one parameter. Today more importance is attached to spreadability than anything else in the evaluation of the consistency of butter. The general consensus is that butter should be spreadable at refrigerator temperatures. There is no suitable method available to measure such a subjective criteria as the spreadability of butter. For this reason, the firmness of butter, which should correlate well with spreadability, was selected as the parameter to be measured. It was recommended that the use of the cone penetrometer, along with other methods, could give good results (13).

Flavor

One of the most important consumer attributes of butter is the pleasing flavor. Butter flavor is made up of many volatile and nonvolatile compounds. Researchers have identified more than 40 neutral volatiles, of which the most prominent are lactones, ethyl esters, ketones, aldehydes, and free fatty acids (14). The nonvolatiles, of which salt (sodium chloride) is the most prominent, contribute to a balanced flavor profile. Diacetyl and dimethyl sulfide also contribute, especially in cultured butter flavor (15).

Ripened or cultured creams use lactic acid bacterial cultures selected for their ability to produce volatile acid and diactyl flavors. The NIZO process was developed to incorporate concentrated lactic starters or to directly inject starter distillates to produce the flavor and characteristics of cultured butter (13). In the UK, Canada, Ireland, and the United States sweet cream, salted butter is primarily used, but in France, FRG, and Scandinavia the taste of cultured butter is preferred (7).

Body and Texture

The physical properties of butter that are noted by the senses are described by butter graders by means of appropriate qualifications of the terms body and texture. The exact meaning of these terms has not been clearly defined. Frequently they are used as if they had the same meaning. Certain properties such as hardness and softness refer to the body of butter, whereas properties such as openness refer to texture. But some of the properties, such as leakiness or crumbliness, lead to confusion. Usually, most body and texture terms are used to describe a defect such as gritty, gummy, sticky, and the like.

Good butter should be of fine and close texture; have a firm, waxy body; and be sufficiently plastic to be spreadable at cold temperatures.

Color

The color of butter may vary from a light creamy white to a dark creamy yellow or orange yellow. Differences in butter color are due to variations in the color of the butter fat, which is affected by the cows' feed and season of the year; variations in the size of the fat globule; presence or absence of salt; conditions of working the butter; and the type and amount of natural coloring added.

Butter colorings are oil soluble and most often are natural annatto, an extraction of the seeds of the tropical tree *Bixa orellana,* or natural carotenes, extractions from various carotene-rich plants. Because they are oil soluble, colorings are added to the cream to obtain the most uniform dispersion.

PROCESSING

Milk and Cream Separation

The most basic and oldest processing method is cream separation. Ancient people are known to have used milk freely. It is probable that they used the cream that rose to the top of milk that had been held for some time in containers, although there is little in ancient literature to suggest that such use was common. It is well established that in early times butter was produced by churning milk.

The principal questions concerning separation in a butter-manufacturing facility are the choice of cream fat content and the choice of the separation technique, milk separation before or after pasteurization, temperature of separation, and regulation of the fat content. Separation of cream from milk is possible because of a difference in specific gravity between the fat and the liquid portion, or serum. Whether separation is accomplished by gravity or centrifugal methods, the result is dependent on this difference (16).

Crystallization

The crystal structure of fat and the resulting physical properties of butter made by both conventional and alternative processes has received considerable study. When conventionally churned or Fritz butter is made, most of the milk fat is contained within the fat globule in cream during the cooling and crystallization process. The fat globule provides a natural limit to the growth of fat crystals. Cooling and holding of cream is normally carried out overnight, and thus sufficient time exists for the crystallization process to approach equilibrium (17).

The principles of crystallization of plastic fats in the type of equipment used for margarine manufacture have been described (18). It is important to develop small fat crystals that remain substantially discrete and do not form a strong interlocking structure. Small crystals (eg, 5-μm diameter) have a greater total surface area than large crystals and will bind water and free liquid fat by adsorption more effectively (17). Large crystals impart a gritty texture to the product. When fats are cooled rapidly in a scraped surface heat exchanger, fat crystallization commences but the fat is substantially supercooled on exiting. If crystallization is then permitted to continue under quiescent conditions, crystals will grow together and form a lattice structure. The product will thus be hard and brittle, and may tend to leak moisture. If, however, crystallization is permitted to occur under agitated conditions (eg, for 1–3 min in a pin worker), the formation of small, independent crystals will be favored and the product will have a fine, smooth texture. If crystallization under agitated conditions is permitted to continue for too long, the product will be too soft for most patting or bulk filling operations, and it is likely to be too soft and greasy at warm room temperatures.

Neutralization

When lactic acid has developed in the raw, unpasteurized cream by microbial activity to a degree considered excessive, neutralizer may be added to return the cream acidity to a desirable level. Sodium carbonates have been found suitable in practice for batch neutralization. For continuous neutralization by pH control, sodium hydroxide is more suitable. These chemicals should be of food grade.

Fractionation

Fractionation by thermal crystallization, steam stripping, short-path distillation, by supercritical fluid extraction, or crystallization using solvents can achieve fat alterations of significance to the dairy industry. Milk fat is a mixture of triglycerides with a range of molecular weights, degrees of unsaturation, melting points, and other physical properties. Milk fat is an important component of most dairy products, but it has been consumed traditionally for the most part as butter.

Fractionation of butter fat for the selective removal of cholesterol has been recently reported using supercritical fluid extraction (19), steam stripping or deodorization (20), short-path molecular distillation (21), and absorption processes (16). Various enzymatic investigations are under way to isolate cholesterol reductase (22) and to insert the gene for cholesterol reductase into lactic and acid bacteria biogenetically (23), resulting in the reduction of the dairy fat cholesterol to its major by-product, coprostanol.

Heat Treatment

The heat treatment of cream plays a decisive role in the butter-manufacturing process and the eventual quality of the butter. It is important that milk and cream be handled in the gentlest possible way to avoid mechanical damage to the fat, a serious problem in continuous manufacture (Fritz process) of butter (24). Cream is pasteurized or heat treated for the following reasons: to destroy pathogenic microorganisms and reduce the number of bacteria; to deactivate enzymes, to liquify the fat for subsequent control of crystallization, and to provide partial elimination of undesirable volatile flavors.

Batch Butter Manufacture

Today batch processing is not used to any extent for the production of large quantities of butter. Batch systems are still encountered in small butter plants, primarily in less industrially developed countries. Continuous systems are more efficient and cost effective for large outputs; batch systems have low capital intensity.

The processing of cream by a batch churn requires filling to approximately 30–50% capacity at a cream temperature of 4.4–12.8°C. Cream temperature will vary depending on seasonably of the cream, the butter characteristics desired, and the desired rate of fat inversion. Churning is accomplished by rotation of the churn at approximately 35 rpm until small butter granules appear. The process usually requires about 45 min for coalescence of the fat globules and clean separation of the buttermilk so that it can be drained (25). The granules may be washed with cold water to remove surface buttermilk. Salt is then added with water if the fat content standardization is required. The butter is worked to ensure uniformity and desirable body and texture characteristics.

Continuous Butter Manufacture

Between 1930 and 1960 a number of continuous processes were developed. In the Alfa, Alfa–Laval, New Way, and Meleshin processes phase inversion takes place by cooling and mechanical treatment of the concentrated cream. In the Cherry–Burrell Gold'n Flow and Creamery Package processes, phase inversion takes place during or immediately after concentration, producing a liquid identical to melted butter, prior to cooling and working. The Alfa, Alfa–Laval, and New Way processes were unsuccessful commercially. The Meleshin process, however, was adopted successfully in the USSR. The Cherry–Burrell Gold'n Flow process appears to have been the more successful of the two American processes (17).

The Fritz continuous butter-making process, which is based on the same principles as traditional batch churning, is now the predominant process for butter manufacture in most butter-producing countries. In the churning process crystallization of milk fat is carried out in the cream, with phase inversion and milk fat concentration taking place during the churning and draining steps. However, because the discovery that cream could be concentrated to a fat content equal to or greater than that of butter, methods have been sought for converting the concentrated or plastic cream directly into butter. Such methods would carry out the principal butter-making steps essentially in reverse order, with concentration of cream in a centrifugal separator, followed by a phase inversion, cooling, and crystallizing of the milk fat (25).

Increased demands on the keeping qualities of butter require that, in addition to careful construction, operation, and cleaning of the milk and cream processing equipment, research to develop machines that will ensure butter production and packing under conditions eliminating contamination and air admixture must be carried out. It has been demonstrated that butter produced under closed conditions has a better keeping quality than butter produced in open systems (26).

There are two classes of continuous processes in use: one using 40% cream, such as the Fritz process, and the other using 80% cream, such as the Cherry–Burrell Gold'n Flow. As much as 85% of the butter in France is made by the Fritz process. In this process 40% fat cream is churned as it passes through a cylindrical beater in a matter of seconds. The butter granules are fed through an auger where the buttermilk is drained and the product is squeeze dried to a low moisture content. It then passes through a second working stage where brine and water are injected to standardize the moisture and salt contents. As a result of the efficient draining of the buttermilk, this

process is suitable for the addition of lactic acid bacteria cultures at this point. The process then becomes known as the NIZO method when the lactic starter is injected (17). Advantages of the NIZO method over traditional culturing are improved flavor development; acid values as a result of lower pH; more flexible temperature treatment of the cream, because culturing and tempering often are accomplished concurrently; and, most important, sweet cream buttermilk is produced.

The Cherry–Burrell Gold'n Flow process is similar to margarine manufacture (27). The process starts with 18.3°C cream that is pumped through a high-speed destabilizing unit and then to a cream separator from which a 90% fat plastic cream is discharged. It is then vacuum pasteurized and held in agitated tanks to which color, flavor, salt, and milk are added. Then this 80% fat–water emulsion, which is maintained at 48.9°C, is cooled by use of scraped surface heat exchangers to 4.4°C. It then passes through a crystallizing tube, followed by a perforated plate that works the butter. Prior to chilling, 5% nitrogen gas is injected into the emulsion.

Although the Meleshin process continues to be in widespread use in the USSR, the use of alternative continuous butter-making processes based on high-fat cream has declined in Western countries during the past 20 years (17). The principal reasons for this decline appear to be the economics and butter quality, particularly when compared to the Fritz process. A Fritz manufacturing process can be installed in existing batch churn factories with almost no modification to cream-handling or butter-packing equipment. The churns could be retained in case the Fritz breaks down. However, very little batch plant equipment could be reused in the alternative systems (ie, Gold'n Flow). When a completely new plant is being bought, the alternative systems still tend to be more expensive, and operational advantages over the Fritz are not significant. Butter from the Fritz process is nearly identical in its physical and flavor characteristics to batch-churned butter, whereas butter produced by the alternative processes tends to be different. These differences may be perceived as defects by the consumer, and manufacturers have been reluctant to alter a traditional product.

There are a number of advantages of the alternative systems compared to a modern Fritz line: (1) The most attractive advantage is the flexibility to produce a wide range of products with fat contents ranging from 30 to 95% butter–vegetable oil blends and the incorporation of fractionated fats. (2) These processes also present the possibility of a number of operational advantages. By use of an efficient centrifuge during the cream concentration stage, fat losses in the buttermilk can be substantially reduced. (3) The composition of the butter can be more accurately controlled, either by including a batch standardization step or by the use of accurate continuous metering systems.

BUTTER PRODUCTS

Butter–Vegetable Oil Blends

In the early 1980s blends of butter and vegetable oil products appeared in the U.S. market. These blends generally were 40% butter and 60% vegetable oil for a total fat content of 80%, within the margarine Standard of Identity and designation. With the increasing popularity of reduced-fat (less than 80%), spreads, starting in the mid-1980s, other blends with butter fat contents of 2–25% were introduced. As noted in Table 1, the full 80% fat blends have declined in preference to the lower fat spreads; the blend spreads have taken a portion of this market share (3).

A number of processes were developed using continuous churns (28) and alternative systems similar to the Cherry–Burrell Gold'n Flow process (29). The major disadvantage to churning, either batch or continuous, was that the resultant buttermilk would be adulterated with some vegetable fat and would be less valuable than standard buttermilk. An advantage of alternative processing systems is their ability to easily accommodate the manufacture of reduced-fat spread blends.

With the popularity of blends in the United States, several products were introduced in Europe and Australia and have subsequently expanded the market for butter fat usage (3). An even earlier entry into the blend category was Swedish Bregott in Europe.

Whipped Butter

Whipped butter is a product produced to improve spreadability; air or nitrogen is incorporated. With the incorporation of gas, the butter's volume increases by approximately 33%. Generally, whipped butter is sold in tub rather than stick form.

Spreadable Butter

In 1937 a process known as the Alnarp method was developed to increase the spreadability of butter made from winter milk fat. The apparent reason for its effectiveness is that it increases the liquid fat content of the milk fat (30).

The consistency of butter is determined by the percentage of solid fat present, which is directly influenced by the fat composition, the thermal treatment given to cream prior to churning, the mechanical treatment given to butter after manufacture, and the temperature at which the butter is held (31). The European butter market demands that butter be softer and more spreadable in winter and harder in summer. With information on the changes in fat composition from gas–liquid chromatography analysis and the use of nuclear magnetic resonance to estimate solid fat, suitable tempering procedures can be selected to modify the fat composition and produce the most acceptable product for the consumer. A spreadable consistency of butter can be achieved by either varying the fatty acid composition or varying heat-step cream-ripening times and temperatures (32).

In efforts to improve the spreading properties of butter in relation to hard butter fat, one alternative put forward is the use of soft fat fractions obtained in the fractionation of anhydrous milk fat. Although several practical methods of fractionation have been presented, the use of soft fat fractions in butter making has not become general practice. This is evidently because fractionation in all cases

significantly raises the cost of the butter produced. In addition, a common problem has been to find suitable uses for the hard fat fractions. Furthermore, in fractionation methods that use solvents or additives, fractionation should be linked to fat refining; and in this process butter also loses its natural food classification. In studies that have used soft butter fat fractions, a substantial softening of the butter has been obtained; however, this butter, like normal butter, hardens as the temperature increases and again decreases. One of the new ways to improve butter spreadability is to add vegetable oil during manufacturing. The best known of these preparations, which cannot be called butter, is the Swedish Bregott, in which 20% of the fat used is vegetable oil. A similar product is made in many other countries, Finland among them, where it has been well received, possibly just because of its better spreadability compared to normal butter.

One industrial process in practice for the fractionation of milk fat is the Tirtiaux system, which is a semicontinuous bulk crystallization process. The Tirtiaux dry fractionation process enables one- or two-step fractionation of butter oil at any temperature from 50 to 2°C. The milk fat fractions thus obtained can be used as such or they can be blended in different proportions for use as ingredients in various food fat formulations or in preparing spreadable butter (33).

Low-Fat Butter

Butterlike products with reduced fat content are manufactured in several countries. Stabilizers, milk and soy proteins, sodium albumin or caseinate, fatty acids, and other additives are used. A product is now available on a commercial scale in the USSR that has the following composition: 45% milk fat, 10% nonfat solids, and 45% moisture. It has a shelf life of 10 days at 5°C (34).

Decreasing the energy content of a diet has clearly been the motive behind those milk fat products in which the fat content is approximately half that of normal butter. Although these products can no longer be called butter according to international standards, they are nonetheless often called low-calorie butter, half-butter, or similar names. A large number of patents have been obtained for these products, because raising the water content to nearly 50% in the manufacture of butterlike spreads requires considerably higher emulsifying properties than the manufacture of normal butter. In addition, the emulsion must often be stabilized with additives. In some countries, a lowfat butter (40% total fat) containing vegetable oil has been designated as Minarine, but Minarine can also be prepared using only butter fat (7).

Healthy Butter

Fractionation by crystallization, supercritical fluid extraction, and other technology are methods being applied commercially to milk fat to create desirable new products, such as decholesterolized butter. The essential purpose of milk fat application development is to adapt products to fit user demands.

Part of the reason for consumer popularity of the butter–vegetable oil mixtures may be seen in the emphasis on saturated animal fats in current nutritional debates as well as the alleged cause-and-effect relation between butter and heart disease. For these reasons, people interested in good nutrition willingly change over to products that contain a certain amount of vegetable oil but also have the natural butter aroma.

Because the physical properties of milk fat influence the rheological properties of dairy products, especially butter, there has been considerable interest in the modification of milk fat by physical and chemical means. Economical fractionation of milk fat into oil and plastic fat fractions will facilitate an increased utilization of milk fat in many food applications, such as chocolate, confectionary and bakery products, and in developing new convenient (spreadable) and dietetic (reduced cholesterol, variable fatty acid composition) butter or butter fat–containing products.

BIBLIOGRAPHY

1. O. F. Hunziker, *The Butter Industry*, 3rd ed., Printing Products Corp., Chicago, 1940.

2. Anonymous, "The World Market for Butter," *International Dairy Federation, Bulletin,* Document **170,** 1984.

3. Land O'Lakes market research, unpublished data.

4. U.S. Department of Commerce, Bureau of the Census, Washington, D.C. *Statistical Abstract of the United States,* 108th ed., 1988.

5. R. Rizek, N. Raper, and K. Tippett, "Trends in U.S. Fat and Oil Consumption," *Journal of the American Oil Chemists' Society* **65**(5), 723 (1988).

6. *Code of Federal Regulations,* Title 7, Part. 58, Subpart P, U.S. Government Printing Office, Washington, D.C., 1983.

7. M. M. Chrysam, "Table Spreads and Shortenings," in T. H. Applewhite, ed., *Bailey's Industrial Oil and Fat Products,* Vol. 3, John Wiley & Sons, Inc., New York, 1985.

8. J. Kisza, K. Batura, B. Staniewski, and W. Zatorski, "Butter Texture and Consistency as a Function of Milk Fat Composition and Method of Manufacture," *Brief Communication* **1**(1), 352 (1982).

9. V. Antila, "The Consistency of Butter and the Possibilities to Manipulate It," *Proceedings of the XXI International Dairy Congress, Moscow* **2,** 159 (1982).

10. P. W. Parodi, "Relationship Between Triglyceride Structure and Softening Point of Milk Fat," *Journal of Dairy Research* **48,** 131 (1981).

11. B. D. Dixon, "Continuous Butter Manufacture: Raw Materials," *International Dairy Federation Bulletin* **204,** 3–5 (1986).

12. B. Schaffer and S. Szakaly, "Heat-Step Cream Ripening for Producing Spreadable Butter in Hungary," *Brief Communication* **1**(1), 337 (1982).

13. G. Van den Berg, "Developments in Buttermaking," *Proceedings of the XXI International Dairy Congress, Moscow* **2,** 153 (1982).

14. T. Siek and J. Lindsay, "Semiquantitative Analysis of Fresh Sweet-Cream Butter Volatiles," *Journal of Dairy Science* **53,** 700 (1970).

15. J. E. Kinsella, "Butter Flavor," *Food Technology* **29**(5), 82–98 (1975).

16. D. H. Hettinga, *Altering Fat Composition of Dairy Products,* CAST Publication, Ames, Iowa, 1991.

17. M. Kimenai, "Continuous Butter Manufacture," *International Dairy Federation Bulletin* **204,** 16 (1986).

18. L. H. Wiedermann, "Margarine and Margarine Oil, Formulation and Control," *Journal of the American Oil Chemists' Society* **55**(11), 823 (1978).

19. D. H. Hettinga, "Processing Technologies for Improving the Nutritional Value of Dairy Products," in Board of Agriculture, ed., *Designing Foods,* National Academy Press, Washington, D.C., 1988, p. 292.

20. R&D Applications, *Cholesterol Reduced Fat: They're Here,* Prepared Foods, Chicago, Ill., 1989, p. 99.

21. Anonymous, *Utilizations of Milk Fat,* IDF Report B-Doc **164,** International Dairy Federation Annual Sessions, Copenhagen, Denmark, 1989.

22. B. J. Spalding, "Combating Cholesterol via Enzyme Research," *Chemish Weekblad,* June 1, 1988, p. 42.

23. S. Harlander, "Biotechnology: Applications in the Dairy Industry," *Journal of the American Oil Chemists' Society* **65**(11), 1727 (1988).

24. M. Schweizer, "Continuous Butter Manufacture: Heat Treatment," *International Dairy Federation Bulletin* **204,** 10 (1986).

25. F. H. McDowall, *The Butter Maker's Manual,* New Zealand University Press, Wellington, 1953.

26. P. Friis, "Advantages of Butter Production in Closed Systems," *Brief Communication* **1**(1), 326 (1982).

27. Anonymous, *Gold'n Flow Continuous Buttermaking Method,* Bulletin **G-493,** Cherry-Burrell Corp., Chicago, 1954.

28. U.S. Pat. 4,425,370 (Jan. 10, 1984), F. Graves (to Land O'Lakes, Inc.).

29. U.S. Pat. 4,447,463 (May 8, 1984), D. Antenore, D. Schmadeke, and R. Stewart (to Land O'Lakes, Inc.).

30. B. Schaffer and S. Szakaly, "The Effect of Temperature Treatment of Cream on the Liquid Fat Ratio," *Brief Communication* **1**(1), 363 (1982).

31. N. Cullinane and D. Eason, "Variation in Percentage Solid Fat of Irish Butter," *Brief Communication* **1**(1), 349 (1982).

32. B. K. Mortensen and K. Jansen, "Lipase Activity in Cream and Butter," *Brief Communication* **1**(1), 334 (1982).

33. U.S. Pat. 4,839,190 (June 13, 1989), J. Bumbalough (Wisconsin Milk Marketing Board, Inc.).

34. S. Gulyayev-Zaitsev, V. Dobronos, J. Berezansky, and N. Zhirnaya, "Low Calorie Butter," *Brief Communication* **1**(1), 328 (1982).

DAVID H. HETTINGA
Land O'Lakes Inc.
Minneapolis, Minnesota

BY-PRODUCT RECOVERY. See DISTILLED BEVERAGE SPIRITS.
BY-PRODUCTS. See WASTE MANAGEMENT AND FOOD PROCESSING.

C

CACAO BEAN. See CHOCOLATE AND COCOA.

CAFFEINE

Caffeine (1,3,7-trimethylxanthine) was isolated from coffee in 1820 (1). "Thein" was isolated from tea in 1827 and later shown to be identical with the coffee isolate. It was subsequently identified in cocoa, maté, kola nuts, and other plants. Its structure was determined in 1875. Caffeine is one of several methylxanthines that occur naturally, primarily in plant matter that is used to prepare beverages. Other methylxanthines that usually accompany it include theobromine and theophylline. The methylxanthines, as dioxypurines, are related to two nucleic acids—adenine and guanine. This is relevant to their biosynthesis and physiological effects. Their relationship to uric acids accounts for their metabolic fate. The structures of caffeine and related compounds are shown in Figure 1.

Total synthesis was first achieved in 1895, starting with dimethylurea. Dimethyluric acid, chlorotheophylline, and chlorocaffeine were intermediates (2).

PRODUCTION

Approximately 4,500 tons of caffeine is required annually in the United States for addition to beverages and for pharmaceutical use. This need is partially met by the decaffeination of coffee. Evaporation of the resulting caffeine-rich solutions or elution from adsorbents used to regenerate the solvents is followed by recrystallization (3). The methylation of theobromine obtained by extraction of cocoa hulls provides most of the remainder of the requirements. Dimethyl sulfate, diazomethane, and methyl iodide can be used to prepare caffeine from xanthine or any of the related monomethyl- or dimethylxanthines (4). Patented synthetic processes are also based on classical organochemical ring closure techniques and do not require the availability of xanthines as starting materials. In 1987 the United States imported 390 tons. Supply barely meets demand and manufacturing capacity was increased by 500 tons in 1989. The price of caffeine in 1989 was approximately $7/lb.

PROPERTIES

Physical Properties

Anhydrous caffeine obtained by crystallization from nonaqueous solvents is a white crystalline substance that melts at 235–237°C. It begins to sublime without decomposition at 120°C at atmospheric pressure and at 80°C under high vacuum. Its crystals are hexagonal and form parallel plates. When crystallized from water, the monohydrate forms long, silky, white needles. It becomes anhydrous at 80°C. Caffeine is soluble to the extent of 0.6% in water at 0°C, 2.13% at 25°C, and 66.7% at 100°C. The pH of dilute solutions is 6.9. In aqueous solution caffeine forms dimers and higher polymers by base stacking (5). It is fairly stable in dilute bases and acids, forming salts with the latter. Caffeine is considerably more soluble in chlorinated solvents—8.67% in methylene chloride and 12.20% in chloroform at 25°C. Its ultraviolet absorption spectrum shows a maximum at 274 nm with no variation over the pH range 2–14.

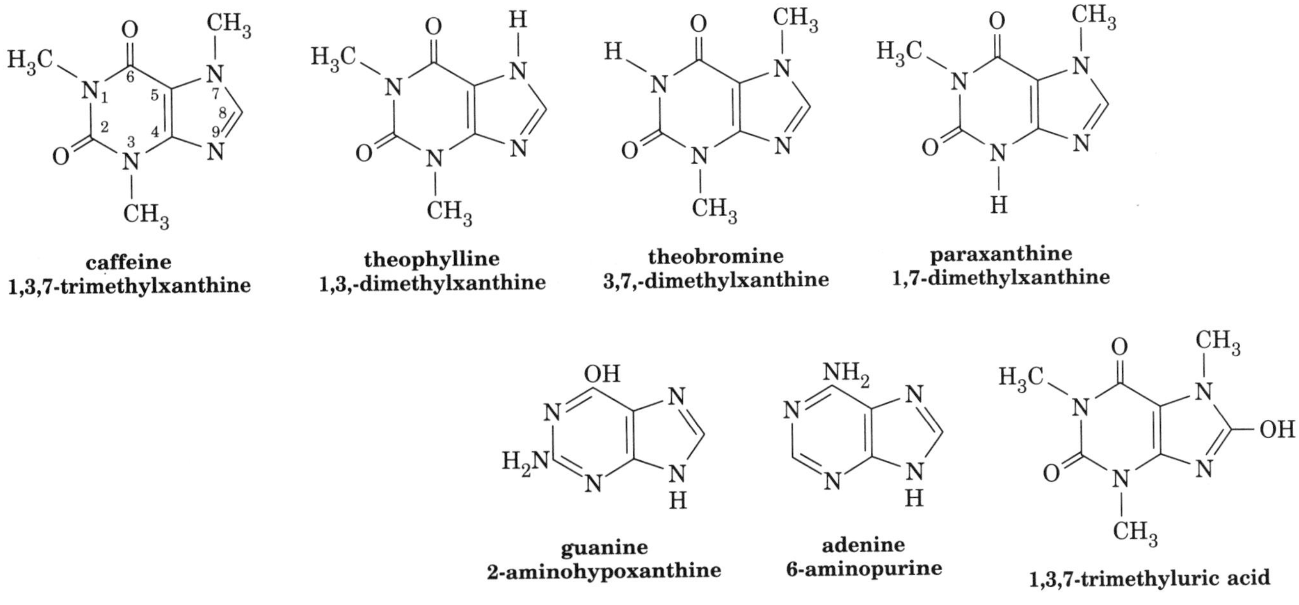

Figure 1. Methyxanthines and related compounds.

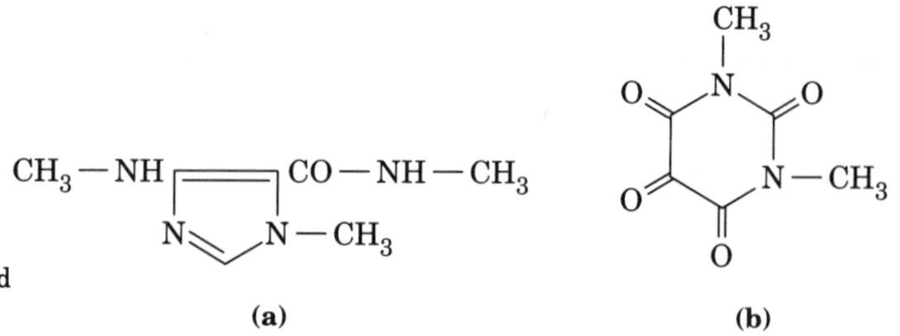

Figure 2. Reaction product of caffeine: (a) caffeidine and (b) dimethylalloxan.

(a)   (b)

Chemical Properties

Complexation occurs with many acid radicals, notably chlorogenate, salicylate, citrate, benzoate, and cinnamate. In plant matter much of the caffeine exists as a chlorogenic acid complex (6). Caffeine is converted to caffeidine by treatment with concentrated NaOH (7). Alloxans are formed by oxidation with chlorine and other oxidants (8). Reaction products are shown in Figure 2.

BIOGENESIS

The biogenesis of caffeine is probably similar in all plants, but with differences in the rates of various steps. It has its origin in the cell purine pool. Precursors include 7-methylxanthosine, 7-methylxanthine, and theobromine (9). Incorporation of adenine into caffeine occurs readily in young tea leaves (10). 7-Methylguanilic acid and 7-methylguanasine may also be part of the biosynthetic path. A 7-methyl-N-nucleoside hydrolase mediates the removal of ribose from 7-methylxanthosine. S-Adenosylmethionine is the methyl donor for the methylation of

the xanthines in reactions catalyzed by N-methyltransferase (11). A possible reaction path is shown in Figure 3.

ANALYSIS

Concern with the physiological effects of caffeine has stimulated research on its analysis in food and biological systems. Modern high-performance liquid chromatographic (hplc) analysis of caffeine in food products requires only the filtration of aqueous extracts and injection onto a suitable hplc column. Detection is based on the measurement of uv absorption at 280 nm (12). Determination of caffeine at the low concentrations present in body fluids is accomplished by direct injection onto reverse-phase hplc columns (13).

OCCURRENCE

In Natural Products

Caffeine is present in various tissues of about 60 plant species, several of which are used to prepare beverages.

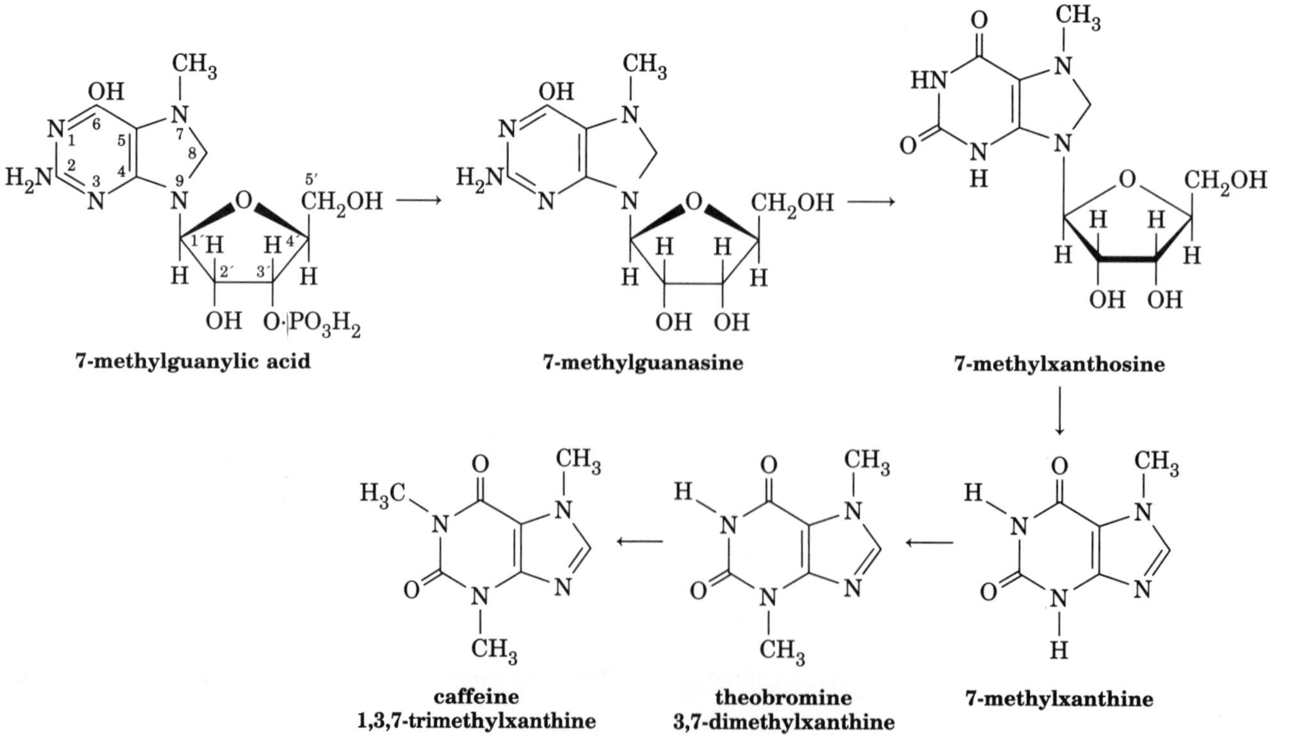

7-methylguanylic acid

7-methylguanasine

7-methylxanthosine

caffeine
1,3,7-trimethylxanthine

theobromine
3,7-dimethylxanthine

7-methylxanthine

Figure 3. Biosynthesis of caffeine.

Table 1. Caffeine in Food Products[a]

Food Product	Caffeine Content, %		Caffeine per Serving, mg	
	Range	Mean	Range	Mean
Coffee, arabica	0.58–1.7	1.2	35–80	60
Coffee, robusta	1.2–3.3	2.2	60–140	120
Coffee, instant	2.2–5.0	3.3	45–105	65
Coffee, decaffinated	0.03–0.09	0.6	2–6	3
Tea, black	1.2–4.6	3.0	20–48	35
Tea, green	1.0–2.4	1.9	17–40	24
Tea, instant	4.0–5.0	4.8	20–30	22
Cocoa, powder	0.08–0.35	0.2	1–8	4
Chocolate milk	0.01–0.025	0.02	2–7	5
Milk chocolate	0.005–0.04	0.02	2–14	6
Dark chocolate	0.02–0.10	0.08	6–28	23
Maté	0.9–2.2	2.0	36–42	40
Guarana paste	3.9–5.8	4.7		
Kola nut	0.8–2.2	1.5		
Cola and pepper drinks	0.08–0.16	0.09	20–30	28

[a] Serving size is 6 oz for hot beverages, 8 oz for cold beverages and 1 oz for chocolate products.

The most important are

coffee beans (*Coffea arabica* and *C. canephora* var. *robusta*),

tea leaves (*Camellia sinensis* vars. *assamica* and *sinensis*),

cocoa beans (*Theobroma cacao*),

maté stems and leaves (*Ilex paraguariensis*),

guarana seeds (*Paullinia cupana*), and

kola nuts (*Cola acuminata* and *C. verticillata*).

Variations in caffeine content result from varietal diversity, climatic changes in the growing areas, and horticultural techniques. In tea, the youngest leaves have the highest concentration (14). High coffee roasting temperatures result in caffeine loss by sublimation (15). The caffeine content of several food products is shown in Table 1.

There is a higher level of caffeine in tea leaves than in coffee beans, but 200 cups of tea are obtained per pound of leaves, whereas only 40–80 cups of coffee are prepared per pound of beans. Increasing amounts of coffee and tea are being decaffeinated.

Caffeine as a Food Additive

In addition to its natural occurrence, caffeine is added to the very widely consumed cola drinks and other soda products. This usage is based on its stimulatory properties and the slight degree of bitterness that it imparts. In 1978 the U.S. Select Committee on GRAS Substances reaffirmed the safety of caffeine as a beverage additive at levels current at that time (10–12 mg/100 mL) but recommended some additional testing. A very few soft drinks have a higher level. A later proposal by the FDA to eliminate its GRAS (generally recognized as safe) status was abandoned. "No Caffeine" colas show the highest rate of growth in the soft drink market.

CAFFEINE IN THE HUMAN DIET

A high proportion of the human population has consumed caffeine for many centuries. Authenticated use of tea in China dates from AD 350. The beverage reached Europe in 1600. Coffee came into use as a beverage in Arabia around AD 1000 and was brought to Europe in the seventeenth century. Cocoa was consumed by the Aztecs in Mexico. In many parts of South America maté has been a source of substantial caffeine intake. World consumption of caffeine in beverages, chocolate products, and medications is approximately 120,000 metric tons annually. In the United States it is approximately 20,000 metric tons or 225 mg/ per person per day.

The consumption by heavy users is of interest because of concern with possible health effects. Mean daily intake (milligrams per kilogram of body weight) in the United States by consumers in the 90th to 100th percentiles varies with age level as follows:

1–5 yr, 4.7
6–11 yr, 3.2
12–17 yr, 2.9
18 yr +, 7.0

Tea and soft drinks are the major caffeine sources for users under the age of 18. Coffee becomes the major source for those over that age (16).

METABOLISM OF CAFFEINE

The caffeine molecule is sufficiently hydrophobic to pass freely through biological membranes. It is completely absorbed from the gastrointestinal tract and rapidly attains peak plasma levels. Caffeine is equally well absorbed from each of its beverage sources. There are no effective barriers to penetration of any tissue, including placenta and fetus. This characteristic prevents efficient excretion of caffeine by the human kidney since it is readily reabsorbed from the renal tubules (17).

The half-life of caffeine in plasma varies not only with species differences, but also with age and condition of the individual. The half-life in rodent plasma is 1–2 h but it is 6 h in that of the healthy adult human. There is variation depending on smoking habits and the use of some medica-

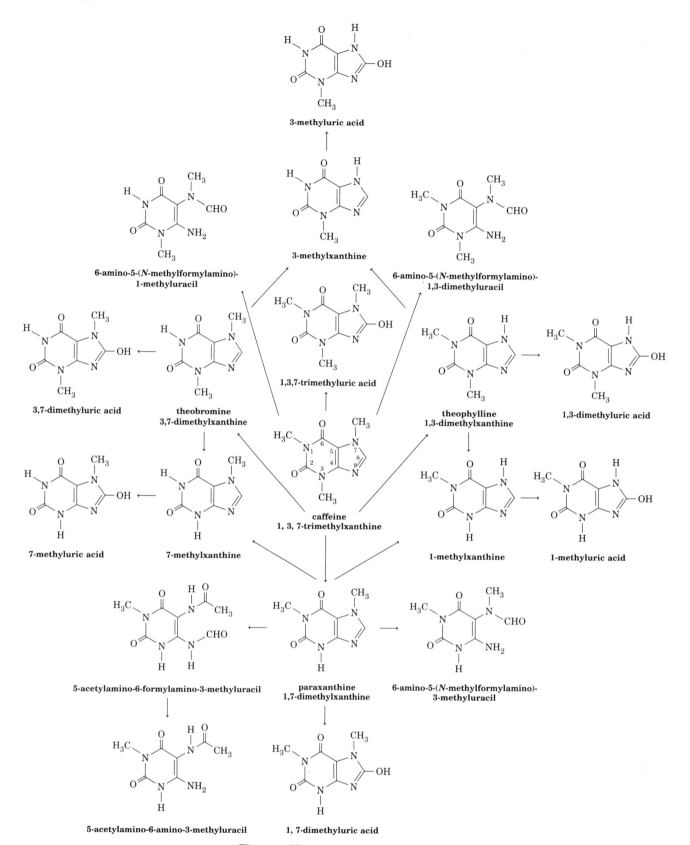

Figure 4. Human metabolism of caffeine.

tions. During pregnancy the half-life is increased to 18 h. The immature liver of the newborn human is limited in its ability to metabolize caffeine, so that its half-life is 3–4 days, similar to that of adults with severe liver damage (17). These differences are significant in extrapolating safety data from animal studies and also in considering diet during pregnancy.

The metabolic pathways of caffeine in those mammalian species that have been studied are similar (17). The same major reactions take place—demethylation, oxidation at the 8 position on the xanthine ring to form uric acids, and ring cleavage between the 8 and 9 positions to form diaminouracils. N-Acetylation also occurs. The three dimethyl- and the three monomethylxanthines that can be formed are all found in human urine after caffeine ingestion, along with all of the corresponding methylated uric acids. Uracils derived from caffeine, theobromine, and paraxanthine, as well as large amounts of acetylated products, are also present. No xanthine is formed. The human metabolic pathway is shown in Figure 4.

Species variations in the metabolic pathways of caffeine were first observed in 1900 (1). In humans about 70% of ingested caffeine is initially converted to paraxanthine, 25% to a mixture of theobromine and theophylline, and about 5% is oxidized without demethylation to form the corresponding uric acid and uracil compounds. Some primates produce theophylline as the predominant initial metabolite. In rats and mice theobromine predominates and a higher proportion of ingested caffeine is oxidized without initial demethylation. In humans, the final product mix is the result of competing reactions, the rates of which vary with gender, dosage, medications in the diet, and individual differences.

Minor amounts of uracils derived from caffeine, theobromine, and paraxanthine also occur. The acetylated product A is believed to be a true metabolite. Product B is probably an artifact of analytical procedures. Averaged values for caffeine metabolites found in human urine are shown in Table 2. Reaction mechanisms are not well known. The biochemical changes take place in the liver. Xanthine oxidase mediates the formation of the uric acid derivatives. It is interesting to note that, in the human infant, theophylline, often used for treatment of apnea, is methylated to caffeine as the first step in its metabolism (1).

PHARMACOLOGY

Effects of Cardiovascular System

Caffeine produces minor transitory increases in blood pressure. Habitual users are less prone to exhibit this effect. Its significance is not known (19). Caffeine has been reported to cause cardiac arrhythmias but there is conflicting evidence (19). There appears to be a caffeine intolerant population that is susceptible to this effect. Caffeine may exacerbate an existing tendency toward arrhythmias. Low dosages may decrease heart rate slightly; high dosages may cause tachycardia in sensitive subjects (20). Cerebral blood flow is decreased and this effect is the basis for its inclusion in drug preparations for the treatment of migraine headaches (21). Some studies indicate a positive correlation between caffeine intake and the development of hypercholesterolemia but many very large efforts, such as the Framingham study, show no correlation between atherosclerotic cardiovascular disease and coffee intake. Other large studies have shown no correlation between coffee drinking and any form of coronary heart disease (20). Disparate results concerning the effects of caffeine on the cardiovascular system probably relate to the size of the studies, lack of control of miscellaneous factors, and, very importantly, to the acceptance of "cups of coffee" as a quantitative measure of caffeine content. Coffee and tea also contain many other physiologically active compounds.

Effects on Central Nervous System

Although caffeine is the stimulatory drug most widely used throughout the world, it is difficult to quantify its behavioral effects. This is primarily due to the differences in the form of its ingestion, ie, coffee, tea, or the pure substance. Another confounding factor is the varying degrees of tolerance that develop among its users. Caffeine affects the cortex, the medulla, and eventually the spinal cord as dosages are increased (21). At a dosage of 2 mg/kg, which generates a peak plasma concentration of $5–10~\mu M$, subjects generally feel more alert and better able to carry out routine tasks after having become bored or fatigued (22). Conversely, omission of a habitual morning dosage of caffeine often results in nervousness, irritability, and poor work performance. Cognitive function does not appear to be improved by caffeine ingestion but may be slightly im-

Table 2. Caffeine Metabolites in Human Urine as Percentage of 5-mg/kg Dosage[a]

Caffeine	1.2	1,3,7-Trimethyluric acid	1.2
Paraxanthine	5.9	1,7-Dimethyluric acid	5.4
Theobromine	1.9	3,7-Dimethyluric acid	
Theophylline	0.9	1,3-Dimethyluric acid	1.8
1-Methylxanthine	18.1	1-Methyluric acid	14.8
7-Methylxanthine	7.7	7-Methyluric acid	1.0
3-Methylxanthine	4.1	3-Methyluric acid	
5-Acetylamino-6-amino-3-methyluracil (A)			3.2
5-Acetylamino-6-formylamino-3-methyluracil (B)			14.7

[a] Ref. 18.
[b] Identified but not determined quantitatively.

paired. The most widely noticed effect of caffeine is the prolongation of the sleep latency period. Its effect on the quality of sleep shows wide individual variation and is dose, time, and age dependent. It is markedly subject to the development of tolerance.

Other Effects

Caffeine produces a mild diuresis in humans and increases the excretion of sodium, potassium, and chloride ions (21). Caffeine increases work output and endurance in long-term exercise regimens because of its ability to promote lipolytic activity. Free fatty acids become available for energy production, thus sparing glycogen (23). The thermogenic properties of caffeine cause an increase in the resting metabolic rate. Consideration has been given to its use in managing obesity (24). Respiratory rate is increased by caffeine at a plasma level of 5 mg/L. The mechanisms for this stimulation are complex and involve many separate effects. They include increased pulmonary blood flow through vasodilation, increased sensitivity of the respiratory centers to carbon dioxide, and improved skeletal muscle contraction (21). This multiple response is utilized in the treatment of infantile apnea.

Mechanism of Physiological Activities

The effects of caffeine on the cardiovascular, neurological, and renal systems and on lipolysis are primarily triggered by the blockage of adenosine receptors through competitive antagonism (25). The structural relationship between caffeine and adenine (and therefore adenosine) was shown previously. Other suggested mechanisms such as the inhibition of phosphodiesterases and modification of calcium metabolism may also be operative but are inadequate to explain these effects at reasonable caffeine concentrations. Adenosine and caffeine show contrasting effects on blood pressure, activity of the central nervous system, urine output, and lipolysis (26). The well known induction of tolerance to caffeine also becomes explicable by this mechanism, since caffeine is known to induce the formation of additional adenosine receptors in rats (25).

Physical Dependence

Addiction to caffeine may develop even from short term (7 days) administration of high dosages. Headache and fatigue are the most common withdrawal symptoms but depression, irritability, anxiety, and vomiting may also occur. Severity of symptoms, which may last for up to one week, range from mild to incapacitating. Many caffeine consumers exhibit no withdrawal symptoms. The physiology of caffeine withdrawal is not well understood (27).

Toxicity

The physiological effects of caffeine become greatly intensified at very high dosages. The LD_{50} of caffeine is similar for many mammals—about 200 mg/kg (28). This would suggest an LD_{50} of about 3.5 g for humans based on normal metabolic weight corrections. It is more generally accepted that the lethal oral dose is about 10 g. Lethal plasma concentration is in the range of 0.5–1.0 mM,

which would require the consumption of about 75 cups of coffee over a very short period of time (17). Lethality resulting from beverage consumption is most unlikely. The few recorded cases of death from caffeine consumption have involved the pure substance or medications containing it.

Caffeine as a Drug

The primary uses of caffeine as a drug are based on its effects on the respiratory, cardiovascular, and central nervous systems. Premature infants are subject to apnea, a transient but potentially dangerous cessation of breathing. Both theophylline and caffeine have been used to control this syndrome. Caffeine decreases apneic episodes and regularizes breathing patterns (21). It has a lower incidence of side effects than theophylline. It is also used for the treatment of bronchiospastic disease in asthmatic patients. Caffeine is used widely in drug mixtures designed for relief from migraine-type headaches because of its vasoconstrictor effects on the cerebral circulation. There are many preparations of this type that are available over the counter that contain 30–200 mg of caffeine per tablet. A common application of caffeine's stimulatory effects on the central nervous system is the use of 100–200-mg tablets to prevent drowsiness when driving or whenever this condition is undesirable. There are about 2,000 nonprescription and about 1,000 prescription drugs containing caffeine.

MUTAGENICITY

Caffeine attains the same concentration in gonadal tissue as in the circulating plasma and may thus reach levels of up to 0.05 mM for very heavy consumers (29). In vitro mutagenicity studies in which positive results are obtained generally involve concentrations at least 100 times greater and are therefore at or above the lethal concentration for humans. At these high concentrations there is inhibition of DNA repair, an effect attributable to its purine analog structure. Caffeine may thus be considered a weak mutagen with no significance for humans.

TERATOGENICITY

Caffeine is teratogenic to rodents at high dosage levels (30). Effects such as cleft palates and ectrodactyly are caused by administration of 100 mg/kg · day or more, but only when given in a single dose by gavage or injection. Increased fetal absorption and decreased fetal weight also occur at these levels, which approach lethality. True teratogenic responses have usually been associated with dosages that result in maternal toxicity. Delayed bone ossification is observed at much lower levels but this effect is rapidly compensated postnatally.

The role of caffeine in the outcome of human pregnancy is generally determined by retrospective questioning. Studies have been based on beverage consumption recall and are therefore inexact with regard to the amount of caffeine ingested or the effects of other beverage components. There is no clear evidence of teratogenic effects due to caffeine (31). Conflicting data exist concerning the relationship between maternal caffeine consumption and infant birth weight (32). Since caffeine crosses the placental barrier, it is possible for high concentrations to occur in the neonate (33). Most medical advice cautions against high caffeine intake during pregnancy.

CARCINOGICITY

The carcinogenic potential of caffeine has been researched intensively in rodents (34). The results of about a dozen well-controlled studies indicate no carcinogenic effects attributable to caffeine.

Epidemiological studies regarding the risk of coffee drinkers of cancers of the bladder, colon, rectum, pancreas, breast, ovaries, and liver have been carried out in different parts of the world (35). Conflicting results have been obtained. In almost all instances, dose–response relationships have not been observed, casting doubt on any causal relationships between the effect and the consumption of coffee. In addition, the fact that tea consumption often produces different effects, sometimes converse, makes it unlikely that caffeine itself is a causative agent of increased cancer risk. Theoretical considerations that suggest a lack of carcinogenic potential for caffeine are based on the fact that it does not alkylate or combine covalently with DNA, it does not undergo metabolic activation, and it lacks some of the structural features of known purine carcinogens (34). Determination of the effect of caffeine on fibrocystic disease of the breast has also been pursued. The weight of evidence indicates no effect.

DECAFFEINATION

Concern with the physiological effects of coffee prompted early consideration of ways to decaffeinate the product. The first commercial plant was established by the firm Kaffee HAG in Germany in 1906 and in the United States a few years later. The U.S. plant was expropriated by the government during World War I, terminating production of decaffeinated coffee in the United States until 1932. Only one brand was available until the mid 1950s and solely as the instant product. Consumption then increased rapidly and in 1988 accounted for 35% of the instant product and 8% of bean and ground coffee. Decaffeinated tea appeared on the market in 1978. In 1988 it accounted for approximately 10% of leaf tea consumed in the United States and 25% of convenience teas.

Coffee

Chlorinated compounds were the first caffeine extractants used. To expedite removal of caffeine, coffee beans are steam treated to bring about swelling for greater solvent permeability and to break down complexation products. In all extraction processes it is necessary to wet the coffee for efficient caffeine removal. Decaffeination is carried out on the green beans to prevent extraction and loss of the aromatic components that are generated during roasting. After solvent removal by steam distillation, the beans are roasted. Trichlorethylene was commonly used as the solvent until it was eliminated in 1977 because of its suspected carcinogenicity. Methylene chloride then became the decaffeinating solvent of choice (36). A residue level of 10 ppm is approved by the USFDA. An unofficial standard providing for the removal of 97% of the original caffeine has come into use for decaffeinated coffee. To eliminate the use of all chlorinated compounds, other solvents are now utilized. Ethyl acetate has been approved for use in the United States and large quantities of decaffeinated coffee are prepared with this solvent. The process has some disadvantages; solids losses are 5–6%. Desolventization is more difficult than with methylene chloride, which has greater volatility.

Decaffeination using water as the only solvent to contact the coffee is also carried out. A concentrated coffee extract containing no caffeine contacts the green beans. Noncaffeine solids are not removed under these conditions. The caffeine-rich extract is then regenerated for use by adsorbing the caffeine on activated carbon (37). Coffee oil expressed from beans is used as a decaffeinating solvent along with other vegetable oils in another operating process. Caffeine is removed from the oils by a liquid–liquid extraction system with water as the second solvent (38).

Supercritical fluid extraction technology has been successfully applied to the decaffeination of coffee (39). The solvent used is carbon dioxide at about 300 bars and 60–90°C. At this pressure the solubility of caffeine is 0.2% at 60°C. Solubility increases as temperature and pressure are raised. The solvent comes in contact with adequately wetted coffee in equipment designed for the necessary high pressures. The extracted caffeine may be removed from the supercritical fluid by reducing the pressure, thereby decreasing its solubility, or by passing the mixture over activated carbon. Techniques that avoid the necessity for repressurizing the gas are preferable. Supercritical processes are capital intensive but result in products of superior quality with no solvent residues.

Tea

The commercial procedures for the decaffeination of tea are, in principle, similar to those used for coffee. For practical reasons it is necessary to utilize black tea rather than free green leaf. Since tea aroma develops during the conversion of fresh leaf to dried black tea, its conservation during decaffeination is desirable. Decaffeinated tea appeared on the market in 1978. The first product was prepared with the use of methylene chloride and the same process evolution took place as with coffee. Decaffeination with supercritical CO_2 became a commercial process in 1984 (40). The resulting caffeine-rich solvent is regenerated with activated carbon. Aroma may be previously stripped with dry supercritical CO_2 or with the moist gas at atmospheric pressure for later addition (41). Caffeine may be selectively adsorbed from tea extracts by resins for the production of decaffeinated instant tea.

BIBLIOGRAPHY

1. M. J. Arnaud, "Products of Metabolism of Caffeine," in P. B. Dews, ed., *Caffeine,* Springer-Verlag, Berlin, 1984, pp. 3–38.

2. E. Fischer and L. Ach, "Synthese des Caffeins," *Chemische Berichte* **28,** 3135–3142 (1895).

3. U.S. Pat. 4,818,552 (Apr. 4, 1989), L. Kaper (to Douwe Egberts).

4. U.S. Pat. 2,534,331 (Dec. 19, 1950), D. W. Woodward (to E. I. du Pont de Nemours & Co., Inc.).

5. A. L. Thakkar and co-workers, "Self-Association of Caffeine in Aqueous Solution: A Nuclear Magnetic Resonance Study," *Journal of the Chemical Society Chemical Communications* **9,** 524–525 (1970).

6. I. Horman and R. Viani, "The Caffeine–Chlorogenate Complex of Coffee, an NMR Study," ASIC, 5th Colloque, Lisbon (1971), pp. 102–111.

7. H. Biltz and H. Rakett, "Kaffeiden und Kaffeiden-carbonsaure," *Chemische Berichte* **61,** 1409–1422 (1928).

8. A. C. Cope and co-workers, "1,3-Dimethyl-5-alkyl Barbituric Acids," *Journal of the American Chemical Society* **63,** 356–358 (1941).

9. E. Looser, T. W. Bauman, and H. Warner, "The Biosynthesis of Caffeine in the Coffee Plant," *Phytochemistry* **13,** 2515–2517 (1974).

10. H. Ashihara and H. Kubota, "Biosynthesis of Purine Alkaloids in Camellia Plants," *Plant and Cell Physiology* **28,** 535–539 (1987).

11. M. F. Roberts and G. R. Waller, "*N*-Methyltransferases and 7-Methyl-*N*-Nucleoside Activity in *Coffee Arabica* and the Biosynthesis of Caffeine," *Phytochemistry* **18,** 451–455 (1979).

12. W. J. Hurst, R. A. Martin, and S. M. Tarka, Jr., "Analytical Methods for the Quantitation of Methylxanthines," in G. A. Spiller, ed., *The Methylxanthine Beverages and Foods: Chemistry, Consumption and Health Effects,* Alan R. Liss, Inc., New York, 1984, pp. 19–26.

13. R. Dorizzi and F. Tagliaro, "Direct HPLC Determination of Theophylline, Theophylline Analogs and Caffeine in Plasma with an ISRP (Pinkerton) Column," *Clinical Chemist* **36,** 566 (1989).

14. M. A. Bokuchava and N. I. Skobeleva, "The Biochemistry and Technology of Tea Manufacture," *CRC Critical Review of Food Science and Nutrition* **12,** 303 (1980).

15. M. Dong and co-workers, "The Occurrence of Caffeine in the Air of New York City," *Atmospheric Environment* **11,** 651–653 (1977).

16. H. R. Roberts and J. J. Barone, "Biological Effects of Caffeine; History and Use," *Food Technology* **37,** 32–39 (1983).

17. R. W. Von Borstel, "Biological Effects of Caffeine; Metabolism," *Food Technology* **37,** 40–45 (1983).

18. D. W. Yesair, A. R. Branfman, and M. M. Callahan, "Human Disposition and Some Biochemical Aspects of Methylxanthines," *Phytochemistry,* **13,** 215–233 (1974).

19. M. G. Meyers, "Effects of Caffeine on Blood Pressure," *Archives of Internal Medicine* **148,** 1189–1193 (1988).

20. T. K. Leonard, R. R. Watson, and M. E. Mohs, "The Effects of Caffeine on Various Body Systems: A Review," *Journal of the American Dietetic Association* **87,** 1048–1053 (1987).

21. M. J. Arnaud, "The Pharmacology of Caffeine," *Progress in Drug Research* **31,** 273–313 (1987).

22. A. Leviton, "Biological Effects of Caffeine; Behavioral Effects," *Food Technology* **37,** 44–46 (1983).

23. E. T. Poehlman and co-workers, "Influence of Caffeine on the Resting Metabolic Rate of Exercise-Trained and Inactive Subjects," *Medical Science and Sports Exercise* **17,** 689–694 (1985).

24. A. G. Dulloo and co-workers, "Normal Caffeine Consumption: Influence on Thermogenesis and Daily Energy Expenditure in Lean and Postobese Human Volunteers," *American Journal of Clinical Nutrition* **49,** 44–50 (1989).

25. R. Von Borstel and R. J. Wurtman, "Caffeine and the Cardiovascular Effects of Physiological Levels of Adenosine," in P. B. Dews, ed., *Caffeine,* Springer-Verlag, Berlin, 1984, pp. 142–150.

26. J. Onrot and co-workers, "The Cardiovascular Effects of Caffeine," *Primary Cardiology* **94,** 104–110 (1984).

27. R. R. Griffiths and P. P. Woodson, "Caffeine Physical Dependence: A Review of Human and Laboratory Animal Studies," *Psychopharmacology* **94,** 437–451 (1988).

28. B. Stavric, "Toxicity to Humans. 2. Caffeine," *Food and Chemical Toxicology* **26,** 645–662 (1988).

29. R. H. Haynes and J. D. B. Collins, "The Mutagenic Potential of Caffeine," in P. B. Dews, ed., *Caffeine,* Springer-Verlag, Berlin, 1984, pp. 231–238.

30. J. G. Wilson and W. J. Scott Jr., "The Teratogenic Potential of Caffeine in Laboratory Animals," in P. B. Dews, ed., *Caffeine,* Springer-Verlag, Berlin, 1984, pp. 165–187.

31. A. Berger, "Effects of Caffeine Consumption on Pregnancy Outcome," *Journal of Reproductive Medicine* **33,** 945–956 (1988).

32. J. Nash and T. V. N. Persaud, "Reproductive and Teratological Risks of Caffeine," *Anatomischer Anzeiger* **167,** 265–270 (1988).

33. W. D. Parsons and A. H. Niems "Prolonged Half-Life of Caffeine in Healthy Term Newborn Infants," *Journal of Pediatrics* **98,** 640–641 (1981).

34. H. C. Grice, "The Carcinogenic Potential of Caffeine," in P. B. Dews, ed., *Caffeine,* Springer-Verlag, Berlin, 1984, pp. 153–164.

35. Committee on Diet and Health, National Research Council, *Diet and Health,* National Academy Press, Washington, D.C., 1989.

36. U.S. Pat. 3,671,263 (June 20, 1972), J. M. Patel and A. B. Wolfson (to Proctor and Gamble Co.).

37. *Tea and Coffee Trade Journal* **160,** 45–47 (1988).

38. U.S. Pat. 4,837,038 (June 6, 1989), J. C. Proudley (to Nestea SA).

39. U.S. Pat. 4,820,537 (Apr. 11, 1989), S. N. Katz (to General Foods Corp.).

40. Ger. Pat. 3,515,740 (May 2, 1985), H. Klima, E. Schütz, and H. Vollbrecht (to SKW Trostberg AG).

41. Eur. Pat. 269,970 (June 8, 1988), E. Schütz and H. Vollbrecht (to SKW Trostberg AG).

HAROLD GRAHAM
Englewood, New Jersey

CANNING. See BEER; CANNING: REGULATORY AND SAFETY CONSIDERATIONS; CRABS AND CRAB PROCESSING; ENERGY USAGE AND FOOD PROCESSING PLANTS; ENERGY USAGE IN THE CANNING INDUSTRY; FOOD PRESERVATION; NUTRITION QUALITY AND FOOD PROCESSING; RETORT POUCH; AND ENTRIES UNDER THERMAL PROCESSING.

CAKE. See BAKING SCIENCE AND TECHNOLOGY.

CALLUS. See entries under GENETIC ENGINEERING.

CANADA. See INTERNATIONAL DEVELOPMENT: CANADA.

CANDY. See CONFECTIONARY.

CANE SUGAR

Cane sugar is the name given to sucrose, a disaccharide produced from the sugarcane plant. It is practically indistinguishable from beet sugar and commands the same price in competitive markets. However, because they come from different plants, the trace constituents are different and can be used to distinguish the two sugars. One effect of the difference is the odor in the package headspace, from which experienced sugar workers can identify the source. Also, the two plants use different photosynthetic pathways, the carbon isotope ratio in the sugars can be used to distinguish sucrose from the two sources. The other major sweetener is corn sugar (glucose), which is a monosaccharide derived from cornstarch. It is available either as a solid or as a syrup (corn syrup). Corn sugar is essentially glucose (dextrose); corn syrup contains mostly glucose but with several percent of residual, incompletely hydrolyzed higher saccharides. Since the commercialization of enzymatic isomerization in 1968, fructose-containing corn syrup has been available. This product, called high fructose corn syrup (HFCS), is essentially equivalent to invert syrup made from sucrose. It can be substituted for invert syrup and even for sucrose in many food products. Other minor sweeteners are maple sugar, sorghum sugar, and palm sugar, all principally sucrose, and honey, whose composition, depends on the source plant.

sucrose

α–D-glucose (dextrose)   **β–D-fructose (levulose)**

All sugar from whatever source is used almost entirely for food. In the United States, only ca 1% of the sugar consumed is used for nonfood purposes. Technology exists to convert sucrose into many other substances by fermentation, esterification, hydrogenation, alkaline degradation, and many others, but the price is prohibitive. As a food, sugar is all energy [kJ (food calories)], and even brown and raw sugar contain virtually no protein, minerals, or vitamins.

Sugar is one of the purest of all substances produced in large volume. Its analysis is approximately: sucrose 99.90%; invert sugars 0.01%; ash (inorganic material) 0.03%; moisture 0.03%; other organic material (mostly polysaccharides) 0.03%. Even raw sugars are >98% sucrose. World production in 1990 was ca 10^8 metric tons.

In the production scheme for cane sugar, the cane cannot be stored for more than a few hours after it is cut because microbiological action immediately begins to degrade the sucrose. This means that the sugar mills must be located in the cane fields. The raw sugar produced in the mills is the item of international commerce. Able to be stored for years, it is handled as a raw material—shipped at the lowest rates directly in the holds of ships or in dump trucks or railroad cars. Because it is not intended to be eaten directly, it is not handled as food. The raw sugar is shipped to the sugar refineries, which are located in population centers. There it is refined to a food product, packaged, and shipped a short distance to the market. In a few places, there is a refinery near or even within a raw-sugar mill. However, the sugar still goes through the raw stage.

There is, however, another category of cane sugar that is not truly refined. It is called direct-consumption, or plantation white, or merely white sugar. White sugar, which should not be confused with refined sugar, is made directly from the cane without going through the raw-sugar stage. It is a little off-white, frequently has a lingering molasses aroma, and is not as pure as truly refined sugar. It is, of course, perfectly sanitary and quite edible, but it does not keep as well as refined sugar and is usually sold only locally at a reduced price. Some sugar factories are improving the quality of the direct-consumption sugars to make them more comparable to refined sugar and thus bypass the refinery.

Still another class of cane sugar is called noncentrifugal sugar or whole sugar. It is made the way sugar was undoubtedly originally made in prehistoric times, by boiling down whole cane juice without the elimination of any impurities. The whole mixture solidifies on cooling and is broken into pieces. It is light to dark brown in color, with a consistency between rock-hard and friable. Much of this sugar is made in India, where it is called *gur*. In other places, this sugar is known as *panela, areado, pile, piloncillo, papelon, chancaca, muscovado, duong-cat, panocha, shakkar*, and *jaggery*.

The principal by-product of cane sugar production is molasses. About 10–15% of the sugar in the cane ends up in molasses. Molasses is produced both in the raw-sugar manufacture and also in refining. The blackstrap or final molasses is about 35–40% sucrose and slightly more than 50% total sugars. In the United States, blackstrap is used almost entirely for cattle feed. In some areas, it is fermented and distilled to rum or industrial alcohol. The molasses used for human consumption is of a much higher grade and contains much more sucrose.

HISTORY

Historians disagree about whether sugarcane is native to India or New Guinea. They do agree that ancient people liked sugarcane and carried it with them in their migrations. It spread throughout much of the South Pacific area, except for Australia, by about 1000 BC. By 400 BC, it was known in Persia, Arabia, and Egypt. The word sugar does not appear in the Bible, but the phrase "sweeter than syrup or honey from the honeycomb" (Psalms 19:11) appears, indicating that although sugarcane was possibly known in the Holy Land in biblical times, only syrups could be obtained from it. In the seventh to tenth centuries AD, the Arabs spread sugarcane throughout their region of influence in the Mediterranean. About the same time, it was carried eastward across the Pacific Ocean to Fiji, Tonga, Samoa, Tahiti, Easter Island, and Hawaii. By the twelfth century, sugarcane had reached Europe, and Venice was the center of sugar trade and refining. Marco Polo reported advanced sugar refining in China toward the end of the thirteenth century. Columbus brought sugarcane to the new world on his second voyage. It spread throughout the Western Hemisphere in the next 200 years, and by about 1750 sugarcane had been introduced throughout the world.

The process for extracting juice from the cane is also very old. In antiquity, the canes were undoubtedly sucked or chewed for their sweet taste. Also, in the ancient past in various places the canes were cut and crushed by heavy weights, ground with circular stones or by a heavy roller running on a flat surface, pounded in a mortar with a pestle, or soaked in water to better extract the sweet juice. The term grinding survives to this day as the name of the process for extracting the cane juice, even though the process no longer involves a true grinding. Parallel rolls, which are used today, were first used in 1449 in Sicily.

The ancient process for obtaining sugar consisted of boiling the juice until solids formed as the syrup cooled. The product looked like gravel; the Sanskrit word for sugar, *shakkara,* has that alternative meaning. Pliny, who traveled widely in the Roman Empire, wrote in AD 77 that sugar was "white and granular." He noted that Indian sugar was more esteemed than Arabian, and that both were used in medicine. By the fourth century AD, the Egyptians were using lime as a purifying agent, and carrying out recrystallization, which is still the main step in refining.

Until quite recently, sugar was strictly a luxury item. Queen Elizabeth I is credited with putting it on the table in the now familiar sugarbowl, but it was so expensive that it was used only on the tables of royalty. Sugar production reached large volume at a reasonable price only by the 18th century.

The development of the sugar industry from the sixteenth century onward is closely associated with slavery, which supplied the large amount of labor used at the time. The low cost of labor and the high price for sugar made many fortunes. The abolition of slavery at various times in different countries between 1761 and 1865 profoundly affected the sugar industry. On the freeing of the slaves, sugar production fell drastically in many producing areas.

The first use of steam power as a replacement for the animal or human power that drove the cane mills occurred in Jamaica in 1768. This first attempt worked only a short time, but steam drive was used successfully a few years later in Cuba. Steam drive for the mills soon spread throughout the world. The use of steam instead of direct firing was soon applied for evaporating the cane juice.

Probably the most essential piece of equipment in the modern process is the vacuum pan invented by Howard (UK) in 1813. This accomplishes the evaporation of water at a low temperature and lessens the thermal destruction of sucrose. The bone-char process for decolorization dates from 1820. The other essential piece of equipment is the centrifuge, which was developed by Weston in 1852 and applied to sugar in 1867 in Hawaii. This machine reduces the time for draining the molasses from the sugar crystals from weeks to minutes by applying a force equal to $1,000g$.

The manufacture of sugar was early understood to be an energy-intensive process. Cuba was essentially deforested to obtain the wood that fueled the evaporation of water from the cane juice. When the forests were gone, the bagasse burner was developed to use the otherwise waste dry cane pulp, bagasse, for fuel. It is to the everlasting credit of the cane-sugar industry that the greatest energy saver of all time was developed in this industry: the multiple-effect evaporator invented by Norbert Rillieux of Louisiana. The 1846 patents of Rillieux describe every detail of the process. This system is now used universally by every industry that has to evaporate water.

The principal analytical methods were developed in the mid-nineteenth century: the polariscope by Ventzke in 1842, the Brix hydrometer in 1854, Fehling's method for reducing sugars, and Clerget's method for sucrose in 1846.

Sugar loaves were for centuries the traditional form in which sugar reached the market. These were formed by pouring the mixture of crystals and syrup into a mold. The molds were kept in the hot room to facilitate the draining of the syrup either through the porous mold or through a hole in the bottom. With a little cooling or drying, the crystals stuck together, forming a convenient marketable loaf of sugar. The sugar loaves required no packaging and were broken up by the user as needed. Only a very small amount of sugar now reaches the market in this form.

The sugar business, like many others, has changed structure in the last 100 years from many small units to few large units. In 1861, the Havemeyer and Elder refinery in Brooklyn was the largest in the world and processed 32 metric tons of sugar daily. In 1981, there were several refineries processing >4000 t daily, and the smallest refineries process 500 t each day. In 1861, there were 1,200 sugar mills in Louisiana. In 1981, there were 25. This consolidation has been repeated throughout the world. The cane and beet farms have also become larger and fewer.

CULTIVATION

Sugarcane is a tropical grass. It is killed by freezing, has thick stems about 2 cm in diameter, and grows very tall—3–5 or even 7 m. Sugarcane only reluctantly forms flowers under ideal conditions. In many areas where cane grows well it never flowers. When it does, the flowers are in the form of tassels containing many tiny blossoms. Each tassel forms many thousands of tiny seeds, called fuzz. Experimental stations for crossbreeding have been set up in those areas where the climate is right for cane to flower. The station at Canal Point, Florida, has a collection of 30,000 varieties of cane for crossbreeding and makes a million (10^6) crosses each year. Since the varieties are all hybrids, seeds will not breed true, but each seed is a new variety. The varieties are designated by the initials of the experiment station that developed them followed by a number. The best-known varieties are those of POJ (Proefstation Ost Java), CP (Canal Point), NCo (Natal selected from canes bred in Coimbatore, India), and Q (Queensland). One variety that today is cultivated all over the world is NCo 310. Selection criteria for new varieties include agronomic characteristics, sugar and fiber content, harvestability, disease and insect resistance, and cold and salt tolerance. A continuing breeding program is essential because all varieties suffer a gradual yield decline. In this baffling phenomenon, the yield from a variety falls off over a period of years. It is interesting to note that some of the original work on tissue culture was done on sugarcane in Hawaii.

The principal commercial cane until the 1920s was the so-called noble cane, *Saccharum officinarium,* with thick stalks and relatively soft rind. However, these canes succumbed to the mosaic virus, which causes a mottling of the leaves and greatly reduced plant vigor. As a result of extensive crossbreeding programs, the experimental stations succeeded in developing multicross hydrids that were resistant to mosaic disease. These new varieties all have a much tougher rind and more fiber and more sugar than the noble canes. The most famous is the miracle cane, POJ 2878, which was the answer to the threat of mosaic disease.

Commercially, cane is grown vegetatively by planting stalks of cane called seed cane or setts. Sugarcane has a bud at each node that is at the base of each leaf. There is also a ring of root primordia. The pieces planted contain one or two buds or may be a whole stalk. When planted, the roots develop and the bud sprouts to form a new plant or stool of the same variety as the cane that was planted.

On harvesting, the stalks are cut off even with the ground; the next day more sprouts start from the buds just below ground level. Thus, a field of sugarcane once planted is self perpetuating. This is the source of one of the economic woes of the whole sugar industry. A high price of sugar results in the planting of many cane fields in different parts of the world. Once started, they continue producing year after year whether the price of sugar is high or low. The cane from the first crop is called plant cane and is always best. Subsequent crops are called stubble or ratoon crops. The ratoon crops have many problems. Gaps appear where the harvester dug too low and pulled the whole stool out by the roots. In other areas, the stalk density becomes too great, resulting in too many spindly stalks. Diseases multiply and weeds get thicker than the cane. In many areas, only two or three ratoon crops are taken off before the whole field is plowed out and started over. However, in some tropical areas where conditions are ideal for growing sugarcane, such as Cuba, fields have remained in production for as long as 20 y.

Good agricultural practice requires the usual close attention to soils, fertilizer, weeds, pests, disease, and water supply. In some cane-growing areas, the land must be drained; in others, irrigated. A common occurrence is for the cane to be knocked down by a thunderstorm or hurricane. The growing tip then turns upward and the stalk grows curved. After several storms the cane field becomes badly tangled, or lodged. Some varieties are more susceptible to lodging, which causes difficult harvesting.

Sugarcane has the very fortunate characteristic that it stores sugar in the stalks at all times. Thus, there is no narrowly defined ripe or mature stage. The sugar content of the stalks increases somewhat with cool weather and decreases again with warm weather. Typically, a cane that had 15% sucrose at its optimum, might have 12% several months before and after the optimum. The harvesting season can therefore be very long. It is nevertheless advantageous to harvest the cane at its highest sucrose content, so considerable attention is paid to the maturity or ripeness of the cane, indicated by a higher sucrose content in the juice in the whole cane stalk. The growing tip has lower sucrose than the rest of the stalk. Tasseling stops growth temporarily and matures the cane. The ripeness can also be controlled to some extent by fertilizer and water supply. Chemical ripeners, which alter the growth of the cane to bring the sucrose content up, can be sprayed on the fields. Because different varieties mature earlier or later in the season, the cane harvest can last 6 mo or more. The shortest harvest season is in Louisiana where freezes occur almost every year and the cane does not start growing again until warm days in spring. In the fall, with the cane only 8-mo-old, harvest is delayed as long as possible so that the cane can grow another day or two and produce a little more sugar. On the other hand, the harvest must be finished by the first freeze, for although frost kills only the growing tip, causing sucrose content to rise, a severe freeze splits open the stalk, and microbial action begins on the next warm day. As a result, the grinding season in Louisiana is compressed to only 2 mo of frenzied activity. At the other extreme, Hawaii normally grinds for more than 11 mo in a deliberate and methodical manner. There, the cane is normally grown for 3 or 4 y. This results in very heavy yields of badly tangled cane. However, the yield per year of growth is the same as for the rest of the world.

RAW SUGAR MANUFACTURE

At the sugar factory, the cane is placed on a carrier that moves it into the factory. The cane is moved from the cane yard or directly from the transport to one of the cane tables. Feed chains on the tables move the cane across the tables to the main cane carrier, which runs at constant speed carrying the cane into the factory. The operator manipulates the speed of the various tables to keep the main carrier evenly filled.

To remove as much dirt and trash as possible, the cane is washed on the main carrier with as much water as is available. This includes recirculated wash water and all the condenser water. Of the order of 1–2% of the sugar in the cane is washed out and lost in the washing, but it is considered advantageous to wash. In areas where there are rocks in the cane, it is floated through a "mud" bath to help separate the rocks. The sugar recovered is normally 10 wt% of the cane. A flow diagram of a raw-sugar factory is shown in Figure 1.

Extraction of the Juice

The juice is extracted from the cane either by milling, in which the cane is pressed between heavy rolls, or by diffu-

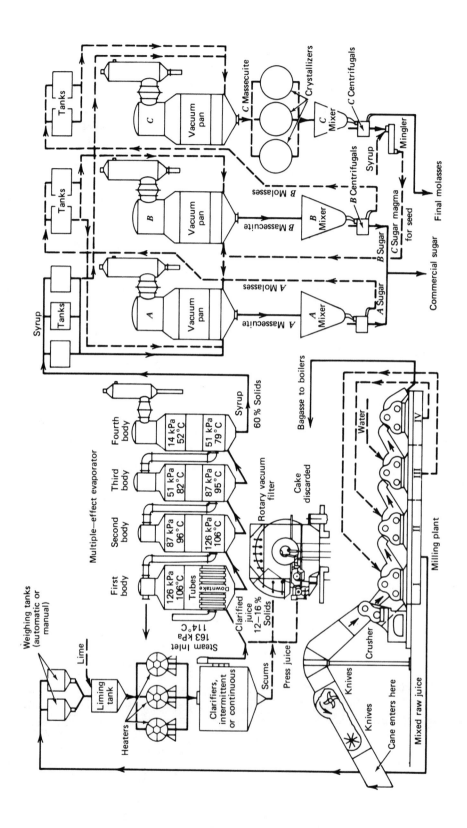

sion, in which the sugar is leached out with water. In either case, the cane is prepared by breaking into pieces measuring a few cm. In the usual system, magnets first remove tramp iron; the cane then passes through two sets of rotating knives. The first set cuts the cane into pieces of 1–2 dm length, splits it up a bit, and also acts as a leveler to distribute the cane more evenly on the carrier. The second set turns faster and combines a cutting and a hammer action. These quite thoroughly cut up and shred the cane into a fluffy mat of pieces a few cm in the largest dimension. In preparing cane for diffusion, it is desirable to break every plant cell. Therefore, cane for diffusion is put through an even finer shredder. No juice is extracted in the knives. In milling, the cane then goes to the crusher rolls, which have large teeth and are widely spaced. These complete the breaking up of the cane to pieces of the order of 1–3 cm. A large amount of juice is removed here.

Milling. The prepared cane passes to a series of mills called a tandem or milling train. These mills are composed of massive horizontal cylinders or rolls in groups of three, one on the top and two on the bottom, in a triangle formation. The rolls are 50—100 cm dia and 1–3 m long and have grooves that are 2–5 cm wide and deep around them. There may be anywhere from 3 to 7 of these three-roll mills in tandem, hence the name. The bottom two rolls are fixed, and the top roll is free to move up and down. The top roll is hydraulically loaded with a force of about 500 t. The rolls turn at 2–5 rpm, and the velocity of the cane through them is 10–25 cm/s. After passing through the mill, the fibrous residue from the cane, called bagasse, is carried to the next mill by bagasse carriers. To achieve good extraction, a system of imbibition is used: bagasse going to the final mill is sprayed with water to extract whatever sucrose remains; the resultant juice from the last mill is then sprayed on the bagasse mat going to the next to last mill, and so on. The combination of all these juices is collected from the first mill and is mixed with the juice from the crusher.

The mills are powered with individual steam turbines. The exhaust steam from the turbines is used to evaporate the water from the cane juice. The capacity of sugarcane mills is 30–300 t of cane per hour.

Diffusion. Diffusion is used universally with sugar beets but is less used with sugarcane. The process in cane is mostly lixiviation (washing) with only a little true diffusion from unbroken plant cells. Because the lixiviation is much faster, great effort is expended in preparing the cane by breaking it so thoroughly that nearly all of the plant cells are ruptured. In many instances, diffusers were added to an already existing mill, and, therefore, the diffuser unit was placed after the crusher rolls. In the diffusers, the shredded cane travels countercurrent to hot (75°C) water. In the ring diffuser, the cane moves around in an annular ring. In tower diffusers, the cane moves vertically, and in rotating drum diffusers, it travels in a spiral. Whatever the apparatus, the juice obtained is much like juice from mills.

Milling achieves ca 95% extraction of the sucrose in the cane, diffusion ca 97% extraction. Diffusion juice contains somewhat less suspended solids (dirt and fiber) and is of higher purity (sucrose as percent of solids). The diffusion plant costs much less and takes much less energy to run. The bagasse from diffusion contains much more water.

Bagasse. The bagasse from the last mill is about 50 wt% water and will burn directly. Diffusion bagasse is dripping wet and must be dried in a mill or some sort of bagasse press. Most bagasse is burned in the boilers that run the factories.

Clarification

The juice from either milling or diffusion is about 12–18 brix (percent solids), 10–15 pol (polarization) (percent sucrose), and 70–85% purity (pol/brix). As dissolved material, it contains in addition to sucrose some invert sugar, salts, silicates, amino acids, proteins, enzymes, and organic acids; the pH is 5.5–6.5. It carries in suspension cane fiber, field soil, silica, bacteria, yeasts, molds, spores, insect parts, chlorophyll, starch, gums, waxes, and fats. It looks brown and muddy, with a trace of green from the chlorophyll.

In the juice from the mill, the sucrose is inverting (hydrolyzing to glucose and fructose) under the influence of native invertase enzyme or an acid pH. The first step of processing is to stop the inversion by raising the pH to ca 7.5 and heating to nearly 100°C to inactivate the enzyme and stop microbiological action. At the same time, a large fraction of the suspended material is removed by settling. The cheapest source of hydroxide is lime, and this has the added advantage that calcium makes many insoluble salts. Clarification by heat and lime, a process called defecation, was practiced in Egypt many centuries ago and remains in many ways the most effective means of purifying the juice. Phosphate is added to juices deficient in phosphate to increase the amount of calcium phosphate precipitate, which makes a floc that helps clarification. When the mud settles poorly, polyelectrolyte flocculants such as polyacrylamides are sometimes used. The heat and high pH serve to coagulate proteins, which are largely removed in clarification.

Diffusion juice contains less suspended solids than mill juice. In many diffusion operations, some or all of the clarification is carried out in the diffusers by adding lime.

The mud from clarification mostly consists of field soil and very finely divided fiber. It also contains nearly all the protein (0.5 wt% of the juice solids) and cane wax. The mud is returned to the fields.

Although clarification removes most of the mud, the resulting juice is not necessarily clear. The equipment is often run at beyond its capacity and the control slips a little, so that the clarity of the clarified juice is not optimum. Suspended solids that slip past the clarifiers will be in the sugar. Clarified juice is dark brown. The color is darker than raw juice because the initial heating causes significant darkening.

Another scheme that makes a better clarification is the sulfitation process. In this method, lime is added as usual, but then sulfur dioxide from a sulfur burner is bubbled through the juice. The precipitate is settled as in the ordinary clarification. The bleaching effect of the sulfite makes a lighter colored sugar. This method is little used in the West but is used extensively in the Orient.

Evaporation

Cane juice has a sucrose concentration of nominally 15 brix (percent). The solubility of sucrose in water is about 72 brix. The concentration of sucrose must reach the solubility point before crystals can start growing. This involves the removal by evaporation of 93% of the water in the cane juice using multiple-effect evaporators, as can be seen in Figure 1.

Because the cane juice contains significant amounts of inorganic ions, including calcium and sulfate, the heating surfaces are quick to scale and require frequent cleaning. In difficult cases, the heating surfaces must be cleaned every few days. This requires shutting down the whole mill or at least one heat-exchange unit while the cleaning is done.

The evaporation is carried on to a final brix of ca 65–68. The juice, after evaporation, is called syrup and is very dark brown, almost black, and a little turbid.

Crystallization. The crystallization of the sucrose from the concentrated syrup is traditionally a batch process. The solubility of sucrose changes rather little with temperature. It is 68 brix at room temperature and 74 brix at 60°C. For this reason, only a small amount of sugar can be crystallized out of a solution by cooling. The sugar must instead be crystallized by evaporating the water. Sucrose solutions up to a supersaturation of 1.3 are quite stable. Above this supersaturation, spontaneous nucleation occurs, and new crystals form. The sugar boiler therefore evaporates water until the supersaturation is ca 1.25 and then seeds the pan. The seeding consists of introducing just the right number of small sugar crystals (powdered sugar) so that, when all have grown to the desired size, the pan will be full. After seeding, the evaporation and feeding of syrup are balanced so that the supersaturation is as high as possible in order to achieve the fastest possible rate of crystal growth, without exceeding 1.3.

The boiling point of a saturated sugar solution at 101.3 kPa (1 atm) is 112°C. Sugar is heat-sensitive and, at this temperature, thermal degradation is too great. The boiling is therefore done under the highest practical vacuum at a boiling point of ca 65°C. The sugar boiler therefore must manipulate the vacuum along with the steam and feed. A proofstick on the vacuum pan allows the contents of the pan to be sampled while under vacuum. When the pan is full, the steam and feed are stopped, the vacuum is broken, and the batch, or strike, is dropped into a receiver below.

A strike is ca 50 t of sugar and it is boiled in 90 min. At the end of this time, the mixture of crystals and syrup, called massecuite, must still be fluid enough to be stirred and discharged from the pan. In practice, about half the sugar in the pan is in crystal form and half remains in the syrup. In this case, the pan yield is said to be 50%. Some very good sugar boilers are able to achieve as much as a 60% yield on first strikes.

In actual practice, there are many more fine details to sugar boiling than is briefly explained here. Because modern pans have been extensively instrumented, automatic boiling can be used.

The sugar-boiling process can be made continuous by continually introducing seed as well as syrup, and also continually removing some of the product sugar. If the continuous boiling is done in one vessel, then the product will always contain crystals of all sizes. The market demand for crystals of a narrow size range can be met by having several vessels in series. In one scheme, all are within one body boiling under the same vacuum. In another scheme, several batch pans are connected in a continuous fashion.

Vacuum Pans

Vacuum pans have a small heating element in comparison to the very large liquor and vapor space above it. The heating element is a chest of vertical tubes called a calandria. The sugar is inside the tubes. There is a large center opening (downcomer) for circulation. A typical pan is shown in Figure 2.

The vacuum pan has a very large discharge opening,

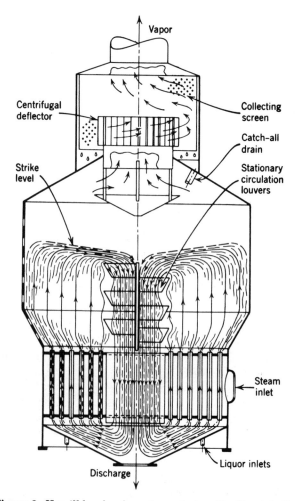

Figure 2. Hamill low-head pan (manufactured by Honolulu Iron Works).

typically 1 m dia. At the end of a strike, the massecuite contains more crystals than syrup and is therefore very viscous. This large opening is required to empty the pan in a reasonable time. At the top or dome of the pan, there are various entrainment separators. The pan may also be equipped with a mechanical stirrer. The pan shown has no such stirrer but relies on natural circulation. The strike is started with the liquor just above the top of the calandria. The strike level cannot be very near the top because vapor space must be allowed for separation of entrainment. In operation, the boiling is very vigorous with much splashing of liquid.

Centrifuging

The massecuites from the vacuum pans enter a holding tank called a mixer that has a very slowly turning paddle to prevent the crystals from settling. The mixer is a feed for the centrifuges. In batch-type centrifuges, the mother liquor is separated from the crystals in batches of about 1 t at a time. A typical batch centrifuge is shown in Figure 3.

The basket is lined with a screen having openings that will retain crystals of the minimum size required. Very fine crystals will go mostly with the mother liquor. The basket is attached to the drive shaft only at the bottom so that discharge devices and wash-water sprays can be mounted through the top. The essential feature of the suspended centrifuge is a universal joint in the centrifugal head, which is inside the brake. If, as is usually the case, the basket is not loaded evenly, this universal joint allows the basket-shaft-sugar mass to rotate around its own center of gravity instead of the center of the shaft. This greatly reduces vibration, which would otherwise shake apart the equipment. The diameter of the basket is ≥1 m, and the top speed is such that a force of ca 1,000g is exerted. This works out to 1,000–2,000 rpm, depending on the diameter. In operation, the loading gate opens, allowing massecuite to fill the basket. The thickness of the massecuite against the wall of the basket is 20 cm. After an initial spin, for most uses, the sugar is sprayed with wash water, sometimes several times. A final spin then allows the greatest possible separation. After this, the centrifuge must be stopped and the sugar plowed out while the centrifuge is turning very slowly. It is then washed, brought up to speed, and the cycle started over. The entire cycle takes as little as 3 min on refined sugars to as long as 2 h on very viscous final strikes. The apparatus has been completely automated.

The starting and stopping of the centrifugal batch consumes a great amount of energy, much of which can be saved by the use of continuous centrifuges (Fig. 4). They consist of a conical basket with a screen as in the basket centrifuge. The point of the cone is down and the feed is at the point. The centrifugal force causes the sugar to crawl up the cone, while the mother liquor passes through. At the top, the sugar flies off and is caught in the outer housing. The sugar crystals are badly broken when they hit the curb, and this limits the use of the continuous centrifuge to sugar that is to be reprocessed. It has its greatest use in final strikes, which are the most difficult of all to centrifuge.

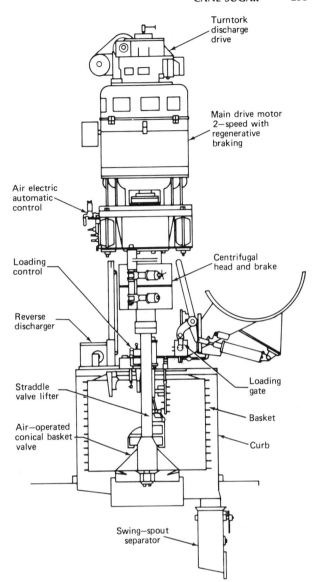

Figure 3. Suspended batch-type automatic centrifuge with reverse-discharge drive.

Boiling Systems

In raw-sugar manufacture, the first strike of sugar is called the A strike, and the mother liquor obtained from this strike from the centrifuges is called A molasses. The pan yield in sugar boiling is about 50%. Because crystallization is an efficient purification process, the product sugar is much purer than the cane juice and the molasses much less pure. As an approximation, crystallization reduces the impurities by a factor of 10 or more in the product sugar. Therefore, almost all the impurities remain in the molasses. Enough molasses accumulates from boiling two first strikes to boil a second strike. The B sugar from the second strike is only half as pure as that from the first strike, but the B molasses is twice as impure. This can go on to a third strike. At this point, $\frac{7}{8}$ of the sugar from the cane juice is in the form of crystals and $\frac{1}{8}$ in the C molasses. In practice, three strikes is about all that can be gotten from cane juice.

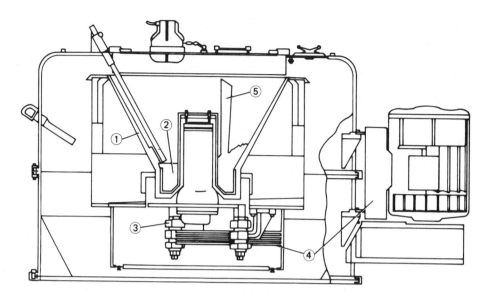

Figure 4. Continuous centrifuge. (1) Stainless steel basket. (2) Loading bowl. (3) Support. (4) Drive. (5) Massecuite side-feed.

Crystallizers

When the steam is turned off at the end of a sugar boiling, evaporation ceases immediately and the mixture of crystals and supersaturated syrup in the pan starts toward equilibrium, which is the point of saturation. In relatively pure sugar solutions, this equilibrium is reached in a few minutes, well before the syrup is removed from the crystals in the centrifuges. However, for low-purity solutions, the crystallization is slower and reaching equilibrium can take a significant amount of time. In the final strike, the time can amount to days, so final strikes are not sent directly to the centrifuges, but instead to crystallizers, holding tanks in which the crystals grow as much as possible and the supersaturation in the molasses is reduced to 1.0. Because the intention in handling the final molasses is to remove as much sugar as possible, advantage is taken of the small temperature coefficient of solubility, and the massecuite is also cooled. The crystallizers are large tanks, some open-top, with a slow-moving stirrer that is sometimes also a cooling coil. At the end of the holding time, the massecuite is warmed slightly as it enters the centrifuge to lower the viscosity and achieve better separation. The limiting factor in exhaustion of molasses is the viscosity. A little more water can always be boiled out, but the molasses must remain fluid enough to run out of the pan, into the centrifuge, and to flow between the sugar crystals on the centrifuge screens.

Storage

The raw sugar, the A and B sugar from the centrifuges, may be slightly dried or cooled but often is not. This sugar goes to the warehouse, eventually to be shipped to a refinery, where it will be warehoused again. Raw sugar can be stored for up to several years. Its chief enemies are heat and moisture. Moisture allows microbiological action to devour the sucrose. High temperature causes an easily detected increase in color long before the sucrose loss can be measured. Sugars at 30°C can be stored for years; at 38°C, for months; and at 45°C, for weeks.

Very High Pol Raw

A recent development has been the production of very high pol raw. If the raw sugar is washed slightly in the centrifuge, and then dried and cooled, the result is a raw sugar with the very high pol (% sucrose) of 99.4. This material is not sticky, is free flowing, and keeps very well. This sugar does not qualify as raw sugar in some tariff rules; therefore, very high pol raw that enters international trade must be "polluted" with something to reduce the pol to the raw-sugar class for that country. This is done by adding some high-test molasses (invert syrup) as the sugar is loaded into the ship. This is, of course, a step backward to conform to regulations.

Direct-Consumption Sugar

Many raw-sugar mills produce direct-consumption sugar by performing a supplemental clarification on the cane juice, as mentioned under clarification, or by filtering it, followed by some decolorization such as with sulfite, hypochlorite, peroxide, or powered carbon. From these improved syrups, one strike of direct-consumption sugar can be obtained; later strikes go into raw sugar.

REFINING

Sugar refineries are located in large cities. They are near seacoasts with harbors and facilities for receiving raw sugar by ship. They thus can receive sugar from anywhere in the world, although each refiner has favorites that suit the refinery, market, or have been the traditional supplier. Refineries are open all year, although the busy season is in the summer.

Refineries are always large. Their capacity is expressed in terms of daily melt. Melt is the sugar term for dissolving, and means the amount of sugar melted or processed each day. The smallest refineries have a daily melt of 450 t, and large ones have as much as 10 times that amount. The yield of refined sugar is nominally 93% of the raw-sugar input.

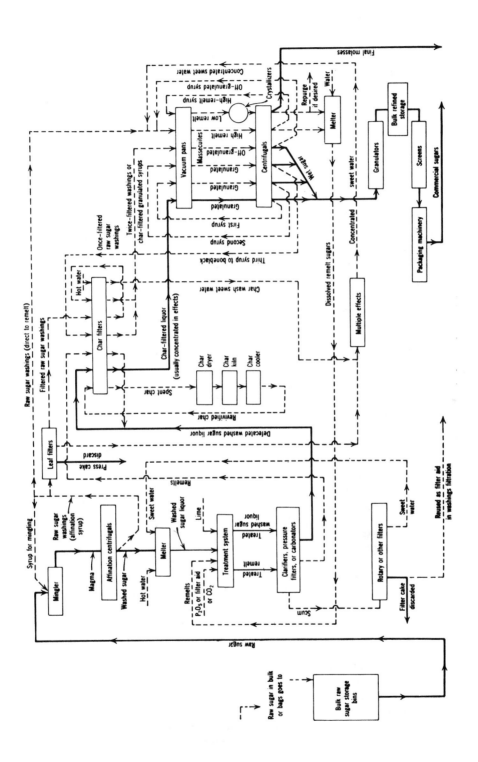

Raw sugar is light to dark brown in color and sticky. The size of the sucrose crystals is ca 1 mm. Refiners would like to have raw sugar that is high in sucrose and of uniform quality; however, they must be prepared to refine anything. Raw sugars are about 98 pol, although they are always described in terms of the equivalent raw value, expressed as 96 pol.

A simplified flow diagram for a typical bone-char refinery is given in Figure 5. Refineries tend to be more complex than this diagram.

Terminology changes slightly in the refinery. In raw sugar, a syrup is a concentrated solution going to the pans. After the boiling, the solution separated from the crystals is molasses. In the refinery, it is liquors that are fed to the pans, and syrups that are separated from the crystals. Local jargon adds to the confusion, with such terms as greens and jets, meaning syrups, and barrel syrup, meaning final molasses.

Affination

The first step in refining is to remove the molasses film from the outside of the raw-sugar crystals. This is done by a washing process known as affination. A syrup that is not quite saturated with sucrose is mingled with the incoming raw sugar in a large trough containing a mixer paddle and scroll. This mixture is then centrifuged and washed in the centrifuge rather more than less. A uniform crystal size is important in raw sugars because a mixture of different sizes or broken crystals does not wash well in the affination centrifuge. The syrup formed is called affination syrup and is used for mingling. The sugar is called washed sugar and is 10 shades lighter in color than the raw sugar. It is estimated that 90% of refining is done in this first step. About 10% of the sugar becomes part of the affination syrup, which thus keeps increasing in volume and is sent to the recovery house. When refining very high pol raw, the affination step can be minimized or skipped completely.

The recovery house is a route through a set of equipment in the same building. It uses the same processes that are used in the main refinery, but in a manner more like a raw-sugar operation. As the name suggests, sugar is recovered in the recovery house, but the main object is to transfer impurities into molasses that contains the least possible amount of sucrose. The recovered sugar is called remelt and is sent back to process.

Melting

The washed sugar is melted in hot water, and usually the pH is adjusted with lime. The washed sugar liquor coming from the melter is adjusted to the operating concentration, usually about 65 brix (percent sucrose). The trend is to operate refineries at higher brix up to 68, because if water is not added, it does not have to be boiled away later. The washed sugar liquor is dark brown and quite turbid and appears much darker than the sugar from which it came. The melter liquor is strained through a plain screen to catch debris in the raw sugar.

Clarification

The object of clarification is the complete removal of all particulate matter. The particles in the sugar come from all sources, eg, field soil and fiber (bagacilio), which escaped clarification in the raw-sugar factory; all microbiological life, including yeasts, molds, bacteria, and their spores; colloids and very high molecular weight polysaccharides; and foreign contaminants such as insect and rodent droppings. The very diversity of the nature of the particulate matter and the wide range of particle sizes makes clarification a difficult and critical step in the refining of sugar. One of three processes is used: filtration, carbonatation, or phosphatation.

Filtration. The most straightforward process is filtration. This removes particles down to the submicrometer size, but not easily. To conserve energy in evaporating water, the sugar liquor is at as high a brix as possible, which means high viscosity. The temperature is raised to reduce the viscosity as much as possible. But, sugar solutions simply cannot be filtered without a filter aid. The first application of diatomaceous earth, or kieselguhr, was to sugar filtration, and sugar remains the largest user of these materials. Without filter aid, the wide range of particle sizes quickly plugs the filtration medium, resulting in greatly reduced, almost zero, flow. With a tight filter aid that achieves good clarity, the flow is still very slow (a faster filter aid does not achieve good clarity). Filtration, as the sole means of clarification, is being replaced by other methods.

Chemical Defecation. To make the liquor more suitable for clarification, it has long been the practice to add various substances that form precipitates in the sugar liquor and that help coagulate the impurities. Today, only two processes are used, carbonatation and phosphatation; both use lime. The addition of lime in the raw sugar process was the first use of defecation. It is applied to both the washed sugar liquor and also to the affination syrup, which carries a heavier load of particulate matter.

Carbonatation. In the carbonatation process a large excess of lime is added. Then carbon dioxide is bubbled through to form a great volume of precipitate, which is removed by filtration. The lime dosage is 0.4–1.2% on melt. It is added as a slurry and is mixed rapidly with the sugar liquor. The pH rises to about 12.6. To avoid destruction of sucrose and reducing sugars at this pH, the holding time in the small mixing tank is kept to 1 min.

The flow then goes to the first saturation tank, where carbon dioxide gas from the flue is bubbled through. (Those kilning their own lime use the carbon dioxide from the lime kiln for this step.) The pH in the first saturator is 9.5 and should be lower for better filterability and higher for better absorption of gas. Several saturation tanks are used: two, three, or four. The pH of the last tank is 7.5, at which point the minimum lime-salts solubility is reached. The holding time is ≤ 1 h in each tank.

In the course of precipitation of the calcium carbonate as calcite, insoluble lime salts are coprecipitated and many other impurities are entangled in the forming precipitate. Considerably more precipitate is formed than is necessary for good clarity. The extra precipitate is formed in order to get good filterability.

Phosphatation. In the phosphate clarification scheme, lime and phosphoric acid (food grade) are added simultaneously with good mixing. The phosphoric acid is added in proportion to the melt at about 0.01–0.02%. The lime is added to bring the pH to 7.8. The calcium phosphate precipitate forms a floc of no particular crystal structure. It is even better at scavenging impurities by entrapment than the carbonate precipitate. It also has the useful property of attaching or entrapping air bubbles. Thus, at the same time that the floc is being formed, some air is injected, mixed, or pumped into the system. Raising the temperature a few degrees also helps tiny air bubbles materialize throughout the liquor. The precipitate then floats to the surface as a scum of 80% organic matter and is scraped off without any filtration. The mixing is very thorough just as the reagents are added, gentle in a floc-development section, and then minimal in the flotation zone.

The phosphate clarifiers are also called frothing clarifiers and have many sizes and shapes with scrapers going forward, backward, and around. Some are heated and some are deep. Sugar is recovered from the scrums by clarifying again. The phosphate system uses only about $\frac{1}{10}$ as much reagents as the carbonate system, and so produces only $\frac{1}{10}$ the scum volume.

No matter what the method of clarification, the clarified liquor is brilliantly clear without any sign of turbidity. It is, however, dark, rather like a cup of weak coffee.

Decolorization

The key process in sugar refining is decolorization. Color is the principal control in every sugar refinery. It is the main property that distinguishes refined sugar from raw sugar. The word color is used loosely. It usually means visual appearance, but in technical sugar work it means colorant, the material causing the color. It can be classified in three groups: plant pigments; melanoidins resulting from the reaction of amino acids with reducing sugars; and caramels resulting from the destruction of sucrose. Many, but not all, compounds in each of these classes have been identified. In sugar work, color refers collectively to the optical sum of all the colorants.

Measurement of color is by light transmission at a wavelength of 420 nm. It is expressed as

$$A = -\log T/bc$$

where A is the attenuation index; T is the transmission; b is the cell depth, cm; and c is the concentration of sugar, g/mL. By dividing by the concentration of sugar, one obtains an index of the amount of colorant relative to the amount of sugar. As many of the colorants are pH sensitive, ie, they show some indicator effect, the color is always measured at pH 7.00. Sucrose concentration must also be kept constant because it affects the refractive index, which in turn affects the light scattering by the residual particulate matter that is always present despite clarification. This is important in measuring the color of refined sugars that have very little color but still have scattering particles. The value of A from the equation is multiplied by 1,000 to give sugar-color units. Raw sugars have a color of 1,000–2,000. Refined sugars have a color of 20 to 40.

Decolorization has traditionally been accomplished by carbon adsorbents, although other processes are available. Bleaching-type decolorizations with hypochlorite, sulfite, or peroxide have largely fallen into disuse because they do not remove the colorant, only camouflage it. Sugars so decolorized do not have good keeping qualities, and these processes are now used only for direct-consumption sugars. One common property of sugar colorant is that much of it is of an anionic nature and, therefore, it is susceptible to ion exchange. Some refineries now use only ion exchange for color removal. The anionic property can also be used to precipitate the color with a suitable cationic material. However, carbon adsorbents remain the principal method of decolorization. Powered carbons can only be used once or twice and are too expensive. Bone char was the standard method, but requires a large investment in plant and uses considerable energy. Granular carbon has a very large capacity for color and is being used in all new installations.

Carbon adsorbents are general adsorbents; they adsorb everything out of the sugar solution, including sugar, with little selectivity. Bone char and granular carbon behave similarly in decolorization of sugar. For the contact with sugar liquor, both are contained in beds called cisterns (also misnamed filters) ca 3 m dia and 7 m tall and holding 30–40 t carbon. The liquor flows downward with a contact time of 2–4 h. The first liquors are water white with a very gradual yellowing. The cistern stays on stream until the color of the liquor becomes too great to be handled by the remainder of the refining process. The decolorization is always greater than 90%. The bone-char cycle is about 4 d; for granular carbon, it is 4 weeks. The first liquors from bone-char treatment are lighter than the first liquors from carbon. However, from the adsorbent point of view, the two systems are different.

Bone char is made by heating degreased cattle bones to about 700°C in the absence of air. It is about 6–10% carbonaceous residue and 90% calcium phosphate from the bone with an open pore structure supplied by the bone. The surface area available to nitrogen is 100 m²/g. The particle size is about 1 mm. Besides being a carbon adsorbent, it has ion-exchange properties that permit removal of considerable ash from the sugar. These same ion-exchange properties result in a buffering effect that keeps the pH of the sugar liquor from falling.

After the decolorization cycle, the sugar is washed out of the bed (sweetened off) and then the bed is washed with cool water to remove as much as possible of the adsorbed inorganic salts (mostly calcium sulfate). The sweet water is of low purity and cannot be recycled. The organic coloring matter is adsorbed so tightly that no amount of washing will remove it. The water is therefore drained from the char and the char moved from the cistern to a kiln where the organic matter is burned off at 500°C, with a little oxygen in the kiln atmosphere to burn away freshly de-

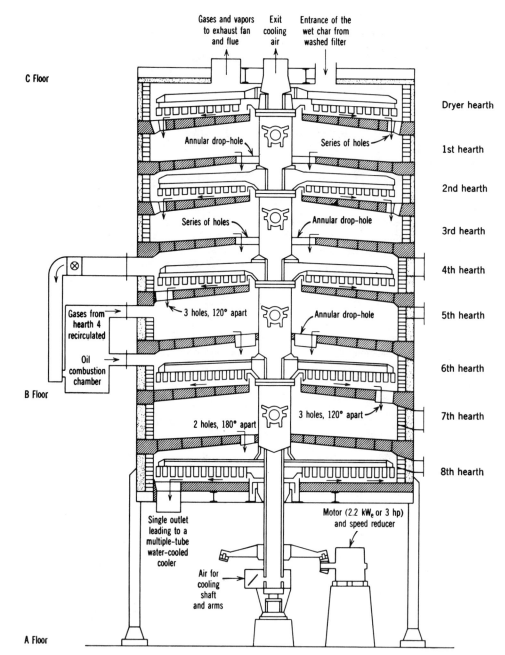

Figure 6. The Herreschoff kiln.

posited carbon and keep the pores open. The kilns used for regeneration of bone char are either pipe kilns in which the char is inside pipes and the flame outside, or Herreschoff-type kilns (Fig. 6), in which the char is in contact with the combustion gases. The Herreschoff kiln is preferred because it offers better control of kiln atmosphere, better heat transfer, and smaller size, hence lower cost. The bone char is then cooled, returned to the cisterns, and settled with sugar liquor. The char loss is ca 0.5% per cycle. The life of bone char is about 200 cycles. The amount of char used to decolorize sugar liquor is expressed as percent char burned (regenerated) on melt. For bone char, it is ca 10%, depending on the raw sugar being processed and the other refining processes.

Granular carbon, originally developed by Pittsburgh Coke and Chemical Co. for use in gas masks, is made from carefully selected mixtures of coal, heated to ca 1,000°C, and steam-activated. It contains more than 70% carbonaceous residue, the remainder being mostly siliceous ash. The surface area available to nitrogen is 1,000 m²/g. The particle size is similar to bone char. Granular carbon has no ion-exchange properties, does not remove ash, and must contain dead-burned magnesite for pH control. The magnesite hydrolyzes very slowly and supplies hydroxide at just the right rate to keep the pH up.

After the sugar-decolorization cycle, the sugar is washed out. The sweet waters are high purity and are sent to the melter. No further washing is used because nothing can be washed out. Kilning is in Herreschoff kilns or rotary kilns at 700°C. The oxygen concentration in the kilns is higher than for bone char to better burn off the heavier loading. The granular carbon is quenched in water and hydraulically returned to the cisterns where it is sweetened on. The cycle is repeated. The carbon loss is ca 5% per cycle, making the life of the carbon ca 20 cycles. The carbon burned on melt is 1% or less.

Powdered carbon is made from a wide variety of starting materials, including wood, coal, agricultural wastes, and black ash from paper mills. All contain organic material that is charred by heat and activated by oxygen or steam. Powdered carbon is introduced into the sugar liquor in an agitated tank with a holding time of 20 min. The mixture is then filtered with the help of filter aid. Powdered carbons are not regenerated. They do a very good job of decolorizing with a minimum of equipment. Use is 0.5% on melt.

Process variations include sending a sequence of liquors of increasing color to the adsorbent. Also, the sugar liquor may be sent through adsorbent in multiple passes, first through the older adsorbent and then through the adsorbent more recently regenerated. Many refineries use two passes. A variation on this scheme is the use of slugged beds of granular carbon. Every so often the flow is stopped, a slug of carbon is hydraulically removed from the bottom of the column, and a fresh slug added to the top. In this system, the liquor flow is upward and the velocity is such that the whole bed is lifted against the top of the column. The whole column may consist of 5–10 slugs, equivalent to 5–10 passes in one column. This approaches continuous adsorption. There is also a continuous adsorption process (CAP) in which the adsorbent is continuously removed from the bottom and added to the top. The flow in this system is such that the bed is expanded but not fluidized or raised.

Ion exchange is also used for decolorization. The chloride form of strong base quaternary ammonium anion exchangers very effectively exchange color for chloride. The chloride that goes into the sugar liquor is very soluble and forms a little more molasses. The ion exchanger is regenerated with brine that acquires a very high BOD (on the order of 10,000 ppm). The ion exchanger gradually becomes fouled with phenolic compounds and can only be cleaned with strong acids, bases, and hypochlorite, but the capacity and life of the ion exchangers makes the process very practical. The significant advantages of ion exchange are in situ regeneration without the use of heat, short contact time, and small size of equipment. Ion exchange is also used for ash removal or deionization of sugar.

The *Talofloc process* takes advantage of anionic properties of the sugar colorant as in ion exchange. In this process, however, the colorants are precipitated by reacting with a quaternary ammonium compound. The compound chosen must be sufficiently soluble in sugar solution to dissolve, but it must also be of just the right size to make an insoluble precipitate with sugar colorant. The compound chosen was dioctadecyldimethylammonium chloride. The precipitated colorant is scavenged from the sugar solution by a phosphate clarification. The dose is 300–500 ppm of the flocculant. The disadvantage of the system is that the additive is used only once and cannot be regenerated.

Crystallization

The color of the washed, clarified, and decolorized liquor going into the crystallization process ranges from water white to slightly yellow. The vacuum pans are the same as were described under Raw Sugar Manufacture, and their operation is the same. They are operated even more carefully to produce crystals of the desired size. Great are is taken to avoid conglomerates and fines. Boiling rate and throughput are important. A new strike of some 50 metric tons must be dropped every 90 min to keep up with the production schedule.

The boiling schemes used in the refinery are more extensive and variable than those used in the raw house. This is because the starting material is of much higher purity. Ordinarily, three, four, or five strikes of refined sugar are obtained. Syrup from the last strike may be used in making speciality syrups or brown sugars. It may also be sent back to decolorization or clarification and recycled.

The refined sugar centrifuges are always batch type because they leave the crystals intact. The centrifuging is easy and the cycles are short.

The drying of the sugar from the centrifuges is done in a rotary dryer using hot air. This dryer is universally misnamed the granulator because by drying in motion, it keeps the sugar crystals from sticking together, or keeps them granular. The hot sugar from the granulator is cooled in an exactly similar rotary drum using cold air.

Conditioning

The sugar from the coolers would appear to be finished, but after a few days storage it becomes wet with water trapped inside the grain because of the very high rate of crystallization and drying. After a few days, this moisture migrates outside the crystal and the sugar is wet again. The moisture is removed by a process known as conditioning, in which the sugar is stored for 4 d with a current of air passing through it to carry away the moisture. In one system, a single silo is used with sugar being continuously added to the top and removed from the bottom, and a current of dry air blowing upward. In another system, the sugar is stored in a number of small bins. It is continuously transferred from bin to bin with dry air blowing around the conveyors that move the sugar.

BIBLIOGRAPHY

This article was adapted from F. G. Carpenter, "Cane Sugar," in M. Grayson, ed., *Kirk-Othmer Encyclopedia of Chemical Technology*, 3rd ed., Vol. 21, Wiley-Intescience, New York, 1983, pp. 878–901.

All references have been removed. For specific citations, refer to the original article.

FRANK G. CARPENTER
Consultant
New Orleans, Louisiana

CANNING: REGULATORY AND SAFETY CONSIDERATIONS

Food canning is an important subject in food processing. In this article, the important safety and regulatory factors to be considered in a food cannery will be discussed. Specifically, this article describes criteria and procedures used by federal inspectors of the U.S. Food and Drug Administration (FDA) to determine if a canning operation is safe. For convenience, the information in this article will be delivered from the perspective of a supervisor who is instructing a field inspector in how to inspect a food cannery. By studying this article, the quality-control officers of a food cannery will have a better understanding of the safety and regulatory issues involved.

BASIC CONSIDERATIONS

At the beginning of an inspection, look into the handling and storage of raw materials, insect and rodent contamination, equipment sanitation, rest rooms, hand-washing facilities, etc. This gives a general idea of the hygienic condition of the cannery.

EQUIPMENT AND PROCESSING

Water Supply

Determine whether the water supply is a public or a private supply; ascertain the source, depth, and type of well; determine the sanitary hazards due to surface contamination or underground contamination; and note if it is chlorinated. It is important to check the chlorine content; acceptable chlorine residual in water varies: 4 ppm for processing and 15–25 ppm for cleanup. Up to 50 ppm are not detectable in the flavor of the finished product.

Raw Materials

Check on wash tanks, soak tanks, flumes, squirrel cage, and spray washers and determine the bactericide used (if any). Determine the efficiency: water pressure at spray washers should be at least 60 lb/in.^{2}.

Check the rate of processing raw stock and the number of operations used; the efficiency regarding adhering dirt, foreign material, and cleaning compound residue; the directions for use of such washing aids to determine if the firm is following manufacturer's instructions about final rinse, amount, etc. How does the firm check cleaning efficiency? What disposition is made of wash water and sort outs?

Report the amount and condition of fresh raw fruits or vegetables being held for processing. Check the holdover from previous day; if refrigerated, determine if the holding temperature is proper and has good air circulation and moisture control. Temperature control should be ±2°F. Check pesticides used by grower for possibility of residue.

Filling Operation

Empty Cans. Check for bad handling practices that could cause dented flanges. Examine the way the cans are cleaned; cans should be inverted to remove gross foreign material and washed while inverted. Hot water wash is best (180°F), cold water next best, and steam or air the least desirable. Jars should be inverted to remove carton dust or have vacuum removal of foreign material.

Determine the nonfood uses of empty containers for possible sources of contamination or adulteration. Check for the possibility of cable burn, which could contribute to defective seams in finished cans. Empty cans beyond the can washer should be removed after production to prevent contamination during cleanup.

Filler. Containers should be ventilated during the filling operation to prevent trapped excess air in the product. Check for possible contamination from ventilator splash that has accumulated and can be reintroduced into the product along with bacteria. Check for the possibility of overhead contamination at filler and on transfer conveyors to the capper.

Exhaust Operation. Treatment with heat, with or without vacuum, to remove air and trapped gases from the container and product is necessary before sealing (185–205°F). Trapped air is a possible cause of internal corrosion.

Seaming Operation (Capping). Check head space control, which is vital for vacuum control and proper processing in continuous cookers for sealed cans. Head space should be generally controlled at $\frac{10}{32}$ in. minimum to $\frac{15}{32}$ in. maximum. Container vacuum absorbs trapped gases, thus the initial vacuum should be higher than the desired finished vacuum. Too little vacuum will not absorb all of the trapped gases and will allow internal corrosion of the container and the spoilage of the product. Proper vacuum reduces the strain on can seams during retorting and prevents ruptures. If the vacuum is too high, some products will deteriorate and thus have a shortened shelf life.

Types of Container Vacuum. There are two types of container vacuum: natural and mechanical. For natural vacuum, the product is filled at +190°F and sealed immediately. On cooling, the product contracts to produce vacuum. Initial head space must be less, because when the product contracts, the head space will increase.

For mechanical vacuum, steam is sprayed into the head space before the cover is sealed to expel air. After sealing, the steam condenses to produce a vacuum. Check for the possibility of contamination from steam condensate, which accumulates in stem line during shutdown. Check the boiler treatment compound used for possible unsatisfactory food additives. This, of course, is significant where live steam or condensate from the steam comes into contact with the product. Carryover of boiler water can usually be noted at the retort operation. A steam pressure cook with boiler water carryover will leave a powdery film on the cans, whereas a water bath cook heated with live steam will show detinning of the cans with boiler water carryover.

Check the capper speed; if it is too high the seams will be unacceptable.

#10 can (603 × 700) 30–33 cans per minute (single head).

#5 can (404 × 700) 40–45 cans per minute (single head).

#1 can (303 × 400) 150–185 cans per minute (four heads).

Note that 603 × 700 indicates can size; thus 603 equals $\frac{3}{16}$ in. diameter and 700 equals 7 in. in height. The second and third digit always indicates the diameter in 16ths of an inch.

Can-Closing Operation

When observing the can-closing operation, use the following guidelines:

1. Can-closing operations may be the cause of a number of problems and must be carefully inspected. Improperly made double seams do not always lead to spoilage, but do represent a potential source of economic loss to the manufacturer and danger to the public. Generally speaking, the code and double seam is formed under rather strained conditions of heat and moisture and is more likely to be subject to defects than the noncode end (can factory end).

2. Note how frequently the firm examines the seams, what type of examination (visual or tear down) is made, and whether records of these examinations are maintained.

3. Visual examination by the canning plant will disclose gross defects in closing operations and should be made at intervals of not more than 30 min on code end seams of cans selected at random from each seaming station. Ask the firm's personnel to provide the records of examinations and to demonstrate the visual inspection operations. To fairly evaluate the operation of a closing machine, cans selected for examination must equal the number of closing heads on the machine so that the cans selected for examination represent ones closed by each closing head of the machine.

4. Determine if the visual examinations of the can seams are made after short or long periods of closing operation shutdown, immediately after can jams, and at the beginning of operations (usually in the first few minutes). If irregularities in seams are found, determine what, if any, action the firm takes.

5. Tear-down examinations of closed cans should be made at frequent intervals (minimum frequency is approximately one can per seaming station every 4 h). Ask the firm's personnel to demonstrate this examination to the inspection team. The examination should include seam width, thickness, length of body and cover hook, seam tightness, a check for cutovers, pressure ridges, droops, lips, and vees. Records of these examinations should be kept by the firm; ask to see them.

Can Handling

Check the capper discharge and the conveyors to the retort basket for dented seams from improper can handling. Check the fill of the retort basket; excess fill will allow can damage in retort when the baskets are stacked. Check handling of cans that fall to the floor, because there is the possibility of bypassing the retort operation. Check the can code legibility; the lid metal could fracture if there is too much pressure at the capper coder. All empty cans must be handled and stored in such a manner as to prevent damage to the can and entry of extraneous material. During the inspection it must be determined if cans are stored and handled under adequate cover, both inside and outside the plant. Find out whether containers are received in bulk, bags, or cardboard containers. Note whether can runways are covered and whether empty cans are washed or cleaned before being filled (if so, is it adequate?) Determine what, if any, precautions have been taken to prevent foreign objects from falling, being thrown, or kicked into open containers (for example, where the can chute goes through the floor). Are the cans used only for the product or are they used for other purposes (ashtrays)? What is the disposition of damaged cans? Generally, cans should not be used for any other purpose except for the product.

Cook Room Operation

In observing cook room operation, use the following guidelines:

1. The cook room operation must be set up in such a manner so that the procedure is systematic and absolute control is maintained. Employees involved in this operation must know the status of each can. All baskets or containers of unretorted and retorted material must be clearly marked. Cans of unknown status must be destroyed. Employees should have information available on the processing of each product and can size.

2. Determine the elapsed time between filling the cans, closing the cans, and retorting the product. A prolonged time between filling and retorting may permit deterioration of the product and the development of large numbers of organisms that may not be destroyed by retorting.

3. There should be a clearly defined procedure for checking and recording the time and temperature of the cook and code or batches involved. Ask to see these records, observe the operation, and determine if there are areas in the operation that are not under absolute control.

4. Check to see if all cans are removed from exhaust boxes, cooling channels, and retorted at the end of the day's operation for the next day's operation.

Retort Operation (Processing)

In observing the retort operation and processing, use the following guidelines:

1. Cans are processed at a high controlled temperature to destroy or inhibit microorganisms considered to be of health significance in the container. Retorting temperatures and times are controlled so that product quality is minimally affected. The autoclave in a bacteriology labo-

ratory is an indispensable piece of equipment, so too is the retort indispensable in the canning industry. Proper equipment and operation for both are absolute necessities. Baskets and trays in retorts must be of a kind and arrangement that do not interfere with venting of air or steam circulation. The National Food Processors Association (NFPA) recommends baskets of strap iron or wire. Perforated baskets should have at least 1-in. holes on 2-in. centers on all sides and the bottom. Dividers in baskets must allow free circulation. Burlap sacks, boards, etc, must not be used. The retort(s) should be connected to a steam line with at least 90 psi.

2. The retort time is based on the initial can filling temperature. Check the lag time from filler to full retort ready for cook. Approximately 3 min is required to fill a retort basket, regardless of can size; therefore, a three-basket retort should be ready for the process 10–15 min after the first basket is started. A longer time period could indicate an unsatisfactory process because the sealed cans may cool down.

3. Vents to allow air removal as the retort heats up should be located at the extreme opposite sides of the retort from the spreader through which steam is admitted. The time and temperature used to vent the retort must be checked by the retort operator.

4. Bleeder valves (to remove any air that may accumulate in the retort during operation due to make-up water in the boiler) on horizontal retorts will be located on top; on vertical retorts, in the lid. Determine where these valves are and if they are used. Bleeder valves must also be on all thermometer wells (and used) if accurate temperature readings are to be obtained.

5. The pressure gauge on each retort must be so located and large enough to make it easily read. Thermometers (the mercury bulb type is preferred) should be placed on each retort so they may be easily read and cleaned. Bleeder valves should be located so as to provide a full flow of steam past the entire length of the thermometer bulb. Accuracy of the thermometer and not a reading from a temperature-recording chart must be used to indicate processing temperature. Recorder readings are easily changed by adjusting the pen; however, the thermometer and records should agree.

6. The inspection team must determine all these factors and actually observe at least one complete cycle of retort operation, being sure to check time and temperature actually used to process a given product against procedures provided by plant management to the retort operator. Check these management procedures against those recommended by NFPA. In case of a line breakdown, where filled or capped cans are held before the process is started, additional time will be required in the retort. Filled cans could become contaminated if held for an extended period uncapped.

7. It is important that proper vent times are adhered to, because trapped air in the retort will form pockets that will act as insulators to prevent proper processing and cause possible spoilage. Check steam spargers for proper steam flow. Vibration could disrupt flow direction, which should be parallel to the bottom of the retort to obtain

tangential flow for maximum efficiency in evacuating air. Check time and temperature controls and records kept to cook. Are Cook-Chex or other similar products used to check the cook?

Cooling of Cans

Use the following guidelines when studying the cooling of cans:

1. Determine whether cans are promptly removed from retorts after processing and if they are air or water cooled. Water used for cooling must be of good sanitary quality. Check the level of free chlorine in the water with chlorine test paper. Samples of cooling water may be taken when deemed necessary. The time required and the temperature to which retorted cans are cooled must be checked. Determine whether or not cans are dry after cooling is completed.

2. Cans are generally cooled to 95–105°F to leave enough heat in the cans to dry them (when water cooled) but not enough to permit the growth of thermophilic organisms. At this temperature the cook is also stopped, preventing overcooking of the product. It must also be remembered that after processing, the sealing compounding the double seam is softened and as it cools, small amounts of cooling water may be drawn into the can. The sealing compound may then set and seal the leak without loss of vacuum. If contaminated cooling water is used, spoilage and possible danger to the consumer may result. When using highly contaminated cooling water, even the heat-closing operation may be taxed beyond its ability to prevent recontamination of the product effectively.

3. Recontamination of the product after retorting may occur in a number of ways. Highly contaminated can exteriors, especially around double seams may cause spoilage. Automatic can handling, frequently at high speed, may cause small deformations and strains on the seam, causing momentary breaks in the seal and pulling in contaminated material; it may result in spoilage or a danger to health. All equipment following can cooling should be checked for accumulation of moisture, grease, and dirt. Look for accessibility to cleaning, frayed belts, speed of operation, and porous material (in or on equipment) coming into contact with the can. These factors tend to complicate the picture and lead to recontamination of the product. Running cans at high speed into dead ends or through short turns in line direction, excessive bumping, and can jams must be avoided. All equipment must be periodically and systematically cleaned.

4. Check the cooling procedure with attention to precautions taken to prevent ruptured seams. Large cans should be cooled under pressure to remove initial heat. Check the chlorine content of the water; for cooling canals there should be 1 ppm residual chlorine at the water overflow discharge. Check rust inhibitors for the possibility of unapproved food additives. Chromates are generally used and impart a yellow color to cooling water. Potassium nitrate is used occasionally, but should be stored in such a manner that it cannot be introduced into the food product.

Storage of Canned Goods

When checking the storage of canned goods, use the following guidelines:

1. Check the stored canned product to determine if it is sufficiently cooled to prevent growth of thermophilic organisms (110°F). The storage area must be clean with no product stored directly on the floor. If the temperature of the warehouse is too high in comparison to the product, the cans may sweat.

2. Rust on the can exteriors may be due to a number of reasons, including improper operation of retort vents, too long of a come-up time in retort, or low-pressure steam containing moisture. Temperature of cans below 100°F coming from the cooler or failure to remove the surface water or chemical composition of the water may cause rust.

3. Examine a representative number of stored canned goods for rust, defective seams, flippers and swells, dents, and leaking cans.

Quality Control

Product. Who is responsible and with how much authority? Who is ultimately responsible? What are the specifications, and who set them up? What are the sampling and testing procedures? Determine the procedure for correcting unsatisfactory conditions. Determine the key to the can codes, legibility, and cycle of change. Determine the policy for marking the codes on shipping cases and invoices.

Container

Seam Tear-Down (Stripping) Examination. Determine the frequency of examinations and the reliability of the examiner and the records. The following is an excerpt from a table used to evaluate can seams. Only the ideal measurement in thousandths of an inch are given here, but in actual practice maximum and minimum dimensions are allowed, usually ±0.006 in. from the value given here. The cover hook will usually be about the same as the body hook; however, variations will be reflected in the W–BH value, which is explained below.

Can Diameter	Width	Thickness	Body Hook	W–BH
303	0.116	0.052	0.080	0.036
404	0.119	0.058	0.080	0.039
603	0.122	0.065	0.082	0.042

Note that the W–BH value, which is the difference between the second operation seam (finished seam) width and the body hook, is used to evaluate the safety margin of a particular seam in regard to the tolerances necessary for the individual components of the seam. Too small a value of W–BH could result in outside cracked seams, whereas too large a W–BH value could result in ruptured seams during the seam distortion accompanying the retort operation, or during handling in automatic equipment.

Sampling. Whether or not in-line samples and finished product samples are to be collected depends on the conditions found during the inspection and on the percent of damaged, defective, or spoiled product found in storage. In any event, the cans selected for a sample must be truly representative of the batch or code and drawn at random. An inadequate sample size or one collected in a nonrandom fashion will probably lead to erroneous concisions as to the extent of abnormalities as well as the cause of spoilage. Use the NFPA table, which gives the number of cans that must be collected to detect given levels of spoilage.

GLOSSARY

Bactericide. Any agent that destroys vegetative bacteria.

Blanch. Precook in live steam or hot water to remove skin, expel air and gases, inactivate enzymes, and arrest flavor changes or wilt for tighter pack (usually at about 145°F).

Body hook. Length of body flange of stripped finished seam.

Cable burn. Can flange and seam damage from can conveyor cable wear.

Can body. Cylindrical shell forming can side; formed from notched body blanks joined and soldered at side seam. Body blanks are notched to reduce the four thicknesses of metal at the side seam to two thicknesses on the can flanges.

Can cover. Formed end or lid with curled edge containing a gasketlike sealing material.

Can flange. Flared rim on open end or ends of formed can body.

Can seam. Referred to as a double seam and consists of five thicknesses of tin plate (seven thicknesses at the lap cross section) interlocked and pressed firmly together in two operations. The first operation rolls from the metal to produce the thicknesses or folds. The second operation rolls flat the folds so the sealing compound will fill all spaces not occupied by metal and act as a gasket and insure tightness.

Cook-Chex. Indicators (cardboard or tape) used on retort baskets or closed cans to determine the minimum time and temperature of the retort operation.

Cover hook. Length of cover curl on a stripped finished seam.

Crossover. The portion of the seam at the lap or body side seam.

Cutover. Sharp fin of the cover formed over the top of the seaming chuck flange, generally at the lap or crossover. A slight sharpness at the crossover only may not indicate a defective seam, but a severe cutover indicates a fracture in an area not protected by a sealant. If the cutover is due to an excessively tight seam, the condition will be continuous; if it is due to machine play or looseness in the machine head or gear train, the condition will occur at irregular intervals and may be overlooked unless regular seam checks are made.

Cut seam. Outer layer of seam perforated. Caused by seam that is too tight.

Droop. A smooth projection of the double seam below the bottom of a normal seam, usually most prominent at the side seam. Severe droops are caused by tight seams or excessive head pressure.

Flume. Channel with a stream of water used to convey material; usually the water is recirculated.

Head space. The unfilled volume of the hermetic container. This volume is important, and for most products it requires positive control.

Jumped seam. Double seam not rolled tight enough adjacent to the lap, caused by jumping of the seaming rolls at the lap. This presents a possibility of seam leakage and is evidenced only on examination of the cover hook of a stripped finished seam by excess wrinkles at the lap section. This condition is sometimes called an overlap.

Lap. Two or more thicknesses of material bonded together such as at the joint formed at the top and bottom of the side seam of the can body flange.

Lip. A sharp projection below the normal seam sometimes referred to as a V droop.

Pin lipping. Point at which two large cover hook wrinkles meet. Metal at this point is forced toward the can body and may form a sharp point, which can cut the can body; leakage results.

Pressure ridge. Formed on the outside of the can body as a result of seaming pressure on the double seam. It should be clearly impressed around the entire inside of the can, but should not be excessively deep on enamel cans to prevent fracturing of the enamel coat.

Squirrel cage washer. Reel-type washer equipped with spray nozzles.

Steam sparger. Spray-type spreader usually constructed of pipes with a series of orifices used to distribute steam into a retort and designed to facilitate removal of air from the retort during venting.

Thickness. Seam dimension referring to the cross-sectional thickness of the double seam.

Width. Seam dimension referring to the overall depth or height of the double seam cross section.

Wrinkle. Waves on cover hook of stripped finished seam that visually indicate the tightness of the seam. Wrinkles should not be more than one-third the cover hook. Lack of wrinkles indicates a tight seam with possibilities of seam fracturing, and wrinkles in excess of one-third indicates a loose seam with possibilities of leakage.

Acknowledgments: This article has been adapted from *FDA's Analyst Operational Manual.*

Y. H. HUI
EDITOR-IN-CHIEF

CANOLA

BACKGROUND

The term canola refers to cultivars of an oilseed crop, known as rapeseed, that is a major source of food and feed throughout the world. The crop has become the world's third most important edible oil source after soybeans and palm (1). The development by plant breeders of low erucic acid and low glucosinolate (double-low) varieties of rapeseed has proved pivotal to the rapid expansion in production and use of rapeseed worldwide. In 1979 the rapeseed industry in Canada adopted the name canola to distinguish those cultivars of *Brassica napus* and *B. campestris* that are genetically low in both erucic acid (seed must yield oil of less than 2% erucic acid) and glucosinolates (oil-free meal must contain less than 30 μmol alkenyl glucosinolate per gram of air-dry meal) (2). As the oil and meal from the double-low cultivars are nutritionally superior to those of earlier-grown varieties, the generic term canola, as it is now defined in the Canadian Food, Feed, and Seed Acts, is used to identify specifically the products derived from these varieties. Canola is no longer a Canadian trademark; regulatory organizations in other countries, including the United States, the United Kingdom, Australia, and Japan, have accepted or are working on the establishment of canola within their existing rapeseed industries.

Canola has become Canada's third most valuable crop after wheat and barley. Its popularity is due to some positive agronomic advantages and marketing alternatives it offers to producers. With an average annual volume of 3.2 million tons, Canada is the leading exporter of seed in the world (3,4); Japan is the major user of Canadian rapeseed/canola. In 1989, with a production at some 5.4×10^5 t of oil, canola counted for 60% of all deodorized fats and oils, 63% of the vegetable oil, and 78% of the salad oils produced in Canada (4); annual exports of canola oil amount to an average of 2.5×10^5 t/yr (1984–1989 period). Among other processed oil and fats, canola has made substantial gains into the shortening and salad and cooking oil segments of the Canadian market during the past decade (3). With respect to the United States, although the current seed production is relatively small, there is a growing demand for canola products; for 1988–1989 alone, economists had forecast consumption for oil and meal in the United States at 2.0×10^5 and 2.7×10^5 t, respectively (5). Canola oil is characterized by a low level of saturated fatty acids and a relatively high content of oleic acid. The recently reported hypocholesterolemic effects in human subjects, when canola oil made up a major portion of the total dietary fat (6), could become an influential factor in further promoting the use of canola products in the world's markets.

Botanical Origin

Although rapeseed production in Canada commenced after World War II, the cultivation of Brassica oilseeds has a long history in Europe and Asia (7). Unlike other oilseed crops, rapeseed is not a product of a single species, but may come from two species of the genus Brassica: *B. na-*

pus and *B. campestris*. The cytogenetic relationships between rapeseed and its close relatives have been discussed extensively in several reports and are important in understanding the origin, evolution, and plant-breeding strategies of the *Brassica* species (8–11). The two rapeseed species, along with *B. juncea* (mustard), commonly referred to as oilseed *Brassicas*, collectively provide more than 13.2% of the world's edible oil supply (11). The small, round *Brassica* seeds contain 40–44% oil (dry weight basis) and produce a high-protein-content (38–41%) oil-free meal. Within the rapeseed species, both spring and winter cultivars exist; the latter are higher yielding than the spring varieties but are less winter hardy than cereals. Winter cultivars of *B. napus* predominate in Northern Europe, whereas spring cultivars of *B. napus* and *B. campestris* are grown in Western Canada. *B. napus* varieties have a generally higher seed yielding potential (15–20%) than those of *B. campestris*, but they require an additional 8–15 frost-free days to mature; in Western Canada, cultivars of *B. napus* require 95–105 days for maturity (12).

Development of Double-Low Cultivars

Rapeseed production in Canada began with the introduction of seed of *B. napus* from Argentina and *B. campestris* from Poland; these materials, highly heterogeneous, constituted the seed stock for the establishment of breeding programs. Although improvements in seed yield and oil content were the first objectives, particular emphasis was also given to oil and meal quality. The elimination of nutritionally undesirable components—erucic acid [found to cause cardiac lipidosis in several animal species (6)] from the oil and sulfur containing glucosinolates from the meal—became a target at the early stages of the breeding programs. Changes in fatty acid composition were made by introducing the inherited trait of low erucic acid into *B. napus* and *B. campestris* adapted lines. Canadian breeders also accomplished the transfer of genes responsible for the low glucosinolate characteristic into rapeseed varieties. These sulfur-containing constituents are hydrolyzed by endogenous myrosinase into thiocyanates, isothiocyanates, and nitriles (Fig. 1) when seed cells are ruptured (eg, during seed crushing). The characteristic odor and flavor of *Brassica* vegetables and condiments (mustard) are largely due to the presence of these compounds. Glucosinolate breakdown products also reduce the palatability and, in nonruminant animals, adversely affect the iodine uptake by the thyroid gland (13). Breeding for lower glu-

cosinolate content of the seed has resulted in nutritional upgrading of rapeseed meal to all classes of livestock and poultry. Further reduction in glucosinolate content is required, particularly of the indolyl glucosinolates (14), if proteins from canola meal are to be used in human food formulations. This has been partly accomplished by extraction processes employed in preparation of protein concentrates and isolates (15). The ultimate solution would be to breed varieties essentially free of both alkenyl and indolyl glucosinolates.

HORTICULTURE—POSTHARVEST HANDLING

Cultivation—Agronomic Considerations

Canola is a cool season crop that requires more available moisture than wheat, as well as cool night temperatures to recover from hot and dry weather. Although it grows well on a variety of soils, it does best on loamy soils that do not crust and impede seedling emergence (16). Good yields are also obtained on light or heavy soils if rainfall, fertilization, and drainage are adequate (12). Canola is moderately tolerant to saline soils and has greater moisture requirements than cereals. Canola also requires more intensive management practices; for a good crop, the need for nitrogen is 20% higher than for a comparable cereal crop. Certified seed is usually used for sowing to ensure absence of mustard or weeds and the quality characteristic of the cultivar.

Canola is grown in Canada on summer fallow or cereal stubble land. From an agronomic viewpoint, the crop fits well as a rotation crop with cereals; wheat or barley following canola is the preferred sequence. Most fields are not cropped with canola more than once every 4 years. This permits for a break in disease and insect pest cycles, while preserving the soil structure. Canola is susceptible to several fungal diseases such as blackleg (*Leptosphaeria maculans*), Sclerotinia, Fusarium, Rhizoctonia, and Alternaria spp.; disease severity varies with year and location. A number of different insects can also cause serious damage to canola (eg, flea beetle). Crop rotation is effective in controlling fungi and insects. Several herbicides are also available for a variety of weed species (12).

Harvest and Storage Practices

The date and method of canola harvesting are important determinants of yield and seed quality (12,16). Immature

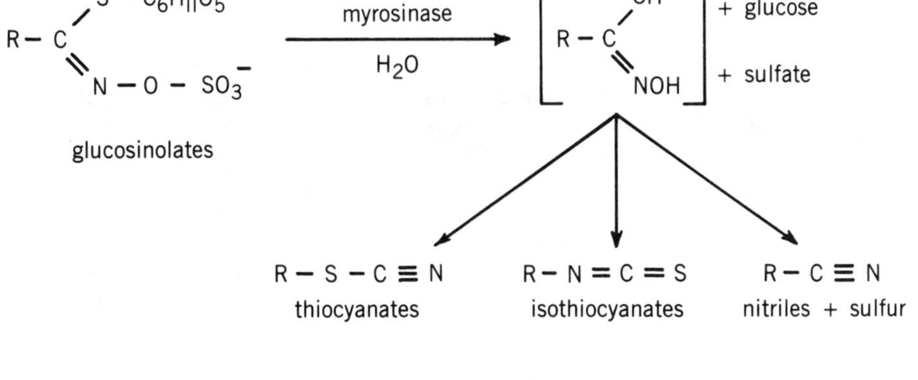

Figure 1. General structure of glucosinolates and their enzymic hydrolysis products.

seeds have higher chlorophyll and free fatty acids contents. On the other hand, if the harvest is delayed, canola is susceptible to shedding, particularly in windy areas. To avoid this, swathing (leaving cut plants to dry in the field) is applied when the majority of the seeds are at a firm doughlike stage (moisture content 35–40%). Swathing facilitates uniform maturation and yields seeds of good color and of high oil and protein contents. Premature swathing (seed moisture >45%) results in reduced yields, whereas late swathing (moisture <20%) increases shattering losses and the possibility of frost damage. Canola is ready to combine when the moisture content is below 15%. Freshly harvested seeds normally require up to 6 weeks to become dormant; storage in dry and cool environments minimizes respiration losses. Canola is usually transported after harvest to dryer–stocker companies and therefrom to oil-processing plants. For storage, silos of various sizes are used, while in the farms the seed is stored in bins. Immature and field-germinated seeds should not be stored for long periods because they deteriorate rapidly. A high percent of damaged or fragmented seeds also gives rise to a rapid increase in free fatty acids (17).

The moisture content of the seed, because of its importance to the growth of microflora and enzyme activity, is the main criterion for safe storage. In the case of oilseeds, including canola, moisture is held by the nonfatty components, thus bringing the critical level of overall seed moisture content much below that (15%) required for safe storage of cereal grains. The maximum moisture content for storage of clean canola seeds in a cool environment (<15°C) is 10.0% (18); others (19) even suggest lower optimum levels of moisture (8–9%) for storage over periods of several months. Above this level, the seeds must be air dried; air temperatures must be kept below 38°C to prevent heat damage. Aeration during storage, although at much smaller airflow rates than drying, is also required to prevent moisture gradients and to cool the stored grain. Figure 2 shows how storage is affected by both temperature and moisture content. Canola is particularly prone to microbial spoilage, with severe losses in seed and oil quality. The rate of spoilage rapidly increases if nonuniform distributions of moisture and temperature exist in the stored grain. An important consideration regarding moisture content of canola is that before solvent extraction of the oil, the seed must be cooked rapidly at temperatures between 80 and 100°C and at water contents between 6 and 10% to inactivate myrosinase: above 10% moisture, hydrolysis of glucosinolates would occur; below 6%, heat inactivation of the enzyme is difficult.

Chlorophyll is another important quality parameter of canola. Besides imparting an undesirable color to the oil, chlorophyll promotes oxidative rancidity in salad oils. Several agronomic practices, including length of growing season and seeding rate, have been shown to affect the chlorophyll level (20). Swathing at 35–37% seed moisture followed by drying is effective in reducing substantially the chlorophyll content of canola seeds (21).

BREEDING CANOLA FOR IMPROVED YIELD AND QUALITY—ROLE OF BIOTECHNOLOGY

Important objectives in canola breeding are both agronomic (seed yield, winter and frost hardiness, disease resistance, early maturity, herbicide tolerance, and resistance to lodging and shattering) and quality (oil and protein content and composition) traits (10,11,22). Although being one of the most desirable attributes, seed yield improvement through selection is a difficult task because of the strong influence from environmental factors.

Oil, being the most valuable component of the seed, is deposited in the form of small lipid droplets in the cytoplasm of embryonal cells. Protein bodies (storage proteins) are also stored in the cytoplasm of these cells (23). In Brassica spp., yellow-seeded cultivars have thinner seed coats and higher oil content than seeds of darker color. It is, therefore, of great interest to producers and crushers to develop by plant breeding yellow-seeded forms or cultivars with increased seed size (24).

Oil quality is determined by the fatty acid composition. Changes in the fatty acid profile can be achieved by modification of the biosynthetic pathways (Fig. 3). Fatty acid biosynthesis in Brassica oilseeds is under enzymic control. Chain elongation and desaturation of the aliphatic chains is carried out by assemblies of enzymes located in various cell compartments; eg, synthesis of oleic acid ($C_{18:1}$) in the developing seed occurs in proplastids, whereas chain elongation by which $C_{18:1}$ is converted to $C_{20:1}$ (eicosenoic acid) and $C_{22:1}$ (erucic acid) occurs at another site (endoplasmic reticulum) in the cell (8,26). Oleic acid thus plays a key role into the biosynthetic pathway because once it is formed, it is transported to the modifying compartment. Changes in single genes can have profound effects on the fatty acid composition of oil. For example, the amount of erucic acid is controlled by a series of alleles (at two loci in B. napus and a single locus in B. campestris) that exert additive action; varying the alleles present, it is possible to have erucic acid content between 1 and 60%. Another oil quality breeding objective of canola has been to reduce

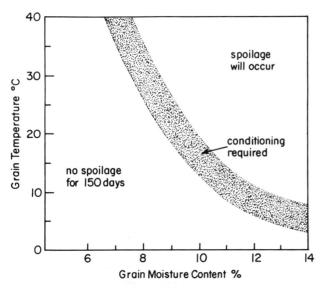

Figure 2. Effects of moisture and temperature on storage of canola (Adapted from Ref. 18).

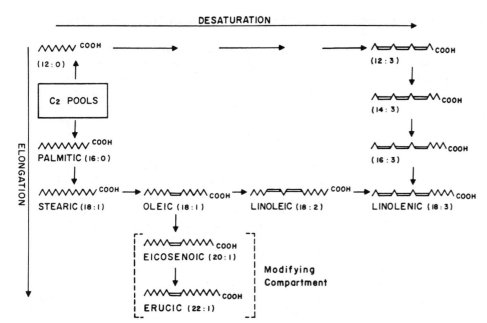

Figure 3. Suggested biosynthetic pathway for fatty acid synthesis in rapeseed (Adapted from Ref. 25 with permission).

the level of linolenic acid (C$_{18:3}$) present from 8–10% to less than 3% for the purpose of improving oxidative stability of the oil. A low-linolenic acid (<3%) Canadian cultivar (Stellar) has been recently introduced (27). The oil from this cultivar exhibited improved stability at 60°C and showed negligible changes in sensory and chemical indices of rancidity, compared to regular canola oil (28); improved odor scores from frying tests were also reported for the low-linolenic canola oil (29). Further modifications in fatty acid composition are being explored; work is being done to develop cultivars with more than 30% linoleic acid and others with higher oleic acid levels. It is feasible to increase the level of palmitic (C$_{16:0}$) and palmitoleic (C$_{16:1}$) acids to about 10–12% in order to make margarines exclusively from low erucic acid rapeseed oils; for such products, the tendency to develop large crystals on storage would be diminished. Overall, it would appear reasonable to expect the development of a broad range of rapeseed oils having fatty acid composition appropriate for specific applications (Table 1; Ref. 25).

In contrast to oil, relatively little attention was paid until now to the amino acid balance of the meal. Indeed, the reduction of glucosinolate levels in the meal has been the primary objective in the breeding programs. Recessive alleles, responsible for the low glucosinolate characteris-

tic, have been introduced from the Polish spring cultivar Bronowski into varieties of *B. napus* and *B. campestris* by backcrossing (8); three to five gene loci are involved in the inheritance of glucosinolate content (22). Additional improvement in the nutritional value and palatability of canola meal would also result if the 1.0–1.5% sinapine and phytic acids present in Brassica oilseeds are reduced or eliminated by breeding.

There is great interest among breeders and producers in developing hybrid cultivars of canola for increased yield. Both cytoplasmic male sterility (CMS) and self-incompatibility are being investigated as means of producing hybrid seeds (10,11,30). For the former, two inbred lines are used: (1) the female parent is male sterile owing to cytoplasmic components (CMS, associated with mitochondria) interacting with the nuclear material and (2) a line carrying nuclear genes that will restore fertility in the hybrid cultivar. The expected level of heterosis (up to 40% seed yield increase) is sufficient to offset the cost associated with the development of an effective cytoplasmic male-sterile, genetic restorer system (10,30). Recent advances made in the production of double-haploid plants from anther and micropore cultures would accelerate the development of hybrid cultivar technology (31).

The Brassica oilseeds have responded to genetic manip-

Table 1. Fatty Acid Composition of Normal and Modified Rapeseed Oil Resulting from Selection[a]

	Fatty Acid							
Oil Type	16:0	16:1	18:0	18:1	18:2	18:3	20:1	22:1
Normal high erucic	3	<1	1	11	13	8	8	50
Selected high erucic	2	<1	1	11	12	9	9	57
Low erucic	4	<1	1	62	20	10	2	<1
Low linolenic	5	<1	2	64	24	<3	1	<1
High 18:2 low 18:3	5	<1	2	55	30	5	1	<1
High palmitic	10	4	1	51	19	13	1	<1

[a] Ref. 25.

ulation better than any other crop. The rapidly growing interest for crop improvement and modification by nonconventional breeding approaches as a result of advances in molecular biology and tissue culture techniques is currently focused on several areas of research (31–33):

1. Production of haploids by culture of reproductive organs to reduce the time required for new cultivar development.

2. Protoplast fusion to transfer desirable genetic traits, particularly those carried by cytoplasmic organelles (chloroplast, mitochondria). Transfer of cytoplasmic traits such as male sterility and herbicide resistance was made possible via protoplast fusion and microinjection techniques (32,34).

3. Selection of cell variants with desirable agronomic characteristics (eg, herbicide resistance, drought tolerance, winter hardiness) after treatment with mutagenic agents and selection pressure of treated cells.

4. Gene transfer using suitable vectors (*Agrobacterium tumefaciens*) or by microinjection techniques.

Consideration is primarily given to genes responsible for herbicide and fungal disease resistance, osmotolerance, manipulation of triglyceride composition, and production of novel high value chemicals (eg, jojoba-type waxes having as precursors eicosenoic and erucic acids). Genes that impart resistance to several herbicides (eg, glyphosphate, the active ingredient in the herbicide Roundup) have been already transferred into canola (33). The impact of genetic engineering and conventional breeding is expected to result in even more valuable and versatile Brassica oilseeds in the future. Specialty oils and by-products for both food and industrial uses are within reach of today's breeders and are expected to influence the production and marketing of rapeseed/canola in the years ahead.

COMPARISON OF CANOLA WITH OTHER OILSEEDS

The reduction of erucid acid in rapeseed oil resulted in a marked increase in oleic acid along with smaller increases in linoleic and linolenic acids. Canola oil has a fatty-acid composition similar to peanut and olive oil with the exception of the low palmitic and higher linolenic acid contents. Canola oil also contains the lowest level of saturated fat (mainly palmitic and stearic acids) and the highest content of unsaturated fat (mainly monounsaturated) of any other edible oil source (Fig. 4). The total quantity of polyunsaturated fatty acids ($C_{18:2}$ and $C_{18:3}$) in canola oil is moderate, because it contains linoleic acid in amounts less than other widely used vegetable oils. The relatively high linolenic acid content of regular canola oil (~10%) makes it susceptible to oxidative rancidity. This problem has been alleviated in the newly developed low linolenic acid cultivars (27). The presence of small amounts of erucic acid (1–2%) in canola oil was suggested as contributing to inhibition of lipoxygenase (35,36); however, the presence of lipoxygenase in canola seeds was recently reported (37). In most respects, there is no obvious physical property distinguishing canola oil from other common vegetable oils. Rapeseed oils (both high and low erucic acid), however, contain a C_{28} sterol, brassicasterol, which is not found in other edible vegetable oils (38); brassicasterol is thus a suitable market to identify Brassica oils in vegetable oil blends.

The protein content of canola meal varies with the variety from which it is produced; it ranges between 36 and 42%. Canola meal is higher in fiber than soybean meal (Table 2). The amino acid composition compares well with that of soybean meal (39). Although the latter has a greater lysine content, canola meal contains more of the sulfur containing amino acids (methionine and cystine). Consequently, the two meals tend to complement each other when used as feedstuffs in rations for livestock and poultry. Canola meal is a richer source of minerals than

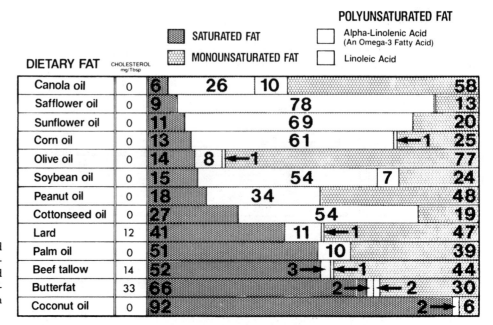

Figure 4. Comparison of fatty acid composition (%) and cholesterol levels of some common edible fats and oils. (Data from Agricultural Handbook No. 8-4 and Human Nutrition Information Service, USDA.)

Table 2. Proximate Analysis and Amino Acid Composition of Canola and Soybean Meals[a]

	Canola Meal		Soybean Meal	
	As Fed, %	In Protein, %	As Fed, %	In Protein, %
Proximate Composition				
Moisture	8.01		11.00	
Crude fiber	11.41		7.30	
Ether extract	3.35		0.80	
Protein (N × 6.25)	37.15		45.21	
Amino Acid Composition				
Alanine	1.73	4.65	1.95	4.31
Arginine	2.26	6.08	3.03	6.71
Aspartic acid	3.03	8.16	5.27	11.66
Cystine	0.90	2.48	0.71	1.61
Glutamic acid	6.46	17.38	8.43	18.65
Glycine	1.92	5.18	2.06	4.55
Histidine	1.04	2.79	1.12	2.48
Isoleucine	1.28	3.44	1.82	4.03
Leucine	2.57	6.92	3.36	7.44
Lysine	2.21	5.95	2.82	6.24
Methionine	0.76	2.10	0.70	1.59
Phenylalanine	1.45	3.90	2.13	4.72
Proline	2.48	6.67	2.27	5.03
Serine	1.69	4.55	2.38	5.27
Threonine	1.68	4.52	1.74	3.85
Tryptophan	0.44	1.19	0.52	1.15
Tyrosine	0.97	2.62	1.33	2.95
Valine	1.64	4.42	1.89	4.18

[a] Ref. 39.

soybean meal (Table 3). However, the presence of higher amounts of phytic acid and fiber reduces the availability of calcium, magnesium, phosphorus, and zinc for poultry. It has also been shown that crude fiber decreases the availability of copper and manganese (39). Nevertheless, canola meal is a better source of available calcium, iron, manganese, phosphorus, selenium, and most vitamins than is soybean meal. On the basis of some data (39), canola meal has been shown to have less metabolizable energy than does soybean meal; on a nutrient content basis alone, canola meal is equivalent to 70–75% of 44% protein content soybean meal for feeding poultry and to about 75–80% of the same for feeding swine and ruminants.

PROCESSING

Processing of edible oils, including canola oil, is designed to produce the purest extract of triglycerides possible. This necessitates removing minor components naturally present in the oil that are detrimental to the quality of the finished product (40).

Cleaning Seeds

Before processing, canola seeds undergo cleaning to reduce the presence of any foreign materials. This material, referred to as dockage, was shown (41) to consist mainly of

damaged canola seed together with weeds. A number of operations are involved, including fanning and sieving mills to remove pods and weed seeds; indent machines to remove any large noncanola seeds; destoners to remove dirt and small stones; and finally gravity tables to eliminate anything not removed in the previous processes. The presence of damaged canola seed has been shown to be detrimental to the quality of the extracted oil and should be reduced as much as possible before oil extraction (42).

Oil Extraction

Extracting oil from canola seeds is carried out with as little chemical alteration to the oil or meal as possible (43). Once cleaned, canola seeds are rolled or flaked to fracture the seed coat and rupture the oil cells. The production of thin flakes, 0.2–0.3 mm thick, is extremely important as a high surface to volume ratio is critical during oilseed processing. To facilitate good oil release, the flakes are cooked at 77–100°C for 15–20 min, to rupture any intact oil cells remaining (44). Flaked and cooked canola seeds generally undergo mild pressing or prepress to reduce the oil content from 42 to 16–20%, while compressing the thin flakes into large cake fragments. Canola cake fragments are solvent-extracted with normal hexane to remove the remaining oil. This is achieved by countercurrent movement of the cells of pressed canola cake and hexane, thus interfacing the oil in the flake or cake with a rich solvent–oil solution (45). The solvent is recovered

Table 3. Mineral and Vitamin Content of Canola and Soybean Meals[a]

	Canola Meal as Fed	Soybean Meal as Fed
Minerals		
Calcium, %	0.68	0.29
Copper, mg/kg	10.4	21.5
Iron, mg/kg	159.0	120.0
Magnesium, %	0.64	0.27
Manganese, mg/kg	53.9	29.3
Phosphorus, %	1.17	0.65
Potassium, %	1.29	2.0
Selenium, mg/kg	1.0	0.1
Zinc, mg/kg	71.4	27.0
Vitamins		
Choline, %	0.67	0.28
Biotin, mg/kg	1.0	0.32
Folic acid, mg/kg	2.3	1.3
Niacin, mg/kg	159.5	29.0
Pantothenic acid, mg/kg	9.5	16.0
Riboflavin, mg/kg	3.7	2.9
Thiamine, mg/kg	5.2	4.5

[a] Ref. 39.

from the oil–hexane solution by a conventional distillation system that ensures the oil and meal are solvent free.

The solvent-extracted oil (43–47% of the total oil) is combined with the prepressed oil (53–56% of the total oil) to form the crude oil fraction, as outlined in Figure 5. The crude oil contains a variety of minor constituents (Table 4)

detrimental to oil quality that are removed by a series of unit processing steps including degumming, alkali refining, bleaching, hydrogenation, and deodorization.

Degumming

Conventional degumming is carried out in most plants by treating the crude oil with either hot water (85–90°C) or steam (2–5%) while mixing intensively from 1 to 30 min (47). This precipitates the water-hydrated phospholipids, which are then removed by centrifugation. The major drawback to this type of degumming process is that it only removes hydratable phospholipids and still leaves 150 to 250 ppm of phosphorus in the oil. To achieve better results, most Canadian processing plants now carry out acid degumming using citric, phosphoric, or malic acid. This is followed by steam, which results in an oil with phosphorus levels of 50 ppm or less. This product is referred to as acid degummed or super degummed oil.

Refining

The crude degummed oil is then subjected to refining, which removes free fatty acids, phospholipids, color

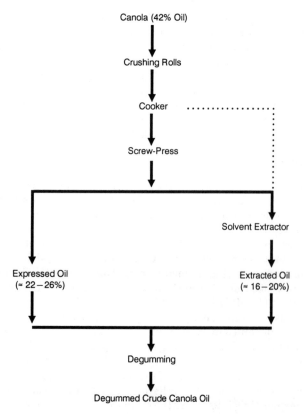

Figure 5. Outline of the primary processing of canola.

Table 4. Minor Constituents in Crude, Degummed Canola Oil[a]

Constituents	Amount
Free fatty acids, %	0.5–0.8
Phospholipids (gums), %	0.5–0.8
Unsaponifiables (sterols and sterol esters), %	0.1–0.8
Color bodies, ppm	20–30
Sulfur compounds, ppm	2–10
Iron, ppm	2
Copper, ppm	5

[a] Ref. 46.

bodies, iron and copper, as well as some sulfur compounds. The major type of refining in Canada is alkali refining, although there is a shift toward physical refining due to the fewer environmental problems associated with the latter process.

Five major steps are involved in alkali refining.

1. Initial pretreatment of the crude degummed oil with phosphoric acid (0.2–0.5% of 85% phosphoric acid) at 40°C for a minimum of 30 min. This conditions or acidifies the nonhydratable phospholipids (gums) and permits their separation during the refining process. In addition to precipitating nonhydratable phosphatides, chlorophyll as well as prooxidant metals such as iron and copper are also removed (48,49).

2. The second and major step in alkali refining, neutralization with sodium hydroxide (8–12%), depends on the free fatty acid content of the oil and the amount of phosphoric acid used during pretreatment. A slight excess is normally added to ensure complete saponification of these fatty acids.

3. The addition of sodium hydroxide in step 2 produces a soap-stock phase that contains precipitated nonhydratable phospholipids as well as free fatty acids and other insoluble impurities. This soap stock is separated from the oil by centrifugation.

4. Any residual soap stock in the oil is further reduced to 50 ppm or less by washing with water at 85–95°C. Citric or phosphoric acid may be added to the washed oil to remove remaining traces of soap stock.

5. During this stage the oil is heated at 105°C with agitation until dry.

Most oil processors use a continuous alkali refining process although batch refining is still used by some processors. The overall result of alkali refining is a marked reduction in free fatty acids, phosphorus, chlorophyll, and sulfur, as summarized in Table 5.

An alternative method to alkali refining is physical refining, which removes free fatty acids from canola oil by steam distillation. This avoids the production of a soap stock and attendant disposal problems. The oil is first acid degummed with citric or phosphoric acid and then bleached to remove phosphatides and trace metals. This method reduces the phosphorus content to less than 5 ppm, as higher levels cause a darkening of the oil during steam distillation. The latter process (distillative deacidification) is carried out in a specially designed deodorizer. Although physical or steam refining is used widely in Europe, its introduction in North America is only recent.

Bleaching

Before hydrogenation or deodorization, canola oil is bleached with acid-activated bleaching clay (at 0.125–2.0%) under vacuum at 100–125°C for 15 to 30 min. This is an adsorptive process attracting polar substances to active sites of clay surface particles. These include pigments such as chlorophylls and carotenes, residual soaps, phospholipids, and trace metals, as well as primary and secondary oxidation products in the oil. Of particular importance is the presence of high levels of chlorophylls, which are extremely detrimental to oil quality and must be removed before deodorization. Unlike carotenes, which are heat bleachable in the deodorizer, chlorophylls are not and must be removed by bleaching (51). Most modern oil-processing plants use a continuous bleaching system.

Hydrogenation

Canola oil, like other edible oils, is hydrogenated to improve oxidative stability and to modify the physical properties of the fat or oil. The basic principle of this method is to add hydrogen at the double bonds of the polyunsaturated fatty acids (those containing two or more double bonds), thus eliminating these sites from reaction with oxygen. Only a small portion of these fatty acids may be affected in partially hydrogenated oil stocks, depending on the type of hydrogenation system used. This process requires a catalyst (usually nickel), which may be poisoned by the presence of low levels of sulfur compounds (3–5 ppm) originating from breakdown products of glucosinolates (52). Once hydrogenated, the oil becomes physically harder and more resistant to oxidation. In Canada, canola oil is selectively hydrogenated for the production of margarines and shortenings.

Selective hydrogenation of canola oil is carried out under conditions of high temperature, low pressure (200°C, 6 psig) and agitation and affects the more unsaturated fatty acids first. This is illustrated by the fatty acid profiles during the course of hydrogenation of soybean oil shown in Fig. 6. All the unsaturated fatty acids are hydrogenated at the same time but at different rates, as indicated by the reaction rate constants. Linolenic acid is hydrogenated 2.3 times faster than linoleic acid, whereas the latter is hydrogenated 12 times faster than oleic acid (53). This method results in the formulation of higher levels of trans fatty acids, steeper solid fat index curves, and higher melting points at lower iodine values. The formation of trans fatty acids changes the physical properties of the oil as a trans double bond is equivalent in physical properties to a saturated single bond fatty acid; eg, cis, trans-linolenic acid is equivalent in physical properties to oleic acid. Hardness of margarine oils is due to the higher lev-

Table 5. Effect of Alkali Refining on the Quality of Canola Oil[a]

Constituents	Crude Degummed	Alkali Refined
Free fatty acids, %	0.4–1.0	0.05
Phosphorus, ppm	150–250	0–5
Chlorophyll, ppm	5–25	0–25
Sulfur, ppm	3–10	2–7

[a] Ref. 50.

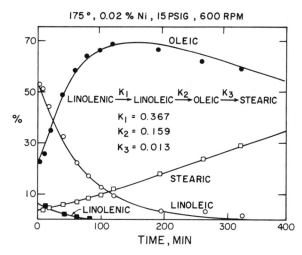

175°, 0.02 % Ni , 15 PSIG , 600 RPM

$$\text{LINOLENIC} \xrightarrow{K_1} \text{LINOLEIC} \xrightarrow{K_2} \text{OLEIC} \xrightarrow{K_3} \text{STEARIC}$$

$K_1 = 0.367$
$K_2 = 0.159$
$K_3 = 0.013$

Figure 6. Time-dependent changes in fatty acid composition during hydrogenation of soybean oil (Adapted from Ref. 53, with permission).

els of trans fatty acids present rather than to a marked decrease in unsaturation of the fatty acids (54). Because different degrees of hydrogenation are often required by the industry, hydrogenation is a batch process.

Deodorization

Any free fatty acids or odiferous or flavor degradation products remaining in the oil are removed by deodorization. This involves steam distillation carried out at very high temperatures (240–270°C) under vacuum (3–8 mmHg). Most plants use a semicontinuous or continuous deodorizing system that is comprised of a large cylindrical tank or shell through which oil is pumped in and passed through a series of trays where it is deaerated and successively deodorized with sparging steam. The oil is then cooled, pumped through a polishing filter, and sparged with nitrogen. The final product is a bland oil that is treated with citric acid (0.005–0.01%) to sequester any trace metals still remaining and other antioxidants to prevent oxidation.

EDIBLE BY-PRODUCTS

Two major edible by-products obtained by processing canola seeds are the oil and meal. Although canola oil remains the primary edible product, meal provides an important source of animal feed.

Canola Oil

The major salad oil in Canada is canola, which accounts for approximately 80% of the total market. It remains clear and liquid under refrigerated conditions and is generally hydrogenated to improve flavor stability by very light or touch hydrogenation to reduce the level of linoleic acid ($C_{18:3}$). In recent years canola oil has been recognized in the United States as a premium quality oil because of the health implications afforded by its unique fatty acid

composition (6). This is attributed to the hypocholesterolemic effect of oleic acid ($C_{18:1}$), a monounsaturated fatty acid present at high levels (>50%) in canola oil, which is recognized for its importance in lowering low-density lipoprotein cholesterol levels (55,56). Because of its low level of saturated fatty acids (<6%), canola oil has become a premium salad oil in North America. The stability of canola oil is comparable to other edible oils, although its high linolenic acid content (8–10%), like that of soybean oil, makes it susceptible to oxidative rancidity (35). Among eight different vegetable oils stored at 60°C for 8–16 days, canola oil had lower peroxide values and volatiles than corn, cottonseed, olive, peanut, safflower, soybean, and sunflower oils (57). A recent study showed a low linolenic acid canola oil product (<2.0%) to be remarkably stable to both heat accelerated and photochemical oxidation (28).

In addition to use as a salad oil, canola oil is used in the production of margarines and shortenings. In North America, margarine is a strictly defined alternative to butter with 80% fat. Although some margarines are made from 100% canola oil, it is generally blended with soybean or palm oil. This is carried out to ensure the development of small crystals (β') essential for smooth creaming properties. The unacceptable sandy mouthfeel (due to large fat crystals) for margarines originates from the $\beta' \rightarrow \beta$ polymorphic transitions; one way to alleviate this problem is by increasing the heterogeneity of acyl chain lengths of the constituent fatty acids in the oil. Canola oil is selectively hydrogenated and the fat and aqueous phases are mixed together in a crystallizer system or votator to form an emulsion. A variety of different soft and stick margarines are available on the North American market. Shortenings are made in a similar manner to margarines, although in this case, the fat is aerated with air or nitrogen (12%) to improve whiteness and opacity.

High Erucic Acid Oil

As discussed earlier, rapeseed oil has been systematically modified by plant breeders from a high erucic acid (HEAR >60%) to a low erucic acid oil (<2%). Nevertheless, HEAR is still permitted in the United States as a stabilizer and thickener component in peanut butter at a maximum level of 2% (58). Superglycerinated, 100% hydrogenated HEAR, has been allowed in cake-mix formulations since 1957 in the United States as an emulsifier in the shortening (58).

Canola Meal

After extraction of the oil and removal of hexane, the meal contains approximately 1.5% residual oil and 8–10% moisture. It may be granulated to uniform consistency and pelletized or sent directly to storage for marketing (59). Gums removed during degumming are usually added back to the meal to a level of 1.5%; some of the acidulated soap stock is added back as well. The final meal produced is used primarily for the animal feed industry. Canola meal has found much wider acceptance in North America owing to a substantial reduction in glucosinolates accomplished by plant breeding.

Canola meal has, as yet, not found wide acceptance in human nutrition despite the high quality of its protein. This is due to a combination of high levels of phytate and polyphenols as well as the presence of some glucosinolates. A dual solvent process has successfully removed all glucosinolates as well as phytate and hulls (15); the resulting meal was reported to be bland, free-flowing, free from glucosinolates, low in polyphenols, and light in color, with a protein content of 50%. Protein isolates containing 90% protein produced from this meal could have considerable potential for use in human foods.

NONEDIBLE BY-PRODUCTS

The majority of rapeseed currently grown in Canada is canola, although there still remains some acreage devoted to high erucic acid rapeseed (HEAR). This is due to the industrial applications of erucic acid and its cleavage products.

Canola Oil

Canola-oil-based printing ink has been found to be superior to petroleum-based ink by reducing the rub problem. Switching to canola-based ink also reduces environmental problems, as 75% of the ink is biodegradable (60). The superior rub properties of canola oil permits both black and color inks to be made with this oil. A number of major Canadian newspapers are now printed with canola-oil-based ink.

High Erucic Acid Oil

Erucic acid, a long chain fatty acid ($C_{22:1}$) containing a single double bond, exhibits high fire and smoke points (271°C). These properties allow erucic acid to withstand high temperatures and still remain a liquid at room temperature (61). Consequently, oils containing high levels of erucic acid are used as lubricants or in lubricant formulations (62). Because of their lubricant properties, HEAR oils have found applications as spinning lubricants in the textile, steel, and shipping industries, as drilling oils, and as marine lubricants (63). HEAR oil is also being used for the clinical treatment of adrenoleukodistrophy, a rare children's disease. Hydrogenated HEAR oils have industrial applications by producing hard, glossy waxes (64).

Erucic Acid Derivatives

Erucic acid and its hydrogenated derivative, behenic acid and erucyl and behenyl alcohols as well as other derivatives including esters, amides, amines, and their salts. These are used extensively in industry as slip, softening, antifoaming, and release agents; emulsifiers; processing aids; conditioners; antistatic agents; stabilizers; and corrosion inhibitors (62).

$$CH_3(CH_2)_7CH=CH(CH_2)_{11}COOH$$

erucic acid

$$CH_3(CH_2)_7CH_2CH_2(CH_2)_{11}COOH$$

behenic acid

Erucamide, the amide derivative of erucic acid, has been used for many years as a processing aid and antiblock agent in plastic films. It facilitates the production of plastic parts by acting as a lubricant and forming a thin layer on the surface of the plastic, thus preventing the sheets from sticking to each other.

$$CH_3(CH_2)CH=CH(CH_2)_{11}CONH_2$$

erucamide

Behenic acid and its esters are used to enhance the performance of a wide range of pharmaceutical, cosmetic, fabric softener, and hair conditioner products. As antifriction coatings they soften and improve the texture and sewability of cotton and synthetic fabrics (62). In addition to the compounds discussed, there are a large number of HEAR-oil-derived alcohols and esters that are used in the cosmetic and pharmaceutical industries.

Erucic acid cleavage products have considerable potential in the production of plastics, resins, and nylons. For example, the oxidative cleavage of erucic acid yields brassylic acid, a 13-carbon dicarboxylic acid, and pelargonic acid, a 9-carbon monocarboxylic acid. A number of long-chain nylons prepared from brassylic acid are important in automotive parts and products (61,65,66).

$$CH_3(CH_2)_6CH=CH(CH_2)_{11}COOH \rightarrow$$

oxidative cleavage

$$CH_3(CH_2)_7COOH + HOOC(CH_2)_{11}COOH$$

pelargonic acid   **brassylic acid**

Alkyl esters of brassylic acid are excellent plasticizers. The corresponding allyl and vinyl esters form polymers and copolymers that may be used in molding compounds, reinforced plastics, laminates, sealants, and coatings (67–70).

STANDARDS AND SAFETY

The standards for crude and refined, bleached, deodorized canola oil, as outlined in Table 6, were established in 1987 by the Canadian General Standard Board (CGSB). Such standards are used for purchasing, consumer protection, health and safety, international trade, and regulatory reference (71). In addition, there is also the Canada Agricultural Products Standard Act, which defines the quality of products, including canola oil (Table 4), for regulating the grading, packing, and marketing of processed canola oil in Canada. These come under the jurisdiction of Agriculture Canada (71).

Canola oil has been approved by Health and Welfare Canada under the Food and Drugs Acts and Regulations (72) as safe for human consumption. The Canadian legislation favors consumption of canola oil over that of rapeseed oil by limiting the amount of C_{22} monoenoic fatty acids such as erucic acid ($C_{22:1}$) permitted in dietary fats. The three sections of this act, outlined in Table 7, refer to this particular restriction. For example, Section B.09.022 refers to a restriction of less than 5% C_{22} monoenoic fatty

Table 6. Canadian General Standards Board Requirements for Canola Oil[a]

A. Crude Canola Oil

Characteristics	Super Degummed	Degummed
Free fatty acids (as oleic acid),		
% by mass, maximum	1.0	1.0
Moisture and impurities, combined		
% by mass, maximum	0.3	0.3
Phosphorus, ppm, maximum	50	200
Chlorophyll, ppm, maximum	30	30
Sulfur, ppm, maximum	10	10
Refined, bleached color,		
Lovibond (133.4 mm cell), maximum	1.5 red	1.5 red
Erucic acid, % by mass, maximum	2.0	2.0

B. Refined, Bleached, and Deodorized Canola Oil

Characteristics	Minimum	Maximum
Free fatty acids (as oleic acid),		
% by mass	—	0.05
Moisture and impurities, combined		
% by mass	—	0.05
Lovibond color (133.4 mm cell)		1.5 red, 15 yellow
Peroxide value, mEq/kg	—	2.0
Cold test, h	12	—
Smoke point, °C	232	—
Unsaponifiable matter, g/kg	—	15
Saponification value, milligrams		
potassium hydroxide per gram oil	182	193
Refractive index (n_D 40°C)	1.465	1.467
Iodine value (Wijs)	110	126
Crismer value, °C	67	70
Relative density (20°C/water 20°C)	0.914	0.920
Erucic acid, % by mass	—	2.0

[a] Ref. 71.

Table 7. Regulations in Canada's Food and Drugs Act Relating to Canola and Rapeseed Oils[a]

B.09.001 (S)	Vegetable fats and oils shall be fats and oils obtained entirely from the botanical source after which they are named, shall be dry and sweet in flavour and odour and, with the exception of olive oil, may contain Class IV preservatives, an antifoaming agent, and β-carotene in a quantity sufficient to replace that lost during processing, if such addition is shown on the label.
B09.022	No person shall sell cooking oil, margarine, salad oil, simulated dairy products, shortening or food that resembles margarine or shortening, if the product contains more than 5 percent C_{22} Monoenoic Fatty Acids calculated as a proportion of the total fatty acids contained in the product.
B07.043	No person shall sell a dressing that contains more than 5 percent C_{22} Monoenoic Fatty Acids calculated as a proportion of the total fatty acids contained in the dressing.
B25.054 (1)	Except as otherwise provided in this Division, no person shall sell or advertise for sale of human milk substitute unless it contains, when prepared according to directions for use, (a) per 100 available kilocalories (i) not less than 3.3 and not more than 6.0 grams of fat, (ii) not less than 500 milligrams of linoleic acid in the form of a glyceride, (iii) not more than 1 kilocalories from C_{22} Monoenoic Fatty Acids.

[a] Ref. 72.

acids in fats and oils in shortenings, salad and cooking oils, as well as table spreads. A similar restriction is cited in Section B.07.043 for salad dressings, while Section B.25.054(a)(iii) limits C_{22} monoenoic acids in infant formula to a maximum of 1% of the calories. In Canada, only 2% erucic acid is now permitted in canola oil, which is much lower than the 5% indicated in the Codex standard for low erucic acid oils by the Codex Alimentarious Commission FAO/WHO 1982 (73). Erucic acid oil with a maximum of 2% was subsequently granted GRAS (generally regarded as safe) status in the United States (1985) by the FDA (74). This oil, now identified as canola oil, has since been recognized by the American Health Foundation as a healthful product in the consumer market. A comprehensive review on the safety and health aspects of canola oil has been published (75), which concluded that, because of its fatty acid composition, canola oil can be a good substitute for saturated fat in diets intended to reduce the risk of hypercholesterolemic and coronary heart disease. Current nutritional recommendations by many health organizations suggest a reduction of the calories derived from fat in the diet to about 30% and no more than 10% of the caloric intake as saturated fat.

FUTURE OF CANOLA

The discovery of cytoplasmic male sterility (CMS) in Brassica spp. a decade ago led to considerable research in developing new hybrids. Because B. napus spp. of canola are mainly self-pollinated, the development of hybrids requires cross-pollination. The CMS system has been perfected and is now used for the production of commercial hybrids. Future research should result in improved hybrids for commercial production. Future breeding will also be focused on developing improved canola lines that mature early, are frost resistant, are resistant to pesticides, are blackleg tolerant, as well as higher yielding. B. napus varieties require long days to flower, so future research will develop lines that mature earlier without any loss in yield (76,77). New high protein lines will also be developed for B. napus varieties, whereas higher yields and oil content are the long-term goals for B. campestris breeding research programs. Biotechnology should also result in new lines with cold temperature hardiness and frost resistance that can germinate at much lower soil temperatures and permit early seeding. As breeders and genetic engineers manage to tailor rapeseed/canola, segregating the crop on the basis of specific characteristics will evolve. Processors may develop contracts with farmers to plant varieties that provide products with the desired end-use properties. In addition to the breeding programs, modifications in canola processing technologies will be sought to improve the finished products while being environmentally safe.

BIBLIOGRAPHY

1. "Fats and Oils Industry Changes," *Journal of the American Oil Chemists Society* **65**, 702–713 (1988).

2. J. K. Daun, "The Introduction of Low Erucic Acid Rapeseed Varieties into Canadian Production," in J. K. G. Kramer, F. D. Sauer, and W. J. Pigden, eds., *High and Low Erucic Acid Rapeseed Oils—Production, Usage, Chemistry and Toxicological Evaluation*, Academic Press, New York, 1983, pp. 161–180.

3. M. Vaisey-Genser and D. F. G. Harris, "Current Markets for Canola Oil", in M. Vaisey-Genser and N. A. M. Eskin, eds., *Canola Oil: Properties and Performance*, Canola Council of Canada, Winnipeg, Man., 1987, pp. 4–9.

4. "Estimate of Production Revised," *Canola Digest* **24** (2), 3 (1990).

5. "Policy Will Affect Canola's Success," *Journal of the American Oil Chemists Society* **66**, 624–625 (1989).

6. B. E. McDonald, "Nutritional Properties," in Ref. 3, pp. 40–45.

7. L. A. Appelqvist, "Historical Background," in L. A. Appelqvist and R. Ohlson, eds., *Rapeseed: Cultivation, Composition, Processing and Utilization*, Elsevier, Amsterdam, 1972, pp. 1–8.

8. R. K. Downey, "The Origin and Description of the *Brassica* Oilseed Crops," in Ref. 2, pp. 1–20.

9. E. S. Bunting, "Oilseed Rape in Perspective," in D. H. Scarisbrick and R. W. Daniels, eds., *Oilseed Rape*, Collins, London, 1986, pp. 1–31.

10. R. K. Downey and G. F. W. Rakow, "Rapeseed and Mustard," in W. R. Fehr, ed., *Principles of Cultivar Development*, Vol. 2, Macmillan, London, 1987, pp. 437–486.

11. R. K. Downey and G. Röbbelen, "*Brassica* Species," in G. Röbbelen, R. K. Downey, and A. Ashri, eds., *Oil Crops of the World: Their Breeding and Utilization*, McGraw-Hill, New York, 1989, pp. 339–362.

12. *Rapeseed-Canola*, Canola Council of Canada, Winnipeg, Man. 1981, pp. 12–14.

13. G. R. Fenwick, R. K. Heaney, and W. J. Mullin, "Glucosinolates and Their Breakdown Products in Food and Food Plants," *CRC Critical Reviews in Food Science and Nutrition* **18**, 123–201 (1983).

14. D. S. Hutcheson, "Development of Improved *B. campestris* Cultivars Essentially Free from Both Alkenyl and Indolyl Glucosinolates," *Canola Research Summary 1985–89*, Canola Council of Canada, Winnipeg, Man., 1989, p. 54.

15. L. L. Diosady, L. J. Rubin, C. R. Phillips, and M. Naczk, "Effect of Alkanol–Ammonia–Water Treatment on the Glucosinolate Content of Rapeseed Meal," *Canadian Institute of Food Science and Technology Journal* **18**, 311–315 (1985).

16. B. Lööf, "Cultivation of Rapeseed," in Ref. 7, pp. 49–59.

17. L. A. Appelqvist and B. Lööf, "Postharvest Handling and Storage of Rapeseed," in Ref. 7, pp. 60–100.

18. *Guide to Farm Practice in Saskatchewan*, University of Saskatchewan, Saskatoon, 1987, pp. 60–61.

19. H. B. W. Patterson, *Handling and Storage of Oilseed, Oils, Fats and Meal*, Elsevier Applied Science, London, 1989, pp. 133–140.

20. K. M. Clear and J. K. Daun, "Chlorophyll in Canola Seed and Oil," in *8th Progress Report on Canola Seed, Oil and Meal Fractions*, Canola Council of Canada, Winnipeg, Man., 1987, pp. 277–279.

21. S. Cenkowski, S. Sokhansanj, and F. W. Sosulski, "The Effect of Drying Temperature on Green Color and Chlorophyll Content of Canola Seeds," *Canadian Institute of Food Science Technology Journal* **22**, 383–386 (1989).

22. B. R. Stefansson, "The Development of Improved Rapeseed Cultivars," in Ref. 2, pp. 143–159.

23. L. Bengtsson, A. Von Hofsten, and B. Lööf, "Botany of Rapeseed," in Ref. 7, pp. 36–48.

24. Canadian International Grains Institute, *Grains and Oilseeds,* 3rd ed., Winnipeg, Man., 1982, pp. 767–769.

25. R. K. Downey, "It's All in the Breeding," *Proceedings of the 19th Annual Convention of the Canola Council of Canada Meeting,* San Francisco, Calif., March 24–26, 1986, pp. 25–36.

26. P. K. Stumpf, "Biosynthesis of Fatty Acids in Higher Plants," in Ref. 11, pp. 38–62.

27. R. Scarth, P. B. E. McVetty, S. R. Rimmer, and B. R. Stefansson, "Stellar: Low Linolenic—High Linoleic Acid Summer Rape," *Canadian Journal of Plant Science* 68, 509–511 (1988).

28. N. A. M. Eskin, M. Vaisey-Genser, S. Durance-Todd, and R. Przybylski, "Stability of Low Linoleic Acid Canola Oil to Frying Temperature," *Journal of the American Oil Chemists Society* 66, 1081–1084 (1989).

29. A. Prevot, J. L. Perrin, G. Laclaverie, Ph. Auge, and J. L. Coustille, "A New Variety of Low-Linolenic Rapeseed Oil; Characteristics and Room-Odor Tests," *Journal of the American Oil Chemists Society* 67, 161–164 (1990).

30. K. F. Thompson and W. G. Hughes, "Breeding and Varieties," in Ref. 9, pp. 32–82.

31. W. A. Keller, "The Application of Biotechnology to Canola Improvement," in Ref. 14, p. 56.

32. W. Keller, "What Biotechnology Offers Canadian Canola Research," in *Proceedings of 18th Annual Convention of the Canola Council of Canada Meeting,* Vancouver, B.C. 1985, pp. 37–44 (1985).

33. "Breeding Canola for Success," *Journal of the American Oil Chemists Society* 65, 1567–1568 (1988).

34. L. Kunimoto, "Biotechnology for the Field and Shelf," in Ref. 32, pp. 45–47.

35. N. A. M. Eskin, "Chemical and Physical Properties of Canola Oil Products," in Ref. 3, pp. 16–24.

36. R. L. Ory and A. J. St. Angelo, "Lipoxygenase Activity in Soybean, Peanut and Rapeseed: Inhibition by Erucic Acid," *Journal of the American Oil Chemists Society* 52, 130A, (1975).

37. A. Khalyfa, S. Kermasha, and I. Alli, "Partial Purification and Characterization of Lipoxygenase of Canola Seed (*Brassica napus* var. Westar)," *Journal of Agricultural and Food Chemistry* 38, 2003–2008 (1990).

38. R. G. Ackman, "Chemical Composition of Rapeseed Oil," in Ref. 2, pp. 85–129.

39. D. R. Clandinin, A. R. Robblee, J. M. Bell, and S. J. Slinger, "Composition of Canola Meal," in R. Salmon and D. R. Clandinin, eds., *Canola Meal for Livestock and Poultry,* Canola Council of Canada, Winnipeg, Man., 1989, pp. 5–7.

40. N. A. M. Eskin and R. Bacchus, "Processing of Canola Oil," in Ref. 3, pp. 25–32.

41. F. W. Hougen, J. K. Daun, and D. C. Durnin, "The Composition and Quality of Canola Dockage," in Ref. 20, pp. 287–297.

42. F. Ismail, N. A. M. Eskin, and M. Vaisey-Genser, "The Effect of Dockage Oil on the Stability of Canola Oil," in *6th Progress Report on Canola Seed, Oil, Meal and Meal Fractions,* Canola Council of Canada, Winnipeg, Man., 1980, pp. 234–239.

43. R. Simmons, "The Effect of Rapeseed Quality on Processing Procedures," in *Proceedings of the International Conference on Science, Technology and Marketing of Rapeseed and Rapeseed Products,* Rapeseed Association of Canada, St. Adele, Quebec, 1970, p. 121.

44. D. L. Beach, "Primary Processing of Vegetable Oils," in J. T. Harapiak, ed., *Oilseed and Pulse Crops in Western Canada,* Western Cooperative Fertilizers Ltd., Calgary, Alta., 1975, pp. 541–550.

45. D. L. Beach, "Rapeseed Crushing and Extraction," in Ref. 2, pp. 181–195.

46. B. F. Teasedale, "Processing of Vegetable Oils," in Ref. 44, pp. 551–585.

47. T. K. Mag, "Canola Processing in Canada," *Journal of the American Oil Chemists Society* 60, 332A–336A (1983).

48. R. Ohlson and C. Svensson, "Comparison of Oxalic Acid and Phosphoric Acid as Degumming Agents for Vegetable Oils," *Journal of the American Oil Chemists Society,* 53, 8–11 (1976).

49. G. R. List, T. L. Mount, K. Warner, and A. J. Heakin, "Steam-refined Soybean Oil. I. Effect on Refining and Degumming Methods on Oil Quality," *Journal of the American Oil Chemists Society* 55, 277–279 (1978).

50. B. F. Teasdale and T. K. Mag, "The Commercial Processing of Low and High Erucic Acid Rapeseed Oils," in Ref. 2, pp. 199–228.

51. H. Niewiadomski, "Progress in the Technology of Rapeseed Oil for Edible Purposes," *Chemistry and Industry* 833–888 (1970).

52. J. K. Daun and F. W. Hougen, "Sulfur Content of Rapeseed Oils," *Journal of the American Oil Chemists Society* 53, 169–171 (1976).

53. R. R. Allen, "Hydrogenation," *Journal of the American Oil Chemists Society* 58, 166–169 (1981).

54. T. J. Weiss, *Food Oils and Their Uses,* AVI Publishing Co., Westport, Conn., 1970.

55. R. S. Mattson and S. M. Grundy, "Comparison of Effects of Dietary Saturated, Monounsaturated and Polyunsaturated Fatty Acids on Plasma Lipids and Lipoproteins in Man," *Journal of Lipid Research* 26, 194–202 (1985).

56. S. M. Grundy, "Comparison of Monounsaturated Fatty Acids and Carbohydrates for Lowering Plasma Cholesterol," *New England Journal of Medicine* 314, 745–748 (1986).

57. J. M. Snyder, E. N. Frankel, and E. Selke, "Capillary Gas Chromatographic Analyses of Headspace Volatiles from Vegetable Oils," *Journal of the American Oil Chemists Society* 62, 1675–1679 (1985).

58. *Federal Register,* 42, 44335–48336 (1977).

59. M. D. Pickard, C. G. Youngs, L. R. Wetter, and G. S. Boulter, "Processing of Canola Seed for Quality Meal," in Ref. 39, pp. 3–4.

60. "What's Black and White and Read All Over?" *Canola Digest* 22(7), 1 (1988).

61. H. J. Nieschlag, G. F. Spencer, R. V. Madrigal, and J. A. Rothfus, "Synthetic Wax Esters and Diesters from Crambe and Limnanthes Seed Oils," *Industrial and Engineering Chemistry Product Research and Development* 16, 202–207 (1977).

62. D. L. van Dyne, M. G. Blase, and K. D. Carlson, *Industrial Feedstocks and Products from High Erucic Acid Oil: Crambe and Industrial Rapeseed,* University of Missouri, Columbia, 1990.

63. H. J. Nieschlag, and I. A. Wolff, "Industrial Uses of High Erucic Oils," *Journal of the American Oil Chemists Society* 41, 1–5 (1971).

64. T. K. Miwa and I. A. Wolff, "Fatty Acids, Fatty Alcohols, Wax Esters and Methyl Esters from *Crambe abyssinica* and *Luna-*

ria annua Seed Oils," *Journal of the American Oil Chemists Society* **40**, 742–744 (1963).

65. J. L. Greene, Jr., R. Perkins, Jr., and I. A. Wolff, "Aminotri-decanoic Acid from Erucic Acid," *Industrial and Engineering Chemistry, Product Research and Development* **8**, 171–176 (1969).

66. R. B. Perkins, Jr., J. J. Roden, A. C. Tanquary, and I. A. Wolff, "Nylons from Vegetable Oils: −13, −13/13 and −6/13," *Modern Plastics* **46**, 136–139 (1969).

67. S. P. Chang, T. K. Miwa, and I. A. Wolff, "Alkyl Vinyl Esters of Brassylic Acid," *Journal of Polymer Science* Part A-1, **5**, 2547–2556 (1967).

68. S. P. Chang, T. K. Miwa, and W. H. Tallent, "Allyic Prepoly-mers from Brassylic and Azelaic Acids," *Journal of Applied Polymer Science* **18**, 319–334 (1974).

69. H. J. Nieschlag, J. W. Hagemann, I. A. Wolff, W. E. Palm, and L. P. Witnauer, "Brassylic Acid Esters as Plasticizers for Poly(vinylchloride)", *Industrial and Engineering Chemistry, Product Research and Development* **3**, 146–149 (1964).

70. H. J. Nieschlag, W. H. Tallent, I. A. Wolff, W. E. Palm, and L. P. Witnauer, "Diester Plasticizers from Mixed Crambe Di-basic Acids," *Industrial and Engineering Chemistry, Product Research and Development* **6**, 201–204 (1967).

71. M. Vaisey-Genser, "Regulations Related to Canola Oil," in Ref. 3, pp. 10–15.

72. Health and Welfare Canada, *The Food and Drugs Act and Regulations,* Supply and Services Canada, Ottawa, 1953, amended to September 1986.

73. FAO/WHO, *Codex Standards for Edible Low Erucic Acid Rapeseed Oil.* Codex Standard 123-1981. Codex Alimentari-ous Commission, Joint Food and Agriculture Organization/World Health Organization Food Standards Programme, April 1982.

74. Food and Drug Administration *Direct Food Substances Af-firmed as Generally Recognized as Safe; Low Erucic Acid Rapeseed Oil.* Federal Food, Drug and Cosmetic Act 21 CRF Part 184, 1555(c), amended 9 Jan. Department of Health and Human Services, Food and Drug Administration, Washing-ton, D.C., 1985.

75. J. Dupont, P. J. White, K. M. Johnston, H. A. Heggtveit, B. E. McDonald, S. M. Grundy, and A. Bonanome, "Food Safety and Health Effects of Canola Oil," *Journal of the American College of Nutrition* **8**, 360–375 (1989).

76. "Contiseed Banks on Hybrids," *Canola Digest* **22**(3), 8 (1988).

77. "Improving Oil/Protein Important at U. of A.," *Canola Digest* **22** (2), 4 (1988).

General References

References 19, 62, 75 are also good general references.

Appelqvist, L. A., and R. Ohlson, *Rapeseed: Cultivation, Composi-tion, Processing and Utilization,* Elsevier Publ. Co., Amster-dam, 1972.

Kramer, J. K. G., Sauer, F. D., and Pigden, W. J. eds., *High and Low Erucic Acid Rapeseed Oils—Production, Usage, Chemis-try and Toxicological Evaluation,* Academic Press, New York, 1983.

Röbbelen, G., R. K. Downey, and A. Ashri, *Oil Crops of the World,* McGraw-Hill, New York, 1989.

Salmon, R. E., and Clandinin, D. R., *Canola Meal for Livestock and Poultry,* Canola Council of Canada, Winnipeg, Man., 1987.

Scarisbrick, D. H., and Daniels, R. W., *Oilseed Rape,* Collins, London, 1986.

Swern, D., *Bailey's Industrial Oil and Fat Products,* 4th ed., John Wiley & Sons, New York, 1985.

Vaisey-Genser, M., and Eskin, N. A. M., *Canola Oil: Properties and Performance,* Canola Council of Canada, Winnipeg, Man., 1987.

C. G. BILIADERIS
N. A. M. ESKIN
University of Manitoba
Winnipeg, Manitoba, Canada

CARBOHYDRATES

Carbohydrates are important components of both natural and processed foods (1–7). Digestible carbohydrates are the principal source of calories for most of the world's population, with starches providing at least 75% of the total caloric intake on a worldwide basis. Nondigestible carbohydrates are the principal component of dietary fi-ber. Carbohydrates are used in processed food formula-tions to improve the body, texture, flavor, appearance, convenience, or stability of the product. Not only do carbo-hydrates have and impart special properties, they are abundant, widely available, inexpensive (low cost per functionality), and safe (nontoxic) (Table 1).

Carbohydrates are almost ubiquitous in food products because they are found in all living cells—those of plants, animals, and microorganisms. Plant material contains more carbohydrate than does animal material. Carbohy-drate molecules make up about three-fourths of the dry weight of plants. They are present in all plant parts eaten by humans (grains, fruits, leaves, stalks, roots, tubers), often as the principal energy storage material, eg, starch and sugar.

Carbohydrates are added in the preparation of many processed food products to improve their sensory attrib-utes. Some are used to provide sweetness. Some influence color and flavor through chemical reactions. Some are used to provide bulk. Others are used to provide texture. In some food products, carbohydrates are added to in-crease viscosity, improve rheological properties, control or hold water, prevent crystallization of water or sugar, sta-bilize emulsions and/or suspensions, or form gels. In others, they provide or improve yet other physical and/or organoleptic properties (Tables 2–4).

FOOD SACCHARIDES

Polysaccharides

About 90% of all the carbohydrate in nature is in the form of polysaccharides, naturally occurring carbohydrate polymers. Polysaccharides are polymer chains of many monosaccharide (sugar) units (6–8). In precise chemical nomenclature, polysaccharides are glycans, polymers composed of glycosyl (monosaccharide) units. Polysac-charides can be linear or branched in any of several ways and composed of a single type of glycosyl unit (a homogly-

Table 1. Principal Food Carbohydrates

Type of Carbohydrate	Source
Monosaccharides	
D-Glucose	Added corn syrups, including high-fructose corn syrups; honey
D-Fructose	Added high-fructose corn syrups, honey
Disaccharides	
Sucrose	Prepared food and beverages, fruits, honey
Lactose	Milk
Maltose	Added corn syrups
Starch Oligo- and Megalosaccharides	
Corn syrup components	Added corn syrups
Maltodextrins	Added maltodextrins
Corn syrup solids	Added corn syrup solids
Soluble Polysaccharides	
Agar	Added gum
Algins	Added gum
Carboxymethylcellulose	Added gum
Carrageenans	Added gum
Beta-glucans	Whole cereal grains
Galactomannans	Legume seeds
Guar gum	Added gum
Gum arabic	Added gum
Hemicelluloses	Vegetables, whole grains, added fiber
Hydroxymethylmethylcelluloses	Added gum
Locust bean gum	Added gum
Methylcelluloses	Added gum
Pectins	Fruits, vegetables, added gum
Starches	Cereal grains, potatoes, vegetables, fruits
Modified food starches	Added starch
Xanthan	Added gum
Insoluble Polysaccharides	
Cellulose	Vegetables, fruits, whole grains, added fiber

Table 2. Selected Food Applications and Functions of Carbohydrates: Low-molecular-weight Carbohydrates

A. Examples:

Sucrose, corn syrups, dextrose, lactose, sorbitol

B. Major Functions:

Sweeten

Control humectancy/water activity

Add body or bulk

Give surface glaze or frost

Glass formation

Provide crispness

Provide viscosity

Impart desirable texture

C. Typical Uses:

Bakery products (biscuits, bread, cookies, crackers, doughnuts, fillings, icings, pretzels, rolls, wafers)

Catsup, chili sauce, tomato sauce

Condensed, sweetened milk

Canned, frozen, and candied fruits

Desserts

Ice creams, water ices, sherbets

Infant foods

Jams, jellies, marmalades, preserves

Juices

Marshmallows and related products

Meat products

Mixes (fillings, frostings, icings, toppings)

Peanut butter

Pickles

Pork and beans

Salad dressings

Soft drinks

Syrups (cordial, fountain, fruit, liqueur, pancake, medicinal)

Table 3. Selected Food Applications and Functions of Carbohydrates: Medium-molecular-weight Carbohydrates

A. Examples:

Maltodextrins, corn syrup solids, low-viscosity gums

B. Major Functions:

Prevent caking
Enhance solubility or dispersibility
Provide body or bulk
Control humectancy/water activity
Impart desirable texture
Bind or strengthen
Carry flavors
Control extrusion expansion
Provide viscosity
Form films and coatings

Inhibit crystallization (ice and sugar)
Control sweetness
Provide easy digestibility
Improve sheen
Disperse fats
Improve mouthfeel
Slow meltdown
Improve freeze–thaw stability
Provide oxygen barrier

C. Typical Uses:

Bakery products
Coffee whiteners
Confectionery products (marshmallows, nougats, hard candies,
    pan coatings)
Cream-type fillings
Dry mixes (beverages, cocktail, cake, doughnut, icing, dressing,
    sauce, soup, gravy)
Frostings
Frozen foods and novelties
Fruit leather

Glazes
Granola bars
Imitation cheeses and sour cream
Infant formulas
Meat spreads
Nut and snack coatings
Pizza, cheese, and other sauces
Sherbets
Soft-serve ice creams
Spice blends
Whipped toppings

Table 4. Selected Food Applications and Functions of Carbohydrates: High-molecular-weight Carbohydrates

A. Examples:

Starches, modified and/or derivatized starches, food gums, cellulose

B. Major Functions:

Adhesion
Anticaking flow aid
Antistaling agent
Add body/bulk
Binding
Clouding agent
Crystallization inhibitor (ice and sugar)
Provide dietary fiber
Dusting
Emulsifying agent
Emulsion stabilizing
Encapsulation
Film forming

Foam forming and strengthening
Gelling
Glazing
Moisture retention
Molding
Processing aid
Protective colloid
Shaping
Suspension stabilizing
Impart texture
Thickening, provide viscosity
Water control
Whipping aid

C. Typical Uses:

Bakery products
Beverages
Breaded products
Chocolate and other syrups
Confectioneries
Dietetic foods
Drink mixes
Extruded foods
Gravies, dry mixes

Ice creams, sherbets, frozen desserts
Jams, jellies, marmalades, preserves
Meat products
Pie fillings
Pet foods
Salad dressings
Soups, soup mixes
Structured foods
Whipped toppings

Table 5. Classification of Some Major Food Polysaccharides by Structure

	Examples
Classification by Shape	
Linear, unbranched	Alginates, starch amylose and its derivatives, carrageenans, pectins, cellulose and cellulose derivatives, gellan gum, curdlan
Branched	
Linear, regular short branches	Guar gum, xanthan, locust bean (carob) gum, scleroglucan
Branch-on-branch structures	Starch amylopectin and its derivates, gum arabic
Classification by Monomeric Units	
Homoglycan (single monosaccharide)	Starch amylose and amylopectin, cellulose, pectins, pullulan, curdlan, scleroglucan
Diheteroglycan (2 monosaccharides)	Guar gum, locust bean (carob) gum, alginates, konjac mannan
Triheteroglycan (3 monosaccharides)	Xanthan, gellan
Tetraheteroglycan (4 monosaccharides)	Gum arabic
Classification by Charge	
Neutral	Guar gum, methylcelluloses, hydroxypropylmethylcelluloses, locust bean (carob) gum, most modified starches, konjac mannan, pullulan, curdlan, scleroglucan
Anionic	Algins, pectins, xanthan,[a] carboxymethylcellulose, gum arabic, carrageenans, agar, gellan gum,[a] curdlan[a]

[a] Due to noncarbohydrate constituents.

can) or from two to six different glycosyl units (a heteroglycan) (Table 5). They generally contain from hundreds to tens of thousands of glycosyl units; some may be even larger.

Amylose, one of the two polysaccharides found in most starches, is an example of a linear, homoglycan (Fig. 1). Its monosaccharide (glycosyl) building block is D-glucose (in the pyranose ring form, α-D-configuration). Therefore, amylose is a glucan.

Monosaccharide Units

All glycosyl units of polysaccharides are cyclic structures (1,4). By far, the most common ring form is the pyranose ring, a saturated six-membered ring composed of five carbon atoms and one oxygen atom. In a few instances (primarily with L-arabinose and D-fructose), glycosyl units occur in a furanose ring, a saturated five-membered ring composed of four carbon atoms and one oxygen atom.

Individual glycosyl rings are joined by acetal bonds given the specific name glycosidic linkages. Like other

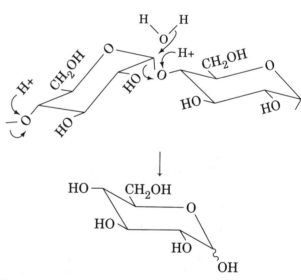

Figure 2. Acid-catalyzed hydrolysis of the glycosidic bonds on both sides of an α-D-glucopyranosyl building-block unit of amylose to produce D-glucopyranose. Ring carbon and hydrogen atoms have been removed for clarity. The disaccharide unit of amylose is a maltosyl unit.

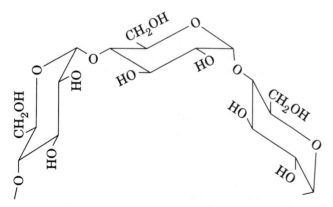

Figure 1. Segment of a molecule of amylose, the linear polysaccharide of starch. Ring carbon and hydrogen atoms have been omitted for clarity.

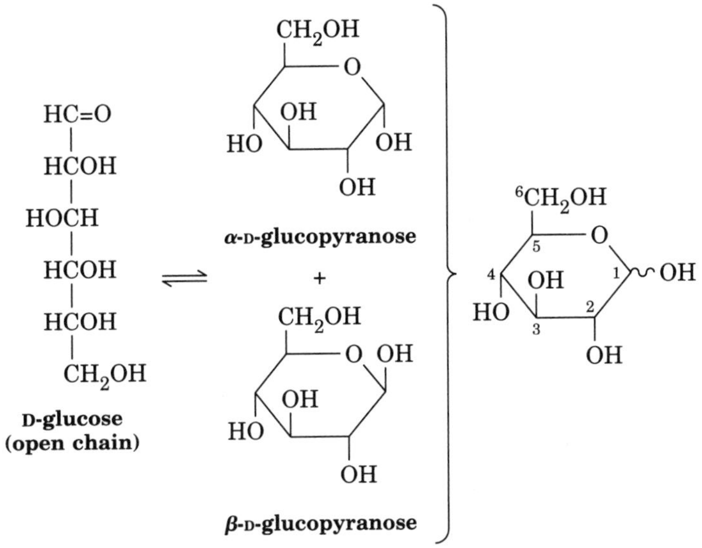

**D-glucose
(open chain)**

α-D-**glucopyranose**

+

β-D-**glucopyranose**

Figure 3. Open-chain (acyclic) and pyranose ring forms of D-glucose. Here, the pyranose ring is presented in the Haworth projection rather than the actual conformational representation depicted in Figures 1 and 2. The structure on the far right represents a mixture of the alpha and beta forms, called anomers. The numbers refer to the carbon atoms in the open-chain form counting from the top (aldehyde group) down.

acetals, glycosidic linkages are sensitive to acids, ie, they undergo acid-catalyzed hydrolysis, often rather easily. The overall process of hydrolytic cleavage of an amylose glycosidic bond is depicted in Figure 2. The product (D-glucose, lower structure) is shown in its most prevalent form, a pyranose ring in the hemiacetal form.

Hemiacetals cleave spontaneously and reversibly in water to an aldehyde group and a hydroxyl group. The hemiacetal form of the pyranose ring is, therefore, the result of an intramolecular reaction.

The structure of D-glucose is often written in the open-chain form, although only very minor amounts (<0.01%) of it ever occur in that form (Fig. 3). The ring forms contain a chiral carbon atom (C-1 in the case of all aldehyde sugars, termed aldoses), termed the anomeric carbon atom, not present in the open-chain form. Thus, there can be two forms of the ring. D Aldoses with the hydroxyl group of C-1 up (in either projection) are in the beta (β) configuration; D aldoses with the hydroxyl group on the anomeric carbon atom down are in the alpha (α) configuration. (For L sugars, the reverse is true.) Hence, amylose is composed of α-D-glucopyranosyl units, ie, they are derived from D-glucose, they are in the pyranose ring form, and they have the alpha configuration. The sugars, other than D-glucose, most abundant in nature and the ring forms in which they are found are given below. The lower two are aldopentoses. L-Rhamnose is a 6-deoxyaldohexose. All others are either aldohexoses or aldohexuronic acids. (See chemistry on page 282)

A linear polysaccharide, such as amylose, has additional hydroxyl groups that can participate in glycosidic linkages. Therefore, polysaccharides can be branched. Representative types of structures are given in Figure 4.

Polysaccharides may contain noncarbohydrate substituent groups attached to the monosaccharide units via ether (eg, methyl), ester (eg, acetate, hydroxybutyrate), or acetal bonds (eg, pyruvic acid).

Depolymerization

Many food products are at least slightly acidic. The combination of acid and heat used during processing and/or

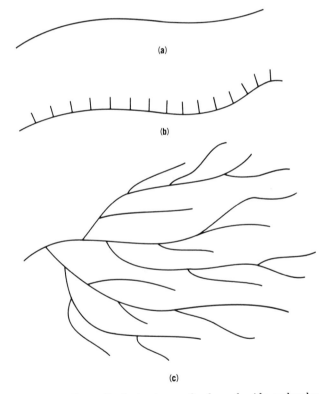

(a)

(b)

(c)

Figure 4. Generalized structures of polysaccharide molecules, which may be (**a**) strictly linear; (**b**) basically linear with short side chains; or (**c**) of a branch-on-branch type. In **b**-type molecules, the side chains may contain from one to four monosaccharide units. These molecules may have regular repeating units so that the branches are regularly spaced along the main chain or they may be irregularly spaced in a random fashion or some other arrangement such as in clusters with gaps between clusters. In **c**-type molecules, the branch points may be clustered or more evenly spaced. Some branches may be long, others short; there may be a spectrum of chain lengths, or there may be medium and short chain length branches. **A** types may be homoglycans or heteroglycans. Linear homoglycans may contain a single type of linkage or different linkages. **B** types are, with few exceptions, heteroglycans. **C** types may be homoglycans or heteroglycans. Both **b** and **c** types must contain at least two different linkages.

α- and β-D-Galactopyranosyl
(α,β-D-Gal$_p$)

3,6-Anhydro-α-D-galactopyranosyl
(3,6-An-α-D-Gal$_p$)

α-D-Galactopyranosyluronic acid
(α-D-Gal$_p$A)

β-D-Glucopyranosyluronic acid
(β-D-Glc$_p$A)

β-D-Mannopyranosyl
(β-D-Man$_p$)

β-D-Mannopyranosyluronic acid
(β-D-Man$_p$A)

α-L-Gulopyranosyluronic acid
(α-L-Gul$_p$A)

L-Rhamnopyranosyl
(L-Rhm$_p$)

β-D-Xylopyranosyl
(β-D-Xyl$_p$)

L-Arabinofuranosyl
(α-L-Ara$_f$)

preparation often effects glycosidic bond cleavage. Sometimes this is desirable; sometimes it leads to undesirable changes. In either case, it is important to control hydrolysis as much as possible. One result of glycosidic bond cleavage is the formation of hemiacetal groups (termed reducing ends) (Fig. 2), that produce aldehydo (carbonyl) groups upon ring opening (Fig. 3) that in turn can react with amino acids and/or proteins to produce color and flavor, an important reaction in the browning of bread crust.

Hydrolysis of polysaccharides is sometimes done deliberately. An example is the controlled breakdown of starch using enzymes or acid (9). Acid-modified starches (thin-

boiling starches, dextrins) are made by a mild treatment of a starch with an acid (10). This is done to reduce paste (molecular dispersion) viscosity, increase gel strength, and reduce the energy required to cook.

Further hydrolysis by acids or enzymes yields two classes of saccharides called maltodextrins and corn syrup solids (from cornstarch). These products are classified by their dextrose equivalent (DE), which is an indication of total reducing sugars calculated as D-glucose, with the DE value of anhydrous D-glucose defined as 100. The DE value is, therefore, inversely related to molecular size (degree of polymerization [DP]). Maltodextrins have a DE of <20, are rather soluble, have a bland taste, and are widely used (Table 3). Dry products with a DE of >20 are classified as corn syrup solids, which are comparatively sweeter.

Specially prepared low-DE starch breakdown products, especially from tapioca and potato starch, produce a fatty mouthfeel and are used as fat replacers or fat-sparing materials in products such as margarine, cheese-type spreads, and spoonable dressings.

More extensive breakdown yields syrups, which are purified, concentrated, aqueous solutions of saccharides with an average DE of 20 or more (11). Corn syrups (those produced from cornstarch) are generally grouped into subclasses. Some syrups will contain as little as 35% maltooligosaccharides. (Oligosaccharides are those saccharides generally containing from 2 to 10 glycosyl units. They are subcategorized according to the number of monosaccharide units present as di-, tri-, tetra-, penta-, etc saccharides. The prefix *malto* indicates a relationship with starch.) These maltooligosaccharides are both linear and branched. The branches are found in the other polysaccharide component of starch, amylopectin, and are formed by the attachment of amylose-like chains to other amylose-like chains at the C-6 hydroxyl group of some α-D-glucopyranosyl units creating a highly branched, branch-on-branch structure. Products with higher degrees of conversion contain progressively higher concentrations of lower molecular weight products and are progressively sweeter.

Enzymes are most often used to make these products of hydrolysis, although combinations of acid-catalyzed and enzyme-catalyzed hydrolysis and complete acid conversion may be used. By using proper conditions, syrups with specific defined compositions can be prepared. An example is high-maltose syrups (Fig. 5).

Because syrups are a class of products with a wide range of compositions, a range of properties can be expected. In general, they are sweet, with the relative sweetness depending on the concentration of low-molecular-weight saccharides, primarily D-glucose and maltose (a disaccharide). They can be concentrated to high dry-substance concentrations without crystallization to give viscous, hygroscopic materials. The hygroscopic character of corn syrups is a useful property in many food products in which they are used.

Complete hydrolysis yields D-glucose. Crystalline glucose (α-D-glucopyranose) is generally sold as dextrose. A solution of D-glucose can also be treated with an enzyme that isomerizes a portion of it into D-fructose. By modern conversion and processing conditions, high-fructose corn syrups (HFCS) that contain a minimum of 42% D-fructose are made. Because the sweetness of D-fructose is greater than that of sucrose, they are used as sweeteners. The ones containing higher concentrations of D-fructose are used to make reduced-calorie products because less is needed, as compared to sucrose, to give equivalent sweetness. Crystalline D-fructose is also obtained from HFCS.

All products of the hydrolysis of starch, like starch itself, are digestible and have caloric/nutritive value. However, reducing sugars can be polymerized in the presence of acid under conditions that remove water, ie, dehydrating conditions. When this reaction is done with D-glucose, D-glucitol (sorbitol), and citric acid, a synthetic, relatively low-molecular-weight, highly soluble, highly branched polymer known as polydextrose results. This material is only partially utilized by humans and is used as a reduced-calorie bulking agent.

Treatment of starch with a specific enzyme produces cyclodextrins, a family of cyclic oligosaccharides, most commonly containing seven α-D-glucopyranosyl units (β-cyclodextrin, cycloheptaamylose). In cyclodextrins, the monosaccharide units are joined by (1→4) linkages to form a ring, the cavity of which is especially useful for the formation of inclusion complexes with certain guest molecules. These stable complexes are potentially useful in a

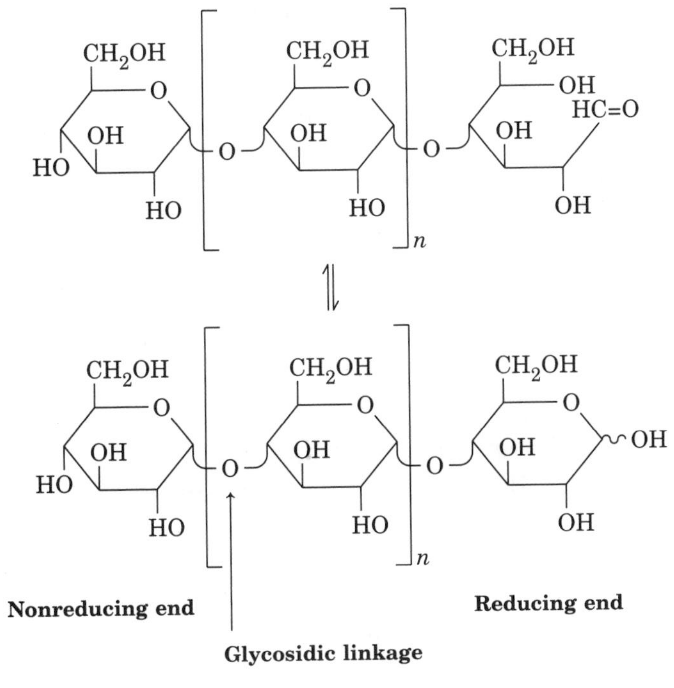

Nonreducing end   **Reducing end**

Glycosidic linkage

Figure 5. The Haworth projection of linear maltooligosaccharides. For maltose, $n = 0$; for maltotriose, $n = 1$; for maltotetraose, $n = 2$; etc.

number of food applications where increased chemical or physical stability, solubility control, or controlled release is desired (12).

Other Mono- and Oligosaccharides (4,6)

Free monosaccharides are rarely found in nature. Exceptions are small amounts of D-glucose and D-fructose in fruits and somewhat larger amounts in honey.

The most abundant disaccharide in nature and in the human diet is sucrose. Sucrose is unique in that the two sugar units, α-D-glucopyranose and β-D-fructofuranose, are joined in a head-to-head manner rather than in a head-to-tail manner as are the constituent glycosyl units of maltose, amylose, and other oligo- and polysaccharides. Because the reducing end-groups of both sugar units are bound up in the glycosidic linkage, sucrose does not contain a hemiacetal group (reducing end) and, hence, is termed a nonreducing sugar. Because the D-fructosyl unit of sucrose is in the furanose ring form and glycosidic bonds involving furanosyl units are much more acid labile than are those involving pyranosyl units, sucrose is very acid sensitive and will undergo hydrolysis easily.

The only other disaccharide to be found abundantly in nature is lactose (milk sugar). Lactose is composed of a β-D-galactopyranosyl unit joined to O-4 of a D-glucosyl unit.

Some tri- and tetrasaccharides other than those made by partial hydrolysis of starch also find their way into the human diet, primarily via ingestion of legume seeds. These oligosaccharides are sucrose molecules modified by the attachment of one or more α-D-galactopyranosyl units. α-D-Galactosidic bonds cannot be split by human digestive enzymes, so these oligosaccharides are utilized only by colonic bacteria that use them for anaerobic fermentation, which produces the flatulence associated with eating beans.

CARBOHYDRATES DERIVED FROM REDUCING SUGARS

Reduction of the carbonyl (aldehyde or ketone) group of sugars produces polyhydroxy alcohols called alditols (also sugar alcohols, polyols, glycitols) (2,4,13). The most common examples of this class of compounds are D-glucitol (sorbitol) made by reduction of D-glucose and xylitol made from D-xylose. Reduction of D-fructose produces a mixture of D-glucitol and D-mannitol.

D-Glucitol (Sorbitol)   **Xylitol**   **D-Mannitol**

Sorbitol (glucitol) and mannitol are generally recognized as safe (GRAS). An important use of sorbitol is as a

humectant. It can extend shelf life in confections and bakery products. Like other alditols, and unlike reducing sugars, it will not undergo Maillard browning and caramelization (see following section). Mannitol can be used as a dusting agent because of its low hygroscopicity. However, most food applications of alditols are in dietetic products.

Alditols are sweet. Xylitol has essentially the same sweetness as sucrose; sorbitol is about half as sweet as sucrose. In chewing gum, polyols provide texture, sweetness, and mouthfeel and reduce the incidence of dental caries.

Products made by hydrogenation of various corn syrups are viscous, hygroscopic, noncariogenic, and sweet (depending on the amounts of sorbitol and maltitol present). Their physical properties are generally similar to those of the syrup from which they are made, usually a high-maltose syrup; but they exhibit a greatly decreased tendency to brown, a decreased tendency to crystallize, reduced fermentability, and slower conversion to D-glucose. The last-named property makes these products of potential use as carbohydrate sources in diets for diabetics.

A series of sorbitol-based nonionic surfactants are used in foods as water-in-oil emulsifiers and defoamers. They are produced by reaction of fatty acids with sorbitol. During reaction, cyclic dehydration as well as esterification (primary hydroxyl group) occurs so that the hydrophilic portion is no longer sorbitol but 1,4-sorbitan. Sorbitan esters can be further modified by reaction with ethylene oxide to produce "ethoxylated sorbitan esters", also FDA approved.

1,4-Anhydro-D-glucitol (1,4-Sorbitan)

D-Glucose can be oxidized to D-gluconolactone/D-gluconic acid (5). D-Gluconolactone, which can hydrolyze spontaneously to gluconic acid in aqueous systems, is used as an acidulant, leavening agent, and sequestrant.

D-Glucopyranose   **D-Glucose**

D-Glucono-1, 5-lactone
(D-Glucono-δ-lactone)   D-Gluconic acid

5-Hydroxymethyl-2-furaldehyde
(HMF)

Maltol
(Enol form)

Isomaltol
(Enol form)

I

REACTIONS OF REDUCING SUGARS IN FOODS

Nonenzymic/Nonoxidative Browning Reactions

Nonenzymic browning reactions are important in foods and involve both carmelization and the Maillard reaction.

Heating of carbohydrates effects a group of thermal dehydration and degradation reactions resulting in the formation of caramel (3). Carmelization is catalyzed by acids and certain salts. In general, heating effects dehydration (formation of anhydro rings and introduction of double bonds). The resulting unsaturated heterocyclic rings can condense to form colors and/or flavors.

Sucrose is commonly used to make caramel colors and flavors. The first step in this process is the hydrolysis of sucrose into its constituent reducing sugars, D-glucose and D-fructose. A variety of products used in food, confectioneries, and beverages are formed by heating sucrose in solution with acids or acidic ammonium salts. Three types of caramel colors are produced commercially: acid-fast caramel, the color used in cola soft drinks; brewer's color for beer; baker's color.

Thermolytic reactions of carbohydrates generally produce unsaturated ring systems that have unique flavors and aromas. The primary product from hexoses is 5-hydroxymethyl-2-furaldehyde (HMF). Maltol and isomaltol contribute to the flavor of freshly baked bread. The furanone I contributes to the flavor of cooked meat and can be used to enhance certain other flavors. Pentoses, less often encountered in foods, yield 2-furaldehyde as the main dehydration product. Carbohydrate degradation products include acetic acid, acetoin, acetol, diacetyl, formic acid, lactic acid, levulinic acid, and pyruvic acid, all of which impart odors and flavors that may be desirable or undesirable. (See chemistry top of next column)

The Maillard reaction is neither completely defined nor understood, except that it occurs between a reducing sugar and a compound with an amino group, generally an amino acid or protein, and produces HMF, reductones, and melanoidin pigments (3,14–16).

Amino acids can react further with dicarbonyl intermediates in the thermal dehydration and Maillard reaction pathways to produce other volatile products, such as aldehydes and pyrazines, that contribute to distinctive aromas and flavors, such as those found in honey, maple syrup, and bread.

Much care is taken to control the kind and amount of desirable and undesirable reactions that occur via heating of carbohydrates in the presence or absence of amino acids and proteins.

DIETARY ASPECTS

Although food preferences and economics play a role in the composition of diets, carbohydrates supply 40–60% of the calories in most diets (United States). These calories come almost exclusively from those carbohydrates that are digestible and/or absorbable (17). The digestible carbohydrates in an average human diet in the United States are starch and related materials such as maltodextrins, sucrose, and lactose. Complete hydrolysis of these saccharides to monosaccharides is required before absorption from the small intestine can take place. Dietary monosaccharides come largely from corn syrup (D-glucose) and high-fructose corn syrup (D-glucose + D-fructose). Lesser amounts are ingested as honey (D-glucose + D-fructose) and sorbitol.

The food we eat also contains considerable amounts of nondigestible polysaccharides. Cellulose is the nondigestible polysaccharide consumed in largest amounts; lesser amounts of pectin and other polysaccharides are also present in plant cell walls. These nondigestible polysaccharides that add fiber (roughage) to diets are found in vegetables, legumes, and cereal grains. Various insoluble (oat, wheat, and corn bran; cellulose; sugar beet fiber, etc) and soluble (gums) materials that are exclusively or primarily polysaccharide in nature are added to prepared foods to improve their physical properties or physiological value. All the food carbohydrates listed in Table 1 are described in more detail in other articles, with the exception of cellulose, which is described below.

CELLULOSE

Cellulose is a polysaccharide that is the principal cell-wall component of higher plants. (18,19) It is composed of β-D-glucopyranosyl units in (1→4) linkages. Humans have no gastric or intestinal enzymes that can digest cellulose, ie, convert it into D-glucose. It, therefore, is classified as dietary fiber and may be added to food products, especially bakery products, to increase fiber levels.

Cellulose in the form of microcrystalline cellulose (which can be labeled cellulose gel) is used to prepare reduced-lipid or lipid-free ice cream- and mayonnaise-like products and low- and no-oil pourable salad dressings. It is also used to make shredded cheese and can be found in barbecue sauces and marshmallow toppings. Microcrystalline cellulose also strengthens and stabilizes foams. It has GRAS status and is widely used as a noncaloric, non-nutritive bulking agent for dietary foods.

BIBLIOGRAPHY

1. R. L. Whistler and J. R. Daniel, in O. R. Fennema, ed., *Food Chemistry,* 2nd ed., Marcel Dekker, New York, 1985, pp. 69–137.

2. D. R. Lineback and G. E. Inglett, eds., *Food Carbohydrates,* AVI Publishing Co., Westport, Conn., 1982.

3. G. G. Birch and L. F. Green, eds., *Molecular Structure and Function of Food Carbohydrates,* John Wiley & Sons, Inc., New York, 1973.

4. W. Pigman and D. Horton, eds., *The Carbohydrates: Chemistry and Biochemistry,* 2nd ed., Academic Press, New York, Vol. IA, 1972.

5. Ref. 4, Vol. IB, 1980.

6. Ref. 4, Vol. IIA, 1970.

7. Ref. 4, Vol. IIB, 1970.

8. G. O. Aspinall, ed., *The Polysaccharides,* Academic Press, New York/Orlando, Vol. 1, 1982; Vol. 2, 1983; Vol. 3, 1985.

9. G. M. A. van Beynum and J. A. Roels, eds., *Starch Conversion Technology,* Marcel Dekker, New York, 1985.

10. R. G. Rohwer and R. E. Klem, in R. L. Whistler, J. N. BeMiller, and E. F. Paschall, eds., *Starch: Chemistry and Technology,* 2nd ed., Academic Press, Orlando, Fla., 1984, pp. 529–541.

11. N. E. Lloyd and W. J. Nelson, in Ref. 10, pp. 611–660.

12. R. B. Friedman and A. R. Hedges, in R. P. Millane, J. N. BeMiller, and R. Chandrasekaran, eds., *Frontiers in Carbohydrate Research-1,* Elsevier Applied Science, London, 1989, pp. 74–82.

13. W. C. Griffin and M. J. Lynch, in T. E. Furia, ed., *Handbook of Food Additives,* 2nd ed., The Chemical Rubber Co., Cleveland, Ohio, 1982, p. 431.

14. C. Erikkson, eds., *Maillard Reactions in Food,* New York, Pergamon, 1981.

15. G. R. Waller and M. S. Feather, eds., *The Maillard Reaction in Foods and Nutrition,* American Chemical Society Symposiam Series, Vol. 215, 1983.

16. M. Fujimaki, M. Namiki, and H. Kato, *Amino-Carbonyl Reactions in Food and Biological Systems,* Kodansha, Tokyo and Elsevier, Amsterdam, 1986.

17. S. Reiser, eds., *Metabolic Effects of Utilizable Dietary Carbohydrates,* Marcel Dekker, New York, 1982.

18. T. P. Nevello and S. H. Zeronian, eds., *Cellulose Chemistry and Its Applications,* Halsted Press, New York, 1985.

19. R. A. Young and R. M. Rowell, eds., *Cellulose,* John Wiley & Sons, Inc., New York, 1986.

J. N. BeMiller
Purdue University
West Lafayette, Indiana

CARBONATED BEVERAGES

Today carbonated beverages are consumed not only by teenagers but by people of all ages. Consumers choose soft drinks for many reasons: taste, refreshment, relaxation, pleasure, sociability, and simply for the product's thirst-quenching capability. The soft drink share of the beverage market keeps growing, often at the expense of other beverages.

INGREDIENTS

The carbonated beverage container labels reveal the product ingredients: carbonated water, sweetener, acid, color, flavor (natural or artificial), and preservatives. In some beverages there may also contain foaming agents, emulsions, caffeine, gum, antioxidants, and anti-foaming compounds. The percentage level of these materials in the drink can vary based on the type of beverage flavor or type of product. Water is the major ingredient in soft drinks and it may range from 86 to 93% of the total volume. The sweetener level can be 8–15% in beverages, with most beverages having a 10–11% sweetener level on a weight-to-volume basis. Carbon dioxide is used at 1.5–5.0 volumes of gas, with principal beverages employing 3.4–3.8 volumes of gas. Citric and phosphoric acids are employed at levels to produce a pH range of 2.5–4.5. In general, the fruit- and ginger ale-type beverages contain citric acid, while cola- and pepper-type beverages use phosphoric acid as the acidulent. Almost all beverages, except the lemon-lime–type, use color. Caramel color produced from sugar-type products are used in cola, pepper, ginger ale, and root beer. The fruit-flavored drinks use one or a combination of the certified red, yellow, and blue colors. The beverage flavors are natural or artificial or a combination. For example, a high-quality grape beverage can be formulated with a combination of artificial flavor and natural grape juice.

Water

Water is an integral part of soft drinks. It is 86–93% of most regular beverages, and the diet drinks contain up to 98% water. For the beverage to have a fresh and wholesome taste, the water used to make the drink must be free of any medicinal, fishy, woody, musty, or other off-flavor. That is, it must be free of contamination with chemicals or microorganisms that can produce an off-flavor or odor in the beverage. Because water is such a high percentage of the total volume of the beverage, objectionable contaminants in the water, such as suspended matter, high alkalinity, and microorganisms, must be eliminated.

It is essential that the water used in soft drinks meet established quality standards. To meet these high standards it must be properly treated. Water-treatment or water-processing procedures used in bottling plants today are the following: (1) The raw or city water is pumped into a specially designed chemical reaction tank; (2) chemical feed pumps deliver precalculated levels of iron, lime, and chlorine into the tank for the chemical reactions that remove suspended matter, reduce alkalinity, and eliminate microorganisms; (3) the water is passed through a sand filter; (4) then it is pumped through an activated carbon bed filter; (5) the water flows through a special micro water polisher prior to entering the carbonator and proceeding through the other steps in beverage preparation. Detailed description of the water-treatment process will not be presented here.

Carbon Dioxide

Carbon dioxide as used in carbonated beverages is a colorless, nontoxic gas with slight pungent or biting odor produced by burning carbon compounds, such as coke, oil, or gas; heating limestone to form lime and carbon dioxide; fermentation to produce alcohol and carbon dioxide; or trapping from gas wells. It is distributed as a liquid in cylinders under high pressure, as a liquid in tank trucks or rail cars under low pressure, or in solid form in insulated containers or trucks at atmospheric pressure.

Carbonation is the process of saturating the beverage with carbon dioxide. The popularity of carbonated drinks is due to the unique taste, zest, and sparkle imparted by the carbon dioxide. The dissolved gas also plays a principal part in inhibiting or destroying harmful bacteria. This fact, however, does not in any way cause the bottlers and canners to relax in maintaining strict sanitary controls throughout the plant.

Gas volume is a measure of carbonation. The gas volume of carbonation of different types of soft drink is as follows:

Gas Volume of CO_2	Drink Classification
≥ 3.5	Ginger ale, lemon–lime (up type), colas, pepper type, and mixes (club soda, tonics, etc)
2.5–3.5	Root beer, grape, lemon, cherry, lime, orange, cream soda, strawberry, and grapefruit

The cola, pepper, and up-type beverages have an average volume of carbon dioxide gas of 3.6.

Sweeteners

Bottlers' sugar is used in the manufacture of carbonated beverages. The nutritive sweeteners employed in carbonated beverages are sucrose ($C_{12}H_{22}O_{11}$) produced from cane or beets and high-fructose corn syrup (HFCS). Sucrose is available in granular form, and since 1950 with larger plants and improved storage and handling systems, liquid sucrose and invert liquid sugar are used as sweeteners in soft drinks. HFCS is currently approved as a nutritive sweetener for use in carbonated beverages. The bottlers' sugar and HFCS sweeteners must meet the technical specifications as established by the Society of Soft Drink Technologists regarding taste, color, odor, ash content, and microbial content. The growth in the use of HFCS nutritive sweetener in carbonated beverages since 1970 has established it as the primary sweetener for soft drinks.

The nonnutritive sweeteners have made it possible to provide diet soft drinks with no or few calories. Since 1983 aspartame (NutraSweet), a methyl ester of two amino acids—phenylalanine and aspartic acid—has been used as the low-calorie sweetener in diet soft drinks. Earlier diet beverages employed saccharine and/or saccharine and cyclamate. Today there are other synthetic sweeteners being tested by the Food and Drug Administration (FDA) that are considered candidates for sweeteners in soft drinks: sucralose, made from sucrose with three OH groups replaced by chlorine atoms; alitame, a dipeptide ester consisting of L-aspartic acid and D-alanine; and acesulfame K.

For the people who are weight conscious and like carbonated beverages, the diet or low-calorie drinks are ideal. The principal diet drinks are similar to the regular nutritive beverages except for the artificial sweeteners employed as replacements for sugar or HFCS.

Flavors

The two general classes of flavors used in soft drinks are (1) natural or fruit juices; essential oils; and extracts and oleoresins from plant leaves, stems, roots, and flowers; and (2) imitation, synthetic, and artificial flavor materials, from extraction processes or chemical synthesis, usually purchased from a reliable flavor company that specializes in marketing flavors for food and beverages.

A flavor selected for a soft drink should meet the following requirements: approval by the FDA, good taste and aroma, good shelf life, ease of mixing, ease of storage and handling, resistance to sunlight and oxygen, resistance to chemical changes in low-pH media, economical, and commercial availability.

The total flavor or taste of a carbonated beverage is a combination of many flavors and other ingredients. In general, these include the materials listed in Table 1.

Colors

The use of certified colors in some beverages gives them more eye appeal and makes them easier to market. Carbonated soft drinks possess quite different color characteristics from still beverages. In many cases, the color for carbonated beverages is combined with the flavor in a product known to the trade as a compound. When color is dissolved with flavor in a compound, solubility problems can occur. Even though there may be enough water in the compound to keep the color in solution under ordinary circumstances, the presence of water-miscible solvents, such as alcohol and propylene glycol, can reduce the solubility of some colors. If a large proportion of such solvents is required to keep flavor oils in solution, it may be impossible to dissolve the required quantity of color.

If solubility problems arise, five alternatives are generally available.

1. Substitute glycerine for propylene glycol or alcohol as a flavor oil solvent.
2. Change the color blend to one that contains primaries with higher solubility.

Table 1. Primary Ingredients Commonly Detected in Carbonated Beverages[a]

Ingredients	Cola	Orange	Grape	Lemon–Lime	Root Beer
Water	x	x	x	x	x
Sugar	x	x	x	x	x
Phosphoric acid	x	—	—	—	—
Citric acid	—	x	x	x	x
Caffeine	x	—	—	—	—
Sodium benzoate	—	x	x	—	x
Carbon dioxide	x	x	x	x	x
Gum acacia	x	x	—	—	x
Caramel color	x	—	—	—	x
FDA colors	—	x	x	—	—
Nutmeg oil	x	—	—	—	x
Methyl anthranilate	—	—	x	—	—
Orange oil	x	x	—	—	—
Lemon oil	x	—	—	x	x
Vanilla–vanillin	x	—	—	—	x
Lime oil	x	—	—	x	—
Cinnamon oil	x	—	—	—	x
Ethyl acetate	—	—	x	—	—
Ethyl alcohol	—	—	x	x	—
Citral	—	x	—	x	—
Kola nut extract	x	—	—	—	—
Ascorbic acid	—	x	—	—	—
Cassia oil	x	—	—	—	x
Clove oil	x	—	—	—	x
Ethyl butyrate	—	—	x	—	—
Methyl salicylate	—	—	—	—	x

[a] Ref. 1.

3. Reduce the strength of the flavor, for example, from 2- to 4-oz goods.

4. Pack the color as a separate unit to be dissolved by the bottler and added to the syrup.

5. Try a darker shade that permits use of less color.

As has been indicated, caramel colors are natural food colorings currently produced by the scientifically controlled caramelization of highly refined corn sugar. Basically, there are three types: acid proof, bakers' and confectioners', and the foaming variety. The carbonated soft drink industry has taken advantage of the eye appeal of caramel, and, in fact, the industry constitutes about 75% of the caramel-coloring market.

To meet the specific requirements of bottlers of acidulated beverages and satisfy the consumer's expectations of a brilliant color, it is necessary to use an acid-proof caramel. These are formulated to remain stable under the acidic conditions commonly encountered in the beverage industry. An inferior quality caramel or the wrong type of caramel may have a tendency to haze and precipitate under such conditions. Equally important to beverage industry users is the acid-proof type of negative colloidal charge. During the early development work on caramel the need for an accurate means of measuring the amount of coloring ability per pound of caramel, generally expressed as tinctorial power (TP), was recognized.

Acid-proof caramels are made either single or double strength, depending on the preference of the individual bottler. Generally, higher acid-proof caramels are used when greater acid stability is required. Frequently, double-strength caramel is used in dietetic drinks where ca-loric value must be considered. Because double-strength, acid-proof caramels contain a minimum of unreacted sugars and only half the normal quantity is needed, they contribute substantially fewer calories to the finished beverage than the single-strength types.

Acids

Citric Acid. Citric acid is by far the most widely used acid in fruit beverages. Its light fruity character blends well with most fruit flavors and quite satisfactorily with lighter flavors such as ginger. It represents 6–8% of the juice of unripe lemons and somewhat less of the juice of fully ripe citrus fruits. Citric acid is the chief acid constituent of currants, cranberries, and citrus fruits. It is associated with malic acid in apples, apricots, blueberries, cherries, gooseberries, loganberries, peaches, plums, pears, strawberries, and raspberries; with isocitric acid in blackberries; and with tartaric acid in grapes and mountain ash berries.

Commercially, citric acid has been recovered from lemon and pineapple juice. Citric acid is used in beverages for its acidic properties, as well as in salts (sodium citrate, calcium citrate) and in other forms for its powerful sequestering action.

Because citric acid occurs naturally in fruits, it is the preferred acidulant for carbonated and still beverages. Citric acid adds tartness and refreshing properties to the drink, often duplicating natural fruit products. It acts as a preservative in syrups and the finished beverage and aids in obtaining the desired bouquet by modifying the sweet flavors. It sequesters harmful metals, which cause haze and speed deterioration of color and flavor. The amount of acid added to soft drinks depends on the flavor and the particular end use intended. Some syrup flavors, such as grape or orange, contain as little as 0.5 oz of citric acid per gallon of syrup, whereas the citric content of certain mixers may be as high as 4 oz/gal. In general, sufficient citric acid should be added to give a final pH of 2.5–4.5.

Anhydrous citric acid is generally used as a 50% solution prepared by dissolving 5 lb of anhydrous acid in enough water to make 1 gal of final solution.

Phosphoric Acid. Phosphoric acid is universally used in cola drinks and probably acidifies more beverages than citric acid. Its use dates to the earliest soft drinks when phosphates of various fruit flavors were popular at the soda fountain. The taste is one of a flat sourness in contrast with the sharp fruitiness of citric and it seems to blend better with most of the nonfruit drinks.

Phosphoric acid is widely used in cola and pepper beverages, as well as in root beer and sarsaparilla drinks. The usual cola contains about 0.05% phosphoric acid and has a pH of 2.3–2.6. Root beer has a pH of about 5.0 and contains 0.01% phosphoric acid.

Other Acids. Other acids used include tartaric acid, fumaric acid, adipic acid, and malic acid.

Preservatives

The preservatives employed in soft drinks include benzoic acid and its sodium and potassium salts and sorbic acid and its sodium, potassium, or calcium salts. The usage levels of these agents in soft drinks range from 0.025 to 0.050 ppm. The percentage of juice and nutritive substances present in the drink, as well as the pH of the finished beverage, play a role in the required level of preservative in the beverage.

Benzoic acid is widely used as a preservative in soft drinks and, in some countries, sorbic acid is also permitted for this purpose. Owing to its unpleasant flavor, sulfurous acid is used only on a small scale.

It has been discovered that the effectiveness of these acids depends largely on the pH value of the environment. They are least effective in a neutral medium and effectiveness increases considerably with decreasing pH. By reducing the pH value from 4.5 to 3.0, the preservative effect of benzoic acid is increased nearly threefold. This means that only in soft drinks with a low pH is the full benefit derived from the addition of a preservative. Although the full benefit of the preservative added is only obtained at a pH of 2.0, in practice, satisfactory results are achieved at pH values between 2.5 and 4.0.

DEVELOPMENT

The research laboratory for carbonated beverages has many functions, one of which is new product development. The steps or stages in new carbonated beverage product development include (1) concept or type of product, (2) feasibility (in regard to legality and compatibility), (3) laboratory formulation, (4) packaging materials (type, specifications, and graphics), (5) ingredient sources and availability, (6) production procedure, (7) production equipment, (8) market research study for consumer profile, (9) product specification, (10) ingredient labeling, (11) product shelf life, (12) microbiological characteristics, (13) food and drug approval of all ingredients, (14) schedule or timetable, (15) marketing plan, and (16) sales and advertising program.

PACKAGES

Packages for carbonated beverages include individual units, packs, cartons, cases, and pallets. These are changing constantly in size, materials, shape, colors, stability, convenience, and cost. Materials for packages include

1. *Glass.* Returnable and nonreturnable bottles of a variety of sizes, shapes, and colors, in 2-, 6-, 8-, 12-, or 24-packs.
2. *Steel.* Plain; tin-, plastic-, and lacquer-coated cans of a variety of sizes; soldered or welded; neck-in, three-piece, and embossed texturized cans; trays, cartons, and pallets.
3. *Aluminum.* Plain, seamless necked, embossed, texturized, pop-top, and screw-top cans; foil-caps; trays, cases, and pallets.
4. *Plastic.* Bottles, packs, crowns, trays, cartons, and composite.
5. *Fiberboard.* Packs, cartons, and trays.
6. *Polystyrene Foam.* Wraparound bottle sleeves and trays.
7. *Waxed Paper.* Individual and multiserving cups at service counters and vending machines.

The continued growth of the carbonated beverage industry may be due as much to improvements in convenience of handling and serving as to any other factor.

PRODUCTION EQUIPMENT

Today the production equipment for bottling and canning of carbonated beverages is engineered to operate efficiently at high speeds as shown in Figure 1. There are a number of units that make up a production line. The equipment required for product manufacture includes water treatment and finished syrup units, as well as those units presented in Figure 2 and Table 2. Each piece of equipment is made of stainless steel and requires proper maintenance and sanitation for efficient production runs.

A flow chart for the production area is shown in Figure 3. The following sequence of stages in this production line (Fig. 2) is adaptable to either bottle or can conveyors or case or carton conveyors.

Stage 1. From warehouse.
Stage 2. To palletizers and warehouse.
Stage 3. Bulk can depalletizer.

The use of a can depalletizer is more economical than receiving empty cans in reshipper cartons or trays. Simi-

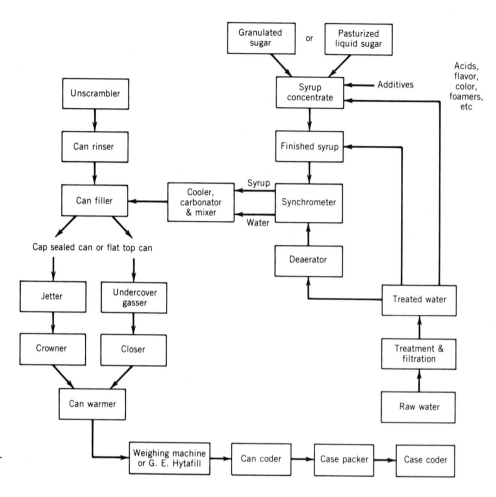

Figure 1. Typical soft drink canning line continuous mix system.

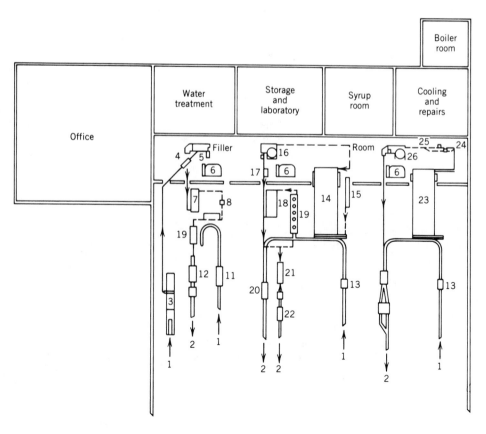

Figure 2. Production area plan (1).

Table 2. Bottling Line Equipment[a]

1. Uncaser
2. Case conveyor
   a. Infeed conveyor
   b. Uncaser to case packer
   c. Full product conveyor
3. Bottle washer
   a. Load table
4. Nonreturnable bottle rinse
   a. Various-size bottle tracks
5. Bottle conveyor
   a. Uncaser to load table
   b. Load table to rinser
   c. Bottle washer to filler
   d. Rinser to filler
   e. Filler to case packer
6. Inspection units
   a. Empty bottle light
   b. Full bottle light
   c. Electronic bottle inspector
   d. Electronic Brix and CO_2 tester
7. Filler
   a. Crowner
   b. Screw capping
8. Carbonation
9. Proportioning equipment
10. Cooling equipment
    a. Water
    b. Syrup

11. Washer heating equipment
    a. Caustic heaters
    b. Steam boiler
12. Case packer
    a. Accumulation table
13. Palletization
    a. Palletizer
    b. Depalletizer

Other Areas Involved

14. Syrup room
    a. Tanks
    b. Pumps
    c. Strainer
    d. CIP system
15. Premix, postmix
    a. Tank washer
    b. Tank filler
    c. Tank or pallet conveyor
16. Liquid sweetener tanks
17. Liquid caustic tanks
18. Water-treating system
19. Evaporative condenser
    a. Cooling compressor
    b. Cooling tower
20. Air compressor

[a] Ref. 1.

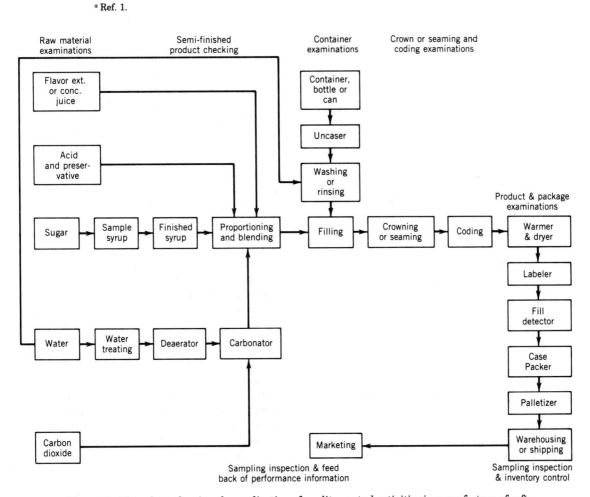

Figure 3. Flow sheet showing the application of quality control activities in manufacture of soft drinks (1).

lar depalletizers may be used for bulk receipt of nonreturnable glass.

Stage 4. Can rinser.

Stage 5. Can filler and closing machine.

Stage 6. Beverage preparation equipment.

Stage 7. Can warmer.

Cans are filled at about 1.1°C and must be warmed above the dewpoint to prevent condensation from damaging paper six-packs and trays or cartons.

Stage 8. Short-fill detector.

Stage 9. Accumulating table.

Accumulating tables are frequently used to collect product when packing equipment malfunctions. The purpose is to prevent lost production. Without the accumulating table, packing equipment stoppage would cause instant filler stoppage.

Stage 10. Six-pack machine.

Stage 11. Tray maker.

The tray maker sets up cartons from flat blanks, adding to the economics obtained by use of a depalletizer (stage 3). An operating speed of 50 cases/min is assumed for this can line.

Stage 12. Tray packer.

Stage 13. Bottle uncaser.

Stage 13 through stage 22 represent a typical line. This line is capable of handling any size bottle, closed with conventional crowns or with aluminum roll-on caps. Nonreturnable bottles bypass the returnable bottle washer. A bottle warmer is provided for nonreturnable bottles for the same reason that stage 7 is provided. Provisions are made to label nonreturnable bottles and to transport them to the loose packer or to the six-pack system. Returnable bottles, which are not labeled and which are packed in rigid wood or plastic cases, bypass the warmer and labelers go directly to the loose packer. An operating speed of 20 cases/min is assumed.

Stages 13 to stage 23 and through stage 27 represent a concept for a high-speed returnable-bottle line for efficient operation based on fewer changeovers than on the typical line. An operating speed of 35 cases/min is assumed.

Stage 14. Washer for returnable bottles.

Stage 15. Rinser for nonreturnable bottles.

Stage 16. Filler for returnable and nonreturnable bottles.

Stage 17. Roll-on capper.

Stage 18. Warmer for nonreturnable bottles.

Stage 19. Labelers for nonreturnable bottles.

Stage 20. Case packer for loose-pack returnable and nonreturnable bottles.

Stage 21. Six-pack machine for nonreturnable bottles.

Stage 22. Case packer for nonreturnable bottles in six-packs.

Stage 23. Washer for returnable bottles.

Stage 24. Electronic empty bottle inspectors.

Stage 25. Bottle combiner.

Stage 26. High-speed returnable-bottle filler–crowner.

Stage 27. High-speed dual-case packer.

PRODUCTION

The production process of a carbonated beverage generally begins with the preparation of the finished syrup. Treated water, metered into the stainless-steel tank, is mixed with the sweetener (granulated or liquid) to a Brix level of approximately 40–55% solids. After the sweetener and water (simple syrup) are mixed, the preservative is added and mixed. The flavor compound and color are blended, and then the acid solution is added into the finished syrup tank. Continuous mixing and blending of the flavored syrup is done throughout the entire process of adding and mixing the various ingredients. After the ingredients are dissolved and properly mixed, a sample of the finished syrup is collected and tested for Brix, color, and acid and tasted when diluted to the finished beverage level.

Second, the finished syrup is blended with the water at the proper ratio, eg, 1 + 5, 1 + 5.5, or 1 + 6 syrup and water, to yield a finished beverage. The predetermined proportion of syrup and water are mixed and delivered to the carbo–cooler. Third, the beverage is cooled and carbonated. The carbonated beverage is then pumped to the filler. Fourth, the clean and commercially sterile container is conveyed to the filler and filled. Then the closure is applied. Fifth, the packaged beverage is conveyed and automatically placed in a case and then palletized prior to the sixth and final step of placing it in the warehouse ready for transfer to retail outlets.

The production or manufacturing process is more complicated than that described here. It involves control of product temperature, gas pressures, equipment speeds, closure application, foam, filler jetting, air removal from filler, and fill level.

Based on this description, one of the primary operations in a beverage plant is production. With good planning, organization, and control, it is possible to establish and maintain an efficient production operation. The chief goal of a beverage production operation is the economic production of quality beverages. To achieve this end, the plant must have a medium for coordinating its activities into a single effort. The production operation in a beverage plant includes the use of workers, materials, and machines. Productivity, measured in output per man-hour, is a constant goal of a production operation.

The production operation in a bottling plant is engineered to produce a variety of beverages. The machinery and equipment used are designed to be operated for long periods of time. However, the processing does not necessarily require around-the-clock activity. The machines and equipment are lined up within the plant according to the sequence of operations, such as the water-treating system, syrup preparation, and the bottling operation. The bottling line includes the conveyor, depalletizer, decaser, bottle sorter, bottle washer, bottle inspector, cooler and carbonator, syrup and water blender, bottle filler, crowner, case packer, palletizer, and forklift truck handling of pallets that are stored in the warehouse. Producing different-flavored beverages on a profitable basis for quality-minded consumers is a challenge to beverage plants of all sizes.

The functions of the production department within the organization include production scheduling; quality control; ordering raw materials and supplies; receiving; inventory control; equipment maintenance; training and supervising production personnel; safety; good manufacturing practices; work assignments; keeping records; noise, air, and solid waste control; budget control; and follow-up and coordination of all these activities. The production operation is also concerned directly with scheduling the syrup preparation, bottling operation, and premix operation.

QUALITY CONTROL

Quality control of beverages is not just limited to a socalled shop-floor inspection. It starts with raw materials, specification, and formulation of a product, process control and product inspection, quality assurance, and performance auditing. It ends when the product reaches the hands of the consumer. Hence, a quality-control program should be operated not only to check the raw materials, maintain their quality within strict limits, and employ continuous inspections during stages of production, but should also maintain the inventory control by rotating finished products in the warehouse according to their shelf life, preserve the plant in a respectable and sanitary condition, insure that the plant and packaged products are in compliance with food laws and regulations, and extend to shipping and transportation of the product.

It is difficult to have a uniform list of activities performed by every quality-control organization because of the differences in size, personalities, level within the company, and technical capabilities. In a large company, for example, a properly organized quality-control group is involved in quality functions along with many other departments, whereas in a small organization, many functions center in an individual or a small group. However, some principal areas that an effective quality-control program should encompass in a bottling plant are listed below and in Figure 3.

Carbonated beverage bottling and canning plants have found that quality control is essential for success. A quality-control laboratory can help control raw materials through setting up specifications and checking each incoming product to insure that it meets established standards, check and approve all packages, improve production quality procedures, improve plant sanitation and good manufacturing properties, create confidence in the quality and uniformity of all beverages packed, and maintain quality records and reports. The savings from improved uniformity in manufacturing conditions and product quality can make the quality-control laboratory self-supporting.

The trend in carbonated beverage production operations today is to increase production speed and automa-tion, improve control instruments, and increase efficiency and productivity. With technological changes, the quality-control laboratory becomes more essential for the bottling and canning plants.

BIBLIOGRAPHY

J. G. Woodroof and G. F. Phillips, *Beverages: Carbonated and Non-Carbonated*, AVI Publishing Co., Westport, Conn., 1981.

G. FRANK PHILLIPS
Dr. Pepper and
Seven-Up Companies, Inc.
Dallas, Texas

CARBONATED WATERS. See FOOD PROCESSING: STANDARD INDUSTRIAL CLASSIFICATION.
CARCINOGEN. See FOODBORNE DISEASES; FOOD MICROBIOLOGY; FOOD TOXICOLOGY; NITROSAMINES; TOXICOLOGY AND RISK ANALYSIS.

CAROTENOIDS

The carotenoid pigments together with other plant and animal pigments were, because of their bright colors, some of the first compounds to be studied. Carotene was first isolated from and named after the carrot in 1831. In 1837, the yellow pigment(s) in autumn leaves were named xanthophyll.

There are four periods of carotenoid research (1). During the first period (1800s) pigments were characterized by measurements of light absorption. During the second period (1900–1927) attempts were made to define the empirical formulas of the isolated carotenoids. The third period (1928–1949) was a time in which some carotenes with provitamin A activity were discovered. Vitamin A and some of the carotenes including β-carotene, were synthesized. The fourth, and most recent, period has witnessed a virtual explosion in the number of known carotenoids. The instrumentation that has made this possible was developed during this time. When the classic book on carotenoids was published in 1948 (2) there were about 80 known carotenoids and of these perhaps only 35 had established structures. Today there are about 600 known carotenoids.

The carotenoids as natural pigments are unique in that they are found in both plants and animals. Animals cannot make the basic 40-carbon structure but are capable of making changes in the hydrocarbon rings and chain. Table 1 shows the distribution of the various plant pigments in food. The carotenoid pigments are important from a food science point of view because the color they impart to various food products and the fact that a number of the pigments have vitamin A activity.

The carotenoids are chemically and biochemically related to the more general class of compounds known collectively as the terpenes and terpenoids. These are compounds composed of repeating isoprenoidlike units. Mevalonic acid has been shown to be an obligate precursor for the biosynthesis of the terpenoids.

The carotenoids have traditionally been classified as carotenes (hydrocarbons) and xanthophylls (containing oxygen in some form). Generally, the hydrocarbons (lyco-

Table 1. Distribution of Natural Coloring Matter

Food	Heme	Carotenoids	Chlorophyll	Betalains	Anthocyanins	Flavonoids	Caramel	Melanins
Red meat	X		—	—	—	—	—	—
Fish	X	X	—	—	—	—	—	X
Eggs	—	X	—	—	—	—	—	—
Crustaceans	—	X	—	—	—	—	—	—
Dairy products	—	X	—	—	—	—	—	—
Green vegetables	—	X	X	—	X	X	—	—
Root vegetables	—	X	X	X	X	X	—	—
Fruits	—	X	X	—	X	X	—	—
Cereal	—	X	—	—	—	—	X	—

pene, β-carotene, etc) are produced first and the xanthophylls are produced at a later stage. However, it has recently been shown that some fish can reverse the process and make retinol from oxygenated carotenoids. Xanthophyll is an old name for lutein.

During the 1960s and part of the 1970s, mevalonic acid was stereochemically labeled with ^{3}H and ^{14}C and was incorporated into various plant tissues. A number of biosynthetic problems were solved with this technique. More recently the chirality of the carotenoids has been studied, highlighted by the discovery of epilutein in the goldfish.

Tables 2 and 3 give the historical development of some yellow and red carotenoids. In most cases the compounds have been totally synthesized.

Today there is much interest in the carotenoids because they are natural pigments and also because of their proposed role in medical applications.

Every three years an informal group meets in Europe or the United States and the plenary and session lectures are published (4–10). A book on the technical aspects of the carotenoids (11) and a book on vitamin A deficiency (12) have also been published.

The carotenoids are the ultimate sources of vitamin A in the diet, and the lack of retinol or provitamin A has led to nutritional blindness. A number of countries have been listed as having a potentially serious vitamin A deficiency problem. In some well-studied Southeast Asian countries it has been estimated that half a million children will go blind because of a lack of vitamin A. When one extrapolates to other countries in Asia, Africa, and Latin America, the mortality and morbidity are staggering.

Finally, there is a body of evidence that suggests that some of the carotenoids themselves have medical applications. These include the deposition of carotenoid pigments in the skin of people who do not develop protective pigmentation and protection against certain cancers. While we have known of the protection of green plants from photodynamic destruction by the carotenoids, the protection of animals from neoplastic transformation is a recent discovery.

HISTORY

Table 4 lists some important milestones in regard to carotenoid and vitamin A research. These milestones represent a 160-year span of research that can be divided into four periods. The emphasis on the isolation of the carotenoid pigments rather than, for example, the related steroids was undoubtedly made because of the bright colors of these compounds.

It is understandable that much of the research, includ-

Table 2. Early Historical Development of Some Yellow Carotenoids[a]

Period	Lutein	Zeaxanthin	Cryptoxanthin	Physalien	Bixin
1830	Berzelius[b] (1837)				
1850					
1870		Thudichum[b] (1869)			
1890	Willstatter[c] (1907)				
1910					Heiduschka[c] (1917) Kuhn[b] (1928–1931)
1930	Karrer[d] (1930–1933)	Karrer[d] (1930–1933)	Yamamoto[b] (1932)	Kuhn[b] (1929) Kuhn[c] (1933)	Kuhn[c] (1929–33)
1950		Isler and co-workers[e] (1955)	Isler and co-workers[e] (1956)	Isler and co-workers[e] (1955)	Weedon[e] (1951)
1960					
1970					
1980	Meyer[e] (1980)				

[a] Ref. 3.
[b] First mention.
[c] Molecular formula.
[d] Constitution.
[e] Total synthesis.

Table 3. Early Historical Development of Some Red Carotenoids[a]

Period	Lycopene	Astaxanthin	Capsorubin	Torulhodin	Capsanthin
1830					Branconnet[b] (1830)
1850					
1870	Harsten[b] (1873)	Pouchet[b] (1876)			
1890					
1910	Willstatter[c] (1910)				Zechmeister[d] and co-workers (1927)
1930	Karrer[e] (1929–1931)	Kuhn[d] and co-workers (1933) Kuhn[c] and co-workers (1938) Kuhn[e] and co-workers (1938)	Zechmeister[d] and co-workers (1934) Zechmeister[c] and co-workers (1935) Zechmeister[e,f] and co-workers (1934–35)	Lenderer[b] (1933) Karrer[d] and co-workers (1943) Karrer[e,f] and co-workers (1946)	Zechmeister[e,f] and co-workers (1927–1935) Karrer[e] and co-workers (1927–1935)
1950	Karrer[g] and co-workers (1950)		Karrer[e,f] and co-workers (1956)		
1960	Isler[h] and co-workers (1956)			Isler[g] and co-workers (1959)	
1960			Cooper and co-workers (1962)		Weedon[g] (1972)
1970		Brogu[g,h] and co-workers (1979)			
1980					

[a] Ref. 3.
[b] First mention.
[c] Molecular formula.
[d] Isolation.
[e] Constitution.
[f] Proposed structure.
[g] Total synthesis.
[h] Commercial method.

Table 4. Some Pertinent Dates and Events Leading to Commercially Produced Carotenoids and Major Carotenoid Literature Compilations[a]

1831		Carotene isolated from carrots (Wackenroder)
1906–1911		Column chromatography developed for carotenoids (Tswett)
1907		Empirical formula of carotene established (Willstatter and Mieg)
1913–1915		Fat soluble growth factor "A" recognized in butter (McCollum and Davis, Osborne and Mendel)
1919–1920		Vitamin A activity related to yellow carotenoids of corn (Steenbock and co-workers)
1922		ACS monograph on carotenoids (Palmer)
1928		Carotene gives Carr-Price reaction and cures Vitamin A deficiency in rats (von Euler and co-workers)
1928–1931		Structure of carotene and vitamin A established (Karrer and co-workers)
1930		Carotene demonstrated to be converted to vitamin A (Moore)
1931		Standard developed: 0.6 μg β-carotene = 1 IU vitamin A (League of Nations)
1934		Book on carotenoids (Zechmeister)
1948		Book on carotenoids (Karrer and Jucker)
1950		Synthesis of β-carotene (Karrer and co-workers, Inhoffen and co-workers, and Milas and co-workers)
1952		Book on carotenoids (Goodwin)
1953–1956		Industrial synthesis of β-carotene (Isler and co-workers) Commercial production of β-carotene
1958–1959		Synthesis of β-apo-8'-carotenal (Isler and co-workers and Rüegg and co-workers)
1959		Synthesis of β-apo-8'-carotenoic acid ethyl esters (Isler and co-workers)
1960		Commercial production of β-apo-8'-carotenal
1956–1963		Industrial synthesis of canthaxanthin (Isler and co-workers)
1962		Conversion of astaxanthin to vitamin A in fish (Granguad and co-workers)
1964		Commercial production of canthaxanthin
1965		Demonstration of independent formation of β and ε rings by use of specifically labeled MVA (Goodwin and co-workers)
1971		Book on carotenoids (Isler)
1978		Isolation of 3^1-epi-lutein (Eugster and co-workers)
1979		Synthesis of astaxanthin (Broger and co-workers)
1979		Reverse phase HPLC separation of α and β-carotene (Zakaria)
1980, 1984		Books on biochemistry of plant and animal carotenoids (Goodwin)
1981		Book on carotenoids as colorants (Bauernfeind)
1984		Retinol found in halobacteria (Oesterhelt and co-workers)

[a] Ref. 3.

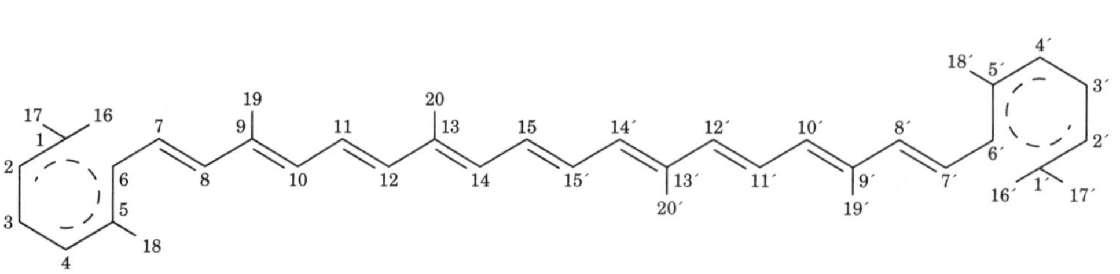

ing chemical synthesis, was motivated by the need for natural coloring matters. Beta-apocarotenal, β-carotene, and canthaxanthin are on the U.S. market as natural coloring matters.

CAROTENOID NOMENCLATURE

In 1948, there were about 35 proven structures of carotenoids (2). Trivial names were used because the structures were not established and the number was small. With more than 600 carotenoids of proven structure now known there is still a need for trivial names, but there is also a need for more precise names that follow a chemical convention.

Under trivial names the carotenoids are generally grouped as carotenes (hydrocarbons) and the oxygen-containing compounds as xanthophylls. Some of the carotenes' names are preceded by a Greek letter (eg, α, β, ε) and others have names denoting the source (lycopene-tomato). Some are a combination such as β-zeacarotene and α-zeacarotene. Here the ring double bond position and the source (maize) are indicated.

Xanthophylls are pigments that contain hydroxyl, carbonyl, epoxide, ester, and carboxylic acid groups. The convention today is to indicate the source and add the suffix xanthin (eg, tunaxanthin). Some of the older trivial names such as lutein denoted their yellow color. Structural features are indicated by the prefix iso, as in isozeaxanthin; the hydroxyl group in zeaxanthin is in the three position and it is in the four position in isozeaxanthin. cis-Lycopene would indicate a departure from the expected all-trans structure.

In recent years an attempt has been made to limit the number of trivial names by building on a known name such as β-carotene. Thus astaxanthin might be 3,3'dihydroxy-4,4'-diketo-β-carotene. The complete numbering system of lycopene is (see chemistry above)

The systematic method of carotenoid nomenclature uses the IUPAC convention (13). The great advantage is that any carotenoid can be exactly named, including chiral centers. The disadvantage is that the name for a complicated compound can take two to three lines, whereas the trivial name is a single word (eg, fucoxanthin). In the IUPAC system (14) the ring configuration is specified as follows: (see chemistry below)

The central chain is assumed to be unsaturated and all-trans. Any departure from this structure is specified. Thus β-carotene is β,β-carotene. The term nor indicates that a methyl is missing and seco indicates a break in the ring structure as 5,6,SECO, β,β-carotene-5,6-dione (semi-β-carotenone). The change in the hydrogenation level to form saturated positions as well as acetylenes and allenes can be designated by hydro and dehydro.

Oxygenated derivatives of carotenoid hydrocarbons are named by use of suffixes and prefixes according to the rules of general chemical nomenclature. The principal group is chosen and is cited by use of a suffix and the other groups are cited as prefixes. The sequence is as follows: carboxylic acid, ester of carotenoid acid, aldehyde, ketone, alcohol, ester of carotenoid alcohol, and epoxide. The following structure illustrates this rule (see chemistry top of page 297)

Apo-carotenoids are found naturally as well as occurring synthetically. The apo designates where the parent compound has been cleaved. Thus β-apo-8'-carotenal should be designated 8'-apo-β-caroten-8'-al.

There are rules for carotenoids with more than 40 car-

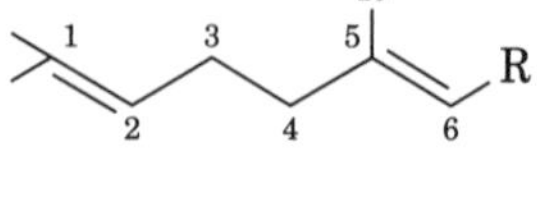

acyclic
ψ (psi)

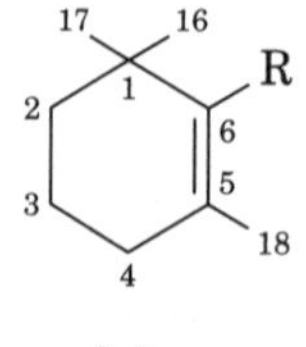

cyclohexene
β (beta)

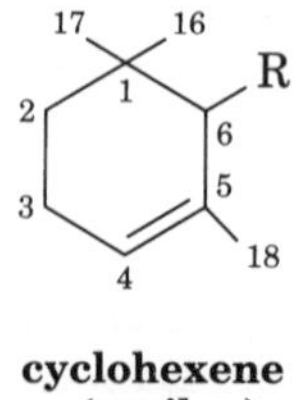

cyclohexene
ϵ (epsilon)

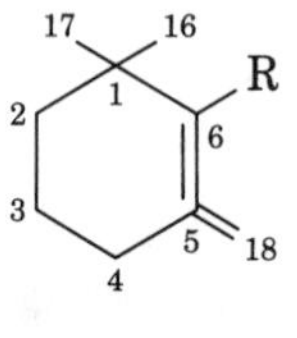

methylene cyclohexane
γ (gamma)

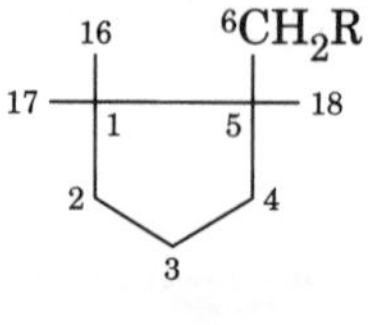

cyclopentane
κ (kappa)

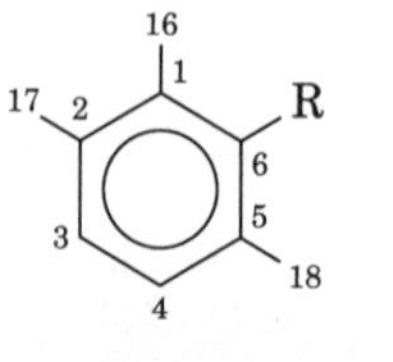

aryl
ϕ (phi)

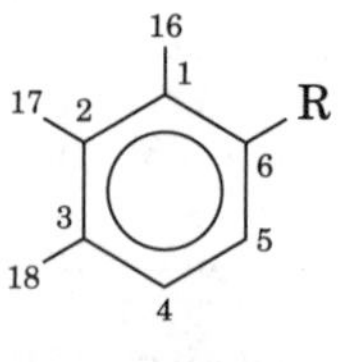

aryl
χ (chi)

3-hydroxy-3-oxo-β, ϵ-carotin-16-oic acid

bons, retro-double bond structures, designated chiral centers, and the use of the primes. This area is covered in more detail elsewhere (13,14).

BIOSYNTHESIS

In 1950 a biosynthetic pathway for the conversion of phytoene to lycopene to β-carotene was proposed (15). While the structures of several of these compounds were not proven at that time, it nevertheless was the first sequential pathway proposal. The cyclization of lycopene to β-carotene was shortly challenged because of a problem in tomatoes ripened at higher temperatures. Above approximately 30°C the red color (lycopene) is inhibited but the cyclic pigments are not. It was argued that, if lycopene is cyclized to β-carotene, both should be inhibited by ripening at higher temperatures (16).

The 1950s and 1960s were intense years in the polyene biosynthetic field because steroid and carotenoid biogenesis followed similar pathways. Thus when mevalonic acid (MVA) was discovered it was a breakthrough for both groups. The use of stereospecific [^{14}C]- and [^{3}H]mevalonic acids incorporated into cell-free systems provided a valuable tool for studying biosynthetic conversions. Acetyl-CoA may be considered the starting point in terpene biosynthesis. As can be seen in Figure 1, two acetyl-CoA molecules condense to form acetoacetyl-CoA. This in turn condenses to form the branched 6-carbon acid β-hydroxy-β-methylglutaryl CoA (HMG-CoA). Through a series of reactions, HMG CoA can also be formed from leucine. HMG CoA is reduced from the dicarboxylic acid to a monoacidic–monohydroxyl compound, MVA. For most systems, MVA once formed cannot be converted back to HMG CoA. MVA in the presence of ATP is converted to MVAP and then to MVAPP. With a third ATP, MVAPP is converted to isopentenyl pyrophosphate (IPP). This step involves a phosphorylation and a decarboxylation. IPP is the key intermediate to the terpenoids, which include rubber, sterols, bile acids, squalene, some sex hormones, ubiquinones, essential oils, the phytol side chain of chlorophyll, and vitamins E and K. It is isomerized to form dimethylallylpyrophosphate (DMAPP). IPP and DMAPP condense to form the C_{10} compound geranyl PP (GPP). The 10-carbon unit is further reacted with IPP to form the C_{15} farnesyl (FPP) and the C_{20} geranylgeranyl (GGPP). FPP can be dimerized to form squalene and the C_{20} GGPP, to form phytoene. The formation of phytoene follows a rather complicated pathway. By analogy to squalene, lycopersene with a saturated 15–15' position was proposed as the first carotenoid and the precursor to phytoene. This became a point of controversy, however. More recent studies have shown that the unlikely reduction to form lycoper-

sene from GGPP and the oxidation to form phytoene from lycopersene does not occur (16–19); see Figure 1. Plants are able to form phytoene and the higher carotenoids but animals can only make changes to preformed carotenoids.

Figure 2 shows the structures in the pathway from phytoene to the oxygenated plant carotenoids. It can be seen that there is a branch of neurosporene leading to either lycopene or α- or β-zeacarotenes. The data on the cyclization of neurosporene or lycopene seem to favor lycopene as the major pathway. In the red tomato, synthesis of β-zeacarotene rules out lycopene as the sole precursor to β-carotene because it is difficult to explain the higher temperature–selective inhibition of lycopene. When tomatoes were treated with DMSO at various stages of maturity (20,21), it was found that the acyclic but not the cyclic carotenoids were inhibited. It was proposed that parallel pathways exist for phytoene to lycopene and phytoene to β-carotene. By the use of a mutant tomato it was found that the DMSO inhibited the formation of lycopene at one stage and both pathways at an earlier stage of maturity. Also, the β-carotene is found in the chloroplast, where as the lycopene is found crystallized in the chromoplasts.

Research has been reported on the use of cell-free homogenates. However, the interpretation of the results of these cell-free systems was difficult because the cell was no longer intact. A number of inhibitors were used. These included mainly nitrogen-containing compounds. One compound, 2-(4-chlorophenyl)triethylamine (CPTA), inhibits the cyclization of lycopene and thus lycopene accumulates. When washing out the CPTA, β-carotene is formed at the expense of lycopene. A number of other inhibitors cause an increase in β-zeacarotene and a decrease in β-carotene. Still other compounds have been found to stimulate β-carotene and β-zeacarotene synthesis. Clearly both pathways are operating and it is possible to "prove" a pathway by the choice of inhibitor, stimulant, or bioregulator (22).

A great deal of information on the biosyntheses of the carotenoids has come from the use of ^{14}C/^{3}H dual-labeled MVA. By use of MVA with ^{3}H in the two and four positions and ^{14}C in the two position, it was found that the ϵ- and β-rings are formed independently. Figure 3 shows the ^{3}H as expected in the rings. Eight carbons and a number of ^{3}H would be found in the entire molecule. Two ^{3}H atoms are found in position four of the β-ring and one in α-carotene. However, if the ϵ-ring is converted to the β-ring, one ^{3}H would be lost. This is not found experimentally. Likewise, the 4-^{3}H from MVA is in the six position: ^{3}H must be lost in the formation of β-carotene but not in the direct formation of α-carotene. Because the 6'-^{3}H is retained in α-carotene, the β-ring is not converted to the ϵ-ring (23). The nonrandomized dimethyl groups in toru-

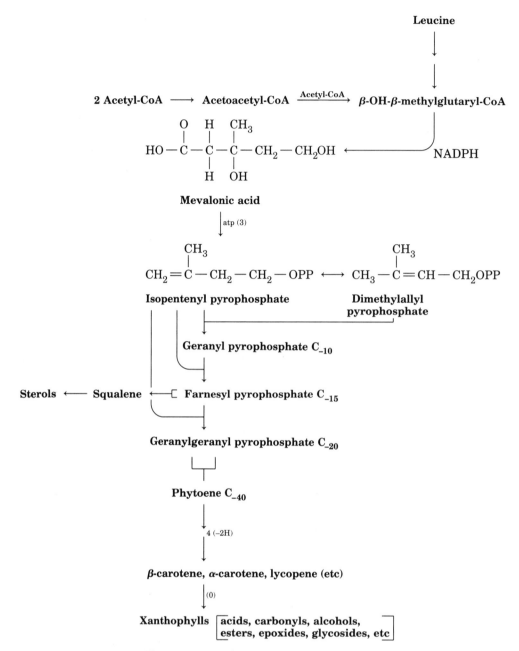

Figure 1. Biosynthetic formation of carotenoids.

larhodin were proved with double-labeled MVA. This result suggested that the interconversion of IPP and DMAPP did not result in the randomization of the C-2 hydrogen from MVA (24).

The ring formation in the red pepper to form capsanthin and capsorubin has been suggested to follow a pathway through a postulated epoxide intermediate (19). It is generally thought that the insertion of oxygen occurs at a later stage in the biosynthetic pathway, and that the epoxide, hydroxyl, and one-carboxyl oxygen came from molecular oxygen (18).

The view had long been held that the direction of biosynthesis is from saturated to unsaturated: from hydrocarbon to the xanthyphylls. Recently, however, it has been shown that some fish can convert astaxanthin to retinol. Bacteria have also been shown to form a series of 45- and 50-carbon carotenoids.

OCCURRENCE IN NATURE

As indicated in Table 1, the carotenoids are widely distributed in foods and are the most widespread group of pigments in nature. They are present without exception in photosynthetic tissue, occur with no definite pattern in nonphotosynthetic tissues such as roots, flower petals, seeds, fruits, and vegetables; and are found sporadically in the Protista (fungi–yeast, and bacteria). The red, yellow, and orange pigments in the skin, flesh, and shell or exoskeleton of some animal species are due to these pigments.

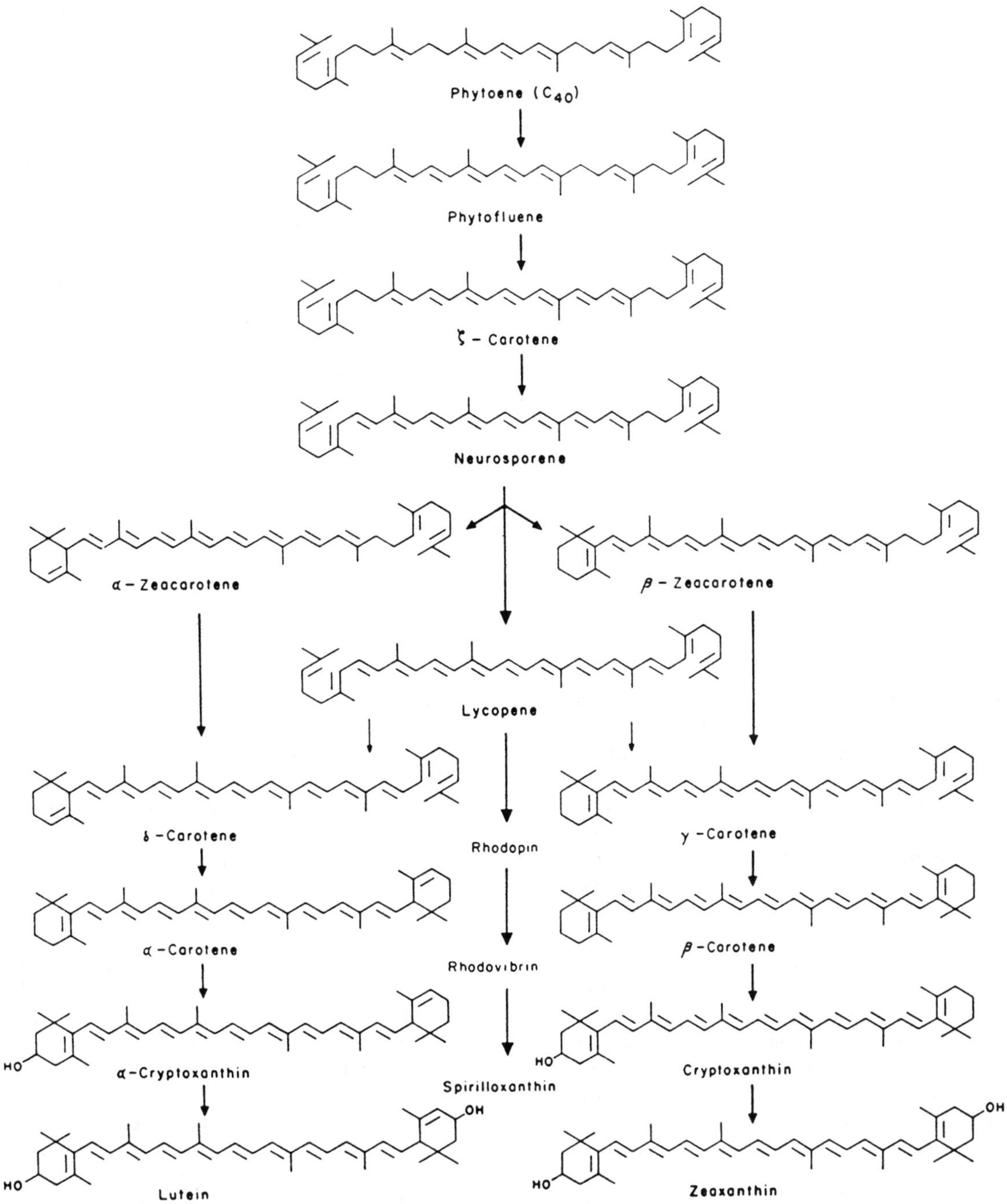

Figure 2. Biosynthetic pathways of some carotenoids.

These would include the lobster, shrimp, salmon, goldfish, and flamingo. People who consume large amounts of carrot or tomato juice may turn orange or red because of an excess intake of these pigments. The carotenoids, which cannot be made by animals, are the precursors to vitamin A but, unlike vitamin A, are not toxic in large doses. Major source's of information on carotenoid occurrences are references 19, 25, 26, and 27.

Generally the concentration of the carotenoids in various foods is low but can vary widely. The cornea of the pheasant's eye narcissus contains 18% β-carotene (dry weight), whereas the concentration in foods is much lower (2). It has been estimated that the annual production of

carotenoids is on the order of 10^8 tons. The major pigment is fucoxanthin, the pigment in marine brown algae. Other major pigments include those found in green leaves, mainly lutein, violaxanthin, and neoxanthin. By contrast, β-carotene occurs widely but in smaller amounts. It is of interest that 3,3'-dihydroxy-α-carotene (lutein) is a major pigment whereas 3,3'-dihydroxy-β-carotene (zeaxanthin) is a minor pigment in nature. The major hydrocarbon carotenoid found in nature is β-carotene, whereas α-carotene is a minor pigment. A typical percentage profile of the xanthophylls of a green plant could be lutein, 40%; violaxanthin, 34%; neoxanthin, 19%; cryptoxanthin, 4%; and zeaxanthin, 2%. Beta-carotene, lycopene, and capsan-

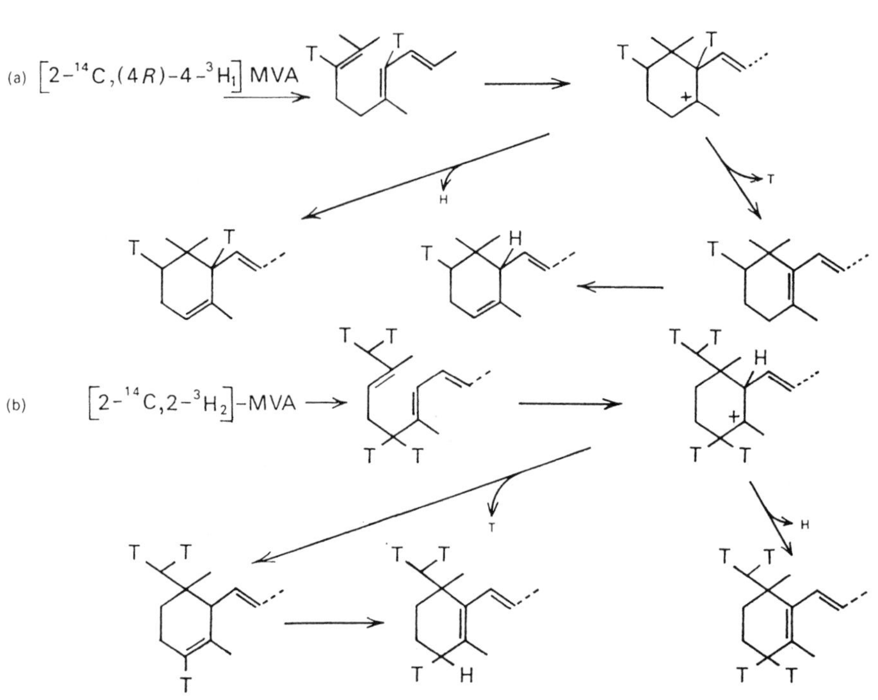

Figure 3. Demonstration of the independent formation of β and ε-rings by the use of specifically labeled species of MVA.

thin can be high in some tissues such as the sweet potato, tomato, and red pepper. By contrast, the carrot is a major source of α-carotene where the concentration can be 30–40%.

The level of carotenoids is often directly related to the quality of food. The lack of pigment, as in white butter or salmon, for example, is a quality defect (28). However, flour is bleached to destroy the carotenoids, and β-carotene is often removed from red palm oil to make it colorless.

The acyclic pigments phytoene, phytofluene, ζ-carotene, neurosporene, and lycopene are often found in carotenoid-producing systems such as higher and lower plants. They are not often found in animal tissue. Of these, lycopene is the most common and may occur in large amounts in the tomato, watermelon, orange, pink grapefruit, and some apricots. Further reactions in products of lycopene lead to the closing of one or both rings. However, in some oranges (Valencia) 3,4-didehydrolycopene is found. There have been a number of acyclic xanthophylls isolated, mainly from the purple photosynthetic bacteria. The 1,2-epoxide has been isolated from tomatoes as a minor pigment. Because phytofluene is fluorescent it must be accounted for in the fluorometric analysis of vitamin A (29).

The alicyclic carotenoids (eg, β-carotene) are common in higher plants, bacteria, fungi, algae, and animals. It has been estimated, on the basis of structure, that some 50–60 alicyclic compounds have potential vitamin A activity (30). Apricots are an excellent source of β-carotene (60%), whereas the level in tomatoes (10%) and peaches (10%) is much lower. Papaya, if it is yellow–orange, is a good source, whereas the red flesh-papaya contains mainly lycopene. When the flesh is green, some plastid pigments are present and thus some β-carotene. Red grapes, pears, figs, red apples, and beet root, although brightly colored, are not good sources of β-carotene. Delta-

carotene has been isolated from certain tomato varieties when apo-poly-*cis*-isomers of β-, α-, and ε-carotene and lycopene have been found.

Vegetables may be excellent sources of the provitamin A carotenoids. Spinach may have up to 4.0 mg/100 g β-carotene and carrots may have as much as 6.0 mg/100 g of β-carotene. Yellow corn (maize) contains small amounts of β-carotene and β-zeacarotene. Sweet potatoes may be purple, white, or bright orange. The two former kinds are low in carotenoids and the latter may have more than 9 mg/100 g. Squash, broccoli, peas, and pumpkin contain relatively large amounts of carotenes. Most vegetable oils are slightly yellow, indicating a small amount of carotenoids. The red palm oil produced in Brazil, Southeast Asia, and parts of Africa may contain large amounts of β- and α-carotene (50 mg/100 g total carotenoids). Egg yolks are a rich source of carotenoids (lutein and zeaxanthin) but relatively poor sources of β-carotene. Green peppers are a better source of carotene than are the red peppers. Wheat flour is low in carotene and this amount is further lowered in the bleaching process.

The hydroxylated carotenoids are very common in plants, where lutein and zeaxanthin and, to a lesser extent, β-cryptoxanthin, are major pigments. Lutein is a common pigment in freshwater fish, whereas tunaxanthin (3,3'-dihydroxy-ε,ε'-carotene) is common in marine fish. Peaches are a good source of β-cryptoxanthin. Many fruits contain xanthophyll esters. In the persimmon the carotenoid alcohol is β-cryptoxanthin. Corn varieties are selected for their high content of zeaxanthin and lutein. These pigments are fed to poultry to color the skin and egg yolks.

The xanthophylls with hydroxyl or keto substitution in the four position are common in various animal tissues. Echinenone (4-keto-β-carotene) is widely found in marine invertebrates and algae, and canthaxanthin (4,4'-keto-β-

carotene) is found in edible mushrooms, blue–green algae, trout, brine shrimp, and flamingos, for example.

Astaxanthin is the pigment of the salmon and the invertebrates Annelida and Crustacea. Recently it has been isolated from red yeast and Adonis flowers (31). Shrimp raised in intensive culture are often blue because of a protein complex with astaxanthin. This complex extends the chromophore resulting in a blue color. This has been mistakenly described as a disease. The condition, however, results in a lower price for the shrimp, particularly on the Japanese market. When cooked, the complex is broken down and the red color is seen. Lobsters can be green, black, red, or blue because of various protein–astaxanthin complexes, all of which are broken down by heat, yielding a red pigment. When treated with alkali, astaxanthin is converted to astacene.

Aromatic carotenoids have recently been isolated from photosynthetic bacteria and some sea sponges. Carotenoid epoxides are fairly common in nature. The so-called violaxanthin cycle is an example of the natural occurrence of epoxidation.

$$\text{zeaxanthin} \rightleftharpoons \text{antheraxanthin} \rightleftharpoons \text{violaxanthin}$$

The usual epoxides are in the five, six position and these are easily converted to the 5,8-epoxide with acid. Where a keto group is in the four position, in-chain epoxides are formed (eg, 9,10-epoxides of canthaxanthin). The 1,2-epoxides of acylic pigments have been reported in tomato leaves. The allenic seaweed pigment fucoxanthin is a natural epoxide, as is the plant allenic epoxide neoxanthin. The cyclopentyl ketones are found in red peppers and are probably formed by a pinacolic rearrangement of the epoxides of zeaxanthin.

The acetylenic carotenoids are found in diatoms, *Euglena,* giant scallops, mussels, and starfish. A number of apo-carotenoids are formed in some fruit, bacteria, orange peel, and fungi. The most well-known apo compound is bixin, the pigment in annatto seeds. Crocin is an apo compound occurring in saffron. In a sense, retinol can be considered a degraded β-carotene. The carotenoid acid torularhodin is a common pigment in Rhodotorula yeast.

EFFECT OF AGRICULTURAL PRACTICES

A number of agricultural practices, particularly those with plants, directly affect the carotenoid content. Thus the nutrient content of edible plants can vary severalfold, up to 20 times (32). The major changes are caused by genetic manipulation. However, a number of factors such as sunlight, rainfall, topography, soils, location, season, fertilization of soils, and maturity may effect carotene content. Thus the nutritional tables give an average value with an understood large variation. A number of countries have developed nutritional tables that, when compared, show a great variation for the same fruit or vegetable (33). In the past, genetic manipulation has been concerned with the color, flavor, texture, and yield. Today greater effort is being made to produce foods that are nutritionally superior. Thus interest is evident in developing not only the color but also the quality of the color (eg, more β-carotene). No agricultural practice has been as effective as

genetic manipulation. High β-carotene tomatoes have been developed, but because of the yellow color were not popular. Sweet potatoes can vary from 0 to 7 mg/100 g in carotene and the carotene from carrots can be increased fivefold. Generally, light has a positive effect on carotene synthesis. Fruit grown in the dark, however, will develop some color (34). Temperature has an effect but this depends on the product. The most pronounced effect was on the ripening of the tomato. In temperatures below 27°C the synthesis of lycopene is favored, whereas a yellow tomato results at a higher temperature. Gamma irradiation also affects the formation of carotene (35).

Most vegetables and fruits increase in carotene during the ripening stage. The carotene content of the carrot increases until a certain maturity level is reached. The β-carotene level falls off with further growth (36). In the case of the red pepper, the conversion from green to red lowers the carotene content. Because of cultural practices and shipping constraints, fruits may be picked green and thus the full color may not develop.

ANALYSIS

A universal extraction and analysis scheme for the carotenoids is not possible because of the wide variation in carotenoid structures. Thus there are numerous methods suggested for the separation of carotenoids from such things as foodstuffs and blood plasma. The classic methods used by most investigators rely on some method of extraction and a chromatographic separation of the carotenoids (37,38). The methods require some sort of size reduction to increase surface area, followed by extraction with a suitable solvent or solvent mixture. Often a bipolar solvent is used to extract nonpolar carotenoids from polar tissue. Ethyl alcohol and acetone are examples of bipolar solvents, and $CHCl_3$–MeOH is an example of a bipolar solvent mixture. The extraction may be followed by saponification, removal of alkali, transfer to a developing solvent, and chromatography under vacuum or pressure. The column may be developed with a gradient of solvents and the pigments separated by elution or cut from the column and extracted. Often rechromatography is required for some bands, followed by verifying chemical reactions. The quantity of each purified fraction is determined by spectroscopy. The method requires several hours, and in some cases artifacts (cis–trans isomers, epoxides) are formed due to the long exposure to solvents, absorbents, light, and oxygen. Although the method does give a separation of provitamin A precursors from inactive carotenoids, it is time consuming and requires skilled technical operators. Thus the method may not be suitable for analysis of a large number of food items for a table of food compositions.

A much simpler method has been developed that chromatographically separates the carotenes from the oxygenated compounds, but it may not separate individual carotenes, their cis isomers, or carotenoid esters (39). Therefore, this method tends to overestimate provitamin A activity, especially if β-carotene represents only a small part of the total. High-performance liquid chromatography (hplc) is clearly the method of choice and the more recent values in the nutritional tables are generally determined by this method. A number of reports have appeared that show the general overestimation of this Association of Official Analytical Chemists (AOAC) method. A modified AOAC method and chromatographic procedures that separated the individual components were used to estimate the provitamin A content of Clingstone peaches (40). The percentage of USRDA for 100 g was much higher by the AOAC method than by the complete separation (60% vs 11% of the RDA for raw, 12% vs 6% for canned).

A comparison of the AOAC method with a stepwise gradient, a saponification step, and hplc has been published (41). The AOAC method and hplc were comparable for green vegetables. For the carrot and pumpkin, the AOAC method was higher for both vegetables than the gradient elution and hplc method. Where fruit xanthophylls were esterified, these compounds were chromatographed with β-carotene and thus the estimation was higher in the AOAC method when the fruit extract was not saponified.

EFFECT OF FOOD PROCESSING

The carotenoids are a group of compounds derived from phytoene that may vary considerably in regard to functional groups, solubility, color, and stability. The carotenoids are altered or partly destroyed by acids, are usually but not always stable in inorganic bases, are destroyed by some enzymes (eg, lipoxygenase), and are usually bleached by light. Most carotenoids are fat soluble and thus are not subject to leaching losses. They are generally fairly stable to the heat involved in the canning operation but are rapidly lost on dehydration because of oxidation (42).

Although the most stable form of the carotenoids is the all-trans form, the cis geometrical isomer can occur on heating, especially in the presence of H⁺. The cis isomer carotenoids are formed naturally in the tangerine tomato. It has been shown that, if β-carotene activity was set at 100%, the neo-β-carotene vitamin A activity was 38%, and the neo-α-carotene activity was 13% (43). The transformation from trans to cis results in a loss of extinction and a shift in absorption maxima to shorter (2–5 mm) wavelengths than the all-trans carotenoid. Cooking causes the formation of cis isomers from all-*trans*-β-carotene (44). Broccoli lost 13% of the all-trans isomer, spinach lost 7%, and the sweet potato lost 32%. A 15% loss of vitamin A was reported, due mainly to a cis–trans isomerization of α- and β-carotene during canning (45). High-temperature–short-time (HTST) processing was shown to have less effect on the carotene in the sweet potato than conventional cooking (46). Heat processing produced enough acid in pineapple to cause the formation of cis isomers (47). This could result in a lighter yellow color.

The formation of 5,6- and 5,8-epoxides of β-carotene in model systems has been studied (48). Diphenylamine, a free-radical inhibitor stopped the loss of β-carotene. It is generally thought that epoxide formation is the initial step in the degradation of the carotenoid. Several reports have been published that show the complexity of the reactions. Stored hydrogenated oils and enzymatic-coupled oxidation of carotenoids result in the formation of epoxides and other products.

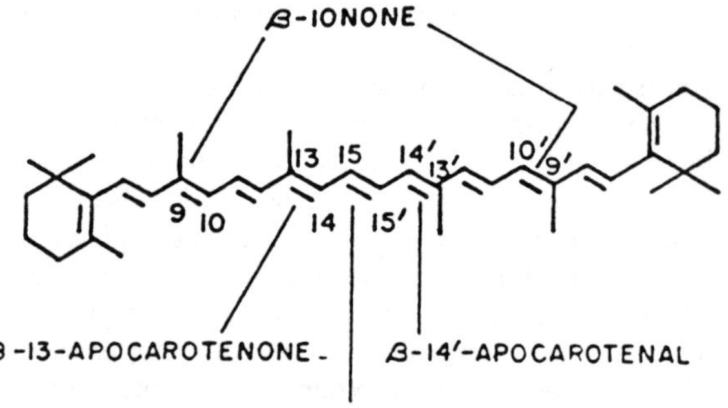

Figure 4. Relationship between the attacking positions by oxygen formation of three carbonyl compounds (51).

Dehydration and size reduction of fruits and vegetables increase surface area and, if products are not protected from light and oxygen, result in poor stability of the carotenoid pigments. The shelf life of the carotenoids in the dry state is shorter than in the wet state (49). Different dehydration methods result in varying losses of carotenes (50). Retention of carotene in carrots varies: fresh (100%) > blanched (95%) > freeze-dried (89%) > air-dried (80%) > and puff-dried (12%). Carotenoid oxidation in foods is generally associated with unsaturated fatty acids and is usually autocatalytic. The oxidation may or may not be enzymatic in nature and is directly related to the available water (A_w), oxygen, heat, and certain metals. Generally, antioxidants are effective in slowing the reaction. The decomposition products from a simulated deodorization of red palm oil were isolated (51). β-15'- and β-14'-apo-carotenals and 13-apo-carotenone were isolated from the reaction mixture (Fig. 4) and a pathway for the reaction was suggested (Fig. 5).

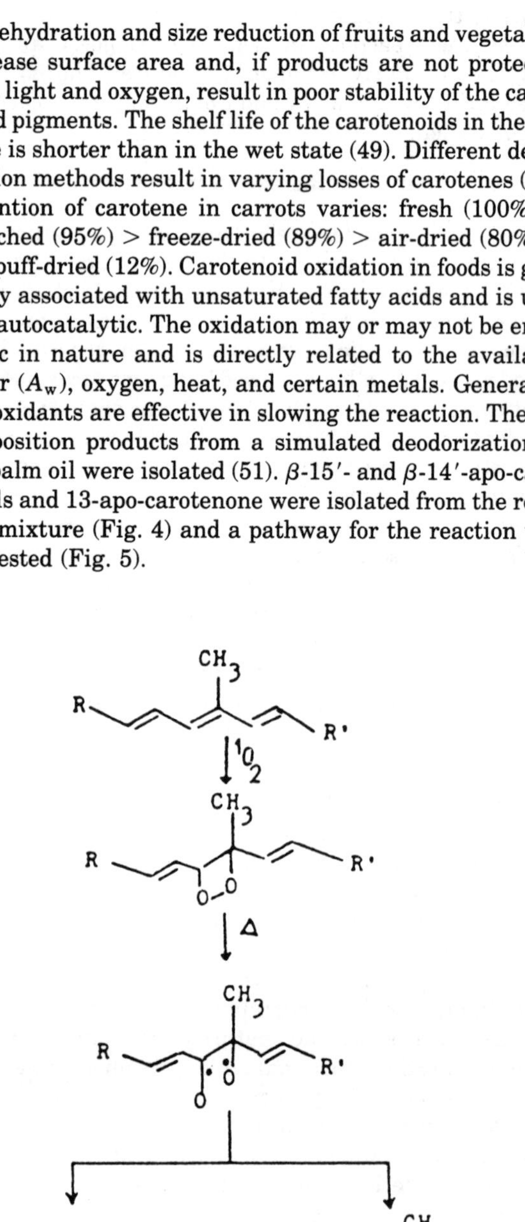

Figure 5. Dioxetane mechanism for the oxidation of β-carotene (51).

A number of studies have reported an apparent increase in the level of carotenes in various canned products over the fresh product. A review of the problem (52) found two reasons for this surprising result. Where soluble solids are leached out of the products during cooking and storage, the fat-soluble carotenes appear to increase in amount. It has also been observed that the fresh control may have an active lipoxygenaselike enzyme that bleaches the carotenoids. When antioxidants were added to the fresh sample before extraction, the vitamin A value in the fresh product increased.

CONVERSION OF CAROTENOIDS TO VITAMIN A

The carotenoids of fruits, vegetables, and animal products are generally fat soluble and are associated with the lipid fractions. In some cases they may be in the crystalline state, as lycopene is in the tomato. During the digestion process, esterases, lipases, and proteases act on the carotenoids, which are solubilized by the bile salts. Within the mucosal cell β-carotene and the other provitamin A carotenoids, are split by carotenoid dioxygenase(s), resulting in an oxidative cleavage. In theory a molecule of β-carotene should yield two molecules of retinal. Retinol is formed by reduction of retinal. It is esterified and transported to the liver for storage. There is some difference of opinion as to whether there is a central cleavage enzyme only or whether there is also a random-splitting enzyme as well, which acts on various places on the chain. The carotenes may also be absorbed, carried by low-density lipoproteins and deposited in the skin, fat, and various organs. Different kinetic profiles were found for β-carotene fed to rats (53). The extent of accumulation, rate of accumulation, time to reach saturation, and turnover time were different for each tissue. Liver, adrenal, and ovary have high concentrations of β-carotene.

Animals differ in the efficiency of the splitting reaction. The rat is generally regarded as very efficient and the mink and the fox relatively inefficient in this reaction. Man is average with a diet : absorption ratio of about 6 : 1. The cat does not have the enzymes necessary to utilize provitamin A carotenes.

Table 5 lists some common carotenoids that are vitamin A precursors. Generally, for a compound to be active it must have an unsaturated central chain and an unsubstituted β-ring. On the basis of structure, of the more than 600 carotenoids listed, the provitamin A compounds would number between 50 and 60.

Alpha-carotene on the above basis would be half as active as β-carotene. The metabolism of α-carotene has been shown to result in liver storage of retinol and α-vitamin A. The latter compound does not combine with retinol-binding protein (RBP). It can induce hypervitaminosis A because it is not delivered to the tissues. There is some confusion in regard to the capacity of the diepoxide to serve as a vitamin A source. The compound is listed as active *in vivo* but not active *in vitro*. Recently, radioactive β-carotene-5,6,5',6'-diepoxide and luteochrome (5,6,5',8'-β-carotene-diepoxide) were fed to rats and [14C]retinol was not detected in the liver (55). The absorption or conversion to retinal is enhanced by bile salts, proteins, lipid, and zinc. It is not clear what effect various

Table 5. Some Carotenoids with Provitamin A Activity

Carotenoid	Activity (%)
β-Carotene	100[a]
Neo-β-carotene-U	38[b]
α-Carotene	50–54[a]
Neo-α-carotene	13[b]
3,4-Dehydro-β-carotene	75[a]
3,4,3,4-Bisdehydro-β-carotene	38[a]
α-Carotene	42–50[a]
7,8-Dihydro-β-carotene (β-zeacarotene)	20–40[a]
β-Carotene-5',6-epoxide	21[a]
α-Carotene-5,6-epoxide	25[a]
β-Carotene-5,6,5',6-diepoxide	Inactive[c]
3-Keto-β-carotene	52[a]
3-Hydroxy-β-carotene	50–60[a]
4-Hydroxy-β-carotene	48[a]
β-Apo-2'-carotenal	Active[a]
β-Apo-8'-carotenal	72[a]
β-Apo-10'-carotenal	Active[a]
β-Apo-12'-carotenal	120[a]
Lycopene	Inactive[a]
Lutein	Inactive[a]
Astaxanthin	Active[d]

[a] Ref. 54.
[b] Ref. 43.
[c] Ref. 55.
[d] Ref. 61.

nonprovitamin A compounds have on the bioavailability of the provitamin A compounds (56). Lycopene is reported to enhance carotene uptake. It was found that nonprovitamin A compounds may decrease the utilization of provitamin A pigments, whereas β-cryptoxanthin and α-carotene promote utilization (57).

The conversion factors for β-carotene and other provitamin A compounds would be as follows for retinol equivalent (RE) and international units (IU):

$$RE = \mu g \text{ retinol} + \frac{\mu g \ \beta\text{-carotene}}{6}$$

$$+ \frac{\mu g \text{ other provitamin A compounds}}{12}$$

$$1 \text{ IU} = 0.3 \ \mu g \text{ retinol}$$
$$= 0.6 \ \mu g \ \beta\text{-carotene}$$
$$= 1.2 \ \mu g \text{ other provitamin A carotenoids}$$

CAROTENOIDS AS FOOD COLORS

It was the color of the carrot that first interested research workers 160 years ago in studying carotenoids. Many years later it was this same color that was the visual means of identification of the compounds separated by chromatography. It is the same today as it has always been, color is one of the means by which the quality of food is identified; people often reject the unfamiliar. Most pigment changes such as the changes in chlorophyll on thermal processing, the bleaching of anthocyanins, the browning of fresh meat, or the off-coloring of flavonoids do not in themselves affect the product. In the case of the destruction of carotene, the loss or change in color would result in

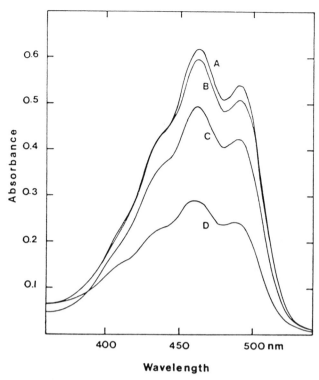

Figure 6. Irradiation of β-carotene with and without antioxidants **(A)** β-Carotene in toluene; **(B)** β-Carotene with 0.1% BHA irradiated for 30 min. with 275-W sunlamp 8–12 in. away; **(C)** same with 0.1% BHT; and **(D)** same without antioxidant.

poorer quality, but also might result in loss of vitamin A activity. Beta-carotene has 11 double bonds in conjugation, which is the reason why this compound is colored yellow–orange. Any disruption in the chromophore, such as that depicted in Figure 5 would result in bleaching. This subject was covered earlier, in the section "Effect of Food Processing (Fig. 6)."

Because of the uncertainty about some of the artificial pigments and because of the delisting of several, (eg, Red No. 2 in the United States, Red No. 40 in the UK) interest has been shown in using carotenoids as food colorants.

We can consider the use of carotenoids as following a direct or an indirect route. As a case of the former, margarine might be colored with β-carotene. Examples of the latter might be the coloring of poultry skin, or eggs, or salmon flesh with the appropriate pigment. Some confusion exists regarding natural and artificial pigments. It can be seen in Figure 7 that the infrared fingerprints for synthetic and natural β-carotene are identical. Many people prefer the term nature-identical to describe the synthetic carotenes. What follows is a brief treatment of the subject of natural coloring matters. Fuller treatments have been published (58,59).

One of the most common natural coloring preparations is obtained by extracting the seeds of the *Bixa orellana* tree, which grows in the tropics. The apo compound bixin is the main component of the oil-soluble annatto preparations, and norbixin is the 6,6′-diapo-dicarboxylic acid (bixin has one methyl ester). Bixin is most stable at pH 8 and shows diminished stability in the 4–8 pH range. In general, annatto extracts have a good shelf life. Annatto oil applications include the coloring of butter, bakery products, and salad oil. Combined with paprika, annatto

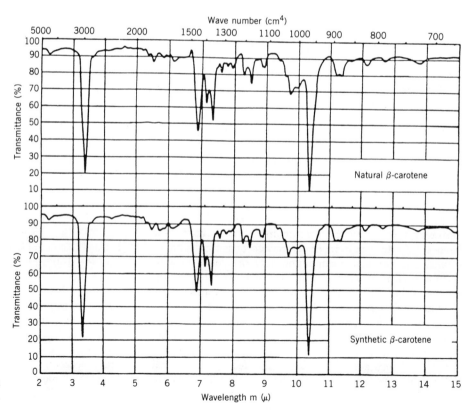

Figure 7. Infrared absorption spectra of natural and synthetic β-carotene (3).

oils are used to color cheese. Water-soluble preparations may be used in ice cream (56).

Saffron consists of the dried stigmas of *Crocus sativus*. It contains crocin, the digentiobioside of crocetin. The preparation is yellow. Saffron is also used as a spice and is widely accepted in such foods as soups, meat products, and cheese. Formerly it was used in cakes and other bakery products.

Tomato extracts are red because of the pigment lycopene. The color of the product will be red to orange, depending on the medium. Lycopene is not stable, and its use is limited.

Carrot extracts, carrot oil, and red palm oil contain large amounts of α- and β-carotene. These are used in fat-based products. They have the advantage of being natural, but have limited use because of the lower price of synthetic β-carotene. Recently, hypersaline algae have been cultured for their β-carotene content.

Oleoresin paprika is an oil extract of paprika. The main carotenoids are capsanthin and capsorubin. The color of products colored by paprika varies from red to pinkish yellow. Oleoresin paprika can be used in salad dressings, sauces, meat products, and processed cheeses. The indirect coloring of trout with oleoresin paprika resulted in a yellow rather than a red hue of the flesh.

The synthetic pigments have a number of advantages of the natural product extracts. While the principal advantage is price, it should not be overlooked that wide variation in quality is not a problem with the synthetic FDC dyes. The major synthetic carotenoids that are marketed include β-carotene, β-apo-8'-carotenal, and canthaxanthin. All occur naturally and only canthaxanthin does not have vitamin A activity.

Beta-carotene is widely encountered in nature. In foods it has been used in butter, margarine, salad oil, popcorn, baked goods, confections and candy, eggnog, coffee whiteners, juices, and soups, to name a few. Beta-apo-8'-carotenal has been isolated from oranges and several natural sources. However, the main source is the chemical synthesis. It is also yellow and its uses are similar to those of β-carotene.

Canthaxanin (rosanthin) (4,4'-diketo-β-carotene) is used where a red color is desired. It was first isolated from the mushroom, but it is also a widely distributed animal pigment. Many crustaceans, especially brine shrimp, and birds, notably the flamingo, contain canthaxanthin. One of its main uses is as a pigmenting agent for salmon. While a red color is obtained, it has been observed that the pigment tends to cook out. Recently canthaxanthin has been used in simulated meats, shrimp, crab, and lobster products.

Synthetic astaxanthin is now available and is approved for use in Europe. To date, it is not approved in the United States. This is the major pigment in a number of fish and shellfish species. The color tends to be more stable than canthaxanthin. When shrimp are raised in intensive culture the blue carotenoprotein is often found. This can be corrected by feeding more green algae or astaxanthin. Marigold flowers contain mainly lutein and are fed to chickens to supply yellow–orange pigments for the skin and egg yolks.

The crystalline carotenoids are not absorbed or metabolized by animals. This problem has been solved by the development of special water-soluble forms marketed to satisfy various product needs. Two approaches have been primarily employed—the production of oil suspensions of micropulverized crystals and the development of emulsions or beadlet forms containing the carotenoid in supersaturated solution or finely colloidal forms in liquid or dry products. These are summarized in Figure 8.

CAROTENOIDS BY CHEMICAL SYNTHESIS

After the elucidation of the structure of β-carotene, lycopene, zeaxanthin, and vitamin A in the early 1930s, much effort was made toward the total synthesis of a number of carotenoids, especially those of economic significance. The

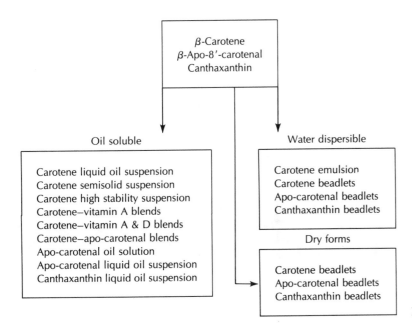

Figure 8. Some carotenoid market forms (3).

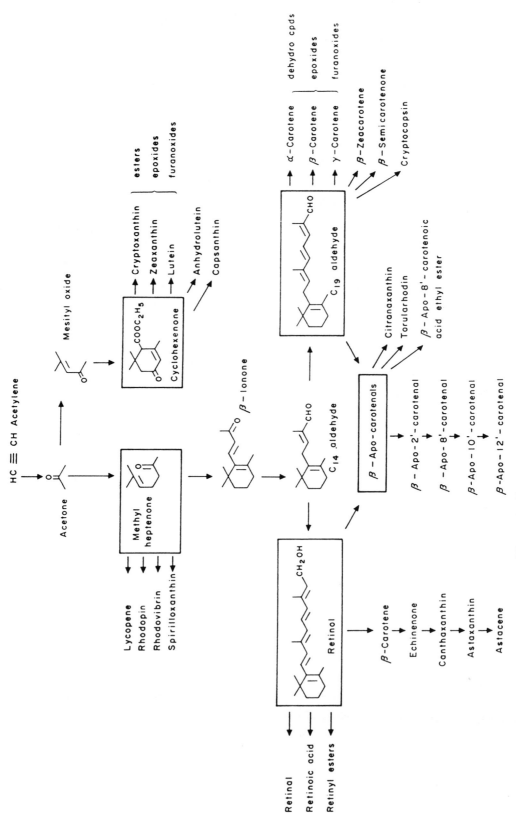

Figure 9. Chemical schemes of carotenoid syntheses (3).

total synthesis of β-carotene was reported in 1950 (60). Since that date, β-carotene, retinol, *cis*-retinoic acid, canthaxanthin, β-apo-8'-carotenal, and, more recently, astaxanthin have been produced by commercial chemical synthesis. A great many other carotenoids have been synthesized but currently they have no commercial significance.

The Roche synthesis of β-carotene starts from β-ionone and is related to the vitamin A synthesis. The β-ionone may be prepared from citral, a constituent of lemongrass oil, or prepared synthetically from acetylene as a starting material. A systematic synthesis of β-carotene and other carotenoids is shown in Figure 9. In the β-carotene procedure, β-ionone is converted to a C-14 aldehyde, which, through further chain-lengthening steps, is converted to C-19 aldehydes. These are joined through a Grignard reaction to form the C-40 diol, which, through allylic rearrangement and dehydration, is converted to 15,15'-dehydro-β-carotene. A partial hydrogenation and rearrangement yields *trans*-β-carotene.

Crystalline carotenoids produced by chemical synthesis are of high purity and uniform color. The majority of carotenoids, when pure, can be obtained in crystalline form. In general, the melting points of crystalline carotenoids are fairly high and quite sharp for the pure compounds. The synthesis of the carotenoid series of compounds employs a variety of organic synthetic techniques.

The literature on the synthesis of the various carotenoids is extensive and the details are certainly beyond this chapter. Reference 25 includes sources for information about synthesis as well as the uses of various methods of analysis.

FUNCTION OF CAROTENOIDS

It is tempting to propose a single universal function for the carotenoids because they are found in such diverse tissues. Failing this, it is tempting to ascribe some function wherever a carotenoid is found. While the function of carotenoids has been proven in some cases, their function, if any, remains to be determined in a large number of organisms. Where the function has been proven, it generally has to do with some aspect of the light-absorbing property of the carotenoid pigment. The critical role of these pigments in photosynthesis has been the best documented. Phototropism and phototaxis responses of plants seem to be related to the light absorption spectra of the carotenoids. Functions in animals are related to the antioxidant and the light-absorbing properties of these pigments. The function in vitamin A formation has been discussed. Because these pigments are found in many reproductive tissues, it has been suggested that they have a role in reproduction. The results in this area are less convincing than those in other areas.

In the photosynthetic process the carotenoids have been shown to be active in the light-gathering process (eg, they absorb light at wavelengths not absorbed by chlorophyll). In *Chlorella*, other green algae, and higher plants the light absorbed by the carotenoids was used at low efficiency. In the diatom and brown algae the energy transfer is comparable to chlorophyll where the main pigment is fucoxanthin. In the photosynthetic process two photo systems are involved: formation of ATP and the reduction process. Carotenes are found in photosystem I and xanthophylls in photosystem II. Beta-carotene rather than the xanthophylls appears to be involved in the energy transfer to the chlorophyll of photosystems I and II.

The role of the carotenoids in photoprotection of photosynthetic bacteria has been well documented. In some studies a mutant which only produced the colorless polyenes phytoene and phytofluene was compared to the wild type *Rhodopseudomonas sphaeroides*. The wild type showed good growth under all conditions of light and oxygen. However the mutant could only grow under anaerobic–light or aerobic–dark conditions.

Under light with O_2, massive killing resulted at room temperature and bacterial chlorophyll was destroyed. If the mutant was grown in the dark with air, chlorophyll disappeared. Where chlorophyll was missing, light and O_2 had no effect.

Diphenylamine (DPA) inhibits the formation of colored carotenoids. When the wild type blue–green bacteria is treated with DPA it is destroyed like the mutant. At low temperatures killing proceeds in the mutant but bacterial chlorophyll is not destroyed.

Corynebacterium poinsettiae is not photosynthetic, and both the wild type and the carotenoid-less mutant can tolerate high intensities of light in the presence of air. When the exogenous dye, toluidine blue, is added as a photosensitizing pigment, the carotenoid-less mutant is killed. The mechanism involves the simultaneous interaction of visible light, a photosensitizing dye, and O_2. Generally, the carotenoids are effective for visible light but have no effect in ultraviolet, gamma, or x radiation.

The primary reactions are listed below:

$$CHL + h_v \rightarrow {}^*CHL \quad \text{excited state}$$

$${}^*CHL \longrightarrow \text{photosynthesis}$$

$$\text{or}$$

$${}^3CHL \text{ triplet excited state intersystem crossing}$$

$${}^3CHL + {}^3O_2 \rightarrow {}^1CHL + {}^1O_2$$

$$\text{or}$$

$${}^3CHL + {}^1CAR \rightarrow {}^1CHL + {}^3CAR$$

$${}^3CAR \rightarrow {}^1CAR \quad \text{harmless decay}$$

$${}^1O_2 + CAR \rightarrow CAR\,O_2 \quad \text{CAR can be regenerated}$$

$$\text{or}$$

$${}^1O_2 + A \rightarrow A\,O_2 \quad \text{photodynamic action}$$

$${}^1O_2 + {}^1CAR \rightarrow {}^3O_2 + {}^3CAR$$

$${}^3CAR \rightarrow {}^1CAR \quad \text{harmless decay}$$

where CHL = chlorophyll; CAR = carotenoid; A = other compounds; and superscript *, 1, and 3 indicate excited, singlet, and triplet states.

As can be seen in these reactions, carotenoids may protect photosynthetic bacteria at various levels by quenching the singlet-excited state of O_2 or the triplet-excited state of chlorophyll. The ground states of oxygen and CHL would be the triplet state. The carotenoids may be the preferred substrate for oxidation or may act in quenching reactive species.

Phototropism is a response of higher plants and some fungi that results in the plant turning to the light. Rival claims for either β-carotene or riboflavin as the active compound have been made based on the action spectra. More recent reports seem to rule out β-carotene (19) as the mediator in the response.

Phototaxis is a response in which an organism such as *Euglena* can move toward the light. The light absorption spectra do not match the spectra of the carotenoids.

The involvement of the carotenoids in reproduction appear to be coincidental to the process of sexual reproduction and of no known significance to the process. The major pigment of the brine shrimp is canthaxanthin. However the female *Artemia* converts all-*trans*-canthaxanthin to *cis*-canthaxanthin during the time of sexual activity.

In animals the principal function of carotenoids is as a precursor to the formation of vitamin A. It is assumed that to have vitamin A activity, a molecule must have one-half of the structure similar to that of β-carotene. Recently it has been shown that astaxanthin can be converted to zeaxanthin in trout when the fish has sufficient vitamin A. Tritiated astaxanthin was converted to retinol in strips of duodenum or everted sacs of trout intestines (61). Astaxanthin, canthaxanthin, and zeaxanthin can be converted to vitamins A and A_2 in guppies (62).

We have become increasingly aware of a mounting body of evidence that suggests that the carotenoids can function in medical applications apart from their role as vitamin A precursors. The symptoms of erythropoietic porphyria can be relieved by large doses of β-carotene (63). This pigment was deposited in the skin of patients who lacked skin pigmentation.

Lately some carotenoids have been found to be effective in the treatment of cancer, particularly cancer of epithelial origin. Other carotenoids such as canthaxanthin and phytoene, with no vitamin A activity, were in some cases found to be as effective as β-carotene.

Some of the first studies were based on epidemiological work. In one such study it was shown that the subjects who consumed diets rich in carotenoids developed fewer lung cancers whether or not they smoked. This study assumed that the active ingredients are the carotenoids contained in fruits and vegetables.

More recently a number of animal studies have shown that β-carotene or other carotenoids can prevent or slow down the growth of skin cancer and other cancerous tumors.

Reviews have been published on the role of the carotenoids in relation to cancer (64,65). Human studies are presently being conducted (66,67).

Acknowledgment. Rhode Island Experimental Station Number 2527.

BIBLIOGRAPHY

1. O. Isler, "Introduction," in O. Isler, ed., *Carotenoids*, Birkhäuser Verlag, Basel, 1971.

2. P. Karrer and E. Jucker, *Carotenoids*, Birkhäuser Verlag, Basel, 1948.

3. J. C. Bauernfeind, "Carotenoids: A Class of Natural Pigments in Foods," in M. S. Peterson and A. H. Johnson, eds., *Encyclopedia of Food Science*, AVI Publishing Co., Inc., Westport, Conn., 1978.

4. *Plenary and Session Lectures, First International Symposium Carotenoids other than Vitamin A*, Butterworth & Co. (Publishers) Ltd., Kent, UK, 1967, p. 278.

5. *Plenary and Session Lectures, Second International Symposium on Carotenoids other than Vitamin A*, Butterworth & Co. (Publishers) Ltd., Kent, UK, 1969.

6. *Plenary and Session Lectures, Third International Symposium on Carotenoids other than Vitamin A*, Butterworth & Co. (Publishers) Ltd., Kent, UK, 1973, p. 130.

7. B. C. L. Weedon, ed., *Plenary and Session Lectures, Fourth International Symposium on Carotenoids-4*, Pergamon Press, Oxford, UK, 1976, p. 243.

8. T. W. Goodwin, ed., *Plenary and Session Lectures, Fifth International Symposium on Carotenoids-5*, Pergamon Press, Oxford, UK, 1979, p. 886.

9. G. Britton and T. W. Goodwin, eds., *Plenary and Session Lectures, Sixth International Symposium on Carotenoids-6*, Pergamon Press, Oxford, UK, 1982.

10. G. Britton, ed., *Plenary and Session Lectures, Seventh International Symposium on Carotenoids-7*, Munich, Pergamon Press, Oxford, UK, 1985.

11. J. C. Bauernfeind, ed., *Carotenoids as Colorants and Vitamin A Precursors. Technological and Nutritional Applications*, Academic Press, Orlando, Fla., 1981, p. 938.

12. J. C. Bauernfeind, ed., *Vitamin A Deficiency and Its Control*, Academic Press, Orlando, Fla., 1986, p. 530.

13. J. B. Davis, "Rodd's Chemistry of Carbon Compounds," in *Supplement IIA/IIB*, Elsevier, Amsterdam, The Netherlands, 1974, pp. 192–358.

14. "Nomenclature of Carotenoids (Rules Approved 1974," *Journal of Pure and Applied Chemistry* **41**, 406–431 (1975).

15. J. W. Porter and R. E. Lincoln, "I. Lycopersicon Selections Containing a High Content of Carotenes and Colorless Polyenes; II. The Mechanism of Carotene Biosynthesis," *Archives of Biochemistry and Biophysics* **27**, 390–403 (1950).

16. T. W. Goodwin, "Biosynthesis of Carotenoids," in T. W. Goodwin, ed., *Chemistry and Biochemistry of Plant Pigments*, Academic Press, Orlando, Fla., 1965.

17. B. H. Davies and R. F. Taylor, "Carotenoid Biosynthesis—The Early Steps," *Pure and Applied Chemistry* **47**, 211–221 (1976).

18. G. Britton, "Later Reactions of Carotenoid Biosynthesis," *Pure and Applied Chemistry* **47**, 223–236 (1976).

19. T. W. Goodwin, *The Biochemistry of the Carotenoids, Vol. 1 Plants*, 2nd. ed., Academic Press, Orlando, Fla., 1980, p. 377.

20. L. C. Raymundo, A. E. Griffiths, and K. L. Simpson, "Effect of Dimethyl Sulfoxide (DMSO) on the Biosynthesis of Carotenoids in Detached Tomatoes," *Phytochemistry* **6**, 1527–1532 (1967).

21. L. C. Raymundo, A. E. Griffiths, and K. L. Simpson, "Biosynthesis of Carotenoids in the Tomato Fruit," *Phytochemistry* **9**, 1239–1245 (1970).

22. M. Elahi, R. W. Glass, T -C. Lee, C. O. Chichester, and K. L. Simpson, "Effect of CPTA Analogs and Other Nitrigenous Compounds on the Biosynthesis of the Carotenoids in *Phycomyces blakesleeanus* Mutants," *Phytochemistry* **14**, 133–381 (1975).

23. R. J. H. Williams, G. Britton, and T. W. Goodwin, "Biosynthesis of Cyclic Carotenes," *Biochemical Journal* **105**, 99–105 (1967).

24. R. E. Tefft, T. W. Goodwin, and K. L. Simpson, "Aspects of the Stereospecificity of Torularhodin Biosynthesis," *Biochemical Journal* **117**, 921–927 (1970).

25. O. Straub, *Key to Carotenoids,* Birkhaüser Verlag, Basel, 1987, p. 296.

26. T. W. Goodwin, *The Biochemistry of the Carotenoids, Vol. 2 Animals,* 2nd ed., Chapman and Hall, London, 1984, p. 224.

27. B. C. L. Weedon, "Occurrence," in O. Isler, ed., *Carotenoids,* Birkhaüser Verlag, Basel, 1971.

28. J. Ostrander, C. Martinsen, J. Liston, and J. McCullough, "Sensory Testing of Pen-Reared Salmon and Trout," *Journal of Food Science* **41**, 386–390 (1975).

29. J. N. Thompson, P. Erdody, R. Brien and T. K. Murray, "Fluormetric Determination of Vitamin A in Human Blood and Liver," *Biochemical Medicine* **5**, 67–89 (1971).

30. K. L. Simpson and C. O. Chichester, "Metabolism and Nutritional Significance of Carotenoids," *Annual Review of Nutrition* **1**, 351–374 (1981).

31. T. Kamata and K. L. Simpson, "Study of Astaxanthin Diester Extracted from Adonis Aestivalis," *Comparative Biochemistry and Physiology* **86B**, 587–596 (1987).

32. R. S. Harris, "Effects of Agricultural Practices on the Composition of Foods," in R. S. Harris and E. Karmas, eds., *Nutritional Evaluation of Food Processing,* 2nd ed., AVI Publishing Co., Inc., Westport, Conn., 1957, pp. 33–57.

33. S. C. S. Tsou, J. Gershon, K. L. Simpson, and C. O. Chichester, "Promoting Household Gardens of Nutritional Improvement," in V. Tanphaichitri, W. Dahilan, V. Suphakarn, and A. Valyasevi, eds., *Human Nutrition, Better Nutrition Better Life, Proceedings of the 4th Asian Congress on Nutrition,* Aksornsmai Press, Bangkok, 1984.

34. L. C. Raymundo, C. O. Chichester, and K. L. Simpson, "Light Dependent Carotenoid Synthesis in the Tomato Fruit," *Journal of Food and Agricultural Chemistry* **24**, 59–64 (1976).

35. C. N. Villegas, C. O. Chichester, L. C. Raymundo, and K. L. Simpson, "The Effect of Gamma Irradiation on the Biosynthesis of Carotenoids in the Tomato Fruit," *Plant Physiology* **50**, 694–697 (1972).

36. C. Y. Lee, "Changes in Carotenoid Content of Carrots during Growth and Post-Harvest Storage," *Food Chemistry* **20**, 285–293 (1986).

37. B. H. Davies, "Analytical Methods—Carotenoids," in T. W. Goodwin, ed., *Chemistry and Biochemistry of Plant Pigments,* Vol. 2, 2nd ed., Academic Press, Orlando, Fla., 38–165 (1976).

38. E. DeRitter and A. E. Purcell, "Carotenoid Analytical Methods," in Ref. 11, pp. 815–823.

39. W. Horwitz, ed., *Official Methods of Analysis,* Association of the Official Analytical Chemists Washington, D.C., pp. 734–740.

40. S. E. Gebhardt, E. R. Elkins, and J. Humphrey, "Comparison of Two Methods of Determining the Vitamin A of Clingstone Peaches," *Journal of Agricultural and Food Chemistry* **25**, 628–631 (1977).

41. K. L. Simpson and S. C. S. Tsou, "Vitamin A and Provitamin A Composition of Foods," in Ref. 12, pp. 461–478.

42. K. L. Simpson, "Chemical Changes in Natural Food Pigments" in T. Richardson and J. W. Finley eds., *Chemical Changes in Food During Processing,* AVI Publishing Co., Inc., Westport, Conn., 1985, pp. 409–441.

43. L. Zechmeister, "Stereoisomeric Provitamin A," *Vitamins and Hormones* **7**, 57–81 (1962).

44. J. P. Sweeney and A. C. Marsh, "Effect of Processing on Provitamin A in Vegetables," *Journal of the American Dietetic Association* **59**, 238 (1971).

45. A. T. Ohunlesi and C. Y. Lee, "Effect of Thermal Processing on the Stereoisomerization of Major Carotenoids and Vitamin A Value of Carrots," *Food Chemistry* **4**, 311 (1979).

46. W. G. Lee and G. R. Ammerman, "Carotene. Stereoisomerization in Sweet Potatoes as Effected by Rotation and Still Retort Canning Processes," *Journal of Food Science* **39**, 1188 (1974).

47. V. I. Singleton, W. A. Gortnrer, and H. Y. Young, "Carotenoid Pigment of Pineapple Fruit. I. Acid Catalyzed Isomerization of the Pigments," *Journal of Food Science* **26**, 49–52 (1961).

48. A. H. El-Tinay and C. O. Chichester, "Oxidation of β-Carotene: Site of Initial Attachment," *Journal of Organic Chemistry* **35**, 2290–2298 (1970).

49. M. Goldman, B. Horev, and I. Saguy, "Decolorization of β-Carotene in Model Systems Simulating Dehydrated Foods. Mechanism and Kinetic Principles," *Journal of Food Science* **48**, 751–754 (1983).

50. E. D. Della Monica and P. E. McDowell, "Comparison of β-Carotene Content of Dried Carrots Prepared by 3 Dehydration Processes," *Food Technology* **19**, 1597 (1965).

51. J. M. Ouyang, H. Daun, S. S. Chang, and C-T. Ho, "Formation of Carbonyl Compounds from β-Carotene during Palm Oil Deodorization," *Journal of Food Science* **45**, 1214–1217 (1980).

52. C. Y. Lee, K. L. Simpson, and L. Gerber, "Vegetables as a Major Vitamin A Source in Our Diets," *New York's Food and Life Science Bulletin* **126**, 11 (1989).

53. S. S. Shapiro, D. J. Mott, and L. J. Machlin, "Kinetic Characteristics of β-Carotene Uptake and Depletion in Rat Tissue," *Journal of Nutrition* **114**, 1924–1933 (1984).

54. J. C. Bauernfeind, "Carotenoid Vitamin A Precursors and Analogs in Foods and Feeds," *Journal of Agricultural and Food Chemistry* **20**, 456–473 (1972).

55. M. I. Heinonen, "Diepoxides of β-Carotene and Their Provitamin A Activity," M.S. thesis, University of Rhode Island, Kingston, R.I., 1988.

56. B. R. Premachandra, V. N. Vasantharajan, and H. R. Cama, "Availability of β-Carotene from High β-Tomato Hybrids to Vitamin A-Deficient Rats and Utilization of β-Carotene in the Presence of Other Carotenoids," *Indian Journal of Experimental Biology* **16**, 473–477 (1978).

57. C. S. C. Tsou, unpublished data, 1988.

58. H. Klaüi and J. C. Bauernfeind, "Carotenoids as Food Color," in Ref. 11, pp. 48–317.

59. H. Kläui, ed., *NATCOL,* Natural Food Colours Association, Basle, Switzerland.

60. H. Mayer and O. Isler, "Total Synthesis" in O. Isler, ed., *Carotenoids,* Birkhaüser Verlag, Basel, 1971.

61. A. S. Al-Khalifa and K. L. Simpson, "Metabolism of Astaxanthin in the Rainbow Trout (*Salmo gairdneri*)," *Comparative Biochemistry and Physiology* **91B**, 563–568 (1988).

62. J. Gross and P. Budowski, "Conversion of Carotenoids into Vitamin A and A₂ in Two Species of Freshwater Fish," *Biochemical Journal* **101**, 747–754 (1966).

63. M. M. Mathews-Roth, M. A. Pathak, T. B. Fitzpatrick, L. C. Harber, and E. H. Kass, "Beta-Carotene as a Photoprotective Agent in Erythropoietic Protoporphyria," *New England Journal of Medicine* **282**, 1231–1234 (1970).

64. R. Peto, R. Doll, J. D. Buckley, and M. B. Sporn, "Can Dietary β-Carotene Materially Reduce Human Cancer Rates?" *Nature (London)* **290**, 201–208 (1981).

65. M. M. Mathews-Roth, "Carotenoids and Cancer Prevention— Experimental and Epidemiological Studies," *Pure and Applied Chemistry* **51**, 717–722 (1985).

66. N. I. Krinsky, "Function," in Ref. 1, pp. 669–716.

67. M. M. Mathews-Roth, "Carotenoids in Medical Application," in Ref. 11, pp. 755–786.

KENNETH L. SIMPSON
University of Rhode Island
Kingston, Rhode Island

CASEIN AND CASEINATES

Casein is the principal (75–80%) protein of all ruminant milks and occurs in all mammalian milks. The casein concentration in cow's milk is between 2.5 and 3.2%. The per capita consumption of bovine (cow) caseins as milk and dairy products in the United States in 1985 was close to 7.1 kg (15.9 lb) and a further 300 g (0.7 lb) was consumed in the form of sophisticated food products in which casein was an important functional ingredient.

The caseins, by definition, precipitate from cow's milk at a pH of 4.6 at 20°C. In the cow there are four genes that code to give four distinctly different proteins (1). These are modified by intracellular posttranslational phosphorylation and glycosylation and postsecretory proteolysis to give the wide range of proteins observed using high-performance liquid chromatography (hplc) or gel electrophoresis. Analysis of the complementary-DNA sequences that code for the bovine caseins has generally verified the earlier chemical sequence studies (2). Interspecies comparison of the complementary-DNA sequences has shown that they have not been well conserved, except for the signal peptides, the chymosin-sensitive site of κ-casein, the phosphorylation sites, and the 5' noncoding region. This suggests that the detailed molecular structure of these proteins is not important for their biological function, namely nutrition of the mammalian young, as long as the general features of easy enzyme digestibility, coagulability, and mineral-carrying and micelle-forming capabilities are preserved.

HETEROGENEITY

Gel electrophoresis of the caseins, which separates them on the basis of their net charge and size, demonstrates that every milk sample gives either one or two protein bands for each protein type (Fig. 1). Early studies showed that these differences were genetically controlled. It is now realized that during milk synthesis each of the two alleles is expressed, giving rise to the so-called genetic variants. There are likely to be a number of silent variants involving mutations that result in uncharged residue substitutions; these would not be readily identified using current methods of analysis. In addition, there can be varying levels of phosphorylation of the same gene product. In the cow the α_{s1}-casein can carry eight or nine phosphate residues, the protein with nine being previously known as α_{s0}-casein; α_{s2}-casein can be phosphorylated at several different levels. Varying levels of glycosylation only occur to a marked degree on the threonines of κ-casein. The presence of an active enzyme, plasmin, in milk gives rise to several observable degradation products from β-casein. These are known as γ-caseins and as components of the proteose peptone (heat stable and acid soluble) fraction of milk.

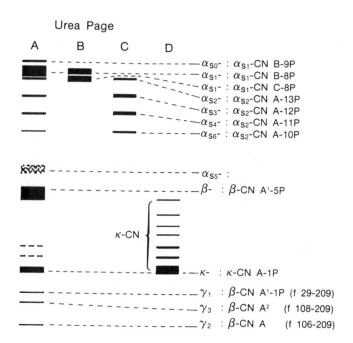

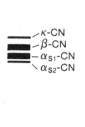

Figure 1. Electrophoretic mobilities. Patterns of the caseins obtained in two types of gel electrophoresis. The left-hand pattern is typical of that obtained in an SDS buffer system, while the right-hand patterns are typical of those obtained by various kinds of alkaline electrophoresis in urea-containing buffers: (**A**) whole casein; (**B**) α s₂-casein showing B and C genetic variants; (**C**) α s₂-casein showing the various levels of phosphorylation; (**D**) κ-casein showing the various glycosylated proteins.

Table 1. Amino Acid Composition of the Principal Caseins

Amino Acid	α_{s1}-Casein B	α_{s2}-Casein A[a]	β-Casein A^2	κ-Casein B
Aspartic acid	7	4	4	3
Asparagine	8	14	5	8
Threonine[b]	5	15	9	14
Serine	8	6	11	12
Phosphoserine	8	11	5	1
Glutamic acid	25	24	19	12
Glutamine	14	16	20	14
Proline	17	10	35	20
Glycine	9	2	5	2
Alanine	9	8	5	15
Cysteine[b]	0	2	0	2
Valine[b]	11	14	19	11
Methionine[b]	5	4	6	2
Isoleucine[b]	11	11	10	13
Leucine[b]	17	13	22	8
Tyrosine[b]	10	12	4	9
Phenylalanine[b]	8	6	9	4
Tryptophan[b]	2	2	1	1
Lysine[b]	14	24	11	9
Histidine[b]	5	3	5	3
Arginine	6	6	4	5
Pyroglutamine	0	0	0	1
Number of residues	199	207	209	169
Molecular weight	23,612	25,228	23,980	19,005[c]
Quantity in skim milk, g/L[d]	10.25	2.74	10.48[e]	3.45

[a] Corresponds to α_{s4}-casein in the earlier nomenclature.

[b] Indispensable or essential amino acids (note that phenylalanine and tyrosine are metabolically interconverted, as are cysteine and methionine).

[c] Excluding any carbohydrate.

[d] Ref. 3.

[e] Including 0.88 g/L of γ-casein.

The ratio of the various protein components changes during the period of a lactation (3) and average concentrations are shown in Table 1. It has also been noted that the quantities of some milk proteins are correlated with their genetic variant (4,5).

NOMENCLATURE

As methods of analysis have improved and greater knowledge of the caseins has been obtained, substantial changes in the naming of the caseins have taken place. Some of the recent recommended changes are listed in Table 2 and more information can be obtained from the various revisions carried out by the American Dairy Science Association (1). Basically, the first part of the name is the gene product and this is followed by the level of phosphorylation and finally by a fragment designation. There is not enough information available to allow adequate naming of the various glycosylated κ-casein species.

α_{s1}-Casein

α_{s1}-Casein is one of the principal protein components of bovine milk (Table 1). It is calcium sensitive (insoluble in 5 mM CaCl$_2$ solution at pH 7.0) and exists in several genetic forms. One of these, the A variant, is missing a 13 amino acid peptide. This affects the functional properties of the casein. α_{s1}-Casein also forms strong self-association complexes as well as complexes with the other caseins. The amino acid composition of the casein is shown in Ta-

ble 1 and the distribution of charged residues and hydrophobicity along the chain is shown in Figure 2. Clearly the phosphates, shown as double-length lines, are clustered in a single region (residues 45–70) of the protein.

Table 2. Recommended Names for the Casein Components of Bovine Milk[a]

Earlier Name(s)	Recommended Name
α_{s0}-Casein	α_{s1}-Casein B-9P
α_{s1}-Casein	α_{s1}-Casein B-8P
α_{s2}-Casein	α_{s2}-Casein A-13P
α_{s3}-Casein	α_{s2}-Casein A-12P
α_{s4}-Casein	α_{s2}-Casein A-11P
α_{s5}-Casein	A dimer of $_{s2}$-caseins
α_{s6}-Casein	α_{s2}-Casein A-10P
β-Casein A^2	β-Casein A^2-5P
β-Casein C	β-Casein C-4P
κ-Casein A	κ-Casein A-1P
γ_1-Casein A^1	β-Casein A^1-1P (f29–209)
γ-Casein	
γ_2-Casein A^{2b}	β-Casein A^2 (f106–209)
γ_3-Casein A[b]	β-Casein A (f108–209)
Proteose peptone components	
5	β-Casein-5P (f1–105)
5	β-Casein-5P (f1–107)
8-Slow	β-Casein-1P (f29–105)
8-Slow	β-Casein-1P (f29–107)
8-Fast	β-Casein-4P (f1–28)

[a] Ref. 1.

[b] These were previously called TS-, R-, and S-caseins.

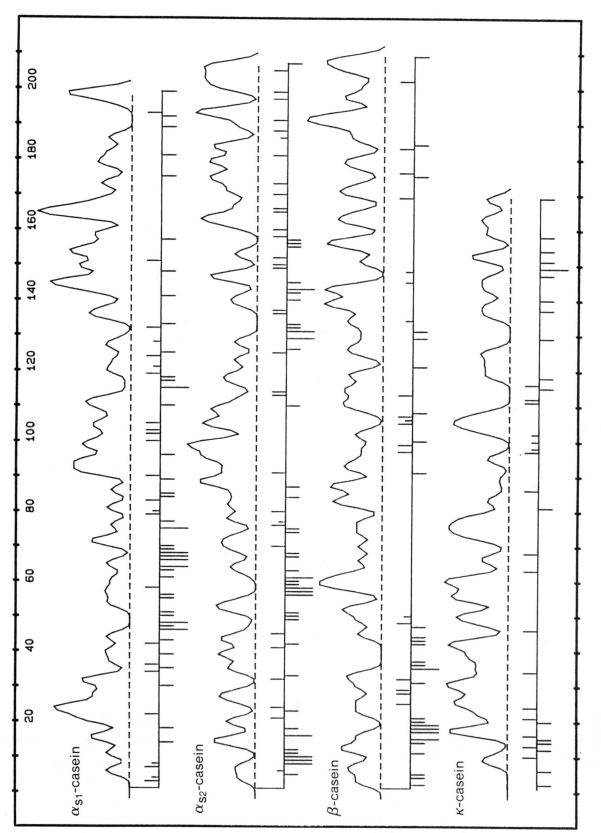

Figure 2. Sequence data for the four major caseins (1) with corrections (2). From the top; α_{s1}-casein B, α_{s2}-casein, β-casein A[2] and κ-casein B. The upper curve in each group is the distribution of smoothed hydrophobicity along the protein chain while the lower section depicts the linear distribution of charged residues. The vertical lines below the base show phosphoserine (longer lines) and glutamic and aspartic acid residues while the lines above the base show histidine (shorter lines) as well as lysine and arginine residues. The charges on the terminal amino and carboxyl groups are also shown.

This region is important for the formation of casein–mineral complexes. The hydrophobic region near residue 25 is important for the functional properties of casein in products such as young cheese.

α_{s2}-Casein

Much less is known about α_{s2}-casein, although many of its properties are similar to those of α_{s1}-casein. It exists with 10–13 phosphorylated residues and as a disulfide-bonded dimer in nonreducing conditions. It is readily hydrolyzed by plasmin.

β-Casein

β-Casein is present in all milks. It forms complexes by a temperature-dependent micellar mechanism with only monomers and large calcium-sensitive polymers present at the same time. This self-association reaction is dependent on the presence of the hydrophobic C-terminal 20 amino acids (Fig. 2, third trace) in the cow. Most of the negatively charged residues are in the N-terminal segment of the protein so that this region is most involved with mineral complexes. There is a large number of structure-breaking proline residues in the protein; this probably keeps the protein in an open form, which is an important characteristic for a protein that is primarily a target for digestive enzymes.

κ-Casein

Although κ-casein is present at low concentrations, it is important in dairy protein chemistry. Not only does it act as a micelle stabilizer in the natural milk system but it also forms disulfide links with the whey proteins when the heat processing of milk denatures them. Hydrolysis by rennet enzymes of the bond between residues 105 and 106 separates the hydrophobic, basic N-terminal region of the protein from the hydrophilic, acidic region (Fig. 2). The consequences of this hydrolysis form the basis for the cheese industry.

MICELLE STRUCTURE

In milk, the caseins are aggregated into large colloidal particles, called casein micelles, containing from 10 to many thousands of casein molecules. In addition to the caseins (in the approximate molecular ratio of $3:3:1:1$ for α_{s1} to β to α_{s2} to κ), there are about 5% by weight of calcium and phosphate and lesser quantities of other small ions. The smaller micelles contain significantly more κ-casein than the larger ones. One group of models for the micelle, based largely on electron microscopic evidence, assumes that the caseins are aggregated into submicelles that are cross-linked with calcium and phosphate. The submicelles with larger quantities of κ-casein tend to take up surface positions. Another group of models gives the minerals a greater role, while a third possibility is that the micelle is a matrix of α_s-caseins with calcium and phosphate; β- and κ-caseins fit into this lattice. Each of these models was based on certain observed behavior of the casein micelle under different conditions and they are, in many ways, not mutually exclusive. The milks of all mammals contain micelles, even though the ratios of the different types of casein vary widely and the levels of serine (and threonine) phosphorylation vary widely.

The effects of acid and heat on milk have important technological implications. Acid causes the minerals to dissociate from the proteins of the micelle. Alkali causes the reverse reaction to occur. Heat has only a minor effect on the proteins of the casein micelle, but causes denaturation of the whey proteins, which become disulfide bonded to the κ-casein on the outer surface of the micelle. As a consequence, the micelles of heated milk behave differently from the native micelles.

COMMERCIAL MANUFACTURE OF CASEIN

Casein has been commercially isolated from milk since 1900 (6). Pasteurized skim milk, produced from the separation of whole cow's milk, is used as the raw material for all edible-grade casein products. Casein is precipitated from the milk and washed and dried according to the scheme outlined in Figure 3. When the casein curd has been precipitated (by any one of three types of precipitating or coagulating agent, as shown in Fig. 3) and the mixture heated, the curd is separated from the whey and is subsequently washed several times with water in vats prior to mechanical dewatering by pressing or centrifuging. The dewatered curd is then dried using fluid-bed driers or pneumatic-conveying driers (the latter with simultaneous grinding). The resulting warm, granular casein is subjected to several dry-processing operations (Fig. 3), including cooling, tempering (to permit equilibration of moisture both within and between particles), milling, sifting, and blending prior to bagging and storage.

Two basic types of casein, acid and rennet, are produced and these are named according to the coagulating agent employed. Three types of acid casein are manufactured commercially: lactic, hydrochloric, and sulfuric casein. In New Zealand, lactic casein is by far the most common product although small quantities of sulfuric casein are also produced for particular requirements. In Europe and Australia, however, hydrochloric acid is commonly employed as a precipitant in casein manufacture because this acid is available as a relatively inexpensive by-product of the chemical industry. Rennet casein is produced from milk that has been clotted by the action of calf rennet (chymosin) or a microbial rennet.

ACID CASEIN MANUFACTURE

In the manufacture of acid casein the following processing operations occur (Fig. 3).

Mineral Acid (Hydrochloric or Sulfuric Acid) Casein

Skim milk (pH 6.6) is pasteurized (72°C, 15 s) and mixed with dilute (0.5 N) acid at a temperature of about 20°C to a pH of approximately 4.6. The mixture is heated (or cooked), for instance, by injection of steam, to a temperature of 50–55°C to aid agglomeration of the casein particles. Following a short period of residence in a cooking line and acidulation vat or tubes, the resultant curd is separated from the whey, washed and dried.

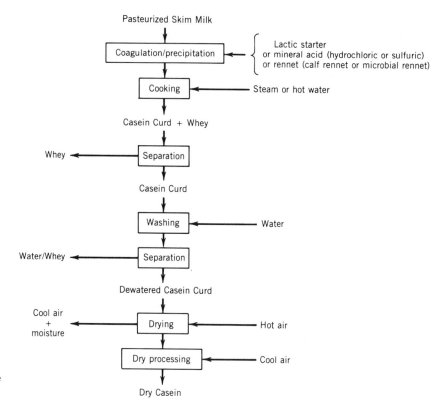

Figure 3. Processing operations used in the production of casein.

Lactic Acid Casein

Pasteurized skim milk is inoculated with cultures of lactic-acid–producing bacteria known as starter (eg, *Lactococcus lactis* subsp. *cremoris,* ≤0.3% of milk volume) at a temperature of 22–26°C. After a period of incubation of some 14–16 h, in which some of the lactose in the milk is fermented to lactic acid by the starter, the pH of the milk is reduced to about 4.6, causing coagulation of the casein. The lactic coagulum is then cooked and processed further in a manner similar to that described for mineral acid casein.

RENNET CASEIN MANUFACTURE

In the manufacture of rennet casein (Fig. 3), calf rennet or microbial rennet (60 RU/mL) is added to skim milk (ratio 1 : 7,000) at a temperature of about 29°C. When rennet casein is intended for nonfood (technical) purposes, it is important that the skim milk be left unpasteurized, because rennet casein plastic produced from pasteurized milk has a darker color than that from unpasteurized milk.

During the renneting of the skim milk, the enzyme specifically cleaves one of the bonds in κ-casein, releasing a glycomacropeptide. This action destabilizes the casein micelles and they form a three-dimensional clot with calcium ions. This process normally takes place in about 30 min at pH 6.6. The clotted milk may then be cooked and the casein processed in a manner similar to that described for acid casein.

Yield

The yield of commercial casein from 100 kg of skim milk is approximately 3 kg.

COMMERCIAL MANUFACTURE OF CASEINATES

Acid casein is insoluble in water. However, it may be dissolved in dilute alkali at pH 6.6 to produce the water-soluble derivatives of casein, known as caseinates. The general scheme for their production is shown in Figure 4. The starting material is either dry acid casein or dewatered acid casein curd. Sodium caseinate, the commonest of the caseinates, is generally produced by adding dilute sodium hydroxide to an aqueous slurry of finely milled acid casein curd or acid casein. Dissolving of the casein is aided by heating (to 60–70°C) and vigorous agitation. The viscous caseinate solution (20 Pa · s at ca 20% total solids and 25°C) is subsequently spray or roller dried prior to blending and bagging.

The high costs of spray drying solutions of sodium caseinate have led to alternative methods for production of granular sodium caseinate (where sodium carbonate and partly dried acid casein curd may be mixed together prior to drying in a casein drier) and extruded sodium caseinate (where acid casein, water, and sodium hydroxide are mixed together in an extruder).

Ammonium and potassium caseinates, formed by reacting acid casein with ammonium hydroxide and potassium hydroxide, have properties generally similar to those of sodium caseinate. In contrast to the translucent, straw-colored solutions of these caseinates, however, cal-

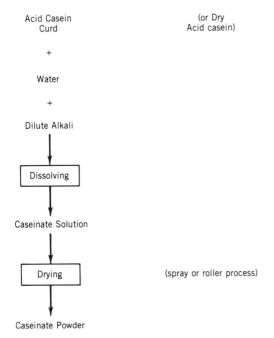

Acid Casein
Curd

(or Dry
Acid casein)

+

Water

+

Dilute Alkali

↓

Dissolving

↓

Caseinate Solution

↓

Drying   (spray or roller process)

↓

Caseinate Powder

Figure 4. Basic steps involved in the manufacture of spray- or roller-dried caseinates from acid casein curd or dry acid casein. Alkali may be sodium hydroxide, calcium hydroxide, potassium hydroxide, or ammonia.

cium caseinate (produced by reacting lime with an aqueous dispersion of acid casein) forms micelles in water, giving an intensely white, opaque, milky solution of relatively low viscosity. Unlike the other caseinates, calcium caseinate solutions may be destabilized on intense heating.

For the vast majority of edible applications, acid casein is converted to caseinates before use. The typical ranges of compositions for spray- or roller-dried caseinates are shown in Table 3.

FOOD ADDITIVE STATUS

Sodium caseinate is listed as generally recognized as safe (GRAS) in the U.S. *Code of Federal Regulation* as a multipurpose GRAS food substance (21 *CFR* 182.1748) and is also approved for use as a binder and extender in imitation sausages, nonspecific loaves, soups, and stews (9 *CFR* 318.7). Other caseinates, such as calcium caseinate, are also generally recognized as safe, although not all caseinates are specified in the GRAS list. All caseinates are currently under review for affirmed as GRAS status, and this action should occur after other higher priority additives are considered. Calcium, potassium, sodium, and ammonium caseinate have been granted prior sanction

Table 3. Typical Compositional Ranges for Caseinate

Protein ($N \times 6.38$), %	88–91
Moisture, %	3.5–6.0
Ash, %	3.7–6.0
Fat, %	0.5–2.0
Lactose, %	0.1–1.0

via clearance for optional use in frozen desserts (21 *CFR* 135.110) and casein and sodium caseinate are GRAS for use as clarifying agents in wine (27 *CFR* 240.1051).

The health aspects of casein and caseinates as food ingredients were reviewed in 1979 by the Federation of American Societies for Experimental Biology (7). It was concluded that there is no evidence to suspect a hazard from the use of casein or caseinates in food or packaging materials at current or expected future levels. At that time, it was recommended that food-grade specifications be established. In 1986, Food Chemical Codex (FCC) specifications were published (8). They are significant in that the commissioner of food and drugs has indicated that the FDA would regard the specifications in the Codex as defining the appropriate food grade within the meaning of the food additive regulations (9).

SPECIFICATIONS AND STANDARDS FOR CASEIN PRODUCTS

Specifications and international standards, in addition to those mentioned above, for casein and caseinates have been published for use by the European Economic Community (10), Food and Agricultural Organization (11), International Dairy Federation (12,13), Argentina (14), Australia (15,16), France (15), New Zealand (17–21), South Africa (15), USSR (22), and the United States (23). These specifications cover grades of edible acid casein, industrial acid casein, edible rennet casein, industrial rennet casein, and sodium caseinate.

USE OF CASEIN PRODUCTS IN FOOD APPLICATIONS

Casein products are incorporated into a wide range of food products for two reasons: nutrition and function.

NUTRITIONAL PROPERTIES OF CASEIN PRODUCTS

Several factors are responsible for the high level of caseinate use in nutritional foods. Caseinates are recognized as having high protein quality (Table 4). Protein quality as measured by the protein efficiency ratio (PER) is used in

Table 4. Nutritional Properties of Caseinates[a]

Protein efficiency ratio (PER)	2.5–2.6
True digestibility (TD)	96
Biological value (BV)	84
Net protein utilization (NPU)	80
Essential amino acid content (g/100 g protein)	
Isoleucine	5.3
Leucine	10.1
Lysine	8.4
Methionine plus cystine	3.5
Phenylalanine plus tyrosine	11.4
Threonine	4.6
Tryptophan	1.3
Valine	6.8
Amino acid score[b]	100

[a] Refs. 24–26.
[b] Relative to the 1985 FAO–WHO–UNO suggested patterns of amino acid requirements for preschool children 2–5 years.

the United States as the determinant in the required protein for U.S. Recommended Daily Allowance (USRDA) labeling. The *Code of Federal Regulations* (27) states "if the protein efficiency ratio of protein is equal to or better than that of casein, the U.S. RDA is 45 grams." If the PER is less than that of casein (2.5), 65 g of protein are required to meet the recommended daily allowance. Casein and caseinate usage in nutritional foods allows for relatively low addition levels to meet desired protein levels. Further advantages of caseinates are that they contain approximately 90% protein and lactose levels in the range of 0.1–0.5%. Relative to vegetable protein concentrates and isolates, commercially manufactured caseinates are bland in flavor.

These four factors (high protein quality, high protein content, low lactose, and bland flavor) have resulted in a high level of caseinate usage in nutritional supplements. The protein can be added as a lower proportion of the formula to produce a clean-flavored product, and the finished product can be specifically designed for lactose-intolerant individuals.

Caseinates are excellent complementary proteins when combined with vegetable proteins. Caseinates contain excess quantities of a number of nutritionally important amino acids. The excess can compensate for the amino acid deficiencies in vegetable proteins. Defined blends of casein and vegetable proteins can produce a nutritionally balanced protein product. For example, wheat gluten alone has an adjusted PER of approximately 1.0. Blending casein with wheat flour at a protein ratio of 1:1 results in a PER of 2.4 (25).

Nutritional foods commonly contain between 7 and 18 g of caseinate per serving, which represents 15–35% of the USRDA for protein. Nutritional foods containing caseinates include enteral products, high-protein beverage powders, fortified cereals, infant formulas, and nutrition bars.

THE FUNCTIONAL PROPERTIES OF CASEIN PRODUCTS

As previously mentioned, acid casein is primarily used in the form of a caseinate because of specific functionality

Table 5. Functional Attributes Imparted to Foods by Selected Casein Products[a]

Property	Functional Attributes
Organoleptic	Flavor, odor, texture, color
Appearance	Opacity, color, grain
Hydration	Solubility, dispersibility, viscosity, gelation
Surfactant	Emulsification, aeration, expansion and stability, gas entrapment during baking
Structural–textural	Elasticity, cohesion, adhesion, texturization, aggregation, viscosity, gelation
pH Stability	Isoelectric precipitation
Heat stability	Ability to retain functionality through high-temperature processing
Rehydration	Ability to be wetted, become hydrated, and function after drying

[a] Refs. 31–33.

Table 6. Relative Functional Properties of Caseinates

Property	Sodium Caseinate	Calcium Caseinate
Solubility	Excellent	Colloidal dispersion
Dispersibility	Poor	Good
Viscosity	Moderate	Low
Water absorption capacity	High	Moderate–low
Heat stability	Excellent	Very good
Whip overrun	Excellent	Very good
Aeration stability	Excellent	Very good

and high solubility or colloidal suspendability requirements for the protein. Rennet casein has found more limited use in the food industry; it is primarily used in imitation cheese, where it imparts thermoplasticity, emulsion stability, and bland flavor.

It is the physical and molecular properties of caseinates that contribute to their ability to provide benefits to, or functionality in, a finished food product (28,29). For example, low levels of sulfur-containing amino acids and high levels of proline are primarily responsible for the open, random structure of casein products; the structure enhances the heat stability of casein and minimizes its gelation or denaturation properties. The structure also results in the ability of the protein to form strong flexible membranes or films and a tendency of the soluble caseinate to bind water tightly. In addition, the high level of hydrophobic amino acid groupings, as well as charged amino acid groupings, results in caseinates with unique amphiphilic character, exhibiting the ability to impart excellent emulsification properties (30).

Functional properties generally attributed to casein and caseinate products are shown in Table 5. Selected caseinates are used to impart one or more of the properties listed.

Caseinates are commercially available in a wide range of functional forms. Processors can modify functionality

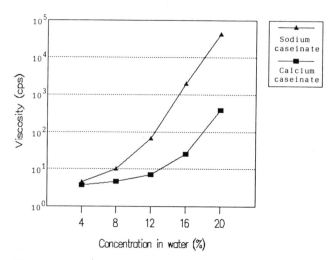

Figure 5. Viscosity comparison of sodium caseinate (Alanate 110) versus calcium caseinate (Alanate 310) as a function of concentration at 20°C.

during caseinate manufacturing using a range of treatments including alkali selection, mineral balance, processing modifications, flavor reduction, drying mode, and subsequent powder treatment (32,34,35). For example, significant differences exist between caseinates neutralized with sodium hydroxide and calcium hydroxide. Table 6 and Figure 5 demonstrate some of the differences between these two products.

Caseinates are widely used throughout the food industry for functional purposes. In general, optimum functionality is obtained with the use levels in the range of 0.5–2.0%. Numerous applications for caseinates have been mentioned in the literature. A thorough listing of applications for casein products has been published (36). Examples include coffee whiteners, soups, sauces, ice cream, salad dressing, formulated meats, bakery glazes, whipped toppings, and yogurt.

BIBLIOGRAPHY

1. W. N. Eigel, J. E. Butler, C. A. Ernstrom, H. M. Farrell, Jr., V. R. Harwalker, R. Jenness, and R. M. Whitney, "Nomenclature of Proteins of Cow's Milk: Fifth Revision," *Journal of Dairy Science* **67**, 1599–1631 (1984).

2. B. Ribadeau-Dumas, "Structure and Variability of Milk Proteins." In C. A. Barth and E. Schlimme, eds., *Milk Proteins, Nutritional, Clinical, Functional and Technological Aspects*, Springer-Verlag, New York, 1988, pp. 112–123.

3. D. T. Davies and A. J. R. Law, "The Content and Composition of Protein in Creamery Milks in South-West Scotland," *Journal of Dairy Research* **47**, 83–90 (1980).

4. F. Grosclaude, "Le polymorphisme génétique des principales lactoprotéines bovines. Relations avec la quantité, la composition et les aptitudes fromagères du lait," *INRA Productions Animales* **1**, 5–17 (1988).

5. D. McLean, E. R. B. Graham, and R. W. Ponzoni, "Effects of Milk Protein Variants and Composition on the Heat Stability of Milk," *Journal of Dairy Research* **54**, 219–235 (1987).

6. J. R. Spellacy, *Casein, Dried and Condensed Whey*. Lithotype Process Co., San Francisco, 1953.

7. Federation of American Societies for Experimental Biology, Life Sciences Research Office, *Evaluation of the Health Aspects of Casein, Sodium Caseinate and Calcium Caseinate as Food Ingredients, SCOGS-96*, Federation of American Societies for Experimental Biology, Bethesda, Md., 1979.

8. Committee on Food Chemicals Codex, *Food Chemicals Codex*, 2nd Suppl. to 3rd ed., National Academy Press, Washington, D.C., 1986.

9. Committee on Codex Specifications, *Food Chemicals Codes*, 3rd ed., National Academy Press, Washington D.C., 1981, p. XXIII.

10. Council of the European Community, "Caseins and Food Grade Caseinates," *Industrie Alimentari* **22**(209), 768–770 (1983).

11. H. W. Kay, "International Code of Principles for Milk and Milk Products. Results of the 18th Session of the FAO/WHO Committee for the Code of Principles," *Milchwissenschaft* **32**, 24, (1977).

12. *International Standard FIL–IDF 45:1969, Compositional Standards for Casein (Edible and Industrial)*, International Dairy Federation, Brussels, 1969.

13. *International Standard FIL–IDF 72-1974, Edible Caseinate: Compositional Standard*, International Dairy Federation, Brussels, 1974.

14. "Codigo Alimentario Argentino" (Argentine Food Code), *Industria Lechera* **61**(669), 14–16 (1981).

15. M. Amariglio, *Analysis Methods and Standards of Quality*, paper presented at the IDF Seminar on Caseins and Caseinates, Paris, May 31 to June 2, 1967, subject 7.

16. A. Bergmann, "Continuous Production of Spray Dried Sodium Caseinate," *Journal of the Society of Dairy Technology* **25**, 89–91 (1972).

17. New Zealand Dairy Board, *Edible Acid Casein—Product Data Sheets CA. 1, CA 2*, New Zealand Dairy Board, Wellington, 1976.

18. New Zealand Dairy Board, *Industrial Acid Casein—Product Data Sheet CA. 3*, New Zealand Dairy Board, Wellington, 1976.

19. New Zealand Dairy Board, *Edible Rennet Casein—Product Data Sheet CA. 4*, New Zealand Dairy Board, Wellington, 1976.

20. New Zealand Dairy Board, *Industrial Rennet Casein—Product Data Sheet CA. 5*, New Zealand Dairy Board, Wellington, 1976.

21. New Zealand Dairy Board, *Spray Process Sodium Caseinate—Product Data Sheet CA. 7*, New Zealand Dairy Board, Wellington, 1976.

22. *Molochnaya Promyshlennost (New Standards)*, (6), 35 (1985).

23. U.S. Department of Agriculture, "U.S. Standards for Grades of Edible Dry Casein (acid)," *Federal Register* **33**(141), 10385 (July 20, 1968).

24. M. Williams, unpublished data, New Zealand Dairy Board, 1987.

25. E. E. Lohrey and M. A. Humphries, "The Protein Quality of Some New Zealand Milk Products," *New Zealand Journal of Dairy Science and Technology* **11**, 147–154 (1976).

26. J. W. G. Porter, "The Present Nutritional Status of Milk Proteins," *Journal of the Society of Dairy Technology* **31**(4), 199–202 (1978).

27. *Code of Federal Regulations*, Title 21, Part 104.20, 1985.

28. J. E. Kinsella, "Milk Proteins: Physiochemical and Functional Properties," *CRC Critical Reviews in Food Science and Nutrition* **21**(3), 197–261 (1982).

29. J. Leman and J. E. Kinsella, "Surface Activity, Film Formation, and Emulsifying Properties of Milk Proteins," *CRC Critical Reviews in Food Science and Nutrition* **28**(2), 115–138 (1989).

30. M. Van den Hoven, "Functionality of Dairy Ingredients in Meat Products," *Food Technology* **40**, 72–77 (1987).

31. C. V. Morr, "Functional Properties of Milk Proteins and Their Uses as Food Ingredients," in P. F. Fox, ed., *Developments in Dairy Chemistry—I. Proteins*. Elsevier Applied Science Publishers, Ltd., Barking, UK, 1982, pp. 375–399.

32. C. V. Morr, "Production and Use of Milk Proteins in Food," *Food Technology* **7**, 39–48 (1984).

33. H. W. Modler, "Functional Properties of Nonfat Dairy Ingredients—A Review Modification of Products Containing Casein," *Journal of Dairy Science* **68**, 2192–2205 (1985).

34. C. Towler, "Conversion of Casein Curd to Sodium Caseinate," *New Zealand Journal of Dairy Science and Technology* **11**, 24–29 (1976).

35. K. J. Kirkpatrick and R. M. Fenwick, "Manufacture and General Properties of Dairy Ingredients," *Food Technology*, (10), 58–65, 1987.

36. C. R. Southward and N. J. Walker, "Casein, Caseinates and Milk Protein Coprecipitates," in I. A. Wolff, ed., *CRC Handbook of Processing and Utilization in Agriculture,* Vol. 1, CRC Press, Inc., Boca Raton, Fla., 1982, pp. 445–552.

MARSHA SWARTZ
NEIL WALKER
New Zealand Milk Products, Inc.
Santa Rosa, California

LAWRIE CREAMER
RAMSEY SOUTHWARD
New Zealand Dairy Research Institute
Palmerston North, New Zealand

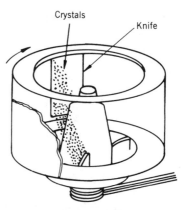

Figure 2. Basket centrifuge: imperforate bowl. Low g force (500–1,000 g) and intermittent feed limit these to applications where moderate solids dryness is needed. Capacity: 100–200 L/min.

CENTRIFUGES: PRINCIPLES AND APPLICATIONS

Centrifuges are used widely in food processing to separate liquids from solids, liquids from liquids, and even to separate two immiscible liquids from the accompanying solids. Centrifuges can be blanketed with inert gas and operate under high temperatures and pressures, as well as discharge centrate under pressure to retard foam and oxidation. They rinse the mother liquor from solids. They can be cleaned in place, and some can be steam sterilized. Centrifuges may also be used to separate large particles from similar small particles when they are operated as classifiers.

Centrifuges were first used in the 1880s to separate milk from cream. They use rotational acceleration to separate heavier phases from lighter phases, similar to a gravity-settling basin. The separating force is measured in units called g's, which are units of acceleration; 1 g = 980 cm/s.^{2}

Centrifuges usually generate at least 1,200 g for low-speed machines and up to 63,000 g for small tubular centrifuges. From a process standpoint, a higher g force results in a higher capacity for a given centrifuge; capacity is measured by volume throughput, degree of solids concentration, clarity of the liquid, and entrainment of one phase in another. Also, because the length of time the fluid is held under high g affects the separation, the volume of fluid in the centrifuge directly affects its capacity. Physically, the separation follows Stokes' law for the settling velocity of a particle in a fluid, except that the accel-

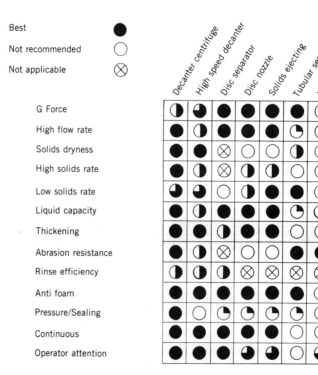

Figure 1. Properties of different centrifuge types.

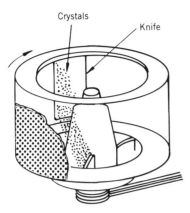

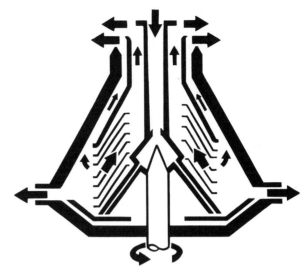

Figure 3. Basket centrifuge: perforate bowl. Same as Figure 2 except that the liquid filters through the solids. Excellent rinsing capability. Very dry cakes if the particle size is large enough to filter well.

eration of the rotational field is substituted for the earth's gravitational field.

Figure 5. Disc stack centrifuge: nozzle version. Same as Figure 4 except that the solids are continuously discharged as a pumpable slurry. Rates as high as 5,000 L/min are possible.

STOKES' LAW

$$v_g = \frac{\Delta\varphi \, dw^2 r}{18 \, \mu}$$

where V_g = settling velocity, w = rotational speed, r = radial position of the particle, d = diameter of particle, $\Delta\varphi$ = difference in the specific gravity of the liquid and solid, and μ = viscosity of fluid. Stokes' law is used to

design separation systems because the faster the settling velocity of a particle, the greater the chance of capture will be. Generally the temperature and the particle size are, to some extent, variables in a system.

For example, in rendering the fat from animal tissues, the amount of fat separated from the tissue increases with increasing temperature. Above 45°C, the residual tissue changes character, and its value as a meat additive drops. For edible rendering, the separation is made at 45°C, for the maximum fat yield with high-value solids. Inedible rendering is done at 90–95°C where the fat yield is at a

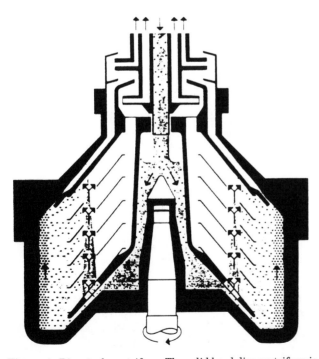

Figure 4. Disc stack centrifuge. The solid bowl disc centrifuge is a solid bowl version liquid–liquid separator operating at 5,000–10,000 g with some models as high as 15,000 g. They can handle a wide range of rates up to 2,400 L/min.

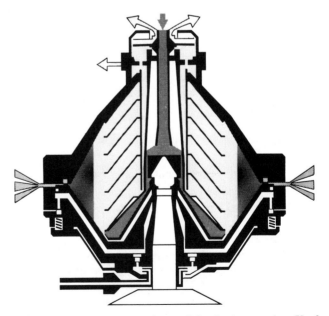

Figure 6. Disc stack centrifuge: solids ejecting version. Used where the solids volume is low, the bowl opens up intermittently to dump the solids as pumpable slurry. Maximum rates 40 L/min solids, 2,500 L/min liquid; g force 5,000–10,000 g.

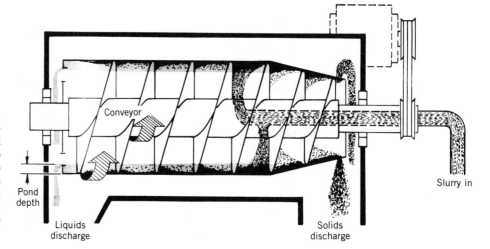

Figure 7. Decanter centrifuges. Solid bowl decanter centrifuges develop 1,000–3,000 *g*. They are used where large volumes of solids are present and the solid phase must be as dry as possible. Rates vary from 8 to 4,000 L/min. Very high speed decanters from 5–10,000 *g* are available but rates are lower, 10–200 L/min.

maximum, because the value of the inedible solids is quite low.

Stokes' law can be generalized as follows. Separation is improved if

1. The difference in specific gravity between the phases is large.
2. The particle size is large.
3. The viscosity is low.
4. The centrifuge speed is high.
5. The feed rate to the centrifuge is low.

Scale-up to estimate between different types of centrifuges is difficult, but within a class (eg, all disk centrifuges) a reasonably accurate scale-up can be developed. Centrifuges are usually scaled up based on *g* times volume or *g* times area. More detailed scale-up is usually left to centrifuge specialists.

SCREENING THE VARIOUS CENTRIFUGE TYPES

The goals of the separation must be defined. Centrifuges in general can:

Separate liquids from liquids.
Separate liquids from solids.
Separate solids from immiscible liquids.
Classify solids by size or density.
Rinse mother liquid from solids.

In brewing, the spent grains must be as dry as possible, the solids rate is high, and the solids are somewhat abrasive. Based on Figure 1, the following relationships can be seen.

	Solids Dryness	High Flow Rate	Abrasion Resistance
Decanter centrifuge	●	●	●
High-speed decanter	●	◑	◑
Perforate basket	●	◑	●

The decanter centrifuge is the clear choice followed by the perforate basket and, a distant third, the high speed decanter. An inquiry of decanter manufacturers would provide much data and reference accounts. The perforate basket centrifuge, in fact, is a filtration device and is not suitable for the application. Most manufacturers have laboratories where tests of new applications can be run at nominal cost. Some manufacturers also rent portable equipment to test on location.

CENTRIFUGE TYPES AND CONCLUSION

The centrifuge has wide and growing applications in the food industry for both liquid–liquid and solid–liquid sepa-

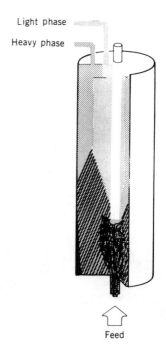

Figure 8. Tubular centrifuges. Manual disassembly to remove solids limits these to applications with little or no solids in the feed. As liquid–liquid separators, they develop much higher *g* force than most other centrifuges; maximum feed rates are 40–80 L/min.

rations. Advances in both speed and capacity have been made in the last few years, which increase yields and capacities. There is extensive literature on both theory and applications to aid the design engineer in equipment selection (Figs. 2–8).

PETER LaMONTAGNE
Sharples, Inc.
Warminster, Pennsylvania

CENTRIFUGATION. See Entries under GENETIC ENGINEERING.

CEREALS, NUTRIENTS, AND AGRICULTURAL PRACTICES

CEREAL GRAIN PRODUCTION AND COMPOSITION

Wheat, rice, and corn were the three largest cereal grains produced in the world in 1987–1988 (Table 1). Average chemical compositions of grains of barley, corn, millet, oats, rice, rye, grain sorghum, triticale, and wheat are listed in Table 2. All cereal grains contain starch as the principal component. Vitamin, mineral, fatty acids and amino acid contents of cereal grains have been published (2). Content and nutritive value of cereal proteins depend both on seed heredity and environment during cultivation and harvest.

WHEAT

Dry and Wet Milling

Of all cereal grains, wheat (*Triticum vulgare*) has the largest production worldwide. Products from dry milling hard wheat are farina, patent flour, first clear flour, second clear flour, germ, shorts, and bran. Similar flour milling processes for soft and durum wheats provide products for many foods that differ from those incorporating hard wheat products. A small amount of the world's wheat is processed to starch and gluten (about 80% of the total wheat flour protein) by wet milling. The baking industry is the largest user of gluten. Alkaline extraction procedures can also yield protein and starch from whole wheat (5) and mill feeds (6).

Nutrient Composition of Wheat Products

Wheat flour, the principal refined product of wheat milling, is the major ingredient in many foods: hard wheat

Table 1. World Production of Cereal Grains in 1987–1988[a]

Grain	Production in Million Metric Tons
Wheat	505
Rice	446
Corn	439
Barley	181
Sorghum	52
Oats	43
Rye	34

[a] Ref. 1.

flour for breads; soft wheat flour for cakes, pastries, quick breads, crackers, and snack foods; and durum semolina for spaghetti, macaroni, and other pasta products. The nutritional value of wheat foods depends mainly on the chemical composition of the refined flour used in their preparation. The average percentage composition of hard wheat milled products is given in Table 3. Wheat averages nearly 13% protein (the range is 7–22%). The lysine content of wheat protein varies between 2.2 and 4.2%, with a mean value of approximately 3.0%. Successful breeding of wheat for better protein quality and quantity could have an important impact on the nutrient content of wheat food products.

CORN

Dry and Wet Milling

Corn (*Zea mays* L.) is processed to provide food ingredients, industrial products, feeds, alcoholic beverages, and fuel ethanol. Wet millers produce starch; modified starch products, including dextrose and syrups (by starch hydrolysis); feed products; and oil (8). Dry millers produce hull, germ, and endosperm fractions that vary in particle size and fat content. Endosperm products—grits, meal, and flour—are the primary products used by the food processors and in consumer markets (9).

Corn germ flour prepared from a commercial dry-milled fraction appears to be a promising fortifying ingredient for the food industry (10). Protein concentrates from normal and high-lysine corns (11) and from defatted dry-milled corn germ (12) are obtained by alkaline extraction of corn or corn germ and by precipitating the extracted proteins by acid.

Nutrient Composition of Corn Products

Typical yields and analyses of dry-milled products from corn have been published (9,13). The discovery of the superior protein quality of *opaque-2* corn (14) makes possible the production of dry-milled products with improved nutritional value. However, prime products of this soft high-lysine corn include no flaking grits, but a considerable amount of table grits, meal, and flour. The International Maize and Wheat Improvement Center (CIMMYT) in Mexico developed commercial corn varieties with high lysine, high yields, traditional appearance, and conventional hardness. These new varieties are called quality-protein maize (QPM). Research is in progress in the United States to produce commercial QPM hybrids.

RICE

Rice (*Oryza sativa* L.) is the second largest cereal grain produced in the world. The whole grain (rough rice) consists of 18–28% hull and the edible portion (brown rice). After the hull is removed, the resulting brown rice is processed by abrasive milling to remove 6–10% by weight as bran polish to produce white or milled rice. Conventional and Japanese rice milling systems consist of precleaning, dehulling, rough rice separation, and whitening (15); the chemical composition of rice and its fractions have been listed (16). Brown rice and white rice are usually con-

Table 2. Average Compositions of Cereal Grains (As-is)[a]

Constituent	Barley (Pearled)	Corn (Field)	Millet (Dehulled)	Oats (Oatmeal)	Rice (Brown)	Rye	Sorghum	Triticale	Wheat[b]
Water, %	10.1	10.4	8.7	8.8	10.4	11.0	9.2	10.5	13.1
Calories per 100 g	352	365	378	384	370	335	339	336	327
Protein, %	9.9	9.4	11.0	16.0	7.9	14.8	11.3	13.0	12.6
Fat, %	1.2	4.7	4.2	6.3	2.9	2.5	3.3	2.1	1.5
Starch, %	57.5	65.0	73.1	50.3	69.2	57.1	65.1	57.5	55.7
Crude fiber, %	0.7	2.9	1.0	1.1	1.3	1.5	2.4	2.6	2.3
Total dietary fiber, %	15.6	8.5	8.5	10.3	3.5	14.3	—[c]	18.3	12.3
Ash, %	1.1	1.2	3.2	1.9	1.5	2.0	1.6	2.2	1.6

[a] Refs. 2–4.
[b] Hard red winter.
[c] Not determined.

sumed as food after cooking. Rice bran, because of rancidity problems, was used exclusively as animal feed until recently. Stabilized rice bran is now commercially available in the United States and is going into ready-to-eat breakfast cereals. It is estimated that 1 million lb/month of stabilized rice bran was used for food in January 1990 in the United States, and this quantity will probably increase by several fold.

OATS

Processing

Oat (*Avena sativa* L.) processing includes cleaning, dehulling, steaming, and flaking (17). Typical products from milling oats are rolled oats, oat hulls, feed oats, mixed grains and seeds, and fines.

Oats are dry milled into break flour, reduction flour, shorts flour, shorts, bran, and hulls (18). Weight distribution and protein contents of dry-milled oat fractions, sequential solvent extraction of proteins from milled fractions, and amino acid compositions of milled fractions and oat extracts were also determined (18). Several commercial high-protein oat cultivars are available. Protein concentrates have been prepared from oat flours of ordinary and high-protein varieties by air classification (19) and by wet-milling procedures (20). These protein products appear to have some promise for food applications.

Nutrient Composition of Oat Products

The nutritional quality of oat protein is good compared with other cereals; rolled oats have a protein efficiency

ratio (PER) of 2.2 compared to casein with a PER of 2.5. Oat groats used for food contain 11–15% protein. However, some dehulled oats from the Near East have protein contents of up to 25%.

BARLEY

Conventional roller milling of barley (*Hordeum vulgare* L.) gives four principal products: flour, tailings flour, shorts, and bran (21). The yields, protein contents, and amino acid composition of roller-milled barley fractions have been reported (22). Barley protein concentrates from normal and high-protein, high-lysine varieties were prepared by an alkaline extraction method (23). Barley and malted barley flours have been fractionated by air classification to yield high-protein flour and starch fractions (24).

A high-lysine and high-protein barley variety, Hiproly, has been described (25). Many other high-lysine barleys have also been developed since 1970. Food uses of barley include parched grain, pearled grain, flour, and ground grain. About half of the total U.S. barley crop is used for malting and the remainder is used for feed.

SORGHUM

Procedures for dry milling of sorghum [*Sorghum bicolor* (L. Moench)] grain range from cracking, which produces a crude product, to debranning and degermination, which yield refined bran, germ, meal, and grit fractions (26). Most of the sorghum that is consumed as food is prepared directly from the whole grain. Sorghum is a major feed grain in the United States.

Table 3. Average Percentage Composition of Dry-Milled Hard Wheat Products (As-is)[a]

Constituent	Wheat	Farina	Patent Flour	First Clear Flour	Second Clear Flour	Germ	Shorts	Bran
Moisture, %	12.0	14.2	13.9	13.4	12.4	10.5	13.5	14.1
Ash, %	1.8	0.4	0.4	0.7	1.2	4.0	4.1	6.0
Protein, %	12.0	10.3	11.0	12.7	13.5	30.0	16.0	14.5
Crude fiber, %	2.5	—[b]	—[b]	—[b]	—[b]	2.0	5.5	10.0
Total dietary fiber, %	12.4	2.6	2.7	—[b]	—[b]	15.1	—[b]	40.5
Fat, %	2.1	0.8	0.9	1.3	1.3	10.0	4.5	3.3

[a] Refs. 3, 7.
[b] Not determined.

Normal sorghum has the lowest lysine content among the cereal grains. However, two floury lines of Ethiopian origin were exceptionally high in lysine at relatively high levels of protein (27). Another high-lysine sorghum, *P721 opaque,* was produced by treating normal grain with a chemical mutagen (28). An alkaline extraction process gives protein concentrates and starch from ground normal and high-lysine sorghums (29). Air classification of flour and horny endosperm from high-lysine sorghum yielded fractions with higher protein contents than the starting flour or horny endosperm (30).

TRITICALE

Triticale (X. *Triticosecale*) is a cross between wheat (*Triticum*) and rye (*Secale*). Most of the food products made from wheat flour, including fermented and nonfermented dough products, can be made successfully from pure triticale flour. Good-quality bread can be made with mixtures of up to 75% triticale flour and 25% wheat flour. Commercial dry milling of triticale is limited to 100% triticale flour in the United States. Protein concentrates and starch have been prepared from ground triticale by an alkaline extraction process (31).

RYE

Rye (*Secale cereale* L.) can be milled into flours and meals for bread, crackers, and snack foods. Ground meat products may contain rye flours as fillers.

BIBLIOGRAPHY

1. U.S. Department of Agriculture, "Agricultural Statistics 1988," U.S. Government Printing Office, Washington, D.C., 1988.

2. D. L. Drake, S. E. Gebhardt, and R. H. Matthews, "Composition of Foods: Cereal Grains and Pasta, Raw, Processed, Prepared," *Agriculture Handbook* Number 8–20, U.S. Department of Agriculture, Washington, D.C., 1989.

3. R. H. Matthews and P. R. Pehrsson, *Provisional Table on the Dietary Fiber Content of Selected Foods,* HNIS/PT–106, U.S. Department of Agriculture, Human Nutrition Information Service, Washington, D.C., 1988.

4. R. A. Olson and K. J. Frey, eds., *Nutritional Qualities of Cereal Grains,* American Society of Agronomy Inc., Crop Science Society of America, Inc., Soil Science Society of America, Inc., Madison, Wis., 1987.

5. Y. V. Wu and K. R. Sexson, "Preparation of Protein Concentrate from Normal and High-Protein Wheats," *Journal of Agricultural and Food Chemistry* **23,** 903–905 (1975).

6. R. M. Saunders, M. A. Conner, R. H. Edwards, and G. O. Kohler, "Preparation of Protein Concentrates From Wheat Shorts and Wheat Mill Run by Wet Alkaline Process," *Cereal Chemistry* **52,** 93–101 (1975).

7. Y. V. Wu and G. E. Inglett, "Effects of Agricultural Practices, Handling, Processing, and Storage on Cereals," in E. Karmas and R. S. Harris, eds., *Nutritional Evaluation of Food Processing,* 3rd. ed., AVI/van Nostrand Reinhold Co., New York, 1988.

8. J. B. May, "Wet Milling: Process and Products," in S. A. Watson and P. E. Ramstad, eds., *Corn: Chemistry and Technology.* American Association of Cereal Chemists Inc., St. Paul, Minn., 1987.

9. R. J. Alexander, "Corn Dry Milling: Processes, Products, and Applications," in Ref. 8.

10. C. W. Blessin, W. J. Garcia, W. L. Deatherage, J. F. Cavins, and G. E. Inglett, "Composition of Three Food Products Containing Defatted Corn Germ Flour," *Journal of Food Science* **38,** 602–606 (1973).

11. Y. V. Wu and K. R. Sexson, "Protein Concentrates from Normal and High-Lysine Corns by Alkaline Extraction: Preparation," *Journal of Food Science* **41,** 509–511 (1976).

12. H. C. Nielsen, G. E. Inglett, J. S. Wall, and G. L. Donaldson, "Corn Germ Protein Isolate—Preliminary Studies on Preparation and Properties," *Cereal Chemistry* **50,** 435–443 (1973).

13. S. A. Watson and P. E. Ramstad, eds., *Corn: Chemistry and Technology,* American Association of Cereal Chemists Inc., St. Paul, Minn., 1987.

14. E. T. Mertz, L. S. Bates, and O. E. Nelson, "Mutant Gene That Changes Protein Composition and Increases Lysine Content of Maize Endosperm," *Science* **145,** 279–280 (1964).

15. H. T. L. van Ruiten, "Rice Milling: An Overview," in B. O. Juliano, ed., *Rice: Chemistry and Technology,* 2nd ed., American Association of Cereal Chemists, Inc., St. Paul, Minn., 1985.

16. B. O. Juliano and D. B. Bechtel, "The Rice Grain and its Gross Composition," in Ref. 15.

17. D. Deane and E. Commers, "Oat Cleaning and Processing," in F. H. Webster, ed., *Oats: Chemistry and Technology,* American Association of Cereal Chemists, Inc., St. Paul, Minn., 1986.

18. Y. V. Wu, K. R. Sexson, J. F. Cavins, and G. E. Inglett, "Oats and Their Dry-Milled Fractions: Protein Isolation and Properties of Four Varieties," *Journal of Agricultural and Food Chemistry* **20,** 757–761 (1972).

19. Y. V. Wu and A. C. Stringfellow, "Protein Concentrates from Oat Flours by Air Classification of Normal and High-Protein Varieties," *Cereal Chemistry* **50,** 489–496 (1973).

20. J. E. Cluskey, Y. V. Wu, J. S. Wall, and G. E. Inglett, "Oat Protein Concentrates from a Wet-Milling Process: Preparation," *Cereal Chemistry* **50,** 475–481 (1973).

21. Y. Pomeranz, H. Ke, and A. B. Ward, "Composition and Utilization of Milled Barley Products. I. Gross Composition of Roller-Milled and Air-Separated Fractions," *Cereal Chemistry* **48,** 47–58 (1971).

22. G. S. Robbins and Y. Pomeranz, "Composition and Utilization of Milled Barley Products. III. Amino Acid Composition," *Cereal Chemistry* **49,** 240–246 (1972).

23. Y. V. Wu, K. R. Sexson, and J. E. Sanderson, "Barley Protein Concentrate from High-Protein, High-Lysine Varieties," *Journal of Food Science* **44,** 1580–1583 (1979).

24. J. R. Voss and C. G. Youngs, "Fractionation of Barley and Malted Barley Flours by Air Classification," *Cereal Chemistry* **55,** 280–286 (1978).

25. A. Hagberg and K. E. Karlsson, "Breeding for High Protein Content and Quality in Barley," in *Symposium: New Approaches to Breeding for Improved Plant Protein,* International Atomic Energy Agency, Vienna, 1968.

26. J. S. Wall and W. M. Ross, *Sorghum Production and Utilization,* Avi Publishing Co., Westport, Conn., 1970.

27. R. Singh and J. D. Axtell, "High Lysine Mutant Gene HL That Improves Protein Quality and Biological Value of Grain Sorghum," *Crop Science* **13,** 535–539 (1973).

28. D. P. Mohan and J. D. Axtell, "Diethyl Sulfate Induced High Lysine Mutants in Sorghum," *Proceedings of the Ninth Biennial Grain Sorghum Research and Utilization Conference,* Lubbock, Tex., Mar. 4–6, 1975.

29. Y. V. Wu, "Protein Concentrate from Normal and High-Lysine Sorghums: Preparation, Composition, and Properties," *Journal of Agricultural and Food Chemistry* **26,** 305–309 (1978).

30. Y. V. Wu and A. C. Stringfellow, "Protein Concentrate from Air Classification of Flour and Horny Endosperm from High-Lysine Sorghum." *Journal of Food Science* **46,** 304–305 (1981).

31. Y. V. Wu, K. R. Sexson, and J. S. Wall, "Triticale Protein Concentrate: Preparation, Composition, and Properties," *Journal of Agricultural and Food Chemistry* **24,** 511–517 (1976).

General References

W. Bushuk, ed., *Rye: Production, Chemistry and Technology,* American Association of Cereal Chemists Inc., St. Paul, Minn., 1976.

J. H. Hulse, E. M. Laing, and O. E. Pearson, *Sorghum and the Millets: Their Composition and Nutritive Value,* Academic Press, Orlando, Fla., 1980.

National Research Council. *Quality-Protein Maize.* National Academy Press, Washington, D.C., 1988.

National Research Council, *Triticale: A Promising Addition to the World's Cereal Grains,* National Academy Press, Washington, D.C., 1989.

Y. Pomeranz, ed., *Wheat: Chemistry and Technology,* 3rd ed., American Association of Cereal Chemists Inc., St. Paul, Minn., 1988.

D. C. Rasmusson, ed., *Barley,* American Society of Agronomy, Inc., Crop Science Society of America, Inc., Soil Science Society of America, Inc., Madison, Wis., 1985.

F. H. Webster, ed., *Oats: Chemistry and Technology,* American Association of Cereal Chemists Inc., St. Paul, Minn., 1986.

Y. Victor Wu
USDA/ARS/NCAUR
Peoria, Illinois

CEREALS SCIENCE AND TECHNOLOGY

Cereals supply the most calories per acre, can be stored safely for a long time, and can be processed into many acceptable products. They are adapted to a variety of soil and climatic conditions; can be cultivated both on a large scale mechanically with a small amount of labor; and on a garden scale almost entirely by human labor. They are excellent sources of energy and relatively good sources of inexpensive protein, certain minerals, and vitamins. More than two-thirds of the world's cultivated area is planted with grain crops. Most developing countries rely on cereals as food sources. Cereals provide more than half of the calories consumed for human energy, and in many developing countries they provide over two-thirds of the total food. Cereals also contribute greatly to production of animal proteins in feeds (1–9).

GRAINS

Wheat

Wheat grows on almost every kind of arable land, from sea level to elevations of 3,000 m, in regions where water is sufficient or areas that are relatively arid; it also thrives in well drained loams and clays. Wheat is a basic food throughout the world. In developing countries, wheat consumption is increasing and accompanies rising living standards. The increase in wheat consumption is at the expense of the more costly rice or the less costly barley or sorghum. Wheat is the most widely cultivated grain crop in temperate areas. Winter wheat may be multicropped prior to soybeans or grown as spring wheat in rotation with barley and oats to reduce loss of soil fertility (5,10–13).

Wheat provides about one-fifth of all calories consumed by humans. It accounts for nearly 30% of worldwide grain production and for over 50% of the world grain trade. It is harvested somewhere in the world nearly every month of the year. The United States, Canada, Australia, Argentina, and France are the main wheat exporters. Wheat production in North America and Western Europe has declined relatively to the rest of the world since 1950. These two areas accounted for 41% of world production in 1950 and 36% in 1981. Eastern Europe and Asia accounted for 48% of the total world production in 1950 and 56% in 1981.

The USSR produces more wheat than any other nation. The main wheat producing areas in Asia are China, India, Turkey, and Pakistan. Wheat is second only to rice as a source of human food in China. Japan is a large wheat importer; Argentina and Australia are major wheat exporters. The United States is the second largest wheat producing country.

Hexaploid wheats are the most widely grown around the world. They can be classified according to

1. Growth habitat: spring wheats are sown in the spring and harvested in late summer; and winter wheats are sown in the fall and harvested early the next summer; winter wheats are preferred because of a generally higher yield.
2. Bran color: white, amber, red, or dark.
3. Kernel hardness: hard or soft; and vitreousness; glassy-vitreous or mealy-starchy-floury. In many European countries all bread wheats (hard and soft) are called soft, and the term hard is reserved for the very hard durum wheats; in the United States the term hard is used for bread wheats; soft is used for wheats to produce cookies, cakes, etc; and durum for wheats to produce alimentary pastes.
4. Geographic region of growth.
5. End use properties: protein range, physical dough properties, breadmaking quality.
6. Variety.
7. Composite designations (14–17).

Corn

Corn (maize) is now the most popular grain used for animal feed in temperate regions. In many tropical areas, it is a basic human food. There are five major types of corn: flint, dent, floury, pop, and sweet corn. Dent varieties, grown most widely in the United States, provide high yields and are used to a large extent as feed grain. Flint corn, with a somewhat higher food value, is common in Europe and South America and is valued for its physical properties and as a raw material in the production of certain Central American foods (tortilla and arepa). Pop, sweet, and floury corn are relatively small food crops (18–21).

The starch in waxy corn is almost entirely amylopectin, in amylomaize up to 70–80% amylose, and in regular corn about 25% amylose and 75% amylopectin. The starches vary widely in their physical properties (gelatinization characteristics, water binding capacity, gel properties, etc) and usefulness in various foods and industrial applications. High-lysine and high-lysine-tryptophan genetic mutants have improved the nutritional properties of corn and are now available as dent or flint corn. The latter is likely to be acceptable in the production of maize based foods.

The United States accounts for over half of the total world corn production and ca 80% of the annual world corn exports. In the United States, most corn is used as feed or seed or in alcoholic beverages. Only ca 10% is used as food (production of starch, corn syrups, breakfast cereals, and various foods). Livestock feed accounts for almost 85% of the total U.S. domestic use. The main importers of United States corn are Western Europe and Japan, primarily for livestock feed.

Rice

Rice is the staple food of about half of the human race, providing over one-fifth of the total food calories consumed by the people of the world. Most rice is produced in the Far East and is primarily consumed within the borders of producing countries. Asia, which has nearly 60% of the world population, produces and consumes about 90% of the world rice production. The United States produces less than 2% of the world crop, but accounts for about 30% of the world rice trade (22,23).

Long-grain rice accounts for ca 50%, medium-grain rice for 40%, and short-grain rice for ca 10% of the United States rice production. California is practically the exclusive producer of short-grain rice. California and Louisiana lead in medium-grain rice production. Growers in Arkansas, Texas, and Mississippi produce mostly long-grain rice.

Barley

Barley is a winter hardy and drought resistant grain. It matures more rapidly than wheat, oats, or rye and is used mainly as feed for livestock and in malting or distilling industries. The two main types of barley, depending on the arrangement of grains in the ear, are two-rowed and six-rowed barley. The former predominates in Europe and in parts of Australia; the latter is more resistant to extreme temperatures and is grown in North America, India, and the Middle East (24,25).

Although barley is grown throughout the world, production is concentrated in the northern latitudes. Since 1950, the world production of barley tripled with most of its increase in Europe. USSR is the largest barley producer. Barley is adaptable to a variety of conditions and is produced commercially in 36 states in the United States. In the United States, the use of barley as a livestock feed is declining, and the use for processing into malt is increasing.

Oats

Oats are grown most successfully in cool, humid climates and on neutral to slightly acidic soil. The bulk of the crop is consumed as animal feed (primarily for horses). Heavy oats (with high groat to hull ratios) are processed into rolled oats, breakfast foods, and oatmeal (26).

The world production of oats is about one-tenth of the world production of wheat and slightly over one-fourth of the world production of barley. The chief oat producing states in the United States are in the north-central Corn Belt and farther northward. About 90% of the oat is left on the farm as a feed, and only a small proportion is processed as food or used for industrial products. The use of oats in food processing is increasing, especially in breakfast cereals.

Sorghum

Sorghum is the fifth most widely grown crop in the world. Sorghum gained significance as a grain as a result of the development of high yielding hybrids. However, sorghum production is <5% of the total grain production. There has been a small increase in the area harvested with sorghum but a substantial increase in production because of a rise in yield. The primary sorghum producing areas are Asia, Africa, and North America. The primary sorghum producing countries are the United States, China, India, Argentina, Nigeria, and Mexico. These countries produce >75% of the world total, with the United States producing ca 30%.

More than 50% of the worldwide production of sorghum is for human consumption. In the United States, sorghum is grown almost entirely as a feed grain for local use or export. Sorghum is produced under a wide range of climatic conditions. It withstands limited moisture conditions and adapts to high temperatures (27,28).

In some varieties of sorghum, the nucellar tissue contains pigments that complicate production of acceptable white sorghum starch. Varieties with no nucellar tissue persisting to maturity could be used in starch production, but the availability of corn has discouraged such attempts. Waxy (starch) sorghums are potentially promising for special food uses. Some sorghums are rich in pigments, notably tannins, which reduce palatability and feed value.

Rye

Rye is characterized by good resistance to cold, pests, and diseases, but it cannot compete with wheat or barley on good soils and under improved cultivation practices. Approximately 90% of rye is produced in Europe (mainly Poland and Germany) and in the USSR. Rye is considered a bread grain in Europe, but its proportion in mixed

Table 1. Total Grains: Production, Consumption, and Net Exports, Selected Years[a]

Region	1960–1961–1962–1963[b]			1969–1970–1971–1972[b]			1983–1984[b]			Rate of Growth of Net Exports, %
	Production	Consumption	Net Exports[c]	Production	Consumption	Net Exports[c]	Production	Consumption	Net Exports[c]	
Developed	316	300	20	403	377	32	459	417	120	8.8
United States	168	140	33	209	169	40	206	180	96	5.2
Canada	24	15	10	35	22	15	48	25	28	5.2
European Community	70	91	−22	93	111	−17	124	118	−11	11.9
USSR & East Europe	180	179		237	340	−2	283	318	−34	22.0
Others	54	54	−1	66	75	−6	82	95	−16	16.9
Developing	210	220	−10	285	305	−19	449	510	−67	9.3
East Asia	20	23	−4	30	38	−9	15	27	−12	6.9
South Asia	98	99	−2	126	128	−2	256	255	−5	5.0
South & Central America	42	42	−1	64	61	−3	92	95	−3	6.5
Middle East	30	35	−4	39	49	−9	49	87	−38	9.5
Africa	16	17	−1	21	23	−2	37	47	−9	12.0
World	806	801		1,051	1,051	—	1,483	1,545		

[a] Ref. 30.
[b] Production in million metric tons.
[c] Numbers with a minus sign are imports.

wheat-rye bread is decreasing. Much of the rye is used as a feed grain and a small proportion in the distilling industry (29).

Total production, consumption, and exports of grains are summarized in Table 1. Figure 1 compares production of various cereal grains in the United States, USSR, and the European Economic Community (30).

PHYSICAL PROPERTIES AND STRUCTURE

Some physical properties of cereal grains are listed in Table 2 (31). Table 3 summarizes approximate grain size and proportions of the principal parts comprising the mature kernels of different cereals (32).

Optimized utilization of cereal grains requires knowledge of their structure and composition. The practical implications of kernel structure are numerous (33,34). They relate to the various stages of grain production, harvest, storage, marketing, and use.

Kernel Structure

The cereal grain is a one-seeded fruit called a caryopsis, in which the fruit coat is adherent to the seed. As the fruit ripens, the pericarp (fruit wall) becomes firmly attached to the wall of the seed proper. The pericarp, seed coats, nucellus, and aleurone cells form the bran. The embryo occupies only a small part of the seed. The bulk of the seed is taken up by the endosperm, constituting a food reservoir.

The floral envelopes (modified leaves known as lemma and palea), or chaffy parts, within which the caryopsis develops, persist to maturity in the grass family. If the chaffy structures envelop the caryopsis so closely that they remain attached to it when the grain is threshed (as in rice and most varieties of oats and barley), the grain is considered to be covered. However, if the caryopsis readily separates from the floral envelopes when the grain is threshed, as with common wheats, rye, hull-less barleys, and the common varieties of corn, these grains are considered to be naked.

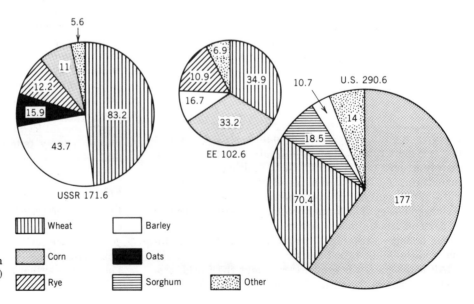

Figure 1. Average grain production for 1980–1984 (million metric tons) (30).

Table 2. Some Physical Properties of Cereal Grains[a]

Name	Length, mm	Width, mm	Grain Mass, mg	Bulk Density, kg/m³	Unit Density, kg/m³
Rye	4.5–10	1.5–3.5	21	695	—
Sorghum	3–5	2.5–4.5	23	1360	—
Paddy rice	5–10	1.5–5	27	575–600	1370–1400
Oats	6–13	1–4.5	22	356–520	1360–1390
Wheat	5–8	2.5–4.5	37	790–825	1400–1435
Barley	8–14	1–4.5	37	580–660	1390–1400
Corn	8–17	5–15	285	745	1310
Bullrush millet	3–4	2–3	11	760	1322
Wild rice	8–20	0.5–2	22	388–775	—
T'ef	1–1.5	0.5–1	0.3	880	—
Findi	1	1–1.5	0.4	790	—
Finger millet	1.5	1.5	—	—	—

[a] Ref. 31.

Wheat. The structure of the wheat kernel is shown in Figure 2 (35). The dorsal side of the wheat grain is rounded, while the ventral side has a deep groove or crease along the entire longitudinal axis. At the apex or small end (stigmatic end) of the grain is a cluster of short, fine hairs known as brush hairs. The pericarp, or dry fruit coat, consists of four layers: the epidermis, hypodermis, cross cells, and tube cells. The remaining tissues of the grain are the inner bran (seed coat and nucellar tissue), endosperm, and embryo (germ). The aleurone layer consists of large rectangular, heavy-walled, starch-free cells. Botanically, the aleurone is the outer layer of the endosperm, but as it tends to remain attached to the outer coats during wheat milling, it is shown in the diagram as the innermost bran layer.

The embryo (germ) consists of the plumule and radicle, which are connected by the mesocotyl. The scutellum serves as an organ for food storage. The outer layer of the scutellum, the epithelium, may function as either a secretory or an absorption organ. In a well-filled wheat kernel, the germ comprises ca 2–3% of the kernel, the bran 13–

17%, and the endosperm the remainder. The inner bran layers (the aleurone) are high in protein, whereas the outer bran (pericarp, seed coats, and nucellus) is high in cellulose, hemicelluloses, and minerals. The germ is high in proteins, lipids, sugars, and minerals; the endosperm consists largely of starch granules embedded in a protein matrix. Grains of rice, barley, oats, rye, and triticale are similar in structure to wheat.

Rice. Rice is a covered cereal; in the threshed grain (or rough rice), the kernel is enclosed in a tough siliceous hull, which renders it unsuitable for human consumption. When this hull is removed during milling, the kernel (or caryopsis), comprised of the pericarp (outer bran) and the seed proper (inner bran, endosperm, and germ), is known as brown rice or sometimes as unpolished rice. Brown rice is in little demand as a food. Unless stored under favorable conditions, it tends to become rancid and is more subject to insect infestation than are the various forms of milled rice. When brown rice is milled further, the bran and germ are removed and the purified endosperms are

Table 3. Approximate Grain Size and Proportions of the Principal Parts Comprising the Mature Kernel of Different Cereals[a]

Cereal	Grain Mass, mg	Embryo, %	Scutellum, %	Pericarp, %	Aleurone, %	Endosperm, %
Barley	36–45 (41)	1.85	1.53	18.3		79.0
Bread wheat	30–45 (40)	1.2	1.54	7.9	6.7–7.0	81–84
Durum wheat	34–46 (41)		1.6	12.0		86.4
Maize	150–600 (350)	1.15	7.25		5.5	82
Oats	15–23 (18)	1.6	2.13		28.7–41.4	55.8–68.3
Rice	23–27 (26)	2–3	1.5	1.5	4–6	89–94
Rye	15–40 (30)	1.8	1.73	12.0		85.1
Sorghum	8–50 (30)		7.8–12.1		7.3–9.3	80–85
Triticale	38–53 (48)		3.7	14.4		81.9

[a] Ref. 32.

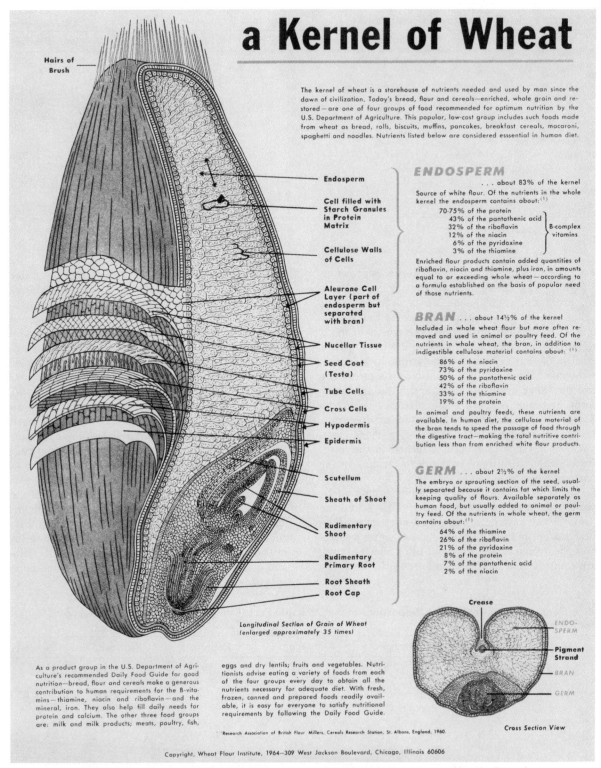

a Kernel of Wheat

Hairs of Brush

The kernel of wheat is a storehouse of nutrients needed and used by man since the dawn of civilization. Today's bread, flour and cereals—enriched, whole grain and restored—are one of four groups of food recommended for optimum nutrition by the U.S. Department of Agriculture. This popular, low-cost group includes such foods made from wheat as bread, rolls, biscuits, muffins, pancakes, breakfast cereals, macaroni, spaghetti and noodles. Nutrients listed below are considered esssential in human diet.

Endosperm

Cell filled with Starch Granules in Protein Matrix

Cellulose Walls of Cells

Aleurone Cell Layer (part of endosperm but separated with bran)

Nucellar Tissue

Seed Coat (Testa)

Tube Cells

Cross Cells

Hypodermis

Epidermis

Scutellum

Sheath of Shoot

Rudimentary Shoot

Rudimentary Primary Root

Root Sheath

Root Cap

Longitudinal Section of Grain of Wheat (enlarged approximately 35 times)

ENDOSPERM

... about 83% of the kernel

Source of white flour. Of the nutrients in the whole kernel the endosperm contains about:[1]

70-75% of the protein
43% of the pantothenic acid
32% of the riboflavin   B-complex
12% of the niacin   vitamins
6% of the pyridoxine
3% of the thiamine

Enriched flour products contain added quantities of riboflavin, niacin and thiamine, plus iron, in amounts equal to or exceeding whole wheat—according to a formula established on the basis of popular need of those nutrients.

BRAN ... about 14½% of the kernel

Included in whole wheat flour but more often removed and used in animal or poultry feed. Of the nutrients in whole wheat, the bran, in addition to indigestible cellulose material contains about:[1]

86% of the niacin
73% of the pyridoxine
50% of the pantothenic acid
42% of the riboflavin
33% of the thiamine
19% of the protein

In animal and poultry feeds, these nutrients are available. In human diet, the cellulose material of the bran tends to speed the passage of food through the digestive tract—making the total nutritive contribution less than from enriched white flour products.

GERM ... about 2½% of the kernel

The embryo or sprouting section of the seed, usually separated because it contains fat which limits the keeping quality of flours. Available separately as human food, but usually added to animal or poultry feed. Of the nutrients in whole wheat, the germ contains about:[1]

64% of the thiamine
26% of the riboflavin
21% of the pyridoxine
8% of the protein
7% of the pantothenic acid
2% of the niacin

Crease

ENDO-SPERM

Pigment Strand

BRAN

GERM

Cross Section View

As a product group in the U.S. Department of Agriculture's recommended Daily Food Guide for good nutrition—bread, flour and cereals make a generous contribution to human requirements for the B-vitamins—thiamine, niacin and riboflavin—and the mineral, iron. They also help fill daily needs for protein and calcium. The other three food groups are: milk and milk products; meats, poultry, fish, eggs and dry lentils; fruits and vegetables. Nutritionists advise eating a variety of foods from each of the four groups every day to obtain all the nutrients necessary for adequate diet. With fresh, frozen, canned and prepared foods readily available, it is easy for everyone to satisfy nutritional requirements by following the Daily Food Guide.

[1] Research Association of British Flour Millers, Cereals Research Station, St. Albans, England, 1960.

Figure 2. A wheat kernel. Longitudinal section (enlarged ca 12×): (**a**) hairs of brush; (**b**) endosperm; (**c**) cell filled with starch granules in protein matrix; (**d**) cellulose walls of cells; (**e**) aleurone cell layer (part of endosperm but separated with bran); (**f**) nucellar tissue; (**g**) seed coat (testa); (**h**) tube cells; (**i**) cross cells; (**j**) hypodermis; (**k**) epidermis; (**l**) scutellum; (**m**) sheath of shoot; (**n**) rudimentary shoot; (**o**) rudimentary primary root; (**p**) root sheath; (**q**) root cap. Cross section: (**a**) crease; (**b**) endosperm; (**c**) pigment strand; (**d**) bran; (**e**) germ (35).

marketed as white rice or polished rice. Milled rice is classified according to size as head rice (whole endosperm) and various classes of broken rice, known as second head, screenings, and brewer's rice, in order of decreasing size.

Corn. The corn grain is the largest of all cereals (Fig. 3) (35). The kernel is flattened, wedge shaped, and broader at the apex than at its attachment to the cob. The aleurone cells contain much protein and oil; they also contain the pigments that make certain varieties appear blue, black, or purple. Two types of starchy endosperms, ie, horny and floury, are found beneath the aleurone layer. The horny endosperm is harder and contains a higher level of protein. In dent corn varieties, the horny endosperm is found on the sides and back of the kernel and bulges in toward the center at the sides. The floury endosperm fills the crown (upper part) of the kernel, extends downward to surround the germ, and shrinks as dent corn matures, causing an indentation at the top of the kernel. In a typical dent corn, the pericarp comprises ca 6%, the germ 11%, and the endosperm 83% of the kernel. Flint corn varieties contain more horny than floury endosperm.

Barley. In barley, the husks are cemented to the kernel and remain attached after threshing. The husks protect the kernel from mechanical injury during commercial malting, strengthen the texture of steeped barley, and contribute to more uniform germination of the kernels. The husks are also important as a filtration bed in the separation of extract components during mashing and contribute to the flavor and astringency of beer. The main types of cultivated covered barley, depending on the arrangement of grains in the ear, are two-rowed and six-rowed barley. The axis of the barley ear has nodes throughout its length; the nodes alternate from side to side. In the six-rowed types of barley, three kernels develop on each node, one central kernel and two lateral kernels. In the two-rowed barley, the lateral kernels are sterile and only the central kernel develops.

The kernel of covered barley consists of the caryopsis and the flowering glumes (or husks). The husks consist of two membranous sheaths that completely enclose the caryopsis. During development of the growing barley, a cementing substance causing adherence is secreted by the caryopsis within the first two weeks after pollination. The husk in cultivated malting barley amounts to ca 8–15% of the grain. The proportion varies according to type, variety, grain size, and climatic conditions. Large kernels have less husk than small kernels. The husk in two-rowed barley is generally lower than that in six-rowed barley. The caryopsis, as in wheat, is a one-seeded fruit in which the outer pericarp layers enclose the aleurone, the starchy endosperm, and the germ. The aleurone layer in barley is at least two cell layers thick; in other cereal grains, except rice, it is one cell layer thick.

Oats. The common varieties of oats have the fruit (caryopsis) enveloped by a hull composed of certain floral envelopes. Naked or hull-less oat varieties are not grown extensively. In light thin oats, hulls may comprise as much as 45% of the grain; in very heavy or plump oats,

they may represent only 20%. The hull normally makes up ca 30% of the grain. Oat kernels, obtained by removing hulls, are called groats.

Sorghum. Sorghum kernels are generally spherical, have a kernel mass of 20–30 mg, and may be white, yellow, brown, or red. Sorghum grains contain polyphenolic compounds, primarily in the outer kernel layers and to some degree in the endosperm. These compounds are colored and impart various colors to certain foods from sorghums. Tannins, present in some sorghum grains, are polyphenols that interact with and precipitate protein. Tannins that are not hydrolyzed by enzymes impart some resistance to sorghum from attack by birds, but they also reduce nutritional value and germinability.

Composition

The chemical composition of the dry matter of different cereal grains, as with other foods of plant origin, varies widely (36). An approximate analysis of important cereal grains is summarized in Table 4. Variations are encountered in the relative amounts of proteins, lipids, carbohydrates, pigments, vitamins, and ash; mineral elements that are present and the quantities of them also vary widely. As a food group, cereals are characterized by relatively low protein and high carbohydrate contents; the carbohydrates consist essentially of starch (90% or more), dextrins, pentosans, and sugars.

The various components are not uniformly distributed in the different kernel structures. The hulls and pericarp are high in cellulose, pentosans, and ash; the germ is high in lipid content and rich in proteins, sugars, and ash components. The endosperm contains the starch and is lower in protein content than the germ and, in some cereals, than bran; it is also low in crude fat and ash components. As a group, cereals are low in nutritionally important calcium; its concentration and that of other ash components is greatly reduced by the milling processes used to prepare refined foods. In these processes, hulls, germ, and

Table 4. Approximate Analysis of Important Cereal Grains, Percentage of Dry Mass[a]

Cereal	Nitrogen	Protein[b]	Fat	Fiber	Ash	NFE[c]
Barley (grain)	1.2–2.2	11	2.1	6.0	3.1	—
(kernel)	1.2–2.5	9	2.1	2.1	2.3	78.8
Maize	1.4–1.9	10	4.7	2.4	1.5	72.2
Millet	1.7–2.0	11	3.3	8.1	3.4	72.9
Oats (grain)	1.5–2.5	14	5.5	11.8	3.7	—
(kernel)	1.7–3.9	16	7.7	1.6	2.0	68.2
Rice (brown)	1.4–1.7	8	2.4	1.8	1.5	77.4
(milled)	—	—	0.8	0.4	0.8	77.4
Rye	1.2–2.4	10	1.8	2.6	2.1	73.4
Sorghum	1.5–2.3	10	3.6	2.2	1.6	73.0
Triticale	2.0–2.8	14	1.5	3.1	2.0	71.0
Wheat (bread)	1.4–2.6	12	1.9	2.5	1.4	71.7
(durum)	2.1–2.4	13	—	—	1.5	70.0
Wild rice	2.3–2.5	14	0.7	1.5	1.2	74.4

[a] Ref. 32.
[b] Typical or average figure.
[c] NFE = nitrogen free extract (an approximate measure of the starch content).

a Kernel of Corn

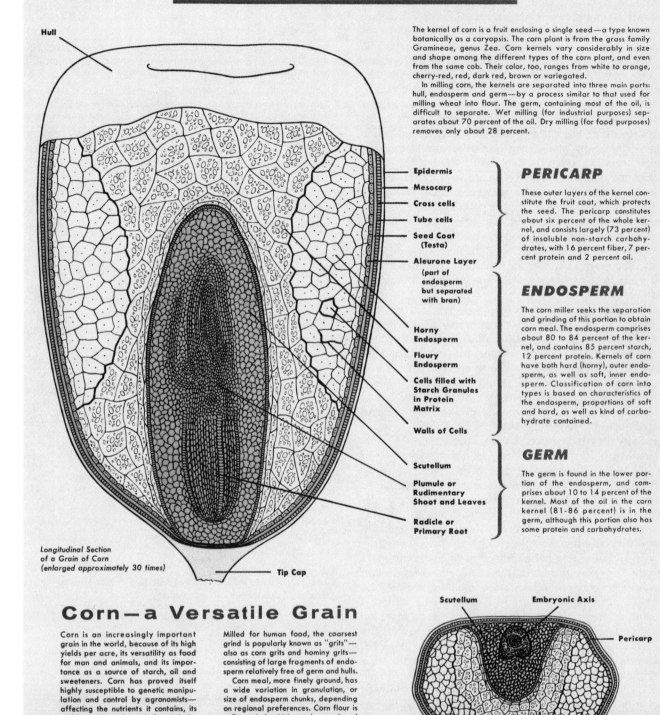

Hull

The kernel of corn is a fruit enclosing a single seed—a type known botanically as a caryopsis. The corn plant is from the grass family Gramineae, genus Zea. Corn kernels vary considerably in size and shape among the different types of the corn plant, and even from the same cob. Their color, too, ranges from white to orange, cherry-red, red, dark red, brown or variegated.

In milling corn, the kernels are separated into three main parts: hull, endosperm and germ—by a process similar to that used for milling wheat into flour. The germ, containing most of the oil, is difficult to separate. Wet milling (for industrial purposes) separates about 70 percent of the oil. Dry milling (for food purposes) removes only about 28 percent.

Epidermis
Mesocarp
Cross cells
Tube cells
Seed Coat (Testa)
Aleurone Layer
(part of endosperm but separated with bran)

Horny Endosperm
Floury Endosperm
Cells filled with Starch Granules in Protein Matrix
Walls of Cells

Scutellum
Plumule or Rudimentary Shoot and Leaves
Radicle or Primary Root

PERICARP

These outer layers of the kernel constitute the fruit coat, which protects the seed. The pericarp constitutes about six percent of the whole kernel, and consists largely (73 percent) of insoluble non-starch carbohydrates, with 16 percent fiber, 7 percent protein and 2 percent oil.

ENDOSPERM

The corn miller seeks the separation and grinding of this portion to obtain corn meal. The endosperm comprises about 80 to 84 percent of the kernel, and contains 85 percent starch, 12 percent protein. Kernels of corn have both hard (horny), outer endosperm, as well as soft, inner endosperm. Classification of corn into types is based on characteristics of the endosperm, proportions of soft and hard, as well as kind of carbohydrate contained.

GERM

The germ is found in the lower portion of the endosperm, and comprises about 10 to 14 percent of the kernel. Most of the oil in the corn kernel (81-86 percent) is in the germ, although this portion also has some protein and carbohydrates.

Longitudinal Section of a Grain of Corn (enlarged approximately 30 times)

Tip Cap

Corn—a Versatile Grain

Corn is an increasingly important grain in the world, because of its high yields per acre, its versatility as food for man and animals, and its importance as a source of starch, oil and sweeteners. Corn has proved itself highly susceptible to genetic manipulation and control by agronomists—affecting the nutrients it contains, its yields, its growth periods, and the climatic conditions in which it will flourish.

The U.S. leads the world in corn production, with a harvest of over 99 million tons (over 4 billion bushels) in 1965, or almost half of the total world production.

Milled for human food, the coarsest grind is popularly known as "grits"—also as corn grits and hominy grits—consisting of large fragments of endosperm relatively free of germ and hulls.

Corn meal, more finely ground, has a wide variation in granulation, or size of endosperm chunks, depending on regional preferences. Corn flour is a fine grind, which can be produced instead of, or together with grits or meal. In processing corn, there is almost no waste—there are uses and demands for all parts of the kernel. Corn can be called the most efficient and economical of the cereal grains.

Scutellum
Embryonic Axis
Pericarp
Horny Endosperm
Floury Endosperm

Cross Section View

Figure 3. A corn kernel. Longitudinal section (enlarged ca 8×): (**a**) hull; (**b**) epidermis; (**c**) mesocarp; (**d**) cross cells; (**e**) tube cells; (**f**) seed coat (testa); (**g**) aleurone layer (part of endosperm but separated with bran); (**h**) horny endosperm; (**i**) floury endosperm; (**j**) cells filled with starch granules in protein matrix; (**k**) walls of cells; (**l**) scutellum; (**m**) plumule or rudimentary shoot and leaves; (**n**) radicle or primary root; (**o**) tip cap. Cross section: (**a**) scutellum; (**b**) embryonic axis; (**c**) pericarp; (**d**) floury endosperm; (**e**) horny endosperm (35).

bran, which are the structures rich in minerals and vitamins, are nearly completely removed.

All cereal grains contain vitamins of the B group, but all are completely lacking in vitamin C (unless the grain is sprouted) and vitamin D. Yellow corn differs from white corn and the other cereal grains in containing carotenoid pigments (principally cryptoxanthin, with smaller quantities of carotenes), which are convertible in the body to vitamin A. Wheat also contains yellow pigments, but they are almost entirely xanthophylls, which are not precursors of vitamin A. The oils of the embryos of cereal grains are rich sources of vitamin E. The relative distribution of vitamins in the kernel is not uniform, although the endosperm invariably contains the least.

The protein contents of wheat and barley are important indexes for the manufacture of various foods. For example, the quantity and quality of wheat protein is critical in breadmaking. Cereal grains contain water soluble proteins (albumins), salt soluble proteins (globulins), alcohol soluble proteins (prolamins), and acid and alkali soluble proteins (glutelins). The prolamins are characteristic of the grass family, and together with the glutelins, comprise the bulk of the proteins of cereal grains. The following are names given to prolamins in proteins of the cereal grains: gliadin in wheat, hordein in barley, zein in maize, avenin in oats, kaffirin in grain sorghum, and secalin in rye.

The various proteins are not distributed uniformly in the kernel. Thus, the proteins fractionated from the inner endosperm of wheat consist chiefly of a prolamin (gliadin) and glutelin (glutenin). The embryo proteins consist of nucleoproteins, an albumin (leucosin), a globulin, and proteoses, whereas in wheat bran a prolamin predominates with smaller quantities of albumins and globulins. When water is added, the wheat endosperm proteins, gliadin, and glutenin, form a tenacious colloidal complex known as gluten. Gluten is responsible for the superiority of wheat over the other cereals for the manufacture of leavened products, because it makes possible formation of a dough that retains the carbon dioxide produced by yeast or chemical leavening agents. The gluten proteins collectively contain 17.55% nitrogen; hence, in estimating the crude protein content of wheat and wheat products from the

determination of total nitrogen, the factor 5.7 is normally employed rather than the customary value of 6.25; the latter value is based on the assumption that proteins contain an average of 16% nitrogen.

As a class, cereal proteins are not as high in biological value as those of certain legumes, nuts, or animal products. Zein, the prolamin of corn, lacks lysine and is low in tryptophan. The limiting amino acid in wheat endosperm proteins is lysine. Biological values of the proteins of entire cereal grains are greater than those of the refined mill products, which consist chiefly of the endosperm; however, American and Western European diets normally include animal products as well as cereals. Under these conditions, the different proteins tend to supplement each other.

The main form of carbohydrate is starch which is the primary source of calories provided by the grains. Most of the carbohydrates are in the starchy endosperm.

The lipids in cereals are relatively rich in the essential fatty acid, linoleic acid. Saturated fatty acids (mainly palmitic) represent <25% of the total fatty acids of most grains.

In summary, cereal grains are a diversified and primary source of nutrients. Their high starch contents make them major sources of calories; they also contribute to our needs for proteins, lipids, vitamins, and minerals. Vitamins and minerals lost during milling to produce refined food products can be, and in many countries are, replaced by nutrient fortification (Table 5). The composition of cereal grains and their milled products makes them uniquely suited for producing wholesome, nutritional, and consumer acceptable foods.

STANDARDS AND CLASSIFICATION

At the turn of the century, many organizations in the United States and in several other grain producing countries tried to develop uniform grain standards. The demand for uniform grades and application of the standards resulted in the introduction during 1903–1916 of 26 bills in the United States Congress.

Those bills and the extensive hearings culminated in

Table 5. FDA Cereal Enrichment Standards[a]

Product	Thiamine, mg/lb	Riboflavin, mg/lb	Niacin, mg/lb	Iron, mg/lb	Calcium,[b] mg/lb	Vitamin D, IU/lb
Flour	2.9	1.8	24	20	(960)	—
Self-rising flour	2.9	1.8	24	20	960	—
Corn grits	2.0–3.0	1.2–1.8	16–24	13–26	(500–750)	(250–1000)
Corn meals	2.0–3.0	1.2–1.8	16–24	13–26	(500–750)	(250–1000)
Rice	2.0–4.0	1.2–2.4	16–32	13–26	(500–1000)	(250–1000)
Macaroni products[c]	4.0–5.0	1.7–2.2	27–34	13–16.5	(500–625)	(250–1000)
Noodle products[c]	4.0–5.0	1.7–2.2	27–34	13–16.5	(500–625)	(250–1000)
Bread[d]	1.8	1.1	15	12.5	(600)	—
US RDA (mg or IU/day)[e]	1.5	1.7	20	18	1000	400

[a] The figures in this table are the nutrient levels to be present in the final products as set forth by the Food and Drug Administration (21 CFR 101.9, 137 and 139). Figures in parentheses are optional. When figures are shown as a range, these indicate min/max values. When one figure is shown, it is the minimum value required. These are 1986 standards.

[b] Calcium is added by the milling company. It cannot be included in the enrichment product.

[c] Enriched pasta products are normally made from flours enriched to these levels. Macaroni products with fortified protein must have the maximum figures shown but no Vitamin D.

[d] If enriched bread is manufactured using at least 62.5% of enriched flour, the bread will meet standards for enriched bread.

[e] U.S. RDA (recommended daily allowances) as published in the 1986 21 CFR 104.20.

the United States Grain Standard Act, which was passed Aug. 11, 1916, and amended in 1940 to include soybeans. This act provides in part for the establishment of official grain standards; the federal licensing and supervision of the work of grain inspectors; and the mechanisms of filing appeals concerning grades assigned by licensed inspectors. The official United States standards for grain cover wheat, corn, barley, oats, rye, sorghum, flaxseed, soybeans, mixed grain, and triticale.

According to United States grain standards, wheat is divided into seven classes on the basis of type and color: hard red spring, durum, red durum, hard red winter, soft red winter, white, and mixed. The division of wheat into classes and subclasses is according to suitability of different wheats for specific uses. Hard red spring, hard red winter, and hard white wheats are valued for the production of flours used in making yeast-leavened breads. Soft red winter, soft white, and white club wheats are especially suited for making chemically leavened baked goods (cookies, cakes). Amber durum wheat is prized for the manufacture of semolina and farina used to produce alimentary pastes (macaroni, vermicelli, etc). Red durum is used primarily as a poultry and livestock feed.

The wheat in each subclass is sorted into a number of grades on the basis of quality and condition. The tests generally used to determine the grade of grain and its conformity to the official United States grain standards include plumpness, soundness, cleanliness, dryness, purity of type, and the general condition of the grain (37,38).

Other cereals are graded in a similar manner. Corn is divided into three classes: yellow, white, and mixed. These class designations apply to dent type corn. If the corn consists of at least 95% flint varieties, the word flint is used. If the corn contains >5% but <95% flint varieties, the corn is designated as flint and dent. Waxy corn is corn of any class consisting of at least 95% waxy corn. For many years moisture content was a grading factor for corn and sorghum. Especially at the beginning of the harvest, moisture content is the single most important quality factor of corn. Broken corn and foreign material (BCFM) is determined by sieving through a $\frac{12}{64}$ in. (4.76 mm) sieve.

Rough rice is divided into three classes: long, medium, and short. There are separate standards for rough, brown, and milled rice. All rough rice is graded on the basis of milled rice yields.

Barley is divided into three classes (with several subclasses): six-rowed (malting, blue malting, and barley), two-rowed (two-rowed malting and two-rowed), and barley. Plumpness is determined by sieving. The designation malting requires inclusion of only varieties that have been approved as suitable for malting purposes. Special grades and designations of oats are: heavy or extra heavy (test weights of 48.9–51.5 kg/hL and >51.5 kg/hL, respectively), thin, bleached, bright, ergoty, garlicky, smutty, tough (moisture between 14–16%), and weevily.

BIBLIOGRAPHY

1. Canadian International Grains Institute, *Grain and Oilseeds, Handling, Marketing, Processing,* 2nd ed., Canadian International Grains Institute, Winnipeg, Canada, 1975.

2. R. C. Hoseney, *Principles of Cereal Science and Technology,* American Association of Cereal Chemists, St. Paul, Minn., 1986.

3. D. W. Kent-Jones and A. J. Amos, *Modern Cereal Chemists,* 6th ed., Food Trade Press, London, UK, 1967.

4. R. A. Olson and K. J. Frey, eds., *Nutritional Quality of Cereal Grains,* American Society of Agronomy, Madison, Wis., 1987.

5. Y. Pomeranz, *Modern Cereal Science and Technology,* VCH Publishers, Inc., New York, 1987.

6. Y. Pomeranz, ed., *Industrial Uses of Cereals,* American Association of Cereal Chemists, St. Paul, Minn., 1973.

7. Y. Pomeranz, ed., *Adv. Cereal Science and Technology,* Vol. I–X, American Association of Cereal Chemists, St. Paul, Minn., 1976–1990.

8. Y. Pomeranz, ed., *Cereals' 78: Better Nutrition for the Worlds' Millions,* American Association of Cereal Chemists, St. Paul, Minn., 1978.

9. Y. Pomeranz and L. Munck, eds., *Cereals: A Renewable Resource, Theory and Practice,* American Association of Cereal Chemists, St. Paul, Minn., 1981.

10. H. Hanson, N. E. Bourlaug, and R. G. Anderson, *Wheat in the Third World,* Westview Press, Inc., Boulder, Colo., 1982.

11. E. G. Heyne, ed., *Wheat and Wheat Improvement,* 2nd ed., American Society of Agronomy, Madison, Wis., 1987.

12. Y. Pomeranz, ed., *Wheat is Unique,* American Association of Cereal Chemists, St. Paul, Minn., 1989.

13. Y. Pomeranz, ed., *Wheat Chemistry and Technology,* 3rd ed., 2 Vol., American Association of Cereal Chemists, St. Paul, Minn., 1988.

14. N. L. Kent, *Technology of Cereals with Special Reference to Wheat,* Pergamon, Oxford, UK, 1960.

15. I. D. Morton, ed., *Cereals in a European Context,* VCH Publishers, Inc., New York, 1987.

16. W. R. Akroyd and J. Doughty, *Wheat in Human Nutrition,* FAO Nutritional Studies No. 23, FAO, Rome, Italy, 1970.

17. J. Holas and J. Kratochvil, eds., *Progress in Cereal Chemistry and Technology,* 2 Vol., Elsevier Science Publishing Co., Inc., Amsterdam, New York, 1983.

18. R. W. Jugenheimer, *Corn: Improvement, Seed Production and Uses,* John Wiley & Sons, Inc., New York, 1976.

19. P. J. Barnes, ed., *Lipids in Cereal Technology,* Academic Press, Inc., Orlando, Fla., 1983.

20. M. N. Leath, L. H. Meyer, and L. D. Hill, *U.S. Corn Industry,* Report No. 479, U.S. Department of Agriculture, Economic Research Service, Washington, D.C., 1982.

21. S. A. Watson and P. E. Ramstad, eds., *Corn: Chemistry and Technology,* American Association of Cereal Chemists, St. Paul, Minn., 1987.

22. B. O. Juliano, ed., *Rice, Chemistry and Technology,* 2nd ed., American Association of Cereal Chemists, St. Paul, Minn., 1987.

23. B. S. Luh, ed., *Rice: Production and Utilization,* AVI Publishing Co., Westport, Conn., 1980.

24. W. G. Heid, *U.S. Barley Industry,* Agr. Econ. Report No. 395, U.S. Department of Agricultural Economics, Statistics and Cooperative Service, Washington, D.C., 1978.

25. D. C. Rasmusson, ed., *Barley,* American Society of Agronomy, Madison, Wis., 1985.

26. F. H. Webster, ed., *Oats, Chemistry and Technology,* American Association of Cereal Chemists, St. Paul, Minn., 1986.

27. H. Doggett, *Sorghum,* Longmans, Green, and Co., London, UK, 1970.

28. L. W. Rooney, D. S. Murthy, and J. V. Mertui, eds., *Interna-*

tional Symposium on Sorghum Grain Quality, Proceedings, INCRASAT Patancheru, A. P. India, 1982.

29. W. Bushuk, ed., *Rye: Production, Chemistry, and Technology,* American Association of Cereal Chemists, St. Paul, Minn., 1987.

30. *Yearbook of Agriculture,* Government Printing Office, Washington, D.C., 1985.

31. H. G. Muller, "Some Physical Properties of Cereals and Their Products as Related to Potential Industrial Utilization," in Y. Pomeranz, ed., *Industrial Uses of Cereals,* American Association of Cereal Chemists, St. Paul, Minn., pp. 20–50, 1973.

32. D. H. Simmonds, "Structure, Composition, and Biochemistry of Cereal Grains," in Y. Pomeranz, ed., *Cereals '78: Better Nutrition for the World's Millions,* American Association of Cereal Chemists, St. Paul, Minn., pp. 105–137, 1978.

33. Y. Pomeranz, *Functional Properties of Food Components,* Academic Press, Inc., Orlando, Fla., 1985.

34. J. E. Kruger, D. R. Lineback, and C. E. Stauffer, eds., *Enzymes and their Role in Cereal Technology,* American Association of Cereal Chemists, St. Paul, Minn., 1987.

35. "From Wheat to Flour," Wheat Flour Institute, Chicago, Ill., 1965.

36. B. Holland, I. D. Unwin, and D. H. Buss, *Cereals and Cereal Products,* Royal Society of Chemistry, Ministry of Agriculture, Fisheries, and Food, London, UK, 1988.

37. C. M. Christensen, ed., *Storage of Cereal Grains and their Products,* 3rd ed., American Association of Cereal Chemists, St. Paul, Minn., 1982.

38. J. L. Multon, ed., *Preservation and Storage of Grains, Seeds, and their By-Products,* Lavoisier Publishing Inc., New York, 1988.

Y. POMERANZ
Washington State University
Pullman, Washington

CHEESE

ORIGIN

The first use of cheese as food is not known. References to cheeses throughout history are widespread. "Cheese is an art that predates the biblical era" (1). The origin of cheese has been dated to 6000–7000 BC and the worldwide number of varieties has been estimated at 500, with an annual production of more than 12 million tons growing at a rate of about 4% (2). This food served as a source of vital nourishment in huts and castles and on journeys; armies also used cheese. It also was a way of preserving food during periods of shortages. A plausible tracing of the origin and development of cheese, as a food, around the world has been published (1).

CLASSIFICATION

Cheeses have been classified in several ways. Several attempts to classify the varieties of cheese have been made. One suggestion consists of scheme that divides cheeses into the following superfamilies based on the coagulating agent.

1. *Rennet Cheeses.* Cheddar, brick, Muenster
2. *Acid Cheeses.* Cottage, Quarg, cream

3. *Heat–Acid.* Ricotta, sapsago
4. *Concentration–Crystallization.* Mysost

A more simple but incomplete scheme, would be to classify cheeses as follows.

1. *Very Hard.* Parmesan, Romano
2. *Hard.* Cheddar, Swiss
3. *Semisoft.* Brick, Muenster, blue, Havarti
4. *Soft.* Bel Paese, Brie, Camembert, feta
5. *Acid.* Cottage, baker's, cream, ricotta

A broad look at cheeses might divide them into two large categories, ripened and fresh.

Ripened

Cheeses can be ripened by adding selected enzymes or microorganisms (bacteria or molds) to the starting milk, to the newly made cheese curds, or to the surface of a finish cheese. The cheese is then ripened (cured) under conditions controlled by one or more of the following elements: temperature, humidity, salt, and time.

Depending on the style of cheese, the ripening can be principally carried out on the cheese surface or the interior. The selection of organisms, the appropriate enzymes, and ripening regime determine the texture and flavor of each cheese type.

Fresh

These cheeses do not undergo curing and are generally the result of acid coagulation of the milk. The composition, as well as processing steps, provides the specific product texture, while the bacteria used to provide the acid usually generate the characteristic flavor of the cheese.

CHEESEMAKING—GENERAL STEPS

1. Milk is clarified by filtration or centrifugation.

2. Depending on the composition of the final cheese, the fat content is standardized in a special centrifuge (separator).

3. Depending on the variety, the milk is pasteurized (generally at about 72°C (161°F) for 16 s) or undergoes any equivalent heat treatment that renders the milk phosphatase negative. Phosphatase is inactivated in the same range as the tuberculin bacilli. The phosphatase test insures that the milk is free of tuberculin bacilli and other pathogens.

4. Before pumping the milk to the cheese vat, the milk fat can be homogenized, if desired. The extent of this homogenization will depend on the cheese variety being made (Fig. 1).

5. A bacterial starter culture is added to the milk, which is at 30–36°C, depending on the cheese being made. This inoculated milk is generally ripened (held at this temperature) for 30–60 min to allow the lactic organisms to multiply sufficiently for their enzyme systems to convert some of the milk sugar (lactose) to lactic acid.

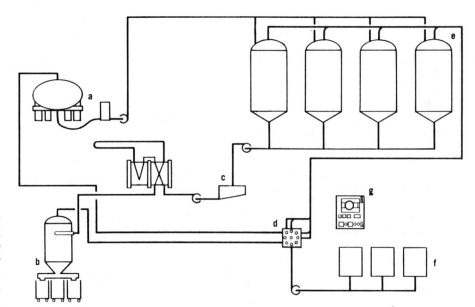

Figure 1. Natural cheese system flow key (**a**) receiving pump/air eliminator; (**b**) cyclone filter; (**c**) balance tank; (**d**) flow direction control; (**e**) cheese vat; (**f**) CIP system; (**g**) CIP control panel. Courtesy of Sherping Systems, Winsted, Minn.

6. After ripening, a milk-coagulating enzyme is added with stirring. The milk in the vat is then allowed to set in a quiescent state.

7. The resulting coagulum is usually ready to be cut 20–30 min after the enzyme addition. For most fresh cheeses little or no coagulating enzyme is added. Coagulation is allowed to occur by the development of lactic acid alone.

8. When the coagulum is firm enough to be cut, a combination of vertical and horizontal stainless steel knives (wires) are pulled lengthwise and crosswise through the coagulum in a rectangular vat. If an automatic enclosed circular vat is used, the cutting is programmed to insure a curd-size range by the speed and timing of the automatic knives.

9. For most varieties a heal time is allowed where no stirring is done for 5–15 min after cutting. This allows the newly cut curd particles to firm up slightly and begin syneresis (excluding whey from the particle during the shrinking process).

10. Following the heal period, heat is applied to the jacket of the vat and gentle stirring begins. For most ripened cheeses the curds are cooked in the whey until the temperature of the mixture reaches 37–41°C, depending on the variety. For some cheeses, for example, Swiss and Parmesan, this temperature can be as high as 53°C. The cheese is generally cooked for 30 min. Fresh cheeses, such as cottage, cream, and Neufchâtel, are cooked at temperatures as high as 51.5–60°C to promote syneresis and provide product stability.

11. Generally, when the cook temperature is reached, a 45–60 min period is allowed to promote curd firming and the releasing of whey. During this stir-out time the contents of the vat are agitated somewhat vigorously.

12. From this point the whey is drained and the treatment of the new curd depends on the nature of the final product.

a. For cheeses such as Cheddar, Monterey Jack, Colby, brick, and mozzarella the curd can be handled in a number of ways. Cheddar can be made by allowing the curd to settle in a rectangular vat, then draining the whey from the curd. When the whey is removed, the curd is trenched down the middle of the vat, lengthwise. The curd is hand cut, turned over, and then piled, at intervals, one slab on another. During this time acid continues to develop and whey continues to be exuded from the curd. When the proper pH is reached (about 5.1–5.3) the curd is milled by machine into finger-size pieces and slowly dry-salted to provide about 1.6–1.8% salt in the final cheese. The salted curds are then packed into 40-lb stainless-steel hoops with porous netting lining the hoops or into 600–700-lb boxes. These containers are generally pressed at 15–20 psi to remove more whey over several hours and then placed into a vacuum chamber while under pressure to fuse the curd particles into a solid mass.

Several methods are used to arrive at a solid cheese block. One process provides for pumping the curds into a tower that allows whey drainage and compacting of the curds by their own weight. At the bottom of the tower a portion of the curd is cut off to fit into a 40-lb hoop, which travels to a press station and, finally, to a packaging station.

In another method, curd and whey are pumped from the vat to a draining-and-matting conveyer (DMC) under controlled temperature and curd depth. When the curd has reached the proper pH, the mat is cut and is automatically salted and transferred onto another conveyer where it is finally carried to a boxing operation (Fig. 2).

There is also a stirred-curd procedure used for mild- and medium-flavored Cheddar as well as Colby, brick, mozzarella, etc. In this method the curd and whey are pumped to a drain table or automatic-finishing vat (AFV) where the whey is drained and the curds are continually stirred to prevent matting. When the

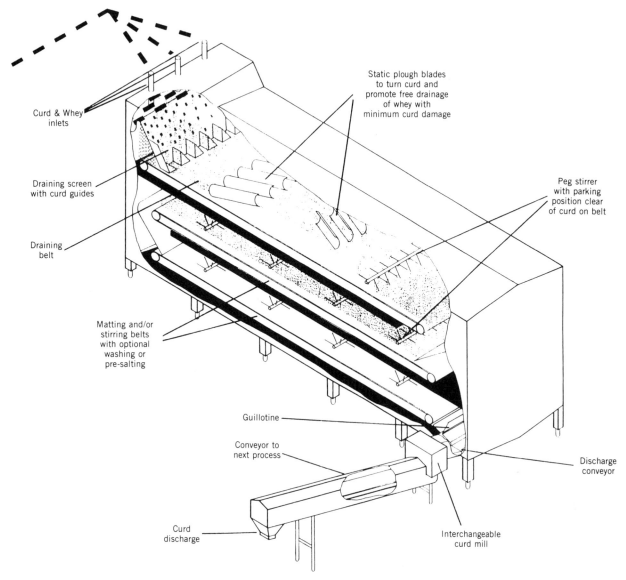

Curd & Whey inlets

Static plough blades to turn curd and promote free drainage of whey with minimum curd damage

Draining screen with curd guides

Draining belt

Peg stirrer with parking position clear of curd on belt

Matting and/or stirring belts with optional washing or pre-salting

Guillotine

Conveyor to next process

Discharge conveyor

Curd discharge

Interchangeable curd mill

Figure 2. Cheese curd draining conveyor. Courtesy of Sherping Systems, Winsted, Minn.

proper pH is achieved the cubes are salted and then boxed as previously described (Fig. 3).

If stirred curd Cheddar is intended for use in process cheese it can be packed into 500-lb drums and used after a short curing period. It should be noted that cheeses such as Monterey Jack and Colby, should not be placed under vacuum to preserve a slight open texture.

b. Blue cheese is made in a similar manner up to this point, but the mold (*Penicillium* sp.) can be added to the starting milk or to the drained curds. This cheese is usually formed in 5-lb hoops to promote an open texture. When the cheese is removed from the hoops it is dry-salted several times over a number of days and then punctured to create air channels in which the mold grows; this develops the characteristic blue veining. This cheese may be salt brined instead to obtain a 4% salt content in the final cheese. The cheese is cured

for 60–120 days at 13°C and 95% humidity to achieve full veining and flavor.

c. Swiss, Parmesan, and other varieties are handled somewhat differently at the whey drainage step. For years these cheeses were manufactured in Europe in wheel-shaped hoops; 20-lb wheels were used for Parmesan and 175–225-lb hoops, for Swiss. The principles are the same so the description here will pertain to most U.S. products. For Swiss cheese the curds and whey are pumped to a large stainless-steel universal vat, which has been equipped with porous side plates, where the curd is allowed to settle evenly at the bottom. Large stainless-steel plates are used to press the curd at a precise depth of the curd mass. This pressing under whey results in a tightly fused cheese, required later for eye development. The whey is drained and the huge curd block is kept under pressure for about 16 h while it continues to develop acidity. Then the 3,000–

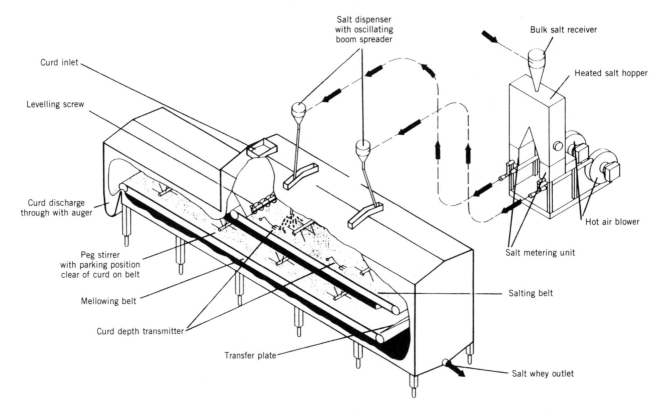

Figure 3. Dual salt applicating and cheese curd mellowing system. Two stage continuous salting and mellowing of milled or granular curd. Courtesy of Sherping Systems, Winsted, Minn.

3,500-lb block is cut into sections of about 180 lb and immersed into 2°C saturated brine for 24 h. The surface of the blocks are then dried. The blocks are packaged and boxed. These boxes are stacked and banded to help keep the block shape as the cheese cures in a hot room at 20–25°C. When the eyes have formed (about 18–28 days) the blocks are transferred to a room at about 2°C to prevent further eye development and held for at least 60 days before cutting into retail sizes. The cheese can be held for longer periods if stronger Swiss flavor is desired.

For Parmesan cheese, the curd is stirred while draining off the whey and then placed into round 20-lb hoops. After pressing, the round cheeses are allowed to surface dry before brining. If salt has not been added during the stirring of the curds, additional brining time is required. The cheeses are held at 13–16°C until the proper weight is reached (moisture reduced) and then vacuum packed in plastic bags with a tight seal. Some factories wax the cheese. This cheese requires a minimum of a 10-mo cure time by federal regulation. Most of the U.S. Parmesan is grated for use as a retail product.

d. Mozzarella and provolone are among those cheeses referred to as pasta filata varieties. Provolone has some lipase enzymes added to produce its characteristic flavor and sometimes is sold naturally smoked or with smoke flavor added. These cheeses have traditionally been further cooked, after whey drainage, in hot water and stretched by hand with subsequent shaping into the familiar forms. Today, however, most moz-

zarella and provolone are mechanically cooked under water and further stretched to provide texture and then shaped in molding machines. The cheeses are then brined to the proper salt level.

e. Cheeses such as Camembert, Brie, Muenster, and Limburger are made in smaller equipment and formed into smaller shapes to allow surface salting or brining and treatment of the cheese surface with selected bacteria or molds to provide the characteristic flavors, and textures.

f. Fresh cheeses such as cottage, queso fresco, and cream are mostly acid in nature and, following cooking of the curds, are treated differently. Cottage curds are washed with filtered water and then combined with a cream dressing. Queso fresco is pressed into loaves after salting and eaten fresh after slicing. Cream cheese curd is separated centrifugally and then standardized in regard to fat, salt, and stabilizers. Stabilizers are added to prevent moisture release. It can then be packaged in small loaves, hot or chilled.

RECENT DEVELOPMENTS

A number of cheese varieties have been manufactured from milk concentrated by ultrafiltration (3,4). Patents have been issued (5) and some procedures have been reduced to practice, but after more than 20 years of research in this area only a minimal amount of the total world production of cheese is made with ultrafiltered milk.

Bactofugation used to remove microbes and spores from cheese milk has sparked some interest, especially in Europe (6). Microfiltration may find significant application in special milk treatment prior to, or during, product manufacture. Significant developments in recombinant biotechnology and fermentation technology may provide specificity previously not available for milk coagulation and cheese flavor development. New cultures, which are resistant to phage, may be on the horizon. New cheese-making technology may allow previously unattainable processes and product control to become reality.

Cheese Starter Cultures

Starter cultures are organisms that ferment lactose to lactic acid and other products. These include streptococci, leuconostocs, lactobacilli, and streptococcus thermophilus. Starter cultures also include propionibacteria, brevibacteria, and mold species of Penicillium. The latter organisms are used in conjunction with lactic acid bacteria for a particular characteristic of cheese, eg, the holes in Swiss cheese are due to propionibacteria, and the yellowish color and typical flavor of brick cheese is due to brevibacterium linens. Blue cheese and Brie derive their characteristics from the added blue and white molds, respectively. Lactic acid-producing bacteria have several functions:

1. Acid production and coagulation of milk.
2. Acid gives firmness to the coagulum, which influences cheese yield.
3. Developed acidity determines the residual amount of animal rennet influencing cheese ripening; more acid curd binds more rennet.
4. The rate of acid development influences dissociation of colloidal calcium phosphate, which then influences proteolysis during manufacture and the rheological properties of cheese.
5. Acid development and production of other antimicrobials control the growth of certain nonstarter bacteria and pathogens in cheese.
6. Acid development contributes to proteolysis and flavor production in cheese.

Types of Culture

There are some impending changes in the nomenclature of starter culture organisms; the nomenclature used here follows present standards (7). Mesophilic cultures are used for cheese types that do not exceed 40°C during cheese making (8). These starters are propagated at 21–23°C and include *Streptococcus lactis* ssp. *lactis*, *S. lactis* ssp. *cremoris*, *S. lactis* ssp. *diacetylactis* and *Leuconostoc cremoris*. These are used for Cheddar, Gouda, brick, Muenster, cream, cottage, and Quarg cheese types. leuconstoc and *S. lactis* ssp. *diacetylactis* ferment citrate to produce carbon dioxide and diacetyl to characterize Edam, Gouda, and cream cheese.

Thermophilic cultures are used in cheese types where curd is cooked to 45–56°C. These starters are propagated at 40–45°C and include *S. thermophilus,* sometimes fecal streptococci, *Lactobacillus helveticus, L. delbrueckii* ssp. *lactis,* and *L. delbrueckii* ssp. *bulgaricus.* These are used for Swiss, Emmentaler, Gruyère, Parmesan, Romano, mozzarella, and Gorgonzola cheeses (9).

Propagation

Cultures are propagated in milk or a medium containing milk components and other nutrients and are heat treated (85°–90°C for 45 min) to render milk free of contaminants. The medium is then cooled to incubation temperature before inoculation. This processing is accomplished in a jacketed stainless-steel vessel provided with agitation and supplied with sterile air pressure. Culture inoculum is available from suppliers in frozen or freeze-dried form. It consists of appropriate mixture of culture strains and is free of contamination. After inoculation, culture is allowed to grow in quiescent state for 12–18 h. At the end of growth the medium is acidic with a pH of 4.5 or lower, depending on the culture. Ripened culture cell population is about 10^9 cfu/g. The ripened culture is cooled to at least 5°C. This form of propagation is called bulk-starter propagation. The culture may be used immediately or held for several days. The growth medium may contain buffering salts or the acid produced may be neutralized and controlled by the addition of ammonia, potassium hydroxide, or sodium hydroxide under controlled conditions. The latter is called pH-controlled propagation.

Starter cultures are susceptible to bacteriophages (viruses), which destroy the cultures and hamper cheese production. Cultures are also sensitive to antibiotics and bacteriocins (proteinaceus bacterial inhibitors). Propagated cultures are used at 1–5% of cheese milk. Highly concentrated (10^{10} cfu/g) frozen cultures are available for direct addition to cheese milk (350 g/5,000 lb). These are called direct vat set (DVS). This form is convenient, but expensive.

Coagulation

Coagulation of milk is essential to cheese making. The coagulation entraps fat and other components of milk. Most proteolytic enzymes can cause milk to coagulate. Rennet (chymosin, EC 3.4.23.4) is widely used for milk coagulation in cheese making (10). It is extracted from the fourth stomach of young calves. Commercial rennets may include blends of chymosin and pepsin (bovine or other animals). Microbial rennets with similar functionality prepared from *M. miehei* are known as Marzyme, Hannilase, Rennilase, and Fromase. Preparations from *M. pusillus* and *Endothia parasiticus* are called Emporase and Sure Curd, respectively. Proteases from plants are known to coagulate milk but are not used in commercial cheese making. Coagulation of milk is also effected by lowering pH to about 4.6 in quiescent state either by fermentation of lactose to lactic acid or by hydrolysis of gamma–delta lactones to produce cottage, Quarg, ricotta, and cream cheese.

Animal and microbial rennets are aspartic proteases with optimum activity under acidic conditions. Their molecular weights range from 30,000 to 38,000. The primary stage of rennet coagulation involves partial hydrolysis of K-casein, the principal stabilizing factor of milk protein, at Phe-105 to Met-106 bond. Destabilization of the residual protein (*para*-casein) occurs in the presence of Ca^{2+} at temperatures >18°C in the secondary, nonenzymatic stage of rennet coagulation. About 30% of calf rennet is retained in the curd before pressing and only 5–8% after pressing (11). The amount of rennet retained is governed by the pH of the curd. Microbial rennets are retained to a lesser extent (3–5%).

During coagulation, linkages between casein micelles are formed and many micelles are joined by bridges, but later these appear to contract, bringing the micelles into contact and causing partial fusion. The firmness of curd is due to an increase in both the number and strength of linkages between micelles. It is suggested that phosphoryl side chains of casein specially β-casein are involved. These may be linked by Ca^{2+} bridges. In cheese, αs-1-casein plays a structural role (12).

Casein aggregation and fusion continues after coagulum cutting and throughout cheese making and early stages of cheese ripening. This process of rennet coagulation of milk is fundamental to the conversion of milk to cheese curd and its manipulation is essential for the control of cheese manufacture (Table 1). Casein aggregation leads to curd formation and syneresis. Increase in acidity and temperature and the addition of calcium chloride accelerates the rate of curd formation and its syneresis. On the other hand, cold storage of milk, homogenization, higher pasteurization temperature, and increased fat content impede the two processes. Acid coagulation of milk, as in cottage-cheese making results in less aggregation of casein than is seen with rennet. During cottage-cheese making the pH of milk casein drops to its isoelectric point resulting in a soft, fragile gel, firm enough to be cut. For large-curd cottage cheese, a very small amount of rennet is added. Essentially all added rennet is inactivated at the cooking temperature of curd (57–60°C).

MICROBIOLOGICAL AND BIOCHEMICAL CHANGES DURING CHEESE MAKING AND CHEESE RIPENING

During cheese manufacture, milk is dehydrated to a given composition, characteristic of a family of cheeses. The distinguishing features of a cheese within a family are a function of smaller but significant deviations from set practices. The following pairs of cheeses represent good examples of the effect of manufacturing variations: Edam and Gouda, brick and Limburger, mozzarella and provolone, and Cheddar and Colby. Rennet curds have unique properties that can be handled to produce high- and low-moisture cheeses without hardening. Limited, but essential, proteolysis of milk protein by rennet enzymes augments the shift of microbial populations that, in turn, are pressed into summoning those metabolic activities that must transform the milk component simply to survive. In doing so, the chemical entities generated interact among themselves and with the microbial population to result in a more flavorful and preserved milk. Both casein and milk fat hydrolysis is needed for cheese flavor development, but the rate and extent of such hydrolysis must be controlled to maintain cheese identity. Different cheese types are discussed below with these comments in mind.

Lactic starter cultures are added to vat milk to give about 10^6–10^7 cfu/mL. The amount and type of starter may vary significantly depending on the type of cheese and characteristics desired. In most cheese types an overnight pH range is 4.95–5.3.

Cheddar-Type Cheese

In Cheddar, Caciocavallo, and Colby types of cheese, about 30–40% of the added culture cells are lost in whey. The cells trapped in the rennet coagulum rapidly multiply

Table 1. Milk Coagulants for Cheese Making

Coagulant	Supplier/Designation			
	Hansen's	Marschall's	Pfizer	SANOFI
Calf	Standard	Calf rennet Cheese rennet	Calf rennet	American rennet[a]
Bovine	Bovine			Beef rennet
Porcine			Marla Set	Pepsin
Calf–porcine	50–50	Chymoset	Econozyme	Quick Set[b]
Bovine–porcine	B–P			
Calf–bovine–porcine				Trizyme
Mucor miehei (modified thermolabile equivalent to Novo Rennilase XL)	Hannilase (HL)		Mor Curd	
Mucor miehei (modified extra thermolabile equivalent to Novo Rennilase XL)	Superlase	Marzyme Supreme	Mor Curd Plus	
Endothia parasiticus			Sure Curd	
Mucor pusillus				Emporase[c]
Microbial, animal	C and P (Chees-zyme and Pep-zyme blend)	M–P (50–50 Hog and *M. miehei*)		Regalase (porcine and Emporase) Beemase (bovine and Emporase)

[a] DFL sells three grades of calf rennet: PD300 (highest chymosin), SF100 (intermediate chymosin), and EL400 (lowest chymosin).
[b] DFL sells three grades of Quick Set.
[c] DFL sells three grades of Emporase.

and ferment lactose to lactic acid. The population of starter organisms may reach in excess of 5×10^8 cfu/g in curd before salting. These bacterial numbers include a 10-fold increase due to milk solids concentration. The cultures multiply only slightly during coagulation and cooking, but growth and acid production accelerate after whey is removed and continues through cheddaring as the starter cells are concentrated in the curd. Acid production will continue until the lactose is depleted. At the time of milling the curd may have a pH of 5.3 and titratable acidity of whey at 0.57% or higher.

The milling operation involves cutting the curd into small pieces to facilitate uniform salt distribution and whey drainage. Enough salt is applied to the curd in two or three applications to yield a salt-in-moisture phase ($S/S + M$) of the cheese of about 4–5%. Due to steady acid production throughout cheese making the casein curd is demineralized with respect to calcium and phosphorus. The rate and extent of acid production governs the residual chymosin, minerals, moisture, and the cheese texture and structure. Higher acid in cheese reflects higher starter populations, possibly resulting in bitter cheese due to total proteinase activity from starter cells.

Cheddar-type curd is dry salted. Addition of salt to the curd is the last step in the manufacture of cheese. It controls microbial growth and activity and the final pH of hooped cheese. At hooping Cheddar cheese may contain 0.6% lactose. Its fermentation to lactic acid depends on $S/S + M$. Culture activity is inhibited by $S/S + M$ of > 5% (13). Salt also helps to expel moisture. Proteolytic activity of chymosin and pepsin on αs-1-casein is stimulated by salt, whereas β-casein proteolysis is inhibited by $S/S + M$ of 5%. Salt appears to affect physical changes in cheese protein, which influence cheese texture, protein solubility and probably protein conformation. Cheddar-type cheese is internally ripened by chymosin in concert with starter proteinases and adventitious lactobacilli. For normal ripening a high starter population must lyse to release proteinases and peptidases affecting cheese flavor, body, and texture development. In most cheese types there is a shift in the population of starter organisms to lactobacilli and in some cases pediococci. It has been demonstrated that in young cheese there are inhibitory and stimulatory factors for lactobacilli. As the cheese ripens, most of the inhibition vanishes. These stimulatory factors for lactobacilli appear to originate primarily in αs-casein (14). In Cheddar cheese, β-casein is not extensively degraded (15). Studies with starters lacking proteinase indicate the importance of starter proteinases (16). Proteolysis and increasing concentration of free amino acids add to the savory taste and provide substrate for sulfur compounds. Many flavor compounds are chemical interactions of microbially derived substrates under conditions of low pH and low oxidation–reduction potential. Numerous compounds, such as hydrocarbons, alcohols, aldehydes, ketones, acids, esters, lactones, and sulfurs are important in cheese flavor. Hydrogen sulfide and methanethiol along with phenylacetaldehyde, phenylacetic acid, and phenethanol are considered key compounds of good Cheddar flavor (17). Lactic acid bacteria are not lipolytic but small increases in butyric and other fatty acids appear desirable.

Swiss, Emmentaler, and Gruyère

These cheese types are made with thermophilic streptococci and lactobacilli to which propionibacteria are added for eye formation. During the early phases of cheese making $S.$ $thermophilus$ multiplies rapidly and utilizes the glucose moiety of lactose to produce L-lactate, leaving behind galactose. Lactobacilli start vigorous acid production after whey drainage when the curd temperature drops to 46–49°C. Acid production and starter bacteria number are greater toward the periphery where the cheese has cooled down. Lactic acid fermentation in cheese is complete within 24 h with no detectable sugars.

Changes During Ripening. Increase in starter organisms population and acidity are halted when the cheese is brined. This is due to exhaustion of fermentable sugar and the low brine temperature. Cheese is cured in a hot room (20°–25°C) for 18–40 days. During this period the added starter organisms disappear and the niche is occupied by propionibacteria, adventitious lactobacilli, and fecal streptococci. Propionibacteria ferment lactate to propionic acid and other flavor compounds according to the reaction: $3CH_3CHOHCOOH \rightarrow 2CH_3CH_2COOH + CH_3COOH + CO_2 + H_2O$.

Eye Formation. The quality of Swiss cheese is judged by the size and distribution of eyes. Swiss cheese eyes are essentially due to carbon dioxide production, diffusion, and accumulation in the cheese body (18). The number and size of eyes depend on carbon dioxide pressure, diffusion rate, body texture, and temperature of cheese. The bulk of protein breakdown products are derived from αs-casein. β-Casein is not greatly proteolysed in Swiss cheese (15). Bacterial proteinases and plasmin (milk proteinase) are considered important in Swiss cheese ripening.

Flavor. Cheese flavor is derived in part from cheese milk. However, the characteristic flavor of Swiss-type cheese comes from the microbial transformation of milk components. These contain water-soluble volatiles (acetic acid, propionic acid, butyric acid, and diacetyl), which give the basic sharpness and general cheesy notes (17). Water-soluble nonvolatile amino acids (especially proline), peptides, lactic acid, and salts provide mainly sweet notes. Oil-soluble fractions (short-chain fatty acids) other than the water-soluble volatiles are important to flavor. Nutty flavor is attributed to alkylpyrazines. Several compounds, eg, ketones, aldehydes, esters, lactones, and sulfur-containing compounds are important. Due to the activity of certain strains of lactobacilli and fecal streptococci, biogenic amines are sometimes found in Swiss cheese.

Camembert Cheese

This is a soft cheese with 48–52% moisture. Milk inoculated with 1–5% mesophilic lactic starter is renneted at about 30°C. When the acidity reaches 0.20–0.23% the curd is dipped, molded, and held at 21°C in rooms with 85–

90% humidity until it reaches pH 4.5–4.6. During storage, yeast and *Geotrichum candidum* appear on the surface. After about a week at about 10°C *Penicillium camemberti* covers the surface and deacidifies the cheese in 15–20 days. This is followed by colonization of the cheese surface by *B. linens* and micrococci. During ripening, the surface pH rises to about 7.0. β-Casein is not extensively degraded. Appearance of some γ-casein suggests plasmin activity. There is αs-casein degradation but less compared to Cheddar cheese. In Camembert, due to extensive deamination, ammonia constitutes 7–9% of the total *N*. Free fatty acids found in large quantities contribute to the basic flavor of cheese and serve as the precursors of methylketones and secondary alcohols. Aldehydes, methylketones, 1-alkanols, 2-alkanols, esters (C$_2$, 4, 6, 8, 10-ethyl 2-phenylethylacetate), phenol, *p*-cresol, lactones, hydrogen sulfide, methanethiol, methyldisulphide, and other sulfur compounds along with anisoles, amines, and other compounds constitute the volatile compounds.

Blue Cheese

Blue cheese and its relatives, Roquefort, Stilton, and Gorgonzola are characterized by peppery, piquant flavors produced by the mold *Penicillium roqueforti*. Milk for cheese making may be homogenized to promote lipolysis in cheese. An acid curd with pH 4.5–4.6 is produced with mesophilic lactic starter and sometimes include *S. lactics* ssp. *diacetylactis*. Gorgonzola is made with *S. thermophilus* and *Lactobacillus bulgaricus*. The blue mold is added to the milk or curd. The molded curd is dry salted, and salt is rubbed on cheese to attain 4–5% salt. After about a week the cheese heads are pierced with needles to aerate the cheese and let carbon dioxide escape and air enter. This promotes mold growth.

Due to high acid and salt, starter population declines rapidly and the blue mold grows throughout the cheese. Salt-tolerant *B. linens* and micrococci appear on the cheese surface in two to three weeks. Due to deacidification of the cheese and extensive proteolysis, the cheese pH rises from 4.5 to 4.7 at 24 h to a maximum of 6.0–6.25 in two to three months.

Molds are both proteolytic and lipolytic, resulting in extensive proteolysis and lipolysis of cheese. In blue cheese both αs- and β-casein are degraded. There is a large accumulation of free amino acids due to extracellular acidic and alkaline endopeptidases (19). *B. linens* and micrococci are also proteolytic and contribute to the overall proteolysis. Occurrence of citrulline, ornithine, γ-aminobutylic acid, and biogenic amines reflect amino acid break-down products. Other volatile compounds such as ammonia, aldehydes, acids, alcohols, and methanethiol are also produced from amino acid metabolism.

Mold-ripened cheeses have a high free fatty acid content, 27–45 meq of acid per 100 g of fat. *Penicillium roqueforti* produces two lipases, one active at pH 7.5–8.0 or higher and the acid lipase active at pH 6.0–6.5. The fatty acids produced are converted into methylketones via ketoacyl–coenzyme A and the β-keto acids (20). Thiohydrolases are also present in cheese, resulting in the accumulation of 2-heptanone. The rate of fatty acid release governs the rate of methylketone formation (21).

Brick Cheese

Brick cheese is representative of a large group of cheeses (Limburger, Muenster, Tilsiter, Bel Paese, and Trappist) ripened with the aid of surface flora made of yeast, micrococci, and coryneform bacteria (*B. linens*). A culture of *S. lactis* ssp. *lactis*, *S. lactis* ssp. *cremoris*, and sometimes in combination with *S. thermophilus* is used to cause a cheese pH of 5.0–5.2. After brining for one or more days cheese is held at about 15°C (60°F) in rooms with a 90–95% RH where yeast (*Mycoderma*) appear on the surface in two or three days followed by micrococci and then *B. linens*. Organisms from the genus *Arthrobacter* have also been isolated from the cheese surface. Cheese ripens in four to eight weeks depending on the moisture and intensity of cheese flavor desired. Film-wrapped cheese has more rounded flavor than waxed cheese.

Growth of yeast lowers the acidity on the surface, making it suitable for micrococci and *B. linens* colonization (22). Sometimes *G. candidum* may also be present. αs-Casein is always hydrolysed but β-casein disappearance was seen in Muenster and not in brick. Yeasts found on Limburger cheese synthesize considerable amounts of pantothenic acid, niacin, and riboflavin (23). Pantothenic acid and *para*-aminobenzoic acid are required by *B. linens* Thus explaining the population sequence. Liberated free amino acids are much higher on the surface than in the interior of cheese. Activities of micrococci appear essential to the development of typical flavor of brick or Limburger cheese. These proteolytic organisms are able to convert methionine to methanethiol. Acetyl methyl disulfide is characteristic of Limburger flavor.

Edam and Gouda

These Dutch cheese varieties are made with *S. lactis* ssp. *lactis*, *S. lactis* ssp. *cremoris*, *Leuconstoc cremoris* and/or *L. lactis*. *Leuconstoc* and *S. lactis* ssp. *diacetylactis* ferment citrate to produce carbon dioxide, diacetyl, and other compounds. The starter organisms reach ~10^9 cfu/g in curd with a pH of about 5.7. The remaining lactose in curd is fermented in an uncoupled state to lower cheese pH to 5.2. Molded cheese forms are placed in 14% brine at 14°C for three to seven days. After brief drying, the cheese is coated with two to three coats of plastic emulsion, which permits slow moisture evaporation. Carbon dioxide production is essential to eye formation. Ripening changes involve proteolysis by rennet enzymes, enzymes of starter bacteria, and, to a small extent, native milk proteinase. Rennet is responsible for extensive degradation of αs-casein to large molecular weight water-soluble peptides. This proteolysis is significant in determining the body characteristics of cheese. Starter organism production of low molecular weight peptide fractions and amino acids are stimulated by rennet action.

Lipolysis occurs to a small degree mainly due to enzymes of starter bacteria liberating free fatty acids from monoglycerides and diglycerides formed by milk or microbial lipases. In well-made Edam and Gouda lactobacilli do not exceed 10^5–10^6/g or certain defects in the cheese are noticed. Lactic acid, carbon dioxide, diacetyl, aldehydes, ketones, alcohols, esters, and organic acids in proper balance are important to flavor. Anethole and bismethylthiomethane are important odor compounds of Gouda cheese (24).

Mozzarella and Provolone

These cheese types may contain 45–60% moisture. *S. thermophilus*, thermophilic *Lactobacillus* sp. or mesophilic lactics with *S. thermophilus* combination starters are used. Sometimes citric acid or other acids are used in place of lactic starters.

The cultures used generally cleave the galactose moiety of lactose in cheese, which, if not utilized, result in dark brown cheese color on baking. In fresh brined cheese there may be 10^8 cfu/g of starter organisms in 1 : 1 ratio of streptococci and lactobacilli. αs-Casein is proteolysed to a lesser degree by rennet enzymes compared to other cheese types, while β-casein is largely intact. This level of rennet proteolysis of milk protein is sufficient to give the melt and stretch characteristics to cheese during hot water kneading at about 57°C. The molded cheese is brined. The pH of cheese should be 5.1–5.4. There is little lipolysis and fatty acids liberation in traditional mozzarella cheese (25). Cheese made with mesophilic starter and *S. thermophilus* tends to have greater protein and fat hydrolysis during storage. Such cheese is difficult to shred and has atypical flavor when aged.

Mozzarella and provolone are manufactured in a similar manner. The former is consumed fresh while the latter may be ripened at 12.5°C for three to four weeks and then stored at 4.5°C for 6–12 months for grating. The ripened cheese have mainly *Lactobacillus casei* and its subspecies. Sometimes leuconstocs are detected in cheese, particularly if no starter was used.

Parmesan and Romano

Parmesan and Romano cheese are made with *S. thermophilus*, *Lactobacillus bulgaricus*, or other species of thermophilic lactobacilli. In addition to rennet, pregastric estrases, or rennet paste may be added to cheese milk for their lipolytic activity. The curd is cooked to 51–54°C, when the whey acidity reaches about 0.2% it is hooped (packed) in round forms. Sometimes salt is added to the curd, which slows the starter and regulates moisture. The cheese at pH 5.1–5.3 is placed in 24% brine for several days.

Compared to other cheese types, the starter populations in the fresh cheese is low. Throughout cheese ripening, 12–24 months, cheese flora seldom exceeds 10^5 cfu/g. Fecal streptococci and salt-tolerant lactobacilli predominate. In these hard grating cheeses αs- and β-caseins are not overly proteolysed compared to other cheese varieties. This is attributed to high cooking temperatures of curd, which inactivates the coagulant, and the high salt in the moisture, which discourages growth of adventitious flora. In ripened cheeses quality varies from location to location. Volatile free fatty acid and nonvolatile free fatty acid (C_4 through C_{18}) concentrations are high in these cheeses, particularly in Romano (25). Butyric acid and minor branched chain fatty acids that occur in milk appear to contribute to the piquant flavor of Parmesan. For a more balanced flavor a concomitant increase in free amino acids

(glutamic acid, aspartic acid, valine, and alanine) has been noted. Too high a free fatty acid level in cheese gives a strong, soapy, undesirable flavor.

Cottage Cheese and Cream Cheese

These are cultured to the isoelectric point of casein (pH ~4.6) followed by heating to separate the cheese solids. Cottage cheese curd is blended with a cultured or uncultured cream to no more than 4% fat. Cream cheese is cultured with lactics and citrate-fermenting cultures. Diacetyl, ethanol, acetone, lactic aid, acetic acid, and other less-characterized compounds are responsible for the flavor of these products.

ACCELERATED CHEESE RIPENING

One of the major costs of cheese is the expense of curing time before desired flavor develops. While some maturation time is inevitable, there are systems available where ripening time is shortened by speeding up proteolysis and lipolysis to generate flavor and modify texture. Elevated temperature (13°C or higher) curing offers the simplest approach to speed up ripening of otherwise normal cheese. Cheese intended for this type of curing must not contain measurable levels of heterofermentative lactobacilli or leuconostocs, because an open-texture defect and off-flavors will develop (26).

Microbial proteinases and gastric esterases have been used with little success to achieve acceptable cheese with uniformity. Activity of these exogenous enzymes is unregulated and may contribute to the detriment of cheese quality. Several unproven systems are available from culture houses.

Additions of partially inactivated starter organisms have been used with mixed results. Presently, this is not economical. Most of the proprietary systems investigated cause a minor to major deviation from characteristic flavor, body, and texture of cheese.

CHEESE DEFECTS

Cheese varieties exist because manufacturing and curing are carried out to preserve characteristic flavor, body, and texture of a given cheese. Attributes of one cheese may prove to be defects in another.

Even though certain molds are used for manufacturing and curing of cheese, their growth on most cheeses is undesirable (27). Gas production with putrid, unclean odors due to nonstarter bacteria constitute a defect in cheese. Moisture accumulation on the surface of hard cheeses permits rind rot due to yeasts, molds, and proteolytic bacteria growth, sometimes accompanied by discoloration. Milk itself can at times contribute weedy, feedy, cowy, barny, and related flavors coming from the cow. Psychrotrophic organism growth in milk can generate rancidity due to lipolysis and bitterness due to proteases from these organisms. Specific defects are discussed below.

Swiss Cheese. Swiss is graded by the size, number, and distribution of eyes; there are many defects of the eyes

(18). *Blind* or cheese with no eyes may be due to inhibition of carbon dioxide production of proprionibacteria by pH of less than 5.0, prolonged brining (high salt), and antibiotics or other inhibitors in milk or cheese.

Overset (pinny or too many eyes) is generally caused by improper acidity, high moisture, entrapment of air in curd, and other factors that prevent knitting of curd before gas production. Early gas is caused by *Enterobacter aerogenes*. Flavor defects, such as stink spots, white spots (28), and excessive amine production, are caused by *Clostridium tyrobutyricum*, *S. faecalis* ssp. *liquefaciens,* and strains of *Lactobacillus buchenerii*, respectively.

Cheddar Cheese. Excessive acidity, below pH 5.0 at three or more days, causes short, crumbly body and a sour, often bitter flavor. Normal body and flavor development may not occur in this cheese. On the other hand, too little acid or pH above 5.3 at three or four days may lead to corky, pasty body with off-flavor development. Open-texture defect may be caused by lactobacillus fermentum, L. brevis, and leuconostocs. Raw milk and rennet preparations have been linked to these defects.

Edam and Gouda. Edam and Gouda are subject to spoilage similar to Cheddar. In these cheeses, brine has been known to contaminate the cheese surface with gas-forming lactobacilli. These may cause the occurrence of phenols, putrid odor, and sulfidelike flavors and excessive production of carbon dioxide. Butyric acid fermentation and carbon dioxide production are due to *C. tyrobutyricum.* Growth of propionibacteria may cause a sweet taste and an open texture due to carbon dioxide.

Brick Cheese. Excessive acidity is a major defect. Coliform bacteria can grow during draining and salting, causing early gas, which if excessive can yield a spongy condition. Sometimes, due to lack of surface smear growth, typical flavor does not develop.

Blue Cheese. Blue cheese may not have sufficient mold growth, leading to lack of flavor and general development of cheese attributes. Excessive growth of mold may cause mustiness and loss of flavor. Mold-ripened cheeses are sensitive to bitterness development due to excessive acidity in the curd followed by abundant mycelium growth and proteases. Surface discoloration by other molds is not uncommon. Excessive slime and undesirable flavors occur due to excessive humidity in curing rooms.

Mozzarella. Mozzarella cheese is generally shredded before use. Softening of this cheese, due to *Lactobacillus casei* strain, is a problem during curing. It is also prone to softening and gas production by thermoresistant strains of *Leuconostoc*.

Cottage Cheese. Cottage cheese has a short shelf life generally due to *Pseudomonas*, yeasts, and molds. Sometimes bitterness and excessive acid development due to surviving starter culture is noticed during refrigerated storage (Tables 2, 3).

For total dollar annual sales and a breakout by cheese type see Tables 4 and 5.

Table 2. General Cheese Composition Ranges

Variety	Moisture, %	Fat (Dry Basis), %	Salt, %
Blue cheese	43.0–45.0	51.0–54.0	3.0–4.0
Brick	40.0–42.0	51.0–54.0	1.5–1.7
Cheddar	36.5–38.0	51.0–55.0	1.5–1.8
Colby	38.0–39.5	51.5–55.0	1.5–1.8
Cream cheese	53.0–54.5	34.0–36.0	1.0–1.2
Edam	35.0–38.0	41.0–47.0	1.6–2.0
Gorgonzola	35.0–38.0	51.0–53.0	3.0–4.0
Granular–stirred curd (CM)[a]	33.0–35.0	51.0–54.0	1.8–2.0
Limburger	46.0–48.0	51.0–54.0	2.0–2.1
Monterey Jack	40.5–42.5	51.0–54.0	1.5–1.7
Mozzarella	54.0–58.0	46.0–50.0	1.8–2.4
LM mozzarella[b]	45.0–48.0	46.0–50.0	1.7–2.2
PS mozzarella[c]	54.0–58.0	34.0–39.0	1.7–2.2
LMPS mozzarella[d]	46.0–50.0	34.0–39.0	1.6–1.9
Muenster	41.5–43.5	51.0–54.0	1.5–1.7
Neufchâtel	61.0–63.0	21.0–25.0	0.8–1.0
Parmesan	29.5–31.5	36.0–39.0	3.5–4.0
Provolone	41.0–44.0	46.0–48.0	1.8–2.2
Romano	32.5–33.5	39.0–41.0	3.5–4.0
Skim curd	47.0–50.0	11.0–15.0	0.9–1.1
Swiss	37.0–39.5	46.0–48.5	0.2–0.6

[a] Cheese for manufacturing.
[b] Low moisture.
[c] Part skim.
[d] Low moisture, part skim.

PROCESS CHEESE

Process cheese is the food produced by grinding and blending various cheese types into a uniform, pliable mass with the aid of heat. Water, emulsifying salts, color, etc are usually added and the hot plastic mass is cooled into the desired form.

Process cheeses were prepared as early as 1895 in Europe, but the use of emulsifying salts was not widely practiced until 1911 when Gerber and Co. of Switzerland invented process cheese. A patent issued to J. L. Kraft in 1916 marked the origin of the process cheese industry in America and describes the method of heating natural cheese and its emulsification with alkaline salts.

Process cheeses in the United States generally fall into one of the following categories (29).

A. Pasteurized blended cheese. Must conform to the standard of identity and is subject to the requirements prescribed by pasteurized process cheese except
   1. A mixture of two or more cheeses may include cream or Neufchâtel.
   2. None of the ingredients prescribed or permitted for pasteurized process cheese is used.
   3. The moisture content is not more than the arithmetic average of the maximum moisture prescribed by the definitions of the standards of identity for the varieties of cheeses blended.
   4. The word process is replaced by the word blended.
B. Pasteurized process cheese.
   1. Must be heated at no less than 65.5°C for no less than 30 s. If a single variety is used the moisture

Table 3. Code of Federal Regulations Cheese Composition Standards

Cheese Type	Legal Maximum Moisture, %	Legal Minimum Fat (Dry Basis), %	Legal Minimum Age
Asiago fresh	45	50	60 days
Asiago soft	45	50	60 days
Asiago medium	35	45	6 months
Asiago old	32	42	1 year
Blue cheese	46	50	60 days
Brick cheese	44	50	—
Caciocavello Siciliano	40	42	90 days
Cheddar	39	50	—
Low-sodium Cheddar	(Same as cheddar but less than 96 mg of sodium per pound of cheese)		
Colby	40	50	—
Low-sodium Colby	(Same as colby but less than 96 mg of sodium per pound of cheese)		
Cottage cheese (curd)	80	0.5	—
Cream cheese	55	33	—
Washed curd	42	50	—
Edam	45	40	—
Gammelost	52	(skim milk)	—
Gorganzola	42	50	90 days
Gouda	45	46	—
Granular–stirred curd	39	50	—
Hard grating	34	32	6 months
Hard cheese	39	50	—
Gruyère	39	45	90 days
Limburger	50	50	—
Monterey Jack	44	50	—
High-moisture Monterey Jack	44–50	50	—
Mozzarella and scamorza	52–60	45	—
Low-moisture mozzarella and scamorza	45–52	45	—
Part-skim mozzarella and scamorza	52–60	30–45	—
Low-moisture, part-skim mozzarella	45–52	30–45	—
Muenster	46	50	—
Neufchâtel	65	20–33	—
Nuworld	46	50	60 days
Parmesan and reggiano	32	32	10 months
Provolone	45	45	—
Soft-ripened cheese	—	50	—
Romano	34	38	5 months
Roquefort (sheep's milk)	45	50	60 days
Samsoe	41	45	60 days
Sapsago	38	(skim milk)	5 months
Semisoft cheese	39–50	50	—
Semisoft, part-skim cheese	50	45–50	—
Skim-milk cheese for manufacturing	50	(skim milk)	—
Swiss and Emmentaler	41	43	60 days

content can be no more than 1% greater than that prescribed by the definition of that variety, but in no case greater than 43%, except for special provisions for Swiss, Gruyère, or Limburger.

2. The fat content must not be less than that prescribed for the variety used or in no case less than 47% except for special provisions for Swiss or Gruyère.

3. Further requirements refer to minimum percentages of the cheeses used.

C. Pasteurized process cheese food.
1. Required heat treatment minimum is the same as pasteurized process cheese.
2. Moisture maximum is 44%; fat minimum is 23%.
3. A variety of percentages are prescribed.

4. Optional dairy ingredients may be used, such as cream, milk, skim milk, buttermilk, and cheese whey.
5. May contain any approved emulsifying agent.
6. The weight of the cheese ingredient is not less than 51% of the weight of the finished product.

D. Pasteurized process cheese spread.
1. Moisture is more than 44% but less than 60%.
2. Fat minimum is 20%.
3. Is a blend of cheeses and optional dairy ingredients and is spreadable at 21°C.
4. Has the same heat treatment minimum as pasteurized process cheese.
5. Cheese ingredients must constitute at least 51%.
6. A variety of percentages are prescribed.

Table 4. Manufacturers' Sales of Cheese[a]

Year	Total ($, Millions)	Annual Percent Change
1967	1,751.8	—
1972	3,094.6	12.1%[b]
1973	3,644.4	17.8
1974	4,504.7	23.6
1975	4,900.5	8.8
1976	5,764.1	17.6
1977	6,073.6	5.4
1978	6,688.5	10.1
1979	7,903.6	18.2
1980	9,415.9	19.1
1981	10,188.0	8.2
1982	10,170.0	−0.2
1983	10,561.7	3.9
1984	10,492.1	−0.7
1985	10,707.5	2.1
1986	11,378.3	6.3
1987	11,232.5	−1.3
1988[c]	11,388.8	1.4
1997[d]	17,644.8	4.6[b]

[a] Ref. 31.
[b] Average annual growth.
[c] Estimate.
[d] Projection.

Advantages of Process Cheese

- Compact, dense body.
- Smooth texture.
- Improved slicing properties without crumbling.
- Smooth melting without phase separation.
- Flavor ranges depending on the cheese used.
- Can control organoleptic and physical properties.
- Versatile in use attributes (cooking cheese, dips, sauces, snacks, etc.).
- Shelf stability (nonrefrigeration, fixed properties).
- Can take on various shapes and forms.

Basic Emulsification Systems for Cheese Processing

Citrates
- Trisodium citrate (most common, used for slices).
- Tripotassium citrate (used in reduced-sodium formulations, promotes bitterness).
- Calcium citrate (poor emulsification).

Orthophosphates
- Disodium phosphate and trisodium phosphate (most common, used for loaf and slices).
- Sodium aluminum phosphate (used in slices).
- Dicalcium phosphate and tricalcium phosphate (poor emulsification, used for calcium ion fortification).
- Monosodium phosphate (acid taste, open texture).

Condensed Phosphates
- Sodium tripolyphosphate (nonmelting).
- Sodium hexametaphosphate (used to restrict melts).
- Tetrasodium pyrophosphate and sodium acid pyrophosphate (minimal usage).

General Rules for Emulsifier Selection

Citrates. Used where firm-textured, close-knit products are desired, such as slices, and when products are refrigerated.

Table 5. Manufacturers' Sales of Cheese by Type

	Natural Cheese		Process Cheese and Related Products		Cottage Cheese		Other Cheese[b]	
	Sales ($, Millions)	Percent Change	Sales ($, Millions)	Percent Change	Sales ($, Millions)	Percent Change	Sales ($, Millions)	Percent Change
1967	829.2	—	562.5	—	218.0	—	142.1	—
1972	1,400.0	11.0[c]	1,134.1	15.1[c]	340.9	9.4[c]	219.6	9.1[c]
1973	1,705.9	21.9	1,363.5	20.2	405.6	19.0	169.4	−22.9
1974	2,458.7	44.1	1,496.6	9.8	456.0	12.4	93.4	−44.9
1975	2,668.7	8.5	1,654.4	10.5	508.7	11.6	68.7	−26.4
1976	3,267.9	22.5	1,859.7	12.4	530.7	4.3	105.8	54.0
1977	2,727.2	−16.5	2,518.5	35.4	545.6	2.8	282.3	66.8
1978	3,104.1	13.8	2,681.4	6.5	588.5	7.9	314.5	11.4
1979	3,949.3	27.2	2,822.0	5.2	729.3	23.9	403.0	28.1
1980	4,821.1	22.1	3,303.4	17.1	840.9	15.3	450.5	11.8
1981	5,225.6	8.4	3,567.9	8.0	856.5	1.9	538.0	19.4
1982	5,625.6	7.7	3,194.3	−10.5	683.2	−20.2	666.9	24.0
1983	5,824.0	3.5	3,325.4	4.1	693.8	1.6	719.5	7.9
1984	5,617.3	−3.5	3,390.1	1.9	748.3	7.9	736.4	2.3
1985	5,664.6	0.8	3,552.6	4.8	738.3	−1.3	752.0	2.1
1986	6,289.8	11.0	3,548.9	−0.1	725.1	−1.8	814.5	8.3
1987	6,208.0	−1.3	3,463.7	−2.4	731.6	0.9	829.2	1.8
1988[d]	6,294.9	1.4	3,529.5	1.9	722.8	−1.2	841.6	1.5
1997[e]	9,826.9	4.7[c]	5,482.8	4.7[c]	1,083.9	3.9[c]	1,251.2	4.2[c]

[a] Ref. 31.
[b] Includes cheese substitutes.
[c] Average annual growth.
[d] Estimate.
[e] Projection.

Phosphates. Used where shelf stability is required, pH is to be controlled for melting, and other physical properties.

Optional Dairy Ingredients

- Milk fat.
- Milk and skim milk.
- Nonfat dry milk (NFDM).
- Whey (lots of lactose, some sweetness).
- Whey protein.
- Buttermilk.
- Skim-milk cheese.

Other Optional Ingredients

- Acidifying agents.
- Salt.
- Artificial coloring.
- Spices or flavorings (other than cheese flavors).
- Mold inhibitors.
- Enzyme-modified cheese (EMC).
- Gums.
- Sweetening agents.

Processing Equipment

- Grinders.
- Transfer of cheese from grinder to blender.
- Blenders.
- Scale hoppers and auger cart.
- Batch cooking.
- Continuous cooking.
- Aseptic processing.
- Surge tanks.
- Flash tanks, scraped surface heat exchange (SSHE).
- Line screens.
- Shear pumps and silverson mixers.
- Specialty equipment.

Forms of Packaging

- Can and cup.
- Jar.
- Pouch.
- Loaf.
- Slice.

Methods of Cooling

- Chill roll.
- Hot pack.
- Fast cooling.
  —Blast coolers.
  —Water spray tunnels.
- Slow cooling.
  —Cased product.
  —Restricted melt properties.

CHEESE ANALOGUES: HISTORY

Cheese analogues are products that resemble natural or process cheeses. They are intended to have the appearance, taste, texture, and nutrition of their counterpart cheeses, but are made without butterfat. The procedures by which they are made are also quite similar to those used for traditional cheeses.

Throughout history, substitute food products have been developed as the result of shortages or the opportunity to reduce costs. With cheese analogues, the impetus was the opportunity to replace valuable butterfat with less-expensive fats made from vegetable oils. Early natural cheese analogues were made by skimming butterfat from whole milk, replacing it with some other fat, then following traditional cheese-making procedures. These products, called filled cheeses, were first made around the turn of the century (30).

In the early 1970s technology was developed in which process cheese analogues could be made by combining dried milk protein, hydrogenated vegetable oil in place of butterfat, emulsifying salts, and other ingredients, then cooking the mixture. These products simulated process American and mozzarella cheeses.

DEFINITIONS

Filled Cheese. A product that simulates a natural cheese, made by substituting the butterfat in fluid whole milk with some other fat, then following traditional natural cheese-making procedures. Typical products are similar to mild Cheddar, Colby, or similar cheeses. A filled cream cheese can also be made by replacing butterfat with vegetable fat following conventional procedures.

Process Cheese Analogue. A product that simulates a process cheese, made by substituting the butterfat in a formulation, combining it with a protein source, emulsifying salts, and other ingredients and cooking the mixture. Typical products are simulations of pasteurized process cheese, pasteurized process cheese food, or pasteurized process cheese spread. Imitation mozzarella cheese is also made with this method.

Cheese analogues are classified as imitation or substitutes depending on their functional and compositional characteristics. An imitation cheese need only resemble the cheese it is designed to replace. Thus the choice of ingredients that can be used is quite extensive. However, a product can generally be called a substitute cheese only if it is nutritionally equivalent to the cheese it is replacing and meets the minimum compositional requirements for that cheese. Thus substitute cheeses will often have higher protein levels than imitation cheeses and be fortified with vitamins and minerals.

COMMERCIAL USE

Since the introduction of these products in the 1970s, cheese analogue consumption in the United States grew to about 208 million lb in 1988, or about 2.2% of the total cheese markets (31). The main incentive was a lower price than natural cheeses. In 1980, for example, the average price of cheese analogues was $0.47/lb less than natural cheese (5). This was due to the use of vegetable oil in analogues, rather than butterfat, as well as imported casein products as the source of dairy protein.

The principal use of cheese analogues has been in sandwich slices or as an ingredient in food products manufactured by food-processing companies or food service operators. In these applications, the analogues can be used as extenders of natural cheeses. Common uses are in frozen pizzas and in food service cheese shreds and cheese sauces. The United States retail market has not grown substantially, because cheese analogues have a lower quality than natural cheeses. Nonetheless, analogues of pasteurized process American cheese and mozzarella shreds are being sold in the grocery stores.

INGREDIENTS IN CHEESE ANALOGUES

Formulations

Cheese analogues all contain a protein source, a fat component, and a variety of minor components chosen to provide the desired flavor, texture, color, nutrition, and keeping quality. Table 6 lists ingredient statements provided by these types of analogue product: a filled Cheddar type, an imitation process cheese, and an imitation cream cheese. The formulations are considerably more complex than for natural cheeses.

PROTEIN SOURCES

Casein Products

The most critical component in cheese analogues is the protein. The cheese analogues that have had commercial success are all based on casein, the protein in milk. In

Table 7. Composition of Casein(ates)

	Acid Casein (Lactic)	Sodium Caseinate	Rennet Casein
Protein ($N \times 6.38$)			
Dry basis, %	96.7	94.7	90.6
As is, %	87.3	90.4	80.6
Ash, %	1.8	3.8	7.8
Moisture, %	9.7	4.5	11.0
Fat, %	1.2	1.1	0.5
Lactose, %	0.1	0.1	0.1
pH (5% at 20°C)	4.6	6.7	7.1

some cases a soluble casein salt, or caseinate, is used. One of three forms of casein is generally used. For each, the starting material is skim milk (Table 7).

Vegetable Proteins

Several patents have been issued for, and papers written about, the use of vegetable proteins such as soy or cottonseed protein in place of caseins (16,32). However, these products have not been commercially successful to date because of many technical problems. These include such important characteristics as flavor, melting properties, gel strength, color, translucence, and mouth feel (such as creaminess or chewiness). The vegetable proteins simply do not function as milk proteins without further modifications, and these developments remain in the laboratory. However, some soybean protein isolates have been used as an extender in cheese analogues to replace up to 30% of the casein.

Vegetable Fats

Fats are essential ingredients in cheese analogues, for they are replacing butterfat in the products. The fats utilized become emulsified and intermixed with protein and

Table 6. Ingredients in Cheese Analogues

Kraft Golden Image Imitation Cheddar Cheese	Kraft Lunchwagon Imitation Pasteurized Process Cheese	American Whipped Products King Smoothee Imitation Cream Cheese
Pasteurized part-skim milk	Whey	Water
Liquid and partially hydrogenated corn oil	Partially hydrogenated soybean oil	Partially hydrogenated vegetable oil (may contain one or more of the following: coconut oil, soybean oil, corn oil)
Cheese culture	Sodium caseinate	Corn syrup solids
Salt	Granular cheese	Sodium caseinate
Enzymes	Food starch, modified	Tapioca flour
Apocarotenal (color)	Sodium citrate	Monodiglycerides
Vitamin A palmitate	Monodiglycerides	Salt
	Potassium chloride	Lactic acid
	Sorbic acid as a preservative	Guar gum
	Magnesium oxide	Monopotassium phosphate
	Zinc oxide	Artificial flavor
	Vitamin B_{12}	Citric acid
	Apocarotenal (color)	Carageenan
	Vitamin A palmitate	Artificial color
	Calcium pantothenate	Sorbic acid (as a preservative)
	Riboflavin	

water and, therefore, contribute to melting and textural characteristics. The fats also must behave in the mouth something like butterfat, which does melt in a characteristic way and influences creaminess, flavor, and other factors. Consequently, fats most often utilized in cheese analogues are partially hydrogenated or mixtures of hydrogenated and unhydrogenated oils that simulate butterfat. These fats are similar to vegetable shortenings and can be made from soybean, corn, cottonseed, palm, or coconut oils or blends of these oils. These blends often have a melting point of about 35–47°C (33).

Other Ingredients

The principal ingredients in cheese analogues are protein, fat, and water. Thereafter, ingredients are added to solubilize the protein, add flavor and color, and provide nutritional supplements. A typical formula for an imitation pasteurized process American cheese is presented below (34).

Water	45.20
Acid casein	25.00
Hydrogenated soybean oil	20.60
Adipic acid	1.30
Na–Ca hydroxide	0.80
Color–flavor blend	0.45
Dry ingredients	
Salt	1.50
Tapioca flour	1.50
Tricalcium phosphate	1.10
Disodium phosphate	1.00
Vitamin mix	0.50
Cheese flavor	0.50
Amino acids	0.31
Dipotassium phosphate	0.14
Sorbic acid	0.10
	100.00%

As can be seen, the formulation of cheese analogues is quite complex.

CHARACTERISTICS

Cheese analogues have both advantages and disadvantages compared to their natural counterparts. To date, the biggest problem is flavor. Casein products have a characteristic off-flavor that needs to be masked, in some manner. Likewise, vegetable proteins do not have characteristic, clean dairy flavors. But there are some offsetting advantages (Table 8).

Table 8. Cheese Analogues Characteristics

Advantages	Disadvantages
Cost of raw materials	Flavor
Keeping quality	No U.S. large source of casein
Nutritional equivalency or superiority	Unnatural image
Varying functional properties	

Analogues generally have better keeping quality than natural cheeses because hydrogenated vegetable oil is less likely to develop rancidity than butterfat. The absence of microbial cultures and their enzyme systems creates stable flavor and texture. Analogues made with vegetable oil also have less cholesterol than natural cheeses and can be formulated to have less saturated fat and fewer calories. The varying functional properties can be incorporated by the utilization of different ingredients.

BIBLIOGRAPHY

1. A. Carlson, G. Hill, and N. F. Olson, *Biotechnology and Bioengineering,* Vol. 29, John Wiley & Sons, Inc., New York, 1987, p. 582.

2. P. F. Fox, *Cheese: Chemistry, Physics and Microbiology,* Vol. 2, Elsevier Applied Science Publishers, Ltd., Barking, UK, 1987.

3. M. L. Green, "Effect of Milk Pretreatment and Making Conditions on the Properties of Cheddar Cheese from Milk Concentrated by Ultrafiltration," *Journal of Dairy Research* **52,** 555–564 (1985).

4. U.S. Pat. 3,914,435 (Oct. 21, 1975), J. L. Maubois and G. Mocquot (to Institut National de la Recherche Agronomique-France).

5. G. Sillen, "Modern Bactofuges in the Dairy Industry," *Dairy Industries International* **52,** 27–29 (1987).

6. D. D. Duxbury, "Breakthrough for Cheese—100% Chymosin Rennet," *Food Processing* **50,** 19–22 (1989).

7. P. H. A. Sneath, ed., *Bergey's Manual of Systematic Bacteriology,* Vol. 2, Williams and Wilkins, Baltimore, Md., 1984.

8. C. Daly, "The Use of Mesophilic Cultures in the Dairy Industry," *Antonie Van Leeuwenhock* **49,** 297–312 (1983).

9. J. Auclair and J. P. Accolas, "Use of Thermophilic Lactic Starters in the Dairy Industry," *Antonie Van Leeuwenhock* **49,** 313–326 (1983).

10. M. L. Green, "Review of the Progress of Dairy Science: Milk Coagulants," *Journal of Dairy Research* **44,** 159–188 (1977).

11. C. A. Ernstrom, "Enzyme Survival during Cheesemaking," *Dairy & Ice Cream Field* **159,** 43–46 (1976).

12. D. J. McMahon and R. J. Brown, "Enzymic Coagulation of Casein Micelles: A Review," *Journal of Dairy Science* **67,** 919–929 (1984).

13. R. C. Lawrence and J. Gilles, "Factors That Determine the pH of Young Cheddar Cheese," *New Zealand Journal of Dairy Science Technology* **17,** 1–14 (1982).

14. K. R. Nath and R. A. Ledford, "Growth Response of *Lactobacillus casei* var *casei* to Proteolysis Products Occurring in Cheese During Ripening," *Journal of Dairy Science* **56,** 710–715 (1973).

15. R. A. Ledford, A. C. O'Sullivan, and K. R. Nath, "Residual Casein Fractions in Ripened Cheese Determined by Polyacrylamide-Gel Electrophoresis," *Journal of Dairy Science* **49,** 1098–1101 (1966).

16. J. Stadhouders, L. Toepoel, and T. M. Wouters, "Cheesemaking With prt⁻ and prt⁺ Variants of N-Streptococci and Their Mixtures. Phage Sensitivity, Proteolysis and Flavor Development during Ripening," *Netherlands Milk Dairy Journal* **42,** 183–193 (1988).

17. D. J. Manning and H. E. Nursten, "Flavor of Milk and Milk Products," in P. F. Fox, ed., *Developments in Dairy Chemistry,*

Vol. 3, Elsevier Applied Science Publishers, Ltd., Barking, UK, 1985.

18. G. W. Reinbold, *Swiss Cheese Varieties, Pfizer Cheese Monographs,* Vol. 5, Pfizer, Inc. New York, 1972.

19. P. Trieu-cuot and J. C. Gripon, "A Study of Proteolysis during Camembert Cheese Ripening Using Isoelectric Focusing and Two Dimensional Electrophoresis," *Journal of Dairy Research* **49,** 501–510 (1982).

20. J. E. Kinsella and D. Hwang, "Biosynthesis of Flavors by *Pencillium roqueforti*," *Biotechnology and Bioengineering* **18,** 927–938 (1976).

21. R. D. King and G. H. Clegg, "The Metabolism of Fatty Acids, Methyl Ketones and Secondary Alcohols by *Pencillium roqueforti* in Blue Cheese Slurries," *Journal of the Science of Food and Agriculture* **30,** 197–202 (1979).

22. W. L. Langhus, W. Y. Price, H. H. Sommer, and W. C. Frazier, "The "Smear" of Brick Cheese and Its Relation to Flavor Development," *Journal of Dairy Science* **28,** 827–838 (1945).

23. M. Purko, W. O. Nelson, and W. A. Wood, "The Associative Action between Certain Yeasts and *Bacterium linens*," *Journal of Dairy Science* **34,** 699–705 (1951).

24. D. Sloot and P. D. Harker, "Volatile Trace Components in Gouda Cheese," *Journal of Agriculture and Food Chemistry* **23,** 356–357 (1975).

25. A. H. Woo and R. C. Lindsay, "Concentration of Major Free Fatty Acids and Flavor Development in Italian Cheese Varieties," *Journal of Dairy Science* **67,** 960–968 (1984).

26. T. Conner, "Advances in Accelerated Ripening of Cheese," *Cultured Dairy Products Journal* **23,** 21–25 (1988).

27. E. M. Foster, F. E. Nelson, R. N. Doetsch, and J. C. Olson, Jr., *Dairy Microbiology,* Prentice Hall, Englewood Cliffs, N.J., 1957.

28. K. R. Nath and B. J. Kostak, "Etiology of White Spot Defect in Swiss Cheese Made from Pasteurized Milk," *Journal of Food Protection* **49,** 718–723 (1986).

29. *Code of Federal Regulations,* Title 21 §100.120.

30. S. K. Gupta, G. R. Patil, and A. A. Patel, "Fabricated Dairy Products," *Indian Dairyman* **39,** 199–208 (1987).

31. U.S. Department of Commerce, Bureau of the Census, International Trade Administration, reprinted in *U.S. Industrial Outlook 1989—Food, Beverages, and Tobacco,* U.S. Government Printing Office, Washington, D.C., Jan. 1989.

32. U.S. Pat. 4,303,691 (Dec. 1, 1981), R. E. Sand and R. E. Johnson (to Anderson, Clayton & Co.).

33. U.S. Pat. 4,822,623 (Apr. 18, 1989), J. L. Middleton (to Universal Foods Corp.).

34. U.S. Pat. 3,922,374 (Nov. 25, 1975), R. J. Bell, J. D. Whynn, G. T. Denton, R. E. Sand, and D. L. Cornelius (to Anderson Clayton & Co.).

General References

F. L. Davies and B. A. Law, eds., *Advances in Microbiology and Biochemistry of Cheese and Fermented Milk,* Elsevier Applied Science Publishers, Ltd., Barking, UK, 1986.

P. F. Fox, *Cheese: Chemistry, Physics and Microbiology,* Vol. 2, *General Aspects,* Elsevier Applied Science Publishers, Ltd., Barking, UK, 1987.

F. Kosikowski, *Cheese and Fermented Milk Foods,* 2d ed. rev., F. V. Kosikowski & Associates Publishers, Brooktondale, N.Y., 1982.

B. H. Webb, A. H. Johnson, and J. A. Alford, *Fundamentals of Dairy Chemistry,* 2d ed., AVI Publishing Co., Westport, Conn., 1974.

G. H. Wilster, *Practical Cheesemaking,* 11th ed., O.S.C. Cooperative Association, Corvalis, Ore., 1959.

K. Rajinder Nuath
John T. Hynes
Ronald D. Harris
Kraft, General Foods, Inc.
Glenview, Illinois

CHEESE RHEOLOGY

The word rheology in the most literal sense means the study of flow. At first sight it appears an unlikely pursuit to apply to the study of cheese. However, one may extend the concept of the flow of a material to include the idea of any change in its shape, under the action of an external agency, which is not instantaneous, and which is not entirely recoverable. These are characteristics of flow (1), and the application to cheese becomes immediately obvious. In fact, in the history of rheology applied to food materials, many of the pioneers found dairy products, and particularly cheese, very suitable materials for their experiments. This followed naturally from the fact that cheese-making was originally a craft industry. Long before rheology developed as a science, the skilled cheese-maker used various tactile tests to estimate the progress of the development of body in the curd and the firmness of the final cheese, using memory as a guide to the consistency of the results. The rheologist has, in more recent years, attempted to apply instrumental techniques and universally adopted physical units to these measurements.

At the same time as the science of rheology was developing, and often in parallel with it, the methods of testing using the human senses have been developed into a separate field of study now known as texture studies (2–4). It is necessary for the complete food rheologist to keep abreast of these developments since food is judged, in the final analysis, by its deformation in the mouth during mastication, a purely sensory judgment. Food scientists devote much effort to correlating the physical and sensory observations. It may well be that in endeavoring to seek parity between instrumental rheological measurements and properties judged subjectively one is asking the wrong question and a question to which there can be no final and complete answer. The real questions which can be asked are: What can instrumental measurements tell us about the properties and in particular the structure of the material on which those properties depend? How can this information be useful to the cheese-maker? How may the instrumental measurements be used as a guide to the consumer's assessment?

FIRST PRINCIPLES

Before proceeding to the detailed study of the application of rheology to cheese, it is pertinent to consider a few basic principles. These will appear obvious to the reader educated in a physical science, but may be less obvious to those from other walks of life. One is taught that matter exists in one of three distinct states: gas, liquid, and solid. This context is mainly concerned with only the latter two. Both are characterized by the fact that any sample has a well-defined volume. Solids have a definite shape which can be altered only by the application of an external agency. Liquids, on the other hand, have no characteristic shape: they take up the shape of the vessel containing them.

The property, then, which distinguishes a solid is its rigidity and this term has been formally adopted by physicists as a measure of the effort required to change the sample's shape. The precise definition will be given later. If the material is a true solid, it will follow that this rigidity is invariant for the material. Furthermore, time does not enter into this, so it is understood that the change is instantaneous and once the effort is removed, the sample will recover its original shape spontaneously. This is, in plain language, the physicist's concept of an elastic solid.

In contrast with the solid, the characteristic by which a layman distinguishes a liquid is its fluidity. This, again, has a precise definition, but in practice it is more convenient to use the inverse concept, which is known as the viscosity. This is a measure of the relation between the effort applied to the liquid and the rate at which it flows. As before, once the effort has been withdrawn, the flow ceases, but the liquid remains where it is—there is no recovery. It is the rate of flow which spontaneously reverts to its original value. This describes the physicist's concept of a viscous liquid.

It is important to keep clearly in mind that the fundamental difference in behavior between solids and liquids involves the dimension of time. Time does not enter at all into the description of the behavior of an elastic solid—only spatial dimensions are involved—whereas time is equally as important as the spatial dimensions in the description of fluid behavior.

So far, only ideal materials have been described. These are seldom encountered in practice and are only of interest to rheologists as reference points, the simplest extremes of material behavior, one may say the "black and white" of classical physics. It is that extensive "gray" area between that is the rheologist's domain. Everyone knows that a piece of cheese, once squeezed, may recover its original shape only slowly and probably not completely. Indeed, some of the softer cheeses may, if sufficient effort be applied, be spread and only recover very little. This leads to a definition of a third category of materials. Any material falling within this gray area, possessing at the same time some of the characteristics of elastic and of viscous behavior, is known as a viscoelastic material.

Yet another class of materials may be mentioned in passing, since they may be encountered in everyday parlance. This is the plastic material. Plasticity may be defined as that property of a material whereby it remains rigid until a certain minimum force is applied, whereupon it deforms. Once this force is removed or falls below the critical value, the material again becomes rigid but remains in the deformed state. There is a simple mathematical model for this type of behavior (5) and the rheologist may be particularly interested in the transition from solid to fluid behavior, known as the yield point; however, the ideal plastic is probably as rare as the ideal solid or liquid.

DEFINITIONS

Before the rheological measurements which have been made on cheese can be discussed it is necessary to define a few of the terms used by rheologists. In the preceding paragraphs the words effort and force have been used in their everyday connotation. The term used by the rheologist is stress, denoted by the Greek letter σ. This is formally defined as the force measured in newtons (N) divided by the area in square meters (m^2) over which it is applied, and is measured in pascals (Pa). A convenient aide-mémoire for the size of these units is that a newton is approximately the weight, under the action of gravity, of a medium-sized apple. The stress may be applied normal to the surface, as in Figure 1a, or tangentially, as in Figure 1b. In either case it is equal to the force divided by the area over which it is applied.

In the case of Figure 1a, which depicts a weight resting on the upper surface of a sample, the deformation which occurs is a change (Δh) in the height (h) of the sample. The fractional change in height, $\Delta h/h$, defines the strain, which is denoted by the Greek letter ε. There is a practical difficulty here: the height is continuously changing during the deformation, so there is some uncertainty over what value should be used for h (6). For small deformations this is relatively unimportant, but when large deformations occur, such as is usual when examining cheese, it is necessary to be clear on this point. For the sake of clarity, the ratio of the change in height to the original height will be referred to as the fractional compression

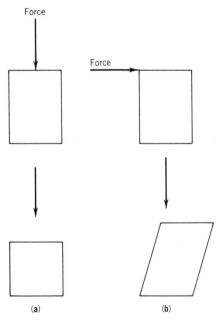

Figure 1. Application of stress to a sample: (a) compressive strain; and (b) shear.

and expressed as a percentage. When the change is related to the varying height, the formula for the strain becomes

$$\varepsilon = \ln(h_0/h_1) \tag{1}$$

This is sometimes called the true strain and will be given in decimal form.

If the stress is applied tangentially, as in Figure 1**b**, the resulting strain is described as a shear. It is defined as the distance through which one plane has traveled relative to another divided by the distance between them and is given the symbol γ. Both the strain and the shear are the quotients of two lengths and are therefore dimensionless numbers.

With these definitions of the physical quantities stress, strain, and shear, it is now possible to define the properties of some ideal materials. The rigidity of a solid, hitherto used loosely, can now be defined formally as the ratio of the tangential stress to the shear produced and is given the symbol n:

$$n = \sigma/\gamma \tag{2}$$

The dimensions of n are evidently the same as the dimensions of the stress, but it is usual to express the modulus of rigidity in newtons per square meter and to reserve the use of pascals for stresses only, thereby preserving an easily recognizable distinction between them.

The second property of an elastic solid which may be defined can be seen by referring again to Figure 1**a**. This is the modulus of elasticity and is the ratio of the stress to the strain. If the material is ideal, it will be isotropic; that is, the properties will be the same in all directions. During compression the volume will remain constant, so that while the height of the sample is changing in the direction of the applied stress, the dimensions in the two directions at right angles will change simultaneously. At any point within the sample only one-third of the stress will be used in causing the change in height. Another third will be used in each of the two other directions normal to this, causing the sample to expand laterally. Hence, for any ideal elastic material the modulus of elasticity will be three times the modulus of rigidity. However, if the material is compressible, some of the stress will be used up in compressing it and the factor 3 relating the modulus of elasticity to that of rigidity falls toward 2. The importance of this is that some authors express their results in terms of elasticity and some in terms of rigidity. An accurate comparison is possible only when some information is available concerning the compressibility of the material. In practice the ideal incompressible material is as unlikely to exist as any other ideal material. Almost all cheese is somewhat compressible, but in general not so compressible that any serious error is introduced by using the factor 3 to convert from one unit to the other.

In the case of liquids there is only one material constant that is of interest. The rate of flow or deformation is now given by the quotient of the strain and the time taken to reach it if the motion is constant, or in the unsteady state by the first derivative with respect to time of the

strain, $\dot{\varepsilon}$. The ratio of the applied stress to the rate of flow is called the viscosity and this is denoted by the Greek letter η, so that

$$\eta = \sigma/\dot{\varepsilon} \tag{3}$$

The unit of viscosity is the pascal-second. A rough guide to its magnitude is that the viscosity of water at room temperature is very nearly $\frac{1}{1000}$ Pa-s.

MODELS OF RHEOLOGICAL BEHAVIOR

In order to provide a theoretical framework within which to discuss the determination of the physical quantities just described it is necessary to consider some models for a viscoelastic body. There are two ways in which this may be visualized. Viscoelasticity may be considered either to be a property intermediate between viscosity and elasticity, or the result of a combination of viscous and elastic behavior occurring simultaneously in the material. In the first case, recalling that the elastic modulus is a function of spatial dimensions only, whereas viscosity is a function also of the first derivative of the strain with respect to time, the suggestion is that viscoelasticity is a function of a fractional derivative, the fraction lying between zero and unity. Although the concept is simple to describe, the mathematical equipment required to convert measurements which are always made in a conventional three-dimensional space and real time to this hypothetical continuum where space and time are no longer independent is formidable. The concept is now of historical interest only, but readers may find some early writers used it to discuss the rheology of cheese.

The alternative hypothesis, that both elastic and viscous behavior occur simultaneously, leads to a number of simple models (7), which require only elementary mathematics. It should be emphasized at this stage that these are only mathematical models, whose purpose is to codify the experimental observations; they cannot be taken to be real physical structures. It will be shown later that it is sometimes possible to relate elements in a model to the existence of particular structural elements within a material. The simplest model for an elastic solid is a spring whose strain or compression ε_s is always proportional to the stress. Writing G for the constant of proportionality to distinguish it from the rigidity or elasticity of any real material,

$$\varepsilon_s = \sigma/G \tag{4}$$

In a similar manner a viscous liquid may be modeled by an infinitely long dashpot whose displacement is always inversely proportional to the viscosity

$$\dot{\varepsilon} = \sigma/\eta \tag{5}$$

There are two ways in which one spring and one dashpot can be combined to give a compound structure. If they are placed in tandem as shown in Figure 2**a**, the stress is continuous throughout the model and the total displace-

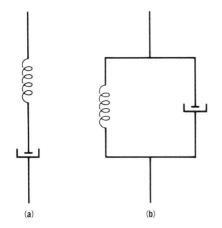

Figure 2. Viscoelastic models: **(a)** Maxwell body; and **(b)** Voigt body.

ment is given by the sum of those of the elements:

$$\varepsilon = \sigma/G + \sigma t/\eta \qquad (6)$$

The first term of the right-hand side expresses the elastic component and indicates that immediately on applying the stress there will be some deformation; thereafter (second term) the deformation will increase with time t. On removal of the stress, the elastic component recovers immediately while the strain due to the dashpot remains. This model is called a Maxwell body. A complete creep and recovery curve of this model is given in Figure 3**a**. It is

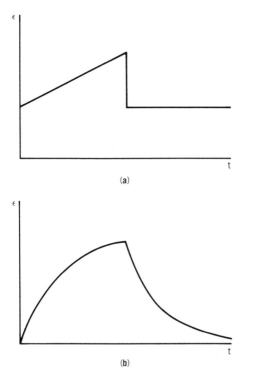

Figure 3. Creep recovery curve: **(a)** Maxwell body; and **(b)** Voigt body.

already apparent that a two-component model represents the rheological behavior of a sample of cheese better than the assumption that it is either a solid or a fluid.

If the two components are combined in the other possible way, so that they share the stress but the strain is common to both, we arrive at a formula for a Voigt body, sometimes known as a Kelvin body:

$$\sigma = G\varepsilon + \frac{\eta\dot{\varepsilon}}{t} \qquad (7)$$

Transforming this equation so that it may be compared with equation 6 gives

$$\varepsilon = \frac{\sigma}{\eta}\left[1 - \exp\left(-\frac{G}{\eta}t\right)\right] \qquad (8)$$

The creep and recovery curve of this model is given in Figure 3**b**. Again, there is some resemblance between this curve and the behavior observed with cheese. The deformation and the recovery are not instantaneous in this case and this is more in line with practical experience, but the model predicts a complete recovery, which is not always observed.

The student of electric circuit theory will be aware of some familiarity with the Voigt and Maxwell models. If one replaces the spring with a capacitor and the dashpot with a resistor, the two models become the two alternative representations of a leaky dielectric. Just as in electrical network theory these binary combinations can be built up into more complicated networks by adding more elements, so the Voigt and Maxwell bodies can be extended. One popular combination is of a Voigt body placed end-to-end with a Maxwell body. This is known as a Burgers body and it does indeed give a creep and recovery curve in which both the deformation and the recovery are retarded to different degrees and in which the recovery is incomplete.

Although the process of building up more complicated models may be continued indefinitely, the expressions for the behavior rapidly become cumbersome. As the number of elements increases it becomes disproportionately difficult to analyze the data, and the accuracy of the parameters extracted decreases (8). Indeed, the precision of any experimental measurements made on cheese may well be insufficient for more than two or three parameters to be identified with any pretensions of accuracy. Only if the rheological behavior can be linked unequivocally to the structure of the sample may the use of complex models be justified. Otherwise, the model should be kept as simple as possible.

EMPIRICAL INSTRUMENTAL TESTS

The earliest applications of rheology to cheese, even before rheology was conceived as a scientific discipline, consisted of making subjective estimations of some of the more obvious properties and trying to describe these in plain language. Often the cheese-maker or the grader, though skillful and competent in judgment, would find it difficult to define precisely the terms being used. One can see that confusion can arise when one tries to distinguish between terms such as consistency and body, firmness and hardness, chewiness and meatiness. The grader's mind may be clear about what is understood by each, but communicating the precise meaning to the outside world is less easy. The difficulty is further compounded by the fact that these words do not translate precisely into other languages, making international communication more difficult. To some extent the development of texture studies, particularly in the United States, has clarified the semantic problem (9), but subjective assessment can never match the precision which can be attained with sophisticated instrumentation. Much of the earlier instrumental rheology sought to replace, or at least to reinforce, subjective judgment with numerical quantities which could be measured reproducibly by any competent technical assistant (10).

The earliest instruments were empirical (3) and, in general, intended to give some indication of the firmness or similar qualities. To this end, they were designed to simulate what the scientist perceived to be the mechanical action of the hand or fingers of the expert during examination of the cheese. They were generally unsophisticated, not too expensive, and made no pretence of measuring any precisely defined physical property. Because of this, their measurements cannot be analyzed and expressed in terms of fundamental rheological parameters as defined above. Nevertheless, some of them may still perform a useful function if their limitations are understood, because they possess the undoubted merit that their measurements are easily reproducible between different laboratories and different times.

One such instrument became known as the Ball Compressor (11). In the course of examining a cheese, one of the actions of any grader is to press either a thumb or a finger into the surface and, from the reaction which is felt, gain an impression of the firmness or springiness of the cheese. The Ball Compressor might in fact be described as a mechanical thumb. It consists of a metal hemisphere which is initially placed on the surface of a whole cheese and allowed to sink into the cheese under the action of a load; the depth of indentation can be measured and recorded after a fixed time. The load can then be removed and the recovery of the cheese observed. The instrument is shown diagrammatically in Figure 4.

Making a number of simplifying assumptions, it is possible to convert a reading obtained with the Ball Compressor into a modulus G (12,13), which may be considered to be the equivalent of the modulus of rigidity, using the formula

$$G = 3M/16(RD^3)^{1/2} \tag{9}$$

where M is the applied load, R the radius, and D the depth of the indentation. This formula has been arrived at assuming that the sample is a homogeneous isotropic elastic solid, the load is static, and there is no friction between the sample and the surface of the hemisphere. It is clear that every one of these conditions is violated. Cheese is far from homogeneous, particularly if it has a rind or has

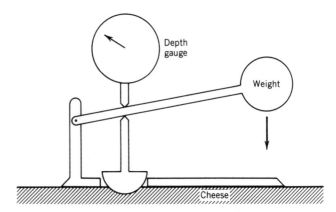

Figure 4. The Ball Compressor (diagrammatic).

been salted by immersion in brine, causing the properties to vary from the surface to the center. It is not isotropic: the method of manufacture of some varieties, such as Cheddar, on which many of the earlier experiments were carried out, or Mozzarella, is intended to produce an oriented structure. Some of the objections, however, would appear to be much less serious in the case of other cheeses where uniformity of texture is the aim of the manufacturer. Cheese is not an elastic solid; it is viscoelastic. The indentation is not instantaneous and may not be completed within the duration of the experiment. Finally, if the load is removed, some indentation remains, showing that some of the energy has been dissipated in causing flow and not stored up as in an elastic solid. In spite of all these limitations, if the test proceeds at least until the indentation becomes so slow that it has virtually ceased, the use of equation 9 gives an indication of the magnitude of the firmness in readily understandable physical units. This enables the reader to compare measurements given in the early literature with those made more recently by more sophisticated means.

The Ball Compressor has been discussed at some length, not only because of its historical importance, but because it has the merit of being a nondestructive means of testing and does not require an operator to undergo any specialized training. Also, it is used on the whole cheese, whereas almost every other rheological test requires that a sample be cut from the whole. Using the Ball Compressor, it has been shown that in a whole cheese such as a Cheddar or a Cheshire weighing some 25 kg there are considerable variations in the firmness over the surface of the cheese (14) and that the firmness differs on the upper and lower sides. This difference is influenced by the frequency with which the cheese is turned in the store during its maturing period, and the time which has elapsed since it was last turned. The implication of this is that it is not possible, whatever instrumental measurement is made, to assign a single number to any property of the cheese, nor is it possible to assess the properties of the whole cheese by means of a measurement made at one local point in that cheese.

Clearly there is a need for some form of nondestructive test and for many purposes there is no reason why it should not be empirical, provided that the implications are understood. The Ball Compressor has the merits of cheapness and simplicity, but the time taken to obtain a representative reading limits its use to the research laboratory. The problem of devising a suitable test is as yet unsolved. It is unfortunate that the application of ultrasonic techniques, which have proved so useful in the nondestructive testing of many engineering materials, have proved unrewarding with cheese (15,16). This is because the dimensions of the cracks and other inhomogeneities in cheese are commensurate with, or sometimes even larger than, the wavelength of the ultrasound and large-scale scattering takes place. It is difficult for a pulse to penetrate the body of the cheese and both the velocity of propagation and the attenuation are more influenced by the scattering than by the properties of the bulk.

One other empirical test deserves to be mentioned on account of its simplicity. This is the penetrometer (17–19). It is not quite nondestructive, but very nearly so, since it only requires that a needle be driven into the body of the cheese; no separate sampling is required. A penetrometer test may take one of several forms. For example, a needle may be allowed to penetrate under the action of a fixed load (17,18), or it may be forced into the cheese at a predetermined rate (19) and the required force measured.

Whichever method is used, let the specific actions be considered. As the needle penetrates the cheese, that part of the cheese immediately ahead of it is ruptured and forced apart. If the needle is thin, the deformation normal to its axis is small, so that the force required to accomplish this may be ignored. On the other hand, the progress of the needle is retarded by the adhesion of its surface to the cheese through which it passes. This may be expected to increase with the progress of the penetration until a point is reached when the restraining force matches the applied load and further penetration ceases. If a suitable diameter for the needle and a suitable weight have been chosen, this test may be completed in a few seconds. The test will be more useful for cheese whose body is reasonably homogeneous on the macroscopic scale, as, for example, the Dutch cheeses. With cheeses such as Cheddar or the blue cheeses the heterogeneities are generally on too large a scale and the penetration becomes irregular. The needle may pass through weaknesses in the structure or even cracks and so give rise to the impression that the cheese is less firm than it really is, or the point of the needle may attempt to follow a line of weakness, not necessarily vertical, and as a result there will be additional lateral forces acting on the needle and its penetration will be arrested prematurely.

Measurements made with a penetrometer cannot be converted theoretically to any well-defined physical unit. Both the cohesive forces within the cheese and the adhesive forces between the cheese and the surface of the needle are a consequence of the forces binding the structure together. One may infer that these are related to its viscoelastic properties but there is no simple theory which attempts to establish these relations. It has been shown experimentally (20) that there is a statistically significant correlation between firmness as measured by the Ball Compressor and by penetration, but this differs among different types of cheese (21). It has also been shown ex-

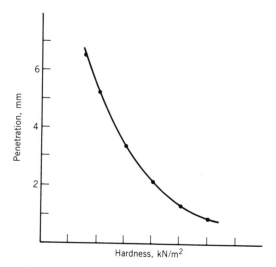

Figure 5. Penetrometer readings compared with compression modulus.

perimentally that a curvilinear relation exists between the resistance to penetration and an elastic modulus of some Swiss cheeses, calculated from the results of a compression experiment (Fig. 5). While this gives some idea of the magnitude of the forces involved, it has little practical application since it refers only to one series of experiments on only one type of cheese. In the absence of a similar experimentally determined relation for the cheese in which one is particularly interested, penetrometer measurements can only be regarded as empirical.

Another test of a similar nature is the use of a cone to penetrate the sample (22). This test was originally developed for use with high-viscosity lubricants. A loaded cone with its apex pointed vertically downward and initially just making contact with the upper surface of the sample is allowed to penetrate the sample until equilibrium is reached. If a suitable combination of the apical angle of the cone and the load is used, the equilibrium is usually reached within a few seconds and the penetration is deep enough to be measured comfortably. The test assumes that the sample behaves as a plastic or Bingham body; that is, when acted upon by small stresses it behaves as a solid, but once a critical stress, the yield stress, is exceeded, it flows as a viscous liquid.

In fact, the test works in reverse. At the commencement the stress is infinite, since the area of application, which is the tip of the cone, is zero. As the cone penetrates, the material is caused to flow laterally and the rate of penetration is controlled by this lateral flow. As the penetration proceeds, the cross-sectional area increases proportionally with the square of the depth of penetration and the stress decreases correspondingly. When the stress no longer exceeds the critical or yield point, motion ceases and an equilibrium is established. It is a matter of contention among rheologists whether a yield value really exists. Some maintain that there is always some flow, albeit minimal, however low the stress. For practical purposes, however, the cone soon appears to be stationary and the depth of penetration can be recorded. This is a simple, quick test and, while it is not entirely nondestructive, it does only a little damage near the surface of the sample. The vertical stress Y may be calculated from the penetration and the angle of the cone:

$$Y = (Mg/\pi h^2)\cot^2\alpha/2 \qquad (10)$$

where α is the apical angle of the cone, h the penetration depth, M the applied load, and g the gravitational constant. This is greater than the yield stress since it does not take into account the stresses involved in causing the sample to flow laterally. An estimate of the yield stress may be obtained by multiplying the equilibrium vertical stress by the factor $\frac{1}{2}\sin\alpha$ (23). It is only an estimate, though, because it takes no account of any friction between the surface of the cone and the cheese.

Another form of penetrometer-type test is that sometimes referred to as the puncture test (24). In place of a needle or cone, a rod is driven into the sample and the resistance to its motion measured. The mechanism of the deformation of the sample is more complicated in this case. At least four principal factors are involved. In the first place, the sample ahead of the rod is compressed. Second, the rod must cut through the sample at its leading edge: hence the name puncture, and this requires a force. Third, there is the frictional resistance between the surface of the rod and the cheese, and finally, there is the force required to set up the lateral flow within the sample. If the sample is large enough compared with the dimensions of the rod so that any compression in directions perpendicular to the motion may be ignored, once a dynamic equilibrium has been set up the first two factors will be constant and the third will increase linearly with the penetration. By using rods of different cross-sectional areas and keeping the perimeter the same it is possible to separate these effects (25,26).

Instead of performing this test on an extensive sample, such as a whole cheese, it is more often carried out on a small sample contained within a rigid box (27–29). When done in this manner, additional forces are called into play. The lateral forces on the rod as the sample is compressed between the rod and the walls of the container now become important. In addition, the compressive force on the leading face of the rod is no longer constant as the distance between this face and the bottom of the container decreases. If the sample is shallow, it is arguable (27,28,30) that this may indeed be the largest single contributory force, and, by ignoring the others in comparison with it, one can calculate a modulus from the ratio of the measured stress to the compression. It is obvious that this will be an overestimate of any true modulus which could have been derived in a simple unrestrained test and that the excess will be arbitrary and unknown. In spite of this the test is useful because it gives an indication of the magnitude of the rigidity of the sample and allows comparisons to be made between similar types of cheese.

PHYSICAL MEASUREMENTS

It has already been pointed out that rheological phenomena take place in real time. The basic rheological experiment consists in applying a known stress for a known time and observing the strain which ensues (31). In this context it is immaterial whether one refers to normal stresses, which result in strains, or to tangential stresses, which result in shears. For the purpose of the argument strain and shear are interchangeable. The generic mathematical equation which connects the variables is

$$\sigma = f(\varepsilon, t)$$

This is the equation of a surface in a three-dimensional diagram. The material is said to be linearly viscoelastic if, for any value of t, the line representing the variation of strain with stress is a straight one. For many substances this is approximately true, at least while the strains and stresses are small. Cheeses vary in this respect. For a linear viscoelastic material the two-dimensional graph of strain versus time will contain all the required rheological information.

The basic graph is the creep curve. If the experiment is prolonged indefinitely, some equilibrium will ultimately be established. If the material is essentially a solid, but one whose deformation is retarded, the strain will eventually become constant. There are obviously two parameters which can be used to denote the material behavior, one based on the ultimate deformation and the other on the rate at which it is (asymptotically) established. On the other hand, if the material is more akin to a liquid, the final equilibrium will be a dynamic one in which a constant rate of strain is established. As before, two parameters are needed to describe the behavior: one, the ultimate flow; the other, the rate of attaining it. The experiment may also be conducted rather differently. Instead of allowing it to proceed ad infinitum, the stress may be removed at a predetermined point. Should the material possess any elastic characteristics, some of the energy used in the deformation will be stored within it; on removal of the stress the strain will decrease again and once more eventually reach an equilibrium. This part of the graph is known as the recovery curve. Again there are two points of particular interest—the amount of recovery and the rate.

So far the description refers to what may be called an ideal viscoelastic material and is an oversimplification of real-life experience. If the material is not linearly viscoelastic, the creep and recovery curve can still be obtained, but a single curve will not include all the rheological information about the sample. A series of such curves is necessary, giving the full three-dimensional representation. It has also been assumed so far that the properties of the material are not altered in any way by the action of the strain itself. In the case of cheese it is very evident that at large strains the whole structure breaks down and cannot recover. Indeed, the stress and strain at the point of onset of this breakdown may be a useful quantity when describing the properties of any cheese, as will be seen later. Even at lower strains there may well be unobserved changes in the structure. These could occur during the early part of a cycle of measurement, in which case they may be revealed by the failure of the recovery curve to follow the expected pattern. However, they may occur during handling of the cheese at any stage between manufacture and sampling. A complete rheological description of a sample really needs not only the experimental curve or curves but a statement of the history of the straining of the sample up to the moment of measurement. This is most likely to be unavailable and it is necessary to presume that the sample is in good condition and to allow that variations in its history are included within the natural variation among samples.

It is often more convenient from the point of view of practical instrumentation to carry out the compression experiment in a different way. Instead of applying a fixed stress and observing the reaction, the sample is subjected to a known rate of strain and the build-up of the stress observed. Again, this may be prolonged indefinitely, if the operation of the instrument will allow it. If the straining is stopped at any point, it is not usually practicable to apply a reverse rate of strain. In this case, the strain is usually held constant and the relaxation of the stress observed. Again, a complete cycle of deformation and relaxation contains all the available information about the rheological properties of the sample. Quite sophisticated instruments have been developed which enable this type of test to be carried out with precision and then record the results, making this now the most widely used test. In principle, the operation of these instruments is simple. The sample is compressed between parallel plates at a known fixed rate and the force required is monitored.

However, before discussing the results of any measurements made on these instruments it is important to consider the exact mode of operation. As the measurement is usually carried out, the sample is in the form of a small cylinder, although other right prisms may be used. This is placed with its axis vertical on one plate and is then compressed along that axis. Since the compression is linear, the height of the sample at any instant after the compression has begun is given by

$$h = h_0(1 - at)$$

where h_0 is the original height of the sample and a denotes the rate at which the two places approach each other, ie,

$$a = -dh/dt$$

Using equation 1 for the strain leads to

$$\varepsilon = -\ln(1 - at) \qquad (11)$$

If the instrument's recorder treats time (or its equivalent linear distance traveled) as one axis, then this is not a true strain axis. As the compression increases from 0 to 100% the true strain increases from zero to infinity. In fact, whatever definition of strain is used (6), other than the linear compression, the time axis is a distortion of the strain axis. Second, as a sample is compressed its volume remains constant, so that the cross-sectional area increases as the height decreases. Assuming that the plates are perfectly smooth, so that there is no lateral friction between them and the end surfaces of the sample, a purely viscous, or equally a perfectly elastic solid, sample would deform uniformly and the cross-sectional area A at any height would vary precisely inversely with the height, ie,

$$A = A_0(h_0/h) \qquad (12)$$

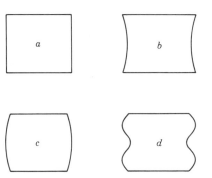

Figure 6. Compression of a cylinder: (*a*) ideal; (*b*) concave; (*c*) barrel; and (*d*) complex.

A viscoelastic material does not deform so simply, but in such a way that the lateral movement is greatest near the ends, so that a concave shape results (32), as shown in Figure 6. In this case the stress is not uniform throughout the sample. When there is some friction between the plates and the sample, the lateral movement of the end layers is restricted and a rotational couple is set up within the sample, using the perimeter of the end surfaces as fulcrum. If the sample is homogeneous or at least cohesive, the effect of this is that the middle of the sample will spread outward, giving rise to barrel-shaped distortion (Fig. 6c). When the rotational forces which develop are sufficient to cause fractures to develop, even though they may be undetected by eye, then with the weakened structure the sample may spread out near the ends, and what is known as mushroom distortion results (33) (Fig. 6). Both barrel and mushroom distortion may occur simultaneously, and a shape such as Figure 6d may be the outcome.

Also, during the compression some liquid, which may be free fat or water, or both, may be squeezed out. This might be thought to lubricate the sample ends to some extent; but it might also ensure closer contact between what is inevitably the slightly rough cut surface, the degree of roughness depending largely upon the type of cheese, and the smooth plates of the instrument. In practice it has been found (34) that the ends of the cheese do not expand as much as would be predicted by the use of equation 12; yet on the other hand it would be quite erroneous to assume that the cross section remained at its original value. The instrument does not measure stress, but the total force transmitted by the end surface to the measuring transducer. The stress within the sample is obtained by dividing the force F by the cross-sectional area. In the case of a perfectly lubricated sample this is given by

$$\sigma = F/A = (F/A_0)(1 - at) \qquad (13)$$

When barrel distortion occurs, the calculation of the true stress is less simple. If one assumes that the material behaves as a perfect elastic solid and that the ends of the sample are firmly bonded to the instrument plates so that they cannot expand laterally, the barreling can be shown to assume a parabolic profile. The actual correction to be applied depends on the height and radius of the uncompressed sample as well as the degree of compression. It is somewhat greater than the correction for a lubricated sample. A force–compression curve is thus not a true representation of a stress–strain curve. Near the origin it is close enough, but the distortion increases as the compression progresses.

If the sample behaves as one of the simple models, it is possible to predict the shape of the stress–strain curve. In the case of the Voigt body the solution is straightforward. The stress is the sum of the elastic and the viscous contributions (eq. 7). From equations 1 and 11

$$\dot{\varepsilon} = ae^{\varepsilon}$$

so that

$$\sigma = G\varepsilon + a\eta e^{\varepsilon} \qquad (14)$$

This is the equation of a curve in which the intercept on the σ axis is proportional to the viscous component and the rate of compression, and the initial slope is proportional to the elastic component. The curve is concave upward throughout. On substituting for σ and ε, equation 14 becomes

$$F = [A_0/(1 - at)][a\eta/(1 - at) - G\ln(1 - at)] \quad (15)$$

This is not so convenient to analyze but there is still an intercept on the F axis proportional to the viscous component and the rate of compression. The solution for the Maxwell body is more complicated. Making the same substitutions as before leads to an infinite series

$$\sigma = Ge^{(G/a\eta)e^{-\varepsilon}}\left[\varepsilon + \sum_{1}^{\infty} \frac{(-G/a\eta)^{i}(1 - e^{-\varepsilon i})}{i.i!}\right] \quad (16)$$

At the commencement of the compression the stress is zero and the initial slope is proportional to the elastic component. Initially the curve is convex upward, but as the compression proceeds a point of inflexion is reached, depending on the ratio of the viscous to the elastic components and the rate of compression. Finally, the curve becomes asymptotic to a line through the origin, with a slope given by the elastic constant.

If the stress–strain curve obtained for a particular sample can be recognized as being similar to one of these two patterns, the appropriate model can be identified and the material constants evaluated. It is seldom that such a simple fit occurs. A material as heterogeneous as cheese is unlikely to conform to the behavior of a two-element model. More sophisticated models may give a better representation, but adding further elements only makes the theory more complex and the analysis that much more difficult. One word of caution must be entered here. The Voigt body is characterized by the finite intercept on the stress axis. The converse is not necessarily true. A finite intercept also arises in the case of any material possessing a yield value, since this is the stress which must be overcome before any deformation takes place.

In the usual way in which compression tests are carried out, the compression is allowed to proceed to a point far beyond that at which any simple theory may be expected to apply. Often it reaches 80%, a true strain of 1.609. Long before this is reached, any structure which may have been present in the original sample of cheese can be seen to have broken down. If it is carried to this extent, only the early part of the test may be considered as measuring the rheological properties of the original sample, while the latter part becomes a test of the mechanical strength of the structure. A typical compression curve is shown in Figure 7. It is often difficult to decide whether there is an intercept on the stress axis, ie, whether there is an instantaneous build-up of stress at the moment the compression begins or a rapid development of stress rising from zero but in a finite though brief time. Nevertheless, it is essential that the distinction be made if the correct model is to be assigned. The instrument itself necessarily has a finite, even if very rapid, acceleration from rest, and the response of the recording mechanism also takes a finite time, so that even if the stress were instantaneous, it would appear as a very steep rise (35).

However, much more serious than any limitations in the instrument is the problem of attaining perfection in the shape of the sample. Usually the dimensions of the sample and the rate of compression are chosen so that the strain rate at the commencement is of the order of 0.01 s^{-1}. In a sample, say 20 mm in height, a lack of parallelism of the two end surfaces of only 0.2 mm means that for the whole of the first second of compression only part of the sample is being compressed. This will inevitably appear on the record as a flow, whatever the real properties of the cheese.

Assuming that any doubts about the shortcomings of either sample or instrumentation may be satisfactorily allayed, it is pertinent to consider the principal features of the compression curve. In the early stages, after the initial rise there is a smooth increase of stress with the strain. Should the compression be halted at this stage, the cheese would recover either completely or in part and a repeat of the curve could then be obtained, the new curve having the same general shape as the original, showing that the internal structure of the cheese had remained more or less intact, although there may have been some

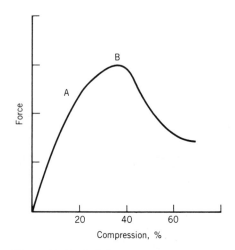

Figure 7. Typical force–compression curve.

rearrangement. Eventually a point is reached, the point A in Figure 7, where the structure begins to break down and the slope becomes noticeably less. The conventional view is that this is the point at which cracks in the structure appear and the cracks then spread spontaneously (18,36). In the case of a hard cheese these cracks may be evident to the eye, but they may at first be so small and so localized that they do not become visible until the compression has proceeded somewhat further. If the cheese has a very homogeneous structure, it is to be expected that these cracks will develop throughout the structure at about the same time and the change of slope will be clearly defined. In a more heterogeneous cheese they may appear over a range of strains and the change of slope will be much more diffuse. Once this region has been passed, the cracks continue to develop at an increasing rate and become more evident, until at point B the rate at which the structure breaks down overtakes the rate of build-up of stress through further compression and a peak is reached. This value is obviously dependent on a balance between the spontaneous failure of the structure and the effect of increasing deformation applied by the instrument to any residual structure. It is a convenient parameter to determine and may be used as a measure of firmness or hardness of the cheese (18). Beyond this point the stress may continue to fall if the failure of the structure becomes catastrophic, until the fragments become compacted into a new arrangement which can take up the stress and this now rises again.

If the compression test is carried on beyond the point at which it is reasonable to attempt to interpret it in terms of an acceptable model, and hence to evaluate any of the customary rheological parameters, it becomes purely empirical. One of the points on the curve most frequently used is the peak (18,37) (point B). This is a kind of yield point and one can obtain from it two useful parameters, the stress required to cause catastrophic breakdown of the structure and the amount of deformation that the cheese will stand before this breakdown occurs. The two are not entirely independent. In a dynamic measurement such as this on an essentially viscoelastic material, the rate of strain as well as the strain itself influences the moment at which the breakdown occurs. For reliable comparisons, all measurements should be made at the same rate of straining. Fortunately, many workers have found it convenient to use similar sizes and rates of compression, so that at least approximate comparisons can be made between their results. The stress measured at point B is in fact sometimes called a yield value, but it should not be confused with the yield stress of a plastic material determined by other methods such as a cone penetrometer. The yield at point B is not a material constant defining the strength of the sample; breakdown has already been occurring at least since point A, and maybe earlier. The peak value at B only indicates the maximum to which the stress rises before the collapse of structure overtakes the build-up of stress in what remains of that structure.

The other point frequently used is the stress at 80% compression. The damage to the structure when the compression reaches this extent is generally so complete that it is unrealistic to regard the stress as applying to the original sample. This is particularly true in the case of hard cheeses such as the typical English and Grana varieties, which crumble long before this degree of compression is reached. Nevertheless, compressions of this order and much greater arise during mastication (38). The stress at 80% can therefore give some indication of the consumer's response to the firmness of the cheese, although it must be said that in mastication the rate of compression is also much greater.

Measurements have been made on four types of hard cheese, Cheddar, Cheshire, Double Gloucester (36), and Leicester (39), which confirm that the value of the stress at the peak point B is dependent on the rate at which the compression is carried out. It was shown that over at least a twentyfold range of the factor a in equation 11, the rate at which the plates approached each other, the peak stress was linearly related to the fourth root of a. This is shown in Figure 8, where straight lines have been drawn through the points for each cheese type. If one bears in mind that this is a destructive test, so that each measurement had to be made on a different sample, the fit of the lines to the experimental points is quite acceptable. This result is, however, purely empirical; as far as can be seen there is no theoretical justification for it and it is limited to those types of cheese which crumble on breakdown. But there is a practical benefit. Using this finding, it is reasonable to reduce measurements made at any practical rate of compression to a standard rate so that the results of workers in different laboratories may be compared.

The investigation was carried one step further on Double Gloucester cheese (36). Not only were the peak values found to obey this fourth-root law but so were the stresses at other degrees of compression; the results are shown in Figure 9. The fact that the relation between stress and rate of deformation is more or less constant before, at, and beyond the yield point may be considered to lend support to the hypothesis that the processes taking place within the cheese are similar throughout the compression. The

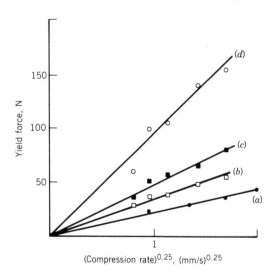

Figure 8. Effect of rate of compression on stress at yield point: **(a)** Double Gloucester; **(b)** Cheshire; **(c)** Leicester; and **(d)** Cheddar.

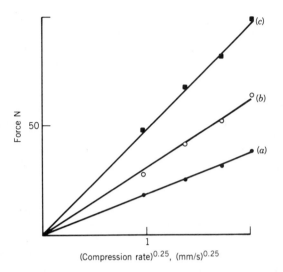

Figure 9. Effect of rate of compression on stress at various compressions: **(a)** 10%; **(b)** 40%; and **(c)** 70%.

stress at any point is the result of a balance between the rate of collapse of structure, ie, the spread of cracks, and the build-up of stress in that structure which remains. If it is assumed that the basic framework within the cheese has at least some structural strength, there will be some build-up of stress before any cracks appear, and the hypothesis predicts that the stress–strain curve will show an initial instantaneous rise due to viscoelastic deformation before any cracks develop, followed by a rising curve, convex upward, as the discrete minute cracks increase, with a pronounced change of slope as they begin to coalesce.

There are alternative explanations which may be advanced for the shape of the curves. In the early stages the curve rises from the origin and is often convex upward. This is characteristic of a Maxwell body. The interpretation could be that cheese behaves as an elastic fluid with a very high viscosity term. Neither the elasticity or the viscosity are readily obtained from the curves, but it is possible to calculate an apparent elasticity from the slope. Sometimes the slope at the origin is taken; more usually, the slope over the middle portion of the rise.

On the other hand, it is arguable that the structure of cheese is basically a solid one, but that even under very small stresses minute cracks begin to appear in that structure (40,41), even though these are far too small to be observed by the naked eye and may be disguised to some extent by the fact that some liquid component from the fat could flow into some of the opening interstices. Experiments on cheese analogs (42) at very low strains have confirmed that the structure does indeed break down well before the compression reaches 1%. It is possible to analyze the curves further on the basis of this hypothesis. Suppose that the cheese has an initial rigid structure giving rise to an elasticity E_0. Then at the instant at which the compression commences the slope of the stress–strain curve $d\sigma/d\varepsilon$ is equal to E_0. If the cheese breaks down continuously by the appearance of cracks, infinitessimal at first but becoming gradually more widespread and larger, the strength of the cheese is progressively reduced, so that at any subsequent instant the elasticity $E = d\sigma/d\varepsilon$

is less than E_0. If the breakdown of the structure is consequent upon the extent of the strain, the equation may be written

$$E = d\sigma/d\varepsilon = E_0[1 - f(\varepsilon)] \qquad (17)$$

This is the equation of a curve through the origin with an initial slope E_0 and subsequently convex upward, as is usually observed.

Pursuing this a little further, the function $f(\varepsilon)$ is a distribution function of the breaking strains of the interparticulate junctions within the cheese. As a trial hypothesis, one may postulate that the distribution function is linear up to the point at which all the junctions are broken. Replacing $f(\varepsilon)$ by a constant c and integrating equation 17 one gets

$$\sigma = E_0\varepsilon - c\varepsilon^2 (\varepsilon < \varepsilon_{\text{critical}}) \qquad (18)$$

This is the equation of a parabola with its apex upward and a maximum stress of $\sigma = E_0^2/4c$ at a strain of $\varepsilon = E_0/2c$. Eventually the situation is reached, at $\varepsilon_{\text{critical}}$, where all the structure is more or less completely destroyed and individual crumbs may move more or less independently as in the flow of a powder. If the cheese is spreadable, the stress required to maintain this flow may be expected to be constant. On the other hand, some further compaction may take place and the stress begin to rise again. In general, the distribution would not be expected to be a linear one. While the argument remains the same, leading to a convex upward curve with a peak and a subsequent trough, algebraic analysis is more involved. However, if the constant c is replaced by a general expression, such as a series expansion in ε, the slope at the origin is still given by E_0 but the parabolic form becomes distorted.

When studying individual curves it is not always possible to determine the slope at the origin with any confidence, particularly since this is the region where the response time of the recorder is most likely to introduce its own distortion. However, rewriting equation 18 as

$$\sigma/\varepsilon = E_0 - c\varepsilon$$

it is possible to construct fresh curves of σ/ε versus ε and these may be easier to extrapolate to zero strain and thereby estimate the value of E_0. The (negative) slope of this curve is then the distribution function of c. In a few cheeses c has been found to be more or less constant up to strains approaching unity (ie, about 60% compression) but with most cheeses the value of c decreases as the strain increases, indicating that the rate of breakdown of structure is actually greatest at the lowest strains. Figure 10 shows a few typical derived curves.

RELAXATION

If the compression of the sample is halted before the sample is completely broken up and then held, the stress which has built up during that compression will decay. The two binary models show characteristically different types of decay behavior. In the Voigt body the stress at

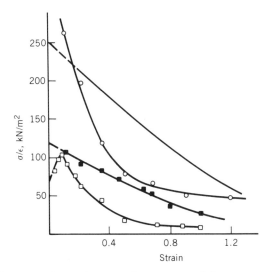

Figure 10. Derived curves of apparent modulus versus strain.

any instant is due partly to the strain and partly to the rate of straining. At the instant of arresting the motion the strain rate falls abruptly to zero, so that its contribution to the stress disappears, leaving only the stress due to the amount of strain, which then remains constant. In the Maxwell body the stress decays progressively, ie, relaxes, falling exponentially to zero if followed indefinitely according to the equation

$$\sigma = \sigma_0 \exp(-Gt/\eta) \qquad (19)$$

where σ_0 is the stress at the instant of halting. The ratio η/G has the dimensions of time and is known as the relaxation time. This is readily determined from the experimental curve, either by redrawing it in logarithmic form or as the time for the stress to fall to $1/e$ of its starting value. The relaxation curve by itself does not allow the elastic and viscous components to be obtained separately. When the simple Maxwell model is evidently insufficient to describe the relaxation behavior, an analysis of the curve may become more difficult. The sample may be described by a more complex model created by combining several simple binary models. There are computational procedures which enable the various constants of the model to be evaluated in such a case, but the precision falls off rapidly as the number of constants increases. The rheologist may sometimes be interested in determining several relaxation times and moduli for a particular sample, especially if these can be assigned to recognizable structural elements in the material, but more often the proliferation of constants does little to clarify the behavior of cheese.

There is an empirical treatment which is sometimes useful in analyzing relaxation data (43,44), particularly in cases where the precision of the data may be in some doubt. If one takes Y_t as the fraction of the stress which has decayed in time t, ie,

$$Y_t = (\sigma_0 - \sigma_t)/\sigma_0$$

where σ_0 is the stress at the commencement of the relaxation and σ_t the stress after time t, it has been found that

many complex viscoelastic materials relax in such a way that the relation

$$1/Y_t = k_1 + k_2/t \qquad (20)$$

holds to a fair degree of approximation; k_1 and k_2 may easily be found. Then $1/k_2$ is the extent to which the stress eventually will decay, while the ratio k_2/k_1 is a measure of the rate of decay. For a perfectly elastic body or a Voigt body $1/k_2$ is zero; there is no relaxation. For a liquid or a Maxwell body it is unity. For more complex models it lies somewhere in between. This is a useful device for deciding whether a simple model will suffice. For a Cheddar cheese $1/k_2$ was found to be in the region of 0.8, so a simple model is not adequate for describing the relaxation behavior of this cheese (43). An alternative treatment has been suggested (45), which, it is claimed, sometimes fits experimental data more closely. This may be derived by considering a binary model of the Maxwell type, but one in which the viscosity associated with the dashpot is not constant but follows a power-law variation with stress. The resulting expression is

$$(\sigma_0/\sigma)^{n-1} = 1 + k(n - 1)t \qquad (21)$$

where n is the exponent of the power law. This is an interesting suggestion: One of the consequences of the power-law variation of viscosity is that it becomes infinite at zero stress, and a Maxwell body with an infinitely viscous dashpot is in fact a solid. Thus this is a model for a material which is solid while at rest but becomes progressively more fluid as the stress increases. This is not too different from the description of cheese given above: a solid material whose structure breaks down progressively when subject to strain.

RHEOLOGY AND STRUCTURE

From the foregoing considerations it will be evident that the rheological properties of any material are consequent upon its structure. The three major constituents of cheese, casein, fat, and water, each contribute to the structure and therefore to the rheological properties in their own specific way. At normal room temperatures the casein is solid, the fat is an intimate mixture of solid and liquid fractions, giving it what may be described as plastic properties, while the water is liquid. The casein forms an open meshlike structure (46–48). Within this mesh is entrapped the fat, which had its origin as the fat globules of the milk. The water is more precisely the aqueous phase, since dissolved in it are the soluble constituents of the milk serum together with any salt which has been added during the cheese-making. Some of the water is bound to the protein and therefore largely immobilized; the remainder is free and fills the interstices between the casein matrix and the fat. This structure is common to all types of cheese. The differences among various types of cheese are brought about by the influence of different manufacturing regimes on that structure. In addition to the characteristic differences arising from different procedures, adventitious variations may also occur within the same type of cheese. These reflect differences in the original milk or in the conditions during the manufacturing or maturing processes.

It is the casein within the cheese which is responsible for its solid nature. The primary structure is a three-dimensional cage whose sides consist of chains of casein molecules (49,50). This provides a structure of considerable inherent rigidity. The chains are not linear (51), but have an irregular, somewhat gnarled structure. This may be deformed elastically under the action of external forces and this elasticity modifies the rigidity of the cage.

During the clotting process these chains have been formed by the joining together of individual casein particles in the serum. This serum surrounds the fat globules so that each cage may be expected to encase at least one globule or cluster of globules (52). The distribution of sizes of these cells will be controlled by the distribution and size of the fat globules. For instance, when the milk is first homogenized, the cheese made from it will have a more uniform distribution of cells than cheese made from fresh milk (53). The complete cheese curd at this stage consists of an aggregate of these cells of casein plus fat and the whole is pervaded by the aqueous phase (52).

If a force is applied to such a structure, the deformation will be primarily controlled by the rigidity of the cage, modified by any elasticity in its structural members. In the absence of the fat and water this would behave simply as any other open-type structure and its deformation would be characterized by a single modulus of rigidity or elasticity. However, the deformation of any of the cells is limited by the fat within it. At very low temperatures the fat would be solid and this would only add to the rigidity. At the normal temperatures at which cheese is matured and used the fat has both solid and liquid constituents and has its own peculiar rheological properties. Any deformation of the casein matrix would also require the fat to deform. At the same time, the water between the fat and the casein acts as a lubricant. As a result, the rigidity of the fat is added to that of the casein in a complex manner and it is this which gives rise to the peculiar viscoelastic properties of the cheese.

Even this is not the whole picture. The final cheese is not merely a continuous aggregate of cells as just described. During manufacture the curd is cut into small pieces at least once to allow any excess serum to drain away. As the serum drains away, the casein matrix shrinks on to the fat globules, making a more compact whole. The granules so formed may then be further distorted, as in the cheddaring process, or may be allowed to take up a random distribution, as in a Cheshire cheese. The final cheese mass is an aggregation of these granules which forms a secondary structure having its own set of rheological properties. Even this may be further modified by subsequent processes such as milling, which gives rise to a tertiary structure, and by pressing, which distorts the whole (54).

This rudimentary account of the factors contributing to the rheological properties applies to any cheese. During the course of manufacture, and subsequently during ripening, the basic structure may be modified by mechanical or thermal treatment, or the casein itself may be acted upon by bacteria and any residual enzymes. These agencies may change the organization of the structure or they may cause contiguous fat globules to coalesce. Finally, water may be lost by evaporation from the surface.

Before discussing the effect of the structure of the cheese on its rheological properties it is appropriate to consider differences which may occur within a single cheese or between cheeses from the same batch which might be expected to be alike. A cheese which matures in contact with the air will develop a pronounced rind and may show considerable variation throughout its body. The surface layers lose moisture more rapidly than the inner portions. This results in a difference in composition, which may be reflected in the rheological properties. It may also affect the progress of the maturation process itself. Another source of variation in some cheeses, particularly some of the larger varieties with long maturation periods, is that they are turned at intervals: The top and bottom layers are subjected to alternate low and high compressive forces due to their own weight, while the middle will have a more uniform treatment. This again may be reflected in the nature of the maturation. An otherwise uniform cheese will show a distribution of firmness as in Figure 11. A practical consequence of this is that measurements made on or near the surface of a large cheese will not be characteristic of the main body of that cheese.

Differences between cheeses may be expected to be greater and the magnitude of the variations will depend to some extent on the method of measurement. Penetration methods, where the instrument only acts locally on a limited quantity of the sample, give greater differences than compression testing in which a larger volume of the sample is involved. Variations of the order of 10% of the mean value of a parameter have been found on different samples from the same batch in a compression test (37); this probably indicates the limit of reproducibility which one may reasonably expect when making measurements on cheese. Differences between batches may be considerably larger. As an extreme example, one experimenter measuring eight different lots of cream cheese obtained readings which ranged from 54 to 251 units (55). Before leaving the topic of variation, one other source which must always be borne in mind is that between laboratories nominally making similar measurements. Because this is not a problem unique to either rheology or cheese, it will be disregarded in what follows.

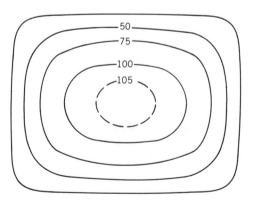

Figure 11. Typical distribution of firmness throughout a mature cheese.

Although it is the interaction of the properties of the principal constituents which gives cheese its viscoelasticity, it is profitable to consider some of the features associated with each of the constituents. First, consider the casein, which, it has been pointed out, gives the cheese its solid appearance. Because the casein forms chains within the spaces around the fat globules, there must necessarily be a minimum amount of casein below which any continuous structure cannot exist. This will depend on the number, size, and size distribution of the fat globules and on the size and size distribution of the casein micelles themselves. Once the quantity of casein required for this minimum structure has been exceeded, any additional casein will serve to strengthen the branches and the junctions. It is to be expected that, irrespective of the type of cheese, there will be a general relationship between the amount of casein present in the cheese and its firmness. Figure 12 shows this relation for some ten different types of hard cheese (29). Naturally, since the data refer to cheeses of very different provenance, there is considerable scatter, but it is possible to draw a regression line through the points and this indicates that about 25% of the weight of the cheese must be casein in order to provide a rigid framework and that above this limit more protein only strengthens it.

The requirement that there be sufficient casein to build a structure around the fat has been clearly shown in measurements on two other varieties of cheese. In some Meshanger cheese (56) it was found that, unless the casein occupied more than about 37% of the volume of the cheese not occupied by the fat, it had virtually no rigidity and behaved like a soft paste. In a similar way, it has been shown quite dramatically (Fig. 13) that in Mozzarella cheese (57) the protein must exceed about 42% of the non-aqueous part of the cheese for any rigidity to exist. Again, this corresponds to a minimum casein content of about 25% for a rigid structure.

Although it has been claimed that the casein matrix gives rigidity to the cheese, there is still a theoretical question which has not been answered. Is the matrix con-

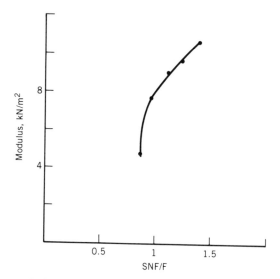

Figure 13. Relation between firmness and protein content in a single cheese. SNF/F = ratio of nonfat to fat solids.

tinuous throughout the whole cheese, so that it may be treated as a solid body, or do the dislocations in the structure which have been introduced by cutting allow it to flow, albeit imperceptibly, however small the stress? The evidence is inconclusive. In theory, an examination of the stress–strain curves at very low strains should decide the issue. If there is a continuous structure, there may be elastic deformation but no flow until a finite stress is reached which is sufficient to rupture some of that structure. Force–compression curves should indicate whether the cheese is of the Maxwell body or the Voigt body type. Commercially available instruments are generally not precise enough in this region, for they are not designed for this purpose. More precise measurements on cheese analogs have already been mentioned as showing that these began to break down at very small strains (42). Measurements of the recovery after compression of a few real cheeses (58) showed that even the smallest strains applied broke down some structure in a Mozzarella cheese, but a Cheddar cheese and a Muenster were able to recover completely (Fig. 14).

Cheddar cheese has also been studied by means of a relaxation experiment (44). Samples were compressed under different stresses, and hence at different rates, to the same ultimate deformation and the subsequent relaxation of the stress was observed. Using Peleg's treatment (43) it was found that the value of $1/k_2$ in equation 20 was dependent on the stress applied. The greater the stress, the more the structure broke down, although only the same strain was reached; the relation was curvilinear, particularly at the lowest stresses (Fig. 15). Extrapolating this curve back to zero stress should indicate whether any breakdown occurred. A zero value of $1/k_2$ would indicate a solid structure. Unfortunately, any extrapolation of this curve would be speculative. It is possible that it would lead to a positive intercept on the vertical axis. If this were so, it would lead to the conclusion that there was no continuous structure throughout that cheese, but to establish this with confidence would require measurements at much lower stresses.

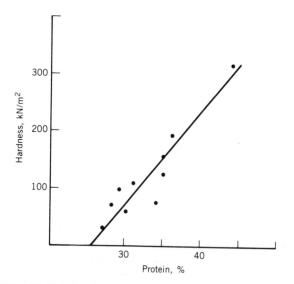

Figure 12. Relation between firmness of cheeses and total protein content.

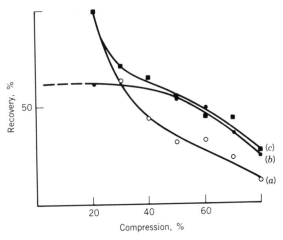

Figure 14. Relation between recovery of cheese and previous compression: (*a*) Cheddar; (*b*) Mozzarella; and (*c*) Muenster.

Another series of experiments which could throw light on this question has been carried out (59). A small sinusoidal vibration was applied to one surface of the cheese sample and the stress transmitted through it observed. If one makes the assumption that the cheese behaves as a Voigt body, a modulus of rigidity can be calculated from the in-phase component of the signal received and a viscosity from the out-of-phase component. For a true Voigt body both of these should be independent of the frequency. In fact, neither was constant with either Cheddar or Gouda cheese, even when the strain was as low as 0.04. The inference is that even at this strain some failure in the rigid structure had taken place. Either internal cracks or slip planes had developed within the cheese (42). However, everyday experience suggests that cheese should have some rigid structure. The stress on the lower layers of a large cheese, such as a Parmigiano or a Cheddar, is of the order of 3 to 4 kPa; yet such a cheese retains its shape more or less indefinitely.

Summarizing the previous paragraphs, it appears that the rheological role of the casein is to provide a continuous elastic framework within the individual granules. It is likely that where the casein chains lie in the granule sur-

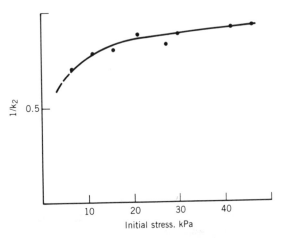

Figure 15. Amount of relaxation after previous stress.

face and are contiguous with chains in neighboring granules they may be held together by physicochemical bonds which develop during maturation and give rise to some rigidity in the aggregation (54). However, these bonds will be sparser than those within the original network, since they form only where chance contacts occur. This results in a weaker secondary structure. These bonds may be broken when a positive strain is applied, giving a Maxwell body type of response, but are strong enough to preserve rigidity under the cheese's own weight. Some further support is given to this view by the fact that individual granules have a much higher modulus than the whole cheese (60).

The role of the other major constituents is more clearly documented than is that of the casein. The fat derived from the original milk is very roughly one third of the total mass. Its rheological properties are very sensitive to temperature changes and, as is to be expected, this sensitivity is imparted to the whole cheese. At 5°C (around normal refrigeration temperatures) many of the glycerides in the milk fat are solid. The proportion of solid fat decreases with rise of temperature, particularly sharply in the 12–15°C region, which is close to the ripening temperature for many varieties. Above this region the proportion of solid decreases further until at around 35°C almost all is liquid. This is near the temperature that a small portion will rapidly attain if it is placed in the mouth and chewed. The ratio of the solid to liquid fat components is the principal factor influencing the rheological properties of the fat (61). Although factors such as breed and species of animal, pasture, and herd management have some effect on this ratio, temperature is by far the most important. Fat also supercools readily, so that the thermal history is almost as important as the absolute value of the temperature itself. During maturation and storage cheese is usually kept at a lower temperature than that at which it was made. Originally most of the glycerides will have been liquid and at the lower temperature these will slowly solidify. Most of the change will take place in the first day or two, but solidification will continue progressively, maybe for several weeks before final equilibrium is reached.

Measurements made on Cheddar, Cheshire (39), Emmentaler (62), Gouda (63), and Russian (64) cheeses at a range of temperatures, using different instruments, all show the same general trend (Fig. 16). It may be mentioned that different instrumental methods give rise to somewhat different results. In general, penetrometers appear to indicate a rather steeper dependence of firmness on temperature than compression tests. Nevertheless, bearing in mind the difference in origin of the cheeses and the limitations in accuracy of the individual measurements, already discussed, this curve shows clearly the influence of the (test) temperature on the firmness of the cheeses.

Differences in the glyceride composition of the original milk fat are also reflected in the final cheese. Using the iodine number as an indication of the ratio of solid to liquid fat at any one temperature, the firmness of Swiss cheese was shown to vary seasonally with the solidity of the fat (65) (Fig. 17). The lower the iodine number, the greater the degree of saturation of the glycerides and the

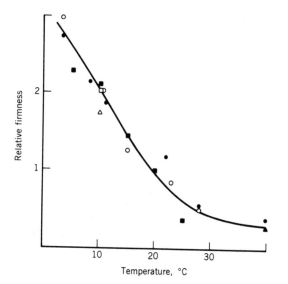

Figure 16. Variation of firmness with temperature: ● Cheddar; ○ Cheshire; ■ Emmental; □ Gouda; and ▲ Russian.

higher the proportion of solid fat. In these particular experiments the cheese produced in December appeared to be anomalous. However, even including the December result, the relation was statistically significant: excluding it, it was highly so and this regression has been drawn on the figure.

The water serves two functions, one purely physical, in that it is a low viscosity liquid, the other in its action as a carrier for the substances dissolved in it which may undergo chemical reactions with the fat or casein. It has already been stated that the water occupies the space not taken up by the casein matrix and the fat. Provided that there is a sufficient quantity, it takes up all of that space. Only near the outer surface, for instance, in the rind, is there likely to be any deficiency. In the case of Swiss-type cheeses, which contain eyes filled with gas, the main body of the cheese is still replete with water. The rheological consequence of this arrangement is that the water acts as

a low-viscosity lubricant between the surfaces of the fat and the casein. The greater the space between these, which is another way of saying the higher the water content of the cheese, the easier the flow of water within these spaces. Hence there should be less restraint on the movement of the casein mesh around the enclosed fat. This freer movement should manifest itself as a lesser resistance to the deformation of the whole cheese and as a greater recovery after deformation.

Taking the latter point first, a study was made of the elastic recovery of samples of some ten varieties of cheese after they had been compressed (29), but only up to a point at which no visible damage to the structure had occurred (Fig. 18). The cheese with the lowest water content, Parmigiano, showed the least recovery, while a Mozzarella with over 50% water showed the most. From the figure it is seen that while there was considerable scatter about the regression line, as was to be expected with cheeses of different origins, a general trend was well established.

The lubrication effect can be more clearly demonstrated with a single variety and within a single cheese. In a freshly made cheese the water is distributed more or less evenly throughout the whole mass. During the maturation period of those cheeses which are allowed to ripen in air, water evaporates from the surface and is only slowly replaced by more water migrating from the interior. A moisture gradient is set up within the cheese. Measurements made on Emmentaler cheese (62) (Fig. 19) showed a very close relation between the water content of successive layers and the firmness as determined by a penetrometer. Similar results have been obtained with Edam cheese (66). In that series of experiments a close relation was observed between the water content of the cheeses and the firmness, although they were made in two batches at different temperatures (Fig. 20).

In its other role as carrier of active ingredients the presence of water is principally seen in the effect on the changes which take place in the casein network. An illustration of this is found in Meshanger cheese (56). This

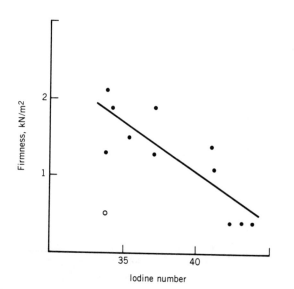

Figure 17. Variation of firmness with saturation of glycerides.

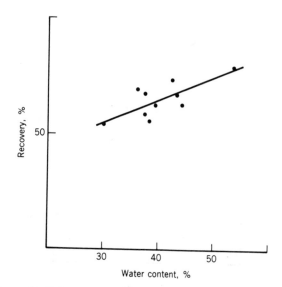

Figure 18. Relation between elastic recovery and moisture content.

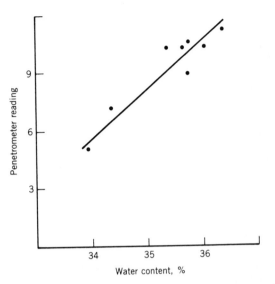

Figure 19. Relation between firmness and moisture content in a single cheese.

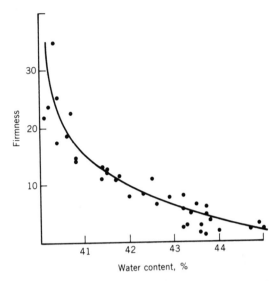

Figure 21. Relation between firmness and moisture content.

cheese has the usual well-organized casein structure when freshly made. As the ripening proceeds, the structure breaks down until it becomes a more or less amorphous loose mass of casein submicelles. Figure 21 shows the relation found between an arbitrary measurement of the firmness and the water content of cheeses from some ten batches during ripening. In the same series of experiments the rate of degradation of the casein was determined for some of the batches over the first few weeks of life. The rate of degradation, calculated as a first-order reaction, showed good agreement with the water content (Fig. 22).

Although protein, fat, and water constitute by far the greatest part of any cheese mass, other constituents cannot be ignored. Salt, when added dry before the cheese is pressed, may sometimes appear as crystals embedded in the fat–protein mass. Then it is not usually present in large enough quantities to make any sensible contribu-

tion to the rheological properties of the whole. Almost all of the salt is present only in solution in the water. There its effect on the properties of the whole cheese is minimal. Any serious contribution of the salt to the rheological properties of the whole cheese is by indirect action (67). A high concentration of salt increases the osmotic pressure, diverting a significant quantity of water from the structural bonds of the casein network (68,69). This may be seen in the effect on the ripening of some samples of Mozzarella cheese (70). When the salt content was low (0.27%) the modulus decreased from 120 to 45 N/m^2 in five weeks, but when salt was present in excess of 1% no change was observed within the same period.

In cheese where the salt is added by immersion in brine there is a further consideration. The inward diffusion of the brine results in a concentration gradient being set up within the cheese. This has a twofold effect. First, the presence of the salt simply excludes some of the water; this in turn reinforces the moisture gradient simulta-

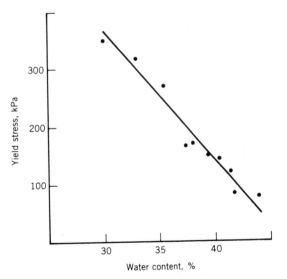

Figure 20. Relation between quasi-modulus and moisture content in a single cheese variety.

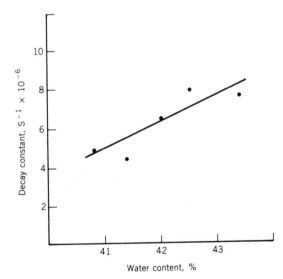

Figure 22. Relation between rate of casein breakdown and moisture content.

neously arising from the evaporation of water from the surface layers. Probably more important is the fact that the diffusion of the salt into the cheese is a slow process (71). In the early stages of ripening the salt concentration in the inner regions is unlikely to be sufficient to limit the protein degradation described above. The presence of more water and an enhanced proteolytic activity affect the rheological properties in the same sense, giving rise to a weaker structure and therefore a less firm cheese. It is therefore probable that the variation in firmness within a single cheese, which, as described earlier, is closely related to the moisture distribution, is accentuated by brining.

Some cheeses also contain a significant quantity of gas. In the Swiss-type cheeses this is concentrated in the eyes; the remainder of the cheese has a close waxy texture and is virtually free of any entrapped gas. In hard and crumbly cheeses the gas may simply be air which has infiltrated into cracks which have developed in the mass. The significance of either eyes or cracks is that measurements made on the whole cheese or on its surface are not a fair representation of the properties of the main body of the cheese. Hidden cracks or eyes will give rise to irregularities when measurements are made by any form of penetrometer and probably account for a substantial amount of the observed scatter. In compression measurements they will give rise to premature breakdown. They also account for the fact that difficulties are encountered when any attempt is made to assess the rheological properties of cheese by studying the propagation of ultrasound through it. The velocity of propagation is more influenced by the scattering at the discontinuities than by the properties of the cheese mass.

SOME EXPERIMENTAL RESULTS

Most of the rheological methods described earlier have been used at one time or another for the study of cheese. Problems associated with penetrometers and tests on the surface have already been indicated. The most satisfactory tests are made on samples cut out specifically from the cheese mass. In this case the sample can be inspected to insure that there are at least no major cracks or other inhomogeneities in it and the dimensions can be precisely determined. The force–compression test using small prepared samples has been widely used for routine measurements. However, in spite of its widespread use, there has been no consensus of opinion on the most suitable operating conditions. Sample size and shape, rate of compression, and even temperature have all varied. A most interesting feature is that, notwithstanding these differences and the number of cheeses which have been examined, the force–compression curves show a remarkably common pattern. This underlines its general usefulness.

There are several important considerations to be taken into account when deciding on the operating conditions. In the first place, the sample should be large enough to be representative of the whole (72); the more heterogeneous the cheese, the larger the sample. On the other hand, it should not be so large that it may contain hidden cracks or irregularities. These two requirements are almost mutually exclusive. As a compromise most workers have taken

samples with linear dimensions of 10 to 25 mm. The preferred shape has usually been a cylinder. It is easier to prepare a rectangular sample with precise dimensions but the symmetry of the cylindrical shape helps to minimize the development of irregular cracks during compression (63). A wide range of compression rates has been used: from 2 mm/min up to 100 mm/min (3.33×10^{-5} to 1.66×10^{-3} m/s). The faster rates tend to obscure the true behavior at the onset of compression unless a recorder of very short response time is available. The strain rate at any instant depends on both the rate of compression and the height of the sample at that instant.

Table 1 summarizes the principal measurements that have been reported on hard cheeses using the force–compression test. In order to produce this table and make the results from different workers comparable, the original curves from the references cited have been transformed into stress–strain curves. As previously mentioned, the instrument in its usual commercially available form produces curves of force versus linear travel. While linear travel is easily converted to true strain (eq. 11), the stress correction requires a knowledge of the actual shape at any instant during compression. Often this is unknown, or at least unreported. Unless special precautions have been taken, the behavior of the sample will be intermediate between that obtaining in the lubricated and in the bonded situations. For the present purposes the simpler correction applying to the lubricated condition has been used throughout, although it is realized that it will not quite compensate for the barreling which occurs. The values were then read from the transformed curves. The modulus given in the column headed E_0 is from the slope at the origin, ie, it refers to the undeformed sample. The entries in the columns headed yield refer to the stress and strain at the peak of the curve.

Table 2 summarizes a number of results which have been obtained by other methods where it has been possible to calculate a modulus from the data. In order to compare the two tables, it will be recalled that Table 1 lists the values of an elastic modulus. It must be most nearly comparable with any quasi-modulus determined by the puncture test or by simple compression. The yield stress measured by a cone penetrometer is more likely to be comparable with the yield value in the force–compression test, since both are influenced by partially broken-down structure. It must be pointed out, however, that lack of adequate documentation makes it difficult to make meaningful comparisons between the results from different laboratories.

VARIATION WITH AGE

The development of firmness during the ripening and subsequent storage of cheese has always been of particular interest to both the practical cheese-maker and the rheologist. During this period all three principal constituents undergo change. Moisture evaporates from the surface of the cheese if this is exposed to the air; additional moisture migrates from the interior to replace that lost and some of this also evaporates. Eventually a moisture gradient is set up, but even the center will be drier than it was originally. The protein in the matrix undergoes progressive change

Table 1. Principal Measurements of Hard Cheeses Using the Force–Compression Test

Cheese	Modulus, E_0, kN/m²	Yield Point		80% Stress, σ_{80}, kPa	Reference
		σ_y, kPa	ε_y		
American loaf	70	25	0.72	46	58
Bel Paese	300	44	1.1		37
Caerphilly	1100	92	0.16	67	36
Cheddar	280	41	0.34	50	22
Cheddar		53	0.25		39
Cheddar		63	0.24		47
Cheddar (mild)	160	40	0.60	78	58
Cheddar (strong)	400	58	0.22	65	58
Cheddar	320	45	0.21	64	73
Cheddar (green)	75	54	0.80	37	47
Cheddar (mature)	170	23	0.20	22	47
Double Gloucester	1600+	139	0.24	75	36
Edam	660	214	0.60		36
German loaf	300	18	ca 1.1		33
Gouda	240	98	0.72	52	36
Lancashire	2000+	105	0.23		36
Leicester	400	56	0.31	38	32
Montasio	40	85	0.80		37
Mozzarella	300	(no peak)			34
Mozzarella	80	14	0.57		58
Muenster	40	16	0.72	59	58
Parmigiano	700	132	0.28		37
Provolone	120	37	0.60		37
Silano	230	26	ca 0.30		57
Processed	195	57	0.39	30	36
Processed	440	71	0.27		37
Processed	60	9	0.21		37

as the available water is reduced and the residual enzymes and bacteria continue to act on it (46,47,75). Furthermore, on the purely physical plane, as the water content decreases, the matrix may shrink or collapse under pressure so that voids do not form and the protein becomes

Table 2. Modulus Measurements Using Various Methods

Cheese	Modulus, kN/m²	Method	Reference
Brie	1.3	Extruder	74
Sbrinz	41	Extruder	74
Chanakh	58	Cone	69
Emmental	40	Cone	69
Emmental	3.5	Cone	22
Gouda	44	Cone	69
Kostroma	54	Cone	69
Lori	40	Cone	69
Cheddar	195	Punch	29
Edam	340	Punch	29
Emmental	220	Punch	29
Gouda	270	Punch	29
Kachkaval	120	Punch	75
Mozzarella	170	Punch	29
Mozzarella	22	Punch	74
Muenster	120	Punch	29
Parmigiano	550	Punch	29
Provolone	130	Punch	29
Cheddar	270–400	Compression	76
Mozzarella	80	Compression	70
Kachkaval	60–100	Ball Compressor	77

denser. Some of the glycerides in the fat slowly crystallize, resulting in a more solid mass of fat.

These changes which take place in the structural components of the cheese are reflected in force–compression curves. Figure 23 shows curves for two cheeses, Cheddar (47) and Gouda (63), when green and when mature. During aging the strain sustainable by the cheese before

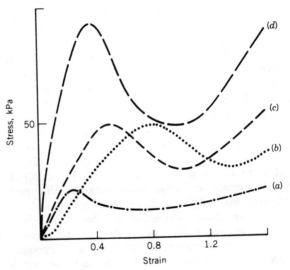

Figure 23. Force–compression curves for young and mature cheeses: (*a*) Cheddar (mature); (*b*) Cheddar (young); (*c*) Gouda (young); and (*d*) Gouda (mature).

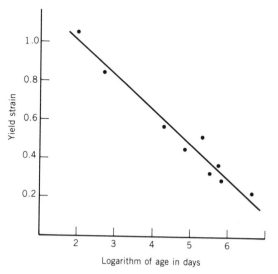

Figure 24. Relation between age and strain at yield point for Cheddar cheese.

breaking down at the yield point (point B in Fig. 7) decreased more or less exponentially with time in both cases. Figure 24 shows this for one batch of Cheddar cheese. At the same time the stress at the yield point diminished. The strength of the matrix had clearly been reduced through aging. In the Gouda cheese the behavior was rather different. Although the yield strain decreased, the stress at this point, ie, the ultimate strength of the matrix, had increased. It is only to be expected that as the actual balance between the different mechanisms of aging of the components varies from one cheese to another, so the paths of change in the whole cheese will vary. Figure 25 shows some of the results that have been reported (47,78,79).

Not only does the firmness of cheese change with age, but also its springiness. The degradation of the protein,

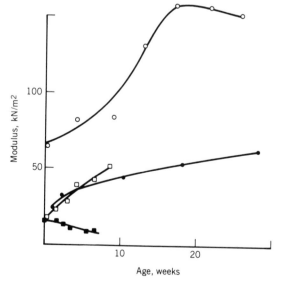

Figure 25. Variation of firmness with age: ● Cheddar; ○ Kachkaval; □ Tamismi and Russian; and ■ unspecified French cheese.

the solidification of the fat, and the reduction of the water available to lubricate the relative motion all tend to reduce the springiness. Springiness is not a simple rheological concept. It implies not only the ability to recover from compression or indentation but also that this will be immediately evident. It does not show itself directly in a force–compression curve; a recovery curve is more appropriate for this purpose. Nevertheless, this decrease of springiness has been remarked upon consistently using subjective observations on a number of cheeses (27,29,47,49,80).

SUBJECTIVE ASSESSMENTS

How the rheological properties measured by instruments, as just described, relate to the grader's or the consumer's assessments of the cheese remains to be considered. The emphasis in this article has been on the measurement of the firmness of cheese which arises from its structure. It should be stressed that meaningful comparisons can only be made between instrumental and subjective methods if the subjective terminology is unambiguous and the instrumental measurements precise. Early experiments on single varieties of cheese showed that the subjective assessment of firmness, which is a fairly simple concept, correlates very highly with measurements made with simple apparatus, such as the Ball Compressor (81). But a simple correlation by itself does not indicate to what extent the instrumental measurement should be used to predict a typical user's appraisal. Nor can the results of experiments on a single variety be extrapolated to other types of cheese without verification. Some authors have arbitrarily assigned a specific consumer quality to a particular instrumental reading. For instance, the quasimodulus calculated from a simple compression test (27), the yield value given by a cone penetrometer (22), and the stress at 80% compression have all been proposed as measures of firmness. The recovery after a limited compression (29,82), arbitrarily decided upon, has been used as an indication of springiness. These intuitive opinions may help visualize the significance of an instrumental measurement and are probably adequate for internal comparisons within a single investigation, but they lack scientific rigor.

The force–compression test, giving as it does a characteristic curve with several readily identifiable features, provides a number of potentially useful parameters. In Figure 26 the values of hardness as assessed by a panel, of a number of different cheeses, versus the stress measured at 80% compression are plotted (73). A logarithmic regression line has been drawn through the points. The standard deviation about this regression line was no greater than that due to the differences among the panel. In this case, then, using the regression line, the instrumental reading can be safely used to predict that panel's assessments. In the same series of experiments the instrumental measurements were carried further. As soon as the compression had reached 80% the compressing plates were withdrawn and the recovered height of the sample was measured. At the same time the springiness was assessed subjectively. As has already been pointed out, springiness involves not

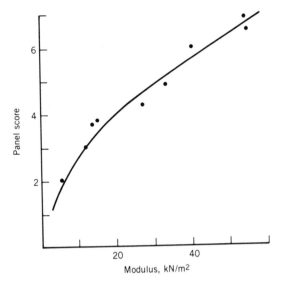

Figure 26. Comparison of subjective and objective measurements of firmness.

only the extent of recovery but also its speed; possibly all individuals will have, subconsciously, their own ideas of the relative contribution of the two to the final judgment. As was to be expected, there was poorer agreement among the panel members in this part of the experiment. Even so, there was a highly significant correlation between the measured recovery and the average panel assessment of springiness.

The relation between user acceptance and instrumental rheological measurement is an ongoing study (82). The few examples cited above show that, if the proper parameter is chosen and the terms clearly understood, rheological measurements can successfully predict users' assessments.

CONCLUSION

Although the study of cheese rheology is as old as rheology itself, only a few cheese varieties have been investigated. These have been drawn almost exclusively from among the firmer varieties. The softer cheeses, of which there are many, have received scant attention. They pose rather different rheological problems because they are less like solids and more akin to stiff pastes or creams; they should, however, be nonetheless interesting to rheologists.

It has been shown, by bringing together diverse observations on the firmer cheeses from a number of workers, that it is possible to build up plausible theories to relate the behavior of the finished cheese to the structure developed during its manufacture. Mathematical models may be posited for the rheological behavior, but with a material as heterogeneous as cheese, these, if they are to be simple enough to be easily handled, are unlikely to give more than an approximate representation of the observed pattern. Using these, or directly from experimental measurements, material constants may be extracted to characterize the cheese. A knowledge of these may be helpful to the cheese-maker, both as a means of controlling the

production regime and as a guide to the acceptability of the product.

This article previously appeared in *Cheese: Chemistry, Physics and Microbiology,"* R. F. Fox, ed., Elsevier Applied Science, New York, 1987. It is printed here with some revision by permission of the publishers.

BIBLIOGRAPHY

1. M. Reiner, *Twelve Lectures in Theoretical Rheology,* North Holland Publishing Co., Amsterdam, 1949, p. 19.
2. A. S. Szeczniak, *Journal of Food Science* **28,** 385 (1963).
3. M. Baron, *The Mechanical Properties of Cheese and Butter,* United Trade Press, London, 1952, p. 29.
4. R. Jowitt, *Journal of Texture Studies* **5,** 351 (1974).
5. R. Houwink, *Elasticity, Plasticity and Structure of Matter,* Dover Publications, New York, 1958, p. 5.
6. M. Peleg, *Journal of Texture Studies* **15,** 317 (1984).
7. J. H. Prentice, *Measurements in the Rheology of Foodstuffs,* Elsevier Applied Science Publishers, London, 1984, p. 32.
8. R. B. Bird, *Annual Review of Fluid Mechanics* **8,** 13 (1976).
9. A. S. Szeczniak and M. C. Bourne, *Journal of Texture Studies* **1,** 52 (1969).
10. J. E. Caffyn, *Dairy Industries* **10,** 257 (1946).
11. J. E. Caffyn and M. Baron, *The Dairyman* **64,** 345 (1947).
12. N. N. Mohsenin, *Physical Properties of Plant and Animal Materials,* Gordon & Breach Scientific Publishers, New York, 1970, p. 288.
13. N. N. Mohsenin, C. T. Morrow, and Y. M. Young, *Proceedings of the 5th International Congress on Rheology,* Vol. 2, 1970, p. 647.
14. C. P. Cox and M. Baron, *Journal of Dairy Research* **22,** 386 (1955).
15. A. D. Konoplev, P. E. Krashenin, and V. P. Tabachnikov, *Trudy Vsesoyuznogo Nauchno-Issledovatel'skogo Instituta* masl. i. syr. Prom. **17,** 40 (1947).
16. S. Poulard, J. Roucou, G. Durrange, and J. Manry, *Rheologica Acta* **13,** 761 (1974).
17. M. Baron, *Dairy Industries* **14,** 146 (1954).
18. P. Eberhard and E. Flückiger, *Schweizerische Milchzeitung* **104,** 24 (1978).
19. V. P. Tabachnikov, V. Ya. Borkov, V. B. Ilyushkin, and I. I. Tetereva, *Trudy Vsesoyuznogo Nauchno-issledovalel'skogo Instituta* masl i. syr. Prom., **27,** 61 (1979).
20. R. Harper and M. Baron, *British Journal of Applied Physics* **2,** 35 (1951).
21. W. G. Wearmouth, *Dairy Industries* **19,** 213 (1954).
22. E. Flückiger and E. Siegenthaler, *Schweizerische Milchzeitung Wissenschaftliche Beilage* **89,** 707 (1963).
23. F. J. Mottram, *Laboratory Practice* **10,** 767 (1961).
24. M. C. Bourne, in P. Sherman, ed., *Food Texture and Rheology,* Academic Press, London, 1979, p. 95.
25. J. M. deMan, *Journal of Texture Studies* **1,** 114 (1969).
26. B. S. Kamel and J. M. deMan, *Lebensmittel-Wissenschaft und Technologie* **8,** 123 (1975).
27. R. Davidov and N. Barabanshchikov, *Molochnaya Promyshlennost* **9**(4), 27 (1950).
28. V. P. Tabachnikov, *Trudy Vsesoyuznogo Nauchno-issledovatel'skogo Instituta* masl. i. syr. Prom. **17,** 84 (1974).
29. A. H. Chen, J. W. Larkin, C. J. Clarke, and W. E. Irvine, *Journal of Dairy Science* **62,** 901 (1979).

30. R. Ramanauskas, S. Urbene, L. Galginaitye, and P. Matsulis, *Trudy Litovskii Filial Vsesoyuznogo Nauchno-Issledovatel'skogo Instituta Maslodel'noi Syrodel'noi Promyshlennosti* **13**, 64 (1979).

31. Ref. 7, p. 9.

32. E. J. V. Carter and P. Sherman, *Journal of Texture Studies* **9**, 311 (1978).

33. P. Sherman and G. Atkin, *Advances in Rheology*, **4**, 133 (1976).

34. E. M. Casirighi, E. B. Bagley, and D. D. Christianson, *Journal of Texture Studies* **16**, 281 (1985).

35. P. W. Voisey and M. Kloek, *Journal of Texture Studies* **6**, 489 (1975).

36. P. Shama and P. Sherman, *Journal of Texture Studies* **4**, 344 (1973).

37. P. Masi, in *Physical Properties of Foods*, Vol. 2, Elsevier Applied Science Publishers, London, 1987, p. 383.

38. M. M. Bryan and D. Kilcast, *Journal of Texture Studies* **17**, 221 (1986).

39. E. Dickinson and I. C. Goulding, *Journal of Texture Studies* **11**, 51 (1980).

40. N. V. Polak, *Dissertation Abstracts International B* **42**, 3178 (1982).

41. R. Jowitt, in P. Sherman, ed., *Food Texture and Rheology*, Academic Press, London, 1979, p. 146.

42. R. J. Marshall, *Journal of the Science of Food and Agriculture* **50**, 237 (1990).

43. M. Peleg, *Journal of Food Science* **44**, 277 (1979).

44. M. Peleg, *Journal of Rheology* **24**, 451 (1980).

45. B. Launay and S. Buré, *DECHEMA Monographien* **77**, 137 (1974).

46. A. M. Kimber, B. E. Brooker, D. G. Hobbs, and J. H. Prentice, *Journal of Dairy Research* **41**, 389 (1974).

47. H. K. Creamer and N. P. Olson, *Journal of Food Science* **47**, 631 (1982).

48. M. V. Taranto, P. J. Wan, S. L. Chen, and K. C. Rhee, *Scanning Electron Microscopy*, 273 (1979).

49. M. Green, A. Turvey, and D. G. Hobbs, *Journal of Dairy Research* **48**, 343 (1981).

50. T. van Vliet and P. Walstra, *Netherlands Milk and Dairy Journal* **39**, 115 (1985).

51. Ref. 7, p. 157.

52. M. L. Green and A. I. Grandison, in P. F. Fox, ed., *Cheese: Chemistry, Physics and Microbiology*, Vol. 1, Elsevier Applied Science Publishers, London, 1987, p. 97.

53. D. B. Emmons, M. Kalab, E. Larmond, and K. J. Lorne, *Journal of Texture Studies* **11**, 15 (1980).

54. M. Kalab, *Milchwissenschaft* **32**, 449 (1977).

55. Z. D. Roundy and W. V. Price, *Journal of Dairy Science* **24**, 135 (1941).

56. L. de Jong, *Netherlands Milk and Dairy Journal* **32**, 1 (1978).

57. P. Masi and F. Addeo, *Advances in Rheology* **4**, 161 (1976).

58. E. M. Imoto, C-H. Lee, and C. K. Rha, *Journal of Food Science* **44**, 343 (1979).

59. S. Taneya, T. Izutsu, and T. Sone, in P. Sherman, ed., *Food Texture and Rheology*, Academic Press, London, 1979, p. 367.

60. E. P. Suchkovae, A. M. Maslov, and N. G. Alekseev, *Izvestiya Vysshikh Uchebnykh Zavedenii Pishchevaya Tekhnologiya* **4**, 113 (1985).

61. H. Jonsson and K. Anderson, *Milchwissenschaft* **31**, 593 (1976).

62. C. Steffen, *Schweizerische Milchwirtschaftliche Forschung* **5**, 43 (1976).

63. J. Culioli and P. Sherman, *Journal of Texture Studies* **7**, 353 (1976).

64. N. Ya. Dykalo and V. P. Tabachnikov, *Trudy Vsesoyoznogo Nauchno-Issledovatel'shoko Instituta* masl. i. syr. Prom., **77**, 85 (1979).

65. E. Flückiger, P. Walser, and H. Hanni, *Deutsche Molkerei Zeitung* **96**, 1524 (1975).

66. C. W. Raadsveld and H. Mulder, *Netherlands Milk and Dairy Journal* **3**, 117 (1949).

67. T. P. Guinee and P. Fox, in P. F. Fox, ed., *Cheese: Chemistry, Physics and Microbiology*, Elsevier Applied Science Publishers, London, 1987, p. 251.

68. R. Ramanauskas, *Proceedings of the 20th International Dairy Congress* 1978, p. E 265.

69. G. G. Khachatryan, K. Zh. Dilanyan, V. P. Tabachnikov, and I. I. Tetereva, *Proceedings of the 19th International Dairy Congress* 1974, 1E, p. 717.

70. M. A. Cervantes, D. B. Lund, and N. F. Olson, *Journal of Dairy Science* **66**, 204 (1983).

71. T. J. Guerts, P. Walstra, and H. Mulder, *Netherlands Milk and Dairy Journal* **34**, 229 (1980).

72. M. Peleg, *Journal of Food Science* **42**, 649 (1977).

73. C-H. Lee, E. M. Imoto, and C. K. Rha, *Journal of Food Science* **43**, 579 (1978).

74. C. S. T. Yang and M. V. Taranto, *Journal of Food Science* **42**, 906 (1982).

75. R. Stefanovic, *DECHEMA Monographien* **77**, 211 (1973).

76. J. C. Weaver and M. Kroger, *Journal of Food Science* **43**, 579 (1978).

77. G. Szabo, *Proceedings of the 19th International Dairy Congress* 1974, 1E, p. 505.

78. S. D. Sakharov, V. P. Tabachnikov, V. K. Nebert, and P. F. Krasheninin, *Trudy Vsesoyuzuogo Nauchno-Issledovatel'skogo Instituta* masl. i. syr. Prom. **18**, 29 (1975).

79. M. Ostojic, D. Miocinovic, and G. Niketic, *Mljekarstvo* **32**, 139 (1982).

80. B. A. Nikolaev and R. M. Abdullina, *Trudy Vologogradskaya* moloch. Inst. **30**(7), 27 (1969).

81. N. Baron and R. Harper, *Dairy Industries* **15**, 407 (1950).

82. A. S. Szczesniak, *Journal of Texture Studies* **18**, 1 (1987).

J. H. PRENTICE
Axminster
Devon, United Kingdom

CHEMICAL COMPOSITION. See FOOD ANALYSIS.

CHEMISTRY. See FOOD CHEMISTRY AND BIOCHEMISTRY.

CHERRIES AND CHERRY PROCESSING

ORIGINS AND BOTANICAL DATA

The cherry is part of the family *Rosaceae* and belongs to the genus *Prunus*. Other members of this genus include apricot, nectarine, almond, peach, and plum.

Sweet Cherry

It is thought that the sweet cherry, *Prunus avium* L., originated in an area bordered by the Caspian and Black Seas.

Theophrastus, the father of botany is given credit for the first writing of cherries ca 300 BC, although the cherry was probably cultivated for several centuries earlier (1). It is a drupe, or fleshy stone fruit that can be separated into two subspecies based upon the texture of the flesh: the *Bigarreaus* with firm flesh and the *Heart* with tender flesh. Only the former are able to tolerate the stress of commercial handling.

The leaf and fruit of the sweet cherry are usually larger than other cherry varieties, but the trees have less foliage. The leaves droop and are supported by long, rigid central spines, called petioles. The petioles may be completely smooth or can be lightly to heavily furred. Double flowered types sometimes develop the white petaled cherry flower and are usually single; its flower buds generally contain one to five blooms. The sweet cherry has both male and female parts. Unlike the red cherries, the trees of one sweet cherry variety usually do not pollinate that same variety. Rather, the trees require cross-pollination from other varieties in order to bear fruit. Red cherry trees cannot be used as cross-pollinizers. The fruit may vary from an ovate to a long heart shape. The flesh of the sweet cherry ranges in color from yellow to dark red to almost black; the skin color can be yellow, variants of mahogany, or purplish black. Within the cherries are elliptical or slightly roundish pits, semiclinging or free from the flesh.

The young tree, to be productive and economically managed for commercial use, must be carefully pruned to develop wide-angled limbs and to control growth to maturity. Sweet cherry trees produce 1,000 lb per acre at eight to nine years, and increase to 7,000–8,000 lb when they are about 14 years old. The trees have an estimated productive life of 30 years.

One of the most important decisions made by the sweet cherry growers is the selection of cultivars to be grown. A number of major factors are considered such as consumer preferences, resistance to disease, cracking, insects, time of fruit maturity, availability of facilities, labor, and markets. The favorite sweet cherry varieties in the Eastern United States are the Windsor, Schmidt, Victor, Napoleon, and Black Tartarian. In the Western United States, the Lambert, Royal Ann, Bing, and Black Tartarian have been utilized for commercial production.

Red Cherry

The red cherry, *P. cerasus* L., is also known as the sour, tart, or pie cherry. It is a drupe fruit and is believed to have originated in the territory between Switzerland and the Adriatic Sea on the west and the Caspian Sea and northward on the east (1). That both sweet and red cherries coexisted in the same area is a certainty as the red cherry is considered to be a hybrid of the sweet cherry (2). For more detailed information on the red cherry's background a comprehensive treatise is available (3).

The principal red cherry varieties are self-fertile. With warming weather, adequate bee activity, required pollen tube growth, and proper nutrition, a tree will produce pollen that can fertilize the pistils of the flowers on the same tree—and fruit will set. Beehives are placed in the orchards to facilitate effective pollination in both the red

and sweet cherry groves. In selecting the proper location for a red cherry orchard, it is necessary to consider which sites are more susceptible to spring frosts. Cherries are very vulnerable to freezing when they are about three to four weeks before full bloom. Even though the buds have only begun to swell, they can be killed at temperatures of 28°F (−2°C) or below. Since hot air rises and cold air descends, orchards located on slightly higher ground usually are less likely to be damaged by frost than trees planted at lower elevations.

In about one third of the orchards on the best sites, substantial crop loss from frost occurs less than once in five years. About one out of every three crops in the average commercial orchard is lost to spring frosts. Secondarily, high levels provide better soil drainage. A site with good air drainage and/or proximity to a large body of water is excellent for successful cherry production. Winter temperature should not go below −15°F (−26°C), and the mean temperature for June, July, and August should be about 60°F (15°C). In the major red cherry producing areas the blossoming period is sometime between mid-April and mid-May, depending on the location. The fruit ripens in late June to early July and may be harvested as late as mid-August in the more northern regions.

Although there are 270 named varieties of red cherries, only a few are grown commercially and therefore, are, well known. Most of the important varieties of red cherries grown in this country are of European origin, coming from France, England, Holland, and Germany. Based upon the color of juice in the fruits, these can be divided into two groups. Those giving a colorless juice are known as Amarelles while those with a darker color and a reddish juice are known as Morellos. Pale and predominantly flattened on the ends, the Amarelles are less acid and sour than the darker varieties. Two outstanding representatives of this type are the Early Richmond and Montmorency varieties. The Morello type has fruit with dark red skin and flesh. The trees vary in size from smaller bushier types to trees as large as most sweet cherry trees. The English Morello is typical of this group in the United States and the Schattenmorello and Pandy are typical in Europe.

Only three varieties of the red cherry are cultivated in quantities in the United States: Montmorency, Early Richmond, and English Morello. The Montmorency cherry represents the overwhelming red cherry variety grown in the United States. It began life in the Montmorency Valley of France. In the United States, it was cultivated under various names for some years and was added to the American Pomological Society's fruit catalogue in 1897. Approximately three quarters of an inch in diameter, the Montmorency is roundish to oblate. Its skin color is light to rather bright red with pale yellow, reddish-tinged flesh. The hardy, vigorous trees are exceptionally productive on good soil; however, yields are highly variable.

PRODUCTION FIGURES

Total worldwide cherry production is close to 1.4 million tons. About 20% of this is produced in North America, 71% in Europe, 7% in Asia, 1.5% in South America, and

Table 1. World Sour Cherry Production Figures[a]

Country	Production (1000) Tons	Year(s)
USSR	450	1970s
United States	150	1986–1990[b]
Germany, West	225	1982–1985
Yugoslavia	137	1982–1983
Hungary	80	1975
Turkey	66	1984–1985
Poland	50	1975
Romania	48	1981–1984
Germany, East	29	1970s
Greece	21	1981–1985
Bulgaria	18	1970s
Canada	11	1981–1984
Belgium	11	1984–1985
Italy	10	1980–1983
Denmark	6	1980–1982
Czechoslovakia	6	1970s
France	5	1981–1983
Austria	4	1984–1985
Norway	0.4	1981–1983

[a] Refs. 4,5.

[b] Ref. 6.

0.5% in Oceania. The United States is among the world's leaders in red cherry production (Table 1). Approximately 75% of the United States crop is grown in Michigan with New York, Utah, and Wisconsin producing most of the remainder. The majority of the crop is further processed and only a small fraction is utilized fresh. The center of production moves west for the United States sweet cherry crop. Washington grew 33%, Oregon 32%, and California 14% of the 1988 crop of 186,200 tons. Michigan also grows sweet cherries and produced 15% of the United States' crop (7).

HARVESTING AND HANDLING

About 40% of the annual sweet cherry crop is sold on the fresh market. Once blemished fruits are removed, the cherries are sorted into several different grades according to size. Highest prices are generally obtained for the largest fruits. Fresh sweet cherries are usually on the market from May through August. Since quality is an all-important factor in the sweet industry, the fruit must be harvested at the proper time. Harvesting too early can result in less color, less pleasing flavor, and smaller size. Although the use of mechanical harvesters is more attractive economically, harvesting by hand protects the fruit from damage and prevents the loss of stems. However, the cost of maintaining a group of harvest workers is extremely high and many times workers are not available. With the advent of more stringent immigration laws, this shortage of workers is becoming more of a problem, not just in the cherry industry, but in agriculture in the United States as a whole.

For many years, red cherry orchards also relied on hundreds of migrant workers, who moved north annually for the harvest. The pickers traversed the orchards with pails or baskets attached to their belts. Picking with both hands, an enterprising worker could harvest over 400 lb of cherries per day from trees with good crops. (Trees eight to twelve years old are in their prime, and may each produce about 100 lb of fruit.) However, the average pickers harvested about 200 lb per day. Owners of individual orchards quite often used members of their own families for picking, especially as the number of migrant workers began to decline. Prior to 1964, over 45,000 workers were necessary to harvest the Michigan red tart cherry crop. After 1964 many growers were forced to begin using mechanical harvesting methods because of the increased cost of hand picking and the scarcity of migrant labor. These were developed by Michigan State University, the U.S. Department of Agriculture, and cooperating commercial companies and growers.

The machines simplify and speed the harvest operation and improve the production for processing, where nearly all cherries are utilized. The percentage of the total crop harvested mechanically rose dramatically, as evidenced in Michigan. In 1964, only 3% of the crop was harvested mechanically; whereas today nearly all of the yield is taken by mechanical methods. Average equipment can harvest 25–30 trees an hour; with machine production, cost can be reduced considerably. The money saved by machine harvesting 100,000 tons of red cherries in Michigan, figuring amortization charges on equipment, was well over $6 million in the early years of its use.

Mechanical harvesting is really a simple process—the tree is vibrated and the cherries fall. One problem with this process is the danger of possible injury to the cherries. Procedures such as special pruning have been developed to modify the trees for machine harvest. The first mechanical limb shaker to be perfected followed a very simple procedure. Usually, two men operated a harvester as large mechanical arms reached up, grasped each primary limb firmly and then shook it vigorously, loosening the cherries. Eventually, it was discovered that shaking the limbs could damage the individual trees. As a result, a trunk shaker was developed and has been proven to be very successful. These sturdy machines shake each trunk, loosening the fruit all over the trees with a few firm movements. Dropping into inclined frames, the fruit rolls onto conveyor belts and then to tanks of cold water. Placing the cherries in cold water promotes firming, which enables the cherries to better handle the pitting process and prevents scald, a color change caused by bruising. It also washes the fruit and removes spray residue. Less than an hour after harvest, the cherries are resting in ice and water in a cooling station within the orchard area.

As a first step in processing, the cherries go through an eliminator, where foreign matter and defective, undersized and immature fruit is removed. An electronic sorter has taken the place of hand sorting. All cherries that pass inspection are given the grade U.S. No. 1, the only standard federal grade for unprocessed cherries. Once graded, the cherries move through mechanical pitters. They are dropped individually into small cups on a revolving cylinder, where the pits are shoved through and out by small cross-shaped plungers. Free of pits, the cherries drop from the cylinder as it moves 360° in rotation, and the pits disappear through the middle of the revolving cylinder. From here, the cherries move along belts and are inspected again for blemishes, loose pits, and foreign matter.

PROCESSING

Canning Red Cherries

Cherries selected for canning, generally, are deposited in either small, consumer sized, enameled cans called 303s which hold 16 oz., or in institutional cans, called No. 10s, which hold 6 lbs 9 oz. No sugar is added to these canned red cherries. Water is added and the cans are given sufficient exhaust in hot water to eliminate air from the can. This reduces corrosion of the metal and produces a good vacuum in the sealed can. Once sealed, the cans are heated from 8–20 min depending on the can size. The cans are then passed through canals of cold water to cool them prior to casing or stacking.

Most of the cherry crop is converted into pie stock (still the most popular use of cherries), jellies, dessert fillings, frozen cherry pies, and other convenience foods. Canned cherry pie filling is prepared by blending pitted red cherries with food starch, sugar, and citric acid in proportions required to give a suitable flavor and consistency to the final canned product. The noncherry ingredients are mixed with a quantity of water and heated to produce a thick gel. Cans are filled with a mixture of the starch gel, red color, and pitted cherries, then sealed and heated at appropriate times and temperatures to achieve commercial sterilization. After cooling, the cans are labeled and cased or bright stacked.

Freezing Red Cherries

Frozen cherries retain a fresher appearance and a more natural flavor than canned cherries. Therefore, freezing has become a more popular method of preservation. In one method, cherries are placed in large containers, predominately 30 lb metal cans or plastic pails. Usually a sugar cap, with a common ratio of one part sugar to five parts cherries, is added to the filled container just before freezing. This helps the cherries to retain color during freezing and thawing. Once the cherries are frozen, the containers must be stored at temperatures below 0°F (−18°C). A comparatively new process used in freezing cherries is the individually quick frozen (IQF) method. This provides the buyer with pitted, frozen cherries, minus the sugar or liquid, which retain the flavor and color of fresh cherries. Because the cherry is frozen quickly, smaller ice crystals are formed in the fruit. This causes less cell disruption and less fluid is lost when the fruit is thawed so that drained weights are higher.

Some frozen cherries are available at the retail level, but most are sold to institutions and food processors. They in turn use the product to make pies, jams and jellies, fruit fillings, toppings, and so forth.

Cherry Juice and Beverages

The red cherry is the only deciduous fruit yielding bright red juice, but its commercial use is currently limited as compared to orange, grape, and other popular juices. It has a pleasing, but strong, flavor and is therefore often blended or diluted with another juice such as apple to better meet the consumer's tastes.

To get a better flavored and colored juice, a better quality cherry should be used. Overripe or spotted fruit produce a juice with a higher benzaldehyde flavor which is characterized as being undesirable. Mushy fruit can yield juice with so much suspended material that filtering it is difficult and it will usually become cloudy during storage. Less mature fruit gives juice which is low in sugar and is therefore sour as well as being poor in color. It has a low density and is lacking in some of the soluble components contributing to flavor (8).

There are three methods of pressing cherries for juice and each has its advantages as well as some disadvantages. Hot pressing may be the simplest procedure. Pitted fruit is heated to about 150°F (66.5°C) and the cherries are pressed before they cool. The resulting juice is strained, chilled, allowed to settle, and then filtered. Yields should be in the 62–68% range for Montmorency cherries. Juice from this method is a deep red color as the heating extracts a greater level of the pigments present. However, the juice's flavor is more like that of canned cherries rather than fresh fruit.

Conversely, the cold pressing method yields a juice that is not as rich in color, but has a flavor more akin to fresh cherries. In this procedure the cherries are coarsely ground or pulped to aid in the extraction of color. The fruit may be ground with the pits as long as they are kept intact. The ground fruit is then pressed and the expressed juice is flash heated, cooled to inactivate enzymes and reduce the microbiological load, clarified, and filtered. Yields should be 61–68%, similar to the hot press method.

Cold pressing frozen cherries combines the best features of the hot and cold press methods. A dark juice is obtained, although it is not quite as dark as that from the hot press, and its flavor resembles the fresh cherries. The cherries are either pitted or crushed and then frozen with or without sugar added. Prior to pressing they are thawed to a temperature of 40–50°F (4.5–10°C). After pressing, the juice is clarified and filtered. Yields from Montmorency cherries average 70–76% (8).

Sweet Cherries

As mentioned earlier, about 40% of the sweet cherry crop is sold fresh; over 50% is used for brining purposes; the majority of the small amount that remains is canned in syrup concentrate ranging from 10–40° Brix. Napoleon (Royal Ann) is the most important sweet cherry variety for canning, but Schmidt, a dark variety, also is canned for dessert. Dark varieties, such as the Bing, Schmidt, Windsor, or the Black Republican, are frozen for commercial distribution, packed with one part sugar to three or four parts cherries. Details of the canning and freezing procedures are very similar to those previously discussed for red cherries.

Brined, Maraschino, and Glace Cherries

Over 50% of the total sweet cherry crop is channeled into brine cherries which in turn are used to produce maraschino and glace cherries. Properly matured, whole cherries are packed in a solution made up of 0.75–1.5% sulfur dioxide and 0.4–0.9% unslaked lime that will bleach the fruits to a uniform white or yellowish-white color and firm them. Originally, the cherries to be brined were stored in huge barrels, with 250 lb of cherries packed tightly in 20 gal of brine. Coopers then sealed the barrels and stacked them for the brining period. As time progressed, the brine cherries were stored in large cypress tanks that could handle over 25,000 lb per tank. Eventually, brining advanced a step further when the swimming pool approach was developed. Deep pits, lined with plastic, hold thousands of pounds of pale, colorless cherries, steeped in brine. Stored in these huge tanks, the cherries soak four to six weeks.

To produce maraschino cherries, the brined and pitted fruit is further bleached using sodium chlorite. Residual bleaching agent is removed by leaching in water for 24–36 h. The cherries are then stored in sodium bisulfite brine for at least two weeks to firm them. The cherries can be dyed either red or green. Added simultaneously with food coloring is a sugar solution that is interchanged five times over a five to seven day span until the cherries achieve the necessary 40–50° Brix (sugar content). Lastly, mint flavor is added to the green cherries and almond flavor to the red.

A step beyond the maraschino is the glace cherry. Glazing or candying of fruit is probably one of the oldest methods of preservation. It is accomplished by dehydrating the fruit with the osmotic pressure difference of a sugar solution. Initially the glacing process utilized honey for this purpose. Instead of a finished Brix of 40–50° as for mara-

schino cherries, a final Brix of 70–75° is achieved by keeping the cherries in the sugar solution for a longer period of time. The glace cherries are then dried to achieve a firm texture. The finished fruit is sensitive to temperature and humidity and should be stored below 50F (10°C) and at 50% RH. The majority of this product is sold to institutional bakers and confectioners for use in fruitcakes, cookies, and candies. Detailed descriptions of all the cherry processes have been reviewed (9,10).

Dried Cherries

Dried cherries are now a commercial reality rather than just a curiosity. Frozen sugared cherries (commonly referred to as 5+1, the ratio of cherries to sugar) that have been frozen in bulk or IQF cherries (unsweetened) are thawed to a temperature of 40–50°F (4.5–10°C). Any juice present after thawing or formed during the first stages of drying is collected and processed separately. The cherries are processed through a multiple stage drier with the first stage being the hottest. Each succeeding stage is cooler until the dried product emerges at room temperature. It is packaged with a nitrogen headstage to preserve its color. The sugar mutes the cherry's natural tartness just enough to leave a pleasant, tangy taste that is compatible with many flavors. The product is still in its introductory phase but it is being increasingly utilized in trail mix-type snacks, candy, and other products requiring a low moisture cherry product.

Table 2. Nutritional Analysis of Fresh Cherries (per 100g of product)[a]

Nutrient and Units	Montmorency Red Cherry	Schmidt (Dark) Sweet Cherry	Emperor Francis (Light) Sweet Cherry
Proximate			
Water, g	82.53	79.75	79.19
Calories, kcal	63	73	74
Protein, g	1.08	0.63	0.79
Lipid, g	0.22	0.14	0.08
Carbohydrate, g	15.78	19.10	19.47
Fiber, g	1.48	0.73	0.57
Ash, g	0.39	0.39	0.47
Minerals			
Calcium, mg	17.2	13.4	17.0
Iron, mg	0.44	0.28	0.34
Magnesium, mg	11.1	10.7	8.4
Phosphorus, mg	20.1	20.3	18.8
Potassium, mg	185.5	200.7	273.3
Sodium, mg	0.02	0	0.04
Zinc, mg	0.12	0.12	0.27
Copper, mg	0.15	0.07	0.11
Manganese, mg	0.10	0.06	0.05
Vitamins			
Ascorbic acid, mg	9.8	7.0	7.4
Thiamin, B_1, mg	0.04	0.03	0.05
Riboflavin, B_2, mg	0.04	0.05	0.06
Niacin, mg	0.19	0.40	0.42
Pantothenic acid, mg	0.16	0.10	0.10
Pyridoxine, B_6, mg	0.05	0	0.04
Folacin, mcg	7.5	4.2	4.2
Vitamin B_{12}, mcg	0	0	0
Vitamin A, IU	850	210	275

[a] Ref. 11.

NUTRITION

A study conducted in 1989 analyzed the nutritional content of both red and sweet cherries (11). The cherries evaluated were both fresh and in various processed forms. This study showed that, among fruits, the red cherry is a good source of Vitamin A, calcium, iron, potassium, and phosphorus, yet is low in sodium and fat. The sweet cherry's nutritional attributes are similar to the red's except its Vitamin A content is not as high. More detailed nutritional information on both the red and sweet cherries is given in Table 2.

BIBLIOGRAPHY

1. U. P. Hedrick, *The Cherries of New York,* New York Agricultural Experiment Station, Geneva, N.Y., 1914.

2. E. J. Olden, and N. Nybom, "On the origin of *Prunus cerasus* L.," *Hereditas* **59**, 327–345 (1968).

3. A. F. Iezzoni, "Sour Cherry Germplasm," in J. N. Moore, ed., *Genetic Resources of Temperate Fruit and Nuts,* International Society of Horticulture Science, Wageningen, The Netherlands, 1989.

4. J. V. Christensen, "Production of Cherries in Western Europe", *Acta Horticulturae* **169**, 15–26 (1985).

5. S. Kramer, "Production of Cherries in the European Socialist Countries", *Acta Horticulturae* **169**, 27–34 (1985).

6. *Red Tart Cherries Crop Statistics & Market Analysis,* Michigan Agricultural Cooperative Marketing Association, Inc., Lansing, Mich., 1990.

7. *Noncitrus Fruits and Nuts Annual,* Agricultural Statistics Board, USDA, Washington, D.C., 1989.

8. D. K. Tressler and M. A. Joslyn, *Fruit and Vegetable Juice Processing Technology,* AVI Publishing Co., Inc., Westport, Conn., 1971.

9. R. E. Marshall, *Cherries and Cherry Products,* Interscience Publishers, N.Y., 1954.

10. J. G. Woodroof and B. S. Luh, *Commercial Fruit Processing,* 2nd ed., Van Nostrand Reinhold Co., Inc., New York, 1986.

11. J. N. Cash, A. Shirazi, and W. C. Haines, *Analysis of Physical and Chemical Properties of Cherries,* Cherry Marketing Institute, Inc., Okemos, Mich., 1989.

Cherry Marketing Institute
Okemos, Michigan

CHEMICAL SENSORS. See SENSORS AND FOOD PROCESSING.
CHEMISTRY. See FOOD CHEMISTRY AND BIOCHEMISTRY.

CHILLED FOODS

The term chilled foods can imply many things to many people. It can range from products that rely on refrigeration as a method of preservation for all or part of a product's life to products that, for marketing purposes only, are retailed through refrigerated display counters. One study observed that most consumers do not appreciate that they are buying chilled foods but consider them to be fresh, frozen, or packeted (1). It is generally accepted by food scientists, however, that chilled foods are refrigerated to retard or prevent deterioration and growth of microorganisms. The following definition from the Institute of Food Science and Technology (U.K.) is widely accepted: "Perishable foods, which to extend the time during which they remain wholesome, are kept within specified ranges of temperature above $-1°C$ and below $+8°C$ (2)".

This article considers technological developments and some issues of concern relating to chilled foods during the 1980s and into the 1990s.

RECENT MARKET TRENDS

The development of the chilled foods market throughout Europe was dynamic during the 1980s, particularly since 1985, resulting in a wide variety of products available to the consumer through catering and retail outlets. Consumer demands and technological and commercial developments have been key factors in the food industry maximizing the unique commercial opportunities presented by the chilled food market. Such a response is likely to continue into the 1990s.

Consumer demands increasingly fall into two main areas: healthy eating and convenience. Healthy eating has become a major media issue; it is often supported by national government-led campaigns and has been accompanied by increasing numbers of vegetarian consumers. Such developments have been to the advantage of chilled foods, which are perceived by consumers as being fresh and less likely to contain additives and preservatives. Indeed, additive-free foods have presented a major market opportunity during this period. In addition, social trends of less time and greater expendable income and increased travel have resulted in less-frequent shopping, increased snack eating, less food preparation in the home but increased eating-out, and the demand for quick-to-prepare and easy-to-use novel, convenient food—a demand fully appreciated by the chilled food industry, resulting in conveniently packed, small-portion, ready meals, often of foreign, exotic recipes. Microwave ownership has increased dramatically; in the UK it has increased from less than 5 to more than 50% home ownership during the 1980s. This has led to the development of many chilled foods, particularly ready meals, specifically intended for microwave cooking or reheating.

Key developments in the food industry (other than technical developments, which will be considered separately) include the concentration of the grocery trade into a small number of major operators, increasingly with multinational operations. This has resulted in the requirement and hence development of sophisticated distribution systems and chilled display facilities to cope with high-turnover, short shelf life chilled foods. In the case of fruit and vegetables, this has resulted in primary producers providing facilities for cooling and storing at chill temperatures on the farm and the sourcing of produce from other parts of the world to ensure a continuity of supply of traditional and exotic products. In addition, intense competition in the marketplace has led to continuous innovation.

As a result of this development, considerable markets have developed throughout Europe, as indicated in Table

Table 1. European Market Value Chilled Food and Drink Products[a]

	Total Markets, All Fresh Food and Drink Products[b]					Total Markets, All Chilled Food and Drink Products[c]				
	1985	1988		1985–1988 Growth	Per Capital	1985	1988		1985–1988 Growth	Per Capital
	Current ECU, Millions		1988 Shares, %	Real Average Annual %[d]	Expenditure, 1988	Current ECU, Millions		1988 Shares, %	Real Average Annual %[d]	Expenditure, 1988 ECUS
Major Countries										
France	37,750	41,685	21.9	2.3	746.11	21,022	23,273	20.6	2.3	416.55
Germany	31,525	35,172	18.5	1.6	574.71	22,858	25,980	23.0	2.2	424.51
Italy	41,422	49,686	26.1	2.1	865.00	19,849	24,464	21.7	3.1	425.91
Spain	17,055	19,457	10.2	0.5	498.90	7,261	8,353	7.4	0.8	214.19
United Kingdom	17,826	29,777	10.9	5.8	363.99	15,496	16,587	14.7	2.8	290.59
Major markets	145,578	166,777	87.7	1.1	616.34	86,487	98,657	87.4	1.3	364.60
Belgium/Luxembourg	4,784	5,173	2.7	1.1	502.70	2,806	3,194	2.8	2.9	310.36
Denmark	2,533	2,832	1.5	1.3	552.00	1,646	1,834	1.6	1.2	357.57
Greece	2,842	2,989	1.6	3.4	298.86	1,386	1,448	1.2	3.2	144.80
Ireland	2,210	1,806	0.9	3.7	510.19	1,117	783	0.7	1.5	221.26
Netherlands	5,772	6,222	3.3	1.0	421.54	5,296	5,672	15.0	0.8	384.27
Portugal	3,433	4,338	2.3	9.0	416.72	1,180	1,277	1.1	3.5	122.63
European Community Total	167,152	190,136	100.0	0.3	585.54	99,918	112,864	100.0	1.1	347.57

[a] Food for Thought, P.O. Box 370, CH-11211, Geneva 1, Switzerland.

[b] Chilled vegetables, liquid milk, cream, yogurt, fresh cheese, natural cheese, chilled desserts, butter, margarine, processed meat, chilled ready meals.

[c] Fresh fruit, fresh vegetables, fresh potatoes, fresh meat, eggs, fresh fish.

[d] Constant food prices and exchange rates.

[e] Current ECUs and exchange rates. ECU, European Community Unit.

1. Chilled food and drink products account for approximately 17% of the total European market. Germany, Italy, France, and the United Kingdom continue to be the major markets for chilled foods and drinks in Europe, accounting for 80% of the market value. This market is still dominated by fresh, ie, unprocessed foods (ca 63%).

The chilled food market, particularly in the UK, developed during the 1980s into a sophisticated product sector. It had been active in terms of new product introductions and, by the end of the decade, offered a wide variety of further-processed/value-added chilled products, based on interesting and sophisticated recipe and ingredient combinations of both traditional and ethnic character. The chilled foods market in the United Kingdom is dominated by own-label branding. Innovation and development of new products has to a large extent been retailer led. New product activity, at the end of the 1980s, is described in Table 2.

TECHNOLOGY OF CHILLED FOODS

Chilling Techniques

Air-Blast Coolers. Air-blast coolers have the advantage of cooling food rapidly and are necessary for highly perishable products. High chilling rates are achieved by high heat transfer coefficients obtained by the increase of air velocity; an optimum velocity of 2–3 m/s is necessary when chilling thick products. The correct airflow pattern around the stacks of food products is important to minimize chilling time. It is also necessary to maintain the relative humidity at high levels for unwrapped moist foods to prevent undue drying.

Air-blast coolers of various capacities are available. Cabinet coolers are relatively small and employ trays in mobile racks. The fixed-air cooler is usually fitted at the top and provides intensive air circulation. Chilling tunnels are designed to provide uniform high-velocity air using a variety of different airflow methods, such as cross flow, longitudinal flow, vertical upward and downward flow. It is generally necessary to have a high cooling effect initially to remove the large amount of heat present in food. A number of different methods are used for supporting food products to be chilled eg, trays on trolleys, wheeled racks, pallets, or hung-on overhead rails. Both cabinet coolers and chilling tunnels are batch systems, which are limited in output and experience air temperature changes during operation. Continuous chilling equipment, however, can cope with much higher product loads, and the air temperatures are relatively constant. Continuous chillers consist essentially of a basic conveyor with either a straight-through longitudinal or spiral configuration, parallel or cross-flow arrangements of airflow are possible with the air velocity set relatively high to ensure rapid chilling. Some typical velocities are 2–3 m/s for carcasses, 8–10 m/s for smaller products, and up to 16 m/s for berry fruits. With the spiral configuration, the air is usually directed upward through the product on a perforated belt. This type of system is ideally suited for formulated meat products and prepared pastry products.

Cold Rooms. Cold rooms are used to maintain food at chilled temperatures but are not suitable for rapid chilling applications. The chilled air is normally delivered at the top of the room; it then flows downward through the stacked product. The warmed air then returns directly to the refrigeration unit. In general, the chilling rate is very low, mainly owing to inadequate air circulation, but it may be intensified by using extra fans.

Cryogenic Chillers. The use of direct-contact refrigerants such as liquid nitrogen and carbon dioxide snow has been growing in popularity in recent years. These cryogens product very rapid chilling. However, control is nec-

Table 2. New Product Activity in the UK Chilled Food Markets during the 1980s

New Product Activity	Description
Fruit and vegetables	Many prepackaged, prepared/ready-to-cook vegetables; exotic varieties of fruits and vegetables increasing; recipe vegetable dishes and filled/topped baked-potato products
Salad products	Modified atmosphere packaging (MAP) of prepared green salad combinations; wide variety of delicatessen-style salads packed in containers, often with added meat
Meat and poultry products	
Fresh meat	MAP packing of some fresh meat cuts; in the late 1980s, development of heat and cooking sauce products—adding value
Cooked meats and paté products	Wide choice of British and Continental prepacked cooked meats and paté products produced in the UK and overseas
Poultry products—raw	Fresh-chilled raw poultry widely available from whole birds to specific cuts; free-range and corn-fed chicken; introduction of fresh game, including duck, pheasant, and pigeon
Further processed/added-value poultry	Several different styles of added-value poultry: cooked/flavored/marinated chicken, crumb- and batter-coated chicken products
Ready meals/recipe dishes	One of the most active new product areas for chilled foods; early recipe-dish developments in 1980–1981, but major activity after 1983, exceeding 150 new product introductions per annum by the end of the decade; the majority of meals are single portion, microwavable
Fish and seafood	MAP packing in the early 1980s, basic and exotic fish and seafood varieties; later came coated (batter and crumb) products; wide variety of fish and seafood recipe-dish products
Pastry and pizza products	Meat, poultry, and vegetable varieties available in a range of presentations: individual and family-sized pies, quiches, pasties, and rolls. Pizza varieties: deep pan, thin and crispy, family, party, individual-sized. Poultry-based savory accompaniment products, usually ethnic in character—samosas, spring rolls, etc
Bakery products	Chilled cream-filled cakes and pastries: choux buns, roulades, and cream slices
Fruit juices and drinks	Pure fruit juices; freshly squeezed juices, often seasonal, and fruit-juice-based drinks
Soups and sauces	
Fresh soups	Up-market recipe styles, microwavable
Pour-over savory sauces	Mainly pasta sauce recipes, microwavable
Savory blessing and toppings	Spoonable/creamy types suitable for salad dressings, toppings, or dips; also oil-based varieties for salad dressing
Sweet sauces	Custard-based with seasonal (Christmas) varieties launched annually
Fresh pasta and cooked rice	Wide variety of fresh pasta; MAP packaging particularly evident; ready-cooked rice varieties for main-meal accompaniments
Dairy products	An active chilled foods sector; European influence on the style of some new product introductions: fromage frais, French-set yogurts, Greek yogurt, continental cheeses
Desserts	Chilled desserts, predominantly individual servings in separate containers or multipacks; dairy-based and cold serve. By the end of the decade family-sized, traditional hot puddings were appearing in the chilled cabinet
Fresh sandwiches	First recorded in the early 1980s, now widely available at all major retailers
Cooking fats and spreads	Active product development of low-fat spread products

essary to prevent freezing. Cryogenic chillers are particularly useful for chilling hot cooked products, where it is desirable to reduce the temperature rapidly.

Hydrocooling. This technique is used extensively for removing the field heat from fruit and vegetable crops, but it is also used for rapidly chilling sous-vide and Capkold-packaged products. Hydrocooling is carried out by flooding, spraying, or immersion in cooled water. The process can be continuous, and cooling times of 20–40 min can be achieved. The main advantages of hydrocooling for produce are that freezing cannot occur, there is no weight loss, and slightly wilted produce may be revived.

Vacuum Cooling. Vacuum cooling is used for leafy crops such as lettuce that have a large surface area and contain plenty of free water. The system is unsuitable for bulky crops or those with a thick, waxy surface. Vacuum cooling is a batch process that relies on the cooling effect of water evaporation under reduced air pressures. Lettuce cools to 2–3°C almost as fast as the vacuum can be established. Vacuum cooling must be controlled to prevent freezing. Some crops may be sprayed with water before cooling to minimize weight loss, as a 1% weight loss occurs for every 5°C the produce is cooled.

Ice-Slush and Ice-Bank Chilling. In these methods, produce is held in containers and cooling air is blown through the stacks. The air is cooled by close contact with water, which is cooled either by melting ice or by a bank of refrigerated tubes. Most crops can be cooled by these methods with the exception of dry onion bulbs. The ice-bank system is most successful and has been specifically designated for both precooling and short-term storage. The benefits of this method are that the cooling rate is better than for a conventional cold room, high humidities are maintained because fine droplets of cold water are carried over with the air, and there is no danger of freezing. For further reading on chilling techniques see Refs. 3–6.

Packaging Materials

The wide diversity of chilled foods available is more than matched by an enormous variety of packaging materials used to present attractive foods in supermarket chill display counters. Table 3 lists the main requirements of a chilled food package (7). The type of food package will determine which of these requirements will need to be satisfied. Having established the requirements of a container for a particular chilled foodstuff, the next step is to

Table 3. Requirements of a Chilled Food Package

Contain the product
Be compatible with food
Nontoxic
Run smoothly on filling lines
Withstand packaging processes
Handle distribution stresses
Prevent physical damage
Possess appropriate gas permeability
Control moisture loss or gain
Project against light
Possess antifog properties
Seal integrity
Prevent microbial contamination
Protect from odors and taints
Prevent dirt contamination
Resist insect or rodent infestation
Be cost-effective
Have sales appeal
Communicate product information
Show evidence of tampering
Easily openable
Tolerate operational temperatures

Table 4. Chilled Food Packaging Materials

Paper
Cardboard
Cellulose and cellulose fiber
Aluminum foil
Metallized film
Glass
Natural casings
Plastics
    Polystyrene (EPS, HIPS)
    Ionomers
    Polyetherimide
    ABS (acrylonitrile–butadiene–styrene)
    PVC [poly(vinyl chloride)]
    PVDC [poly(vinylidene chloride)]
    EVOH (ethylene vinyl alcohol)
    EVA (ethylene vinyl acetate)
    CPET (crystallized polyester)
    PET [poly(ethylene terephthalate)]
    PC (polycarbonate)
    Polyethylene (HDPE, LDPE, LLDPE)
    OPP (oriented polypropylene)
Nylon (polyamide)

ascertain which type of packaging material would provide the necessary properties.

Table 4 lists the principal materials used in the packaging of chilled foods. For any one product, a number of materials are generally used either as separate components or in the manufacture of an individual component.

Paper and Board. These materials are widely used in chilled food packaging. They are easy to decorate attractively and are available with coatings such as poly(vinylidene chloride) (PVDC), silicone, poly(ethylene terephthalate) (PET), polypropylene (PP), or polyethylene (PE) to improve the oxygen, moisture, or grease-barrier properties. Paper and board are complementary to all the other materials, either in the form of labels, cartons, or outer packaging or as laminates with flexible plastics.

Glass. Glass containers are the oldest form of high-barrier packaging, have good axial strength, and may be processed similarly to cans. Their great assets are product visibility and recyclibility. Twist-off container closures make opening simple, while tamper-evident features such as pop-up buttons provide the necessary safety factor. Importantly, new glass technology and plastic sleeves of poly(vinyl chloride) (PVC) or foamed high-impact polystyrene (HIPS) have done a great deal to reduce glass breakage problems.

Aluminum. Pressed aluminum trays have a long history of use for prepared frozen meals and hot take-away foods. They are also used for many chilled ready meals. Their temperature stability makes them ideal for conventional ovens, but there is controversy on their suitability for use in microwave ovens. Although most modern microwave ovens operated under correct conditions can tolerate properly designed aluminum trays, there is still the small possibility that arcing may occur. Hence, aluminum foil containers are not recommended to be used in microwave ovens (8).

Plastics. Rigid plastic containers for chilled foods are largely made from low-density and high-density polyethylene (LDPE, HDPE), polypropylene (PP), high-impact polystyrene (HIPS), acrylonitrile–butadiene–styrene (ABS), and poly(vinyl chloride) (PVC). Other plastics such as polycarbonate (PC) are used in small quantities, but a very important new area of development is that of high-barrier laminates, which have good barrier properties and which are microwavable or dual ovenable/microwavable. Such laminates are typically made by coextrusion processes, using sandwich layers of poly(vinylidene chloride) (PVDC) or ethylene vinyl alcohol (EVOH) to provide the oxygen barrier. Table 5 compares the oxygen transmissions of the high barrier plastics with monolayer plastics.

Rigid plastic containers are available in a wide range of bottle, pot, and other shapes, and because container blow-molding or thermoforming equipment is less expensive than a metal can manufacturing line, food processors have an option to manufacture in-house.

Flexible plastics offer the cheapest form of barrier packaging and may be used to pack perishable chilled foodstuffs under vacuum or modified atmospheres. Because plastics such as PE and PP may be laminated with foil, very high barrier structures may be produced. Alternatively, plastic materials may be combined together in coextruded sheets so as to meet the necessary packaging requirements. The coextrusion of multilayer films is based on the ability to extrude two or more dissimilar materials at the same time, the polymers being delivered in their molten state to the feed block feeding the multilayer structure to the die. This method of operation enables multilayer films of excellent quality to be produced and offers advantages in adapting the relative thicknesses, widths, and positions of the layers.

Before considering the packaging employed for individual categories of chilled foodstuffs, it is appropriate to mention thermoforming as one of the principal methods of achieving rigid or semirigid containers. Usually the lower web of material to be formed into the tray is heated in a closely controlled manner until sufficiently soft and then molded into a die by the application of vacuum from below or pressure from above. The operation may be assisted by

means of a plug that helps to distribute film evenly into the die. After forming, the lower web container is filled with product and transported together with the top web into the sealing station. Here the package is first evacuated or gas-flushed, depending on the properties of the product and then sealed. Finally, the individual packages are cut from the continuous web in the longitudinal and cross-cutting units.

The ability to thermoform in-line does, however, depend on the materials in use. For certain substances, such as crystalized polyester (CPET), the complexities of the process are such that only specialized plastics manufacturers are able to undertake the operation. Consequently, food processors often use premade trays.

Packaging for Food Types

The types of packaging materials used for major chilled food product groups are described as follows. The information is not exhaustive, but a general overview due to the huge range of both chilled foods and packaging materials available.

Chilled Ready Meals. Choice of packaging material for chilled ready meals depends particularly on the requirement for microwavability and dual ovenability together with optimal product presentation. Ultimate barrier performance is perhaps less important, owing to the relatively short shelf lives involved.

The major categories of packaging used for chilled ready meals include aluminum foil, PET-lined paperboard, PET-lined fiber trays, CPET-trays, PC/polyetherimide trays, and glass.

Ovenable paperboard first became available in the 1970s. Essentially, poly(ethylene terephthalate) (PET) is extrusion-coated onto paperboard to give material that is microwavable and ovenable to a temperature of about 220°C. A second generation of materials has now been produced in the United States for coating use that utilizes cheaper application technology. Acrylic latex is spray-coated and produces a moisture evaporation/transmission rate that ensures that the package contents do not dry out during heating, sometimes a problem with PET.

Fiber trays are plastic-molded from wet cellulose pulp slurry and then dried before a top coating of PET is bonded to the inner tray, which may be varied to suit specific purposes, eg, molded fiber trays can be lidded with heat-resistant polyester film and will withstand temperatures from −40°C to 220°C without discoloration or signs of brittleness.

CPET trays are currently being used for perhaps the majority of high-cost dual-ovenable chilled ready meals. Trays are made from polyester granules that are extruded to make a sheet that has an amorphous state; that is where its internal structure is disorganized. In this condition it is as clear as glass. On subjecting the amorphous material to conditions of heat and pressure, a container can be formed that has a crystalline structure, which becomes opaque and highly heat resistant. CPET containers may be closed with polyester film lidding.

CPET containers are recommended for use in conventional ovens up to about 220°C. A higher temperature,

Table 5. Oxygen Transmission Rates for Plastic Packaging Materials

Material[a]	Oxygen Transmission Rate[b]
PVC/EVOH/PVC	0.05
PP/EVOH/PP	0–10
PP/PVDC/PP	1
PET/PVDC	2
PET 4	4
PVC 10	10
HDPE	70
OPP 80	80
LDPE	400

[a] Nominal 500 μm thickness of monolayer or laminate.
[b] Typical values cm^3/m^2/24 h/atm at 25°C and 50% relative humidity.
Abbreviations: PVC, poly(vinyl chloride); EVOH, ethylene vinyl alcohol; PP, polypropylene; PVDC, poly(vinylidene chloride); PET, poly(ethylene terephthalate); HDPE, high-density polyethylene; OPP, oriented polypropylene; LDPE, low-density polyethylene.

240–260°C, is obtainable by using some of the newer materials such as the polycarbonate/polytheimide trays. The trays are usually overwrapped with cardboard sleeves that allow superb presentation by using high-quality graphics and increased pack area for product or promotional information.

Where barrier properties are important and microwave heating only is to be considered, coextruded structures or EVOH, which effectively sandwiches a higher barrier layer of PVDC between polypropylene are used. The coextruded laminate is thermoformed into trays, a process that is technologically demanding in terms of the control necessary to achieve consistent results.

A variety of materials can be used for the packaging of chilled desserts. Yogurt containers are traditionally made from thermoformed or injection-molded polystyrene capped with adhesive-coated foil, although there has been some tendency to move to more sophisticated multilaminate materials if aseptic or clean-fill packaging is considered for longer shelf life.

Milk, Butter, and Margarine. Nonreturnable packages such as blow-molded HDPE bottles or PE-coated paperboard gable-top packages are most often used for packaging and distributing milk under refrigeration in Europe. Returnable glass bottles are still employed to a great extent in the UK.

Butter is usually packed in thin aluminum foil supported by paper or in paper itself. Margarine is, however, usually packed in semirigid plastic tubs. These are generally thermoformed but occasionally, especially for larger sizes, are injection-molded. Typical materials include ABS, PP, and multilayer films.

Cheese and Slicing Meats. Packaging requirements for both cheese and slicing meats are similar. The laminates used should prevent moisture loss and exclude oxygen, which may lead to oxidative deterioration. The most common packaging methods used are vacuum packaging in premade pouches, vacuum packaging on thermoform-fill-seal machinery, and gas flushing with either premade pouches or reel-fed machinery.

The types of material used include PVDC, which exhibits good clarity, and structures composed typically of PET/PE, cellulose film/PE, or nylon/PE. PVDC is also extensively used in the preparation of cold meats in chub form. Meat emulsion is filled into clipped tubular casings,

which are then cooked to pasteurize the product and set the protein into sliceable form before chilling.

Fresh Meats, Fish, Poultry, and Pasta. Traditionally, chilled fresh meats, fish, and poultry for retail sale were packaged, and indeed still are, in EPS trays covered with PVC cling film. However, modified atmosphere packaging (MAP) has recently become a very popular method of enhancing the visual appearance and extending the shelf life of these perishable chilled products (9). The MAP concept consists of modifying the atmosphere surrounding a food product, thereby retarding the biochemical, enzymatic, and microbial deterioration rates. The gas mixture used in MAP must be chosen to meet the needs of the specific food product, but for nearly all products this will be some combination of CO_2, O_2, and N_2. Typically used gas mixtures are given in Table 6.

Figure 1 illustrates a typical modified atmosphere pack based (10). Typically, the lidding film consists of 60 μm PVDC/12 μm PET antifog coating and the tray consists of 350 μm PVC/PE. Alternatively, HDPE can be used for the tray. If higher-barrier trays are necessary, then EVOH/foam polystyrene or EVOH/HIPS laminations can be used (11).

Fresh and Prepared Fruit, and Vegetables. Packaging of fresh produce must taken into account the postharvest respiration of plant tissue. The depletion of oxygen and accumulation of carbon dioxide are natural consequences of respiration, which is a complex process that is affected

Table 6. Gas Mix for Modified Atmosphere Packaging

Commodity	% O_2	% CO_2	% N_2
Poultry		25	75
Meat, red	60–85	40–15	
Meat, cured		20–35	65–80
Meat, cooked		25–30	75–70
Fish, white	30	40	30
Fish, oily		40	60
Bakery, nondairy		20–70	30–80
Bakery, dairy			100
Pasta, fresh			100
Cheese, hard		0–70	30–100
Cheese, mold ripened			100
Fruit and vegetables	2–5	3–10	85–95

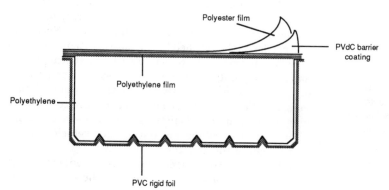

Figure 1. Construction of a modified atmosphere pack.

by intrinsic factors such as produce size, severity of preparation, and variety, maturity, and type of tissue (12).

Modified atmospheres are developed whenever fresh produce is packaged in sealed plastic film. MAP has a wide range of useful benefits such as delayed maturation of vegetables and retardation of ripening, delayed softening, reduced microbial spoilage, reduced enzymatic browning, and alleviation of physiological disorders and chilling injury. Packaging film of correct intermediary permeability must be chosen for produce so that a desirable equilibrium-modified atmosphere of 2–5% O_2 and 3–8% CO_2 is established. Undesirable anaerobiosis will be initiated if produce is sealed with film of insufficient permeability, whereas, if produce is sealed within film of excessive permeability, little or no atmosphere modification will result. The exact equilibrium modified atmosphere attained will obviously depend on the product's intrinsic respiration rate but will also be greatly influenced by several external parameters such as temperature, initial microbial load, packaging film and equipment, relative humidity, fill weight, pack volume, film surface area, and degree of illumination. These parameters need to be optimized for each commodity so that the full benefits of MAP are realized (12).

Various packaging materials are used for the retail packaging of fresh produce. Plastic materials used primarily for the containment function include perforated and open-weave bags and thermoformed and molded punnets and trays. Physical damage is minimized by the use of foamed polystyrene trays, and firm, but gentle holding in position is achieved by means of highly elastic cling films consisting of thin LDPE and ethylene vinyl alcohol (EVOH) ionizers of plasticized PVC. Permeable plastics used for pillow-packs and lidding materials include, LDPE, PVC, EVA, butadiene–styrene copolymer, oriential polypropylene (OPP), and cellulose acetate microporous and microperforated films.

Sous-Vide and Capkold Products. The sous-vide method involves vacuum packing of food before pasteurization, chilling, refrigerated storage, and subsequent reheating. The Capkold method involves pasteurization of pumpable food before packaging, chilling, refrigerated storage, and subsequent reheating. Both these cook-chill methods use patented flexible multilayer barrier plastic pouches or casings. These multilayer plastics are designed to resist physical abuse and extreme temperature changes and hence can withstand pasteurization, chilling, and reheating temperatures.

TEMPERATURE CONTROL AND MONITORING

The importance of adequate temperature control to prevent or retard deterioration of chilled foods cannot be overstressed. Careful monitoring and control of the temperature of chilled foods throughout their life is therefore critical. The range of requirements for temperature monitoring is extensive; it includes primary production sites such as the farm or slaughterhouse; receipt of raw materials to production sites; chill storage areas; preparation and holding areas; cooking and chilling operations; transport, including vehicles and storage depots; retail display; and also during reheating and on service counters. Measurements are required not only in food itself, but frequently in the environment in which a food is held, eg, air temperature in a chilled store or vehicle or the temperature of the chilling or heating medium. This will ensure correct functioning of the equipment used to control food temperature. In addition, the range of temperatures requiring monitoring can be very wide—from frozen storage to cooking temperatures. The demands placed on temperature-measuring equipment are therefore considerable.

A wide variety of temperature-monitoring devices are available. Appropriate equipment should be used based on the application, the user, and the restraints of the equipment available, such as its accuracy, resolution, speed of response, functionality, and cost. For most practical purposes, the technologies used are platinum-resistance thermometry, thermistors, thermocouples, and bimetallic strips. Liquid-in-glass thermometers are not recommended for use in the food industry because of the likelihood of breakage and contamination of food. All equipment should be maintained in good working order and regularly calibrated against traceable standards.

A wide variety of electronic thermometers that can be either hand-held or fixed have become available for taking spot-check measurements. In recent years, however, there has been an increasing demand for continuous monitoring of temperature to indicate the temperature history of products or environments. Original equipment used for such purposes typically used bimetallic strips to sense temperature and a pen for recording on a moving chart, frequently a circular chart recorder. Such devices have limited accuracy and resolution and interpretation is difficult. These devices have now been largely superseded by equipment using platinum resistance thermometer, thermistors, and thermocouple and automatic recording and logging of both temperature and time. These devices are located at the site of temperature monitoring and can store data for long periods of time (up to months depending on the frequency of data recording). Microprocessors are incorporated into the devices to store data that can subsequently be interfaced with a personal computer. This enables data to be presented in a variety of forms, including temperature traces. Sophisticated analysis can be undertaken if wished.

Temperature sensors may be linked to remote sites, enabling continuous display of temperature in central control rooms at some distance from the point of measurement. This has the advantage of allowing temperatures in several separate areas to be monitored from a central site under the control of a person who has the responsibility for ensuring corrective action is undertaken. Such systems can be further refined by the incorporation of audible or visible alarms that are activated by electrical signals from the temperature sensor at predetermined default settings and may even directly trigger the call out of engineers. Smaller, more rugged data loggers are continually being developed for monitoring the time–temperature history of chilled cabinets or refrigerated vehicles, where concealment of the monitor is important.

Systems are also being developed using radio waves for signal transmission to a control center, where the temperature information is displayed. This has real time-saving benefits in setting up as such a system uses no wiring from the sensors themselves. Therefore, a chilled store can be fitted with this type of monitoring system in only a few hours. Another interesting development with potential used for monitoring temperature in the chilled food industry is the use of thermal-imaging cameras that take a temperature picture of the object under view, allocating different colors to different areas of temperature within the field of view. Such images can be stored and translated into actual temperature readings when interfaced with a computer.

Although the technology of equipment used for measuring temperatures is advancing rapidly to suit the needs of the chilled food industry, most of the devices currently available are capable of sensing temperatures only at one or at most a few specific points within a product or its environment. Such measurements give no information regarding the rest of the food within a consignment, environment, or even a single pack of food. Considerable care and expertise is therefore required to choose points of temperature measurements that give meaningful results. Two approaches are generally adopted. Within temperature-controlled areas such as cold stores, distribution vehicles, and display cabinets, the temperature of the circulated cold air is recorded. Measurements are usually made at the coldest point where air leaves the cooling apparatus and the warmest point where air returns to the cooling apparatus. Temperatures are assessed against known parameters to indicate correct functioning of the equipment. In addition, temperatures of products are frequently monitored at positions that may be either representative of a

consignment or considered to be the warmest, to ensure that effective temperatures control of the product is achieved. Effective temperature control is not only dependent on correct functioning of refrigeration equipment, but also on such factors as the temperature of the food delivered into storage areas, the way in which it is stacked (to ensure that cold air circulates freely to all parts of the consignment), the insulation properties of the storage area, the frequency of opening doors in these areas, the thermal properties of the food, and the effect of packaging (13). Good temperature control is thus achieved by a variety of methods, not only temperature monitoring, but also education and training of the work force and management and supervision. Guidance on temperature control and monitoring is given in several of the documents listed in Table 7.

SAFETY CONSIDERATIONS

Low-Temperature Pathogens

Foods that have been freshly produced and/or stored may contain a wide variety of microorganisms present as part of the normal food microflora or owing to contamination of the food during processing. Foods may be considered microbiologically unacceptable as a result of growth of either spoilage or pathogenic microorganisms. Spoilage microorganisms cause an unacceptable change in the product, whereas pathogens render the food unsafe. In food production, matters relating to safety must take precedence over those of quality.

A number of pathogenic bacteria are capable of causing

Table 7. Guidelines and Codes of Practice and Chilled Foods

Guidelines for Good Hygienic Practice in the Manufacture of Chilled Foods	Chilled Food Association, UK, 1989
Guidelines for Cook Chill and Cook Freeze Catering System	Department of Health, UK, 1989
A Basic Guide to Vacuum (Sous-vide) Cooking	Dorset Institute, UK, 1988
Food and Drink Manufacture—Good Manufacturing Practice. A Guide to its Responsible Management, 2nd ed.	Institute of Food Science and Technology, UK, 1989
Guidelines for the Handling of Chilled Foods, 2nd ed.	Institute of Food Science and Technology, UK, 1990
Recommended Conditions for Land Transport and Perishable Foodstuffs	International Institute of Refrigeration, France, 1988
Good Hygienic Practice Guidelines for the Modified Atmosphere Packaging of Vegetable Products	Official Bulletin of the Republic of France, 1988
Guidelines for the Handling of Fish Packed in a Controlled Atmosphere	Seafish Industry Authority, UK, 1988
The Transport of Perishable Foodstuffs	Shipowners Refrigerated Cargo Research Association, UK, 1985
Guidelines on Sous-Vide Catering Systems	Sous-Vide Advisory Council, UK, 1990
Guidelines of Good Hygienic Practices for Prepared Refrigerated Meals	Syndicat National des Plats Préparés, France, 1989
Guidelines for Retail Operations	The Retail Consortium, UK, 1989
Guidelines for the Handling and Distribution of Chilled Foods	Transfigaroute (UK) and The National Cold Storage Federation, 1989

foodborne illness (14). These can be divided into infective organisms, which invade the body (eg, *Salmonella* sp.), and intoxicative organisms, which produce toxins (eg, *Clostridium botulinum*). It must be noted that foods may appear fit to eat despite the presence of large numbers of pathogens and toxins may even remain in foods in the absence of viable microorganisms. Lack of spoilage therefore cannot be used to indicate safety.

The effect of chill storage on microorganisms is to increase the lag period before growth commences and to reduce the rate of, or totally inhibit, growth. Consequently, the time taken for pathogens (or spoilage microorganisms) to grow is increased compared with storage at higher temperatures. This time becomes increasingly pronounced as the temperature approaches the minimum for growth of a microorganisms (15). Also, as the minimum growth temperature is approached, organisms are often more susceptible to the effects of other preservative factors of the food, such as pH and water activity. Although microbial growth will be prevented by temperatures below that for growth, the microorganism may survive and subsequently grow if the temperature increases. It is therefore important that appropriate temperatures are maintained throughout the life of the product.

It was recognized in 1936 that storage of foods at low temperatures greatly retarded or prevented growth of the foodborne pathogens recognized at the time (16). The investigators recommended that foods should be stored below 50°F (10°C) and preferably at 39°F (ca 4°C) to prevent growth of pathogenic bacteria or toxin development. Many pathogenic bacteria are unable to grow at temperatures below 5°C; these include the most common food-poisoning microorganisms: *Salmonella, Campylobacter, Clostridium perfringens,* and *Staphylococcus aureus* (17). It is now recognized, however, that some pathogenic microorganisms are capable of growing at chill temperatures and are termed psychrotrophs. These include *Listeria monocytogens, Yersinia enterocolitica,* and *Aeromonas hydrophila,* which may grow at temperatures of −1 to +1°C. Also, some spore-forming pathogens may grow at temperatures of 3–5°C, including some strains of *Clostridium botulinum* and *Bacillus cereus.*

Listeria Monocytogenes. *Listeria monocytogenes* is a bacterium that may invade the human body and cause disease. The nature and extent of the disease are dependent on the age, sex, physiological, and immunological status of the host. Certain groups are at increased risk of disease (eg, pregnant women, the elderly, and the immunocompromised) and with these groups, mortality may be high (18). *L. monocytogenes* is common in the environment and may be isolated from a wide variety of foods (19). Foods implicated in outbreaks of disease include coleslaw (20), pasteurized milk (21), and soft cheeses (22,23).

In several of these outbreaks, growth during chill storage was considered to have been significant. *L. monocytogenes* is of particular concern in chilled foods, as it has been reported to grow at temperatures as low as −0.4°C. Chill storage cannot therefore prevent growth; however, many of the processes involved in food processing will be effective against *L. monocytogenes* (24). The presence of *L.*

monocytogenes in many foods has been attributed to postprocess contamination; therefore, hygiene control in the factory environment is important.

Yersinia Enterocolitica. *Yersinia enterocolitica* is another invasive microorganism causing human disease characterized by diarrhea, abdominal pain, fever, and vomiting. Death is rare. Several chill foods have been implicated in outbreaks of disease, including chocolate milk (25), tofu (26), and pasteurized milk (27). Like *L. monocytogenes, Y. enterocolitica,* has been found in a wide variety of foods including meat, poultry, fish, dairy products, and vegetables (28,29), although many of these strains are not those commonly associated with disease.

The minimum growth temperature for *Y. enterocolitica* has been reported as −1°C (30), and growth can occur during storage of chilled foods. Control of *Y. enterocolitica* should therefore be focused on its elimination from raw materials and the prevention of postprocess contamination. Further details on factors affecting *Y. enterocolitica* can be obtained from Ref. 31.

Aeromonas Hydrophila. Although there has been no conclusively proven foodborne outbreaks of human disease caused by *Aeromonas hydrophlia,* there is considerable circumstantial evidence that it is pathogenic (32). This organism has been isolated from a wide range of food types, including some chilled modified atmosphere products (33,34).

A study of the growth of four strains of *A. hydrophila* in chicken broth reported minimum growth temperature of −0.1 to +1.4°C (35). Other factors that affect growth have been summarized in Ref. 36.

Psychrotrophic Spore-Forming Bacteria. The spore-forming bacteria capable of growth at chill temperatures (some strains of *C. botulinum* and *B. cereus*) are significant as they produce heat-resistant spores. These may survive some of the thermal processes applied to foods and then grow during refrigerated storage.

Human botulism is caused by toxins of *C. botulinum.* It is now recognized that some strains of *C. botulinum* (nonproteolytic strains of types B, E, and F) are capable of growth and toxin production at temperatures of 3.3–5°C in inoculated beef stew and fresh herring (37,38). Reference 39 reviews the literature in relation to growth of clostridia at low temperatures.

Bacillus cereus may act as an agent of foodborne illness causing either diarrhea or vomiting. Studies on isolates from humans have reported minimum growth temperatures of 10–12°C (40). Psychrotrophic strains have, however, been isolated from pasteurized milk and cream and are able to grow at 1–4°C (41). At present, it is not clear whether these strains are of public health significance, although they can be of importance in determining the shelf life of pasteurized dairy products.

In conclusion, chill storage alone cannot be relied on to prevent growth of pathogenic microorganisms. The use of low temperatures will, however, reduce the rate of growth. This effect becomes increasingly significant as the minimum growth temperature is approached. Also, chill temperatures for storage will interact with other preservation systems present in the food (eg, pH, water activity, preservatives), and the combined effect may be used to prevent growth of pathogens.

Pasteurization

For precooked chilled and cook-chill foods that heat process given is critical to product safety and stability as it kills microorganisms and inactivates enzymes that may cause product deterioration. The term pasteurization is used to describe a wide range of heat treatments that are mild by comparison with the botulinum cook used to achieve commercial sterility in the canning industry; a process with a lethal effect on microorganisms equivalent to the effect of heating at 121°C for 3 min (42). The process temperatures used in pasteurization can vary over a wide range of about 60–105°C (43). Such mild treatments are used to attain high sensory quality, avoiding the overprocessed qualities often associated with canned foods. To maintain safety and extend shelf life the heat treatment is often used in combination with other preservative factors such as pH, water activity, modification of pack atmosphere (including vacuum), as well as low-temperature storage.

As temperatures rise above the optimum for growth of an organism, they exert an increasingly inhibitory effect until lethality occurs. Organisms vary in their ability to withstand heat; the heat resistance is traditionally described in terms of D and Z values. Thermal destruction of organisms (vegetative cells and spores) follow a logarithmic order of death. The time required to destroy 90% of the organisms, ie, to reduce the population by 1 log, is termed the D value. If D values obtained at different heating temperatures are plotted against temperature, a loga-

rithmic order results and the reciprocal of the slope is termed the Z value, ie, the number of degrees (C or F) that cause a tenfold change in D value (44). These values are specific to the organism tested and the substrate in which it was tested, as the lethal effect of heat can be enhanced, eg, by the presence of acid (45) or diminished, eg, by the presence of sugar or fat (46). Examples of typical heat-resistance data are given in Table 8.

Within heterogeneous foods, such as a ready meals, a variety of different environments are likely to exist and organisms may not be randomly distributed throughout the food. Furthermore, variation is to be expected in the heat received among different packs within a batch of food undergoing processing owing to the ability of heat to penetrate throughout the load. During process evaluation it is therefore important to determine the heating effect achieved at the slowest heating point ie, that point which receives the lowest heat process.

Pasteurization processes used in the production of chilled foods, if carefully designed and operated, should result in the destruction of vegetative forms of microorganisms, including food-poisoning bacteria, but will not eliminate microbial spores. Spore-forming organisms are therefore of concern in precooked chilled food, particularly the pathogens *C. botulinum*, *C. perfringens*, and *B. cereus*. Prompt and rapid chilling and maintenance of adequately low temperatures during storage and distribution are required to prevent surviving spores germinating and growing in products (see previous section). Contamination of products after processing may result in the presence of pathogenic and/or spoilage organism in a product and is of particular concern in ready-to-eat products are given no further heat treatment before consumption.

Pathogenic bacteria such as *C. botulinum* and *S. aureus* produce toxins in foods that are responsible for food

Table 8. Examples of Typical Heat-Resistance Data of Organisms Applicable to Pasteurized Products

Organism	Heating Substrate	Heating Temperature, °C	D Value, minimum	Z Value (C)
Escherichia coli	Broth	56	4.5	4.9
Zygosacchararomyces bailii	Broth	60	8.1	5.0
Pseudomonas fluorescens	Broth	60	3.2	7.5
Streptococcus faecalis	Fish	60	15.7	6.7
Staphylococcus aureus	Pea soup	60	10.4	4.6
Salmonella senftenberg	Pea soup	60	10.6	5.7
Lactobacillus plantarum	Tomato juice	70	11	12.5
Listeria monocytogenes	Carrot	70	0.27	6.7
Clostridium botulinum:				
Nonproteolytic type B	Buffer (pH 7.0)	82.2	32.3	9.7
type F	Water	80	3.3	9.4
Clostridium pasteurianum	Buffer (pH 4.5)	95	3.95	
Clostridium butyricum	Buffer (pH 7.0)	85	23	
Clostridium tyrobutyricum	Buffer (pH 7.0)	90	18	
Byssochlamys fulva	Grape drink	93	5.0	7.8
Bacillus cereus	Buffer (pH 7.0)	100	8.0	10.5
Bacillus licheniformis	Not stated	100	13.5	
Bacillus polymyxa	Buffer (pH 7.0)	100	18.0 ca	
Bacillus coagulans	Buffer (pH 4.0)	98.9	9.5	
Bacillus subtilis	Buffer (pH 6.8)	121	0.57	9.8

[a] Ref. 43.

poisoning. These toxins are comparatively resistant to heat (47,48) and will be largely unaffected by pasteurization or reheating. Considerable care must therefore be taken to prevent toxin production prior to and following heat processing.

Chemical Contaminants

Chilled foods, in common with all foodstuffs, are susceptible to contamination from external or environmental sources. Various chemical contaminants released into the environment by industrial activity invariably find their way into the food chain (49,50). It should be emphasized, however, that most substances detected are only found at very low levels, generally well below those levels at which current medical knowledge indicates potential hazards might exist. Also, legal limits are set by government agencies for the maximum amounts of contaminants allowed, and these safety limits are continuously under review.

Chemical contaminants of particular concern include pesticides, metals, radioactivity, chlorinated compounds, plastic components, and nitrates. The use of all pesticides is controlled, but concern has been expressed over residues in food resulting from incorrect application, peculiar climatic/horticultural conditions, or environmental contamination due to persistent compounds such as the organochlorine-based compounds (eg, DDT and Lindane) (51,53). Recent concern has been expressed over various compounds, particularly daminozide (Alar) and the dithiocarbamate group of fungicides, both of which break down to carcinogenic compounds on heating. Nuvan, used for treating fish in fish farms, has also been implicated with cancer. Other examples of residues causing concern due to high levels being found in foods include the fumigant bromomethane, the fungicide carbendazim, and the sprouting inhibitor tecnazene.

The toxicity of heavy metals is well known, particularly lead, mercury, cadmium, and arsenic (54,56). Trace amounts of most of these elements can be found naturally in the environment, but industrial activity has caused pollution and increased their levels. Several major poisoning outbreaks have been attributed to metals, eg, mercury in shellfish and cadmium in rice, both in Japan. The cumulative effect of lead is also well known. Moves to reduce use of leaded gasoline, the major source of lead pollution, must be encouraged.

Several nuclear accidents, weapons testings, and disposal procedures have caused an elevation in the amount of radioactivity in the environment. Isotopes such as strontium-90, caesium-134 and 137, and iodine-131 have been found in foods, milk and milk products being particularly susceptible (57).

Dioxins and polychlorinated biphenyl compounds (PCBs) are chlorinated compounds that are extremely hazardous. They have been manufactured for some time and have found their way into the environment. Dioxins are by-products associated with the burning of chlorine-containing waste or with the reaction of chlorine with organic compounds. As such, they have also become universal contaminants and have been found in foods, water, paper products, and animal tissues (58,59).

The use of plastic packaging materials can give rise to migration of certain plastic components into foodstuffs. Particularly at risk are fatty foods and alcoholic drinks, which have a higher solubility for the components and materials undergoing heating (60). It is therefore important to specify the correct form of plastic for the foodstuff and application.

The rising level of nitrate in the environment is mainly due to run-off from agricultural use of nitrogenous fertilizers. Particular attention should be paid to the quality of water used in food production, as this may contain nitrate at unacceptable levels (61). Certain foods also contain high nitrate levels, eg, cured meats, broad-leaf green vegetables, and celery.

Problems associated with the carry over of veterinary compounds in meats are also known. Antibiotics, artificial growth hormones, and boosted natural hormone levels have all been detected. Of most concern is the carry over of antibiotics, as this can lead to resistance in bacteria; however, there is little data on the effect of hormones at the levels found in meats.

In conclusion, many substances may inadvertently contaminate foods, it is therefore imperative that food producers maintain rigorous standards of quality control for all supplies of food and water to ensure that the final products produced are wholesome.

The Consumer

As already stated, many chilled foods rely on maintenance of strict temperature control throughout their life to prevent or retard growth of microorganisms that may cause food poisoning or spoilage. Adequate temperature control is required at all stages of a product's life, including transport and storage by the consumer, often considered to be the weakest link in the chill chain. A survey in the UK showed that only 11% of consumers considered the home to be a possible source of food poisoning and 53% or less had an understanding of the basic food-poisoning hazards (inadequate thawing, reheating, cross contamination, cooking, refreezing, and storage of food at room temperature) (62). When questioned about their refrigerator, 63% did not adjust the temperature dial and only 6% had ever measured the temperature.

A recent survey of domestic refrigerators recorded air temperatures up to $+14.6°C$, approximately half the temperatures above 5°C (63), the maximum temperature recommended for storage of chilled foods in the home (64). Temperatures of 18–38°C were recorded in chilled foods that had been transported in the trunk of a car for 1 h (65). These foods took ca 5 h to achieve a temperature of 7°C when placed in a refrigerator. Studies at the Campden Food and Drink Research Association (CFDRA) (unpublished data) showed that most chilled foods are unrefrigerated for ca 2 h during shopping trips. Considerable variation in the length of time for which chilled foods are stored in refrigerators in the home, with some being stored for up to 46 days has been noted (65).

Labeling on food packaging advises consumers to keep refrigerated and eat within a specified number of days (frequently 2) of purchase. European Community legislation requires labeling with use by or best before (for longer-life chilled products) accompanied by the latest date on which the product should be eaten. Concern exists that consumers may therefore be able to purchase foods within a short time from production and thus the potential exists for products to spend a large proportion of their life in relatively uncontrolled storage conditions within the home. It is to be hoped that the large-scale media attention given to food issues in recent years, together with the campaigns and promotions conducted by government and the food industry, will promote a greater awareness and understanding of the responsibilities of the consumer.

QUALITY ISSUES

Sensory Characteristics

Foods are only purchased and consumed if they are perceived to be of satisfactory quality. Judgment of quality is on the basis of appearance, flavor, and texture of the food. These sensory characteristics are the result of the physical and chemical structure of the product; hence any change in the physicochemical properties or the microbiological flora that may occur during chill storage can give rise to sensory changes. There may be a number of reasons for such changes in the different sensory attributes of chilled foods.

Appearance may be altered through discoloration, changes in water content, or physical separation. Discoloration itself may be due to a number of mechanisms, ie, enzymatic browning (apple slices) (66), oxidation of heme pigments (raw meat) (67), senescence (watercress) (68), or migration of color from one product to another (tomato on toppings). Food may lose or gain moisture, thus effecting texture and appearance. Moisture loss can cause loss of turgidity in cut salad vegetables, giving a limp appearance and reducing textural crispness; it also contributes to the dulling of uncooked poultry. Migration of moisture from the filling to the pastry of a product such as quiche can cause staining of the latter and give a soggy/sticky consistency. Specific foods show other visual changes during chill storage: Certain gels and milk products, including yogurts and cottage cheese, exhibit syneresis in which water is excluded from the product, giving a separated aqueous layer.

Although chill storage reduces the chemical and microbiological activity in foods, these still occur and can give rise to a range of stored flavors. In addition, many unprocessed fruits and vegetables lose flavor intensity during storage. Dependent on the type of food, these stored flavors (which may sometimes first be detected by odor) can be developed through cell breakdown and autolysis, which gives rise to stale cabbage note (sulfur-containing compounds) in Brassicae (69) and the gamey/livery flavor in uncooked poultry (70). Fat or lipid oxidation is characterized by a typical old/stale flavor detected in cooked, chill-stored chicken and potatoes (warmed-over flavor) (71). In some foods, the presence of light increases the rate of this reaction any may even initiate the development of staleness (72).

Many of the odors/flavors that develop during chill storage are due to microbial activity. Such flavors are frequently accompanied by visible growth: These include the sour/cheesey flavors associated with the activity of lactic acid bacteria, the sweet/ammoniacal odor found in meat and poultry associated with growth of pseudomonads (73), and the musty/damp flavor associated with mold growth found in a range of foods (74). Storage under certain low-oxygen conditions (modified atmospheres) is sometimes used in association with chill conditions: this can give rise to anaerobic conditions and a winey flavor in certain fruits and vegetables (75) or a cheesy flavor in ducks (76).

Although the development of a sensory change results from other reactions in the food, flavor or textural change may precede the detection of chemical or physical changes or marked microbial growth. The stale flavor associated with lipid oxidation may be perceived by a sensory panel before the oxidation products can be measured chemically (72). This is largely due to the acute sensitivity of humans to certain chemicals.

Similarly, a winey flavor can develop in fruit stored under anaerobic conditions before the appearance of mold (75). Conversely, in aerobically stored poultry the development of the typical off-odor closely parallels the growth of pseudomonads (73).

Chemical Changes

Chilling as a means of preservation is dependent on the temperature sensitivity of chemical and biochemical (enzyme catalyzed) reactions. Generally, lowering the temperature to chill temperatures reduces the likelihood of a reaction occurring and the rate at which it proceeds. By holding products at chill temperatures, changes in food chemistry are reduced and the shelf life is extended. There are exceptions, however, and some products, such as tomatoes, apples and particularly tropical and subtropical fruits, are prone to chill injury, ie, physiological damage that occurs as a result of exposure to nonfreezing temperatures. Chilling injury causes loss of quality through poor ripening, pitting, collapse of structural integrity, the development of off-flavors, and rotting. Methods to alleviate the development of these symptoms include manipulation of storage conditions (eg, temperature cycling), exogenous chemical treatments (eg, application of phospholipids), and selection or genetic manipulation to prevent chill sensitivity (77,78).

Quality issues in chilled foods focus on color, flavor, and texture, and, in the consumer's eye, their close approximation to fresh or freshly prepared foods. Changes in food chemistry that affect these sensory characteristics, and in some instances the nutritional status of the product, are dependent on the intrinsic factors of the product, eg, the moisture content, enzymes and substrates available, pH, and the influence of extrinsic factors on them, eg, handling of raw materials, processing, conditions during storage.

A knowledge of the factor or factors most critical to the perception of quality and an understanding of the food chemistry involved can be used to improve quality or extend shelf life. For example, an important factor in the assessment of the quality of uncooked meats is color, bright red indicating freshness, darker and brown colors being undesirable (67). By using conditions that favor the formation of oxymyoglobin (the form of myoglobin responsible for the red color), ie, high-oxygen tension (60–80% O_2; 40–20% CO_2) and suppressing the activity of oxygen-utilizing enzymes by holding at chill temperatures, the retention of freshness (ie, red color) can be extended for up to 1 week (79).

Interactions between intrinsic factors occur in fresh vegetables prepared ready for use as a result of chopping, trimming, and shredding. Small pieces of tissue are more susceptible to moisture loss and wilting, and the cut surfaces frequently became discolored owing to the enzymes and substrates usually separated in intact tissues being brought together. Use of humidified atmospheres can reduce water loss and wilting, and treatments of cut surfaces with chemical dips such as ascorbic acid retard oxidation and discoloration (80).

Migration of ingredients such as water, oils, coloring, and enzymes from one component of a product to another can be the cause of physicochemical changes (eg, the thinning of mayonnaise in coleslaw and the translucent appearance of cabbage) (81), adverse color changes (eg, migration of coloring from one layer of a trifle to another), and flavor changes (eg, on pizza, lipases from unblanched green peppers attacking fatty acids in cheese) (82).

A quality issue believed to be related to the oxidation of lipids in chilled ready-cooked meats for reheating before serving, particularly poultry and pork, is the development of flavors during chill storage that consumers describe as left-over reheated, more commonly termed warmed-over flavor (71). These flavors develop rapidly after cooking, being detected after only 48 h of chill storage. Current attempts at control involve the use of antioxidants, both natural and synthetic. However, a better understanding of the problem is needed to enable a more targeted approach.

Microbiological Spoilage

Many microbial species may be present as part of the natural flora of raw materials used in the production of chilled foods (17). The preservative mechanisms used in foods are selected for their ability to restrict/inhibit microbial growth but are rarely effective against all the organisms that may be present in a food. Organisms that are unaffected by the preservative mechanisms used during processing and handling of a food are those that will cause subsequent spoilage.

The means by which microorganisms produce spoilage are varied and depend on the organism present and the substrate on which it is growing. Different organisms utilize different substrates and metabolic pathways for growth and can often change the type of metabolism used to suit the available nutrients and conditions. For example, the spoilage process of aerobically stored meats involves the successive utilization of glucose, lactic acid, ammonia, and gluconic acid by different groups of microorganisms until finally *Pseudomonas fragi* generates additional nutrients to support its growth by proteolytic activity, resulting in the nitrogenous and sulfurous off-odors (eg, hydrogen sulfide, ammonia, and trimethylamine) associated with overt spoilage (83).

Overt spoilage occurs when the sensory characteristics of a product become unacceptable. The sensory characteristics involved are flavor, odor, and texture–visual appearance. Compounds causing off-odors and flavors are primarily volatile, and volatile compounds resulting from the microbial metabolism of protein are distinctive causes of spoilage. Souring is another frequently encountered type of microbial spoilage, caused by organisms that form acids from the metabolisms of carbohydrates. Textural changes can be caused either by the presence of microorganisms themselves (eg, slime on meat) or products of metabolism (eg, polysaccharides causing rope in milk and dairy products) or by the subsequent effects of microbial by-products such as enzymes (eg, soft rots on fruits and vegetables) or gas (eg, holes in hard cheese). Visual spoilage of microbial origin can take a variety of forms, including discoloration, pigmentation, gas disruption, surface growth, cloudiness, and rotting. Table 9 shows some of the types of spoilage found in chilled food and their causes.

CONTROL OF CHILLED FOODS

Manufacturing Control

A number of factors influence the special requirements for control during the manufacture and distribution of chilled foods. These include the perishability of raw materials, the minimal processing often used to maximize sensory quality, the potential for spoilage and for food poisoning of perishable products, and the requirement for adequate temperature control throughout product life. Many organizations have produced guidelines and codes of practice concerned with the production, storage, and distribution of chilled foods; some of these are listed in Table 7. The key factors affecting safety and quality at each stage of production and the approaches to control used in commercial operations are discussed.

Raw Materials. The quality of incoming ingredients is important in determining subsequent quality of chilled foods. The use of poor-quality materials is also a major means of introducing potential safety and stability problems to the production site. Ingredients with high microbial loading may then provide sources of infection to subsequent batches of production. The variety of organisms introduced on raw materials will be great and will include species capable of causing subsequent problems. The introduction of such organisms will be largely unavoidable, eg, salmonellae and clostridia on raw meat or spores on dried goods. Control of incoming ingredients is essential, however, to minimize the microbial loading on ingredients before processing, deterioration, and spread of these organisms within the processing environment. Controls taken include vetting of suppliers to ensure the best microbiological standards are achieved, specification of temperature of storage and of delivery (where appropriate), maximum age of ingredients at the time of delivery, and microbiological checks (where appropriate).

Raw Material Storage. On receipt, goods should be transferred to appropriate stores with the minimum of

Table 9. Some Microbiological Spoilage Problems in Chilled Foods

Problem	Cause	Food	Microorganism
Off Odors and Flavors			
Nitrogeneous	Hydrogen sulfide, ammonia trimethylamine	Fresh meat and dairy products	*Pseudomonas* sp, *Acinetobacter* sp, *Moxarella* sp
		Heat-processed meat	*Clostridium* sp
Souring	Acids: lactic, acetic, butyric	Meat, dairy, fruits and vegetables	*Lactobacillus* sp, *Streptococcus* sp, *Brocothrix thermosphacta*, *Bacillus* sp, coliforms
Earthy	Geosmin	Mushroom, water	*Actinomycetes*
		Meat	Cyanobacteria, Streptomycetes
Fruity	Lipase, esterase activity	Milk	*Pseudomonas fragi*
Potato	2-Methoxy-3-isopropyl pyrazine	Surface-ripened cheese	*Pseudomonas* sp
Textural Spoilage			
Slime	Polysaccharides	Meats	*Pseudomonas fragi*
Ropiness	*Polysaccharides*	Milk	*Alcaligenes* sp
Bittiness	Polysaccharides	Milk	*Bacillus cevens*
Rotting	Enzyme activity: pectinases, cellulose, xylanose	Fruits and vegetables	*Erwinia* sp, *Clostridia* sp, yeasts and molds
Visual Spoilage			
Bubbles	Gas formation	Cottage cheese, coleslaw	Coliforms
Holes	Gas formation	Hard cheese	Coliforms
Discoloration	Fluorescent pigment	Meat, egg	*Pseudomonas* sp
	Black pigment	Dairy products	*P. nigrifaciens*
	Surface growth	Potato salad	*Pichia membrane-faciens*

delay to prevent further microbial growth and deterioration. Stores in which temperature and/or humidity controls are necessary should be monitored and fitted with alarms. All food should be stored under hygienic conditions that prevent contamination by microorganisms, foreign bodies, or chemicals. Physical separation of products that are to be cooked from those already, or not to be, cooked is essential to prevent cross contamination. Separation is also important where aromatic ingredients are to be stored to prevent flavor transfer. Efficient stock rotation is also necessary to ensure materials are used as quickly as possible.

Preparation. Ingredients should be prepared with the minimum delay to prevent deterioration and microbial growth. This can be aided by the use of a temperature-controlled environment. Preparation of raw materials should occur in an area separated from those for handling cooked foods to prevent cross contamination. Should processing be held up, facilities for cold storage of prepared ingredients must be available.

Processing. The cooking process is a critical factor controlling the safety and stability of cooked chilled foods. The efficiency of the process in reducing the microbial load is dependent on several factors: the initial microbial load, the species present, and the temperature and time of the process given, which are in turn affected by the physical dimensions and thermal properties of the food itself. Heat processing results in an exponential decrease of microbial

numbers; thus, the greater the initial loading, the greater the heat process required to reduce numbers to a predetermined level. In addition, the ability of organisms to withstand heating is very variable and varies among species and the state of the organism (eg, vegetative cells/spores). Spores are more resistant than vegetative cells, and some species, eg, clostridia and streptococci, are more resistant than others eg, salmonellae, *E. coli*, and yeasts.

Selection of processing parameters may be determined by factors other than the microflora of a product, eg, sensory parameters. Processing regimes used in most cooked chilled foods will not produce sterile product. Knowledge of the processing parameters and the intrinsic microflora of raw materials will enable a judgment to be made as to the organisms likely to be remaining in a product after cooking. Thereafter, the potential for outgrowth of surviving organisms will be determined by properties of the food itself due to its formulation (eg, acidity, water activity, preservatives) and the storage conditions (eg, temperature, headspace gas, time).

Chilling. Chilling should occur as soon after cooking as possible to prevent the possibility of microbial growth and determination in the product. For the same reasons, chilling should occur as rapidly as possible. Several factors influence the chilling rate of foods, ie, thermal properties of the food, dimensions, packaging, method of chilling, and food temperature before chilling. Consideration must be given to each of these to ensure that specifications are achieved.

It is essential at this stage and in all subsequent stages that foods do not become recontaminated, eg, by cross contamination from raw materials, utensils, etc, or by food-handlers, particularly as competitive flora have been largely removed during cooking and thus introduced pathogens may be able to proliferate in the absence of competing organisms. Cross contamination has been shown to be one of the major causes of food-poisoning outbreaks.

Many manufacturers now use a system of segregating factories into high- and low-risk areas and operations, to minimize the opportunity for contamination of sensitive (high-risk) product after processing. High-risk areas are specially designed to high standards of hygiene and physically segregated from low-risk areas. Within such areas special operating practices exist relating to personnel, materials, equipment, and environment to prevent product contamination. Such practices include use of filtered air, controlled temperature environment, restricted access to personnel who must wear special protective clothing and use special foot baths and hand-washing facilities, dedication of equipment to the area, removal of outer packaging materials before entry to the area, and rigid cleaning schedules.

Subsequent Distribution and Storage. A variety of possible distribution and storage systems exist. In manufacturer/retail systems the chill chain frequently involves storage at the site of production, distribution to central storage depots where loads are broken down and reformulated, followed by further distribution to retail stores. Additional storage depots may be used in complex, long-distance chill chains. Within cook-chill catering systems several types of system may be used, including storage at the site of production and distribution to the end-service point, distribution to a secondary (satellite) storage area and subsequent distribution to the point of service.

Whatever the system operated, it is essential that the temperature of the food is controlled and does not rise above specified limits at any time. Monitoring systems must exist to ensure that correct temperatures are maintained at all times. There are several areas of potential failure to maintain adequate product temperature. The point of transfer of foods, eg, transfer into and out of stores, service points, displays, etc, where foods may be left in loading bays, is of particular concern, especially at staff breaktimes and in hot weather. External packs of food are the most readily susceptible to temperature fluctuation. Temperature fluctuation can result in condensation problems that can cause surface-water films and enhance microbial growth and deterioration. Other problems may occur due to equipment failure resulting from incorrect stacking of loads preventing adequate circulation of cooled air or failure of personnel to operate equipment correctly. Continuous temperature-recording devices, together with frequent monitoring of these, are important, and built-in alarm systems are highly recommended.

It is essential that an efficient system of stock rotation, involving adequate date marking of products, is used to ensure that product reaches the consumer with the prescribed shelf life still remaining.

Packaging. In some cook-chill systems, eg, sous-vide and Capkold, food is subsequently packaged into closed packaging with or without modification of the headspace. Where the headspace is modified, eg, by pulling a vacuum or gas flushing, it is important to ensure that the correct atmosphere is achieved and maintained, as this is being used to achieve microbial control. Packaging films of suitable permeability to gases in the pack and atmosphere need to be selected. Of particular concern is failure of the integrity of the pack caused either by failure during sealing/closure or by subsequent damage to packs. Any holes in the packaging will cause alteration in the pack atmosphere. Most importantly, these also allow a means of entry for organisms into the pack, including food-poisoning organisms that may be able to grow rapidly in the absence of competitive microflora destroyed during cooking. Pack failure before chilling may allow cooling fluids to be drawn into a pack as the contents contract. Where cool water is used, a good supply of potable water should be ensured.

Shelf Life

Control of the potential problems of food poisoning and spoilage results from the control of both temperature and time. Control of time minimizes the opportunity for deterioration and growth of microorganisms and is achieved by restricting shelf life.

The shelf life of the product may be described as that time for which, under normal conditions, a product remains wholesome. The shelf life may be limited by a product becoming unsafe or developing unacceptable sensory qualities or by commercial considerations. In order to establish the shelf life of a product is is important to determine that change which is the first to render the product unwholesome/unacceptable. Some of the changes that result in spoilage of chilled foods have been discussed previously.

Determining shelf life requires consideration of every aspect of manufacture, storage, distribution, and consumer handling. Much information is required on the conditions to which the product is likely to be exposed during its life and the effect these are likely to have on the product itself. It is particularly important to appreciate the potential variation that may occur in a product from batch to batch and the effect that this may have on the shelf life. The ultimate shelf life must take account of all the variations that are likely to occur.

Determination of shelf life requires considerable resource and expertise and understanding of the manufacturing process and therefore is usually undertaken by the manufacturer. Many experimental approaches are used in determining shelf life, including HACCP, challenge testing, and predictive modeling, as well as storage trials. Advice on determining shelf life of chilled foods is given in Ref. 84.

Although setting shelf life is usually undertaken by the industry itself, in certain circumstances shelf life is prescribed in official guidance or legislation. In the UK, the Department of Health recommends a maximum shelf life of 5 days (at 3°C) for cook-chill foods in catering systems. In France, legislation stipulates various maximum shelf lives for different products, depending on the heat process (pasteurization value) given.

Hazard Analysis (HACCP and HAZOP)

Formalized procedures for hazard analysis were first developed in the nuclear chemical industries. The concept has since been extended to other industries. During the 1960s, the Pillsbury Corporation, with the National Aeronautics and Space Administration and the U.S. Army, adopted the concept of Hazard Analysis Critical Control Point (HACCP) to assure food safety for the U.S. space program. Since that time, the concept has gained wide acceptance for use in the food industry in the United States and Europe.

HACCP is a structured approach to the identification and control of hazards and provides a more structured and critical approach than traditional inspection and quality-control procedures. The emphasis is removed from final product testing to the control of raw materials and process operations within the factory environment. The system is proactive and has the potential to identify areas of concern where failure has not yet been experienced and is therefore particularly useful for new operations. Although widely used for food-poisoning hazards, analyses can be made for other hazards such as foreign bodies or chemical contaminants or for spoilage problems or control of waste.

The HACCP approach consists of describing assessments of potential hazards associated with all stages of the food manufacturing operation from acquisition of raw material through to product sale and consumption. A detailed flow diagram of all these stages is produced, and each stage is considered in turn to identify potential problems that may arise, taking into account realistic process deviations that may occur, the characteristics of the product, and its use. Stages of the process that need to be controlled to assure safety (or product stability) are identified as critical control points (CCP). Following the hazard analysis, effective control options are devised to control the identified hazard. The results form the basis for the development and management of quality assurance and quality control programs. Resources can be concentrated into areas identified as critical (85,86).

Recently, application of the Hazard and Operability Studies (HAZOP) concept originally developed in the chemical industry has been recommended for use in the food industry (87). This is a creative and interactive rule-based technique to identify existing and potential hazards. Each stage in the process is examined to determine the process intent, and deviations from the intention are proposed following a systematic series of guide wards. Possible causes for each deviation are then considered and assessed as to whether they are likely to occur and what consequences would result if they did occur. As with HACCP, CCPs are identified and controls devised.

Both approaches require the use of a multidisciplinary team to ensure that knowledge and expertise are available to cover all aspects of the process. Commitment is required to derive the full benefits of the analysis and to ensure its subsequent implementation and success within the factory environment.

Predictive Microbiology

The application of HACCP/HAZOP systems can be further enhanced by the use of mathematical models that predict the changes that may occur in a food. This article is only concerned with those changes that are of microbiological significance, although models can be used for other factors. The use of mathematical models in microbiology is not new, as these have been used for many years in thermal-process microbiology to ensure the production of ambient-stable, high water activity (A_w) food products.

Recently, there has been an upsurge of interest in the use of mathematical models to predict growth of pathogenic and spoilage organisms and its consequences. The models may be broadly divided into those that are probabilistic and those that are kinetic. The former indicate the overall probability of growth for specific bacteria under set conditions and have been used to predict toxin production by *C. botulinum* (88) and *S. aureus* (89). Kinetic models that will predict the rate and extent of growth for spoilage and pathogenic bacteria are likely to be more useful. Such models are based on factors that may be used to characterize the microbial growth curve. The two factors most commonly used are the lag period before which growth starts and the generation time, which is a measure of the rate of growth. A number of mathematical equations using these factors have been used and include the Arrhenius equation (90), a modified Arrhenius equation (91), polynomial models (92), the Ratkowsky (square-root) equation (93), and the Schoolfield (nonlinear Arrhenius) equation (94). The choice of model type is still under debate and until recently few comparative studies have been published.

Most models were initially developed for the effect of temperature on microbial growth, but many have now been extended to include other factors such as pH, A_w, and salt (92,95). Such models will allow predictions of microbial growth to be made under a wide range of conditions. This is in contrast to the challenge-testing procedures that have been traditionally used to determine growth of microorganisms in foods, as the results obtained from these are relevant only to the specific conditions tested. Also, these studies tend to be cumbersom, time-consuming, and expensive.

A number of factors must be considered in the use of models to predict microbial growth. First, it is important to ensure that the factor being modeled is the factor of most importance, eg, the growth of spoilage microorganisms may not be the limiting factor for shelf life. Second, the model used must be accurate over the range of conditions of concern. The Schoolfield and Ratkowsky equations predict growth well over most of the temperature range for growth; however, the latter is less accurate as temperatures approach the minima for microbial growth (96). Obviously, this is of great significance for chill foods. Third, good quality data are required for the construction of appropriate mathematical models to ensure that subsequent predictions are accurate. Fourth, models should not be extrapolated beyond the limits of the original data as this may result in poor predictions. Finally, most current work on model development is based in microbiological growth media. Clearly, the models from this work must be evaluated using real foods.

The use of mathematical models to predict microbial growth offers significant time and cost benefits compared with conventional procedures. They may be usefully incorporated in HACCP/HAZOP programs and will be of great benefit in ensuring the microbiological quality and safety of foods in a cost-effective manner.

Quality Systems for the Chilled Food Industry

Formal quality systems encompassed in the internationally recognized standards EN 29000/ISO 9000/BS 5750 introduce a disciplined and formalized approach to the management of quality and production processes, regardless of the products or service being provided (97,98). It has now been applied very successfully in a wide variety of production and service industries throughout the world. Within the food industry, 50–60 processing sites in the UK are currently registered under the scheme, but no other European food manufacturer has achieved this status yet. At the present time, ISO 9000 schemes are being developed in many areas of the industry, including some chilled-food producers.

Adopting a structured and pragmatic approach ensures that staff at all levels within the organization know and understand exactly what is expected of them and how to achieve their goals. This is done by provision of detailed outlines of objectives and requirements in each production area, supported by written, detailed instructions and by tailored training programs. Significant improvements can be achieved in production levels. It must be borne in mind that a simple increase in output is not necessarily a positive achievement: If the goods do not meet specifications and are therefore subject to rejection or reworking, an increase in successful production cannot be claimed. By implementing ISO 9000, the likelihood of attaining the desired level of success is greatly increased. Planned, systematic internal audits ensure not only that the detailed quality plans are being adhered to but, where lapses do occur, that the auditing and corrective action procedures provide agreed courses of action that support and encourage the work force rather than allowing them to think they are being criticized for doing wrong. This in turn boosts staff morale, encouraging an ever more responsible approach to their daily work.

Implementation requires time and effort in developing a well-designed and carefully constructed ISO 9000-compatible system. Development of the system can be achieved comfortably within 12 months in any operation of less than 400–500 production staff. Payback times are seldom more than a year and can be as little as a few weeks in some cases.

Once the step has been taken and registration under the scheme achieved, the standards must be maintained. A system of 6-monthly surveillance visits is an obligatory part of the administration of the scheme. Defaulters can be struck off the nationally circulated register of approved companies.

Challenge Testing

Challenge testing is a procedure frequently used by the food industry to determine the limits of stability, to assess the risk of food poisoning of a product, and to determine which components of a food are responsible for its preservation. It can be defined as the laboratory simulation of what can happen to a product during manufacture, distribution, and subsequent handling. In its simplest form challenge testing can involve the storage of product, or different formulations of a product, under realistic and abuse conditions, as is used in determining shelf life. Microbiological challenge testing involves the inoculation of a product with relevant microorganisms and/or holding the product under a range of controlled environmental conditions (99). As such, challenge testing provides information on the conditions that allow/inhibit growth of a particular microorganism, but it does not provide information on the likelihood of organisms being present in the product. An essential part of designing a challenge test is therefore the choice of relevant organisms used to inoculate the product. These should be those with known tolerance to the food and storage conditions and therefore likely to cause spoilage or food poisoning in the product under test.

Challenge testing can include a variety of experimental approaches in addition to the storage trials already noted. The earliest approach used in the canning industry was inoculated pack studies. In these, suspensions of highly heat-resistant spores of a known spoilage organism, *C. sporogenes,* were used to challenge a process in order to determine the processing condition required to reduce possible contamination with *C. botulinum* to acceptable limits. Spore suspensions introduced into the product were usually contained in glass spheres that could be retrieved after processing for analysis. Recently, other forms of inoculation have been developed to suit particular processes, eg, the incorporation of organisms into alginate particles for use in processes where product is pumped or mixed (100), or the inoculation of filter-paper strips to determine the effectiveness of decontamination of packaging (101). Methods of introduction of microorganisms to a product and their subsequent retrieval for analysis must be suited to the particular product and process. The preparation of the microorganisms is also critical, as different handling procedures before inoculation may influence the results obtained, eg, cultures of *L. monocytogenes* pretreated at 4°C grow to high levels in chilled broth much more quickly than those pretreated at 30°C (15).

LEGISLATION

Legislative control of the production and distribution of foods varies among different European countries, although protection of the health of the consumer and the promotion of food safety are often the objectives. Most countries operate a system whereby government has a separate ministry concerned with either agriculture or public health responsible for initiating policy and legislation. Many details of food safety are contained in subordinate legislation, eg, in regulations, administrative procedures, and instructions to local government. Food-control services and inspectorates are usually the responsibility of regional/local authorities, which frequently operate in an autonomous manner, varying in the implementation and interpretation of regulations. In response to the expansion of the chilled food market, legislation is being implemented to ensure the safety of new foods and, increasingly, process techniques. With the advent of the single European Market in 1992, legislation is increasingly approached from an international basis through such groups as the European Economic Community (EEC), Council for Mutual Economic Assistance (CMEA), and the FAO/WHO Codex Alimentarus Commission. Once ratified by member states, such legislation takes precedence over national regulations for intracommunity trade.

Table 10 lists existing and proposed community legislation of particular relevance to temperature control of chilled foods. In addition to the directives listed, 25 countries, including all major European countries, the United States, and the USSR have ratified the Agreement on the International Carriage of Perishable Foodstuffs (the ATP Agreement) and most member states of the European Community consider this to be mandatory. In this agreement, storage temperatures are recommended for frozen foods and the following temperatures are recommended for chilled foods: 7°C for fresh meat; 6°C for meat products, butter, eggs; 4°C for poultry, rabbit, game, milk and dairy products, partly cooked meat products, ready meals, sauces, egg products, and fresh cakes and sweets; 2°C for fresh fish. These are, however, currently being reconsidered. Testing and certification of distribution vehicles

with regard to insulation and refrigeration performance is also covered by this agreement.

European Community legislation exists or is proposed covering aspects pertinent to the production and sale of all food, not specifically chilled foods. This includes such aspects as labeling, food composition, food additives, consumer protection, and packaging materials.

FUTURE DEVELOPMENT

The preceding sections have shown major developments in the marketplace and the technology of the production of chilled foods. There is no reason to suspect that these developments should not continue to occur. Consumer awareness and media attention to problems, real and potential, in the food chain are, however, likely to increase, with a possibility of a knock-on effect on the market. This is likely to be associated with increasing legislation and powers of enforcement for control and inspection agencies. It therefore follows that the food industry is likely to be subject to much closer scrutiny and control than has hitherto existed. Within the food industry, closer self-control through the use of documented quality-management systems are likely to arise in the continuing battle to attain and maintain the highest standards of safety and quality and to be seen to be so doing.

Acknowledgments: The author would like to thank her colleagues for their considerable assistance in preparing this article, particularly Sandara Bond, Helen Churchill, Brian Day, Nerys Griffiths, David J. Rose, Patrick Stewart, Gary Tucker, and Stephen Walker.

Table 10. Existing and Proposed European Community Legislation on Temperature Control on Chilled Foods

Legislation	Requirements
Directive 88/657/EEC (mince and meat in small pieces)	Storage and transport at +2°C; temperature recorders required in vehicles
Directive 88/658/EEC (meat products)	Preparation/cutting rooms to be at 12°C; temperature to be indicated on label; storage and transport at that temperature; prepared meals to be cooled to 10°C within 2 h of cooking
Directive 64/433/EEC (fresh meat)	7°C for carcases; 3°C for cuts and offal; 12°C for cutting rooms
Directive 89/437/EEC (egg products)	Storage and transport at 4°C
Directive 85/397/EEC (heat-treated milk)	Raw milk cooled to 8°C for short or 6°C for longer storage; Transport at < 10°C, on arrival cool to 5°C; meat-treated milk— storage and transport, 6°C
Proposed regulation on fish hygiene	Fresh fish at temperature of melting ice; temperature recording in cold stores
Proposed regulation on hygiene of any animal product not covered by other directives	No specific temperatures; temperature recording in cold stores
Proposed regulation on rabbit and game	Rabbit meat and small game, 4°C; large game, 7°C
Proposed regulations on rendering or animal fat	Raw materials storage, 7°C; greaves storage, 7°C

BIBLIOGRAPHY

1. M. Cowling, "Social Trends—Their Direction and the Implications for Chilled Foods," *Proceedings of a Seminar on Chilled Foods,* Stratford-on-Avon, UK, May 7–9, 1985, pp. 9–16, Campden Food and Drink Research Association, Chipping Campden, UK, 1985.

2. *Guidelines for the Handling of Chilled Foods,* 2nd ed., Institute of Food Science and Technology (UK), London, 1990.

3. A. Ciobanu, G. Lascu, V. Bercescu, and L. Nicolescu, *Cooling Technology in the Food Industry,* Abacus Press, Tunbridge Wells, UK, 1976.

4. W. Gosney, *Principals of Refrigeration,* Cambridge University Press, UK, 1982.

5. E. Hallowells, *Cold and Freezer Storage Manual,* 2nd ed., AVI Publishing Co., Westport, Conn., 1980.

6. Trent Regional Health Authority, *The Cook-Chill System: An Appraisal of Equipment and Consumables,* Green Belfield-Smith Consultants, London, 1988.

7. B. Turtle, "Cost Effective Food Packaging," *World Packaging Directory,* Cornhill Publications Ltd., London, 1988, pp. 67–72.

8. R. Goodard, "Packaging for the Microwave," *Food Manufacture* **61**, 29–33, 1986.

9. B. Day, "Extension of Shelf-Life of Chilled Food," *European Food and Drink Review* **4**, 47–56, 1989.

10. R. Pearson, "The Quiet Revolution," *Plastics Today* No. 29, Imperial Chemical Incorporated, Wellwyn Garden City, UK 1988.

11. "The Impact of Materials and Gas on Chilled Food," *Food Engineering* **61**, 62–63, 1989.

12. B. Day, "Optimization of Parameters for Modified Atmosphere Packaging of Fresh Fruit and Vegetables," *CAP '88,* Schotland Business Research Inc., Princeton, N.J., 1989, pp. 147–170.

13. R. Heap, "Design and Performance of Insulated and Refrigerated ISO Intermodal Containers," *International Journal of Refrigeration* **12**, 137–145, 1989.

14. M. Doyle, ed., *Foodborne Bacterial Pathogens,* Marcel Dekker, New York, 1989.

15. S. Walker, P. Archer, and J. Banks, "Growth of *Listeria monocytogenes* at Refrigeration Temperatures," *Journal of Applied Bacteriology* **68**, 157–162, 1990.

16. S. Prescott and L. Geer, "Observations on Food Poisoning Organisms under Refrigeration Conditions," *Refrigeration Engineering* **32**, 211–212, 282–283, 1936.

17. S. Walker and M. Stringer, "Microbiology of Chill Foods," in *Chilled Foods—the State of the Art* (In press).

18. B. Ralovich, "Epidemiology and Significance of Listeriosis in the European Countries," in *Listeriosis: Joint WHO/ROI Consulation on Prevention and Control.* Ved-Med, Berlin, 1987, pp. 21–55.

19. R. Brackett, "Presence and Persistence of *Listeria monocytogenes* in Food and Water," *Food Technology* **42**, 162–164, 1988.

20. W. Schlech and co-workers, "Epidemic Listeriosis—Evidence for Transmission by Food," *New England Journal of Medicine* **308**, 203–206, 1983.

21. D. Fleming and co-workers, "Pasteurized Milk as a Vehicle of Infection in an Outbreak of Listeriosis," *New England Journal of Medicine* **312**, 404–407, 1985.

22. S. James, "Listeriosis Outbreak Associated with Mexican-Sytle Cheese—California," *Morbidity and Mortality Weekly Report* **34**, 357–359, 1985.

23. B. Bannister, "*Listeria monocytogenes* Meningitis Associated with Eating Soft Cheese," *Journal of Infection* **15**, 165–268, 1987.

24. S. Walker, "*Listeria monocytogenes*: An Emerging Pathogen," *Food Technology International Europe* 237–239, 1990.

25. R. Black and co-workers, "Epidemic *Yersinia enterocolitica* Infection Due to Contaminated Chocolate Milk," *New England Journal of Medicine* **298**, 76–79, 1978.

26. C. Tacket and co-workers, "An Outbreak of *Yersinia enterocolitica* Infections Caused by Contaminated Tofu," *Journal of Epidemiology* **121**, 705–711, 1986.

27. C. Tacket and co-workers, "A Multistate Outbreak of Infections Caused by *Yersinia enterocolitica* Transmitted by Pasteurized Milk," *Journal of the American Medical Association* **251**, 483–486, 1984.

28. M. Greenwood and W. Hooper, "*Yersinia* spp. in Foods and Related Environments," *Food Microbiology* **2**, 263–269, 1985.

29. A. Gilmour and S. Walker, "Isolation and Identification of *Yersinia enterocolitica*-like bacteria," *Journal of Applied Bacteriology* Suppl. **65**, 213S–236S, 1988.

30. S. Walker, P. Archer, and J. Banks, "Growth of *Yersinia enterocolitica* at Chill Temperatures in Milk and Other Media," *Milchwissenschaft* (In press).

31. S. Walker, "*Yersinia enterocolitica*: A Review of Foodborne Pathogen," *Campden Food Preservation Research Association Technical Bulletin No. 63,* Chipping Campden, UK, 1987.

32. A. Stelma, "*Aeromonas hydrophila,*" in M. Doyle, ed., *Foodborne Bacterial Pathogens,* Marcel Dekker, New York, 1989.

33. C. Fricher and S. Tompsett, "*Aeromonas* spp. in Foods: A Significant Cause of Food Poisoning?," *International Journal of Food Microbiology* **9**, 17–23, 1989.

34. S. Enfors, G. Molin, and A. Ternstrom, "Effect of Packaging Under Carbon Dioxide, Nitrogen or Air on the Microbial Flora of Pork Stored at 4°C," *Journal of Applied Bacteriology* **47**, 197–208, 1979.

35. S. Walker and M. Stringer, "Growth of *Yersinia enterocolitica* and *Listeria monocytogenes* at Chill Temperatures," *Campden Food and Drink Research Association Technical Memorandum* No. 503, Chipping Campden, UK, 1988.

36. S. Palumbo and R. Buchanan, "Factors Affecting Growth or Survival of *Aeromonas hydrophila* in foods," *Journal of Food Safety* **9**, 37–51, 1988.

37. C. Schmidt, R. Lechowich, and J. Folmazzo, "Growth and Toxin Production by Type E *Clostridium botulinum* Below 40°F," *Journal of Food Science* **26**, 626–634, 1961.

38. D. Cann, B. Wilson, G. Hobbs, and J. Shewan, "The Growth and Toxin Production of *Clostridium botulinum* in Certain Vacuum Packed Fish," *Journal of Applied Bacteriology* **28**, 431–436, 1965.

39. Y. Roberts and G. Hobbs, "Low Temperature Characteristics of *Clostridia,*" *Journal of Applied Bacteriology* **31**, 75–88, 1968.

40. J. Goepfert, W. Spira, and H. Kim, "*Bacillus cereus* Food Poisoning: A Review," *Journal of Milk Food Technology* **35**, 213–227, 1972.

41. D. Coghill and H. Juffs, "Incidence of Psychrotrophic Spore Forming Bacteria in Pasteurized Milk and Cream Products and Effect of Temperature on Their Growth," *Australian Journal of Dairy Technology* **3**, 150–153, 1979.

42. Department of Health and Social Security and co-workers, *Food Hygiene Codes of Practice No. 10: The Canning and Low Acid Foods: A Guide to Good Manufacturing Practice*, HMSO, London, 1982.

43. "Guidelines to the Types of Food Products Stabilized by Mild Heat (Pasteurization) Treatments," *Campden Food and Drink Research Association Technical Manual No. 27*, Chipping Campden, UK (In preparation).

44. C. Stumbo, "Death of Bacteria Subject to Moist Heat," in C. Stumbo, ed., *Thermobacteriology in Food Processing*, Academic Press, New York, 1965.

45. A. Russell, "Inactivation of Bacterial Spores by Thermal Processes (Moist Heat)," in A. Russel, ed., *The Destruction of Bacterial Spores*, Academic Press, London, 1982.

46. J. Gaze, "The Effect of Oil on the Heat Resistance of *Staphylococcus aureus*," *Food Microbiology* 3, 277–283, 1985.

47. M. Woodburn, E. Somers, J. Rodrigues, and E. Schantz, "Heat Inactivation Rates of Botulinum Toxins, A. B. E and F in Some Foods and Buffers," *Journal of Food Science* 44, 1658–1661, 1979.

48. S. Tatini, "Thermal Stability of Exterotoxins in Foods," *Journal of Milk Food Technology* 39, 432–438, 1976.

49. B. Dunn and H. Stich, eds., *Carcinogens and Mutagens in the Environment*, Vol. 1, CRC Press, Boca Raton, Fla., 1982.

50. *Royal Commission on Environmental Pollution*, 9th report, HMSO, London, 1983.

51. D. Gunn and J. Stevens, eds., *Pesticides and Human Welfare*, Oxford University Press, Oxford, UK 1976.

52. Ministry of Agriculture Fisheries and Food/Health and Safety Executive, *Pesticides 1988*, (Reference book 500), HMSO, London, 1988.

53. "Report of the Working Party on Pesticides Residues (1982–1985)," *Ministry of Agriculture Fisheries and Food: Food Surveillance*, Paper No. 16, HMSO, London, 1986.

54. C. Reilly, ed., *Metal Contamination in Food*, Applied Science, London, 1980.

55. M. Webb, ed., "Chemicals in Food and Environment," *British Medical Bulletin* 31 (3), 1975.

56. F. Schmidt and A. Hildebrandt, "Health Evaluation of Heavy Metals in Infant Formula and Junior Foods," Springer-Verlag, New York, 1983.

57. T. Twomey, "Radioactivity and its Measurement in Foodstuffs," *Dairy Food Sanitation* 7, 452–457, 1987.

58. Department of Environment, "Dioxins in the Environment," *Pollution Paper* No. 27, HMSO, London, 1989.

59. "PCB's in Food and Human Tissue," *Ministry of Agriculture Fisheries and Food: Food Surveillance*, Paper No. 13, HMSO, London, 1983.

60. R. Ashby and C. Vom Bruck, eds., *Food Additives and Contaminants* 5, (Suppl. 1), 1988.

61. "Nitrate and Drinking Water," *Technical Report*, No. 27, European Chemical Industry Ecology and Toxicology Centre, Brussels, 1988.

62. Ministry of Agriculture Fisheries and Food, *Food Hygiene Report on a Consumer Survey*, HMSO, London, 1988.

63. S. Rose, S. Steadman, and R. Brunskill, "A Temperature Survey of Domestic Refrigerators," *Campden Food and Drink Research Association Technical Memorandum*, No. 577, Chipping Campden, UK, 1990.

64. Ministry of Agriculture Fisheries and Food and Department of Health and Social Security, *The Recipe for Food Safety: How to Prevent Food Poisoning*, HMSO, London, 1988.

65. S. James and J. Evans, "Temperatures in the Retail and Domestic Chilled Chain," *Proceedings of Cost 91bis Concluding Seminar on the Effects of Processing and Distribution on the Quality and Nutritive Value of Foods*, October 2–5, 1989, Gothenburg, Sweden.

66. L. Meyes, ed., *Food Chemistry*, Reinhold Publishing, New York, 1960.

67. R. Lawrie, "The Eating Quality of Meat," in *Meat Science*, 4th ed., Pergamon Press, Oxford, UK, 1986, pp. 169–208.

68. V. Arthey, *Quality of Horticultural Products*, Butterworths, London, 1975.

69. L. Bedford, "Shelf Life of Fresh Brussel Sprouts," *Campden Food and Drink Research Association, Technical Memorandum*, No. 578, Chipping Campden, UK, 1990.

70. J. Jones, G. Mead, N. Griffiths, and B. Adams, "Influence of Packaging of Changes in Chill-Stored Turkey Portions," *British Poultry Science* 23, 25–40, 1982.

71. A. St. Angelo and M. Bailey, eds., *Warmed Over Flavour in Meat*, Academic Press, London, 1987.

72. H. Chan, G. Levitt, and N. Griffiths, "Light Induced Flavour Deterioration. The Effect of Exposure to Light on Pork Luncheon Meat Containing Erythrosine," *Journal of the Science of Food and Agriculture* 28, 339–344, 1977.

73. E. Barnes and C. Impey, "The Shelf-Life of Eviscerated and Uneviscerated Chicken Carcasses Stored at 10°C and 4°C," *British Poultry Science* 16, 319–327, 1975.

74. J. Robinson, K. Browne, and W. Burton, "Storage Characteristics of Some Vegetables and Soft Fruit," *Annals of Applied Biology* 81, 399–408, 1975.

75. K. Browne, J. Geeson, and C. Dennis. "Effects on Harvest Date and CO_2 Enriched Storage on the Shelf-Life of Strawberries," *Journal of Agriculture Science* 59, 197–204, 1984.

76. E. Barnes, C. Impey, and N. Griffiths, "The Spoilage Flora and Shelf-life of Duck Carcasses Stored in Oxygen Permeable and Impermeable Films," *British Poultry Science* 20, 491–500, 1979.

77. R. Jackman and co-workers, "Chilling Injury. A Review of Quality Aspects," *Journal of Food Quality* 11, 253–278, 1988.

78. R. Wills and co-workers, "Physiological Disorders" in ed., *Postharvest: An Introduction to the Physiology and Handling of Fruit and Vegetables*, 3rd ed., BSP Professional Books, Oxford, UK, 1989, pp. 73–87.

79. D. Kropf, "Effects of Retail Display Conditions on Meat Colour," *Reciprocal Meat Conference Proceedings* 33, 15–32, 1980.

80. J. Geeson and T. Brocklehurst, "A Fresh Approach to Salad," *Food Processing* 58, 41–42, 44, 1989.

81. A. Tunaley, G. Brownsey, and T. Brocklehurst, "Changes in Mayonnaise-Based Salads during Storage," *Lebensmittel Wissenschaft Technologie* 18, 220–224, 1985.

82. T. Labuza and M. Schmidl, "Accelerated Shelf-Life Testing of Foods," *Food Technology* 39, 57–134, 1985.

83. C. Gill, "Microbial Interactions with Meats," in M. Brown, ed., *Meat Microbiology*, Applied Science Publishers, Barking, UK, 1982, pp. 225–264.

84. "Evaluation of Shelf-Life for Chilled Foods," *Campden Food and Drink Research Association Technical Manual*, No. 28, Chipping Campden, UK, 1990.

85. International Commission on Microbiological Specifications for Foods, *Microorganisms in Foods 4: Application of the Hazard Analysis Critical Control Point (HACCP) System to*

Ensure Microbiological Safety and Quality, Blackwell Scientific, Oxford, 1988.

86. "Guidelines to the Establishment of Hazard Analysis Critical Control Point (HACCP)," *Campden Food and Drink Research Association Technical Manual,* No. 19, Chipping Campden, UK, 1987.

87. T. Mayes and D. Kilsby, "The Use of HAZOP Hazard Analysis to Identify Critical Control Points for the Microbiological Safety and Food," *Food Quality Preference* 1, 53–57, 1989.

88. T. Roberts, A. Gibson, and A. Robinson, "Prediction of Toxin Production by *Clostridium botulinum* in Pasteurized Pork Slurry," *Journal of Food Technology* 16, 337–355, 1981.

89. C. Genigeorgis, M. Savoukidis, and S. Martin, "Initiation of Staphylococcal Growth in Processed Meat Environments," *Applied Microbiology* 21, 940–942, 1971.

90. W. Reichardt and R. Morita, "Temperature Characteristics of Psychrotrophic and Psychrophilic Bacteria," *Journal of General Microbiology* 128, 565–568, 1982.

91. K. Davey, "A Predictive Model for Combined Temperature and Water Activity on Microbial Growth During the Growth Phase," *Journal of Applied Bacteriology* 67, 483–488, 1989.

92. A. Gibson, N. Bratchell, and T. Roberts, "Predicting Microbial Growth: Growth Responses of Salmonellae in a Laboratory Medium as Affected by pH, Sodium Chloride and Storage Temperature," *International Journal of Food Microbiology* 6, 155–178, 1988.

93. D. Ratkowsky and co-workers, "Model for Bacterial Culture Growth Rate Throughout the Entire Biokinetic Temperature Range," *Journal of Bacteriology* 154, 1222–1226, 1983.

94. R. Schoolfield, P. Sharpe, and C. Magnuson, "Non-linear Regression of Biological Temperature-Dependent Rate Models Based on Absolute Reaction-Rate Theory," *Journal of Theoretical Biology* 88, 719–731, 1981.

95. J. Brougall and C. Brown, "Hazard Analysis Applied to Microbial Growth in Foods: Development and Application of Three-dimensional Models to Predict Bacterial Growth," *Food Microbiology* 1, 13–22, 1984.

96. C. Adair, D. Kilsby, and P. Whittall, "Comparison of the Schoolfield (Non-linear Arrhenius) Model and the Square Root Model for Predicting Bacterial Growth in Foods," *Food Microbiology* 6, 7–18, 1989.

97. European Standards EN 29000–EN 29004, *Quality Systems,* Parts 1–4, European Committee for Standardization, Brussels, 1987.

98. British Standard 5750, *Quality Systems,* Parts 1–4, British Standards Institution, London, 1987.

99. S. Rose, ed., "Guidelines for Microbiological Challenge Testing," *Campden Food and Drink Research Association,* Technical Manual No. 20, Chipping Campden, UK, 1987.

100. K. Brown, C. Ayres, J. Gaze, and M. Newman, "Thermal Destruction of Bacterial Spores Immobilized in Food/Alginate Particles," *Food Microbiology* 1, 187–198, 1984.

101. S. Leaper, K. Bloor, and M. Stringer, "Resistance of Bacterial Spores in Relation to Chemical Sterilization Methods for Packaging Materials," *Campden Food and Drink Research Association,* Technical Memorandum No. 453, Chipping Campden, UK, 1987.

SALLY A. ROSE
Campden Food and Drink
Research Association
Gloucestershire, United
Kingdom

CHITOSAN. See EDIBLE FILMS AND COATINGS.
CHLOROPHYLL. See PHOTOSYNTHESIS.

CHOCOLATE AND COCOA

The name *Theobroma cacao,* food of the gods, indicating both the legendary origin and nourishing qualities of chocolate, was bestowed on the cacao tree by Linnaeus in 1720. All cocoa and chocolate products are derived from the cocoa bean, the seed of the fruit of this tree. Davila Garibi, a contemporary Mexican scholar, has traced the derivation of the word from basic root words of the Mayan language to its adoption as chocolate in Spanish (1).

The terms cocoa and cacao are often used interchangeably in the literature. Both terms describe various products from harvest through processing. In this article, the term cocoa will be used to describe products in general and the term cacao reserved for botanical contexts. Cocoa traders and brokers frequently use the term raw cocoa to distinguish unroasted cocoa beans from finished products and this term is used to report statistics for cocoa bean production and consumption.

STANDARDS FOR COCOA AND CHOCOLATE

In the United States, chocolate and cocoa are foods standardized by the U.S. Food and Drug Administration (FDA) under the Federal Food, Drug and Cosmetic Act. The current definitions and standards resulted from prolonged discussions between the U.S. chocolate industry and the FDA. The definitions and standards originally published in the *Federal Register* of Dec. 6, 1944, have since been only slightly revised.

The Codex Alimentarious Committee on Cocoa Products, established by the Food and Agricultural Organization (FAO) of the United Nations in 1963, has also drafted product standards. The draft standards for several products, including sweetened chocolate and milk chocolate, have been sent to member governments for acceptance. A draft standard becomes an official Codex standard when it has been accepted by a sufficient number of countries.

There are several major differences between the current U.S. standards for chocolate and cocoa products and those prepared by Codex, including a difference in the use of the word chocolate. According to U.S. standards, chocolate refers only to the material prepared from grinding cocoa beans (see below), also called chocolate liquor. According to Codex standards, chocolate designates a category of products as well as specific products. The FDA has already stated its intention to adopt the Codex food standards (2).

COCOA

Production

Worldwide cocoa bean production has increased significantly over the past 10 years from approximately 1,500,000 t in the 1977–1978 crop year to more than

Table 1. Production of Raw Cocoa, Thousands of Metric Tons

	1977/1978	1987/1988	1988/1989 Estimate
Ivory Coast	304	665	780
Brazil	283	402	345
Ghana	268	188	305
Malaysia	23	222	240
Nigeria	207	150	160
Indonesia	4	49	65
Other	423	524	507
Total	1,512	2,200	2,402

2,400,000 t today. The production share by country has also changed dramatically in the last 10 years. The big gainers are Malaysia, Indonesia, and the Ivory Coast. The large gains in Malaysia and Indonesia have helped to diversify production and partially shield the market from adverse weather-induced supply shocks. The big losers in production share have been Ghana, Nigeria, and Brazil. Table 1 lists production statistics for these countries.

Malaysian and Indonesia beans are viewed by the market as inferior beans. However, trade groups from these countries are working with chocolate manufacturers around the world to improve quality. The use and acceptance of these beans are increasing and most market experts believe that these two countries will continue to become bigger factors in the world cocoa market.

Consumption

The United States is the largest consumer of cocoa beans in the world. However, its share has been declining from approximately 21% in 1965 to 13% in 1988. Table 2 gives the annual tonnage of cocoa beans imported into the United States and other leading countries.

Another trend to note is that an increasing share of cocoa beans are processed at the origin country in which the cocoa was produced. The share of origin-processed cocoa beans has grown from 16% in 1965 to 33% in 1988.

Marketing

Most of the cocoa beans and products imported into the United States are done so by New York and London trade houses. The New York Sugar, Coffee and Cocoa Exchange provides a mechanism by which both chocolate manufacturers and trade houses can hedge their cocoa bean transactions.

COCOA BEANS

Significant amounts of cocoa beans are produced in about 30 different localities. These areas are confined to latitudes 20° north or south of the equator. Although cocoa trees thrive in this very hot climate, young trees require the shade of larger trees such as banana, coconut, and palm for protection.

A cocoa tree produces its first crop in three to four years and a full crop after six to seven years. A full-grown tree can reach a height of 12–15 m but is normally trimmed to 5–6 m to permit easy harvest. Cocoa pods are harvested twice a year, once in May and again in October or November.

Fermentation (Curing)

Prior to shipment from producing countries most cocoa beans undergo a process known as curing, fermenting, or sweating. These terms are used rather loosely to describe a procedure in which seeds are removed from the pods, fermented, and dried. Unfermented beans, particularly from Haiti and the Dominican Republic, are used in the United States.

The age-old process of preparing cocoa beans for market involves specific steps that allegedly promote the activities of certain enzymes. Various methods of fermentation are used to the same end.

Fermentation plays a major role in flavor development of beans by mechanisms that are not well understood (3). Because freshly harvested cocoa beans are covered with a white pulp, rich in sugars, fermentation begins almost immediately on exposure to air. The sugars are converted to alcohol and, finally, to acetic acid, which drains off, freeing the cotyledon from the pulpy mass. The acetic acid and heat formed during fermentation penetrate the skin

Table 2. Grind of Raw Cocoa, Thousands of Metric Tons

	1965	1970	1975	1980	1985	1988
United States	285	266	208	142	205	249
UK	102	82	72	65	91	100
FRG	157	126	151	158	207	218
Netherlands	118	115	119	133	167	213
France	63	40	37	48	42	38
Eastern Europe and USSR	147	190	282	218	243	226
Producing Countries	216	286	366	527	625	647
Total	1,335	1,357	1,471	1,510	1,838	1,961

Table 3. Main Varieties of Cocoa Beans Imported into the United States

Africa	South America	West Indies	Other
Accra (Ghana)	Arriba (Ecuador)	Trinidad	New Guinea
Ivory Coast	Venezuelean	Grenada	Malaysia
Lagos	Bahia (Brazil)	Sanchez (Dominican Republic)	Samoa
Nigeria			
Fernando Po			
Sierra Leone			

or shell, killing the germ and initiating chemical changes within the bean that play a significant role in the development of flavor and color. During this initial stage of fermentation, the beans acquire the ability to absorb moisture, which is necessary for many of the chemical reactions that follow.

Commercial Grades

Most cocoa beans imported into the United States are one of about a dozen commercial varieties that can be generally classified as criollo and forastero. Criollo beans have a light color, a mild, nutty flavor, and an odor somewhat like sour wine. Forastero beans have a strong, somewhat bitter flavor and various degrees of astringency. The forastero varieties are more abundant and provide the basis for most chocolate and cocoa formulations. Table 3 shows the main varieties of cocoa beans imported into the United States. The varieties are usually named for the country or port of origin.

Bean Specification

Cocoa beans vary widely in quality necessitating a system of inspection and grading to ensure uniformity. Producing countries have always inspected beans for proper curing and drying as well as for insect and mold damage. Only recently, however, has a procedure for grading beans been established at an international level. This ordinance, reached primarily through the efforts of FAO, has been adopted by Codex as the model ordinance for inspection and grading of beans. It classifies beans in two principal categories according to the fraction of moldy, slaty, flat, germinated, and insect-damaged beans (4).

Cocoa beans are sometimes evaluated in the laboratory to distinguish and characterize flavors. The procedure usually consists of the following steps: beans are roasted at a standardized temperature for a specific period of time; shelled, usually by hand; and ground or heated slightly to obtain chocolate liquor. The liquor's taste is evaluated by a panel of experts who characterize and record the particular flavor profile. The Chocolate Manufacturer's Association of the United States recently formed a committee to standardize this laboratory evaluation (4).

Blending

Most chocolate and cocoa products consist of blends of beans chosen for flavor and color characteristics. Cocoa beans may be blended before or after roasting, or nibs may be blended before grinding. In some cases finished liquors are blended. Common, or basic beans, are usually African or Brazilian and constitute the bulk of most blends. More

expensive beans from Venezuela, Trinidad, Ecuador, etc are added to impart specific characteristics. The blend is determined by the end use or type of product desired.

MANUFACTURE OF COCOA AND CHOCOLATE PRODUCTS

The cocoa bean is the basic raw ingredient in the manufacture of all cocoa products. The beans are converted to chocolate liquor, the primary ingredient from which all chocolate and cocoa products are made. Figure 1 depicts the conversion of cocoa beans to chocolate liquor and, in turn, to cocoa powder, cocoa butter, and sweet and milk chocolate, the chief chocolate and cocoa products manufactured in the United States.

Chocolate Liquor

Chocolate liquor is the solid or semiplastic food prepared by finely grinding the kernel or nib of the cocoa bean. It is also commonly called chocolate, baking chocolate, or cooking chocolate. In Europe, chocolate liquor is often called chocolate mass.

Cleaning. Cocoa beans are imported in the United States in 70-kg bags. The beans can be processed almost immediately or stored for later use. They are usually fumigated prior to storage.

The first step in the processing of cocoa beans is clean-

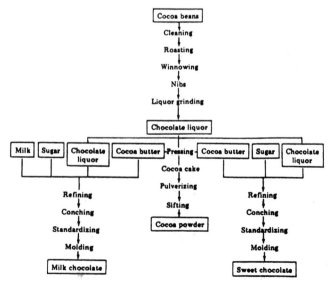

Figure 1. Flow diagram of chocolate and cocoa production.

ing. Stones, metals, twigs, twine, and other foreign matter are usually removed by passing beans in a large thin layer over a vibrating screen cleaner. Large objects are retained as the beans fall through a lower screen. The second screen removes sand and dirt that have adhered to the beans. Strategically placed magnets are commonly used to remove small pieces of metal.

Roasting. The familiar chocolate flavor is primarily developed during roasting, which promotes reactions among the latent flavor precursors in the bean. Good flavor depends on the variety of bean and the curing process used. The bacterial or enzymatic changes that occur during fermentation presumably set the stage for the production of good flavor precursors.

Although flavor precursors in the unroasted cocoa bean have no significant chocolate flavor themselves, they react to form highly flavored compounds. These flavor precursors include various chemical compounds such as proteins, amino acids, reducing sugars, tannins, organic acids, and many unidentified compounds.

The cocoa bean's natural moisture combined with the heat of roasting cause many chemical reactions other than flavor changes. Some of these reactions remove unpleasant volatile acids and astringent compounds, partially break down sugars, modify tannins and other nonvolatile compounds with a reduction in bitterness, and convert proteins to amino acids, which react with sugars to form flavor compounds, particularly pyrazines (5). To date, more than 300 different compounds, many of them formed during roasting, have been identified in the chocolate flavor (6).

Roasting is essentially a cooking process, developed by craftsmen who were guided principally by their senses of smell and taste and also to some extent by their knowledge of how beans of differing degrees of roast behaved in the subsequent winnowing, grinding, pressing, and conching processes; the ease and efficiency with which these processes can be performed is strikingly affected by the degree of roast. Roasting conditions can be adjusted to produce different types of flavor. Low, medium, full, and high roasts can be developed by varying time, temperature, and humidity in the roaster. Low roasts produce mild flavors and light color; high roasts produce strong flavors and dark color (7).

Roasters have evolved from the coke-fired rotary-drum type to continuous-feed roasters. It is usual to roast cocoa beans with the shell still on, however, other methods of roasting include nib roasting (where the shell is first removed by a rapid or moist heating step) and liquor roasting. The newer nib and liquor roasters are designed to subject the cocoa to more uniform heat conditions in addition to minimizing the loss of cocoa butter to the shell. Roasting times vary from 30 to 60 minutes. Actual temperature of the bean in the roaster is difficult to measure but probably ranges from as low as 70°C to as high as 180°C.

Winnowing. Winnowing, often called cracking and fanning, is one of the most important operations in cocoa processing. It is a simple process that involves separating the nib, or kernel, from the inedible shell. Failure to re-

Table 4. Analyses of Cocoa Shell from Roasted Cocoa Beans

Component	Shell Percentage
Water	3.8
Fat	3.4
Ash	
Total	8.1
Water soluble	3.5
Water insoluble	4.6
Silica, etc	1.1
Alkalinity (as K_2O)	2.6
Chlorides	0.07
Iron (as Fe_2O_3)	0.03
Phosphoric acid (as P_2O_5)	0.8
Copper	0.004
Nitrogen	
Total nitrogen	2.8
Protein nitrogen	2.1
Ammonia nitrogen	0.04
Amide nitrogen	0.1
Theobromine	1.3
Caffeine	0.1
Carbohydrates	
Glucose	0.1
Sucrose	0.0
Starch (Taka-diastase method)	2.8
Pectins	8.0
Fiber	18.6
Cellulose	13.7
Pentosans	7.1
Mucilage and gums	9.0
Tannins	
Tannic acid (Lowenthal's method)	1.3
Cocoa-purple and cocoa-brown	2.0
Acids	
Acetic (free)	0.1
Citric	0.7
Oxalic	0.32
Extracts	
Cold water	20.0
Alcohol (85%)	10.0

move the shell results in lower quality cocoa and chocolate products, more wear on nib grinding machines, and lower efficiency in all subsequent operations.

Because complete separation of shells and nibs is virtually impossible, various countries have established maximum allowable limits of shell in finished products. The maximum in the United States is 1.75% of shell by weight. However, U.S. manufacturers average from 0.05 to 1%.

The analyses of cocoa shell is given in Table 4 (8). In the United States, shells are often used as mulch or fertilizer for ornamental and edible plants or for animal feed. Recent studies demonstrate that cocoa shells incorporated into the diets of ruminants can improve their appetites. This phenomenon is attributed to the theobromine content of shells.

Grinding. The final step in chocolate liquor production is the grinding of the kernel or nib of the cocoa bean. The nib is a cellular mass, containing about 50–56% cocoa fat (cocoa butter). Grinding liberates the fat locked within the cell wall while producing temperatures as high as 110°C.

Nibs are usually ground while they are still warm after roasting. The original horizontal three-tier, stone mills and vertical-disk mills have been replaced by modern horizontal-disk mills, which have much higher outputs and are capable of grinding nibs to much greater fineness. Two modern machines in particular account for a large percentage of liquor grinding. One uses a pin mill mounted over a roller refiner. The pin mill grinds the nibs to a coarse, but fluid, liquor. The liquor is delivered to a roll refiner that reduces the particle size to a fine limit. The second type is a vertical- or horizontal-ball mill. Coarsely ground nib is fed to the base of a vertical cylinder that contains small balls in separate compartments. A central spindle causes the balls to rotate at high speeds, grinding the liquor between them and against the internal wall of the cylinder (9).

Cocoa Powder

Cocoa powder (cocoa) is the food prepared by pulverizing the material remaining after part of the fat (cocoa butter) is removed from the chocolate liquor. The U.S. chocolate standards define three types of cocoa based on fat content. These are (1) breakfast, or high-fat, cocoa containing not less than 22% fat; (2) cocoa, or medium-fat cocoa, containing less than 22% fat but more than 10%; and, (3) low-fat cocoa, containing less than 10% fat.

Cocoa powder production today is an important part of the cocoa and chocolate industry because of increased consumption of chocolate-flavored products. Cocoa powder is the basic flavoring ingredient in most chocolate-flavored cookies, biscuits, syrups, cakes, and ice cream. It is also used extensively in the production of confectionary coatings for candy bars.

Manufacture. When chocolate liquor is exposed to pressures of 34–41 mPa (5,000–6,000 psig) in a hydraulic press and part of the fat (cocoa butter) is removed, cocoa cake (compressed powder) is produced. The original pot presses used in cocoa production had a series of pots mounted vertically one above the other. These have been supplanted by horizontal presses that have 4–24 pots mounted in a horizontal frame. The newer presses are capable of complete automation, and by careful selection of pressure, temperature, and time of pressing, cocoa cake of a specified fat content can be produced.

Cocoa powder is produced by grinding cocoa cake. Cocoa cake warm from the press breaks easily into large chunks, but is difficult to grind into a fine powder. Cold, dry air removes the heat generated during most grinding operations. Because the finished cocoa powder still contains fat, great care must be taken to prevent the absorption of undesirable odors and flavors.

Commercial cocoa powders are produced for various specific uses and many cocoas are alkali treated or Dutched to produce distinctive colors and flavors. The alkali process can involve the treatment of nibs, chocolate liquor, or cocoa with a wide variety of alkalizing agents (10).

Cocoa powders not treated with alkali are known as cocoa, natural cocoa, or American-processed cocoa. Natural cocoa has a pH of 5.4–5.8 depending on the type of cocoa beans used. Alkali-processed cocoa ranges in pH from 6.0 to as high as 8.5.

Cocoa Butter

Cocoa butter is the common name given to the fat obtained by subjecting chocolate liquor to hydraulic pressure. It is the main carrier and suspending medium for cocoa particles in chocolate liquor and for sugar and other ingredients in sweet and milk chocolate.

The FDA has not legally defined cocoa butter, and no standard exists for this product under the U.S. Chocolate Standards. For the purpose of enforcement, the FDA defines cocoa butter as "the edible fat obtained from sound cocoa beans either before or after roasting." Cocoa butter as defined in the *U.S. Pharmacopeia* is the fat obtained from the roasted seed of *Theobroma cacao Linne*. The Codex Committee on Cocoa and Chocolate Products defines cocoa butter as "the fat produced from one or more of the following: cocoa beans, cocoa nibs, cocoa mass (chocolate liquor), cocoa cake, expeller cake or cocoa dust (fines) by a mechanical process and/or with the aid of permissible solvents" (11). It further states that "cocoa butter shall not contain shell fat or germ fat in excess of the proportion in which they occur in the whole bean."

Codex has also defined the various types of cocoa butter in commercial trade (11). Press cocoa butter is defined as fat obtained by pressure from cocoa nib or chocolate liquor. In the United States, this is often referred to as prime pure cocoa butter. Expeller cocoa butter is defined as the fat prepared by the expeller process. In this process, cocoa butter is obtained directly from whole beans by pressing in a cage press. Expeller butter usually has a stronger flavor and darker color than prime cocoa butter and is filtered with carbon or otherwise treated prior to use. Solvent-extracted cocoa butter is cocoa butter obtained from beans, nibs, liquor, cake, or fines by solvent extraction (qv), usually with hexane. Refined cocoa butter is any of the above cocoa butters that have been treated to remove impurities or undesirable odors and flavors.

Composition and Properties. Cocoa butter is a unique fat with specific melting characteristics. It is a solid at room temperature (20°C), starts to soften around 30°C and melts completely just below body temperature. Its distinct melting characteristic makes cocoa butter the preferred fat for chocolate products.

Cocoa butter is composed mainly of glycerides of stearic, palmitic, and oleic fatty acids. The triglyceride structure of cocoa butter has been worked out and is as follows: trisaturated, 3%; monounsaturated–oleodistearin, 22%; oleopalmitostearin, 57%; oleodipalmitin, 4%; diunsaturated–stearodiolein, 6%; palmitodiolein, 7%; and triunsaturated, triolein, 1% (12,13). Although there are actually six crystalline forms of cocoa butter, four basic forms are generally recognized as gamma, alpha, beta prime, and beta. The γ form, the least stable, has a melting point of 17°C. It changes rapidly to the α form, which melts at 21–24°C. At normal room temperature the β' form changes to the β form, melting at 27–29°C, and finally, the β form is reached. It is the most stable form with a melting point of 34–35°C (14).

Table 5. Properties and Composition of Cocoa Butter[a,b]

Characteristics	Press Cocoa Butter	Expeller Cocoa Butter	Refined Cocoa Butter
Refractive index, n_D 40°C	1.456–1.458	1.453–1.459	1.453–1.462
Melting behavior			
Slip point, °C	30–40	30–34	30–34
Clear melting point, °C	31–35	31–35	31–35
Free fatty acids			
(mol % oleic acid)	0.5–1.75	0.5–1.75	0.0–1.75
Saponification value			
(mg KOH/g fat)	192–196	192–196	192–196
Iodine value (Wijs)	33.8–39.5	35.6–44.6	35.7–41.0
Unsaponifiable matter			
(petroleum ether % m/m)	Not more than 0.35%	Not more than 0.40%	Not more than 0.50%

[a] Ref. 16.
[b] Contaminants not to exceed 0.5 mg/kg of arsenic, 0.4 mg/kg of copper, 0.5 mg/kg of lead and 2.0 mg/kg of iron.

Because cocoa butter is a natural fat, derived from different varieties of cocoa beans, no single set of specifications or chemical characteristics can apply. Attempts have been made to define the physical and chemical parameters of the various types of cocoa butter (15) (Table 5).

Cocoa Butter Substitutes and Equivalents

In the last 25 years, many fats have been developed to replace part or all of the added cocoa butter in chocolate-flavored products. These fats fall into two basic categories commonly known as cocoa butter substitutes and cocoa butter equivalents. Neither can be used in the United States in standardized chocolate products, but they are used in small amounts, usually up to 5% of the total weight of the product, in some European countries.

Cocoa butter substitutes do not chemically resemble cocoa butter and are compatible with cocoa butter only within specified limits. Cocoa butter equivalents are chemically similar to cocoa butter and can replace cocoa butter in any proportion without deleterious physical effects (16–17). Cocoa butter substitutes and equivalents differ greatly with respect to their method of manufacture, source of fats, and functionality, and are produced by several physical and chemical processes (18,19).

For example, cocoa butter substitutes are produced from lauric acid fats such as coconut, palm, and palm kernel oils by fractionation and hydrogenation; from domestic fats such as soy, corn, and cotton seed oils by selective hydrogenation; or from palm kernel stearines by fractionation. Cocoa butter equivalents can be produced from palm kernel oil by fractional crystallization, from glycerol and selected fatty acids by direct chemical synthesis, or from edible beef tallow by acetone crystallization.

Cocoa butter substitutes of all types enjoy widespread use in the United States chiefly as ingredients in chocolate-flavored products. Cocoa butter equivalents are not widely used because of their higher price and limited supply. At present the most frequently used cocoa butter equivalent in the United States is that derived from palm kernel oil but a synthesized product is expected to be available in the near future.

Sweet and Milk Chocolate

Most chocolate consumed in the United States is consumed in the form of sweet chocolate or milk chocolate. Sweet chocolate is chocolate liquor to which sugar and cocoa butter have been added. Milk chocolate contains these same three ingredients and milk or milk solids.

The U.S. definitions and standards for sweet chocolate are quite specific (20). Sweet chocolate must contain at least 15% chocolate liquor by weight and must be sweetened with sucrose, or mixtures of sucrose, dextrose and corn syrup solids in specific ratios. Semisweet chocolate and bittersweet chocolate, although often referred to as sweet chocolate, must contain a minimum of 35% chocolate liquor. The three products, sweet chocolate, semisweet chocolate, and bittersweet chocolate are often called simply chocolate or dark chocolate to distinguish them from milk chocolate. Table 6 gives some typical formulations for sweet chocolates (6).

Sweet chocolate can contain milk or milk solids up to 12% maximum, nuts, coffee, honey, malt, salt, vanillin and other spices and flavors as well as emulsifiers. Many

Table 6. Typical Formulations for Sweet (Dark) Chocolates

Ingredients	Percent in Formulation		
	No. 1	No. 2	No. 3
Chocolate liquor	15.0	35.0	70.0
Sugar	60.0	50.4	29.9
Added cocoa butter	23.8	14.2	—
Lecithin	0.3	0.3	—
Vanillin	0.9	0.1	0.1
Total fat	*32.0*	*33.0*	*37.1*

Table 7. Typical Formulations for Milk Chocolate

Ingredients	Percent in Formulation		
	No. 1	No. 2	No. 3
Chocolate liquor	11.0	12.0	12.0
Dry whole milk	13.0	15.0	20.0
Sugar	54.6	51.0	45.0
Added cocoa butter	21.0	21.6	22.6
Lecithin	0.3	0.3	0.3
Vanillin	0.1	0.1	0.1

different kinds of chocolate can be produced by careful selection of bean blends, controlled roasting temperatures, and varying amounts of ingredients and flavors (21).

The most popular chocolate in the United States is milk chocolate. The U.S. Chocolate Standards state that milk chocolate shall contain no less than 3.66 wt % of milk fat and not less than 12 wt % of milk solids. In addition, the ratio of nonfat milk solids to milk fat must not exceed 2.43 : 1 and the chocolate liquor content must be not less than 10% by weight. Some typical formulations of milk chocolate and some compositional values are shown in Table 7 (6).

Production. The only principal difference in the production of sweet and milk chocolate is that in the production of milk chocolate, water must be removed from the milk. Many milk chocolate producers in the United States use dry milk powder. Others condense fresh whole milk with sugar and either dry it, producing milk crumb, or blend it with chocolate liquor and then dry it, producing milk chocolate crumb. These crumbs are mixed with additional chocolate liquor, sugar, and cocoa butter later in the process (22).

Mixing. The first step in chocolate processing is the weighing and mixing of ingredients. Today, this is for the most part a fully automated process carried out in a batch or continuous-processing system. In batch processing, all the ingredients for one batch are automatically weighed into a mixer and mixed for a specific period of time. The mixture is conveyed to storage hoppers directly above the refiners. In the continuous method, ingredients are metered into a continuous kneader, which produces a constant supply to the refiners (23). The continuous process requires very accurate metering and rigid quality-control procedures for all raw materials.

Refining. The next stage in chocolate processing is refining. This is essentially a fine grinding operation in which the coarse paste from the mixer is passed between steel rollers and converted to a drier powdery mass. This process breaks up crystalline sugar, fibrous cocoa matter, and milk solids.

Tremendous advances have been made in the design and efficiency of roll refiners. The methods currently used for casting the rolls have resulted in machines capable of high output and consistent performance. The efficiency of the newer refiners has also been improved by hydraulic control of the pressure between the rolls and thermostatic control of cooling water to the rolls.

Modern five-roll refiners with 2-m-wide rollers can process 2200 kg of paste per hour.

Particle size is extremely important to the overall quality of sweet and milk chocolate. Hence the refining process, which controls particle size, is critical. Fine chocolates usually have no particles larger than 25 or 30 microns. This is normally accomplished by passing the paste through refiners more than once. However, smooth chocolates can be produced with only a single pass through the refiners if the ingredients are ground prior to mixing.

Conching. After refining, chocolate is subjected to conching, a step critical to the flavor and texture development of high-quality chocolates. Conching has long been, and still is, one of the less-satisfactorily explained parts of the chocolate-making process. This is at least in part because it can embrace a wide range of phenomena, including the relatively simple process of reliquification of a newly refined chocolate paste and the complex and often-controversial processes of flavor development, gloss development, and agglomerate reduction, modification of the melting quality, and so on. Conching is a mixing–kneading process allowing moisture and volatile components to escape while smoothing the chocolate paste. It is well known that the name *conche* derives from the seashell shape of the first really effective conching machine, which consisted of a tank with curved ends and a granite bed on which the chocolate paste from the refiners was slowly pushed back and forth by a granite roller. The longitudinal conche, development of which is commonly attributed to Rudolph Lindt of Switzerland in 1879, is still used and many experts consider it best for developing subtle flavors.

Conching temperatures for sweet chocolate range from 55 to 85°C and from 45 to 55°C for milk chocolate. Higher temperatures are sometimes used for milk chocolate if caramel or butterscotch flavors are desired (24).

Several other kinds of conches are also used today. The popular rotary conche can handle chocolate paste in a dry stage direct from the refiners (25). The recently developed continuous conche actually liquifies and conches in several stages and can produce up to 3,600 kg of chocolate per hour in a floor area of only 34 m².

The time of conching varies from a few hours to many days. Many chocolates receive no conching. Nonconched chocolate is usually reserved for inexpensive candies, cookies, and ice cream. In most operations, high-quality chocolate receives extensive conching for as long as 120 hours.

Flavors, emulsifiers, or cocoa butter are often added

during conching. The flavoring materials most commonly added in the United States are vanillin, a vanillalike artificial flavor, and natural vanilla (26). Cocoa butter is added to adjust viscosity for subsequent processing.

Several chemical changes occur during conching including a rise in pH and a decline in moisture as volatile acids (acetic) and water are driven off. These chemical changes have a mellowing effect on the chocolate (27).

Standardizing. In standardizing or finishing, emulsifiers and cocoa butter are added to the chocolate to adjust viscosity to final specifications. Lecithin (qv) is by far the most common emulsifier in the chocolate industry (6). It is a natural product, a phospholipid, possessing both hydrophilic and hydrophobic properties. The hydrophilic groups of the lecithin molecules attach themselves to the water, sugar, and cocoa solids present in chocolate. The hydrophobic groups attach themselves to cocoa butter. This reduces both the surface tension between cocoa butter and the other materials present and the viscosity. Less cocoa butter is then needed to adjust the final viscosity of the chocolate.

The amount of lecithin required falls within a narrow range of about 0.2–0.6% (28). It can have a substantial effect on the amount of cocoa butter used, reducing the final fat content of chocolate by as much as 5%. Because cocoa butter is usually the most costly ingredient in the formulation, the savings to a large manufacturer can be substantial.

Lecithin is usually introduced in the standardizing stage, but can be used earlier in the process. Some lecithin is often added during mixing or in the later stages of conching. The addition at this point has the added advantage of reducing the energy necessary to pump the product to subsequent operations because the product viscosity is reduced.

Viscosity control of chocolate is quite complicated because chocolate does not behave as a true liquid owing to the presence of cocoa particles. This non-Newtonian behavior has been best described using the Casson flow rela-

tion (29,30). When the square root of rate of shear is plotted against the square root of shear stress for chocolate, a straight line is produced (29). With this Casson relationship method two values are obtained, Casson viscosity and Casson yield value, which describe the flow of chocolate. The chocolate industry was slow in adopting the Casson relationship but this method now prevails over the simpler MacMichael viscometer and instruments such as Brookfield and Haake are now replacing the MacMichael.

Tempering. Tempering follows conching and standardizing in the processing of chocolate. The state or physical structure of the fat base in which sugar, cocoa, and milk solids are suspended is critical to the overall quality and stability of chocolate. Production of a stable fat base is somewhat complicated because the cocoa butter in chocolate exists in several polymorphic forms. Tempering is the process of inducing satisfactory crystal nucleation of the liquid fat in chocolate.

The reason why tampering is required at all is that cocoa butter has a tendency to solidify in an unsatisfactory form if left to itself, usually going into an unstable form if it is cooled rapidly, but going into an equally unacceptable superstable form if cooled too slowly. This commonly happens, for instance, when a chocolate turns gray or white after being left in the sun, the coarse, white fat crystals that can form in the slowly cooled center of a very thick piece of chocolate are similarly in a superstable form (this is known in the industry as fat bloom).

The reason why nucleation tempering of the still-molten fat is necessary in the case of chocolate is that the cocoa butter can solidify in a number of different physical forms.

Control of the polymorphic forms in cocoa butter is further complicated by the presence of other fats such as milk fat. "The fat in a chocolate can be likened to the mortar between the bricks in a mason's wall. The solid particles in a well conched chocolate bed down better than the solids in a coarsely refined and poorly mixed one" (31).

A stable crystalline form for chocolate depends primar-

Table 8. Variations in Theobromine and Caffeine Content of Various Chocolate Liquors

Country of Origin	Theobromine, %	Caffeine, %	Total, %	Theobromine to Caffeine Ratio
New Guinea	0.818	0.329	1.147	2.49:1
New Guinea	0.926	0.330	1.256	2.8:1
Malaysia	1.05	0.252	1.302	4.17:1
Malaysia	1.01	0.228	1.238	4.45:1
Brazil (Bahia)	1.21	0.183	1.393	6.61:1
Nigeria (Main Lagos)	1.73	0.159	1.889	14.9:1
Nigeria (Light Lagos)	1.23	0.137	1.367	8.99:1
Dominican Republic (Sanchez)	1.57	0.177	1.757	8.93:1
Dominican Republic (Sanchez small)	1.25	0.261	1.511	4.77:1
Africa (Fernando Po)	1.47	0.064	1.534	23.2:1
Mexico (Tabascan)	1.41	0.113	1.523	12.4:1
Trinidad	1.24	0.233	1.473	5.30:1
Maximum	1.73	0.330	1.889	23.2:1
Minimum	0.818	0.064	1.147	2.49:1

Table 9. Theobromine and Caffeine Content of Finished Chocolate Products

Product	Theobromine, %	Caffeine, %
Baking chocolate	1.38	0.092
Chocolate flavored syrup	0.24	0.014
Cocoa, 15% fat	1.46	0.250
Dark sweet chocolate	0.41	0.078
Milk chocolate	0.19	0.018

ily on the method of cooling the fat present in the liquid chocolate. To avoid the grainy texture and poor color and appearance of improperly cooled chocolate, the chocolate must be tempered or cooled down to form cocoa butter seed crystals (32). This is usually accomplished by cooling the warm (44–50°C) liquid chocolate in a water-jacketed tank that has a slowly rotating scraper or mixer. As the chocolate cools the fat begins to solidify and form seed crystals. Cooling is continued to around 26–29°C, during this time the chocolate becomes more viscous and if not quickly processed further it will become too thick to process.

In another method of tempering, solid chocolate shav-

Table 10. Composition of Cocoa Beans and Products Made Therefrom, Whole Weight Basis in Percent

	Number of Sample	Total Solids	Total Protein[a]	Cocoa Protein[b]	Milk Protein[c]	Fat	Ash	Total Carbohydrates[d]
			Whole Cocoa Beans					
Ghana	1	92.9	10.1	10.1	—	47.8	2.7	30.3
	2	94.0	9.8	9.8	—	51.6	2.6	28.0
	3	94.5	10.2	10.2	—	46.4	2.9	33.0
	4	94.7	10.3	10.3	—	46.3	3.1	33.0
Bahia	1	94.0	10.0	10.0	—	49.3	2.7	30.0
	2	94.1	10.2	10.2	—	48.6	2.7	30.6
	3	95.1	10.2	10.2	—	48.2	2.7	32.0
	4	94.9	10.2	10.2	—	48.4	2.7	31.6
			Chocolate Liquor					
Natural	1	98.4	9.4	9.4	—	56.2	2.4	28.5
	2	98.5	9.5	9.5	—	54.1	2.6	30.5
	3	98.9	10.1	10.1	—	57.0	2.4	27.4
	4	98.5	10.2	10.2	—	55.1	2.6	28.6
Dutch	1	98.6	9.2	9.2	—	55.4	3.8	28.5
	4	99.2	9.4	9.4	—	56.0	3.8	28.1
			Cocoa					
Natural	1	96.3	18.4	18.4	—	12.8	4.6	56.9
	2	96.2	18.4	18.4	—	16.4	4.8	52.9
	3	97.4	19.8	19.8	—	12.7	4.5	56.5
Dutch	1	97.1	17.5	17.5	—	12.0	8.3	55.9
	2	97.4	18.3	18.3	—	14.3	7.4	53.7
			Sweet Chocolate					
	1	99.6	3.4	3.4	—	35.1	1.0	59.4
	2	99.3	3.8	3.8	—	36.5	1.0	57.3
	4	99.5	3.6	3.6	—	35.0	1.0	59.2
			Milk Chocolate					
12% whole milk solids	1	99.2	4.2	1.0	3.2	34.7	0.9	59.2
	2	99.5	4.3	1.1	3.1	30.2	1.0	63.8
	3	99.6	4.5	1.1	3.4	32.3	0.9	61.6
	4	99.5	4.0	1.4	2.6	29.6	1.0	64.6
20% whole milk solids	1	98.8	6.6	1.3	5.2	34.4	1.5	56.1
	2	99.5	6.5	1.2	5.2	33.1	1.4	58.3
	3	99.4	6.8	1.4	5.4	30.5	1.5	60.4

[a] Total protein = milk protein + cocoa protein.
[b] Cocoa protein = (total nitrogen − milk nitrogen) × 4.7.
[c] Milk protein = milk nitrogen × 6.38.
[d] Total carbohydrate by difference using cocoa N × 5.63.

Table 11. Amino Acid Content of Cocoa and Chocolate Products

	Whole Beans[a]	Chocolate Liquor[b] Natural	Chocolate Liquor[b] Dutch	Cocoa[c] Natural	Cocoa[c] Dutch	Sweet Chocolate[d]	Milk Chocolate[a] 12% MS	Milk Chocolate[a] 20% MS
				mg/g				
Tryptophan	1.2	1.3	—	—	—	0.6	—	—
Threonine	3.5	3.9	3.6	7.7	8.0	1.5	1.8	2.8
Isoleucine	3.3	3.8	4.0	7.0	7.4	1.4	2.2	3.5
Leucine	5.3	6.0	6.3	11.5	11.3	2.3	3.8	6.1
Lysine	4.8	5.1	5.1	8.7	8.3	1.9	2.4	3.9
Methionine	0.7	1.1	0.9	2.0	1.7	0.4	0.9	1.4
Cystine	1.4	1.1	1.0	2.1	2.1	0.4	0.3	0.4
Phenylalanine	4.1	4.9	5.3	9.9	9.7	1.7	2.2	3.6
Tyrosine	2.6	3.5	3.6	7.8	8.0	1.2	1.9	3.0
Valine	5.1	5.8	6.3	11.1	10.9	2.1	2.7	4.3
Arginine	5.0	5.3	5.1	11.3	11.3	1.9	1.2	1.9
Histidine	1.6	1.7	1.7	3.4	3.0	0.6	0.7	1.0
Alanine	3.8	4.3	4.1	8.7	8.4	1.5	1.4	2.3
Aspartic acid	9.2	10.0	9.8	19.1	18.3	3.9	3.5	5.5
Glutamic acid	12.8	14.1	14.1	28.0	26.2	5.7	8.5	13.7
Glycine	4.0	4.4	4.5	8.3	8.5	1.6	1.0	1.6
Proline	3.4	3.7	3.9	7.6	7.5	1.4	3.4	5.5
Serine	3.7	4.1	4.0	6.8	8.2	1.5	1.9	3.0
Total AA Recovered[f]	*75.5*	*84.1*	*83.3*	*162.0*	*158.8*	*31.6*	*39.8*	*63.4*

[a] Whole beans = 48% fat, 5% moisture, 10% shell.

[b] Chocolate liquor = 55% fat.

[c] Cocoa = 13% fat.

[d] Sweet chocolate = 35% chocolate liquor, 35% total fat.

[e] 12% MS milk chocolate = 12% whole milk solids, 10% liquor, 32% total fat; 20% MS milk chocolate = 20% whole milk solids, 13% liquor, 33% total fat.

[f] Total AA recovered = sum of individual amino acids.

ings are added as seed crystals to liquid chocolate at 32–33°C. This is a particularly good technique for a small confectionery manufacturer that does not produce its own chocolate. However, the shavings are sometimes difficult to disperse and may cause lumps in the finished product (21).

Molding. The final stage in the processing of chocolate is molding. The three basic methods of molding are block, shell, and hollow molding; block molding predominates (21). Chocolate, either plain or mixed with nuts, raisins, or other ingredients, is deposited in molds, allowed to cool, and removed from the molds as solid pieces. Shell molding

Table 12. Fatty Acid Composition of Raw Cocoa Beans and Cocoa Butter

	Number of Sample	Fatty Acid[a] 14:0	16:0	18:0	18:1	18:2	20:0
		Cocoa Beans					
Ghana	1	0.16	28.31	34.30	34.68	2.55	—
	2	0.53	30.20	31.88	33.55	3.84	—
	3	0.19	31.72	32.57	32.82	2.70	—
	4	0.23	31.50	32.39	33.06	2.82	—
Bahia	1	0.15	29.29	31.70	35.24	3.62	—
	2	0.12	26.68	32.06	37.90	3.24	—
	3	0.25	33.99	28.80	33.62	3.34	—
	4	0.19	30.91	30.37	35.22	3.31	—
		Natural Cocoa Butter					
	1	0.15	27.08	32.64	35.61	3.63	0.89
	2	0.19	27.68	32.64	35.03	3.63	0.83
	3	0.14	28.42	32.55	34.71	3.23	0.95
	4	0.14	27.29	32.41	35.36	3.70	1.10
		Dutch Cocoa Butter					
	1	0.16	27.23	32.69	35.54	3.31	1.07
	2	0.15	26.63	34.24	34.68	3.52	0.78
	4	0.15	26.47	33.53	35.45	3.40	1.00

[a] Expressed as mole percent and calculated from peak areas of the gas chromatograms.

Table 13. Vitamin Content of Cocoa Beans and Chocolate Products,[a] Whole Weight Basis

	Number of Sample	B_1	B_2	Pantothenic Acid	Niacin	B_6
				mg/100 g		
Whole Cocoa Beans						
Ghana	1	0.21	0.16	0.24	0.19	0.22
	2	0.17	0.18	0.35	1.07	0.21
	3	0.19	0.18	0.57	0.91	0.18
	4	0.16	0.15	0.32	0.52	0.01
Bahia	1	0.14	0.18	0.34	0.46	0.61
	2	0.17	0.18	0.35	1.13	0.16
	3	0.13	0.27	0.61	1.00	0.16
	4	0.16	0.16	0.38	0.81	0.09
Chocolate Liquor						
	1	0.08	0.17	0.20	0.88	0.09
	2	0.11	0.16	0.27	1.02	0.20
	3	0.08	0.15	0.17	1.01	0.16
	4	0.05	0.11	0.15	0.29	0.02
Cocoa						
	1	0.05	0.19	0.33	1.34	0.17
	2	0.13	0.23	0.35	1.53	0.17
	3	0.15	0.22	0.32	1.37	0.24
Milk Chocolate						
	1	0.07	0.10	0.37	0.14	0.02
	2	0.11	0.24	0.37	0.38	0.02
	3	0.07	0.16	0.45	0.21	0.07
	3a	0.10	0.25	0.61	0.24	0.08
	4	0.15	0.33	0.32	1.11	0.20

[a] Vitamins A and C—negligible amounts present.

is a very complicated process. Chocolate is deposited into metal or polycarbonate molds and, by a reversal process, a layer of liquid chocolate remains clinging to the mold's inner surface. After this layer of chocolate has cooled, it is filled with a confection, such as caramel, fondant crème, or more liquid chocolate. The molds used in hollow molding are divided in two halves and connected by a hinge. Chocolate is deposited in one half. The mold is then closed and rotated so that the entire mold is coated. Easter eggs and other hollow-chocolate products are produced by this process.

THEOBROMINE AND CAFFEINE

Chocolate and cocoa products, like coffee, tea, and cola beverages, contain alkaloids (1). The predominant alkaloid in cocoa and chocolate products is theobromine, although significant amounts of caffeine may be present, depending on the origin of the beans. Published values for theobromine and caffeine content of chocolate vary widely, mainly because of natural differences in various beans and differences in methodology. This latter problem has been alleviated by the recent introduction of high-pressure liquid chromatography (hplc), which has greatly improved the accuracy of analyses. Hplc values have been

published for theobromine and caffeine in a number of chocolate liquor samples (33) (Table 8). Of the 12 varieties tested, the ratio of theobromine to caffeine varied widely from 2.5 : 1 for New Guinea liquor to 23.2 : 1 for that obtained from Fernando Po. Total alkaloid content, however, remained fairly constant, ranging from 1.5 to 1.89%. The theobromine and caffeine contents of several finished

Table 14. Tocopherols of Chocolate of Cocoa Beans and Chocolate Products

	Total Tocopherol	Alpha Tocopherol
	mg/100 g	
Bahia–Ghana beans	10.3	1.0
Liquor, natural	10.9	1.1
Liquor, Dutch	10.0	0.8
Cocoa butter, natural	19.2	1.2
Cocoa butter, Dutch	18.7	1.1
Cocoa, natural	2.3	0.2
Cocoa, Dutch	2.2	0.2
Dark chocolate	6.0	0.7
Milk chocolate, 12% milk	5.6	0.7
Milk chocolate, 20% milk	6.3	0.7

Table 15. Mineral Element Content of Cocoa and Chocolate Products (by Atomic Absorption Spectrophotometry)[a]

Product	Ca	Fe	Mg	P[b]	K	Na	Zn	Cu	Mn
					mg/100 g				
Raw Accra nibs	59.56	2.50	232.16	385.33	626.70	11.98	3.543	1.930	1.600
Raw Bahia nibs	52.73	2.45	229.11	383.33	622.55	13.55	3.423	1.940	2.060
Natural cocoa	115.93	11.34	488.51	7716.66	1448.56	20.12	6.306	3.620	3.770
Dutch cocoa	111.41	15.52	475.98	7276.00	2508.58	81.14	6.370	3.610	3.750
Chocolate liquor	59.39	5.61	265.23	3996.66	679.61	18.89	3.530	2.050	1.850
12% milk chocolate	106.41	1.23	45.56	159.00	156.64	80.09	0.773	1.020	0.282
20% milk chocolate	174.00	1.40	52.26	207.96	346.33	115.40	1.240	0.126	0.139
Dark chocolate	26.33	2.34	93.70	142.90	302.53	18.63	1.500	0.432	0.345

[a] Data from duplicate analyses of each of three samples. Mean values.
[b] Total phosphorus—ash below 550°C (AOAC procedure).

chocolate products as determined by hplc are presented in Table 9.

NUTRITIONAL PROPERTIES OF CHOCOLATE PRODUCTS

Chocolate and cocoa products supply proteins, fats, carbohydrates, vitamins, and minerals. The Chocolate Manufacturers' Association of the United States, in McLean, Va., during the period 1973 to 1976 completed a nutritional analysis of a wide variety of chocolate and cocoa products representative of those generally consumed in the United States. The analyses were conducted in Philip Keeney's laboratory at the Pennsylvania State University and complete nutritional data for the various products analyzed are given in Tables 10–15. Where possible, data on more than one sample of a given variety or type of product are presented.

BIBLIOGRAPHY

"Chocolate and Cocoa," in M. Grayson, ed., *Kirk-Othmer Encyclopedia of Chemical Technology*, Vol. 5, 2nd ed., John Wiley & Sons, Inc., New York, 1979, pp. 363–402.

1. W. T. Clarke, *The Literature of Cacao*, American Chemical Society, Washington, D.C., 1954.
2. B. Siebers, *Manufacturing Confectioner* 57(8), 52 (1977).
3. Cocoa, Chocolate and Confectionery Alliance in association with the Office International du Cacao et du Chocolat, *Report of the Cocoa Conference*, Cocoa, Chocolate and Confectionery Alliance, London, 1968.
4. C. E. Taneri, *Manufacturing Confectioner* 52(6), 45 (1972).
5. G. A. Reineccius, P. G. Keeney, and W. Weissberger, *Journal of Agricultural Food Chemistry* 20(2), 202 (1972).
6. L. R. Cook, *Chocolate Production and Use*, Magazines for Industry, New York, 1972.
7. H. R. Riedl, *Confectionery Production* 40(5), 193 (1974).
8. A. W. Knapp and A. Churchman, *Journal of the Society of Chemical Industry, London* 56, 29 (1937).
9. A. Szegvaridi, *Manufacturing Confectioner* 50, 34 (1970).
10. H. J. Schemkel, *Manufacturing Confectioner* 53(8), 26 (1973).
11. Food and Agriculture Organization of the United Nations and the World Health Organization, *Report of Codex Committee on Cocoa Products and Chocolate*, 10th sess. Codex Alimentarious Commission, Geneva, 1974.
12. T. P. Hilditch and W. J. Stainsby, *Journal of the Society of Chemical Industry, London* 55, 95T (1936).
13. M. L. Meara, *Journal of the Chemical Society,* Part 3, 2154 (1949).
14. S. J. Vaeck, *Manufacturing Confectioner* 40(6), 35 (1960).
15. Codex Standards on Cocoa Products and Chocolate, Food and Agriculture Organization of the United Nations, World Health Organization, Rome, Italy, Vol. VII, 1st ed., 1981.
16. J. Robert Ryberg, *Cereal Science Today* 15(1), 16 (1970).
17. K. Wolf, *Manufacturing Confectioner* 57(4), 53 (1977).
18. P. Kalustian, *Candy Snack Industry* 141(3) (1976).
19. B. O. M. Tonnesmann, *Manufacturing Confectioner* 57(5), 38 (1977).
20. "Cacao Products," *Code of Federal Regulations*, no. 21, part 14, Office of the Federal Register, National Archives and Records Administration. Apr. 1, 1974.
21. B. W. Minifie, *Chocolate, Cocoa and Confectionery: Science and Technology*, AVI Publishing Co., Westport, Conn., 1970.
22. B. Christiansen, *Manufacturing Confectioner* 56(5), 69 (1976).
23. H. R. Riedl, *Confectionery Production* 42(41), 165 (1976).
24. L. R. Cook, *Manufacturing Confectioner* 56(5), 75 (1976).
25. E. M. Chatt, "Cocoa," in Z. I. Kertesz, ed., *Economic Crops*, Vol. 3, Interscience Publishers, New York, 1953, p. 185.
26. H. C. J. Wijnougst, *The Enormous Development in Cocoa and Chocolate Marketing Since 1955*, H. C. J. Wijnougst, Mannheim, FRG, 1957, p. 161.
27. J. Kleinert, *Manufacturing Confectioner* 44(4), 37 (1964).
28. R. Heiss, *Twenty Years of Confectionery and Chocolate Progress*, AVI Publishing Co., Westport, Conn., 1970, p. 89.
29. E. H. Steiner, *International Chocolate Review* 13, 290 (1958).
30. N. Casson, *British Society of Rheology Bulletin* ns 52, (Sept. 1957).
31. M. G. Reade, unpublished data, UK.
32. W. N. Duck, *Journal of Agricultural Food Chemistry* 20(2), 22 (1972).
33. W. R. Kreiser and R. A. Martin, *Journal of the Association of Official Analytical Chemists* 61(6) (1978).

B. L. ZOUMAS
J. F. SMULLEN
Hershey Foods Corporation
Hershey, Pennsylvania

CHOLECALCIFEROL. See entries under VITAMINS.

CHOLESTEROL AND HEART DISEASE

At this stage it is not an exaggeration to state that practically every adult American is familiar with the terms cholesterol and heart disease. This article does not go into all the arguments surrounding this important subject. Instead, the discussion centers on facts disseminated by public health authorities of the U.S. government. The first part of this article presents some established scientific information about dietary cholesterol and fat and their relationship to heart disease. The second part describes some basic medical approaches recommended by health authorities in dealing with certain patients with clinical disorders of the heart and circulation, with a special emphasis on cholesterol, fat, and related factors.

FACTS ABOUT BLOOD AND DIET CHOLESTEROL

What Is "Blood Cholesterol"? For That Matter, What Is Cholesterol?

Pure cholesterol is an odorless, white, waxy, powdery substance. It cannot be tasted or seen in foods. Cholesterol is found in all foods of animal origin and is part of every animal cell. The body uses cholesterol to make essential body substances such as cell walls and hormones, as well as for various other functions. Even if no cholesterol was consumed in the diet the liver would manufacture enough for the body's needs.

Cholesterol is like other fatlike substances in that it will not mix with water. Therefore, to carry cholesterol and fat lipid in the blood, the body wraps them in protein packages. This combination is called a lipoprotein. Blood cholesterol is found in all the major lipoproteins, including the low-density lipoproteins (LDLs) and the high density lipoproteins (HDLs).

How Is Blood Cholesterol Measured and How Are the Results Expressed?

To measure blood cholesterol level, a small blood sample is taken and the amount of cholesterol is determined in a laboratory. The cholesterol level is expressed as milligrams per deciliter (mg/dL). The average blood cholesterol level for middle-aged men and women in the United States is about 215 mg/dL). This means that the cholesterol found in a deciliter of liquid (which is one-tenth of a liter or approximately one-tenth of a quart) weighs 215 mg. For comparison, 28,350 mg = 1 oz.

Why Should One Care about Cholesterol?

High blood cholesterol is one of three main controllable risk factors for coronary heart disease. A risk factor is a habit, trait, or condition in a person that is associated with an increased chance (or risk) of developing a disease. The other two main controllable risk factors for coronary heart disease are high blood pressure and cigarette smoking. Any one of these risk factors increases an individual's chance of developing heart disease, and all three together may greatly increase heart disease risk, perhaps by 10 times or more. Obesity and diabetes are other risk factors. Being a male or having a family history of premature

heart disease will also add to an individual's risk of heart disease.

Genetic and animal studies have shown that elevated levels of blood cholesterol, whether caused by genetic defects or dietary excesses, lead to early development of hardening of the arteries and coronary heart disease. Scientific studies of large population groups (epidemiologic studies) have shown that people with high blood cholesterol have more chance of developing coronary heart disease than do people with lower levels of cholesterol, and that the chances of developing coronary heart disease increase in proportion to the amount the cholesterol is elevated, especially for values over 200 mg/dL. In the United States, people with a blood cholesterol of 240 mg/dL or higher have more than two times the risk of developing heart disease as do those with a level of under 200 mg/dL and more than half of U.S. adults have levels over 200 mg/dL. Recently, blood cholesterol levels for adults have been classified as

1. desirable (less than 200 mg/dL);
2. borderline-high (200 to 239 mg/dL); and
3. high (240 mg/dL and above).

These categories apply to all adults over age 20, regardless of age or sex, and are part of medical guidelines defined by the Adult Treatment Panel of the National Cholesterol Education Program (NCEP) in October 1987. In adults, a total blood cholesterol level above 240 mg/dL warrants medical attention to help bring it down. However, levels above 200 mg/dL also increase the risk of heart disease and may require further evaluation, depending on whether other heart disease risk factors are present. When persons are evaluated for borderline-high blood cholesterol levels, other factors that increase their risk status for coronary heart disease are low HDL cholesterol levels (below 35 mg/dL); advanced hardening of the arteries in the head, legs, feet, hands, or arms; angina or other evidence of blockages in the arteries serving the heart; or a previous heart attack. These factors are considered in addition to the main heart disease risk factors mentioned earlier.

A physician can assess a person's risk for heart disease, offer advice on how to make dietary changes that are generally sufficient to lower blood cholesterol to an acceptable level, and monitor progress toward cholesterol reduction. Persons with very high blood cholesterol levels might also be prescribed a cholesterol-lowering drug.

What Is Coronary Heart Disease and How Important Is It to the Average American?

Almost 30% of the nearly 2 million deaths in this country each year are the result of coronary heart disease. Most coronary heart disease is due to blockages in the arteries that supply blood to the heart muscle. Fat and cholesterol, circulating in the blood, are deposited in the inner walls of the arteries. Over the years, scar tissue and other debris build up as more fat and cholesterol are deposited. The arteries become narrower and narrower, much as old water pipes build up scaly mineral deposits. This process is known as atherosclerosis. When one or more of the arter-

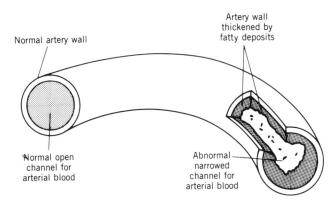

Normal artery wall

Artery wall thickened by fatty deposits

Normal open channel for arterial blood

Abnormal narrowed channel for arterial blood

Figure 1. Atherosclerosis

ies is seriously narrowed, and generally when an obstructing blood clot forms at a site of narrowing, the result is a heart attack (see Fig. 1).

What Factors Influence Blood Cholesterol Level?

Diets high in saturated fat and cholesterol play a major role in the high levels of blood cholesterol found in millions of Americans. Saturated fat is the key dietary factor raising blood cholesterol levels. In contrast, dietary cholesterol has a smaller effect on blood cholesterol levels. Obesity, primarily due to an intake of calories that exceeds the needs of the body, can also lead to high cholesterol values.

In some persons, inherited tendencies affect blood cholesterol levels. When adults try to lower their blood cholesterol by dietary changes, the size of the reduction will be influenced, in some people, by whether or not they are high or low responders, ie, whether their body tends to make big or small changes in blood cholesterol levels in response to dietary changes. This response rate is somewhat genetically determined. Only about 1 in every 500 adults has an inherited tendency to have very high blood cholesterol levels. However, even among these persons, dietary changes can do much to bring the high levels down. Most people can lower their blood cholesterol levels by following a diet that is lower in saturated fat and cholesterol.

Age and sex also influence blood cholesterol levels. In the United States, blood cholesterol levels in men and women start to rise at about age 20. The average blood cholesterol levels in women before menopause (45–60 yr) are lower than those of men of the same age. After menopause, however, the average cholesterol level of women usually increases to a level higher than that of men. In men, blood cholesterol levels off around age 50, and the average blood cholesterol level declines slightly after age 50. Use of oral contraceptives can increase blood cholesterol levels in some women, as can pregnancy. However, blood cholesterol should return to normal 20 weeks after childbirth.

What Are LDL and HDL?

LDL and HDL refer to two types of lipoproteins, packages of fat, cholesterol, and protein that are made by the body

to transport fat and cholesterol through the blood. LDLs are the low-density lipoproteins that contain the greatest amounts of cholesterol and may be responsible for depositing cholesterol in the artery walls. For that reason they are sometimes known as "bad" cholesterol. HDLs are the high-density lipoproteins. HDLs contain the greatest amounts of protein and small amounts of cholesterol. They are believed to take cholesterol away from cells in the artery wall and transport it back to the liver for reprocessing or removal from the body. Researchers have noted that persons with higher levels of HDL have less heart disease. Thus HDLs have become known as the "good" cholesterol. Women generally have higher levels of HDL; this may explain in part why they have fewer heart attacks than do men. In some cases, but certainly not in all cases, some women may even have a high total cholesterol due to a high HDL level.

What Can Be Done to Raise Levels of HDLs?

Higher levels of HDL usually are found in people who exercise regularly, do not smoke, and stay at desirable weight. Therefore, exercising regularly, quitting smoking, and losing weight (if overweight) are ways to raise HDL levels. Adopting these practices is also good for general health.

Should Blood Cholesterol Be Checked?

Blood cholesterol should be checked periodically as part of a physical examination. It is recommended that all persons age 20 and over have their blood cholesterol measured at least once every 5 yr. Cholesterol measurement usually requires that a blood specimen be obtained by drawing blood from the arm. The sample is then sent to a laboratory for analysis. Sometimes, to get a complete picture of the cholesterol fractions in the blood, the physician will ask that the report be made in terms of the amount of cholesterol carried on the low-density lipoproteins (LDL cholesterol) and on the high-density lipoproteins (HLD cholesterol).

In screening programs, blood cholesterol may often be measured using a drop of blood obtained by pricking a finger, an almost painless procedure. The tiny fingerstick blood sample is then analyzed by a portable machine that gives a cholesterol value within minutes. Such methods provide approximate values for blood cholesterol levels. However, because of variations due to differences in sample handling techniques, types of machines used, and training of volunteer personnel, high or borderline-high values should be rechecked by a second sample taken by a physician.

Are There Benefits from Reducing Blood Cholesterol?

The benefits of lowering blood cholesterol are real. It will slow the fatty buildup in the arteries and in some cases even reverse the process. Lowering blood cholesterol also definitely reduces risk of a heart attack and of death caused by a heart attack.

In a study of coronary bypass patients, reported in 1987, substantial reduction of blood cholesterol led to slowing of the atherosclerotic process and, in some cases, to its partial reversal. In the beginning of the study, all the men had fatty blockages in key arteries serving the heart that were severe enough to require a bypass operation. Among persons who lowered their blood cholesterol markedly, there were fewer new fatty deposits and fewer increases in size of existing deposits than in the group who did not lower their blood cholesterol as much. In addition, there was some shrinkage of fatty deposits associated with marked lowering of blood cholesterol levels during the 2 yr of the study.

Another major research study proved that people who have high blood cholesterol and who reduce it also reduce their risk of heart attack. At the beginning of this study, all persons had blood cholesterol levels in the high category but did not have any heart disease. After 6 yr, among persons who lowered their high blood cholesterol substantially, there were fewer heart attacks or deaths from heart attacks than in the group who did not lower their levels as much. In fact, among the adults in this study, for each 1% reduction in the number of heart attacks. This means that, if blood cholesterol levels are reduced by 10 to 15%, the risk of heart attack drops by 20 to 30%.

The results of these two major studies and the overwhelming body of scientific evidence support the concept that high blood cholesterol increases risk of coronary heart disease and that lowering the high blood cholesterol levels, especially LDL cholesterol, decreases the risk of coronary heart disease. This same evidence led the National Heart, Lung, and Blood Institute to launch the Na-

tional Cholesterol Education Program (NCEP) to help educate people about reducing high blood cholesterol.

What Are Saturated Fat and Unsaturated Fat and Where Are They Found in Foods?

The fat we eat contains fatty acids that are saturated or unsaturated, terms that refer to the chemical structure of the fatty acids in the fat molecules. Unsaturated fatty acids are further divided into two kinds—monounsaturated fatty acids and polyunsaturated fatty acids. Food fats contain a mixture of the three kinds of fatty acids. When a fat contains a large proportion of saturated fatty acids, it is said to be a saturated fat. Alternatively, the same fat can be called highly saturated or high in saturates. When a fat contains a large proportion of polyunsaturated fatty acids, it is often called a polyunsaturated fat, but it can also be said to be high in polyunsaturates or highly polyunsaturated. Similarly, when a fat or oil contains a large proportion of monounsaturated fatty acids, it is often called a monounsaturated fat. However, it is also said to be highly monounsaturated or high in monounsaturated fatty acids.

The major sources of saturated fatty acids in the diet are the fats in meats and dairy products. Beef fat and butterfat are rich in saturated fatty acids, as can be seen in Table 1. Butterfat is the fat found in milk, cheeses, cream, ice cream, and other products made from milk or cream. However, dairy products that are low in fat are

Table 1. Fats and Oils: Differences in Fatty Acids Are Important

	Polyunsaturated Fatty Acids, %[a]	Monounsaturated Fatty Acids, %[a]	Total Unsaturated Fatty Acids, %[b]	Saturated Fatty Acids, %[a]
Vegetable Oils and Shortening				
Safflower oil	75	12	86	9
Sunflower oil	66	20	86	10
Corn oil	59	24	83	13
Soybean oil	58	23	81	14
Cottonseed oil	52	18	70	26
Canola oil	33	55	88	7
Olive oil	8	74	82	13
Peanut oil	32	46	78	17
Soft tub margarine[c]	31	47	78	18
Stick margarine[c]	18	59	77	19
Household vegetable shortening[c]	14	51	65	31
Palm oil	9	37	46	49
Coconut oil	2	6	8	86
Palm kernel oil	2	11	13	81
Animal Fats				
Tuna fat[d]	37	26	63	27
Chicken fat	21	45	66	30
Lard	11	45	56	40
Mutton fat	8	41	49	47
Beef fat	4	42	46	50
Butter fat	4	29	33	62

[a] Values are given as percent of total fat.

[b] Total unsaturated fatty acids = polyunsaturated fatty acids + monounsaturated fatty acids. The sum of total unsaturated fatty acids + saturated fatty acids will not add to 100% because each item has a small amount of other fatty substances that are neither saturated nor unsaturated. The size of the other category will vary.

[c] Made with hydrogenated soybean oil + hydrogenated cottonseed oil.

[d] Fat from white tuna, canned in water, drained solids.

also lower in saturated fat. In contrast, the fat from poultry or fish is, in general, more unsaturated than beef fat or butterfat. A few vegetable fats are quite high in saturated fatty acids. These are coconut oil, palm kernel oil, palm oil, and the cocoa fat found in chocolate. These four are not available for consumers to purchase but are often used in commercial baked goods and other processed foods.

Other vegetable oils, although composed of all three types of fatty acids, are rich in polyunsaturated or monounsaturated fatty acids. Vegetable oils with the highest amounts of polyunsaturates are safflower oil, sunflower oil, corn oil, soybean oil, and cottonseed oil. Monounsaturated fatty acids are found in large amounts in olive oil, canola oil, and peanut oil. A few foods from plants (nuts or avocado, eg) do contain substantial total fat, although their fat is largely in an unsaturated form. Hydrogenated vegetable oils and solid vegetable shortenings are lower in polyunsaturated fatty acids and higher in both saturated and monounsaturated fatty acids than the unhydrogenated versions of the same oils.

What Is Dietary Cholesterol and Where Is It Found in Foods?

Dietary cholesterol is the cholesterol found in the foods we eat. Although it is not visible to the eye, it is found in all foods of animal origin, including meat, fish, poultry, and dairy products. Because cholesterol is not the same as fat, a food may contain substantial cholesterol but only a moderate amount of saturated fat (eg, an egg yolk). Foods of plant origin have no cholesterol. These include vegetables, fruits, grains (which are made into cereals and flours), nuts and seeds, and vegetable oils.

How Do Dietary Fats and Cholesterol Influence Blood Cholesterol Levels?

In the typical American diet, the saturated fat content is the strongest contributor to raising blood cholesterol. The cholesterol in foods also contributes, but to a much lesser extent than saturated fat. Polyunsaturated fats will lower blood cholesterol, but only about half as much as saturated fats will raise it. In other words, if eating a given amount of saturated fat will raise your blood cholesterol by 10%, the same amount of polyunsaturated fat will lower blood cholesterol, but only by about 5%. Unsaturated fats in general (including both monounsaturates and polyunsaturated), when substituted for saturated fat, will lower blood cholesterol levels.

Does the Average American Eat Too Much Fat and Cholesterol?

Fat is a major source of calories in the American diet, contributing about 35 to 40% of the total caloric intake. For comparison, in Japan where heart disease is uncommon, the typical diet contains only about 25% fat. The average American eats some 350 to 450 mg of cholesterol each day.

For adults with high blood cholesterol levels, the National Cholesterol Education Program's Adult Treatment Report recommends a reduction in daily fat intake to less than 30% of calories (with less than 10% of calories from saturated fat, no more than 10% from polyunsaturated fat, and 10 to 15% from monounsaturated fat) and a reduction in dietary cholesterol to less than 300 mg/day. Such dietary changes will help lower blood cholesterol and reduce the overall risk of heart disease.

Can Blood Cholesterol Levels Be Lowered?

Many studies have shown that blood cholesterol can be lowered by dietary changes, and, on average, a 10 to 15% reduction in blood cholesterol can be achieved. Some people will do even better. Depending on the initial level, on how much eating habits are changed, and on the body's response, this can translate, over several months, into a blood cholesterol reduction of 30 to 55 mg/dL. The higher the initial blood cholesterol level, the greater the overall reduction that can be expected. Also, people whose diets are high in saturated fat will probably see larger reductions than persons whose diets are low in saturated fat. Because saturated fat raises blood cholesterol more than anything else in the diet, the most effective way to lower blood cholesterol is to eat less saturated fat.

How much blood cholesterol is lowered depends on how much saturated fat and cholesterol are eliminated from the diet and on how consistently a low-saturated fat, low-cholesterol eating style can be maintained. The size of blood cholesterol reduction also depends, in some persons, on whether they are high or low responders, that is whether their body tends to respond to dietary changes with big or small changes in blood cholesterol.

For anyone who is overweight, reduction of weight will often lower blood cholesterol, especially LDL cholesterol. Reducing dietary fats, the most concentrated source of calories, is essential in weight reduction. All fats, whether they are saturated or unsaturated, are a rich source of calories. One gram of fat provides about 9 calories for a gram of protein or carbohydrate. For persons who are of desirable weight, blood cholesterol can be lowered by cutting down on saturated fat and cholesterol; by replacing saturated fatty acids with polyunsaturated fatty acids, monounsaturated fatty acids, and complex carbohydrates; and by monitoring the daily intake of calories so that weight remains constant.

How Is Blood Cholesterol Lowered?

Dietary changes that work together to reduce total fat, especially saturated fat, and cholesterol will work to lower blood cholesterol levels in most people. A number of approaches have been proven to help. For persons whose blood cholesterol is too high, the following dietary changes are recommended.

1. Eat less total fat
   by eating less fat and oil,
   by eating fewer high-fat foods (high-fat foods often contain large amounts of saturated fat).
2. Eat less saturated fat
   by eating fewer foods high in saturated fat,
   by replacing saturated fat with unsaturated fat when possible.

3. Eat less cholesterol by choosing fewer or eating smaller amounts of high cholesterol foods.
4. Eat more complex carbohydrates (starch and fiber). Foods high in complex carbohydrates are usually low in fat and contain no cholesterol
5. Lose weight, if overweight, by decreasing the number of calories taken in and increasing the number of calories used (exercise).

These steps are consistent with the Dietary Guidelines for Americans that have been developed by the U.S. Department of Agriculture (USDA) and the U.S. Department of Health and Human Services (USDHS). Four of the seven dietary guidelines include advice to eat a variety of foods, to maintain desirable weight, to eat foods with adequate starch and fiber, and to avoid too much fat, saturated fat, and cholesterol. Eating a variety of foods will help supply essential nutrients in the diet. The other three dietary guidelines advise avoiding too much sugar, avoiding too much sodium, and moderation in drinking alcoholic beverages.

For persons with high blood cholesterol, some key points to remember and some practical steps for following the five dietary changes that help to lower blood cholesterol levels are given next.

To eat less total fat, key points to remember are:

- Within any food category, there are high-fat and low-fat items. Examples are given in Table 2.
- Sausage and most processed luncheon meats are high in fat and saturated fat.
- Cream, sour cream, ice cream, butter, and many cheeses are high in fat and saturated fat.

Some practical ideas are:

- Choose low-fat items when selecting foods.
- Choose fish, poultry, lean cuts of meat, and eat moderate portions.
- Trim fat from meat and remove skin from poultry before cooking and eating.
- Choose low-fat dairy products, such as skim or low-fat milk, low-fat yogurt, low-fat cheeses, sherbet, or ice milk, instead of high-fat dairy products.
- Bake, roast, or broil foods, instead of frying.
- Add less fat and shortening to foods when cooking.

To eat less saturated fat, key points to remember are:

- Steps that reduce total fat can also work to reduce saturated fat.
- Most animal fats generally contain high proportions of saturated fat (as shown in Table 1) whereas the fat in chicken and fish contains higher proportions of polyunsaturated fatty acids.
- The vegetable oils from palm kernel, coconut, and palm, as well as cocoa fat, contain large proportions of saturated fat.
- Vegetable oils with the highest proportions of polyunsaturated fat are safflower, sunflower, corn, soybean, and cottonseed oils.

- Many margarines are lower in saturated fat and higher in unsaturated fats than butter.

Some practical ideas are:

- Choose low-fat dairy products, lean cuts of meat, chicken, and fish.
- Choose low-fat baked goods made with oils high in unsaturated fat and low in saturated fat.
- When using fats and oils, use only small amounts and replace those high in saturated fat with items high in polyunsaturates or high in total unsaturates.
- Read the nutrient section of food labels to choose items that are low in saturated fatty acids.

To eat less cholesterol, a key point to remember is:

- Cholesterol is found in high amounts in organ meats (liver, kidney, sweetbread, brain) and egg yolks.

A practical idea is:

- Eat only moderate portions of high cholesterol foods, or choose them less often.

To eat more complex carbohydrates, a key point to remember is:

- Vegetables, fruits, cereal grains, dried peas and beans, rice, and pasta contain complex carbohydrates, little or no saturated fat, and no cholesterol.

A practical idea is:

- Choose foods high in complex carbohydrates more often and use them in place of high fat foods.

To help lose weight, if overweight, a key point to remember is:

- Fats are high in calories. Fat and oils supply 9 cal/g as compared to protein and carbohydrates, which supply only 4 cal/g.

Some practical ideas are:

- Reduce the total amount of fat eaten each day, to help reduce caloric intake.
- Increase daily physical activity.

How Much Fat and Cholesterol Are Contained in Basic Foods and Where Can More Information about This Be Found?

Foods differ in the amount of saturated fat and cholesterol they contain. Table 2, the fat and cholesterol comparison chart, gives values for a few basic foods that are grouped into categories. Within each category, there are higher and lower fat items. The examples are meant to illustrate these differences and not to endorse or slight any one food.

The chart shows the fat and cholesterol in a specified amount of food. Values for dairy products and oils and fats

Table 2. Fat and Cholesterol Comparison Chart

Example	Item	Saturated Fatty Acids, g	Total Fat, g	Cholesterol, mg
Beef	Top round, lean only, broiled	2.2	6.2	84
100 g (3½ oz)	Ground lean, broiled medium	7.3	18.5	87
	Beef prime rib, meat, lean and fat, broiled	14.9	35.2	86
Processed Meats	Dutch loaf, pork and beef	6.4	17.8	47
100 g (3½ oz)	Sausage smoked, link, beef and pork	10.6	30.3	71
	Bologna, beef	11.7	28.4	56
	Frankfurter, beef	12.0	29.4	48
	Salami, dry or hard, pork, beef	12.2	34.4	79
Pork	Ham steak, extra lean	1.4	4.2	45
100 g (3½ oz)	Pork, center loin, lean only, braised	4.7	13.7	111
	Pork, spareribs, lean and fat, braised	11.8	30.3	121
Poultry	Chicken broilers or fryers, roasted:			
100 g (3½ oz)	Light meat without skin	1.3	4.5	85
	Light meat with skin	3.1	10.9	84
	Dark meat without skin	2.7	9.7	93
	Dark meat with skin	4.4	15.8	91
	Chicken skin	11.4	40.7	83
Fin fish	Cod, Atlantic, dry heat cooked	0.1	0.7	58
100 g (3½ oz)	Perch, mixed species, dry heat cooked	0.2	1.2	115
	Snapper, mixed species, dry heat cooked	0.4	1.7	47
	Rockfish, Pacific, mixed species, dry heat cooked	0.5	2.0	44
	Tuna, bluefin, dry heat cooked	1.6	6.3	49
	Mackerel, Atlantic, dry heat cooked	4.2	17.8	75
Mollusks	Clam, mixed species, moist heat cooked	0.2	2.0	67
100 g (3½ oz)	Mussel, blue, moist heat cooked	0.9	4.5	56
	Oyster, eastern, moist heat cooked	1.3	5.0	109
Crustaceans	Crab, blue, moist heat cooked	0.2	1.8	100
100 g (3½ oz)	Lobster, northern, moist heat cooked	0.1	0.6	72
	Shrimp, mixed species, moist heat cooked	0.3	1.1	195
Liver and organ meat	Chicken liver, cooked, simmered	1.8	5.5	631
	Beef liver, braised	1.9	4.9	389
100 g (3½ oz)	Pork brains, cooked	2.2	9.5	2,552
Eggs				
(1 yolk = 17 g)	Egg yolk, chicken, raw	1.7	5.6	272
(1 white = 33 g)	Egg white, chicken, raw	0	trace	0
(1 whole = 50 g)	Egg, whole, chicken, raw	1.7	5.6	272
Nuts and seeds	Chestnuts, European, roasted	0.4	2.2	0
100 g (3½ oz)	Almonds, dry roasted	4.9	51.6	0
	Sunflower seed kernels, dry roasted	5.2	49.8	0
	Pecans, dry roasted	5.2	64.6	0
	Walnuts, English, dried	5.6	61.9	0
	Pistachio nuts, dried	6.1	48.4	0
	Peanut kernels, dried	6.8	49.2	0
	Cashew nuts, dry roasted	9.2	46.4	0
	Brazil nuts, dried	16.2	66.2	0
Fruits	Peaches, raw	0.010	0.09	0
100 g (3½ oz)	Oranges, raw	0.015	0.12	0
	Strawberries, raw	0.020	0.37	0
	Apples, with skin, raw	0.058	0.36	0
Vegetables	Cooked, boiled, drained			
100 g (3½ oz)	Potato, without skin	0.026	0.10	0
	Carrots	0.034	0.18	0
	Spinach	0.042	0.26	0
	Broccoli	0.043	0.28	0
	Beans, green and yellow	0.064	0.28	0
	Squash, yellow, crookneck	0.064	0.31	0
	Corn	0.197	1.28	0
	Avocado, raw, without skin or seed:			
	Florida origin	1.74	8.86	0
	California origin	2.60	17.34	0
Grains and legumes	Split peas, cooked, boiled	0.054	0.39	0
	Red kidney beans, cooked, boiled	0.07	0.5	0
100 g (3½ oz)	Oatmeal, cooked	0.19	1.0	0

Table 2. (*continued*)

Example	Item	Saturated Fatty Acids, g	Total Fat, g	Cholesterol, mg
Milk and cream	Skim milk	0.3	0.4	4
1 cup (8 fl oz)	Buttermilk (0.9% fat)	1.3	2.2	9
	Low fat milk (1% fat)	1.6	2.6	10
	Whole milk (3.7% fat)	5.6	8.9	35
	Light cream	28.8	46.3	159
	Heavy whipping cream	54.8	88.1	326
Yogurt and sour	Plain yogurt, skim milk	0.3	0.4	4
cream	Plain yogurt, low fat (1.6%)	2.3	3.5	14
1 cup (8 fl oz)	Plain yogurt, whole milk	4.8	7.4	29
	Sour cream	30.0	48.2	102
Soft cheeses	Cottage cheese, low fat (1% fat)	1.5	2.3	10
1 cup (8 fl oz)	Cottage cheese, creamed	6.0	9.5	31
	Ricotta, part skim	12.1	19.5	76
	Ricotta, whole milk	18.8	29.5	116
	American processed spread	30.2	48.1	125
	Cream cheese	49.9	79.2	250
Hard cheeses	Mozzarella, part skim	22.9	36.1	132
(8 oz)	Mozzarella, whole milk	29.7	49.0	177
	Provolone	38.8	60.4	157
	Swiss	40.4	62.4	209
	Blue	42.4	65.1	170
	Brick	42.7	67.4	213
	Muenster	43.4	68.1	218
	American processed	44.7	71.1	213
	Cheddar	47.9	75.1	238
Vegetable oils and	Canola oil	14.8	218.0	0
shortening	Safflower oil	19.8	218.0	0
1 cup (8 fl oz)	Sunflower oil	22.5	218.0	0
	Corn oil	27.7	218.0	0
	Olive oil	29.2	216.0	0
	Soybean oil	31.4	218.0	0
	Margarine, regular soft tub[a]	32.2	182.6	0
	Margarine, stick or brick[a]	34.2	182.6	0
	Peanut oil	36.4	216.0	0
	Household vegetable shortening[a]	51.2	205.0	0
	Cottonseed oil	56.4	218.0	0
	Palm oil	107.4	218.0	0
	Coconut oil	188.5	218.0	0
	Palm kernel oil	177.4	218.0	0
Animal fats	Chicken fat	61.2	205.0	174
1 cup (8 fl oz)	Lard	80.4	205.0	195
	Mutton fat	96.9	205.0	209
	Beef fat	102.1	205.0	223
	Butter	114.4	183.9	496

[a] Made with hydrogenated soybean oil + hydrogenated cottonseed oil.

are all given for a one-cup volume. Values for meats, poultry, seafood, nuts, fruits, and vegetables are given for 100 g, which is equal to $3\frac{1}{2}$ oz. However, grams of fat actually eaten will depend on the portion size used for a meal as well as the type of ingredients selected.

It is important to remember that prepared dishes, which are made from a combination of basic foods, will contain amounts of fat related to the fat-containing ingredients, especially the high-fat ones. Addition of fat during frying or basting will also add to the fat content of the final meal. Prepared foods include recipes made at home, takeout or fast-food, restaurant food, and manufactured, prepackaged items.

Fat and cholesterol values for only a few items are listed in Table 2. To see how other foods rank in choles-terol and saturated fat, more extensive lists should be consulted. Commercially prepared foods may have values available from the manufacturers or listed on the labels.

MEDICAL APPROACHES

Increased blood cholesterol levels, more specifically increased levels of LDL cholesterol, are causally related to an increased risk of coronary heart disease (CHD). Coronary risk rises progressively with cholesterol level, particularly when cholesterol levels rise above 200 mg/dL. There is also substantial evidence that lowering total and LDL cholesterol levels will reduce the incidence of CHD.

Two approaches can be used to lower blood cholesterol

levels. The first is a patient-based approach that seeks to identify persons at high risk who will benefit from intensive intervention efforts. The goal here is to establish criteria that define the candidates for medical intervention and to provide guidelines on how to detect, set goals for, treat, and monitor these patients over time. The second approach, the population (public health) strategy, aims to shift the distribution of cholesterol levels in the entire population to a lower range. These two approaches are complementary and, together, represent a coordinated strategy aimed at reducing cholesterol levels and coronary risk.

Initial Classification by Total Blood Cholesterol

Serum total cholesterol should be measured in all persons 20 yr of age and over at least once every 5 yr; this measurement may be made in the nonfasting state. Levels below 200 mg/dL are classified as desirable blood cholesterol, those 200 to 239 mg/dL as borderline-high blood cholesterol, and those 240 mg/dL and above as high blood cholesterol. The cut point that defines high blood cholesterol (240 mg/dL) is a value above which risk of CHD rises steeply and corresponds approximately to the 75th percentile for the adult U.S population. The cut points recommended in this article are uniform for adult men and women of all ages (see Table 3).

Along with cholesterol testing, all adults should also be evaluated for the presence of other CHD risk factors, including hypertension, cigarette smoking, diabetes mellitus, severe obesity, and a history of CHD in the patient or of premature CHD in family members. Patients with other risk factors should be given other forms of preventive care as appropriate.

Patients with desirable blood cholesterol levels (<200 mg/dL) should be given general dietary and risk-reduction educational materials, and advised to have another serum cholesterol test within 5 yr. Patients with cholesterol levels of 200 mg/dL or greater should have the value confirmed by repeating the test; the average of the two test results is then used to guide subsequent decisions. Patients with high blood cholesterol (≥240 mg/dL) should

undergo lipoprotein analysis, as should those with borderline-high blood cholesterol (200–239 mg/dL) who are at high risk because they have definite CHD or two other CHD risk factors; male sex is considered a risk factor for the purpose of estimating risk status. Patients with confirmed borderline-high blood cholesterol levels who do not have CHD or two other risk factors do not need further evaluation and active medical therapy; they should be given the dietary information designed for the general population and re-evaluated after 1 yr.

Some experts believe that patients in the borderline-high blood cholesterol group who have one other risk factor (eg, hypertension) or are of younger age (20–39 yr) should also undergo lipoprotein analysis. Although this is not recommended here as a general approach for the borderline-high blood cholesterol group, it is clear that individualized clinical judgment and patient management is appropriate for this group.

Classification by LDL Cholesterol

Once someone is identified as requiring lipoprotein analysis, the focus of attention should shift from total cholesterol to LDL cholesterol. The ultimate objective of case finding and screening is to identify persons with elevated LDL cholesterol levels. Similarly, the specific goal of treatment is to lower LDL cholesterol levels. Hence, the level of LDL cholesterol will serve as the key index for clinical decision making about cholesterol-lowering therapy (see Table 4).

Lipoprotein analysis involves measurement of the fasting levels of total cholesterol, total triglyceride, and HDL cholesterol. From these values, LDL cholesterol is calculated as follows:

$$LDL\ cholesterol = total\ cholesterol - HDL\ cholesterol - (triglyceride/5)$$

Levels of LDL cholesterol of 160 mg/dL or greater are classified as high-risk LDL cholesterol, and those 130 to 159 mg/dL as borderline-high-risk LDL cholesterol. Patients with high-risk LDL cholesterol levels and those

---

Table 3. Initial Classification and Recommended Follow-up Based on Total Cholesterol

Classification	
<200 mg/dL	Desirable blood cholesterol
200–239 mg/dL	Borderline-high blood cholesterol
≥240 mg/dL	High blood cholesterol

Recommended Follow-up	
Total cholesterol <200 mg/dL	Repeat within 5 yr
Total cholesterol 200–239 mg/dL	
Without definite CHD or two other CHD risk factors (one of which can be male sex)	Dietary information and recheck annually
With definite CHD or two other CHD risk factors (one of which can be male sex)	Lipoprotein analysis; further action based on LDL cholesterol level
Total cholesterol ≥240 mg/dL	

Table 4. Classification and Treatment Decisions Based on LDL Cholesterol

	Classification
<130 mg/dL	Desirable LDL cholesterol
130–159 mg/dL	Borderline-high-risk LDL cholesterol
≥160 mg/dL	High-risk LDL cholesterol

Dietary Treatment		
	Initiation Level	Minimal Goal
Without CHD or two other risk factors[a]	≥160 mg/dL	<160 mg/dL[b]
With CHD or two other risk factors[a]	≥130 mg/dL	<130 mg/dL[c]

Drug Treatment		
	Initiation Level	Minimal Goal
Without CHD or two other risk factors[a]	≥190 mg/dL	<160 mg/dL
With CHD or two other risk factors[a]	≥160 mg/dL	<130 mg/dL

[a] Patients have a lower initiation level and goal if they are at high risk because they already have definite CHD, or because they have any two of the following risk factors: male sex, family history of premature CHD, cigarette smoking, hypertension, low HDL cholesterol, diabetes mellitus, definite cerebrovascular or peripheral vascular disease, or severe obesity.
[b,c] Roughly equivalent to total cholesterol <240 mg/dL (b) or <200 mg/dL (c) as goals for monitoring dietary treatment.

with borderline-high-risk LDL cholesterol who have definite CHD or two other risk factors (one of which can be male sex) should have a complete clinical evaluation and then begin cholesterol-lowering treatment. A basic principle adopted in this article is that the presence of other risk factors or definite CHD warrants initiating treatment at lower LDL cholesterol levels and the setting of lower LDL cholesterol treatment at lower LDL cholesterol levels and the setting of lower LDL cholesterol treatment goals. In this scheme, a low level of HDL cholesterol (below 35 mg/dL) is considered another risk factor (like hypertension) that will affect the assessment of overall coronary risk and in this way influence clinical decisions about treatment.

The clinical evaluation should include a complete history, physical examination, and basic laboratory tests. This workup will aim to determine whether the high LDL cholesterol level is secondary to another disease or a drug and whether or not a familial lipid disorder is present. The patient's total coronary risk and clinical status, as well as age and sex, should be considered in developing a cholesterol-lowering treatment program (see Table 4).

Dietary Treatment

Medical treatment begins with dietary therapy. The minimal goals of therapy are to lower LDL cholesterol to levels below the cut points for initiating therapy, ie, to below 160 mg/dL or to below 130 mg/dL if definite CHD or two CHD risk factors are present. Ideally, even lower levels of LDL cholesterol should be attained, if possible, to achieve a further reduction in risk.

Although the goal of therapy is to lower LDL cholesterol, most patients can be managed during dietary therapy on the basis of their total cholesterol levels. This has the advantage of avoiding the additional costs and the need for fasting blood involved in the measurement of LDL cholesterol levels. Serum total cholesterol levels of 240 and 200 mg/dL correspond roughly to LDL cholesterol levels of 160 and 130 mg/dL, respectively. Thus, the monitoring goals during dietary therapy are to lower the serum total cholesterol level to below 240 mg/dL for patients with an LDL cholesterol goal of <160 mg/dL, or to below 200 mg/dL for patients with an LDL cholesterol goal of <130 mg/dL.

The general aim of dietary therapy is to reduce elevated cholesterol levels while maintaining a nutritionally adequate eating pattern. Dietary therapy should occur in two steps, the step-one and step-two diets, that are designed to progressively reduce intakes of saturated fatty acids and cholesterol and to promote weight loss in patients who are overweight by eliminating excess total calories. The step-one diet should be prescribed and explained by the physician and his or her staff. This diet involves an intake of total fat less than 30% of calories, saturated fatty acids less than 10% of calories, and cholesterol less than 300 mg/day. The step-two diet consists of a reduction in saturated fatty acid intake to less than 7% of calories and in cholesterol to less than 200 mg/day. The step-one diet calls for the reduction of the major and obvious sources of saturated fatty acids and cholesterol in the diet; for many patients this can be achieved without a radical alteration in dietary habits. The step-two diet requires careful attention to the whole diet so as to reduce intake of saturated fatty acids and cholesterol to a minimal level compatible with an acceptable and nutritious diet. Involvement of a registered dietitian is very useful, particularly for intensive dietary therapy such as the step-two diet.

After the step-one diet is started, the serum total cholesterol level should be measured and adherence to the diet should be assessed at 4 to 6 wk and at 3 mo. If the total cholesterol monitoring goal is met, then the LDL cholesterol level should be measured to confirm that the LDL goal has been achieved. If this is the case, the patient enters a long-term monitoring program and is seen quarterly for the first year and twice yearly thereafter. At these visits total cholesterol levels should be measured and dietary and behavior modifications should be reinforced.

If the cholesterol goal has not been achieved with the step-one diet, the patient should generally be referred to a registered dietitian. With the aid of the dietitian, the patient should progress to the step-two diet or to another trial on the step-one diet (with progression to the step-two diet if the response is still not satisfactory). On the step-two diet total cholesterol levels should again be measured and adherence to the diet should be assessed after 4–6 wk and at 3 mo of therapy. If the desired goal for total cholesterol (and for LDL cholesterol) lowering has been attained, long-term monitoring can begin. If not, drug therapy should be considered. A minimum of 6 mo of intensive dietary therapy and counseling should usually be carried out before initiating drug therapy; shorter periods can be considered in patients with severe elevations of LDL cholesterol (>225 mg/dL) or with definite CHD. Drug therapy should be added to dietary therapy, not substituted for it.

Drug Treatment

Drug therapy should be considered for an adult patient who, despite dietary therapy, has an LDL cholesterol level of 190 mg/dL or higher if the patient does not have definite CHD or two CHD risk factors (one of which can be male sex). If the patient does have definite CHD or two other risk factors, then drug therapy should be considered at LDL cholesterol levels of 160 mg/dL or higher. The goals of drug therapy are the same as those of dietary therapy: to lower LDL cholesterol to below 160 mg/dL or to below 130 mg/dL if definite CHD or two CHD risk factors are present. These are minimal goals; if possible, considerably lower levels of LDL cholesterol should be attained.

Individualized clinical judgment is needed for patients who do not meet these criteria for drug therapy but who have not attained their minimal goals of dietary therapy. These patients include those without definite CHD or two risk factors whose LDL cholesterol levels remain in the range of 160 to 190 mg/dL and patients with CHD or two risk factors in the range of 130 to 160 mg/dL on adequate dietary therapy. In general, maximal efforts should be made in this group to achieve lower cholesterol levels and lower CHD risk by means of nonpharmacologic approaches. Consideration should also be given to the use of low doses of bile-acid sequestrants in these patients, especially in males. Moreover, many experts believe that patients with definite CHD should receive drug therapy if their minimal LDL cholesterol goal (<130 mg/dL) has not been reached.

The drugs of first choice are the bile-acid sequestrants (cholestyramine, colestipol) and nicotinic acid. Both cholestyramine and nicotinic acid have been shown to lower CHD risk in clinical trials, and their long-term safety has been established. However, these drugs require considerable patient education to achieve effective adherence. Nicotinic acid is the preferred drug in patients with concurrent hypertriglyceridemia (triglycerides ≤250 mg/dL), because bile-acid sequestrants tend to increase triglyceride levels.

A new class of drugs, to be considered after the bile-acid sequestrants and nicotinic acid, to the HMG CoA reductase inhibitors (eg, lovastatin). These drugs are very effective in lowering LDL cholesterol levels, but their effects on CHD incidence and their long-term safety have not yet been established.

Other available drugs include gemfibrozil, probucol, and clofibrate. Gemfibrozil and clofibrate are fibric acid derivatives; they are primarily effective for lowering elevated triglyceride levels, but are not FDA approved for routine use in lowering cholesterol.

After starting drug therapy, the LDL cholesterol level should be measured at 4 to 6 wk, and then again at 3 mo. If the response to drug therapy is adequate (ie, the LDL cholesterol goal has been achieved), then the patient should be seen every 4 mo, or more frequently when drugs requiring closer follow-up are used, in order to monitor the cholesterol response and possible side effects of therapy. For long-term monitoring, serum total cholesterol alone can be measured at most follow-up visits, with lipoprotein analysis (and LDL cholesterol estimation) carried out once a year.

If the response to initial drug therapy is not adequate, the patient should be switched to another drug, or to a combination of two drugs. The combination of a bile-acid sequestrant with either nicotinic acid or an HMG CoA reductase inhibitor has the potential of lowering LDL cholesterol levels by 40 to 50% or more. The combination of colestipol and nicotinic acid has been shown to beneficially influence coronary atherosclerotic lesions. For most patients, the judicious use of one or two drugs should be able to provide an adequate LDL cholesterol-lowering effect.

Drug therapy is likely to continue for many years or for a lifetime. Hence, the decision to add drug therapy to the regimen should be made only after vigorous efforts at dietary treatment have not proven sufficient. The patient must be well informed about the goals and side effects of medication and the need for long-term commitment. In the ideal treatment setting, the management of high-risk cholesterol levels would call on the expertise of a variety of professionals. The office nurse or physician's assistant can help greatly in promoting adherence to dietary and drug therapy. A registered dietitian can be of great value in dietary therapy. The pharmacist can help to provide counseling and promote adherence with drug therapy. Consultation with a lipid specialist is useful for patients with unusually severe, complex, or refractory lipid disorders.

This article has been adapted from the following:

1. *National Cholesterol Education Program. Report of the Expert Panel on Detection, Evaluation, and Treatment.* NIH Publication No. 89-2925. Washington, D.C., USDHHS, PHS, NIH, 1989. (Source of Tables 3 and 4.)
2. *Facts about Blood Cholesterol.* NIH Publication No. 88-2696. Washington, D.C., USDHHS, PHS, NIH, 1987. (Source of Tables 1 and 2 and Figure 1.)

Y. H. Hui
Editor-in-chief

CHROMATOGRAPHIC ENRICHMENT. See Corn and corn products.

CHROMATOGRAPHY, HIGH PERFORMANCE LIQUID

Chromatography is a technique in analytical chemistry in which a mixture of chemical substances is separated by its distribution between two or more phases. Chromatography is a separation technique and not an identification procedure as is often believed. It is a reference method where known authentic chemicals are used to compare to unknowns. Therefore without pure standard chemicals, it is difficult to determine the nature of unknown compounds with chromatography alone. For structural determination, compounds are often isolated after chromatographic separation for further characterization with ultraviolet, visible, infrared, nuclear magnetic resonance, and mass spectroscopy.

Liquid chromatography is a branch of chromatography where the mechanism of separation involves the chemistry between a solid and/or liquid stationary phase and a liquid mobile phase. Chemical compounds are separated by differences in their interaction between the stationary phase and the mobile phase. The stationary phases for liquid chromatography are usually in the form of a piece of paper, a thin layer of solid on a glass plate, or solid particulate material packed inside a column. Paper chromatography, thin layer chromatography, and column chromatography are all various forms of liquid chromatography. The advantages of column chromatography over paper and thin layer techniques are mainly speed, automation, quantitation, and preparative separations.

Gas chromatography is a technique which has been developed primarily for analysis of relatively volatile, oil-based compounds. The sample is in equilibrium between the stationary phase and a carrier gas. Based on the strength of interaction with the stationary phase, sample components elute at varied retention times.

In the last 20 to 30 years, column liquid chromatography has divided into two principal branches—open columns or low to medium pressure systems and high pressure or high performance systems. The distinction between the two branches is mainly the equipment used and the particle size of the stationary phase. As column packing material reduces in size, high pressure pumps are required to force the mobile phase through the column. In low pressure liquid chromatography, typical particle sizes are ≥40 microns. This allows the use of gravity or low-cost pumps for solvent/eluent delivery. Also the matrix and columns are bought separately and column packing is done in the laboratory itself, as opposed to prepacked columns or cartridges for high performance liquid chromatography (HPLC).

Open column chromatography lends itself to sample preparation and large scale processes due to the variety of available matrixes and their low cost. In HPLC, the particle size of the packing material is in the range of 3 to 10 μ in diameter. HPLC columns are generally smaller than open column chromatography columns because of their high power of resolution, and analytical nature. The columns are made of stainless steel with length ranges from 2 to 30 cm and internal diameters from 2 to 10 mm. The packing materials are spherical or irregular in shape and they usually have a rather narrow particle size distribution. Some of these materials are packed into stainless steel columns under extremely high pressure; several thousand psi (350 bar 35 MPa) is not uncommon. Special equipment is required for packing HPLC columns. The apparatuses are expensive and column packing techniques vary considerably depending on the characteristics of individual matrixes. As a result, most HPLC chromatographers buy columns that are already prepackaged with the desired material. At the present time, there are a few hundred types of HPLC columns available in the marketplace. Each has its own unique chemistry and application.

In order to achieve the highest resolution possible with these smaller particle packing materials, specialized HPLC equipment is required. A complete HPLC system consists of 5 components: (1) pump, (2) injector, (3) column, (4) detector, and (5) an integrator or recorder. Figure 1 shows a modern HPLC system.

An HPLC pump is best made of an inert and corrosion resistant material to accommodate all types of liquid mobile phases eg, acids, bases, salts, alcohols, ketones, chlorinated hydrocarbons, ethers, amines, etc. The parts that are in contact with the mobile phase are generally made of 316 stainless steel, titanium, ruby, sapphire, or high technology plastic polymers, eg, Teflon (Dupont), Kel-F(3M) and PEEK (ICI). HPLC pumps deliver the liquid eluents at constant flow rates with very low pulsation and often under high pressures (ca 1,000–3,000 psi). The volume of liquid delivered must be very reproducible from run to run. They are designed to operate as programmed and can be controlled from a computer.

HPLC injectors are also constructed of the inert materials mentioned previously. Thoughtful design has allowed samples to be introduced to HPLC columns under high pressure. For qualitative analysis a syringe can be used to vary the amount of sample injected for any one analysis. For quantitative analysis, fixed volume injectors are recommended as they provide more reproducible results. Sample volume is kept to a minimum, usually less than 50 microliters, to reduce mixing and diffusion. For routine analysis, automatic samplers are widely used. Hundreds of samples can be processed in the absence of an operator and a good HPLC system should give a relative standard deviation of 0.5 to 1.0 for most assays. To mini-

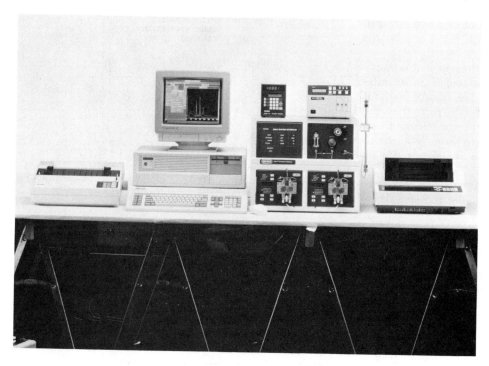

Figure 1. Modern HPLC system. Courtesy of Bio-Rad Laboratories, Richmond, Calif.

mize mixing and diffusion, all tubing, connectors, and adaptors used between the injector and the detector are precision engineered with minimal dead volumes. For example, the inner diameter of the tubing is less than 0.01 in. in diameter. Hence all samples and mobile phases must be filtered through 0.45 μ filters to eliminate particulate matter which otherwise could cause blockage and/or damage to the system's moving parts.

The HPLC column is the heart of separation. Depending on the interaction between the sample and the stationary phase, HPLC is classified into 5 major techniques: (1) size exclusion chromatography; (2) reversed phase chromatography; (3) normal phase chromatography; (4) ion exchange chromatography; and (5) affinity chromatography. Often the mechanism of separation is not well defined and in many cases several forces are working at the

same time. The choice of an HPLC column is mainly based on the experience of the analyst and published data in the literature. A well equipped HPLC methods development chemist will have 40 to 50 new columns in the laboratory at any given moment. This would allow a method to be developed quickly. An experienced HPLC chromatographer would only take a few hours to work out the parameters for resolving an average sample but difficult separations take weeks or even months to perform.

The fourth component in the HPLC system is the detector. The most widely used detector is the fixed or variable wavelength uv/visible spectrophotometer. Most molecules absorb uv or visible light. Examples are: proteins, amino acids, lipids, vitamins, antioxidants, preservatives, color additives, and non-nutritional sweeteners. As substances pass through the spectrophotometer, absorbance is con-

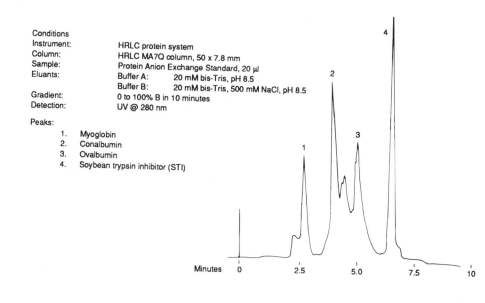

Conditions
Instrument: HRLC protein system
Column: HRLC MA7Q column, 50 x 7.8 mm
Sample: Protein Anion Exchange Standard, 20 μl
Eluants: Buffer A: 20 mM bis-Tris, pH 8.5
               Buffer B: 20 mM bis-Tris, 500 mM NaCl, pH 8.5
Gradient: 0 to 100% B in 10 minutes
Detection: UV @ 280 nm

Peaks:
    1.   Myoglobin
    2.   Conalbumin
    3.   Ovalbumin
    4.   Soybean trypsin inhibitor (STI)

Figure 2. Analysis of a protein standard. Courtesy of Bio-Rad Laboratories, Richmond, Calif. (3).

verted into electrical outputs. This voltage response is directly proportional to the absorbance, hence quantitative analysis can be easily performed.

For compounds that do not absorb uv or visible light, a refractive index (RI) detector can be used. Carbohydrates, alcohols, and many polymers are often studied by the RI detector which measures the difference in refractive index between the mobile phase filled reference cell and the effluent from the HPLC column passing through the sample cell. In general, the RI detector is approximately 10 times less sensitive than a uv/Vis monitor. The limit of RI detection is 100 nanograms of sucrose. Other widely used detectors are fluorescence, conductivity, electrochemical, and radioactivity monitors.

The final component in the HPLC system is the recorder or integrator. A detector sends its output in the form of an electrical signal to the recorder or integrator. With the chart paper moving, the pen moves up and down as substances enter and exit the detector's flow-through cell. In this way a chromatogram is generated on the chart paper. A chromatogram is a plot of detector signal output versus time. The area under each peak is directly proportional to the amount of a specific substance present in the original sample.

Integrators can calculate peak areas and/or peak heights for quantitative analysis. Computer softwares are

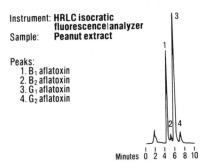

Figure 4. Aflatoxin analysis. Courtesy of Bio-Rad Laboratories, Richmond, Calif. (5).

available for data storage and further data manipulation eg, reintegration, baseline editing, integration of negative peaks, and peak summation. As computer software becomes more complex, HPLC chromatographers often spend a great proportion of their time learning how to manipulate the software for optimal data output.

In the last 20 years, HPLC has become the first method of choice by the analytical chemist, due to a number of advantages over other separation techniques.

1. *Separation.* A variety of packing material and solvents for any application, including thermally unstable and nonvolatile samples.

2. *Quantitation.* HPLC is precise and accurate.

3. *Sensitivity.* High sensitivity is achieved by specific detection methods.

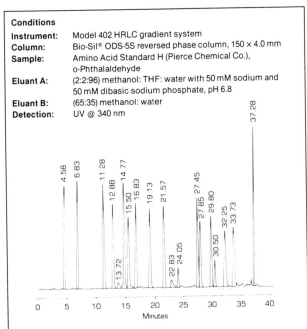

Conditions
Instrument: Model 402 HRLC gradient system
Column: Bio-Sil® ODS-5S reversed phase column, 150 × 4.0 mm
Sample: Amino Acid Standard H (Pierce Chemical Co.),
o-Phthalaldehyde
Eluant A: (2:2:96) methanol: THF: water with 50 mM sodium and
50 mM dibasic sodium phosphate, pH 6.8
Eluant B: (65:35) methanol: water
Detection: UV @ 340 nm

The procedure used for automated pre-column OPA/AA analysis begins with a 10 μl injection of OPA reagent at zero flow rate. Next, a 10 μl sample of Amino Acid Standard H is injected at zero flow rate. The amino acids are at a concentration of 2.5 μmoles/ml in 0.12 N HCl. The OPA and amino acids are then mixed in a column packed with glass beads at 0.1 ml/min for 1 minute. When derivatization is complete, the flow rate is ramped up to 1.5 ml/min. The derivatized amino acid mixture is then analyzed chromatographically, using a gradient method. Detection is performed with a UV monitor at 340 nm. Total time for this procedure is approximately 45 minutes.

Figure 3. The procedure used for automated pre-column OPA/AA analysis. Courtesy of Bio-Rad Laboratories, Richmond, Calif. (4).

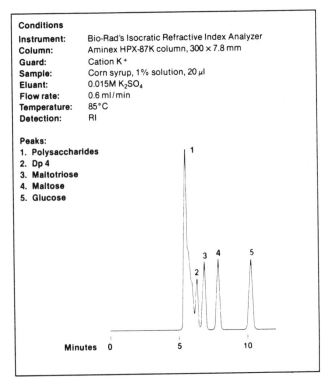

Conditions
Instrument: Bio-Rad's Isocratic Refractive Index Analyzer
Column: Aminex HPX-87K column, 300 × 7.8 mm
Guard: Cation K+
Sample: Corn syrup, 1% solution, 20 μl
Eluant: 0.015M K₂SO₄
Flow rate: 0.6 ml/min
Temperature: 85°C
Detection: RI

Peaks:
1. Polysaccharides
2. Dp 4
3. Maltotriose
4. Maltose
5. Glucose

Figure 5. Analysis of corn syrup on the Aminex HPX-87K column. Courtesy of Bio-Rad Laboratories, Richmond Calif. (6).

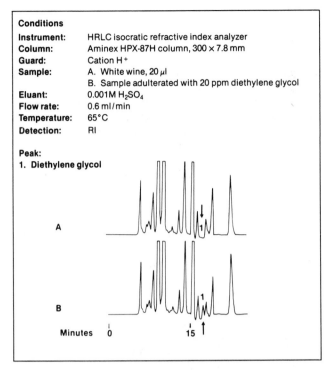

Conditions

Instrument:	HRLC isocratic refractive index analyzer
Column:	Aminex HPX-87H column, 300 × 7.8 mm
Guard:	Cation H⁺
Sample:	A. White wine, 20 µl
	B. Sample adulterated with 20 ppm diethylene glycol
Eluant:	0.001M H_2SO_4
Flow rate:	0.6 ml/min
Temperature:	65°C
Detection:	RI

Peak:
1. Diethylene glycol

Figure 6. Diethylene glycol. Courtesy of Bio-Rad Laboratories, Richmond, Calif. (7).

4. *Sample.* Technique is simple, easy to learn, and with very little sample preparation.

5. *Speed.* Analysis time is short and several components can be analyzed for a single operation.

6. *Automation.* HPLC is reliable for unattended operation.

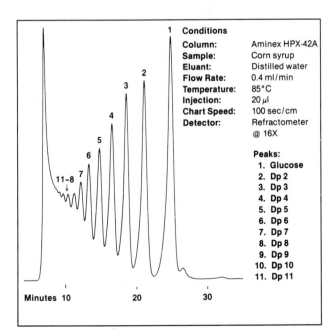

Conditions

Column:	Aminex HPX-42A
Sample:	Corn syrup
Eluant:	Distilled water
Flow Rate:	0.4 ml/min
Temperature:	85°C
Injection:	20 µl
Chart Speed:	100 sec/cm
Detector:	Refractometer @ 16X

Peaks:
1. Glucose
2. Dp 2
3. Dp 3
4. Dp 4
5. Dp 5
6. Dp 6
7. Dp 7
8. Dp 8
9. Dp 9
10. Dp 10
11. Dp 11

Figure 7. Single HPX-42A column separation. Courtesy of Bio-Rad Laboratories, Richmond, Calif. (8).

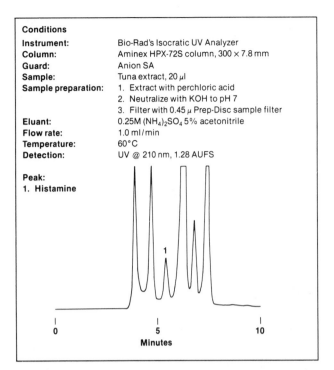

Conditions

Instrument:	Bio-Rad's Isocratic UV Analyzer
Column:	Aminex HPX-72S column, 300 × 7.8 mm
Guard:	Anion SA
Sample:	Tuna extract, 20 µl
Sample preparation:	1. Extract with perchloric acid
	2. Neutralize with KOH to pH 7
	3. Filter with 0.45 µ Prep-Disc sample filter
Eluant:	0.25M $(NH_4)_2SO_4$ 5% acetonitrile
Flow rate:	1.0 ml/min
Temperature:	60°C
Detection:	UV @ 210 nm, 1.28 AUFS

Peak:
1. Histamine

Figure 8. Histamine in tuna. Courtesy of Bio-Rad Laboratories, Richmond, Calif. (9).

7. *Cost.* Cost per analysis is lower than other techniques due to increased automation.

8. *Preparative.* Easily scaled up for production of a specific quantity of a desired substance.

9. *Nondestructive.* Detection is such that purified components are recovered for further study.

10. *Methods.* Methods are available or easily developed for most compounds.

With all the advantages mentioned above, more and more HPLC methods are appearing as the official methods of analysis in regulatory agencies. Some examples are as follows: determination of aflatoxins in cottonseed, saccharides in corn syrup, and sorbitol analysis (1,2).

Some applications pertinent to the food industry are ion exchange of proteins (Fig. 2); reversed phase of amino acids (Fig. 3); normal phase of aflatoxins (Fig. 4); and ion moderated partition of saccharides, diethylene glycols ogliosaccharides, and histamines (Figs. 5–8).

BIBLIOGRAPHY

1. *Official Methods of Analysis,* 14th ed., AOAC, Arlington, Va., 1984.

2. *USP XXI–NF XVI,* U.S. Pharmacopoeia, Rockville, Md., 1984.

3. *Bio-Rad Technical Bulletin,* No. 1410, Bio-Rad Laboratories, Richmond, Calif.

4. *Bio-Rad Technical Bulletin,* No. 1415, Bio-Rad Laboratories, Richmond, Calif.

5. *Bio-Rad Technical Bulletin,* No. 1296, Bio-Rad Laboratories, Richmond, Calif.

6. *Bio-Rad Technical Bulletin,* No. 1564, Bio-Rad Laboratories, Richmond, Calif.

7. *Bio-Rad Technical Bulletin*, No. 1278, Bio-Rad Laboratories, Richmond, Calif.

8. *Bio-Rad Technical Bulletin*, No. 1157, Bio-Rad Laboratories, Richmond, Calif.

9. *Bio-Rad Technical Bulletin*, No. 1312, Bio-Rad Laboratories, Richmond, Calif.

VICTOR CHU
JIM HEARD
Bio-Rad Laboratories
Richmond, California

CHROMOSOME. See entries under GENETIC ENGINEERING.
CIGUATOXIN, See MARINE TOXINS.
CITRIC ACID CYCLE, See CARBOHYDRATES; FOOD UTILIZATION.
CIVILIZATION. See HISTORY OF FOODS.

CITRUS INDUSTRY

ORIGIN AND HISTORY OF CITRUS

A brief synopsis of the comprehensive description of the origin and history follows: species of the genus *Citrus* are believed to have originated in the tropical and subtropical regions of Asia and the Malay Archipeligo. The citron was known to Europeans since ca 310 BC. The sweet orange (*Citrus sinensis* Osbeck) was introduced into southern Europe about 1400 AD. The new world received its first citrus in 1493 when Columbus landed on Haiti in his second voyage. Ponce de Leon was likely to have introduced the first citrus to Florida when he initially landed there in 1513. In California, it was not until the establishment in the early 1700s of the first missions that citrus was grown there (1).

Northern cities of the United States began to receive shipments of Florida oranges packed in barrels in the early 1800s, with grapefruit to follow in the 1880s. Today, both fresh and processed forms of citrus are shipped from producing areas to nonproducing areas in the world.

CULTIVARS OF COMMERCIAL IMPORTANCE

Worldwide, the sweet orange is the most important commercial group of citrus. In the United States, the Valencia (Fla., Calif., Ariz.); Hamlin, Pineapple, and Parson Brown (Fla.); and navel (Calif.) are the chief sweet orange culti-

vars. In Brazil, the largest sweet orange growing country in the world, the Pera, Hamlin, Natal, and Valencia are the chief cultivars grown. Other sweet oranges of importance in the world include the Shamouti orange (Israel) and pigmented or blood oranges (Mediterranean area of Europe, especially Spain).

There are several types of mandarins (tangerine-like citrus), perhaps the most important being the Satsuma mandarin grown primarily in Japan and Spain. Other citrus of note in this highly diversified group, with most being grown in Florida, include Temple and Page orange; Dancy, Robinson and Sunburst tangerine; Honey tangerine (formerly called Murcott); Orlando and Minneola tangelo.

The most important grapefruit of commerce and produced primarily in Florida is the Marsh seedless and its pigmented forms, Marsh Pink and Ruby Red; the white seedy grapefruit cultivar, Duncan, is of less importance today.

Lemons, grown in the United States primarily in California, are of chiefly the Eureka and Lisbon cultivars. Italy and the United States are the two major producing countries. The two major lime cultivars are the small Indian, West Indian, Mexican or Key (as known in Fla.), and the larger Persian or Tahiti.

PRODUCTION

Bearing acreage and production figures for the United States and the major producing states for oranges and grapefruit for the 1987–1988 citrus season appear in Table 1 (2). Value in U.S. dollars of all U.S. orange production for the 1987–1988 season was $1,322 million; for grapefruit, $364 million (3). An overview of world citrus production and exports of fresh citrus by country is shown in Table 2 (4). Data presented are for countries that produced greater than one million metric tons of citrus for the 1987–1988 season. Other countries that produced citrus in significant quantity but less than one million metric tons in 1987–1988 in order of decreasing volume were South Africa, Cuba, Australia, Greece, Venezuela, Lebanon, Algeria, Cyprus, Tunisia, USSR, and Uruguay. China has gained the fourth spot in total citrus production by country behind Brazil, United States, and Spain by quadrupling its production in the past decade. Current production trends in the United States and around the world are discussed by area grown by numerous authors (5).

Table 1. Bearing Acreage and Production of Citrus for the U.S. and its Principal Producing States for the 1987–1988 Citrus Season[a–c]

	Florida		California		Texas		Arizona		United States	
	Acres	Tons	Acres	Tons	Acres	Tons	Acres	Tons	Acres	Tons
Oranges	382.9	6.2	172.0	2.1	9.1	0.1	10.6	0.1	574.6	8.5
Grapefruit	106.8	2.3	20.6	0.3	14.0	0.1	6.0	0.1	147.4	2.8
P.C.F.[d]	523.1	9.0	249.7	3.1	23.1	0.2	36.5	0.3	832.4	12.6

[a] Ref. 2.
[b] Acreage is in thousands.
[c] Production in millions of short tons. One short ton = 0.9078 metric tons.
[d] Principal citrus fruits.

Table 2. World Production and Export Data for Fresh Citrus for the 1987–1988 Citrus Season[a–c]

Production Area	Total Citrus		Production			
	Production	Exports	Oranges	Grapefruit	Tangerines	Lemons/Limes
World	60.8	7.4	42.3	4.5	6.7	5.5
Northern Hemisphere[d]	42.3	6.5	26.5	4.0	6.0	4.5
United States	11.4	1.0	7.9	2.5	0.3	0.8
Mediterranean	14.2	4.7	8.9	0.6	2.7	2.0
Egypt	1.6	0.2	1.3	—	0.1	0.2
Israel	1.1	0.5	0.6	0.3	0.1	—
Italy	2.5	0.2	1.5	—	0.3	0.7
Morocco	1.3	0.6	0.9	—	0.3	—
Spain	4.3	2.4	2.4	—	1.2	0.6
Turkey	1.2	0.2	0.6	—	0.3	0.2
China	4.0	0.1	3.4	0.2	0.3	0.1
Japan	3.3	—	0.4	—	2.9	—
Mexico	2.7	—	1.7	0.1	0.2	0.7
Southern Hemisphere[d]	17.6	0.8	15.0	0.5	1.1	1.0
Argentina	1.4	0.1	0.6	0.2	0.2	0.4
Brazil	12.3	0.1	11.3	0.1	0.6	0.3

[a] Ref. 4.

[b] Data in millions of metric tons. One metric ton = 1.1016 short tons.

[c] Only countries producing greater than 1 million metric tons are included.

[d] The northern hemisphere harvest started in the autumn of 1987; the southern hemisphere harvest began and ended in 1988.

There is a detailed accounting of citrus production practices including propagation, planting, weed control, soils, fertilizing, pruning, irrigating, and frost protection (6). Another source discusses the different facets of crop protection including diseases and injuries, registration, certification and indexing, regulatory measures, vertebrate pests, biological control, and nematodes (7).

Commercial citrus is usually propagated by grafting a desired cultivar onto a suitable rootstock. Selection of rootstock to use as the basis for a budded citrus tree is of prime importance because of the horticultural characteristics influenced by the rootstock, namely: tree vigor and size, depth of rooting, cold tolerance, adaptation to certain soil conditions, disease and nematode resistance, cold hardiness and fruit yield, size, texture, quality and maturity date (8). The authors present a table listing 12 rootstocks and a ranking for each on the above-named horticultural characteristics for Florida conditions. For trees planted in Florida during 1987, the major rootstocks were sour orange, Carrizo, Cleopatra, Swingle, and Volkameriana (9).

ANATOMY

Botanically, citrus fruit is classified as a berry, developing from an ovary with axile placentation. Citrus fruits are structurally quite similar internally but may vary considerably in shape and size. Most oranges are round, grapefruit and mandarins somewhat flat, while lemons and limes are usually oblong. Limes are generally the smallest of the popular citrus fruits followed in increasing size order by lemons, mandarins, oranges, and grapefruit. Citrus cultivars, especially oranges and grapefruit, are often broken down into seedy and nonseedy (or nearly seedless) cultivars. Examples of seedy cultivars include Honey tangerine (Murcott), Temple (orange), Pineapple orange, and Duncan grapefruit; seedless (or nearly seedless) include

Tahiti (Persian) lime, Satsuma (mandarin), Hamlin, Valencia and navel oranges, and Marsh grapefruit (white and pink forms). The major parts of the citrus fruit are labelled in the photograph of the equatorial section of a Duncan grapefruit (Fig. 1). Citrus juice is located in the cells within the vesicles (often called juice vesicles or juice sacs for this reason). However, when a citrus fruit is squeezed, other components may also be expressed along

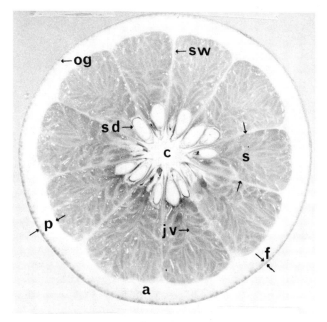

Figure 1. Equatorial section of a Duncan grapefruit showing the various common parts: (**a**) albedo; (**c**) core; (**f**) flavedo; (**jv**) juice vesicle; (**og**) oil gland; (**p**) peel; (**sd**) seed; (**s**) segment; (**sw**) segment wall. Unlabeled photograph courtesy of L. G. Albrigo and R. D. Carter, Lake Alfred, Fla.

Table 3. Proximate Composition of Raw Citrus Fruit Juices (g/100 g)[a]

Fruits	Water	Food Energy, kcal	Protein, N x 6.25	Total Lipid, fat	Carbohydrate, total	Fiber	Ash
Oranges	88.30	45	0.70	0.20	10.40	0.10	0.40
Grapefruit	90.00	39	0.50	0.10	9.20	—	0.20
Tangerines[b]	87.60	44	0.63	0.19	11.19	0.33	0.39
Lemons	90.73	25	0.38	0.00	8.63	—	0.26
Limes	90.21	27	0.44	0.10	9.01	—	0.24

[a] Ref. 10.

[b] Raw tangerines, edible portion.

with the juice including seeds or seed fragments, pulpy material including that from cell walls (primarily of a pectic nature), juice vesicles, segment walls, flavedo and albedo, and core.

PROXIMATE COMPOSITION

The proximate composition of the major commercial raw citrus juices is listed in Table 3 (10).

CHEMICAL COMPOSITION OF CITRUS JUICES

Carbohydrates

The chief ingredient (after water) in orange, grapefruit, and mandarin-type citrus juices is soluble carbohydrates or sugars. However, for mature lemons and limes, citric acid is the main soluble solid. The soluble solids fraction of the carbohydrates in orange juice, for example, is made up primarily of the reducing sugars glucose and fructose and the disaccharide sucrose in an approximate ratio of $1:1:2$ (11). Traces of other sugars have been found in citrus juice but have not been found important to the overall flavor of the juice. Quite different from most citrus fruit is the lemon which has about 90% of its sugar as reducing sugars. Total sugars: (a) vary considerably by cultivar ranging from <1% in Key limes to >10% in most orange cultivars; (b) constitute about 60–80% of the total soluble solids found (75–80% for Florida sweet oranges); (c) tend to remain more or less constant in juices of grapefruit with advancing maturity; but (d) tend to increase in juice of oranges with advancing maturity with most of the increase being due to sucrose; and (e) are significantly higher in the stylar-end half of citrus fruit than in the stem-end half.

Pectin

Pectin is the term used for a group of complex carbohydrate derivatives/complex polysaccharides known as pectic substances. Technically, pectin is composed of anhydrogalacturonic acid units that exist in a chain-like combination with each unit connected through the 1,4 glycosidic linkages forming a polygalacturonic acid (12). The high molecular weight pectic substances occur in most citrus fruits rather abundantly as a major component of primary cell wall and the middle lamella. Up to 30% of the albedo of orange (dry weight basis) may be pectin, and grapefruit peel may be ca 3.5% pectin (fresh weight basis). However, concentrations in the juices are relatively small, eg, 0.01–0.13% in orange juice, and 0.3% in grapefruit juice sacs or pulp as normally eaten on a fresh weight basis (13).

Pectin is involved in juice quality primarily because of its ability to impart the positive attributes of body or viscosity and cloud (cloudy appearance) to juices. Improper heat treatment of juices allowing less than total inactivation of pectinesterase enzyme may lead to quality problems involving pectin. In concentrate, pectin-induced gelation may occur causing nonpourable product and mixing problems; in single-strength juices, the enzyme may be involved in a series of reactions resulting in clarification or loss of cloud (settling out of pectins and other particulate material). Finally, pectin is considered as fiber, thus linking citrus juices in with certain health benefits resulting from fiber consumption.

Organic Acids

The chief organic acid found in citrus juice is citric acid with considerably lower amounts of malic and succinic acids. Acidity of citrus juice is generally reported as percent anhydrous citric acid. In addition to the above named acids, citrus peel has oxalic and malonic acids. Juices of orange cultivars generally have the least amount of acid, 0.5–1.0%; grapefruit, 1–2%; and lemons and limes, 5–7%. Actual weighted averages for percent acid for all Florida oranges and grapefruit received at processing plants for the 1987–1988 season were 0.82 and 1.14, respectively (14).

Flavor descriptions of citrus cultivars are often keyed to acidity levels, thus the designations, eg, sweet orange and sour lemon. Because of the high acid content of lemon and lime juice, consumption of these juices is in a diluted form such as lemonade or limeade, or as garnishes.

The pH values of the various citrus cultivars generally follow the acidity levels with orange juice having the highest pH at about 3.5, grapefruit 3.0, and lemons and limes 2.2 (15).

Essential Oils

The oil sacs in the peel of citrus fruits contain the bulk of citrus essential oils with much less being found in the juice sacs. Of greatest commercial significance are the oils extracted from orange, grapefruit, lemon, lime, and tangerine fruit.

Peel oil content (expressed as lb oil/ton of fruit and as percent oil in the fruit) of 13 citrus cultivars generally from 2–3 seasons of date has been reported (16). Ranges for means of percent oil in fruit for Hamlin orange was 0.37–0.45; Pineapple orange, 0.44–0.84; Valencia orange, 0.60–0.97; Marsh grapefruit, 0.29–0.34; Dancy tangerine, 0.73–0.84; Persian lime, 0.34–0.43; and lemon, 0.62–0.88. A compilation was made in 1977 of 161 volatile essential oil components identified in the oils of orange, grapefruit, tangerine, lemon, and lime, the result of 58 separate investigations (17). Orange oil alone contained 112 of these components. Since that time, numerous other constituents have been discovered in citrus oils primarily because of the use of more sophisticated analytical techniques.

D-limonene is by far the most abundant component of citrus oils, comprising, eg, over 95% by weight of orange oil with the numerous other hydrocarbons identified appearing in trace amounts (18). The oxygenated aldehydes and esters are considered important contributors to orange flavor, and to the characteristic flavors of the other citrus cultivars as well. Oxygenated alcohols are abundant in citrus oils, but few ketones are noted. Percentages of aldehydes, esters, alcohols, and acids present in terpeneless citrus oils have been reported (19). Nonvolatile constituents comprising substantial portions as coumarins and flavonoids in citrus oils range from ca 1% in orange to 7% or more in lime and grapefruit (20). Whereas total yield of oil decreases with increasing maturity, individual components may increase significantly such as the important grapefruit flavor-contributing ketone, nootkatone, that was found to increase more than 12-fold in Duncan grapefruit over a seven month Florida season (21).

Pigments

Citrus peel color is due primarily to orange-colored carotenoids and green-colored chlorophyll, but only in limes is the green color acceptable in temperate climates. The color of immature citrus peel of most cultivars is green and eventually turns to different shades of yellow and orange with carotenoid synthesis and chlorophyll degradation at advancing maturity. Cool temperatures play an important role in the color break from green to yellow/orange, thus a significant quantity of tropical citrus is green when mature. Although several carotenoids have been identified in citrus peel, the most important in oranges and tangerines is reddish-orange β-citraurin (22).

β-cryptoxanthin is the main contributor to the orange color in orange juice (23). The other main carotenoids in orange and tangerine juice are α-carotene, β-carotene, Zeta carotene, α-cryptoxanthin, lutein Zeaxanthin, and two yellow pigments occurring in the largest amounts of all the carotenoids, antheraxanthin and violaxanthin. The main pigment in Florida pink and red grapefruit juices is lycopene with somewhat less β-carotene (24).

Enzymes

While pectinesterase and its various forms are probably the most important enzymes in citrus because of potential quality problems (gelation in concentrates, and cloud loss in single-strength juices) associated with this group of en-

zymes, numerous other enzymes have been identified as occurring in citrus. There is a comprehensive listing of enzymes reported in citrus with comments on each for reactions or substrate, source, and properties. The list includes 18 oxidoreductases, 9 transferases, 7 hydrolases, 9 lyases, 3 isomerases, and 3 ligases (25).

Amino Acids

Free amino acids account for ca 70% of the nitrogen in citrus juices and constitute ca 0.1% w/w for oranges and grapefruit (26). There is a table listing amino acid composition of citrus juices from various sources (27). Proline predominates in the juices of most citrus cultivars with somewhat lesser quantities reported for asparagine, aspartic acid, arginine, and gamma amino butyric acid.

Lipids

The bulk of lipids in citrus fruit occurs in the seeds. The largest lipid component in citrus fruits is fatty acids, the five major ones comprising between 85–93% of all fatty acids being palmitic, palmitoleic, oleic, linoleic, and linolenic (28). A listing was made of fatty acids as percentage in citrus juices by type of fatty acid (saturated, monounsaturated, diunsaturated, and triunsaturated), and by cultivar (29). The phospholipid fraction of Florida and California orange juices is composed primarily of phosphatidyl choline and phosphatidyl ethanolamine with lesser amounts of phosphatidyl acid, phosphatidyl serine, and phosphatidyl inositol (30).

Minerals

Potassium is the most abundant mineral of citrus juices. Other major elements reported in California oranges and lemons are Ca, Cl, Fe, N, Na, P, and S; minor elements included Al, B, Mn, Si, Sr, with the least amounts of Ag, As, Ba, Bi, Co, Cr, Cu, Li, Mo, Ni, Pb, Sn, Ti, V, Zn, and Zr (31). Ranges of contents of 32 trace metals in orange juices from Florida, California, Brazil, and Mexico have been presented (32). Current data for Na, K, Ca, Fe, P, Mg, Zn, and Cu on more than 500 commercial processed orange and grapefruit juice samples representing the five major juice products produced in Florida and surveyed on a regular basis over two years' time have also been presented (33).

Flavonoids

Flavonoid constituents having a basic chemical structure of C_6–C_3–C_6 have been reviewed (34). Flavonoids are abundant in citrus, having remarkable taste and nutritional properties. They are useful as taxonomic markers and yield valuable by-products. Flavonoids occur as flavanones, flavones, and anthocyanins, the latter of minor importance in citrus and occurring only in blood oranges. The principal flavanones are naringin, isosakuranetin, eriodictyol, and hesperetin. Naringin (5,7,4-trihydroxyflavanone 7-β-neohesperidoside) is of importance as a highly significant bitter constituent in grapefruit juice. Other flavonoids described in detail include 10 additional flavonone glycosides, 2 flavanone aglycones, 14 flavone glycosides, 7 C-glycosylflavones, 22 flavone aglycones, and 5 anthocyanins (35).

Limonin

Of the many chemically related triterpene derivatives called limonoids, limonin is by far the most well known and important because of its role in sometimes producing excessive bitterness in grapefruit and navel orange juices, and to a lesser extent, certain other citrus juices as well. Limonin is a highly oxygenated triterpine derivative whose structural features include a furan ring, two lactone rings, a five-membered ether ring, and an epoxide (36). Limonin becomes bitter when limonate A-ring lactose, the nonbitter metabolically-active form of limonin, undergoes acid-catalyzed lactonization as a result of disruption of fruit tissues such as in juicing. Heating, as in juice pasteurization also promotes this reaction. Limonin levels in citrus juices generally decrease with increasing maturity. Seeds are the major source of limonoids in citrus. The other major limonoid in addition to limonin in citrus is deacetylnomilin.

FRESH CITRUS FRUIT

Harvesting

Fruit is harvested primarily by hand throughout the world. Fruit to be used for processing is generally pulled off with a slight twist of the wrist. Fruit that will be used in the fresh market (particularly specialty-type fruit such as tangelos and tangerines) are often clipped off the tree. Small clippers are used to cut the stem close to the fruit. Some mechanical harvesting has been experimentally employed in the United States in harvesting certain citrus for processing, but not fruit destined for the fresh market because of the often excessive damage caused to the fruit. Machines are generally either of two types: those with picking heads to detach fruit directly (such as an auger spindle); or noncontact machines which rely on fruit removal by shaking a limb (80–90% removal efficiency for Hamlin and Pineapple oranges), the trunk or foliage (37); pulsating air blasts shake fruit from trees with 85–95% efficiency when used in conjunction with abscission chemicals (38).

Research is underway at the University of Florida, Gainesville on a robotic picking device using a combination of sensor technology and computers. Harvested fruit allowed to fall to the ground may be used for processing but not for fresh packing because of excessive fruit damage. For fresh market, especially, pickers most often place fruit into a picking bag or other suitable device. When the container is full, the picker empties it carefully into a durable pallet box holding, eg, ca 900 lb (408.2 kg) of oranges. The boxes are mechanically loaded onto a flatbed truck for transportation to the packinghouse. Fruit which has been detached and dropped to the ground is collected by hand for eventual depositing into trucks and hauled in bulk to the processing plant.

In general, citrus fruit is one of the most durable fresh commodities. However, damage during harvest may cause oloecellosis or bruising, stem-end rind breakdown, blos-

som-end clearing in grapefruit, stylar-end breakdown in limes, and red-colored lesions as a result of superficial wounds (39). Weather related disorders on citrus fruit include water spot, zebra skin of tangerines, sunburn, freeze injury, wind scar, and endoxerosis or internal decline especially in lemons (40).

The term, packinghouse eliminations, is used for harvested fruit which for some reason does not meet grade standards for fresh utilization, but generally is all right for processing. Fruit going to Florida packinghouses were surveyed for grade-lowering factors and found to have 30–32% of the eliminations due to wind injury; 20–23% due to melanose, a fungal disease; 10–20% due to poor color; 12–15% due to scale insects; and 4–5% due to rust mites (41).

Maturity Standards

Most citrus producing countries have both internal and external quality standards with the bulk of these standards being involved with fruit maturity or, rather, lack thereof. Maturity standards are often complicated and depend mainly on cultivar and time of year. The grade standards in the United States are under the authority of the United States Department of Agriculture (USDA) and the different state agencies, eg, for Florida, the Florida Department of Agriculture and Consumer Services and the Florida Department of Citrus.

Perhaps the most important internal quality factor considered in standards is that the juice of each type of citrus must meet certain minimum ratios of degrees Brix (or percent total soluble solids) to percent acid for acceptable palatability. If this ratio is too low, the juice will be too tart or sour. Other factors often considered are total soluble solids (°Brix), total acid, and juice content.

External color and its role in consumer acceptance is the most important of the external grade standards. Other external factors often included in standards are texture, discoloration, blemishes, similar varietal characteristics, form, and firmness as well as maturity and uniformity of fruit size and pack (42).

Marketing

In the United States, per capita consumption of fresh citrus in 1987 was 28.16 lb (12.77 kg). Oranges constituted just over half of this amount at 14.29 lb (6.48 kg); grapefruit, 6.66 lb (3.02 kg); tangerines, 3.51 lb (1.59 kg); lemons, 2.57 lb (1.17 kg); and limes, 0.65 lb (0.29 kg) (43). Whereas California markets the bulk of its citrus for fresh consumption (67% of its oranges and 56% of its grapefruit), Florida processes ca 93% of its oranges and 57% of its grapefruit (44). The United States exported 16.3 and 25.5 million 4/5 bu (0.029 m³) of fresh oranges and grapefruit, respectively, during the 1987–1988 citrus season (45). Japan took the bulk of the exported citrus: 6.0 and 13.3 million 4/5 bu (0.029 m³) of oranges and grapefruit, respectively. Other large importers of U.S. fruit are Canada and Hong Kong (primarily oranges); France, Canada, Taiwan, and The Netherlands (primarily grapefruit). Whereas Florida is the chief supplier of fresh grapefruit for export in the United States, California is the chief supplier of oranges.

An estimation in 1986 (46), indicated 80% of fresh citrus fruit in the United States was packed in fiberboard boxes holding 32.0–41.9 lb (14.5–19.0 kg) net wt of fruit; 14% in fiberboard bagmaster containers holding 39.9–50.0 lb (18.1–22.7 kg) net wt of fruit packaged in consumer sized bags; and 6% in miscellaneous containers including 2% in wirebound boxes (mostly Florida tangerines).

Maintaining Quality in Fresh Citrus

When certain citrus fruits are mature internally but still are greenish or patchy green externally, consumers generally believe the fruit to be immature, thus it is desirable to degreen the fruit. Degreening may be accomplished through use of either ethylene, or color enhancement utilizing a dye or a combination of the two treatments. In Florida, degreening with ethylene is generally at 1–5 ppm in a room maintained for 36 h or less between 84.2–86.0°F (29–30°C), 90–96% relative humidity, and having at least one change of air per hour (47). Use of much longer holding periods such as 72 or 96 h are not uncommon at times. Degreening procedures vary with citrus growing area, cultivar, fruit condition, etc. Color enhancement of the peel may be used in certain cases. In Florida, for example, for degreened (or naturally pale) oranges, Temples and tangelos having a final yellow appearance, these fruit, under government supervision, may be flooded in a tank containing Citrus Red No. 2 for orange color enhancement, with 2 ppm residue being permitted following rinsing (48). California and many foreign countries do not allow coloring of citrus.

Preparation of fruit for the fresh market may involve a number of operations that include washing, grading, sizing, waxing, and fungicide application. In the washing operation, natural waxes are lost and rind permeability is increased allowing moisture loss, decrease in sugar content, flavor, texture, quality, and ascorbic acid (49). Waxing of fruit tends to minimize the negative side effects of washing plus enhancing the appearance of the rind. Many types and formulations of wax and waxlike materials are used around the world including carnauba and candelilla waxes, beeswax, microcrystalline waxes (complexes of hydrocarbons having branched chains), shellac, polyethylene and wood rosins (50). A table is also presented listing 13 film formers used in citrus fruit-coating formulations and the appropriate United States Food and Drug Administration (U.S. FDA) citations for approval of each (51).

Postharvest decay control techniques have been reviewed (52). Common postharvest fungal diseases of citrus are stem-end rot (*Diplodia natalensis* or *Phomopsis citri*) and green mold (*Penicillium digitatum*), with sour rot (*Geotrichum candidum*) and anthracnose (*Colletrotrichum gloeosporioides*) important primarily for mandarin-type citrus fruit. Some of the more extensively used fungicides for decay control are sodium ortho-phenylphenate (SOPP) applied as a foam or as a soak or drench, thiabendazole (TBZ), and benomyl (the latter two compounds applied in water, by either suspension in water emulsion wax or by dissolution in a hydrocarbon-soluble wax). Diphenyl as a vapor-phase fumigant is the most effective and common of the fumigants used for decay control.

For the control of several potentially harmful species of fruit flies and their eggs, larvae and pupae in fruit destined for export, fumigation with methyl bromide or cold treatment is used since the recent disallowance of ethylene dibromide (EDB) as a fumigant. Cold treatment in the United States prescribed for the disinfestation of citrus of the Caribbean fruit fly is 34.0°F (1.1°C) for 17 days (53).

Unlike many fruits such as apples that may continue to ripen and increase in quality after harvest, citrus fruits severed from the tree begin to lose the quality attained up to the time of harvest. Cold storage may increase the time of marketability of most fresh citrus by several weeks. Typical storage conditions for various citrus fruits are outlined in a table including information on temperature, relative humidity, estimated storage life, chilling injury symptoms (if held too cold), and typical storage diseases for each cultivar (54). Typical storage conditions are 50–60°F (10–15.6°C) for grapefruit; 40°F (4.4°C) for tangerines; 32–34°F (0 to 1.1°C) for Florida and Texas oranges; and 38–48°F (3.3–8.9°C) for California oranges. Estimated cold storage life varies greatly by cultivar, eg, 2 weeks for tangelos, 6–8 weeks for sweet oranges, to a maximum of 20 weeks for green lemons. Controlled atmosphere storage which works well for apples, eg, to significantly prolong storage life, is generally not feasible in the case of citrus.

There has been limited success in use of wrapping individual fruit with plastic film. Advantages of film wrapping have been primarily to extend storage life by reducing water loss.

Production and Marketing of Freshly Squeezed Citrus Juice

In citrus producing areas around the world, there has always been some marketing of freshly squeezed citrus juices, most often at roadside stand-type operations. Now, primarily in Florida, there is a viable fresh juice industry utilizing several million 90-lb (40.8 kg) boxes of oranges but much less of grapefruit. Juice from mature fruit is typically packed into quart-size (32 fl oz or 946 ml) polyethylene or other type of plastic bottles which are shelf-dated for a maximum storage time at cool temperatures of up to 17 days (55). Juice from about one dozen average-sized oranges will equal a quart. Ascorbic acid retention was ca 91–93% for freshly squeezed orange juice packed in polyethylene bottles after two weeks of storage at 29–40°F (−1.7 to 4.4°C); for grapefruit juice under similar circumstances, the percentage retention was 86–88 (55). For orange juice, there are both U.S. federal (56) and Florida (57) regulations. Certain supermarkets throughout North America are also engaged in extracting juice from fresh fruit and marketing the juice, primarily in-store. Some freshly squeezed juice is also being marketed as frozen juice, generally in liter polyethylene bottles. This product may be thawed in ca 20 min in a microwave oven with negligible loss of flavor and ascorbic acid.

PROCESSED CITRUS PRODUCTS

General Descriptions

Recommended international standards have been drawn up by the United Nations Food and Agricultural Organization/World Health Organization Codex Alimentarius Commission in 1972 for orange, grapefruit, and lemon juices (58) and concentrated orange juice (59) with certain amendments for each issued in 1976. Some of the most important provisions made for the various citrus juices are the recommended standards for soluble solids exclusive of added sugar of not less than 10% mass/mass (m/m) in orange juice, 9% m/m in grapefruit juice, and 6% m/m in lemon juice. For the concentrated orange juice, an important provision is ". . . that the soluble orange solids shall be not less than 11% m/m (exclusive of added sugars) . . ." in the product obtained by reconstituting the concentrate (59).

The citrus processing industry in the United States is highly regulated through rather complex federal and state laws. At the federal level Standards of Identity are issued by the U.S. FDA and United States Standards for Grades are put forth by the USDA. States engaged in citrus production have their own laws, eg, for Florida, "Florida Citrus Fruit Laws, Chapter 601, Florida Statutes" (60) overseen by the state legislature, and "Official Rules Affecting the Florida Citrus Industry Pursuant to Chapter 601, Florida Statutes" (57) overseen by the Florida Department of Citrus.

"United States Standards for Grades of Orange Juice" (61) include the products: canned orange juice, frozen concentrated orange juice, reduced acid frozen concentrated orange juice, concentrated orange juice for manufacturing, canned concentrated orange juice, dehydrated orange juice, pasteurized orange juice, and orange juice from concentrate. The key in these standards is a 100-point grading system for quality in which flavor, color, and absence of defects are taken into account. Flavor and color can each attain 40 points while absence of defects can attain 20 points. Minimum points for Grade A color and flavor is 36; for Grade B, 32. Minimum points for absence of defects for Grade A is 18; for Grade B, 16. The total points required for minimum U.S. Grade A is 90; for U.S. Grade B, 80. However, there is a limiting rule: the lowest score of any single factor (flavor, color or defects) determines the grade, even though the total score for the product might place it in a higher grade. Other quality factors which

may be considered are: appearance, reconstitution (for concentrated products), coagulation, and separation.

"United States Standards for Grades of Grapefruit Juice" (62) include the products: grapefruit juice, grapefruit juice from concentrate, frozen concentrated grapefruit juice, concentrated grapefruit juice for manufacturing, and dehydrated grapefruit juice. Grapefruit juices are also graded on a 100-point system, however, unlike orange juices, flavor constitutes 60 points, while color and absence of defects constitute 20 points each. Grade A requires a minimum of 54 points for flavor, 48 for Grade B; and a minimum of 18 points for Grade A color and absence of defects, 16 for Grade B for each factor. Total points required for minimum Grade A is 90; for Grade B, 80. The limiting rule also applies for grapefruit juices.

Flavor is graded by trained USDA inspectors when federal inspection is requested by a processing plant or continuously as is required in the state of Florida. Color of orange juice is determined by plant personnel and by federal inspectors objectively through use primarily of a HunterLab D45-2 citrus colorimeter (Hunter Associates Laboratory, Inc., Reston, Va). Other tristimulus colorimeters recently made available for measurement of orange juice color are the HunterLab LabScan, MacBeth Model 1500 Color-Eye (Kollmorgen Corp, Newburgh, N.Y.) and the Minolta Chroma Meter II/CR100 (Minolta Camera Co., Ltd., Osaka, Japan) (63). No machine is available as yet for measurement of color of grapefruit juice which may vary naturally from bone white, through various shades of amber, pinks and reds, even to ripe-tomato red.

The major analytical factors most often considered in United States processed citrus juice standards include minimum °Brix, minimum and maximum values for °Brix to percent acid ratio, and maximum percent by volume of recoverable oil. Data in Table 4 list °Brix and °Brix to percent acid ratio values for U.S. Standards for Grades of selected processed citrus products. For some of these products, there are certain important allowances made by law, eg, for pasteurized orange juice, if it is necessary to adjust soluble solids to achieve standardization of product quality, up to 25% soluble solids from concentrated orange juice may be added with a proper label declaration, such

Table 4. Minimum U.S. Standards for °Brix and °Brix to Percent Acid Ratio for Selected Processed Orange Juice (OJ) and Grapefruit Juice (GJ) Products[a]

	°Brix		°Brix to Percent Acid Ratio	
	Grade A	Grade B	Grade A	Grade B
Canned OJ	10.5	10.0	Depends on °Brix[b]	
Froz. conc. OJ (recon.)	11.8	11.8	11.5:1–12.5:1[c]	10.0:1
Conc. OJ for mfg[d]	11.8	11.8	8.0:1	8.0:1
Pasteurized OJ	11.0	10.5	11.5:1–12.5:1[c]	10.5:1
OJ from concentrate	11.8	11.8	11.5:1–12.5:1[c]	11.0:1
GJ	9.0	9.0	8.0:1	7.0:1
GJ from concentrate	10.0	10.0	8.0:1	7.0:1
Froz. conc. GJ (recon.)	10.6	10.6	9.0:1	7.0:1
Conc. GJ for mfg[d]	—	—	6.0:1	5.5:1

[a] Refs. 61, 62.
[b] At <11.5 °Brix, minimum ratio is 10.5:1 for Grade A, 9.5:1 for Grade B; at 11.5 °Brix or more, minimum ratio is 9.5:1 for Grade A or B.
[c] Depends on where manufactured.
[d] mfg = manufacturing.

as prepared in part from concentrated orange juice. Similarly, in the manufacture of grapefruit juice, concentrated grapefruit juice may be added to adjust soluble solids content, but such added concentrate cannot contribute more than 15% of the grapefruit juice soluble solids in the finished food. Finally, in order to improve the color of light-colored orange juice (such as Hamlin juice), up to 10% by volume of highly-colored *citrus reticulata,* or hybrids thereof (mandarin types of orange) may be legally added in the manufacture of frozen concentrated orange juice.

Florida imposes some requirements on products manufactured there beyond those required by federal standards. For example, Florida requires for frozen concentrated orange juice: a 13:1 minimum ratio of °Brix to percent acid for both U.S. Grade A and B instead of 11.5–12.5 for U.S. Grade A and 10 for Grade B depending on the fruit source outside of Florida; a maximum of 12% by volume of sinking pulp (no requirement out-of-state); no pulpwash (the product obtained from washing the pulp following juice extraction and finishing) may be used, whereas outside the state the amount of pulpwash recovered from the fruit used to make the juice (ca 3–8%) can be incorporated into the final product; continuous inspection by the USDA is required, whereas continuous inspection is not required out-of-state. There are no federal regulations for the bitter constituents, limonin and naringin in the two products, grapefruit juice and frozen concentrated grapefruit juice, however, there are Florida regulations limiting these constituents (57). Essentially, for Florida-packed canned and chilled grapefruit juice from August 1 to December 1 of each season, juice must contain <5.0 ppm limonin by high performance liquid chromatography, or <600 ppm naringin by the Davis Test (64); for frozen concentrated grapefruit juice from August 1 to December 1 of each season, Grade A must have <5.0 ppm limonin or <600 ppm naringin while Grade B must have <7.0 ppm limonin or <750 ppm naringin (65).

A processed citrus product of great importance in Japan is canned Satsuma mandarin sections produced through a series of automatic and mechanical operations. The grading, scalding, peeling and separating of segments, chemical treatment for the removal of segment membranes, can filling, and pasteurization processes are described by Itoo (66).

Products that have lost the importance they once had, especially in the United States are canned and chilled citrus sections packed in juice, water, or light sugar/dextrose syrup. Most of the canned sections currently packed are from grapefruit. Chilled citrus sections, which are not heat processed and must be kept chilled, are generally packed with the covering liquid containing less than 0.1% of preservatives as sodium benzoate and potassium sorbate. Some of this glass pack is institutional gallon size. Packs may be as either orange or grapefruit sections; a mixture of the two; or various mixtures of orange, occasionally grapefruit, pineapple, maraschino cherry, and shaved coconut, the latter mixture being known as ambrosia. The product is designed for cool marketing temperatures. The advantages of chilled citrus section products are their convenience to the user, the relatively good firmness or texture of the sections, and attractiveness of the product. A disadvantage is that a significant number of consumers can detect the preservative as a sharp and disagreeable off-flavor.

There are grade standards for both dehydrated orange and grapefruit juices, but production of these products has never been great. At least one Florida plant currently manufactures dehydrated citrus juices. There are generally flavor problems of a heated/processed type associated with the products that limit consumer acceptance.

A relatively small volume of citrus juices are manufactured as blends of citrus juices with one another such as orange-grapefruit, or lemon-lime, and as blends of citrus juices with other fruit juices, eg, grapefruit-pineapple and orange-strawberry-banana.

Production Volumes

Brazil leads the world in frozen concentrated orange juice production with 244.8 million gal (926.6 mil ℓ) equivalent 42° Brix being produced during the 1987–1988 season (67), while for the same season Florida produced 171.2 mil gal (648.0 mil ℓ) (68) for second place. In recent years, chilled types of ready-to-serve orange juice have surpassed retail sales in the United States of frozen concentrated orange juices. For 1988, 428 mil gal (1,620.0 mil ℓ) of chilled orange juice were sold at the retail level versus 353 mil gal (1,336.1 mil ℓ) of frozen concentrate (reconstituted basis of 11.8° Brix) (67). The 1987–1988 season average retail price per single-strength equivalent gal of all types of Florida orange juice was $3.97 ($1.05/L). There were canned Florida orange juice sales of ca 16 mil gal (60.6 mil ℓ) for 1988. Florida produced 31.9 mil gal (120.7 mil ℓ) of frozen grapefruit concentrate during the 1988–1989 season (68). Sales in millions of gallons (followed by millions of liters in parentheses) of the various forms of grapefruit juice produced in Florida for 1988 were: frozen concentrated (reconstituted basis of 10.6° Brix)—9.8 (37.1); chilled types—35.6 (134.7); canned single-strength—26.6 (100.7); while the average retail price for all forms of Florida-produced grapefruit juice was $4.42 per single-strength equivalent gallon ($1.17/ℓ) (67).

Whereas Florida processed about 93% of its orange crop in the 1986–1987 season amounting to 4,987,000 short tons (4,523,209 metric tons), California during the same season processed about 27% of its oranges amounting to 600,000 short tons (544,200 metric tons); total United States processed orange production was 5,625,000 short tons (5,101,875 metric tons) (69). For grapefruit during the 1986–1987 season, Florida processed 58% or 1,227,000 short tons (1,112,889 metric tons) of its crop while California processed 34% or 101 short tons (91.6 metric tons); for all of the United States, 1,369,000 short tons (1,241,683 metric tons) were processed (70); California processed 53% or 437 short tons (396.4 metric tons) of its lemon crop, Arizona 67% or 181 short tons (164.2 metric tons), for all of the United States 57% or 618 short tons (560.5 metric tons) (71).

Manufacture

As is the case for citrus marketed fresh, citrus destined for processing must also meet minimum maturity standards. For example, for Florida, in addition to meeting minimum °Brix to percent acid ratio requirements, there must be no mixing of immature fruit with mature fruit; fruit not passing initial maturity tests may be regraded or separated then retested; if fruit continues to fail to meet minimum maturity standards, fruit is then condemned and destroyed.

Some major factors to consider in fruit prior to harvest, in addition to those of economic significance, should include: °Brix or percent soluble solids and the ratio of the °Brix to percent acid in the juice; the color of juice in oranges; the flavor; in grapefruit, (especially in Florida) the limonin and naringin (bitter constituents) content; and finally, the processed product destined to be made from the fruit.

In the United States, citrus is generally trucked in bulk from the grove directly to the processing plant. Fruit may or may not be placed in large wooden holding bins. Fruit may undergo initial grading (prior to washing) where debris, green fruit, overripe fruit, decayed fruit, etc are removed. Citrus fruit including lemons are washed using roller brush washers with suitable detergents, carefully graded again, sized, and in general made ready for the extractors. There are many extractor types available in the world, but the two main ones are made in the United States—the FMC In-Line (FMC Corp., Lakeland, Fla.) (Fig. 2) and the Brown International Corp. (Covina, Calif.) rotary (Fig. 3). With the FMC machine, peel oil as an oil/water emulsion is separated from the juice stream as is the peel and pulp. For the rotary machine, the peel oil can be recovered from the flavedo which is shaved off of the reamed-out halves following extraction. However, the current trend for oil recovery is to use another Brown International Corp. machine which can deoil the whole fruit prior to juice extraction by puncturing the flavedo causing the oil to be released from the cells.

A 90 lb (40.8 kg) box of Florida oranges yielded on average for the 1987–1988 season 50.14 lb (22.74 kg) of juice; for the same season, an 85 lb (38.6 kg) box of grapefruit yielded 43.84 lb (19.88 kg) of juice (72). Yield from grapefruit increases with increasing maturity over the season with a slight reduction toward the end of the season. Yield for oranges (in Fla.) also increases over the season and then falls off towards the end, however, three major cultivars are involved over the season. Setting the severity of extraction pressure to control yield is an extremely important step: too much can hurt juice quality very significantly while too little may adversely affect the dollar return to the grower. Thus yield is generally monitored by law.

Following extraction, juice is typically run through a screw-type finisher where excess pulp, rag, seeds, and peel parts are screened out. Citrus juice finishers may be set to allow varying quantities of pulp to remain in the juice. This is accomplished primarily through use of screens having perforations of varying sizes with 0.020 in. (0.508 mm) diam being most common. Finished orange juice is 50–60% of the raw fruit weight (73).

Deoiling, if necessary, and deaeration may be accomplished continuously using a heated product in a vacuum,

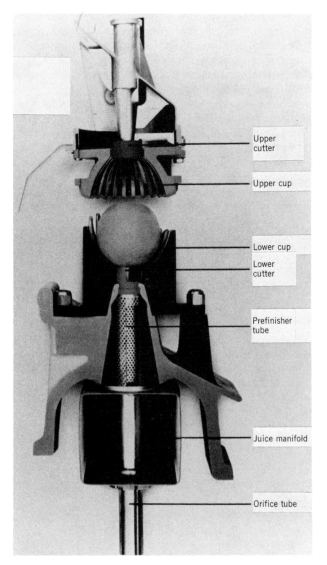

Upper cutter

Upper cup

Lower cup

Lower cutter

Prefinisher tube

Juice manifold

Orifice tube

Figure 2. A cross section of an FMC Corporation In-Line citrus fruit extraction head which in operation allows for internal extraction of juice and juice cells from a whole fruit without halving the fruit. Photograph courtesy of FMC Corporation, Citrus Machinery and Services Division, Lakeland, Fla.

Figure 3. Brown International Corporation Model 700 citrus fruit extractors in a commercial operation. Photograph courtesy of Brown International Corporation, Covina, Calif.

eg, for canned single-strength orange juice, 160°F (71°C) at 21 in. vacuum (74). Plate, tubular, or steam infusion heat exchangers may be used. Up to about 90% of the peel oil may be removed in the deoiling process.

Citrus juices destined for manufacture into processed products undergo heat treatment to inactivate pectic enzymes and destroy spoilage microorganisms. Some representative processing times and temperatures follow: a treatment often used in the production of Florida orange concentrate is 210°F (99°C) for 6 s (75); for reconstituted concentrated orange juice to be remanufactured into orange juice from concentrate, 1 s at 185°F (85°C) or 16 s at 165°F (73.9°C) (76); for lemon juice, 30 s at 180°F (82°C) (77); for canned orange juice, flash processing at 235–265°F (113–129°C) for a fraction of a min (74). The required heat treatment depends on the container selected and the product's storage environment. Both tubular and plate-type pasteurizers are used in the industry. Due to the extreme sensitivity of citrus juice flavor to excessive heat, juices are cooled as rapidly as possible following processing. If the juice is made into hot-pack, then the filled cans are inverted for a short period of time to thoroughly sterilize the inside of the can, generally cooled to somewhat less than 100°F (37.8°C) to allow drying of the can to prevent rusting, and stored from about 40°F (4.4°C) to ambient conditions.

Cold-packing of citrus juice requires aseptic conditions: commercially sterile cooled product; sterilized containers; aseptic hermetic sealing; sterilized closure; and microorganism-free filling atmosphere. The several methods currently available for aseptic packaging of fruit juices including citrus juices are reviewed in detail (78). An example of the complexity of packaging material used in some of the aseptic packaging processes is that presently used in Tetra Pak and Pure-Pak packages (by layer from the inside out): polyethylene, surlyn, aluminum foil, polyethylene, paper, printing ink, polyethylene (79). Typical temperatures and times for commercial sterilization of aseptically packaged high acid foods (such as citrus juices) are 200–205°F (93–96°C) for 15–30 s (80). Unfortunately, aseptic packaging of 100% citrus juices in portion packs and marketed at ambient temperatures result in juices having browning and excessive heated/processed aroma and taste (81), a situation which negates to a large extent the big advantages of being able to keep aseptically packed juices at ambient temperature, and in convenient portion-pack boxes.

Ready-To-Serve Chilled Citrus Juice

Ready-to-serve (RTS) orange juice because of its convenience has become more popular in the United States in recent years than frozen concentrated orange juice. Carter (73) has described in detail the production, packaging and distribution of reconstituted Florida orange juice. In the United States during 1988, the bulk (68.5%) of the RTS juice was packed into plastic-lined fiberboard cartons, 26.5% into plastic (generally polyethylene) containers, and 5% into glass containers (82). As to container sizes, 73.6% of RTS orange juices were marketed in one-half gal (64 fl oz or 1.89 ℓ) containers, 10.5% in 1 gal (128 fl oz or 3.78 ℓ) containers, 7.7% in 96 fl oz (2.84 ℓ) containers, 5.4%

in 1 qt (32 fl oz or 0.95 ℓ) containers with 2.3% being packed in miscellaneous-sized containers. Pasteurized orange juice is being marketed in ever increasing quantities in the United States and in 1988 commanded 21.9% of all RTS orange juice consumed, however, the major RTS juice is still orange juice from concentrate. These chilled types of juice in cartons or plastic, marketed at about 32–36°F (0–2.2°C), are shelf-dated at generally 5–8 weeks. A major U.S. marketer of processed citrus juice recently introduced a 96 fl oz (284 L) specially constructed multi-layer oxygen-barrier plastic container which is being used for pasteurized orange juice shelf-dated at about three months. Grapefruit juice is also marketed as chilled juice in cartons, plastic, and glass. Orange or grapefruit juice packed in glass with metal caps and kept at cool temperatures have shelf lives in excess of one year. The quality of the product held at cool temperatures is better than that held at higher temperatures.

Frozen Concentrated Orange Juice

Frozen concentrated orange juice (FCOJ) is the Cinderella product which opened up the U.S. market to large consumption of citrus juice beginning in the late forties. Scientists MacDowell, Moore, and Atkins of the Florida Citrus Commission in the mid-1940s discovered that by adding some fresh orange juice (6–24%) (sometimes called cut back juice) back to a bland but otherwise excellent quality orange concentrate made under vacuum at low temperature, a product resulted that could be frozen for long periods of time while retaining excellent quality in the reconstituted juice (83). The discovery allowed year-round consumption of convenient, reasonably priced orange juice having much the same quality as freshly squeezed juice.

The evaporator in common use currently utilizes steam and is the very efficient thermally accelerated short time evaporation (TASTE) evaporator which typically has eight stages and seven effects plus a flash cooler. Carter (84) has listed the advantages of a TASTE evaporator over recirculation-type evaporators: improved juice quality, 85% reduction in clean-up time, 66% reduction in steam requirements, low power requirements, 85% reduction in space, and 18% savings in construction costs. A grouping of three modern TASTE evaporators in commercial use is illustrated in Figure 4. With each increase in stage, there is a corresponding increase in °Brix and vacuum as the water is evaporated off. The different effects are the result of combining two or more stages. The juice temperature in the initial stages may be about 212°F (100°C) allowing destruction of most yeast, mold, bacteria, and all enzymes, while the temperature in the final concentrate may be about 50°F (10°C). Evaporators are rated on the amount of water-evaporating capacity per h with an evaporating capacity of 79,920 lb (36,000 kg) per h being common. Processing plants may have several evaporators totalling up to ca one million lb water removal capacity per h. Concentrate is run into cold-wall tanks, where blending, if desired, of other concentrates, essence, cut-back juice or folded cold-pressed peel oil (oil that has been concentrated by distillation with loss of terpenes mostly such as d-limonene) may be made. Blending may be done at a later

Figure 4. A grouping of three thermally accelerated short time evaporation (TASTE) evaporators is pictured at an interior central Florida citrus processing plant. Photograph courtesy of Gulf Machinery Company, Safety Harbor, Fla.

time also. Any of these natural additives may be used to enable the processor to help standardize the product. Currently, FCOJ made in the United States is mostly a blend of orange concentrates representing different cultivars and different sources of supply. By far the largest source outside the U.S. is Brazil which, for example, in 1987 exported 408,583 short-tons (370,900 metric tons) of 65° Brix frozen orange concentrate to the United States (85) with a significant quantity of this product being utilized in FCOJ manufactured in Florida.

Recovery of orange essence (aqueous and oil-phase mixtures of volatile compounds) during the concentrating process, if recovered, is generally carried out at the beginning of the evaporation process at which time the essence is distilled off from about 25% of the juice being evaporated (86). Essence may be concentrated (or folded) using a series of condensers, then added during the blending process described above to the concentrate, or stored under nitrogen or carbon dioxide at cool temperatures for use in the same citrus product or other products including orange juice from concentrates later on. An advantage to packers who use essence is that it allows for considerable standardization of product year-round. Probably well over 50% of FCOJ produced in Florida is essence-add product. In general, natural essence when added to juice imparts a somewhat flowery aroma and taste to the juice different

from that of freshly extracted juice. Essence quality may vary greatly depending on a number of factors such as method of manufacture, cultivar, fruit maturity, year, and storage conditions. In addition, the levels at which essence is added back to juice is critical—too little accomplishes nothing while too much may cause an overpowering essency aroma and taste.

Cut-back juice, if used, is now usually heat-treated to inactivate pectinesterase, cooled rapidly and added to the concentrate at a rate of ca 10% by volume to return significant fresh flavor to the rather bland concentrate. In some cases, freshly squeezed juice, pasteurized or not, may be frozen into large blocks, and stored at 0°F (−18°C) or lower for later use.

If cold-pressed peel oil is used to enhance the quality of the raw concentrate, a minimum amount (0.005% by vol, reconstituted basis) may be added initially to have alpha-tocopherol, a naturally-occurring antioxidant in the oil, available to help lessen oxidation in the concentrate. When final blending of flavor enhancers (including essence if used), pulp, other juices, etc is being done, enough good quality oil is added to allow about 0.015–0.017% by volume in the reconstituted juice.

In the United States and Brazil, primarily, much of the concentrated orange juice (generally 60–68° Brix) is stored in operations known as tank farms. Stainless steel tanks of 100,000–150,000 gal (378,500–567,750 ℓ) capacity are themselves in huge insulated rooms at temperatures from about 9–20°F (−12.8 to −7.6°C). Product may be stored very well for a year or longer using this method. Careful inventory of the contents of each tank allows dispensing of product as desired using fully automated systems. There is still a great quantity of concentrate in the world's citrus producing areas that is stored (at 0°F or lower) and transported in 55-gal (208.2 ℓ) drums having polyethylene liners.

The marketplace in recent years has seen several variants of the typically blended commercial FCOJ, eg, reduced acid FCOJ (a blend of regular FCOJ with juice having its acidity lowered but not significant quantities of ascorbic acid) through use of anionic ion exchange, the blend having a maximum °Brix to percent acid ratio of 21.0 and a maximum of 26.0; high-density FCOJ (primarily for the institutional market) having °Brixes greater than 42° such that instead of the usual three volumes of water being used for reconstituting, eg, four volumes of water are used; FCOJs having bits or pieces of pulp added; FCOJ manufactured from 100% Valencia oranges as a premium pack; and FCOJ-like product, calcium-supplemented frozen orange concentrate beverage, in which about 20% of the U.S. Recommended Daily Allowance for this mineral is supplied in the form of the lactate, the carbonate, calcium hydrate or the tribasic phosphate (87).

A recent application of membrane technology to the concentration of orange juice has resulted in a product superior to conventional FCOJ in a small scale commercial operation of a large Florida citrus processor. The two-step process removes pulp, enzymes, and bacteria in a filtering step and removes water by reverse osmosis in a second step. The filtrate from the first step, accounting for only ca 5% of the original juice, is pasteurized and combined with the filtrate or serum concentrate of the second step. Present production costs are estimated at ca 7 times the cost of conventional product.

There are numerous container sizes and types for retailing FCOJ. For 1988, the most common size in the United States accounting for 74.2% of sales was the 12 fl oz (354 mL) composite fiberboard, plastic-lined can fitted with metal ends and easy-open zip-tops (82), which reconstitutes with three volumes of water to 48 fl oz (1,416 mL). Other can sizes were the 16 fl oz (473 mL) with 16.8% of sales, while the remaining 9% of sales was made up primarily of 6 fl oz (177 mL) and 32 fl oz (946 mL) sizes.

Since 1966, there has been a futures market in the United States for U.S. Grade A concentrated orange juice for manufacturing. Citrus futures are traded on the New York Cotton Exchange where a futures contract is equal to 15,000 lb of orange solids. Trading in frozen orange concentrate is regularly conducted for deliveries in January, March, May, July, September, and November.

NONPROCESSING FACTORS THAT CAN AFFECT QUALITY OF PROCESSED CITRUS JUICES

Four of the major factors: rootstock, cultivar, maturity, and year (season) have been extensively reviewed (88). Some other nonprocessing factors include freezing weather; soil type; production area; yield; climatic conditions; fertilizer, cultivation, pruning and spray practices; and probably other factors as well.

In Florida and elsewhere, citrus grown on rough lemon rootstock is generally lower in quality than that grown on other commercial rootstocks. In California, juices from navel oranges grown on trifoliate-orange rootstock were of excellent quality, while sour-orange and sweet orange rootstocks produced juices of only fair quality (89).

Each citrus cultivar produces a juice having its own unique qualities. And, of course, different cultivars of the same citrus species produce juices having more-or-less similar qualities. The world's most important species, sweet orange (*Citrus sinensis*) and its cultivars, such as mature Hamlin, Pineapple, and Valencia, all possess delicious, relatively sweet (very slightly tart), yellow/yellow-orange juice of excellent processing quality. However, there are differences—the Valencia has a distinctly superior flavor to Hamlin and is somewhat superior to Pineapple (90); Hamlin juice lacks somewhat in body as compared with Valencia especially and to a lesser extent, Pineapple; Hamlin juice has significantly more ascorbic acid than does Valencia and somewhat more than Pineapple; Pineapple juice possesses a notable fruity aroma somewhat more than Valencia and significantly more than Hamlin. There is a table (91) listing 32 citrus cultivars with comments from the literature regarding sensory flavor of each cultivar.

In the normal course of the ripening process of citrus fruit, internal flesh of the immature fruit is generally coarse or ricey textured, very acidic and in some cases as in grapefruit, quite bitter. As ripening progresses, the flesh becomes more juicy, acidity decreases, bitterness especially in the case of grapefruit decreases, fruity top

notes appear that are typical of the cultivar, and in general the fruit becomes edible and fit for processing. Should the fruit remain on the tree too long, an overripe condition manifests itself generally in the form of ricey or dry textured flesh, bland or stale juice flavor, and if seeds have sprouted, a deleterious off-flavor which is especially evident in grapefruit. Grapefruit enjoy a long period of time during which harvest of mature fruit may be accomplished—in Florida, eg, this period may extend from late September–June or about three-fourths of a year, while most sweet orange seasons last only about three months. Satsumas apparently retain peak quality for only a short time (92) and if not picked in 10–14 days after peak maturity, the fruit become puffy, rough in appearance, and the juice insipid in taste.

In the event of damaging natural freezes (temperatures below 29°F or −1.7°C for several h) in citrus-producing regions, quality of citrus may be hurt depending on the severity of the freeze. Freeze-damaged mature fruit generally cannot be marketed fresh. However, they may be utilized as processed fruit. Seriously damaged fruit must be harvested shortly after the freeze in order to be used. Because of the freeze-damage, primarily to the rind, extraction pressures should be less than would be normally used. Excess pressure could cause an excess of harsh tasting components to be extracted from the albedo and flavedo into the juice.

Improper harvesting, fruit transport, and storage could result in quality loss and fruit loss in the final processed product. Plugging (removal of an area of peel around the stem while harvesting) allows microbiological decay to occur and the introduction of foreign material. Fruit stored in bulk in trailers during hot and humid days will deteriorate in quality before processing.

PROCESSING FACTORS THAT CAN AFFECT THE QUALITY OF CITRUS JUICE PRODUCTS

Some of these factors have already been discussed such as the importance of applying proper extraction and finisher pressures; maintaining proper levels of good quality peel oil, cutback juice or essence in the juice; and processing fully mature fruit so as to eliminate problems associated with immature fruit such as excessive tartness, green off-flavors and excessive bitterness, especially as may occur in grapefruit.

Proper processing temperature and time will allow for destruction of both enzymes and potential spoilage microorganisms as has already been discussed. However, excessive processing can lead to development of undesirable heated/processed off-aroma and taste, and browning of the juice. Navel orange juice is not manufactured into 100% juice products and is used in juice blends because of the formation of excessive bitter limonin which develops from the nonbitter precursor limonin monolactone (93) once the juice is extracted and allowed to age for a short while or until heated. The research efforts on methods to reduce limonin in processed grapefruit juice have been recently reviewed (94). Several current projects center around adsorption methods for debittering citrus juices using various resins such as polystyrene divinylbenzene copolymers. Other methods to reduce limonin include enzymatic processes which act to hydrolyze the bitter compounds, and use of supercritical carbon dioxide. Florida has set limits on both of the major bitter constituents, limonin and naringin, occurring in processed grapefruit juice as discussed earlier. Incorporation of air into juice during any step of the processing operation should be avoided as citrus juice is subject to oxidation with possible development of citrus-oxidized types of off-flavor and some loss of ascorbic acid.

The potential problem of inadequate orange color in orange juice has been briefly discussed earlier. The major method currently used to avoid color problems is to blend yellowish juice such as occurs in Hamlin orange with more orangish juice such as Valencia. Of course, relatively small quantities, eg, 5%, of highly colored citrus juices such as that from Honey tangerine or Mikan (having color scores in excess of 43), when added to a yellowish juice can increase the juice color from $\frac{1}{2}$–1 full color score point.

STORAGE AND MARKETING FACTORS WHICH CAN AFFECT THE QUALITY OF PROCESSED CITRUS JUICE PRODUCTS

Excessive storage temperatures for each of the various products can be highly damaging to product quality. Canned citrus juices allowed to reach elevated temperatures of 100°F (38°C) will develop severe heated, scorched, and metallic off-flavors in addition to browning. Marketing of chilled juices at temperatures above 36–38°F (2.2–3.3°C) will result in greatly reduced shelf life brought on by increased microbiological activity and oxidation. Storage temperatures for FCOJ or frozen concentrated grapefruit juice above 0°F (−17.8°C) for many months or above 20°F (6.7°C) for a much shorter length of time can result in product developing stale, old, oxidized, etc type of off-flavors. Transport of bulk citrus juice or concentrate by truck over long distances for long durations of time require strict monitoring to make sure the product is at a safe temperature during the journey.

MICROBIOLOGY OF PROCESSED CITRUS PRODUCTS

Because of the low pH values/generally high acidities of citrus juices, growth of pathogens such as *Salmonella, Shigella, Clostridium,* etc are not a problem. However, even though most types of bacteria cannot grow in citrus juice, *Leuconostoc, Lactobacillis* and various genera of yeasts can grow, and at low temperatures, and can produce off-aromas and tastes (95). Several species of *Lactobacillis* will grow at relatively high temperatures. Spoilage due to (1) *Leuconostoc* or *Lactobacillis* bacteria generally manifests itself in production of diacetyl (buttermilk off-flavor) without gas formation; (2) yeast usually results in product having a fermented taste with production of gas (carbon dioxide); and (3) mold generally results in a filamentous growth on the product surface. Since some microorganisms survive processing, care must be taken during the life of the various citrus products to

discourage growth of these microorganisms. Of course, elimination of spoiled/decayed fruit initially followed by good manufacturing practices with proper sanitation will aid greatly in maintaining manageable very small numbers of microorganisms in product. Orange serum agar is used extensively in the microbiological examination of citrus products resulting in reliable total plate counts of microorganisms that can cause spoilage problems, the counts usually reported as colonies/ml. Normal microorganism counts of 45° Brix orange concentrate will vary from 25,000–75,000 per ml with 300–500 per ml for chilled or pasteurized juices (96). Another source (97) indicates that some manufacturers of orange juice from concentrate endeavor to keep total microorganism counts per ml of reconstituted juice below 500 with less than one-half of these yeasts.

CITRUS BY-PRODUCTS AND SPECIALTY PRODUCTS

Distilled Citrus Oil

Basic chemistry of citrus essential oils has been discussed in an earlier section. An in-depth 1971 report of Florida citrus oils exists as a technical bulletin (98). The most important use of peel oil is as an additive to the juice (or similar juice) from which it was extracted for flavoring purposes, eg, approximately 0.015% by volume in reconstituted FCOJ, and probably should not be considered as a by-product when used in this capacity. However, citrus essential oils have a multitude of uses as one of the most popular natural flavoring ingredients in numerous foods manufactured worldwide. Production of d-limonene, primarily from citrus peel press liquor d-limonene (stripper oil), primarily ranged from about 11–15 million lb (5.0–6.8 million kg) in Florida since the 1981–1982 season and was 14.5 million lb (6.6 million kg) for 1987–1988 (99). Some uses of d-limonene are as a solvent, especially as a substitute for turpentine and certain chlorinated solvents which have been found to be toxic; a substitute for certain chlorofluorocarbons which have been incriminated in the destruction of the ozone layer; carbon source for various syntheses of compounds, such as synthetic resins; penetrating oil; a base for manufacture into flavorings such as l-carvone and l-menthol.

Essences

Annual U.S. production of natural orange essence (aroma or aqueous fraction) was estimated at 10 million lb (4.5 million kg) in 1983 while that of orange essence oil was about 700,000 lb (317,460 kg) (86). Essence produced during the orange juice concentration process has been discussed in the previous section on FCOJ. Most of the essence produced is added back to FCOJ or orange juice from concentrate. However, essence aroma and essence oil are also used elsewhere such as in the manufacture of drinks, and certain other applications where citrus aroma/taste is required.

Dried Pulp and Meal, and Molasses

Nearly 700,000 tons (634,900 metric tons) of dried pulp and meal and 30,000 tons (27,210 metric tons) of molasses were produced in Florida during the 1987–1988 season for cattle feed supplements, with 508,000 tons (460,756 metric tons) or almost 73% of the pulp and meal being exported (99). Commonly, to produce these products, the mixture of peel, pulp, rag, and seed remaining following juicing is pressed (reducing the load to the dryer) resulting in press cake and pressed liquor. The press cake is sent to a single-pass rotary drum dryer where it is dried to about 10% final moisture (100). This product can either be bagged and shipped in this form or pelletized, which in this latter form reduces the bulk density considerably thus reducing storage and shipping space. The pressed liquor is generally concentrated to 72° Brix molasses using a waste heat evaporator (100). The molasses may be marketed as such, but most often it is added back to the citrus pulp for economic considerations.

Fermented Products

A number of fermented products such as alcohol, wines, vinegar, citric acid, lactic acid, riboflavin, feed yeast, 2,3-butylene glycol, and methane may be produced from sound, cull, or waste citrus (101). Citrus wines have never achieved the high popularity of grape wines due primarily to flavor problems still unsolved. Fermentation of citrus molasses to ethyl alcohol for use primarily as neutral spirits for brandy is a viable industry. Quality vinegars may be made (102) possessing some of the characteristic flavors of the citrus cultivars used to make them.

Pectin

Citrus pectin chemistry was discussed in an earlier section. Swisher and Swisher (103) have reviewed methods of citrus pectin production, types and uses. Citrus peel, from which the peel oil has been removed, and pulp to a lesser degree are the major sources for pectin extraction. Prime sources of citrus pectin is the albedo layer of lemon and lime peel, with grapefruit being somewhat less desirable and orange least desirable. Pectin may be manufactured using a number of processes, all of which are complicated. Pectins extracted from citrus peel are high-methoxyl, rapid-set pectins requiring both sugar and acid for gel formation used in the manufacture of jams, preserves, and marmalades. Low-methoxyl pectins not requiring sugar to gel may be made from the high-methoxyl pectin by selective partial deesterification. An overview of the subject with numerous formulae using citrus pectin for jams, preserves, marmalades, apple butter, mincemeat, etc has been presented (104).

Pulpwash Solids

Citrus juice and soluble solids remaining in the finisher pulp is generally recovered in a process in which the pulp is washed with water to allow a leaching of the soluble solids. Sometimes this product made from oranges is known as water extracted soluble orange solids (WESOS) or just pulpwash. About 3–8% additional solids may be recovered by this method, however, the quality of the product is generally bitter and of poor flavor and color quality with few characteristics of normally extracted juice (105). For this reason, Florida does not allow WESOS

to be used in any of its products, but instead it may be used in drink bases. Orange pulpwash produced outside the state of Florida may be used in the preparation of both FCOJ and concentrated orange juice for manufacturing, provided it is the pulpwash removed from the juice used in making those particular concentrates at the time of manufacture. For the 1985–1986 season in Florida, 15.5 million lb (7.0 million kg) of WESOS were produced from early-mid season fruit (Hamlin and Pineapple cultivars) and 8.3 million lb (3.8 million kg) from late season Valencias, or a total of 23.8 million lb (10.7 million kg), and estimated at that time to be worth (for the Valencies only) $0.89/lb ($1.96/kg) or ca 50% that of the regular juice price (106).

Miscellaneous By-Products and Specialty Products

Numerous other products are manufactured from citrus other than those already discussed including dried or frozen stabilized pulp or juice sacs; citrus jellies and marmalades; brined and candied peel; canned gel-type products such as jellied sauces using pulpy citrus juices or gelled products prepared from whole, broken, or crushed citrus sections gelled in citrus juices; citrus purees; frozen citrus juice bars; citrus syrups made from pulp-free citrus juice bases; and citrus juice-based, low-pulp thirst quenchers having added sodium and calcium ions. In the UK and countries with strong English ties, nonalcoholic beverages known as squashes and cordials are popular. These products are essentially mixtures of fruit juices with sugar syrup intended for consumption after dilution with water (107). However, the different names may take on different meanings depending on country, eg, in British and Indian usage, the term squash is used to describe a beverage containing suspended fruit material and cordial describes a clarified product, while in Australia both types of product are called cordials. The most popular are citrus-based orange, lemon, grapefruit or lime, with sulfur dioxide, benzoic acid or sorbic acid added as preservative (108). Many of the above-mentioned citrus by-products and specialty products plus several others are described in some detail in a technical bulletin (109).

Juice vesicle separation technology recently perfected in Japan is used for production of citrus products in that country. Two methods of separation are used on the peeled fruit, one utilizing air and water nozzles and another involving nitrogen freezing then shattering of the fruit. Both methods produce a high percent of intact vesicles. Juice products containing 30% whole Mikan juice vesicles and Florida grapefruit vesicles in flavored gelatin packaged in a plastic simulated half grapefruit peel have been popular in Japan.

NUTRITIONAL ASPECTS, ESPECIALLY VITAMINS AND MINERALS

Citrus and its products are nearly universally accepted for their pleasing distinctive flavors with large quantities consumed, and in many areas such as the United States, often on a daily basis. It is indeed fortunate that water-soluble ascorbic acid (Vitamin C) occurs in citrus juices in quantity, thus supplying hundreds of millions of people with a natural source of this important vitamin that is required daily. To put the enormity of this statement in perspective, with 797 million gal (3,017 million L) of orange juice being consumed in the United States during 1988 (82), Americans ingested along with the juice 2,249,109 lb (1,020,004 kg) of ascorbic acid assuming the average 6 fl oz serving contained 100% of the U.S. Recommended Daily Allowance (U.S. RDA) (33) of 60 mg.

Of course, there is more to the nutritional value of citrus juices and its products than just ascorbic acid. Citrus supplies liquid (mostly water), calories (mainly from the carbohydrates present), good amounts of thiamine, folic acid and potassium, minor amounts of magnesium, phosphorus and copper, but generally insignificant amounts of protein, Vitamin A, riboflavin, niacin, calcium and iron. Citrus also has insignificant quantities of fat and sodium, the latter being especially beneficial to millions of people on low-sodium diets for medical reasons. Previous discussions have covered some of the nutritional aspects of citrus in the sections on proximate analysis, carbohydrates, pectin, amino acids, lipids, and minerals. Data in Table 5 present nutrition information primarily on vitamins and minerals of representative orange and grapefruit juice products.

Ascorbic acid loss due to processing factors is minimal, thus for fresh juice, values for this vitamin and other nutrients are similar to that found in the processed products shortly after packing. Ascorbic acid in chilled-type of processed products packed in fiberboard cartons, polyethylene, or polystyrene plastic bottles generally lose about 1.4–2.0% per day (112) due to oxidation. Single strength juice packed in glass retains about 87% of the initial ascorbic acid after one year at 40°F (4.4°C), with progressively less at increasing storage temperatures, eg, 67% after eight months at 80°F (26.7°C). FCOJ packed in fiber cans with polyethylene liners and aluminum ends retains 91.5% of the initial ascorbic acid after one year at −5°F (−20.5°C) (112). Some additional data on ascorbic acid stability follow: (1) freshly extracted orange and grapefruit juices lose only ca 2% of their original ascorbic acid at 70°F (21.1°C) after three days (113); (2) ascorbic acid loss

in freshly squeezed juice and reconstituted FCOJ was about the same after a week in a home refrigerator (114); (3) the loss of ascorbic acid in whole oranges during their marketing life is not expected to be greater than 10% of the original quantity (114).

Citrus products are a good source of thiamine which was found to be very stable with only a 16–17% loss in canned single-strength orange juice reported after the product was stored for 18 months at 80.6°F (27°C) (114).

Citrus fruit, especially orange juice, is a good and stable source of folate protected from oxidation by the presence of ascorbic acid; unlike other nutritional sources of this vitamin, the folate in orange juice is not subjected to destruction caused by cooking of foodstuff; it occurs in the metabolically active and most available form, N-5 methyl tetrafolate (115).

With pectin now being considered as a factor in dietary fiber, citrus pectin has taken on a new nutritional significance in this subject. Pectin has long been associated with treatment of diarrhea and in recent years, the ability of dietary fiber to lower serum cholesterol levels in humans has been well established. Baker (116) has reviewed studies on the hypocholesterolemic effect that pectin has on serum cholesterol levels in human subjects.

In 1936, Rusznyak and Szent-Gyorgyi (117) discovered that the flavone fraction of lemons had therapeutic effects on abnormal capillary permeability and fragility in humans. Much research has transpired since that time in this area as reviewed by Robbins (118), who also included a list of 60 flavonoids isolated from citrus species. In 1977, Robbins (119) made the discovery that the fully methoxylated flavones, especially tangeretin and nobiletin, and hepta-methoxyflavone were effective in preventing the adhesion of red blood cells, a discovery of note, since the prevention of red blood cell adhesion has been associated with a lessening of some forms of heart disease.

Krehl (120) has discussed the relationships of ascorbic acid, folic acid, thiamine, riboflavin, potassium, sodium, pectin, dietary fiber, and citrus bioflavonoids found in citrus with health and disease.

Table 5. Nutritional Data Expressed as Percent U.S. Recommended Daily Allowance (U.S. RDA) for Reconstituted Frozen Concentrated Orange Juice (FCOJ) and Reconstituted Frozen Concentrated Grapefruit Juice (FCGJ) Based on a Standard 6 fl oz (177 mL) Serving[a]

Nutrients	U.S. RDA (mg)	U.S. RDA, % FCOJ	FCGJ		U.S. RDA (mg)	U.S. RDA, % FCOJ	FCGJ
Calories	—	90	80	Minerals			
Vitamins				Calcium	1000	<2	<2
Vitamin A[b]	5000	2	<2	Iron	18	<2	<2
Vitamin C	60	120	100	Phosphorus	1000	2	2
Thiamine	1.5	8	4	Magnesium	400	4	4
Riboflavin	1.7	2	2	Zinc	15	<2	<2
Niacin	20	<2	2	Copper	2	4	2
Vitamin B-6	2	4	4	Sodium[c]	—	0	0
Folic acid	0.4	20	2	Potassium[c]	—	350	250
Pantothenic acid	10	2	2				

[a] Refs. 110, 111.

[b] Expressed in international units.

[c] No U.S. RDA established for these nutrients; values given expressed as mg/6 fl oz serving.

ADULTERATION OF CITRUS PRODUCTS AND METHODS OF DETECTION

Citrus juice products may be adulterated in a number of ways: dilution with water; addition of sugars (especially from beet medium invert which is difficult to detect); acids; pulpwash; preservatives, especially sodium benzoate and potassium sorbate; artificial or other color ingredients, especially turmeric and annatto; juices from other citrus cultivars, eg, grapefruit juice to orange juice; juices from other commodities, especially apples or pears; use of processed juice in fresh juice; and perhaps other forms of adulteration as well. Methods for the detection of adulteration in processed citrus products have been reviewed (121). These include the physical methods of chromatography, isotope ratios (eg, carbon isotope ratio to detect adulteration with certain sugars and oxygen isotope ratio to distinguish presence of processed juice in fresh juice), optical rotation, visible and ultraviolet absorption spectrometry (to determine dilution, addition of pulpwash, preservatives, natural turmeric and annatto and artificial FD and C 5 and 6 food colors, and ascorbic acid), fluorescence (to detect addition of pulpwash, turmeric and potassium sorbate), ash and mineral analyses (classical methodology), and Brix or soluble solids. Chemical methods include: amino acids, organic acids (especially citric acid), phenolics (a wide variety of compounds ranging from fat-soluble coumarins and psoralins to water-soluble glycosides of the flavanones and flavones), sugars, lipids, vitamins (especially β-carotene added to improve color), minerals (especially total ash and alkalinity of the ash), and volatiles. The final class of methods is that using biological assays. Two specific methods (122) include: an immune assay for orange juice that was able to distinguish lemon juice from orange juice, was sensitive to commercial orange juice products, and was insensitive to common food additives; and an assay using *Lactobacillus plantarum* whose growth in the orange juice assay system is proportional to the amount of juice present.

CITRUS CULTIVAR IMPROVEMENT PROGRAMS

In most of the citrus growing areas of the world, plant breeding programs are underway to produce new citrus cultivars, and/or cultivars having certain new/improved fruit characteristics and/or other horticultural characteristics. In Florida, there have been several recent releases of new cultivars including Sunstar, Midsweet, and Gardner midseason sweet orange cultivars maturing from January–March released in 1987 (123); an orange hybrid, Ambersweet, released in 1989 having good color, relatively good cold hardiness, good ease of peeling, and maturing before Christmas (124); a citrus hybrid, Fallglo, which is a large fruit (with highly colored juice) maturing from late October–late November (125); and Flame (known as Henderson in Texas), a red grapefruit having good external and internal quality. New citrus varieties for California were recently reviewed (126). The grapefruit cultivars, Rio Red, introduced in 1984, which has characteristics similar to those of Ruby Red, except for a

much deeper color pigmentation which lasts all through the season, and Ray Ruby a cultivar exhibiting improved pigmentation over that of Ruby Red have recently come from Texas.

In Florida, there are plant breeding studies to produce a nearly seedless Foster pink grapefruit having reduced acid naturally; seedless or nearly seedless Duncan grapefruit and Pineapple orange; and more intense-colored pink/red grapefruit. Two of the methods used to obtain seedless fruit are: use of tissue culture techniques with standard pollination which results in triploid citrus (citrus having three of each set of chromosomes); and use of irradiation techniques in efforts to produce useful variant forms or mutations resulting in, eg, seedless fruit and earlier maturation. In addition, there are ongoing studies to breed in cold hardiness to certain cultivars, resistance to a number of diseases, and perhaps other factors as well. Similar programs or less-involved programs are in place in many other areas both in the United States and the world including California, Texas, Japan, Australia, Israel, Spain, France, Italy, Argentina, South Africa, and Brazil (for the last-named country, essentially a cultivar selection program).

BIBLIOGRAPHY

1. H. J. Webber, W. Reuther, and H. W. Lawton, "History and Development of the Citrus Industry," in W. Reuther, H. J. Webber, and L. D. Batchelor, eds., *The Citrus Industry*, Vol. 1, Division of Agricultural Science, University of California, Berkeley, 1967, pp. 1–39.

2. Florida Department of Agriculture and Consumer Services, *Florida Agricultural Statistics, Citrus Summary, 1987–88,* Florida Agricultural Statistics Services and USDA, ARS and National Agricultural Statistics Services, Orlando, Fla., 1989, pp. 5–7.

3. Ref. 2, pp. 14, 16.

4. Food and Agricultural Organization of the United Nations, "Citrus Fruit, Fresh and Processed, Annual Statistics," Rome, Italy, 1989, pp. 27–31.

5. W. F. Wardowski, S. Nagy, and W. Grierson, eds., *Fresh Citrus Fruits*, AVI Publishing Co., Inc., Westport, Conn., 1986.

6. W. Reuther, ed., *The Citrus Industry*, Vol. III, *Production Technology*, Division of Agricultural Science, University of California, Berkeley, 1973.

7. W. Reuther, E. C. Calavan, and G. E. Carman, eds., *The Citrus Industry*, Vol. IV, *Crop Protection*, Division of Agricultural Science, University of California, Berkeley, 1978.

8. L. K. Jackson, J. M. Bulger, and R. M. Davis, "Citrus Basics: Citrus Rootstocks," *The Citrus Industry* **66**(9), 18–23 (1985).

9. J. F. Fisher, "Castle Charts Change in Rootstock Use," *The Citrus Industry* **69**(7), 42–45 (1988).

10. U.S. Department of Agriculture, *Composition of Foods: Fruits and Fruit Juices, Raw, Processed, Prepared*, Agriculture Handbook 8–9, USDA, 1982, pp. 130, 154, 159, 184, 276.

11. A. L. Curl, and M. K. Veldhuis, "Composition of Sugars in Florida Valencia Orange Juice," *Fruit Products Journal* **27**, 342, 343, 361 (1948).

12. S. Nagy, P. E. Shaw, and M. K. Veldhuis, eds., *Citrus Sci-*

ence and Technology, Vol. 1, *Nutrition, Anatomy, Chemical Composition, and Bioregulation,* AVI Publishing Co., Inc., Westport, Conn., 1977, p. 112.

13. S. Nagy and J. A. Attaway, eds., *Citrus Nutrition and Quality,* American Chemical Society Symposium Series 143, American Chemical Society, Washington, D.C., 1980, pp. 111, 124.

14. Florida Department of Agriculture and Consumer Services, *1987–88 Season Annual Report,* Fruit and Vegetable Inspection Division, Winter Haven, Fla., 1988, p. 11.

15. J. F. Kefford and B. V. Chandler, *The Chemical Constituents of Citrus Fruits,* Academic Press, New York, 1970, p. 31.

16. J. W. Kesterson, R. Hendrickson, and R. J. Braddock, *Florida Citrus Oils,* Bulletin 749, Institute of Food Agricultural Sciences, University of Florida, Gainesville, 1971, pp. 158, 159.

17. Ref. 12, pp. 430–435.

18. Ref. 12, p. 436.

19. R. J. Braddock and J. W. Kesterson, "Quantitative Analysis of Aldehydes, Esters, Alcohols, and Acids from Citrus Oils," *Journal of Food Science* **41**, 1007–1010 (1976).

20. Ref. 12, pp. 442, 443.

21. J. F. Fisher and H. E. Nordby, "Two New Coumarins from Grapefruit Peel Oil," *Tetrahedron* **22**, 1489–1493 (1966).

22. Ref. 13, p. 135.

23. Ref. 13, p. 141.

24. S. V. Ting and E. J. Deszyck, "The Internal Color and Carotenoid Pigments of Florida Red and Pink Grapefruit," *Proceedings of the American Society of Horticultural Science* **71**, 271–277 (1958).

25. Ref. 12, pp. 248–252.

26. R. L. Clements and H. V. Leland, "An Ion Exchange Study of the Free Amino Acids in the Juices of Six Varieties of Citrus," *Journal of Food Science* **27**, 20–25 (1962).

27. Ref. 13, p. 403.

28. Ref. 12, p. 267.

29. Ref. 12, p. 268.

30. C. E. Vandercook, H. C. Guerrero, and R. L. Price, "Citrus Juice Characterization. Identification and Estimation of the Major Phospholipids," *Journal of Agricultural and Food Chemistry* **18**, 905, 907 (1970).

31. J. J. Birdsall, P. H. Derse, and L. J. Tebly, "Nutrients in California Lemons and Oranges. II. Vitamin, Mineral, and Proximate Composition," *Journal of the American Dietetic Association* **38**, 555–559 (1961).

32. Ref. 13, pp. 363–392.

33. P. J. Fellers, S. Nikdel, and H. S. Lee, "Nutrient Content and Nutrition Labeling of Several Processed Florida Citrus Juice Products," *Journal of the American Dietetic Association* **90**(8) 1079–1084 (1990).

34. Ref. 12, pp. 397–426.

35. Ref. 12, pp. 407–419.

36. Ref. 13, pp. 63–82.

37. Ref. 5, p. 214.

38. Ref. 5, p. 216.

39. Ref. 5, pp. 367–370.

40. Ref. 5, pp. 363–365.

41. W. Grierson, "Causes of Low Packouts in Florida Packinghouses," *Proceedings of the Florida State Horticultural Society* **71**, 166–170 (1958).

42. Ref. 5, p. 34.

43. U.S. Department of Agriculture, *Fruit and Tree Nuts, Situation and Outlook Yearbook,* USDA, Economic Research Services, TFS-246, Washington, D.C., 1988, p. 38.

44. Ref. 2, p. 16.

45. Ref. 2, p. 38.

46. Ref. 5, p. 398.

47. Ref. 13, p. 131.

48. Ref. 13, p. 145.

49. Ref. 5, p. 381.

50. Ref. 5, pp. 383, 384.

51. Ref. 5, p. 382.

52. Ref. 13, pp. 193–224.

53. U.S. Department of Agriculture, *Plant Protection and Quarantine Treatment Manual,* USDA, Animal and Plant Health Inspection Services, Washington, D.C., 1976.

54. Ref. 5, p. 494.

55. P. J. Fellers, "Shelf Life and Quality of Freshly Squeezed, Unpasteurized, Polyethylene-Bottled Citrus Juice," *Journal of Food Science* **53**(6), 1699–1702 (1988).

56. United States Food and Drug Administration, *Code of Federal Regulations,* Title 21, Parts 100–169, Section 146.135, "Orange Juice," 341, 342, 1984.

57. Florida Department of Citrus, *Official Rules Affecting the Florida Citrus Industry, Pursuant to Chapter 601, Florida Statutes,* Chap. 20–64, Florida Department of Citrus, Lakeland, 1975.

58. Food and Agriculture Organization of the United Nations, "Recommended International Standards for Grapefruit and Lemon Juices Preserved Exclusively by Physical Means," Joint FAO/WHO Food Standards Programme, Codex Alimentarius Committee, FAO, Rome, Italy, 1972.

59. Food and Agriculture Organization of the United Nations, "Recommended International Standards for Concentrated Apple Juice and Concentrated Orange Juice Preserved Exclusively by Physical Means," Joint FAO/WHO Food Standards Programme, Codex Alimentarius Committee, FAO, Rome, Italy, 1972.

60. Florida Department of Citrus, *Florida Citrus Fruit Laws, Chapter 601, Florida Statutes,* Florida Department of Citrus, Lakeland, 1982.

61. U.S. Department of Agriculture, "United States Standards for Grades of Orange Juice," USDA, Agricultural Marketing Service, Fruit and Vegetable Division, Processed Products Branch, Washington, D.C., 1983.

62. U.S. Department of Agriculture, "United States Standards for Grades of Grapefruit Juice," USDA, Agricultural Marketing Service, Washington, D.C., 1983.

63. B. S. Buslig and C. J. Wagner, Jr., "General Purpose Tristimulus Colorimeters for Color Grading Orange Juice," *Proceedings of the Florida State Horticultural Society* **97**, 74–76 (1984).

64. Ref. 57, Chap. 20–64, pp. 2, 3.

65. Ref. 57, Chap. 20–64, p. 10.

66. S. Nagy, P. E. Shaw, and M. K. Veldhuis, eds., *Citrus Science and Technology,* Vol. 2, *Fruit Production, Processing Practices, Derived Products, and Personnel Management,* AVI Publishing Co., Inc., Westport, Conn., 1977, pp. 600–605.

67. R. M. Behr, M. G. Brown, and G. F. Fairchild, "Florida Citrus Outlook Update for the 1988–89 Season," *The Citrus Industry* **70**(4), 17, 21, 22 (1989).

68. Ref. 14, p. 17.

69. Ref. 43, p. 22.

70. Ref. 43, p. 27.

71. Ref. 43, p. 30.

72. Ref. 14, p. 11.

73. R. D. Carter, *Reconstituted Florida Orange Juice: Production, Packaging, Distribution,* Florida Department of Citrus, Lakeland, 1985, p. 4.

74. A. Lopez, *A Complete Course in Canning and Related Processes,* Book III, *Processing Procedures for Canned Food Products,* The Canning Trade, Inc., Baltimore, Md., 1987, p. 226.

75. Ref. 73, p. 8.

76. Ref. 73, p. 10.

77. Ref. 74, p. 224.

78. A. Lopez, *A Complete Course in Canning and Related Processes,* Book II, *Packaging, Aseptic Processing, Ingredients,* The Canning Trade, Inc., Baltimore, Md., 1987, p. 160–211.

79. Ref. 7, 8, p. 179.

80. Ref. 78, p. 164.

81. Consumers Union of U.S., Inc., "Orange Juice," in *Consumer Reports,* Consumers Union of U.S., Inc., Mount Vernon, N.Y., 1987, pp. 76–80.

82. P. Mittal, ed., *A. C. Nielsen Food Index Annual Summary, Feb. 13, 1988,* Florida Department of Citrus Economic and Market Research Department, Lakeland, 1988.

83. U.S. Patent 2,453,109 (Nov. 9, 1948) L. G. MacDowell, E. L. Moore, and C. D. Atkins (to the United States of America as represented by the Secretary of Agriculture).

84. R. D. Carter, "New Evaporator Boosts Concentrated Orange Juice Production," *Food Manufacture* **40**, 48, 49 (1965).

85. Ref. 4, p. 52.

86. J. D. Johnson and J. D. Vora, "Natural Citrus Essences," *Food Technology* **37**(12), 92, 93 (1983).

87. R. D. Carter, "Some Recent Advances in the Citrus Processing Industry in Florida," in R. Goren and K. Mendel, eds., *Citriculture,* Vol. 4, Balaban Publishers, Rehovot, Israel, 1989, pp. 1697–1702.

88. P. J. Fellers, "Citrus: Sensory Quality as Related to Rootstock, Cultivar, Maturity and Season," in H. Pattee, ed., *Evaluation of Quality of Fruits and Vegetables,* AVI Publishing Co., Inc., Westport, Conn., 1985, pp. 83–128.

89. G. L. Marsh, "Bitterness in Navel Orange Juice," *Food Technology,* **7**, 145–150 (1953).

90. P. J. Fellers, "Relation of Processing, Variety and Maturation to Flavor Quality and Particle Size Distribution in Florida Orange Juice, I, Flavor Considerations," *Proceedings of the Florida State Horticultural Society,* **88**, 351–353 (1975).

91. Ref. 88, pp. 102–105.

92. F. P. Lawrence, *Citrus Fruit for the Dooryard,* Bulletin 166B, University of Florida, Agricultural Extension Service, 1962, p. 6.

93. V. P. Maier and G. D. Beverly, "Limonin Monolactone, the Nonbitter Precursor Responsible for Delayed Bitterness in Certain Citrus Juices," *Journal of Food Science* **33**, 488–492 (1968).

94. P. J. Fellers, "A Review of Limonin in Grapefruit Juice, Its Relationship to Flavour and Efforts to Reduce It," *Journal of the Science of Food and Agriculture* **49**, 389–404 (1989).

95. Ref. 13, p. 312.

96. Ref. 13, p. 315.

97. Ref. 73, p. 58.

98. J. W. Kesterson, R. Hendrickson, and R. J. Braddock, *Florida Citrus Oils,* Technical Bulletin 749, Institute of Food and Agricultural Science, University of Florida, Gainesville, 1971.

99. Florida Citrus Processors Association, *Statistical Summary, 1987–88 Season,* Florida Citrus Processors Association, Winter Haven, Fla., 1989, p. 1D.

100. Ref. 66, p. 370.

101. Ref. 66, p. 337.

102. R. R. McNary and M. H. Dougherty, *Citrus Vinegar,* University of Florida Agricultural Experimental Station, Bulletin 622, 1960.

103. Ref. 66, pp. 309–315.

104. Sunkist Growers, *Citrus Pectin: Preservers Handbook,* Sunkist Growers Products Department, Ontario, Calif., 6th Ed., 2nd printing, 1956.

105. P. J. Fellers, "Sensory Flavor Characteristics of Water-Extracted Soluble Orange Solids Produced in Florida," *Journal of Food Science* **50**, 739–743, 753 (1985).

106. M. Brown, "An Economic Evaluation of Inclusion of Water Extracted Soluble Orange Solids (WESOS) from Valencia Oranges in FCOJ and Concentrated Orange Juice for Manufacturing. A Report to the WESOS Workshop," Florida Department of Citrus, Lakeland, (Feb. 3, 1987).

107. Ref. 66, p. 346.

108. Ref. 66, pp. 346, 348.

109. J. W. Kesterson and R. J. Braddock, *By-Products and Specialty Products of Florida Citrus,* Bulletin 784, Institute of Food and Agricultural Science, University of Florida, Gainesville, 1976.

110. United States Food and Drug Administration, "Nutrition Labeling of Food," in *Code of Federal Regulations,* Title 21, Food and Drug Administration; Parts 100–169, Section 101.9, pp. 17–24, Apr. 1988.

111. S. E. Gebhardt, R. Cutrufelli, and R. H. Matthews, "Composition of Foods: Fruits and Fruit Juices—Raw, Processed, Prepared," *Agriculture Handbook No. 8–9,* USDA, Human Nutrition Information Service, Washington, D.C., 1982.

112. O. W. Bissett and R. E. Berry, "Ascorbic Acid Retention in Orange Juice as Related to Container Type," *Journal of Food Science* **40**, 178–180 (1975).

113. Ref. 13, p. 13.

114. Ref. 13, p. 15.

115. Ref. 13, pp. 26, 27.

116. Ref. 13, p. 120.

117. I. Rusznyak and A. Szent-Gyorgyi, "Vitamin P–Flavonols as Vitamins," *Nature* **27**, 138, 1936.

118. Ref. 13, pp. 43–59.

119. R. C. Robbins, "Stabilization of Flow Properties of Food with Phenylbenzo-8-pyrone Derivatives (Flavonoids)," *International Journal of Vitamins Nutrition Research* **47**(4), 373–382 (1976).

120. W. A. Krehl, *The Role of Citrus in Health and Disease,* Rose Printing Co., Tallahassee, Fla., 1976, pp. 19–37.

121. Ref. 13, pp. 395–421.

122. Ref. 13, pp. 412, 413.

123. C. J. Hearn, "The Performance of Sunstar, Midsweet, and Gardner Oranges," *Proceedings of the Florida State Horticultural Society* **101**, 33–36 (1989).

124. C. J. Hearn, "Ambersweet: Promising Orange Hybrid Released by USDA," *Citrus Industry,* **70**(3), 56, 61 (1989).

125. C. J. Hearn, "The Fallglo Citrus Hybrid in Florida," *Proceedings of the Florida State Horticultural Society* **100**, 119–121 (1988).

126. E. M. Nauer, D. J. Gumpf, T. L. Carson, and J. A. Bash, "New Varieties for California," *Citrograph,* **73**(8), 152–158 (1988).

General References

References 5, 13, 15, 73, and 120 are good general references.

G. Hassee, ed., *The Orange, A Brazilian Adventure 1500–1987,* Coopercitrus Industrial Frutesp S. A., Duprat and Iobe Publishing Co., Sao Paulo, Brazil, 1987.

S. Nagy, J. A. Attaway and M. E. Rhodes, eds., *Adulteration of Fruit Juice Beverages,* Marcel Dekker, Inc., New York, 1988.

S. Nagy, P. E. Shaw, and M. K. Veldhuis, eds., *Citrus Science and Technology,* Vol. I, II, AVI Publishing Co., Inc., Westport, Conn., 1977.

J. B. Redd, C. M. Hendrix, and D. L. Hendrix, *Quality Control Manual for Citrus Processing Plants,* Book 1, Intercit, Inc., Safety Harbor, Fla., 1986.

W. Reuther, and co-authors, *The Citrus Industry,* Vol. I–V, Division of Agricultural Science, University of California, Berkeley, 1967.

W. B. Sinclair, ed., *The Orange, Its Biochemistry and Physiology,* Division of Agricultural Science, University of California, Riverside, 1961.

W. B. Sinclair, ed., *The Grapefruit, Its Composition, Physiology and Products,* Division of Agricultural Science, University of California, Riverside, 1972.

U.S. Department of Agriculture, *Chemistry and Technology of Citrus, Citrus Products, and By-Products,* ARS, USDA, Agricultural Handbook No. 98, Washington, D.C., rev., 1962.

PAUL J. FELLERS
Florida Department of Citrus
Lake Alfred, Florida

CLEANING-IN-PLACE (CIP)

From the time of its early use in the 1950s, the practice of cleaning-in-place (CIP) for cleaning of plants processing potable liquids and other products such as ice cream and butter has become widespread and is now considered an established cleaning technique (1,2). The technique of CIP stands for cleaning of the tanks, pipelines, processing equipment, and process lines by circulation of water and chemical solutions (hereafter referred to as solution) through them. The term CIP or cleaning-in-place emphasizes that the technique does not require dismantling of pipelines or equipment, which was the case with manual cleaning. Manual cleaning was extremely time-consuming and expensive, and often the level of hygiene (bacteriologic cleanliness) achieved through it was low and inconsistent (2). Introduction of CIP, which became inevitable in the face of economic pressures to increase throughput, increasing cost of labor, scarcity of labor, and technical developments by equipment manufacturers and detergent chemists, alleviated the problems associated with manual cleaning (2,3). Three forms of energy—chemical, kinetic, and thermal—are generally needed for

any cleaning operation. In comparison to manual cleaning, higher temperatures and stronger chemical (detergent) concentrations can be used in CIP. Solution temperatures up to 88°C and detergent pH up to 13 can be used in CIP (4). The equipment design and the properties of the material to be cleaned impose limits on temperature and strength of solution. Lower temperatures may have to be used if product residue (soil on product contact surface) becomes harder to clean at higher temperatures. Manual brushing, which contributes a great deal in manual cleaning, is totally eliminated in CIP and, in some sense, is replaced by the kinetic energy of turbulent flow in pipelines or impingement of jets in vessels (1,4). The circulation time is also a factor in cleaning, and an increase in the time, within a limit, improves the cleaning achieved. Circulation time from 5 min to 1 h is used in practice (5). One of the most important though less noticeable advantages of CIP over manual cleaning is the fact that CIP provides freedom in plant design from the severe limitation of keeping the plant manually cleanable and hence helps in development of new processes and ideas (3). The use of CIP has also made it possible and convenient to automate the cleaning operations in a plant where it was impossible in the case of manual cleaning. The sequence or cycle of operations in any cleaning are the same regardless of cleaning technique (5). A normal cleaning sequence consists of prerinsing with water, washing with detergent solution, and postrinsing with clean water. In addition, there may be a disinfection (sanitizing) stage followed by a final water rinse. Prerinsing is an important stage. It should be started as soon as possible and should continue until the discharging water is free from product residue (soil).

REQUIREMENTS FOR CIP

General Requirements

Only a plant that is properly designed to be cleaned by CIP can be cleaned by CIP efficiently (4). The design of a plant that is to be cleaned in place and the design of the equipment and system that is to apply the CIP technique should be compatible. The CIP system should be designed as an integral part of the processing plant; modifying the plant later for CIP may pose problems (4). All product contact surfaces must be accessible to the pre- and postrinse water and solution. The material that comes in contact with cleaning solution must be able to withstand the solution at its concentration and temperature. For this reason, most of the construction material is stainless steel.

Requirements for Pipelines

There should be no crevice condition, especially in the pipe joints. Crevice condition can lead to bacterial trap, which may be impossible to clean by CIP (6). As far as possible, pipelines should have welded joints. Proper drainage of prerinse water is important in CIP system to avoid any dilution or cooling of solution. To provide the drainage, all piping should have a minimum fall (pitch or slope) of 1 : 100. Pipe work should also have good support to prevent the pipes from sagging. Any sags in the pipeline would prevent complete drainage and put strain on the pipe joints. Uncleanable dead pockets must be avoided inside the pipeline. This may not be achieved, if CIP is added to the plant as an afterthought instead of being an integral part (2). A mean flow rate of 1.5 m/s is normally recommended, but a mean solution velocity of 1.0 m/s may be sufficient in some cases. In practice, there is not much gain in exceeding the mean velocity beyond 2.0 m/s (4). Volume flow rate would depend on the diameter of the pipe. Higher flow rates create a higher hydraulic pressure drop, hence power requirements of circulating pumps could be considerably larger. Abrupt changes in the diameter of pipes that disturb the flow and reduce the cleaning efficiency should be avoided (6).

Requirements for the Vessels and Tanks

The cleaning-in-place of storage tanks or vessels is performed by spraying the cleaning solution onto the surface of the vessel through pressure spray devices located inside the vessel. The spray devices may be rotating, oscillating, or fixed. The fixed device, which is in the form of a perforated ball, is used most commonly. The fixed or static spray ball device does not have any moving parts and therefore gives trouble-free operation with minimal maintenance. Rotating or oscillating devices may wear and give a distorted spray pattern. They may also jam or stick in one place with the consequence of incomplete cleaning of the vessel. A single rotating or oscillating jet, however, can clean a larger-sized vessel than can a single static spray ball. The installation of the spray device should result in a spray pattern that must always cover all parts of the vessel, including probes, agitators, and areas shadowed by them. If required, more than one static spray ball should be used for total coverage in the vessel. Suitable filters should be used to prevent blockage of the spray device. Spray devices should be run at designed pressure and throughput. Too high a pressure can cause atomization of the solution and too low a pressure will reduce the force of jet impingement—both resulting in unsatisfactory cleaning. Permanently installed spray devices are commonly used, but removable spray devices may be preferred in certain special circumstances. Adequate venting of vessels is extremely important to avoid the collapse of the vessel due to a vacuum created during in-place-cleaning when a cold-water rinse immediately follows a wash period with hot detergent. The vessel and the supply and return CIP lines to it should have adequate drainage, otherwise undrained prerinse water can dilute and cool the solution, hence resulting in unsatisfactory cleaning.

Cleaning Circuits

Factory installations as a whole are generally divided into a number of circuits that can be cleaned at different times by CIP technique. It is a usual practice to group pipelines, vessels, and special equipment such as heat exchangers and evaporators into different cleaning circuits because of their different cleaning requirements with respect to flow rates, pressures, and chemicals (1). The product residue deposit (soil) should be of the same kind in a circuit, and all components of the circuit must be available for cleaning at the same time. Hot and cold lines of the plant may be placed in different circuits. After the introduction of automation, it has become practical to clean some parts of the plant while production continues in the adjacent areas. In such circumstances, it is necessary to prevent contamination of product by the cleaning solution. In recent years, double seat valves with the internal leakage drains being used to safeguard against such contamination (2).

TYPES OF CIP SETS/SYSTEMS

CIP sets can be categorized as centralized, local, and satellite or decentralized. In a centralized system, the various CIP circuits are connected by a network of pipes to one or two central CIP stations or units, which consist of all necessary equipment for storage and monitoring of cleaning fluids (water for rinse and solution). Large capacity water and detergent tanks (13,600 to 27,700 L) are used in centralized CIP (4). The system may work well except in the case of large processing plants where long pipe runs require large pump capacities and excessive energy use owing to heat losses. Also, the longer pipelines may contain some water after the prerinse operation that can dilute the detergent solution (7). Overcoming this problem increases chemical consumption. Another disadvantage of a centralized system is the fact that total reliance for cleaning the plant is placed on one or two central stations or units (4). In case of failure or malfunctioning of the central units, the whole plant may have to be shut down. In contrast to the centralized system, the local system requires a greater number of smaller tanks and pumps and shorter pipe runs. The system is more reliable. If one local unit fails or malfunctions, the cleaning of the plant not served by that local unit can still be achieved. The local system would require a greater number of heating units and detergent-strength (concentration) controllers (4). The satellite system is a combination of central and local systems. In the satellite system, there are central solution tanks and the local units draw the required volume of solution using properly sized pipes from the central tanks. The heating of the solution is arranged locally (4). Large modern food-processing plants generally employ a satellite system (7) because of the advantages of saving energy, water, and detergent.

The CIP systems are also classified as single-use, reuse, or multiuse systems, depending on whether the same cleaning solution is used for one, many, or few cycles of cleaning (2). The single-use system is most suited for cleaning heavy soil loads (such as in thermal processing equipment) or cleaning of small plants (8). The system uses the minimum amount of detergent needed for cleaning a circuit and discharges the solution to the sewer after one cleaning cycle is over. The system may be wasteful of detergent and energy and can cause effluent problems (4). However, it is simple in installation and operation. By contrast, the reuse system is complex in installation and operation. The reuse system provides for the reclamation and reuse of the cleaning solution and final rinse water—

the latter is used as prerinse for the next cleaning cycle. More detergent may be added to the solution between the cleaning cycles to counteract the loss of detergency due to expending chemical energy to remove the soil. The solution is discharged to the sewer when it becomes very dirty. The reuse systems require greater capital expenditure but offer savings on volume of water and detergent solutions and energy used (4). Greater benefits can be derived from the reuse systems in plants that have light soil loads and use large-diameter pipe circuits (especially in the brewing industry) (8). The multiuse system is a compromise between the single-use and reuse systems. The final rinse water and solution are used for a few cleaning cycles before discharging to the sewer. In terms of capital and operating costs and complexity of the installation and operation, it is between single-use and reuse systems. Many factors: type and size of plant to be cleaned, soil loading, range and type of chemicals available, pressure drop in pipeline circuits, and pressure and throughput required for vessel spray device, should be considered before selecting the type and size of the CIP system.

CONTROL SYSTEM

Control systems for operating CIP must try to perform CIP operations in a precise sequence. In the early CIP control systems, pumps and valves were controlled manually and time was kept using a stopwatch. Overdependence on the human element prevented the precise sequencing in early control systems (4,9). As the CIP control system evolved with time, the use of electromechanical relays became common. The relays were used to perform the automatic sequencing, but the system had an inherent ability to fail at the wrong time (9). The present-day CIP control systems are microprocessor-based. The development of inexpensive and reliable microprocessors has accelerated the use of microprocessor-based control systems in CIP. Control systems for CIP have improved a great deal in the past 40 years from the unreliable to a very reliable and flexible microprocessor-based controls where cycle time, chemical concentration, and temperature are fully adjustable to suit individual cleaning needs (10).

The center of any microprocessor-based control system is a process controller containing a central processing unit (CPU), memory, and an input/output (I/O) interface. Input–output devices provide the operator access to the control system. The CPU accesses the operator-oriented equipment (such as pushbuttons, pilot lights, rotary switches, thumbwheel switches, LED readouts, CRT screens, and printers) and process-oriented equipment (such as pump motors, valve solenoids, pressure sensors, level probes, flow meters, and temperature switches) through the I/O devices. An operating logic, which may originate as a verbal description or an algorithm, describes the interaction and coordination among the operator-oriented equipment, the process-oriented equipment, and the control system. The operating logic or computer program can be prepared from the algorithm. The CPU executes the operating logic or the program. Because the controller's actions are determined by the computer program, any configuration of equipment and/or program sequence can be implemented and controlled by such a control system. The application of microprocessor-based control systems has helped in making the modern food-processing plants and associated CIP systems completely automatic (7,9,10). With precisely controlled operations, a more sophisticated CIP unit design has become possible, resulting in increased savings on energy, water, and cleaning chemicals (10). The microprocessor-based control systems have provided new opportunities for modifying and improving CIP operations and concepts.

Even very reliable modern control systems can malfunction; therefore, regular routine checks and maintenance are necessary (1,4).

ENERGY AND COST CONSIDERATIONS

Heat energy required for CIP operations constitutes 99% of the energy requirement (11). Heat energy needed is related to cleaning temperature and hence energy savings can be achieved by reducing the cleaning temperature. However, this would necessitate an increase in the detergency requirement of the solution and/or cleaning time. The most energy efficient method, therefore, may not be the most cost effective method. Cost and energy calculations are required in several plants for combinations of time, temperature, and strength of detergent to achieve a standard level of bacteriologic cleanliness before any meaningful conclusion can be reached (11).

Type of plant to be cleaned, CIP system used, mechanical components, control system, building construction, and plant installation would all influence the capital cost associated with CIP (12). However, it is the operating cost that becomes the major factor in making decisions about the CIP system to be used (12). Cleaning chemicals, water supply, required heat energy, operating labor, effluent discharge, maintenance, and electric power would all contribute to the operating cost. The satellite or decentralized system may be the best option in economic terms (12).

BIBLIOGRAPHY

1. D. A. Timperley, "Cleaning in Place (CIP)," *Journal of the Society of Dairy Technology* **42**, 32–33 (1989).

2. N. P. B. Sharp, "CIP System Design and Philosophy," *Journal of the Society of Dairy Technology* **38**, 17–21 (1985).

3. J. R. Franklin, "The Concept of CIP," Proceedings of a seminar in Melbourne, Australia, June 3–4, 1980.

4. W. Barron, "A Practical Look at CIP," *Dairy Industries International* **49**, 34–34 (1984).

5. R. K. Guthrie, *Food Sanitation,* AVI Publishing Co., Inc., Westport, Conn., 1980.

6. J. Farmer, "Engineering Design for CIP," in Ref. 3.

7. "Cleaning of Dairy Equipment," *Dairy Handbook,* Alfa–Laval, S-221 03 Lund, Sweden; 19??, pp. 307–318.

8. P. F. Davis, "Single Use, Re-use and Multi-use CIP systems," in Ref. 3.

9. A. L. Foreshew, "Control Systems for Operating CIP," in Ref. 3.

10. J. Hyde, "State-of-the-Art CIP/Sanitation Systems," *Dairy Record* **84**, 101–105 (1983).

11. C. C. Sillett, "Energy Aspects of Cleaning-in-Place," *Journal of the Society of Dairy Technology* **35**, 87–91 (1982).

12. G. F. Taylor, "The Economics of CIP," in Ref. 3.

D. S. JAYAS
P. SHATADAL
S. CENKOWSKI
University of Manitoba
Winnipeg, Manitoba, Canada

CLONING. See entries under GENETIC ENGINEERING.
COBALAMIN. See entries under VITAMIN (VITAMIN B-12)

COFFEE

Coffee was originally consumed as a food in ancient Abyssinia and was presumably first cultivated by the Arabians in about 575 AD (1). By the sixteenth century, it had become a popular drink in Egypt, Syria, and Turkey. The name coffee is derived from the Turkish pronounciation *kahveh* of the Arabian word *gahweh*, signifying an infusion of the bean. Coffee was introduced as a beverage in Europe early in the seventeenth century and its use spread quickly. In 1725, the first coffee plant in the Western Hemisphere was planted on Martinique in the West Indies. Its cultivation expanded rapidly and its consumption soon gained the wide acceptance it enjoys today.

MODERN COFFEE PRODUCTION

Commercial coffees are grown in tropical and subtropical climates at altitudes up to ca 1,800 m; the best grades are grown at high elevations. Most individual coffees from different producing areas possess characteristic flavors. Commercial roasters obtain preferred flavors by blending—mixing the varieties before or after roasting. Colombian and washed Central American coffees are generally characterized as mild, winey-acid and aromatic; Brazilian coffees as heavy body, moderately acid, and aromatic; and African robusta coffees as heavy body, neutral, slightly acid, and slightly aromatic. The premium coffee blends contain higher percentages of Colombian and Central American coffees.

ECONOMIC IMPORTANCE OF COFFEE

Coffee has been a significant factor in international trade for about 175 years. It is among the leading agricultural products in international trade along with wheat, corn, and soybeans. The total world exportable production of green coffee in the 1989–1990 growing season is approximately 71.7 million bags (Table 1). This compares with a total of about 95 million bags, the difference being internal consumption.

The United States imports from producing countries in 1988–1989 totaled about 16.7 million bags with about 75% of this from countries in the Western Hemisphere (Table 2).

Table 1. Exportable Production of Green Coffee 1989–1990[a]

Principal Countries	Million Bags[b]
Brazil	16.5
Colombia	10.8
Indonesia	5.4
Ivory Coast	4.4
Mexico	3.2
Uganda	3.1
Guatemala	2.9
Costa Rica	2.5
El Salvador	2.2
Kenya	1.9
Honduras	1.4
Ethiopia	1.2
India	0.9
World Total	71.7

[a] Ref. 2.
[b] Millions of 60-kg bags.

In September 1989 producer and consumer members of the International Coffee Organization agreed to discontinue the International Coffee Agreement. This agreement was intended to establish export–import quotas for green coffee to stabilize pricing. The short-term result of no agreement is expected to result in lower green coffee prices. Prices under $2.00/kg of green coffee are expected initially. This compares with a peak of about $6.60/kg in April 1977 following a disasterous frost in Brazil in 1975.

PROCESSING AND PACKAGING

Green Coffee Processing

The coffee plant is a relatively small tree or shrub, often controlled to a height of 2–3 m, belonging to the family *Rubiaceae*. *Coffea arabica* accounts for 69%; *Coffea robusta*, 30% and *Coffea liberica* and others, 1% of world production. Each of these species includes several varieties. After the spring rains, the plant produces white flowers. About six months later, the flowers are replaced by fruit, approximately the size of a small cherry. The ripe fruit is red or purple.

The outer portion of the fruit is removed by curing: yellowish or light green seeds, the coffee beans, remain. They are covered with a tough parchment and a silvery

Table 2. Net Imports of Green Coffee 1988[a]

Principal Countries	Million Bags[b]
United States	16.7
FRG	8.4
France	5.4
Japan	5.1
Italy	4.2
Netherlands	2.7
UK	2.4
Spain	2.4
World Total	67.7

[a] Ref. 2.
[b] Millions of 60-kg bags.

Table 3. Typical Analyses of Green Coffee Types[a]

Variety	H$_2$O	Oil	Total Nitrogen	Ash	Caffeine	Chloro-genic Acid[a]	Trigonelline	Protein	Reducing Sugar	Sucrose	Total Carbo-hydrate
African robusta	11.5	7.0	2.5	3.8	2.06	4.7	0.76	11.4	0.40	4.2	35.0
Colombian arabica	13.0	13.7	2.1	3.4	1.10	4.1	0.94	10.5	0.17	7.2	34.1
Brazilian arabica	11.0	14.3	2.2	4.1	1.01	4.1	1.24	11.1	0.27	7.1	32.0

[a] Ref. 3.
[b] Chlorogenic acid values vary somewhat with the method of analysis used.

skin known as the spermoderm. Each cherry normally contains two coffee beans.

Curing is effected by either the dry or wet method. The dry method produces so-called natural coffees; the wet method, washed coffees. The latter coffees are usually more uniform and of higher quality.

Dry curing is used in most of Brazil and in other countries where water is scarce in the harvesting season. The ripe cherries are spread on open drying ground and turned frequently to permit thorough drying by the sun and wind. Sun drying usually takes two to three weeks depending on weather conditions. Some producing areas use, in addition to the sun, hot air, indirect steam, and other machine-drying devices. When the coffee cherries are thoroughly dry, they are transferred to hulling machines that remove the skin, pulp, parchment shell, and silver skin in a single operation.

In wet curing, freshly picked coffee cherries are fed into a tank for initial washing. Stones and other foreign material are removed. The cherries are then transferred to depulping machines that remove the outer skin and most of the pulp. However, some pulp mucilage clings to the parchment shells that encase the coffee beans. Fermentation tanks, usually containing water, remove the last portions of this pulp. Fermentation may last from 12 h to several days. Because prolonged fermentation may cause development of undesirable flavors and odors in the beans, some operators use enzymes to accelerate the process.

The beans are subsequently dried either in the sun, in mechanical dryers, or in combination. Machine drying continues to gain popularity in spite of higher costs because it is faster and independent of weather conditions. When the coffee is thoroughly dried, the parchment is broken by rollers and removed. Further rubbing removes

the silver skin to produce ordinary green unroasted coffee, containing about 12–14% moisture.

Coffee prepared by either the wet or dry method is machine graded into large, medium, and small beans by sieves, oscillating tables, and airveyors. Damaged beans and foreign matter are removed by handpicking, machine separators, electronic color sorters, or a combination of these techniques. Commercial coffee is graded according to the number of imperfections present, black beans, damaged beans, stones, pieces of hull, or other foreign matter. Processors also grade coffee by color, roasting characteristics, and cup quality of the beverage.

Chemical Composition of Green Coffee

Coffee varies in composition according to the type of plant region from which it comes, altitude, soil, and method of handling the beans. As shown in Table 3, differences are greater between species, eg, arabica vs robusta (African) than within the same species grown in different regions, eg, Colombian vs Brazilian arabicas.

The lower oil, trigonelline, and sucrose contents are typical of robusta beans as is the higher caffeine content. Green coffee contains little reducing sugar but a considerable quantity of carbohydrate polymers. The polymers are mainly mannose with varying percentages of glucose, arabinose and galactose.

Table 4 summarizes the polysaccharide composition of different sources of green coffee (4).

Effects of Roasting on Major Components

Green coffee has no desirable taste or aroma; these are developed by roasting. Many complex physical and chemical changes occur during roasting including the obvious

Table 4. Polysaccharide Analysis of Green Coffee Beans, Polysaccharide Content[a] (wt %[b])

Variety	Arabinose	Mannose	Glucose	Galactose	Total
India robusta	4.1	21.9	7.8	14.0	48.2
Ivory coast robusta	4.0	22.4	8.7	12.4	48.3
Sierra Leone robusta	3.8	21.7	8.0	12.9	46.9
El Salvador arabica	3.6	22.5	6.7	10.7	43.5
Colombia arabica	3.4	22.2	7.0	10.4	43.0
Ethiopia arabica	4.0	21.3	7.8	11.9	45.0

[a] Expressed as anhydro-sugars.
[b] Dry basis.

Table 5. Average Composition of Green and Roasted Coffee

Constituents	Green, Percent Dry Basis	Roasted, Percent Dry Basis[a]
Hemicelluloses	23.0	24.0
Cellulose	12.7	13.2
Lignin	5.6	5.8
Fat	11.4	11.9
Ash	3.8	4.0
Caffeine	1.2	1.3
Sucrose	7.3	0.3
Chlorogenic acid	7.6	3.5
Protein (based on nonalkaloid nitrogen)	11.6	3.1
Trigonelline	1.1	0.7
Reducing sugars	0.7	0.5
Unknown	14.0	31.7
Total	100.0	100.0

[a] Not corrected for dry-weight roasting loss, which varies from 2 to 5%.

change in color from green to brown, and a large increase in bean volume. As the roast nears completion, strong exothermic reactions produce a rapid rise in temperature, usually accompanied by a sudden expansion, or puffing of the beans, with a volume increase of 50–100%. However, this behavior varies widely among coffee varieties because of differences in composition and physical structure.

Table 5 shows the most significant and well-established chemical changes that occur in green coffee as a result of roasting. The principal water-soluble constituents of green coffee are protein, sucrose, chlorogenic acid, and ash, which together account for 70–80% of the water-soluble solids. Most sucrose disappears early in the roast. Reducing sugars are apparently formed first and then react rapidly so that the total amount of sugar decreases as the roast nears completion. The sugar reactions, dehydration and polymerization, form high molecular weight water-soluble and water-insoluble materials. The formation of carbon dioxide and other volatile substances as well as the loss of water, account for most of the 2–5% dry-weight roasting loss.

Roasting essentially insolubilizes the proteins, which constitute 10–12% of green coffee, and 20–25% of the fraction soluble in cold water. The flavor and aroma of roasted coffee are probably due in large part to breakdown and interaction of the amino acids derived from these proteins. Analyses of the amino acids present after acid hydrolysis in both green and the corresponding roasted coffee show marked decreases in arginine, cysteine, lysine, serine, and threonine in Colombian and Angola robusta types after

roasting. The amounts of glutamic acid and leucine for both coffee types, and in the robusta, phenylalanine, proline, and valine increase with roasting (5). Cysteine is the probable source of the many sulfur compounds found in the coffee aroma.

About 15–40% of the trigonelline is decomposed during roasting. Trigonelline is a probable source of niacin, which reportedly increases during roasting, and of the potent aromatic nitrogen ring compounds, such as pyridine, found in roasted coffee aroma. However, the pyrazines, oxazoles, and thiazoles, also components of the coffee aroma, are probably products of protein breakdown. Caffeine is relatively stable, and only small amounts are lost by sublimation during roasting.

The chlorogenic acids 3-caffoylquinic acid (cholorogenic acid), 4-caffeoylguinic acid, cryptochlorogenic acid, and 5-caffeoylquinic acid (neochlorogenic acid) occur at least in part as the potassium caffeine chlorogenate complex. They decompose in direct relationship to the degree of roast. Table 6 shows the changes that occur in these acids during roasting.

Apparently, cholorogenic acids modify and control reactions that occur during the roast and are particularly important to the decomposition of sucrose.

Glycerides of linoleic and palmitic acids along with some glycerides of stearic and oleic acids, make up the 7–16% fat content of coffee. Some cleavage of glycerides and some loss of unsaponifiables occur during roasting. Table 7 details these losses (7).

Aroma

A combination of advanced chromatography and mass spectra analytical techniques have advanced the identification of volatile flavor components in roasted coffee to a total of 670. Although present in minute quantities they are extremely significant to the balance of flavor in a cup of coffee. An historical account of this work has been published (8). A summary of the volatile components by chemical class is in Table 8.

Freshly roasted ground coffee rapidly loses its fresh character when exposed to air, and within a few weeks develops a noticeable stable flavor. The mechanism for the development of staling is not known but is believed to be caused by an oxidation reaction that can be catalyzed by increasing levels of moisture (9).

ROASTING TECHNOLOGY

The main processing steps in the manufacture of roasted coffee are blending, roasting, grinding, and packaging.

Table 6. Changes in Chlorogenic Acids during Roasting, Percent[a]

Acid	Santos				Colombian			
	Green	Light Roast	Medium Roast	Dark Roast	Green	Light Roast	Medium Roast	Dark Roast
Chlorogenic acid	5.56	2.90	1.96	1.11	3.77	2.74	2.16	0.93
Neochlorogenic acid	0.88	1.59	1.02	0.63	0.60	1.53	1.16	0.49
Isochlorogenic acid	0.41		0.24					

[a] Ref. 6.

Table 7. Characteristics of Oil from Green and Roasted Coffee, Percent

Variety	Green			Roasted		
	Oil	FFA[a]	Unsaponifiables	Oil	FFA[a]	Unsaponifiables
Santos	12.77	0.78	6.56	16.05	2.70	6.10
Robusta (Indonesia)	9.07	1.00	6.40	11.27	1.99	5.65

[a] As oleic; FFA = free fatty acid.

Green coffee is shipped in bags weighing from 60 to 70 kg. Prior to processing, the green coffee is dumped and cleaned of string, lint, dust, hulls, and other foreign matter. Coffees from different varieties or sources are usually blended before or after roasting.

Roasting is usually by hot combustion gases in rotating cylinders. The bean charge absorbs heat at a fairly uniform rate and most moisture is removed during the first two-thirds of this period. As the temperature of the coffee increases rapidly during the last few minutes, the beans swell and unfold with a noticeable cracking sound, like that of popping corn, indicating a reaction change from endothermic to exothermic. This stage is known as development of the roast. The final bean temperature, 200–220°C, is determined by the blend, variety, or flavor development desired. A water or air quench terminates the roasting reaction. Most, but not all, of any added water is evaporated from the heat of the beans.

Theoretically, about 315 kJ (300 Btu) is needed to roast 0.45 kg of coffee beans. However, the air-recirculation rate determines the thermal efficiency achieved. Older roasters that do not use recirculation may have an efficiency rate as low as 25%, requiring as much as 1260 kJ (1195 Btu). Roasters with complete air recirculation have an efficiency rate of 75% or more, requiring about 420 kJ (400 Btu).

Most roasters are equipped with temperature controls and automatic quenches that control the roasting operation. The bean temperature, correlated to the color of ground coffee measured by a photometric reflectance instrument, determines the quench end point of a roast.

Conventional roasting by hot combustion gases in ro-

tating cylinders requires 8–15 min. Fast rotating technologies patented in the 1980s develop roast flavor in under 5 min with some roasting times in 1–3 min (10,11).

These fast rotating technologies are the basis of high-yield coffees. Fast roasting is achieved by increasing the heat transfer into the bean through fluidization of the bed of beans with significantly higher quantities of heated air than used in conventional roasting. The increased rate of heat input results in about a 20% increase in extractable soluble solids in conventional household brewers. Because the heat transferred to the beans is carried by significantly higher quantity of air the temperature of the air can be lower, ie, 250–310°C compared to 425–490°C. It is further claimed that the lower roasting temperature results in higher aroma retention. The principles of fast roasting can be utilized in batch or continuous roasters as with conventional roasting times.

Air must be circulated through the beans to remove excess heat before the finished and quenched roasted coffee is conveyed to storage bins. Residual foreign matter, such as stones and tramp iron, which may have passed through the initial green coffee cleaning operation, must be removed before grinding. This is accomplished by an air lift adjusted to such a high velocity that the roasted coffee beans are carried over into bins above the grinders, and heavier impurities left behind. The coffee beans flow by airveying or gravity to mills where they are ground to the desired particle size.

Grinding

Roasted coffee beans are ground to improve extraction efficiency in the preparation of the beverage. Particle size distributions ranging from about 1,100 μm average (very coarse) to about 500 μm average (very fine) are tailored by the manufacturer to the various kinds of coffee makers used in households, hotels, restaurants, and institutions.

Most coffee is ground in mills that use multiple steel cutting rolls to produce the most desirable uniform particle size distribution. After passing through cracking rolls, the broken beans are fed between two more rolls, one of which is cut or scored longitudinally, the other, circumferentially. The paired rolls operate at controlled speeds to cut, rather than crush, the coffee particles. A second pair of more finely scored rolls, installed below the main grinding rolls and running at higher speeds, is used for finer grinds. A normalizer section is then used to distribute particles uniformly.

Packaging

Most roasted and ground coffee sold directly to consumers in the United States is vacuum packed in metal cans; 0.45, 0.9, or 1.35 kg are optional, although other sizes have

Table 8. Aromatic Components of Roasted Coffee

Hydrocarbons	51
Alcohols	19
Aldehydes	28
Ketones	70
Acids	20
Esters	30
Lactones	8
Amines	21
Sulfur thiols	13
Phenols	44
Furanes	108
Thiophenes	26
Pyrroles	74
Oxazoles	28
Thiazoles	27
Pyradines	13
Pyrazines	79
Others	11
Total	670

been used to a limited extent. After roasting and grinding, the coffee is conveyed, usually by gravity, to weighing-and-filling machines that achieve the proper fill by tapping or vibrating. A loosely set cover is partially crimped. The can then passes into the vacuum chamber, maintained at about 3.3 kPa (25 mm Hg) absolute pressure, or less. The cover is clinched to the can cylinder wall and the can passes through an exit valve or chamber. This process removes 95% or more of the oxygen from the can. Polyethylene snap caps for reclosure are placed on the cans before they are stacked in cardboard cartons for shipping. A case usually contains 10.9 kg of coffee, and a production packing line usually operates at a rate of 250–350 0.45-kg cans per minute.

Vacuum-packed coffee retains a high-quality rating for at least one year. The slight loss in fresh roasted character that occurs is due mostly to chemical reactions with the residual oxygen in the can and previous exposure to oxygen prior to packing (9).

Coffee vacuum packed in flexible, bag-in-box packages has gained wide acceptance in Europe and the United States. The inner liner, usually a preformed pouch of plastic-laminated foil is placed in a paperboard carton that helps shape the bag into a hard brick form during the vacuum process (12). The carton also protects the package from physical damage during handling and shipping. This type of package provides a barrier to moisture and oxygen as good as that of a metal can.

Inert gas flush packing in plastic-laminated pouches, although less effective than vacuum packing, can remove or displace 80–90% of the oxygen in the package. These packages offer satisfactory shelf life and are sold primarily to institutions.

Some coffee in the United States, and an appreciable amount in Europe, is distributed as whole beans, which are ground in the stores or by consumers in their homes. Whole-bean roasted coffee remains fresh longer than unprotected ground coffee and retains its fresh roasted flavor for several days longer than ground.

The packaging of roasted whole beans in bags requires the use of a one-way valve to allow carbon dioxide gas to escape and prevent air from entering package. Carbon dioxide gas is a reaction product of roasting.

Modified Coffees

Coffee substitutes, which include roasted chicory, chickpeas, cereal, fruit, and vegetable products, have been used in all coffee-consuming countries. Although consumers in some locations prefer the noncoffee beverages, they are generally used as lower-cost beverage sources, rather than as coffee. The coffee shortage created by a severe Brazilian frost in July 1975 resulted in increased use of these materials as extenders with coffee in the United States, Canada, and Europe in the late 1970s.

Chicory is harvested as fleshy roots, which are dried, cut to a uniform size, and roasted. Chicory contains no caffeine and on roasting, develops an aroma compatible with that of coffee. It gives a high yield, about 70%, of water-soluble solids with boiling water and can also be extracted and dried in an instant form. Chicory extract has a darker color than does normal coffee brew (13). The growing technology for the processing and use of roasted

cereals and chicory is evidenced by the introduction on the market of coffees extended with these materials.

INSTANT COFFEE

Instant coffee is the dried water-extract of roasted, ground coffee. Although used in army rations during the Civil War, instant coffee did not become a popular consumer item until after World War II. Improvements in manufacturing methods and product quality as well as a trend toward the use of convenience foods, account for this rise in popularity. The patent literature on instant coffee products and processes dating back to 1865 is extensive.

Beans for instant coffee are blended, roasted, and ground as they are for regular coffee. Roasted, ground coffee is then charged into columns called percolators through which hot water is pumped to produce a concentrated coffee extract. The extracted solubles are dried, usually by spray or freeze drying, and the final powder is packaged in glass jars at rates of up to 200 jars per minute.

Blends

Coffee is blended to achieve desired flavor characteristics. The concepts used to blend green coffees for regular roasted coffee may also be applied to blends for instant coffee. Most soluble coffee blends contain Brazilian, Central American, Colombian, and African robusta coffees. Some soluble coffees are manufactured in producing countries for export.

Roasting

The batch- or continuous-type roasters used for roasted and ground coffee are also used for instant coffee. The degree of roast can be varied somewhat, depending on the varieties of coffees and the blend composition, to develop the desired characteristics of flavor and aroma.

Grind

Grind is adjusted to suit the type of percolation used and is generally coarser than the regular grind for vacuum- or bag-packed coffee (14). Coarser products avoid the development of excessive pressures in the percolator hydraulic system. The amount and distribution of very fine particles must also be controlled as they may interfere with uniform extraction. However, a grind that is too coarse necessitates longer time periods and higher temperatures for adequate extract concentrations and solubles yields.

Extraction

Commercial extraction equipment and conditions have been designed to obtain the maximum soluble yield and acceptable flavor. In most processes, the water-soluble components in roasted coffee are first extracted with boiling water at atmospheric pressure. Additional solubles are then removed by pressure extraction at higher temperatures, thus hydrolyzing hemicelluloses and other components of the roasted coffee to water-soluble materials.

The factors influencing extraction efficiency and product quality are: (1) grind of coffee, (2) temperature of water fed to the extractors and temperature profile through the system, (3) percolation time, (4) ratio of coffee to water, (5) premoistening or wetting of the ground coffee, (6) design of extraction equipment, and (7) flow rate of extract through the percolation columns (14).

Cylindrical percolators with height-to-diameter ratios range from 7 : 1 to 4 : 1 are common. They are usually operated in series as semicontinuous units of 5–10 percolators, with the water flowing countercurrent to the coffee. The ground coffee may be steamed or wetted with water or coffee extract; this supposedly improves extraction (15). Feed water temperatures range from 154 to 182°C, and unless the columns are heated, the temperature drops so that the extract effluent will have cooled to 60–82°C. The effluent extract temperatures may be reduced by water cooling in plate heat exchangers to minimize flavor and aroma loss prior to drying.

The extract is removed from the percolators and stored in insulated tanks until dried. The extract solubles yield is calculated from the extract weight and soluble solids concentration as measured by specific gravity or refractive index. Yield is controlled directly by adjusting the weight of soluble solids removed and depends primarily on the properties of the coffee, operating temperatures, and percolation time. Soluble yields of 24–48% on a roasted coffee basis are possible. Robusta coffees give yields about 10% higher than arabica coffees (14).

High solubles concentrations are desirable to reduce evaporative load in drying and provide good flavor retention. Percolate concentrations are usually maintained in the range of 20–30% soluble solids for good flavor quality. Some processors concentrate solubles by vacuum evaporation prior to drying. The concentrated percolate may also be clarified by centrifuging prior to drying so that the dry product will be completely free of insoluble fine particles.

The flavor of instant coffee can be enhanced by recovering and returning some of the natural aroma lost during processing. The aroma constituents may be collected from coffee grinders or percolator vents, or may be obtained by concentrating the percolate. Many patents have been issued in the last 10 years on the separation, collection, and transfer of aroma from roasted coffee to instant coffee.

Drying

The following factors are important criteria for good instant coffee drying processes: (1) minimum loss or degradation of flavor and aroma, (2) free-flowing particles of desired uniform size and shape, (3) suitable bulk density for packaging requirements, (4) desirable product color, and (5) moisture content below 4.5% Operating costs, product losses, capital investment and other economic aspects must be considered in selecting the drying process.

Spray Drying

This process is most often used for drying instant coffee. Atomization is usually obtained with pressure nozzles. Selection of nozzles and nozzle combinations is based on properties of the extract pressures used, desired particle size, and bulk density and capacity requirements. The flow of hot air is usually concurrent with the atomized extract spray. Most processors prefer to use low inlet air temperatures (200–260°C) for best flavor quality, outlet air is 107–121°C. Spray dryers are usually constructed of stainless steel and must be provided with adequate dust collection systems, such as cyclones or bag filters (14).

The particles of dried instant coffee are collected from the conical bottom of the spray dryer through a rotary valve and conveyed to bulk storage or packaging bins. Processors may screen the dry product to obtain a uniform particle-size distribution.

Agglomeration

Most instant coffees have been marketed in a granular-appearing form since the mid-1960s. Previously, most soluble coffees were marketed as small spherically shaped particles.

The granular appearance is usually achieved by fusing small spray-dried particles with steam and a low level of heat in a tower similar to a spray-drying tower. Other methods using a continuous belt are described.

Freeze Drying

Freeze drying became commercially important in the United States in 1964, although it was introduced a few years earlier in Europe.

Freeze drying occurs at much lower temperatures than spray drying. Sublimation of ice crystals to water vapor under a very high vacuum [about 67 Pa (500 μm Hg) or lower] removes most of the water. Heat input is controlled to give maximum end-point temperatures between 38 and 49°C (16). Drying times are significantly longer than for spray drying (16). Freeze-dried coffees use high-quality coffee and have better retention of volatile aromatics than do spray-dried coffees and are thus considered quality instant coffees.

Packaging

In the United States, instant coffee for the consumer market is usually packaged in glass jars containing from 56 to 340 g of coffee. Larger units for institutional, hotel, restaurant, and vending-machine use are packaged in bags and pouches of plastic or paper. In Europe, instant coffee is packaged in glass jars, and, frequently, plug-closure metal containers with foil liners.

Protective packaging is primarily required to prevent moisture pickup. The flavor quality of regular instant coffee changes very little during storage. However, the powder is hygroscopic and moisture pickup can cause caking and flavor impairment. Moisture content should be kept below 5%.

Recently, many instant coffee producers in the United States have incorporated natural coffee aroma in coffee oil in the powder. These highly volatile and chemically unstable flavor components necessitate inert-gas packing to prevent aroma deterioration and staling from exposure to oxygen.

DECAFFEINATED COFFEE

Decaffeinated coffee was first developed on a commercial basis in Europe about 1900. The basic process is described in a 1908 patent (18). Green coffee beans are moisturized by steam or water to a moisture content of at least 20%.

The added water and heat separate the caffeine from its natural complexes and aid its transport through the cell wall to the surface of the beans. Solvents are then used to remove the caffeine from the wet beans.

Up to the 1980s man-made organic solvents were commonly used. The caffeine is removed either by direct contact of solvent with the beans or by contact with a secondary water system that has previously removed the caffeine from the beans (12). In either case additional steaming or stripping is used to remove solvent from the beans. The beans are dried to their original moisture content of about 10–12% prior to roasting.

In the 1980s decaffeination processes were commercialized making use of solvents that occur in nature or can be made from substances that occur in nature. The use of these processes are the basis of positioning a coffee product as naturally decaffeinated.

Carbon dioxide under supercritical conditions is a very specific solvent for caffeine. This is based on a 1970 patent of Studiengesellschaft Kohler of Mulheim, FRG. Subsequent patents have been issued disclosing and claiming the use of this technology in a commercial process (17). Fats and oils, including oil from roasted coffee, are disclosed and claimed for a decaffeination process (19). Edible esters, including ethyl acetate, which is present in coffee, are also disclosed and claimed for a decaffeination process. A process using direct water contact with green beans is also in use. This contact removes caffeine as well as some noncaffeine solids. The noncaffeine solids, containing flavor precursors, are reabsorbed on green beans prior to drying and roasting.

In all of the above decaffeination processes prewetting of the green beans is necessary and drying afterward is needed prior to roasting. These steps, in addition to caffeine removal cause changes in the beans that affect roast flavor development.

The degree of decaffeination as claimed on the product is based on the caffeine content of the starting material and the time–temperature process conditions used by the manufacturer to achieve a desired end point.

Roasted decaffeinated coffee is vacuum packed as ground or whole beans for consumer use. The roasted decaffeinated coffee can also be made into a soluble coffee by methods previously described. Soluble coffee can also be made by extracting nondecaffeinated roasted coffee and then removing caffeine from the extract of the roasted coffee containing caffeine, by using many of the solvents described previously.

Roasted and ground decaffeinated coffee is vacuum packed or made into instant coffee by methods previously described. Decaffeinated coffee represented about 23% of cups of coffee consumed in 1989 in the United States (20).

BIBLIOGRAPHY

This article was adapted and updated from the article "Coffee" by A. Stefanucci, W. P. Clinton, and M. Hamell in M. Grayson, ed., *Kirk-Othmer Encyclopedia of Chemical Technology*, Vol. 6, 3rd ed., John Wiley & Sons, Inc., New York, 1979, pp. 511–522.

1. W. A. Ukers, *All About Coffee*, 2nd ed., Tea & Coffee Trade Journal, New York, 1935, pp. 1–3.

2. Foreign Agricultural Service, *USDA World Coffee*, July, 1989.

3. A. Stefanucci and K. Sloman, *Internal Report*, Technical Center, General Foods Corp.

4. A. G. W. Bradbury and co-workers, "Polysaccharides in Green Coffee," paper presented at the Association Scientifique Internationale du Cafe, 12th colloquium, Montreux, Switzerland, 1987, p. 266.

5. H. Thaler and R. Gaigal, *Zeitschrift fuer Lebensmittel-Untersuchung und-Forschung* **120**, 449 (1963).

6. J. R. Feldman and co-workers, *Journal of Agriculture and Food Chemistry* **17**(4), 733 (1969).

7. L. Gariboldi and co-workers, *Journal of the Science of Food and Agriculture* **15**, 619 (1964).

8. I. Flament, "Research on the Aroma of Coffee," paper presented at the A.S.I.C. 12th colloquium, 1987, p. 146.

9. W. Clinton, "Evaluation of Stored Coffee Products," paper presented at the A.S.I.C. 9th colloquium, London, 1980, p. 273.

10. U.S. Pat. 4,169,164 (Sept. 25, 1979), M. Hubbard (to Hills Bros. Coffee Co.).

11. U.S. Pat. 4,737,376 (Apr. 12, 1988), L. Brandlein and co-workers (to General Foods Corp.).

12. U.S. Pat. 2,309,092 (Jan. 26, 1943), N. E. Berry and R. H. Walters (to General Foods Corp.).

13. R. J. Clarke and R. Macrae, eds., *Coffee*, Vol. 5, *Related Beverages*, Elsevier Applied Science, Publishers, Ltd., Barking, UK.

14. H. Foote, M. Sivetz, *Coffee Processing Technology*, Vols. 1, 2, Avi Publishing Co., Westport, Conn., 1963.

15. U.S. Pat. 3,549,380 (Dec. 22, 1970), J. M. Patel and co-workers (to Procter & Gamble).

16. U.S. Pat. 3,438,784 (Apr. 15, 1969), W. P. Clinton and co-workers (to General Foods Corp.).

17. U.S. Patent 4,820,537 (Apr. 8, 1989) S. Katz (to General Foods Corp.).

18. U.S. Pat. 897,763 (Sept. 1, 1908), J. F. Meyer and co-workers (to Kaffee-Handels-Aktien-Gesellschaft).

19. U.S. Pat. 4,465,699 (Aug. 14, 1984) F. A. Pagliaro and co-workers (to Nestlé, SA).

20. International Coffee Organization, *Coffee Drinking Study, U.S.A.*, Winter, 1989.

General References

Reference 2 is a good general reference.

R. J. Clarke and R. Macrae, eds., *Coffee*, Vols. 1–6, Elsevier Applied Science Publishers, Ltd., Barking, UK, 1985–1988.

M. Sivetz, *Coffee Processing Technology*, Vol. 2, AVI Publishing Co., Westport, Conn., 1963.

WILLIAM P. CLINTON
Monsey, New York

COLE CROPS. See VEGETABLE PRODUCTION.

COLLAGENS

Collagens are a family of proteins that are widely distributed in vertebrates and invertebrates and are the most abundant proteins in animals (1). Macromolecular assemblies of collagen support and hold the body together. The protein of bone is collagen; tendon, which connects muscle to bone, is collagen; and deep layers of skin contain collagen. In food, collagen contributes to the texture of meat and meat products. It is used as stabilizer in frozen dairy products, as gelatin in desserts, and as casings for sausages. Collagen is also the principal component of leather and is used as glues and as binders for emulsions in photographic films.

STRUCTURE

Collagen molecules are unique in their composition. Like other proteins, they are high-molecular weight polymers composed of amino acids joined by peptide bonds. Collagen has a special amino acid sequence, a repeating Glysine-X-Y where the X residue is frequently proline and the Y residue is frequently hydroxyproline (2,3). This repeating sequence forms a left-handed helical conformation (4) of the collagen chain. Three chains of collagen monomer combine to form the collagen molecule. Collagen molecules contain a fibrous domain that is made up of three left-handed helices fitted together to form a right-handed triple helix. The length of the triple-helical portion of the molecule varies with the type of collagen and is from 100 to 450 nm in length (3). Collagen molecules also contain globular *N*- and *C*-terminal domains. The exact composition and structure of the *N*- and *C*-terminal globular domains are collagen-type dependent (3) and are referred to as the telopeptides.

The three chains may or may not be identical. If they are different, the chains are coded on separate genes (3). Collagens undergo extensive posttranslational modification such as hydroxylation of proline residues to make hydroxyproline (necessary for collagen stability at body temperature) in the fibrous domain and the formation of hydroxylysine in both the fibrous and the telopeptide domains (3). Hydroxyproline and hydroxylysine are amino acids almost unique to collagens. Hydroxyproline and proline restrict the chain to the left-handed collagen helix (4), and hydroxylysine forms part of the covalent intermolecular and intramolecular cross-links that stabilize fibrils (macromolecular forms of collagen formation (3)).

CLASSIFICATION

Collagens are classified into three groups depending on the macromolecular form and type of cross-links present. The banded (or striated) fibrous group (group I) consists of Types I, II, and III (3). Recently, Types V and XI have been included based on their cross-links (3) and their genomic structure (5), which is the pattern of the exons (coding DNA sequences) and introns (noncoding DNA sequences) in the triple-helix coding region of the particular collagen gene. Type I collagen predominates in vertebrates (6) and is found in bone, skin, tendon, dentine, and muscle (3). Type II is found in cartilage, disk, and vitreous humor (3). Type III is usually associated with Type I in skin (7), muscle (3), and supporting connective tissue in internal organ systems (7). Type V collagen is found in tissues associated with Type I collagen, and Type XI collagen is found in tissues associated with Type II collagen (5).

The nonfibrous collagen (group II) consists of Type IV collagen that in the collagenous component of nonfibrous sheets, called the basement membrane, that underlie the epithelial and endothelial cells surrounding muscle and nerve and that provide the filtration properties of glomeruli (3). Type IV collagen has its triple-helical domain interrupted by nonglycine residues; it assembles as a tetramer and the macromolecular assembly is described as a chicken-wire structure rather than a fibril (3). The nonglycine portions, which interrupt the triple-helical domains, are theorized to add to the flexibility of the basement membrane structure, and the tetramer is stabilized by disulfide and lysino–alanino cross-links in the *C*-terminal end and by disulfide cross-links in the globular *N*-terminal end.

The microfibrillar collagens (group III) are Types VI, VII, IX, and X. Their macromolecular assembly is in the form of fine fibers, and they do not have the striated or banded appearance typical of group I fibers (3). Type VI collagen has been reported to link cell surfaces with the extracellular matrix and to form an interconnecting mesh work between collagen and elastin. Type VII collagen is found as anchoring fibrils for skin basement membrane and was first isolated from the placenta. These two collagens (Types VI and VII) are matrix-associated collagens. Types IX and X are cell-associated collagens and are found in cartilaginous tissue. Type IX is known to be the core protein for glycosaminoglycans.

Different criteria have been used to classify the collagens for groups II and III (6). Group II (the nonfibrous) collagens do not laterally aggregate as do the group I collagens and, therefore, Types VI, VII, and VIII are classified (6) as group II collagens along with Type IV. Type VIII collagen is found in bovine aorta and possibly as part of the basement membrane of the eye (2). Group III collagens have chain molecular weights of <95,000 and only Types IX and X are included in this group (6).

There are other types of collagen that have not been classified into groups. Type XII collagen is found in the bovine periodontal ligament as well as in chicken tendon (8), and a recent report (9) describes the characterization of a portion of another collagen type, Type XIII, which was isolated from the basement membrane of the bovine anterior lens capsule. Collagens have also been isolated from nonvertebrates such as annelids and insects (3).

Collagen genes for the individual chains of Types I (7) and IV (10), and the pro $\alpha1$(II) (7), pro $\alpha1$(III), pro $\alpha2$(V), and pro $\alpha1$(XI) (5) chains have been isolated, characterized, and, in some cases, cloned. The transcription of the collagen gene (DNA to RNA) is regulated by a particularly complex system that is probably cell specific (11).

BIOSYNTHESIS

Collagens are made on ribosomes and are then modified by cotranslational and posttranslational processing. Each collagen gene product undergoes slightly different processing steps depending on the type of collagen. The processing steps outlined here are for the $\alpha1$(I) chain, which eventually combines with itself and one $\alpha2$(I) chain to form Type I collagen, the most abundant type of collagen (Fig. 1). Each named chain is both chemically and geneti-

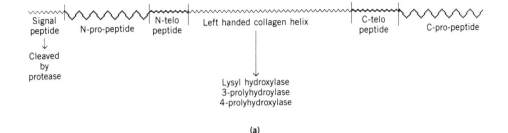

Signal
peptide

N-pro-peptide

N-telo
peptide

Left handed collagen helix

C-telo
peptide

C-pro-peptide

↓

Cleaved
by
protease

Lysyl hydroxylase
3-prolyhydroylase
4-prolyhydroylase

(a)

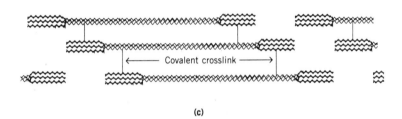

OH*   OH* OH   OH   OH   OH   OH* OH*

N-pro-peptide

OH*
N-telo
peptide

[May contain hydrohylysine]

Triple helix collagen fiber
[contains both hydrohylysine and hydrohyproline]

OH*
C-telo
peptide

[May contain hydrohylysine]

C-pro-peptide
[contains interchain S-S bonds]

* - Site for formation of intermolecular covalent crosslink
via lysine and hydroxylysine

Cleaved
by
N-propeptidase

Lysyl oxidase

Cleaved
by
C-propetidase

(b)

← Covalent crosslink →

(c)

Figure 1. Post-translational assembly of type I collagen: (**a**) pre pro collagen chain; (**b**) pro collagen molecule; and (**c**) collagen fibril. Courtesy of Elizabeth D. Strange.

cally distinct even though there is considerable homology among the different chains even between species.

Collagens are translated from the mRNA (which also must be processed to remove the introns or noncoding sequences) starting with the N-terminal end. The first series of amino acids is called the signal peptide and allows the collagen to penetrate the membranes of the rough endoplasmic reticulum (where the protein is manufactured from the RNA). After the collagen has been secreted into the lumen of the rough endoplasmic reticulum, the signal peptide is cleaved by a protease. The next series of amino acids to be translated forms the aminopropeptide, which has several functions. It is involved in fibrillogenesis, the building of the ultimate structure of collagen in bone or connective tissue (6), the control of formation of the collagen fiber, and the feedback control of collagen synthesis. Along with the C-terminal propeptide, it also prevents the formation of collagen fibrils (macromolecular assemblages of collagen molecules) within the cell (3).

After the aminopropeptide is translated, the N-telopeptide domain of the collagen chain is made. This portion of the preprocollagen molecule is the first section made which will not eventually be cleaved in subsequent processing and is a site for the eventual formation of the covalent cross-links that stabilize the collagen molecule. Other cross-link sites are in the helical and the C-telopeptide domains (3), which, respectively, are the next sections of the preprocollagen molecule to be synthesized.

As the collagen molecule is being translated on the ribosomes of the endoplasmic reticulum, a series of three enzymes are hydroxylating the prolines and lysines as they are added to the growing collagen chain. In addition, galactose is added to hydroxylysine while the polypeptide chains are assembled on the ribosomes and the galactose can be glucosolated. The C-terminal propeptide chain also contains an asparagine residue that is glycoslated with a high mannose oligosaccharide.

Although the preprocollagen chains form on the ribosomes and the cotranslational modifications of the amino acids take place from the N-terminal to the C-terminal, the eventual assembly of the triple-helical collagen molecule takes place from the C-terminal end to the N-terminal end after the collagen has been released from the ribosome. Cross-links between the collagen chains are formed. In the C-terminal propeptide, intrachain and interchain disulfide linkages are formed. The interchain disulfide linkages are between the two α1(I) chains and between one of the α1(I) chains and the α2(I) chain.

After the triple helix has been assembled, it goes to the Golgi apparatus and is then secreted. The final assembly of the collagen outside the cell requires the proteolytic cleavage of both the N-terminal and C-terminal propeptides. The rate at which propeptides are cleaved varies with the type of collagen even within the same structural groupings.

Group I collagens assemble into a quarter-stagger pattern (7) with the *N*-telopeptide of one triple-helical collagen overlapping the *C*-telopeptide of another triple-helical collagen molecule. The procollagen *N*-domain helps align the triple-helical molecules during fibril formation. During extracellular processing these procollagen peptides are cleaved (3). The spaces that have less proteinaceous material lead to the characteristic classic banding pattern with a polarized repeat (D) of 65–70 nm seen in electron microscopy of negatively stained group I collagens (7) and found by x-ray diffraction.

CROSS-LINKS

At least two types of covalent cross-links are formed extracellularly between the telopeptide regions of one collagen molecule and the triple helix of another collagen molecule in the assembled fibril. Lysine and hydroxylysine are enzymatically oxidized. The oxidized lysines (oxidized lysine is an aldehyde instead of an amine) or hydroxylysines react spontaneously with each other or with unoxidized lysines and hydroxylysines. The reaction between two oxidized lysines (to form allysine aldol) is largely intramolecular and connects the *N*-telopeptide regions of the monomers. The reaction between the oxidized lysine and hydroxylysine (to form dehydrohydroxylysinonorleucine) is largely intermolecular and connects the *N*-telopeptide of one chain of one molecule with the helical domain of another molecule and connects the *C*-telopeptide with another site in the helical domain of another collagen molecule. These linkages are acid labile, easily reduced, and form the majority of cross-links in fetal collagen. These covalent linkages change as the animal matures to form more stable types of cross-links. Maturation affects the solubility and denaturation behavior of collagen. The alteration of the cross-links as collagen ages is extremely important in the texture of meat products. It makes the texture of meat from relatively young animals more desirable than the meat from older animals. Meat preparation steps that involve an acid treatment, for example, marinating, take advantage of the acid lability of the immature collagen cross-links. The cross-links stabilize the collagen molecule and give the strength and resistance to deformation that is characteristic of connective tissue.

MEAT

Figure 2 shows the architectural structure of muscle connective tissue (3). Connective tissue is mostly aggregated collagen with a small amount of other molecular species such as complex carbohydrates, proteins, and fats. The tendon (Type I collagen) (3) attaches the muscle to the bone so that work generated by muscle fibers can be translated into movement. Epimysial (Type I and III collagens) connective tissue surrounds the muscle; muscle fiber bundles are, in turn, surrounded by perimysial (Type I and III collagens) connective tissue; and the muscle fibers are surrounded by endomysial (Type I, III, IV, and V collagens) connective tissues. Both the myofibrillar and connective tissues contribute to the texture of meat; the connective

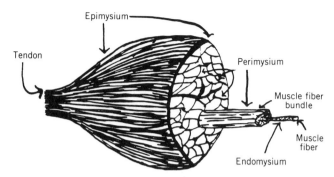

Figure 2. Muscle connective tissue. Courtesy of Elizabeth D. Strange.

tissue contribution is more subtle than the myofibrillar contribution.

Studies have concentrated on the role that the intramuscular connective tissue (ie, perimysium and endomysium) plays in the texture of muscle. Correlations of the total amount of collagen with relative tenderness of muscle have been demonstrated (12). The solubility of collagen also correlates with relative tenderness (13), and solubility is affected by collagen type and by age of the animal (14).

Excessive collagen (or connective tissue) can have an adverse effect on the acceptability of restructured meat products that are made by reducing the size of the muscle pieces and then reforming them into a steaklike product. In this kind of product, the myofibrillar structures are from a variety of muscles and the larger macromolecular connective tissue aggregates (tendon and epimysium) are not trimmed from the starting material. Different connective tissue structures have differing effects on hedonic perception of texture (15). Tendon is most objectionable, epimysium less so. Added perimysium and endomysium have little effect on the overall acceptability of the restructured product.

Collagen is also present in emulsion products such as frankfurters and high-collagen content by-products are sometimes added for economic reasons. In emulsion products, the finely chopped fat is stabilized by a coating of solubilized protein. Actomyosin is a hydrophilic protein that favors an oil-in-water emulsion, whereas collagen is a more hydrophobic protein that favors a water-in-oil emulsion. The addition of the starting materials that have a high collagen content destabilizes emulsions and causes fat pockets and water losses when products are heated to temperatures that cause gelation (unraveling of the triple helix and melting) of the collagen as well as undesirable texture changes. Successful meat emulsion products that contain high-collagen content by-products can be made if the amount of by-product added is carefully controlled (3).

Another aspect of the addition of high-collagen content by-products is the dilution of the protein quality of the product. Collagen is an incomplete protein. It lacks the essential amino acid tryptophan and is low in all the other essential amino acids. However, collagen is an excellent source of proline, which can become the limiting amino acid during times when collagen is being rapidly laid down by the body, such as during wound healing (3).

CASINGS

Denatured collagen is used to make two different types of food products: casings for sausages and gelatin. Casing collagen is usually made from the hides of mature cattle. The collagen is treated with alkali and decalcified, which opens up the fiber bundles, destroys some of the hydrogen bonding, deamidates the amide amino acids, destroys cross-links, and lowers the isoelectric point (pI) of the collagen. The prepared collagen is ground, acid swollen, extruded, and dried; it is then used as sausage casing. The extent of the alterations is critical to the success of the product (3).

GELATIN

Gelatin is made from collagen that has been solubilized with minimal cleavage of the peptide bonds. Class A gelatin is made from the hide of young animals and is acid extracted. It has pI of 6–9 and carries a net positive charge in all food systems. Class B gelatin is made from bone and hide of more mature animals and is alkali extracted. It is fully deamidated, has a pI of 5, and carries a net positive charge in acidic food systems and a net negative charge in neutral foods. Class A gelatin has a higher proportion of Type III collagen and class B gelatin contains some Type II collagen. The molecular weight distribution of the proteins of class A gelatin is lower than that of class B gelatin. Gelatin gels by the partial reformation of the triple-helicial forms of the native collagen. The collagen in gelatin gels is not in exact register (triple helices are formed but they will be a variety of lengths) or as extensively aggregated as native collagen. This results in a lower melt temperature, from 60°C for native collagen to 40°C for gelatin (3).

Gelatin acts as a stabilizer for ice cream and frozen desserts. It prevents the formation of large ice crystals by increasing the viscosity of the ice cream mix. This imparts a desired smooth texture and firm body to the frozen product (16).

Some basic aspects of collagen must be elucidated. The chemical nature of the mature cross-links and the mechanisms for control of the biosynthesis of collagen are active areas of research. The solutions to these and other questions on collagen will have an impact on both the medical and food fields.

BIBLIOGRAPHY

1. H. J. Swatland, *Structure and Development of Meat Animals,* Prentice-Hall, Inc., Englewood Cliffs, N.J., 1984, p. 43.

2. D. A. D. Parry, "The Molecular and Fibrillar Structure of Collagen and Its Relationship to the Mechanical Properties of Connective Tissue," *Biophysical Chemistry* **29**, 195–209 (1988).

3. A. M. Pearson, T. R. Dutson, and A. J. Bailey, eds., *Advances in Meat Research, Vol. 4. Collagen as a Food,* AVI Publishing Co., Westport, Conn., 1987.

4. G. N. Ramachandran, "Stereochemistry of Collagen," *International Journal of Peptide and Protein Research* **31**, 1–16 (1988).

5. M. Bernard, H. Yoshioka, E. Rodriguez, M. van der Rest, T. Kimura, Y. Ninomiya, B. R. Olsen, and F. Ramirez, "Cloning and Sequencing of Pro-α1(XI) Collagen DNA Demonstrates That Type XI Belongs to the Fibrillar Class of Collagens and Reveals That the Expression of the Gene Is Not Restricted to Cartilagenous Tissue," *Journal of Biological Chemistry* **263**, 17159–17166 (1988).

6. E. J. Miller and S. Gay, "The Collagens: An Overview and Update," in L. W. Cunningham, ed., *Methods in Enzymology,* Vol. 144. *Structural and Contractile Proteins, Part D, Extracellular Matrix,* Academic Press, Inc., Orlando, Fla., 1987.

7. K. L. Piez and A. H. Reddi, eds., *Extracellular Matrix Biochemistry,* Elsevier Science Publishing Co., Inc., New York, 1984, Chapts. 1–3.

8. B. Dublet, E. Dixon, E. de Miguel, and M. van der Rest, "Bovine Type XII Collagen: Amino Acid Sequence of a 10 kDa Pepsin Fragment from Peridontal Ligament Reveals a High Degree of Homology with the Chicken α1(XII) Sequence," *Federation of European Biochemical Societies Letters* **233**, 177–180 (1988).

9. R. Dixit, M. W. Harrison, and S. N. Dixit, "Characterization of 26k Globular Domain of a New Basement Membrane Collagen," *Connective Tissue Research* **17**, 71–82 (1988).

10. S. L. Hostikka and K. Tryggvason, "The Complete Primary Structure of the α2 Chain of Human Type IV Collagen and Comparison with the α1(IV) Chain," *Journal of Biological Chemistry* **263**, 19488–19493 (1988).

11. P. Bornstein and J. McKay, "The First Intron of the α1(I) Collagen Gene Contains Several Transcriptional Regulatory Elements," *Journal of Biological Chemistry* **263**, 1603–1606 (1988).

12. E. Dransfield, "Intramuscular Composition and Texture of Beef Muscles," *Journal of the Science of Food and Agriculture* **28**, 833–842 (1977).

13. N. D. Light, A. E. Champion, C. Voyle, and A. J. Bailey, "The Role of Epimysial, Perimysial and Endomysial Collagen in Determining Texture in Six Bovine Muscles," *Meat Science* **13**, 137–149 (1985).

14. D. E. Goll, R. W. Bray, and W. G. Hoekstra, "Age-Associated Changes in Muscle Composition. The Isolation and Properties of a Collagenous Residue from Bovine Muscle," *Journal of Food Science* **28**, 503–509 (1963).

15. E. D. Strange and R. C. Whiting, "Effects of Added Connective Tissues on the Sensory and Mechanical Properties of Restructured Beef Steaks," *Meat Science,* **27,** 61–74 (1990).

16. W. S. Arbuckle, *Ice Cream,* 2nd ed., AVI Publishing Co., Westport, Conn., 1972, pp. 98–99.

Elizabeth D. Strange
United States Department of Agriculture
Philadelphia, Pennsylvania

COLLOID MILLS

A colloid mill is a device used in the preparation of emulsions and dispersions. The name implies that it can generate colloidal size droplets for the disperse phase, but in reality the colloid mill produces emulsions with droplets in the size range of 1–5 micrometers. It is sometimes used for dispersing solid particulates throughout the continuous phase, but it does not grind particles, rather it deagglomerates and disperses the solids (1).

There are many different colloid mills commercially available. In general, the basic components and principles of operation are the same. The basic parts of a colloid mill are the rotor and stator. The rotor turns at high revolutions per minute (rpm), in close proximity to a fixed stator. The fluid being processed is drawn into the small gap between the rotor and stator because of the centripetal or centrifugal action (depending on the design) and is subjected to hydraulic shear forces (1,2). The colloid mill can generate shear levels of 10 million reciprocal seconds because of the large velocity gradient generated in the liquid layer between the rotor and stator (3).

The intensity of shear produced depends on the speed of the rotor; the clearance between the rotor and stator; and the rheological properties of the material being processed (1,3). The shape of the rotor or stator may also have an effect on the product, and some colloid mills have the faces of the rotor and stator roughened by concentric or radial corrugations (4). The diameter of the rotor may range from 2–21 in. and the peripheral speed of the rotor can vary from 10,000 to 18,000 feet per minute (fpm) depending on the rpm and diameter of the rotor (1,5). The gap between the rotor and stator may be set at .001–.050 in. depending on the model, many mills have adjustable gaps. The greatest shear on the product will occur at the smallest gap. However, when the gap is reduced, the flow rate is also reduced.

The output of the mill is inversely proportional to viscosity, but the efficiency of the mill seems to increase with the viscosity (3). There will be a temperature rise in the product as it passes through the mill and the amount of heat input will depend on the length of time the product resides in the working area. The most common food applications of the colloid mill are the processing of mayonnaise and salad dressing.

BIBLIOGRAPHY

1. E. K. Fischer, *Colloidal Dispersions,* John Wiley & Sons, Inc., New York, 1950, pp. 338–346.
2. E. A. Hauser, *Colloidal Phenomena,* The MIT Press, Cambridge, Mass., 1954, p. 63.
3. P. Walstra, "Formation of Emulsions," in P. Becher, ed., *Encyclopedia of Emulsion Technology,* Vol. 1, Marcel Dekker, Inc., New York, 1983.
4. P. Becher, *Emulsions: Theory and Practice,* 2nd ed., Van Nostrand Reinhold Co., Inc., New York, 1965.
5. H. Bennett, H. L. Bishop, Jr., and M. F. Wulfinghoff, *Practical Emulsions,* Vol. 1, Chemical Publishing Co., Inc., New York, 1968.

WILLIAM D. PANDOLFE
APV Gaulin Inc.
Wilmington, Massachusetts

COLOR AND FOOD

IMPORTANCE OF FOOD COLOR

The role that color plays in the reaction to food is so automatic that it may be taken for granted. This does not

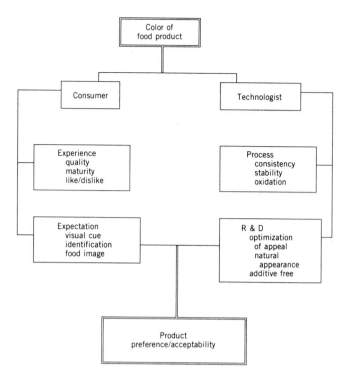

Figure 1. Food color considerations for the consumer and for the food technologist.

make that role any less important it merely decreases awareness. The color of a food has important considerations for both the consumer and the technologist. These considerations are quite different yet they are interrelated as shown in Figure 1.

The psychological effects of color have long been recognized. A room decorated in red exudes warmth and may increase pulse and respiration rates and even increase vivacity. Whereas a room decorated in blue or green is cool and peaceful encouraging concentration and relaxation. But darkening that blue can turn it into a subdued and even depressing atmosphere. The color of food also has its psychological aspects but these are less of mood swings than they are of learned associations. The color of a food not only sends a message of expectation but can also provide clues as to the condition of that food; a yellow peach is ripe, brown strawberry jam is old. Food of an unnatural color raises a barrier that most people have difficulty overcoming. Witness the red-fleshed potato that cooks up blue. Most people given the chance will avoid tasting it even when assured it tastes similar to the familiar white varieties. In contrast a recently introduced yellow-fleshed variety, reminiscent of buttery mashed potatoes, is gaining in popularity. In a very real sense the appearance of a food acts as a gate keeper and the old saying "We eat with our eyes" is not far off the mark. Experience and memory play important roles in food assessment. This is easily demonstrated by mismatching gelatin samples for color and flavor. Even a familiar flavor such as orange is difficult to recognize when colored red. This is because red is associated with red fruits, thus a red-colored orange gelatin sends the wrong visual cue. This is of such proportion that it is difficult to rely on the messages sent by the taste buds and olfactory receptors.

In practice consumers do not isolate their sensory perceptions of a food but combine them in obtaining a total assessment for a food. This association is sufficiently strong that when asked to separate them there may still be a strong influence of one sensory perception on another. Yet the visual sense is both sophisticated and sensitive, it is only that experience his confirmed what the eye sees, which has caused the interaction of the senses to become unconscious.

The technologist has an interest in the color of food for several reasons. One is the need to maintain a uniformity of color over production runs. A second is the avoidance of color changes brought about by chemical reactions occurring during processing or storage life of the product. A third is the optimizing of color and appearance in relation to consumer preference. A fourth is the maintenance of a color in accordance with consumer experience and expectations. The need to maintain color uniformity over production runs varies with the food and especially with the packaging used. For example, the need is greater for fruit juices packaged in clear containers than for those packaged in individual drinking boxes where very little of the juice is actually seen. The color problems of processed green vegetables are examples of color changes arising from unstable pigment reactions. The retailing practices of fresh meat, particularly beef, have been limited for many years by the need to maintain meat pigments in the bright cherry red of oxymyoglobin because of the strong effect on consumer acceptability.

For many new products the product developer must consider the eye appeal of the product, and this should be given serious consideration in consumer acceptability ratings in any optimization procedure undertaken. The need to be aware of dynamic consumer reactions in regard to color and appearance is paramount. What is preferred in the color of a product is influenced by experience and availability and changes over time. The replacement of nitrites for curing meat products is complicated by the fact that the nitrites contribute to the characteristic color of cured meat. A satisfactory replacement has yet to be found. To support the deletion of nitrites from the process, consumers must exhibit sufficient concern about their diets that they would be willing to accept brown cured meat products or at least products that appeared different from those to which they are accustomed.

COLOR THEORY

A thorough review of color theory is beyond the scope of this article; however, several references have been published (1–3). The portion of the electromagnetic spectrum that comprises visible light falls between 380 and 750 nm, bordered by ultraviolet light on the low end and infrared light on the high end. The visible spectrum is represented by the colors seen in the rainbow with blue existing at wavelengths less than 480 nm; green roughly 480–560 nm; yellow, 560–590; orange 590–630; and red, at wavelengths longer than 630 nm (1). Purple is achieved by mixing blue and red and is considered to be a nonspectral color. If white light is dispersed by a prism a spectrum is obtained representing all the visible colors at appropriate wavelengths. The relative power or energy (power multiplied by time) emitted at these wavelengths can be plotted to produce the spectral power distribution curve of the light source. A group of light sources, black bodies, change from black through red to white when heated, the color produced being dependent on the temperature reached. This is referred to as color temperature. Tungsten filaments are close approximations of black bodies, although their color temperature is not exactly equal to their actual temperatures. Real daylight and fluorescent lights do not approximate black bodies.

When light falls on an object it may be reflected, absorbed, or transmitted or a combination of these may occur. When light passes through a material essentially unchanged it is transmitted. Wherever light is slowed down, as occurs at the boundary of two materials, the light changes speed and the direction of the light beam changes slightly (refractive index). The change in direction is dependant on the wavelength and accounts for the dispersion of light into a spectrum by a prism. The change in refractive index results in some light being reflected from the object. Light that is absorbed by the material is lost as visible light. If light is absorbed completely the object appears black and opaque. If some of the light is not absorbed but transmitted the object appears colored and transparent. Lambert's law is always true in the absence of light scattering but not all materials follow Beer's law.

When light interacts with matter it may travel in many directions. The sky is blue because of light scattered by molecules of air. Scattering caused by larger particles produces the white of clouds, smoke, etc. Scattering results in light being diffusely reflected from an object. Translucent materials transmit part of the light and scatter part of the light, if no transmittance occurs, the object is said to be opaque. The color of the object is dependent on the amount and kind of scattering and the absorption occurring. The amount of light that is scattered is dependent on the difference between the refractive indexes of the two materials. The boundary between two materials having the same refractive index cannot be seen because no light scattering occurs. Large particles scatter more light than small particles until the particle size approaches the size of the wavelength, at which point scattering decreases. Many foods are not completely reflecting or transmitting materials, which introduces an element of empiricism into an attempt to measure the color (4). With these foods absorption and scattering are factors affecting visual judgments. The absorption coefficient K and the scatter coefficient S in the Kubelka–Munk equations have been used in attempts to deal with this (5, 6).

The chemist tends to think of color rather simplistically as being determined by the amount of light absorbed at a specified wavelength. But that is not how color is seen, for it is the light that is reflected from an object that determines color. The appearance of an object may vary over the entire object because of the angle of viewing and the light falling on the object. Appearance attributes have been divided into two categories: color and geometric or spatial attributes (3). Color refers to the light reflected from the object to provide a portion of the spectrum as well as white, gray, black, or any intermediate. Geometric attributes result from the spatial distribution of the light from the object and are responsible for the variation in perceived light over a surface of uniform color, such as gloss or texture. The mode in which the eye is operating affects the visual evaluation. There are three modes: the illuminant mode, in which the stimulus is seen as a source of light; the object mode, in which the stimulus is an illuminated object; and the aperture mode, in which the stimulus is seen as light. The aperture mode may be thought of as a lighted window where first the light is seen, but at a closer range the object mode takes over and the contents of the room can be seen. The aperture mode views an object through a smallish aperture removing the effects of spatial distribution of light. This method is usually used for visual color-matching experiments. However, it is the object mode that is of practical importance in assessing the appearance of a food. The everyday evaluations of color, haze, clarity, gloss, and opacity are made by observing what the object does to the light falling on it.

The human eye is an incredibly sensitive and discriminating sensor; it can detect up to 10,000,000 different colors (4). The basic units of sight are the eye, the nervous system, and the brain. The cone light receptors located in the fovea of the retina of the eye are responsible for the ability to see color. It is generally agreed that there are three types of cone receptor responding to red, green, or blue and that these responses are converted in nerve-signal–switching areas within the eye and the optic nerve to opponent-color signals as proposed by Meuller. Three opponent-color systems, black–white, red–green, and yellow–blue, were first proposed by Herring (3). The other type of light receptor, the rods, are also located in the retina; they increase in density as the distance from the fovea increases. The rods are responsible for black-and-white vision, the ability to see in dim light. Rods do not contribute to color vision. Approximately 8% of the population, predominately men, have abnormal color vision. For all humans the mode of presentation including viewing angle, source of illumination, and background, affect the color perceived by the viewer.

COLOR MEASUREMENTS

Color may be measured colorimetrically or spectrophotometrically. A colorimeter is an instrument designed to reproduce the physiological sensation of the eye (4). They are designed to match colors (the sample) by blending amounts of red, green, and blue light.

The most important system used for color measurement was adopted in 1931 by the Commission International de l'Éclariage and is known as the CIE system (1). Its purpose was to derive numbers that provided a measure of a color seen under a standard source of illumination by a standard observer. CIE Source A represented incandescent light and a color temperature of 2,854 K, CIE Source B simulated noon sunlight, and CIE Source C simulated overcast-sky daylight. These light sources were supplemented by the CIE in 1965 and illuminant D$_{65}$ with a color temperature of 6,500 K is now widely used. In 1931 the CIE adopted the standard observer as having color vision representative of the average of that of the population having normal color vision. These data were used to

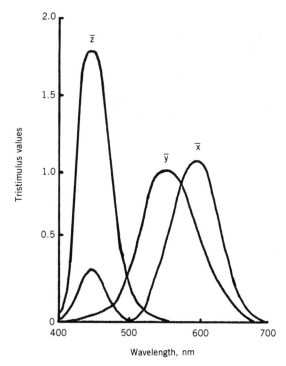

Figure 2. Curves for 2° standard observer expressed as tristimulus values x, y, and z.

obtain CIE standard-observer curves for the visible spectrum for the tristimulus values r, g, and b, for a set of red, green, and blue primaries (Fig. 2) (1). With this system negative amounts of light at the lower wavelengths are required to match some reds. This was considered undesirable and a mathematical transformation of these data was made to produce the unreal primaries X, Y, and Z. It is these transformed tristimulus values of the equal-energy spectrum colors in X, Y, and Z, defining the 2° 1931 CIE standard observer, that are now used. This has the advantage that the tristimulus value of Y provides information on the color's lightness.

The CIE tristimulus values of a color are calculated by multiplying the relative power of a standard illuminant, the reflectance of the object, and the standard observer functions of x, y, and z. Tables are available to make these calculations but todays' instruments provide direct readouts. Tristimulus Y is limited to 100 for a perfectly white object but there is no limit for X and Z. Fluorescent samples can exceed 100 for all three. To plot the color on the 1931 CIE chromaticity diagram, chromaticity coordinates are calculated by the equations:

$$x = \frac{X}{X + Y + Z}$$

$$y = \frac{Y}{X + Y + Z}$$

$$z = \frac{Z}{X + Y + Z}$$

By convention x and y are plotted on the diagram that features the horseshoe-shaped spectrum locus. Dominant

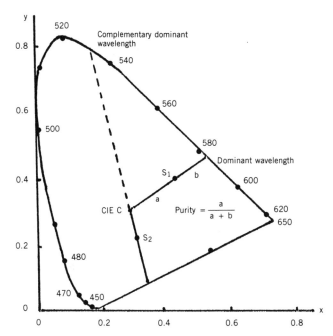

Figure 3. CIE chromaticity diagram showing dominant wavelength and purity.

wavelength is obtained by drawing a line from the illuminant through the sample to the spectrum locus as shown in Figure 3 and represents visual hue. Purity is obtained by the equation $a/(a + b)$ and represents visual chroma or saturation. Samples located nearer the spectrum locus are more saturated. Colors placing below the illuminant at the horseshoe base are considered nonspectral colors and a complimentary dominant wavelength is calculated by drawing a line from the sample through the illuminant to the spectrum locus. Purity is determined as for spectral colors. The horseshoe spectrum locus represents two di-

mensions of color and the third is represented by Y as shown in Figure 4. The plane of the horseshoe rises as the color becomes lighter. Spectrum colors on the spectrum locus have a low lightness factor.

The three dimensionality of color is an important concept for measuring the color of a food. Examination of the wavelength distribution along the spectrum locus makes it clear that color is not equally visually spaced with this system. The elliptical nature of the color spaces prove troublesome when trying to match or establish specifications for color. Much attention has been devoted to providing more uniform color spaces and L^*, a^*, and b^* are frequently calculated to do this using CIELAB (1). Color-matching computers are now available and are much used in textile and other industries. It may be important to determine whether a color difference exceeds the just noticeable difference (JND). Computers have removed the computational labor of the more accurate calculations of differences.

The Hunter L, a, and b color scale is an opponent-type system and has the advantage of being more uniform than the CIE system. It has proved popular for measuring food products. This scale was developed in 1958, allowing computation and direct readout from the dials of the colorimeter (3). Readout is as L_L, a_L, and b_L and the functions are related to the CIE X, Y, and Z as:

$$L_L = 10(Y)^{1/2}$$

$$a_L = \frac{17.5(X - Y)}{(Y)^{1/2}}$$

$$b_L = \frac{7.0(Y - Z)}{(Y)^{1/2}}$$

The a and b chromaticity dimensions are expanded so that intervals on the scale for a and b approximate the visual correspondences to those of the 100-unit lightness scale

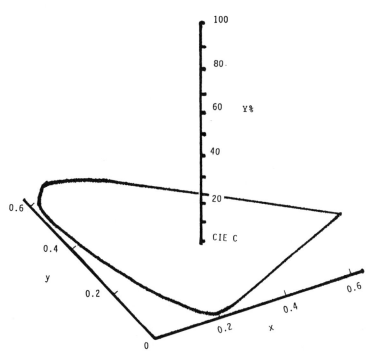

Figure 4. Third dimension of color represented by Y function rising from the plane of the chromaticity diagram.

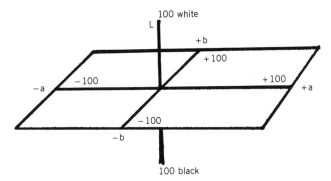

Figure 5. Hunter L a b color scale.

L_L. This color scale is based on Hering's theory that the cone receptors in the eye are coded for light–dark, red–green, and yellow–blue signals. It is argued that a color cannot be both red and green at the same time therefore a is used to represent these; $-a$ represents greenness and $+a$ represents redness. Similarly, $-b$ represents blueness and $+b$ represents yellowness. Lightness is represented by L_L (Fig. 5). Hue is calculated as the angle going counterclockwise from $+a$ and hue angle is calculated as $\tan^{-1} b/a$. Chroma is calculated as $(a^2 + b^2)^{1/2}$. Color difference is calculated as $\Delta E = [(L_1 - L_2)^2 + (a_1 - a_2)^2 + (b_1 - b_2)^2]^{1/2}$, but this value gives no indication in which dimension the difference lies. The Hunter color scale was first developed for use with reflectance and later adapted for transmission measurements. Confusion arises with luminosity of dark liquids such as dark fruit juices. This prompted adaptation and revision of scales for use at low luminosity levels (4). These scales are useful when the ratio of absorbance to transmittance is high.

Colorimeters are instruments designed to reproduce the physiological sensation of the human eye. Their design is based on how the eye sees color. In simplest terms light is projected through red, green, and blue filters in front of a lens and the colored beam projected on a screen. A filter of unknown color is projected on the same screen and the amount of red, green, and blue light is adjusted to match the unknown color. In practice a set of glass filters with transmission curves shaped like those of the standard observer are used. A photocell and meter take a reading of the light reflected from an object through each of these filters and X, Y, and Z values are obtained. Spectrophotometers provide a measure of color in terms of the reflection curve integrated over the visible spectrum, modified by the illuminant source and the standard observer. Needless to say these instruments are more expensive than colorimeters. Choice of instrument depends on the task to be performed. Spectrophotometers are required for computing color mixtures. If the task is color grading or specifications pass–fail assessments colorimeters are more rapid and precise for this purpose (7). Developments with minicomputers have taken much of the tedium out of color measurement and allowed for specialized applications.

SAMPLE PRESENTATION

Most instruments treat foods subjected to measurement as having opaque, matte, and uniform surfaces. In practice, foods seldom meet these requirements but deviate from ideal conditions of flatness, uniformity, opaqueness, diffusion, and specularity in ways that affect the measurement. This introduces an element of empiricism into any attempt to instrumentally assess the color of a food and sample preparation, and presentation becomes important. Color measurements are made to obtain repeatable numbers that correspond to visual assessments of the color of the food. Repeatability requires standardized procedures for sample preparation.

Many foods require a container of optical material thus introducing an element of gloss in contrast to the ideal matte surface. A uniform sample works best but grinding, mixing, milling, and blending are all processes that affect the light scattering properties of the sample. This may be to the extent that the sample no longer represents visual evaluation of the product, ie, crushed potato chips. The presence of water or other liquids in the mixture decreases the light scattering of the sample. Foods that consist of pieces that do not fit tightly together result in light from the source becoming trapped and thus reducing the light scattering. These foods are best measured with an aperture larger than the incident light beam to maximize collection of the light that has been trapped and diffused beyond the normal-size aperture. Liquids that depend on the transmission of light require optical cells of a thickness that will maximize the color difference. Translucent samples measured by reflection also pose the problem of light being trapped because it is reflected from within the food not from the surface. Light so trapped is not collected to be included in the measurement and a large aperture is useful.

Because most foods do not represent ideal conditions it is recommended that a second measurement be made after turning the sample 90° and the two readings be averaged. In devising a presentation technique that works well for the measurement it is well to remember to check that it retains a true relationship with the visual assessment of the product in its original form.

DATA INTERPRETATION

For quick assessments the simplest evaluation that will do the job is desired. Three-dimensional data may be considered too complex and difficult to deal with. It is important that any color measurement used to assess a food relates to what the eye sees and this should go further to determining those aspects of color that are important to consumers' assessment of quality. Any reduction of data needs to take these factors into account. Consumer assessments are generally obtained in less-controlled conditions, which are useful in determining the breadth of the problem. An intermediate stage under more—controlled conditions may be necessary in translating consumer results into instrumental measurements. However, using company personnel, closely involved with the technical side of the operation, as a substitute for consumers is risky because they are not representative of the consuming population.

Generally it is advisable to go beyond the basic CIE X, Y, and Z or Hunter L, a, and b instrument readouts to get a true picture of the color of the food. Lightness, hue, and chroma or saturation dimensions should be examined to obtain an indication of how the color of the samples measured differ. If the samples differ in only one dimension that dimension may be enough to provide a useful assessment of the product. However it should be remembered that lightness varies with both hue and chroma.

Some caution is required in using color difference calculations. The general ΔE calculation is affected by the equation selected for the calculation and by the geometry of the instrument on which the data were collected. Therefore, a suitable equation should be selected and used consistently. In addition, the ΔE calculation does not provide information about the nature of the difference. Calculating differences for the components as either CIELAB ΔL^*, Δa^*, and Δb^* or ΔH for hue, ΔL, or ΔC for chroma, is likely to be more informative. If these are to be used for establishing tolerances they must be checked against visual judgments. It is important to note that while numbers may be convenient, color acceptions or rejections in practice are not number based but made on visual assessments.

The availability of computer-equipped instruments has removed much of the tedium from calculating color-measurement results and made readily available data in different forms. This should not be allowed to lull researchers into a false sense of security. It is still important to use well-prepared, representative samples for measurement. It still takes time to think about the problem, to evaluate, and to understand the results obtained.

COLOR MEASUREMENT OF SPECIFIC FOODS

An extensive review of the color measurement of various foods has been published (8). This discussion highlights treatments of color measurement that represent newer developments or deal with conditions that present real problems in color measurement in attempting to relate measured color to what the eye sees.

Orange juice is a food for which color is considered important enough to represent a large proportion of the grading system yet it is difficult to measure because of translucency that may be compounded by the presence of particles. One approach to color measurement of this type of product has been the use of Kubelka–Munk equations developed for reflectance at a single wavelength (5), use with colorimeters is strictly an empirical adaptation. It has been observed that the effect of different rates of transmittance change relative to concentrations; <1.0 produced large changes in measured lightness, hue, and chroma (7). The falloff in lightness and chroma observed at higher concentrations resulted from smaller change in transmittance at concentrations >2.0. It was concluded that for strongly colored scattering materials in dilute suspension, instrumental measurement was inadequate because it did not measure what the human vision perceived as appearance (7). This is because the instrument measures intensity of back-scattered light over a limited angle whereas human vision is stimulated by internally scattered light emerging multidirectionally from the suspension in addition to that which is reflected directly. For purposes of grading, the color of orange juice is measured using the Citrus Colorimeter (3). This instrument has two scales: citrus redness (CR); and a subsidiary citrus yellow (CY), which are used to calculate a color-score equivalent to a visual color score.

The color of fresh meat is of interest because of the importance of the degree of oxygenation of the surface to consumer acceptability and also because of the incidence of dark cutting beef and pale, soft exudate pork. More traditional approaches have been to determine relative concentrations of the myoglobin, oxymyoglobin, and metmyoglobin pigments spectrophotometrically at appropriate wavelengths (9). The problem of measuring meat color is made difficult because of the variability in concentration of the heme pigment myoglobin, the condition of the cut surface that may have undergone a degree of desiccation, the chemical state of the myoglobin, and the light-scattering properties of the muscle pigments (7). The muscle of a freshly slaughtered animal is dark and translucent in appearance becoming lighter and more opaque as the pH falls and the glycogen is converted to lactic acid. Because of the degree of translucency inherent in fresh meat, the Kubelka–Munk equations have been explored as a means of measuring meat color (7, 9). The scatter coefficient has been used to measure color as a condition of fresh meat; however, the range of scatter is too large to estimate pigment concentration accurately (7). Scattering data were supplemented with data for L, hue angle, and chroma to produce typical values for conditions of fresh meat. It has been pointed out that derived formulas and multiple regression equations for expressing meat color suffer in accuracy because most people tend to think of how the viewed samples differ from a mental image of ideal meat color (10).

A mathematical approach rather than a visual approach has been used for estimating physical color parameters L^*, a^*, and b^* for red and tawny ports based on consumer data (11). Consumers ratio-scaled 15 blends of port for redness, brownness, and intensity of color and then provided ratings for their ideal port. Assessors were separated on the basis of those who preferred tawny and those who preferred ruby port. Deviations of the samples from the ideal were regressed against L^*, a^*, b^*, hue angle, and chroma. Three simultaneous equations including port preference, hue angle, and chroma were established for relating sensory information with physical information. These data were further treated by the method of inverse simultaneous estimation to obtain estimates for color parameters for ideal ports. However, as with all ideal assessments, caution must be exercised because of the large degree of error. In this case the ideal fell outside the range of blends, a common occurrence with ideal data. Although the study does demonstrate that appropriate use of mathematics and data collection can increase efficiency in relating sensory and physical color data.

ROLE OF COLORANTS

Because the color of a food is used by consumers to aid in forming a judgment about that food, it is not surprising that food processors and manufacturers consider colorants important adjuncts. The use of colorants to enhance the appearance of a food is not a recent development. Spices and condiments have been used since early civilization.

However, the use of colorants has not been without its more unscrupulous aspects such as the propensity to use them to mask inferior products in the eighteenth and nineteenth centuries. The use of copper sulfate or copper fittings in making pickles to obtain an appetizing green appearance has lingered to more recent times. Candies, wines, liquors, and even flours have all been subjected to harmful coloring practices (12). As a result the use of colorants in foods came under scrutiny in the late nineteenth century, and the use of food colors has been controlled at the federal level in the United States since an act of Congress in 1886 allowed butter to be colored. This act recognized the need to provide the marketplace with products having both an attractive and a consistent appearance. By 1900 a wide range of foods such as jellies, catsup, alcoholic beverages, milks, ice creams, candy, sausages, pastries, etc were allowed to be colored. Both the substances allowed and the amounts specific to a food have been regulated where coloring has been permitted.

Colorants allowed in foods include those from both natural and synthetic sources. Natural colorants as a rule are more expensive, less stable, and possess lower tinctorial power. In addition, they are frequently present as mixtures in the host materials and vary with season, region, variety, etc. Disadvantages associated with natural colorants have been identified as low yields; color instability resulting from effects of pH, light, heat, and freezing; and possible association with other properties that may be undesirable (13). However, they do present the advantage of being perceived as safer than synthetic colors. Pigments to be used as colorants must be extracted from the host materials and prepared in some form, such as a dried powder or beadlets, which increases the ease of use in a food product (12). A breakthrough in the use of natural pigments occurred with the synthesization of the carotenoids β-carotene, β-apo-8′-carotenal, and canthaxanthin. But even here considerable confusion exists in the interpretations by regulatory bodies. In the United States these synthetic duplications of natural extracts must be identified as an artificial color, whereas in Canada they must be labeled a natural color.

The safety of synthetic dyes continues to be questioned. In 1938 the FDA evaluated synthetic dyes and determined that 15 dyes met the criteria for use in food, drugs, and cosmetics, and these were identified as FD&C colors. Since then the list of FD&C colors has been reduced to seven. The most controversial deletion was FD&C Red Dye No. 2, or amaranth, which was banned in the United States based on concerns regarding teratogenic and reproduction effects raised by USSR studies in 1970 and later confirmed by an in-house FDA study. This dye was also under fire at the time as contributing to behavioral problems in children. However the Canadian Health Protection Branch, having assessed the same data, retained amaranth as a colorant on the basis of lack of carcinogenic evidence and inadequate experimental control used in the FDA study (12). Sweden, Denmark, the FRG, Japan, and nine European Economic Community countries also retained it. In contrast the FDA gave approval in 1974 for FD&C Red Dye No. 40, and it is now the only general-purpose red color certified in the United States. Canada's Health Protection Branch did not allow FD&C Red Dye

No. 40 known as Allura Red, for use in that country. A result of these discrepancies in regulations between countries is an added complexity between trading partners and to global trade.

A quick assessment of the commodity use of synthetic colors indicates that a large proportion are used in fruits and fruit juices and nonalcoholic beverages. These are foods where water solubility and stability are requirements and suitable natural colorants have yet to be developed.

EFFECT OF COLOR ON SENSORY PERCEPTIONS

The effect of color on sensory perceptions has been well demonstrated. Color can influence people's perception of the basic tastes and affect their ability to distinguish threshold levels (14). Presenting the basic tastes in colored solutions raised the thresholds of the test subjects. Thresholds for sour presented in green or yellow solutions were higher than those presented in red solutions, which in turn were higher than those presented in colorless solutions. This is not entirely accounted for by color association with certain foods, for example, both green and yellow sour and nonsour foods are common. In the case of bitter taste, green and yellow presentations resulted in higher concentrations being required for detection than for colorless or red presentations. It was concluded that red color was not associated with bitter flavor (14). In contrast to sour and bitter, sweet thresholds were lower for green-colored solutions. Thresholds for colored salt solutions were no different than those for colorless solutions. This was attributed to the wide variety of foods associated with a salty taste.

The levels of basic tastes in most foods are well above threshold levels and taste–color associations vary with specific foods. Color was shown to be a distraction in correctly identifying the sweeter sample in a paired comparison test of pear nectar, but it was not confirmed that the color green lowers the perception of sweetness, a result obtained in an earlier study (15). It was concluded that the effect of color varied considerably between individuals. The intensity of red color has been shown to affect a subject's perception of sweetness in beverages and solutions (16). A red color in fruit-flavored beverages was associated with an increase in perceived sweetness although the sucrose level was constant. Sweetness of darker red solutions was reported to be perceived as being 2–10% greater than the uncolored reference when actual sucrose concentrations were 1% less. There is an interrelationship between color and flavor, for example, the introduction of blue color to cherry- and strawberry-flavored drinks markedly decreased perception of tartness and flavor (17). The addition of yellow color decreased perception of sweetness to a lesser extent, and the addition of red increased sweetness. This study introduces the possibility of using color to reduce the concentration of sucrose required in a beverage. The perception of sweetness seems to be more affected by color than the perception of salt. It has been observed that color had no influence on salt perception of chicken broth colored red or yellow (18). Neither was flavor preference affected by color.

An investigation into the effect of color on aroma, flavor, and texture of foods, showed that appropriately colored foods were perceived to contain greater intensity of aroma and flavor and as having a higher quality, with aroma being the most affected (19). Perception of texture was not affected by color. Another study found that fruit-flavored beverages were correctly identified when their color was visible but when their color could not be seen dramatic reductions in correct identification occurred (20). Only 70% of the tasters correctly identified grape; 50%, lemon–lime; 30%, cherry; and 20%, orange. When the study was expanded to include combinations of color, flavor, colorless, and flavorless in beverages, flavor identification was color cued, and in the case of the colorless beverages a spread of flavors was identified. Flavorless beverage colored red was identified as strawberry by 22% of the tasters; orange-colored, as orange by 22%; and green-colored, as lime by 26%. Identification as having no flavor ranged from 41% for colored beverages to 48% for the colorless beverage. These same researchers observed that when white cake was varied in lemon flavor and yellow color that flavor intensity was most affected by color when no lemon flavoring was added, but the effect was nonlinear. Intensity of flavor was most affected between no added color and the lowest level. Color has been shown to have a strong influence on the assessment of aroma and flavor for port wines (21). These ratings were shown to be influenced by visual cues when these were made available. Examination of assessments by individual assessors showed assessors differed in the degree of this influence. These researchers reported that trained panelists were able to remove color bias from evaluation of Bordeaux wines when instructed to do so.

That color plays a role in the acceptability of many foods cannot be denied. One study varied the color of cherry- and orange-flavored beverages above and below normal levels. It was observed that the overall liking peaked around the normal color (20). It was also noted that overall acceptability was more closely related to flavor than to color acceptability, suggesting that color cannot substitute for flavor. This same study demonstrated a significant color-flavor interaction for overall acceptability of lemon-flavored cakes. More intense flavors were liked less at higher color concentrations. As flavor levels increased, overall acceptability peaked at lower color levels. These findings suggest that intensities reach a point where they exceed the accustomed expectation and are, therefore, liked less. The interrelationships between pleasantness, color, and sucrose concentration have been observed (16). The interaction was evident in the lower pleasantness rating given to the darker red intensity beverage. This may have resulted because the beverage was perceived as sweeter and, therefore, as less pleasant or the red color may not have been associated with the tasters' expectations of a desirable red color. One study examined the perceived pleasantness of beverages varying in sweetness and color and concluded that the most intense color or sweetest flavor did not always produce the optimum product formulation (17). Color and appearance have been shown to have important influences on overall acceptability of wines but the degree varies with the individual assessor and the study (21).

MARKETING IMPLICATIONS

Any research and development technologist should be aware of the marketing implications related to the color of a food as they affect product development and quality control. The technologist must bear in mind that it is what is perceived by the consuming public that controls sales not necessarily the truth of the perception.

A 1985 survey conducted by *Good Housekeeping* magazine, asked respondents to indicate whether nutrition or appetite appeal was the more-important selection criterion for food purchases; 45% stated appetite appeal (22). Color certainly plays a considerable role in appetite appeal and this is one reason that foods are colored. Yet some consumers consider the addition of color as perpetrating deception and in Europe there have been attempts to ban the use of colorants (13). It has been reported that the lack of artificial colorings and the presence of natural color rank high as positive food attributes, whereas the use of artificial color was ranked as negative (22). However, 35–40% of respondents considered artificial color as somewhat acceptable. More than 50% of the respondents considered coal tar color derivatives to be unacceptable. Apparently, the use of colorants for cosmetic reasons, once considered to enhance acceptability, may now be a detractor because of the pressure to reduce the use of additives in processed foods as a result of prevailing consumer attitudes and safety concerns. It should also be remembered that experiences change over time and that this influences acceptability. The combination of experience and prevailing consumer concerns can combine to bring about changes in acceptability. The Spy variety of apple was once prized as a cooking apple but it has gradually been superseded by other varieties for processing and retail sales. When cooked, the Spy apple produces a very intense yellow color. This bright yellow color is less preferred by today's consumers than the duller browner color of the cooked product of more familiar varieties. Part of this is undoubtedly that the intense color is unexpected and unfamiliar but the lower preference may also result from the fact that the intense yellow, although natural, has become suspect.

Although within the marketing profession there has been much exploration of the effects of color, particularly in the packaging, on the sales of food products, little research of this nature has been published. Package design and color are important factors in attracting a buyer at the point of purchase. That packaging becomes part of a familiar product has been witnessed by any shopper who has had to search for an often-purchased product because the color of the package has been changed. Marketers are concerned about product image and thus both the product and the package must contribute to this image. Both can become dated and in an industry that experiences a short life for many of its products this can be of concern for the long-lived products. Black has not generally been considered a desirable color for food packages but with the popularity of art deco design, sophisticated black packages and labels are beginning to appear. Marketers are quick to take advantage of color association with consumer concerns such as that of green with environmental concerns. One Canadian food retailer has capitalized on this to develop an array of G.R.E.E.N. products, which have been promoted as being environmental friendly. This sort of use of color with food is usually beyond the purvey of the technologist, but such marketing decisions may result in formulation requests to research and development.

It can be concluded that color has an important association with food, which extends far beyond the technologist's concern with measurement. An examination of all the facets as they affect any product is becoming increasingly important in coping with the greater degree of sophistication and competitiveness occurring in the food industry.

BIBLIOGRAPHY

1. F. W. Billmeyer, Jr., and M. Saltzman, *Principles of Color Technology*, 2nd ed., John Wiley & Sons, Inc., New York, 1981.

2. D. B. Judd, and G. Wyszecki, *Color in Business, Science, and Industry* John Wiley & Sons, Inc., New York, 1963.

3. R. S. Hunter, *The Measurement of Appearance*, John Wiley & Sons, Inc., New York, 1975, pp, 44, 81, 167.

4. F. J. Francis, "Colour and Appearance as Dominating Sensory Properties of Foods," in, G. G. Birch, J. G. Brennan, and K. J. Parker, eds., *Sensory Properties of Foods*, Elsevier Applied Science Publishers, Ltd., Barking, UK, 1977, p. 27.

5. E. A. Gullett F. J. Francis, and F. M. Clydesdale, "Colorimetry of Foods: Orange Juice," *Journal of Food Science* **37**, 389 (1972).

6. D. B. MacDougall, "Colour in Meat" in G. G. Birch, J. G. Brennan, and K. J. Parker, eds., *Sensory Properties of Foods*, Applied Science Publishers Ltd., 1977, p. 59.

7. D. B. MacDougall, "Instrumental Assessment of the Appearance of Foods," in A. A. Williams and R. K. Atkin, Eds., *Sensory Quality in Foods and Beverages*, Ellis Horwood Ltd., 1983, p. 121.

8. F. J. Francis, and F. M. Clydesdale, *Food Colorimetry: Theory and Applications*, AVI Publishing Co., Inc., Westport, Conn., 1975.

9. F. M. Clydesdale, and F. J. Francis, "Color Measurement of Foods. 28. The Measurement of Meat Color," *Food Product Development* **5**, 87 (1971).

10. B. A. Eagerman, F. M. Clydesdale, and F. J. Francis, "Determination of Fresh Meat Color by Objective Methods," *Journal of Food Science* **42**, 707 (1977).

11. A. A. Williams, C. R. Baines, and M. S. Finnie, "Optimization of Colour in Commercial Port Blends," *Journal of Food Technology* **21**, 451 (1986).

12. Institute of Food Technologists' Expert Panel on Food Safety & Nutrition and the Committee on Public Information, *Food Colors. Scientific Status Summary, Institute of Food Technologists*, Chicago, Ill., 1980.

13. D. Blenford, "Natural Food Colours," *Food Flavourings, Ingredients and Processing* **7**(7), 19 (1985).

14. J. A. Maga, "Influence of Color on Taste Thresholds," *Chemical Senses and Flavor* **1**, 115 (1974).

15. R. M. Pangborn and B. Hansen, "The Influence of Color on Discrimination of Sweetness and Sourness in Pear-Nectar," *American Journal of Psychology* **76**, 315 (1963).

16. J. Johnson and F. M. Clydesdale, "Perceived Sweetness and Redness in Colored Sucrose Solutions," *Journal of Food Science* **47**, 747 (1982).

17. J. L. Johnson, E. Dzendolet, R. Damon, M. Sawyer, and F. M. Clydesdale, "Psychophysical Relationships Between Perceived Sweetness and Color in Cherry-Flavored Beverages," *J. Food Protect.* **45**, 601 (1982).

18. S. R. Gifford and F. M. Clydesdale, "The Psychophysical Relationship Between Color and Sodium Chloride Concentrations in Model Systems," *Journal of Food Protection* **49**, 977 (1986).

19. C. M. Christensen, "Effects of Color on Aroma, Flavor and Texture Judgements of Foods," *Journal of Food Science* **48**, 787 (1983).

20. C. N. DuBose, A. V. Cardello, and O. Maller, "Effects of Colorants and Flavorants on Identification, Perceived Flavor Intensity, and Hedonic Quality of Fruit-Flavored Beverages and Cake," *Journal of Food Science* **45**, 1393 (1980).

21. A. A. Williams, S. P. Langron, C. F. Timberlake, and J. Bakker, "Effect of Colour on the Assessment of Ports," *Journal of Food Technology* **19**, 659 (1984).

22. K. W. McNutt, M. E. Powers, and A. E. Sloan, "Food Colors, Flavors and Safety: A Consumer Viewpoint," *Food Technology* **40**(1), 72 (1986).

ELIZABETH A. GULLETT
University of Guelph
Guelph, Canada

COMPLEXATION. See CAFFEINE; TEA.
COMPOSTING. See MEAT INDUSTRY PROCESSING WASTES: CHARACTERISTICS AND TREATMENT

COMPUTER APPLICATIONS IN THE FOOD INDUSTRY

BACKGROUND AND DEFINITIONS

Computers are general-purpose machines that process data according to sets of instructions they are given. Computer science is the study, design, and application of electronic and mechanical devices in data manipulation and problem solving (1). The applications of computers range from recreational games to large database systems. The term hardware refers to the machinery or equipment that forms the computer system. Hardware includes items such as the central processing unit (CPU), disk drive, keyboard, and printer. The CPU consists of a control unit that extracts and executes information from the computer's memory and an arithmetic–logic unit that performs arithmetic calculations. In large main-frame computers the CPU comprises a number of circuit boards. In a personal computer the CPU is contained on a chip called the microprocessor.

Software refers to the program or instructions that tell the hardware how to operate and what tasks to perform. In the early days of computers it was common to have programs written specifically for a particular company. Software packages designed for a variety of uses are currently available to the public. However, it is still common if not essential to employ specialists in the implementation and operation of these packages. Types of commercially available software include database management systems, inventory, payroll and accounts receivable programs, spreadsheets, and word processors. A computer can move data into its main memory called random access memory (RAM).

After processing, the data can be sent to peripheral devices such as disk drives, monitors, or printers. Major features of computer systems include the operating system, type of CPU, clock speed, memory capacity, and communications ability. The operating system is the main control program for the computer. In small computers the operating system loads the application program into memory for execution. In multitasking computers the operating system is responsible for the concurrent operation of one or more programs. A major function of the operating system is to keep track of the data on the disk.

The CPUs commonly used in personal computers are the 8088, 8086, 80286, 80386, and 80486. The 8088 uses an 8-bit data bus, while the 8086 uses a 16-bit data bus. These microprocessors were introduced in 1978. The 80286 was introduced in 1984; it is a multitasking 16-bit CPU that is up to 20 times faster than the 8088. The 80386 was introduced in 1986; it is a multitasking 32-bit CPU that is up to 35 times faster than the 8088. The 80486 is a high-speed version of the 80386.

The clock speed dictates the speed of the CPU. The clock circuit uses a quartz crystal to generate a uniform electrical frequency delivering a steady stream of pulses to the CPU. Clock speeds are expressed in megahertz (million cycles per second).

A computer's memory is made up of RAM chips. RAM chips hold the program being used and the data being processed. They are functional only as long as the electrical power is on. The amount of RAM determines the size of program and the amount of data that a computer can manage.

With the proper device such as a modem or a cable and a communications program, two computers can exchange data. With a local area network (LAN), groups of personal computers can share software and peripheral devices such as printers. Main-frame computers can communicate with many terminals through separate machines called communications controllers.

The decrease in cost and increase in availability and power of computers has resulted in their incorporation in nearly all phases of the food industry. Computers are well suited for repetitive calculations. One of the earliest applications of computers in food processing was linear programming to obtain optimum utilization of resources. Typical problems included production planning and least cost formulations. These techniques continue to be widely used (2). Major areas of use in food processing include information processing, process control, data acquisition, and computer-aided manufacturing and design.

FOOD INDUSTRY APPLICATIONS

Information Systems

Business applications for computers such as word processing, databases, and spreadsheets are commonly used in the food industry. Data from remote and varied locations can be combined for analysis and report generation. Spreadsheet programs are commonly used for accounting purposes. A spreadsheet creates a two-dimensional matrix of rows and columns. The rows and columns intersect to form cells. When cells are assigned values, formulas can be developed to perform calculations on the values. Relational spreadsheets store data in a central database and formulas in the spreadsheet. This allows the data to be easily updated for multiple users of the spreadsheet. A database is a collection of electronically stored, interrelated files. Database management systems control the organization, storage, and retrieval of database information. Numerous word processors are available. Some are specialized for science and engineering applications. These are able to produce complex equations or chemical structures. Most word processors are designed to accomplish the functions of a typewriter. The ability to edit, move, and copy text; to merge files; and to check spelling has made word processing an efficient method of document preparation.

Process Control

Process control systems monitor signals from various sensors, perform analyses on the information, and provide the proper control. In the past, processes were often controlled by pneumatic analogue controllers. The processes were monitored from remote-control panels and various values recorded. Either alarms would alert an operator of deviations or hard-wire relay logic was used to control motors or other devices.

Control is now accomplished with a programmable logic controller (PLC). Data can be stored in and accessed from the PLC by other hardware for report preparation. Many of the sensors used in the food industry such as those measuring movement or temperature are analogue or continuous. These require an analogue to digital converter to change the signal into binary coded form for the computer. Computerized control of basic processes such as weighing can be accomplished using PLC equipment that can weigh and record various ingredients. With PLC-based systems the controller's function is to keep certain parameters constant.

The more advanced process control systems adjust continuously to provide the best possible settings for the existing process and materials. For example, in chocolate tempering where cooling rate, flow, and shear forces affect the quality of the chocolate, changes in raw materials necessitate adjusting the process conditions. Computer systems can provide the necessary control to produce a product with optimum quality (3).

Computerized systems are used to automate various phases of food processing. Boiler efficiency is improved with microprocessor controls that optimize air and fuel ratios. Sensors can detect preliminary cavitation impulses in a pump and automatically adjust the variable speed drive to compensate. The values for pH can be maintained or adjusted depending on the acid production of a bacterial culture. For retort processing of canned foods, a computerized control system can provide consistent control of retort temperature and pressure. All aspects of the system can be monitored including cooking, venting, and cooling. If a temperature deviation is unacceptable, the system can automatically recalculate a new process time and make all necessary adjustments (4).

In another example, a dairy processing plant can choose to program the computer system to require the input of certain data such as weight, quality checks, antibiotics, and driver identification before the unloading pump will start. After unloading begins, the computer starts the flow meter, sampler, and cooling media. Proper valves will open and tank levels will be monitored. Milk will flow to the processing operations such as standardizing, pasteurizing, homogenizing, and filling while critical parameters such as temperature and pressure are being monitored by the computer. Bottles filled and any product lost are recorded (5).

Computer-integrated manufacturing (CIM) combines office and accounting functions with process control systems in the plant (Fig. 1). CIM involves interfaces, materials control, data communication and local area networks, warehouse and inventory control, supervisory and monitoring control, and statistical process control. Examples of CIM include activities such as reading gauges, preparing reports, filing and upgrading forms, calculations, and storage of information. CIM is implemented through a communications architecture that interfaces electronic control equipment with data-collection equipment.

Computerized management systems require automation and optimization of unit operations, which are then tied into a total plant control system. Planning and control systems examine factors such as quality-control specifications, shelf life, schedule control, and raw material and finished product inventory. High-level systems have been developed so that complete process control strategies can be configured without conventional language programming. Information from computers on the manufacturing line tells management the amount being produced and tells the warehouse operators how much is on hand. The office personnel can determine the demand for orders immediately. The amount of paperwork decreases when production, financial, and warehousing data are all on the same terminal.

Computer-aided management allows a company to use materials as soon as they arrive and move them quickly through the process. With this type of control, massive warehouse stocks of finished goods can be reduced and finished product can be shipped out immediately. Creating a situation where information is exchanged between administrative and process control activities allows the implementation of just-in-time (JIT) manufacturing. Separate activities such as customer orders and production scheduling are integrated into a total system.

Computer-controlled workstations along with integrated materials handling stations enabling the production of various related products are known as flexible manufacturing systems (FMS) (6). Dedicated, continuous equipment offers high volume, low unit cost, and little flexibility. However, batch control systems are most commonly used in the food industry. The use of FMS for optimizing batch processing operations is complimentary to consumer demands for lower volume specialty foods. Currently, FMS applications are used mainly for nonfoods. Likely applications in the food industry are packaging, filling, and thermal processing.

Laboratory Automation

Laboratory automation systems (LAS) and laboratory information management systems (LIMS) have been developed to reduce errors and increase efficiency of laboratory analyses. Rapid access to previous analyses, data manipulation, and report preparation are valuable features of LIMS. Laboratory management systems are principally concerned with data handling. Instrument data can be recorded and stored for analysis and reporting. Many devices include both data acquisition and control features.

Microprocessors are standard components of many laboratory instruments. Spectroscopy curves and chromato-

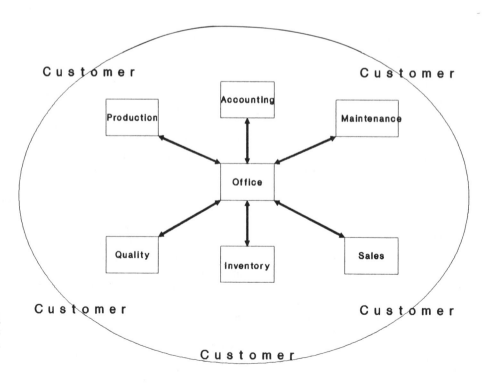

Figure 1. Computer integrated manufacturing unites all vital company functions through a common database.

grams of unknown samples can be compared to standard references. Computer-aided analysis of protein electropherograms provides an array that can be compared to a library of standard patterns to determine those having the highest similarity (7).

Computer programs that assist in chemical solution preparation are useful in laboratory operations. Programs are able to perform calculations for dilutions, fortifications, and titrations. Databases containing information on microbial properties and characteristics can be used in identifying unknown cultures. Computer-assisted identification of lactic acid bacteria on meat can be accomplished using a probability matrix (8).

Robots are devices that imitate human actions. Laboratory robotics applications include nutrient analysis, toxin and residue analysis, additive detection, and microbial analysis. One system currently in use in several laboratories consists of a centrally located robotic arm surrounded by various workstations. Specific applications include fat and microbial analysis.

Sensory Evaluation

Computer-based sensory evaluation systems are able to develop questionnaires, administer sensory tests, and collect sensory data. Menu-driven software packages are designed to automate the collection and analysis of sensory data. Some systems have been developed that use touch screen hardware, software, and interface designed to perform sensory evaluation tasks such as test design, scorecard preparation, and data analysis (9). Computerized quantitative descriptive analysis (QDA) may improve freedom of response in making judgments about samples. Computerization may also eliminate human error associated with recording data. Variations of responses for computerized QDA has been shown to be smaller than or equal to manual procedures (10). Computer-based statistical programs are used in the analysis of sensory evaluation panel results, analysis of variance being the most common.

Artificial Intelligence

Artificial intelligence (AI) is a group of techniques that simulate human reasoning. AI and robotics are mainly used in automotive and other hard-goods industries. Artificial intelligence encompasses the areas of natural language processing, image processing or machine vision, expert systems, and robotics. Natural language programs allow requests or commands to be given in the language spoken by the user. Aside from language translation, natural language processing is being used to develop database searches and manipulations from natural language questions. Image processing or machine vision is used for quality inspection and for guidance of robot arms. Image processing combines microprocessors with video, laser, or x-ray technology to inspect products and packaging on production lines. In the food industry image processing has been applied to detection of slicing defects, color of baked goods, dough thickness, surface texture and size, missing or damaged labels, faulty closures, defective pieces, and extraneous matter such as glass or bone fragments (11).

Computer pattern recognition analysis of orange juice has been successful in detecting adulteration (12). An image processing program has been developed for detecting stress cracks in corn kernels using a commercial vision system (13). Estimation of carcass composition in live cattle using a picture processing system gives results comparable to traditional prediction techniques.

X-ray systems can be used to inspect opaque containers. Infrared light that is absorbed by glass, labels, or caps can indicate chipped finishes and missing caps or labels. Gamma ray systems are able to measure levels of fluids and other materials in a container. The technology allows interfacing with process controls to detect as well as correct defects. Expert systems manipulate a set of facts obtained from a human expert to solve problems normally solved by that expert (14). Examples of successful expert systems include programs to determine and correct problems with large hydrostatic retorts, schedule facilities and personnel, identify the source of defects in Cheddar cheese, and assist in label preparation.

Robotic devices are most prominent in packaging or material handling devices. The use of robots in laboratories was discussed above.

Computer Modeling

Models are sets of mathematical equations that represent an actual set of operations. Values can be substituted for variables in the equations to predict the success of the set of operations. Various models have been developed to help predict economic returns on automation. These models include plant capacity, insurance, taxes, sales revenue, capital investment, wages, and benefits.

Although computer modeling in the design of molecular structures has been used for pharmaceutical and basic chemistry applications, its use in food ingredient research is more recent. Models for sweet molecules can be determined by superimposing known sweet compounds on a computer screen and noting similarities in factors such as shape and charge. Other potential uses for computer modeling for food ingredients include fat and starch research and developing new bulking agents.

Many applications that are not understood well enough to use a mathematical model may be modeled by simulation techniques. Using simulation programs, new or altered processes and marketing schemes can be tried before actual implementation. Simulation programs allow investigation of numerous alternatives at a greatly reduced expense. Simulation techniques use objects and activities to develop models.

Computer models of sterilization processes can predict the temperature at the point of slowest heating in specified containers as a function of retort temperature (15). Computer-aided design, or computer-aided manufacturing, has been used in process and packaging design. Computer models can be used in the prediction of shelf life based on moisture changes, cost of materials, shock resistance, prediction of stress, design of molds, and design of cartons or cases. Computer generated pictures can be used to design, redesign, or modify packaging size and shape to meet a specific requirement such as a store's shelf space. Carcasses can be evaluated and classified by scanning the loin area with a video camera followed by computer analysis of the resulting data (16).

Quality Assurance

The standard tools of statistical process control include flow charts, histograms, and control charts. Numerous software packages have been developed to perform statistical procedures and chart generation. These programs are particularly useful for plotting charts such as X-bar and R charts and provide immediate information about the state of the process. Programs have also been used in the selection of sampling plans and operational characteristic curves (17). These programs require the user to specify the alpha level or producer's risk, acceptable quality level, beta level or consumer's risk, lot tolerance proportion defective, and the lot size. Double sampling plans are also available. The type of distribution such as poisson, binomial, hypergeometric, or normal can be selected. Portable data loggers have been useful for measuring and recording temperature at remote locations.

Existing commercial spreadsheets can be used for quality functions. For example, a spreadsheet containing quality data for malting barley performs several calculations and executes a selection index based on the importance of each trait to overall malt quality (18). Computer-assisted quality control systems allow presentation of pertinent production and quality-control data at the time and location from which it is generated. With computerized systems, trends can be rapidly observed and corrected.

ADVANTAGES AND DISADVANTAGES

The principal advantage of computer technology in food science and processing is speed and accuracy. Software-reconfigurable machinery allows greater levels of flexibility so that short processing runs can be made and a variety of products can be produced to meet market demands. Computer systems that plan materials flow permit smaller inventories and lower costs. Major benefits of CIM are flexibility, production control, lower manufacturing costs, and improved product quality. Computer assisted quality control programs allow operators to recognize trends toward a nonconformance situation more quickly. Interdepartmental communications are more rapid and decisions more accurate. Advantages of robotics systems over manual operations include speed, better precision, and improved worker safety. Labor savings are the major justifications cited for a robot investment. However, another advantage often overlooked is reduction in mundane tasks making better use of highly trained technicians and scientists. Use of microcomputer controlled robots removes the human factor as a source of error. Scheduling can be based on orders and quickly revised as needed.

Disadvantages in using computers include the loss of productive time during the learning period and human data handling errors. Programmable logic controllers represent a great improvement over older, relay logic networks. Various hardware and software systems can collect data from the controller, perform analyses, and generate appropriate reports. Unfortunately, factory automation cannot usually compensate for a poorly designed product or process or for ineffective management.

FUTURE DEVELOPMENT

In the area of information systems, future database programs will be better able to manage multiple information forms such as data, pictures, graphics, and voice. The success of voice as an information form will depend on natural language developments. Digital communications will replace analogue signals so that sensors can keep up with high-speed operations. Continued efforts will be made in simplifying and reducing costs of process control by integrating the computer functions within the various operations of the company. CIM areas that will be emphasized in the future include computer-aided design, bar coding and automated data collection, and computer-aided manufacturing. The use of modular hardware will allow plant personnel more flexibility to make changes. In the future, remote laboratory analysis may be reduced with the placement of robotics in the processing area to provide direct process control feedback.

BIBLIOGRAPHY

1. A. Freedman, *The Computer Glossary*, 4th ed., The Computer Language Company, Inc., Point Pleasure, Pa., 1989.
2. G. L. Kerrigan and J. P. Norback, "Linear Programming in the Allocation of Milk Resources for Cheese Making," *Journal of Dairy Science* **69**, 1432 (1986).
3. K. L. Bibbo and A. C. Warren, "Automation and Systems Control and Their Effect on Productivity and Product Quality," *Food Technology* **40**(7), 43 (1986).
4. R. J. Swientek, "Computerized Control System Optimizes Retort Processes," *Food Processing* **46**(10), 73 (1985).
5. C. Honer, "Computerizing the Dairy Plant," *Dairy Record,* **85**(5), 81 (1984).
6. J. T. Clayton, "Flexible Manufacturing Systems for the Food Industry," *Food Technology* **41**(12), 66 (1987).
7. J. C. Autran and P. Abbal, "Wheat Cultivar Identification by a Totally Automatic Soft Laser Scanning Densitometry and Computer-Aided Analysis of Protein Electropherograms," *Electrophoresis* **9**(5), 205 (1988).
8. B. Doring, S. Ehrhardt, F. K. Luke, and U. Schillinger, "Computer-Assisted Identification of Lactic Acid Bacteria from Meats," *Systematic and Applied Microbiology* **11**(1), 67 (1988).
9. R. L. Winn, "Touch Screen System for Sensory Evaluation," *Food Technology* **42**(11), 68 (1988).
10. A. J. King and A. Morzenti, "Response Freedom in Computerized and Manual Modes of Sensory Scoring," *Food Technology* **42**(10), 150 (1988).
11. J. Mans, "Sensors: Windows into the Process," *Prepared Foods* **157**(3), 81 (1988).
12. G. A. Perfetti, F. L. Joe, Jr., T. Fazio, and S. W. Page, "Liquid Chromatographic Methodology for the Characterization of Orange Juice," *Journal of the Association of Official Analytical Chemists* **71**(3), 469 (1988).
13. S. Gunasekaran, T. M. Cooper, A. G. Berlage, and P. Krishnan, "Image Processing for Stress Cracks in Corn Kernels," *Transactions of the ASAE* **30**(1), 266 (1987).

14. M. R. McLellan, "Introduction to Artificial Intelligence," *Food Technology* **43**(5), 120 (1989).

15. P. T. Kelly and P. S. Richardson, "Computer Modelling for the Control of Sterilization Processes," *Technical Memorandum Campden Food Preservation Research Association* No. **459,** Campden Food Preservation Research Assoc., Chipping Campden, UK, 1987.

16. Ger. Pat. DD259, 346 (1988), C. Liebe, W. Konig, R. Osterloh, and T. Frouse.

17. G. Pappas and V. N. M. Rao, "Computer Codes and Their Application. Development of a Sampling Plan for Specified Risks by Microcomputers," *Journal of Food Processing and Preservation* **11**(4), 339 (1987).

18. J. A. Clancy, and S. E. Ullrich, "Analysis and Selection Program for Malt Quality in Barley by Microcomputer," *Cereal Chemistry* **65**(5), 428 (1988).

General References

G. Hettich, "Pyramid Power, The Basics of Food Automation," *Prepared Foods,* **158**(4) 132 (Apr. 1989).

I. Saguy, *Computer-Aided Techniques in Food Technology,* Marcel Dekker, Inc., New York, 1983.

A. A. Teixeira and J. E. Manson, "Computer Control of Batch Retort Operations With On-Line Correction of Process Deviation," *Food Technology* **36**(4), 84 (1982).

A. A. Teixeira and C. F. Shoemaker, *Computerized Food Processing Operations,* Van Nostrand Reinhold Co., Inc., New York, 1989.

ROBERT L. OLSEN
Schreiber Foods, Inc.
Green Bay, Wisconsin

CONDENSERS. See DISTILLATION: TECHNOLOGY AND ENGINEERING; FREEZING SYSTEMS FOR THE FOOD INDUSTRY.

CONFECTIONERY. See FOOD PROCESSING: STANDARD INDUSTRIAL CLASSIFICATION.

CONFECTIONS

HISTORY AND PRODUCT CATEGORIES

History

Confections date back more than 3,000 years to the time of the Egyptians; hieroglyphics have been found depicting the preparation of sweets. Romans also possessed a sweet tooth and recent archaeological explorations at Herculaneum uncovered a store, thought to be a confectionery kitchen, equipped with various early confectionery implements (ie, pots, pans, and formers), many of which are somewhat similar to those currently used. During this early time period, confections consisted of products prepared with various dried or fresh fruits, nuts, spices, and herbs. Honey was the main confectionery sweetener.

With the coming of the Middle Ages, sugar-based (sucrose) confections arose first in Europe but quickly spread. While initially sold only by apothecaries, their sweet taste created a consumer demand that quickly outgrew supply. Through the discovery of sugarcane cultivation and further refinement by the Persians, supply grew, but not un-

til the late fourteenth or early fifteenth centuries, when the Venetians started importing sugar from Arabia, did its widespread use begin and sugar confections were made available to the common people.

In the fifteenth and sixteenth centuries, to meet the growing consumer demand, confectioners were offering a variety of hand-prepared high-boiled sweets, molded sweets, and long-established fruit- and nut-containing products.

With the availability of specialized confectionery equipment (Industrial Revolution period), annual world production and consumption grew to its current high of 107.3 million metric tons (8.2 million tons for the U.S. market solely—1988–1989 totals—Economic Resources: Amstar Corporation, American Sugar Division).

United States confectionery, per capita, consumption continues to increase with growth rates of 5–10% per year expected. When viewed from a worldwide perspective (of industrialized countries), the United States rated number 13 in sweets consumption in 1986; averaging approximately 19 lb per person per year. What will the future bring, is not really known. Space limitation does not permit a detailed analysis of the future development of this popular category of food products.

Product Categories

Confections or candies can be roughly divided into two types: amorphous, or noncrystalline, and crystalline. Products categorized in these types are

Amorphous (Hard Chewy Homogeneous, Noncrystalline)	Crystalline (Crystal Structure)
Caramels	Chocolates
Taffys	Creams
Toffees	Fudge
Brittles	Fondant
	Nougats
	Pressed sweets (lozenges)

Subgroups within each category can be easily established, based on each formula's moisture and fat levels, as well as processing techniques (see Figures 1, 2 and Table 1 concerning confectionery products' relative moistures).

Common to almost all confections is the use of either cane or beet sugar (the disaccharide sucrose) as their principal constituent. Other ingredients and sweeteners commonly used are corn syrup, dextrose (corn sugar) invert sugar, honey, starch, molasses, colors, and flavors. These ingredients are used in both dry and liquid forms.

INGREDIENTS

Corn Syrup

The terms glucose syrup and corn syrup are often used interchangeably, with the following definition: "a purified concentrated solution of nutritive saccharides obtained from starch and having a DE of 20 or more (2)." In the United States and Canada glucose syrup made from corn

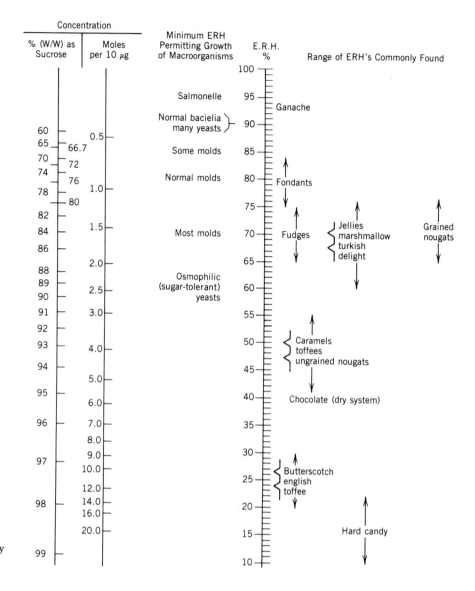

Figure 1. Equilibrium relative humidity of some sugar confectionery products.

starch is called corn syrup. Dextrose equivalent (DE) is defined as the percent reducing sugars, calculated as dextrose on a dry substance basis.

In 1935 only 42 DE acid-converted syrup was available, and it remains a principal raw material for confections today. When only acid hydrolysis is used, DE is adequate to define the carbohydrate composition of a syrup. It is the carbohydrate composition that controls the functional properties and, therefore, the choice of syrup to be used. The corn syrup industry has defined 17 functional properties of corn syrups as they relate to DE. These are shown in Figure 3. It must be remembered that in any multicomponent system, it is rare for a single ingredient to be solely responsible for a functional property, however, Figure 3 serves as a good guide in corn syrup choice. As the width of the arrow increases, so does the property in question.

In the 1950s high-maltose syrups were introduced. These are produced by first converting the starch to a DE of about 20 and then treating it with a malt enzyme, which selectively hydrolyzes the starch to the disaccharide maltose (DP-2). These syrups are used almost exclu-

sively in hard candy because of their less-hygroscopic nature.

While an acid-converted and acid–enzyme–converted (high-maltose) syrup can have the same DE, this term no longer defines the saccharide composition and the functional properties. It is, therefore, necessary to define a syrup by its type of conversion and the amount of the individual saccharides (DP-1, DP-2, etc). Table 2 illustrates the dramatic differences between type of conversion, DE, and saccharide composition of three types of syrup-conversion technique. The significant differences, for example, between acid and high-maltose syrup each at 42 DE in confections are that the high-maltose product has better heat and color stability, better flavor transfer, slower moisture absorption, and reduced browning.

High-Fructose Corn Syrup

The highest conversion corn syrup is about 95 DE and contains 92% dextrose on a dry substance basis. Functionally, this product is similar to dextrose. It also serves as

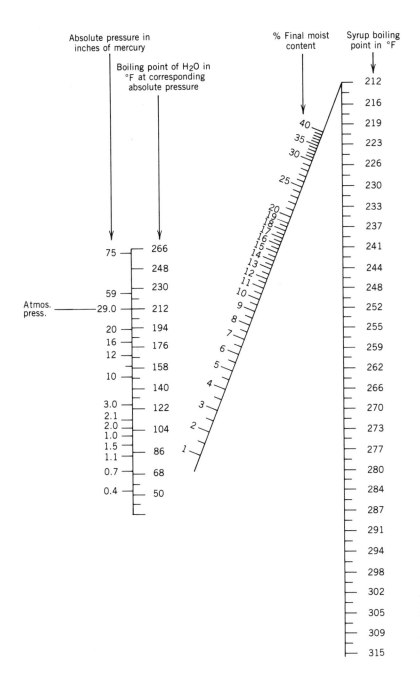

Absolute pressure in
inches of mercury

Boiling point of H₂O in
°F at corresponding
absolute pressure

% Final moist
content

Syrup boiling
point in °F

Figure 2. General data graph. Nomograph for determining the moisture content of sugar solutions at various temperatures and pressures. Note: this chart refers to 100/30 sugar:corn syrup ratio, but is still reasonably accurate for other proportions.

the starting material for the production of high-fructose corn syrup (HFCS). The 95 DE syrup is treated with an enzyme that converts the dextrose to fructose. This process is known as isomerization. The functional properties of this product are similar to honey or total invert sugar. A typical saccharide composition is

Fructose	42%
Dextrose	50%
DP-2	6%
DP-3 and higher	2%

Additional high-fructose products containing 55, 90, and 100% fructose are made from starch syrups by enrichment techniques of the isomerized syrups. Principal advantages of these products are high sweetness, solubility, and hy-

groscopicity. The most-recent commercial corn syrups, the HFCSs, were introduced in the 1970s and came into widespread use in the 1980s, particularly in soft drinks. Dry HFCS was recently announced as a new product from several suppliers. It is encapsulated to prevent moisture pickup and is available in 42, 55, 90, and 95% fructose levels.

Corn Syrup Solids

By removal of about 97% of the water in corn syrup, corn syrup solids are produced. Commercially available products range in DE from 20 to 64 and include the high-maltose types. The functional properties are identical to their liquid counterpart. Their main advantages include ease of handling, suitability for low-moisture products,

Table 1. Boiling Point vs. Solids for Sucrose and Corn Syrups[a]

Boiling Point, °F	NSE 62 D.E.	Sucrose	Reg. 42 D.E.
213	—	25.5	—
214	38.0	42.0	46.0
215	46.0	49.5	53.0
216	51.5	54.5	57.5
217	56.5	58.5	60.5
218	60.0	62.0	64.0
219	63.0	65.0	67.0
220	65.5	67.5	69.5
221	67.5	69.5	71.5
222	69.5	71.5	73.5
223	71.0	73.0	75.0
224	72.5	74.5	76.5
225	74.0	76.0	78.0
226	75.0	77.0	79.0
227	76.0	78.0	80.0
228	77.0	79.0	81.0
229	78.0	80.0	82.0
230	79.0	81.0	83.0
231	79.5	81.5	83.5
232	80.5	82.5	84.5
233	81.5	—	85.0
234	82.5	84.0	85.5
235	83.0	84.5	86.0
236	83.5	85.0	86.5
237	84.0	85.5	87.0
238	84.5	86.0	87.5
239	—	86.5	—
240	86.0	87.0	88.0
241	—	87.5	—
242	87.0	88.0	89.0
243	—	—	—
244	—	—	—
245	88.5	89.5	90.5
250	90.0	91.0	92.0
255	—	93.0	—
260	93.5	94.0	94.5
265	94.5	95.0	95.5
270	95.0	95.5	96.0
275	95.5	96.0	96.5
280	96.0	96.5	97.0
285	96.5	97.0	97.5
300	97.5	98.0	98.5
320	98.0	98.5	99.0

[a] Ref. 1.

ability to increase solids without cooking, and ease of blending with dry ingredients.

Dextrose (d-Glucose)

Dextrose is by far the most abundantly occurring sugar in nature. It is found free, in honey, fruits, and berries. The presence in starch has been discussed and it makes up one-half of the sucrose molecule.

It is generally conceded that Kirchoff's work in 1811 was the forerunner of the starch hydrolyzate industry. Kirchoff's process involved heating a suspension of potato starch with sulfuric acid. Samples of the evaporated syrup, the solidified dry syrup, and the dextrose crystals separated from the syrup were shown to the Academy Imperiale des Sciences de St. Petersburg. For his discov-

ery, Kirchoff was awarded a lifetime pension and was decorated with the Order of St. Anne by the Russian czar.

Not until more-efficient methods for crystallization of dextrose hydrate were developed did pure crystalline dextrose become a principal industrial product. By careful control of concentration, agitation, temperature, and amount of seed crystals, a practical process for the manufacture of pure dextrose hydrate was developed in the early 1920s (3).

Today dextrose is made by an enzyme that essentially hydrolyzes the corn starch into a syrup rich in dextrose (92%). The syrup is purified, evaporated, and crystallized, yielding dextrose hydrate, the most common commercial product. The anhydrous form is made by crystallization at higher temperatures. Powdered, granular, liquid, and coproducts combined with maltodextrins represent the available forms of dextrose.

The functional properties of dextrose in confections have been defined as flavor, humectant, compressibility, hygroscopicity, and sweetness (3) and these properties find use in jellies, gums, marshmallows, pastes, and panned and pressed products. Anhydrous dextrose is permitted, with limitations, in chocolate products and may be used as desired in compound coatings.

Dextrose has a cooling effect when dissolved in the mouth and this property, along with its compressibility and flow characteristics, has resulted in its use in direct compression tableting. For this purpose, coarse grades and the coproducts made with maltodextrins are used.

Hydrogenated Corn Syrup and Sorbitol

Dextrose and the saccharides of corn syrup are reactive materials. When treated with hydrogen under pressure, at elevated temperatures and with a catalyst, a class of products known as polyols are formed when corn syrup is treated; the resulting products are known as hydrogenated corn syrups. If dextrose is the starting material, the product is sorbitol.

These products have a high resistance to color development in storage or when heated and offer good moisture retention. The dry form of sorbitol has a cooling effect when eaten, similar to dextrose hydrate. Certain forms find application in direct compression tableting, panning, high- and low-boiled sweets, and diabetic chocolate.

The principal use for sorbitol and some hydrogenated corn syrups has been in the formulation of sugarless products. The term sugarless is construed to mean that the product is noncariogenic: it must not be interpreted to mean low in calories. While its metabolic pathway may offer some physiological advantages, the caloric value is similar to carbohydrates, that is, 4 cal/g. Depending on the assumed daily consumption of products formulated with these polyols, a laxative warning may be necessary.

Honey

Honey was probably the original sweetener, possibly predating written history. Hannibal's march across the Alps, according to history, was made possible, in part, because his troops carried rations of a honey almond nougat, which was high in caloric value and compact in size. Honey is naturally high in invert sugar, comparatively

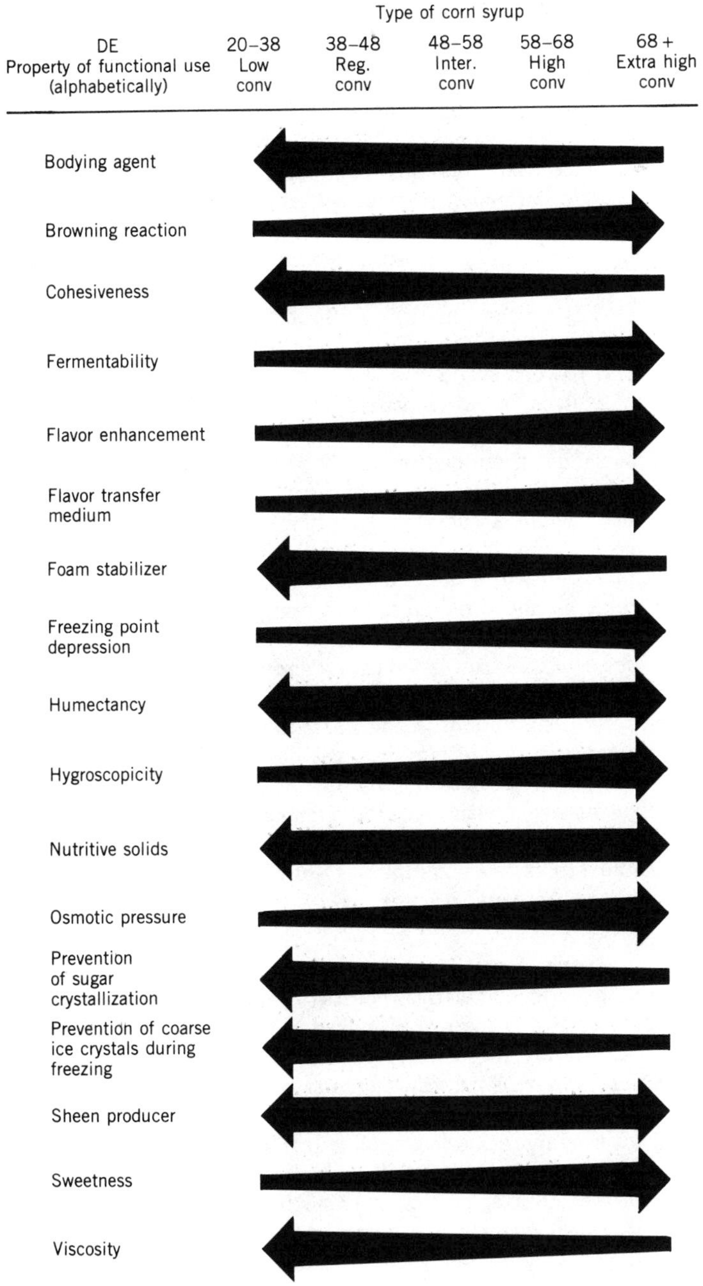

Type of corn syrup

DE Property of functional use (alphabetically)	20–38 Low conv	38–48 Reg. conv	48–58 Inter. conv	58–68 High conv	68 + Extra high conv

Bodying agent

Browning reaction

Cohesiveness

Fermentability

Flavor enhancement

Flavor transfer medium

Foam stabilizer

Freezing point depression

Humectancy

Hygroscopicity

Nutritive solids

Osmotic pressure

Prevention of sugar crystallization

Prevention of coarse ice crystals during freezing

Sheen producer

Sweetness

Viscosity

Figure 3. Properties and functional uses of corn syrups (2).

Table 2. Saccharide Distribution in Starch Syrups vs. Type of Conversion[a]

Syrup Type	D.E. Value	Amount of Constitutent, %						
		DP-1[b]	DP-2	DP-3	DP-4	DP-5	DP-6	DP > 6
Acid conversion	30	9	10	12	9	7	6	47
	36	14	12	10	9	7	6	42
	42	19	14	12	10	8	7	30
	55	31	18	13	10	7	5	16
Acid–enzyme conversion	51	20	31	18	6	4	3	18
	64	37	31	11	5	4	3	13
	70	43	30	7	5	3	2	10
High maltose	36	5	28	12	4	3	2	46
	42	6	44	13	3	2	2	30
	48	9	52	15	2	2	2	18

[a] Ref. 2.
[b] DP refers to degree of polymerization of dextrose, where DP-1 is the monosaccharide dextrose, DP-2 is disaccharide, etc.

high in cost, and variable in flavor. In confections, it is now primarily used as a flavor enhancer although it also retards sucrose crystallization. The flavor depends on the source of nectar; clover and orange blossom honeys are usually preferred for their delicate flavor. Honeys from alfalfa, mesquite, and buckwheat are stronger in flavor and must be carefully used. Honey usually contains 16–18% water and 82–84% solids; it varies in color from amber to water-white. USDA standards for grades exist as does a subsidy program, which has resulted in large surpluses in storage.

Molasses

Today molasses is basically a coproduct of the sugar-refining process. That which is first separated from sugar liquor is light in color and milder in flavor than molasses produced in later stages of the process. The darker, stronger molasses can also impart a bitter flavor to candy if used at too high a level. Because molasses contains a large amount of invert sugar it acts to inhibit sucrose crystallization. Powdered molasses has been available for about 20 years and has a limited use in confections.

Maple Sugar

Maple sugar is produced in northern United States and Canada by tapping the sugar maple tree (*Acer saccharum*), collecting the sap, and evaporating most of the water. The crystallized maple sugar contains slightly more than 80% sucrose and has a distinctive flavor. Because quantities are limited and prices are high, maple sugars and syrups have largely been supplemented by artificial flavors. Nevertheless, they still find a use in fudges, creams, and in combination with walnuts.

PRODUCT CHARACTERISTICS

By proper ingredient selection and processing methodology, a variety of confections can be produced. Following the amorphous, noncrystalline and crystalline categories, basic product characteristics can be formed for each type of product.

Noncrystalline
Hard candy

Brittle

Caramel

Toffee

Licorice

Jellies and gums

Crystalline
Fondants and creams

Fudge: beater and kettle types

Nougats

Marshmallows

Pralines

Pressed candy (tablets)

Marzipan and pastes

Panned candies (*dragées*)

Noncrystalline

Hard Candy. The objective in the production of this confection is to use ingredients and conditions that will allow the finished product to be formed into a shape with no apparent evidence of crystalline form, commonly called a glass. Examples are fruit drops, clear mints, butterscotch, lollipops, candy canes, and most cough drops.

The most common ingredients and ratios used are sucrose (liquid or dry), which makes up 50–70% of the carbohydrate, and 42 DE acid-converted corn syrup, which makes up the remaining 30–50%. Invert sugar has also been used as a crystallization inhibitor, but is not as effective as corn syrup.

Normal cooking procedure is to combine the sucrose and corn syrup and heat, usually with the aid of vacuum, to reduce the moisture to 2% or less. Special steam pressure cookers through which syrup passes continuously are used when a constant supply is required. The mass is cooled somewhat and colors, flavors, and acidulants are mixed in while the mass is still plastic. These colors and flavors may be mixed to form multicolored confections for sticks, suckers, starlights, and other novelty designs. If the cooked syrup is fed into a special spinning machine a threadlike glass is formed known as cotton candy.

While still pliable, the warm mass is formed by cutting, stamping, or pressing into the desired shape. A satinlike finish may be obtained by pulling the plastic mass on rotating arms with repeated overlapping. Partial crystallization results with a short spongy texture; candy canes and sticks are examples. Cooling and packaging completes the usual process.

Brittle. Brittles are essentially hard candy with ingredients added at critical points to change their eating characteristics. They are prepared in much the same way as hard candy, ie, sucrose and corn sugar are boiled to evaporate water; often raw peanuts or almonds are included (some confectioners prefer to add previously roasted nuts after evaporation is completed). The addition of bicarbonate of soda puffs the batch, which is then cooled and cut or broken. Butter crunch is basically the same, with much less corn syrup or invert and considerably more butter for flavor. It is virtually the same as English toffee.

Caramel. The manufacture of caramel resembles hardcandy making except that milk and fat are added to the sweetener blend. Historically, invert sugar in small amounts has been included with the sucrose and 42DE corn syrup. Some confectioners, however, have converted to 64 DE corn syrup and eliminated the invert sugar. Sweetened, condensed, or evaporated milks are usually employed and the fats may be either butter or vegetable oil, preferrably emulsified with the milk or with a milk-syrup mix before cooking. The cooking may be batch or continuous, depending on the manufacturer's needs.

The final moisture content of caramel is higher than hard candy, usually 10–12% and the texture is plastic at normal temperatures. The reaction of the sugars with the milk proteins, known as the Maillard reaction, results in the caramel flavor and color. The aforementioned plasticity allows caramel to be formed by extrusion or deposited as sheets, ribbons, ropes, or individual patties.

Toffee. Toffee, as it is known worldwide, is merely a high-cooked caramel deriving its characteristic flavor from heating the milk protein in the presence of the sugars to a high degree (the Maillard reaction, or special caramelization process). Butter is sometimes added, which partially breaks down with age, lending a distinctive flavor that is attractive. Brown sugar goes well with this kind of treatment and sometimes sodium bicarbonate is used to change the alkaline composition of the syrup.

Toffee is often confused with Taffy. Taffy is generally a low-cost, chewy candy containing a high percentage of corn syrup and very little fat or milk solids. It is cooked just high enough to hold its shape but low enough to soften when held in the mouth. Most taffy is aerated either by pulling or through the addition of a frappé.

In this country, a very popular confection is known as English toffee. An example of this type is the Heath Bar or Brown and Haley's Almond Roca. Nearly all manufacturing retailers feature a version of this toffee, although each manufacturer uses a little different approach. Essentially, it is straight granulated white sugar, butter, and nut meats (usually almonds) cooked to a high temperature 155°C (310°F). Sometimes it is cut in bite size pieces and coated with milk chocolate, or slabbed in sheets, coated with milk chocolate, and dusted with chopped nut meats. If made properly, this candy has an unusually crunchy texture and appealing flavor and is widely accepted in the United States. The origin of the name English toffee is unknown; it is not featured to any great extent in the UK.

Licorice. Licorice is extracted by means of water from the root of the *Glycyrrhiza glabra* plant and concentrated to become licorice block. Generally, flour makes up about 33% of the product; various sugars (sucrose, corn syrup, molasses, invert), about 50%; licorice, 3–6%; and water, about 17–18%. A small amount of gelatin is also used. In the United States the most popular forms are ropes, bite-size pieces, and sugar-coated centers, such as Good and Plenty or Ferrara's Licorice Lozenges.

Jellies and Gums. These confection types, considered to be synonymous by consumers, have the common general characteristic of being chewy, which can be varied from rather soft to quite hard. Orange slices, gumdrops, gummy bears, jelly beans, fruit snacks, and jujubes represent some of the product types.

Whether jellies or gums, these candies all contain sucrose with corn syrup and/or invert sugar. Generally, the higher the corn syrup concentration, the chewier the product, although the use of 64 DE corn syrup results in softer, more tender candy. Hard gumdrops made with gum arabic usually have high levels of sucrose while jelly bean centers may contain twice as much corn syrup as sucrose.

To impart the chewy property, starch, gelatin, agar, gum arabic, pectin, and, in some cases, combinations are used to form the gel. Each of these ingredients impart specific gelling properties and have specific requirements for use in a given formula. Gelatin, for example, must first be soaked in water. It is added to the sweeteners after cooking to the desired solids and cooling somewhat because, if boiled, its gel properties are reduced. Typical use levels are 6–9%. Agar or gum arabic are dispersed in water, the sweeteners added, and the ingredients cooked to

the required solids. Fruit acids, if desired, must be added after the cook to prevent destruction of the agar. Normal amounts of fruit acids range from 1 to 2%, although gum arabic may be as much as 40% in hard gumdrops. Pectin requires the presence of sugars and fruit acids for gel formation and use is 1.5–4%. Starch is normally gelatinized in boiling water where the sweetener concentration is about 50%. Higher concentrations of sugars will prevent complete gelation of the starch. Techniques of cooking starch under pressure allow for reduced water and offer savings in time, space, and energy. Starch levels range from 7 to 14%.

For most items, a typical formula consists of about equal parts of sucrose and corn syrup. Flavors, colors, and acidulants are added after the cooking. The finished cook averages about 70–75% solids (pressure cooking, 80% solids) depending on the gelling material. The cooked material may then be cast into starch, plastic, or metal molds or formed into slabs and dried to 80–88% solids. Further processing and packaging normally follows.

Crystalline

Fondants and Creams. To many consumers the terms fondant and cream, when referring to candy, are synonymous. To the old-time confectioner they are distinctly different; fondants being more solid and used in preparing items such as peppermint patties, while creams are softer and used in making chocolate-covered products. In this article they will be treated as one, because they are prepared from the same ingredients, often used interchangeably, and differ only in degree of hardness and crystal size.

Fondants and creams are nothing more than mixed sugars contained in two phases, solid and liquid. The solid phase is composed of sucrose crystals held in a saturated solution of sucrose and other sugars.

Virtually all fondants are prepared from a blend of sucrose and other sugars, invert or corn syrup (the latter is preferred and far more widely used). One exception is the fondant used in the making of cordial cherries. A very coarse fondant is used for this purpose, prepared from sucrose only, allowing the added enzyme invertase to act during an initial storage period of four to seven days after manufacture. The result is liquefaction of the sucrose with the desired cordial effect.

The amount of corn syrup in normal fondant production varies from 0 to 40%, depending on the final use of the product. The sucrose, plus water to dissolve it, and the corn syrup are heated to a temperature of ca 116–119°C (240–246°F), again depending on the desired consistency. After boiling, the syrup is cooled to ca 37.8°C (100°F) without agitation, to create a supersaturated state. Agitation is then initiated, which causes crystallization to occur in the form of tiny crystals, usually no larger than 15 μ (0.0006 in.). The fondant of 12–13% moisture may then be stored for several days or used immediately.

In large, modern manufacturing operations, the process is a continuous one, utilizing water-jacketed equipment and incorporating other ingredients such as fruit, flavors, butter, colors, etc at the point of agitation. The product is then generally held for a period of only a few hours before being formed into centers or deposited into shells or starch.

Many small confectioners do not have the facilities for preparing their own fondant, either because of space or cost. They rely on dry fondant products, which are offered by their sugar suppliers, that require only the addition of water to achieve the desired consistency.

The candies commonly known as butter creams are fondant based and contain varying amounts of butter, ranging from a low of about 5% to a high of almost 20%. The higher the butter content the richer the taste and the shorter the shelf life. A very few confectioners also use real cream, incorporating it into the sucrose–corn syrup mix before boiling; this provides a somewhat distinctive flavor and slightly different body to the finished fondant.

Some fondant products contain no butter or fat of any kind and are known as water fondants. They are usually found in lower-priced items or contain other ingredients that are not compatible with fat. In this category are recent developments in liquor-containing candies, now permitted in 10 states. These are usually water fondants and total alcohol content is normally less than 5%.

Fondants and creams are the staple of the boxed-chocolate industry. Every manufacturer, whether large or small, considers its creams to be the leader and the basis on which business will rise or fall.

Fudge. Fudge is hybrid confection, a cross between a caramel and a cream. Perhaps the first fudge was made by someone accidently graining a batch of caramel. Ingredients are sucrose, corn syrup, milk (in various forms), and fat (also in various forms). These are the basics to which egg white, salt, nut meats, flavorings, etc are added. Most fudges are improved by the addition of up to 20% brown sugar. Processing is varied to obtain the most desirable crystal structure compatible with flavor and type.

Beater Fudge. Beater fudge (4) is made by incorporating the milks, fats, sucrose, and corn syrup and cooking under agitation to approximately 114°C (237°F). It is cooled slightly on a flat-bed beater equipped with plows and beaten while still warm. Other ingredients (chocolate, frappé, nut meats, salt, flavoring, etc) are added at intervals during the beating process. When a grain or first indication of crystallization appears, the batch is removed to boards, trays, or pans where it is allowed to set. This process requires considerable skill but results in a superior product. With the use of a continuous fondant machine and a screw-type mixing conveyor at the discharge end of the fondant maker, the procedure can become a continuous operation, which is practical, but the result is a more coarsely grained fudge.

Kettle Fudge. Generally, in this method the sucrose, corn syrup, milks, and fats are cooked in a kettle to about the same temperature or slightly higher than the beater fudge. The kettle is set aside undisturbed until the contents cool to about 93°C (200°F).

Then fondant and the other ingredients are added (including chopped chocolate liquor if desired) and mixed in. The kettle, which may contain 100 lb or more of product, is given a blast of heat, which facilitates the scraping out of the kettle when emptying. The heat also imparts a sheen to the surface of the fudge, which is now formed into a slab on a table to be marked into squares after cooling. A

continuous method can also be devised by using a caramel cooking plant where the flow of hot liquid caramel empties into a screw-type mixing conveyor and the other ingredients, including the fondant, are incorporated.

Nougats. Although the origin of traditional nougat is unknown, the town of Montélimar in southern France is given credit for it. Originally it consisted of honey, whipped egg whites, and nuts, probably almonds. Today there are a great variety of nougats in the United States, but they fall into only two categories, chewy or grained. The main difference is the sucrose-to-corn syrup ratio; chewy nougats will have about equal amounts of each while grained (or short) nougats will contain sucrose at approximately twice the level of corn syrup. Although most formula books specify 43 DE corn syrup, the largest nougat manufacturer in the United States uses 64 DE corn syrup to achieve a more tender product.

All nougat processes encompass two steps: first, the preparation of frappé (sometimes referred to as mezzeta) and, second, the preparation of the cooked syrup. Frappé is a mixture of an aerating agent and syrup. The aerating agents in use today are egg whites (usually dried), soy albumen, and a special milk protein. The ingredients are whipped to a light foam by a high-speed mixer and the cooked syrup—132°C (270°F) for a short nougat, 135–141°C (275–285°F) for a chewy nougat—is slowly added, followed by other ingredients such as fruit, nuts, vegetable fat, or milk powder, all with as little mixing as possible to retain the foam structure. Sometimes powdered sucrose is added to short nougat at the last stage to induce the formation of tiny crystals. Moisture content approximates 9–11% for a short nougat and 5–7% for a chewy one.

In the United States short textured nougats are far more popular as evidenced by Milky Way, 3 Musketeers, and Snickers candy bars. The first two include a chocolate ingredient and the third incorporates peanut butter in the nougat portion.

Marshmallows. Marshmallow resembles nougat but has a higher moisture content, usually 15–20%, and contains no fat. Gelatin is the predominant gelling agent although egg albumen is sometimes included as well. Infrequently gum arabic, agar, or pectin are used. Sucrose, invert sugar, and 43 DE and 64 DE corn syrups are the sweeteners normally used, although sorbitol has also been used in small amounts.

Most marshmallow found in today's supermarkets is produced by a continuous-extrusion method where air under pressure is forced into the cooked syrup ingredients. The product is then extruded as ropes into a bed of starch where it dries, followed by cutting, cooling, and packaging. An extremely high strength gelatin is essential to this process. Where the marshmallow is to be coated with chocolate, it is often cast into starch molds. No matter the process, marshmallow requires a higher concentration of flavor because the high moisture and occluded air dilute the taste impression.

Pralines. Pralines are difficult to define because the term means different things in different parts of the United States and has been frequently misused in efforts to upgrade the image of certain products. The traditional Southern praline is made from pecans, sucrose, brown sugar, corn syrup, maple syrup, and butter or cream. Both grained and noncrystalline pralines are known, but the tender, grained product is the most widely accepted. The methods of manufacture vary but are based in principle on depositing the cooked, grained syrup onto the pecans. Old-time candy makers placed the pecans inside metal rings and covered them with the praline syrup. The finished praline is very sweet and quite fragile, requiring protective packaging if it is to be shipped.

Pressed Candy (Tablets). Sweeteners such as sucrose, dextrose, and sorbitol may be formed into unique shapes by the use of high pressure and punches and dies to form the shape. Examples are products such as peppermint Life Savers, Certs, and Breath Savers.

The process usually consists of dry blending the sweeteners (95–98%) with a color, to coat the crystals, and the desired flavor. If the flavor is a fruit type, a fruit acidulant such as citric or malic acid may also be added at a level of 0.5–3.0%. Finally, a release agent such as magnesium or calcium stearate is blended at a level of about 1% to prevent sticking to the punches and to aid release from the dies. The mix is fed into a press fitted with punches and dies. Under a pressure of several tons, the mix is compressed and formed into the desired shape then discharged to a conveyor and packaged.

Not all sweeteners perform equally in a tableting press. The compressibility of the sweetener is often improved by the supplier in the crystallization process or by addition of corn syrups or malto dextrins.

Marzipan and Pastes. A large range of candies may be classified as pastes. These generally consist of flavored mixtures using sucrose, corn syrup, and a suitable binder such as gelatin or edible gum. Shredded coconut or pastes of other nuts are frequently added. Marzipan is generally regarded as an almond paste and with it and other nut pastes binders are not usually added. Pastes are mixed in a kneading machine; then for marzipan and other such candies, the nuts are coarsely ground on rollers and mixed in. When fully blended, the pastes are formed into bars or small pieces by extrusion and cutting or by rolling out on tables and cutting.

Panned Candy (*Dragées*). The process of coating nuts, fruit (dry or preserved), caramels, nougat, chocolate, or certain kinds of pastes with sugar or chocolate is known as panning (5). The customary equipment for this procedure consists of a motor-driven, revolving, openmouthed pan usually of oval or pear shape, but sometimes round. The center material is placed in the pan and when the pan rotates, the centers tumble over one another, providing an excellent mixing and increased surface area. The process may be classified into two categories, soft panning and hard panning.

Soft panning consists of alternate additions of a syrup (usually a sucrose–corn syrup blend), to make the center material sticky, and dry sucrose, which adheres and dries on the surface. The process is continued until the desired thickness is reached. Then a finishing syrup, which may contain color and flavor, is applied to provide a smooth surface. After finishing, the candy may be held for a period of 12–24 h, then polished and packaged. The best example of this kind of candy is jelly beans.

Hard panning is much the same process, but instead of applying powdered sucrose to dry the centers, dry air is blown over the surface of the candy between liquid applications. After a thin bonding coat of gum arabic and flour (or other material) has been applied, subsequent applications are of liquid sucrose only. If the air is cold, the process is called cold panning; if hot, it is termed hot panning. Heat may also be applied by means of gas-fired burners immediately below the revolving pan. The result of hard panning techniques is a continuum of thin layers of fine sucrose crystals, which are extremely smooth and crisp. Jordan almonds are a typical example.

Sugar-roasted nuts are a specialty item prepared by cooking in sucrose. That is, an approximate 60% sucrose and water solution is brought to a boil, and a quantity of raw nut meats equal in weight to the weight of the sucrose in solution is added. They are stirred by hand or by machine over a gas fire. Or they may be tumbled in a revolving kettle over a gas fire.

When the temperature reaches about 130°C (265°F), enough moisture has been boiled off so that the sucrose crystallizes and forms a crystal coating around each nut. The continued heat eventually melts this coating, resulting in a sucrose syrup surrounding each nut. This, when cooled, forms a sugar glaze. Butter or vegetable oil is added both for flavor and as a lubricant. Sometimes sanding sugar is used for separation.

A high degree of skill is required for this work, and the sucrose used must be pure so that there is nothing inherent that could retard the graining process. There are many variations of this confection depending on the ratio of sucrose-to-nut meats as well as the flavorings used, such as cinnamon and honey.

CHOCOLATE PRODUCTS

Candies are often combinations of several confections. This is especially true in chocolate products where the chocolate is used as a coating (enrobed confections) on centers (6,7).

Chocolate is one of the most delightful creations of man and accounts for the largest volume of confectionery products. It has evolved over the years from an astringent watery suspension of crushed cocoa beans to its current state of smooth mellow taste, texture, and richness. Chocolate products are enjoyed by millions of people throughout the world and are constantly changing forms to please consumers. Chocolate is eaten as straight chocolate or as coating (also known as couverture) on candy bars, cookies, ice cream, nuts, and more. The flavor of chocolate is developed by combining chocolate liquor with sugar and other ingredients such as milk, lecithin, and flavorings.

In the United States, chocolate is subject to federal standards of identity, which prescribe rather rigidly the quantitative elements and consequently, some of the qualitative aspects as well (8) (Table 3).

Milk chocolate, which makes up approximately 60% of all U.S. chocolate, usually contains 45–50% sugar (but may be as low as 40% or as high as 55%, per standards).

Table 3. Typical Category Formulations for Chocolate Products

Product	Percent
Milk Chocolate	
Chocolate liquor	10.0
Cocoa butter	21.0
Whole milk powder	13.0
Sugar	55.47
Lecithin	0.3–0.5
Vanillin	0.03
Total	*100.00*
Milk Chocolate Compound	
Cocoa powder (10–12% fat)	8.0–12.0
Vegetable fat (palm kernel oil, mp 35–39°C)	28.0–32.0
Sugar	45.47–55.67
Nonfat milk powder	8.0–10.0
Lecithin	0.3–0.5
Vanillin	0.03
Total	*100.00*
Dark Chocolate	
Chocolate liquor	16.0
Cocoa butter	25.07
Sugar	58.4
Lecithin	0.5
Vanillin	0.03
Total	*100.00*

Sweet chocolate will also have approximately 45–50% sugar content, but can go as high as 60–62% in some low-cost dark chocolates. Strong-flavored bittersweet chocolates are usually formulated with as low as 25–30% sugar.

Not all products that appear to be chocolate to the casual observer meet these standards, and by regulation these products may not be labeled as chocolate. They are generally referred to as confectioners' coatings or compound coatings and have standards of their own with such lengthy names as sweet chocolate and vegetable fat (other than cocoa butter) coating. Many have been developed for specific uses where real chocolate is inappropriate, ie, coatings for ice cream bars and novelties. Because they may also be less expensive they have gained considerable use as coatings for candy bars and baked goods. Products whose labels bear the statements chocolate flavored or chocolatey can be expected to contain this type of coating with a generally less satisfying flavor.

Also included in this category are groups of products referred to in the chocolate industry as pastel, or colored, coatings. They are not truly chocolates, by standard of identity, but many of the higher quality formulations contain significant amounts of cocoa butter, giving them excellent eating qualities. One example is white chocolate. This, of course, is a misnomer because it contains none of the characteristic chocolate liquor or flavor. It is, rather, a blend of cocoa butter, sugar, milk solids, and vanilla or vanillin flavorings.

When cocoa butter is not used in a formulation, a vegetable fat is usually substituted, and the resulting product is considered to be a compound coating. Various melting points of vegetable fats are used, thereby affecting the final melting point of the coating. This is especially help-ful in the preparation of summer coatings, or coatings designed for use in warm climatic conditions or locales.

Regardless of fat source, these products contain sugars at levels equal to or exceeding that of real chocolates. Because of the higher levels of sugars and the increase in sweetness, specific flavorings are usually added to help retard or lessen the excessive sweetness impact.

Of growing interest to some American consumers are the specialty items referred to as dietetic or sugar-free chocolates and for which there are currently no standards. In almost all cases, the caloric content is not significantly less than that of normal chocolate; the actual benefit is improved dental care by caries reduction. In these products the sugar (sucrose) is replaced by one or more substances such as sorbitol, mannitol, or xylitol. Sorbitol is by far the most widely used and is approximately one-half to two-thirds as sweet as sucrose. Artificial sweeteners may be added to bring the perceived sweetness up to acceptable levels. Saccharin is the artificial sweetener of choice today but the use of aspartame and others is not far away. Sorbitol is frequently used; however, it produces a mild laxative effect and care should be exercised in consuming products that contain sorbitol. Xylitol also has been used for the same purpose, and although it is now commercially available, it is even more costly. Its use as a food additive has already been discussed. Figures are not available, but this segment of the chocolate products market is slowly growing in volume.

CHEWING GUM

One final category of confections should also be included to complete this review—chewing gum. In addition to being a sticky goop to step on or something to find under a desk in class, chewing gum is really an interesting, unique product. Sometimes it is described as a nearly homogenous solid–liquid physical mixture emulsion. It is made up of sugar, gum base, water, corn syrup, color, flavor, and sometimes acids. The gum contains all these ingredients mixed together with those that make up the gum base (ie, rubber, wax, fats, oils, and emulsifiers).

Chewing gum or chewing something (roots, animal skins, sweet grasses, tree barks, etc) dates back to prehistoric times; tree resins were the most popular of the chewables. For centuries the Greeks have chewed mastic gum (in Greek it's pronounced *mas-tee-ka*). This is the resin obtained from the bark of the mastic tree, a shrublike plant found mainly in Greece and Turkey. Grecian women especially favored chewing mastic gum because it cleaned their teeth and sweetened their breath.

Closer to home, the colonists of New England learned from native Americans to chew the gumlike resin that formed on spruce trees when the bark was cut. Lumps of spruce gum were sold in the eastern United States during the early 1800s, making it the first commercial chewing gum in that country. In about 1850, sweetened paraffin wax became popular and eventually exceeded spruce gum in popularity.

Chewing gum continued to evolve, and modern chewing gum had its beginning in the late 1860s when chicle was brought to the United States and tried as a chewing gum ingredient. Chicle comes from the milky juice (latex) of the sapodilla tree, which grows in the tropical rain forests of Central America, especially in the Yucatán Peninsula. Until the recent development of synthetic poly(vinyl acetate) resins and synthetic gum bases, chicle had served as the foundation of most commercial chewing gum operations.

From the prehistoric to current times, chewing gum has grown in popularity and now is one of the largest confectionery categories, second only to chocolates. Chewing gum's popularity is twofold: (1) taste and (2) physiology (the act of chewing is a pleasure that almost everyone enjoys). It relaxes the nerves and muscles, moistens and freshens the mouth, cleans the breath and eases tensions that can cause restlessness and fidgetiness. And recent studies have indicated that chewing gum can help keep teeth clean. Chewing stimulates saliva, which helps neutralize traces of acids from fermented food that may cause tooth decay, and the gum base and sugar act as an abrasive to help remove dental plaque.

Next to chocolates, chewing and bubble gum consist of the largest confectionery sales category representing more than 18% of the consumer sales dollar market (9). Annual sales are estimated at in excess of $2.5 billion. An interesting note, based on these figures, is that approximately 200 pieces or sticks of gum are purchased for every man, woman, and child in the United States each year.

Basic Gum Composition

In general all gums are basically similar, consisting of approximately 25% gum base, 25% corn syrup (80% solids, 43 Bé), and 50% sugar. Depending on the type of product desired, gum base used, ingredients selected, and product characteristics, these levels are adjusted and modified.

General Formulation

Bubble Gum	Percent
Bubble gum base	15.5–16.7
Corn syrup, 44.5–45.0 Bé, 42 DE	23.0–24.5
Sugar	To make 100
Flavor	0.4–1.0
Color	0.03
Acidulant (to taste)	0.1–0.3
Glycerin	0.5

Stick Gum	Percent
Gum base	20.0–22.0
Corn syrup, 44.5–45.0 Bé, low DE	17.0
Sugar	To make 100
Flavor (usually mint oil)	0.6–0.7
Glycerin	0.5
Salvage or trim (rework)	5.0

Sugarless Gum	Percent
Sugarless gum base	25.0–33.0
Mannitol	0.0–5.0
Sorbitol	To make 100
Sorbitol solution (70% solids) or lycasin	15.0–23.0
Xylitol	0.0–8.0
Glycerin	1.0–6.0
Flavor (dry and liquid)	1.0–2.0

Ingredients

Gum Base. Gum base is the ingredient that separates chewing and bubble gums from all other confections. It is the rubbery substance the properties of which allow it to be chewed for hours without any substantial changes. Present gum bases contain synthetic rubbers, plasticizers, softeners, fillers, emulsifiers, and a variety of other ingredients used to impart the unique characteristics desired.

While these ingredients are present to some degree in all bases, generally the major ingredients are latex rubbers, 10–20%; fillers (sometimes politely called texturizers—either talc or calcium carbonate), 30–60%; softeners and plasticizers (fats and waxes), 10–20%; and others, less than 10%. Most gum bases also contain BHA and BHT as stabilizer–antioxidants.

It is important to note that there are really only two types of gum base: chewing gum and bubble gum. And the difference between these two is obvious—the ability to form and blow bubbles. Bubble gum bases usually contain higher levels of rubber to allow this stretching and bubble forming.

Sugars. The most common types of sugars used in gum manufacture are sucrose (common sugar) and dextrose monohydrate. The dextrose is used because it is lower in cost in some countries. In general, dextrose is a negative to the chewing character of the gum and also the processing character.

Corn Syrup. The most common Baumé for corn syrup in chewing gum and bubble gum is 45°, 84–85% solids. The reason for this is that it allows a high level of corn syrup to be used. This gives strength and body to the gum for processing, and in most countries corn syrup is lower in price than sugar.

Soft bubble gums are usually made with 43 Bé corn syrups, but other specialty syrups are also gaining acceptance. For example, high DE syrups are sometimes used for added sweetness as well as cost reduction. High maltose types help control crystallization. Just like other confections, the selection of the proper corn syrup can affect the characteristics of the resulting gum.

Flavors and Plasticizers. It is important to know that the flavors do more than just give taste to the gum product. The flavor is a plasticizer for the gum base and will have an effect on the softness or firmness of the gum.

Generally, highly concentrated essential oil flavors are the best plasticizers for the gum base. Normal candy flavors do not function well and if used, another plasticizer must be added to the gum to achieve a good chew character.

Because the flavors plasticize the gum base, the ratio of flavor to gum base is very important. This ratio is adjusted to give the chewing characteristics desired, and for bubble gums, the bubbling characteristics desired. Flavor is added in low levels, as compared to the sugar, corn syrup, and gum base, yet its effects on the final product is pronounced.

Gum Processing

Gum processing is rather simple, but because of the many interactions possible, a number of factors can affect the final gum's quality. Starting with the gum mixer (some of which will hold almost a ton of finished gum) the gum base is added and mixed. In mixing gum the sequence of ingredient addition also plays an important role in the gum characteristics. A rule of thumb in gum formulation is What goes in last, comes out first.

Recommended for gum is three 5-min cycles as follows. The gum base (presoftened), hard gum salvage, corn syrup, color, and plasticizers (if used) are mixed for 5 min. Then one-half of the sugar and the glycerin are added and mixed for 5 min. Finally, the soft gum salvage and the other half of the sugar are added and mixed. The flavor is added after the sugar is incorporated 2–3 min before the end of the cycle. The total mixing time for the third cycle is 5 min.

The total mixing time is 15 min. Shorter cycles on the two sugar additions can be done if the mixing goes quickly. If the batch size is large, poor mixing will result or longer mixing times will be required. If the mixer is not capable of handling this two times sugar addition cycle, then a three times, by one-third, sugar cycle with mixing intervals of 3 min each can be tried.

The finished gum, which looks similar to stiff cookie dough, is then removed from the mixer and allowed to cool slightly before being formed or sheeted. In the case of stick gum, the gum dough is extruded into a sheet that, through the use of a series of sizing rollers, is compressed into a thin, wide ribbon of gum. Each sizing roller is set closer together than the previous pair, and gradually the gum's thickness is reduced. To aid in processing and to reduce sticking to the rollers, powdered sugar is usually slowly applied on top of the gum during this process. Finally, the ribbon of gum is passed through a set of scoring rollers, which cut the gum into the desired size and shape.

Although processing is nearly complete, the gum matrix is still somewhat unstable and conditioning is required to develop the desired chewing qualities and shelf life. Depending on the type of gum produced, conditioning can take a few minutes to 8–12 h. During conditioning the gum's environment is controlled to optimize character development, usually conditions of 26.7°C (80°F) and 40–50% RH are used.

For soft bubble gums or chunk gums (Bubblicious, Hubba Bubba, Bubble Yum) the gum is processed similarly, however rather than sheeting (flattening out) through sizing rollers, this type of gum is extruded into a rope form, which is passed through a conditioning tunnel to cool and develop texture. (Something similar to a long snake, one, it is hoped, that does not break.) After either operation, the gum is ready to be packaged.

Packaging

Packaging plays an important part in this product's shelf life, quality, and salability. Gum, just like other confections, is sensitive to heat, moisture, and, for some flavors, oxidation. To increase gum's shelf life and maintain quality standards, exotic packaging materials and machines are used. These packaging machines are exotic not only because of the speeds that they run at (some over 1,800 rpm) but also because they form air tight packages.

BIBLIOGRAPHY

1. Confectionery Industry Archives, general industry knowledge, various publishers—various authors.
2. J. M. Newton, *Products of the Wet Milling Industry in Foods,* symposium proceedings of Corn Processing Company, Clinton, Iowa, Washington, D.C., 1970.
3. E. R. Kooi, "Dextrose and Starch Syrups," in M. Grayson, ed., Kirk-Othmer *Encyclopedia of Chemical Technology,* Vol. 6, 2d ed., John Wiley & Sons, Inc., New York, 1965.
4. H. Knechtel, unpublished data, Knechtel Research Sciences, Inc., 1980–1986.
5. A. Meiners and H. Joike, *Handbook for the Sugar Confectionery Industry,* Nos. 1 and 2, Silesia-Essenzenfabrik Gerhard Hanke KG, Norf, FRG, 1969.
6. B. W. Minifie, *Chocolate, Cocoa, and Confectionery: Science and Technology,* 3rd ed., Van Nostrand Reinhold Co., Inc., New York, 1989.
7. S. T. Beckett, ed., *Industrial Chocolate Manufacture and Use,* Van Nostrand Reinhold Co., Inc., New York, 1988.
8. Frederick J. Bates & Associates, eds., *Polarimetry, Saccharimetry and the Sugars,* circular C440, National Bureau of Standards, Washington, D.C., 1942.
9. Distributor Electronic Billing Systems, eds., *DEBS Confectionery Marketing Report,* written commentary, Distributor Concepts, Ann Arbor, Mich., 1987–1988.

ROBERT F. BOUTIN
Knechtel Research Science, Inc.
Skokie, Illinois

CONSERVES. See FRUIT PRESERVES AND JELLIES
CONTAMINANTS, ENVIRONMENTAL. See FOOD TOXICOLOGY
CONTAMINATION. See FREEZING SYSTEMS FOR THE FOOD INDUSTRY

CONTROLLED ATMOSPHERES FOR FRESH FRUITS AND VEGETABLES

Controlled-atmosphere (CA) storage is a technique for maintaining the quality of fresh fruits and vegetables in an atmosphere that differs from normal air with respect to the concentrations of oxygen (O_2), carbon dioxide (CO_2), and/or nitrogen (N_2). The desired compositions of the atmosphere for storing commodities are usually obtained by initially increasing CO_2 or decreasing O_2 levels in a gastight storage room or container. Sometimes, the addition of carbon monoxide (CO) or removal of ethylene (C_2H_4) may also be beneficial.

Modified atmosphere (MA) is a condition similar to CA, but with less or no active control of the gas concentrations. In MA, the O_2 level is reduced and the CO_2 level is increased at a rate determined by the respiration rate of the commodity, the storage temperature, and the permeability of the container and film wrap to the gases. Judicious selection of the commodity, the package dimensions, and the package material will insure establishment and maintenance of the desired atmosphere under specified storage temperatures.

TYPES OF CA STORAGE

During the past 60 years, tremendous progress has been made in the technology of CA storage. The commercial application of CA during transit and storage has received considerable attention since 1960, resulting in the development of different methods of establishing and maintaining CA. These include regular CA; short-term, high-CO_2 treatment; rapid CA; low-oxygen CA; low-ethylene CA; and low-pressure, or hypobaric, storage.

Short-Term High-CO_2 Treatment

Short-term, high-CO_2 treatment was originally developed to maintain the firmness of Golden Delicious apples (1). Subsequently, it was found that pears and other fruits and vegetables also benefit from this treatment (2,3). This treatment involves the exposure of fruit to 10–20% CO_2 for four to seven days prior to adjustment of the atmosphere to regular CA concentrations. Carbon dioxide injury of the fruit skin may occur if moisture has condensed on the surface of the fruit. This high-CO_2 treatment gives excellent results in maintaining the quality of Golden Delicious apples and Anjou pears (1,2,4).

Rapid CA

Rapid CA is a strategy that shortens the time between harvest and establishment of the desirable CA conditions (5). The faster the CA conditions are attained after harvest, the better the fruit quality can be maintained, providing that the cooling rate in storage is not adversely affected by the rapid loading of the room. To achieve the objectives of rapid CA, the storage room should be filled and sealed within three days or less of harvest.

Low-Oxygen CA

Low-oxygen CA has recently received increased attention; not only because it markedly retards fruit softening, but also because it greatly reduces the development of storage scald and breakdown of apples and pears (6). In regular CA, the recommended O_2 concentrations are usually 2% or higher. It has been found that O_2 levels between 1 and 1.5% are even more effective in extending the storage life of some fruits and vegetables (7). Careful monitoring to maintain the precise O_2 level is essential to avoid damage due to anaerobic respiration.

Low-Ethylene CA

In low-ethylene CA, ethylene is scrubbed from the CA room to improve storage quality of the fruit. Removal of ethylene from the storage atmosphere results in retardation of ripening, retention of flesh firmness, and reduction of the incidence of superficial scald of apples (8). It is generally recognized that the concentration of ethylene should be maintained below 1 ppm to obtain the beneficial effect of low-ethylene CA. Storage life of several apple varieties such as Empire and Bramley's Seedling can be extended by this technique (9,10).

Low-Pressure Hypobaric, Storage

Low-pressure, or hypobaric, storage consists of storing fruits and vegetables at below-normal atmospheric pressure. Enhanced diffusion of gases under reduced-pressure facilitates the loss of CO_2 and ethylene from the commodity and reduces the O_2 gradient between the inside and outside of the commodity. The partial pressures of O_2 is directly related to the absolute pressure of air (11). Thus the O_2 concentration is equivalent to 0.55% at 20 mm Hg. Ethylene inside the fruit is also reduced proportionally. Therefore, this storage technique combines the advantages of low O_2 storage and low-ethylene storage. Ripening can be inhibited and storage life prolonged when fruit are stored under hypobaric conditions.

BENEFICIAL EFFECTS OF CA STORAGE

Beneficial effects of CA storage include reduction of respiration, scald, decay, discoloration, and internal breakdown, inhibition of ethylene production and ripening, and retention of firmness, flavor, and nutritional quality.

The rate of respiration of fresh fruits and vegetables has been shown to be reduced by low O_2 or high CO_2 (12). The lower respiration rate indicates that CA has an inhibitory effect on the overall metabolic activities of stored commodities. A slower rate of utilization of carbohydrates, organic acids, and other reserves usually leads to prolonging the life of the produce.

Ethylene production of fresh fruits and vegetables is suppressed by low O_2 and/or elevated concentrations of CO_2. The biosynthesis of ethylene in plant tissues requires the presence of O_2 (13,14). When O_2 is absent or when plant tissue is under low O_2 atmosphere, ethylene biosynthesis is inhibited. The time of onset of ethylene production in apples is inversely related to the O_2 concentration in storage, and the maximum rate of production is directly related to the O_2 level (15). High CO_2 inhibits ethylene action on ripening (16), which in turn inhibits the autocatalytic production of enzymes involved in ethylene biosynthesis, including 1-aminocyclopropane-1-carboxylic acid (ACC) synthase and ethylene-forming enzymes (17). These inhibitory effects on ethylene action and production by CA consequently lead to the delay of the ripening process.

The loss of organic acids in apples and pears during storage is reduced by CA (18,19). This reduced loss of organic acids in CA fruits is probably due to an increase in CO_2 fixation, an inhibition of respiratory metabolism, and a lower consumption of acids under CA (19, 20).

A slower decline in carbohydrates under CA has been reported in sugar beet, Chinese cabbage, apricots, and peaches (21,22,23). CA storage is also beneficial in the retention of ascorbic acid and amino acids in several fresh fruits and vegetables (3,22,24,25).

EFFECTS ON PHYSIOLOGICAL DISORDERS

Scald is one of the most serious physiological disorders in apples, pears, and many other fruits during and after storage. Storage of fruits in CA, particularly in low O_2 atmosphere, has been shown to reduce the susceptibility of fruits to scald (6,26). Russet spotting of lettuce can be caused by exposure to ethylene or warm shipping or storage temperatures (27). This physiological disorder can be substantially reduced by low O_2 atmosphere (28,29). The development and spread of necrotic spots on the outer leaves of cabbage was largely prevented by low O_2 atmosphere but not by high-CO_2 treatment (30). The incidence and severity of vein streaking of cabbage leaves was also reported to be reduced in CA storage (31).

Controlled atmosphere reduces chilling injury in certain sensitive crops, while it aggravates or has no effect in other crops. Holding zucchini squash in low O_2 alleviated chilling injury during storage at 2.5°C (32,33). Conditioning grapefruit with short-term prestorage treatment of high CO_2 (40%) at 21°C reduced subsequent brown staining and rind pitting, two symptoms of chilling injury, at 1°C (34). Intermittent exposure of unripe avocados to 20% CO_2 reduced chilling injury at 4°C (35). Addition of CO_2 to the storage atmospheres was also effective in reducing chilling-induced internal breakdown and retaining ripening ability of peaches (36,37). Controlled atmosphere has also been reported to reduce the severity of chilling injury symptoms in okra (38). However, increasing the level of CO_2 or reducing O_2 concentration can accentuate symptoms of chilling injury in cucumbers, bell peppers, and tomatoes (39,40).

Some other physiological disorders have also been reported to be aggravated by CA conditions. The development of a brown discoloration in and around the core and adjacent cortex tissue of apples and pears has been associated with high CO_2 (41,42). White core inclusions in kiwifruit can also be induced by elevated CO_2 in combination with ethylene (43). The severity of brown stain on lettuce

has also been found to increase with increasing CO_2 levels (44).

CONCLUSION

Controlled atmosphere storage has withstood the test of time and has been proven to be effective for maintaining quality and extending the storage life of a number of horticultural crops. The rate of deterioration of fresh produce is often greatly retarded by CA storage. Each type of fruit or vegetable has its own specific requirement and tolerance for atmosphere modification. The optimum CA condition for each commodity is continually being revised and improved. Table 1 presents the most recent recommendations for O_2 and CO_2 levels, the most suitable storage temperatures, feasible storage periods, and pertinent remarks for some commodities.

The maintenance of CA storage requires continual monitoring of the gases and temperature to prevent any deviation from the recommended conditions. Injuries induced by low O_2 or high CO_2 often lead to tissue discoloration or off-flavor and loss of market value. Proper maturity, the internal condition of the fruit at harvest, and the speed at which storage atmospheres are established are some of the key factors affecting the success of CA storage.

It should be emphasized that proper CA storage should always be accompanied by good temperature control. CA storage is considered to be a supplement—not a substitute—for proper refrigeration and careful handling.

Table 1. Summary of Controlled Atmosphere Storage Requirements for Fruits and Vegetables

Commodity Cultivar	Beneficial Concentration (%)		Suitable Temperature		Approximate Storage Period[a]	Remarks	Reference
	O_2	CO_2	°C	°F			
Fruits of Temperate Zone							
Apple (*Malus domestica*)							
Boskoop	1.5–2	<1.5	4	39	5–7 mo	Sensitive to low temperature breakdown	45
Bramley's Seedling	2	6	3–4	37–39	7 mo	CA reduces bitter pit	45
Cortland	2–3	5	2	36	4–6 mo		46
Cox Orange Pippin	1–3	<1	3	37	4–6 mo	2% O_2 first week then 1.00–1.25% O_2	45
Empire	1.5–3	1–5	2	36	5–7 mo		45
Fuji	2–2.5	1–2	0	32	7–8 mo	CA reduces scald	45
Gala	1–2	1–5	0	32	5 mo		45
Golden Delicious	1–3	1–5	0	32	7–11 mo	Rapid CA is beneficial	45
Granny Smith	1–2	1–3	0	32	7–9 mo	CA reduces scald	45
Jonathan	1–3	1–6	0–3	32–37	4–7 mo	CA reduces Jonathan spot	45
McIntosh	1.5–3	1–5	3	37	7–9 mo		45
Newtown	3	5–8	2–4	36–39	8 mo	Susceptible to low temperature injury	
Northern Spy	2–3	2–3	0	32	8 mo		46
Red Delicious	1–3	1–3	0	32	8–10 mo	Susceptible to scald	45
Rome Beauty	2–3	2–5	0	32	6–8 mo		45
Spartan	1.5–2.5	1–2	0	32	6–8 mo		45
Stayman	2–3	2–5	0	32	7–8 mo		46
Worcester Pearmain	3	5	1	34	6 mo		45
Apricot (*Prunus armeniaca*)	2–3	2–3	−0.5–0	31–32	7 wk	CA delays ripening	47

Table 1. (*continued*)

Commodity Cultivar	Beneficial Concentration (%)		Suitable Temperature		Approximate Storage Period[a]	Remarks	Reference
	O$_2$	CO$_2$	°C	°F			
Blackberry (*Rubus* sp.)	5–10	15–20	−0.5–0	31–32	1 wk	Prompt cooling is important	47
Black currant (*Ribes nigrum*)	—	25–50	2	36	4 wk	50% first week, then 25% CO$_2$	48
Blueberry (*Vaccinium* sp.)	5–10	15–20	−0.5–0	31–32	2–3 wk	Prompt cooling is important	47
Cherry, sweet (*Prunus avium*)	3–10	10–15	−1–0	30–32	4 wk	High CO$_2$ reduces decay	47
Cranberry (*Vaccinium macrocarpon*)	1–2	0–5	3	37	2–4 mo		47
Fig (*Ficus carica*)	5–10	15–20	−1–0	30–32	2 wk		47
Grape (*Vitis vinifera*)	2–5	1–3	−1–0	30–32	1–6 mo		47
Kiwifruit (*Actinidia chinensis*)	1–2	3–5	0	32	3–5 mo	CA delays ripening	47
Nectarine (*Prunus persica*)	1–2	3–5	−0.5–0	31–32	6–9 wk	CA reduces internal breakdown	47
Peach (*Prunus persica*)	1–2	3–5	−0.5–0	31–32	6–9 wk	Cultivars differ in response	47
Pear, European (*Pyrus communis*)							
Anjou	0.5–2	0.5–2	−1	30	7–9 mo	CA reduces scald	49
Bartlett	1–2	1–2	−1	30	3–5 mo	Rapid cooling recommended	49
Bosc	1–3	0.5–1	−1	30	4–6 mo	Optimum maturity is critical	49
Comice	2	2–3	−1	30	5–7 mo		49
Conference	2	2	−1	30	4–6 mo		49
Packham's Triumph	2–3	1–2	−0.5	31	8 mo	Optimum maturity is critical	49
Passe Crassane	3–4	5–7	−1	30	6–7 mo	Tolerant to high CO$_2$	49
Pear, Asian (*Pyrus serotina* and *Pyrus bretschneideri*)							
Nijiseiki (20th Century)	3	1	0	32	9–12 mo		49
Tsu Li	1–2	3	0	32	6–8 mo		49
Ya Li	3–4	2	0–3	32–37	6–8 mo		49
Persimmon (*Diospyros kaki*)	3–5	5–8	−1–0	30–32	4 mo		47
Plum (*Prunus domestica*)	1–2	0–5	−0.5–0	31–32	4–5 mo	CA delays ripening	47
Rasberry (*Rubus idaeus*)	5–10	15–20	−0.5–0	31–32	1 wk	Prompt cooling is important	47
Strawberry (*Fragaria* sp.)	5–10	15–20	−0.5–0	31–32	1 wk	Commercial use during transport	47

Subtropical and Tropical Fruits

Commodity Cultivar	O$_2$	CO$_2$	°C	°F	Storage Period	Remarks	Reference
Avocado (*Persea americana*)	2–5	3–10	10	50	3–6 wk	CA reduces chilling injury	47
Banana (*Musa* spp)	2–5	2–5	14	58	1–6 mo	CA delays ripening	47
Grapefruit (*Citrus paradisi*)	3–10	5–10	13	55	6–8 wk	CA reduces pitting	47
Lemon (*Citrus limon*)	5–10	0–10	13	55	1–6 mo	CA reduces decay	47
Lime (*Citrus aurantifolia*)	5–10	0–10	13	55	6–8 wk	CA retards degreening	47
Mango (*Mangifera indica*)	3–5	5–10	13	55	5 wk	CA delays ripening	47
Olive (*Olea europaea*)	2–3	0–1	7	45	2 mo		47
Orange (*Citrus sinensis*)	5–10	0–5	7	45	8–12 wk	Storage life varys with cultivars	47
Papaya (*Carica papaya*)	3–5	5–10	12	54	2–3 wk		47
Passion fruit (*Passiflora edulis*)	5	5	7–10	45–50	6 wk		50
Pineapple (*Ananas comosus*)	2–5	5–10	10	50	4 wk		47

Table 1.   *(continued)*

Commodity Cultivar	Beneficial Concentration (%)		Suitable Temperature		Approximate Storage Period[a]	Remarks	Reference
	O$_2$	CO$_2$	°C	°F			
Vegetables							
Artichoke (*Cynara scolymus*)	2–3	2–3	0	32	1 mo	CA decreases discoloration	51
Asparagus (*Asparagus officinalis*)	Air	10–14	2	36	3 wk	High CO$_2$ is beneficial	51
Bean, lima (*Phaseolus limensis*)	>5	10–35	5–7	40–45	7–10 d	Shelled only	52
Bean, snap (*Phaseolus vulgaris*)	2–3	4–7	8	46	2 wk	CA reduces color loss	51
Broccoli (*Brassica oleracea italica*)	1–2	5–10	0	32	1 mo	CA maintains green color	51
Brussels sprouts (*Brassica oleracea gemmifera*)	1–2	5–7	0	32	3–5 wk	CA reduces yellowing	51
Cabbage (*Brassica oleracea, capitata*)	2–3	3–6	0	32	6–8 mo	Large scale commercial use	51
Cantaloupe (*Cucumis melo*)	3–5	10–20	8	46	2–3 mo	CA reduces ripening	51
Cauliflower (*Brassica oleracea botrytis*)	2–3	3–4	0	32	1 mo		51
Celery (*Apium graveolens*)	1–4	3–5	0	32	3 mo		51
Chinese cabbage (*Brassica campestris*)	1–2	0–5	0	32	4–5 mo	CA reduces leaf abscission	51
Corn, sweet (*Zea mays*)	2–4	5–10	0	32	2 wk	CA reduces sugar loss	51
Cucumber (*Cucumis sativus*)	1–4	0	12	54	3 wk	CA reduces yellowing	51
Leek (*Allium porrum*)	1–6	5–10	0	32	4 mo		51
Lettuce, head (*Lactuca sativa*)	1–3	0	0	32	3–4 wk		51
Mushroom (*Agaricus bisporus*)	Air	10–15	0	32	1–2 wk	CO$_2$ retards cap opening	51
Onion, dry (*Allium cepa*)	0–1	0–5	0	32	8 mo	Tolerant to low O$_2$	51
Onion, green (*Allium cepa*)	1	5	0	32	2 mo		53
Parsley (*Petroselinum crispum*)	8–10	8–10	0	32	3 mo		51
Pepper, sweet (*Capsicum annuum*)	2–5	0	12	54	3 wk		51
Radish (*Raphanus sativus*)	1–2	2–3	0	32	4 mo		51
Spinach (*Spinacia oleracea*)	7–10	5–10	0	32	3 wk		51
Tomato (*Lycopersicon esculentum*)	3–5	2–3	12	54	4–6 wk		51

[a] The approximate storage periods are compiled by the authors based on all the information available.

BIBLIOGRAPHY

1. H. M. Couey and K. L. Olsen, "Storage Response of Golden Delicious Apples Following High Carbon Dioxide Treatment," *Journal of the American Society of Horticulture Science* **100**, 148–150 (1975).

2. C. Y. Wang and W. M. Mellenthin, "Effect of Short Term High CO$_2$ treatment on Storage of d'Anjou Pear," *Journal of the American Society of Horticulture Science* **100**, 492–495 (1975).

3. C. Y. Wang, "Effect of Short Term High CO$_2$ Treatment on the Market Quality of Stored Broccoli," *Journal of Food Science* **44**, 1478–1482 (1979).

4. H. M. Couey and T. R. Wright, "Effect of a Prestorage CO$_2$ Treatment on the Quality of d'Anjou Pears After Regular or controlled Atmosphere Storage," *HortScience* **12**, 244, 245 (1977).

5. O. L. Lau, "Storage Responses of Four Apple Cultivars to a 'Rapid CA' Procedure in Commercial Controlled Atmosphere Facilities," *Journal of the American Society of Horticulture Science* **108**, 530–533 (1983).

6. W. M. Mellenthin, P. M. Chen, and S. B. Kelly, "Low Oxygen Effects on Dessert Quality, Scald Prevention, and Nitrogen Metabolism of d'Anjou Pear Fruit During Long Term Storage," *Journal of the American Society of Horticulture Science* **105**, 522–527 (1980).

7. D. G. Richardson and M. Meheriuk, eds., *Controlled Atmospheres for Storage and Transport of Perishable Agricultural Commodities,* Timber Press, Beaverton, Oreg., 1982.

8. F. W. Liu, "Conditions for Low Ethylene CA Storage of Apple: A Review," in S. M. Blankenship, ed., *Controlled Atmospheres for Storage and Transport of Perishable Agricultural Commodities,* Horticulture Report 126, North Carolina State University, Raleigh, 1985.

9. G. D. Blanpied, "Low Ethylene CA Storage for Empire Apples," in Ref. 8.

10. C. J. Dover, "Effects of Ethylene Removal during Storage of Bramley's Seedling," in Ref. 8.

11. D. R. Dilley, P. L. Irwin, and M. W. McKee, "Low Oxygen, Hypobaric Storage and Ethylene Scrubbing," in Ref. 7.

12. J. B. Biale, "Respiration of Fruits," in W. Ruhland, ed., *Encyclopedia of Plant Physiology,* Vol. 12, Springer-Verlag, New York, 1960.

13. E. Hansen, "Quantitative Study of Ethylene Production in Relation to Respiration of Pears," *Botanical Gazette* 103, 543–558 (1942).

14. D. O. Adams, and S. F. Yang, "Ethylene Biosynthesis: Identification of 1-Aminocyclopropane-1-Carboxylic Acid as an Intermediate in the Conversion of Methionine to Ethylene," *Proceedings of the National Academy of Sciences of the United States of America* 76, 170–174 (1979).

15. M. Knee, "Physiological Responses of Apple Fruits to Oxygen Concentrations," *Annals of Applied Biology* 96, 243–253 (1980).

16. S. P. Burg and E. A. Burg, "Molecular Requirements for the Biological Activity of Ethylene," *Plant Physiology* 42, 144–152 (1967).

17. S. F. Yang, "Biosynthesis and Action of Ethylene," *HortScience* 20, 41–45 (1985).

18. P. H. Li and E. Hansen, "Effects of Modified Atmosphere Storage on Organic Acid and Protein Metabolism of Pears," *Proceedings of the American Society for Horticultural Science* 85, 100–111 (1964).

19. D. A. Kollas, "Preliminary Investigation of the Influence of Controlled Atmosphere Storage on the Organic Acid of Apples," *Nature (London)* 204, 758–759 (1964).

20. N. Allentoff, W. R. Phillips, and F. B. Johnston, "A ¹⁴C Study of Carbon Dioxide Fixation in the Apple. II. Rates of Carbon Dioxide Fixation in the Detached McIntosh Apple," *Journal of the Science of Food and Agriculture* 5, 234–238 (1954).

21. M. T. Wu, B. Singh, J. C. Theurer, L. E. Olson, and D. K. Salunkhe, "Control of Sucrose Loss in Sugarbeet During Storage by Chemicals and Modified Atmosphere and Certain Associated Physiological Changes," *Journal of the American Society of Sugar Beet Technology* 16, 117–127 (1970).

22. C. Y. Wang, "Postharvest Responses of Chinese Cabbage to High CO_2 Treatment and Low O_2 Storage, *Journal of the American Society for Horticultural Science* 108, 125–129 (1983).

23. B. N. Wankier, D. K. Salunkhe, and W. F. Campbell, "Effects of CA Storage on Biochemical Changes in Apricot and Peach Fruits," *Journal of the American Society for Horticultural Science* 95, 604–609 (1970).

24. J. N. McGill, A. I. Nelson, and M. P. Steinberg, "Effects of Modified Storage Atmospheres on Ascorbic Acid and Other Quality Characteristics of Spinach," *Journal of Food Science* 31, 510–517 (1966).

25. T. Suzuki, N. Kubo, S. Haginuma, S. Tamura, and H. Yamamoto, "Changes in Free Amino Acid Content During Controlled Atmosphere Storage of Lettuce, Chinese Cabbage and Cucumber," *Report of National Food Research Institute (Japan)* 38, 46–55 (1981).

26. C. R. Little, H. J. Taylor, and F. McFarlane, "Postharvest and Storage Factors Affecting Superficial Scald and Core Flush of Granny Smith Apples," *HortScience* 20, 1080–1082 (1985).

27. W. J. Lipton and J. K. Stewart, *An Illustrated Guide to the Identification of Some Market Disorders of Head Lettuce,* U.S. Department of Agriculture, Marketing Research Report, U.S. Government Printing Office, Washington, D.C., 950 (1972).

28. W. J. Lipton, *Market Quality and Rate of Respiration of Head Lettuce Held in Low Oxygen Atmospheres,* U.S. Department of Agriculture, Marketing Research Report, U.S. Government Printing Office, Washington, D.C., 777 (1967).

29. C. S. Parsons, J. E. Gates, and D. H. Spalding, "Quality of Some Fruits and Vegetables after Holding in Nitrogen Atmospheres," *Proceedings of the American Society for Horticultural Science* 84, 549–556 (1964).

30. H. Bohling and H. Hansen, "Storage of White Cabbage (*Brassica oleracea* var. Capitata) in Controlled Atmospheres," *Acta Horticulturae* 62, 49–54 (1977).

31. L. S. Beard, "Effect of CA on Several Storage Disorders of White Cabbage," in Ref. 8.

32. F. Mencarelli, W. J. Lipton, and S. J. Peterson, "Responses of Zucchini Squash to Storage in Low O_2 Atmospheres at Chilling and Nonchilling Temperatures," *Journal of the American Society for Horticultural Science* 108, 884–890 (1983).

33. C. Y. Wang and Z. L. Ji, "Effect of Low Oxygen Storage on Chilling Injury and Polyamines in Zucchini Squash," *Scientia Horticulture* 39, 1–7 (1989).

34. T. T. Hatton and R. H. Cubbedge, "Conditioning Florida Grapefruit to Reduce Chilling Injury During Low Temperature Storage," *Journal of the American Society for Horticultural Science* 107, 57–60 (1982).

35. P. Marcellin and A. Chaves, "Effects of Intermittent High CO_2 Treatment on Storage Life of Avocado Fruits in Relation to Respiration and Ethylene Production," *Acta Horticulture* 138, 155–162 (1983).

36. R. E. Anderson, C. S. Parsons, and W. L. Smith, *Controlled Atmosphere Storage of Eastern-Grown Peaches and Nectarines,* U.S. Department of Agriculture, Marketing Research Report, U.S. Government Printing Office, Washington, D.C.

37. N. L. Wade, "Physiology of Cold Storage Disorders of Fruits and Vegetables," in J. M. Lyons, D. Graham, and J. K. Raison, eds., *Low Temperature Stress in Crop Plants,* Academic Press, Orlando, Fla., 1979.

38. Y. Ilker and L. L. Morris, "Alleviation of Chilling Injury of Okra," *HortScience* 10, 324 (1975).

39. I. L. Eaks, "Effect of Modified Atmosphere on Cucumbers at Chilling and Nonchilling Temperatures," *Proceedings of the American Society for Horticultural Science* 67, 473–478 (1956).

40. A. A. Kader, "Biochemical and Physiological Basis for Effects of Controlled and Modified Atmospheres on Fruits and Vegetables," *Food Technology (Chicago)* 40, 99–104 (1986).

41. R. M. Smock and A. Van Doren, "Controlled Atmosphere Storage of Apples," *Cornell University Agricultural Experimental Station Bulletin,* 762 (1941).

42. E. Hansen, "Reactions of Anjou Pears to Carbon Dioxide and Oxygen Content of the Storage Atmosphere," *Proceedings of the American Society for Horticultural Science* 69, 110–115 (1957).

43. M. L. Arpaia, F. G. Mitchell, A. A. Kader, and G. Mayer,

"Effects of 2% O_2 and Varying Concentrations of CO_2 with or without C_2H_4 on the Storage Performance of Kiwifruit," *Journal of the American Society for Horticultural Science* **110**, 200–203 (1985).

44. P. E. Brecht, A. A. Kader, and L. L. Morris, "The Effect of Composition of the Atmosphere and Duration of Exposure on Brown Stain of Lettuce," *Journal of the American Society for Horticultural Science* **98**, 536–538 (1973).

45. M. Meheriuk, "Controlled Atmosphere Recommendations for Apples," in J. Fellman, ed., *Proceedings of the Fifth International CA Research Conference,* Washington State University, Wenatchee, Washington, 1989.

46. R. E. Hardenburg, A. E. Watada, and C. Y. Wang, *The Commercial Storage of Fruits, Vegetables, and Florist and Nursery Stocks,* U.S. Department of Agriculture, Agricultural Handbook **66** (1986).

47. A. A. Kader, "A Summary of CA Requirements and Recommendations for Fruits Other than Pome Fruits," in Ref. 45.

48. W. H. Smith, "The Use of Carbon Dioxide in the Transport and Storage of Fruits and Vegetables," *Advances in Food Research* **12**, 95–146 (1963).

49. D. G. Richardson and M. Meheriuk, "Controlled Atmosphere Recommendations for Pears," in Ref. 45.

50. J. S. Pruthi, "Physiology, Chemistry, and Technology of Passion Fruit," *Advances in Food Research* **12**, 203–282 (1963).

51. M. E. Saltveit, "A Summary of Requirements and Recommendations for the Controlled and Modified Atmosphere Storage of Harvested Vegetables," in Ref. 45.

52. A. L. Ryall and W. J. Lipton, *Handling, Transportation, and Storage of Fruits and Vegetables,* Vol. 1, AVI Publishing Co., Westport, Conn., 1979.

53. H. W. Hruschka, *Storage and Shelf Life of Packaged Green Onions,* U.S. Department of Agriculture, Marketing Research Report, U.S. Government Printing Office, Washington, D.C. **1015** (1974).

CHIEN YI WANG
ALLEY E. WATADA
USDA/ARS
Beltsville, Maryland

COOKIES. See BISCUITS AND CRACKERS.

CORN AND CORN PRODUCTS

CORN

Origin and History

Corn (*Zea mays*) was probably gathered from the wild for millennia before being domesticated about 7,000 years ago. Early Americans spread this food staple from what is today southern Mexico to as far south as Argentina and northward to Canada. Much later, Columbus brought corn from the New World back to Spain and within two generations, it had been spread around the world (1). There are 6 types of corn being grown: pod, dent, flint, flour, sweet, and pop. While a prehistoric pod corn may have been the forerunner of all modern varieties, pod-type corns have little value today except for being grown as ornamental curiosities. Genetic differences cause kernal color to range from black to blue to red or white, but yellow is the most commonly found color.

Corn kernels contain an embryo (a tiny dormant corn plant) and endosperm (starch granules in a protein matrix) that are encased within a proteinaceous aleurone layer and a cellulosic outermost protective pericarp (seed coat). Yellow dent corn grown in the Corn Belt of the United States, in Canada, and in much of Europe was developed in North America after the Revolutionary War. Hybridization has made dent varieties the most productive corn found anywhere in the world. The kernels contain vitreous endosperm at their sides and back with the remainder being floury endosperm. When the kernel dries during maturation, the floury endosperm shrinks, causing the top (crown) of the kernel to collapse or dent.

The kernel retains its rounded shape when dried because nearly all the endosperm in flint corn is vitreous. Flint corn is most popular in South America and Northern Europe. Popcorn is a special type of flint corn that was selectively bred by American Indians.

One of the oldest types of corn, the floury variety can be traced back to the early Americas. Because the endosperm is nearly all soft starch, it was easily ground and provided a staple food for the Indians. Lacking vitreous endosperm, the kernels tend to shrink uniformly when dried. Once widely grown, floury corn today is popular only in the Andean region of South America.

In most varieties of corn, sugar produced by photosynthesis is rapidly converted to starch as a source of future energy for the embryo. In sweet corn varieties, this rate of conversion is slowed or prevented, resulting in a delicious vegetable that is consumed fresh, canned, or frozen.

Breeding/Genetics

Since prehistoric times, man has cultured corn to promote qualities he felt improved the crop for his needs. Through carefully controlled hybridization, the composition, yield, resistance to disease (and to a lesser degree, insects), and time to maturity have been modified. Table 1 shows that composition can be changed dramatically.

Besides the quantities of starch, protein, and oil, the composition of these components can be modified as well. Normal corn starch contains 26–28% amylose (linear starch molecules) and 72–74% amylopectin (branched starch molecules). High amylose varieties containing as much as 85% amylose and waxy varieties that are almost all amylopectin have been developed. Protein nutritional quality has also been improved by development of high lysine strains. The properties of the starch component can be altered dramatically and are the subject of many patents (2).

The composition and properties of components are determined by a genetic code. Changes in this code through hybridization is a lengthy process; the variability shown in Table 1 is the result of more than 87 years of effort at the University of Illinois. Developments in genetic engineering and tissue culture techniques offer hope of reducing this development time. Methods to produce genetically transformed plants include the direct transfer of DNA by microinjection, electroporation, DNA entry aided by polyethylene glycol, infection of plant tissue with Agro-

Table 1. Corn Composition Changes Possible by Breeding[a]

	Typical Corn	Composition Variability Possible by Breeding	
Component	Content, %	Minimum, %	Maximum, %
Starch	72	35	80
Protein	9.8	4.0	32
Oil	4.8	0.2	22

[a] Ref. 3.

bacterium engineered to carry foreign DNA into plants, and use of a particle gun to inject DNA-coated microprojectiles (4). Fertile corn plants have been produced from protoplasts cultured in-vitro, making recovery of genetically manipulated progeny produced via tissue culture at least theoretically possible (5,6).

Present Production (Volumes/Techniques) and Economics

Every continent, except Antarctica, grows corn; 40% of the present world crop is produced in the United States. In the 1987–1988 crop year, 12 states (Iowa, Ill., Nebr., Minn., Ind., Ohio, Wis., Mo., S. Dak., Mich., Kans., and Tex. in order of production) produced 157.5 million metric tons (6.2 billion bushels) that was 88% of the United States and 36% of the world's crop (7). Yield is influenced by many factors, including climate, pest control, planting density, and fertilization. Yield in the United States has increased from about 1.5 metric tons/hectare in the 1930s to about 7.5 metric tons/hectare today. In 1985, a test plot produced 23.2 metric tons/hectare and yields approaching 40 metric tons/hectare are considered possible; corn is the most productive of the major food crops. Approximate yields, production, and consumption data are shown in Table 2.

A crop of corn is always maturing somewhere in the world. It grows from north latitude 58° to south latitude 40° and from below sea level to altitudes of 4,000 meters. It is adapted to areas with fewer than 10 in. of rainfall and regions having more than 400 in. Early varieties, that have been adapted to cold climates, mature in as little as 60 days. Late varieties grown in the tropics need nearly a year to reach maturity. Corn can grow to as little as 60 cm in height or as tall as 6.5 meters. Ears of corn can be as small as your thumb for some popcorn varieties or as large as 60 cm as those grown in the Jala Valley of Mexico (1).

A century ago, all corn was harvested by hand and stored on the cob in drying cribs. Today, one machine that picks and shells in the field can do the work of 50 men. After harvesting, kernels are dried to less than 15% moisture content to maintain grain quality and prevent long term storage spoilage. Drying must be done carefully to allow the use in wet milling (8,9).

As with most major agricultural commodities, corn economics are tied to governmental policy. These policies, designed to achieve domestic objectives, generally force local prices above free market levels. A worldwide network of grain exchanges provides instantaneous response to perceived changes in supply, demand, and harvest estimates throughout the growing season. Seasonally produced commodities like corn generally are least expensive at harvest and rise in price to compensate for storage charges.

Physical Properties/Quality

Typical analysis of dent corn is shown in Table 3. Corn grain quality is defined by governmental agencies. The United States Department of Agriculture grading scale (Table 4) represents qualities sought by consumers around the world. Insects can adversely affect the market value of stored corn. Infestation must be avoided by sanitation/fumigation of storage areas. Mechanical damage lowers the value of corn in many applications. Breakage of the seed hull or damage to the tip cap renders the kernels susceptible to fungal invasion that can make the corn unfit for nearly every use. The seed corn industry has perhaps the highest quality requirements, followed closely by beverage alcohol, dry milling, and snack food producers. The wet milling process can tolerate less than Grade 1 quality because of the low-pH, sulfurous acid steeping that precedes milling; although significant heat damage must be avoided.

Table 2. Worldwide Corn Production Data

	Early 1950s		Mid 1970s		Late 1980s		
Country	Yield[a]	Production[b]	Yield	Production	Yield	Production	Consumption[c]
World	1.5	130	2.7	335	3.6	460	460
United States	2.4	70	5.2	140	7.3	200	150
China	1.3	15	2.5	45	3.4	74	75
Brazil	1.3	6	1.5	16	1.7	25	NA
USSR	1.4	7	3.0	10	3.0	14	27
S. Africa	1.0	3	1.5	9	1.8	8	NA

[a] Yield in metric tons/hectare.
[b] Production in millions of metric tons/year.
[c] Consumption in millions of metric tons/year.

Table 3. Component Composition of Typical Dent Corn Composition Shown as Percent of Dry Matter, % d.b

Part	Percent of Kernel	Composition of Kernel Parts					
		Starch	Protein	Oil	Sugar	Ash	Other
Whole kernel	100	72	9.8	4.5	1.9	1.5	10.3
Endosperm	82	86	9	1	0.6	0.3	3.1
Germ	12	8	19	30	11	10	22
Hull & tip cap	6	7	4	1	1	1	86

CORN USE IN FEED

While corn was originally grown as food, the single largest use today is as feed for farm animals. In the United States, 80% of the domestic corn usage is for this purpose. Swine consume ca one-third; beef cattle eat one-fourth; and dairy cattle and poultry each account for about one-fifth of the feed corn. Corn is an important feed ingredient because it supplies the energy component and a large portion of the protein input to the animal's diet. It has been a dominant force in feed manufacture, consistent in quality and composition, and never presenting its customers with a shortage of supply.

Also, the co-products of the various industries that use corn to produce beverage alcohol, starch, corn sweeteners, corn oil, and dry milled products provide the concentrated protein and vitamin content of corn, making them valuable feed ingredients as well. Corn gluten feed is a source of additional protein in feeds compounded for beef cattle and dairy herds. The fiber content is readily converted by these ruminants. The xanthophylls in corn gluten meal provide the coloration in chicken and eggs so desired by many United States consumers. The distilling industry provides distillers dried grain while dry millers produce high-fiber, high-calorie hominy feed.

WET MILLING

Process Description

The United States wet millers buy ca 15% of the corn used in the U.S. (worldwide, wet millers consume about 10% of the corn used). There are two dozen corn wet mills in the United States ranging in capacity from 600–10,000 metric tons/day. Shelled corn is shipped to wet millers by truck, rail, or barge. A process flowsheet is shown as Figure 1. After cleaning to remove coarse material, ie, cobs, and fines (broken corn, dust, etc), the corn is steeped in a sulfurous acid solution to soften the corn and render the starch granules separable from the protein matrix that envelopes them. About 7% of the kernel's dry substance is leached out during this step, forming protein-rich steepwater, a valuable feed ingredient and fermentation adjunct.

The softened kernels are coarsely ground (first grind) to release the germs. Because of their high oil content, the germs are lighter than the starch, protein, and fiber fractions and can be separated in hydrocyclones. The germs are washed free of remaining starch, dried, and the valuable corn oil is removed by expelling, solvent extraction, or a combination of both. The spent germs are a valuable feed ingredient.

The starch-protein-fiber slurry is subjected to an intense milling to release additional starch from the fiber. The fiber is then wet-screened from the starch-protein slurry, washed free of starch, and dried to form the major component of corn gluten feed—a valuable feed ingredient. The best fiber can be additionally purified to become corn bran, a dietary fiber ingredient that has been shown to lower serum cholesterol and triglycerides (10). The starch-protein slurry is separated into its component parts, again taking advantage of the density difference between the heavier starch and the lighter protein (gluten) particles. Separation is usually done with combinations of disk-nozzle centrifuges and banks of hydrocyclones.

The protein fraction is filtered and dried to become high (60%) protein content corn gluten meal. The starch slurry can be dewatered and dried to produce regular corn starch. Dry starch can be sold as is or heat treated in the presence of acid catalysts to produce dextrins. Or, it is chemically modified before dewatering and drying to produce modified starches used in food and industrial applications. Lastly, it can be hydrolyzed to produce corn sweeteners.

Starch

Corn starch is a principal ingredient in many food products, providing texture and consistency, as well as energy. More than half of the corn starch sold is used in industrial applications, primarily in paper, textile weaving, adhesives, and coatings. Starch is a polymer consisting of α-linked anhydroglucopyranose units. Two forms exist: amylose is an essentially linear molecule in which the anhydroglucopyranose units are linked almost exclu-

Table 4. USDA Grading Standards for Corn Maximum Limits, %

Grade[a]	Minimum Test Weight Per bu, lb	Broken Corn and Foreign Material	Damaged Kernels	
			Total	Heat Damaged
1	56	2	3	0.1
2	54	3	5	0.2
3	52	4	7	0.5
4	49	5	10	1.0
5	46	7	15	3.0

[a] A sixth, or sample grade, is either lower quality than Grade 5, or shows significant contamination by odor or noncorn material(s).

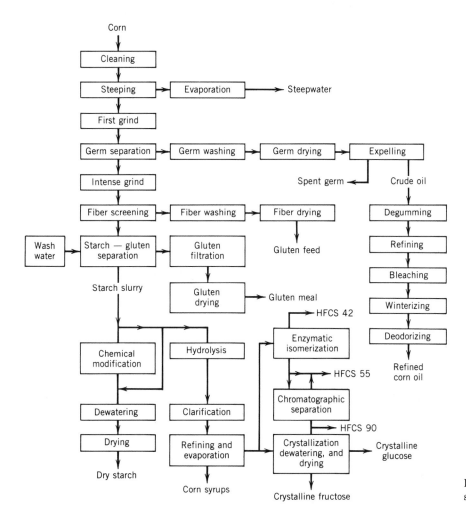

Figure 1. Corn wet milling process flow sheet.

sively via α-1,4 bonds. Amylopectin is a much larger, branched molecule (the mol wt is ca 1,000 times greater than amylose) (11); α-1,4 linkages predominate but there is a significant number of α-1,6 linkages that result in the branched structure. The structure of these molecules is shown in Figure 2. While the ratio of amylose to amylopectin is quite consistent in normal corn varieties, it varies considerably when starch is obtained from either waxy or high amylose varieties (Table 5).

When heated in the presence of water to 62–72°C, normal starch granules swell, forming high viscosity pastes or gels. This process is called gelatinization. Starch from normal corns form characteristic firm, opaque gels because of the amylose fraction. The linear molecules align on cooling after gelatinization in a process called retrogradation, forming a thick, rubbery mass. The bushy amylopectin molecules in waxy starch cannot align to form such a mass, resulting in softer, translucent salve-like gels.

Table 5. Starch Composition from Various Corns

Corn Type	Amylose Content, % dB	Amylopectin Content, % dB
Normal	26–28	72–74
Waxy	0–2	98–100
High amylose	55–85	15–45

High amylose starches are difficult to fully gelatinize and provide little viscosity unless cooked above the boiling point of water. Brabender Viscoamylograph viscosity curves of these three corn starches are shown as Figure 3. These vastly different characteristics are further enhanced by modification of the native granules, resulting in starches with a wide range of properties for industrial and food applications.

Corn Starch-Based Sweeteners

Acid or enzyme catalysts can be used to break the linkages between the anhydroglucopyranose units in the starch molecule with the addition of a molecule of water at the break site. This process, called hydrolysis, produces a variety of corn-based sweeteners. The first sweetener from starch (arrowroot, not corn) was produced in Japan in the ninth century. By the nineteenth century, starch sugars were being produced in Europe and the United States. Because glucose is not as sweet as sucrose, products made from corn syrups are not cloyingly sweet, allowing delicate flavors to reach the palate. However, enzymatic isomerization of glucose to fructose produces high fructose corn syrups (HFCS) that are as sweet as sucrose syrups, thus allowing corn sweeteners to replace sugar in liquid applications (such as soft drinks).

To produce sweeteners from corn starch, the starch is

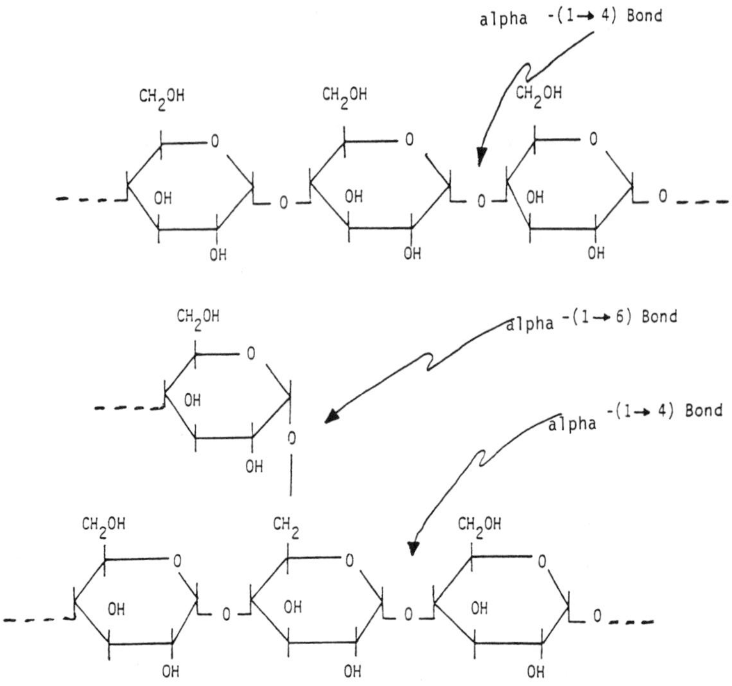

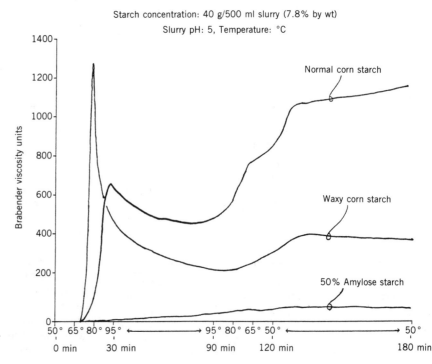

Figure 2. Chemical structure of amylose (top) and amylopectin (bottom) fractions of corn starch.

gelatinized in the presence of a catalyst under conditions that promote hydrolysis. Acid is usually used to make slightly converted or low dextrose equivalent (DE) syrups. Enzymatic conversion, using thermostable alpha-amylases (for liquefaction), beta-amylases (for maltose production), and glucoamylase (for high glucose content) is widely practiced to produce a full range of saccharide compositions. After the desired degree of hydrolysis is achieved, insolubles are separated by centrifugation or filtration (or both). Soluble impurities are removed using activated carbon or ion exchange (either singly or in com-

bination) before evaporation of the purified syrup to the desired solids concentration. Pure glucose is obtained by crystallization from highly converted starch syrups. Highly converted glucose syrups can also be enzymatically isomerized to 42% (dry basis) fructose content, significantly increasing their sweetness. Further increases in fructose content are possible using chromatographic enrichment (12). Pure fructose is produced by crystallization from syrups enriched to fructose contents above 90% (dry basis). A variety of corn starch-based sweeteners are described in Table 6.

Figure 3. Brabender Viscoamylograph viscosity curves for various corn starches.

Table 6. Characteristics of Various Corn Starch-Based Sweeteners

Characteristic	Type I Syrup	Type II Syrup	Type III Syrup	Type IV Syrup	HFCS 42	HFCS 55	Crystalline Product	
							Glucose	Fructose
DE[a]	27–30	42–43	70	95	NA	NA	99.8	NA
Glucose, % dB	9–10	19–20	43–47	92	52	40	99.6	—
Maltose, % dB	8–9	14	27–30	4	—	—	—	—
Maltotriose, % dB	9–10	12	5–7	1	—	—	—	—
Maltotetraose, % dB	7–8	9–10	4–5	1	—	—	—	—
Higher polymers	63–67	44–46	14–17	2	6[b]	5[b]	0.4[b]	0.4[d]
Fructose	—	—	—	—	42	55	—	99.6
Sweetness[c]	20–30	25–45	40–70	55–85	80–110	85–120	60–90	100–140

[a] Dextrose equivalent—a measure of reducing sugar content.
[b] Nonglucose content.
[c] Relative to sucrose when tasted at 10% solids solution; data show the range of taste panel results reported in the literature.
[d] Nonfructose content.

Corn Oil

The crude corn oil recovered from the germs consists of a mixture of triglycerides, free fatty acids, phospholipids, sterols, tocopherols, waxes, and pigments (13). Refining removes the substances that detract from the quality, resulting in a nearly pure (99%) triglyceride stream. The first refining step is degumming; 1–3% water is added and the dense, hydrated gums are removed by centrifugation. The degummed oil is then refined. Treatment with dilute (12–13 wt%) NaOH forms water soluble soaps of the free fatty acids, allowing centrifugal separation. An alternative physical refining process steam strips the volatile components, primarily free fatty acids.

Pigments are removed by sorption on bleaching clay that is then separated from the oil by filtration. The oil is then winterized by cooling to ca 4°C, precipitating the waxes that are then filtered from the oil. Winterization is not required if the oil is to be hydrogenated. Deodorization, a steam stripping process similar to physical refining, removes volatile impurities, resulting in an oil with lighter color and improved oxidative stability.

Corn oil's flavor, color, stability, retained clarity at refrigerator temperatures, polyunsaturated fatty acid composition, and vitamin E content make it a premium vegetable oil. The major uses are frying or salad applications (50%) and margarine formulations (35%).

Other Co-Products

The solubles removed during steeping are concentrated to ca 50% dry substance by evaporation. Most of this high protein material is added to corn gluten feed before drying. Some is sold directly as a feed ingredient and it is also used as a nutrient in fermentations. Steepwater has also shown value as a growth stimulant for chickens and has demonstrated a synergistic effect with fertilizer when applied to certain plants (14,15). When combined with an insecticide, steepwater has been shown to act as a repellent to bees (thus protecting them) while attracting undesirable insects.

Its sponge-like properties and protein content make spent corn germ meal an ideal carrier of nutrient supplements and medications in animal feeds. However, most of

this material is combined with the fiber and steepwater streams and dried to form corn gluten feed. Corn gluten meal has the highest protein content (60%) of the co-product streams. Its high digestibility coupled with carotene and xanthophyll content make it an ideal chicken feed ingredient. It is also used to formulate ruminant feeds (high bypass protein value) and its low residue content make it desirable in pet foods.

DRY MILLING

General

The United States dry millers buy ca 3% of the corn used in the U.S. There are 71 dry corn mills in 22 states in the United States (16). They tend to be smaller than wet mills; while the largest U.S. dry mill has a capacity of 1,750 metric tons/day, most grind fewer than 250 metric tons/day. About one-fifth of the corn ground is white corn, the remainder being yellow dent. Dry millers use three processes; tempering-degerming, stone-grinding or nondegerming, and alkali-cooking. These processes are diagrammed in Figure 4.

Tempering-Degerming

The most prevalent process is tempering-degerming. The first step is to dry clean the corn as in wet milling. Sometimes a secondary wet cleaning process is added to remove surface dirt. The corn is moistened (tempered) to a water content of ca 20% to loosen the hull from the endosperm and to make the germ pliable. The corn is then coarsely ground through a special machine to crack the endosperm, free the germ, and release the hull. The fines are removed by sifting and the hulls are separated by air aspiration.

The remaining mixture of germ (whole and broken) and endosperm pieces is separated into narrowly sized fractions by sieving. Clean, heavier endosperm can be separated from the lighter germs by air flotation. The germ fraction (and the part of the endosperm that could not be separated) are then milled between rolls spaced slightly closer than the particles; the soft germ particles squeeze through without damage and spring back to their original

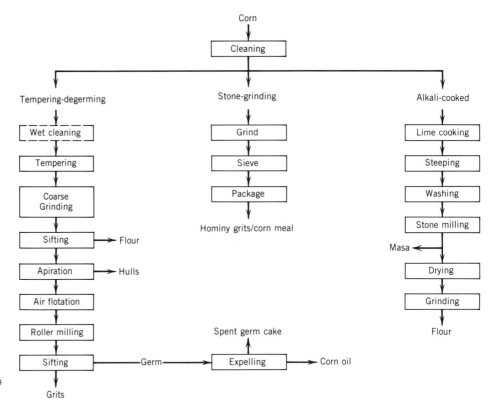

Figure 4. Corn dry milling process flow sheet.

size while the brittle endosperm is crushed to smaller pieces. Germ particles can then be separated from the endosperm by screening.

Oil is expelled from the germ and the spent material is either mixed with the hulls to form hominy feed or sold as food-grade corn germ flour. Food-grade corn bran is also sold as a dietary fiber ingredient. The various size fractions of endosperm have a variety of applications; the largest pieces (flaking or hominy grits) are used to produce corn flakes while smaller sizes have many food and industrial uses.

Speciality Mills

Several corn meal products are purchased by the U.S. Government to provide nutrition to people in the Third World via the food donation program (P.L. 480) that include: enriched corn meal, soy fortified corn meal or corn-soy blend (CSB), corn soy milk (CSM), and sweetened and unsweetened instant corn soy milk (ICSM). Industrial uses are a growing market for dry-milled products. Regular and chemically modified corn flours provide an inexpensive starch source for foundry core binders and adhesives in building materials and charcoal briquettes, and substrate for pharmaceutical fermentations. Production of corn-based sweeteners from dry-milled grits has been reported (17).

Stone-Ground and Alkaline-Cooked

Whole ground or stone-ground corn (usually white corn) produces hominy grits and corn meal, ingredients for corn bread, muffins, hush puppies, and spoon bread. The increasing popularity of Mexican foods in the United States

has created a demand for masa (dough) produced by the alkaline-cooked (nixtamalization) process. Corn is cooked in a boiling lime solution for up to one h and then steeped for up to an additional 12 h. Washing removes the excess alkali and loosened hulls and the remaining wet endosperm/germ mixture (nixtamal) is stone-ground to produce masa which can be formed into finished products, ie, tortillas or dried, screened, and reformulated into flours for later use.

FOOD USES FOR CORN

Corn is used directly and ingredients produced from corn are widely used for food in Asia, Africa, North and South America, and parts of the USSR. Much of the corn consumed in the United States is in the form of ready-to-eat breakfast cereals.

Corn Flakes

The corn flake production process has remained unchanged for decades. Flaking grits are pressure cooked with enough flavored syrup to raise the grit moisture to ca 33%. The starch gelatinizes but there is insufficient water to allow complete granule disruption; the grits appear translucent but do not lose their form. The grits are dried to ca 20% moisture and flaked through high pressure rolls. They are then toasted, cooled, and packaged to retain crispness until eaten. Each flake is formed from a single large piece of endosperm. Extrusion processes have been developed to produce corn flakes from smaller endosperm particles but the flavor and texture differs from flakes made in the traditional manner (18).

Sweet Corn

Sweet corn is popular in the United States, its country of origin; a natural mutant was first grown by North American Indians. It differs from dent corn in genetic makeup, accumulating sugar at the expense of starch production. Corn breeders have concentrated not only in developing varieties that have high, stable sugar contents, but have worked to improve insect and disease resistance as well. Because sweet corn is consumed directly, factors such as a long, tight husk as an insect barrier and reduced silk to minimize bacterial and fungus growth must be considered in strain selection. The thin-hulled, high sugar content hybrids preferred by consumer's palates require careful harvesting and the seeds need more care in processing and planting than their sturdier field corn cousins. Most of the sweet corn seed is grown by six producers in the optimum climate that exists in Treasure Valley, Idaho.

Other Corn-Based Foods

Degerminated corn flour, meal, and grits are used to make a variety of other breakfast foods including expanded or puffed shapes, gruels, and porridges. They are also ingredients in both fermented and unleavened breads (tortilla, arepa, etc), steamed products (eg, tamales, couscous, and dumplings), and snacks (chips, expanded shapes, and fritters). A favorite North American snack that is spreading around the world is popped corn. Beverages, both alcoholic and nonalcoholic, are produced from corn grits, corn starch, and corn syrups. Beers range from the clear yeast-fermented pale lager brews made with corn grits in the United States to thick, opaque, Bantu or Kaffir beers that are soured with an Acetobacter or Lactobacillus fermentation before yeast is added. Parched corn is a popular food ingredient around the world.

FERMENTATION

Corn has long been used as a source of substrate for fermentations, primarily for the production of alcohol. The high productivity and worldwide availability of corn makes it an ideal starting commodity for such activity. The starch must first be converted to a fermentable saccharide, usually done by enzymes either external to or concurrently with the fermentation process. The United States has an annual production capability of 3.9×10^9 liters of fuel-grade ethanol (19). Originally considered a gasoline extender but now used as an oxygenating agent and octane enhancer, production is predicted to double between 1988 and 1992 (20). Corn also serves as a raw material for the manufacture of food-grade acids (citric, lactic, acetic, etc), enzymes, amino acids, vitamin C, and a variety of antibiotics.

USES FOR BIOMASS (STOVER, CORNCOBS)

While the kernel is the most valuable part of the corn plant, a lot of other biomass is also produced. For each 100 kg of grain, ca 45 kg of stalk and 18 kg of corncob is produced. Corncobs are converted into furfural, and ground for use as carriers for agricultural chemicals and vitamins, amino acids, and medications in feeds as well as absorbents for chemical and oil spills. However, most corncobs remain in the field because of the development of the sheller-picker. Because seed corn is harvested on the cob to minimize kernel damage, only seed corn producers have a reliable source of cobs. Forcing corncob products to bear the cost of separate collection and storage has resulted in occasional shortages, price increases, and searches for alternatives.

Cornstalks have been evaluated as a fiber source for the production of paper and board products. The stalk residue has never been found to be cost competitive with conventional fiber sources. Cornstalk juice can contain as much as 30% impure sugar and has been used to produce molasses and other sweeteners since the 1700s. Stalks have been used for feed, but low digestibility limits their nutritional benefits, even when fed to ruminants. Alkaline treatment has been reported to improve the feed value (21). However, the improvement of soil condition by plowing under stalks and cobs remains the biggest use today.

BIBLIOGRAPHY

1. P. C. Mangelsdorf, *Corn, Its Origin Evolution and Improvement*, Belknap Press, Cambridge, Mass., 1974, p. 2.

2. Examples include: U.S. Pat. 4,107,338 (Aug. 15, 1978), G. L. Tutor, J. C. Fruin, and James L. Helm (to Anheuser-Busch, Inc.). U.S. Pat. 4,615,888 (Oct. 7, 1986), J. Zallie, R. Trimble, and H. Bell (to National Starch and Chemical Corp.). U.S. Pat. 4,798,735 (Jan. 17, 1989), R. B. Friedman, D. J. Gottneid, E. J. Faron, F. J. Pustek, and F. R. Katz (to American Maize Products Company).

3. J. W. Dudley, unpublished data, University of Illinois, Urbana, June 28, 1989.

4. R. L. Phillips, "Maize Genetic Engineering Update," *1988 Scientific Conference Proceedings,* Corn Refiners Association, Inc., Washington, D.C.

5. R. D. Shillito, G. K. Carswell, C. M. Johnson, J. J. DiMaio, and C. T. Harms, "Regeneration of Fertile Plants from Protoplasts of Elite Inbred Maize," *Bio/Technology* **7**, 581–587 (June, 1989).

6. L. M. Prioli and M. R. Söndahl, "Plant Regeneration and Recovery of Fertile Plants from Protoplasts of Maize (*Zea mays* L.)," *Bio/Technology* **7**, 589–594 (June, 1989).

7. *1988 Corn Annual,* Corn Refiners Association, Inc., Washington, D.C., p. 21.

8. R. B. Brown, G. N. Fulford, T. B. Daynard, A. G. Meiering, and L. Otten, "Effect of Drying Method on Grain Corn Quality," *Cereal Chemistry* **56** (6), 529–532 (1979).

9. R. B. Brown, G. N. Fulford, L. Otten, and T. B. Daynard, "Note on the Suitability for Wet Milling of Corn Exposed to High Drying Temperatures at Different Moisture Contents," *Cereal Chemistry* **58** (1), 75, 76 (1981).

10. L. Earll, J. M. Earll, S. Naujokaitis, S. Pyle, K. McFalls, and A. M. Altschul, "Feasibility and Metabolic Effects of A Purified Corn Fiber Food Supplement," *Journal of the American Dietetic Association* **88** (8), 950–952 (1988).

11. J. J. M. Swinkels, *Starke* **37** (1), 3 (1985).

12. L. Hobbs, F. W. Schenck, J. E. Long, F. R. Katz, "Corn-Based Sweeteners," *Cereal Foods World* **31** (12), 852–869 (1986).

13. R. A. Reiners, "Corn Oil in Products of the Corn Refining

Industry" in *Foods Seminar Proceedings,* Corn Refiners Association, Inc., Washington, D. C., 18–23 (1978).

14. J. M. Russo and V. Heiman, "The Value of Corn Fermentation Condensed Solubles as a Growth Stimulant for Chickens," *Poultry Science* **38** (1) (1959).

15. U.S. Pat. 4,976,767 (Dec. 11, 1990), A. M. Kinnersley and W. E. Henderson (to CPC International).

16. Ref. 7, p. 7.

17. K. Kroyer, *Die Starke* **18** (10), 311–316 (1966).

18. S. A. Watson, P. E. Ramstad eds., *Corn: Chemistry and Technology,* American Association of Cereal Chemistry, Inc., St. Paul, Minn., 1987, p. 404.

19. J. G. Joyce, *Investor News* **10** (3), 1, 4 (1988).

20. J. M. Murray, *The Current Status and Future Prospects for Food and Industrial Products From Corn,"* Wisconsin Corn Growers Association, Madison, 1988, p. 4.

21. U. I. Oji, D. N. Mowat, and J. E. Winch, "Alkali Treatment of Corn Stover to Increase Nutritive Value," *Journal of Animal Science* **44** (5), 798–802 (1977).

FRED W. SCHENCK
Corn Products,
CPC International
Argo, Illinois

CORONARY HEART DISEASE. See CHOLESTEROL AND HEART DISEASE.

CORROSION AND FOOD PROCESSING

MATERIALS OF CONSTRUCTION

The "Stainless Steels"

Many people refer to this range of alloys as austenitic stainless steels, 18/8s, 18/10s, etc, without a full appreciation of what is meant by the terminology. It is worth devoting a few paragraphs to explain the basic metallurgy of stainless steels.

It was in 1913 that Harry Brearley discovered that the addition of 11% chromium to carbon steel would impart a good level of corrosion and oxidation resistance, and by 1914 these corrosion resisting steels had become commercially available. It was Brearley who pioneered the first commercial use of these steels for cutlery, and it was also he who coined the name "stainless steels." For metallurgical reasons, which are outside the scope of this article, these were known as ferritic steels because of their crystallographic structure. Unfortunately, they lacked the ductility to undergo extensive fabrication and furthermore, they could not be welded. Numerous workers tried to overcome these deficiencies by the addition of other alloying elements and to produce a material where the ferrite was transformed to austenite (another metallurgical phase) that was stable at room temperature. Soft stainless steels that were ductile both before and after welding were developed in Sheffield, England (then the heart of the British steel industry) exploiting scientific work undertaken in Germany. This new group of steels was based on an 18% chromium steel to which nickel was added as a second alloying element. These were termed the austenitic stainless steels. The general relationship between chromium and nickel necessary to maintain a

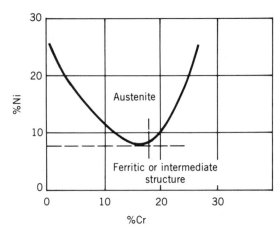

Figure 1. Graph showing the various combinations of chromium and nickel that form austenitic stainless steels (ref. 1).

fully austenitic structure is shown diagrammatically in Figure 1. It will be seen that the optimum combination is 18% chromium, 8% nickel—hence the terminology 18/8s.

Probably the next major advance in the development of stainless steels was the discovery that relatively small additions of molybdenum had a pronounced effect on the corrosion resistance, greatly enhancing the ability to withstand the effects of mineral acids and other corrodents such as chloride solutions. Needless to say, from these early developments, there has been tremendous growth in production facilities and the number of grades of stainless steel available. Table 1 lists some of the more commonly available grades, while Figure 2 illustrates how the basic 18/8 composition is modified to enhance specific physical or chemical properties.

In spite of the plethora of stainless steels available, grades 304 and 316 have, and continue to be, the workhorses for fabrication of dairy and food processing equipment.

Although 316 stainless steel offers excellent resistance to a wide range of chemical and nonchemical environments, it does not offer immunity to all. In the case of the food industry, these are notably anything containing salt, especially low-pH products. There was, therefore, a demand by industry to develop more corrosion-resistant materials and these are finding increasing use in the food industry for certain specific processing operations.

Super Stainless Steels and Nickel Alloys

The super stainless steels are a group of alloys that have enhanced levels of chromium, nickel, and molybdenum, compared to the conventional 18/8s. The major constituent is still iron; hence the classification under the "steel" title. Still further increases in the three aforementioned alloying elements result in the nickel alloys. (The classification of an alloy is generally under the heading of the major constituent.)

There are a large number of these alloys but those of primary interest to the food industry are shown in Table 2 together with their composition. In general terms, it will be noted that the increase in nickel content is accompanied by an increase in chromium and molybdenum. As

Table 1. Composition of Some of the More Commonly Used Austenitic Stainless Steels. Unless Indicated Otherwise, All Values Are Maxima.

Alloy	UNS No.	Composition, %						
		Carbon	Manganese	Silicon	Chromium	Nickel	Molybdenum	Others
302	S30200	0.15	2.00	1.00	16.0–18.0	6.0–8.0		Sulfur 0.030 Phosphorus 0.045
304	S30400	0.08	2.00	1.00	18.0–20.0	8.0–10.5		Sulfur 0.030 Phosphorus 0.045
304L	S30403	0.03	2.00	1.00	18.0–20.0	8.0–12.0		Sulfur 0.030 Phosphorus 0.045
316	S31600	0.08	2.00	1.00	16.0–18.0	10.0–14.0	2.0–3.0	Sulfur 0.030 Phosphorus 0.045
316L	S31603	0.03	2.00	1.00	16.0–18.0	10.0–14.0	2.0–3.0	Sulfur 0.030 Phosphorus 0.045
317	S31700	0.08	2.00	1.00	18.0–20.0	11.0–15.0	3.0–4.0	Sulfur 0.030 Phosphorus 0.045
317L	S31703	0.03	2.00	1.00	18.0–20.0	11.0–15.0	3.0–4.0	Sulfur 0.030 Phosphorus 0.045
321	S32100	0.08	2.00	1.00	17.0–19.0	9.0–12.0		Sulfur 0.030 Ti ≮ 5 × Carbon
347	S34700	0.08	2.00	1.00	17.0–19.0	9.0–13.0		Sulfur 0.030 Cb + Ta ≮ 10 × Carbon

stated previously, this element is particularly effective in promoting corrosion resistance.

Just like insurance, you get only what you pay for, and generally speaking, the higher the corrosion resistance, the more expensive the material. In fact, the differential between type 304 stainless steel and a high-nickel alloy may be as much as 20 times, depending on the market prices for the various alloying elements that fluctuate widely with the supply and demand position.

Aluminum

High-purity grades of aluminum (±99.5%) and its alloys still are preferred for some food and pharmaceutical appli-

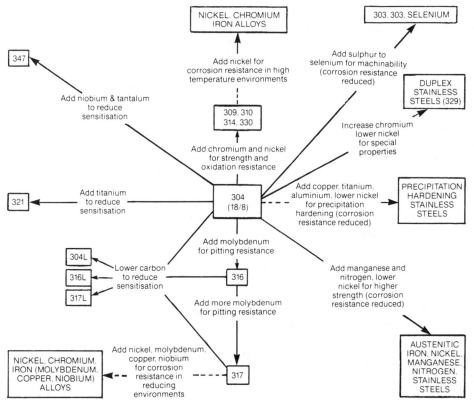

Figure 2. Outline of some compositional modifications of 18/8 austenitic stainless steel to produce special properties (ref. 2).

Table 2. Composition of Some of the More Commonly Used Wrought Super Stainless Steels and Nickel Alloys, Unless Indicated Otherwise, All Values are Maxima

Alloy	UNS No.	Composition, %						
		Carbon	Silicon	Manganese	Chromium	Nickel	Molybdenum	Others
904L	N08904	0.02	0.70	2.0	19.0–21.0	24.0–26.0	4.2–4.7	Cu 1.2–1.7
Avesta 254 SMO	S31254	0.02	0.80	2.0	19.5–20.5	17.5–18.5	6.0–6.5	Cu 0.5–1.0, N 0.18–0.22
Incoloy 825	N08825	0.05	0.50	1.0	19.5–23.5	38.0–46.0	2.5–3.5	Al 0.2, Ti 0.6–1.2
Hastelloy G-30	N06030	0.03	0.08	1.5	28.0–31.5	Bal.	4.0–6.0	Co 5.0, Cu 1.0–2.4, Cb + Ta 0.3–1.5, Fe 13.0–17.0
Inconel 625	N06625	0.10	0.50	0.50	23.0–28.0	Bal.	8.0–10.0	Co 1.0, Fe 5.0, Al 0.4, Ti 0.4
Hastelloy C-276	N10276	0.02	0.08	1.0	14.5–16.5		15.0–17.0	W 3.0–4.5, V 0.35

cations because of the reasonable corrosion resistance of the metal. This resistance is attributable to the easy and rapid formation of a thin, continuous, adherent oxide film on exposed surfaces. This oxide film, in turn, exhibits a good corrosion resistance to many foodstuffs, and it is reported that fats, oils, sugar, and some colloids have an inhibitory or sealing effect on these films (ref. 3).

As aluminum salts formed by corrosion are colorless, tasteless, and claimed to be nontoxic, the metal is easy to clean, inexpensive, and light and has a high thermal conductivity. It still is used quite extensively in certain areas of food manufacture and distribution. However, in recent years, the claim of nontoxicity is being questioned as a high dietary incidence has been implicated in Alzheimer's disease (senile dementia) with compounds of aluminum (aluminosilicates) being found in the brain tissue of sufferers (ref. 4). However, the case is far from proven, and it is not clear if the increased levels of aluminosilicates are due to a high intake of aluminum per se or other factors such as a dietary deficiency of calcium.

For many years, aluminum was used extensively for containment vessels in the diary and brewing industry, and it was Richard Seligman who founded the then Aluminium Plant and Vessel Company (now APV plc) to exploit the technique of welding this material for the fabrication of fermenting vessels in the brewing industry (ref. 5). Many of these original vessels are still in use in some of the smaller, privately owned breweries in the United Kingdom.

Although large fermenting vessels and storage tanks now tend to be fabricated from stainless steels, there is still widespread use of aluminum for beer kegs, beer cans, and a miscellany of small-scale equipment where the resistance of aluminum is such that it imparts no change or modification of flavor, even after prolonged storage.

While still used for holding vessels and some equipment when processing cider, wines, and perry, prolonged contact is inadvisable because of the acidity of the sulfites employed as preservatives for these products—inadvisable, that is, unless the surface of the metal has been modified by anodizing or has been protected with a lacquer.

In the manufacture of preserves, aluminum is still employed for boiling pans, the presence of sugar appearing to inhibit any corrosion. In the field of apiculture, it has even been used for making prefabricated honeycombs, which the bees readily accepted.

Extensive use is made of aluminum and the alloys in the baking industry for baking tins, kneading troughs, handling equipment, etc.

In other areas of food manufacture and preparation, the use of aluminum extends virtually over the whole field of activity—butter, margarine, table oils, and edible fats, meat and meat products, fish and shellfish, certain sorts of vinegar, mustards, spices; the list is almost endless.

No mention has so far been made of the application of this metal in the dairy industry, and indeed it still has limited application mostly in the field of packaging, eg, bottle caps, wrapping for cheese, butter, carton caps for yogurt, cream.

It will be appreciated that the uses of aluminum in the food industry so far mentioned have tended to be for equipment used in batch operation, hand utensils, and packaging. There are probably three major factors that have mitigated against its more widespread use, not only in the dairy industry but in brewing and many other branches of food processing.

1. Modern, highly automated plants operating on a continuous or semicontinuous basis employ a wide variety of materials of construction. Because of the position in the electrochemical series (to be discussed later), aluminum and its alloys are susceptible to galvanic corrosion when coupled with other metals.

2. The commercial availability of stainless steels and their ease of fabrication, strength, ease of maintenance, appearance, and proven track record of reliability.

3. The fact that since modern plants operate on a semicontinuous basis with much higher levels of fouling, cleaning regimes require strongly alkaline detergents to which aluminum has virtually zero corrosion resistance.

Copper and Tinned Copper

Copper and tinned copper were used extensively in former times because of their excellent thermal conductivity (8 times that of stainless steel), ductility, ease of fabrication, and reasonable level of corrosion resistance. However, the demise of copper as a material of construction is largely attributable to the toxic nature of the metal and its catalytic activity in the development of oxidative rancidity in fats and oils. Even at the sub-part-per-million level, copper in vegetable oils and animal fats rapidly causes the development of off-flavors. In equipment where high levels of liquid turbulence are encountered, eg, plate heat exchanger or high-velocity pipe lines, copper is subject to erosion. Nevertheless, there is an area of the beverage industry where copper is still the only acceptable material of construction, ie, pot stills for Scotch and Irish whiskey production. It is also used in the distillation of the spirits such as rum and brandy. Much old copper brewing equipment such as fermenting vessels and wort boilers is still in use throughout the world, and an interesting observation is that even though the wort boilers in modern breweries are fabricated from stainless steel, they are still known as "coppers" and UK craftsmen fabricating stainless steel are still known as coppersmiths.

Titanium

There are certain areas of the food industry, especially in equipment involving heat transfer, where stainless steels are just not capable of withstanding the corrosive effects of salty, low-pH environments. Food processors are increasingly accepting the use of titanium as an alternative, in the full knowledge that it offers corrosion immunity to the more aggressive foodstuffs and provides a long-term solution to what was an on-going problem with stainless steels. Titanium is a light metal, the density of which is almost half that of stainless steels. Although relatively expensive (6–7 times the cost of stainless steel), being a low-density material offsets this price differential for the raw material by almost half. It is ductile and fabricable using normal techniques, although welding it does require a high degree of expertise.

Other Metals

Tin, in the form of tin plate, is used extensively in the canning industry, where its long-term corrosion resistance to a wide range of food acids makes it a material par excellence for this purpose.

Cadmium, used as a protective coating for carbon steel nuts and bolts, was favored at one time. However, the high toxicity of the cadmium compounds has come under increasing scrutiny from many health regulatory bodies and now cadmium-plated bolting is not permitted in food factories. Indeed, Denmark and Sweden have totally banned the import of cadmium-plated components into their countries, and many other countries are likely to follow suit.

Lead and lead-containing products are generally not acceptable for food contact surfaces, although some codes of practice permit the use of lead-containing solder for capillary pipeline joints on water supplies and service lines.

SELECTING MATERIALS OF CONSTRUCTION (Ref. 6)

Designing equipment is a multidiscipline exercise involving mechanical engineers, materials–corrosion engineers, stressing experts, draftsmen, etc. The corrosion engineer has an important role in this team effort, namely, to ensure the materials specified will offer a corrosion resistance that is just adequate for all the environmental conditions likely to be encountered during normal operation of the equipment. A piece of equipment that prematurely fails by corrosion is as badly designed as one in which the materials have been overspecified. Unfortunately, all too often the functional requirements for a piece of equipment are analyzed in a somewhat arbitrary manner and all too often, the basic cost of the material tends to outweigh other equally important considerations.

Figure 3 shows the primary criteria that must be considered in the initial selection process.

- *Corrosion Resistance.* For any processing operation, there will be a range of materials that will offer a corrosion resistance that is adequate (or more than adequate) for a particular job. When considering corrosion resistance, the operational environment is the obvious one, but the other point must be whether the material will also offer corrosion resistance to the chemicals used for cleaning and sanitizing.
- *Cost.* Many of the materials originally considered will be eliminated on the grounds of their high cost. For example, there is no point in considering a high-nickel alloy when a standard 300 series stainless steel at a lower cost will be perfectly satisfactory.
- *Availability.* Availability is a less obvious feature of the material selection process. Many steel producers will require a minimum order of, say, 3 tons for a nonstandard material. Clearly, the equipment manufacturer is not going to buy this large quantity when the job that is to be done may only require the use of one ton of material.
- *Strength.* Strength is a factor that is taken into account at the design stage but, as with all the others, cannot be considered in isolation. For example, many of the new stronger stainless steels, although more expensive on a ton-for-ton basis than conventional stainless steels, are less expensive when considered on a strength/cost ratio.
- *Fabricability.* There is little point in considering materials that are either unweldable (or unfabricable) or can be welded only under conditions more akin to a surgical operating theater than a general engineering fabrication shop.
- *Appearance.* Appearance may or may not be an important requirement. Equipment located outside

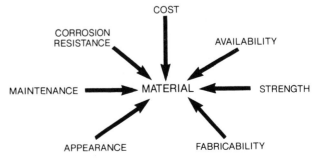

Figure 3. Materials selection criteria.

must be resistant to environmental weathering and therefore may require the application of protective sheathing, which could double the basic material cost.

- *Maintenance.* Is the equipment to be essentially maintenance-free or is some maintenance, such as periodic repainting, tolerable? How long will the equipment operate without the need for major servicing?

When all these interrelated criteria have been considered, the long list of possible starters will have been reduced to maybe one or two. Also, somewhere through the selection process some of those materials initially rejected because, for example, of their high cost, may have to be reconsidered because of other factors.

TYPES OF CORROSION

Defining Corrosion

Before embarking on a discussion of the various forms of corrosion, it is worthwhile considering exactly what corrosion is. There are several definitions of corrosion. For example, Fontant (Ref. 7) defines it as extractive metallurgy in reverse using the diagram, shown in Figure 4, to illustrate the point.

A more general and descriptive definition is "it is the deterioration or destruction of a material through interaction with its environment." This covers all materials of construction including rubber and plastics as well as metal. However, the primary object of this article is to deal with corrosion of metals, in particular stainless steels, and how this corrosion can be classified.

There are two basic forms of corrosion—wet corrosion and dry corrosion. Dry corrosion is concerned with the oxidation of metals at high temperature and clearly outside the scope of this text. Wet corrosion occurs in aqueous solutions or in the presence of electrolytes and is an electrochemical process. It should be noted that the "aqueous" component of the system may be present in only trace quantities, eg, present as moisture; the classical example is the corrosion of steel by chlorine gas. In fact, steel is not corroded by chlorine since steel is the material used for storing liquid chlorine. However, in the presence of even

trace quantities of moisture, chlorine rapidly attacks steel and, for that matter, most metals.

The corrosion of metals involves a whole range of factors. These may be chemical, electrochemical, biological, metallurgical, or mechanical, acting singly or conjointly. Nevertheless, the main parameter governing corrosion of metals is related to electrochemistry. Electrochemical principles therefore are the basis for a theoretical understanding of the subject. In fact, electrochemical techniques are now the standard method for investigating corrosion although the standard "weight loss" approach still provides invaluable data. It is not proposed to discuss in depth the electrochemical nature of corrosion but should further information be required, several excellent texts are available (refs. 7,8).

Forms of Corrosion

Wet corrosion can be classified under any of eight headings, namely:

- Galvanic or bimetallic corrosion
- Uniform or general attack
- Crevice corrosion
- Pitting corrosion
- Intergranular corrosion
- Stress corrosion cracking
- Corrosion fatigue
- Selective corrosion (castings and free-matching stainless steels)

Galvanic Corrosion. When two dissimilar metals (or alloys) are immersed in a corrosive or conductive solution, an electrical potential or potential difference usually exists between them. If the two metals are electrically connected, then, because of this potential difference, a flow of current occurs. As the corrosion process is an electrochemical phenomenon and dissolution of a metal involves electron flow, the corrosion rates for the two metals is affected. Generally, the corrosion rate for the least corrosion resistant is enhanced while that of the more corrosion resistant is diminished. In simple electrochemical terms, the least resistant metal has become anodic and the more resistant cathodic. This, then, is galvanic or dissimilar metal corrosion.

The magnitude of the changes in corrosion rates depends on the so-called electrode potentials of the two metals; the greater the difference, the greater the enhancement or diminution of the corrosion rates. It is possible to draw up a table of some commercial alloys that ranks them in order of their electrochemical potential. Such a table is known as the galvanic series. A typical one as shown in Table 3 is based on work undertaken by the International Nickel Company (now INCO Ltd.) at their Harbor Island, NC, test facility. This galvanic series relates to tests in unpolluted seawater, although different environments could produce different results and rankings. When coupled, individual metals and alloys from the same group are unlikely to show galvanic effects that will cause any change in their corrosion rates.

The problem of dissimilar metal corrosion, being rela-

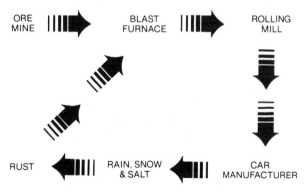

Figure 4. Extractive metallurgy in reverse.

Table 3. The Galvanic Series of Some Commercial Metals and Alloys in Clean Seawater

Noble or cathodic ↑

Platinum
Gold
Graphite
Titanium
Silver

Chlorimet 3 (62 Ni, 18 Cr, 18 Mo)
Hastelloy C (62 Ni, 17 Cr, 15 Mo)

18/8 Mo stainless steel (passive)
18/8 stainless steel (passive)
Chromium stainless steel 11–30% Cr (passive)

Inconel (passive) (80 Ni, 13 Cr, 7 Fe)
Nickel (passive)

Silver solder

Monel (70 Ni, 30 Cu)
Cupronickels (60–90 Cu, 40–10 Ni)
Bronzes (Cu–Sn)
Copper
Brasses (Cu–Zn)

Chlorimet 2 (66 Ni, 32 Mo, 1 Fe)
Hastelloy B (60 Ni, 30 Mo, 6 Fe, 1 Mn)

Inconel (active)
Nickel (active)

Tin
Lead
Lead–tin solders

18/8 Mo stainless steel (active)
18/stainless steel (active)

Ni–resist (high-Ni cast iron)
Chromium stainless steel, 13% Cr (active)

Cast iron
Steel or iron

Active or anodic ↓

2024 aluminum (4.5 Cu, 1.5 Mg, 0.6 Mn)
Cadmium
Commercially pure aluminum (1100)
Zinc
Magnesium and magnesium alloys

Figure 5. Galvanic corrosion of 304 stainless steel initiated by a 316 weld deposit. Note the large pit associated with the weld splatter.

it is most unusual for the potential difference between molybdenum and nonmolybdenum containing stainless steels to be sufficient to initiate galvanic corrosion, the environmental factors in this particular case were obviously such that corrosion was initiated (Fig. 5). As stated, this is somewhat unique and it is not uncommon for 316 welding consumables to be used for welding 304 stainless steel with no adverse effects. As a practice, however, it is to be deprecated and the correct welding consumables should always be employed.

Not all galvanic corrosion is bad; indeed, galvanic corrosion is used extensively to protect metal and structures by the use of a sacrificial metal coating. A classic example is the galvanizing of sheet steel and fittings, the zinc coating being applied not so much because it doesn't corrode, but because it does. When the galvanizing film is damaged, the zinc galvanically protects the exposed steel and inhibits rusting. Similarly, sacrificial anodes are fitted to domestic hot water storage tanks to protect the tank.

Uniform or General Attack. As the name implies, this form of corrosion occurs more or less uniformly over the whole surface of the metal exposed to the corrosive environment. It is the most common form of corrosion encountered with the majority of metals, a classic example being the rusting of carbon steel. Insofar as the corrosion occurs uniformly, corrosion rates are predictable and the necessary corrosion allowances built into any equipment. In the case of stainless steels, this form of corrosion is rarely encountered. Corrodents likely to produce general attack of stainless steel are certain mineral acids, some organic acids, and high-strength caustic soda at concentrations and temperatures well in excess of those ever likely to be found in the food industry. The same remark applies to cleaning acids such as nitric, phosphoric, and citric acids, but not for sulfuric or hydrochloric acids, both of which can cause rapid, general corrosion of stainless steels. Hence, they are not recommended for use, especially where corrosion would result in a deterioration of the surface finish of process equipment.

The behavior of both 304 and 316 stainless steels when subjected to some of the more common acids that are en-

tively well understood and appreciated by engineers, is usually avoided in plant construction and in the author's experience, few cases have been encountered. Probably the most common form of unintentional galvanic corrosion is on service lines where brass fittings are used on steel pipelines—the steel suffering an increase in corrosion rate at the bimetallic junction.

One of the worst bimetallic combinations is aluminum and copper. An example of this is in relation to aluminum milk churns used to transport whey from Gruyere cheese manufacture (in Switzerland), where copper is used for the cheesemaking vats and the whey picks up traces of this metal. The effect on the aluminum churns which are internally protected with lacquer that gets worn away through mechanical damage is pretty catastrophic.

Another, somewhat unique, example of galvanic corrosion is related to a weld repair on a 304 stainless-steel storage vessel. Welding consumables containing molybdenum had been employed to effect the repair, and although

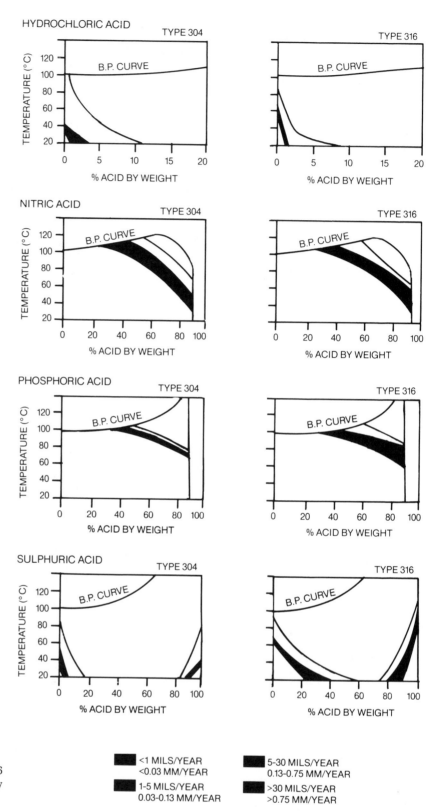

Figure 6. Corrosion resistance of 304 and 316 stainless steels to mineral acids (reproduced by permission of British steel Plc).

countered in the food industry is graphically illustrated in Figure 6. These isocorrosion graphs, ie, lines that define the conditions of temperature and acid concentration that will produce a constant corrosion rate expressed in mils (0.001 in.) or mm loss of metal thickness per year, are used extensively by corrosion engineers in the material selection process when the form of corrosion is general attack.

They are of no value whatsoever when the corrosion mode is one of the other forms that will be defined, such as pitting or crevice corrosion.

Crevice Corrosion. This form of corrosion is an intense local attack within crevices or shielded areas on metal surfaces exposed to corrosive solutions. It is characteristi-

Figure 7. Crevice corrosion at the interplate contact points of a heat-exchanger plate.

cally encountered with metals and alloys that rely on a surface oxide film for corrosion protection, eg, stainless steels, titanium, aluminum.

The crevices can be inherent in the design of the equipment (eg, plate heat exchangers) or inadvertently created by bad design. Although crevice corrosion can be initiated at metal-to-metal surfaces (see Fig. 7), it is frequently encountered at metal to nonmetallic sealing faces. Any nonmetallic material that is porous and used, for example, as a gasket, is particularly good (or bad!) for initiating this form of attack. Fibrous materials that have a strong wicking action are notorious in their ability to initiate crevice attack. Similarly, materials that have poor stress relaxation characteristics, ie, have little or no ability to recover their original shape after being deformed, are also crevice creators, as are materials that tend to creep under the influence of applied loads and/or at elevated temperatures. Although used for gasketing, PTFE suffers both these deficiencies. On the other hand, elastomeric materials are particularly good insofar as they exhibit elastic recovery and have the ability to form a crevice-free seal. However, at elevated temperatures, many rubbers harden and in this condition, suffer the deficiencies of many non-elastomeric gasketing materials.

Artificial crevices can also be created by the deposition of scale from one of the process streams to which the metal is exposed. It is necessary, therefore, to maintain food processing equipment in a scale-free condition, especially on surfaces exposed to service fluids such as services side hot/cold water, cooling brines, which tend to be overlooked during plant cleaning operations.

Much research work has been done on the geometry of crevices and the influence of this on the propensity for the initiation of crevice corrosion (ref. 9). However, in practical terms, crevice corrosion usually occurs in openings a few tenths of a millimeter or less and rarely is encountered where the crevice is greater than 2 mm (0.08 in.).

Until the 1950s, crevice corrosion was thought to be due to differences in metal ion or oxygen concentration within the crevice and its surroundings. While these are factors in the initiation and propagation of crevice corrosion, they are not the primary cause. Current theory supports the view that through a series of electrochemical reactions and the geometrically restricted access into the crevice, migration of cations, chloride ions in particular, occurs. This alters the environment within, with a large reduction in pH and an increase in the cations by a factor of as much as 10. The pH value can fall from a value of, say, 7 in the surrounding solution to as low as pH 2 within the crevice. As corrosion is initiated, it proceeds in an autocatalytic manner with all the damage and metal dissolution occurring within the crevice and little or no metal loss outside. The confined and autocatalytic nature of crevice corrosion results in significant loss of metal under the surface of site of initiation. As a result, deep and severe undercutting of the metal occurs (see Fig. 8). The time scale for initiation of crevice corrosion can vary from a few hours to several months and, once initiated, can progress very rapidly. Stopping the corrosion process can

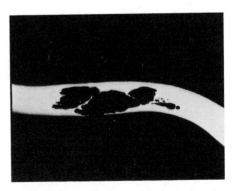

Figure 8. Photomicrograph of a section through a site of crevice corrosion. Note the deep undercutting which is typical of chloride-induced attack on stainless steel.

be extremely difficult as it is necessary to remove all the trapped reactants and completely modify the occluded environment. The difficulty of attaining this will be appreciated by reference to Figure 8, where the entrance to the corroded region is only 0.5 mm (0.020 in.).

While methods for combating the onset of crevice corrosion can be deduced from the foregoing text, a reiteration of some of the more important precautions is not out of place, viz:

- Good-quality, crevice-free welded joints are always preferable to bolted joints
- Good equipment design (well-designed gasket sealing faces) that avoids unintentional crevices and does not permit the development of stagnant regions
- Frequent inspection of equipment and removal of surface deposits
- Use of good-quality rubber gaskets rather than absorbent packings
- Good gasket maintenance; replacement when hardened or damaged

However, certain pieces of equipment are by virtue of their design highly creviced. In such cases, it is necessary to recognize the potential corrosion risk and select the materials of construction that will resist the initiation of crevice corrosion by the environment. Similarly, cleaning and sanitizing regimes must be developed to avoid the onset of attack.

In the case of stainless steels, although there are several ionic species that will initiate the attack, by far the most common are solutions containing chloride. The presence of salt in virtually all foodstuffs highlights the problem. Low pH values also enhance the propensity for initiation of attack.

Other environmental factors such as temperature and the oxygen or dissolved air content of the process stream all play a role in the corrosion process.

Because the presence of oxygen is a prerequisite for the onset of crevice corrosion (and many other forms of attack), in theoretical terms complete removal of oxygen from a process stream will inhibit corrosion. In practice, however, this is difficult to achieve. Only in equipment where complete and effective deaeration occurs, such as a multiple effect evaporator operating under reduced pressure, will the beneficial effect of oxygen removal be achieved.

Stainless steels containing molybdenum (316, 317) have a much higher resistance to crevice corrosion than do alloys without this element (304, 321, 347). The higher the molybdenum content, the greater the corrosion resistance. For particularly aggressive process streams, titanium is often the only economically viable material to offer adequate corrosion resistance.

Pitting Corrosion. As the name implies, pitting is a form of corrosion that leads to the development of pits on a metal surface. It is a form of extremely localized but intense attack, insidious insofar as the actual loss of metal is negligible in relation to the total mass of metal that may be affected. Nevertheless, equipment failure by per-

Figure 9. Pitting corrosion of a stainless-steel injector caused by the presence of hydrochloric acid in the steam supply.

foration is the usual outcome of pitting corrosion. The pits can be small and sporadically distributed over the metal surface (Fig. 9) or extremely close together, close enough, in fact, to give the appearance of the metal having suffered from general attack.

In the case of stainless steels, environments that will initiate crevice corrosion will also induce pitting. As far as the food industry is concerned, it is almost exclusively caused by chloride containing media, particularly at low pH values.

Many theories have been developed to explain the cause of initiation of pitting corrosion (ref. 10), and the one feature they have in common is that there is a breakdown in the passive oxide film. This results in ionic migration and the development of an electrochemical cell. There is, however, no unified theory that explains the reason for the film breakdown. Evans (ref. 11) for example, suggests that metal dissolution at the onset of pitting may be due to a surface scratch, an emerging dislocation or other defects, or random variations in solution composition. However, propagation of the pit proceeds by a mechanism similar to that occurring with crevice corrosion. Like crevice corrosion, the pits are often undercut and on vertical surfaces may assume an elongated morphology due to gravitational effects (Fig. 10).

The onset of pitting corrosion can occur in a matter of days but frequently requires several months for the development of recognizable pits. This makes the assessment of the pitting propensity of a particular environment very difficult to determine, and there are no short-cut laboratory testing techniques available. Methods and test solutions are available to rank alloys; the best known and most frequently quoted is ASTM Standard G48 (ref. 12), which employs 6% ferric chloride solution. Another chemical method involving ferric chloride determines the tem-

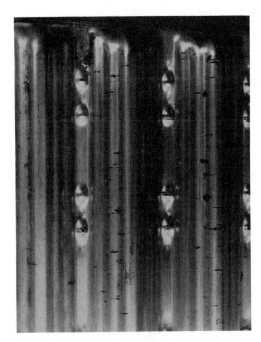

Figure 10. Elongated pitting attack on a 316 stainless-steel heat exchanger plate.

Figure 11. Scanning electron micrograph of the surface of sensitized stainless steel showing preferential attack along the grain boundaries.

perature at which the solution will cause pitting within a 24-h test period, the results being expressed as the critical pitting temperature or CPT (ref. 13). However, as stated, both of these methods are used to rank the susceptibility of a range of alloys rather than define the performance of a material in a service environment. Electrochemical methods have also been used.

As with crevice corrosion, alloy composition has a profound effect on the resistance of a material to pitting attack. Greene and Fontana (ref. 14) summarized the effect of various elements as shown in Table 4.

Intergranular Corrosion. A fact not often appreciated is that metals and alloys have a crystalline structure. However, unlike crystalline solids such as sugar or salt, metallic crystals can be deformed or bent without fracturing; in other words, they are ductile. In the molten state, the atoms in a metal are randomly distributed but on cooling and solidification, they become arranged in crystalline form. Because crystallization occurs at many points in the solidification process, these crystals or grains are randomly orientated and the region where they meet are grain boundaries. In thermodynamic terms, the grain

boundaries are more susceptible to corrosion attack because of their higher free energy, although in practice the difference in free energy of the grain boundaries and the main crystals or grains in a homogeneous alloy are too small to be significant. However, when the metal or alloy has a heterogeneous structure, preferential attack at or adjacent to the grain boundaries can occur. This is intergranular corrosion as shown in Figure 11.

When austenitic stainless steels are heated to and held in the temperature range of 600–900°C (1100–1650°F), the material becomes sensitized and susceptible to grain boundary corrosion. It is generally agreed that this is due to chromium combining with carbon to form chromium carbide, which is precipitated at the grain boundaries. The net effect is that the metal immediately adjacent to the grain boundaries is denuded of chromium and instead of having a composition of, say, 18% chromium and 8% nickel, it may assume an alloy composition where the chromium content is reduced to 9% or even lower. As such, this zone depleted in chromium bears little similarity to the main metal matrix and has lost one of the major alloying elements on which it relied for its original corrosion

Table 4. The Effect of Alloying on Pitting Resistance of Stainless-Steel Alloys

Element	Effect on Pitting Resistance	Element	Effect on Pitting Resistance
Chromium	Increases	Molybdenum	Increases
Nickel	Increases	Nitrogen	Increases
Titanium/ Niobium	No effect in media other than ferric chloride	Sulfur (and selenium)	Decreases
Silicon	Decrease or increase depending on the absence or presence of molybdenum	Carbon	Decreases if present as grain boundary precipitates

resistance. Indeed, the lowering of corrosion resistance in this zone is so great that sensitized materials are subject to attack by even mildly corrosive environments.

As supplied from the steel mills, stainless steels are in the so-called solution annealed condition, ie, the carbon is in solution and does not exist as grain boundary chromium carbide precipitates. During fabrication where welding is involved, the metal adjacent to the weld is subjected to temperatures in the critical range (600–900°C/1100–1650°F) where sensitization can occur. As such, therefore, this zone may be susceptible to the development of intergranular carbide precipitates. Because the formation of chromium carbides is a function of time, the longer the dwell time in the critical temperature zone, the greater the propensity for carbide formation. Hence, the problem is greatest with thicker metal sections due to the thermal mass and slow cooling rate.

By heating a sensitized stainless steel to a temperature of 1050°C (1950°F), the carbide precipitates are taken into solution and by rapidly cooling or quenching the steel from this temperature, the original homogeneous structure is reestablished and the original corrosion resistance restored.

The first stainless steels were produced with carbon contents of up to 0.2% and as such, were extremely susceptible to sensitization and in-service failure after welding. In consequence, the carbon levels were reduced to 0.08%, which represented the lower limit attainable with steel making technology then available. Although this move alleviated the problem, it was not wholly successful, particularly when welding thicker sections of the metal. Solution annealing of the fabricated items was rarely a practical proposition and there was a need for a long-term solution. It was shown that titanium or niobium (columbium) had a much greater affinity for the carbon than chromium and by additions of either of these elements, the problem was largely overcome. The titanium or niobium carbides that are formed remain dispersed throughout the metal structure rather than accumulating at the grain boundaries.

Grade 321 is a type 304 (18 Cr, 8 Ni) with titanium added as a stabilizing element, while grade 347 contains niobium. By far, the most commonly used is 321, grade 347 being specified for certain chemical applications.

Modern steelmaking techniques such as AOD (air–oxygen decarburization) were developed to reach even lower levels of carbon, typically less than 0.03%, to produce the "L" grades of stainless steel. These are commercially available and routinely specified where no sensitization can be permitted. With these advances in steel making technology, even the standard grades of stainless steels have typically carbon levels of 0.04–0.05% and, generally speaking, are weldable without risk of chromium carbide precipitates at metal thicknesses of up to 6 mm (¼in). Above this figure or where multipass welding is to be employed, the use of a stabilized or "L" grade is always advisable.

Stress Corrosion Cracking. One of the most insidious forms of corrosion encountered with the austenitic stainless steels is stress corrosion cracking (SCC). The morphology of this type of failure is invariably a fine filamentous crack that propagates through the metal in a trans-

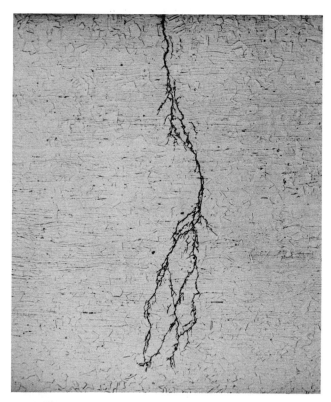

Figure 12. Photomicrograph of a typical stress corrosion crack showing its highly branched morphology and transgranular propagation.

granular mode. Frequently, the crack is highly branched as shown in Figure 12, although sometimes it can assume a single-crack form. Factors such as metal structure, environment, and stress level have an effect on crack morphology. The disturbing feature of SCC is that there is virtually no loss of metal and frequently, it is not visible by casual inspection and is only apparent after perforation occurs. Some claim that as much as 50% of the failures of stainless steel are attributable to this cause.

Another characteristic of SCC in stainless steels is that once detected, repair by welding is extremely difficult. Crack propagation frequently occurs below the surface of the metal, and any attempt to weld repair results in the crack opening up and running ahead of the welding torch. The only practical method of achieving a satisfactory repair is to completely remove the affected area with a 15–25 cm (6–9 in.) allowance all around the area of visible damage and replace the section. Even then, there is no guarantee that the damaged zone has been entirely removed. In most cases, there are three prerequisites for the initiation of SCC.

- *Tensile Stress.* This may be either residual stress from fabricating operations or applied through the normal operating conditions of the equipment. Furthermore, it has been observed that a corrosion pit can act both as a stress raiser and a nucleation site for SCC.

- *Corrosive Species.* Although there are a number of ionic compounds that will act as the corrodent, in the food industry this invariably is the chloride ion.

High-strength caustic soda at elevated temperatures will also induce SCC, but the concentrations and temperatures required are well in excess of those ever likely to be encountered. Furthermore, the crack morphology is inter- rather than transgranular. pH also plays a role and generally speaking, the lower the pH the greater the propensity for SCC.

- *Temperature.* It generally is regarded by many that a temperature in excess of 60°C (140°F) is required for this type of failure, although the author has seen examples occurring at 50°C (122°F) in liquid glucose storage vessels.

In the absence of any one of these prerequisites, the initiation of SCC is eliminated. Therefore, it is worth considering the practical approach to its elimination from equipment.

Figure 13 is a diagrammatic representation of the effect of stress on "time-to-failure." As will be seen, by reducing the stress level below a certain critical point, the "time-to-failure" can be increased by several orders of magnitude. On small pieces of equipment, residual stress from manufacturing operations can be removed by stress-relief annealing, which, in the case of stainless steels, is the same as solution annealing. For large pieces of equipment such as storage vessels this approach is clearly impractical. Applied stress is very much a function of the operational conditions of the equipment and only by reducing the stress level by increasing the thickness of the metal can this be reduced. However, this too is a somewhat impractical and uneconomic approach. Some (ref. 15) claim that by placing the surface of the metal under compressive rather than tensile stress, by shot peening with glass beads, the problem of SCC can be minimized or eliminated. This too is not a practical proposition for many items of food processing equipment.

As for the matter of corrosive species, it is questionable if anything can be done about elimination. With foodstuffs, for example, this invariably will be the chloride ion, a naturally occurring or essential additive.

Similarly, little can be done in respect to the temperature as this is going to be essential to the processing operation.

As shown by Copson (ref. 16), the tendency for iron–chromium–nickel alloys to fail by SCC in a specific test medium (boiling 42% magnesium chloride solution) is related to the nickel content of the alloy. Figure 14 shows this effect, and it is unfortunate that stainless steels with a nominal 10% nickel have the highest susceptibility to failure. Increasing the nickel content of the alloy results in a significant increase in the time-to-failure, but of course, this approach incurs not only the increased cost of the nickel but also the added penalty of having to increase the chromium to maintain a balanced metallurgical structure. The more effective approach is by reducing the nickel content of the alloy.

A group of stainless steels have been developed that exploit this feature, and although their composition varies from producer to producer, they have a nominal composition of 20/22% chromium, 5% nickel. Molybdenum may or may not be present, depending on the environment for which the alloy has been designed. These alloys differ from the austenitic stainless steels insofar as they contain approximately 50% ferrite; hence their designation, austenitic–ferritic, or more commonly, duplex stainless steels. It is only with the advent of modern steel making technology, particularly in relation to the lower carbon levels that can be achieved, that these alloys have become a commercially viable proposition. Because of their low carbon content, typically 0.01–0.02%, the original problems associated with welding ferritic stainless steels and chromium carbide precipitation have been overcome. The

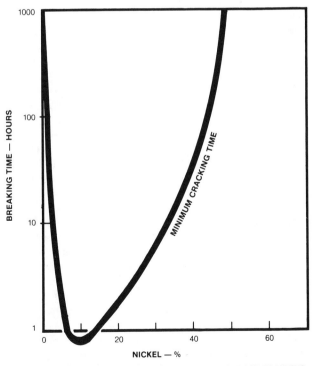

STRESS CORROSION CRACKING OF IRON-NICKEL-CHROMIUM ALLOYS: COPSON CURVE. (DATA POINTS OMITTED FOR CLARITY)

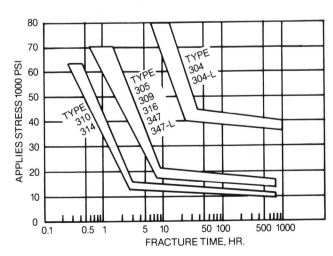

Figure 13. Composite curves illustrating the relative resistance to stress corrosion cracking of some commercial stainless steel in a specific test solution.

Figure 14. Stress corrosion cracking of iron, chromium, nickel alloys—the Copson curve. Data points have been omitted for clarity.

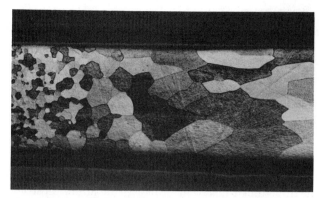

Figure 15. Photomicrograph of a weld deposit on a ferritic stainless steel. Compare the size of grains in the weld with those in the parent material on the left.

alloys are almost twice as strong as the austenitic stainless steels and are ductile and weldable. From a general corrosion standpoint, they are comparable with, or marginally superior to, their 300 series equivalent; but from an SCC standpoint in test work and from field experience, they offer a resistance orders of magnitude better.

Also now available are fully ferritic stainless steels such as grade 444, which contains 18% chromium and 2% molybdenum. This alloy contains carbon at the 0.001% level and therefore does not suffer the problems of welding that were encountered with the original ferritic steels. Furthermore, stabilizing elements such as titanium and niobium are also alloying additions that minimize the tendency for intergranular chromium carbide formation. The main disadvantage of these materials is their susceptibility to grain growth during welding (Fig. 15), which makes them extremely sensitive to fracture even at room temperature. Welding sections thicker than 3 mm ($\frac{1}{8}$ in.) are not regarded as a practical proposition, and, therefore, their use tends to be limited to tubing.

Corrosion Fatigue. Fatigue is not a form of corrosion in the accepted sense as there is no loss of metal, but can be associated with other forms of localized attack. Because pure fatigue is an in vacuo phenomenon, a more correct term is corrosion fatigue or environmental cracking, which is the modern expression and takes into account cracking where the corrosive factor has played a major role on the crack morphology.

The primary cause of corrosion fatigue is the application of fluctuating pressure loads to components that, while of adequate design to withstand normal operating pressures eventually fail under the influence of cyclic loading. The components can be of extremely rigid construction such as a homogenizer block or of relatively light construction such as pipework. There are many potential sources of the fluctuating pressure, the most common of which are positive-displacement pumps (eg, homogenizer or metering pumps), rapid-acting on–off valves that will produce transient pressure peaks, frequent stop–start operations, dead-ending of equipment linked to a filling machine, etc.

Generally speaking, fatigue cracks are straight, without branching and without ductile metal distortion of the material adjacent to the crack. The one characteristic of

Figure 16. Fatigue cracking of a 1-in.-diameter stainless steel bolt showing the characteristic conchoidal striations on the crack face.

fatigue cracks is that the crack face frequently has a series of conchoidal markings that represent the stepwise advance of the crack front (see Fig. 16). Although, as stated, corrosion fatigue cracks are generally straight and unbranched, where fatiguing conditions pertain in a potentially corrosive environment, the influence of the corrosive component may be superimposed on the cracks. This can lead to branching of the cracks, and in the extreme case, the crack may assume a highly branched morphology which is almost indistinguishable from stress corrosion cracking. It is only when the crack face is examined under high-power magnification that it is possible to categorize the failure mode. An example of this is shown in Figure 17, which has all the features of stress corrosion cracking but the scanning electron micrograph (Fig. 18) clearly shows the stepwise progression of the crack front.

The site for initiation of corrosion fatigue is frequently a discontinuity in the metal section of the component. This may be, for example, a sharp change in diameter of a shaft where inadequate radiusing of the diametric change results in a high cycle stress level through shaft rotation and thus an initiation point for fatigue. A corrosion pit, which under cyclic loading will have a high stress-level association, can also act as a nucleation site (epicenter) for fatigue failure. Because of the corrosive element, it is sometimes difficult to establish whether the pit and associated corrosion were the initiating mechanism for the cracking or the result of corrosion superimposed on a crack, the fracture face of which will be in an "active" state and therefore more susceptible to corrosion processes.

The whole subject of fatigue and corrosion fatigue is complex. However, as far as food processing equipment is concerned, avoidance of fatigue failure is best achieved by avoiding pulsing and pressure peaks. This requires the use of well-engineered valving systems and avoiding the

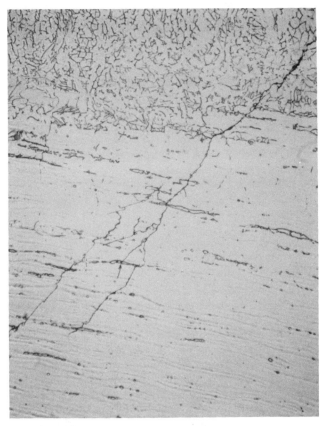

Figure 17. Photomicrograph of a fatigue crack showing all the features of stress corrosion cracking. Compare this with Figure 12.

use of positive-displacement pumps. Where this is impractical, provision should be made to incorporate pulsation dampers to smooth out the pressure peaks and minimize the risk of fatigue failures.

Selective Corrosion

Corrosion of Castings. There are a number of stainless-steel components found in food processing plants such as pipeline fittings and pump impellers that are produced as castings rather than fabricated from wrought material—notably, the cast equivalents of grade 304 (CF8) and 316 (CF8M). Although the cast and wrought materials have similar, but not identical, compositions with regard to their chromium and nickel contents, metallurgically they have different structures. Whereas the wrought materials are fully austenitic, castings will contain some ferrite or more terminologically correct, δ (delta) ferrite in the basic austenitic matrix. The ferrite is necessary to permit welding to the castings, to avoid shrinkage cracking during cooling from the casting temperature, and to act as nucletion sites for the precipitation of chromium carbides, which will invariably be present as it will not always be possible to solution anneal the cast components.

The nominal ferrite level is usually 5–12%. Below 5%, cracking problems may be experienced while above 12%, the ferrite tends to form a continuous network rather than remain as isolated pools.

Because the crystallographic structures of the ferrite and austenite differ (austenite is a face-center cubic and the ferrite, body-center cubic), the ferrite has, thermodynamically speaking, a higher free energy, which renders it more susceptible to attack, particularly in low-pH chloride-containing environments such as tomato ketchup and glucose syrups. Although the problem is not so severe when the ferrite occurs as isolated pools, when present as a continuous network, propagation of the corrosion occurs along the ferrite, with the austenite phase being relatively unaffected (see Fig. 19). Because the products of corrosion are not leached out from the corrosion site and are more voluminous than the metal, corroded castings frequently assume a blistered or pockmarked appearance. The common environments encountered in the industry that produce preferential ferrite attack are the same as those causing stress corrosion cracking. Therefore, this form of damage is frequently also present (see Fig. 20).

Depending on the method used to make the casting, the surface of the castings can be chemically modified, which reduces the corrosion resistance. Small components are usually cast by either the shell molding process or produced as investment castings. In the shell molding process, the sand forming the mold is bonded together with an organic resin that carbonizes when the hot metal is

Figure 18. Scanning electron micrograph of the fracture face of the crack shown in Figure 17. Note the stepwise progression of the crack.

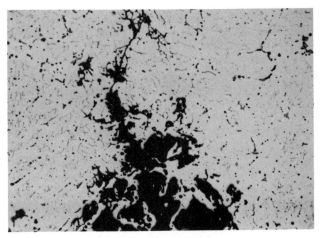

Figure 19. Photomicrograph of a section of cast 316 (CF8M) showing preferential attack of the ferrite phase.

Figure 20. Corroded pump impeller (above) and valve body (below) caused by tomato ketchup. Note the pocklike corrosion sites.

Figure 21. Free machining and nonfree machining bolts after immersion in a mildly acidic environment.

poured. This results in the metal adjacent to the mold having an enhanced carbon level with the formation of intergranular carbide precipitates and hence, a susceptibility to intergranular attack and other forms such as crevice, pitting, and stress corrosion cracking. Methods of overcoming this include solution annealing or machining off the carburized skin of metal.

With investment castings, the mold is made of zircon sand (zirconium silicate), and fired at a high temperature to remove all traces of organic material and wax that is used as a core in the moldmaking process. They do not, therefore, have this carburized layer and offer a much better resistance to surface corrosion.

Free-Machining Stainless Steels. Stainless steels are notoriously difficult to machine, especially turning, not so much because they are hard, but because the swarf tends to form as continuous lengths that clog the machine and to weld to the tip of the machine tools. One method of overcoming this is to incorporate a small amount, typically 0.2%, of sulfur or selenium in the alloy. These elements react with the manganese to form manganese sulfide or selenide, which, being insoluble in the steel, forms as discrete pools in castings or as elongated, continuous stringers in, say, wrought bar. The effect of the sulfide inclusions is to cause the material to form chips rather than long strings of swarf when being machined. Both manganese sulfide and selenide have virtually zero corrosion resistance to dilute mineral acids or other corrosive media. Thus, the free-cutting variants have a much lower corrosion resistance than their designation would imply. Indeed, some believe that the addition of sulfur to a type 316 material will offset the beneficial effect of the alloying addition of molybdenum. As stated, the sulfide inclusions will occur in castings as discrete pools, and therefore there will not be a continuous corrosion path. However, one of the products of corrosion in acidic media will be hydrogen sulfide, which has a profound effect on the corrosivity of even dilute mineral acids, causing attack of the austenitic matrix.

In the case of wrought materials, in particular bar stock, the sulfide inclusions are present as semicontinuous stringers and can suffer so-called end-grain attack in mildly corrosive media.

Stainless-steel nuts and bolts, which are produced on automatic thread-cutting machines, are invariably made from free-cutting materials. Figure 21 illustrates the difference in corrosion resistance of a bolt made from this and a nonfree machining 316. Both bolts were exposed to the same mildly acidic environment.

When specifying materials of construction, quite clearly due cognizance of this difference must be recognized. Any components turned from bar stock which are likely to come into contact with potentially corrosive environments should always be specified in 316 and not in the free machining, sulfur-containing variant.

CORROSION OF SPECIFIC ENVIRONMENTS

From a corrosion standpoint, the environments likely to be encountered in the food industry that may cause premature equipment failure may be classified under main headings:

Noncorrosive
Mildly corrosive
Highly corrosive
Service fluids
Alkaline detergents
Acidic detergents
Sanitizing agents

Noncorrosive Foodstuffs

In general terms, natural foodstuffs such as milk, cream, natural fruit juices, and whole egg do not cause corrosion problems with 304 or 316 stainless steels. Prepared foodstuffs to which there is no added salt such as yogurt, beer, ice cream, wines, spirits, and coffee also fall within this classification. For general storage vessels, pipelines, pumps, fittings or valves, grade 304 is perfectly satisfactory. However, for plate heat exchangers that are highly creviced and therefore prone to crevice corrosion, grade 316 frequently is employed. This offers a higher degree of protection against some of the more acidic products such as lemon juice, which may contain small quantities of salt and also provides a higher level of integrity against corrosion by service liquids and sanitizing agents.

It is quite common to use sulfur dioxide or sodium bisulfite for the preservation of fruit juices and gelatin solutions and in such cases, storage vessels always should be constructed from 316. Although the sulfur dioxide is noncorrosive at ambient temperature in the liquid phase, as a gas contained in significant quantities within the head space in a storage tank it tends to dissolve in water droplets on the tank wall. In the presence of air, the sulfurous acid that forms is oxidized to sulfuric acid at a concentration high enough to cause corrosion of 304 but not of 316.

Mildly Corrosive Foodstuffs

This category of foodstuffs covers products containing relatively low levels of salt and where pH values are below 7. Examples include glucose/fructose syrups and gelatin, the production of which may involve the use of hydrochloric acid. For storage vessels, pipelines, fittings, and pumps, grade 316 has established a good track record, and boiling pans in this grade of steel are perfect for long and satisfactory service. The corrosion hazards increase however in processing operations involving high temperatures and where the configuration of the equipment is such as to contain crevices, especially when the product contains dissolved oxygen. For example, multistage evaporators operating on glucose syrup will usually have the first stage, where temperatures may approach 100°C (212°F), constructed in a super stainless steel such as 904L. Subsequent effects where temperatures are lower and where the product has been deoxygenated may be fabricated in grade 316 stainless steel.

As previously indicated, it is common practice to use sulfur dioxide as a preservative in dilute gelatin solutions during storage prior to evaporation. In some cases, excess hydrogen peroxide will be added to neutralize the sulfur dioxide immediately before concentration. This can have a catastrophic effect on the 300 series stainless steels and indeed on even more highly alloyed metals such as 904L, because of the combined effect of the chlorides present with the excess hydrogen peroxide. Because of this, it is a more acceptable practice to make the peroxide addition after rather than before evaporation.

Gelatin for pharmaceutical end use is subject to UHT treatment to ensure sterility. This will involve heating the gelatin solutions to 135°C (285°F) and holding at that temperature for a short period of time. Plate heat exchangers are used extensively for this duty. Although plates made from 316 stainless steel give a reasonable life of typically 2–3 years, a corrosion resistant alloy with an enhanced-level molybdenum is to be preferred.

Highly Corrosive Foodstuffs

The list of foodstuffs falling in this category is almost endless—gravies, ketchups, pickles, salad dressings, butter, and margarine—in fact, anything to which salt has been added at the 1–3% level or even higher. Also within this category must be included cheese salting brine and other brines used in the preservation of foodstuffs that undergo pasteurization to minimize bacterial growth on food residues remaining in the brine. *Note:* Although these brines are usually too strong to support the growth of common organisms, salt resistant strains (halophiles) are the major problem.

Low-pH products containing acetic acid are particularly aggressive from a corrosion standpoint, but selection of materials for handling these products depends to a great extent on the duty involved.

When trying to define the corrosion risk to a piece of equipment handling potential corrodents, several factors come into play. While temperature, oxygen content, chloride content, and pH are the obvious ones, less obvious and equally important is contact time. All three main forms of corrosion induced in stainless steels (crevice, pitting, and stress corrosion cracking) have an induction period before the onset of corrosion. This can vary from a few hours to several months depending on the other operative factors. In a hypothetical situation where stainless steel is exposed to a potentially corrosive environment, removal of the steel and removal of the corrodents will stop the induction and the status quo is established. On repeating the exposure, the induction period is the same. In other words, the individual periods the steel spends in contact with the corrodent are not cumulative and each period must be taken in isolation.

When the contact period is short, temperatures are low and a rigorous cleaning regime is implemented at the end of each processing period, 316 stainless steel will give excellent service. However, where temperatures are high and contact periods are long, the corrosion process may be initiated. This is especially so in crevices such as the interplate contact points on a plate heat exchanger where, albeit at a microscopic level, corrodents and corrosion products are trapped in pits or cracks. Geometric factors may prevent the complete removal of this debris during cleaning and under such circumstances, the corrosion process will be ongoing.

Because of the perishable nature of foodstuffs, storage is rarely for prolonged periods or at high temperatures, and regular, thorough cleaning tends to be the norm. The one exception to this is buffer storage vessels for holding "self-preserving" ketchup and sauces. For such duties, an alloy such as 904L, Avesta 254 SMO, or even Inconel 625 may be required.

While all the foregoing applies to general equipment, the one exception is plate heat exchangers. Their highly creviced configuration and the high temperatures employed render them particularly susceptible. Plates made from grade 316 have a poor track record on these types of duties. Even the more highly alloyed materials do not offer complete immunity. The only reasonably priced material that is finding increased usage in certain areas of food processing is titanium.

The fact that butter and margarine have been included in this group of corrodents requires comment. Both these foodstuffs are emulsions containing typically 16% water and 2% salt. A fact not often appreciated is that the salt is dissolved in the water phase, being insoluble in the oil. From a corrosion standpoint, therefore, the margarine or butter may be regarded as a suspension of 12% salt solution and as such is very corrosive to 316 stainless steel at the higher processing temperatures. The only mitigating feature that partly offsets their corrosivity is the fact that the aqueous salty phase is dispersed in an oil rather than the reverse and that the oil does tend to preferentially wet the steel surface and provide some degree of protection. However, the pasteurizing heat exchanger in margarine rework systems invariably has titanium plates as the life of 316 stainless steel is limited and has been known to be as little as 6 weeks.

Corrosion by Service Fluids

Steam. Because it is a vapor and free from dissolved salts, steam is not corrosive to stainless steels. Although sometimes contaminated with traces of rust from carbon steel steam lines, in the author's experience no case of corrosion due to industrial boiler steam ever has been encountered.

Water. The quality and dissolved solids content of water supplies varies tremendously, with the aggressive ionic species, chloride ions, being present at levels varying between zero, as found in the lakeland area of England, to several hundred parts per million, as encountered in coastal regions of Holland. It is also normal practice to chlorinate potable water supplies to kill pathogenic bacteria with the amount added dependent on other factors such as the amount of organic matter present. However, most water supply authorities aim to provide water with a residual chlorine content of 0.2 ppm at the point of use. Well waters also vary in composition depending on the geographical location, especially in coastal regions where the chloride content can fluctuate with the rise and fall of the tide.

What constitutes a "good" water? From a general user viewpoint, the important factor is hardness, either temporary hardness caused by calcium and magnesium bicarbonates that can be removed by boiling, or permanent hardness caused by calcium sulfate that can be removed by chemical treatment. While hardness is a factor, chloride content and pH probably are the most important from a corrosion standpoint.

What can be classed as a noncorrosive water supply from a stainless-steel equipment user? Unfortunately, there are no hard-and-fast rules that will determine whether corrosion of equipment will occur. As has been repeatedly stated throughout this article, so many factors come into play. The type of equipment, temperatures, contact times, etc all play a role in the overall corrosion process. Again, as has been stated before, the most critical items of equipment are those with inherent crevices—evaporators and plate heat exchangers among others. Defining conditions of use for this type of equipment will be the regulatory factor. Even then, it is virtually impossible to define the composition of a "safe" water, but as a general guideline, water with less than 100 ppm chloride is unlikely to initiate crevice corrosion of type 316 stainless steel while a maximum level of 50 ppm should be used with type 304.

Cooling tower water systems are frequently overlooked as a potential source of corrodents. It must be appreciated that a cooling tower is an evaporator, and although the supply of makeup water may contain only 25 ppm chloride, over a period of operation this can increase by a factor of 10 unless there is a routine bleed on the pond.

Water scale deposits formed on heat-transfer surfaces should always be removed as part of the routine maintenance schedule. Water scale deposits can accumulate chloride and other soluble salts that tend to concentrate, producing higher levels in contact with the metal than indicated by the water composition. Furthermore, water scale formed on a stainless-steel surface provides an ideal base for the onset of crevice corrosion.

As stated previously, potable water supplies usually have a residual free chlorine content of 0.2 ppm. Where installations have their own private wells, chlorination is undertaken on site. In general terms, the levels employed by the local water authorities should be followed and over-chlorination avoided. Levels in excess of 2 ppm could initiate crevice corrosion.

Cooling Brines. Depending on the industry, these can be anything from glycol solutions, sodium nitrate/carbonate, or calcium chloride. It is the latter which are used as a 25% solution that can give rise to corrosion of stainless steel unless maintained in the ideal condition, especially when employed in the final chilling section of plate heat exchangers for milk and beer processing. However, by observing certain precautions, damage can be avoided.

The corrosion of stainless steels by brine can best be represented as shown in Figure 22. It will be noted that an exponential rise in corrosion rate with reducing pH occurs in the range pH 12–7. The diminution in number of pits occurring in the range pH 6–4 corresponds with the change in mode of attack, ie, from pitting to general corrosion. It will be seen that ideally the pH of the solution should be maintained in the region 14–11. However, calcium chloride brine undergoes decomposition at pH values higher than 10.6:

$$CaCl_2 + 2NaOH \rightarrow Ca(OH)_2 \downarrow \ + 2NaCl$$

Scale deposition occurs and heat-transfer surfaces become fouled with calcium hydroxide scale. Furthermore, the scale that forms traps quantities of chloride salts that cannot be effectively removed and remain in contact with the equipment during shutdown periods. This is particularly important in equipment such as plate heat exchangers that are subjected to cleaning and possibly hot-water sterilization cycles at temperatures of 80°C (176°F) or higher.

The other aspect, nonaeration, is equally important. Air contains small quantities of carbon dioxide that form a slightly acidic solution when dissolved in water. This has the effect of neutralizing the buffering action of any alkaline components in the brine:

$$2NaOH + CO_2 \rightarrow Na_2CO_3 + H_2O$$
$$Na_2CO_3 + CaCl_2 \rightarrow 2NaCl + CaCO_3 \downarrow$$

or

$$Ca(OH)_2 + CO_2 \rightarrow CaCO_3 \downarrow + H_2O$$

Therefore, the pH of the brine decreases and assumes a value of about pH 6.5, which is the region where pitting

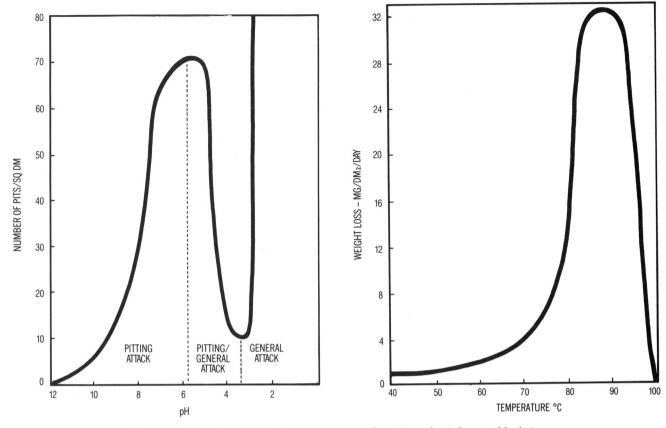

Figure 22. The effect of pH and temperature on the pitting of stainless steel by brine.

incidence is highest. Furthermore, scale deposits of calcium carbonate are laid down on heat-transfer surfaces, creating the problems referred to above.

The precautions to be observed when using brine circuits are

- Ensure correct pH control and maintain in the range pH 9.5–10.
- Eliminate aeration. In particular, make certain that the brine return discharge line is below the surface in the storage tank during running and that the method of feeding brine from the tank does not cause vortexing with resultant air entrainment. Baudelot type evaporators cause aeration of the chilling liquor and should never be used on brine circuits.
- When cleaning and sterilizing the brine section of a pasteurizer, flush out all brine residues until the rinse water is free of chloride. As an added precaution, it is advisable to form a closed circuit and circulate a $\frac{1}{4}$–$\frac{1}{2}$% caustic soda or sodium metasilicate solution to ensure that any brine residues are rendered alkaline.
- In plate heat exchangers and similar equipment, make sure that stainless-steel brine section components remain free of scale.
- When shutting down the plant after a cleaning run, it is advisable to leave the section full with a dilute ($\frac{1}{4}$%) caustic solution. Before startup, this should be drained and residues flushed out prior to reintroducing brine.

When operating conditions prevailing in a plant do not permit such a disciplined cleaning, operating, and shutdown procedure, only two materials can be considered for the brine section of a plate heat exchanger. These are Hastelloy C-276 or titanium. The corrosion resistance of both of these materials is such that cleaning and sanitizing of the product side of the heat exchanger can be carried out without removing the brine. Although both are more expensive than stainless steels, especially Hastelloy C-276, the flexibility of plant operation which their use permits could offset their premium price.

Alkaline Detergents

Supplied to food processing plants either as bulk shipments of separate chemicals or as carefully preformulated mixtures, the composition of alkaline detergent formulations can vary widely in accordance with individual preference or the cleaning job to be done. The detergents do, however, generally include some or all of the following compounds:

> Sodium hydroxide
> Sodium polyphosphate
> Sodium metasilicate
> Sodium carbonate

Additionally, it is not uncommon to find that a selection of sequestering agents such as EDTA and any of the many available wetting agents also may be present in the formulations.

None of these compounds are corrosive to stainless steels at the concentrations and temperatures used by the food industry for cleaning. Indeed, 316 stainless steel is unaffected by concentrations of sodium hydroxide as high as 20% at temperatures up to 160°C (320°F). They can, therefore, be used with impunity at their usual maximum concentration of 5%, even in ultra-high-temperature (UHT) operation where temperatures can rise to 140°C (284°F).

Companies have reported that some of these preformulated alkaline detergents cause discoloration of the equipment.

The discoloration starts as a golden yellow, darkening to blue through mauve and eventually black. It has been established that this discoloration is caused by the EDTA sequestering agent, which complexes with traces of iron in the water. It then decomposes under certain conditions of pH and temperature to form an extremely fine film of hydrated iron oxide, the coloration being interference colors that darken as the film thickness increases. Although the film is not aesthetically pleasing, it is in no way deleterious and removing it by conventional cleaning agents is virtually impossible.

Some alkaline detergents are compounded with chlorine release agents such as sodium hypochlorite, salts of di- or trichlorocyanuric acid that form a solution containing 200–300 ppm available chlorine at their usage strength. Although the high alkalinity reduces the corrosivity of these additives, generally speaking they should not be employed on a regular basis at temperatures exceeding 70°C (160°F).

Acidic Detergents

Alkaline detergents will not remove the inorganic salts such as milkstone and beer-stone deposits frequently found in pasteurizers. For this, an acidic detergent is required and selection must be made with regard to their interaction with the metal. As was shown earlier, sulfuric and hydrochloric acids will cause general corrosion of stainless steels. Although it could be argued that sulfuric acid can be employed under strictly controlled conditions because stainless steels, especially grade 316, have a very low corrosion rate, its use could result in a deterioration of surface finish. This, in corrosion terms, is an extremely low rate but from an aesthetic viewpoint, is undesirable.

Acids such as phosphoric, nitric, and citric, when used at any concentration likely to be employed in a plant cleaning operation, have no effect on stainless steels and can be used with impunity. Three cautionary notes are worthy of mention:

- It is always preferable to use alkaline cleaning before the acid cycle to minimize the risk of interacting the acid with any chloride salts and, therefore, minimize the formation of hydrochloric acid.
- It is inadvisable to introduce an acid into a UHT sterilizing plant when it is at full operating temperature (140°C/285°F) as part of a "clean-on-the-run" regime.
- Nitric acid, which is a strong oxidizing agent, will attack certain types of rubber used as gaskets and seals. As a general guideline, concentrations should not exceed 1% and temperature 65°C (150°F), al-

though at lower concentrations, the upper recommended temperature is 90°C (195°F).

Another acid that is finding increasing use in the food industry for removing water scale and other acid-soluble scales is sulfamic acid. Freshly prepared solutions of up to 5% concentration are relatively innocuous to stainless steels but problems may arise when CIP systems incorporating recovery of detergents and acids are employed. Sulfamic acid will undergo hydrolysis at elevated temperatures to produce ammonium hydrogen sulfate

$$NH_2SO_2OH + H_2O \rightarrow NH_4HSO_4$$

which behaves in much the same way as sulfuric acid. In situations where the use of this acid is contemplated, prolonged storage of dilute solutions at elevated temperature is inadvisable, although at room temperature the hydrolysis is at a low rate.

Sanitizing Agents

Terminology for the process of killing pathogenic bacteria varies from country to country. In Europe, disinfection is preferred; in America, sanitizing. Rgardless, the term should not be confused with sterilization, which is the process of rendering equipment free from all live food spoilage organisms including yeasts, mold, thermophilic bacteria, and most importantly, spores. Sterilization with chemicals is not considered to be feasible and the only recommended procedure involves the circulation of pressurized hot water at a temperature of not less than 140°C (285°F).

For sanitization, while hot water (or steam) is preferred, chemical sanitizers are extensively used. These include noncorrosive compounds such as quaternary ammonium salts, anionic compounds, aldehydes, amphoterics, and potentially corrosive groups of compounds that rely on the release of halogens for their efficacy. By far, the most popular sanitizer is sodium hypochlorite (chloros), and this is probably the one material that has caused more corrosion in food plants than any other cleaning agent. For a detailed explanation of the corrosion mechanism, the reader is referred to an article by Boulton and Sorenson (ref. 17) that describes a study of the corrosion of 304 and 316 stainless steels by sodium hypochlorite solutions. It is important, therefore, if corrosion is to be avoided, that the conditions under which it is used are strictly controlled. For equipment manufactured from grade 316 stainless steel, the recommended conditions are:

- Maximum concentration—150 ppm available chlorine
- Maximum contact time—20 min
- Maximum temperature—room temperature that is well in excess of the minimum conditions established by Tastayre and Holley to kill pseudomonas aeruginosa (ref. 18).

In addition, several other precautions must be observed:

- Before introducing hypochlorite, equipment should be thoroughly clean and free of scale deposits. Organic residues reduce the bactericidal efficiency of the disinfectant and offer an artificial crevice in which stagnant pools of hypochlorite can accumulate.

- It is imperative that acidic resides be removed by adequate rinsing before introducing hypochlorite solutions. Acid solutions will react with hypochlorite to release elementary chlorine, which is extremely corrosive to all stainless steels.

- The equipment must be cooled to room temperature before introducing hypochlorite. In detergent cleaning runs, equipment temperature is raised to 80–85°C (176–185°F), and unless it is cooled during the rinsing cycle, a substantial increase in temperature of the disinfectant can occur. An important point, frequently overlooked, is that a leaking steam valve can cause a rise in the temperature of equipment even though it theoretically is shut off.

- After sanitizing, the solution should be drained and the system flushed with water of an acceptable bacteriological standard. This normally is done by using a high rinse rate, preferably greater than that used in the processing run.

While these comments relate specifically to the santizing of plate heat exchangers, similar precautions must be taken with other creviced equipment. Examples include manually operated valves that should be slackened and the plug raised to permit flushing of the seating surface. Pipeline gaskets also should be checked frequently to make sure that they are in good condition and not excessively hardened. Otherwise they will fail to form a crevice-free seal over their entire diameter. Where it is not possible to completely remove hypochlorite residues such as in absorbent gland packing materials, hot water is preferred.

All the foregoing relate specifically to sodium hypochlorite solutions but other sanitizing agents that rely on halogen release, such as di- and trichlorocyanuric acid, should also be used under strictly controlled conditions, including such factors as pH.

Iodophors also are used for sanitizing equipment. These are solutions of iodine in nonionic detergents and contain an acid such as phosphoric to adjust the pH into the range at which they exhibit bactericidal efficacy. This group of sanitizers is employed where hot cleaning is not necessary or on lightly soiled surfaces such as milk road tankers, and farm tanks. Extreme caution should be exercised with this group for, although used at low concentrations (50 ppm), prolonged contact with stainless steel can cause pitting and crevice corrosion. Furthermore, in storage vessels that have been partially filled with iodophor solutions and allowed to stand overnight, pitting corrosion in the head space has been observed as a result of iodine vaporizing from the solution and condensing as pure crystals on the tank wall above the liquid line. Another factor is that iodine can be absorbed by some rubbers. During subsequent processing operations at elevated temperatures, the iodine is released in the form of organic iodine compounds, especially into fatty foods, which can cause an antiseptic taint. The author knows of one dairy that used an iodophor solution to sanitize a plate pasteurizer to kill an infection of a heat resistant spore-forming organism.

The following day, there were over 2000 complaints of tainted milk. CIP cleaning cycles did not remove the antiseptic smell from the rubber seals, and complete replacement with new seals was the only method of resolving the problem.

Another sanitizing agent that is assuming increasing popularity, especially in the brewing industry because of its efficacy against yeasts, is peracetic acid. As such, peracetic acid will not cause corrosion of 304 or 316 stainless steels, and the only precautionary measure to be taken is to use a good quality water containing less than 50 ppm of chloride ions for making up the solutions to their usage concentration. Because of the strongly oxidizing nature of some types of peracetic acid solutions, deterioration of some types of rubber may occur. A recent survey undertaken by the IDF for the use of peracetic acid in the dairy industry (ref. 19) found few corrosion problems reported. The general consensus of opinion was that it permitted greater flexibility in the conditions of use, compared with sodium hypochlorite, without running the risk of damage to equipment.

For comprehensive information on the cleaning of food processing equipment, albeit primarily written for the dairy industry, the reader is referred to the British Standards Institute publication BS 5305 (ref. 20).

CORROSION BY INSULATING MATERIALS

Energy conservation now is widely practiced by all branches of industry, and the food industry is no exception. For example, in the brewing industry, wort from the wort boilers is cooled to fermentation temperature, and the hot water generated in the process is stored in insulated vessels (hot-liquor tanks) for making up the next batch of wort. An area of corrosion science that is receiving increased attention is the subject of corrosion initiated in stainless steels by insulating materials. At temperatures in excess of 60°C (140°F) these can act as a source of chlorides that will induce stress corrosion cracking and pitting corrosion of austenitic stainless steels.

Among the insulating materials that have been used for tanks and pipework are

Foamed plastics—polyurethane, polyisocyanurate, phenolic resins, etc
Cellular and foamed glass
Mineral fiber—glass wool, rock wool
Calcium silicate
Magnesia
Cork

All insulating materials contain chlorides to a lesser (10 ppm) or greater (1.5%) extent. The mineral based insulants such as asbestos may contain them as naturally occurring water soluble salts such as sodium or calcium chloride. The organic foams, on the other hand, may contain hydrolysable organochlorine materials used as blowing agents (to form the foam), fire retardants such as chlorinated organophosphates, or chlorine containing materials present as impurities. Insulation manufacturers are becoming increasingly aware of the potential risk of chlorides in contact with stainless steels and are making serious efforts to market a range of materials that are essentially chloride-free.

A point not frequently appreciated is that even though many of the insulating materials contain high levels of chloride, in isolation they are not corrosive.

Corrosion, which is an electrochemical process, involves ionic species, and in the absence of a solvent (water) the chloride or salts present in the insulation cannot undergo ionization to give chloride ions. Therefore, they are essentially noncorrosive. Similarly, where organochlorine compounds are present, water is necessary for hydrolysis to occur with the formation of hydrochloric acid or other ionisable chloride compounds.

The main problem, therefore, is not so much the insulant but the interaction of the insulant with potential contaminants to release corrosive species. Under ideal conditions, that is, if the insulating material could be maintained perfectly dry, the chloride content would not be a critical factor in the material specification process. Unfortunately, it must be acknowledged that even in the best regulated installation, this ideal is rarely (if ever) achieved and therefore, thought must be given to recommended guidelines.

Any material that is capable of absorbing water must be regarded as a potential source of chloride. Although the chloride content of the insulant may be extremely low (25 ppm), under extractive conditions when the insulation becomes wet and concentration effects come into play, even this material may cause the initiation of stress corrosion cracking. The chloride content of the contaminating water cannot be ignored. Even a "good-quality" water with a low chloride content (say, 30 ppm), if being continuously absorbed by the insulant, forms a potential source of chlorides that, through evaporation, can reach very significant levels and initiate the corrosion mechanism.

The more absorptive the insulant, the greater the risk, and materials such as calcium silicate and certain types of foams must be regarded as least desirable. It is interesting to note that one of the least absorptive insulants from those listed is cork, and in the experience of the author, there have been no reported cases of this insulant causing problems with stress corrosion cracking of stainless steels. Of course, it could be argued that the bitumen used as an adhesive to stick the cork to the vessel walls has to be applied so thickly that the bitumen forms an impermeable barrier preventing contact of the contaminated water with the stainless-steel substrate. Irrespective of the protection mechanism, the net effect is that cork has an extremely good "track record." Unfortunately, due to the cost, cork is now rarely used.

No hard-and-fast rules can be laid down for specifying the acceptable maximum tolerable chloride content of an insulating material. Any specification must take into account what is commercially available as well as all the other factors such as price, flammability, and ease of application. To specify zero chloride is obviously impractical and even a figure of 10 ppm may be difficult to achieve in commercially available products. As a general compromise, 25 ppm maximum is considered to be technically and commercially feasible while minimizing the potential source level of chloride corrodent.

The primary function of the insulant is to provide a thermal barrier between the outside of the vessel and the environment. It does not provide a vapor or moisture barrier, and provision of such protection must be regarded as equally important in the insulating process. As mentioned previously, the use of bitumen as an adhesive for cork insulation must provide an extremely good water barrier, although there are a wide range of products marketed specifically for this purpose. These include specially developed paints of undefined composition and zinc free silicone alkyd paints as well as silicone lacquer. These silicone based products are particularly appropriate because of their inherent low water permeability and also their stability at elevated temperatures. Aluminum foil has also been successfully used as a water–vapor barrier between the stainless steel and the insulant. There is little doubt that the foil provides an extremely good barrier but at laps between the sheets of foil, ingress of water can occur unless sealing is complete, the achievement of which is most unlikely. Furthermore, it is likely that there will be tears in the foil, providing yet another ingression path for water. However, it is believed that aluminum foil has a role additional to that of a barrier. As it is, from an electrochemical aspect, "anodic" to stainless steel, it will provide galvanic protection of the steel in areas where there is a "holiday" in the foil, thus inhibiting corrosion mechanisms.

When insulation material becomes wet, the insulating efficiency shows a dramatic fall-off. In fact, the thermal conductivity of wet insulation will approach that of the wetting medium, and it is ironic that the thermal conductivity of water is among the highest known for liquids. It is imperative that from the standpoint of preventing moisture ingression to minimize corrosion risks and maintain insulation efficiency, the insulation is externally protected from rain and water. There is a variety of materials available for this such as aluminum sheeting, plastic-coated mild steel, and spray applied polyurethane coatings. Much of the value of the protection will be lost unless particular attention is taken to maintaining a weathertight seal at the overlaps in individuals sheets of cladding. There are many semiflexible caulking agents that can be employed for this. One that is particularly effective is the RTV silicon rubber, which exhibits extremely good weather resistance and long-term reliability. Nevertheless, maintenance is required, especially around areas of discontinuity such as flanged connections and manway doors. Operators are beginning to realize that routine maintenance work on insulation is as important as that on all other items of plant and equipment.

It will be appreciated from all the foregoing that insulation of a vessel or pipeline is a composite activity with many interactive parameters. It is impossible to lay down specification rules, as each case much be viewed in the light of requirements. These notes must, therefore, be regarded as guidelines rather than dogma, and for further information, an excellent publication by the American Society for Testing Materials (ref. 21) is essential reading.

CORROSION OF RUBBERS

Many involved in the field of corrosion would not regard the deterioration of rubber as a corrosion process. The author's opinion differs from this viewpoint as "rubber undergoes deterioration by interaction with its environment."

General

Rubber and rubber components form an essential part of food processing equipment—joint rings on pipelines, gaskets on heat exchangers, and plate evaporators. Although natural rubber was the first material to be used for manufacturing these components, nowadays they are made almost exclusively from one of the synthetic rubbers listed in Table 5.

Unlike metals and alloys, which have a strictly defined composition, the constituents used in the formulation of rubbers are rarely stipulated. More often, they will reflect the views and idiosyncrasies of the formulator on how to achieve the desired end product. The important constituents of a rubber are

- Basic Polymer. Largely determines the general chemical properties of the finished product.
- Reinforcing Fillers. These are added to improve the mechanical properties and will invariably be one of the grades of carbon black—or if a white rubber is required, mineral fillers such as clays or calcium silicate.
- Vulcanizing Agents. These cross-link the basic polymer and impart rubberlike properties that are maintained at elevated temperatures.

- Antioxidants. To stabilize the rubber against oxidative degradation, hardening or softening, after prolonged operating periods at elevated temperatures.
- Processing Aids. Which facilitate the molding of the rubber.
- Plasticizers. To modify the mechanical properties.

A complicating factor to be considered when formulating rubber for food-contact surfaces is the acceptability (or nonacceptability) of the compounding ingredients. Some countries, notably Germany and the USA, have drawn up lists of permitted ingredients (refs. 22, 23), while other countries regulate the amount of material that can be extracted from the finished article by various test media.

Invoking these regulations may impose limits on the in-service performance of a rubber component, which could be a compromise, exhibiting desirable properties inferior to those achievable were it for a nonfood application. For example, the resistance to high pressure steam of some rubbers can be enhanced by using lead oxide as an ingredient. Obviously, such materials could not be contemplated for any food contact application.

Corrosion by Rubber

The majority of rubbers and formulating ingredients have no effect on stainless steels even under conditions of high temperature and moisture. There are two notable exceptions, namely, polychloroprene and chlorosulfonated polyethylene. Both of these contain chlorine, which under the influence of temperature and moisture, undergo hydrolysis to produce small quantities of hydrochloric acid. In contact with stainless steel, this represents a serious corrosion hazard causing the three main forms of attack.

When specifying a rubber component, it is easy to avoid these two polymers but a fact not often appreciated is that many of the rubber adhesives are produced from one of these polymers. That is the reason why manufacturers of heat exchangers and other equipment, which necessitates sticking the rubber onto metal, specify what type of adhesive should be used. Many DIY adhesives and contact adhesives are formulated from polychloroprene, and it is not

Table 5. Some Synthetic Rubbers

Rubber	Common Trade Names	Basic Structure
Polychloroprene	Neoprene	Poly (2-chlor-1,8 butadiene)
	Perbunan C	
	Butachlor	
	Nairit	
SBR (styrene butadiene rubber)	Buna S	Copolymer of styrene and 1,4 butadiene
	Plioflex	
	Intol	
	Krylene	
Nitrile	Buna N	Copolymer of acrylonitrile and 1,4 butadiene
	Chemigum N	
	Paracril	
	Perbunan	
	Hycar	
EPDM (ethylene propylene diene methylene)	Nordel	Copolymer of ethylene and propylene with a third monomer such as ethylidene norbonene, cyclopentadiene
	Royalene	
	Vistalon	
	Dutral	
	Keltan	
	Intolan	
Butyl	GR-1	Copolymer of isobutylene and isoprene
	Bucar	
	Socabutyl	
Silicone	Silastic	Poly(dimethyl siloxane), poly(methyl vinyl siloxane), etc
	Silastomer	
	Sil-O-Flex	
Fluoroelastomer	Viton	Copolymer of hexafluoropropylene and vinylidene fluoride
	Technoflon	
	Fluorel	

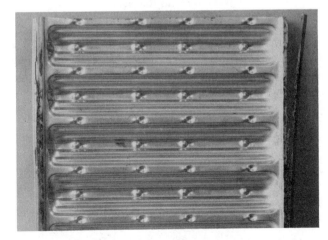

Figure 23. Catastrophic failure of the gasket groove of a heat-exchanger plate by stress corrosion cracking caused by the use of a polychloroprene based adhesive.

Table 6. Performance of Rubbers in Some Environments Found in the Food and Beverage Industries

	SBR	Medium Nitrile	Natural Rubber	Poly-chloroprene	Butyl	EPDM	Silicone[a]
Products							
Whole milk	E	E	E	E	E	E	E
Beer, wines, spirits	G–F	E	F	E	E	E	E
Fats, oils, cream	F	E	F	G	P	G	E
Sauces	E	E	F[b]	E	E[b]	E	E
Salad dressings	F	E	F	G	P	G	E
Fruit drinks and juices[d]	E	E	G	E	E	E	E
Cleaning Agents							
Sodium hydroxide[e]	G	E	G	G	E	E	E
Sodium carbonate[f]	E	E	E	E	E	E	E
Sodium hypochlorite[g]	G	G	G	G	E	E	E
Nitric acid[h]	F	F	P	F	G	G	G
Phosphoric acid[i]	E	E	E	E	E	E	E
Quaternary ammonium compounds[c]	E	E	E	E	G	G	G

[a] Depending on the type of basic polymer.
[b] Depending on the fat/oil content.
[c] Used as an aqueous 1% solution.
[d] The performance of a rubber will be affected by the presence of essential oils.
[e] All strengths up to 5% at maximum operating temperatures of the rubber.
[f] Sodium carbonate and other detergent/additives, eg, sodium phosphate, silicate.
[g] Sodium hypochlorite as used at normal sterilizing concentration—150 mg/L.
[h] Nitric acid as used at normal cleaning strength of ½–1%.
[i] Up to 5% strength.

unknown for a maintenance engineer having to stick gaskets in a heat exchanger to get the supply of rubber cement from the local hardware shop. The results can be catastrophic, as shown in Figure 23. Similarly, many self adhesive tapes used as a polychloroprene-based adhesive and direct contact of these with stainless steels should be avoided.

Corrosion of Rubber by Environments

From the standpoint of food processing, the environments likely to interact with rubber are classified under the fol-

lowing headings:

- Foodstuffs containing no fat or a low level of fat, eg, milk

- Fatty products: butter, cream, cooking oils, shortenings

- Alkaline detergents

- Acid detergents

- Sanitizing agents

Table 7. Suggested Limits of Concentration and Temperature for Peracetic acid in Contact with Rubber

Peracetic Acid Concentration (Active Ingredient)	Temp., °C	Medium Nitrile	Butyl (Resin-Cured)	EPDM	Silicon	Fluoro-Elastomer
0.05% (500 mg/L)	20	R[a]	R	R	R	R
	60	R	R	R	R	R
	85	(?)	R	R	R	(?)[b]
0.10% (1 g/L)	20	R	R	R	R	R
	60	NR[c]	R	R	NR	R
	85	NR	(?)	R	NR	NR
0.25% (2.5 g/L)	20	R	R	R	(?)	R
	60	NR	R	R	NR	R
	85	NR	NR	R	NR	NR
0.5% (5 g/L)	20		R	R	NR	R
	60	NR	R	R	NR	R
	85	NR	NR	R	NR	NR

[a] R—Little effect on the rubber.
[b] (?)—Possibly some degradation.
[c] NR—Not recommended as significant degradation may occur.

Unlike the corrosion of metals, which is associated with oxidation and loss of metal, rubber deterioration usually takes other forms. When a rubber is immersed in a liquid, it absorbs that liquid or substances present in it to a greater or lesser degree. The amount of absorption determines whether the rubber is compatible with the environment. The absorption will be accompanied by changes in mass, volume, hardness, and tensile strength. For example, immersing and oil resistant rubber in vegetable oil may produce a change in volume of only 2–3%, whereas a non-oil-resistant rubber may swell by 150% or more. Such a volumetric change will be accompanied by a large reduction in the tensile strength and a high degree of softening.

Broadly, speaking, a rubber should not exhibit a volumetric or weight change greater than 10% nor a hardness change of more than 10 degrees (International Rubber Hardness Degrees—IRHD or Shore A) to be classed as compatible. For general guidance, data presented in Tables 6 and 7 (19) indicate the compatibility of rubbers with some food industry environments. But for more information, the reader should refer to one of the national or international test procedures (refs. 24–26).

BIBLIOGRAPHY

Adapted from: C. T. Cowan, "Materials Science and Corrosion Prevention for Food Processing Equipment," *APV Corrosion Handbook,* March 1990, 44 pp.
Copyrighted APV Crepaco, Inc. Used with permission.

1. N. Warren, *Metal Corrosion in Boats,* London Stanford Maritime, 1980.
2. A. J. Sedricks, *Corrosion of Stainless Steels,* Wiley, New York, 1979.
3. The British Aluminium Co., Ltd., *Aluminium in the Chemical and Food Industries,* 1951. (London: The British Aluminium Co., Ltd.)
4. G. Ferry, *New Scientist,* 27 Feb. 1986, 23.
5. G. A. Dummett, *From Little Acorns, A History of the APV Company,* Hutchison Benham, London, 1981.
6. C. T. Cowan, "Corrosion Engineering with New Materials," *The Brewer* 163–168 (1985).
7. M. G. Fontana, *Corrosion Engineering,* McGraw-Hill, New York, 1986.
8. L. L. Shrier, *Corrosion,* Newnes-Butterworth, London, 1976.
9. J. W. Oldfield and B. Todd, Transactions of the Institute of Marine Engineering (C), 1984, Conference 1, 139.
10. J. W. Oldfield, *Test Techniques for Pitting and Crevice Corrosion of Stainless Steels and Nickel Based Alloys in Chloride-Containing Environments,* NiDi Technical Series 10 016, 1987.
11. U. R. Evans, *Corrosion* **7,** 238 (1981).
12. ASTM G48-76, *Standard Methods for Pitting and Crevice Corrosion Resistance of Stainless Steels and Related Alloys by Use of Ferric Chloride Solution,* American Society for Testing Materials, Philadelphia.
13. R. J. Brigham, *Materials Performance* **13,** 29 (Nov. 1974).
14. N. D. Greene and M. G. Fontana, *Corrosion* **15,** 25t (1959).
15. M. Woelful and R. Mulhall, *Metal Progress,* 57–59 Sept. 1982.
16. M. R. Copson, "Effect of Composition on SCC of Some Alloys Containing Nickel," in T. Rhodin, ed., *Physical Metallurgy of Stress Corrosion Fracture,* Interscience Publishers, New York, 1989.
17. L. H. Boulton and M. M. Sorensen, *New Zealand Journal of Dairy Science & Technology* **23,** 37–49 (1988).
18. G. M. Tastayre and R. A. Holley, Publication 1806/B. Agriculture Canada, Ottawa, 1986.
19. IDF Bulletin 236, *Corrosion by Peracetic Acid Solutions,* International Dairy Federation, 41 Square Vergote, 1040 Brussels, Belgium.
20. BS 5305 Code of practice for cleaning and disinfecting of plant and equipment used in the dairying industry, British Standards Institute, London.
21. ASTM Special Technical Publication 880, Corrosion of Metals under Thermal Insulation, American Society for Testing Materials, Philadelphia.
22. Sonderdruck aus Bundesgesundheitsblatt, 22.1.79, West Germany.
23. USA *Code of Federal Regulations,* Title 21, Food and Drugs Section 177.2600. Rubber Articles Intended for Repeated Use.
24. BS 903 Part A16, Resistance of Vulcanized Rubbers to Liquids, British Standards Institution, London.
25. ASTM D. 471, Testing of Rubbers and Elastomers. Determination of Resistance to Liquids, Vapors, and Gases, American Society for Testing Materials, Phildelphia.
26. ISO 1817, Vulcanized Rubbers—Resistance to Liquids—Methods of Test.

APV CREPACO
Lake Mills, Wisconsin

COTTONSEED OIL. See OILSEEDS AND THEIR OILS.

CRABS AND CRAB PROCESSING

Crabs play an important part in the United States fishing industry, ranking fourth in quantity (pounds landed) and third in value in 1988. The quantity and value of the major crab fishes in the United States in 1988 are shown in Table 1 (1). Variations in processing methods for crabs are related to the species and the seasonal characteristics that affect yield, ease of meat removal, and color. Speed in handling the product from time of harvesting to freezing or canning is a most important process requirement (2,3). Quality-control procedures generally emphasize sanitation, food-regulatory requirements, and compliance with end product specifications of the processor (4).

KING CRAB

Nomenclature

Three species of king crab harvested in the North Pacific and Bering Sea are of commercial importance. The most important species is red king crab (*Paralithodes camtschatica*). Blue king crab (*P. platypus*) is caught in commercial quantities primarily near the Pribilof Islands. Brown or golden king crab (*Lithodes aequispina*) are found in deeper waters at the edge of the continental shelf.

Table 1. Landings and Values of U.S. Crab Species in 1988

Crab Species	Quantity (million pound)	Value (million dollars)	Average Ex-Vessel Price (dollar/pound)
Blue	218.7	84.4	0.39
Chesapeake	77.8		
South Atlantic	54.2		
Gulf	77.8		
Middle Atlantic	8.9		
Dungeness	47.4	54.8	1.15
Washington	18.1		
Alaska	10.3		
Oregon	9.4		
California	9.6		
King	21.0	84.2	4.01
Snow	146.3	137.1	0.94

King crabs are actually not true crabs like dungeness, tanner, or blue crab, but are more closely related to an order of hermit crabs. The legs of true crabs are jointed forward, whereas the legs of king crabs are jointed to fold behind its body (5).

Harvesting

The world's most-extensive king crab populations dwell in Alaskan waters, where 15-year-old male king crabs may grow to a weight of 24 lb and a leg span of 6 ft. Most of the commercial catch consists of 7- or 8-year-old crabs that weight about 6 lb. They are captured by means of large pots consisting of iron frames about 6 ft^2 and 3 ft high covered with synthetic webbing. These pots may weigh 300–700 lb a piece. The pots are baited with fresh fish (herring is the most preferred) and then dropped to the bottom, a 60–300 ft depth, with a heavy line and marker buoy. The pots are lifted every few days and the male crabs of legal size are kept. The crabs are held alive on the vessel in a well with circulating seawater. At the docks or on the deck of a processor vessel the live crabs are transferred to seawater tanks to await processing. Weak crabs may be sorted out and processed immediately, but generally all dead and injured crabs are discarded.

Processing

The crab is butchered by use of a stationary iron blade. The back, or carapace, is removed and the crab is split into halves, each with legs attached, called sections. A quick shake will jar out viscera that cling to the body cavities. Revolving nylon brushes are generally used to clean off gills and viscera.

Batch cooking is common, although a continuous cooker requires less labor, and cooking is done without the delay of filling a basket. It also has automatic steam controls and a fixed rate of travel through the tank, insuring identical cooking. For crab planned for frozen sales, the sections are cooked for 22–28 min in either boiling fresh water or seawater. Cook times will vary slightly, but cooking should be continued to insure that the internal temperature of the crabmeat reaches at least 88°C (190°F).

For canned or retorted products, crab need only be cooked to make meat removal easy. A two-stage cook of 20 min at 60°C (140°F), rinsing the sections well with a spray of water followed by cooking in boiling water for 12 min has been proposed (6). This prevents a blue discoloration in canned Kegani crab, because the crabmeat is coagulated during the precook while the blood (hemocyanin) can be washed free of the meat. Regardless of cooking method, cooked crabmeat should be picked, and thermally processed as rapidly as possible to insure an excellent canned product (7). Normally, canned crabmeat is packed in 307 × 113 C-enamel cans, sealed in a vacuum, retorted for 50 min at 116°C (240°F), then water cooled (8).

For freezing, the sections are cooled in cold running water after cooking. Ideally, chilled water is used to cool the sections to at least 4°C (40°F). Usually, sections are then frozen either in brine or air-blast freezers. The majority of Alaskan king crab production is brine frozen. The sections are packed with the legs stacked in the center of baskets, shoulders out, and held in liquid brine (22–23%) at a temperature of −18 to −20°C (0 to −5°F) for 30–60 min. Internal temperature should reach at least (15°F). The sections are then removed from the brine and dipped in cold fresh water to rinse the salt from the section and to form a protective glaze on the surface of the crab. Blast freezing freezes the product in air by blowing refrigerated air (less than −30°C, or −20°F) over it. Proponents of blast freezing contend the method results in a product with a lower sodium content. Brine freezing, if carefully controlled will be similar to blast freezing and some processors believe it results in a moister product.

Product Forms

Almost all of Alaska's king crab is sold frozen. Frozen king crab is packed in a variety of product forms including both in-shell products and extracted crabmeat.

Bulk Clusters (Sections). These are packs of shoulders, legs, and claws, commonly packed in 80-lb, random-weight cartons. The pack should average about 75% legs and 25% shoulder–body meat by weight.

Legs and Claws. Legs and claws are available in a number of forms including both whole legs and claws or split legs and claws. King crab have six walking legs, two of which have claws. Packs of legs and claws should be in equivalent proportion to the live crab and should include the whole leg from shoulder to walking tip or claw. As a general rule, the larger the leg the higher the price.

King Crab Meat. The body, leg, and claw meats are removed by shaking or by blowing with water under pressure. After the shoulder–body meats are removed, each leg is broken just below the uppermost joint and the two connective tendons that run through the leg meat are removed by pulling the remaining leg sections free. This step is repeated for each leg segment allowing the leg meats to be removed intact. For larger industrial processes, the legs are divided at the joints, in some cases, by use of a band saw, and fed into two large rubber rollers. The rollers are adjusted for proper clearance and are continuously rotated so that the shell is squeezed as it passes through and the meat is forced out. Meat yield averages about 20% but may be 26% or more by weight of the live crabs (9,10).

The meats are washed, sorted, inspected for shell and debris, and packed into cartons or trays for freezing. King crab meat is available in a number of forms.

Merus Meat. The red-colored merus meat is the largest segment of a king crab's leg and the most expensive. It is available in both blocks and individually quick frozen (IQF) in poly bags.

Fancy Meat. The most common pack of king crab meat and consists of a mixture of merus meat (25–35%), shoulder–body meat (40–50%), and red leg meat (30–35%). The blocks of meat are packed in triple layers beginning with the merus meat, then white meat, finally red leg meat, and then frozen. The higher the percentage of leg meat in the block, the higher the price.

Salad Meat. Salad meat is a cheaper frozen pack of small chunks of both red and white meat.

Rice Meat. The cheapest form of king crab meat is rice meat, which consists of uniformly shredded body and leg meats.

King crab, if adequately protected from dehydration, can be held for six months or more at −18 to −23°C (0– −10°F).

SNOW CRAB

Nomenclature

Commonly referred to as tanner crab, the two crab species *Chionoecetes bairdi* and *C. opilio* were relatively unknown in the United States until the king crab stocks began to decline. Today these crab species are marketed as snow crab and have emerged as the most valued of the crabs harvested in the United States (1). *C. bairdi*, the larger and thus more desirable of the two snow crab species, is

caught only in the North Pacific and Bering Sea. The smaller species, *C. opilio*, is caught in the North Atlantic as well. Pacific *C. opilio* is caught primarily by Alaskan fisherman while Atlantic *C. opilio*, which is slightly smaller, is caught by Canadians. Canadian snow crab may be marketed as queen crab (5).

Snow crabs are members of the spider crab family. In the North Pacific, they occur broadly from the southeast coast of Alaska westward along the coast throughout the Aleutian Islands and the Bering Sea. Snow crabs are caught in the winter and spring by the same boats that fish for king crab in the fall. Although only half the size of king crabs, snow crabs are still larger than most other species of crab. The largest, *C. bairdi*, occasionally exceed 3 lb and have a leg span of 2 ft.

Alaskan snow crab are fished and processed in approximately the same manner as king crab and by the same fisherman and processors. Snow crab is almost always marketed frozen and the product forms are similar to king crab, ie, clusters, claws, legs, and meats (fancy, merus, salad, and shredded). The yield of picked meats from snow crab is slightly lower than for king crab and average 17%, ranging from 12 to 25% (9,11). Like fancy meat from king crab, frozen blocks of fancy snow crab meat consist of three layers: red meat (45%), white meat (35–45%), and shreds (10–20%). Snow crab has the same handling characteristics as king crab. Like king crab, care should be taken to insure the product is well protected by proper glazing and packaging to prevent dehydration and subsequent quality loss during frozen storage.

A snow crab product that is popular in the Japanese market is called green crab, which is raw frozen snow crab sections. It is important to handle, process, and freeze the snow crab quickly and carefully because there is a tendency for bluish to black discoloration to develop in the sections. Although there is no general agreement on the cause(s) for the discoloration (12), it may be the result of the enzymatically mediated oxidation and polymerization of phenolic compounds naturally present in crab (2). Thus antioxidants, sulfites, and chelating agents have been used to block the bluing discoloration. However, speed or quick processing is the best insurance in preventing the bluing, which will still occur and even develop at a faster rate after the frozen raw crab section is thawed. If crab are not cooked properly to destroy these enzymes, the crab should be further processed quickly or consumed soon after thawing.

DUNGENESS CRAB

Nomenclature

Dungeness (*Cancer magister*) crabs inhibit sandy and grassy bottoms along the Pacific Coast from California to the Aleutian Islands. They are found at various depths below the intertidal zone, but are usually fished commercially offshore in depths from 12 to 120 ft. Dungeness crabs can grow to a weight of over 3 lb; however, the commercial catch usually ranges between 1.25 and 2.25 lb. They are harvested with circular iron pots covered with wire mesh, 3 ft in diameter and constructed with two entrance tunnels on the sides. Pots are hauled up every day or two, the legal size male crabs are placed in seawater wells in the boat either dry or flooded, and the crabs are delivered alive at the plant. Only live, vigorous crabs are processed (13).

Product Forms

Dungeness crabs are marketed in a variety of forms, although the current trend is more toward whole crabs, both live and cooked.

Live Crab. Live Dungeness are commonly air freighted in 50-lb wetlock boxes (14). As with snow crab, the live crab may be packed in sawdust that has been saturated with water.

Whole-Cooked Crab. For Dungeness crab planned for fresh sales, whole crab are cooked for 20–25 min in salted (4–5%) boiling water in stainless-steel batch or continuous cooker. Adding citric acid (0.1–0.2%) to the saltwater may enhance the color of the cooked crab by bleaching the surface of the shell, creating a white color. The hot crabs are cooled in cold water and, if the crab are to be sold fresh, packed in ice immediately. Properly handled, the shelf life should be up to seven days. If not sold fresh, the crab is frozen as soon after cooking as possible. Brine freezing has become very popular, but any rapid freezing system can be as good if care is taken to avoid any loss of quality from dehydration.

Meat. Dungeness crab sections intended for picking are prepared in the same manner as king crab and generally cooked in fresh (unsalted) boiling water for 10–12 min in either stainless-steel batch or continuous cookers. After cooling the sections in cold water, the meat is removed from the crab section by hand. The steps in meat extraction have been well documented (15). Generally while holding the legs of a section with one hand the body is pressed with the free hand against a stainless-steel table and then the key bone is removed. This step frees the body meat, which can be dislodged from the shell by striking the section and hand vigorously against the side of an aluminum or stainless-steel pan. Then the shell associated with the body is broken free of the legs just below the joint of each leg. Using a small metal mallet and anvil the top section of each leg is hit to crack the shell. The leg meat is then removed by striking the leg against the pan. This step is repeated at each joint to remove the leg meat. Body and leg meat are usually kept separate. The meat is placed in a strong salt brine (25% sodium chloride solution) to separate the bits of shell from the meat. The meat floats to the top of the tank and is conveyed out of the tank, where it is immediately inspected under uv light (shell pieces will fluoresce) then washed with cold water sprays to remove excess salt.

The body and leg meats are weighed separately and packed in about equal proportions in #10 C-enamel cans holding 5 lb net weight of drained crabmeat or in #2 cans holding 1 lb of meat. The cans are sealed under a low vacuum, frozen in a blast freezer, and stored −18–−23°C (0–−10°F) or colder. Frozen Dungeness crab meat stores well only if protected by suitable packaging against dehydration and oxidation, and stored at −23°C (−10°F) or lower. Under these conditions it stores well for six months.

BLUE CRAB

Nomenclature

Blue crabs (*Callinectes sapidus*), are found along the Atlantic and Gulf coasts from Massachusetts to Texas. Essentially a shallow-water crab, they live in bays, sounds, and channels near the mouths of coastal rivers. Blue crab, when fully grown average 5–7 in. across the carapace. The top of the shell is generally green and the bottom is a whitish shade. The name blue crab comes from the coloring of the legs, of which there are five pairs. The hind legs, flattened at the outer ends, are employed for swimming. At times, using their hind swimming legs, blue crab can swim beautifully through the water with great speed and ease.

Harvesting

Catching methods depend on season, regional preferences, and state regulations. Crabs are taken by crab pots, trotlines, dredges, scrapes, and dip nets. The trotline is a long length of rope with pieces of bait attached at intervals. It is laid on the bottom, ends anchored and marked with buoys. To collect the catch, the boat is run along the line, forcing it to pass over a roller attached to the boat. As the boat moves forward, the crabs cling to the bait until they reach the surface where they are caught with a dip net and placed in a basket or barrel. Fishing boats are usually small. Fishing is usually a one- or two-person operation, employing inexpensive gear. The pots used for blue crab are cubical in shape, 2 ft on each side, and covered with chicken wire. The pot consist of two parts: the crab enter the lower or bait compartment and then pass into the upper or trap compartment through a slit in the partition. The pots are lifted daily and the crabs are removed from the top compartment.

Processing

At the plant, the crabs are weighed and dumped into baskets for cooking in vertical retorts. Cooking conditions vary based on type of equipment used, but most plants use steam at 121°C (250°F) for 9–20 min (4,16). A few plants use boiling water for 15–20 min. After cooling, many processors deback the cooked crabs, wash, and refrigerate the crabs overnight before picking the meat.

The meat is picked primarily by hand and divided into three categories: the lump meat, which is the large muscle controlling the swimming legs; the regular or flake meat, which is the remainder of the meat from the body; and claw meat. Total meat yields consist of 50% flake meat, 25% lump meat, and 25% claw meat and varies from 12 to 17% of the weight of the whole crab.

Blue crab meat does not freeze and store well; therefore, only a small volume of frozen meat is produced. Most of the meat is packed in hermetically sealed cans or plastic bags and sold in the fresh chilled form. To extend the marketing period for the chilled product from several days to several weeks, a heat pasteurization process was developed and has been used successfully for 1-lb sealed cans of meat, which must be stored and distributed at 0°C (32°F) or as near to that as practicable (16).

A popular and important product is the soft-shell crab, a product common only in the blue crab industry. Soft-shell crabs are crabs that have just molted. They are considered a delicacy and bring higher prices. The entire body of a soft-shell crab may be eaten after cooking. Because the optimal period for shipping soft crabs is so short, crab fisherman of the Atlantic and Gulf coasts are by necessity close observers of the molting cycle. As the molt approaches, the peelers move from their normal foraging areas to shallow beds of marine grasses near the shoreline. The peelers are collected in pots or with trawllike devices called scrapes and held in peeler floats, large floating containers. Immediately after molting the then-soft crab is removed, graded for size, and shipped in a refrigerated container to market. Occasionally, soft-shell crabs are frozen. For freezing, the crabs may be held in refrigerated storage for two or three days, then are killed, eviscerated, washed, wrapped individually with parchment, packed in a carton, and frozen (4).

HANDLING TIPS FOR CRAB

A common problem with processed crab is a discoloration called bluing although the actual color may range from light blue to blue-gray to black. Despite almost a century of study, general agreement on the causes of or cures for bluing in crab has not been reached (12). However, a scheme for bluing relating the discoloration to the enzymatically mediated oxidation and polymerization of phenolic compounds naturally present in crab has been proposed and can be used to illustrate the importance of quick and proper handling of crab (2).

Only crabs that are alive and in good condition should be processed. Crabs that have been held in a cooler (1°–3°C) even for up to four days are edible if properly cooked. However, the chances of bluing occurring in the cooked crab are greatly increased even if proper cooking times and temperatures are used because the compounds causing the bluing have already been formed (2). Certainly, care must be taken to cook the crab properly, because underprocessing will not destroy the enzymes responsible for the bluing. Molting is a factor in the bluing reaction, because the concentration of phenolic compounds responsible for bluing is high as they are involved in the formation of the new shell (sclerotization). So during times of molting, extra care must be used in handling and processing crab to prevent bluing.

If the crabmeat is to be thermal processed, the cooking time and temperatures of the raw crab are not as critical. However, as soon as the crab is cooked, the meat from the shell must be removed and thermal processed. Regardless of how long the live crab are cooked, any delay in thermal processing the crabmeat will result in a higher incidence of bluing. Using frozen crab sections to produce canned crabmeat will also increase the risk of bluing. Whole crab or crab sections should be thoroughly cooked before freez-ing to reduce the risk of bluing. Finally, avoid any contact of copper or iron with the crab, because these metals can greatly intensify the bluing discoloration.

BIBLIOGRAPHY

1. M. C. Holliday and B. K. O'Bannon, ed., in *Fisheries of the United States, 1988,* Current Fishery Statistics No. **8800,** U.S. Commerce, National Oceanic and Atmospheric Administration, National Marine Fisheries Service, Washington, D.C., 1990.
2. J. K. Babbitt, "Blueing Discoloration of Dungeness Crabmeat," in R. E. Martin, and co-workers, eds., *Chemistry and Biochemistry of Marine Food Products,* AVI Publishing Co., Inc., Westport, Conn., 1982, p. 423.
3. E. B. Dewberry, "Speed Is the Essence," *Food Processing Industry* **39,** 50–53 (1970).
4. J. A. Dassow, "Preparation for Freezing and Freezing of Shellfish," in D. K. Tressler and C. F. Evers, eds., *The Freezing Preservation of Foods,* 4th ed., Vol. 3, AVI Publishing Co., Inc., Westport, Conn., 1968, p. 266.
5. P. Redmayne, "Crab: King, Snow, and Dungeness," *Seafood Leader* **2**(4), 72 (1982).
6. I. Osakabe, "Low Cook Produces High Quality in Kegani Crab," *Pacific Fisherman* **55**(12), 48 (1957).
7. J. K. Babbitt, D. K. Law, and D. L. Crawford, "Effect of Precooking on Copper Content, Phenolic Content and Blueing of Canned Dungeness Crabmeat," *Journal of Food Science* **40,** 649 (1975).
8. National Food Processors Association, *Processes for Low-Acid Canned Foods in Metal Containers,* Bulletin **26-L,** NFPA, Washington, D.C., 1976.
9. C. Crapo, B. Paust, and J. K. Babbitt, *Recoveries and Yields from Pacific Fish and Shellfish,* Marine Advisory Bulletin No. **37,** Alaska Sea Grant College Program, Fairbanks, 1988.
10. R. A. Krzeczkowski, R. D. Tenney, and C. Kelley, "Alaska King Crab: Fatty Acid Composition, Carotenoid Index and Proximate Analysis," *Journal of Food Science* **36,** 604 (1971).
11. R. A. Krzeczkowski, and F. E. Stone, "Amino Acid, Fatty Acid and Proximate Composition of Snow Crab (*Chionoecetes bairdi*)," *Journal of Food Science* **39,** 386 (1974).
12. D. D. Boon, "Discoloration in Processed Crabmeat. A Review," *Journal of Food Science* **40,** 756 (1975).
13. J. K. Babbitt, *Improving the Quality of Commercially Processed Dungeness Crab,* Oregon State University Extension Marine Advisory Program **SG65,** Corvallis, 1981.
14. H. J. Barnett, R. W. Nelson, and P. J. Hunter, "Shipping Live Dungeness Crabs by Air to Retail Market," *Commercial Fisheries Review* **31**(5), 21 (1969).
15. B. Engesser, *Crab Meat Extraction,* a training film, College of Engineering, Oregon State University, Corvallis, 1972.
16. T. M. Miller, N. B. Webb, and F. B. Thomas, *Technical Operations Manual for the Blue Crab Industry,* University of North Carolina Sea Grant Publication **UNC-SG-74-12,** Special Scientific Report No. **28,** Raleigh, 1974.

JERRY K. BABBITT
USDC/NMFS
Kodiak, Alaska

CRACKERS. See BISCUITS AND CRACKERS.
CRUSTACEANS. See SHELLFISH.
CRYOGENICS. See FOOD FREEZING AND FREEZING SYSTEMS.

CRYSTALLIZATION. See CORN AND CORN PRODUCTS; EVAPORATORS: TECHNOLOGY AND ENGINEERING; HEAT EXCHANGERS: FOULING; HONEY ANALYSIS; HYDROGENATION; OLIVES; OILSEEDS AND THEIR OILS; PALM OIL; SOYBEANS AND SOYBEANS PROCESSING; SYRUPS; and entries under FATS AND OILS.

CULTURAL NUTRITION

A LITERAL MELTING POT

Most of us can easily answer the question: What are your favorite foods? Generally, we respond with tastes developed during childhood. Basic food habits usually prove resistant to change.

Despite their food habits, Americans have a wide variety of food tastes. This variety is due partly to the significant mobility of Americans, partly to a wide variety of choices, and partly to such trends as eating at restaurants and eating ethnic. It is considered stylish to sample the cuisines of other countries.

At home, though, we usually prefer less variation. Our food patterns tend to follow family history—and especially what our parents liked and disliked. Where an ethnic community is established, food habits can become readily apparent. Especially in major cities, it is not uncommon to find neighborhood grocery stores that have shelves of food catering to ethnic tastes of a significant segment of the surrounding community.

Personal food patterns become especially important when we are old and when we are ill. Success in encouraging such patients to eat often depends on including their special preferences in their diets. This article addresses food patterns of people in the United States of different regions, ethnic origins, cultural backgrounds, religious backgrounds, and life-styles.

Regional Differences

For many years, obvious regional food patterns could be found in the United States. Those pronounced differences have faded considerably as techniques for processing, storing, and transporting foods have advanced. National advertising has had its effects, as has the mobility of our population.

In certain parts of the country, though, remnants of regional food preference persist. Such preferences often reflect availability that has spawned popularity. For example, seafood enjoys popularity in coastal areas because of ready availability, lower cost that reflects lower transportation expense, and local industry advertising. Preference is thus influenced by climate, geography, and economics.

In California, the state's agricultural richness affects preference, as do ethnic demographics—characteristics of human populations. California's large quantities of fruits and vegetables, the popularization of the salad as a part of the meal, and the large numbers of Mexican-Americans, Orientals, and Italians all combine to give California its distinct cuisine.

In the southern states, corn, fish, and rice are often used. Hot breads, such as biscuits or corn bread, accompany meals. Green leafy vegetables are cooked with a piece of fatback (pork back fat), a regional practice that yields a consumable liquid known as pot liquor. Before mechanical refrigeration was developed, milk and cheese were consumed only in small quantities; storage and shipping were not possible. The national consumption of dairy products has shown an increase since the 1940s and 1950s.

ETHNIC, CULTURAL, AND RELIGIOUS INFLUENCES

Each ethnic, cultural, and religious group has certain characteristic food patterns or preferences. Sometimes these characteristics remain unique to the group, but more often they become a part of our nation. In that regard the United States can be described as a melting pot, both figuratively and literally. The remaining part of this article explores such influences.

EUROPEANS

Many of the foods we have come to think of as typically American were in fact brought to us from European countries. Norwegian, Swedish, and Danish immigrants brought a greater use of milk, numerous cheeses, cream, and butter. In their native countries, Western Europeans relied on fish, shellfish, and vegetables. Especially popular were potatoes, dark breads, eggs, cheese, beef, pork, and poultry.

Central Europeans favored potatoes, rye flour, wheat, pork, sausage, and cabbage. Cabbage was especially favored and was eaten raw, cooked, or salted as sauerkraut. Common vegetables included turnips, carrots, squash, onion, beans, and greens.

Italian in the United States have perpetuated many favorite foods of their ancestors and have concurrently made them basics of the American diet. The huge sale of pizza and pasta products and the number of Italian restaurants confirm this fact. There are nearly as many pizza stands in the United States as there are hamburger and fried chicken businesses. Daily, the typical Italian diet features pasta made from hard-wheat dough. Its innumerable forms are seasoned with sauces, onions, tomatoes, cheese, peppers, and meat. Bread usually accompanies the meal.

Differences exist between northern and southern Italian foods. The northern area of Italy is more industrialized, and meat, root vegetables, and dairy products are more popular there. Southern Italy's preference tends toward fish, spices, and olive oil. The southern Italian diet can benefit from more green vegetables, eggs, fruits, meat, and milk. The last two food items can especially benefit the nutrient intake of children.

Experienced dietitians and nurses know that hospitalization can be especially difficult for Italians who have a strong sense of family and a strong preference for traditional foods. Institutional diets that cater to Italian patients can both reduce feelings of isolation and increase nutritional intake.

NATIVE AMERICANS

It may come as a surprise that more than half the plant foods you eat—including corn, potatoes, squash, pumpkin, tomatoes, peppers, beans wild rice, cranberries, and cocoa—come from North, Central, and South American Indians. The Native American diet also used acorns, wild fruit, fish, wild fowl, small and large game, and seafood. The food preservation methods that produce jerky and pemmican were introduced to the frontier Americans by the Native Americans. Corn, which has become a world staple, is a significant example of their culture's contribution.

Diet varied from tribe to tribe depending on geographic location. Historically, the natural resources of the tribe's home area determined such occupations as hunting, fishing, agriculture, and herding. Native Americans who lived near the coast used large amounts of seafood. Those in the Northwest were and still are great fishermen, and from Alaska to the Columbia River one of the most prized items is salmon. A favorite method of preserving fish and meat is smoking; racks of drying fish can be seen along the river banks where Native Americans live. Along the New England coast, shellfish is a favorite, especially clams. A favored method of cooking is open-air baking of clams and fish over hot coals. Sometimes seafood is baked in the sand. The item is wrapped in seaweed or similar materials and placed in sand kept hot by a fire.

In the past, the Plains Indians depended more on game, fowl, and freshwater fish. Native vegetables and fruits were generously used and cultivated if the group was a stationary, agricultural one. Many vegetables that originated in North and South America have become part of the American and world diet. Because little food storage was possible, nutritional life was characterized by the extremes of bounty and scarcity. Despite such fluctuations in availability, the historical Native American diet was undoubtedly more nutritional than is their diet today.

With the introduction of trading-post products such as lard, sugar, coffee, and canned meat and canned milk, the American Indian diet began to assimilate European food customs, leaving little today that can be distinguished as a Native American food pattern.

Native American food specialties that can be found today include clambakes, fry bread (biscuit dough fried in fat), and wajupi, a pudding made from berries, sugars, and cornstarch. Northern California Indians have revived the use of acorns in preparing soup and flour, methods of preparing eels, and stews combining meat, vegetables, and berries. Often these traditional foods play a part in tribal ceremonies. As cultural renewal has taken place among Native Americans, and as the Native American heritage is becoming better understood, interest in traditional Native American foods is being revived.

The nutritional weakness of the Native American diet is a low intake of milk by children after weaning. Vitamin C intake is also low, and the greater use of fruits, vegetables, and lean meat will improve the diet.

The present generation of Native Americans may be consuming too much sugar and soft drinks. Sound nutritional education is needed. Extensive alcohol abuse among some adult males has also hurt the nutritional status of these people.

BLACK AMERICANS

Among middle- and upper-income blacks, food selections are similar to those of their white counterparts. Among poor families, many of whom have come from or still live in the southern states, the traditional eating pattern is determined to a large extent by the cost and availability of foods. However, there are many food items preferred by black families, rich or poor.

The popularization of the term soul food indicates the effort to recognize the uniqueness of black American food habits. One characteristic black food habit is the use of hominy grits, especially for breakfast with some form of pork. Corn bread, muffins, or biscuits are preferred to yeast bread. There are differences between the dietary habits of southern blacks and northern blacks, and one of these differences is in the customary use of greens.

Among southern blacks, fresh greens such as mustard, turnip, and collard are popular and are usually served with pork such as fatback or salt pork. In the north, fresh greens are less available, and a much wider variety of frozen greens are used. Other vegetable favorites include fresh corn, lima beans, cabbage, sweet potatoes, and squash. The use of fresh fruit emphasizes oranges, watermelon, and peaches (Table 1).

Black-eyed peas and corn are well liked, as indicated earlier; a mixture of black-eyed peas and rice called Hoppin John is a favorite southern dish. A favorite method of preparing green vegetables is to boil them with a small amount of fat pork; the resultant juice (pot liquor) is also used in the meal. Potatoes (white and sweet) and yams are popular, and sweet potatoes are often made into pies.

Carbohydrate is generally obtained from grits, potatoes, and rice, though black-eyed peas and dried beans are also used. The last-named provide some protein, as does fried fish, especially when locally caught. Other meats frequently used include poultry, pork (both cured and fresh), and wild game. All parts of an animal are used, especially among poor families. Dishes made from a slaughtered pig, for instance, include chitlings (the small intestines, cleaned and fried crisp); hog maws (the stomach lining); boiled pig's feet; and neckbone, jaws, snout, and head, which are cooked and made into scrapple. Frying, stewing, and barbecuing are the favored methods of preparation. Sweets such as molasses, pastries, cakes, and candy are especially popular.

Milk consumption, which is often low, should be encouraged, especially among children. Buttermilk and ice cream are well liked by black Americans, although there is little use of cheeses. The use of more citrus fruits should be encouraged.

As the economic status of black people improves, they eat more meat and their diet improves generally. Because their typical diet is high in fat and carbohydrate and low in protein, iron, and vitamin C, an increase in consumption of lean meat, milk, and citrus fruits will help balance their nutrient intake.

Table 1. Characteristic Black Americans Food Choices[a]

Protein Foods	Milk and Milk Products	Grain Products	Vegetables	Fruits	Other
Meat	Milk	Rice	Broccoli	Apples	Salt pork
Beef	Fluid	Corn bread	Cabbage	Bananas	Carbonated
Pork and ham	Evaporated,	Hominy grits	Carrots	Grapefruit	beverages
Sausage	in coffee	Biscuits	Corn	Grapes	Fruit drinks
Pig's feet, ears, etc	Buttermilk	Muffins	Green beans	Nectarines	Gravies
Bacon	Cheese	White bread	Greens	Oranges	
Luncheon meat	Cheddar	Dry cereal	Mustard	Plums	
Organ meats	Cottage	Cooked cereal	Collard	Tangerines	
Poultry	Ice cream	Macaroni	Kale	Watermelons	
Chicken		Spaghetti	Spinach		
Turkey		Crackers	Turnips, etc		
Fish			Lima beans		
Catfish			Okra		
Perch			Peas		
Red snapper			Potatoes		
Tuna			Pumpkins		
Salmon			Sweet potatoes		
Sardines			Tomatoes		
Shrimp			Yams		
Eggs					
Legumes					
Kidney beans					
Red beans					
Pinto beans					
Black-eyed peas					
Nuts					
Peanuts					
Peanut butter					

[a] California Department of Health, *Nutrition During Pregnancy and Health*, revised, 1975.

It is in preparation that the essence of the term soul food can be found. It is not necessarily the unique foods but rather the care in preparation that is emphasized. Food preparation offers the opportunity to minister to the physical health and well-being of those who will consume the food. The opportunity to bring happiness, health, and love through food preparation lies at the figurative and literal heart of the customary black dietary habits.

JEWS

Orthodox Jews follow Old Testament and rabbinical dietary laws. Conservative Jews distinguish between meals served in the home and those outside it. Reformed Jews do not observe dietary regulations. The country of origin can further influence a Jewish family's dietary practices.

Foods used according to strict Jewish laws are referred to as Kosher. Those who follow those dietary laws eat only animals that are designated as clean and are killed in a ritualistic manner. The method of slaughter minimizes pain and maximizes blood drainage. Blood, the symbol of life, is strictly avoided by soaking the meat in cold water, draining it, salting it, and rinsing it three times. Permitted meats include poultry, fish with fins and scales, and quadruped animals that chew the cud and have divided hooves. (Thus, pork cannot be eaten.) The hindquarter of quadruped animals must have the hip sinew or the thigh vein removed.

Kosher laws additionally require separation of meat and milk. Milk and its products must be excluded from meals involving meat but can be consumed before the meat meal. If a milk meal is to be consumed, meat and its products are excluded from the milk meal and for 6 hours thereafter. Usually two milk meals and one meat meal are eaten each day. A Kosher household will maintain two sets of dishes, utensils, and cooking supplies, one for meat meals and one for milk meals.

Certain foods, viewed as neutral, are referred to as pareve. Fruits, uncooked vegetables, eggs, and clean fish, because they are recognized as neutral, can be consumed with either meat or milk meals.

No food can be cooked or heated on Saturday, the Sabbath, so the evening meal preceding the Sabbath tends to the most substantial of the week. Both chicken and fish are served at that meal. Any foods eaten on the Sabbath must be cooked on a preceding day.

On Jewish holidays, symbolic foods are eaten. For example, Passover, a spring festival, commemorates the flight of the Jews from Egypt. Passover celebration lasts 8 days, and only unleavened bread, called matzo, is permitted. An Orthodox household will, therefore, use separate utensils for preparing unleavened bread to avoid any contact with leavening. Other holidays include Rosh Hashanah (Jewish New Year), Sukkoth (fall harvest), Chanukkah (the festival of lights), Purim (the arrival of spring), and Yom Kippur (the day of atonement) when fasting occurs.

MEXICAN-AMERICANS

People of Mexican descent make up a large part of the population of the Southwest. They therefore reside close to their native country and maintain a close tie to traditional Mexican foods. A lack of refrigeration in such warmer climates creates a scarcity of meat, milk, eggs, vegetables, and fruits. For lower-income families, the expense of these nutrient-rich foods can be prohibitive.

Items basic to the Mexican diet include dry beans, chili peppers, tomatoes, and corn. The corn is ground, soaked in lime water, and baked on a griddle to make tortillas, the popular flat breads that can be rolled and filled with ground beef and vegetables to make tacos. Tacos provide nutrients from all four food groups, although the user of wheat tortillas reduces the calcium obtained from lime-soaked corn tortillas. Many foods are fried, and beans are even refried until they absort all the fat.

Nutritionists encourage retention of the basic Mexican diet with only minor modifications. Because of iron, vitamin A, and calcium deficiencies, use of lean meat is recommended in taco fillings, dark green and yellow vegetables, fruits, and milk—dried if fresh milk is too expensive—are also recommended. The milk is especially important for children. One Mexican dietary practice that is particularly encouraged is the practice of limiting the amount of sweets consumed (Table 2).

PUERTO RICANS

The Puerto Rican diet has similarities to the dietary patterns of other Caribbean Islands but is noteworthy because so many Puerto Ricans emigrate to the United States.

Traditional emphasis is on beans, rice, and starchy root vegetables (and plantains) known as viandas. Rice and beans eaten together are a high-quality, complementary protein combination. The beans are often boiled and then cooked with sofrito, a mixture made from tomatoes, green peppers, onions, garlic, salt pork, lard, and herbs. Viandas are good sources of B vitamins, iron calories, and, in some cases, vitamin C. Salt codfish is more popular than fresh fish. Chicken, beef, and pork are favored. Coffee with 2 to 5 oz of milk per cup—cafe con leche—is drunk several times a day, contributing to an otherwise low consumption of milk. Fruits prove an especially popular food owing to availability, with such native fruits as papaya, mango, and acerola (the West Indian cherry, which is extremely high in vitamin C) being joined by bananas, oranges, and pineapple.

The mainstays of the Puerto Rican diet may be hard to find or expensive in the United States. And because Puerto Ricans come from an agrarian background and thus often lack the job skills for a highly industrialized society, they often find themselves in the lowest socioeconomic classes in the United States, living in crowded urban conditions with poor cooking and refrigeration facilities and unable to feed their families as well as they did in Puerto Rico. Consequently, malnutrition is not uncommon among their children. Also, Puerto Rican children born in the United States may have adopted favorite mainland foods such as hamburgers, hot dogs, canned spaghetti, and cold cereals instead of the traditional Puerto Rican food prepared by their mothers. If these children have picked up the mainland habit of snacking on nutrient-light foods, they may be more poorly nourished than people adhering to the relatively inexpensive but nutritionally adequate traditional diet of Puerto Rico.

Nutritionists encourage Puerto Ricans living in northern mainland cities to become familiar with different fruits, which are inexpensive when they are in season or canned; to use more milk, cheese, and inexpensive cuts of meat; and to substitute canned tomatoes for fresh tomatoes, which are expensive when not in season.

Table 2. Characteristic Mexican-American Food Choices[a]

Protein Foods	Milk and Milk Products	Grain Products	Vegetables	Fruits	Other
Meat	Milk	Rice	Avocados	Apples	Salsa (tomato-
Beef	Fluid	Tortillas	Cabbage	Apricots	pepper-onion
Pork	Flavored	Corn	Carrots	Bananas	relish)
Lamb	Evaporated	Flour	Chilies	Guavas	Chili sauce
Tripe	Condensed	Oatmeal	Corn	Lemons	Guacamole
Sausage (chorizo)	Cheese	Dry cereals	Green beans	Mangos	Lard (manteca)
Bologna	American	Cornflakes	Lettuce	Melons	Pork cracklings
Bacon	Monterey jack	Sugar coated	Onions	Oranges	Fruit drinks
Poultry	Hoop	Noodles	Peas	Peaches	Kool-aid
Chicken	Ice cream	Spaghetti	Potatoes	Pears	Carbonated
Eggs		White bread	Prickly pear	Prickly pear	beverages
Legumes		Sweet bread	cactus leaf	cactus fruit	Beer
Pinto beans		(pan dulce)	(nopales)	(tuna)	Coffee
Pink beans			Spinach	Zapote	
Garbanzo beans			Sweet potatoes	(or sapote)	
Lentils			Tomatoes		
Nuts			Zucchini		
Peanuts					
Peanut butter					

[a] California Department of Health, *Nutrition During Pregnancy and Health,* revised, 1975.

MIDDLE EASTERN PEOPLE

Middle East people—Greeks, Iranians, Arabs, Turks, Armenians, and Lebanese—tend to be farmers. Dietary emphasis is on crops and animals raised—cattle, sheep, goats, chickens, ducks, geese, grains, fruits, and vegetables. Lamb holds the position of the favored meat, and wheat products, grains, and rice are the major energy sources. Popular dairy foods, derived from sheep, goats, and camels, are sour milk, including yogurt, fermented milk, and sour cream.

Lentils and beans may be boiled or stewed with tomatoes, onions, and olive oil; they may be eaten alone or mixed with other foods. A favorite combination is seasoned chick-peas mixed with bulgur (wheat that has been steamed, cracked, and fried) and spices and then fried in fat.

Traditional vegetables include okra, squash, tomatoes, onions, leeks, peppers, spinach, brussel sprouts, cabbage, peas, green beans, dandelion greens, eggplant, artichokes, and olives. Grape leaves, used either fresh or canned, are stuffed with rice, bulgur, meat, and seasoning.

Fruits grown in these warm climates are used extensively. Some common ones are dates, figs, melons, cherries, oranges, apricots, and raisins.

Common Middle Eastern cereals include rice, wheat, and barley. Typical breads are flat, thin, and round and may be baked outdoors. Olive oil and butter from sheep's and goat's milk are used generously. Turkish coffee, a favorite beverage in the Middle East, is a strong, dark drink containing the crushed coffee bean that is served with a generous amount of sugar. The Middle Eastern diet is a good one, contributing adequate nutrients. But Middle Eastern people in the United States may have difficulty obtaining the foods if they cannot afford them, as they are usually expensive.

ASIANS

The United States is also home to many people of Asian heritage. The diets of the Chinese, Japanese, and Filipino are discussed below.

Chinese

China is a vast country with many different regional foods. In the United States, most of the Chinese on the West Coast come from southern China and adhere to a Cantonese diet. This type of cookery uses little fat and subtle seasonings. Beef, pork, poultry, and all kinds of

Table 3. Characteristic Chinese Food Choices[a]

Protein Foods	Milk and Milk Products	Grain Products	Vegetables	Fruits	Other
Meat	Flavored milk	Rice	Bamboo shoots	Apples	Soy sauce
Pork	Milk (cooking)	Noodles	Beans	Bananas	Sweet and
Beef	Ice cream	White bread	Green	Figs	sour sauce
Organ meats		Barley	Yellow	Grapes	Mustard sauce
Poultry		Millet	Bean sprouts	Kumquats	Ginger
Chicken			Bok choy	Loquats	Plum sauce
Duck			Broccoli	Mangos	Red bean paste
Fish			Cabbage	Melons	Tea
White fish			Carrots	Oranges	Coffee
Shrimp			Celery	Peaches	
Lobster			Chinese cabbage	Pears	
Oyster			Corn	Persimmons	
Sardines			Cucumbers	Pineapples	
Eggs			Eggplant	Plums	
Legumes			Greens	Tangerines	
Soybeans			Collard		
Soybean curd (tofu)			Chinese broccoli		
Black beans			Mustard		
Nuts			Kale		
Peanuts			Spinach		
Almonds			Leeks		
Cashews			Lettuce		
			Mushrooms		
			Peppers		
			Potatoes		
			Scallions		
			Snow peas		
			Sweet potatoes		
			Taro		
			Tomatoes		
			Water chestnuts		
			White radishes		
			White turnips		
			Winter melons		

[a] California Department of Health, *Nutrition During Pregnancy and Health*, revised, 1975.

seafoods are well liked. In Chinese cooking, all parts of the animal are utilized, including organs, blood, and skin. Rice is the predominant cereal used by the Cantonese, both at home and in the United States. Northern Chinese use more bread, noodles, and dumplings, which are prepared from wheat, corn, and millet (Table 3).

The Chinese diet is varied, containing eggs, fish, meat, soybeans, and a great variety of vegetables. Many green, leafy vegetables unfamiliar to Americans, such as leaves of the radish and shephard's purse, are enjoyed. Sprouts of bamboo and beans are incorporated into some dishes, giving a distinctive flavor and texture. Many Chinese recipes call for mushrooms and nuts. Eggs from ducks, hens, and pigeons are widely used—fresh, preserved, and pickled. Soy sauce is used both in preparation and serving. The high salt content of this flavorable condiment is a problem for Chinese patients whose salt intake may be restricted by the physician.

Chinese cookery is unique in many ways. Food is quickly cooked over the heat source, usually cut into small pieces so that the short cooking period is adequate. Vegetables retain their crispness, flavor, and practically all their nutrients; this method of cookery, which has been widely adopted, is quite beneficial. When vegetables are cooked in water, the liquid is also consumed.

Probably the greatest weakness, in the Chinese diet is a low intake of milk and milk products, a consequence of lactose intolerance. However, a higher consumption of meat and the frequent use of soybeans prevent calcium and protein deficiency. In addition, for undefined reasons, Chinese children born in the United States have a higher tolerance for milk.

Japanese

Japanese people who have emigrated to the United States, especially the older people, tend to follow the food habits of their homeland rather closely. Typical of the Japanese food pattern is much use of fish, both fresh and saltwater. Methods of preparation vary. Fish may be eaten raw or deep-fat fried, dried, or salted. Many kinds of vegetables may be prepared with meat or eaten separately. Protein intake comes from meat, fish, eggs, legumes, nuts, and, most popular of all, soybean curd (tofu), which is sold in many American grocery stores. Eggs are prepared in many ways and are well liked.

The basic cereal is rice, although since World War II wheat has been consumed. Milk and cheese are limited in the traditional Japanese diet, mainly because of lactose intolerance. Japanese born in the United States have a greater tolerance for dairy products such as milk and cheese, and they also eat more fruits. It is important that

Table 4. Characteristic Japanese Food Choices[a]

Protein Foods	Milk and Milk Products	Grain Products	Vegetables	Fruits	Other
Meat	Milk	Rice	Bamboo shoots	Apples	Soy sauce
Beef	Ice cream	Rice crackers	Bok choy	Apricots	Nori paste
Pork	Cheese	Noodles (whole	Broccoli	Bananas	used to
Poultry		wheat noodle	Burdock root	Cherries	season rice)
Chicken		called soba)	Cabbage	Grapefruits	Bean thread
Turkey		Spaghetti	Carrots	Grapes	(konyaku)
Fish		White bread	Cauliflower	Lemons	Ginger (shoga;
Tuna		Oatmeal	Celery	Limes	dried form
Mackerel		Dry cereals[b]	Cucumbers	Melons	called denishoga)
Sardines (dried			Eggplants	Oranges	Tea
form called			Green beans	Peaches	Coffee
mezashi)			Gourd (kampyo)	Pears	
Sea bass			Mushrooms	Persimmons	
Shrimp			Mustard greens	Pineapples	
Abalone			Napa cabbage	Pomegranates	
Squid			Peas	Plums (dried	
Octopus			Peppers	pickled plums	
Eggs			Radishes (white	called umeboshi)	
Legumes			radish called	Strawberries	
Soybean curd			daikon; pickled	Tangerines	
(tofu)			white radish		
Soybean paste			called takawan)		
(miso)			Snow peas		
Soybeans			Spinach		
Red beans			Squash		
(azuki)			Sweet potatoes		
Lima beans			Taro (Japanese)		
Nuts			sweet potato)		
Chestnuts			Tomatoes		
(kuri)			Turnips		
			Water chestnuts		
			Yams		

[a] California Department of Health, *Nutrition During Pregnancy and Health,* revised, 1975.
[b] Nisei only.

the Japanese in this country eat enriched, converted, or whole-grain rice, which contains more nutrients than unenriched rice with the husks removed. To avoid excessive nutrients loss, they should also be told to refrain from washing the rice repeatedly before preparing it (Table 4).

Filipinos

Filipino food patterns are similar to those of the Chinese and Japanese. The basic cereal is rice, and the principal sources of protein are fish, meat, eggs, legumes, and nuts. Meat is prepared by roasting, frying, or boiling; the fish used may be dried.

A large variety of vegetables and fruits is found in the diet. Vegetables are usually boiled or pan-fried, and some are used as salad dishes. Most fruits are eaten raw, but they are sometimes used in cooking (Table 5).

Again, because of lactose intolerance, Filipinos have a low intake of milk and milk products. The consumption of good calcium and protein sources should be encouraged, especially among the young. Growing children may also have a problem with low caloric intake. Larger servings of food should be encouraged, especially more meat; food preparers should also be instructed about the use of enriched rice without prewashing.

THE ESKIMOS: MEETING OF TWO CULTURES

For years, the Eskimos of the far north have maintained their culture on a diet consisting almost entirely of meat and fish. This culture has fascinated nutritionists, as it runs counter to all that is known about the necessity of a balanced diet.

Caribou, whale, and seals from the staples of the traditional diet, with walrus, fish, and birds being of minor importance. Edible plants and berries are scarce. Although it had been believed that Eskimos consumed the stomach contents of the animals, that assumption is debatable. There is agreement the Eskimo diet is very high in protein and fat and very low in carbohydrate.

The high protein content provides calories essential to an active life and additionally contributes to blood glucose, an essential metabolic fuel for the nervous system. What carbohydrate has been available has been obtained through synthesis of glycogen from muscles. A by-product of the traditional diet has been the virtual nonexistence of diabetes. This disease has, however, been on the increase since the introduction of sugar to the culture.

A unique characteristic of Eskimo nutrition is the utilization of fatty acids from oils as energy. Tissues can adapt to the utilization of ketone bodies instead of glucose as energy source. The ketone bodies form in the liver and are

Table 5. Characteristic Filipino Food Choices[a]

Protein Foods	Milk and Milk Products	Grain Products	Vegetables	Fruits	Other
Meat	Milk	Rice	Bamboo shoots	Apples	Soy sauce
Pork	Flavored	Cooked cereals	Beets	Bananas	Coffee
Beef	Evaporated	Farina	Cabbage	Grapes	Tea
Goat	Cheese	Oatmeal	Carrots	Guavas	
Deer	Gouda	Dry cereals	Cauliflower	Lemons	
Rabbit	Cheddar	Pastas	Celery	Limes	
Variety meats		Rice noodles	Chinese celery	Mangos	
Poultry		Wheat noodles	Eggplants	Melons	
Chicken		Macaroni	Endive	Oranges	
Fish		Spaghetti	Green beans	Papayas	
Sole			Leeks	Pears	
Bonito			Lettuce	Pineapples	
Herring			Mushrooms	Plums	
Tuna			Okra	Pomegranates	
Mackerel			Onions	Rhubarb	
Crab			Peppers	Strawberries	
Mussels			Potatoes	Tangerines	
Shrimp			Pumpkins		
Squid			Radishes		
Eggs			Snow peas		
Legumes			Spinach		
Black beans			Sweet potatoes		
Chick peas			Tomatoes		
Black-eyed peas			Water chestnuts		
Lentils			Watercress		
Mung beans			Yams		
Lima beans					
White kidney beans					
Nuts					
Peanuts					
Pili nuts					

[a] California Department of Health, *Nutrition During Pregnancy and Health*, revised, 1975.

characteristic of high-fat, low-carbohydrate diets. Eskimos thus provide a human example of a unique metabolic phenomenon.

As we saw in the discussion of vitamins, we should expect scurvy among Eskimos. But their culture has escaped scurvy because they eat fish and meat either raw or only slightly cooked. What little vitamin C is available is not destroyed by oxidation in the cooking process.

Vitamins A, D, E, and K are obtained in sufficient quantities from the traditional Eskimo diet, as are B vitamins. The diet is low in calcium, a problem that surfaces in bone fragility among Eskimos. Children are somewhat protected by a traditionally long nursing period.

As nontraditional dietary habits have gained popularity in the Eskimo culture, our problems have become theirs. Heart disease, diabetes, tooth decay, and hypertension are now on the increase.

PEOPLE AND THEIR LAND

As you have probably determined from these discussions of culture and nutrition, our food habits are very much influenced by our geographic location. Relationships between people and their land can be profound, and they have given rise to fields of knowledge ranging from cultural geography to literary analysis focusing on agrarian mythology. Although it has been said that we are what we eat, perhaps we should also say that we eat where we are.

BIBLIOGRAPHY

This article has been adapted from Y. H. Hui, *Principles and Issues in Nutrition*, Jones and Bartlett Publishers, Inc., Boston, 1985. Used with permission.

General References

L. Bakery, ed., *The Psychobiology of Human Food Selection,* Van Nostrand Reinhold, New York, 1982.

L. K. Brown, and K. Mussell, eds., *Ethnic and Regional Foodways in the United States. The Performance of Group Identity,* University of Tennessee Press, Knoxville,

C. A. Bryant, and co-workers, 1985, *The Cultural Feast: An Introduction to Food and Society,* West Publishing Co., St. Paul, Minn., 1985.

P. Farb, and G. Armelagos, *Consuming Passions: The Anthropology of Eating,* Houghton Mifflin, Boston, 1980.

P. Fieldhouse, *Food and Nutrition: Customs and Culture,* Methuen, New York, 1985.

Food and Nutrition Board, *What is America Eating?* National Academy Press, Washington, D.C., 1986.

Food and Nutrition Board, National Research Council, National Academy of Sciences, *Recommended Dietary Allowances,* 10th ed., National Academy Press, Washington, D.C., 1989.

N. Jerome, and co-eds., *Nutritional Anthropology,* Redgrave Publishing, Pleasantville, N.Y., 1980.

F. E. Johnston, ed., *Nutritional Anthropology,* Alan R. Liss, New York, 1987.

P. G. Kittler, and K. Sucher, *Food and Culture in America,* Van Nostrand Reinhold, New York, 1989.

A. W. Logue, *The Psychology of Eating and Drinking,* W. H. Freeman, New York, 1986.

J. M. Newman, *Melting Pot: An Annotated Bibliography and Guide to Food and Nutrition Information for Ethnic Groups in America,* Garland, New York, 1986.

W. Root, and R. DeRochement, *Eating in America: A History,* Eco Press, New York, 1976.

D. Sanjur, *Social and Cultural Perspectives in Nutrition,* Prentice-Hall, Englewood Cliffs, N.J., 1982.

Y. H. HUI
Editor-in-chief

CULTURED MILK PRODUCTS

Cultured milk products are produced by the lactic fermentation of milk using various bacterial cultures. Some products may also have other fermentations taking place, eg, alcohol. These fermentations lead to the coagulation of milk and the production of typical cultured milk product flavor.

Fermented milk products originated in the Near East and then spread to parts of Southern and Eastern Europe. The earliest forms of fermented products were developed by accident by nomadic tribes who carried milk from cows, sheep, camels, or goats. Under warm storage conditions, milk coagulated or clabbered due to the production of acid end products by lactic bacteria. Fortunately the predominant bacteria were lactic types and, therefore, helped to preserve the product by suppressing spoilage and pathogenic bacteria. Man evidently enjoyed the refreshing tart taste of his discovery and began to handle milk so that this preserving action would be encouraged.

Milk and curdled milk products are mentioned throughout the Old Testament dating back as far as 4000 BC: "He then brought some curds and milk" (Gen. 18 : 8); "I'm thirsty, he said. Please give me some water. She opened a skin of milk, and gave him a drink" (Judg. 4 : 19); "He asked for water, and she gave him milk; in a bowl fit for nobles she brought him curdled milk." (Judg. 5 : 25); "He will eat curds and honey" (Isa. 7 : 15). There is also remarkable pictorial evidence that the custom of keeping milk in containers for later consumption was already a craft systematically practiced by the Sumerians around 2900 BC (1). Through applied scientific principles and advances in manufacturing technology, these early products have developed into a highly diversified group of foods that are popular throughout the world.

Actual figures for world consumption of fermented milk products are not known. It's not known how much of the estimated 1 trillion lb of cows milk produced per year is manufactured into cultured milk products. The International Dairy Federation (IDF) has been collecting data on consumption of fermented milk products since 1966. However, the IDF is composed of only 34 member countries and of those 34 only 28 are supplying data to the IDF on fermented milk consumption. Bulgaria, the entire Middle East, and the Near East are not included in these figures, all of which consume large amounts of fermented products. Table 1 shows consumption figures as recorded by the IDF. The Northern European countries consume the greatest amount of fermented milk products on a per capita basis with Finland leading the way consuming 39.3

Table 1. Total Annual Per Capita Consumption (in Kilograms of Fermented Milks) in Countries for Which Data Are Available[a]

Country	Total Consumption			1985 Consumption		
	1975	1980	1985	Yogurt	Buttermilk	Other
Finland	35.4[c]	41.0	39.3	9.4	1.9	28.0
Sweden	19.9	24.0	27.3	5.4	0.03	21.9
The Netherlands	24.7	27.3	26.6	18.1	8.5	—
Denmark	13.0[c]	26.7	23.9	8.0	8.4	7.5
India		3.7[d]	22.8	4.0	18.8	—
Iceland	1.7[d]	5.7[d]	20.3	6.2	14.1	—
Switzerland	10.9	14.8	17.4	16.2	1.2	—
Israel	14.1	14.3	15.9	6.8[b]	—	9.1
Norway	14.6	15.1	14.0	3.1	—	10.9
France	7.8	9.3	12.7	12.7[e]	—	—
FRG	6.1[c]	10.1	11.1	7.9	2.1	1.1
Austria	7.3[c]	9.8	11.0	6.6	2.2	2.2
Czechoslovakia	3.0[c]	7.3	9.5	2.5	3.9	3.1
Belgium	5.1[c]	7.7	8.4	6.0	2.3	0.06
Japan	2.5	2.4	7.9	2.9	—	5.0
USSR	7.2	6.2	7.5	—	—	7.5
Spain	—	—	5.5	5.5	—	—
United States	0.9[c]	3.1	3.8	1.8	2.0	—
Ireland	—	—	3.4	3.4[e]	—	—
UK	1.6	2.8	3.1	3.1	—	—
Canada	0.7[c]	2.3	3.1	2.5	0.6	—
Italy	—	—	2.9	1.6	—	1.3
Australia	1.0	1.8	2.8	2.8	—	—
Hungary	—	—	2.5	1.1	0.1	1.3
Chile	—	1.4	2.5	2.5	—	—

[a] Ref. 1.

[b] Includes jellified milks.

[c] Buttermilk not included (2–4 kg for Denmark; about 2 kg for United States).

[d] Yogurt only.

[e] Includes products other than yogurt.

kg/yr. India (22.8 kg/yr) and Israel (15.9 kg/yr) are also leading consumers. Compared to others, the United States (3.8 kg/yr) consumes very little on a per capita basis. Although the United States is not a major consumer of fermented milk products, its consumption figures are rapidly increasing, leading to advances in the United States fermented product manufacturing industry.

Throughout the world there are great differences in cultured products. These differences are due to variations in the cultures used (Table 2) and manufacturing principles. Products are most often classified as traditional or nontraditional; however, products can be classified according to culture medium (milk and cream), manufacturing procedure, further processing (packaging and addition of fruits, vegetables, meat, fish, or grains), end use (baking and consumption), and microbial action (type of bacteria or yeast and temperature). Traditional fermented milk products have a long history and are known and made all over the world. Their manufacture is crude and relies on ill-defined procedures and nonstandard cultures that lead to inconsistencies in product characteristics (1). The production of nontraditional products is based on sound scientific principles and leads to the production of products with consistent characteristics. Cultures and manufacturing methods have been standardized to produce the highest quality possible. Despite differences, most modern, or nontraditional, cultured products use the following basic manufacturing steps: (1) culture or starter preparation;

Table 2. Relationship between Type of Lactic Culture and Geographical Area[a]

Type	Culture(s) Used	Region
I	*Streptococcus* and *Leuconostoc* sp.	Norway, Sweden, Finland, Iceland
II	*Lactobacillus* sp.	Bulgaria, Japan
III	*Streptococcus* and *Lactobacillus* sp.	Egypt, Iraq, Lebanon, Syria, Turkey, India
IV	*Streptococcus* and *Lactobacillus* sp. and yeasts	USSR, Lebanon, Finland

[a] Ref. 2.

Table 3. Fermented Foods and Geographical Location of Production[a]

Name	Type	Location	Bacteria
Acidophilus	Dairy	Europe, North America	*Lactobacillus acidophilus Bifidobacterium bifidum*
Bulgarian buttermilk	Dairy	Europe	*Lactobacillus bulgaricus*
Buttermilk	Dairy	North America, Europe, Middle East, North Africa, Indian Subcontinent, Oceania	*Streptococcus cremoris, S. lacticus, S. lacticus* ssp. *diacetylactis, Leuconostoc cremoris*
Filmjolk	Dairy	Europe	*S. cremoris, S. lacticus, S. lacticus* ssp. *diacetylactis, L. cremoris, Alcaligenes viscosus, Geotrichum candidum*
Flummery	Cereal and dairy	Europe, South Africa	Naturally present lactic bacteria
Ghee	Dairy and miscellaneous	Indian Subcontinent, Middle East, South Africa, Southeast Asia	*Streptococcus, Lactobacillus,* and *Leuconostoc* sp.
Junket	Dairy	Europe	*Streptococcus* and *Lactobacillus* sp.
Kefir	Dairy	Middle East, Europe, North Africa	*Streptococcus, Lactobacillus,* and *Leuconostoc* sp.; *Candida kefyr, Kluyveromyces fragilis*
Kishk	Dairy and cereal	North Africa, Middle East, Europe, Indian Subcontinent, East Asia	*Streptococcus, Lactobacillus,* and *Leuconostoc* sp.
Kolatchen	Cereal and dairy	Middle East, Europe	*S. lacticus, S. lacticus* ssp. *diacetylactis, S. cremoris Saccharomyces cerevisiae*
Koumiss	Dairy	Europe, Middle East, East Asia	*S. lacticus, L. bulgaricus, Candida kefyr, Torulopsis*
Kurut	Dairy	North Africa, Middle East, Indian Subcontinent, East Asia	*Lactobacillus* and *Streptococcus* sp., *Saccharomyces lactis, Penicillium*
Lassi	Dairy	Indian Subcontinent, East Asia, Middle East, North Africa, South Africa, Europe	*S. thermophilus, L. bulgaricus,* sometimes yeast
Prokllada	Dairy	Europe	*Streptococcus* and *Lactobacillus* sp.
Sour cream	Dairy	Europe, North America, Indian Subcontinent, Middle East	*S. cremoris, S. lacticus, S. lacticus* ssp. *diacetylactis*
Yakult	Dairy	East Asia	*Lactobacillus casei*
Yogurt	Dairy	Worldwide	*S. thermophilus, L. bulgaricus*

[a] Ref. 3.

(2) treatment of product, such as pasteurization, separation, and homogenization; (3) inoculation with bacterial culture; (4) incubation; (5) agitation; (6) cooling; and (7) packaging. Table 3 lists the world's principal cultured milk products including type, location, and bacterial culture used.

MICROORGANISMS

A culture (starter) is a controlled bacterial population that is added to milk or milk products to produce acid and flavorful substances that characterize cultured milk products. Cultures may be a single strain of bacteria or a combination of cultures (mixed strain). Mixed strain cultures may be used to enhance the production of specific flavors or characteristics and, therefore, must be compatible and balanced. Some cultures will be antagonistic to each other while others will act in a symbiotic relationship. Table 4 shows characteristics of bacteria that are commonly used in the manufacture of cultured milk products.

The proper production of lactic acid is critical in fermented product manufacture. Lactic acid is not only responsible for the refreshing tart flavor of cultured products but is responsible for the destabilization of the milk protein structure (casein micelle) that allows the milk protein to coagulate, thus contributing to the products' body and texture characteristics. The culture may also contribute other flavorful compounds (Table 5). The major flavor compounds, other than lactic acid, encountered in fermented milk products are acetaldehyde, diacetyl, ethanol, carbon dioxide, and acetic acid. These flavor compounds are responsible for the unique flavor characteristics of various products, and, therefore, great care is taken to promote their production in certain products. For example, proper yogurt flavor depends on the production of acetaldehyde while proper cultured buttermilk flavor depends on the production of diacetyl. Products such as kefir and koumiss also undergo alcoholic fermentation with final levels of ethanol reaching 1–2.5%.

NUTRITION

The primary nutritional benefits of cultured products are due to their compositional makeup. All milk products are

Table 4. Bacteria Involved in Fermented Milk Foods, Their Temperature Requirements and Acid Production as Titratable Acidity (TA)[a]

Bacteria	Standard Temperature for Incubation, °C	General Maximum TA Produced in Milk, %
Leu. citrovorum	20	0.1–0.3
S. lactis	22	0.9–1.0
S. cremoris	22	0.9–1.0
S. durans	31	0.9–1.1
S. thermophilus	38–45	0.9–1.1
L. acidophilus	38–45	1.2–2.0
L. bulgaricus	43–47	2.0–4.0

[a] Ref. 4.

excellent sources of high-quality proteins (casein and whey), calcium, and vitamins (especially riboflavin and other B vitamins). Fermented milk products are also good for individuals who suffer from lactose intolerance or lactose maldigestion (the inability to digest lactose properly). People with this condition suffer digestive upsets due to an inability to properly break down the milk sugar lactose. During fermentation, lactose is broken down by lactic bacterial cultures, which makes fermented milk products more readily digestible for lactose-intolerant individuals.

Early in the twentieth century Metchnikoff proposed

Table 5. Starter Cultures and Their Principal Metabolic Products[a]

Starter Organisms	Important Metabolic Products
Mesophilic Bacteria (Optimum Growth Temperature of 20–35°C)	
Str. cremoris	Lactic acid
Str. lactis	Lactic acid
Str. lactis ssp. *diacetylactis*	Lactic acid, diacetyl
Leuc. cremoris	Lactic acid, diacetyl
Lb. acidophilus	Lactic acid
Lb. breves	CO_2, acetic acid, lactic acid
Lb. casei	Lactic acid
Thermophilic Bacteria (Optimum Growth Temperature of > 35°C)	
Str. thermophilus	Lactic acid, acetaldehyde
Lb. bulgaricus	Lactic acid, acetaldehyde
Lb. lactis	Lactic acid
Bifidobacteria	Lactic acid, acetic acid
Yeast	
Saccharomyces cerevisiae	Ethanol, CO_2
Candida (Torula) kefir	Ethanol, CO_2
Kluveromyces fragilis	Acetaldehyde, CO_2

[a] Ref. 2.

that the consumption of cultured milk products such as yogurt results in the prolongation of life. Metchnikoff based his claims on the ability of lactic bacteria to prevent putrefactive processes in the digestive tract. Although lactic cultures do produce antimicrobial effects from the production of acids, hydrogen peroxide, and antibiotics such as nisin, scientific evidence does not support Metchnikoff's claims of prolonged life due to the consumption of cultured products. However, there is some scientific evidence that the addition of alternative organisms to cultured products such as *Lactobacillus acidophilus* and *Bifidobacterium* sp. may have therapeutic effects in the lower digestive system. Many of these organisms are believed to be able to pass through the upper digestive system and then produce beneficial effects in the lower intestine. At the present time a great deal of research is being done to determine the actual nutritional and health properties of fermented milk products. This research will help to answer the question of whether cultured milk products actually do contain therapeutic properties.

BIBLIOGRAPHY

1. M. Kroger, J. A. Kurmann, and J. L. Rasic, "Fermented Milks—Past, Present, and Future," *Food Technology* **43**, 92–99 (1989).

2. F. L. Davies and B. A. Law, *Advances in the Microbiology and Biochemistry of Cheese and Fermented Milk*, Elsevier Applied Science Publishers, Ltd., Barking, UK, 1984.

3. G. Campbell-Platt, *Fermented Foods of the World*, Butterworth & Co. (Publishers) Ltd., Kent, UK, 1987.

4. F. V. Kosikowski, *Cheese and Fermented Milk Foods*, Edwards Brothers, Inc., Ann Arbor, Mich., 1977.

JOHN A. MCGREGOR
Louisiana State University
Baton Rouge, Louisiana

CULTURED SYSTEMS. See EEL; MEAT STARTER CULTURE; VEGETABLES, PICKLING AND FERMENTING.

CUTTING. See MEAT SLAUGHTERING AND PROCESSING EQUIPMENT.

D

DAIRY FLAVORS

Dairy products are very ancient foods and are one of the most important classes of foods in our diet. The basic dairy product, milk, is not a very stable food. Today's practice allows us to heat process it to keep it stable from microbiological attack for various periods of time depending on the nature of the processing conditions to which the milk is subjected. The commercial practice of heat processing (pasteurization) has only been in place for the last 60 or so years. In ancient times milk was allowed to spoil in a controlled way so that it would remain eatable.

The development of ghee (butter) from buffalo milk, an early practice of controlling the spoilage of milk, has been traced to India of perhaps about 4,000 years ago. The method of butter making was to appear in Europe at a much later date. The first commercial creamery in the United States was built by Alanson Slaughter in 1881 in Orange County, New York.

Cheese from milk also has a prehistoric origin. Cheese was an excellent way to preserve the important nutrients of milk. Because of the many ways of stabilizing milk into cheese, many forms of cheese with a variety of flavor and taste profiles have been developed. Recently, a USDA bulletin indicated over 800 varieties and the (United States Food and Drug Administration) U.S. FDA has established standards of identity for 30 different type of cheeses. Typically, cheeses are named after the immediate region where they were originally produced.

At first the manufacture of dairy products, including milk, milk products, butter, and cheese, was based on historical practice and the art of the producers. It took the great scientific revelation of Louis Pasteur working in France between 1857 and 1876 to relate the manufacturing process to the new world of the microorganism and fermentation. The use of pure culture techniques began around 1870, and the dairy and cheese industry was transformed from an industry of inconsistent product quality and quantity. The various organisms responsible for the development of the flavor, taste, and texture of dairy and cheese products were discovered and used in day-to-day operation of the manufacturing plants. More recent scientific advances in the area of organic, analytic, and biochemistry have allowed for the identification of those chemical components that develop during the manufacturing steps and contribute to the flavor profiles that we associate with these products.

Besides the well-known nutritional properties of these products, they are very popular owing to their unique flavor, taste, and textual attributes.

OVERVIEW OF DAIRY PRODUCTS

Dairy products may be defined as food products based on milk developed through a variety of manufacturing methods and fluid milk itself. There are several animal sources of milk, including the cow, water buffalo, goat, and sheep as the commonly used sources.

In 1985, the United States produced an estimated $31.8 billion worth of dairy products (Table 1). The largest segment (31%) of this industry is packaged fluid milk products. The commercial value of these products in terms of their ratio of value added to shipments is well below the food industry's average of 35%. The U.S. Bureau of Census data for 1985 indicated that 140,000 people are employed in the United States in the manufacture of dairy products.

In discussing the nature of the flavors of these products it is natural to organize the discussion based on the types of products manufactured. Based on the major groups of products, there are four distinct areas that will be discussed. They are fluid milk and milk products (dried and canned), milk-derived products (yogurt and similar types of fermented milk products and ice cream), butter, and cheeses.

Milk and Milk Products

These products are minimally processed. Typically they may be reduced in fat, have the fat suspended (homogenized), or heat treated. The heat-processing step may vary in its time/temperature relationship to assure a safe product. Although the heat treatment can have some influence on the taste and flavor of fresh milk, there is a much greater relationship to the time of year, nature of the animals' feed, and the species of animal from which the milk is derived.

Milk is a very complex substance based on fats, sugars (mainly lactose), proteins (such as lactalbumin), and salts (such as calcium phosphate). The taste of milk is mainly affected by the change in this mixture and the degree to which the fat is emulsified (homogenized) in a continuous phase with the aqueous phase. The sensory perception determines the physical nature of the emulsion, with a slightly salty/sweet taste from the salts and lactose. Although some slight flavor changes can be detected in even

Table 1. Dairy Products in the United States[a]

Product Class	Dollar Value Millions	Growth Rate, % 1982–1985	Added Value, %
Butter	1,839	−7	6.2
Natural cheese	5,665	1	17.3
Process cheese	3,763	10	17.3
Dried milk	3,492	3	30.8
Canned milk	1,771	19	30.8
Ice cream	3,882	18	29.9
Fluid milk	10,046	10	25.2
Cottage cheese, Yogurt	1,359	25	
Total	31,817	9	

[a] Ref. 1.

529

Table 2. Free Fatty Acids in Milk[a]

Fatty Acid (Carbon No.: Unsaturation)	% of Fat		
	Fresh Milk	Rancid Milk	Butter
4:0	7.23	6.03	1.32
6:0	3.40	3.42	0.65
8:0	1.37	2.07	0.83
10:0	5.16	3.98	1.41
12:0	4.89	3.98	4.22
14:0	10.02	9.58	11.02
16:0	23.61	26.09	22.69
18:0	9.98	13.65	12.04
18:1	29.78	27.05	38.40
18:2	4.55	4.14	4.66
18:3	—	—	2.69

[a] Ref. 4.

the mildest of heat-treated milk, the flavor defects are generally considered to be notes associated with good quality, pasteurized milk.

Of the various constituents of the milk, the lipid (fat) fraction has the greatest effect on milk's flavor. It is also the precursor of many chemical components considered important in all types of dairy flavors. The fatty acid composition of the milk lipids is also very complex and unique among food products. More than 60 fatty acids have been reported in cow's milk (2). Quantities of butyric acid and caproic, caprilic, decanoic (fatty acids with 6, 8, and 10 carbon atoms) acids are unique to milk and of great importance in the development of flavors in products based on milk. Oxidation of these lipids gives rise to some key flavor components. For example, the oxidation of unsaturated octadecadienoic acid leads to the formation of 4-cis-heptenal, which has been identified as the cream-like flavor component in milk and butter (3). The free fatty acid distribution of milk and butter is shown in Table 2.

The heat treatment of milk is divided into two types of processing: pasteurization in which the milk is heated for 15 s at 72°C and ultra high-temperature treatment (uht), where it is heated at 135–150°C for a few seconds. The flavor changes that occur during the heat treatment may be formed from three distinct chemical mechanisms:

- Degradation of thermally labile precursor found in milk.
- Reaction between the sugars and proteins (Maillard reactions).
- The release or formation of sulfur-containing components.

Degradation of Precursors. The triglycerides (lipids) in milk fat are formed from complex as well as simple fatty acids of many types, including substituted acids such β-keto acids and hydroxy acids. These acids originate in the diet of the animal and also may be the result of normal metabolism. Methyl ketones are formed from the β-keto acids (5). Although they are formed in fresh milk, maximum yield occurs at prolonged elevated heating (140°C for 3 h). The liberation of methyl ketones is generally below their individual odor thresholds, yet they contribute to

the aroma of milk because of their synergistic interaction due to their combined concentration (6). Methyl ketones are formed from the decarboxylation of the β-keto acids, whereas ring closure of δ- and τ-hydroxy acids form the corresponding lactones.

The Maillard Reaction. Reaction between the lactose and the milk protein can lead to many different compounds of flavor significance. These reactions are referred to as the Maillard reaction. Discussion of the Maillard reaction is complex and outside the scope of this article, but its basic chemical pathway is the reaction between aldose and ketone sugars and α-amino acids resulting in the formation of aldosylamines or ketosylamines. These materials then undergo further rearrangement, resulting in the creation of many key flavoring materials, such as acetyls, furans, pyrroles, aldols, and pyrones. The normal heat treatment of milk does not appear to produce enough of the Maillard reaction product to be a flavor contributor. However, the cooked flavor of milk is noted when it is heated to above 75°C (7). At that point many Maillard reaction products start to contribute to the flavor of the milk. Further heating above the cooked-flavor range produces a caramelization flavor, due to the intense amount of Maillard reaction products being formed and the liberation of many sulfur-containing chemicals.

Formation of Sulfur-Containing Compounds. Dimethyl sulfide is produced from the S-methyl methionine sulfonium salt (from the vegetable matter in the animal's diet) that is decomposed on heating (8). At high concentrations this chemical produces a malty or cowy flavor defect in the milk (9). The flavor from the caramelization of milk has found great use in various culinary schools of fine food preparation.

Flavor Defects in Milk. Caramelization is a wanted defect in many food applications; yet, there are some defects that are not wanted. Extended storage of dry milk causes the stale off-flavor where the release to the fatty acid contributes to the further development of methyl ketones above their threshold value, but below the quantity seen in certain cheeses; for example, the blue family of cheeses (10).

The oxidation of milk causes the cardboardy or cappy defect that is evaluated as the milk being metallic, tallowy, oily, or fishy. The oxidation of the unsaturated fatty acids leads to the methyl ketones and series of aldehydes. Very low levels of cis-4-heptenal, hexenal, and 1,3 octenone contribute to the oxidized flavor defect (11).

The lipolysis of the milk by lipases remaining active after the heat treatment can cause rancidity of the milk. This is due to the increase of both the short-chain fatty acids and the increased acidity of the milk. In normal milk the total free acid content is about 360 mg/kg, but in rancid milk it rises to 500–1500 mg/kg (4).

One final major flavor defect is known as the sunlight defect. It is believed to be produced from the degradation of methional (a common amino acid found in milk) to several sulfur-containing compounds, including mathanethiol, dimethyl sulfide, and dimethyl disulfide.

Table 3. Fat Content of Cow's Milk[a]

Type of Fats	Range of Occurrence, %
Triglycerides	97.0–98.0
Diglycerides	0.25–0.48
Monoglycerides	0.015–0.038
Keto acid glycerides	0.85–1.28
Aldehydo glycerides	0.011–0.015
Glyceryl ethers	0.011–0.023
Free Fatty Acids	0.10–0.44
Phospholipids	0.2–1.0
Cerebrosides	0.013–0.066
Sterols	0.12–0.4

[a] Ref. 13.

These contribute to a flavor characterized as burnt or cabbagey. The extent of the flavor defect is related to the length and strength of exposure to light (12).

Milk's Basic Flavor. Milk's basic flavor, then, is due mainly to the taste components found in the milk as noted in Table 3. The fats, sugars, salts, and protein all contribute to give milk a subtle, but enjoyable, flavor and taste. Unless the milk has been mistreated or subjected to high temperature for prolonged time, no significantly predominate flavor is noted. In contrast, products made from milk have characteristic and, in some cases, strongly predominate aroma and tastes.

Cultured Dairy Products

Among the first products of milk to be developed were cultured products. Cultured products are produced through out the world, as seen in Table 4. Originally, fermentation occurred naturally to milk set aside for later use. The spoilage organisms in the milk produced the proper conditions for the preservation of the cultured product. The lowering of the pH of the milk by the organisms allows for the preservation of the product and contributes to its characteristic aroma and taste.

The preparation of these cultured products became an art that survives today. The consumption of these products varies around the world, with the northern European countries consuming the greatest amount (~20 kg/y/person); and the United States consuming the least (~1.2 kg/y/person) (14), However, the United States is a 300-fold increase over the 1966 figure. This, no doubt, reflects the healthy eating trend in the United States prompted by television commercials and magazine ads listing the health benefits of cultured milk products.

The basic fermentation is due to lactic acid bacteria. Historically, this was fermentation done by a mixed or unknown culture or back-slop, but modern technology has developed the pure culture with its predictable performance potentials.

The major flavor and taste component in these products is the lactic acid produced by the fermentation of the sugars by the bacteria. Although this flavor is considered acceptable in many societies, it is rejected completely in others. To increase the commercialization of these products, manufacturers have introduced product that contain added sugars, fruit preserves, and flavors. These innovations have also contributed to increased consumption, particularly in the United States.

During the fermentation the lactose in the milk is transformed via pyruvic acid to lactic acid. The small amount of lactose or its hydrolysis products galactose and glucose is below the threshold for detecting sweetness, but the lactic acid may account for more than 2% of the mass of the product and; by that, contribute a significant flavor and taste effect. The fermentation pathway may favor both enantimer formation (L(+) and D(−) lactic acid), as depicted in Figure 1. The preferred metabolite, in flavor terms, in the L(+) or a harsh off-flavor will be observed (15).

The basic microorganisms used are the lactobacilli, streptococci, and leuconostocs. The major difference in their flavor is the amount of lactic acid produced and the perception of the acid taste. One exception is the product from the USSR called Kifur, which undergoes a secondary fermentation to create ethyl alcohol. This product is also known as the champagne of milk (16).

Yogurt Flavor. Yogurt flavor is also influenced by acetaldehyde, another product of lactose from the fermentation of pyruvic acid by *Lactobacillus bulgaricus*. In yogurt the acetaldehyde level differs greatly with the type of mi-

Table 4. Cultured Milk Products

Product Type	Fermentation	Organism
Yogurt (U.S.)	Moderate acid	*Streptococcus thermophilus* *Lactobacillus bulgaricus*
Buttermilk (U.S.)	Moderate acid	*Streptococcus cremoris* *Streptococcus lactis*
Acidophilus milk (U.S.)	High acid	*Lactobacillus acidophilus*
Bulgarican milk (EUR)	High acid	*Lactabacillus bulgaricus*
Yakult (Japan)	Moderate acid	
Dahi (India)	Moderate acid	*Streptococcus diacetyllactic*
Leben (Egypt)	High acid	*Streptococci, Lactobacilli, yeast*
Kifur (USSR)	High alcohol	*Lactobacillus caucascus* *Leuconostoc* spp

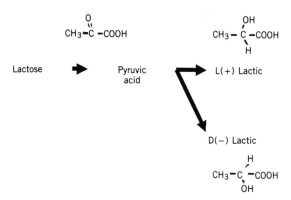

Figure 1. Lactic acid fermentation.

croorganism(s) used (17). There is no agreement on the acetaldehyde values that give an optimum flavor effect (18). The value of 23–41 ppm is given in some reports, whereas the value of 13–16 ppm is considered by others to be the optimum level (19,20).

One further compound of flavor interest in cultured milk products is diacetyl. It is formed from the citrate present in milk by the action of Streptococci, Leuconostocs, and other organisms that can metabolize citrate. (21) This component strongly contributes to the acid cream butter flavor usually at level of 1–2 ppm (22). It is diacetyl that is a significant contributor to butter's flavor.

Butter and Cream

We have discussed the nature of milk flavor and the cultured products derived from milk. The production of butter from milk involves the concentration of the milk fat in the milk by mechanical separation. The resultant product is a water in oil emulsion that has a mild characteristic flavor. During this process the exogenous enzymes (those enzymes existing outside the cell) can break down the triglycerides to free fatty acids; with improper handling this may occur at a high level, leading to a flavor effect known as rancidity. It is the short-chain free fatty acids (eg, butyric acid) that are responsible for the development of rancidity in butter. The role of other free fatty acid directly on flavor seems small, except the deca- and dodecanoic acids, which appear in sweet cream butter at levels exceeding their flavor threshold (23). Because the lipases responsible for the hydrolysis of the triglycerides are very heat stable, their activity continues even at low temperature during refrigerated or frozen storage.

Rancidity. Rancidity is the major flavor defect in butter. The most useful index of rancidity is related to the amount of butyric acid found in the butter. A rancidity score developed by the University of Wisconsin correlates sensory data with individual free fatty acids concentration by stepwise regression and discriminatory analysis to index the relative rancid flavor intensities in butter (24). This is a very useful flavor index for the evaluation of butter.

Lactone Formation. Lactones have also been considered a major flavor attribute in butter and cream. They are

generated from the small amount of δ-hydroxy acids occurring in milk fat. On hydrolysis these compound may spontaneously lactonize to form the corresponding δ-lactones. The τ-lactones also may be formed from unsaturated C18 fatty acids via hydration and β-oxidation. The amount of lactones formed is related to the type of feed, the season, the type of breed, and the stage of lactation of the animal. (25) Estimates of the amount of lactone that can create a flavor effect is reported at 7–85 ppm (26,27).

Heated-Milk Defects. Heated milk fat increases the level of lactones found in butter (28). Although this is considered a defect in butter or cream, it is a desired effect when cooking. The flavor developed is characterized as a coconut-type note, which has been observed in heated butter used to flavor popcorn. Several patents have been issued to commercial companies for the use of lactones or lactone precursors for the flavoring of butter analogues, including margarine.

Imitation Products. Although margarine is not a dairy product, it bears mention here as it is produced in great quantity throughout the world as a substitute for butter. The process goes back to the Napoleonic times as a commercial product. The flavors used in these products generally are composed of diacetyl and lactones or lactone precursors for the butter flavor impression. To various degrees of sophistication, all commercially available margarine has these materials added to the vegetable fat/water emulsion.

Ice Cream. One other product produced from milk and butter is ice cream. Ice cream represents a congealed product produced from cream, milk, skim milk, condensed milk, butter, or a combination of these ingredients, plus the addition of sugar, flavorings, fruits, stabilizers and colors. Again, the basic contribution of the dairy ingredients to the flavor is the fat and lactose, with those aromatic components mentioned in the foregoing discussion and associated with butter or cream flavor. Typically, the aromatic portion of the flavor is greatly modified by the addition of flavors and extracts.

Most of these basic dairy products, like butter, have flavor profiles that are simple. The real flavor masterpieces of the dairy industry are the many varieties of cheese, with their extremes in flavor types and intensities.

Cheese

The many varieties of cheese represent many centuries of the cultivation of the art of cheese-making. Many different, probably accidentally developed, flavors, tastes, and textures have been developed. The products of cheesemaking are the result of many types of microorganisms (yeast, bacteria, and mold) working together with the basic milk source (typically cow's, goat's or sheep's milk) and the knowledge of the control of fermentation. The variety of methods is great and, so too, as has been noted, are the varieties of cheese produced throughout the world. We will not attempt to discuss all the types known, but focus our flavor discussion on those well-known cheeses and the basic science of their formation.

Chemistry. The basic chemistry or biochemistry of cheese-making is the slow degradation of the macromolecules (protein and lipid material) to small components of flavor (aroma and taste) value. Proteolysis (breakdown of the protein by enzymes) during cheese ripening and aging has a major influence on peptide formation. Peptides have various taste characteristics based on their polarity. The basic taste sensations of sour/sweet and salt/bitter and the flavor-enhancing property known as umami (the flavor-enhancer taste associated with monosodium glutamate, MSG) are all taste attributes of peptides and amino acids.

It has been observed that the acid- (sour), hydrophobic- (bitter), and glutamic/aspartic-rich (umami) peptides are the key taste peptides and amino acids in aged cheese (29). The hydrophilic (sweet) and salt-based peptides contribute very little, as they are overwhelmed by the basic taste attributes of the cheese. The natural sweetness from lactose and the saltiness due to the salt (sodium chloride) added to all cheese to produce a selective environment and the chemistry needed to develop the specific cheese character play a more significant flavor role than do the peptides that have salty or sweet taste character.

The salt content of selected cheeses is listed in Table 5. Note that the Roquefort or blue-cheese types have the highest salt content and those of the Swiss types have the lowest. This is reflected in the basic salt taste of these cheeses. As we will see, the volatile chemicals formed from the amino acid, peptides, lipids, and sugar are the backbone of the characterizing flavors for the various cheese.

Types of Cheeses. The family of cheeses generally recognized are

- Fresh soft low-fat cheese
- Cream/Neufchâtel
- Gouda/Muenster
- Port du Salut
- Brie/Camembert
- Blue cheeses
- Goat cheeses
- Emmentaler
- Cheddar
- Parmesan/Romano
- Whey/ricotta

Processing. The basic processing involves the progressive dehydration and working of the coagulated and separated curds of milk to form a homogeneous compact mass. The initial coagulation is achieved by using an added enzyme, rennin, or by lactic acid fermentation, or both, depending on the desired character of the final product. The green cheese is then ripened via the addition of certain microorganisms that will accomplish the breakdown of the protein, lipids, and sugars to the small molecular weight components that contribute to the cheesey flavors found in the final product.

Flavor Attributes of Cheese by Families. The less cheesey flavored cheeses are those with the least amount of ripening: fresh soft low-fat, higher fat cream/Neufchâtel, and the whey/ricotta type. Their flavors are primarily due to the components contributing to milk, cream, and butter flavor. They should be low in flavor intensity or they would be considered to have a flavor defect.

The next group in terms of flavor strength consist of the Swiss (Emmentaler), German (Muenster), and Dutch (Gouda) cheeses. The importance of peptides was recognized in studies with Swiss cheese (30). Results indicated that the small peptide that interacted with calcium and magnesium gives the cheese a sweet flavor, with the small free peptides and amino acid contributing a slight brothy-nutty flavor. An important reaction in Emmentaler is the conversion of pyruvic acid to propionic acid by propionibacterium. Propionic acid is found in Emmentaler in levels of up to 1%, and it is considered the basis of the sweet taste of that cheese (31). Gouda has a defect called catty taint that is due to the formation of 4-mercato-4-methylpentaone-2 (32).

Cheddar cheese is by far the most consumed cheese in the United States. It is produced by a unique process, cheddaring, introduced in 1857 to repress the growth of spoilage organisms during cheese-making. Cheddaring involves the piling and repiling of blocks of warm cheese curd in the cheese vat for ca 2 h. During this period the lactic acid increases rapidly to a point where coliform bacteria are destroyed.

The basic taste of cheddar is due to the peptide fraction and, in particular, to the umami and bitter components. The development of acetic acid during ripening contributes to the sharpness of aged cheddar. The release of the acetic acid and amino acids continue during the ripening process. The medium chain fatty acids (C6 to C18) that are released during the ripening process have also been shown to be important in cheddar cheese flavor character (33).

Many other studies over a 10-year period indicated that it is necessary to produce the right level of acidity to provide a reducing environment in the finished cheese so that there may be a release of sulfur-containing compounds. The lost of these compounds, in particular, methyl mercaptan, results in a cheese with little cheddar character (34–36). Methanethiol has also been shown to be a significant factor in aged cheddar (37–41). Methanethiol has a very fecal-like aroma with a threshold of detection reported at 0.02 ppb (42).

It has been suggested that acetylpyrazine and 2-methoxy-3-ethylpyrazine are important in aged cheddar (43). The possible microbial origins of pyrazines became evident when it was reported that dimethylpyrazine and

Table 5. Salt Content of Certain Cheeses

Cheese Variety	Salt (NaCl), % w/w
Roquefort	4.1–5.0
Feta	2.4–4.4
Parmesan	2.1–3.5
Edam	1.7–1.8
Cheddar	1.6–1.7
Emmentaler	0.9–1.0

Table 6. Summary of the Flavor Chemistry of Dairy Products

Product	Process	Chemistry	Major Flavor Contributors
Milk	Homogenization and mild heat	Natural components	Balance of fats, salts, and lactose (milk sugar)
Defects	Cooked >75°C	Maillard reactions	Aldehydes and pyrazines
	Prolonged cooking	Caramelization	Sulfur-containing components
	Oxidation	Oxidation	Aldehydes
	Sunlight exposure	Sulfur reactions	Methanethiol
	Rancidity	Lipolysis	Butyric acid
Cream	Fat separation	Enzymolysis	Methyl ketones (low level) diacetyl and lactones
Butter	Further fat separation	Enzymolysis	Diacetyl and butyric acid
Cultured products	Fermentation	Lactic acid formation	L(+) Lactic acid
Yogurt	Fermentation		Acetaldehyde
Cheeses	Salting		NaCl
	Aging	Proteolysis	Peptide formation
Swiss types	Fermentation	Propionibacteria	Propionic acid
Cheddar types	Cheddaring	Lactic acid fermentation	Lactic and acetic acid
		Maillard reactions	Acetyl pyrazine and other pyrazines
Blue types	Fermentation	Penicillium	Methyl ketones (high amounts)

2-methoxy-isopropylpyrazine were isolated from milk contaminated by *Pseudomonas taetrolens* (44). Further alkylpyrazines have been isolated from mold ripened cheese using *Penicillium caseicolum* cultures (45). The overall flavor of good cheddar is a balance between the free fatty acids and the other flavor components noted above.

The blue cheese family has been found to contain from 13 to 37 times the amount of free fatty acids found in cheddar. This is due to the lipolysis of the fat by the lipases generated from the use of the mold *Penicillium roqueforti* (46). Roquefort is made from sheep milk and does not have the high levels of free fatty acids seen in the blue cheese made from cow's milk. Part of the flavor difference between the two is due to the lower concentration of propionic acid (4:0) in Roquefort and the relative larger amount of C8:0 and C10:0 free fatty acids.

The major aroma character of the blue cheeses is due to the high levels of methyl ketones found in the cheese. These components have been isolated in milk and are assumed to be the product of the free fatty acid. The importance of the methyl ketone in creating an imitation blue cheese flavor has been well reported (47–52).

Secondary alcohol may be produced from the methyl ketones when they are generated in high amounts driving the equilibrium from the methyl ketones and to the secondary alcohols. Although the secondary alcohols have a similar flavor profile to the methyl ketones, their intensity and quantity in the cheese is less, therefore, making only a small contribution to the complete flavor (52).

CONCLUSION

The flavor of dairy products are extremely varied and represent a great range of chemistry, biochemistry, and microbiology. A summary of the major dairy products and

their associated significant flavor values is given in Table 6. It summarizes the nature of the components contributing to the flavor and taste characters of the particular material. Science has yet to identify completely the components responsible for the overall subtle aromas and taste found in dairy products, but Table 6 represents a good basic start to an understanding of the complexities of dairy flavors.

BIBLIOGRAPHY

1. J. M. Connor, *Food Processing,* Lexington Books, Lexington, Massachusetts, 1988, pp. 91–107.
2. S. F. Herb, P. Magidma, F. E. Lubby, and R. W. Riemenscheider, "Fatty Acids of Cow Milk (II). Compounds by Gas-Liquid Chromatography Aided by Fractionation," *Journal of the American Oil Chemists' Society* 39, 142–149 (1962).
3. P. H. Haverkamp-Begemann and J. C. Koster, "4 Cis-heptenal—cream flavor component of butter," *Nature* 202, 552–523 (1964).
4. J. A. Knitner and E. A. Day, "Major Free Fatty Acids in Milk," *Journal of Dairy Science* 48, 1557–1581 (1965).
5. S. Patton and G. W. Kurtz, "A Note on The Thiobarbituric Acid Test for Milk Lipid Oxidation," *Journal of Dairy Science,* 38, 901 (1955).
6. J. E. Langler and E. A. Day, "Development and Flavor Properties of Methyl Ketone in Milk Fat," *Journal of Dairy Science* 47, 1291–1296 (1964).
7. D. V. Josephson and F. J. Doan, "Cooked Flavor in Milk, Its Sources and Significance," *Milk Dealer* 29, 35–40 (1939).
8. T. W. Keenan and R. C. Lindsay, "Evidence for a Dimethyl Sulfide Precursor in Milk," *Journal of Dairy Science* 51, 122–128 (1968).
9. S. Patton, D. A. Forss and E. A. Day, "Methyl Sulfide and the Flavor of Milk," *Journal of Dairy Science* 39, 1469–1473 (1956).

10. O. N. Parks and S. Patton, "Volatile Carbonyl Compounds in Stored Dry Whole Milk," *Journal of Dairy Science* **44**, 1–9 (1961).

11. H. T. Badings and R. Neeter, "Recent Advances in the Study of Aroma of Milk and Dairy Products," *Netherlands Milk and Dairy Journal* **34**, 9–30 (1980).

12. R. L. Bradley, Jr., "Effect of Light on Alteration of Nutritional Value and Flavor of Milk—A Review," *Journal of Food Protection* **43**, 314–320 (1980).

13. F. E. Kurtz, *Fundamentals of Dairy Chemistry,* The AVI Publishing Co., Inc., Westport, Conn. 1965, pp. 135–180.

14. International Dairy Foundation, *Bulletin 119,* International Dairy Foundation, Brussels, Belgium, 1980.

15. G. Kielwein and U. Dawn, "Occurrence and Significance of D(−)-Lactic Acid in Fermented Milk Products with Special Consideration of Yogurt Based Mixed Products," *Deutsche Molkerei Zeitung* **100**, 290–293 (1979).

16. K. M. Shahani, B. A. Friend, and R. C. Chandan, "Natural Antibiotic Activity of Lactobacillus acidophilus and bulgaricus," *Cult. Dairy Prod. J.* **18**, 1519 (1983).

17. R. K. Robinson, A. Y. Tamine, and L. W. Chubb, "Yogurt-Perceived Regulatory Trends in Europe," *Dairy Ind. Int.* **41**, 449–451 (1976).

18. M. Groux, "Component of Yogurt Flavor," *Le Lait* **53**, 146–153 (1973).

19. H. T. Badings and R. Neeter, "Recent Advances in the Study of Aroma Compounds of Milk and Dairy Products," *Netherlands Milk and Dairy Journal* **34**, 9–30 (1980).

20. M. L. Green and D. J. Manning, "Development of Texture and Flavor in Cheese and Other Fermented Products, *Journal of Dairy Research* **49**, 737–748 (1982).

21. E. B. Collins, "Biosynthesis of Flavor Compounds by Microorganisms," *Journal of Dairy Science* **55**, 1022–1028 (1972).

22. V. K. Steinholt, A. Svensen, and G. Tufto, "The Formation of Citrate in Milk," *Dairy Research Institute* **155**, 1–9 (1971).

23. E. H. Ramshaw, "Volatile Components of Butter and their Relevance to Its Desired Flavor," *Australian Journal of Dairy Technology* **29**, 110–115 (1974).

24. A. H. Woo and R. C. Lindsay, "Statistical Correlation of Quantitative Flavor Intensity Assessments of Free Fatty Acid Measurements for Routine Detection and Prediction of Hydrolytic Rancidity of Flavor in Butter," *Journal of Food Science* **48**, 1761–1766 (1983).

25. P. S. Dimick, N. J. Walker, and S. Patton, "Occurrence and Biochemical Origin of Aliphatic Lactones in Milk Fat—A review," *Journal of Agricultural Food Chemistry* **17**, 649–655 (1969).

26. J. E. Kinsella, S. Patton, and P. S. Dimick, "The Flavor Potential of Milk Fat. A Review of Its Chemical Nature and Biochemical Origins," *Journal of the American Oil Chemistry Society* **44**, 449–454 (1967).

27. R. Ellis and N. P. Wong, "Lactones in Butter, Butter Oil, and Margarine," *Journal of the American Oil Chemistry Society* **52**, 252–255 (1975).

28. G. Urbach, "The Effect of Different Feeds on the Lactone and Methyl Ketone Precursors of Milk Fat," *Lebensmittel Wissenschaft Technologie* **15**, 62–69 (1982).

29. K. H. Ney, "Recent Advances in Cheese Flavor Research," in G. Charalambous and G. Inglett, eds., *The Quality of Foods and Beverages,* Vol. 1, Academic Press Inc., New York, 1981, pp. 389–435.

30. S. L. Biede, A Study of the Chemical and Flavor Profiles of Swiss Cheese, Doctoral Dissertation, Department of Food Science, Iowa State University, Ames, Iowa, 1977.

31. K. H. Ney, "Bitterness and Gel Permeation Chromatography of Enzymatic Protein Hydrolysates," *Fette Seifen Anstrichmittel* **80**, 323–325 (1978).

32. H. T. Badings, "Causes of Ribes Flavor in Cheese," *Journal of Dairy Science* **50**, 1347–1351 (1967).

33. D. D. Bill and E. A. Day, "Determination of the Major Free Fatty Acids of Cheddar Cheese," *Journal of Dairy Science* **47**, 733–738 (1964).

34. T. J. Kristoffersen and I. A. Gould, "Cheddar Cheese Flavor, II., Changes in Flavor Quality and Ripening Products of Commercial Cheddar Cheese During Controlled Curing," *Journal of Dairy Science* **43**, 1202–1209 (1960).

35. B. A. Law, "New Methods for the Controlling Spoilage of Milk and Milk Products," *Perfumer & Flavorist* **7**, 9–13 (1983).

36. T. J. Kristoffersen, I. A. Gould, and G. A. Puruis, "Cheddar Cheese Flavor. III. Active Sulfhydryl Group Production During Ripening," *Journal of Dairy Science* **47**, 599–603 (1964).

37. L. M. Libbey and E. A. Day, "Methyl Mercaptan as a component of Cheddar Cheese," *Journal of Dairy Science* **46**, 859 (1963).

38. D. J. Manning, "Cheddar Cheese Micro Flavor Studies II. Relative Flavor Contribution of Individual Volatile Components," *Journal of Dairy Research* **46**, 531–539 (1979).

39. D. J. Manning, "Headspace Analysis of Hard Cheese," *Journal of Dairy Research* **46**, 539–545 (1979).

40. D. J. Manning, "Cheddar Cheese Aroma—The Effects of Selectively Removing Specific Classes of Compound from Cheese Headspace," *Journal of Dairy Research* **44**, 357–361 (1977).

41. D. J. Manning, "Sulfur Compounds in Relation to Cheddar Cheese Flavor," *Journal of Dairy Research* **41**, 81–87 (1974).

42. R. G. Buttery, D. G. Guadaghi, L. C. Ling, R. M. Seifert, and W. Lipton, "Additional Volatile Components of Cabbage, Broccoli and Cauliflower," *Journal of Agricultural Food Chemistry* **24**, 829–832 (1976).

43. W. A. McGugan, "Cheddar Cheese Flavor, A Review of Current Progress," *Journal of Agricultural Food Chemistry* **23**, 1047–1050 (1975).

44. M. E. Morgan, "The Chemistry of Some Microbially Included Flavor Detects in Milk and Dairy Foods," *Biotechnology and Bioengineering* **18**, 953–965 (1976).

45. T. O. Bosset and R. Liardon, "The Aroma of Swiss Gruyere Cheese, III. Relative Changes in the Content of Alkaline and Neutral Volatile Components During Ripening," *Lebensmittel Wissenchaft Technologie* **18–85**, 178 (1985).

46. D. Anderson and E. A. Day, "Quantitative Analysis of the Major Free Fatty Acids in Blue Cheese," *Journal of Dairy Science* **48**, 248–249 (1965).

47. B. K. Dwivedi and J. E. Kinsella, "Continuous Production of Blue-Type Cheese Flavor by Submerged Fermentation of Penicillium Roqueforti," *Journal of Food Science* **39**, 620–622 (1974).

48. R. C. Jolly and F. V. Kosikowski, "Flavor Development in Pasteurized Milk Blue Cheese by Animal and Microbial Lipase Preparations," *Journal of Dairy Science* **58**, 846–852 (1975).

49. M. Godinho and P. F. Fox, "Ripening of Blue Cheese. Influence of Salting Rate on Lipolysis and Carbonyl Formation," *Milchwissenschaft* **36**, 476–478 (1981).

50. T. Kanisawa and H. Itoh, "Production of Methyl Ketone Mixture from Lipolyzed Milk Fat by Fungi Isolated from Blue Cheese," *Nippon Shokuhin Kogyo* **31**, 483–487 (1984).

51. J. H. Nelson, "Production of Blue Cheese Flavor via Sub-

merged Fermentation by Penicillium Roqueforti," *Journal of Agricultural Food Chemistry* **18**, 567–569 (1970).

52. E. A. Day and D. F. Anderson, "Gas Chromatographic and Mass Spectral Identification of Natural Components of the Aroma Fraction of Blue Cheese," *Journal of Agricultural Food Chemistry* **13**–14, 2 (1965).

CHARLES H. MANLEY
Takasago International
Corporation, USA, Inc.
Teterboro, New Jersey

DAIRY INGREDIENTS: APPLICATIONS IN MEAT, POULTRY, AND SEAFOODS

Meat and dairy products have several things in common. Both are appreciated from organoleptic and nutritional points of view. They have been traditionally regarded as two of the main protein sources for humans. In a modern diet, meat contributes about 35% of the protein intake, and milk about 25% (1).

Meat and dairy industries, furthermore, have a common interest in maintaining a positive image for protein of animal origin. However, in recent years there has been criticism about the fat content of meat and dairy products, and in some countries nutritional councils have recommended substituting vegetable protein for animal protein in the diet. In response to this criticism, meat and dairy manufacturers have succeeded in introducing low-fat products. A whole new range of meat and dairy foods is available, enabling the consumer to decide on the fat level desired in the diet. In addition to providing protein and fat, meat and dairy products contribute to a great extent to the intake of micronutrients. Meat is very important for the intake of iron and certain vitamins such as thiamin and riboflavin, while dairy foods are critical for the intake of calcium and vitamins.

Liquid or solid dairy products are widely used in the meat industry. They are mostly used with the objective to improve taste or eye appeal, eg, cheese topping on burgers, cordon bleu entrees, and cheese franks. Fresh dairy cream is often used in pâtés or liver products and in many seafood products. These applications are not within the scope of this article on functionality. The two most important ingredients from milk that find application in meat products are milk sugar (lactose) and predominantly milk protein. Milk protein causes functional advantages, while lactose is mostly used for taste improvement.

This article reviews and updates the application of these milk derivates in meat, poultry, and fish products. Scientific backgrounds will be highlighted.

PROCESSED MEAT PRODUCTS

Processed meat products are all meat-based products that need more than just cutting from the carcass. It is estimated that ca 25% of the worldwide meat production is being processed (Table 1). In these processed meats dairy ingredients can be used as functional ingredients.

Table 1. Production of Meat in Different Areas of the World (in 1,000 tons) 1987[a]

Area	Total Meat[b]
European community	29,985
United States	26,787
USSR	18,997
Total world	159,669

[a] Ref. 2.
[b] Beef, veal, pork, poultry, and sheep expressed as gross carcass weight.

Meat products can be classified in many different ways. Usually they are categorized as noncomminuted, coarsely comminuted, and finely comminuted. The degree of comminution determines to a large extent the way of applying dairy ingredients (see Table 2).

DAIRY INGREDIENTS IN MEAT PRODUCTS

The two most important ingredients from milk that find application in meat are predominantly milk protein and to some extent the milk sugar or lactose.

Lactose

Water binding in coarse and noncomminuted meats such as pumped ham is improved by extraction and swelling of meat proteins. For optimum extraction of protein from meat, usually phosphate and salt are used. The concentration of these ingredients is limited for organoleptic reasons. The effective salt and phosphate concentration on water and so the ionic strength can be increased by increasing the total solids of the brine mixture. For this purpose lactose is often used. Lactose has advantageous effects on

Taste. Lactose has the capacity of masking the bitter aftertaste of salts and phosphates.

Stability. Lactose improves water binding and sliceability and enhances the cured color.

Yield. Lactose is therefore used in liver products, cooked hams, and cooked sausages. In sterilized products the addition of lactose is less desirable because of the browning reaction.

Table 2. Classification of Meat Products

Degree of Comminution	Examples
None	Hams (cooked, smoked, dry), bacon, cooked loins, shoulders, reformed hams
Coarse	(Semi)dry sausages (salami, cervelat), fresh sausages (bratwurst), hamburgers, meat patties, meatballs, chicken nuggets
Finely	Frankfurters, hot dogs, polony, pariser, bologna, mortadella, liver sausages/spreads, meat loaves

Table 3. Ingredients in Final Product

Ingredient	Percent of Final Product
Phosphates	0.3
Nitrited salt	1.8
Lactose	1.2
Powdered egg white	0.3
Milk protein	Varying from 0–1.5
Sodium ascorbate	0.05

Lactose as an integrated part of whey powder or skim milk powder (smp) is being used in many meat products. The protein from smp also contributes to some extent to the stability, but as this is present in micellar form, it is not optimum for emulsifying properties. Also, it is known that calcium may negatively influence the binding properties of meat proteins (3). Therefore Ca-reduced smp has found quite some application in meat products.

Milk Protein

The application of milk protein in the form of sodium caseinate has found broad acceptance in the meat industry (4). The functional properties of milk protein can be explained from its molecular structure (5) (Fig. 1).

Due to the high proline and the low sulfuric amino acids content, caseinates will have a random coil structure with a low percent helix. As a consequence, caseinates will show no heat gelation and denaturation and will have a high viscosity in solution.

Caseinates have a high electrical charge and have several very hydrophobic groups. This makes them perfect emulsifiers, with a strong preference for interfaces fat/water or air/water. The high charge makes them perfectly soluble in water.

Caseinates build strong, flexible membranes that will hardly be influenced by heat. Whey proteins show a different pattern (Fig. 2). They are low in proline and have many S–S bonds, leading to a globular, strongly folded and organized structure. Whey proteins are sensitive to heat. During heating they unfold and, depending on pH and concentration, they will build intermolecular disulfide bonds resulting in gelation. The globular form is also the reason for the low viscosity in solution.

Whey proteins have a rather high hydrophobicity, which is more evenly distributed than in caseinate. Because of this structure, the application possibilities of whey proteins in meat products are very limited and will not be discussed in this article.

FINELY COMMINUTED MEAT PRODUCTS

The various finely comminuted meat systems, such as frankfurters, Vienna sausages, mortadellas, luncheon meat, chicken sausages, liver sausages, and pâtés, are commonly indicated as meat emulsions (6).

These products are generally prepared by chopping or grinding meat with water or ice, salt, and frequently phosphates to a kind of cold meat slurry, then forming the matrix in which the animal fatty tissue and possibly connective tissue material is dispersed.

In general, the final temperature during processing does not exceed 15–18°C. Often flavorings, binders, or other additives are mixed in. Having this in mind, it will be clear that hardly any of the abovementioned finely comminuted products will show structures that are equal to true and simple emulsions as defined in physical chemistry.

The stability of finely comminuted meat products is—apart from production technology, formulation, the use of binders, etc—mainly determined by the quality of the meat cuts used (6,7).

Structure of Lean Meat

Lean meat generally contains about 20% protein, which can be divided into:

30–35% Sarcoplasmic proteins.
50–55% Myofibrillar or structural proteins (myosin, actin).
15–20% Stromal proteins (collagen, elastin).

The essential unit of all muscle is the fiber (8) (Fig. 3). Each fiber is surrounded by a membrane, the sarcolemma, and consists of a great number of parallel ordered myofibrils (8). In turn, each myofibril represents a similarly parallel-ordered structure of very thin protein threads, the actin- and the myosin filaments (Fig. 4).

Within the sarcolemma, surrounding the myofibrils, is the meat juice or sarcoplasma. This meat juice contains the water soluble meat proteins (wsp), consisting of enzymes, the red meat color myoglobin, and structured bodies. It is easily extracted on pressing, freezing/thawing, or chopping the meat. The myofibrillar proteins in warm slaughtered meat are present as free actin- and myosin filaments and are soluble in brine. Therefore, they are called the salt soluble proteins (ssp). In chilled and matured meat, however, the actin and myosin have gone into a reaction to form the complex actomyosin (7,8). This actomyosin is only partly soluble in brine, depending on the condition of the meat, age of the animal, (pre)slaughter conditions, anatomic location, mechanical treatment, etc.

The nonsoluble part of the myofibrillar proteins is able to swell (take up water). In this postrigor condition the water- and fat-holding capacity of the meat are considerably lower than in warm slaughtered meat.

Meat Emulsions

As indicated in the introduction, hardly any of the abovementioned finely comminuted meat products is a true emulsion. In meat emulsion there are at least four different phase systems at the same time:

1. A true aqueous solution of salt, phosphates, nitrite, sugars, etc and a colloidal solution (sol) of sarcoplasmic (water-soluble) proteins and salt-solubilized myofibrillar proteins. This solution is the continuous phase of three different dispersed systems at a time.

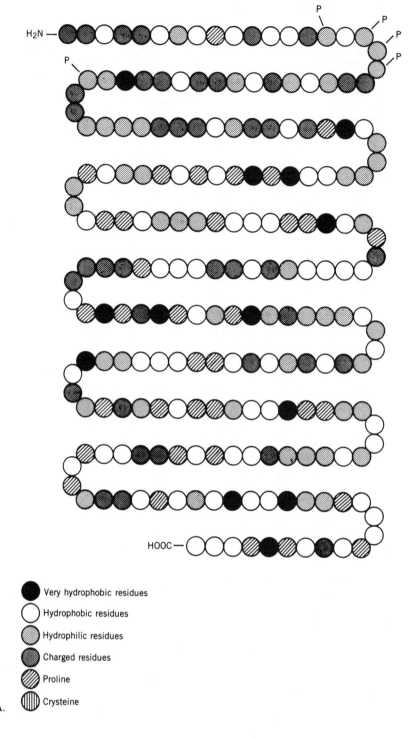

Figure 1. Amino acid sequence of β-casein A.

2. A suspension of undissolved proteins and fatty tissue.

3. An emulsion containing emulsified fat droplets.

4. A foam; during the comminution process some air is corporated.

This system has to be stabilized against heat treatments such as pasteurization or sterilization, against smoking, cooling, frying, reheating, vacuum treatments, etc—whatever physical stress is put upon the products during conserving, storing, and commercializing.

These stabilization properties are important, not only from the standpoint of production practices, but also from the standpoint of cost and quality characteristics, such as texture, consistency, bite, tenderness, juiciness, appearance, and palatability of the finished sausages.

The mechanism for the binding and stabilization phenomenon still is not completely understood, and there are several conflicting viewpoints expressed in the literature (8–15). It is, however, generally agreed on that the myofibrillar proteins are mainly responsible for this stability (8–10,16,17).

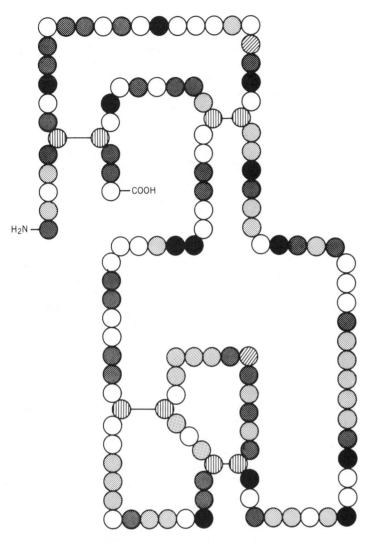

Very hydrophobic residues

Hydrophobic residues

Hydrophilic residues

Charged residues

Proline

Cysteine

Figure 2. Amino acid sequence of α-lactalbumin B.

Stability of the Aqueous Matrix

The stability of the matrix is, apart from the effect of binders, determined by the water holding and gelling capacity of the meat proteins, and, more in particular, of the myofibrillar part. In relation to the stability of fresh sausage batters, the water-binding capacity of the nonsolubilized myofibrillar proteins is decisive. Changes in the net-charge of the proteins, resulting in either attraction or repulsion of the filaments, are responsible for shrinkage and swelling of the protein network and run parallel with the decrease or increase, respectively, in water holding (18).

This mechanism of water binding may be compared to

that of a sponge. The more salt-soluble proteins are solubilized, the better the water binding will be.

Stability of the Suspended and Emulsified Fat

Animal fatty tissue consists of a cellular network in which the fat is enclosed. The network is built up of connective tissue and water. As long as these cell walls are intact, only minor fat separation will take place. However, on chopping the fatty tissue an increasing number of fat cells is damaged and more and more fat is set free. This free fat should be stabilized to prevent fat separation from the product. The origin of the fat (pork, beef, mutton, poultry, etc) and the anatomic location determine the amount of

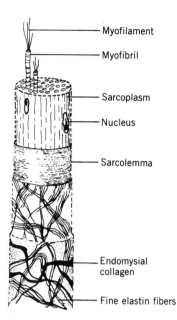

Figure 3. Sketch of a muscle fiber.

free fat during comminution and as a consequence the application of the fatty tissue.

Most beef, mutton, and chicken fat, as well as pork flare fat are hard to stabilize in finely comminuted meat products, unless they are pre-emulsified with a nonmeat protein such as sodium caseinate (13,19,20).

Pork fats other than flare fat (back fat, shoulder fat) are softer and consequently can be used directly in the production of the meat emulsion, because there is more fat in a liquid state.

Particularly in finely comminuted meat products, the interaction between the free fat and the meat proteins plays an important role in the stabilization of both fat and water (14).

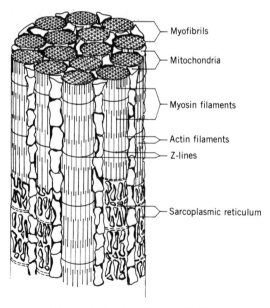

Figure 4. Section of a muscle fiber.

The salt-soluble myofibrillar meat proteins have excellent emulsifying properties and are quickly preferentially absorbed in the fat/water interface (21). The sarcoplasmic proteins are relatively unimportant in this respect. In addition to their gelling capacity and their role in water binding, this emulsification is another reason to aim at a sufficient extraction of these myofibrillar proteins during first stage of chopping.

However, when the myofibrillar meat protein enters the fat/water interface, the protein structure is altered and, as a result, the myofibrillar protein is no longer capable of gel formation and water binding (8). This means that the consumption of ssp for the emulsification of free fat goes at the expense of the water binding.

Milk Proteins in Meat Emulsions

Milk protein, type sodium caseinate, is a perfect emulsifying protein, which is very strongly attracted by the fat/water interface (12). If this type of milk protein gets the opportunity to surround the free-fat particles during sausage manufacture and before the myofibrillar (ssp) proteins do this, the latter are saved for denaturation in the interface.

Research has proved that milk protein, type sodium caseinate, indeed is better and more quickly absorbed in the oil/water interface when emulsification takes place when both milk proteins and meat proteins are present in the continuous phase (Fig. 5).

In this way the abovementioned loss in gelling capacity is avoided. As milk protein is doing the fat binding, the meat proteins can use their full power in water binding and texture formation. In this way sodium caseinate contributes directly to a better fat binding and indirectly to an improved water binding and texture formation (20–22).

Milk proteins can be applied in three ways:

1. Addition of the powder at the beginning of the comminution process. When milk protein is used in this way, it is important that the addition takes place just before the addition of water/ice. In that way the protein is optimally hydrated; this hydration is very important for its functionality.

2. Addition as a jelly. The milk protein is predissolved in water in a bowl chopper or colloid mill. A common ratio is 1 part milk protein dissolved in 6 parts water.

3. In the form of a pre-emulsion. These pre-emulsions consist of milk protein, fat, and water. Technologically difficult to stabilize fats can be used perfectly by this pre-emulsifying technique (eg, beef suet and flare fat). A common ratio of milk protein : fat : water is 1 : 5 : 5. Addition levels of these emulsions to meat emulsions can vary from 10 to 25% maximum.

Temperature control during the production of finely comminuted meat products is very important; when temperatures at the end of comminution exceed 15–18°C, the stability of the final product is affected very negatively. This means that there is little tolerance in processing.

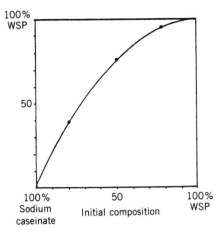
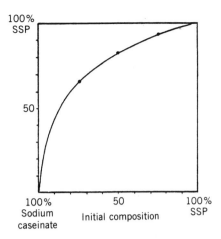

Figure 5. The preferential adsorption of milk protein, sodium caseinate over water soluble protein (WSP) and salt soluble protein (SSP), in a fat/water/protein emulsion; (a) the initial composition of the protein soluble mixtures is shown on the abscissae; (b) the composition of the protein solution after removal of the emulsified fat is shown on the ordinates.

A unique feature of milk proteins is their capacity to widen the processing tolerance as far as temperatures are concerned. This is illustrated in Figure 6.

SECTIONED AND FORMED MEAT PRODUCTS

Principles of Production

Sectioned and formed products are made by mechanically working meat pieces to disrupt the normal muscle cell structure (16). This produces a creamy, tacky exudate on the surface of the meat pieces. When the product is heated during thermal processing, this exudate binds the meat pieces together; during this heating process the solubilized salt-soluble proteins (mainly myosin) are denaturated and form a gel (23–27). The mechanical action of tum-

bling and massaging primarily affects external tissues to produce the surface exudate. Some internal tissue disruption also occurs, which explains the enhanced tenderness, brine penetration and distribution, and improved water-holding capacity. The protein exudate is not only produced by the mechanical action during massaging and tumbling, but also by the synergistic effect of addition of salt combined with alkaline phosphates, which improves yields and maximizes myofibrillar protein solubilization.

Optimum product quality is achieved with the addition of brine, which produces a final product containing 2 to 3% salt and 0.3 to 0.5% phosphate.

Dairy Ingredients

In the production of reformed hams both lactose and milk proteins (sodium caseinate) are very useful ingredients. Lactose can be used in levels varying from 0.5 to 2%. When lactose is used, it serves as a water binder, thereby increasing the yield. It has the capacity of masking the bitter aftertaste of phosphates, giving the final product a more delicate flavor. Milk protein or milk protein hydrolysates can be used in levels varying from 0.8 to 1.6% by dissolving them in the brines used for injection of the meat pieces. Milk protein strongly affects the binding between the meat pieces, thereby an improved sliceability is obtained and they have a positive effect on yield. The effects on yield after cooking as function of milk protein percentage in the final product is presented in Figure 7.

INTERACTION OF MILK PROTEINS WITH MEAT PROTEINS

The results of application research indicate that there is a specific interaction between milk protein and meat proteins. To study this effect we have to make a classification of proteins present in meat. Meat protein can be divided into different groups according to their specific function and solubility in different solvents (Table 4).

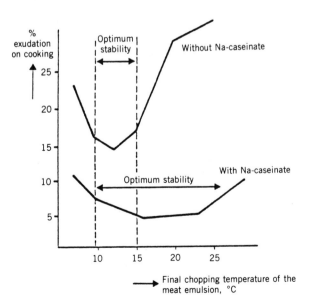

Figure 6. Influence of Na-caseinate on the stability of a meat emulsion in dependence of the final chopping temperature.

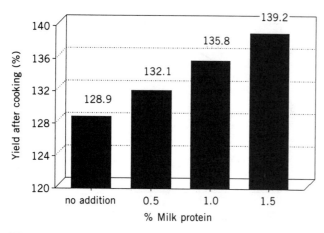

Figure 7. Milk proteins in reformed cooked ham (60% injection).

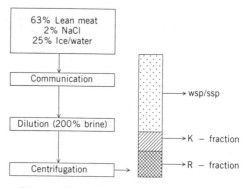

Figure 8. Fractionation of meat proteins.

To study the effect of interaction between milk protein and salt-soluble meat proteins, lean meat was extracted according to the extraction procedure shown in Figure 8. Sixty-three percent lean meat (*M. semimembranosis*) is extracted with 2% salt and 35% ice/water in the bowl chopper. After dilution with 200% brine (2% salt), the lean meat slurry is centrifuged.

After centrifugation there are three distinct layers in the centrifuge tube:

Upper layer: contains wsp (sarcoplasmic proteins) and the salt-soluble myofibrillar proteins.

K-fraction: contains nonsolubilized and swollen myofibrillar proteins, mainly actomyosin.

R-fraction: contains connective tissue materials.

The gellification of wsp/ssp proteins was tested in the following solution:

3% ssp/wsp proteins.
3% NaCL.
2% milk protein.

The equipment for these tests is the gelograph (Fig. 9).

The principle of this gelograph is as follows: A protein solution that is placed in a waterbath is slowly heated (1°C/min). During this heat treatment the viscosity/gel strength of the solution is continuously measured by means of an oscillating needle; this equipment operates in a nondestructive way; this means the gel structure that is formed during the heating of the solution is not destroyed because of minimal oscillating of the needle. In this way it

is possible to imitate the gelling of meat proteins in pasteurized meat products (eg, a cooked ham).

An example is a product containing 60% lean meat, which contains about 12% total meat protein. Roughly 50% of this protein is myofibrillar proteins. This is about 6%. Suppose that 50% of total myofibrillar proteins are extracted during massaging/tumbling. This means that the concentration of ssp will be around 3%. This is the same concentration as used in the model experiments. Results of gelograph experiments are presented in Figures 10 and 11.

Figure 10 shows the gel strength of a 3% wsp/ssp as a function of temperature. The temperature starts at 20°C (68°F) and is increased at a rate of 1°C/min; during this temperature increase the gel strength is measured as milligels. At 44°C (110°F) there is a sudden increase in gel strength; this is caused by the aggregation of myosin molecules (24). At higher temperatures the viscosity decreases again. The reason for this phenomenon is not yet exactly known (23,26,27).

At a temperature of 57°C there is again an increase in gel strength. With further heating of the solution to 75°C the gel strength reaches a constant value of ± 150 m gels. Cooling down the solution results in a final gel strength of ± 200 m gels. The same experiments are done in the presence of 2% milk proteins and 2% soy protein isolate. The results are presented in Figure 12. A conclusion from this figure—addition of 2% milk protein (which has no gellifying capacity itself) results in a much higher gel strength of a heated meat protein solution.

This result supports the theory that there is a synergistic effect of milk proteins on the gellifying capacity of salt-soluble meat proteins, either milk proteins play a role in the cross-linking of myosin molecules or they absorb a significant amount of water, which results in a higher net concentration of myosin molecules. Other nonmeat proteins only partially have this specific function (eg, soy protein isolate) or do not have it at all.

Table 4. Classification of Meat Proteins

Protein	Percent	Role	Solubility	Gellifying Properties	Emulsifying Properties
Sarcoplasmic	30–35	Metabolic	Water	Weak	Weak
Stromal	15–20	Connective	Insoluble	No	Weak
Myofibrillar	50–55	Contractile	Salt soluble	Strong	Strong

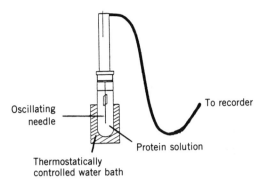

Figure 9. The gelograph.

POULTRY

Poultry Consumption

Whereas the consumption of red meat is stable or even decreasing, poultry consumption has grown impressively over the last decade. The main reasons for the increase of consumption of poultry are a healthy image (lean, low in fat and cholesterol), increase in the amount of further-processed poultry meat (convenience), and relatively cheap source of animal protein.

The consumption of poultry per capita in different areas of the world is given in Table 5.

Poultry Rolls and Poultry Bolognas

Poultry rolls and poultry bolognas are in appearance very similar to their red meat counterparts. They can be manufactured from materials such as trimmings, skin, fat, and mechanically deboned poultry (mdp). The basic difference between a poultry roll and a poultry bologna is that the bologna sausage contains curing salt and is finely emulsified (20). Poultry rolls normally have a certain ratio of fine meat emulsion to meat chunks.

It is important to create an even color appearance within the roll, ie, the color of the surrounding fine emulsion must match the color of the meat chunks. To obtain this even and white color, it is recommended that milk proteins be used in both basic components: part in the fine emulsion and part mixed into the meat chunks.

Some chicken or turkey bolognas are made from 100% mechanically deboned poultry. From a technologic point of view the poultry meat emulsion is not very different from the red meat emulsion. Sometimes poultry bolognas can have a dry mouthfeel. This can be corrected by the addition of approximately 10% chicken skin/fat emulsion. Such a pre-emulsion is made of 1 part milk protein (sodium caseinate), 5 parts chicken skin, 5 parts chicken fat, 6 parts water, and 2 parts ice.

Without the assistance of such a prestabilized fat/skin emulsion, fatting out often occurs when the total fat content reaches 20% or higher. The addition of milk protein as a dry powder to such products (addition level generally between 1.0 and 1.5%) improves the consistency of the final product as well; generally firmer and better sliceable products are obtained.

Chicken Nuggets and Patties

Since their introduction in the early 1980s, these poultry items have become very popular. Initially, a chicken nugget was a solid piece of slightly marinated breast meat that was battered and breaded (20).

Increased demand has forced the poultry processor to use such parts as thigh meat, trimming, mechanically deboned meat, and chicken skin. The target in combining all these ingredients in a nugget or a patty is to reach the appearance, flavor, and bite of the original product as closely as possible. Important in the stabilization of the final product is the binding of the meat pieces, the water holding capacity, and the stability of the fat.

Again, some basic rules in meat technology are very important: for optimal binding of the meat pieces, it is necessary to grind the meat before use; this creates relatively small meat pieces with a large surface area. During the mixing stage, when salt and phosphates are added, it is important that sufficient meat proteins are extracted and become hydrated in the continuous phase (28–30). On

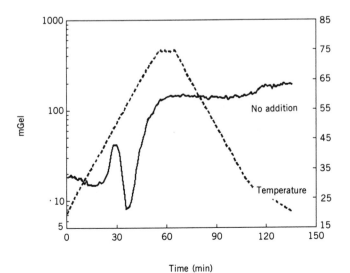

Figure 10. Results of an experiment on gelograph meat protein (no addition).

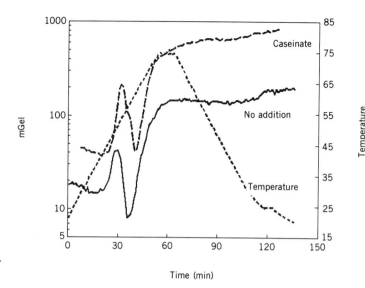

Figure 11. Results of an experiment on gelograph meat protein (no addition, caseinate).

frying, this continuous phase will gellify and cement the meat pieces together. When milk protein is being used in these systems, they boost the meat protein gelation, resulting in higher yields and better consistency of the final products.

Products based on lean meat exclusively tend to be dry and tough after frying. By incorporation of prestabilized chicken fat/skin emulsion (usually 10%), flavor, consistency, and juiciness can be improved.

Whole Muscle Poultry Products

One of the most popular food items in this group of poultry products is breast of turkey. Essentially the manufacturing process comes very close to the production process of reformed hams. Important differences are: lower injection levels are applied (up to 25%); lower salt concentrations are being used; no curing ingredients are being used, because a white color is desired; and the tumbling process is generally much shorter and less intense than that practiced in ham manufacture.

For the production of breast of turkey both milk protein (type sodium caseinate) and milk protein concentrates are used frequently. Usage levels vary from 0.8 to 1.5% (calculated on the final product). Yield increases from 3.0 to 6.0% can be reached. At the same time the consistency, binding, and sliceability of the product are improved. One of the most important characteristics of milk proteins in comparison to, eg, carbohydrate-based stabilizers is the fact that milk proteins do not affect the original flavor and palatability of the final product.

SEAFOOD

Since the growing interest in food and health in general, seafood items have become very popular over the last 10 years. The major reasons for the impressive increase in seafood consumption are sea foods are low in fat, high in protein, contain high percentages of polyunsaturated fatty acids, and most sea foods are relatively cheap.

Although fish and meat are two completely different

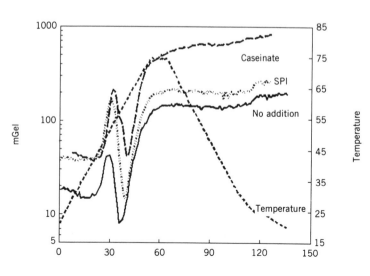

Figure 12. Results of an experiment on gelograph meat protein (no addition, caseinate, SPI).

Table 5. Per Capita Consumption of Poultry, 1982–1987 (kg)[a]

Area	1982	1983	1984	1985	1986	1987
United States	29.3	29.9	30.5	32.0	33.1	35.6
European community	15.5	15.6	15.6	15.9	16.2	16.9
USSR	9.6	9.4	9.5	10.3	10.7	11.2
Japan	11.1	11.4	11.8	12.2	12.9	13.6

[a] Ref. 2.

sources of protein, there is a striking similarity from a technologic point of view. Muscle fibers in seafood are almost identical to fibers in red meat (31). A remarkable difference is the very low content of collagen protein in fish-meat. This is the main reason for its softer texture and bite compared to red meats.

Because of the similarity between red meat and fish-meat structure a lot of processing technologies from the red meat industry can be transferred to processed seafood. This is especially true for fish sausages, fish nuggets, and fish burgers.

Canned Tuna

Canned tuna, but eg, also canned salmon, mackerel, and sardines, has been on the market for years as preserved fish with a long shelf life. For a good heat transfer and maintenance of the natural appearance and taste during sterilization, it is absolutely necessary that the fish is canned with a liquid as heat transfer medium. Until recently the liquid used was usually oil, but as a result of changes in the food consumption pattern and health consciousness of people, an aqueous solution has been applied more frequently in recent times. The switch from oil to an aqueous solution is naturally associated with some changes in product characteristics, and it is apparent in practice that milk protein hydrolysate is a perfect ingredient for stabilizing the canned product in respect to flavor, juiciness, and yield.

In the preparation of canned tuna and other species of fish the starting point is usually precooked fish. After steam-cooking the fish is cooled and bones, skin, and the dark fish flesh are removed carefully and the fish is suitable for canning. A formula for water-packed tuna is given in Table 6.

Milk Protein Functionality

A light color, mild flavor, juicy texture, and good water penetration during sterilization are extremely important characteristics of premium quality canned tuna. Milk protein hydrolysate promotes the juiciness of the fish struc-

ture, gives a mild taste, and has an excellent effect on the water retention of the tuna during sterilization. Adding 10% milk hydrolysate, as indicated in the brine formula, improves the drain weight of the tuna by 6 to 8%. This means that milk proteins not only effect the quality of the final product, but have important economic advantages as well (30).

Table 6. Water-Packed Tuna

Component	Percent of Total
Steam-cooked tuna	75
Brine	25
Brine formula	
Water	86
Salt	4
Milk protein hydrolysate	10

BIBLIOGRAPHY

1. R. Hermus and H. Albers, "Meat and Meat Products in Nutrition," *Proceedings of 32nd European Meeting of Meat Research Workers,* Gent, Belgium, 1986.

2. ZMP, Bilanz 1987, Zentrale Markt- und Preisberichtstelle für Erzeugnisse der Land-, Forst- und Ernährungswirtschaft GmbH, Bonn, FRG, 1988.

3. F. M. W. Visser, "Dairy Ingredients in Meat Products", *Proceedings of the International Dairy Federation Seminar: Dairy Ingredients in the Food Industry,* Luxembourg, May 19–21, 1981.

4. S. S. Gulayev-Zaitsev, "The Use of Caseinates in Foods," in Organizing Committee of the XXII International Dairy Congress, ed., *Milk the Vital Force,* D. Reidel Publishing Company, Dordrecht, Holland, 1986.

5. P. Walstra and R. Jenness, *Dairy Chemistry and Physics,* John Wiley & Sons, Inc. New York, 1984.

6. P. V. Farrant, "Muscle Proteins in Meat Technology", in P. F. Fox and J. J. Cordon, eds., *Food Proteins,* Applied Science Publishers Ltd., New York, 1982.

7. R. Hamm, *Kolloidchemie des Fleisches,* Paul Parey, Berlin and Hamburg, 1972.

8. J. Schut, "Meat Emulsions'" in S. Friberg, ed., *Food Emulsions,* Marcel Dekker Inc., New York, 1976.

9. P. M. Smith, "Meat Proteins: Functional Properties in Comminuted Meat Products," *Food Technology* 4, 116–121 (1988).

10. N. L. King and J. J. MacFarlane, "Muscle Proteins," in A. M. Pearson and T. R. Dutson, eds., *Advances in Meat Research, Restructured Meat and Poultry Products,* Vol. 3, Van Nostrand Reinhold Co., New York, 1987.

11. J. Schut, F. Visser, and F. Brouwer, "Microscopical Observations During Heating of Meat Protein Fractions and Emulsions Stabilized by These," *Proceedings of 25th European Meeting of Meat Research Workers,* Budapest, Hungary, Aug. 27–31, 1979, Paper 6.12.

12. J. Schut, "Zur Emulgierbarkeit von Schlachtfetten bei der Herstellung von Brühwurst," *Fleischwirtschaft* 48, 1201–1204 (1968).

13. J. Schut and F. Brouwer, "The Influence of Milk Proteins on the Stability of Cooked Sausages," *Proceedings of 21st European Meeting of Meat Research Workers,* Aug. 31–Sept. 5, Bern, Switzerland, 1975, pp. 80–82.

14. J. Schut, F. Visser, and F. Brouwer, "Fat Emulsification in

Finely Comminuted Sausage: A Model System," *Proceedings of 24th European Meeting of Meat Research Workers,* Kulmbach, FRG, 1978, paper W 12.

15. S. A. Ackerman, C. Swift, R. Carroll, and W. Townsend, "Effects of Types of Fat and of Rates and Temperatures of Comminution on Dispersion of Lipids in Frankfurters," *Journal of Food Science* **36,** 266–269 (1971).

16. L. Knipe, "Sectioned and Formed Meat Products," *Proceedings Annual Sausage and Processed Meat Short Course,* Ames, Iowa, 1982, Res. Paper, 2113 E-7.

17. P. A. Morrisey, D. M. Mulvihill, and E. O'Neill, "Functional Properties of Muscle Proteins", in B. J. F. Hudson, ed., *Development in Food Proteins,* Vol. 5, Elsevier Applied Science, London, 1987.

18. J. E. Kinsella, "Relationships between Structure and Functional Properties of Food Proteins," in P. F. Fox and J. J. Gordon, eds., *Food Proteins,* Applied Science Publishers Ltd., New York, 1982.

19. R. Hamm and J. Grambowska, "Proteinlöslichkeit und Wasserbindung unter den in Brühwurstbräten gegebenen Bedingungen, Part II," *Fleischwirtschaft* **58,** 1345–1348 (1978).

20. H. Hoogenkamp, *Milk Protein. The Complete Guide to Meat, Poultry, and Sea Food,* DMV Campina bv., Veghel, Holland, 1989.

21. J. Schut and F. Brouwer, "Preferential Adsorption of Meat Proteins during Emulsification," *Proceedings of 17th European Meeting of Meat Research Workers,* Bristol, UK, 1971.

22. A. S. Bawa, H. Z. Orr, and W. R. Usborne, "Evaluation of Synergistic Effects Obtained in Emulsion Systems for the Production of Wieners, *Journal of Food Science Technology* **5,** 285–290 (1988).

23. A. Asghar, K. Samejima, and T. Yasui, "Functionality of Muscle Protein in Gelation Mechanisms of Structured Meat Products," *CRC, Critical Reviews in Food Science and Nutrition* **22,** 27–106 (1985).

24. E. A. Foegeding and T. C. Lanier, "The Contribution of Non Muscle Proteins to Texture fo Gelled Muscle Protein Foods," in R. D. Philips and J. W. Fixley, eds., *Protein Quality and the Effects of Processing* Marcel Dekker Inc., New York, 1989.

25. B. Egelandsdal, K. Fretheim, and O. Honbitz, "Dynamic Rheological Measurements on Heat Induced Myosin Gels: An Evaluation of the Methods Suitable for the Filamentous Gels," *Journal of the Science of Food and Agriculture* **37,** 944–954, 1986.

26. M. K. Knight, "Interaction of Proteins in Meat Products: Part III, Functions of Salt Soluble Protein, Insoluble Myofibrillar Protein and Connective Tissue Protein in Meat Products," Research report no. 630, Leatherhead Food R. A., Leatherhead, UK, 1988.

27. G. R. Ziegler and J. C. Acton, "Mechanisms of Gel Formation by Proteins of Muscle Tissue," *Food Technology* **5,** 77–82 (1984).

28. K. Samejima, M. Ishioroshi, and T. Yasui, "Relative Roles of the Head and Tail Portions of the Molecule in Heat Induced Gelation of Myosin," *Journal of Food Science* **46,** 1412–1418 (1981).

29. J. A. Dudziak, E. A. Foegeding, and J. A. Knopp, "Gelation and Thermal Transitions in Post Rigor Turkey Myosin/Actomyosin Suspensions," *Journal of Food Science* **53,** 1278–1281 (1988).

30. D. G. Huber and J. H. Regenstein, "Emulsion Stability Studies of Myosin and Exhaustively Washed Muscle from Adult Chicken Breast Muscle," *Journal of Food Science* **53,** 1282–1293 (1988).

31. J. Jongsma, *Milk Proteins in Processed Sea Food,* DMV Campina bv., Veghel, Holland, 1987.

MARTIEN VAN DEN HOVEN
BEN VAN VALKENGOED
Creamy Creations
Rijkevoort, The Netherlands

DAIRY INGREDIENTS FOR FOODS

Milk and its products have been consumed by many generations of people of Western, Middle Eastern, Indian, and some African cultures. The nutritional value of dairy materials, properly handled, has been proven by widespread use, and their safety is assured by processing procedures that have been introduced and refined since the time of Pasteur. The great bulk of dairy products today recognize traditions of simple processes and physical separations, avoiding chemicals and thereby preserving the natural properties and the very desirable connotations of traditional products.

The more recently developed food-processing industries have generated requirements for sophisticated performance of dairy materials as ingredients for both familiar and fabricated foods. Milk production does not require the use of pesticides and hormones. Well-managed herds can produce milk with contaminant levels far below the requirements of regulatory authorities. Active components that directly inhibit the processes of digestion, metabolism, and assimilation of nutrients are not present in milk.

The nature of milk and its components is probably better characterized than the properties of any other basic food. Consequently, the ability to transform milk and its individual components into appropriate food ingredients is extensive, as demonstrated by Figure 1.

FORMS OF DAIRY INGREDIENTS

Dairy ingredients are provided in many different forms, which can be grouped according to the effect of the process on the products.

1. Compositionally complete or only partially rearranged ingredients (eg, removal of water and/or fat).

Liquid milk and cream are available, but it is normally more convenient for the food processors to use products from which the water has been removed (as in the manufacture of whole milk powder and butter). The centrifugal removal of cream results in the manufacture of nonfat milk powder. These processes have been industrialized on a very large scale in order to minimize costs. Although they are inherently simple, sophisticated knowledge and process control are required to achieve consistent product performance as may be specified.

2. Products in which a desired component of milk is prepared (eg, milkfat, protein).

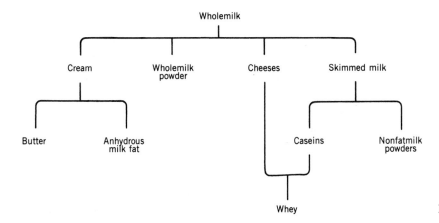

Figure 1. The Cascade of dairy materials.

Complete removal of water and nonfat solids from cream will result in anhydrous milkfat that is usable in the same manner as other oils but that is prepared without requiring a complex series of extractive procedures in its manufacture. The major protein of milk, casein, is obtained by simple isoelectric or enzymic precipitation followed by subsequent washing with water and drying.

3. Products obtained by the reassembly, or rearrangement, of the individual components.

At its simplest level the compositional adjustment of milk ingredients is readily achieved. For example, the control of fat levels in cheese products and milk powders is widely practiced; increased levels of whey solids in infant-food products are provided because they are nutritionally desirable; the recombination of powdered ingredients at locations that are far distant from the point of milk production has provided international availability of milk and its products.

4. Products in which the ingredients have been modified for specific purposes.

Judicious application of heat modifies the properties of milk components and can result in a range of milk powders tailored to satisfy specific industrial requirements (1). Addition of food-grade alkali to casein results in its solubilization (2). Physical modification by processes such as homogenization, to provide homogeneity, and agglomeration of powder particles to provide improved dispersibility, is of considerable value. More extensive modifications of dairy materials have been intensively researched. New processes are becoming available, and products that result from these newer technologies hold considerable promise in meeting the increasingly sophisticated requirements of food formulators. Transformations of milk components are being achieved. Lactose can be converted to lactulose (3), protein is being transformed into its hydrolysates (4), and milkfat can be deodorized and decolorized.

Extensively modified materials have been prepared, but the market has not yet taken up all the possibilities; the great bulk of commercial usage in the food industry lies with products that are well known and are based on simple processes.

THE INGREDIENT RANGE

The composition of dairy ingredients varies both naturally (because of their biological origin) and by design (through the manufacturing procedures) but can, in general terms, be set out as in Table 1.

Control of Component Ratios and Performance

A wide variety of performance requirements are imposed by the needs of food formulation. Products that contain various combinations of milk components are available in traditional, widely used forms and also in forms that have been recently modified for food-industry use. The composition of some available product groups is provided in Table 2.

Table 2 reviews the products that have been traditionally available and that are widely recognized. More recently, the specific needs of the food industry have been met by the development of milk powders for milk recombination plants. Recombiners require particular properties such as heat stability for the manufacture of recombined evaporated milks. Similarly, nonfat milk powders that provide manageable viscosity characteristics in the production of sweetened condensed milks are also required. Manufacture by the appropriate use of heat and by control of the mineral balance of the milks during processing are the essential manufacturing processes. An improvement in the dispersibility of powdered products has been achieved by progressively improved drier design so that powder particles can be agglomerated without damage to the flavor or nutritional properties of the product. More recently, effective coating of the particles with natural materials that impart hydrophillic properties has been achieved. The reduction of mineral levels by the processes of ion exchange and electrodialysis are widely practiced (15,16). This has allowed, in particular, the development of infant foods with low osmotic pressure and with managed mineral profiles that appropriately meet nutritional needs for nutrient balance.

Powders that contain milk fat are manufactured by means that ensure that stale flavors, due to oxidation developed during storage, do not arise in less than 12 months. Fat-containing powders that are intended for dispersion in water require particular care in manufacture (17).

Table 1. Typical Composition of Dairy Products

Product	Content, % dry basis				Moisture content, %
	Fat	Protein	Lactose	Mineral	
Milk	32	27	35	5	87
Cheese (cheddar)	53	38	2[a]	6[b]	33
Products with concentrated fat levels					
Cream	87	5	7	1	55
Butter	98	1	1	<0.5	16
Anhydrous milkfat (AMF)	100	<0.1	<0.1	<0.1	<0.1
Products with concentrated protein levels					
Nonfat dry milk	1	40	50	9	3.5
Casein (acid)	1	96	<0.1	2	11
Caseinate	1	95	<0.1	4	4
Whey powder	1	15	76	8	3.4
Whey protein concentrate (WPC)[a]	3	37	54	6	4
Whey protein concentrate[b]	5	79	12	4	4
Whey protein isolate	1	96		3	4
Isolated products					
Lactose			100		5
Minerals	1	8	1	83	10

[a] As lactate.
[b] Includes salt.

High-fat powders, such as cream and butter powders, which are particularly useful in the bakery industry, require appropriate formulation to ensure that free-flow properties are retained without loss of functional performance (18).

PRODUCTS IN WHICH MILKFAT IS THE DOMINANT COMPONENT

The use of cream, butter, and anhydrous milkfat (AMF), represent around 30% of the milkfat that is produced on

Table 2. Combinations of Milk Components

Product Class	Performance Required by Food Formulator	Process of Manufacture	Reference
Whole milk	A natural liquid	Pasteurized for safety	5
	A stable liquid	Sterilized for microbial stability	5
	A stable liquid with dairy flavor	UHT[a] sterilized and aseptically packaged to protect from oxygen and light	6
Compositionally altered milks	A powder for dry-blend purposes	Dehydrate by roller or spray drier	1
	Fat reduced or zero fat contribution	Remove milk fat centrifugally	7
	Protein enrichment	Protein added before drying	
	Sweetened and stable	Sugar addition, appropriate heat, crystalization concentration	8
	Cocoa crumb as chocolate ingredient	Codried cocoa and sweetened condensed milk	9
Modified milk	Reduced mineral levels (especially for infant foods) or mineral profile adjustment	Ion exchange and/or electrodialysis (of whey generally)	10
	Heat-stable powder	Mineral adjustment and appropriate heat control	11
	A powder for nutritional improvement in bread without loaf volume depression	Appropriate heat control	12
	Sweet but without sugar addition (especially for dietetic uses)	Hydrolysis of lactose by lactase	13
	A powder with high dispersibility	Agglomeration plus surface-coating with hydrophilic materials	14

[a] UHT, ultra-high temperature.

dairy farms worldwide. The 6.8 million tons used annually means that milkfat is the third most widely used fat in worldwide commerce. The nature of fat on presentation to the manufacturing process is pure and wholesome so that the use of extensive clean-up processes and chemical preservative procedures are unnecessary. Milkfat has a desirable and prestigious flavor, so the maintenance of that flavor is of special importance. For this reason, refrigerated storage of the products of milkfat is common. Because of a significant level of naturally occurring materials that have antioxidant activity, and the widespread use of stainless steel, which prevents contamination of milk with pro-oxidants, extended shelf life can be achieved without recourse to added antioxidents.

The manufacturing processes that are used are inherently simple physical procedures such as churning and dehydration. The product range is shown in Figure 2.

Rearrangement of the textural properties of milkfat is achieved in a number of ways (19):

1. Mechanical and thermal procedures are often used simultaneously in scraped surface heat exchangers and pin-working machines and result in plasticization. Butter-type products with controlled physical properties are the result.

2. Processes known as dry fractionation consist of crystallization from melted fat after controlled cooling followed by separation of the liquid from solid phase by vacuum filtration or centrifugation without coming into contact with any contaminating process aid. These processes yield products that contain increased solid fat levels that are particularly useful for croissants and puff pastry (20).

3. Compounding of milkfat with vegetable oil to modify spreadability properties of spreadable products results in a number of products that are successful at the retail level and are attractive to the catering industry. Commonly, these products employ 50% milkfat, 40% liquid vegetable oil to increase the pro-

portion of fat liquid at 5°, and 10% hard stock to provide standup properties.

The intensity of the yellow color of milkfat is, to some extent, controllable. The natural color of milk fat is carotene, a precurser of vitamin A, and is not implicated in any concern for health. Selection of the milkfat origin according to breed of cow, feed source to the herd, and season of production will provide a range of color properties. More extensive color modification can be achieved by use of steam deodorization processes that are commonly employed in the oils industry. It yields a white milkfat that is finding extensive use in the manufacture of white cheeses and coffee whiteners and holds promise in other foods.

Experimental investigation into the extensive modification of milkfat by complex processes such as interesterification and hydrogenation has not, so far, yielded products having adequately improved benefits to justify the cost of the processes. Such approaches would simultaneously harm both the distinctive flavor and the naturalness of milkfat (Table 3).

MILK PROTEINS

The products that contain milk proteins are derived from skimmed (nonfat) milk (Fig. 3). The proteins of milk have a well-balanced amino acid profile, having been designed by nature for the rapid early phase of growth that is necessary for infant mammals. By comparison with a reference protein, which has been described by a FAO/WHO group of experts, milk proteins meet all the requirements for essential amino acids. Whey proteins are particularly rich in the sulfur-containing amino acids (methionine and cystine), so they serve a valuable role complementing the amino acid profile of proteins of vegetable origin in food formulations that require high nutritional quality (30).

A physiological role for milk proteins in human nutrition is based on the activity of the immunoglobulins, lac-

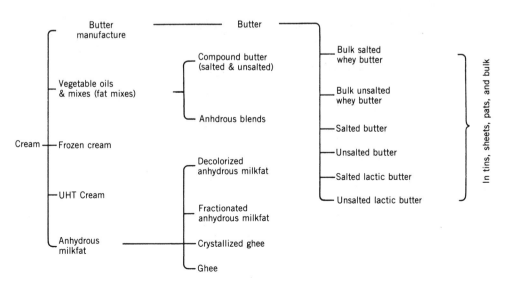

Figure 2. The products derived from cream.

Table 3. Milk Fat Ingredients

Product Class	Performance Requirement by Food Formulator	Process of Manufacture	Reference
Whole milk			
Cream	Natural flavor and liquid, whippable	Centrifugal separation	7
Butter	Solid or spreadable form	Churning	21
	Spreadability	Ammix process	22
Component: Anhydrous milk fat	Strong dairy flavor with no interfering components	Dehydrate and wash	23
Rearrangement of composition	Control of melting characteristics while retaining natural flavor performance	Selection of natural properties	24
		Fractional crystalization and separation	25
		Inclusion of nondairy oils into churning process (compound butter)	
	Greater spreadability or increased levels of polyunsaturated fatty acids	Inclusion of polyunsaturated oils before churning or by recombining	26
		Introduce protected polyunsaturated oils to feed of the cows	27
Modifications	Light color	Destruction of carotene by steam deodorization	
	Deep color	Select milkfat by origin	28
	Melting-point control	Fractional crystalization and separation	29

toferrin, and other proteins that are present in small amounts (31). The particular amino acid profile of whey proteins has been proposed as a fundamental property that improves immunocompetence of humans (32).

The proteins in milk are recoverable by long-practiced separation procedures such as by very specific enzymatic precipitation (as in the case of rennet casein), heat precipitation (as in lactalbumin), and by isoelectric precipitation (as in acid caseins). The most widely used of these is the last-named (33).

Recent technological advances have resulted in the manufacture of soluble whey protein concentrates that are produced by the membrane process of ultrafiltration of whey. Functional whey protein concentrates that contain up to about 80% protein are commercially available.

Products that provide the combined benefits of casein and whey protein have been available in the form of copre-

cipitates, which are manufactured by control of temperature, acidity level, and calcium level (33). More recently developed technology using shifts in pH has resulted in new products that combine strong functionality with high PER levels (34,35).

Specific functional performance has been engineered into caseinates, which have been solubilized from acid casein by reaction with alkali (2). Arising from a precise understanding of the influence of minerals, a growing comprehension of the modifications achievable by controlled management of enzymatic reaction, a number of highly specific protein products have been created. Also of considerable importance to the food formulator is the fact that the protein component of a food should not contribute unwelcome flavors. Adsorption technology has been used to enhance the already bland flavor of casein products. The use of casein (and caseinate) as an ingredient in foods

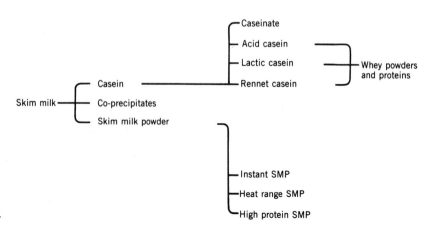

Figure 3. The products derived from skim milk.

is very extensive, as it provides strong functional performance, high levels of protein relative to milk, competitive cost, and strong nutritive quality (36).

Similar product development has been applied to whey protein concentrates (WPCs) with the result that enhanced gelation performance and products with unusual heat stability have been achieved. The selection of an appropriately functional WPC for a particular use in food formulation can offer substantial benefits (37).

Very high purity levels of whey protein have been achieved on the industrial scale by ion exchange processes that produce isolates, but the cost penalty that is incurred means that only the very highest-priced market applications can be considered for such products.

The enzymatic hydrolysis of proteins may be used to produce products of low or zero allergenicity. This technique is receiving intensive scientific attention and is likely to result in products that will have considerable

impact in improving the health of infants, and possibly others, who have specific nutritional requirements (Table 4).

CHEESE AND CULTURED PRODUCTS

Cheese and cultured milk products generally are consumed in their native form, but they also can provide a nutritionally valuable ingredient resource with especially useful flavor and compositional characteristics that are well regarded by consumers (48). The cheesemaker can provide a graduated series of flavor, texture, and compositional properties to suit a variety of needs. Consequently cheese types vary widely, as shown in Figure 4.

Natural cheeses are manufactured by controlled fermentation, which influences both acidity and flavor, and which in combination with controlled composition, influ-

Table 4. Milk Protein

Product Class	Performance Requirement of Food Formulation	Process of Manufacture	Reference
Whole proteins			
Coprecipitate	Balanced amino acid profile	Precipitated under managed conditions of heat, acid, and calcium concentration	38
Isolates	Strong functional performance	pH shifts and isoelectric precipitation	34,35
Components			
Casein-acid	Broad functional performance, bland, nutritionally strong	Isoelectric precipitation, multiple washing procedure	33,36
Casein-rennet	Nutritionally strong, including bioavailable calcium, stable bland flavor, stretch properties for analogue cheese	Enzymic precipitation and multiple washing	39
Whey protein			
Lactalbumin	Amino acid balance complementary to vegetable proteins No interfering function	Heat precipitation and spray dried or ring dried	40
WPC,[a] low protein level	Milk powder alternative at economical price	Ultrafiltration to around 35% protein content	
WPC, high protein level	Soluble, bland, heat setting	Ultrafilter and diafilter to around 75% protein content	41
Isolates	Pure proteins	Ionic adsorption	42
Compounding	Specific functional performance	Introduce vegetable proteins in blends or by interreaction	43
Modification Caseinates			
Sodium	Solubility, natural emulsification, foam, fat binding, water holding Bland flavor	Treat with sodium alkali and spray dry Carbon asorption	36,44
Calcium	Colloidal protein, viscosity control	Treat with calcium alkali and spray dry	45
Potassium	Soluble and functional with low sodium	Treat with potassium alkali	
Various	Special performance (eg, high foam, low viscosity, gelation)	Various proprietary procedures	46
Whey proteins, WPC	Special performance (eg, strong gels, heat stability)	Various proprietary procedures	47
Hydrolysates	Nutritional performance, reduced allergenicity	Enzymic modification	4

[a] WPC, whey protein concentrate.
[b] WPC, whey protein concentrate

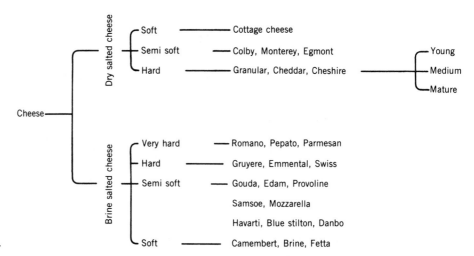

Figure 4. Cheese types.

ences texture (49). Texture control extends from gels, through spreadable pastes, to firmly structured semisolids. Processed cheese, which is made by melting the natural cheese and then dissolving the protein so that a new emulsion is established, extends the range of textural properties and forms of presentation that can be offered.

High-melt properties that can provide intact cheese pieces in, eg, heated sausages have been achieved. Stretch properties as may be required for pizza, and melt characteristics as may be required for crust appearance, can be built-in during the manufacturing process.

Cheese flavors are often desired characteristics, and they range widely. They include fresh and carbohydrate-derived flavor as in lactic cottage cheese and in creamy harvati; protein-derived flavor as in matured cheddar and camembert; and fat-derived flavors as in parmessan and blue cheeses. Further flavor extension by inclusion of natural materials such as spices and herbs can be used with both natural and processed cheeses. Flavor contributions are particularly significant to cheese sauces, dressings, and parmessan toppings for European-style foods.

Particulate forms of cheese, prepared by grating, dicing, or drying, provide convenience for incorporation into food formulations. Varied flavor intensity (such as by enzyme treatment) can also be built into powdered cheeses. Powdered yogurts are available, for coating and as formulation ingredients. These can be used, along with cream cheese, cottage cheese, and quarg for a range of cheesecakes and similar products.

New technologies, such as the use of ultrafiltration of milk before cheesemaking, have already yielded commercially successful forms of feta cheese with increased shelf life and promise to further extend the range of product composition that will be created (Table 5).

LACTOSE

The sugar of milk is lactose, a disaccharide that has one of the lowest levels of sweetness contributed by an carbohydrate. Because it is approximately 25% as sweet as sucrose, it can be used in food formulations without domi-

Table 5. Cheeses

Product Class	Performance Requirement of Food Formulation	Process of Manufacture	Reference
Whole			
Natural	Flavor and texture selection, strong nutritional contribution Stretch and melt control	Fermentation, controlled acidity and heat, controlled maturation	50
Powders	Flavor enhancement	Selection of type and maturity, emulsify and spray dry	51
Particulate	Sprinkling (onto pizza)	Shredding or dicing	
Compositionally modified			
Skim milk cheese	Protein source	Fat-reduced milk	
Cream cheese	Fat source flavor contribution and texture	Cream addition, culturing	
Modified cheese	Flavor enhanced Stretch control	Enzymic modification Acidity management	53
Processed cheese	Specific texture requirements, flavor modification	Emulsification. Use of nondairy ingredients	53
	Specific shape	eg, Slices	54

Table 6. Lactose

Product Class	Performance Requirement	Process	Reference
Component	Low sweetness, reduction of water activity, enhanced browning. Promotion of calcium and phosphorous absorption (especially in infant feeding), dispersing agent	Concentration crystalization and particle size control	55,56,57,59
Modified	Sweet syrup	Hydrolysis by lactase, concentration	13,60
Transformed	New materials	Various	55,61

nating the natural flavors of other food components. Its solubility and sweetness vary with temperature and concentration, and these properties may require control in some forms of use (55).

Lactose is prepared by the processes of concentration and multiple crystalization followed by management of crystal size and properties. Lactose is available both at high levels of purity (including pharmaceutical quality) and, at lower cost, as the major component of dry whey powder or dry whey permeates (55).

When used in the baking industry, it acts as a reducing sugar to promote the Maillard reaction, which increases the browning of the crust. At higher temperatures it will caramelize and so contribute to flavor.

In formulated powdered products lactose crystals are slow to take up moisture and consequently minimize the likelihood of caking and lumping. This same property is also a valuable aid to tableting.

Nutritionally, lactose has been shown to promote calcium and phosphorous absorption, which is especially useful in infant feeding preparations (56). It protects against destabilization of the caseinate complex during the drying process. Its presence can also maintain biological activity in the preparation of enzyme products. In brewing and baking applications, it is not fermented by conventional strains of yeast so its contribution to sweetness and color is not destroyed by the biological processes of those industries.

The hydrolysis of lactose yields a sweet syrup that contains glucose and galactose, which may have nutritional advantages in some dietary applications. Lactose-hydrolyzed syrups from permeates and wheys are commercially available and are being used in confectionery and ice cream (Table 6).

THE NUTRITIONALLY VALUABLE MINOR COMPONENTS OF MILK

The new technologies that have arisen from the biotechnology revolution are permitting the extraction of a number of biologically active materials. Some of these biologically active components are present in large proportions in the colostrum milk but decline to low levels during the lactational period. The extraction of lactoferrin, which is an iron-binding protein, has been achieved commercially by sophisticated procedures of chromatography. This protein has been shown to be inhibitory to a number of pathogenic microorganisms, plays a role in the immunological process, and also provides a highly available source of iron. The enzyme lactoperoxidase has also been extracted from dairy sources and is capable of acting to destroy microbes in a completely safe manner that would provide an alternative to the use of antimicrobial agents.

Natural antibodies are also present in very low concentration in unprocessed milk, but traditional procedures have resulted in their inactivation. Procedures are now available that allow for preparation of protein products having biological activity, and there is evidence that disease control is achievable (62). Currently these products are targeted toward the problems of animal health. Scientific enquiry is targeted toward the improvement of human health. Wide-ranging implications for a new range of foods that have physiological functions exist and may provide a new wave of the future (63,64).

The recognition of the value of calcium naturally present in milk and cheese in meeting the nutritional needs of aging people, particularly women, has resulted in a demand for pure mineral ingredients of natural origin, fine particle size, and clean flavor. The natural calcium content of milk is being recovered by precipitation processes for direct incorporation into foods (Table 7).

PREPARED INGREDIENTS FOR FOODS

Sophisticated food ingredients have been specifically designed to provide physical functional performance for particular food systems. Strong functional properties are provided by dairy-based ingredients.

Casein protein has an open structure and has separate areas of hydrophobic and hydrophyllic nature along the protein molecule. Consequently, it diffuses readily to interfaces and has powerful emulsion-forming end stabilizing properties. Whey proteins unfold on heating and so, similarly, expose hydrophillic areas of the polypeptide chains. They also gel and can provide natural thickening and stabilizing function. The properties of caseinates and coprecipitates can be tailored by control of ionic content to provide a range of viscosities, solubilities, foaming and whipping (67), and surface activity (68). Measurements of the function of milk proteins have proved difficult to relate to the performance of these proteins in the complex environment of food systems (69), but considerable progress has been made (70,71). Manufacturing techniques

Table 7. Nutritionally Valuable Components in Minor Concentrations

Product Class	Performance Requirement of Food Formulations	Process of Manufacture	Reference
Minerals	Natural, origin, low flavor, bioavailable	Precipitation	
Biologically active			
Lactoferrin	Microbiological inhibition by iron binding. Nutritionally available form of iron	Sophisticated chromatography	65
Lactoperoxidase	Natural antimicrobial system	Sophisticated chromatography	65
Immunoglobulins	Natural antibody active against a wide range of diseases	Specialized ultrafiltration	66

that control the function of proteins are being practiced (72).

The functional properties of milkfat vary because of the effects on the secretory process in the mammary glands of cows by the breed of the cow and by the type of feed that she receives and that changes seasonally. Levels of technological control by temperature control during processing, texturizing by physical means, and fractionation by controlled crystalization minimize these variations (19). Milkfats tailor-made for specific purposes are available (73). The primary role of milkfat, when incorporated into a food, is the contribution that it makes to flavor, texture, and consumer acceptability (74).

The benefits that dairy ingredients provide to food formulators differ according to the particular needs of each industry. For confectionery products, milkfat is compatible with cocoa butter in that it becomes part of the continuous fat phase and it contributes to smooth flavor and texture of milk chocolate. Caramel flavor is best developed from sweetened condensed milk by utilizing the browning properties of lactose and protein. Proteins, especially caseins, enhance moisture retention by candy and control the quantity of free and bound water (75). Hydrolyzed milk proteins act as whipping agents for frappés and marshmallows (76).

For bakery products, milk powders improve crust color, resilience, and structural strength of cakes. Whey powders improve tenderness and shortness. Whey powder and skim milk powder, when used in dough for cookies and biscuits, reduces the tendency of the dough to tear and produces smooth, even browning. Milk proteins at levels up to 20% improve the nutritional value of cookies end biscuits because they contain high levels of the essential amino acids that are deficient in soy and wheat flour. Butter imparts distinctive flavor to butter cookies and croissants (77,78).

For meat products, milk proteins offer improved appearance, increased yield, and economy. The powerful emulsion capacity of sodium caseinate is particularly valuable in comminuted meat products. Whey protein concentrates of high gel strength are important in reformed ham. Lactose will mask salts, phosphates, and bitter aftertaste while providing a low sweetness profile in liver products, cooked hams, and cooked sausages (79,80).

A very wide variety of proprietary dairy ingredients are manufactured for convenience of use in the food industry. Specific attention is paid to the physical form of powders to ensure dispersibility, mixibility, and low dust levels. Butter is available as flakes and in sheet form. Cheese may be shredded, diced, or sliced. Flavor enhancement, particularly for cheese and milk powder, is offered (81).

CONCLUSION

The remarkably wide range of products that has been obtained from milk demonstrates the intensity of technological attention that has been devoted to dairy materials over a long period of time. The increasing interest in progressively more sophisticated materials that will be required by the food industry is opening up an increasing range of possibilities for the future. While the opportunities for sophisticated materials are exciting, the great bulk of dairy ingredient purchases are based on the recognition of value for money and a clear understanding of the increasing range of uses for products that can be produced by relatively simple processes from a pure and natural raw material.

Acknowledgment: The authors gratefully acknowledge the valuable assistance of the staff of the New Zealand Dairy Research Institute during the preparation of this paper.

BIBLIOGRAPHY

This article has been adapted and used with permission from K. J. Kirkpatrick and R. M. Fedwick, "Manufacture and General Properties of Dairy Ingredients," *Food Technology* **41** (10), 58–65 (1987).

1. M. E. Knipschildt, "Drying of Milk and Milk Products," In R. K. Robinson, ed., *Modern Dairy Technology,* vol. 1, Elsevier Applied Science Publishers, London, 1986, p. 131.
2. C. Towler, "Conversion of Casein Curd to Sodium Caseinate," *New Zealand Journal of Dairy Sciences and Technology* **11** (24), (1976).
3. U.S. Pat. 4,273,922 (1981) K. B. Hicks.
4. R. J. Knights, "Processing and Evaluation of the Antigenic-

ity of Protein Hydrolysates," in F. Lifshitz, ed., *Nutrition for Special Needs in Infancy,* Marcel Dekker, New York, 1985.

5. M. J. Lewis, "Advances in the Heat Treatment of Milk," in Ref. 1, p. 131.

6. International Dairy Federation, *New Monograph on UHT Milk,* Document No. 133, International Dairy Federation, Brussels, Belgium, 1981.

7. C. Towler, "Developments in Cream Separation and Processing, in Ref. 1.

8. B. H. Webb, "Condensed Products," In B. H. Webb and E. O. Whittier, eds., *By-products from milk,* 2nd ed., AVI Publishing Co., Westport, Conn., 1970, p. 83.

9. B. H. Webb, 1970b. "Miscellaneous Products," in Ref. 8, p. 285.

10. J. G. Zadow, "Utilization of Milk Components: Whey," in ref. 1, p. 93.

11. K. J. Kirkpatrick, "Raw Material Selection for Recombined Evaporated Milk Products." In *Proceedings of IDF Seminar on Recombination of Milk and Milk Products,* Document No. 142. International Dairy Federation, Brussels, 1982, p. 91.

12. E. J. Guy, "Bakery Products," in Ref. 8, p. 197.

13. R. R. Mahoney, "Modification of Lactose and Lactose-containing Dairy Products with Beta-Galactosidase," *Developments in Dairy Chemistry* 3, 69 (1985).

14. H. G. Kessler, "Drying-Instantizing," in *Food Engineering and Dairy Technology,* Verlag A. Kessler, Germany, 1981.

15. B. T. Batchelder, "Electrodialysis Applications in Whey Processing," in *Proceedings of International Whey Conferences.* Whey Products Institute and International Dairy Federation, 1986.

16. H. Jonsson and S-O. Arph, "Ion Exchange for Demineralization of Cheese Whey," in Ref. 15.

17. U.S. Pat. 4,737,369 (1988) I. A. Suzuka, and K. Mori.

18. F. G. Kieseker, J. G. Zadow, and B. Aitkin, "Further Developments in the Manufacture of Powdered Whipping Creams," *Australian Journal of Dairy Technology* 34, 112 (1979).

19. E. Frede, "Technological and Analytical Aspects of Milk Fat Modification," in *Conference Proceedings Food Ingredients Europe 1989,* Expoconsult Publishers, Maarsen, The Netherlands, 1989, pp. 55–61.

20. L. Eyres, "Milkfat Product Development, *Lipid Technology* 1(1), 12 (1989).

21. R. A. Wilbey, "Production of Butter and Dairy-based Spreads," Ref. 1, p. 93.

22. H. T. Truong and D. S. Munroe, "The Quality of Butter Produced by the "Ammix" Process," *Brief Communications,* 21st International Dairy Congress, vol. 1, p. 341.

23. A. Fjaervoll, "Anhydrous Milkfat, Manufacturing Techniques and Future Applications," *Dairy Industries* 35, 424 (1970).

24. M. W. Taylor and R. Norris, "The Physical Properties of Dairy Spreads," *New Zealand Journal of Dairy Science and Technology* 12, 166 (1977).

25. Austr. Pat. 431,955 (1968), I. T. H. Olsson.

26. M. M. Chrysam, "Table Spreads and Shortenings" in T. H. Applewhite, ed., *Bailey's Industrial Oil and Fat Products,* vol. 3, 1985, p. 1.

27. A. D. Fogerty and A. R. Johnson, "Influence of Nutritional Factors on the Yield and Content of Milkfat. Protected Polyunsaturated Fat in the Diet," Document No. 125, International Dairy Federation, Brussels, 1980, p. 96.

28. A. R. Keen, "Seasonal Variation in the Colour of Milkfat from Selected Herds and Two Dairy Plants," *New Zealand Journal of Dairy Science and Technology* 19, 263 (1984).

29. A. E. Thomas, "Fractionation and Winterization: Processes and Products," in Ref. 26, p. 1.

30. E. Renner, "Milk Proteins," in *Milk and Dairy Products in Human Nutrition,"* W-GmbH, Volkswirtschaftlicher Verlag, Munich, Germany, 1983, pp. 90–115.

31. L. Hambraeus, "Importance of Milk Proteins in Human Nutrition: Physiological Aspects," in T. E. Galesloot and B. J. Tinvbergen, eds. *Milk Proteins 1984,* Pudoc, Wageningen, Germany, 1985, p. 63.

32. G. Bounous and P. A. L. Kongshavn, "Influence of Protein Type in Nutritionally Adequate Diets on the Development of Immunity," in M. Friedman, ed., *Absorption and Utilization of Amino Acids,* vol. 2, CRC Press, Inc., Boca Raton, Fla., 19 , pp. 219–233.

33. C. R. Southward and N. J. Walker, "Casein Caseinates and Milk Protein Coprecipitates," in *CRC Handbook of Processing and Utilization in Agriculture,* vol. 1, 1982, p. 445.

34. U.S. Pat. 4,376,072 (1982), P. B. Connolly.

35. U.S. Pat. 4,519,945 (1985), H. A. W. E. M. Ottenhof.

36. C. R. Southward, "Uses of Casein and Caseinates," in P. F. Fox, ed., *Developments in Dairy Chemistry,* vol. 4, *Functional Milk Proteins,* Applied Science Publishers, London, 1989, pp. 173–244.

37. M. J. Rockell, "Selecting the Correct WPC for Your Food Applications via a Knowledge of the Functional Properties of WPC's," in *Food Ingredients Europe, Conference Proceedings,* Expaconsult Publishers, Maarssen, The Netherlands, 1989, p. 51.

38. L. L. Muller, "Manufacture of Casein, Caseinates and Coprecipitates," in Ref. 36, vol. 1, 1982, p. 315.

39. C. R. Southward and N. J. Walker, "The Manufacture and Industrial Use of Casein," *New Zealand Journal of Dairy Technology,* (1980).

40. B. P. Robinson, J. L. Short, and K. R. Marshall, "Traditional Lactalbumin—Manufacture, Properties, and Uses," *New Zealand Journal of Dairy Science and Technology* 11, 114 (1976).

41. K. R. Marshall, "Proteins, Industrial Isolation of Milk Proteins: Whey Proteins" in Ref. 36, vol. 1, 1982, p. 339.

42. D. E. Palmer, "Recovery of Protein from Food Factory Wastes by Ion Exchange," in P. E. Fox and J. J. Congson, eds., *Food Proteins,* Applied Science Publishers, New York, 1981, p. 341.

43. U.S. Pat. 4,486,343 (1984), N. J. Walker and P. B. Connolly.

44. A. Bergman, "Continuous Production of Spray Dried Sodium Caseinate," *Journal of the Society of Dairy Technology* 25, 89 (1972).

45. J. Roeper, "Preparation of Calcium Caseinate from Casein Curd," *New Zealand Journal of Dairy Science and Technology* 12, 182 (1977).

46. U.S. Pat. 4,126,607 (1978), W. C. Easton.

47. J. E. Kinsella, "Proteins from Whey: Factors Affecting Functional Behaviour and Uses," *Proceedings of IDF Seminar, Atlanta, Ga.,* International Dairy Federation, Brussels, Belgium, 1986, p. 87.

48. J. C. Dillon, "Cheese in the Diet," in Ref. 50, pp. 499–511.

49. M. E. Johnson, "Cheese Chemistry," in *Fundamentals of Dairy Chemistry,* 3rd ed., by Van Nostrand Reinhold Co., New York, 1988, pp. 634–654.

50. A. Eck, *Cheesemaking, Science and Technology,* Lavoisier Publishing, Inc., New York, 1986.

51. T. I. Hedrick, "Spray Drying of Cheese," in *Proceedings of Second Marschall International Cheese Conference,* 1981, p. 76.

52. L. Talbott, "The Use of Enzyme Modified Cheeses for Flavouring Processed Cheese Products," in Ref. 51, p. 81.

53. A. Meyer, *Processed Cheese Manufacture, Benckiser-Knapsack GMBH, 1970.*

54. J. H. Carne, "Sliced Process Cheese Production," In *Proceedings from the First Marschall International Cheese Conference,* 1979, p. 305.

55. V. H. Holsinger, "Lactose," in Ref. 49, pp. 279–342.

56. S. G. Coton, T. R. Poynton, and D. Ryder, "Utilisation of Lactose in the Food Industry," *Bulletin of International Dairy Federation,* Document 147, p. 23 (1982).

57. E. Renner, "Lactose," in Ref. 30, pp. 154–171.

58. T. A. Nickerson, "Lactose" in Ref. 8, p. 356.

59. P. A. Morrissey, "Lactose: Chemical and Physico-Chemical Properties," *Developments in Dairy Chemistry* 3, 1 (1985).

60. J. Rothwell, "Uses for Dairy Ingredients in Ice Cream and Other Frozen Desserts," *Journal of the Society of Dairy Technology* 37, 119 (1984).

61. L. A. W. Thelwall, "Developments in the Chemistry and Chemical Modification of Lactose," *Developments in Dairy Chemistry* 3, 35 (1985).

62. C. O. Tacket, G. Losonosky, H. Link, Y. Hoang, P. Guesry, H. Hilpert, and M. M. Levine, "Protection by Milk Immunoglobulin Concentrate Against Oral Challenge with Enterotoxigenic *Escherichia coli,*" *New England Journal of Medicine* 318, 1240 (1988).

63. S. Dosako, "Application of Milk Proteins to Functional Food," *Japan Food Science* 27(1), 25–34 (1988).

64. A. S. Goldman, "Immunologic Supplementation of Cow's Milk Formulations," in *Bulletin No 244,* International Dairy Federation, Brussels, Belgium, 1990.

65. B. Reiter, "The Biological Significance of the Non-immunoglobulin Protective Proteins in Milk," *Developments in Dairy Chemistry* 3, 281 (1985).

66. Jap. Pat. 60–75433 (1985) Y. Minami.

67. H. W. Modler, "Properties of Non-fat Dairy Ingredients—A Review," *Journal of Dairy Science* 68, 2195–2205 (1985).

68. J. Leman and J. E. Kinsella, "Surface Activity, Film Formulation, and Emulsifying Properties of Milk Proteins," *Critical Reviews in Food Science and Nutrition* 28(2), 115–138 (1989).

69. P. D. Patel and J. C. Fry, *The Search for Standard Methods for Assessing Protein Functionality.*

70. C. V. Morr, "Utilization of Milk Proteins as Starting Materials for Other Foodstuffs," *Journal of Dairy Research* 46, 369–376 (1979).

71. C. V. Morr, "Functional Properties of Milk Proteins and Their Use as Food Ingredients," Chapter 12, p. 375, in Ref. 36, vol. 1, Proteins, p. 375.

72. H. W. Modler and J. D. Jones, Selected Processes to Improve the Functionality of Dairy Ingredients. *Food Technology* 41(10), 114–117 (1987).

73. D. Illingworth, J. C. Lloyd, and R. Norris, "Tailor-made Fats from Milkfat—the New Zealand Experience," in *Fats for the Future II, International Conference of New Zealand Institute of Chemistry,* IUPAC, 1989.

74. M. I. Gurr and P. Walstra, "Fat Content of Dairy Foods in Relation to Sensory Properties and Consumer Acceptability," *International Dairy Federation Bulletin,* No. 244, 44–46 (1990).

75. L. B. Campbell and S. J. Pavlasek, "Dairy Products as Ingredients in Chocolates and Confections," *Food Technology* 41(10), 78–85 (1987).

76. L. Munksgaard and R. Ipsen, "Confectionery Products," in *Dairy Ingredients for the Food Industry 3,* International Dairy Federation, Brussels, Belgium, 1989.

77. W. B. Sanderson, "Cakes, Cookies, & Biscuits," in *Dairy Ingredients for the Food Industry,* International Dairy Federation, Brussels, Belgium, 1985.

78. R. O. Cocup and W. B. Sanderson, Functionality of Dairy Ingredients, *Bakery Products Food Technology* 41(10), 86–90 (1987).

79. J. M. G. Lankveld, "Meat and Meat Products," in *Dairy Ingredients for the Food Industry 2,* International Dairy Federation, Brussels, Belgium, 1987.

80. van deu Hoven, "Functionality of Dairy Ingredients in Meat Products," *Food Technology* 41(10), 72–77 (1987).

81. "Product Update—Ingredients from and for Dairy Products," *Food Technology* 43(6), 108–122 (1989).

ROBIN M. FENWICK
K. J. KIRKPATRICK
New Zealand Dairy Board
Wellington, New Zealand

DEHYDRATION

Dehydration (or drying) is defined as the application of heat under controlled conditions to remove the majority of the water normally present in a food by evaporation (or in the case of freeze-drying by sublimation). This definition excludes other unit operations that remove water from foods (eg, mechanical separations, membrane concentration, evaporation, and baking) as these normally remove much less water than dehydration. The main purpose of dehydration is to extend the shelf life of foods by a reduction in water activity. This inhibits microbial growth and enzyme activity, but the product temperature is usually insufficient to cause inactivation. The reduction in weight and bulk of food reduces transport and storage costs and, for some types of food, provides greater variety and convenience for the consumer. Drying causes deterioration of both the eating quality and the nutritive value of the food. The design and operation of dehydration equipment aim to minimize these changes by selection of appropriate drying conditions for individual foods. Examples of commercially important dried foods are sugar, coffee, milk, potato, flour (including bakery mixes), beans, pulses, nuts, breakfast cereals, tea, and spices.

THEORY

Dehydration involves the simultaneous application of heat and removal of moisture from foods. Factors that control the rates of heat and mass transfer are described elsewhere in the encyclopedia. Dehydration by heated air or heated surfaces is described in this article. Microwave, dielectric, radiant, and freeze-drying are described in other entries.

Psychrometrics

The capacity of air to remove moisture from a food depends on the temperature and the amount of water vapor already carried by the air. The content of water vapor in air is expressed as either absolute humidity—the mass of water vapor per unit mass of dry air (in kilograms per kilogram), termed moisture content in Fig. 1)—or relative humidity (RH) (in percent)—the ratio of the partial pressure of water vapor in the air to the pressure of saturated water vapor at the same temperature, multiplied by 100. Psychrometry is the study of the interrelationships of the temperature and humidity of air. These properties are most conveniently represented on a phychrometric chart (Fig. 1).

The temperature of the air, measured by a thermometer bulb, is termed the dry-bulb temperature. If the thermometer bulb is surrounded by a wet cloth, heat is removed by evaporation of the water from the cloth and the temperature falls. This lower temperature is called the wet-bulb temperature. The difference between the two temperatures is used to find the relative humidity of air on the psychrometric chart. An increase in air temperature, or reduction in RH, causes water to evaporate from a wet surface more rapidly and therefore produces a greater fall in temperature. The dew point is the temperature at which air becomes saturated with moisture (100% RH). Adiabatic cooling lines are the parallel straight lines sloping across the chart, which show how absolute humidity decreases as the air temperature increases.

Sample Problems 1. Using the psychrometric chart (Fig. 1), calculate the following.

1. The absolute humidity of air that has 50% RH and a dry-bulb temperature of 60°C.
2. The wet-bulb temperature under these conditions.
3. The RH of air having a wet-bulb temperature of 45°C and a dry-bulb temperature of 75°C.
4. The dew point of air cooled adiabatically from a dry-bulb temperature of 55°C and 30% RH.
5. The change in RH of air with a wet-bulb temperature of 39°C, heated from a dry-bulb temperature of 50°C to a dry-bulb temperature of 86°C.
6. The change in RH of air with a wet-bulb temperature of 35°C, cooled adiabatically from a dry-bulb temperature of 70°C to 40°C.

Solution to Sample Problems 1.

1. 0.068 kg/kg of dry air. Find the intersection of the 60°C and 50% RH lines and then follow the chart horizontally right to read the absolute humidity.
2. 47.5°C. From the intersection of the 60°C and 50% RH lines, extrapolate left parallel to the wet-bulb lines to read the wet-bulb temperature.
3. 20%. Find the intersection of the 45°C and 75°C lines and follow the sloping RH line upward to read the % RH.
4. 36°C. Find the intersection of the 55°C and 30% RH

lines and follow the wet-bulb line left until the RH reaches 100%.

5. 50–10%. Find the intersection of the 39°C wet-bulb and the 50°C dry-bulb temperatures and follow the horizontal line to the intersection with the 86°C dry-bulb line; read the sloping RH line at each intersection (this represents the changes that take place when air is heated before being blown over food.
6. 10–70%. Find the intersection of the 35°C wet-bulb and 70°C dry-bulb temperatures and follow the wet-bulb line left until the intersection with the 40°C dry-bulb line; read sloping RH line at each intersection (this represents the changes taking place as the air is used to dry food; the air is cooled and becomes more humid as it picks up moisture from the food.

Water Activity

Deterioration of foods by microorganisms can take place rapidly, whereas enzymatic and chemical reactions take place more slowly during storage. In either case, water is the single most important factor controlling the rate of deterioration. The moisture content of foods can be expressed either on a wet-weight basis:

$$m = \frac{\text{mass of water}}{\text{mass of sample}} \times 100 \tag{1}$$

$$m = \frac{\text{mass of water}}{\text{mass of water + solids}} \times 100 \tag{2}$$

or a dry-weight basis (1):

$$M = \frac{\text{mass of water}}{\text{mass of solids}} \tag{3}$$

The dry-weight basis is more commonly used for processing calculations, whereas the wet-weight basis is frequently quoted in food composition tables. It is important, however, to note which system is used when expressing a result. Dry-weight basis is used throughout this article unless otherwise stated.

A knowledge of the moisture content alone is not sufficient to predict the stability of foods. Some foods are unstable at a low moisture content eg, peanut oil deteriorates if the moisture content exceeds 0.6%), whereas other foods are stable at relatively high moisture contents (eg, potato starch is stable at 20% moisture) (2). It is the availability of water for microbial, enzymatic, or chemical activity that determines the shelf life of a food, and this is measured by the water activity (A_W) of a food. Examples of unit operations that reduce the availability of water in foods include those that physically remove water (dehydration, evaporation, and freeze- drying or freeze concentration) and those that immobilize water in the food (eg, by the use of humectants in intermediate-moisture foods and by formation of ice crystals in freezing). Examples of the moisture content and A_W of foods are shown in Table 1. The effect of reduced A_W on food stability is shown in Table 2.

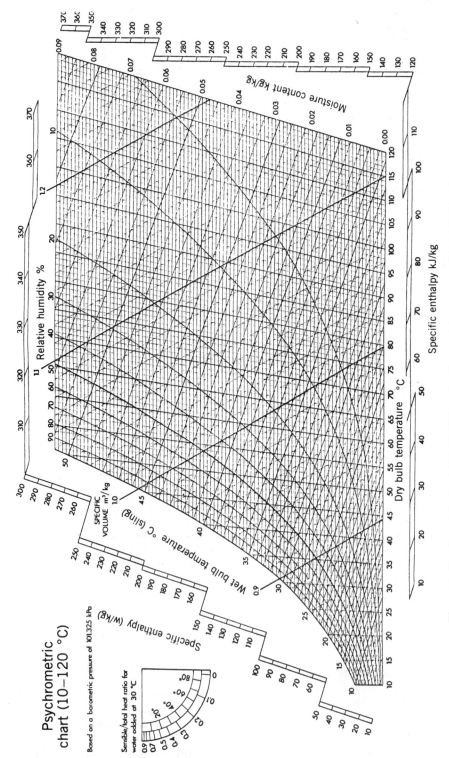

Figure 1. Psychometric chart (10–120°C based on barometric pressure of 101.325 kPa. Courtesy of Chartered Institution of Building Services Engineers.

Table 1. Moisture Content and Water Activity of Foods[a]

Food	Moisture Content, %	Water Activity	Degree of Protection Required
Ice (0°C)	100	1.00[b]	
Fresh meat	70	0.985	
Bread	40	0.96	Package to prevent moisture loss
Ice (−10°C)	100	0.91[b]	
Marmalade	35	0.86	
Ice (−20°C)	100	0.82[b]	
Wheat flour	14.5	0.72	
Ice (−50°C)	100	0.62[b]	Minimum protection or no packaging required
Raisins	27	0.60	
Macaroni	10	0.45	
Cocoa powder		0.40	
Boiled sweets	3.0	0.30	
Biscuits	5.0	0.20	Package to prevent moisture uptake
Dried milk	3.5	0.11	
Potato crisps	1.5	0.08	

[a] Refs. 2–4.
[b]

Water in food exerts a vapor pressure. The size of the vapor pressure depends on

1. The amount of water present.
2. The temperature.
3. The concentration of dissolved solutes (particularly salts and sugars) in the water.

Water activity is defined as the ratio of the vapor pressure of water in a food to the saturated vapor pressure of water at the same temperature:

$$A_W = \frac{P}{P_0} \qquad (4)$$

where P (Pa) is vapor pressure of the food and P_0 (Pa) the vapor pressure of pure water at the same temperature. A_W

Table 2. The Importance of Water Activity in Foods

A_W	Phenomenon	Examples
1.00		Highly perishable fresh foods
0.95	Pseudomonads, bacillus, *Clostridium perfringens*, and some yeasts inhibited	Foods with 40% sucrose or 7% salt; cooked sausages, bread
0.90	Lower limit for bacterial growth (general), salmonella, *Vibrio parahemolyticus*, *Clostridium botulinum*, lactobacillus, and some yeasts and fungi inhibited	Foods with 55% sucrose, 12% salt; cured ham, medium-age cheese. Intermediate-moisture foods ($A_W = 0.90$–0.55)
0.85	Many yeasts inhibited	Foods with 65% sucrose, 15% salt; salami, mature cheese, margarine
0.80	Lower limit for enzyme activity and growth of most fungi; *Staphlococcus aureus* inhibited	Flour, rice (15–17% water) fruit cake, sweetened condensed milk, fruit syrups, fondant
0.75	Lower limit for halophilic bacteria	Marzipan (15–17% water), jams
0.70	Lower limit for growth of most xerophilic fungi	
0.65	Maximum velocity of Maillard reactions	Rolled oats (10% water), fudge, molasses, nuts
0.60	Lower limit for growth of osmophilic or xerophilic yeasts and fungi	Dried fruits (15–20% water), toffees, caramels (8% water), honey
0.55	Deoxyribonucleic acid becomes disordered (lower limit for life to continue)	
0.50		Dried foods, spices, noodles
0.40	Minimum oxidation velocity	Whole egg powder (5% water)
0.30		Crackers, bread crusts (3–5% water)
0.25	Maximum heat resistance of bacterial spores	
0.20		Whole milk powder (2–3% water), dried vegetables (5% water), cornflakes (5% water)

is related to the moisture content by the Brunauer-Emmett-Teller (BET) equation

$$\frac{A_W}{M(1 - A_W)} = \frac{1}{M_1 C} + \frac{C - 1}{M_1 C} A_W \qquad (5)$$

where A_W is the water activity, M the moisture as percentage dry weight, M_1 the moisture (dryweight basis) of a monomolecular layer, and C a constant (2).

A proportion of the total water in a food is strongly bound to specific sites (eg, hydroxyl groups of polysaccharides, carbonyl and amino groups of proteins, the hydrogen bonding). When all sites are (statistically) occupied by adsorbed water the moisture content is termed the BET monolayer value. Typical examples include gelatin (11%), starch (11%), amorphous lactose (6%), and whole spray-dried milk (3%). The BET monolayer value therefore represents the moisture content at which the food is most stable. At moisture contents below this level, there is a higher rate of lipid oxidation and, at higher moisture contents, Maillard browning and then enzymatic and microbiological activities are promoted.

The movement of water vapor from a food to the surrounding air depends on both the moisture content and composition of the food and the temperature and humidity of the air. At a constant temperature the moisture content of food changes until it comes into equilibrium with water vapor in the surrounding air. The food then neither gains nor loses weight on storage under those conditions. This is called the equilibrium moisture content of the food, and the relative humidity of the storage atmosphere is known as the equilibrium relative humidity. When different values of relative humidity versus equilibrium moisture content are plotted, a curve known as a water sorption isotherm is obtained (Fig. 2).

Each food has a unique set of sorption isotherms at different temperatures. The precise shape of the sorption isotherm is caused by differences in the physical structure, chemical composition, and extent of water binding within the food, but all sorption isotherms have a characteristic shape, similar to that shown in Figure 2. The first part of the curve, to point A, represents monolayer water, which is very stable, unfreezable, and not removed by drying. The second, relatively straight part of the curve

(AB) represents water absorbed in multilayers within the food and solutions of soluble components. The third portion (above point B) is free water condensed within the capillary structure or in the cells of a food. It is mechanically trapped within the food and held only by weak forces. It is easily removed by drying and easily frozen, as indicated by the steepness of the curve. Free water is available for microbial growth and enzyme activity, and a food that has a moisture content above point B on the curve is likely to be susceptible to spoilage.

The sorption isotherm indicates the A_W at which a food is stable and allows predictions of the effect of changes in moisture content on A_W and hence on storage stability. It is used to determine the rate and extent of drying, the optimum frozen storage temperatures, and the moisture-barrier properties required in packaging materials.

The rate of change in A_W on a sorption isotherm differs according to whether moisture is removed from a food (desorption) or it is added to dry food (adsorption) (Fig. 2). This is termed a hysteresis loop. The difference is large in some foods, eg, rice, and is important for example in determining the protection required against moisture uptake.

Mechanism of Drying

When hot air is blown over a wet food, heat is transferred to the surface, and latent heat of vaporization causes water to evaporate. Water vapor diffuses through a boundary film of air and is carried away by the moving air (Fig. 3). This creates a region of lower water vapor pressure at the surface of the food, and a water vapor pressure gradient is established from the moist interior of the food to the dry air. This gradient provides the driving force for water removal from the food.

Water moves to the surface by the following mechanisms:

1. Liquid movement by capillary forces.
2. Diffusion of liquids, caused by differences in the concentration of solutes in different regions of the food.
3. Diffusion of liquids, which are adsorbed in layers at the surfaces of solid components of the food.
4. Water vapor diffusion in air spaces within the food caused by vapor pressure gradients.

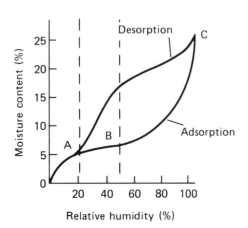

Figure 2. Water sorption isotherm.

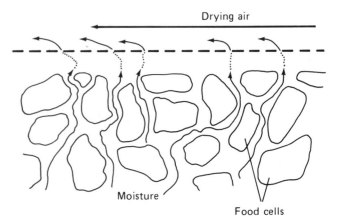

Figure 3. Movement of moisture during drying.

Foods are characterized as hygroscopic and nonhygroscopic. Hygroscopic foods are those in which the partial pressure of water vapor varies with the moisture content. Nonhygroscopic foods have a constant water vapor pressure at different moisture contents. The difference is found by using sorption isotherms.

When food is placed in a drier, there is a short initial settling down period as the surface heats up to the wet-bulb temperature (*AB* in Fig. 4a). Drying then commences and, provided that water moves from the interior of the food at the same rate as it evaporates from the surface, the surface remains wet. This is known as the constant-rate period and continues until a certain critical moisture content is reached (*BC* in Figs. 4a and b). In practice, however, different areas of the surface of the food dry out at different rates and, overall, the rate of drying declines gradually during the constant-rate period. Thus the critical point is not fixed for a given food and depends on the amount of food in the drier and the rate of drying.

The three characteristics of air that are necessary for successful drying in the constant rate period are

1. A moderately high dry-bulb temperature.
2. A low RH.
3. A high air velocity.

The boundary film of air surrounding the food acts as a barrier to the transfer of both heat and water vapor during drying. The thickness of the film is determined primarily by the air velocity. If this is too low, water vapor leaves the surface of the food and increases the humidity of the surrounding air, to cause a reduction in the water vapor pressure gradient and the rate of drying (Similarly, if the temperature of the drying air falls or the humidity rises, the rate of evaporation falls and drying slows).

When the moisture content of the food falls below the critical moisture content, the rate of drying slowly decreases until it approaches zero at the equilibrium moisture content (ie, the food comes into equilibrium with the drying air). This is known as the falling-rate period. Nonhygroscopic foods have a single falling-rate period (*CD* in Fig. 4a and b), whereas hygroscopic foods have two periods. In the first period, the plane of evaporation moves inside the food, and water diffuses through the dry solids to the drying air. It ends when the plane of evaporation reaches the center of the food and the partial pressure of water falls below the saturated water vapor pressure. The second period occurs when the partial pressure of water is below the saturated vapor pressure and drying is by desorption.

During the falling-rate period, the rate of water movement from the interior of the food to the surface falls below the rate at which water evaporates to the surrounding air. The surface therefore dries out. This is usually the longest period of a drying operation and, in some foods, eg, grain drying, where the initial moisture content is below the critical moisture content, the falling-rate period is the only part of the drying curve to be observed. During the falling-rate period, the factors that control the rate of drying change. Initially, the important factors are similar to those in the constant-rate period, but gradually the rate of mass transfer becomes the controlling factor. This depends mostly on the temperature of the air and the thickness of the food. It is unaffected by both the RH of the air (except in determining the equilibrium moisture content) and the velocity of the air. The air temperature is therefore controlled during the falling rate period, whereas the air velocity and temperature are more important during the constant-rate period. In practice, foods may differ from these idealized drying curves owing to shrinkage, changes in the temperature and rate of moisture diffusion in different parts of the food, and changes in the temperature and humidity of the drying air.

The surface temperature of the food remains close to the wet-bulb temperature of the drying air until the end of the constant-rate period, owing to the cooling effect of the evaporating water. During the falling-rate period the amount of water evaporating from the surface gradually decreases but, as the same amount of heat is being supplied by the air, the surface temperature rises until it reaches the dry-bulb temperature of the drying air. Most

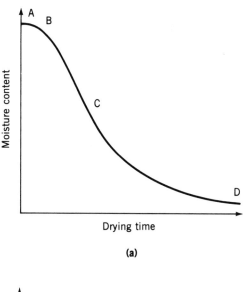

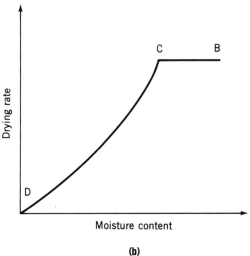

Figure 4. (**a**) and (**b**) Drying curves. The temperature and humidity of the drying air are constant and all heat is supplied to the surface by convection.

heat damage to food therefore occurs in the falling rate period.

Calculation of Drying Rate

The rate of drying depends on the properties of the drier (the dry-bulb temperature, RH and velocity of the air, and the surface heat transfer coefficient) and the properties of the food (the moisture content, surface-to-volume ratio, and the surface temperature and rate of moisture loss). The size of food pieces has an important effect on the drying rate in both the constant- and the falling-rate periods. In the constant-rate period, smaller pieces have a larger surface area available for evaporation, whereas, in the falling-rate period, smaller pieces have a shorter distance for moisture to travel through the food. Other factors that influence the rate of drying include:

1. The fat content of the food (higher fat contents generally result in slower drying, as water is trapped within the food).
2. The method of preparation of the food (cut surfaces lose moisture more quickly than losses through skin.
3. The amount of food placed into a drier in relation to its size (in a given drier, faster drying is achieved with smaller quantities of food).

The rate of heat transfer is found using

$$Q = h_s A \ (\theta_a - \theta_s) \qquad (6)$$

The rate of mass transfer is found using

$$-m_c = K_g A (H_s - H_a) \qquad (7)$$

Because during the constant-rate period, an equilibrium exists between the rate of heat transfer to the food and the rate of mass transfer in the form of moisture loss from the food, these rates are related by

$$- m_c = \frac{h_c A}{\lambda} \ (\theta_a - \theta_s) \qquad (8)$$

where Q (J/s) is the rate of heat transfer, h_c W/m^2/°K the surface heat transfer coefficient for convective heating, A (m^2) the surface area available for drying, θ_a (°C) the average dry-bulb temperature of drying air, θ_s(°C) the average wet-bulb temperature of drying air, m_c (kg/s) the change in mass with time (drying rate), K_g (kg/m^2/s) the mass transfer coefficient, H_s (kilograms of moisture per kilogram dry air) the humidity at the surface of the food (saturation humidity), H_a (kilograms of moisture per kilogram dry air) the humidity of air, and λ (J/kg) the latent heat of evaporation at the wet-bulb temperature.

The surface heat transfer coefficient (h_c) is related to the mass flow rate of air using the following equations: for parallel airflow,

$$h_c = 14.3 \ G^{0.8} \qquad (9)$$

and for perpendicular airflow,

$$h_c = 24.2 \ G^{0.37} \qquad (9)$$

Where G (kg/m^2/s) is the mass flow rate of air.

For a tray of food in which water evaporates only from the upper surface, the drying time is found using

$$- m_c = \frac{h_c}{\rho \lambda x} \ (\theta_a - \theta_s) \qquad (11)$$

where ρ (kg/m^3) is the bulk density of food and x (m) the thickness of the bed of food.

The drying time in the constant-rate period is found using

$$t = \frac{\rho \lambda x (M_i - M_c)}{h_c (\theta_a - \theta_s)} \qquad (12)$$

where t (s) is the drying time, M_i (kg/kg of dry solids) the initial moisture content, and M_c (kg/kg of dry solids) the critical moisture content.

For water evaporating from a spherical droplet in a spray drier, the drying time is found using

$$t = \frac{r^2 \rho_1 \lambda}{3 \ h_c \ (\theta_a - \theta_s)} \frac{M_i - M_f}{1 + M_i} \qquad (13)$$

where ρ_1 (kg/m^3) is the density of the liquid, r (m) the radius of the droplet, M_f (kg/kg of dry solids) the final moisture content.

In the falling-rate period, moisture gradients change throughout the food and the temperature slowly increases from the wet-bulb temperature to the dry-bulb temperature as the food dries. The following equation is used to calculate the drying time from the start of the falling-rate period to the equilibrium moisture content using a number of assumptions concerning, eg, the nature of moisture movement and the absence of shrinkage of the food:

$$t = \frac{\rho x (M_c - M_e)}{K_g (P_s - P_a)} \ln \left(\frac{M_c - M_e}{M - M_e} \right) \qquad (14)$$

where M_e (kg/kg of dry solids) is the equilibrium moisture content, M (kg/kg of dry solids) the moisture content at time t from the start of the falling-rate period, P_s (mmHg) the saturated water vapor pressure at the wet-bulb temperature, and P_a (mmHg) the partial water vapor pressure.

The velocity of the air needed to achieve fluidization of spherical particles is calculated using

$$v_f = \frac{(p_S - p) g}{\mu} \frac{d^2 \ \varepsilon^3}{180 (1 - \varepsilon)} \qquad (15)$$

where v_f (m/s) is the fluidization velocity, ρ_s (kg/m^3 the density of the solid particles, ρ (kg/m^3) the density of the fluid, g (m/s^2) the acceleration due to gravity, μ (N·s/m^2) the viscosity of the fluid, d (m) the diameter of the particles, ε the voidage of the bed.

Formulas for foods of other shapes are described in Ref. 5. The minimum air velocity needed to convey particles is found using:

$$v_e = \sqrt{\left[\frac{4d(\rho_s - \rho)}{3C_d\rho}\right]} \qquad (16)$$

where v_e (m/s) is the minimum air velocity and C_d ($= 0.44$ for Re $= 500$–$200,000$) the drag coefficient.

Sample Problem 2. Peas with an average diameter of 6 mm and a density of 880 kg/m³ are dried in a fluidized-bed drier. The minimum voidage is 0.4 and the cross-sectional area of the bed is 0.25 m². Calculate the minimum air velocity needed to fluidize the bed if the air density is 0.96 kg/m³ and the air viscosity is 2.15×10^{-5} N·s/m².

Solution of Sample Problem 2. From equation 15

$$v_F = \frac{(880 - 0.96)9.81}{2.15 \times 10^{-5}} \frac{(0.006)^2(0.4)^3}{180(1 - 0.4)}$$

$$= 8.5 \text{ m/s}^1$$

Derivations of these equations are described in Refs. 6–8.

Sample Problem 3. A conveyor drier is required to dry peas from an initial moisture content of 78% to 16% moisture (wet-weight basis) in a bed 10 cm deep with a voidage of 0.4. Air at 85°C with a relative humidity of 10% is blown perpendicularly through the bed at 0.9 m/s. The drier belt measures 0.75 m wide and 4 m long. Assuming that drying takes place from the entire surface of the peas and there is no shrinkage, calculate the drying time and energy consumption in both the constant and the falling-rate periods. (Additional data: The equilibrium moisture content of the peas is 9%, the critical moisture content 300% (dry-weight basis), the average diameter 6 mm, the bulk density 610 kg/m³, the latent heat of evaporation 2300 kJ/kg, the saturated water vapor pressure at wet-bulb temperature 61.5 mmHg, and the mass transfer coefficient 0.015 kg/m²/s).

Solution to Sample Problem 3. In the constant-rate period, from equation 10:

$$h_c = 24.2(0.9)^{0.37}$$
$$= 23.3 \text{ W/m}^2/°\text{K}$$

From Fig. 1 for $\theta_a = 85°C$ and RH $= 10\%$,

$$\theta_s = 42°C$$

To find the area of the peas,

$$\text{Volume of a sphere} = \frac{4}{3}\pi r^3$$

$$= 4 \times 3.142(0.003)^3$$

$$= 339 \times 10^{-9} \text{ m}^3$$

$$\text{Volume of the bed} = 0.75 \times 4 \times 0.1$$

$$= 0.3 \text{ m}^3$$

$$\text{Volume of peas in the bed} = 0.3(1 - 0.4)$$

$$= 0.18 \text{ m}^3$$

$$\text{Number of peas} = \frac{\text{volume of peas in bed}}{\text{volume each pea}}$$

$$= \frac{0.18}{339 \times 10^{-9}}$$

$$= 5.31 \times 10^5$$

$$\text{Area of a sphere} = 4\pi r^2$$

$$= 4 \times 3.142(0.003)^2$$

$$= 113 \times 10^{-6} \text{ m}^2$$

and

$$\text{Total area of peas} = 5.31 \times 10^5 \times 113 \times 10^{-6}$$

$$= 60 \text{ m}^2$$

From equation 8:

$$\text{Drying rate} = \frac{23.3 \times 60}{2.3 \times 10^6}(85 - 42)$$

$$= 0.026 \text{ kg/s}$$

From a mass balance

$$\text{Volume of peas} = 0.18 \text{ m}^3$$
$$\text{Bulk density} = 610 \text{ kg/m}^{-3}$$

Therefore,

$$\text{Mass of peas} = 0.18 \times 610$$

$$= 109.8 \text{ kg}$$

$$\text{Initial solids content} = 109.8 \times 0.22$$

$$= 24.15 \text{ kg}$$

Therefore,

$$\text{Initial mass water} = 109.8 - 24.15$$

$$= 85.6 \text{ kg}$$

After constant-rate period, solids remain constant and

$$\text{Mass of water} = 96.6 - 24.15$$

$$= 72.45 \text{ kg}$$

Therefore,

$$(85.6 - 72.45) = 13.15 \text{ kg water lost}$$

at a rate of 0.026 kg/s

$$\text{Drying time} = \frac{13.15}{0.026} = 505 \text{ s} = 8.4 \text{ min}$$

Therefore,

$$\text{Energy required} = 0.026 \times 2.3 \times 10^6$$
$$= 6 \times 10^4 \text{ J/s}$$
$$= 60 \text{ kW}$$

In the falling-rate period,

$$\text{RH} = \frac{P_A}{P_o} \times 100$$

$$10 = \frac{P}{61.5} \times 100$$

Therefore,

$$P = 6.15 \text{ mmHg}$$

The moisture values are

$$M_c = \frac{75}{25} = 3$$

$$M_f = \frac{16}{84} = 0.19$$

$$M_e = \frac{9}{91} = 0.099$$

From Equation 14,

$$t = \frac{(3 - 0.099)\, 610 \times 0.1}{0.015\, (61.5 - 6.15)} \ln\left(\frac{3 - 0.099}{0.19 - 0.099}\right)$$

$$= 737.7 \text{ s}$$

$$= 12.3 \text{ min}$$

From a mass balance, at the critical moisture content, 96.6 kg contains 25% solids = 24.16 kg. After drying in the falling-rate period, 84% solids = 24.16 kg. Therefore,

$$\text{Total mass} = \frac{100}{84} \times 24.16$$

$$= 28.8 \text{ kg}$$

and

$$\text{Mass loss} = 96.6 - 28.8$$

$$= 67.8 \text{ kg}$$

Thus,

$$\text{Average drying rate} = \frac{67.8}{737.7}$$

$$= 0.092 \text{ kg/s}$$

$$\text{average energy required} = 0.092 \times 2.3 \times 10^6$$
$$= 2.1 \times 10^5 \text{ J/s}$$
$$= 210 \text{ kW}$$

Drying Using Heated Surfaces

Heat is conducted from a hot surface, through a thin layer of food, and moisture is evaporated from the exposed surface. The main resistance to heat transfer is the thermal conductivity of the food. Knowledge of the rheological properties of the food is necessary to determine the thickness of the layer of food and the way in which it should be applied to the heated surface. Additional resistances to heat transfer arise if the partly dried food lifts off the hot surface. Equation 17 is used in the calculation of drying rates.

$$Q = UA(\theta_a - \theta_b) \tag{17}$$

where

U = overall heat transfer coefficient (W/m²/K)

θ_a = temperature of hot surface (°C)

θ_b = temperature of food (°C)

Sample Problem 4. A single-drum drier 0.7 m in diameter and 0.85 m long operates at 150°C and is fitted with a doctor blade to remove food after three-fourths of a revolution. It is used to dry a 0.6-mm layer of 20% w/w solution of gelatin, preheated to 100°C, at atmospheric pressure. Calculate the speed of the drum required to produce a product with a moisture content of 4 kg of solids per kilogram of water. (Additional data: The density of gelatin feed is 1020 kg/m³, and the overall heat transfer coefficient is 1200 W/m²/K; assume that the critical moisture content of the gelatin is 450%, dry weight basis.

Solution to Sample Problem 4. First,

$$\text{Drum area} = \pi dl$$
$$= 3.142 \times 0.7 \times 0.85$$
$$= 1.87 \text{ m}^2$$

Therefore,

$$\text{Mass of food on the drum} = (1.87 \times 0.75)\, 0.0006 \times 1020$$
$$= 0.86 \text{ kg}$$

From a mass balance (initially the food contains 80% moisture and 20% solids),

$$\text{Mass of solids} = 0.86 \times 0.2$$
$$= 0.172 \text{ kg}$$

After drying, 80% solids = 0.172 kg.
Therefore,

$$\text{Mass of dried food} = \frac{100}{80} \times 0.172$$
$$= 0.215 \text{ kg}$$

Mass loss $= 0.86 - 0.215$

$$= 0.645 \text{ kg}$$

From equation 17,

$$Q = 1200 \times 1.87 \, (150 - 100)$$

$$= 1.12 \times 10^5 \text{ J/s}$$

$$\text{Drying rate} = \frac{1.12 \times 10^5}{2.257 \times 10^6} \text{ kg/s}$$

$$= 0.05 \text{ kg/s}$$

and

$$\text{Residence time required} = \frac{0.645}{0.05}$$

$$= 13 \text{ s}$$

As only three-quarters of the drum surface is used, 1 rev should take $(100/75) \times 13 = 17.3$ s. Therefore, speed = 3.5 rev/min.

EQUIPMENT

Most commercial driers are insulated to reduce heat losses, and they recirculate hot air to save energy. Many designs have energy-saving devices that recover heat from the exhaust air or automatically control the air humidity (9,10). Computer control of driers is increasingly sophisticated (11) and also results in important savings in energy. The criteria for selection of drying equipment and potential applications are described in Table 3. The relative costs of different drying methods are reported as follows (12): forced-air drying, 198; fluidized-bed drying, 315;

drum drying, 327; continuous vacuum drying, 1840; freeze-drying, 3528. Relative energy consumption (in kWh/kg of water removed) are as follows: roller drying, 1.25; pneumatic drying, 1.8; spray drying, 2.5; fluidized-bed drying, 3.5 (13).

Hot-Air Driers

Bin Driers (Deep-Bed Driers). Bin driers are cylindrical or rectangular containers fitted with a mesh base. Hot air passes up through a bed of food at relatively low speeds (eg, 0.5 m^3/s/m^2 of bin area). These driers have a high capacity and low capital and running costs. They are mainly used for finishing (to 3–6% moisture content) after initial drying in other types of equipment. Bin driers improve the operating capacity of initial driers by taking the food when it is in the falling-rate period, when moisture removal is most time consuming. The deep bed of food permits variations in moisture content to be equalized and acts as a store to smooth out fluctuations in the product flow between drying stages and packaging. However, the driers may be several meters high, and it is therefore important that foods are sufficiently strong to withstand compression at the base and to retain an open structure to permit the passage of hot air through the bed.

Cabinet Driers (Tray Driers). These driers consist of an insulated cabinet fitted with shallow mesh or perforated trays, each of which contains a thin (2–6 cm deep) layer of food. Hot air is circulated through the cabinet at 0.5–5 m/s per square meter tray area. A system of ducts and baffles is used to direct air over and/or through each tray, to promote uniform air distribution. Additional heaters may be placed above or alongside the trays to increase the rate of drying. Tray driers are used for small-scale production (1–20 tons/day) or for pilot-scale work. They have low capital and maintenance costs, but compared to driers

Table 3. Characteristics of Driers

Type of Drier	Solid	Liquid	Initial Moisture Content Moderate to High	Initial Moisture Content Low	Size of Pieces Heat-sensitive	Size of Pieces Small	Size of Pieces Intermediate to Large	Size of Pieces Mechanically Strong	Drying Rate Required Moderate to Fast	Drying Rate Required Slow	Final Moisture Content Required Moderate	Final Moisture Content Required Low
Bin	*			*			*	*		*		*
Cabinet	*		*				*		*	*	*	
Conveyor	*		*				*		*	*	*	
Drum		*	—	—		—	—	—	*			*
Foam mat		*	—	—	*	—	—	—	*			*
Fluid bed	*		*			*		*	*		*	
Kiln	*		*				*			*	*	
Pneumatic	*			*		*		*	*			*
Rotary	*		*			*		*	*			*
Spray		*	—	—	*	—	—	—	*		*	*
Trough	*		*		*	*			*		*	
Tunnel	*		*				*		*	*	*	*
Vacuum band		*	—	—	*	—	—	—	*		*	*
Vacuum shelf	*	*	*		*	*			*	*	*	*
Radiant	*			*	*				*			*
Microwave or dielectric	*			*	*				*	*		*
Solar (sun)	*		*				*				*	

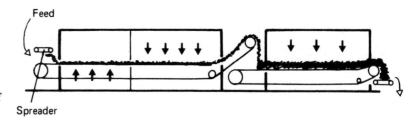

Figure 5. Two-stage conveyor drier. Courtesy of Proctor and Schwartz Inc.

that have more sophisticated control they produce more variable product quality.

In developing countries the high capital investment for sophisticated driers often cannot be justified, but there is a need for better-quality products than those produced by sun or solar drying. A small drier consisting of a 60 kW gas or kerosine heater/blower unit and a cabinet fitted with 15 mesh trays has been developed to meet this need (14). In operation, air is passed across the food in each tray. Trays are loaded at the top of the drier and unloaded at the base using metal fingers to move the tray stack. The drier therefore operates semicontinuously and with counter-current airflow. Typical tray loadings are 5 kg, and the tray change cycle is 15–20 minutes.

Conveyor Driers (Belt Driers). Continuous conveyor driers are up to 20 m long and 3 m wide. Food is dried on a mesh belt in beds 5–15 cm deep. The airflow is initially directed upward through the bed of food and then downward in later stages to prevent dried food from blowing out of the bed. Two- or threestage driers (Fig. 5) mix and repile the partly dried shrunken food into deeper beds (to 15–25 cm and 250–900 cm in three-stage driers). This improves uniformity of drying and saves floor space. Foods are dried to 10–15% moisture content and then transferred to bin driers for finishing. This equipment has good control over drying conditions and high production rates. It is used for large-scale drying of foods (eg, fruits and vegetables are dried in 2–3.5 h at up to 5.5 tons/h). It has independently controlled drying zones and is automatically loaded and unloaded, which reduces labor costs. As a

result, it has largely replaced the tunnel drier in many applications.

A second application of conveyor driers is foam-mat drying in which liquid foods (eg, fruit juices) are formed into a stable foam by the addition of a stabilizer and aeration with nitrogen or air. The foam is spread on a perforated belt to a depth of 2–3 mm and dried rapidly in two stages by parallel and then counter-current airflows (Table 4). Foam drying is approximately three times faster than drying a similar thickness of liquid. The thin, porous mat of dried food is ground to a free-flowing powder that has good rehydration properties. The rapid drying and low product temperatures result in high-quality product. However, a large surface area is required for high production rates, so capital costs are therefore high.

Fluidized-Bed Driers. Metal trays with mesh or perforated bases contain a bed of particulate foods up to 15 cm deep. Hot air is blown through the bed (Fig. 6), causing the food to become suspended and vigorously agitated (fluidized). The air thus acts as both the drying and the fluidizing medium and the maximum surface area of food is made available for drying. A sample calculation of the air speed needed for fluidization is described in sample problem 2. Driers may be batch or continuous in operation; the latter are often fitted with a vibrating base to help move the product. Continuous cascade systems, in which food is discharged under gravity from one tray to the next, employ up to six driers for high production rates.

Fluidized-bed driers are compact and have good control over drying conditions, relatively high thermal efficien-

Table 4. Advantages and Limitations of Parallel Flow, Counter-Current Flow, Center-Exhaust, and Cross-Flow Drying

Type of Air Flow	Advantages	Limitations
Parallel or cocurrent type: Food → Airflow →	Rapid initial drying. Little shrinkage of food. Low bulk density. Less heat damage to food. No risk of spoilage	Low moisture content difficult to achieve as cool, moist air passes over dry food
Counter-current type: Food → Airflow ←	More economical use of energy. Low final moisture content as hot air passes over dry food	Food shrinkage and possible heat damage. Risk of spoilage from warm moist air meeting wet food
Center-exhaust type: Food → Airflow → ↑ ←	Combined benefits of parallel and counter-current driers but less than cross-flow driers	More complex and expensive then single-direction air flow
Cross-flow type: Food → Airflow ↑ ↓	Flexible control of drying conditions by separately controlled heating zones, giving uniform drying and high drying rates	More complex and expensive to buy, operate, and maintain

Figure 6. Fluidized-bed drying. Courtesy of Petrie and McNaught Ltd.

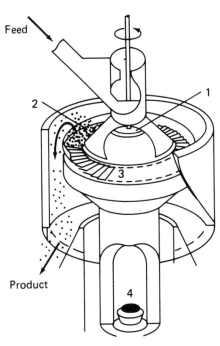

Figure 7. Torbed drier: (**1**) rotating disc distributor to deliver raw material evenly into processing chamber, (**2**) rotating bed of particles, (**3**) fixed blades with hot gas passing through at high velocity, (**4**) burner assembly. Courtesy of Torftech Ltd.

cies, and high drying rates. In batch operation, products are mixed by fluidization, this leads to uniform drying. In continuous driers, there is a greater range of moisture content in the dried product; bin driers are therefore used for finishing. Fluidized-bed driers are limited to small particulate foods that are capable of being fluidized without excessive mechanical damage (eg, peas, diced or sliced vegetables, grains, powders, or extruded foods). These considerations also apply to fluidized-bed freeze-driers and freezers.

A development of the fluidized-bed drier, named the Torbed drier, has potential applications for drying particulate foods. A fluidized bed of particles is made to rotate around a torus-shaped chamber, by hot air blown directly from a burner (Fig. 7). The drier has very high rates of heat and mass transfer and substantially reduced drying times. It is likely that some products (eg, vegetable pieces) would require a period of equilibration to allow moisture redistribution before final drying. The drier operates semicontinuously under microprocessor control and is suitable for agglomeration and puff drying in addition to roasting, cooking, and coating applications.

Kiln Driers. These driers are two-story buildings in which a drying room with a slatted floor is located above a furnace. Hot air and the products of combustion from the furnace pass through a bed of food up to 20 cm deep. These driers have been used traditionally for drying apple rings or slices in the United States, and hops or malt in Europe. There is limited control over drying conditions, and drying times are relatively long. High labor costs are incurred by the need to turn the product regularly, and by manual loading and unloading. However, the driers have a large capacity and are easily constructed and maintained at low cost.

Pneumatic Driers. In pneumatic driers, powders or particulate foods are continuously dried in vertical or horizontal metal ducts. A cyclone separator is used to remove the dried product. The moist food (usually less than 40% moisture) is metered into the ducting and suspended in

hot air. In vertical driers the airflow is adjusted to classify the particles; lighter and smaller particles, which dry more rapidly, are carried to a cyclone more rapidly than are heavier and wetter particles that remain suspended to receive the additional drying required. For longer residence times the ducting is formed into a continuous loop (pneumatic ring driers) and the product is recirculated until it is adequately dried. High-temperature short-time ring driers are used to expand the starch-cell structure in potatoes or carrots to give a rigid, porous structure, which enhances subsequent conventional drying and rehydration rates. Calculation of air velocities needed for pneumatic drying is described in equation 16.

Pneumatic driers have relatively low capital costs, high drying rates and thermal efficiencies, and close control over drying conditions. They are often used after spray drying to produce foods that have a lower moisture content than normal (eg, special milk or egg powders and potato granules). In some applications the simultaneous transportation and drying of the food may be a useful method of materials handling.

Rotary Driers. A slightly inclined rotating metal cylinder is fitted internally with flights to cause the food to cascade through a stream of hot air as it moves through the drier. Airflow may be parallel or counter-current (Table 4). The agitation of the food and the large area of food exposed to the air produce high drying rates and a uniformly dried product. The method is especially suitable for foods that tend to mat or stick together in belt or tray driers. However, the damage caused by impact and abrasion in the drier restrict this method to relatively few foods (eg, sugar crystals and cocoa beans).

Spray Driers. A fine dispersion of preconcentrated foods is first atomized to form droplets (10–200 μm in diameter) and sprayed into a current of heated air at 150–300°C in a large drying chamber. The feed rate is controlled to produce an outlet air temperature of 90–100°C, which corresponds to a wet-bulb temperature (and product temperature) of 40–50°C. Complete and uniform atomization is necessary for successful drying, and one of the following types of atomizer is used.

1. Centrifugal Atomizer. Liquid is fed to the center of a rotating bowl (with a peripheral velocity of 90–200 m/s). Droplets, 50–60 μm in diameter, are flung from the edge of the bowl to form a uniform spray (Fig. 8**a**).
2. Pressure Nozzle Atomizer. Liquid is forced at a high pressure (700–2000 kPa) through a small aperture.

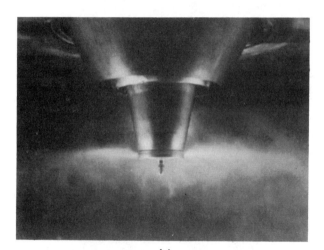

(a)

(b)

Figure 8. Atomizers: (**a**) centrifugal atomizer; (**b**) two-fluid nozzle atomizer (15). Courtesy of Elsevier Applied Science.

Droplet sizes are 180–250 μm. Grooves on the inside of the nozzle cause the spray to form into a cone shape and therefore to use the full volume of the drying chamber.

3. Two-Fluid Nozzle Atomizer. Compressed air creates turbulence, which atomizes the liquid (Fig. 8**b**). The operating pressure is lower than the pressure nozzle, but a wider range of droplet sizes is produced.

Both types of nozzle atomizer are susceptible to blockage by particulate foods, and abrasive foods gradually widen the apertures and increase the average droplet size. Studies of droplet drying, including methods for calculating changes in size, density, and trajectory of the droplets are reported in Refs. 10, 16, and 17.

Rapid drying takes place (1–10 s) because of the very large surface area of the droplets. The temperature of the product remains at the wet-bulb temperature of the drying air, and there is minimum heat damage to the food. Airflow may be co- or counter-current (Table 4). The dry powder is collected at the base of the drier and removed by a screw conveyor or a pneumatic system with a cyclone separator. There are a large number of designs of atomizer, drying chamber, air heating, and powder collecting systems (10,18). The variations in design arise from the different requirements of the very large variety of food materials that are spray dried—eg, milk, egg, coffee, cocoa, tea, potato, ground chicken, ice cream mix, butter, cream, yogurt and cheese powder, coffee whitener, fruit juices, meat and yeast extracts, encapsulated flavors (19), and wheat and corn starch products. Spray driers may also be fitted with fluidized bed facilities to finish powders taken from the drying chamber.

Spray driers vary in size from small pilot-scale models for low-volume high-value products (eg, enzymes and flavors) to large commercial models capable of producing 80,000 kg of dried milk per day (20) (Fig 9). The main advantages are rapid drying, large-scale continuous production, low labor costs, and simple operation and maintenance. The major limitations are high capital costs and the requirement for a relatively high feed moisture content to ensure that the food can be pumped to the atomizer. This results in higher energy costs (to remove the moisture) and higher volatile losses. Conveyor-band driers and fluidized bed driers are beginning to replace spray driers, as they are more compact and energy efficient (21).

The bulk density of powders depends on the size of the dried particles and on whether they are hollow or solid. This is determined by the nature of the food and the drying conditions (eg, the uniformity of droplet size, temperature, solids content, and degree of aeration of the feed liquid). Instant powders are produced by either agglomeration or non-agglomeration methods. Agglomeration is achieved by remoistening particles in low-pressure steam in an agglomerator, and then redrying. Fluidized-bed, jet, disc, cone, or belt agglomerators are described in Ref. 22. Alternatively, straight-through agglomeration is achieved directly during spray drying. A relatively moist powder is agglomerated and dried in an attached fluidized bed drier. Nonagglomeration methods employ a binding agent, (eg, lecithin), to bind particles. This method was

Figure 9. Spray drier. Courtesy of De Melkindustrie Veghel.

previously used for foods with a relatively high fat content, (eg, whole milk powder) but agglomeration procedures have now largely replaced this method (23).

Trough Driers (Belt-Trough Driers). Small, uniform pieces of food, (eg, peas or diced vegetables) are dried in a mesh conveyor belt that hangs freely between rollers to form the shape of a trough. Hot air is blown through the bed of food, and the movement of the conveyor mixes and turns it to bring new surfaces continually into contact with the drying air. The mixing action moves food away from the drying air, and this allows time for moisture to move from the interior of the pieces to the dry surface. The moisture is then rapidly evaporated when the food again contacts the hot air. The drier operates in two stages, to 50–60% moisture and then to 15–20% moisture. Foods are finished in bin driers. These driers have high drying rates (eg, 55 min for diced vegetables, compared with 5 h in a tunnel drier), high energy efficiencies, good control, and minimal heat damage to the product. However, they are not suitable for sticky foods.

Tunnel Driers. Thin layers of food are dried on trays, which are stacked on trucks programmed to move semicontinuously through an insulated tunnel. Different designs use one of the types of air flow described in Table 4. Food is finished in bin driers. Typically, a 20-m tunnel

contains 12–15 trucks with a total capacity of 5000 kg of food. This ability to dry large quantities of food in a relatively short time (5–16 h) made tunnel drying widely used, especially in the United States. However, the method has now been largely superseded by conveyor drying and fluidized-bed drying as a result of their higher energy efficiency, reduced labor costs, and better product quality.

Sun and Solar Drying. Sun drying (without drying equipment) is the most widely practiced agricultural processing operation in the world; more than 250,000,000 tons of fruits and grains are dried by solar energy per annum. In some countries foods are simply laid out on roofs or other flat surfaces and turned regularly until dry. More sophisticated methods (solar drying) collect solar energy and heat air, which in turn is used for drying. Solar driers are classified into three categories (4):

1. Direct natural-circulation driers (a combined collector and drying chamber).
2. Direct driers with a separate collector.
3. Indirect forced-convection driers (separate collector and drying chamber).

Both solar and sun drying are simple, inexpensive technologies, in terms of both capital input and operating costs. Energy inputs and skilled labor are not required. Sun drying is therefore the preferred option in developing countries that have a suitable dry season after harvesting food crops such as paddy or maize. The major disadvantages of sun drying are poor control over drying conditions; lower drying rates than in artificial driers; dependence on sunlight, which causes cessation of drying operations at night or during rain; and contamination of the dried product by dust, etc. Each of these factors contributes to a more variable and generally lower quality product than that produced by artificial driers.

Solar driers aim to improve product quality by providing greater control over drying conditions, protection from rain or dust, and higher drying rates. However, they have a relatively small capacity for drying the large bulk of crops at harvest time when compared to sun drying. In addition, the higher capital investment may not result in a higher income from improved quality of the dried crop. More valuable crops such as herbs and spices offer better potential for solar drying owing to the smaller quantities involved and the increased income from improved quality. However, the dependence of solar driers on sunlight and the better control over drying conditions achieved in artificial (fuel-fired) driers again limit the potential for solar drying.

Heated-Surface Driers. Driers in which heat is supplied to the food by conduction have two main advantages over hot-air drying: (1) It is not necessary to heat large volumes of air before drying commences, and the thermal efficiency is therefore high; (2) Drying may be carried out in the absence of oxygen to protect components of foods that are easily oxidized.

Typically, heat consumption is 2000–3000 kJ/kg of water evaporated compared with 4000–10,000 kJ/kg of water evaporated for hot-air driers. However, foods have low thermal conductives that become lower as the food dries. There should therefore be a thin layer of food to conduct heat rapidly, without causing heat damage. Foods may shrink during drying and lift off the hot surface, therefore introducing an additional barrier to heat transfer. Careful control is necessary over the rheological properties of the feed slurry to minimize shrinkage and to determine the thickness of the feed layer.

Drum Driers (Roller Driers). Slowly rotating hollow steel drums are heated internally by pressurized steam to 120–170°C. A thin layer of food is spread uniformly over the outer surface by dipping, spraying, spreading, or auxiliary feed rollers. Before the drum has completed 1 rev (within 20 s/3 min), the dried food is scraped off by a doctor blade that contacts the drum surface uniformly along its length. Driers may have a single drum (Fig. 10a) or double drums (Fig. 10b) or twin drums. The single drum is widely used as it has greater flexibility, a larger proportion of the drum area available for drying, easier access for maintenance, and no risk of damage caused by metal objects falling between the drums.

Drum driers have high drying rates and high energy efficiencies. They are suitable for slurries in which the particles are too large for spray drying. However, the high capital cost of the machined drums and heat damage to sensitive foods from high drum temperatures have caused a move to spray drying for many bulk dried foods. Drum drying is used to produce potato flakes, precooked cereals,

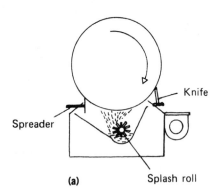

(a)

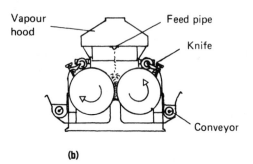

(b)

Figure 10. Drum driers: (**a**) single drum; (**b**) double drum. Courtesy of APV Mitchell Ltd.

molasses, some dried soups, and fruit purees, and whey or distillers' solubles for animal feed formulations.

Developments in drum design to improve the sensory and nutritional qualities of dried food include the use of auxiliary rolls to remove and reapply food during drying, the use of high- velocity air to increase the drying rate, and the use of chilled air to cool the product. Drums may be enclosed in a vacuum chamber to dry food at lower temperatures, but the high capital cost of this system restricts its use to high-value heat-sensitive foods.

Vacuum Band and Vacuum Shelf Driers. A food slurry is spread or sprayed onto a steel belt (or band) that passes over two hollow drums within a vacuum chamber at 1–70 mmHg. The food is dried by the first steam-heated drum, and then by steam-heated coils or radiant heaters located over the band. The dried food is cooled by the second water-cooled drum and removed by a doctor blade. Vacuum shelf driers consist of hollow shelves in a vacuum chamber. Food is placed in thin layers on flat metal trays that are carefully made to ensure good contact with the shelves. A particular vacuum of 1–70 mmHg is drawn in the chamber and steam or hot water is passed through the shelves to dry the food.

Rapid drying and limited heat damage to the food make both methods suitable for drying heat-sensitive foods. However, care is necessary to prevent the dried food from burning onto trays in vacuum shelf driers, and shrinkage reduces the contact between the food and heated surfaces of both types of equipment. Both have relatively high capital and operating costs and low production rates.

Vacuum-band shelf driers are used to produce puff-dried foods. Explosion puff drying involves partially drying food to a moderate moisture content and then sealing it into a pressure chamber. The pressure and temperature in the chamber are increased and then instantly released. The rapid loss of pressure causes the food to expand and develop a fine porous structure. This permits faster final drying and rapid rehydration. Sensory and nutritional qualities are well retained. The technique was first applied commercially to breakfast cereals and now includes a range of fruit and vegetable products.

EFFECTS ON FOODS

The effect of A_W on microbiological and selected biochemical reactions is shown in Fig. 11 and Table 2. Almost all microbial activity is inhibited below $A_w = 0.6$; most fungi are inhibited below $A_w = 0.7$; most yeasts are inhibited below $A_w = 0.8$; and most bacteria below $A_w = 0.9$. The interaction of A_w with temperature, pH, oxygen, and carbon dioxide, or chemical preservative has an important effect on the inhibition of microbial growth. When any one of the other environmental conditions is suboptimal for a given microorganism, the effect of reduced A_w is enhanced (Fig. 11). This permits the combination of several mild control mechanisms that result in the preservation of food without substantial loss of nutritional properties or sensory properties (Table 5).

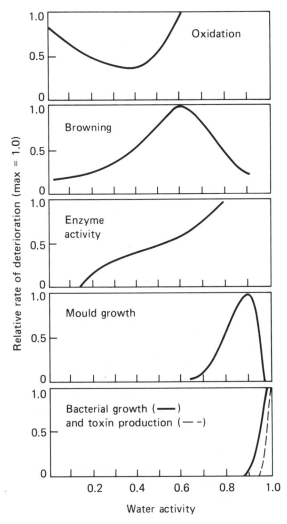

Figure 11. Effect of water activity on microbial, enzymic and chemical changes to foods (7). Courtesy of Marcel Dekker.

different foods. However, in general, a low A_w restricts the mobility of reactants and browning is reduced. At a higher A_w, browning reaches maximum. Water is a product of the condensation reaction in browning, and at higher moisture levels, browning is inhibited by end-product inhibition. At high moisture contents, water dilutes the reactants and the rate of browning falls (Fig. 11).

Oxidation of lipids occurs at low A_w values owing to the action of free radicals. Above the BET monolayer value, antioxidants and chelating agents (which sequester trace metal catalysts) become soluble and reduce the rate of oxidation. At higher A_w values the catalytic activity of metals is reduced by hydration and the formation of insoluble hydroxides but, at high A_w values, metal catalysts become soluble and the structure of the food swells to expose more reactive sites (Fig. 11).

Texture

Changes to the texture of solid foods are an important cause of quality deterioration. The nature and extent of pretreatments (eg, the addition of calcium chloride to blancher water), the type and extent of size reduction, and peeling each affect the texture of rehydrated fruits and vegetables. In foods that are adequately blanched, loss of texture is caused by gelatinization of starch, crystallization of cellulose, and localized variations in the moisture content during dehydration, which set up internal stresses. These rupture, compress, and permanently distort the relatively rigid cells, to give the food a shrunken, shrivelled appearance. On rehydration the product absorbs water more slowly and does not regain the firm texture associated with the fresh material. There are substantial variations in the degree of shrinkage with different foods.

Drying is not commonly applied to meats in many countries owing to the severe changes in texture compared with other methods of preservation. These are caused by aggregation and denaturation of proteins and a loss of water-holding capacity, which leads to toughening of muscle tissue.

The rate and temperature of drying have a substantial effect on the texture of foods. In general, rapid drying and high temperatures cause greater changes than do moderate rates of drying and lower temperatures. As water is removed during dehydration, solutes move from the interior of the food to the surface. The mechanism and rate of

Enzymatic activity virtually ceases at A_w values below the BET monolayer value. This is due to the low substrate mobility and its inability to diffuse to the reactive site on the enzyme. Chemical changes are more complex. The two most important things that occur in foods that have a low A_w are Maillard browning and oxidation of lipids. The A_w that causes the maximum rate of browning varies with

Table 5. Interaction of A_W, pH, and Temperature in Selected Foods

Food	pH	A_W	Shelf Life	Notes
Fresh meat	>4.5	>0.95	Days	Preserve by chilling
Cooked meat	>4.5	0.95	Weeks	Ambient storage when packaged
Dry sausage	>4.5	<0.90	Months	Preserved by salt and low A_w
Fresh vegetables	>4.5	>0.95	Weeks	Stable while respiring
Pickles	<4.5	0.90	Months	Low pH maintained by packaging
Bread	>4.5	>0.95	Days	Preserved by heat and low A_w in crust
Fruitcake	>4.5	<0.90	Weeks	Preserved by heat and low A_w
Milk	>4.5	>0.95	Days	Preserved by chilling
Yogurt	<4.5	<0.95	Weeks	Preserved by low pH and chilling
Dried milk	>4.5	<0.90	Months	Preserved by low A_w

movement are specific for each solute and depend on the type of food and the drying conditions used. Evaporation of water causes concentration of solutes at the surface. High air temperatures (particularly with fruits, fish, and meats) cause complex chemical and physical changes to the surface and the formation of a hard, impermeable skin. This is termed case hardening. It reduces the rate of drying and produces a food with a dry surface and a moist interior. It is minimized by controlling the drying conditions to prevent excessively high moisture gradients between the interior and the surface of the food.

In powders, the textural characteristics are related to bulk density and the ease with which they are rehydrated. These properties are determined by the composition of the food, the method of drying, and the particle size of the product. Low-fat foods (eg, fruit juices, potatoes, and coffee) are more easily formed into free-flowing powders than are whole milk or meat extracts. Powders are instantized by treating individual particles so that they form free-flowing agglomerates of aggregates, in which there are relatively few points on contact (Fig. 12). The surface of each particle is easily wetted when the powder is rehydrated, and particles sink below the surface to disperse rapidly through the liquid. These characteristics are respectively termed wettability, sinkability, dispersibility, and solubility. For a powder to be considered instant, it should complete these four stages within a few seconds.

The convenience of instantized powders outweighs the additional expense of production, packaging, and transport for retail products. However, many powdered foods are used as ingredients in other processes, and these are required to possess a high bulk of density and a wider range of particle sizes. Small particles fill the spaces between larger ones and thus exclude air to promote a longer storage life. The characteristics of some powdered foods are described in Table 6.

Flavor and Aroma

Heat not only vaporizes water during drying but also causes loss of volatile components from the food. The extent of volatile loss depends on the temperature and solids concentration of the food and on the vapor pressure of the volatiles and their solubility in water vapor. Volatiles that have a high relative volatility and diffusivity are lost at an early stage in drying. Fewer volatile components are lost at later stages. Control of drying conditions during each stage of drying minimizes losses. Foods that have a high economic value due to their characteristic flavors (eg, herbs and spices) are dried at lower temperatures.

A second important cause of aroma loss is oxidation of

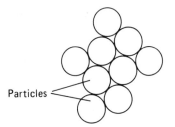

Figure 12. Agglomerated powder.

Table 6. Bulk Density and Moisture Content of Selected Powdered Foods[a]

Food	Bulk Density kg m³	Moisture Content, %
Cocoa	480	3–5
Coffee (ground)	330	7
Coffee (instant)	330	2.5
Coffee creamer	470	3
Cornstarch	560	12
Egg, whole	340	2–4
Milk, powdered, skimmed	640	2–4
Milk, instant, skimmed	550	2–4
Salt, granulated	960	0.2
Sugar, granulated	800	0.5
Wheat flour	450	12

[a] Refs. 24, 25.

pigments, vitamins, and lipids during storage. The open, porous structure of dried food allows access of oxygen. The rate of deterioration is determined by the storage temperature and the water activity of the food.

In dried milk the oxidation of lipids produces rancid flavors owing to the formation of secondary products, including δ-lactones. Most fruits and vegetables contain only small quantities of lipid, but oxidation of unsaturated fatty acids to produce hydroperoxides, which react further by polymerization, dehydration, or oxidation to produce aldehydes, ketones, and acids, causes rancid and objectionable odors. Some foods, (eg, carrots), may develop an odor of violets produced by the oxidation of carotenes to β-ionone. These changes are reduced by vacuum- or gas-packing, low storage temperatures, exclusion of ultraviolet or visible light, maintenance of low moisture contents, addition of synthetic antioxidants, or preservation of natural antioxidants.

The technical enzyme glucose oxidase is also used to protect dried foods from oxidation. A package that is permeable to oxygen but not to moisture and that contains glucose and the enzyme is placed on the dried food inside a container. Oxygen is removed from the headspace during storage. Milk powders are stored under an atmosphere of nitrogen with 10% carbon dioxide. The carbon dioxide is absorbed into the milk and creates a small partial vacuum in the headspace. Air diffuses out of the dried particles and is removed by regassing after 24 h. Flavor changes, due to oxidative or hydrolytic enzymes, are prevented in fruits by the use of sulfur dioxide, ascorbic acid, or citric acid, by pasteurization of milk or fruit juices, and by blanching of vegetables.

Other methods used to retain flavors in dried foods include

1. Recovery of volatiles and their return to the product during drying.
2. Mixing recovered volatiles with flavor- fixing compounds, which are then granulated and added back to the dried product, eg, dried meat powders.
3. Addition of enzymes, or activation of naturally occurring enzymes, to produce flavors from flavor precursors in the food (eg, onion and garlic are dried

under conditions that protect the enzymes that release characteristic flavors). Maltose or maltodextrin are used as a carrier material when drying flavor compounds.

Color

Drying changes the surface characteristics of food and hence alters the reflectivity and color. Chemical changes to carotenoid and chlorophyll pigments are caused by heat and oxidation during drying. In general, longer drying times and higher drying temperatures produce greater pigment losses. Oxidation and residual enzyme activity cause browning during storage. This is prevented by improved blanching methods and treatment of fruits with ascorbic acid or sulfur dioxide. For moderately sulfured fruits and vegetables the rate of darkening during storage is inversely proportional to the residual sulfur dioxide content. However, sulfur dioxide bleaches anthocyanins, and residual sulfur dioxide is an important cause of color deterioration in stored dried fruits and vegetables.

The rate of Maillard browning in stored milk and fruit products depends on the water activity of the food and the temperature of storage. The rate of darkening increases markedly at high drying temperatures, when the moisture content of the product exceeds 4–5%, and at storage temperatures above 38°C (26).

Nutritive Value

Large differences in reported data on the nutritive value of dried foods are due to wide variations in the preparation procedures, the drying temperature and time, and the storage conditions. In fruits and vegetables, losses during preparation usually exceed those caused by the drying operation. For example, losses of vitamin C during preparation of apple flakes are reported to be 8% during slicing, 62% from blanching, 10% from pureeing, and 5% from drum drying (27).

Vitamins have different solubilities in water, and, as drying proceeds, some (eg, riboflavin) become supersaturated and precipitate from solution. Losses are therefore small (Table 7). Others, (eg, ascorbic acid) are soluble until the moisture content of the food falls to very low levels and react with solutes at higher rates as drying proceeds. Vitamin C is also sensitive to heat and oxidation. Short drying times, low temperatures, and low moisture and oxygen levels during storage are necessary to avoid large losses. Thiamin is also heat sensitive, but other water-soluble vitamins are more stable to heat and oxidation, and losses during drying rarely exceed 5–10% (excluding blanching losses).

Oil-soluble nutrients (eg, essential fatty acids and vitamins A, D, E, and K) are mostly contained within the dry matter of the food, and they are not therefore concentrated during drying. However, water is a solvent for heavy-metal catalysts that promote oxidation of unsaturated nutrients. As water is removed, the catalysts become more reactive, and the rate of oxidation accelerates (Fig. 12). Fat-soluble vitamins are lost by interaction with the peroxides produced by fat oxidation. Losses during storage are reduced by low oxygen concentrations and storage temperatures and by exclusion of light.

The biological value and digestibility of proteins in most foods does not change substantially. However, milk proteins are partially denatured during drum drying, and this results in a reduction in solubility of the milk powder, aggregation, and loss of clotting ability. A reduction in biological value of 8–30% is reported, depending on the temperature and residence time (30). Spray drying does not affect the biological value of milk proteins. At high storage temperatures and at moisture contents above approximately 5%, the biological value of milk protein is decreased by Maillard reactions between lysine and lactose. Lysine is heat sensitive, and losses in whole milk range from 3–10% in spray drying and 5–40% in drum drying (31).

The importance of nutrient losses during processing depends on the nutritional value of a particular food in the diet. Some foods, eg, bread and milk, are an important source of nutrients for large numbers of people. Vitamin losses are therefore more significant in these foods than in those that either are eaten in small quantities or have low concentrations of nutrients.

In industrialized countries, the majority of the population achieve an adequate supply of nutrients from the mixture of foods that is eaten. Losses due to processing of one component of the diet are therefore insignificant to the long-term health of an individual. In one example, complete meals that initially contained 16.5 μg of vitamin A lost 50% on canning and 100% after storage for 18

Table 7. Vitamin Losses in Selected Dried Foods[a]

Food	Loss, %						
	Vitamin A	Thiamin	Vitamin B$_2$	Niacin	Vitamin C	Folic Acid	Biotin
Fruits[b]	6	55	0	10	56		
Fig (sun-dried)	—	48	42	37	—	—	—
Whole milk (spray-dried)	—	—	—	—	15	10	10
Whole milk (drum-dried)	—	—	—	—	30	10	10
Pork		50–70					
Vegetables[c]	5	<10	<10				

[a] Refs. 28,29.

[b] Fruits: mean loss from fresh apple, apricot, peach, and prune.

[c] Vegetables: mean loss from peas, corn, cabbage, and beans (drying stage only).

months. Although the losses appear to be significant, the original meal contained only 2% of the recommended daily allowance (RDA), and the extent of loss is therefore of minor importance. The same meal contained 9 mg of thiamin and lost 75% after 18 months' storage. The thiamin content is 10 times the RDA, so adequate quantities therefore remained. Possible exceptions are the special dietary needs of preterm infants, pregnant women, and the elderly. In these groups there may be either a special need for certain nutrients or a more restricted diet than normal. These special cases are discussed in detail in Refs. 32, 33 and 34.

Reported vitamin losses during processing give an indication of the severity of each unit operation. However, such data should be treated with caution. Variation in nutrient losses between cultivars or varieties can exceed differences caused by alternative methods of processing. Growth conditions, or handling and preparation procedures before processing, also cause substantial variation in nutrient loss. Data on nutritional changes cannot be directly applied to individual commercial operations, because of differences in ingredients, processing conditions, and equipment used by different manufacturers.

REHYDRATION

Rehydration is not the reverse of drying. Texture changes, solute migration, and volatile losses are each irreversible. Heat reduces the degree of hydration of starch and the elasticity of cell walls and coagulates proteins to reduce their water-holding capacity. The rate and extent of rehydration may be used as an indicator of food quality; those foods that are dried under optimum conditions suffer less damage and rehydrate more rapidly and completely than poorly dried foods.

NOMENCLATURE

A Area
A_w Water activity
C Constant, drag coefficient
d Diameter
G Mass flow rate of air
g Acceleration due to gravity (9.81 m/s^2)
H Humidity
h Surface heat transfer coefficient
K Mass transfer coefficient
L Length
M Moisture content (dry-weight basis)
m Mass, mass flow rate, moisture content (wet-weight basis)
P Pressure, vapor pressure
Q Rate of heat transfer
r Radius
Re   Reynolds Number
t Time
U Overall heat transfer coefficient
v_f Velocity, air velocity needed for fluidization
x Thickness, depth
ε Voidage of fluidized bed

θ Temperature
λ Latent heat
μ Viscosity
τ Constant = 3.142
ρ Density

Acknowledgments: Grateful acknowledgment is made for information supplied by the following: APV Mitchell Dryers Ltd, Carlisle, Cumbria, CA5 5DU, UK; Petrie and McNaught Ltd, Rochdale, UK; Zschokke Wartmann Ltd, Stahlrain, CH-5200 Brugg, Switzerland; Proctor and Schwartz Inc., Glasgow, Scotland; Torftech Ltd, Mortimer, Reading, Berkshire, UK; Unilever, London EC4P 4BQ, UK; Chartered Institute of Building Services Engineers, 222 Balham High Road London, SW12 985, UK; De Melkindustrie Veghel.

BIBLIOGRAPHY

Reproduced with permission from P. Fellows, *Food Processing Technology*, Ellis Horwood Limited, Chichester, UK, 1988.

1. M. J. Lewis, *Physical Properties of Foods and Food Processing Systems,* Ellis Horwood, Chichester, West Sussex, UK, 1987.
2. C. Van den Berg, "Water Activity," in D. MacCarthy, ed., *Concentration and Drying of Foods,* Elsevier Applied Science, Barking, Essex, UK, 1986, pp. 11–36.
3. J. A. Troller and J. H. B. Christian, *Water Activity and Food,* Academic Press, London, 1978.
4. B. Brenndorfer, L. Kennedy, C. O. Oswin-Bateman, and D. S. Trim. *Solar Dryers,* Commonwealth Science Council, Commonwealth Secretariat, Pall Mall, London, 1985.
5. D. Kunii and O. Leverspiel, *Fluidisation Engineering,* John Wiley & Sons, New York, 1969.
6. M. Loncin and R. L. Merson, *Food Engineering—Principles and Selected Applications,* Academic Press, New York, 1979.
7. M. Karel, "Dehydration of Foods," in O. R. Fennema, ed., *Principles of Food Science. Part 2. Physical Principles of Food Preservation,* Marcel Dekker, New York, 1975, pp. 309–357.
8. C. W. Hall, *Dictionary of Drying,* Marcell Dekker, New York, 1979.
9. P. E. Zagorzycki, "Automatic Humidity Control of Dryers," *Chemical Engineering Progress* April, 66–70 (1983).
10. K. Masters, *Spray Drying.* Leonard Hill, London, 1972, pp. 15, 144, 160, 230, 545–586.
11. K. Grikitis, "Dryer Spearheads Dairy Initiative," *Food Process* (1986).
12. S. F. Sapakie and T. A. Renshaw, "Economics of Drying and Concentration of Foods," in B. M. McKenna, ed., *Engineering and Food,* Vol. 2, Elsevier Applied Science, London, 1984, pp. 927–938.
13. C. Tragardh, "Energy and Energy Analysis in Some Food Processing Industries," *Lebensmittel Wissenschaft Technologie* **14,** 213–217 (1981).
14. A. Axtell, A. B. Bush, and M. Molina, *Try Drying It. A Case Study of Small Industry Driers,* IT Publications, London, 1991.
15. K. Masters, "Recent Developments in Spray Drying," in S. Thorne, ed., *Developments in Food Preservation,* Vol. 2, Applied Science, London, 1983, pp. 95–121.
16. S. E. Charm, *Fundamentals of Food Engineering,* 3rd ed., AVI Publishing Co., Westport, Conn., 1978, pp. 298–408.

17. P. J. A. M. Kerkhof and W. J. A. H. Schoeber, "Theoretical Modelling of the Drying Behaviour of Droplets in Spray Driers," in A. Spicer, ed., *Advances in Preconcentration and Dehydration of Foods,* Applied Science, London, 1974, pp. 349–397.

18. O. G. Kjaergaard, "Effects of Latest Developments on Design and Practice of Spray Drying. In Ref. 17, pp. 321–348.

19. H. B. Heath, "The Flavour Trap," *Food Flavour Ingredients* **7**, 21, 23, 25 (1985).

20. M. Byrne, "The £8M Drier," *Food Manufacture* September, 67, 69 (1986).

21. J. C. Ashworth, "Developments in Dehydration," *Food Manufacture* December. 25–27, 29 (1981)

22. H. Schubert, (1980) "Processing and Properties of Instant Powdered Foods," in P. Linko, Y. Malkki, J. Olkku, and J. Larinkari, eds., *Food Process Engineering,* Vol. 1, *Food Processing Systems,* Applied Science, London, 1980, pp. 675–684.

23. U.S. Pat. 4,490,403 (1983) J. Pisecky, J. Krag, and Ib. H. Sorensen.

24. B. K. Watt and A. L. Merrill, *Composition of Foods,* Agriculture Handbook 8. U.S. Department of Agriculture, Washington D.C., 1975.

25. M. Peleg, "Physical Characteristics of Food Powders," in M. Peleg and E. B. Bagley, eds., *Physical Properties of Foods,* AVI Publishing Co., Westport, Conn., 1983, pp. 293–323.

26. C. H. Lea, "Chemical Changes in the Preparation and Storage of Dehydrated Foods," in *Proceedings of Fundamental Aspects of Thermal Dehydration of Foodstuffs,* Aberdeen, March 25–27, 1958, Society of Chemical Industry, London pp. 178–194.

27. F. Escher and H. Neukom, "Studies on Drum-drying Apple Flakes," *Trav. Chim. Aliment. Hyg.* **61**, 339–348 (1970) (in German).

28. B. A. Rolls, "Effect of Processing on Nutritive Value of Food: Milk and Milk Products, in M. Rechcigl, ed., *Handbook of the Nutritive Value of Processed Foods,* Vol. 1, CRC Press, Boca Raton, Fla., 1982, pp. 383–389.

29. D. H. Calloway, "Dehydrated Foods," *Nutrition Reviews* **20**, 257–260 (1962).

30. W. Fairbanks and H. H. Mitchell, "The Nutritive Value of Skim-milk Powders with Special Reference to the Sensitivity of Milk Proteins to Heat," *Journal of Agricultural Research,* **51**, 1107–1121 (1935).

31. B. A. Rolls and J. W. G. Porter, "Some Effects of Processing and Storage on the Nutritive Value of Milk and Milk Products," *Proceedings of the Nutrition Society* **32**, 9–15 (1973).

32. A. E. Bender, "The Nutritional Aspects of Food Processing," in A. Turner, ed., *Food Technology International Europe,* Sterling, London, 1987, pp. 273–275.

33. R. R. Watson, *Nutrition for the Aged.* CRC Press, Boca Raton, Fla., 1986.

34. D. Francis, *Nutrition for Children,* Blackwell, Oxford, UK, 1986.

General References

References 7, 10, and 12 are good general references.

Hall, C. W., *Dictionary of Drying,* Marcel Dekker, New York, 1979.

Spicer Kjaergaard, O. G., "Effects of Latest Developments on Design and Practice of Spray Drying," in A. Spicer, ed., *Advances in Preconcentration and Dehydration of Foods.* Applied Science, London, 1974, pp. 321–348.

P. FELLOWS
Intermediate Technology
Development Group
Rugby, United Kingdom

DEGUMMING. See CANOLA; HYDROGENATION; OLIVES; OILSEEDS AND THEIR OILS; PALM OIL; SALAD OILS; SOYBEANS AND SOYBEAN PROCESSING; and entries under FATS AND OILS.
DESICCATION. See FREEZING SYSTEMS FOR THE FOOD INDUSTRY.
DETERGENCY. See DISINFECTANTS.

DETERGENTS

A detergent is any material that cleanses or provides surface active properties. Detergents assist in the removal of unwanted soils from surfaces. The use of detergents in food-processing facilities is an important component of their sanitation program. Food-processing plants perform regular sanitation within their operations for two primary reasons: to provide a clean and aesthetically appealing environment and to eliminate microbial organisms that cause food spoilage and health problems for the consumer. Detergents selected for use in food-processing areas should be products that are free rinsing, leaving no residual toxins that might contaminate the food. Detergents used in food-processing areas should also leave no residual odor, after rinsing, that might serve to conceal malodors caused by microbial activity in those areas.

The Food and Drug Administration (FDA) publishes a listing of materials that are acceptable for use in the food environment (1). These include ingredients for food and substances present on food contact surfaces, along with parameters for their use. The U.S. Department of Agriculture (USDA) follows the FDA guidelines when approving detergents and other products acceptable for use in meat- and poultry-processing plants.

TYPES OF DETERGENT

Detergents may be either solvent-based or water-soluble products. Water-soluble detergents may be acidic, neutral, or alkaline. Solvent-based detergents are not widely used in food-processing areas. Many solvents have toxic hazards associated with them, possess strong persistent odors and are not readily biodegradable. The few solvent detergents that are acceptable in food-processing plants tend to be either readily biodegradable or are volatile. Most volatile solvents are also flammable. Because of all the liabilities that are associated with solvent-based detergents, their use is limited to specialty applications where other detergents do not provide adequate cleaning. Acid detergents are generally used to remove mineral deposits such as hard-water scale. These products are also effective in removing coagulated protein films. Specialized applications use acid detergents for microbial control. A product formulation that combines an acid with an anionic surfactant makes an effective sanitizer; as do iodophors, which are acid detergents that release iodine in solution. Neutral detergents (pH between 5.0 and 9.0) are generally synergistic surfactant blends. Some neutral products also contain solvents to compliment the cleaning action of the surfactants. Because of their pH neutrality these detergents are widely used for hand scrubbing. Neutral detergents tend to be most effective on oils, greases, and fats because of their excellent ability to emulsify and disperse these types of soil. A specialized application in this category is the use of quaternary ammonium chloride formulations as sanitizers. Alkaline detergents are effective on organic soils including fats, proteins, and carbohydrates. These detergents are used extensively to clean food-processing facilities. They work by solubilizing, saponifying, peptizing, or hydrolyzing the soils to be removed.

DETERGENT PROPERTIES

Detergents assist the cleaning process by performing one or more of the following: wetting or penetration, dispersing or deflocculating, suspension, emulsification, peptizing, dissolving, chelating or sequestering, rinsing, and sanitizing (2,3). Wetting refers to the detergent's ability to penetrate through and come in contact with all soils and equipment surfaces, thus loosening the soil and facilitating its removal. A dispersant promotes the breakdown of large soil particles into smaller particles and inhibits the formation of larger soil particles by holding them uniformly distributed in the detergent bath. When soil particles are uniformly distributed throughout the detergent bath, without any precipitation to the bottom or flotation to the surface, they are in suspension. An emulsifier breaks down fat and oil into tiny particles and disburses them uniformly throughout the water phase where they are held in suspension. (This is an oil-in-water emulsion.) Peptizing is the partial solubilization and colloidal dispersion of protein by an alkaline detergent. When a soil is converted into a form that is completely soluble in water it will dissolve. Chelation is the process of chemically removing minerals such as calcium, magnesium, or iron from water. This is done by forming a soluble complex with the mineral thus preventing their precipitation from the detergent bath. This enhances the rinsability of the detergent solution and prevents interference of those minerals with the detergent's cleaning ability. Rinsing is the ability to flush the detergent solution and soil from the surfaces being cleaned without leaving any residue. Sanitizing reduces the total microbial population present on the cleaned surfaces. Sanitizers work most efficiently on cleaned surfaces free of any soil. The ideal detergent would possess the following characteristics: rapid and complete solubility in water and complete removal of all soil from the surfaces being cleaned. It would rinse free of all residue and provide germicidal activity. Finally it would be inexpensive to use and safe to handle. To date no universal product possessing all these characteristics exists; hence the plethora of products available in the marketplace.

DETERGENT COMPONENTS

Most detergents are blends of surfactants and builders, although they may be strictly surfactant blends or blends of builders. Surfactants are classed as anionic, nonionic, or cationic. Anionic surfactants are negatively charged molecules and are the most widely used in cleaning. Nonionic surfactants have no charge and are excellent oil and grease solubilizers and emulsifiers. Cationic surfactants are positively charged molecules and are not used to any significant degree as cleaners. Their use is primarily in specialty applications such as fabric softeners and sanitizers.

In aqueous solution, most solid soils are negatively charged minerals (4). If cationic surfactants are present, an electrostatic attraction occurs that is not favorable to soil removal. The negative charge of the anionic surfactants, on the other hand, creates an electrostatic repulsion that facilitates removal of these soils and prevents their redeposition. Uncharged oily soils are most readily removed by solubilization in micelles or by emulsification with nonionic surfactants (4). The solubilization process is most effective when the nonionic surfactant concentration is above the critical micelle concentration (CMC). Also the effectiveness can be further optimized by maintaining a temperature just below the cloud point of the surfactant. Builders are all the other ingredients that are formulated into a detergent to enhance the performance of the surfactants. Detergents that contain no surfactants rely on saponification to produce surfactants *in situ*. Some builders found in detergents are alkalies, acids, phosphates, silicates, chelants, chlorine, defoamers, antiredeposition agents, abrasives, and corrosion inhibitors (5–7). By a judicious selection of builders, detergents can be formulated to enhance one or more of the following properties: emulsification, saponification, hydrolysis, removal of mineral deposits, wetting and penetration, dispersing, antiredeposition, peptizing, water softening and antiscaling, defoaming, and sanitizing.

CLEANING

Regular cleaning is essential to maintaining sanitary conditions in food-processing facilities. A good cleaning program should schedule regular cleaning of all interior building surfaces (floors, walls, and ceilings), all processing equipment, removal of garbage, trash, and other wastes, leaving clean rinsed surfaces that are free of any chemical residues that might contaminate the food (8). The basic steps of the cleaning process are the physical removal of the gross soil load, detergent cleaning, rinsing, and sanitizing. The physical removal of gross soil is accomplished by such methods as squeegeeing, shoveling, scraping, rinsing, or any other mechanical means appropriate to the soil and the surface being cleaned. By reducing the soil load the detergents can work more efficiently. The detergent cleaning process loosens the remaining soil. Rinsing flushes the soil from the surface and prevents its redeposition. Sanitizing inhibits microbiological growth on the cleaned surfaces. Three classes of soil have been characterized: soils bound to the surface by oils or greases, soils absorbed onto the surface, and scale deposits. These soils are separated from surfaces by three possible mechanisms: mechanical action, chemical alteration of the nature of the soil, or reduction of the energy binding the soil to the surface by reduction of the surface tension through wetting and penetration (9). The procedure used for cleaning and the type of detergent selected depend on a number of factors: the type and condition of the soil, the type of surface being cleaned, the condition and temperature of the water supply, the cleaning equipment available, the time alloted for cleanup, and the cost of the sanitation program. The ideal water used in sanitation would be free of microorganisms, clear, colorless, noncorrosive, and free of minerals. Other factors to effective cleaning are time, temperature, concentration of detergent, and mechanical action. In general, cleaning is enhanced with longer time, higher temperature, higher concentrations of detergent, and increased mechanical action. However, this generalization needs to be tempered by practical considerations and limitations of the clean-up project. For example, protein starts to coagulate above 55°C (10). In general the following criteria are used to select an appropriate detergent for cleaning:

Heavy soils, cooked on soils	High alkaline cleaner
Fats and proteins	Chlorinated cleaner
Hand scrubbing, soft metals	Light duty or neutral cleaner
Remove mineral deposits, brighten equipment	Acid cleaner

There are four primary procedures used to apply the detergent cleaners with an infinite variety of variations on these methods. Cleaning in place (CIP) is used primarily on enclosed equipment and uses pumps to circulate the detergent solution through the equipment. This process normally entails the use of an alkaline cleaner followed by an acid cleaner. These cleaners typically are low-foaming detergents, because foam hinders the cleaning process when the solutions are pumped through pipes and spray balls to clean large tanks and other parts of the system. Foaming and high-pressure spraying techniques are the most common methods for cleaning the exposed surfaces of equipment, conveyors, walls, and floors. Foaming is a modification of the high-pressure application where compressed air is injected into the detergent stream to produce foam. Foam is popular for use on vertical surfaces because it holds the detergent onto those surfaces for several minutes before draining off. Soak tank cleaning is used on trays, molds and other portable equipment that can be readily removed to one location for cleaning by soaking in a detergent bath. Boil out is a special soak tank procedure where the equipment to be cleaned is filled with a detergent solution and heat is applied until cleaning is effected. Hand scrubbing is used where none of the other procedures are appropriate, or to supplement inadequate cleaning by those methods. This technique combines the chemical action of the detergent with the mechanical action of wiping or scouring the soil from the surface.

DISPOSAL

Disposal of detergents, as with all chemicals, comes under the purview of federal, state, and local laws and regulations. Food processors should maintain close contact with their waste-disposal facilities to keep abreast of current requirements for disposal in their respective locales. Generally, spent detergents can be flushed down the drain. However, care does need to be taken to ensure that this waste solution does comply with the requirements of the local wastewater treatment plant. Usually their primary concerns are the pH and the phosphate levels in the discharged waste. Avoid using detergents containing phosphates where stringent phosphate disposal levels exist. The pH can be readily neutralized, if necessary, before sending the solution out to the wastewater treatment plant. In the event of a detergent spill, several options exist. One would be to neutralize and dilute the detergent with water and dispose as wastewater. Another would be to contain the spill, absorb onto an appropriate substrate, if a liquid, and dispose in a landfill in accordance with all applicable laws and regulations. Certain solvent cleaners present special disposal problems unique to them, and special care must be taken to comply with the disposal regulations for those products.

FOOD SAFETY

In general, food products must be segregated from the clean-up area to avoid contamination. There are a few specialized cases where detergents are used to process edible food products. These special applications are stringently regulated to insure the safety and wholesomeness of the resulting food product. However, where this is not the case, the food either must be removed from the clean-up area or otherwise covered or segregated to protect it from any possible contamination during the clean-up process. Also, the containers of detergents should be stored in an area segregated from the food-processing and storage areas, again to avoid the possibility of contamination.

WORKER SAFETY

Many detergents used to clean food processing plants are strongly alkaline, strongly acid, release chlorine, or present other safety hazards to the clean-up personnel. These personnel should familiarize themselves with the products they are using and the safety hazards associated with them by studying the material safety data sheet (MSDS), the product label, and any other information the detergent supplier has made available. Pay special attention to the first aid measures that have been recommended. Acids and alkalies can cause both chemical and thermal burns. Chlorine causes respiratory distress. Many chemicals, such as quaternary ammonium chloride compounds, can be sensitizing agents or allergens. Even mild detergents, because of their ability to solubilize oils and fats, can cause dryness, chapping, and cracking of the skin. Appropriate, safety measures need to be taken by clean-up personnel to avoid injury. The purpose of safety equipment is to insure a safe air supply for breathing and to protect the skin from direct contact with any hazardous materials being used. Generally the first aid step to take, after eye or skin exposure, is to rinse with copious quantities of water; for breathing difficulties, remove to fresh air. See a doctor if distress persists. Furthermore, in the area of eye safety, contact lenses should never be worn when working with potentially hazardous chemical products. In the event of any chemical splattering in the eye, the contact lens makes it very difficult to adequately flush the eye. When mixing chemical products such as detergents, they should always be added to water. Furthermore, use cold water where the chemical addition produces heat of reaction as, for example, adding a strong alkaline detergent to water, which can generate sufficient heat to boil water. A healthy, clean, and safe working environment should be the goal of every food-processing plant.

BIBLIOGRAPHY

1. *Code of Federal Regulations,* Title 21 §172, U.S. Government Printing Office, Washington, D.C., 1986.
2. M. E. Parker and J. H. Litchfield, *Food Plant Sanitation,* Reinhold Publishing Corp., New York, 1962, p. 229.
3. N. G. Marriot, *Principles of Food Sanitation,* 2nd ed., Van Nostrand Reinhold, Co., Inc., New York, 1989, pp. 79–81.
4. D. Myers, *Surfactant Science and Technology,* VCH Publishers, Inc., New York, 1988, pp. 315, 316.
5. Ref. 2, pp. 230–250.
6. Ref. 3, pp. 81–91.
7. R. K. Guthrie, *Food Sanitation,* 2nd ed., AVI Publishing Co., Inc., Westport, Conn., 1980, p. 162.
8. Ref. 7, p. 146.
9. Ref. 3, p. 74–77.
10. Ref. 3, p. 91.

General References

Anonymous "Developing a Total Cleaning/Sanitation Program," *Food Processing* **48,** 112,113 (1987).

M. J. Banner, "Keeping It Clean," *Beverage World,* **106,** 41–49 (1987).

W. G. Cutler and E. Kissa, eds., *Detergency Theory and Technology,* Marcel Dekker, Inc., New York, 1987.

E. A. Erten and R. F. Ellis, "Efficient Cleaning Practices Safeguard Product Quality," *Food Processing* **49,** 104–106 (1988).

D. R. Karsa, *Industrial Applications of Surfactants,* The Royal Society of Chemistry, Burlington House, London, 1987.

National Restaurant Association, *Sanitation Operations Manual,* (monogram) 1979.

State Water Quality Board, *Detergent Report, A Study of Detergents in California,* State Department of Water Resources, State Department of Public Health, (monogram) 1965.

G. K. York, "A Guide to Food Plant Sanitizers," *Food Processing* **48,** 118–120 (1987).

Terry L. McAninch
Birko Corporation
Westminster, Colorado

DIETETICS

PREVENTING AND CURING DISEASES

Diet and nutrition may be related to disease in two basic ways.

1. *Therapy.* Therapy includes special dietary and nutritional care and perhaps even cures for people with certain forms of illness. This is the field of therapeutic nutrition. Many questions of health care concern this field. For instance, how should a patient who is recovering from an accidental ingestion of cleansing detergent be fed? What should a burn victim eat to assist in the recovery process? How should a patient with a kidney problem or diabetes be fed to control or reduce the symptoms? And can a disease be cured through a dietary regimen or the administration of some nutrients? Can mental illness be cured with a special diet? Can vitamin C cure cancer?

2. *Prevention.* The knowledge that certain dietary practices or substances may produce specific clinical disorders may be used to prevent certain diseases. For example, eating too much salt may cause high blood pressure. One of the principal questions in preventive nutrition is can specific clinical diseases be prevented by adopting certain lifelong dietary habits? For example, can hypertension be avoided by reducing the consumption of salt? Many health practitioners are now paying increasing attention to such questions and trying to impart knowledge of long-term preventive measures to their patients.

The recent scientific development in basic and clinical nutrition has clearly established two important facts.

1. A proper nutrition and dietary intake over a period of time can protect from various illness.
2. Scientifically applied, nutrition and diet can become an important part of a clinical treatment plan for patients suffering from different disorders.

The next few sections will provide some familiar examples to illustrate the role of nutrition and diet in patient care, both in prevention and treatment.

Obesity

There is absolutely no doubt that people with a weight 10–20 lb over their ideal weight run a higher risk of heart disease, high blood pressure, diabetes, and so on. There are many things that can be done to lose and maintain weight. The appropriate route is a wholesome eating habit throughout life. Thus by being careful about the quality and quantity of food eaten, certain diseases can be avoided.

Heart Diseases

Heart diseases can be prevented and treated through diet. For example, for people in normal health, the risk of getting heart disease can be reduced by eating an appropriate amount of fat, cholesterol, and fiber. This preventive approach is also supported by treatment success. When blood chemistry profiles confirm that a patient is at risk for coronary heart disease, that risk can be reduced by following a diet low in fats and cholesterol. A 60-year-old man recovering from a heart attack can reduce the risk of having another one by modifying his dietary intake with the help of a dietitian or a nutritionist.

Diabetes

Diabetics benefit from special diets. Many of the painful symptoms of diabetes can be controlled by various means, especially by a diet with defined quantities of fat, carbohydrate, and protein. This is a classic example of the successful treatment of a disease by manipulating the diet.

Genetic Disorders

Some infants are born unable to use certain components of dietary protein. One such disorder is known as phenylketonuria. Without special care, these children do not grow properly. The use of a special diet permits them to lead normal lives. This treatment procedure can be supplemented by a preventive approach. If the pregnant woman is willing to follow a special diet, the disease will be less severe in the child after birth and, in some cases, the disorder does not show up in the child at all. This is an unusual case of preventing a childhood disease by dietary manipulation during the prenatal stage.

Surgery

Proper nutritional and dietary management before surgery can decrease complications and improve recovery. This type of dietary preventive treatment is extremely important in patient care. After the operation, an aggressive dietary regimen will definitely hasten recovery and increase the quality of life. It must be emphasized that nutritional therapy is a small part of the management plan. In this case it is assumed that the surgery is successful, the medication protects against infection, and that other surgical risks are minimal.

Intestinal Disorders

The elderly have a higher tendency to develop a special problem of the large intestine: diverticular disease. The victims develop pouches along the intestinal walls. It is certain that the cause of this problem in some patients is a deficient intake of dietary fiber throughout adulthood. A preventive measure is to consume an adequate amount of fiber when young. Clinical evidence shows that in some of these patients, the intestinal pouches can be eliminated by ingesting a calculated amount of fiber. Most medical textbooks prescribe this as one treatment method for some of the patients with the disorder.

PATIENT CARE AND THE HEALTH TEAM

Patient Care and Management

In practice, the two approaches, therapy and prevention, may be used simultaneously, depending on the patient's needs and particular disease. If well-nourished patients have good appetites and eat well, they are fed a regular diet to maintain nutritional status and permit a speedy recovery. The calories and protein may be somewhat higher than normal, but the diet does not require any special modification. On the other hand, some patients need special diets to control a disease, to correct nutrient deficiencies, and to provide life support in the overall management plan of the attending physician.

Especially in a hospital setting, diet therapy is rarely the only component of a treatment or cure. However, diet therapy is almost always part of a coordinated clinical management program. For example, the appropriate treatment for a kidney patient on hemodialysis has many components: the proper use of the artificial kidney; a specific dietary regimen; a program of drugs, including vitamin and mineral supplements; diagnostic tests; psychotherapy; and patient education and compliance. Although diet therapy is undoubtedly a science, the practice of clinical dietetics is also an art. Because eating is important, there are intense human factors to be considered in translating therapeutic ideas into diets that patients will readily follow. The practitioner must be prepared to deal with the patient's emotions, cultural and socioeconomic background, and many other psychological factors.

Nutrition is one of the youngest branches of medical science. Its role in achieving and maintaining good health is both preventive and therapeutic. For nearly a century, experience in clinical medicine has shown that nutrition and diet therapy can positively influence the coarse of recovery in patients with certain pathological conditions. Rapid advances in nutritional science have increased the role of diet therapy in patient care in the last two decades.

As more information becomes available about particular diseases, their management by diet therapy becomes more scientific, exact, and unique. This trend is leading to specialization in specific dietary care and patient management. Together with an increasing patient population, this specialization has increased the frequency of the team approach to inpatient, outpatient, and community health care. By coordinating the knowledge and experience of the members, the team can provide better diagnosis, treatment, and overall patient care. Many experienced clinicians firmly believe that a health team treats the whole person rather than the disease or the symptoms.

In a health team approach, several professionals share responsibility for therapeutic feeding. Traditionally, the physician diagnoses the patient's condition and prescribes a diet order, the dietitian develops detailed meal plans for the patient, and the nurse serves the foods and monitors the patient's responses.

The dietitian and the nurse both play an important role in the feeding care of a patient. The dietitian works with the patient and family to develop a dietary plan that fits into the clinical management program of the health team.

The dietitian works closely with the nurse, who, in the current health care system, has a direct role in making sure that the patient is fed and that the team is fully informed of the patient's eating patterns. With the introduction of total parenteral nutrition, the nurse is assuming more responsibilities in the nutritional care of patients. The nurse is in a unique position to educate the patient about food and nutrition and to lead the patient toward wholesome eating habits after discharge.

Those who work closely with patients everyday have especially good opportunities to tailor meals to varying individual tastes and needs. Eating is a highly personal activity, capable of creating feelings ranging from extreme pleasure to disgust or pain. Whether the nutritional focus is on prevention or therapy, health care professionals must see their clients as individual people with their own feelings about eating, rather than as insensitive food-ingesting machines.

Apart from the professionals discussed, other specialists may be called in as needed. More details about each member of the health team are provided in the next several sections.

The Doctor. Although the extent of a physician's participation in the nutritional and dietary care of a patient varies with the circumstances, such as the types of patient and facility and the availability of trained personnel and nutrient formulas, only the doctor is authorized to perform some basic duties. In general, the doctor diagnoses the illness or condition of the patient and decides if special nutritional or dietary care will benefit the patient. If the doctor believes that such care would be helpful, he or she recommends a prescription for the nutritional and dietary needs of the patient. This recommendation is implemented in a number of ways, as discussed below.

The Dietitian. While the doctor suggests the general framework for a patient's dietary management, the dietitian usually develops the specific dietary plan. Circumstances where this may not be the case are discussed later.

In the United States, a dietitian who has completed specified educational and other requirements can become registered with a private organization, the American Dietetic Association (ADA). This person can then use the title Registered Dietitian (R.D.). A registered dietitian practices dietetics for healthy and sick people. On the other hand, a clinical dietitian, also a registered dietitian, is a member of a health care team and usually practices dietetics in a clinical setting. In most states, if a health care institution, for example, a hospital, provides health services, it must include one or more dietitians on its staff.

In practice, however, many small and underbudgeted hospitals do not have a full-time dietitian. In some cases, ordinary clinical dietetics may be planned by the attending physician and implemented by the nurse. Many of these hospitals may either employ a part-time dietitian as a consultant or contract for the services of professional dietitians in nearby larger hospitals or cities when necessary.

Many hospitals have a small dietetic department where the dietitian attends to food purchasing, storage,

and preparation and personnel management as well as prescribed dietary orders. In such a situation, although the dietitian implements the doctor's diet prescription, the nurse does much of the food serving, patient monitoring, and evaluation. In some hospitals the food service operation is divided into different departments, such as dietetics, food purchasing and planning, kitchen staff, and personnel management. Here other dietitians or dietetic aides assist the chief dietitian. Within such a setting, the dietitian assumes a large part of the practice of clinical dietetics, although nurses still play important roles because of their constant contact with the patients.

In addition to having fully staffed dietetic departments, many university teaching hospitals and large medical centers employ nutritionists, some of whom provide outpatient care. A number of highly regarded hospitals and medical centers have established a sophisticated nutrition support service, which integrates the activities of a number of health professionals. These health professionals give optimal nutrition and dietary care to the patient in an effort to prevent and manage any risk of morbidity and mortality caused by an unsatisfactory nutritional status. Institutions vary in the degree of responsibility they give to dietitians who provide nutritional and dietary care to patients, although dietitians are the best-trained personnel to administer this aspect of patient care.

The Nurse. Under almost any form of personnel management, nurses play extremely important roles in clinical dietetics. They coordinate the activities of doctors, dietitians, and patients, including such concerns as the diet prescription, food service, meal serving, and patient response. The nurse arranges the food trays, helps patients in feeding, answers questions, and provides additional beverages. The nurse can also observe patients directly, recording the amounts of fluid, food, and nutrient supplements consumed. Measuring the amount of urine voided is also an important index. The patient's response to the dietary regimen is best evaluated by the nurse because of the nurse's constant presence.

Enteral and parenteral feedings are administered and monitored by the nurse. Proper administration of these feedings requires the nurse to have a good working knowledge of the nutritional bases of commercial or hospital-prepared products. In the case of intravenous fluids and electrolytes, the doctor's prescription is implemented by the nurse only, who also monitors the patient's response. The nurse's observations and records help the dietitian or nutritionist in planning other aspects of the patient's nutritional and dietary needs. Health professionals are aware that drugs profoundly affect a person's nutritional status. Nurses know firsthand what, how much, and when drugs are taken by the patient. If a drug causes nausea or drowsiness, the nurse may understand why a meal is not eaten.

If the patient is eating a regular diet, the dietitian or doctor will not be directly concerned about the patient's nutritional and dietary needs. In this circumstance, the nurse is the sole source of nutritional and dietary information or advice. The patient's medical record is frequently reviewed and studied by the nurse. The nurse should relate all information regarding body weight, food consumption, medications administered, and overt clinical signs to the dietitian and physician at the earliest opportunity.

The nurse has the best opportunity to teach a patient principles of nutrition. In many cases, after the patient is discharged, the nurse is the only contact with the patient, especially if a particular diet is part of the home care regimen.

In many health clinics the nurse frequently assumes the roles of nutritionist and dietitian as well as nurse. With the increasing popularity of nurse practitioners, some nurses even assume certain aspects of the physician's role. Because of the nurse's importance in the nutritional and dietary care of the patient, some people have argued that nurses need to receive a stronger background in nutrition and dietetics.

Other Health Professionals. The total care plan for a patient with any illness usually has many facets that involve all types of health professional. Apart from doctors, nurses, and dietitians, they include pharmacist, physical therapist, and social worker. More details can be obtained from the literature of each particular profession.

This article has been adapted from Y. H. Hui, *Human Nutrition and Diet Therapy* Copyrighted 1983, Jones and Bartlett Publishers, Inc. Used with permission.

BIBLIOGRAPHY

General References

Y. H. Hui, *Human Nutrition and Diet Therapy,* Jones and Bartlett, Boston, (1983).

American Dietetic Association, *Manual of Clinical Dietetics,* American Dietetic Association, Chicago, Ill., 1988.

V. Aronson, *The Dietetic Technician: Effective Nutrition Counseling,* Van Nostrand Reinhold Co., Inc., New York, 1985.

C. K. Carlson, "Dietetics in Physical and Rehabilitative Medicine," *ADA Newsletter* **6**(2), II, (1987).

T. M. Crump, ed., *Nutrition and Feeding of the Handicapped Child,* Little, Brown & Co., Boston, 1987.

C. C. Doak and co-workers, *Teaching Patients with Low Literacy Skills,* Lippincott, Philadelphia, 1985.

S. Escott-Stump, *Nutrition and Diagnosis-Related Care,* 2nd ed., Lea & Febiger, Philadelphia, 1988.

P. A. M. Hodges and C. E. Vickery, *Effective Counseling Strategies for Dietary Management,* Aspen, Rockville, Md., 1989.

R. S. Hurley and co-workers, "Preprofessional Education," in J. K. Kane, *Exploring Careers in Dietetics and Nutrition,* Rosen Publishing Group, New York, 1988.

C. E. Lang, ed., *Nutritional Support in Critical Care,* Aspen, Rockville, Md., 1987.

Mayo Clinic, *Mayo Clinic Diet Manual,* B. C. Dekker, Philadelphia, 1988.

J. L. Ometer and co-workers, "Quality Assurance, II. Application of Oncology Standards Against A Levels of Care Model," *Journal of the American Dietetic Association* **82**(2), 132 (1982).

J. L. Ometer and co-workers, "Quality Assurance, I. A Levels of Care Model," *Journal of the American Dietetic Association* **82**(2), 129 (1982).

F. R. Smith, "Patient Power," *American Journal of Nursing* **85**(11), 1260 (1985).

M. C. Sundberg, *Fundamentals of Nursing with Clinical Procedures,* Jones and Bartlett, Boston, 1989.

M. L. Wahlqvist and J. S. Vobecky, eds., *Patient Problems in Clinical Nutrition: A Manual,* John Libbey & Co., London, 1987.

Y. H. HUI
Editor-in-chief

DIGESTION. See FOOD UTILIZATION.
DIGESTIVE TRACT. See FOOD UTILIZATION.

DISINFECTANTS

DEFINITIONS

Disinfection originally referred to the destruction of the germs of disease. In modern use, a disinfectant is an agent capable of destroying a wide range of microorganisms, but not necessarily bacterial spores. Sterilization is the process of destroying all forms of life, including bacterial spores. Sanitation is a more general term referring to those factors that improve the general cleanliness of our living environment and aid in the preservation of health. A sanitizer is a chemical substance that reduces numbers of microorganisms to an acceptable level. This term is widely used in the United States and it is virtually synonymous with the term disinfectant. In the food industry, sanitation and sanitizers are used to refer to processes and agents, respectively, for cleaning inanimate (work) surfaces. The term hygiene (from *Hygieia*, the Greek goddess of health) is frequently used to refer to personal cleanliness, hence hand hygiene. A soil is an organic or inorganic residue on equipment and other work surfaces. A soiled surface is cleaned by washing with a detergent. Detergency does not imply anything more than assisting in cleaning a surface; microorganisms are reduced only to the extent that they are physically removed from the surface (1).

Disinfectants may be characterized on the basis of their mode and range of antimicrobial action. A biocide is a disinfectant that kills all forms of life, whereas a biostat is a disinfectant that inhibits the growth of all forms of life. Many biocides are biostats at low concentrations. Bactericides and bacteriostats are agents active against bacteria. The prefixes fungi-, spori-, and viru- indicate action against fungi, spores, and viruses, respectively. The term germicide is sometimes used to describe disinfectants that destroy disease-causing organisms.

A sanitation program refers to the schedule and procedures for cleaning and sanitizing a food-processing plant or food-handling facility, as well as requirements for personal hygiene. Procedures for sanitation of inanimate surfaces and equipment involve a two- or three-step process. In each case the initial step is removal of gross soil. In a three-step process, the initial rinse is followed by a detergent wash, rinse, and sanitizing. In a two-step process a combined detergent–sanitizing agent is used, followed by a rinse.

CHEMICAL DISINFECTING AGENTS

Lister used phenol (carbolic acid) as a germicide in 1867. Although it is the parent compound of chemical disinfection, its use today is limited to substituted phenols, eg, the *bis*-phenols used in germicidal soaps. Both chlorine (in the form of hypochlorite) and phenol were used to deodorize waste materials in the early 1800s, before Pasteur established the germ theory of infection and putrefaction. The use of disinfectant chemicals began in clinical surroundings in the late nineteenth century. In 1908 the first large-scale use of chlorine (chloride of lime) in water purification started in Chicago, and its use for this purpose spread rapidly. Yet disinfectants were not readily accepted in food production until the 1940s, when hypochlorite treatment was permitted in the dairy industry as an alternative to steam sterilization. The range of antimicrobial chemical agents has been comprehensively reviewed (2).

Heavy metal ions such as Hg^{2+}, Cu^{2+} and Ag^+ are universal biocides, but they are generally too toxic to use on food equipment. Alcohols such as ethanol and isopropanol and others with chain length up to 8–10 carbon atoms are biocidal except for bacterial spores. Water is essential for the antimicrobial action of alcohols, and 60–70% (v/v) is the most effective concentration. The alkylating agents formaldehyde and ethylene oxide are active gaseous disinfectants with a broad spectrum of biocidal activity, including bacterial spores. The sporicidal activity is probably due to their ability to penetrate into the spores and the fact that they do not require water for their action. Formaldehyde is too toxic for use with food equipment because it is absorbed onto surfaces as a reversible polymer (paraformaldehyde) that slowly depolymerizes to give free formaldehyde as a food contaminant. Ethylene oxide is the most reliable substance for gaseous disinfection of dry surfaces, especially heat-sensitive materials such as plastics (3).

Halogens

The halogens include fluorine, chlorine, bromine, and iodine, all of which are extremely active oxidizing agents. Fluorine is the most reactive, but it is too toxic, irritant and corrosive to be used as a disinfectant (2). Bromine is also too toxic and irritant for widespread use, but it has some applications in mixed halogen compounds (4). Chlorine is an effective disinfectant (Table 1) through the action of hypochlorous acid (HOCl). Hypochlorous acid is unstable and it is normally stabilized by adding caustic soda to form sodium hypochlorite (NaOCl). HOCl can be formed by any of the following reactions:

Chlorine $\qquad Cl_2 + H_2O \underset{OH^-}{\overset{H^+}{\rightleftharpoons}} HOCl + H^+ + Cl^-$

Hypochlorite $\quad NaOCl + H_2O \underset{OH^-}{\overset{H^+}{\rightleftharpoons}} HOCl + Na^+ + OH^-$

Chloramine $\quad NH_2Cl + 2H_2O \underset{OH^-}{\overset{H^+}{\rightleftharpoons}} HOCl + NH_4^+ + OH^-$

At acid pH, solutions have greater germicidal activity because greater amounts of HOCl are formed; however, for product stability a pH > 8 is used. Sodium hypochlorite (NaOCl) in commercial liquid form contains 10–14% available chlorine. Calcium hypochlorite [$Ca(OCl)_2$] is a powder with about 30% available chlorine. Chlorine-based sanitizers are normally used at concentrations that yield 100–200 ppm available chlorine with contact times of 3–30 min. Organic chlorines such as chloramine-T and isocyanuric acid derivatives are more stable than hypo-

Table 1. Advantages and Disadvantages of Different Classes of Sanitizers[a]

	Hypochlorite	Iodophor	QAC	Acid Anionics
Antimicrobial Spectrum				
Gram-positive bacteria	+	+	+	+
Gram-negative bacteria	+	+	−	+
Spores	+	−	−	−
Viruses	+	−	−	+
Low toxicity	+	+	+	
Nonirritating to skin	−	+	+	
Noncorrosive	−[b]	+	+	
Nonstaining	+	−[c]	+	+
Stability during storage	−	+	+	+
in hard water	+	+	−	+
with organic matter	−	±	+	+
Use at high temperatures	+	−	+	+
Mineral–protein films				
removed	−	+	−	+
Leaves no residues	+	±	−	−
Leaves no flavors or odors	−	−	+	+
Detergent–sanitizer	−	±	+	+
Self-indicating	±[d]	+	−	−
Low cost	+	±	−	

[a] Advantageous attribute = +, disadvantageous attribute = −.
[b] Only a problem if misused.
[c] Only some plastic materials and painted walls; staining of skin is temporary.
[d] By odor of available chlorine.

chlorites in the presence of organic matter, they are less irritant and less toxic, and they release chlorine more slowly.

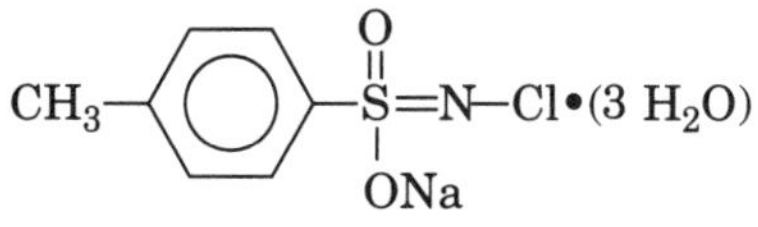

Chloramine-T

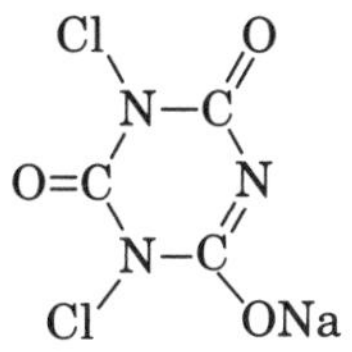

Isocyanuric acid derivatives

Chloramines have 25–30% available chlorine, but they are weaker bactericides than hypochlorites except that they are more active at pH > 10. Isocyanuric acid products are active over the pH range 6–10. Both of these organic chlorines can be formulated as detergent–sanitizers, but they are relatively expensive (1).

Iodine is too insoluble in water to be used as a disinfectant; however, if it is complexed with a nonionic surface active agent, the resulting iodophor can be used as a detergent–sanitizer (Table 1). Diatomic iodine (I_2) is highly bactericidal; hypoiodous acid (HOI) and the hypoiodite ion (IO^-) are less bactericidal; the iodate ion (IO_3^-) is inactive.

$$I_2 + H_2O \underset{H^+}{\overset{OH^-}{\rightleftharpoons}} HOI + H^+ + I^- \underset{H^+}{\overset{OH^-}{\rightleftharpoons}} I^- + IO_3^-$$

Iodophors are generally used at 12.5–25.0 ppm; at these concentrations they are self-indicating, ie, the presence of available iodine is indicated by the amber color of the solution. They must be used at < 50°C to avoid the release of toxic iodine vapor. Iodophors are formulated with an acid, usually phosphoric acid, because they are more active at pH 3–5. They also solubilize and remove mineral and protein films from food-processing equipment (4).

Surface Active Agents (Surfactants)

The synthetic detergents in this group fall into three categories:. anionics, nonionics, and cationics. The anionics are weakly bactericidal against Gram-positive bacteria. The nonionics are not germicidal, but they are good surfactants. The cationics, known as quaternary ammonium compounds, quats or QACs, are bactericidal. They are ammonium salts with some or all of the hydrogen atoms substituted with alkyl or aryl (R) groups, for example benzalkonium chloride. These agents can be used as skin antiseptics and as food equipment sanitizers.

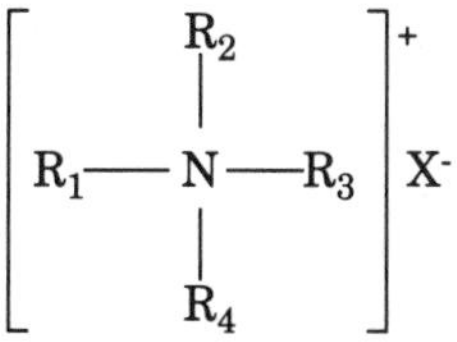

General Structure of QACs

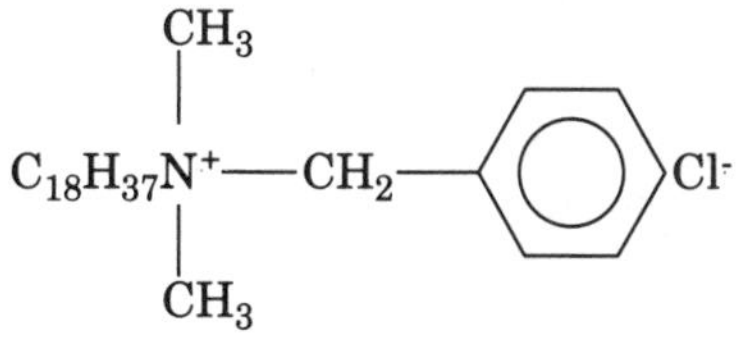

Benzalkonium chloride

Their properties are summarized in Table 1. Many Gram-negative bacteria of significance in food spoilage are not adversely affected by QACs; in fact, some pseudomonads grow in dilute solutions of QACs (5). The combination of QACs with nonionic detergents gives another type of detergent–sanitizer that affords good contact with the surface that requires disinfection. Cationic sanitizers are inactivated by soaps and phospholipids because they are oppositely charged (1).

Phenols

Many phenolic compounds are strong germicides, but their potential for use in the food industry is limited by their odor and the possibility of causing off-flavors in foods. Their action involves cell lysis. Depending on concentration, phenols are either bactericidal or bacteriostatic. They have only limited activity against viruses, and they are not sporicidal. Halogenation of phenols increases their activity 3–30 times. Substitution in the *para* position is more effective than in the *ortho* position, substitution with two halogen atoms is more effective than one, and bromophenols are more active than chlorophenols (6). Chloroxylenol or *para*-chloro-*meta*-xylenol (PCMX), the active ingredient of Dettol, is a phenolic compound that can be used as a skin germicide.

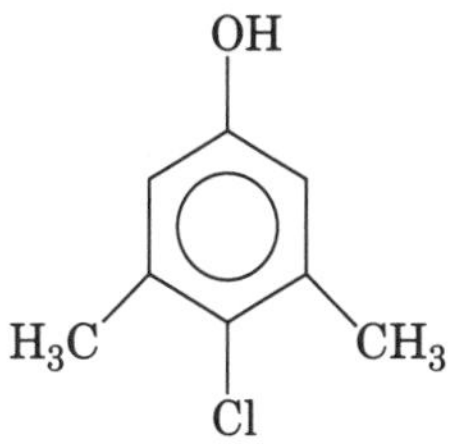

Chloroxylenol (PCMX)

Many halogenated *bis*-phenols (or phenylphenols) have considerable activity against bacteria and fungi, but they have low activity against pseudomonads (2). They are used as clinical disinfectants and in germicidal soaps, eg, hexachlorophene.

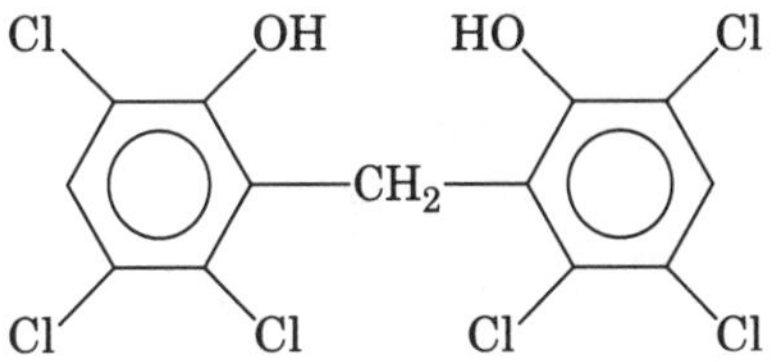

Hexachlorophene

Hexachlorophene is not very volatile and lacks the unpleasant odor of phenols. It is more active against Gram-positive than Gram-negative bacteria. It is bacteriostatic for *Staphylococcus aureus* at extremely low concentrations (0.05 μg/mL) and requires a suitable quenching agent to inactivate residues of the disinfectant so that its bactericidal activity can be accurately assessed (2). It was widely used as a skin antiseptic marketed as pHisoHex and in a wide range of over-the-counter (OTC) pharmaceutical and personal hygiene products. It can be absorbed through inflamed and infant skin with the possibility of serious systemic toxicity (7). As a result, OTC products have generally been limited to 0.75% hexachlorophene. Higher concentrations are sold with medical prescription.

Restrictions on the use of hexachlorophene resulted in the development of a range of other disinfectants for use in germicidal soaps. For example, Irgasan DP 300, also known as triclosan, 2-4-4′trichlor-2′-hydroxy diphenyl ether is active against Gram-positive and Gram-negative organisms.

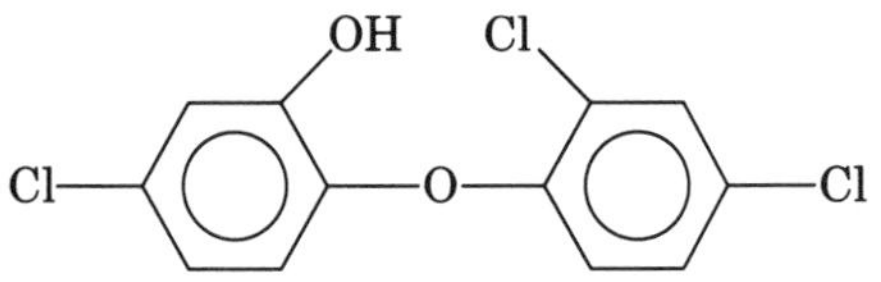

Triclosan

Other Antimicrobial Agents

Chlorhexidine is one of a family of N^1,N^5-substituted biguanides (2). Its use was licensed in the UK and Canada, but not in the United States.

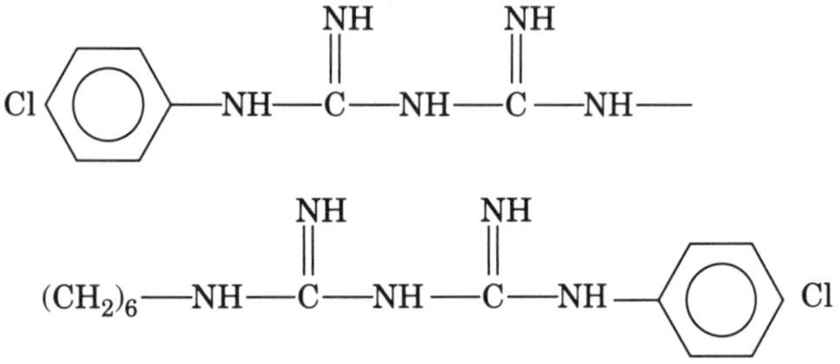

Chlorhexidine

Chlorhexidine is active against Gram-positive and Gram-negative bacteria; it has limited antifungal activity, but it is not active against acid-fast bacilli, bacterial spores, or viruses. However, some resistant pseudomonads can contaminate aqueous solutions of this compound. It is not compatible with anionic compounds. The digluconate salt is freely soluble in water; however, the diacetate and dihydrochloride salts are only poorly soluble. Alcoholic chlorhexidine or a 4% chlorhexidine detergent (Hibiscrub) are highly effective skin germicides (8).

The salicylanilides and carbanilides are families of antimicrobial chemicals (9). One of the more popular disinfectants among these chemicals is 3,4,4′-trichlorocarbanilide (TTC). It has been incorporated into soaps to give an antimicrobial product.

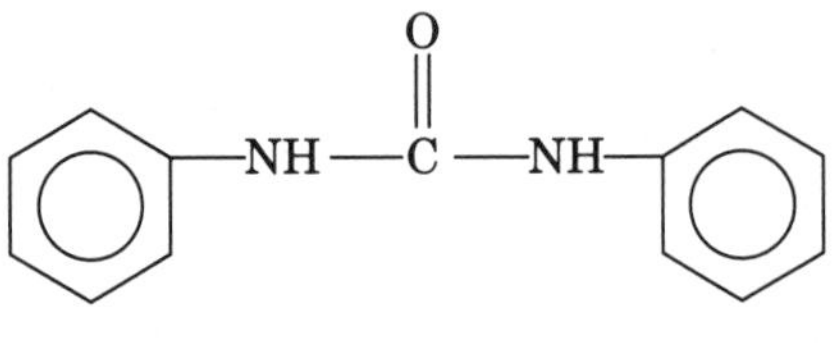

Carbanilides

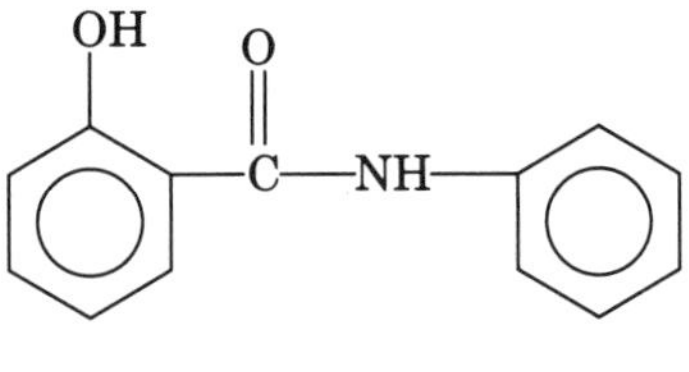

Salicylanilides

PHYSICAL DISINFECTING AGENTS

In addition to chemical compounds, heat and ultraviolet (uv) light can be used to disinfect food contact surfaces. Heat, particularly moist heat, is the most reliable and most widely used method of destroying all forms of microbes (2). The efficacy of heat treatment with steam or hot water (80°C or above) depends on the temperature achieved and the time of exposure. Temperatures above 60°C progressively kill microbial cells. Vegetative cells may be killed or injured. Injured cells can recover in an appropriate environment, for example, in foods. Heat injury has been described in two ways: the extension of the lag phase of growth (10) and the inability to grow on selective media, for example, on salt-containing media for *Staphylococcus aureus* (11) or violet red bile agar for coliform bacteria (12). Because heat cannot be universally applied it is necessary to rely on chemical agents for sanitation programs.

Radiation is an alternative to gaseous disinfection. Gamma (cobalt-60) and electron-accelerated β-ray (<10 MeV) irradiation have good penetrating ability and, therefore, have the potential for use not only in disinfection, but also in food processing. Ultraviolet radiation has a wavelength between 210 and 328 nm, with maximum antimicrobial activity between 240 and 280 nm. Ultraviolet radiation is low energy, so it does not penetrate foods. It is absorbed by glass and plastics, but it can be used for surface disinfection (2). *Micrococcus radiodurans* and bacterial and mold spores are highly resistant to ionizing and uv irradiation. The primary effect of radiation on living cells is by action on DNA. Radiation resistance is genetically determined. Sublethally damaged cells can recover by photoreactivation or by excision and recombination events, which give rise to mutations (2).

EVALUATION OF DISINFECTANTS

A number of laboratory tests have been developed to evaluate the efficacy of disinfectants (2). The Rideal–Walker and Chick–Martin methods of determining the phenol coefficient with *Salmonella typhi* were developed in the early 1900s. Although the Chick–Martin test includes an organic soil, the phenol coefficient tests are artificial in concept, have poor reproducibility and phenol is an unreliable control disinfectant (1). The improved Kelsey–Sykes capacity test attempted to resolve the deficiencies of the phenol tests by using at least four organisms in preliminary screening tests, the most resistant organism being selected for further testing. The disinfectants are prepared in a standardized hard water, with or without a standard soil. Recovery broths are prepared with a neutralizer against the disinfectant. This test also has its shortcomings, but it is used as the official test for disinfectants in the UK (13).

Several tests have been designed to simulate in-use conditions by preparing an air-dried film of microorganisms on an appropriate surface, such as stainless steel, with or without an organic soil. The Lisboa test was developed for testing sanitizers for dairy equipment. Stainless steel tubes are contaminated and disinfected using standardized procedures and neutralized to inactivate any residual disinfectant; the number of surviving organisms is then determined (1). Various modifications of this in-use testing have been developed (2,14). Methods of the Association of Official Analytical Chemists (AOAC) include use-dilution carrier techniques with *Salmonella choleraesuis*, *Staphylococcus aureus*, or *Pseudomonas aeruginosa* contaminated onto polished stainless-steel cylinders. Tests for fungicidal and sporicidal activity are also specified (15).

Tests for evaluating skin germicides under in-use conditions have been even more difficult to standardize (16), and there are no official methods comparable to the AOAC use-dilution technique for evaluating disinfectants for use on inanimate surfaces. Multiple-basin techniques measure the rate of mechanical removal of microorganisms from the skin. Testing of skin disinfectants must take into account the transient and residual microflora of the skin, the time of exposure, methods of sampling the skin after washing, and methods of contaminating skin with a transient microflora appropriate to the use environment (17). Depending on the germicidal agent used, it may be necessary to neutralize the rinse solution. With skin disinfectants it is appropriate to test for a residual (or substantive) effect, ie, improved results with repetitive use of the same agent compared with nongermicidal soaps. Occlusion and expanded flora tests (in which agents are applied topically to the skin, covered with a plastic film, and sealed with impermeable plastic tape) represent alternative approaches to testing skin disinfectants (18).

Comparisons of antimicrobial soaps by various standardized techniques have given results that question the in-use efficacy of most products tested. Studies have indicated that under the specific test conditions, only an iodophor product (0.75%) and 4% chlorhexidine gluconate significantly reduced the number of microorganisms released from hands (17,19,20). Other products, including intermediate strength iodophor products, 2% chlorhexidine, Irgasan DP 300, *para*-chloro-*meta*-xylenol (PCMX), trichlorocarbanilide or tribromosalicylanilide, under in-use conditions of testing were no better than a nongermicidal soap.

MODULAR SYSTEMS OF DISINFECTION

Food industries that process liquid products, such as milk, beer, and soft drinks, use CIP systems for disinfection. The principles are similar to those for manual cleaning, but mechanical force generated by the velocity of liquid flow through the system is relied on to remove food soils.

CIP offers many advantages over manual sanitizing programs, including lower labor costs, more economic operation, better sanitation, faster cleanup and reuse of equipment, less dismantling and reassembly, and greater safety of use and operation (1). The CIP systems can be based on single or multiple use of cleaning solutions. Multiple-use systems are usually automatic, and solutions are recovered according to a preset program and stored in holding tanks for reuse. A typical CIP program is as follows (1):

1. Prerinse (5 min) with cold water.
2. Alkali detergent wash (15 min at 80°C).
3. Intermediate rinse (3 min) with cold water from mains.
4. Cold hypochlorite solution (10 min).
5. Final rinse (3 min) with cold water from mains.

For large tanks that would be uneconomical to fill with cleaning fluids, a permanent or portable spray system is fitted to the vessel. The spraying devices should allow every part of the inside of the vessel to come in contact with the cleaning solutions. Good drainage must be insured to avoid accumulation of fluids and residues on equipment surfaces.

Mechanical aids for plant sanitation include pressurized steam, high-pressure water jets, compressed air, and ultrasound. All of these cleaning aids require specialized equipment and have specific uses. High-pressure steam can remove debris and sterilize. High-pressure water jets also remove debris from inaccessible parts of machinery; however, such inaccessibility should be avoided in the design of food equipment and machinery. Compressed air removes dry powder, dust, and soil, but it spreads, rather than eliminates, the soil. Vacuum cleaners are preferable to compressed air for removing dry solids and dust. Ultrasound is used to clean small or sensitive items of equipment that are otherwise difficult to clean. It requires immersion of the objects in an ultrasound tank for exposure to ultrasonic vibrations, which remove soils by cavitation.

Foam sanitation is an efficient system for cleaning walls, floors, equipment with large contact surfaces, and immovable food-handling equipment. A foaming agent is added to the detergent formulation to produce a thick, long-lasting foam. This gives the cleaning agent a long contact time with the soiled surfaces. This form of cleaning requires a special pressure-generating system. The foam must be removed and the bactericidal agent applied. Foam systems give a good visible awareness of the sanitation process.

LEGISLATION

The food legislation of most countries requires that food is handled in a sanitary manner. The details and standards differ between countries. In the United States the principal federal legislation governing the handling of foods is the Food, Drug and Cosmetic Act of 1938, as amended. This covers all foods except meats and poultry, which are covered by various food inspection acts. These acts are implemented through the *Code of Federal Regulations,* which is revised and published annually. Each State can pass its own laws, provided they respect the relevant federal legislation. Emphasis is placed on good manufacturing practice (GMP) regulations that outline the minimum acceptable standards for processing and storage of food. The U.S. Department of Agriculture (USDA) has legal control over meat and poultry inspection. In the United States, sanitizers for use in food establishments are defined as pesticides and they are regulated by the Environmental Protection Agency under the Federal Insecticide, Fungicide and Rodenticide Act (FIFRA) of 1978. Antimicrobial agents for application on human bodies fall under the regulatory authority of the Food and Drug Administration.

In Canada food legislation is enacted through the federal Food and Drugs Act of 1953, as amended. The Health Protection Branch (HPB) of Health and Welfare Canada is responsible for administering the act. Other agencies also have responsibility for food inspection, including Agriculture Canada, and Fisheries and Oceans Canada. Disinfectants for use in food premises are deemed to be drugs and require a Drug Identification Number (DIN) issued by HPB.

In the UK, food legislation is based on the Food and Drugs Act of 1955, which is the enabling legislation for regulations governing food hygiene and is controlled by the Department of Health and Social Security or the Ministry of Agriculture, Fisheries and Food (MAFF). Enforcement of the regulations is the responsibility of Environmental Health Officers employed by elected local authorities.

In Europe, the EEC has mandatory standards that automatically become law in member countries. In a broader sense, the Codex Alimentarius program of the Food and Agriculture and World Health Organizations (FAO/WHO) has the intention of establishing international agreements on food standards that would safeguard health and encourage good handling practices for foods.

BIBLIOGRAPHY

1. P. R. Hayes, *Food Microbiology and Hygiene*, Elsevier Applied Science Publishers, Ltd., Barking, UK, 1985.

2. A. D. Russell, W. B. Hugo, and G. A. J. Ayliffe, eds., *Principles and Practice of Disinfection, Preservation and Sterilisation*, Blackwell Scientific Publications, London, 1982.

3. C. R. Phillips, "Gaseous Sterilization," in S. S. Block, ed., *Disinfection, Sterilization and Preservation*, 2nd ed., Lea & Febiger, Philadelphia, Pa., 1977.

4. J. R. Trueman, "The Halogens," in W. B. Hugo, ed., *Inhibition and Destruction of the Microbial Cell*, Academic Press, Orlando, Fla., 1971.

5. J. A. Toller, *Sanitation in Food Processing*, Academic Press, Orlando, Fla., 1983.

6. R. F. Prindle, "Phenolic Compounds," in S. S. Block, ed., *Disinfection, Sterilization and Preservation*, 3rd ed., Lea & Febiger, Philadelphia, Pa., 1983.

7. R. A. Chilcote and co-workers, "Hexachlorophene Storage in a Burn Patient Associated with Encephalopathy," *Pediatrics* **59**, 457–459 (1977).

8. E. J. L. Lowbury and H. A. Lilly, "Use of 4% Chlorhexidine Detergent Solution (Hibiscrub) and Other Methods of Skin Disinfection," *British Medical Journal* i, 510–515 (1973).

9. H. C. Stecker, "The Salicylanilides and Carbanilides," in Ref. 3.

10. H. Jackson and M. Woodbine, "The Effect of Sublethal Heat Treatment on the Growth of *Staphylococcus aureus*," *Journal of Applied Bacteriology* **26**, 152–158 (1963).

11. M. E. Stiles and L. D. Witter, "Thermal Inactivation, Heat Injury, and Recovery of *Staphylococcus aureus*," *Journal of Dairy Science* **48**, 677–681 (1965).

12. L. A. Roth, M. E. Stiles and L. F. L. Clegg, "Reliability of Selective Media for the Enumeration and Estimation of *Escherichia coli*," *Canadian Institute of Food Science and Technology Journal* **6**, 230–234 (1973).

13. B. Croshaw, "Disinfectant Testing—With Particular Reference to the Rideal–Walker and Kelsey–Sykes Tests," in C. H. Collins, M. C. Allwood, S. F. Bloomfield, and A. Fox, eds., *Disinfectants: Their Use and Evaluation of Effectiveness*, Academic Press, Orlando, Fla. 1981.

14. R. M. Blood, J. S. Abbiss, and B. Jarvis, "Assessment of Two Methods for Testing Disinfectants and Sanitizers for Use in the Meat Processing Industry," in Ref. 13.

15. Association of Official Analytical Chemists, *Official Methods of Analysis of the Association of Official Analytical Chemists*, 15th ed., AOAC, Arlington, Va., vol. 1, pp. 135–142, 1990.

16. G. A. J. Ayliffe, J. R. Babb and H. A. Lilly, "Tests for Hand Disinfection," in Ref. 13.

17. A. Z. Sheena and M. E. Stiles, "Efficacy of Germicidal Hand Wash Agents in Hygienic Hand Disinfection," *Journal of Food Protection* **45**, 713–720 (1982).

18. R. R. Marples and A. M. Kligman, "Methods for Evaluating Topical Antibacterial Agents on Human Skin," *Antimicrobial Agents and Chemotherapy,"* **2**, 8–15 (1974).

19. A. Z. Sheena and M. E. Stiles, "Efficacy of Germicidal Hand Wash Agents against Transient Bacteria Inoculated onto Hands," *Journal of Food Protection* **46**, 722–727 (1983).

20. A. Z. Sheena and M. E. Stiles, "Low Concentration Iodophors for Hand Hygiene," *Journal of Hygiene* **94**, 269–277 (1985).

Michael E. Stiles
University of Alberta
Edmonton, Canada

DISTILLATION. See Distilled beverage spirits; Food fermentation.

DISTILLED BEVERAGE SPIRITS

HISTORY

The history of distilled spirits goes back into antiquity. Scientists have unearthed pottery in Mesopotamia depicting fermentation scenes dating back to 4200 BC, a small wooden model of a brewery from about 2000 BC is on display at the Metropolitan Museum of Art in New York City, and Aristotle mentions a wine that produces a spirit. The first real distiller was probably a Greek–Egyptian alchemist who in the first or second century AD, in an attempt to transmute base metal into gold, boiled some wine in a crude still. The discovery of ardent spirits that resulted from this effort was looked on with awe. It was kept a secret for centuries.

The technique of distillation probably came from the Egyptians who had been interested in alchemy since the pre-Christian era. At a later time the Arabians gained this knowledge from the Egyptians. Distillation was introduced into Western Europe either through Spain about AD 1150, or by the crusaders who learned about it from the Moslems in the 12th and 13th centuries. Distilled spirits were probably known in Ireland and Scotland before the 12th century, but actually it was not until then that there is a recorded history of distilled spirits in Europe. The first written evidence is in the description by Master Salernus who died in AD 1167. However, for another three centuries distilled spirits were regarded only as a rare and costly medicine called *aqua vitae*, "the water of life."

The first treatise on distillation was written by a French chemist, Arnold de Villeneuve, sometime before 1311, and was printed in 1478 in Venice. A Spaniard, Raymond Lully, was also instrumental in spreading the knowledge of distillation through Europe in the late 13th century. *Liber de Arte Distillandi*, by Hieronimus Brunswick, a well-known author of medical works, was printed in 1500 in Strasbourg. A more comprehensive book by Ryff, another medical author, appeared in 1556 in Frankfurt am Main. Both works contain elaborate chapters on herbs and their distillates with indications of their uses as medicines.

Monks had been producing fermented liquors on a substantial scale since AD 800, and were the first to practice the new art of distillation because churchmen were the only one capable of reading and understanding the treatises on distillation. Gradually the knowledge of *aqua vitae* spread widely, and during the 16th and 17th centuries the production of distilled spirits became a full-scale industry. The popularity of distilled spirits in Europe grew during this era and may best be traced through court records.

The word *whisky* comes from the Gaelic word *uisge-beatha*, or *usquebaugh*, as the Irish called it, meaning "the water of life." *Usquebaugh*, supposedly the Celtic form of whisky, was actually a cordial made with aniseed, cloves, nutmeg, ginger, caraway seeds, raisins, licorice, sugar, and saffron. The real whiskey of the Irish was called potheen, reputedly a formidable drink, full of heavy-bodied flavors resulting from simple distillations.

The early methods of distillation used the alembic, a simple closed container to which heat was applied. The vapors were transferred through a tube to a cooling chamber in which they were condensed. Alembics made of copper, iron, or tin were preferred. Other metals were said to have an adverse effect on the distillate. Occasionally, glass and potter's earth were used. From this developed stills consisting of clay or brick fireboxes into which the copper pots of the still were fitted. Direct heat was applied to the body or cucurbit containing the fermented mixture, and the vapors rose into the head, passed through a pipe (the crane neck) to the worm tub, where a copper coil was immersed in a barrel of cool water. Variations and improvements of this technique resulted in pot stills, some of which are still in use in the production of malt whiskies, notably in Scotland, and brandies, in France.

During the latter part of the 18th century, great strides were made in the development of distillation equipment, mostly in France. Because brandy (from fruit) was the principal distillate in France, the French had an advantage over the British and Germans, who were both hampered by the thick mashes obtained from grains. Argand invented the preheater, and in 1801, Edouard Adam designed the prototype of a charge still. He used egg-shaped vessels to hold the alcoholic liquid through which vapors from the kettle passed to the condenser. In the early 19th century several stills were patented in France and England, and in 1830 Aeness Coffey, in Dublin, developed his continuous still. The Coffey still (Fig. 1) is composed of two columns: in one, the fermented mash is stripped of its

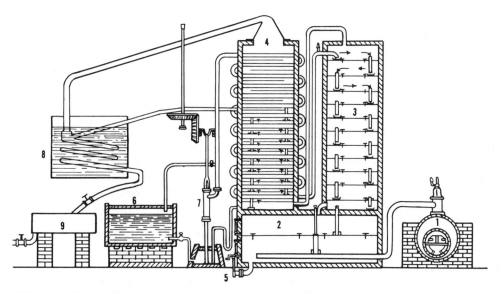

Figure 1. Coffey still. 1. Boiler. 2. Spent mash chamber. 3. Stripping column (analyzer). 4. Rectifying column. 5. Residual mash (stillage) outlet. 6. Fermented mash (feed) 7. Feed pump. 8. Condenser. 9. Product tank.

alcohol and, in the other, the vapors are rectified to a high proof (94–96%). Almost all of the fundamentals of distillation had been recognized and incorporated into the Coffey still. Later developments do not differ fundamentally from the stills of Cellier-Blumenthal and Coffey.

The early history of distillation in the United States is as poorly documented as it is elsewhere. No one knows who made the first rum, colonial New England's most important product; no one knows when or where the first rye whisky was made, or with any certainty who made the first bourbon in Kentucky. The Indians may possibly have known distillation. They had fermented drinks from maple syrup, corn, ground acorns, chestnuts, and chinquapins. Columbus reported an Indian drink made from the marrow of maguey. The fermented sap of maguey makes pulque, and distilled pulque becomes mezcal. Whether or not Columbus tasted mezcal is conjectural. However, the Spanish and Portugese brought the still.

The first recorded beverage spirits made from grains (corn and rye) were distilled on Staten Island in 1640 by William Kieft, the director general of the Dutch Colony of New Netherland. During the early colonial days, fermented beverages made from sugar-bearing fruits and vegetables were very popular. Pumpkins, maple sugar, parsnips, peaches, pears, apples, currants, grapes, and elderberries provided a ready source. The more aristocratic colonists preferred imported wines. However, in the 18th century colonial drinking customs changed to distilled drinks, especially rum. Rum was made in Barbados as early as 1650. In colonial America, the earliest reference is in the records of the General Court of Massachusetts in May 1657. By 1750, there were 63 distilleries. One of the first acts of the Continental Congress was to establish a ration of rum for soldiers and sailors. Brandy, which had been made in the colonies as early as 1650, and gin, which was popular with the poorer classes in England, never achieved the popularity of rum in this country. Nevertheless, rum was not a native drink, because the raw materials had to be imported. But rum starts with an inexpensive by-product of the sugar industry, molasses, and so could be the least-expensive distilled spirit. There are sufficient sugars left after most of the cane sugar has been removed to serve as the source of alcohol for rum.

With the decline of the three-cornered trade, the revolution, and the movement of settlers from the coastal areas inland, the rum industry slowly declined. Americans turned to rye, a successful crop in Pennsylvania and Maryland, and to corn, which grew in the South and West. George Washington is reported to have made rye whisky. His distillery had an excellent reputation for making fine liquor under the supervision of James Anderson, a Scotsman. In 1798, Washington's net profit was £83 with an inventory of 587 L (155 gal) still unsold.

Canada's first distillery produced rum from molasses in Quebec in 1769, but whisky did not develop until the 19th century. By 1850, there were some 200 distilleries operating in Ontario alone. Today there are only 27 of which 2, Joseph E. Seagram & Sons, Ltd., and Hiram Walker & Sons, Ltd., have been in business continuously since 1857 and 1858, respectively.

The Whisky Rebellion in 1791 resulted in the rise of Kentucky as the greatest whisky-producing state in the union. In 1810, Kentucky which had achieved statehood in 1792 with over 70,000 inhabitants, many of them Scotch–Irish and German farmer immigrants with distiller experience, had 2,000 stills out of 14,000 in the United States producing well over 7,500,000 L (2 million gal) of whisky annually. Many counties of Kentucky claim the first production of bourbon. However, it is generally agreed that the first bourbon whisky (genuine, old-fashioned, handmade, sour-mash bourbon) was made by a Baptist minister, the Reverand Elijah Craig, in 1789 at Georgetown in Scott County which was part of Bourbon County, one of the Virginia counties that made up the Kentucky territory, from which Craig's corn whisky got its name. In order to distinguish this corn whisky from Pennsylvania rye, it was called Kentucky bourbon, with a mashing formula of at least 51% corn grains. Until 1865, few Kentucky distilleries produced more than 160,000 L (1,000 bbl) per year. Most of them were small, with an annual production of only 8,000–80,000 L (50–500 bbl).

The method used in Kentucky of making whisky, with the exception of malt preparation, was in most respects similar to that used in Scotland and Ireland for hundreds of years. Ground corn and rye meal were scalded in tubs somewhat larger than barrels, stirred with paddles, and allowed to cool and sour overnight. Then malt made from rye, corn, or barley was added for conversion of the starch to a fermentable grain sugar. Yeast was added and the mash allowed to ferment for 72–96 h. A simple, single-chambered copper still was used to separate the spirits from the mash. When redistilled, the product was called double distilled. In 1819, New Orleans received some 750,000 L (200,000 gal) per month of this product. By the time it had floated down the Ohio and Mississippi on flatboats, the hot sun and boat movement had aged it.

Before the Civil War, not much attention was paid to aging, even though it was recognized that whisky left in charred oak barrels took on a golden color and some mellowness. Originally, whisky was sold in its natural white state or artificially colored to resemble the amber glint of brandy. No one bothered with brand names, whisky was whisky, as everyone knew, and not too much was made of the wide variation in palatability. The hunter or riverman who drank raw white whisky was not particular about quality.

After the Civil War, the rise in taxes made storage in bond desirable. Many family names, always important in Kentucky lore, now became associated with distinctive whiskies. As the industry grew, it engaged in bitter controversy over what was whisky. As a result, in 1909 during President Taft's administration, whisky was finally defined as any volatile liquor distilled from grain, and standards of identity based on current manufacturing processes for various whisky types, such as rye and bourbon, were established (1). In 1906, the Pure Food and Drug Act was passed requiring a statement of manufacturing process and materials on the label. By 1920 when prohibition began, blended whiskies comprised 70% of the whisky market.

Prohibition brought with it evils that were greater than those it was designed to prevent. During Prohibition, consumption of spirits increased from 530 to 750 million L (140 to 200 million gal) annually. In 1930, the Prohibition

Enforcement Bureau estimated that the production of moonshine was more than 3 billion L (800 million gal) per year. In December 1933, Prohibition was repealed and the distilling industry began to catch up with developments in bacteriology, chemistry, engineering, and sanitation. Many of the small producers were absorbed by the major organizations, and capital was provided for new equipment and inventory.

Per capita consumption continued to decline; it was 10.26 L (2.71 gal) in 1864, 6.47 L (1.71 gal) in 1922–1930, and only 4.96 L (1.31 gal) in 1960. However, the growth in population accounted for an increase in total consumption from 550 million L (145 million wine gal) in 1940 to over 1.6 billion L (420 million gal) in 1976. In World War II the industry offered its facilities to the government for war production, even before the United States entered the war. From Pearl Harbor to V-J Day, almost 3 billion L (750 million gal) of 190° proof (95%) alcohol was produced for the war effort in 127 distilling plants.

TAXATION, GOVERNMENT REGULATIONS

The distilling industry has always been taxed heavily, not only in the United States, but throughout the world. On the one hand, governments have laid an unduly heavy burden on the legal producers of distilled beverage spirits. On the other hand, dishonest men have evaded legal obligations and brought the industry into disrepute. Moreover, in many instances, dry interests have attempted to further their cause by advocating high taxes. For example, in England, by 1730, the laws were so complicated and onerous that they all but destroyed the industry, and in 1743 Parliament completely revised the regulations. Whenever taxes were too high, illicit distilling flourished.

Alcoholic beverages were subject to regulations as early as the Babylonian Code of Hammurabi, ca 2000 BC, which contained provisions for the quality, sale, and use of fermented liquors. In the Magna Carta, a clause provided a standard of measurement for the sale of ale and wine. With the increase in consumption of distilled spirits in Europe, governments became increasingly interested in the tax revenues, and in 1643, Parliament imposed the first tax on distilled spirits.

The first liquor tax in the United States (2 guilders on each half-vat of beer) was imposed in 1640 by William Kieft, director general of New Netherland. The Molasses Act of 1733 and the Sugar Act of 1763, imposing a heavy tax on French and Dutch rum and molasses and leaving the more expensive British products free, provoked protest and action against its enforcement.

In 1791, Alexander Hamilton, in testing the strength of the new federal government, imposed an excise tax to be collected by revenue officers assigned to each district. This tax was fiercely contested and was repealed in 1800 during Jefferson's administration and, except for the years between 1812 and 1817, as a war measure, there was no further excise tax on domestic beverage spirits until 1862.

In 1975, the combined federal excise tax and the average state tax (32 licensed states) amounted to $13.13 per proof gal; federal taxes alone, $10.50 per proof gal. Table 1 lists the federal excise tax rates for various years. The

Table 1. Federal Excise Tax Rates on Distilled Spirits

Year	Rate[a], Dollars	Year	Rate[a], Dollars
1812–1817	0.02 (0.09)	1934 (Jan.)	0.53 (2.00)
1862	0.05 (0.20)	1938 (July)	0.59 (2.25)
1846 (July)	0.16 (0.60)	1940 (July)	0.79 (3.00)
1864 (Dec.)	0.53 (2.00)	1941 (Oct.)	1.06 (4.00)
1894	0.29 (1.10)	1942 (Nov.)	1.59 (6.00)
World War I	0.61 (2.30)	1944 (Apr.)	2.38 (9.00)
1919	1.69 (6.40)	1951 (Nov.)	2.77 (10.50)
1933 (Dec.)	0.29 (1.10)	1985 (Oct.)	3.30 (12.50)

[a] Per liter (gal).

federal excise tax was raised in 1985 to $12.50 per proof gal, ie, a gallon of liquid containing 50% by volume of ethyl alcohol (100° proof). Prior to that, and since 1951, it had been $10.50 per proof gal.

Even though the metric system for containers has been adopted, determination of the excise tax remains on a tax-gallon basis. The Bureau of Alcohol, Tobacco, and Firearms has established an official conversion factor: 1 L = 0.26417 U.S. gal (Treasury Decision A.F.F. 39, Jan. 21, 1977).

In the United States, the revenue from excise taxes on distilled spirits has become a substantial part of income realized by the three levels of government: federal, state, and local. In 1863, with a $0.20 rate, the federal government collected over $5 million; in 1960, $3,090 million; and in 1975 the total revenue amounted to $6,277 million, of which 60% represented the federal government's share, collected at a rate of $10.50 per tax gallon. In 1988, the federal government collected $4.0 billion from distilled spirits and another $2.1 billion from wine and beer. State and local governments collected $7.5 billion in 1988, of which $3.1 billion came from distilled spirits and $4.4 billion from wine and beer.

Supervision over the production of distilled beverage spirits is maintained by the Bureau of Alcohol, Tobacco, and Firearms, Department of the Treasury, which succeeded the Alcohol and Tobacco Tax Division (ATTD) of the Internal Revenue Service in July 1972. (The ATTD was called the Alcohol Tax Unit from the repeal of prohibition to 1952.) This organization's enforcement agents are charged with the responsibility of eliminating illicit operations.

Prior to Prohibition, revenue agents and deputy collectors investigated illegal liquor operations, made arrests, seizures, etc, along with income tax and other miscellaneous work. On January 16, 1920, the effective date of Prohibition, federal prohibition agents took over the duties. An act of Congress created the Bureau of Prohibition in the Treasury Department on April 1, 1927. The Prohibition Reorganization Act, effective July 1, 1930, terminated the Bureau of Prohibition and created the Bureau of Industrial Alcohol in the Treasury Department, responsible for the permissive provisions of the act, and the Bureau of Prohibition in the Department of Justice, responsible for the enforcement of the penal provisions of the National Prohibition Act. These two were merged on December 6, 1933, as the Alcohol Tax Unit, Internal Revenue Department, and on March 15, 1952 it was named the

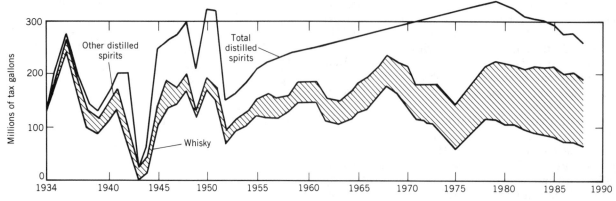

Figure 2. Production of distilled spirits from 1934 to 1988.

Alcohol and Tobacco Tax Division of the Internal Revenue Service.

PRODUCTION AND CONSUMPTION

In 1810, according to government records, Pennsylvania produced 24.5 million L (6.5 million gal); Indiana 83,000 L (22,000 gal); and Kentucky more than 7.5 million L (2 million gal). In 1917, apparently in anticipation of Prohibition, over 1.1 billion L (300 million gal) of beverage spirits were produced, of which 225 million L (60 million gal) were whisky. In the year 1930, the Prohibition Enforcement Bureau estimated that the illicit production amounted to 3 billion L (800 million gal). Figure 2 gives the production of distilled spirits from 1934 to 1988. Table 2 shows U.S. consumption of liquor from 1949 to 1988.

STANDARDS OF IDENTITY

Distilled alcoholic beverages are usually characterized by their geographical origin, type of material used in production, and standard of quality, as evaluated by organoleptic

analysis (taste and bouquet). Secretiveness was a way of life for ancient distillers; producers today operate according to their own methods. As it is not possible to correlate taste and bouquet with chemical composition, evaluation becomes a matter of organoleptic analysis.

Over the centuries legends, traditions, and to some extent political influences were important factors in the production and identification of potable distilled liquors. Since most governments had an economic interest because of the great source of revenue realized, they established standards for distilled alcoholic beverages, generally in keeping with the customs prevailing in their country. As a result, geographical identification has become accepted, and each country respects the identity and exclusiveness of the other's products.

Within each category of product wide variations in flavor can be caused by (1) types of material and their proportions, (2) methods of preparation, (3) selection of yeast types, (4) fermentation conditions, (5) distillation processes, (6) maturation techniques, and (7) blending. Since the alcoholic and water components are insignificant factors in flavor intensity or palatability, distillers are primarily interested in the more flavorful constituents, the

Table 2. U.S. Liquor Consumption by Type, Percent[a]

Type	1949	1960	1966	1976	1980	1984	1988
Blends	66.2	31.3	24.3	11.2	8.8	6.7	5.8
Straight	8.7	25.4	23.9	15.3	12.7	12.1	11.9
Bonds	5.4	4.1	2.4	0.9	1.6	0.31	0.30
Scotch	4.4	8.1	10.4	13.6	10.5	9.5	8.1
Canadian	2.7	5.2	6.9	12.2	11.7	12.2	11.6
Other	0.4	0.3	0.3	0.3	0.12	0.14	0.18
Total whisky	*87.8*	*74.4*	*68.2*	*53.5*	*45.5*	*40.9*	*37.7*
Gin	7.1	9.3	10.5	9.8	9.7	9.5	8.1
Vodka	0.0	7.8	10.4	18.0	21.5	22.8	24.3
Cordials	2.2	3.8	4.3	6.8	8.6	10.0	11.7
Brandy	1.3	2.6	3.2	3.9	4.2	5.0	5.2
Rum	1.3	1.6	2.2	4.2	6.6	7.8	8.4
Other	0.3	0.5	1.2	3.8	3.9	3.9	4.6
Total non-whisky	*12.2*	*25.6*	*31.8*	*46.5*	*54.6*	*59.1*	*62.3*
Total consumption, million L (wine gal)	641.5 (169.5)	888.3 (234.7)	1,169.2 (308.9)	1,621.8 (428.5)	1,692 (447)	1,669 (441)	1,427 (377)

[a] Refs. 2, 3.

Table 3. Congeneric Content of Various Types of Distilled Alcoholic Beverages, mg/L (ppm) at 80° proof (40%)[a]

| Component | Whisky | | | | | Cognac Brandy |
| | U.S. | Canadian | Scotch | Bourbon | | |
				Straight	Blended	
Fusel oil[b]	664	464	1,144	1,624	1,560	1,544
Total acids[c]	240	160	120	552	540	288
Esters[d]	136	112	136	448	344	328
Aldehydes[e]	21.6	23.2	36	54.4	43.2	60.8
Furfural	2.6	0.88	0.88	3.6	7.2	5.4
Total solids	896	776	1,016	1,440	1,272	5,584
Tannins	168	144	64	416	384	200
Total congeners, wt/vol %	0.116	0.085	0.160	0.292	0.309	0.239

[a] Determinations were made according to the official methods of analysis of the Association of Official Agricultural Chemists.
[b] Determined by the Komarowsky colorimetric method.
[c] As acetic acid.
[d] As ethyl acetate.
[e] As acetaldehyde.

so-called congeners, substances that are generated with the alcohol during the fermentation process and in the course of maturation. To produce a palatable product, it is, therefore, necessary to select the proper configuration of these constituents (congeners). In consideration of items 1–7 above, this cannot be accomplished by production techniques alone; therefore, the majority of alcoholic beverages are blended to provide uniformity, balanced bouquet, and palatability. To illustrate the variations in flavor constituents an analysis of some types is given in Table 3.

Whiskies

Although brandy, rum, and gin are substantial items in world markets, whiskies are by far the leading distilled alcoholic beverages, with those from Canada, Scotland, and the United States accounting for most of the sales. Irish whiskey, a distinctive product of Ireland, although not enjoying a large volume of sales, does have good distribution as a specialty item. The Irish use the spelling whiskey, the Scotch and Canadians use whisky, and U.S. citizens use both, although U.S. regulations use the spelling whisky.

Canadian. Canadian whisky is manufactured in compliance with the laws of Canada regulating the manufacture of whisky for consumption in Canada and containing no distilled spirits less than three years old. Canadian whiskies are premium products usually bottled at six years of age or more, and because they are blended, they are not designated as straight whiskies. They are light bodied and, although delicate in flavor, they nevertheless retain a distinct positive character. The Canadian government exercises the customary rigid controls in matters pertaining to labeling and in collection of the excise tax. However, it sets no limitations as to grain formulas, distilling proofs, or special types of cooperage for the maturation of whisky.

The major cereal grains (corn, rye, and barley malt) are used, and their proportions in the mashing formula remains a distiller's trade secret; otherwise, the process is substantially the same as is found in the distilleries of the United States. Because Canadian distillers are not faced with artificial proof restrictions in their distillation procedures, they are able to operate batch and continuous distillation systems under conditions that are optimum for the separation and selection of desirable congeners.

White oak casks (189 L, 50 U.S. gal) are used in the maturation process. A substantial amount of Canadian whiskies are stored in preused cooperage from other whiskies, which should lower the cost. Again, the proportions of new and matured cooperage used for maturation are each distiller's trade secret.

Scotch. Although Scotch whisky has enjoyed a worldwide reputation for its unique smoky flavor and high standard of quality, consumers know very little about it. Not much information is revealed by government regulations, which only specify the use of cereal grains and a minimum requirement for aging in oak casks; in Britain,

> spirits described as Scotch Whisky shall not be deemed to correspond to that description unless they have been obtained by distillation in Scotland from a mash of cereal grain saccharified by the diastase of malt and have been matured in warehouse in cask for a period of at least three years.

In the United States,

> Scotch Whisky is a distinctive product of Scotland, manufactured in Scotland in compliance with the laws of Great Britain regulating the manufacture of Scotch Whisky for consumption in Great Britain, and containing no distilled spirits less than three years old.

This minimum age requirement is greatly exceeded by the Scotch distillers. For example, nothing under 4 years of age is included in their exports to the United States, and for the most part, 6-, 8-, 10-, and 12-yr minimum ages are featured in their brands. Most Scotch brands are blends of grain whiskies and numerous distinctive malt whiskies are produced by more than 100 distilleries in four major areas of Scotland. Malt whiskies are characterized by their location.

Even though there are many distilleries and, no doubt,

slight variations in their production methods (that is how single whiskies acquire the characteristics attributable to each specific plant) there are definite processes generally used. A basic knowledge of these traditional methods, still in use today, is needed to fully appreciate the quality concept inherent in Scotch whisky.

As in Canada, no government limitations are placed on production and maturation techniques. The Scotch distillers are guided by their production experience developed over many centuries and by consumer reactions.

The outstanding taste characteristic of Scotch, its subtle smoky flavor, is due to the techniques used in the production of malt whiskies. Malted barley, dried over peat fires, is the only grain ingredient in the mash. The kind and the amount of peat used in the fires determines the intensity of flavor in the final product. The aroma of the burning peat, known as peat reek, is absorbed by the barley malt. This smoky flavor is carried through to the final distillate and becomes a characteristic of single malt whisky. Peat is heather, fern, and evergreen that have been subjected to aging and compression processes over the centuries.

The dried malted barley is ground to a grist and allowed to hydrolyze in a mash tub. After conversion is completed, the liquid portion, or wort, is drained off, cooled, and placed in a fermenter. After fermentation, the separation of malt whisky from the fermented wort takes place in a batch distillation system, a copper kettle with a "worm," or spiral tube, leading from its head. The size and shape of these pot stills exert a definite influence on the character of the whiskies. Another critical factor is the selection of the product during that portion of the distillation cycle that will produce the desired flavors. The first portion in the cycle is referred to as foreshots and the last as feints (heads and tails in the United States). The middle portion, after further distillation, becomes high wines and is subsequently reduced to maturation proof for storage in oak casks. The final distillation proof is in the 140–160° (70–80%) range.

The grain whiskies used in Scotch brands are produced in a manner similar to the production techniques used in Canada and in the United States. Corn (referred to in the UK as maize), rye, and barley malt are the ingredients. The proportions again depend on the individual distiller. Because delicate flavors are desired, the distillation proof is around 180–186° (90–93%). The distillation system is basically a Coffey still composed of two columns.

The grain whiskies are generally aged in matured oak casks of 190-L capacities not unlike U.S. and Canadian barrels. Some malt whiskies acquire other distinctive qualities by being matured in oak casks that were previously used for sherry.

Whereas the materials, geographic location, and production processes are responsible for the uniqueness of Scotch whisky, it is the skill of the blender that achieves the quality of the final product. As many as 20 and sometimes more, different malt and grain whiskies are married to produce a brand. Of course, the formula is a well-guarded secret.

Irish. Irish whiskey is manufactured either in the Republic of Ireland or in Northern Ireland in compliances with their laws regulating the manufacture of Irish whiskey for home consumption and containing no distilled spirits less than three years old. Like Scotch, Irish whiskeys are blends of grain and malt whiskeys. Unlike Scotch, Irish whiskey does not have the unique smoky taste because the barley malt is not dried over peat fires. Because whiskey has been known in Ireland as long as in Scotland, it is not surprising that their production techniques are closely related. One variation, however, is the use of some small grain, mostly barley, in addition to barley malt in the production of malt whiskey. The production process involves four basic steps: brewing (mashing), fermentation, distillation, and maturation. Mashing takes place in a Kieve (mash tun), a circular metal vessel with two bottoms; the upper is perforated, which permits the wort (converted grain starch or maltose) to be filtered to the underback and sent on to the washbacks (fermenters) where the inoculation with and action by yeast produces the whiskey. The wash (fermented mash) is then distilled in a pot still (batch), producing a low wine of about 100° proof (50%), which in turn is redistilled to produce first shots (fore shots) from which the final product or whiskey of about 140–150° proof (70–75%) is distilled. Irish whiskey brands are generally considered to be more flavorful and heavier bodied than Scotch blended whiskies.

United States. Distilled spirits for beverage purposes in the United States are characterized specifically as to type, materials, composition, distillation proofs, maturation proofs, storage containers, and the extent of the aging period. The federal government also requires that a detailed statement of the production process be filed for each type and any subsequent improvements or changes must be filed and approved before being placed into operation. In addition, a generalized application of the regulations is made to establish the identity of products where the intensity of flavor may not conform to an arbitrary organoleptic evaluation based on chemical analysis. As a result, in spite of extensive manufacturing facilities and know-how available for the production of a wide range of whiskies, the distiller is restrained within narrow limits and does not enjoy the degree of latitude available to the Canadian and Scotch distillers. U.S. regulations, by the Bureau of Alcohol, Tobacco, and Firearms of the Treasury Department, are very specific and more limiting than those of Canada or the UK. Title 27, *Code of Federal Regulations*, Subpart C, paragraph 5.22 *et seq.* sets the standards of identity for all distilled spirits. Those that are a factor in the U.S. market are given here.

Whisky is an alcoholic distillate from a fermented mash of grain produced at less than 190° proof in such manner that the distillate possesses the taste, aroma, and characteristics generally attributed to whisky, stored in oak containers, and bottled at not less than 80° proof.

Neutral Spirits or *alcohol* are distilled spirits produced from any material at or above 190° proof.

The requirement of distilling above 190° proof (95%) gives a distiller an opportunity to take advantage of the physical relationship that exists between water, alcohol, and congeners that are produced during fermentation. The sophisticated equipment utilized by progressive distillers permits a technique known as selective distillation, whereby the distiller can remove all congeners or retain those that are deemed desirable.

Grain spirits are neutral spirits distilled from a fermented mash of grain and stored in oak containers.

Because grain spirits have delicate flavors, they must be stored in oak barrels previously seasoned through the storage of whisky or grain neutral spirits. It is important that the barrel is compatible with the flavor intensity of the grain neutral spirits; otherwise, the woody character of the barrel would overwhelm the light flavor of the grain neutral spirits and thus prevent proper development during aging. Grain neutral spirits are produced on continuous and batch distillation systems. Each produces a number of distillates having a low flavor intensity that, when stored in barrels, develop in flavor in the same manner as whisky (4).

The batch system, related to the pot still, is simpler in concept and offers an opportunity to use the heart-of-the-run principle for the production of grain neutral spirits. Sometimes referred to as a time-cycle distillation system, it involves three stages: (1) the heads (aldehydes) are removed, (2) the product is removed, and (3) the residual distillate (tails) remaining in the kettle is removed for subsequent redistillation in the continuous system. The batch system is composed of a large kettle with a capacity of up to 190,000 L (50,000 wine gal) with a vapor pipe leading to a product-concentrating column having as many as 55 bubble-cap plates. The large capacity of the kettle is important in maintaining product uniformity. Straight whisky, produced in the normal manner in the whisky column, is pumped into the kettle, indirect steam heat is applied through a coil within the kettle, and the alcoholic vapors rise into the product column where they are refined. The grain neutral spirits thus produced are reduced in proof with deionized water to between 110 and 160° proof (55 and 80%), put in oak barrels, and placed in government bonded warehouses for storage.

Vodka is neutral spirits so distilled, or so treated after distillation with charcoal or other materials, as to be without distinctive character, aroma, taste, or color.

This definition has come under fire recently because some vodkas on the market claim to have some flavor. Since the 1950s drinking patterns in the United States have become more diversified. Vodka has moved up from negligible sales in 1949 to 24% in 1988. The fact that it can be mixed with any flavored substance seems to be the reason for its wide acceptance.

Rye whisky, bourbon whisky, or malt whisky is whisky produced at not more than 160° proof from a fermented mash of not less than 51% rye, corn, or malted barley respectively, and stored at not more than 125° proof in charred new oak containers.

Corn whisky is whisky produced at not more than 160° proof from a fermented mash of not less than 80% corn. It may or

may not be stored in oak containers, but only in used or uncharred ones.

In producing bourbon whisky, eg, the mashing formula must contain at least 51% corn and the remaining ingredients (49%) generally are proportioned between rye grains and barley malt. Each distiller selects a preferred formula. Very little bourbon with 49% small grains is produced. The most popular proportions are 60% corn, 28% rye, 12% barley malt (referred to as 40% small grains), 70–18–12 (30% small grains), and 75–13–12 (25% small grains). Although some bourbons use as much as 15% barley malt, the general practice in the industry is to use 12% barley malt for all bourbon production.

Because rye and barley malt produce more intensive flavors than corn, the formula with the greater small-grains proportion will produce a bourbon with more body, provided, of course, that the same distillation techniques are used.

Straight whisky may be any of the whiskies in the preceding two paragraphs that have been stored in the prescribed oak containers for two years or more. A straight whisky may further be identified as *bottled in bond*, provided it is at least four years of age, bottled at 100° proof (50%), and distilled at one plant by the same proprietor. A bottled-in-bond whisky may contain homogeneous mixtures of whiskies, provided they represent one season, or if consolidated with other seasons, the mixture shall be the distilling season of the youngest spirits contained therein, and shall consist of not less than 10% of spirits of each such season.

Light whisky is whisky produced in the United States at more than 160° proof, and stored in used or uncharred new oak containers; and also includes mixtures of such whiskies. If light whisky is mixed with less than 20% by volume of 100° proof (50%) straight whisky the mixture shall be designated Blended Light Whisky.

Blended whisky is a mixture which contains straight whisky or a blend of straight whiskies at not less than 20 percent on a proof gallon basis, and, separately, or in combination, whisky or neutral spirits. A blended whisky containing not less than 51 percent on a proof gallon basis of one of the types of straight whisky shall be further designated by that specific type of straight whisky; for example, "blended rye whisky."

Scotch whisky is whisky which is a distinctive product of Scotland, manufactured in Scotland in compliance with the laws of the United Kingdom regulating the manufacture of Scotch whisky for consumption in the United Kingdom: such product is a mixture of whiskies, such mixture is "blended Scotch whisky."

Irish whisky is whisky which is a distinctive product of Ireland, manufactured either in the Republic of Ireland or in Northern Ireland, in compliance with their laws regulating the manufacture of Irish whisky for home consumption: if such product is a mixture of whiskies, such mixture is "blended Irish whiskey."

Canadian whisky is whisky which is a distinctive product of Canada, manufactured in Canada in compliance with the laws of Canada regulating the manufacture of Canadian whisky for consumption in Canada: if such product is a mixture of whiskies, such mixture is "blended Canadian whisky."

GINS

Gin is a product obtained by original distillation from mash, or by redistillation of distilled spirits, or by mixing neutral spirits, with or over juniper berries and other aromatics, or with or over extracts derived from infusions, percolations, or maceration of such materials, and includes mixtures of gin and neutral spirits. It shall derive its main characteristic flavor from juniper berries and be bottled at not less than 80° proof.

France de La Boe, a 17th-century professor of medicine at Leyden University, The Netherlands, is credited with being the originator of the botanical-flavored spirits known as gin. Because his product's primary flavor was due to the essential oils extracted from juniper berries, he gave it the French name *jenièvre*, which appeared later as the Dutch *geneva* and finally as the English *gin*. Gin produced exclusively by original distillation or by redistillation may be further designated as distilled. Gin derives its main characteristic flavor from juniper berries. In addition to juniper berries, other botanicals may be used, including angelic root; anise; coriander; caraway seeds; lime, lemon and orange peel; licorice; calamus; cardamom; cassia bark; orris root; and bitter almonds. The use and proportion of any of these botanicals in the gin formula is left to the producer, and the character and quality of the gin depends to a great extent on the skill of the craftsman in formulating the recipe. The more skilled producers formulate their aromatic ingredients on the basis of the essential oil content in the raw materials to assure a greater degree of product uniformity.

To expose the essential oils, the ingredients are reduced to a granular form and then immersed directly into the kettle (pot), which is filled with grain neutral spirits at approximately 100° proof (50%). A vapor-phase extraction may also be used. In this case, the botanical mixture is placed on trays or in baskets in the head of the kettle where the alcoholic vapors passing by extract the essential oils and rise to the condenser.

It is important that the grain spirits be as neutral as possible (devoid of congeners) to avoid undesirable flavors. In addition to the kettle, some gin stills have a refinement section (as many as six plates) above the kettle for flavor stability and enrichment. Indirect steam heat is applied and the various essential oils are distilled over during the entire distillation cycle. The first (heads) and last (tails) portions of the cycle are not included in the product. Only the heart of the run is used, representing approximately an 85% recovery of the original alcohol concentration and varying with the type of product desired. Some distillers, to avoid thermal decomposition of the delicate flavors and to acquire a degree of softness, conduct the distillation under reduced pressure at a temperature of about 57°C. London Dry Gin, for example, is produced in this manner.

British and Canadian regulations permit and recognize the use of maturation techniques for gin. Gins stored in special oak casks acquire a pale golden hue and a unique dryness of flavor. Although distillers are permitted to store gins in the United States for further flavor development, the federal government does not permit any reference to aging to appear on the label.

Holland Gin, characterized by its high flavor intensity derived mostly from juniper berries and cereal grains (corn, rye, and barley malt), is produced by immersing the botanical mixture directly into the grain mash before distillation or by extracting the essential oils from the botanical mixture with the heavy distillate (high wines) from a fermented mash of grain, consisting of corn, rye, and barley malt. Consequently, the flavors produced during fermentation become flavor components of the final product. Compound gin is a mixture of grain spirits and essential oil extracts from botanicals. It does not undergo any distillation procedure.

Brandies

Brandy is an alcoholic distillate from the fermented juice, mash, or wine of fruit, or from the residue thereof, produced at less than 190° proof in such manner that the distillate possesses the taste, aroma, and characteristics generally attributed to the product, and bottled at not less than 80° proof.

The most important category of brandy is fruit brandy, distilled solely from the juice or mash of whole, sound, ripe fruit or from standard grape, citrus, or other fruit wine. Brandy derived exclusively from one variety of fruit is so designated. However, a fruit brandy derived exclusively from grapes may be designated as brandy without further qualification, and unless the product is specifically identified, the term brandy always means grape brandy. Brandy is subject to a distillation limitation of 170° proof (85%). If distilled over 170° proof (85%), it must be further identified as neutral brandy. A minimum of two years of maturation in oak casks is required, otherwise the term immature must be included in the designation. Although the age is not indicated on the label, brandies are normally aged from three to eight years.

Brandies are produced in batch or continuous distillation systems. The pot still or its variation is universally used in France, whereas in the United Stated both systems are employed. The batch system produces a more flavorful product, the continuous system a lighter, more delicate flavor.

The history of brandy can be said to be the history of distillation, because in the distant past it was the distillation of wine in crude stills that produced *aqua vitae*. In the ensuing evolution, many areas of Europe and of the United States became renowned for their brandies. Perhaps the most popular brandy comes from the Cognac region of France, in the Department of Charente and Charente Inférieure. As such, it enjoys an exclusive identity, Cognac, under which no other brandy may be labeled. Cognac is produced in the traditional pot stills by small farmers and sold to the bottlers who age the brandies in limousin oak casks. When the brandies reach maturity, they are skillfully blended for marketing under their own brand name.

Cognac is a blend of some Grande Champagne, Petite Champagne, Borderies, and Fins Bois, the proportions of each being a well-kept secret. To further characterize Cognac, the bottle is labeled: E, especial; F, fine; V, very; O, old; S, superior; P, pale; X, extra; C, Cognac; eg, VSOP means very superior old pale, and is considered to be a better-quality product.

Another well-known brandy of Frances is Armagnac, produced in southern France. Armagnac is distilled from wines in a continuous system using two pot stills in series. Armagnac is considered to be more heavy-bodied and drier than Cognac. Brandies are distilled in almost every wine region of France; they are called *eau de vie*, exported as French Brandy, never as Cognac.

In the United States, California produces almost all of the grape brandy. Generally, it is a well-integrated operation, the cultivation of the grapes, the making of the wine, the distilling, aging, bottling, and the marketing of the brandy being done by the same firm. Usually a continuous multicolumn distillation system is employed. Of the total U.S. brandy consumption of approximately 56 million L (15 million gal) California brandies account for over 80%.

Blended applejack is accorded a special classification as a mixture that contains at least 20% of apple brandy (applejack) on a proof basis, stored in oak containers for not less than two years, and not more than 80% of neutral spirits on a proof gallon basis and bottled at not less than 80° proof (40%). Another class of beverage spirits, flavored brandy, is a brandy to which natural flavoring materials have been added with or without the addition of sugar; it is bottled at no less than 70° proof (35%). The name of the predominant flavor appears as part of the designation, ie, blackberry flavored brandy; cherry flavored brandy, etc. Such a flavored brandy may contain up to 12.5% of wine derived from the particular fruit corresponding to the labeled flavor.

Certain areas in Europe and in South America are well known for their specialty brandies, such as Spanish brandies, distilled from Jerez sherry wine; the fragrant, fruity Portuguese brandies, distilled from port wine; the pleasant and flowery muscat bouquet of Pisco brandy from Peru; Kirschwasser brandy, with its almond undertone flavor, distilled from a fermented mash of small black cherries, which grow along the Rhine Valley in Germany and Switzerland; and Slivovitz, the plum brandy, which is produced in Hungary and in the Balkan countries.

Cognac or Cognac (grape) brandy, is grape brandy distilled in the Cognac region of France, which is entitled to be so designated by the laws and regulations of the French Government.

Rums

Rum is an alcoholic distillate from the fermented juice of sugar cane, sugar cane syrup, sugar cane molasses, or other sugar cane by-products, produced at less than 190° proof in such manner that the distillate possesses the taste, aroma and characteristics generally attributed to rum, and bottled at not less than 80° proof.

Blackstrap molasses is the most common raw material for the manufacture of rum. Otherwise, the same basic factors that produce different whiskies are responsible for the flavor varieties of rums. The type of yeast, fermentation environment, distillation techniques and systems, the maturation conditions, and not least the blending skill all contribute to the final character and quality of rum. Blackstrap molasses varies in composition according to origin owing to the environment and to some extent on the processing of cane (a greater recovery of sugar is usu-

ally reflected in a lower concentration of residual sugar in the molasses). A typical composition of Puerto Rican blackstrap molasses is as follows:

pH	4.3
Density (% solids)	89°Brix
Nonfermentable solids	28.7%
Sucrose	33.8%
Reducing sugars	26.9%
Total sugars	*60.7%*

Rums are characterized as light bodied, of which the Puerto Ricans are the best known, and full bodied, which come from Jamaica and certain other islands of the West Indies. Light rums are distilled on multicolumn continuous distillation systems over a proof range of 160–180° (80–90%). They are matured in oak casks that are reused for rum storage. Age may, but need not, be stated on the labels.

Rum has been produced in Jamaica for more than 200 yr as some plantation crop records indicate. A typical small plantation would produce a ton of sugar and 41,000 L (9,000 imperial gal) of rum. Today, only a few large facilities exist and the law requires that a distillery may only be operated in conjunction with a sugar refinery. As in every area where distilled spirits are produced, terminology and local practice play a major role in determining the identity and the flavor characteristics of the product; eg, molasses acid results from a natural fermentation that produces alcohol and subsequently an organic acid mixture. Some of this is used in the wash (fermenting mash) for the production of certain flavorful rums. Dunder is the term used for stillage, the material left after the beverage spirits are removed by distillation. Dunder is also used as an ingredient in the fermenting mash, as much as 40 vol %, to provide buffering action, and flavor development.

Jamaican and other full-bodied rums are distilled between 140 and 160° proof (70 and 80%) in pot stills. They are matured in large casks of 422.4 L (111.6 gal) called puncheons. Unlike the light-bodied rums, which use cultured yeast for inoculation, the Jamaican rums rely on natural fermentation, sometimes referred to as wild fermentation. In this method the mash is inoculated by the yeast that is present in the air and in the raw material. Time of fermentation may vary from 2 to as much as 11 days, depending on the desired flavor characteristics.

Puerto Rican rums are generally labeled as white or gold label. The latter is a little more amber in color and has a more pronounced flavor. Although the rums produced in various areas are not considered distinctive types, they do retain their local characteristics and their names may not be applied to rum produced in any other place than the particular region indicated in the name.

Venezuela and to some extent Mexico are major producers of rum. The former requires a minimum two-year aging period and in the latter the amount of production is closely regulated by the government and keyed to the availability of cane and the need for industrial alcohol. In all other respects, the light-bodied rums are similar to those produced in Puerto Rico. A limited amount of more flavorful rums for use in blending is produced in both countries. Continuous and batch distillation systems are used and thus offer a capability of producing rums with varying degrees of flavor intensities and individual characteristics.

Tequila

Tequila, a distinctive product of Mexico, is an alcoholic distillate produced in Mexico from the fermented juice of the heads of *Agave Tequilana Weber* (blue variety), with or without additional fermentable substances, distilled in such a manner that it possess the taste, aroma, and characteristics generally attributed to tequila. This is Mexico's most popular distilled spirits drink.

The mezcal azul (blue mezcal), the primary source for tequila, is cultivated and usually propagated from 2-yr-old sprouts obtained from 7-yr-old mezcal plants. After 8–12 yr the plants are matured; the trimmed heads, referred to as pine apples because of their appearance and weighing 36–59 kg (80–130 lb) each, are transported to the distillery. In the State of Jalisco, up to 20,000 t of mezcal heads are harvested annually. The heads contain (unsteamed): moisture, 62%; total solids, 38%; fiber, 11%; inulin, 20%; and ash, 2.5%, and they have a pH of 5.5.

The juice from the mezcal heads is first extracted in masonry ovens with a capacity of approximately 40 t for 9–24 h at approximately 93°C. The length of the steaming period is critical for the acid hydrolysis of the inulin to monosaccharides. During a 12-h cooling period, some additional juice (*mieles de escurrido*, drained molasses) is recovered. The mezcal heads are now dark brown, soft in texture, with a taste similar to maple syrup. Residual juice is removed by shredding the steamed heads, compressing the strips between roller mills, and finally washing the strips (bagasse) to recover all of the sugary syrup.

The mezcal juice from the steaming ovens, the roller mills, and the bagasse washes is pumped into fermenters of 3,800–7,500 L (1,000–2,000 gal) capacity made of masonry or of local pine wood. Nitrogen nutrients are added to facilitate fermentation. The Mexican government also permits the addition of piloncillo, a brown sugar, up to 30 wt % of the fermentable sugar in the mezcal heads after steaming. After a fermentation period of about 42 h, the alcoholic concentration is 4.5 vol% of the fermenter mash. When piloncillo is not used, the alcoholic yield is lower and the fermentation takes longer.

The fermented mash is pumped to a copper pot still of about 1,100 L (300 gal) capacity provided with a steam coil within the kettle for heating and a condenser for cooling the vapors. The intermediate of the first distillation called ordinario is collected at 28° proof (14%) and redistilled in a slightly larger pot still. This distillation cycle can be controlled to yield a product of approximately 106° proof (53%). The residual distillate from this process is combined with the fermented mash, starting a new cycle in the primary distillation system.

Tequila, as consumed in Mexico, is unaged and usually bottled at 80–86° proof (40–43%). However, some producers do age tequila in seasoned 190-L (50-gal) white oak casks imported from the United States. In aging, tequila becomes golden in color and acquires a pleasant mellowness without altering its inherent taste. Tequila aged one year is identified as Anejo; aged as much as two to four years, it is identified as Muy Anejo. The annual production of tequila in Mexico is around 20 million L (5.2 million U.S. gal), the major portion of which is produced by 28 plants located in the county of Tequila.

Cordials and Liqueurs

Cordials and liqueurs are the same, with the former term being American and the latter European. They are obtained by mixing or redistilling neutral spirits, brandy, gin, or other distilled spirits with or over fruits, flowers, plants, or pure juices therefrom, or other natural flavoring materials, or with extracts derived from infusions, percolations, or maceration of such materials. Cordials must contain a minimum of 2.5 wt% of sugar or dextrose, or a combination of both. If the added sugar and dextrose are less than 10 wt% then it may be designated as dry. Synthetic or imitation flavoring materials cannot be included in United States cordials, nor can they be designated as distilled or compound.

Cordials were known in ancient Egypt and Athens, but commercial production was started in the Middle Ages when alchemists, physicians, and monks, among others, were searching for an elixir of life. From this activity many well-known cordials were developed, such as Benedictine and Chartreuse, both derived from aromatic plant flavors and bearing the names of the monasteries where they were first prepared.

A great variety of cordials are available encompassing a wide spectrum of flavors from fruits, peels, leaves, roots, herbs, and seeds. Organoleptic attainment, however, becomes a matter of experience and skill in the selection of botanicals and in the extraction and formulation of flavors. Although these elements are carefully guarded secrets, the producer must rely on three basic processes, namely maceration, percolation, and distillation, or any combination thereof. Maceration involves the steeping of the raw materials in the spirits, usually in a vat, to impart the desired aroma, flavor, and color. The liquid is then drawn off and provides the base for further processing. Percolation is accomplished by recirculating the spirits through a percolator containing the raw materials. As the spirits seep down through the raw material, the desired constituents are extracted, which will give the proper aroma, flavor intensity, and color. The distillation method is similar to that used in gin production. The ingredients are either immersed in the beverage spirits or placed in trays or pans in the head of the still. The rising vapors extract the essential flavors, which are then condensed and discharged as a colorless liquid. This distillate contains the basic flavor that is used for further processing.

Cordials are characterized and marketed according to their generic names; eg, anisette (aniseed), crème de menthe (peppermint), triple sec (citrus fruit peel), slow gin (sloe berries), and by their trade names (proprietary brands) of which Benedictine and Chartreuse are well-known examples.

THE MANUFACTURING PROCESS

Any material rich in carbohydrates is a potential source of ethyl alcohol, which for industrial purposes is obtained by the fermentation of materials containing sugar (molasses), or a substance convertible into sugar, such as the starches. In the production of distilled spirits for beverage purposes, however, cereal grains are the principal types of raw material used. Any reference to alcohol in beverages is always to ethyl alcohol (C_2H_5OH). Although other alcohols may be present, they are referred to as higher alcohols, fusel oils, or by their specific name.

The chemical composition of grain varies considerably and depends to a large extent on environmental factors such as climatic conditions and the nature of the soil. Another variable is the malt (sprouted or germinated grain) used. Malt is generally understood to be germinated barley, unless it is further qualified as rye malt, wheat malt, etc. The purpose of malting is the development of the amylases, the active ingredients in malt. Amylases are enzymes of organic origin, which change grain starch into the sugar, maltose. Besides providing the means of converting the grain starch into sugar, the malt also contributes to the final flavor and aroma of the distillate. Malting techniques may produce a malt of such unusual character that it may indeed furnish the outstanding characteristics of the final product, as in Scotch whisky. Figure 3 shows the process flow sheet of a modern beverage spirits plant.

Grain Handling and Milling

The beverage distilling industry utilizes premium cereal grains. Each distiller supplements government grain standards with personal specifications, especially in regard to the elimination of grain with objectionable odors, which may have developed during storage or kiln drying at the elevators. Hybrid corn, usually of the readily available dent variety and plump rye, developed from Polish strains (Rosen), are used for beverage alcohol production. Modern distilleries use Airveyor unloading systems, others the traditional power shovel in conjunction with screw conveyors and bucket elevators. Even though the grain has been subjected to a cleaning process at the elevator, it is passed over receiving separators, a series of vibrating screens which sift out the foreign materials. Air jets and dust collectors remove any light material and magnetic separators remove metallic substances.

Milling breaks the outer cellulose protective wall around the kernel and exposes the starch to the cooking and conversion processes. Distillers require an even, coarse meal without flour. Milling is accomplished by three methods: (1) in roller mills, using pairs of corrugated rolls (breaks), run sharp to sharp (projections facing projections); (2) in hammer mills, where a series of revolving hammers within a close-fitting casing and rotating at 1,800–3,600 rpm shear the grain to a meal, which is removed by suction through a screen, different for various types of grain; and (3) in attrition mills (not widely used) where the grain is ground by two counterrotating disks (1,200–2,000 rpm). This mill is not entirely satisfactory because of excessive flour produced. A three-break roller mill (three pairs of corrugated rolls, arranged vertically) with rolls 23 cm in diameter × 76 cm long has a capacity of 3.5 m^3 (100 bu) of corn per hour; a hammer mill 61 cm in width, 10.6 m^3 (300 bu) per hour; and an attrition mill

with 41 cm grinding plates, 3.5 m^3 (100 bu) per hour, on the basis that 25% of the ground corn remains above the no. 12 mesh (1.68 mm) screens.

Mashing

The mashing process, consists of cooking, ie, gelatinization of starch, and conversion (saccharification), ie, changing starch to grain sugar (maltose). Cooking can be carried out at atmospheric or higher pressure in a batch or a continuous system. For whisky production, batch cooking at atmospheric pressure is widely used although some batch pressure cooking is practiced. For grain neutral spirits production, both batch and continuous systems are used under pressure. After cooling conversion is accomplished in the cooking vessel by the addition of barley malt meal to the cooked grain. Some distillers pump the mash immediately to a converter for the necessary holding time and thus make the cooking vessel available for the next cook. The converted mash is cooled and pumped to the fermenters.

Distillers vary mashing procedures, but generally conform to basic principles, especially in the maintenance of sanitary conditions. The cooking and conversion equipment is provided with direct or indirect steam, propeller or rake-type agitation, and cooling coils or a barometric condenser. Mashing procedures for rye, corn, and malt grains are described below.

Rye. In the preparation of a bourbon mash, rye is generally subjected to the corn-cooking process. However, rye undergoes liquefaction at a much lower temperature than corn, which avoids thermal decomposition of critical grain constituents adversely affecting the final flavor of the distillate. For that reason, many distillers mash rye separately.

Water is drawn at the rate of 35 L/m^3 (28 gal/bu) and rye and malt meal are added. The mash is slowly heated to 54°C and held for approximately 30 min. Proteolytic enzymes, active at 43–46°C, aid in reducing the viscosity, and the optimum temperature for β-amylase is 54°C. The mash is then heated to 63–67°C and held for 30–45 min to ensure maximum conversion. The mash (pH 6.0) is then cooled to the fermenting temperature 20–22°C. This process of converting small grains is called infusion mashing.

Corn. Although the starch in corn grains converts rather easily, higher cooking temperatures are necessary to make the starch available. Usually malt is not added at the beginning, but to reduce viscosity, premalt of 0.5% may be added before cooking, preferably at around 66°C. Thin stillage (the residual dealcoholized fermented mash from the whisky distillation process) is added by some producers to adjust pH to 5.2–5.4. For cookers operating at atmospheric pressure, a mashing ratio of 95–115 L (25–30 gal) of slurry (grain, water, and stillage mixture) per 0.03 m^3 (1 bu) and a holding time of 30 min at 100°C are preferable. The mash is cooled to 67°C and malt is added. Primary conversion, the saccharification taking place during conversion, is in the order of 70–80% of the available starches. The remainder of the conversion to fermentable sugar takes place during the fermentation process and is

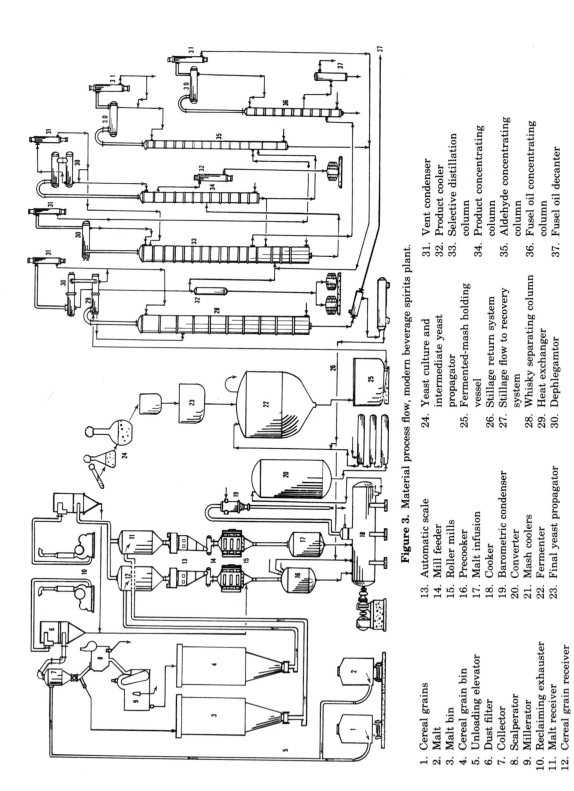

Figure 3. Material process flow, modern beverage spirits plant.

1. Cereal grains
2. Malt
3. Malt bin
4. Cereal grain bin
5. Unloading elevator
6. Dust filter
7. Collector
8. Scalperator
9. Millerator
10. Reclaiming exhauster
11. Malt receiver
12. Cereal grain receiver

13. Automatic scale
14. Mill feeder
15. Roller mills
16. Precooker
17. Malt infusion
18. Cooker
19. Barometric condenser
20. Converter
21. Mash coolers
22. Fermenter
23. Final yeast propagator

24. Yeast culture and
    intermediate yeast
    propagator
25. Fermented-mash holding
    vessel
26. Stillage return system
27. Stillage flow to recovery
    system
28. Whisky separating column
29. Heat exchanger
30. Dephlegamtor

31. Vent condenser
32. Product cooler
33. Selective distillation
    column
34. Product concentrating
    column
35. Aldehyde concentrating
    column
36. Fusel oil concentrating
    column
37. Fusel oil decanter

referred to as secondary conversion. For batch cooking under pressure only 65–83 L (17–22 gal) of water are drawn, and the maximum temperature is 120–152°C. In continuous pressure cooking, water is drawn at a ratio of 30 L/m³ (24 gal/bu) of meal and sufficient thin stillage is added to adjust the pH to 5.2–5.4. The mash is pumped through the continuous pressure cooker, where it is exposed to temperatures of 170–177°C for 2–6 min, and then into a flash chamber where it is cooled immediately by vacuum to the malting (conversion) temperature of 63°C. A malt slurry is continuously introduced and the mixture proceeds through the water cooling system to the fermenters.

Fermentation

In fermentation the grain sugars (largely maltose), produced by the action of malt enzymes (amylases) on gelatinized starch, are converted into nearly equal parts of ethyl alcohol and carbon dioxide. This is accomplished by the enzymes in yeast. Yeast multiples by budding, and a new cell is produced about every 70 min. Although yeasts of several genera are capable of some degree of fermentation, *Saccharomyces cerevisiae* is almost exclusively used by the distilling industry. It has the ability to reproduce prolifically under normal growth conditions found in distilleries, has a high fermentation rate and efficiency, and can tolerate relatively high alcohol concentrations (up to 15–16 vol %). A great variety of strains exist and the characteristics of each strain are evidenced by the type and amount of congeners the yeast is capable of producing.

Alcoholic fermentation is represented by the following reactions:

$$C_{12}H_{22}O_{11} \xrightarrow{\text{maltase}} C_6H_{12}O_6 \rightarrow 2\ C_2H_5OH + 2\ CO_2$$

| maltose | dextrose | ethyl alcohol | carbon dioxide |

A fermentation efficiency of 95% is obtained based on the sugar available. Of the starch converted to grain sugar and subsequently subjected to fermentation, 5–6% is consumed in side reactions. The extent and type of these reactions depend on: (1) yeast strain characteristics, (2) the composition of the wort, and (3) fermentation conditions such as the oxidation–reduction potential, temperature, and degree of interference by bacterial contaminants.

Secondary products formed by these side reactions largely determine the characteristics and organoleptic qualities of the final product. In the production of whisky, the secondary products (known as congeners) formed and retained during the subsequent operations include a number of aldehydes, esters, higher alcohols (fusel oils), some fatty acids, phenolics or aromatics, and a great many unidentified trace substances (Table 3). In the production of grain neutral spirits, the congeners are removed from the distillate in a complex multicolumn distillation system. Some distillers, however, retain a small portion of the low-boiling esters in the distillate when the grain neutral spirits are to be matured. Fermentation of grain mashes is initiated by the inoculation of the set mash with 2–3 vol %

of ripe yeast prepared separately (see below) and followed by three distinct phases.

1. Prefermentation involves rapid multiplication of yeast from an initial 4–8 million/mL to a maximum of 125–130 million/mL of the liquid and an increasing rate of fermentation.
2. Primary fermentation is a rapid rate of fermentation, as indicated by the vigorous "boiling" of the fermenting mash, caused by escape of carbon dioxide. During this phase secondary conversion takes place, ie, the changing of dextrins to fermentable substances.
3. Secondary fermentation is a slow and decreasing rate of fermentation. Conversion of the remaining dextrins, which are difficult to hydrolyze, takes place.

The degree of conversion, agitation of the mash, and temperature directly affect the fermentation rate. Fermenter mash set at a concentration of 144 L (38 gal) of mash per bushel of grain 25.4 kg (56 lb) will be fermented to completion in two to five days, depending on the set and control temperatures. The set temperature (temperature of the mash at the time of inoculation) is largely determined by the available facilities for cooling the fermenting mash. If cooling facilities are adequate, temperatures of 27–30°C may be employed; otherwise, the set temperature must be low enough to ensure that the temperature will not exceed 32°C during fermentation. When no cooling facilities are provided, the inoculation temperature must be below 21°C. Excessive temperatures during the prefermentation phase retard yeast growth and stimulate the development of bacterial contaminants which are likely to produce undesirable flavors.

In the production of sour mash whisky U.S. federal regulations require that a minimum of 25 vol % of the fermenting mash must be stillage (cooled, screened liquid recovered from the base discharge of the whisky separating column, pH 3.8–4.1). In addition to producing a heavier-bodied whisky, this procedure provides the distiller with an economical means of adjusting the setting pH (4.8–5.2) to inhibit bacterial development. It also provides buffering action during the fermentation cycle, which is important because secondary conversion does not take place if the fermenting pH drops below 4.1 in the immediate stages. Thin stillage also provides a means of diluting the cooker mash to the proper fermenter mash concentration, 3,220–3,870 L/m³ (30–36 gal/bu) of grain for the production of spirits, and 4,080–4,850 L/m³ (38–45 gal/bu) for making whisky. This concentration gives about a 12–16% soluble solids in the fermenting liquid, within the range used in beer fermentations but lower than that in wine fermentation.

Preparation of the yeast involves a stepwise propagation, first on a laboratory scale and then on a plant scale to produce a sufficient quantity of yeast for stocking the main mash in the fermenters. A strain of yeast is usually carried in a test tube containing a solid medium (agar slant). A series of daily transfers, beginning with the removal of some yeast from the solid medium, are made into successively larger flasks containing liquid media—

diamalt (commercial malt extract) diluted to 15–20° Balling, malt extract, or strained sour yeast mash—until the required amount of inoculum is available for the starter yeast mash, called a dona. After one day's fermentation, the dona is added to a yeast mash normally composed of barley malt and rye grains, and representing approximately 2–3.5 wt % of the total grain mashed for each fermenter.

The yeast mash is generally prepared by the infusion mashing method and then soured (acidified) to a pH of 3.9–4.1 by a 4–8-h fermentation at 41–54°C with *Lactobacillus delbrucki*, which ferments carbohydrates to lactic acid. Satisfactory souring can be induced with an inoculum of approximately 0.25% of culture per volume of mash. The water-to-meal ratio of 2,580–3,000 L/m³ (24–28 gal/bu) attains a yeast mash balling of approximately 21°. Before inoculation with yeast, the soured mash is pasteurized to 71–87°C to curtail bacterial activity, then cooled to the setting temperature of 20–22°C. The sour mash medium offers an optimum condition for yeast growth and also has an inhibitory effect on bacterial contamination. In 16 h the yeast cell count reaches a 150–250 million/mL. Some distillers use the sweet yeast method for yeast development. In this instance the lactic acid souring is not included and the inoculation temperature is usually above 26.6°C to insure rapid yeast growth.

Distillation

Distillation separates, selects, and concentrates the alcoholic products of yeast fermentation from the fermented grain mash, sometimes referred to as fermented wort or distillers beer. In addition to the alcohol and the desirable secondary products (congeners), the fermented mash contains solid grain particles, yeast cells, water-soluble proteins, mineral salts, lactic acid, fatty acids, and traces of glycerol and succinic acid. Although a great number of different distillation processes are available, the most common systems used in the United States are (1) the continuous whisky separating column, with or without an auxiliary doubler unit for the production of straight whiskies; (2) the continuous multicolumn, system used for the production of grain neutral spirits; and (3) the batch rectifying column and kettle unit, used primarily in the production of grain neutral spirits that are subsequently stored in barrels for maturation purposes. In the batch system, the heads and tails fractions are separated from the product resulting from the middle portion of the distillation cycle.

Although most modern plants have various capacity whisky stills available, a whisky separating column is usually incorporated into the multicolumn system, thus acquiring a greater range of distillation selectivity, ie, the removal or retention of certain congeners. For example, absorptive distillation involving the addition of water to the upper section of a column in the whisky distillation system is a method of controlling the level of heavier components in a product. In the beverage distillation industry, stills and auxiliary piping are generally fabricated of copper, although stainless steel is also used. All piping that conveys finished products is tin-lined copper, stainless steel, or glass.

The whisky column, a cylindrical shell that is divided into sections and may contain from 14 to 21 perforated plates, spaced 56–61 cm apart. The perforations are usually 1–1.25 cm in diameter and take up about 7–10% of the plate area. The vapors from the bottom of the still pass through the perforations with a velocity of 6–12 m/s. The fermented mash is introduced near the top of the still, and passes from plate to plate through down pipes until it reaches the base where the residual mash is discharged. The vapor leaving the top of the still is condensed and forms the product. Some whisky stills are fitted with entrainment removal chambers and also with bubble-cap plate sections (wine plates) at the top to permit operation at higher distillation proofs. Because whisky stills made of copper, especially the refinement section, supply a superior product, additional copper surface in the upper section of the column may be provided by a demister, a flat disk of copper mesh. The average whisky still uses approximately 1.44–1.80 kg/L (12–15 lb of steam/proof gal) of beverage spirits distilled. Steam is introduced at the base of the column through a sparger. Where economy is an important factor, a calandria is employed as the source of indirect heat. The diameter of the still, number of perforated and bubble-cap plates, capacity of the doubler, and proof of distillation are the critical factors that largely determine the characteristics of a whisky.

The basic continuous distillation system for the production of grain neutral spirits usually consists of a whisky separating column, an aldehyde column (selective distillation column), a product concentrating column (sometimes referred to as an alcohol or rectifying column, from which the product is drawn), and a fuel oil concentrating column. In addition, some distillers, to secure a greater degree of refinement and flexibility, may include an aldehyde concentrating column (heads concentrating column) or a fuel oil stripping column. Bubble-cap plates are used throughout the system (except in the whisky column, which may have some bubble-cap plates).

This distillation system offers a wide range of flexibility for the refinement of distilled beverage spirits. Figure 3 shows a five-column, continuous distillation system for the production of grain neutral spirits. A fermented mash (generally 90% corn and 10% barley malt) with an alcohol concentration of approximately 7 vol % is pumped into the whisky column somewhere between the 13th and 19th perforated plate for stripping. The residual mash is discharged at the base and pumped to the feed recovery plant; the overhead distillate [ranging in proof from 105 to 135° (52.5 to 67.5%)] is fed to the selective distillation column (also called the aldehyde column), which has over 75 bubble-cap plates. The main stream [10–20° proof (5–10%)] from the selective distillation column is pumped to the product concentrating column. A heads draw (aldehydes and esters) from the condenser is pumped to the heads concentrating column (also called the aldehyde concentrating column), and a fuel oil and ester draw is pumped to the fuel oil concentrating column. The product is withdrawn from the product concentrating column.

Some accumulation of heads, at the top of the product concentrating column, are removed at the condenser, and transferred to the aldehyde concentrating column, where the heads from the system are removed for disposal. The fuel oil concentrating column removes the fuel oil from the system. Figures 4, 5, and 6 show the distribution and

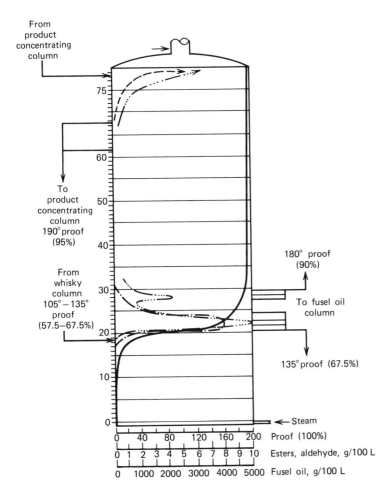

From
product
concentrating
column

75

60

To
product
concentrating
column
190° proof
(95%)

50

40

From
whisky
column
105°–135°
proof
(57.5–67.5%)

30

20

10

0

180° proof
(90%)

To fusel oil
column

135° proof (67.5%)

Steam

| 0 | 40 | 80 | 120 | 160 | 200 | Proof (100%) |

| 0 1 2 3 4 5 6 7 8 9 10 | Esters, aldehyde, g/100 L |

| 0 1000 2000 3000 4000 5000 | Fusel oil, g/100 L |

Figure 4. Selective distillation column.

concentration of alcohols and congeners through the selective distillation column, the product concentrating column, and the heads concentrating column.

By-Products. The discharge from the base of the whisky column is called stillage and contains in solution and in suspension substances derived from grain (except the starch, which has been fermented), and from the mashing and fermentation processes. The suspended solids are recovered by screening and then subjected to a pressing and a drying operation (dehydrating), usually in rotary, steam-tube dryers. The liquid portion, called thin stillage, is concentrated by a multieffect evaporator to a syrup with a solid content of 30–35%. This concentrate can be mixed and dried with the previously screened-out solids, or it can be dried separately, on rotary-drum dryers. These by-products are known as distillers' dried grains and distillers' solubles and are used by the feed industry to fortify dairy, poultry, and swine formula feeds. Distillers' feeds are rich in proteins (24–35%), fat (8%), choline, niacin, and other B-complex vitamins and contain vital fermentation growth factors that produce good growth responses in livestock and poultry.

Maturation

In the United States, the final phase of the whisky production process, called maturation, is the storage of beverage spirits (which at this point are colorless and rather pun-

gent in taste) in new, white oak barrels, whose staves and headings are charred. The duration (years) of storage in the barrel depends on the time it takes a particular whisky to attain the desirable ripeness or maturity. The staves of the barrel may vary in thickness from 2 to 2.85 m. The outside dimensions of a 190-L (50-gal) barrel are approximately: height, 0.88 m (34.5 in.) and diameter at the head, 0.54 m. Whisky storage warehouses vary in construction from brick and mortar types, single and multiple floors (as many as six) having capacities up to 100,000 barrels of 190-L (50-gal) capacity, to wooden sheet-metal-covered buildings (called iron clads), generally not exceeding a capacity of 30,000 barrels of 190-L (50-gal) capacity. It is customary to provide natural ventilation. The whisky in storage is subject to critical factors that determine the character of the final product. The thickness of the stave, depth of the char (controlled by regulating the duration of the firing, 30–50 s), temperature and humidity, entry proof, and, finally, the length of storage impart definite and intended changes in the aromatic and taste characteristics of a whisky. These changes are caused by three types of reaction occurring simultaneously and continually in the barrel: (1) extraction of complex wood constituents by the liquid, (2) oxidation of components originally in the liquid and of material extracted from the wood, and (3) reaction between the various organic substances present in the liquid, leading to the formation of new congeners.

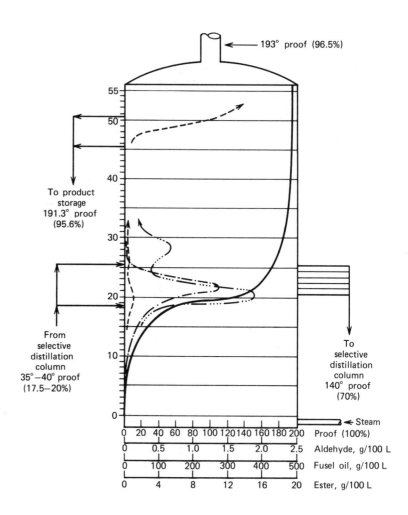

Figure 5. Product concentrating column.

Comprehensive studies of changes occurring during the maturation of whisky have been published by a number of investigators (4–11). Although these reports contain detailed information on the concentration of various congeners throughout the course of the maturation period, little is revealed of the interrelationships between various congeners except for the nonvolatile groups, such as solids, nonvolatile acids, tannin, and color. Notable among the observed changes are increases in the concentrations of acids, esters, and solids. These studies reveal that throughout the maturation period a linear relationship exists between the increase in acids and esters and the increase in dissolved solids; this suggests that acid and ester formation is dependent on some precursor that is extracted from the barrel at the same rate as the bulk of the solid fraction (9).

In 1962 the U.S. Treasury Department increased the maximum allowable entry proof (storage in barrels) for straight whisky from 110 to 125° (55 to 62.5%). The industry engaged in a study in December 1985 to determine the flavor development, chemical composition, and evaporation loss rates of distillates at six different proofs from 109 to 155° (54.5 to 77.5%) over a planned 8-yr period, and the study was continued an additional 4 yr. Evaporation losses averaged approximately 3.0% a year over the 12-year period.

In addition to chemical analysis, they applied regressive analysis for most congeners listed in Table 4. Regression analysis illustrates the functional relationship between congeners. Except for fusel oil, there was an appreciable continuing development of congreners through 12 yr. Esters, the most important flavor constituent in whisky, developed linearly over this time period. Later studies, using radioactive carbon for direct monitoring, determined the congeners derived from ethanol during whisky maturation. Esters (ethyl acetate) are formed in a system that is affected by changes in concentration of acetic acid, ethanol, and water, ie, a loss of water by evaporation; an increase in acetic acid results in an increase of ethyl acetate.

The maturing progress of a type of bourbon as measured by the principal ingredients is illustrated in Tables 4 and 5. It is evident that the congeners amount to only about 0.5–0.75% of the total weight. Yet it is this small fraction that determines the quality of the final product. Thus matured whiskies vary widely in taste and aroma because of the wide variation in congener concentration. For example, esters may vary from 8.0 to 18.0 g/100 L, and aldehydes from 0.7 to 7.6 g/100 L. Nevertheless, there is no correlation between chemical analysis and quality, ie, taste and aroma. Only the consumer can detect the fine variations and thus evaluate the quality of whiskies.

ANALYSIS

Analytical results are expressed in terms of one component for each chemical class; ie, acetic acid for acids, acet-

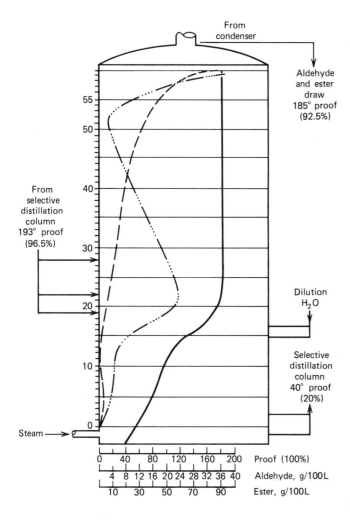

Figure 6. Aldehyde (heads) concentrating column.

Table 4. **Chemical Analysis**[a] **of Bourbon Whiskies Matured at Six Proofs from 109 to 155° (54.5 to 77.5%), g/100 L at 100°**
Proof (50%)

Age, yr	Proof	Color[b]	pH[c]	Solids	Fixed Acids	Volatile Acids	Esters	Fusel Oil	Aldehydes	Furfural	Tannins
0	109	nd[d]	nd	0	0	1.6	4.8	125	0.3	0.06	0
1	106	0.666	4.2	94	6	29	11	134	3.1	1.02	31
	112	0.678	4.2	99	7	27	12	134	3.1	1.03	31
	122	0.686	4.3	90	7	23	11	132	3.1	0.95	29
	132	0.652	4.3	70	6	21	11	128	2.9	0.84	26
	143	0.688	4.2	71	5	21	11	130	2.6	0.78	25
	155	0.652	4.2	63	5	18	11	129	2.6	0.75	23
2	107	0.793	4.1	124	7	31	16	147	3.5	1.16	37
	113	0.810	4.2	105	7	32	18	141	3.5	1.10	37
	123	0.848	4.3	107	7	27	16	143	3.2	1.12	36
	133	0.785	4.1	103	6	28	17	136	2.9	0.96	32
	143	0.807	4.1	92	6	23	16	145	2.5	0.82	30
	152	0.764	4.2	85	5	22	15	142	2.3	0.80	29
3	106	1.036	4.1	149	9	36	21	148	4.5	1.12	45
	112	1.013	4.2	145	9	36	21	146	4.2	1.14	43
	123	1.031	4.1	130	8	32	22	151	3.8	1.09	41
	132	0.928	4.2	118	7	29	19	153	3.5	0.89	36
	142	0.917	4.1	107	7	27	21	142	3.2	0.78	35
	151	0.845	4.1	88	6	25	21	149	2.8	0.71	31
4	106	1.086	4.1	177	12	42	23	148	5.2	1.16	52
	112	1.081	4.1	162	11	40	22	145	5.3	1.16	49
	122	1.081	4.0	152	10	33	21	152	4.9	1.11	45
	131	1.000	4.0	137	9	32	22	148	5.1	0.96	42
	140	0.977	4.1	124	8	28	20	146	4.2	0.88	40
	150	0.910	4.1	110	8	26	20	148	4.0	0.84	36

Table 4. (continued)

Age, yr	Proof	Color[b]	pH[c]	Solids	Fixed Acids	Volatile Acids	Esters	Fusel Oil	Aldehydes	Furfural	Tannins
5	106	1.229	4.1	194	12	43	30	157	6.1	1.20	49
	112	1.260	4.1	182	12	41	31	157	6.0	1.19	50
	122	1.180	4.0	164	10	34	26	155	5.7	1.14	45
	131	1.086	4.1	144	9	30	27	157	5.4	1.02	40
	140	1.102	4.0	131	8	30	28	155	5.3	0.96	38
	149	1.004	4.1	114	7	27	28	159	4.9	0.94	34
6	107	1.215	4.1	186	12	40	33	159	5.6	1.25	49
	113	1.284	4.1	185	12	40	33	163	5.5	1.15	49
	123	1.284	4.0	176	11	34	32	159	5.1	1.30	47
	132	1.155	4.0	153	10	33	32	163	5.1	1.05	42
	141	1.155	4.0	138	10	28	31	163	4.8	1.00	40
	150	1.056	4.0	122	8	27	29	166	4.7	1.00	36
7	107	1.456	4.1	214	13	51	37	157	6.1	1.34	57
	113	1.409	4.1	209	13	49	35	160	6.1	1.34	54
	123	1.409	4.1	190	12	46	36	155	5.7	1.29	51
	132	1.284	4.0	169	10	37	35	164	5.8	1.15	45
	141	1.215	4.0	152	10	32	33	162	5.2	1.08	42
	149	1.137	4.0	135	8	29	35	162	5.3	1.03	40
8	107	1.509	4.1	223	13	50	42	148	6.8	1.38	56
	113	1.523	4.1	218	13	49	43	159	6.8	1.35	57
	122	1.409	4.1	198	12	38	38	155	6.6	1.29	52
	131	1.276	4.0	175	10	34	37	153	5.9	1.13	60
	140	1.284	4.0	57	10	32	40	151	6.1	1.06	60
	149	1.168	4.0	147	9	27	39	152	5.4	1.08	58
9	108	1.509	4.0	225	14	50	48	164	8.7	1.45	57
	113	1.538	4.1	226	13	50	50	160	7.9	1.40	58
	122	1.432	4.0	204	12	38	43	171	7.8	1.27	53
	131	1.284	4.1	179	9	37	42	155	6.2	1.14	45
	140	1.301	4.0	164	9	33	45	159	6.7	1.06	44
	148	1.168	4.1	143	8	28	41	160	6.4	1.06	40
10	108	1.509	4.1	240	15	57	57	182	8.3	1.37	58
	113	1.482	4.1	231	13	52	54	173	8.4	1.33	55
	124	1.523	4.0	206	12	41	54	168	8.2	1.27	53
	132	1.367	4.0	183	11	41	41	175	8.1	1.15	47
	140	1.377	4.0	171	10	37	37	176	7.1	1.03	43
	147	1.215	4.0	152	9	32	32	169	7.1	1.00	41
11	108	1.699	4.0	252	14	60	60	166	8.2	1.37	64
	113	1.721	4.0	242	13	56	56	168	7.7	1.37	62
	123	1.658	3.9	219	12	49	57	179	7.8	1.34	59
	131	1.509	3.9	192	12	42	54	171	7.7	1.21	53
	139	1.398	3.9	181	10	38	51	178	6.7	1.08	50
	146	1.319	4.0	162	9	34	51	166	6.7	1.05	47
12	107	1.770	4.0	270	21	68	65	189	9.1	1.41	67
	112	1.745	4.0	258	19	52	62	187	9.3	1.37	65
	123	1.699	3.9	231	17	46	59	187	8.5	1.30	61
	130	1.538	3.9	211	15	41	62	187	9.4	1.20	55
	138	1.409	3.9	190	14	36	58	184	8.0	1.09	52
	145	1.319	3.9	168	10	33	54	185	7.8	1.07	47

[a] Analyses performed annually on samples reduced to 100° proof (50%). Acids as acetic acid, esters are ethyl acetate, aldehydes as acetaldehyde, and tannins as tannic acid (11).

[b] Absorbance at 430 nm at 100° proof (50%).

[c] At 100° proof (50%).

[d] Not determined.

aldehyde for aldehydes, etc (Tables 4 and 5). Individual constituents can, of course, be determined by more refined techniques. Until the advent of gas chromatography, tedious chemical procedures were generally required for separation of the congeners. Distillation, countercurrent distribution, and other physical methods affected concentration of various congener groups. These concentrates reacted to form chemical derivatives that were then sub-

jected to further separation. Paper and column chromatography have been used extensively for this work. After separation of the components, identification is possible by infrared and mass spectrometry.

With the high sensitivity (flame and β-ray ionization detection) and efficiency offered by gas chromatographic techniques, a qualitative profile of alcoholic distillates can be obtained without prior treatment of the sample. Flame

Table 5. Changes Taking Place in Bourbon Whisky Stored in Wood[a]

Age, yr	Range	Proof	Extract[b]	Acids[b]	Esters[b]	Aldehydes[b]	Furfural[b]	Fusel Oil[b]	Color
New	Average	101.0	26.5	10.0	18.4	3.2	0.7	100.9	0.0
	Maximum	104.0	161.0	29.1	53.2	7.9	2.0	171.3	0.0
	Minimum	100.0	4.0	1.2	13.0	1.0	Trace	71.3	0.0
								42.0	
1	Average	101.8	99.4	41.1	28.6	5.8	1.6	110.1	7.1
	Maximum	103.0	193.0	55.3	55.9	8.6	7.9	173.4	10.9
	Minimum	100.0	61.0	24.7	17.2	2.7	Trace	58.0	5.4
			54.0	10.4	10.4			42.8	4.6
2	Average	102.2	126.8	45.6	40.0	8.4	1.6	110.1	8.6
	Maximum	104.0	214.0	61.7	59.8	12.0	9.1	197.1	11.8
	Minimum	100.0	81.0	25.5	24.4	5.9	0.4	86.2	6.9
			78.0	23.5	11.2			42.8	5.7
			78.0	23.5	11.2			42.8	5.7
4	Average	104.3	151.9	58.4	53.5	11.0	1.9	123.9	10.8
	Maximum	108.0	249.0	73.0	80.6	22.0	9.6	237.1	14.8
	Minimum	100.0	101.0	40.0	28.2	6.9	0.8	95.0	8.6
			92.0	40.0	13.8			43.5	7.4
6	Average	107.9	185.1	67.1	64.0	/9	1.8	135.3	13.1
	Maximum	116.0	287.0	81.0	83.9	23.3	9.5	240.0	17.5
	Minimum	102.0	132.0	53.6	36.4	7.7	0.9	98.1	12.0
			127.0	45.0	17.9				9.8
8	Average	111.1	210.3	76.4	65.6	12.9	2.1	143.5	14.2
	Maximum	124.0	326.0	91.4	93.6	28.8	10.0	241.8	20.9
	Minimum	102.0	152.0	64.1	37.7	8.7	1.0	110.0	12.3
			141.0	53.7	22.1			47.6	10.5

[a] Ref. 5.

[b] Grams per 100 L of 100° proof (50%) spirit.

ionization uses a hydrogen flame for the combustion of organic substances to produce electrons and negative ions, which are collected on an anode. The resulting electrical current is proportional to the amount of material burned. β-Ray ionization uses beta particles emitted from a source such as strontium-90 to ionize the carrier gas and its components. The measure of the electrical current resulting from the collection of electrons on the anode is used in the determination or detection of the substance. This may augment sensory evaluations. A positive identification of the separated congeners can be made by infrared and mass spectrometers. Nmr may offer possibilities in the identification of congeners present in alcoholic beverages.

PACKAGING

Because distilled spirits are packaged under federal government supervision and the product is then distributed to the various states in compliance with laws and regulations, many factors must be considered normally not involved in glass packaging. First, the product represents a high value, because it includes the federal tax at the time of bottling. In the United States, a bottle of whisky retailing at $5.44 is taxed at $2.88. Obviously, care must be taken to avoid losses. Second, in addition to dealing with cases, bottles, cartons, closures, and labels, the distiller must apply to each bottle a federal strip stamp indicating that the excise tax has been paid and also apply bottle stamps of the decal type to indicate identification or tax payment in the seven states that require these in their system of control. With the heavy investment in product taxes and the necessity of applying state stamps, the dis-

tiller's ability to build up a substantial case goods inventory to provide immediate service to his customers in those states having this requirement is limited. Third, because this is a licensed industry, in addition to the normal record keeping necessary for efficient operations and control, federal and state records are required.

Federal regulations prescribe and limit the standards of fill (size of containers). On October 1, 1976, the adoption of the metric system permitted distillers to incorporate the new sizes into their operations as circumstances allow. Once the metric size is adopted for a brand, the distiller may not revert to the former standard.

Now, only the metric size will be permitted, such as 1.75 L (59.2 fl oz); 1.00 L (33.8 fl oz); 750 mL (25.4 fl oz); 500 mL (16.9 fl oz); 200 mL (6.8 fl oz); and 50 mL (1.7 fl oz). Individual states likewise limit the number of sizes that can be distributed within their borders and need not adopt all the sizes made available by the federal government.

The most common packaging operation utilized in the industry is the straight-line system with a line speed from 120 to 200 bottles/min. Along with the general progress in packaging, the distilling industry is moving in the direction of automatic-line operation with variable frequency control systems keyed to the fill and labeling operations. This is only possible, however, when volume on certain sizes is substantial. Most distillers are faced with mechanical equipment changes, requiring approximately four hours per line, to handle various sizes of bottles. In addition to material specifications for quality control on all supplies, built-in quality-control inspection systems are included on the bottling lines. Likewise, the quality-control department, independent of bottling operations, takes random samples for evaluation purposes.

Table 6. U.S. Commercial Exports of Liquor by Class and Country of Destination, 1988 (in proof gal)[a,b]

Country of destination	Grand Total	Whiskey			Nonwhiskey						
		Bourbon	Other	Total	Rum	Brandy	Gin	Vodka	Cordials	N.E.S.[c]	Total
Argentina	34,262	0	34,262	34,262	0	0	0	0	0	0	0
Australia	1,880,227	1,495,182	367,801	1,862,983	0	0	0	180	637	16,427	17,244
Austria	71,267	63,728	7,183	70,911	0	0	0	0	0	346	346
Bahamas	130,889	5,506	9,490	14,996	54,295	0	0	0	15,470	46,128	115,593
Bahrain	2,204	1,255	493	1,748	0	0	0	0	0	456	456
Barbados	6,517	0	343	343	0	0	0	1,212	0	4,962	6,174
Belgium	220,961	30,626	32,539	63,165	21,690	0	16,681	71,243	47,636	546	157,796
Belize	15,589	826	0	826	4,930	1,243	0	490	0	8,094	14,763
Benin	1,617	1,617	0	1,617	0	0	0	0	0	0	0
Bermuda	46,082	982	1,169	2,151	0	0	0	35,217	3,926	4,788	43,931
Bolivia	10,354	0	0	0	10,354	0	0	0	0	0	10,354
British Pacific Islands	124	0	124	124	0	0	0	0	0	0	0
British Virgin Islands	87,843	242	1,266	1,510	33,547	3,489	1,075	13,471	237	34,514	56,333
Brazil	12,391	12,916	5,475	18,391	0	0	0	0	0	0	0
Bulgaria	7,418	7,418	0	7,418	0	0	0	0	0	0	0
Canada	1,387,417	143,404	825,530	968,934	28,548	38,759	10,177	6,651	261,050	73,288	418,433
Cayman Islands	40,628	773	970	1,743	1,902	0	228	6,130	1,367	29,258	38,885
Chile	38,491	17,575	0	17,575	12,324	0	0	6,264	2,328	0	20,616
China (Taiwan)	43,647	7,682	15,991	23,573	2,593	0	2,0489	13,463	774	1,196	20,074
Colombia	39,036	0	629	629	0	2,182	1,560	7,917	10,453	16,345	38,457
Costa Rica	3,793	2,501	367	2,368	0	0	0	0	0	925	925
Cyprus	2,543	677	608	1,375	704	0	0	464	0	0	1,168
Czechoslovakia	378	378	0	378	0	0	0	0	0	0	0
Denmark	52,062	34,656	3,739	38,395	0	0	0	13,667	0	0	13,667
Djibouti	118	0	118	118	0	0	0	0	0	0	0
Dominican Republic	58,507	604	12,771	13,375	1,156	5,917	3,277	2,588	6,764	26,430	45,132
Ecuador	57,459	0	6,582	6,582	23,766	199	0	13,253	0	13,657	50,277
Egypt	2,468	176	2,088	2,266	0	0	0	0	0	202	202
El Salvador	20,875	0	9,806	9,806	5,559	878	797	2,489	1,352	0	11,009
FRG	2,161,915	1,902,920	131,900	2,124,820	3,475	7,437	0	4,675	0	11,505	27,093
Finland	16,206	6,039	0	6,039	7,136	481	0	1,109	441	0	9,167
French Indian Ocean Area	272	0	130	130	0	0	0	0	142	0	142
France	531,228	361,684	130,700	492,392	0	0	0	0	3,770	35,000	32,536
French West Indies	24,414	742	756	1,496	8,275	0	0	1,296	2,000	13,345	22,916
Ghana	1,727	1,727	0	1,727	0	0	0	0	0	0	0
Gibralter	174	174	0	174	0	0	0	0	0	0	0
Greece	95,460	55,340	28,973	84,313	3,967	2,861	0	1,529	2,219	571	11,147
Guatemala	4,932	268	3,668	3,936	0	675	0	0	0	321	996
Guinea	3,925	0	0	0	0	0	760	0	0	3,166	3,925
Guyana	802	0	457	457	0	0	0	0	0	345	345
Haiti	26,168	16,693	11,356	28,049	0	0	0	119	0	0	119
Honduras	2,448	525	949	1,474	0	0	0	974	0	0	974
Hong Kong	127,458	30,749	36,238	66,937	0	3,525	5,554	40,059	2,627	3,706	60,471
Iceland	83,926	1,325	1,423	2,748	4,783	301	1,791	70,327	3,911	0	81,178
Indonesia	6,893	0	0	0	0	931	0	5,902	0	0	6,893
Ireland	28,042	923	10,749	11,672	0	0	0	0	4,793	11,577	16,370
Israel	32,547	7,423	1,742	9,185	190	0	760	7,749	4,548	10,135	23,362
Italy	380,956	298,529	73,339	389,868	2,153	0	0	1,020	3,738	4,177	11,088
Jamaica	32,132	1,160	10,982	12.142	3,648	0	1,016	792	1,845	12,689	19,990
Japan	4,384,397	3,586,977	375,296	3,962,273	53,381	12,606	19,458	43,997	8,020	284,662	422,124
Kenya	662	331	331	662	0	0	0	0	0	0	0
Korea, Republic of	23,763	11,353	3,971	15,324	0	0	0	626	311	7,500	8,439
Lebanon	11,173	571	2,660	3,231	3,960	0	732	0	1,902	1,348	7,942
Liberia	2,735	802	1,093	1,895	840	0	0	0	0	0	840
Malaysia	18,830	11,767	3,050	14,817	0	0	0	4,013	0	0	4,013
Malta	1,471	465	685	1,130	0	0	0	0	0	341	341
Mauritania	2,030	0	0	0	0	0	0	0	0	2,030	2,030
Mexico	817,843	26,548	306,140	332,688	189,841	226,782	1,096	1,632	14,230	51,574	485,155
Netherlands Antilles	385,364	54,755	18,053	72,808	115,977	4,251	6,444	22,206	26,794	136,884	312,550
Netherlands	603,668	259,749	120,678	380,427	90,339	24,245	2,587	26,487	27,433	52,100	823,241
New Caledonia	2,327	1,049	736	1,785	0	0	0	0	0	542	542
New Zealand	153,843	118,808	28,877	147,685	2,946	0	0	2,235	975	0	6,159
Nigeria	2,161	2,161	0	2,161	0	0	0	0	0	0	0
Norway	271,482	2,979	14,607	17,586	0	0	2,101	240,355	5,440	0	253,896
Other Pacific Islands	1,262	674	588	1,262	0	0	0	0	0	0	0

Table 6. (*continued*)

Country of destination	Grand Total	Whiskey			Nonwhiskey						
		Bourbon	Other	Total	Rum	Brandy	Gin	Vodka	Cordials	N.E.S.[c]	Total
Pakistan	1,612	667	0	667	0	0	0	0	0	945	945
Panama	124,300	9,963	4,766	14,731	33,226	516	0	20,032	4,683	51,110	109,569
Papua New Guinea	2,074	2,074	0	2,074	0	0	0	0	0	0	0
Paraguay	19,067	11,864	5,578	17,442	0	0	0	0	0	1,625	1,625
Peru	2,905	410	0	410	0	0	0	0	0	2,495	2,495
Philippines	42,704	5,692	5,298	10,990	8,719	8,816	5,360	2,973	0	6,846	31,714
Poland	31,223	1,778	4,539	6,317	6,206	0	16,190	0	0	2,510	24,906
Portugal	7,184	6,076	1,108	7,184	0	0	0	0	0	0	0
Singapore	119,635	29,547	18,581	58,126	0	0	0	14,649	0	56,858	71,507
South Africa	45,760	16,673	14,074	30,747	0	0	0	5,702	1,023	7,683	15,013
Spain	88,519	59,250	29,025	88,275	0	0	0	0	244	0	244
Sri Lanka	1,521	190	1,179	1,359	0	152	0	0	0	0	152
Sweden	227,895	19,228	12,425	31,653	9,014	0	0	132,722	54,506	0	196,242
Switzerland	40,348	25,132	15,003	40,135	0	213	0	0	0	0	213
Thailand	49,674	5,691	9,726	15,417	7,496	0	0	23,625	0	3,136	34,257
Trust Territories of the Pacific	12,463	972	1,162	2,134	1,077	0	0	9,019	0	233	10,329
Trinidad and Tobago	6,880	0	0	0	0	0	0	0	0	6,880	6,880
Tunisia	693	0	0	0	0	0	0	500	193	0	693
Turkey	710	710	0	710	0	0	0	0	0	0	0
Turks Island	2,822	0	0	0	0	164	0	1,121	197	1,340	2,822
United Arab Emirates	5,959	2,183	2,651	4,834	0	0	0	0	0	1,125	1,125
UK	550,726	219,172	228,412	447,584	58,030	8,325	0	13,797	1,925	21,066	103,142
Uruguay	17,391	4,675	1,494	6,159	0	0	0	0	0	11,222	11,222
USSR	489	242	0	242	0	0	0	0	247	0	247
Venezuela	114,626	21,311	65,763	87,074	0	0	0	0	25,860	1,692	27,552
Western Samoa	374	0	0	0	0	0	0	374	0	0	374
Yugoslavia	21,012	1,663	8,876	10,539	0	0	0	0	10,473	0	10,473
Total	*16,055,409*	*9,126,999*	*3,089,213*	*12,216,212*	*814,117*	*360,008*	*99,692*	*902,381*	*567,084*	*1,095,915*	*3,639,197*

[a] Ref. 12.

[b] Proof gallon equals a standard U.S. gallon of 231 in.³ at 60°F containing 50% ethyl alcohol by volume (100° proof).

[c] N.E.S. = not elsewhere specified.

EXPORTS OF DISTILLED SPIRITS

The United States is well known throughout the world for its world-class whiskies and these are exported to many lands. Tastes differ widely in other countries. In some, like Germany and Japan the major export by far is bourbon whisky. In others, like Belgium, it is vodka. In all, bourbon, the prototypical U.S. whisky, is the largest distilled spirit export. Table 6 shows exports to some 92 countries.

GLOSSARY

Balling. A measure of the sugar concentration in a grain mash, expressed in degrees and approximating percent by weight of the sugar in solution.

Bushel. A distillers bushel of any cereal gain is 25.4 kg (56 lb).

Congeners. The flavor constituents in beverage spirits that are responsible for its flavor and aroma and that result from the fermentation, distillation, and maturation processes.

Feints. The third fraction of the distillation cycle derived from the distillation of low wines in a pot still. This term is also used to describe the undesirable constituents of the wash which are removed during the distillation of grain whisky in a continuous patent still (Coffey). These are mostly aldehydes and fusel oils.

Foreshots. The first fraction of the distillation cycle derived from the distillation of low wines in a pot still.

Fusel oil. An inclusive term for heavier, pungent-tasting alcohols produced during fermentation. Fusel oil is composed of approximately 80% amyl alcohols, 15% butyl alcohols, and 5% other alcohols.

Grain whisky. An alcoholic distillate from a fermented wort derived from malted and unmalted barley and maize (corn), in varying proportions, and distilled in a continuous patent still (Coffey).

Heads. A distillate containing a high percentage of low-boiling components such as aldehydes.

High wines. An all-inclusive term for beverage spirit distillates that have undergone complete distillation.

Low wines. The term for the initial product obtained by separating (in a pot still) the beverage spirits and congeners from the wash. Low wines are subjected to at least one more pot still distillation to attain a greater degree of refinement in the malt whisky.

Malt whisky. An alcoholic distillate made from a fermented wort derived from malted barley only, and distilled in pot stills. It is the second fraction (heart of the run) of the distillation process.

Proof. In Canada, the UK, and the United States, the alcoholic concentration of beverage spirits is expressed in terms of proof. The United States statutes define this standard as follows: proof spirit is held to be that alcoholic liquor that contains one-half its volume of alcohol of a specific gravity of 0.7939 at 15.6°C; ie, the figure for proof is always twice the alcoholic content by volume. For example, 100° proof means 50% alcohol by volume. In the UK as well as Canada, proof spirit is such that at 10.6°C weighs exactly twelve-thirteenths of the weight of an equal bulk of distilled water. A proof of 87.7° would indicate an alcohol concentration of 50%. A conversion factor of 1.142 can be used to change British proof to U.S. proof.

Proof gallon. A U.S. gallon of proof spirits or the alcoholic equivalent thereof; ie, a U.S. gallon of 231 in.^{3} (3,785 cm^3) containing 50% of ethyl alcohol by volume. Thus a gallon of liquor at 120° proof is 1.2 proof gal; a gallon at 86° proof is 0.86 proof gal. A British and Canadian proof gallon is an imperial gallon of 277.4 in.^{3} (4,546 cm^3) at 100° proof (57.1% of ethyl alcohol by volume). An imperial gallon is equivalent to 1.2 U.S. gal. To convert British proof gallons to U.S. proof gallons, multiply by the factor 1.37. Since excise taxes are paid on the basis of proof gallons, this term is synonymous with tax gallons.

Single whisky. The whisky, either grain or malt, produced by one particular distillery. Blended Scotch whisky is not a single whisky.

Spirits. Distilled spirits including all whiskies, gin, brandy, rum, cordials, and others made by a distillation process for nonindustrial use.

Tails. A residual alcoholic distillate.

Wash. The liquid obtained by fermenting wort with yeast. It contains the beverage spirits and congeners developed during fermentation.

Wine gallon. Measure of actual volume; U.S. gallon (3.7845118 L) contains 231 in.^{3} (3,785 cm^3); British (Imperial) gallon contains 277.4 in.^{3} (4,546 cm^3).

Wort. The liquid drained off the mash tun and containing the soluble sugars and amino acids derived from the grains.

BIBLIOGRAPHY

1. *Proceedings Before and By Direction of the President Concerning the Meaning of the Term Whisky*, U.S. Government Printing Office, Washington, D.C., 1909.

2. *Business Week* (Feb. 17, 1962).

3. *Business Week* (Mar. 21, 1977).

4. S. Baldwin and co-workers, *Journal of Agricultural and Food Chemistry* **15**, 381 (1967).

5. C. A. Crampton and L. M. Tolman, *Journal of the American Chemical Society* **30**, 97 (1908).

6. A. J. Liebmann and M. Rosenblatt, *Industrial and Engineering Chemistry* **35**, 994 (1943).

7. A. J. Liebmann and B. Scherl, *Industrial and Engineering Chemistry* **41**, 534 (1949).

8. P. Valaer and W. H. Frazier, *Industrial and Engineering Chemistry* **28**, 92 (1936).

9. M. C. Brockmann, *Journal of the Association of Official Agricultural Chemists* **33**, 127 (1950).

10. G. H. Reazin and co-workers, *Journal of the Association of Official Agricultural Chemists* **59**(4), 770 (1976).

11. S. Baldwin and A. A. Andreasen, *Journal of the Association of Official Agricultural Chemists* **53**(4), 940 (1974).

12. Bureau of the Census, U.S. Department of Commerce, *Distilled Spirits Council of the U.S.*, Aug. 1989.

General References

Annual Statistical Revue (1976), Distilled Spirits Council of the US, Inc., Washington, D.C., 1977.

A. Barnard, *The Whiskey Distilleries of the United Kingdom*, London, 1887.

H. Barron, *Distillation of Alcohol*, Joseph E. Seagram & Sons, Inc., New York, 1944.

C. S. Boruff and L. A. Rittschol, *Agric. Food Chem.* **7**, 630 (1959).

Canada, *Its History, Products, and Natural Resources*, Department of Agriculture of Canada, Ottawa, 1906.

A. Cooper, *The Complete Distiller*, London, 1760.

H. G. Crowgey, *Kentucky Bourbon, The Early Years of Whiskey Making*, The University Press of Kentucky, Lexington, 1971.

Distillers Feed Research, Distillers Feed Research Council, Cincinnati, Ohio, 1971.

G. A. DeBecze, "Alcoholic Beverages, Distilled," *Encyclopedia of Industrial Chemical Analysis*, Vol. 4, Interscience Division of John Wiley & Sons, Inc., New York, 1967, pp. 462–494.

The Excise Act, 1934, Department of National Revenue, Ottawa, Canada, 1947, Chap. 52, pp. 24–25.

G. Foth, *Handbuch der Spiritusfabrikation*, Verlag Paul Parey, Hamburg, Germany, 1929.

G. Foth, *Die Praxis des Brennereibetriebs*, Verlag Paul Parey, Hamburg, Germany, 1935.

A. Herman, E. M. Stallings, and H. F. Willkie, "Chemical Engineering Developments in Grain Distillery," *Trans. Am. Inst. Chem. Eng.* **31**(4), 1942.

A. J. Liebmann, "Alcoholic Beverages," in R. E. Kirk and D. F. Othmer, eds., *Encyclopedia of Chemical Technology*, Vol. 1, Interscience Publishers, New York, 1947, pp. 228–303.

A. McDonald, *Whisky*, Glasgow, UK, 1934.

Methods of Analysis, 12th ed., Association of Official Agricultural Chemists, New York, 1975.

G. W. Packowski, "Alcoholic Beverages, Distilled" in A. Standen, ed., *Kirk-Othmer Encyclopedia of Chemical Technology*, Vol. 1, 2nd ed., Interscience Division of John Wiley & Sons, Inc., New York, 1963, pp. 501–531.

G. W. Packowski, "Distilled Beverage Spirits," in M. Grayson, ed., *Kirk-Othmer Encyclopedia of Chemical Technology*, Vol. 3, 3rd ed., Wiley Interscience, New York, 1978, pp. 830–869.

D. R. Peryam, *Ind. Qual. Control* **11**, 17 (1950).

Regulations, Distilleries and Their Products, Circular ED 203, Department of National Revenue, Ottawa, Canada, Mar. 30, 1961.

Regulations under the Federal Alcohol Administration Act, Alcohol, Tobacco Products and Firearms (27 CFR) ATF P 5100.8 (12/77), U.S. Government Printing Office, Washington, D.C.

J. Samuelson, *The History of Drink*, London, 1880.

E. D. Unger, H. F. Willkie, and H. C. Blankmyer, "The Development and Design of a Continuous Cooking and Mashing System for Cereal Grains," *Trans. Am. Inst. Chem. Eng.* **40**(4), 1944.

H. F. Willkie, *Beverage Spirits in America—A Brief History*, The Newcomen Society in North America, Downington, Pa., 1949.

H. F. Willkie and J. A. Proschaska, *Fundamentals of Distillery Practice*, Joseph E. Seagram & Sons, Inc., New York, 1943.

F. B. Wright, *Distillation of Alcohol,* London, 1918.

H. Wustenfeld, *Trinkbranntweine und Liquöre,* Berlin, Germany, 1931.

C. L. Yaws, J. R. Hopper, "Methanol, Ethanol, Propanol and Butanol, Physical and Thermodynamic Properties," *Chemical Engineering* 119 (June 7, 1976).

Joseph L. Owades
Consultant
Sonoma, California

DISTILLATION: TECHNOLOGY AND ENGINEERING

Although the use of distillation dates back in recorded history to about 50 BC the first truly industrial exploitation of this separation process did not occur until the 12th century, when it was used in the production of alcoholic beverages. By the 16th century, distillation also was being used in the manufacture of vinegar, perfumes, oils, and other such products.

As recently as 200 years ago, distillation stills were small, of the batch type, and usually operated with little or no reflux. With experience, however, came new developments. Tray columns appeared on the scene in the 1820s, along with feed preheating and the use of internal reflux. By the latter part of that century, considerable progress had been made by Hausbrand in Germany and Sorel in France, who developed mathematical relations that turned distillation from an art into a well-defined technology.

Distillation today is a widely used operation in the petroleum, chemical petrochemical, beverage, and pharmaceutical industries. It is important not only for the development of new products but also in many instances for the recovery and reuse of volatile liquids. Pharmaceutical manufacturers, for example, use large quantities of solvents, most of which can be recovered by distillation with a substantial savings in cost and a reduction in pollution.

Although one of the most important unit operations, distillation unfortunately is also one of the most energy-intensive operations. It easily is the largest consumer of energy in petroleum and petrochemical processing and as such, must be approached with conservation in mind. It is a specialized technology in which the correct design of equipment is not always a simple task.

DISTILLATION TERMINOLOGY

To provide a better understanding of the distillation process, the following briefly explains the terminology most often encountered.

Solvent Recovery

The term solvent recovery often has been a somewhat vague label applied to the many and very different ways in which solvents can be reclaimed by industry.

One approach when an impure solvent contains both soluble and insoluble particles is to evaporate the solvent from the solids. This requires the use of a small forced-circulation-type evaporator that combines a heat exchanger, external separator, and vacuum system with a special orifice that causes back pressure in the exchanger and arrests vaporization until the liquid flashes into the separator. Although this will recover a solvent, it will not separate solvents if two or more are present.

A further technique is available to handle an airstream that carries solvents. By chilling the air by means of vent condensers or refrigeration equipment, the solvents can be removed from the condenser. Solvents also can be recovered by using extraction, adsorption, absorption, and distillation methods.

Solvent Extraction

Essentially a liquid/liquid process where one liquid is used to extract another from a secondary stream, solvent extraction generally is performed in a column somewhat similar to a normal distillation column. The primary difference is that the process involves two liquids instead of liquid and vapor.

During the process, the lighter (ie, less dense) liquid is charged to the base of the column and rises through packing or trays while the more dense liquid descends. Mass transfer occurs, and a component is extracted from one stream and passed to the other. Liquid/liquid extraction sometimes is used when the breaking of an azeotrope is difficult or impossible by distillation techniques.

Carbon Adsorption

The carbon adsorption technique is used primarily to recover solvents from dilute air or gas streams. In principle, a solvent-ladened airstream is passed over activated carbon and the solvent is adsorbed into the carbon bed. When the bed becomes saturated, steam is used to desorb the solvent and carry it to a condenser. In such cases as toluene, for example, recovery of the solvent can be achieved simply by decanting the water/solvent two-phase mixture that forms in the condensate. Carbon adsorption beds normally are used in pairs so that the airflow can be diverted to the secondary bed when required.

On occasion, the condensate is in the form of a moderately dilute miscible mixture. The solvent then has to be recovered by distillation. This would apply especially to ethyl alcohol, acetone-type solvents.

Absorption

When carbon adsorption cannot be used because certain solvents either poison the activated carbon bed or create so much heat that the bed can ignite, absorption is an alternative technique. Solvent is recovered by pumping the solvent-ladened airstream through a column countercurrently to a water stream that absorbs the solvent. The air from the top of the column essentially is solvent-free whereas the dilute water/solvent stream discharged from the column bottom usually is concentrated in a distillation column. Absorption also can be applied in cases where an oil rather than water is used to absorb certain organic solvents from an airstream.

Azeotropes

During distillation, some components form an azeotrope at a certain stage of the fractionation and require a third component to break the azeotrope and achieve a higher percentage of concentration. In the case of ethyl alcohol and water, for example, a boiling mixture containing less than 96% by weight ethyl alcohol produces a vapor richer in alcohol than in water and is readily distilled. At the 96% by weight point, however, the ethyl alcohol composition in the vapor remains constant, ie, the same composition as the boiling liquid. This is known as the azeotrope composition. Further concentration requires use of a process known as azeotropic distillation. Other common fluid mixtures that form azeotropes are formic acid/water, isopropyl alcohol/water, and isobutanol/water.

Azeotropic Distillation

In a typical azeotropic distillation procedure, a third component such as benzene, isopropyl ether, or cyclohexane is added to an azeotropic mixture such as ethyl alcohol/water to form a ternary azeotrope. Because the ternary azeotrope is richer in water than the binary ethyl alcohol/water azeotrope, water is carried over the top of the column. The azeotrope, when condensed, forms two phases. The organic phase is refluxed to the column whereas the aqueous phase is discharged to a third column for recovery of the entraining agent.

Certain azeotropes such as the *n*-butanol/water mixture can be separated in a two-column system without the use of a third component. When condensed and decanted, this type of azeotrope forms two phases. The organic phase is fed back to the primary column, and the butanol is recovered from the bottom of the still. The aqueous phase, meanwhile, is charged to the second column, with the wa-

ter being taken from the column bottom. The vapor from the top of both columns is condensed, and the condensate is run to a common decanter (Fig. 1).

Extractive Distillation

This technique is somewhat similar to azeotropic distillation in that it is designed to perform the same type of task. In azeotropic distillation, the azeotrope is broken by carrying over a ternary azeotrope at the top of the column. In extractive distillation, a very high boiling compound is added and the solvent is removed at the base of the column.

Stripping

In distillation terminology, stripping refers to the recovery of a volatile component from a less volatile substance. Again, referring to the ethyl alcohol/water system, stripping is done in the first column below the feed point where the alcohol enters at about 10% by weight and the resulting liquid from the column base contains less than 0.02% alcohol by weight. This is known as the stripping section of the column. This technique does not increase the concentration of the more volatile component but rather, decreases its concentration in the less volatile component.

A stripping column also can be used when a liquid such as water contaminated by toluene cannot be discharged to a sewer. For this pure stripping duty, the toluene is removed within the column while vapor from the top is decanted for residual toluene recovery and refluxing of the aqueous phase.

Rectification

For rectification or concentration of the more volatile component, a top section of column above the feed point is

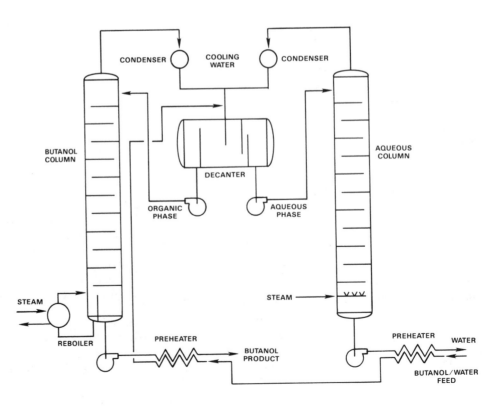

Figure 1. System for recovering butanol from butanol/water mixture.

required. By means of a series of trays and reflux back to the top of the column, a solvent such as ethyl alcohol can be concentrated to more than 95% by weight.

Batch Distillation

When particularly complex or small operations require recovery of the more volatile component, there are batch-distillation systems of various capacities. Essentially a rectification-type process, batch distillation involves pumping a batch of liquid feed into a tank where boiling occurs. Vapor rising through a column above the tank combines with reflux coming down the column to effect concentration. This approach is not very effective for purifying the less-volatile component.

For many applications, batch distillation requires considerable operator intervention or, alternatively, a significant amount of control instrumentation. Although it is more energy intensive than a continuous system, steam costs generally are less significant on a small operation. Furthermore, it is highly flexible and a single batch column can be used to recover many different solvents.

Continuous Distillation

The most common form of distillation used by the chemical, petroleum, and petrochemical industries is the continuous-mode system. In continuous distillation, feed constantly is charged to the column at a point between the top and bottom trays. The section above the feed point rectifies the more volatile component while the column section below the feed point strips out the more volatile component from the less volatile liquid. In order to separate N components with continuous distillation, a minimum of $N - 1$ distillation columns is required.

Turndown

The turndown ratio of a column is an indication of the operating flexibility. If a column, for example, has a turndown ratio of 3, it means that the column can be operated efficiently at 33% of the design maximum throughput.

SYSTEM COMPONENTS: COLUMNS, INTERNALS, INSTRUMENTATION, AND AUXILIARY EQUIPMENT

The following briefly defines the many components required for a distillation system and the many variations in components that are available to meet different process conditions.

Column Shells

A distillation column shell can be designed for use as a free-standing module or for installation within a supporting steel structure. Generally speaking, unless a column is of very small diameter, a self-supporting column is more economical. This holds true even under extreme seismic 3 conditions.

There are distillation columns built of carbon steel, 304 stainless steel, 316 stainless steel, Monel, titanium, and Incoloy 825. Usually, it is more economical to fabricate columns in a single piece without shell flanges. This technique not only simplifies installation but also eliminates danger of leakage during operation. Columns more than 80 feet long have been shipped by road without transit problems.

Although columns of more than 3-ft diameter normally have been transported without trays to prevent dislodgement and possible damage, recent and more economical techniques have been devised for factory installation of trays with the tray manways omitted. After the column has been erected, manways are added and, at the same time, the fitter inspects each tray.

With packed columns of 20-in. diameter or less that use high-efficiency metal mesh packing, the packing can be installed before shipment. Job-site packing, however, is the norm for larger columns. This prevents packing from bedding down during transit and leaving voids that would reduce operating efficiency. Random packing always is installed after delivery except for those rare occasions when a column can be shipped in a vertical position. Access platforms and interconnecting ladders designed to Occupational Safety and Health Administration (OSHA) standards also are supplied for on-site attachment to free-standing columns.

Installation usually is simple because columns are fitted with lifting lugs and, at the fabrication stage, a template is drilled to match support holes in the column base ring. With these exact template dimensions, supporting bolts can be preset for quick and accurate coupling as the column is lowered into place.

Column Internals

During recent years, the development of sophisticated computer programs and new materials has led to many innovations in the design of trays and packings for more efficient operation of distillation columns. In designing systems for chemical, petroleum, and petrochemical use, full advantage can be taken of available internals to assure optimum distillation performance.

Tray Devices

Although there are perhaps five basic distillation trays suitable for industrial use, there are many design variations of differing degrees of importance and a confusing array of trade names applied to their products by tray manufacturers. The most modern and commonly used devices are sieve, dual-flow, valve, bubble-cap, and baffle trays—each with its advantages and preferred usage. Of these, the sieve- and valve-type trays currently are most often specified.

For a better understanding of tray design, Figure 2 defines and locates typical tray components. The material of construction usually is 14 gauge, with modern trays adopting the integral truss design, which simplifies fabrication. A typical truss tray is shown in Figure 3. For columns less than 3 ft in diameter, it is not possible to assemble the truss trays in the column. Trays therefore must be preassembled on rods into a cartridge section for loading into the column. Figure 4 shows this arrangement in scale-model size.

The hydraulic design of a tray is a most important factor. The upper operating limit generally is governed by

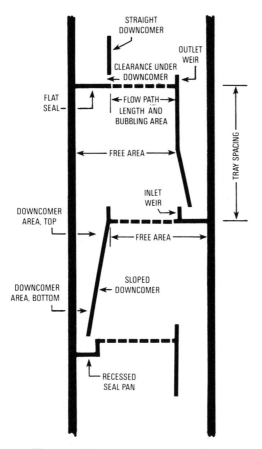

Figure 2. Tray component terminology.

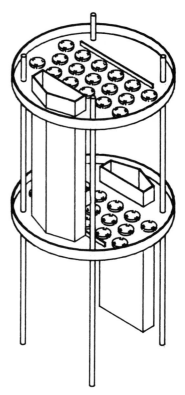

Figure 4. Cartridge tray assembly.

the flood point, although, in some cases, entrainment also can restrict performance. By forcing some liquid to flow back up the column, entrainment reduces concentration gradients and lowers efficiency. A column also can flood by down-comer backup when tray design provides insufficient down-comer area or when the pressure drop across the tray is high. When the down-comer is unable to handle all the liquid involved, the trays start to fill and pressures increase. This also can occur when a highly foaming liquid is involved. Flooding associated with high tray pressure drops and small tray spacing takes place when the re-

quired liquid seal is higher than the tray spacing. Down-comer design also is particularly important at high operating pressures due to a reduction in the difference between vapor and liquid densities.

The lower limit of tray operation, meanwhile, is influenced by the amount of liquid weeping from one tray to the next. Unlike the upward force of entrainment, weeping liquid flows in the normal direction and considerable amounts can be tolerated before column efficiency is significantly affected. As the vapor rate decreases, however, a point eventually is reached when all the liquid is weeping and there is no liquid seal on the tray. This is known as the dump point, below which there is a severe drop in efficiency.

Sieve Tray. The sieve tray is a low-cost device that consists of a perforated plate that usually has holes of $\frac{3}{16}$ in. to 1 in. diameter, a down-comer, and an outlet weir. Although inexpensive, a correctly designed sieve tray can be comparable to other styles in vapor and liquid capacities, pressure drop, and efficiency. For flexibility, however, it is inferior to valve and bubble-cap trays and sometimes is unacceptable for low liquid loads when weeping has to be minimized.

Depending on process conditions and allowable pressure drop, the turndown ratio of a sieve tray can vary from 1.5 to 3 and occasionally higher. Ratios of 5, as sometimes claimed, can be achieved only when the tray spacing is large, available pressure drop is very high, liquid loadings are high, and the system is nonfoaming. For many applications, a turndown of 1.5 is acceptable.

It also is possible to increase the flexibility of a sieve

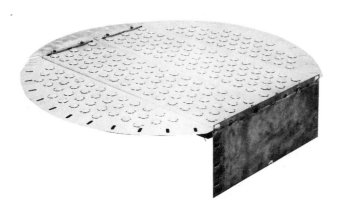

Figure 3. Typical tray of integral truss design.

tray for occasional low throughput operation by maintaining a high reboil and increasing the reflux ratio. This may be economically desirable when the low throughput occurs for a small fraction of the operating time. Flexibility, likewise, can be increased by the use of blanking plates to reduce the hole area. This is particularly useful for initial operation when it is proposed to increase the plant capacity after a few years. There is no evidence to suggest that blanked-off plates have inferior performance to unblanked plates of similar hole area.

Dual-Flow Tray. The dual-flow tray is a high hole area sieve tray without a down-comer. The liquid passes down through the same holes through which the vapor rises. Because no down-comer is used, the cost of the tray is lower than that of a conventional sieve tray.

In recent years, use of the dual-flow tray has declined somewhat because of difficulties experienced with partial liquid/vapor bypassing of the two phases, particularly in larger diameter columns. The dual-flow column also has a very restricted operating range and a reduced efficiency because there is no cross flow of liquid.

Valve Tray. Although the valve tray dates back to the rivet type first used in 1922, many design improvements and innumerable valve types have been introduced in recent years. A selection of modern valves as illustrated provide the following advantages (Fig. 5):

1. Throughputs and efficiencies can be as high as sieve or bubble-cap trays.
2. Very high flexibility can be achieved and turndown ratios of 4 to 1 are easily obtained without having to resort to large pressure drops at the high end of the operating range.
3. Special valve designs with venturi-shaped orifices are available for duties involving low pressure drops.
4. Although slightly more expensive than sieve trays, the valve tray is very economical in view of its numerous advantages.
5. Because an operating valve is continuously in movement, the valve tray can be used for light-to-moderate fouling duties. Valve trays can be successfully used on brewery effluent containing waste beer, yeast, and other materials with fouling tendencies.

Bubble-Cap Tray. Although many bubble-cap columns still are in operation, bubble-cap trays rarely are specified today because of high cost factors and the excellent performance of the modern valve-type tray. The bubble cap, however, does have a good turndown ratio and is good for low liquid loads.

Baffle Tray. Baffle trays are arranged in a tower in such a manner that the liquid flows down the column by splashing from one baffle to the next lower baffle. The ascending gas or vapor, meanwhile, passes through this curtain of liquid spray.

Although the baffle-type tray has a low efficiency, it can be useful in applications where the liquid contains a high fraction of solids.

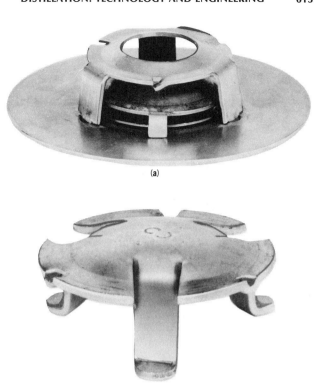

(a)

(b)

(c)

Figure 5. (a) Special two-stage valve with lightweight orifice cover for complete closing; (b) and (c) two typical general purpose valves which may be used in all types of services.

Packings

For many types of duties, particularly those involving small-diameter columns, packing is the most economical tower internal. One advantage is that most packing can be purchased from stock on a cubic-foot basis. In addition, the mechanical design and fabrication of a packed column is quite simple. Disadvantages of packing include its unsuitability for fouling duties, breakage of ceramic packing, and, according to some reports, less predictive performance, particularly at low liquid loads or high column diameters.

The most widely used packing is the random packing, usually Rashig rings, Pall rings, and ceramic saddles. These are available in various plastics, a number of different metals and, with the exception of Pall rings, in ceramic materials. Although packings in plastic have the advantage of corrosion resistance, the self-wetting ability of some plastic packing, such as fluorocarbon polymers,

Figure 6. Segment of high efficiency metal mesh packing.

sometimes is poor, particularly in aqueous systems. This considerably increases the HETP when compared with equivalent ceramic rings.

High-efficiency metal mesh packing as shown in Figure 6 has found increasing favor in industry during recent years. One type uses a woven wire mesh that becomes self-wetting because of capillary forces. This helps establish good liquid distribution as the liquid flows through the packing in a zig-zag pattern.

If properly used, high efficiency packings can provide HETP values in the range 6–12 in. This can reduce column heights, especially when a large number of trays is required. Such packings, however, are very expensive, and each application must be studied in great detail.

With random and, in particular, high-efficiency packing, considerable attention must be given to correct liquid distribution. Certain types of high-efficiency packing are extremely sensitive to liquid distribution and should not be used in columns more than 2 ft in diameter. Positioning of these devices and the design of liquid distribution and redistribution are important factors that should be determined only by experts.

Instrumentation

One of the most important aspects of any distillation system is the ability to maintain the correct compositions from the columns by means of proper controls and instrumentation. Although manual controls can be supplied, this approach rarely is used today in the United States.

Manual control involves the extensive use of rotameters and thermometers, which, in turn, involves high labor costs, possible energy wastage, and, at times, poor quality control. Far better control is obtained through the use of pneumatic or electronic control systems.

Pneumatic Control Systems. The most common form of distillation column instrumentation is the pneumatic-type analogue control system. Pneumatic instruments have the advantage of being less expensive than other types and, because there are no electrical signals required, there is no risk of an electrical spark. One disadvantage is the need to ensure that the air supply has a very low dew point (usually −40°F) to prevent condensation in the loops.

Electronic Control Systems. Essentially, there are three types of electronic control systems:

1. Conventional electronic instruments.
2. Electronic systems with all field devices explosion proof.
3. Intrinsically safe electronic systems.

The need to have a clear understanding of the differences is important. Most distillation duties involve at least one flammable liquid that is being processed in both the vapor and liquid phases. Because there always is the possibility of a leak of liquid or vapor, particularly from pump seals, it is essential for complete safety that there be no source of ignition in the vicinity of the equipment. Although many instruments, such as controllers and alarms, can be located in a control room removed from the process, all local electronic instruments must be either explosion proof or intrinsically safe.

With explosion-proof equipment, electrical devices and wiring are protected by boxes or conduit that will contain any explosion that may occur. With intrinsically safe equipment, barriers limit the transmission of electrical energy to such a low level that it is not possible to generate a spark. As explosion-proof boxes and conduits are not required, wiring costs are reduced.

For any intrinsically safe system to be accepted for insurance purposes, FM (Factory Mutual) or CSA (Canadian Standards Association) approval usually must be obtained. This approval applies to a combination of barriers and field devices. Therefore, when a loop incorporates such instruments from different manufacturers, it is essential to ensure that approval has been obtained for the combination of instruments.

Auxiliary Equipment

In any distillation system, the design of auxiliary equipment such as the reboiler, condenser, preheaters, and product coolers is as important as the design of the column itself.

Reboiler. Although there are many types of reboilers, the shell-and-tube thermosyphon reboiler is used most frequently. Boiling within the vertical tubes of the ex-

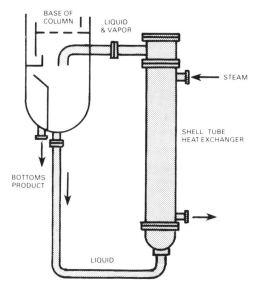

Figure 7. Typical shell and tube thermosyphon reboiler arrangement.

changer produces liquid circulation and eliminates the need for a pump. A typical arrangement is shown in Figure 7.

For certain duties, particularly when the bottoms liquid has a tendency to foul heat transfer surfaces, it is desirable to pump the liquid around the heat exchanger. Because boiling can be suppressed by use of an orifice plate at the outlet of the unit, fouling is reduced. The liquid being pumped is heated under pressure and then is flashed into the base of the column where vapor is generated.

An alternative approach is the use of a plate heat exchanger as a forced-circulation reboiler. With this technique, the very high liquid turbulent flow that is induced within the heat exchanger through the use of multiple corrugated plates holds fouling to a minimum. Meanwhile, the superior rates of heat transfer that are achieved reduce the surface area required for the reboiler.

Condensers. Because most distillation column condensers are of shell-and-tube design, the processor has the option of condensing on either the shell or tube side. From the process point of view, condensation on the shell side is preferred because there is less subcooling of condensate and a lower pressure drop is required. These are important factors in vacuum duties. Furthermore, with cooling water on the tube side, any fouling can be removed more easily.

Tube-side condensation, on the other hand, can be more advantageous whenever process fluid characteristics dictate the use of more expensive, exotic materials. Capital cost of the unit then may be cut by using a carbon steel shell.

Preheaters/Coolers. The degree to which fluids are aggressive to metals and gasketing materials generally determines the selection of plate or shell-and-tube preheaters and product coolers. If fluids are not overly aggressive toward gasket materials, a plate heat exchanger is an

extremely efficient preheater because a very close temperature approach may be achieved. Added economy is realized by using heat from the top and bottoms product for all necessary preheating.

Very aggressive duties normally are handled in a number of tubular exchangers arranged in series to generate a good mean temperature difference. The use of multiple tubular units obviously is more expensive than a single plate heat exchanger but is unavoidable for certain solutions such as aromatic compounds.

Vent Condenser. It is normal practice on distillation systems to use a vent condenser after the main condenser to minimize the amount of volatiles being driven off into the atmosphere. Usually of the shell-and-tube type, the vent condenser will have about one-tenth the area of the main unit and will use a chilled water supply to cool the noncondensible gases to about 45–50°F.

Pumps. Because most distillation duties involve fluids that are highly flammable and have a low flash point, it is essential that explosion-proof (Class 1, Group D, Division 1) pump motors be supplied. Centrifugal pumps generally are specified as they are reliable and can provide the necessary head and volumetric capacity at moderate costs.

PACKAGED DISTILLATION SYSTEMS

For distillation systems of moderate size, it often can be economical to fabricate and supply columns, heat exchangers, tanks, pumps, and other elements as a fully preassembled package. This technique was used for the distillation unit as pictured in Figure 8 during fabrication as a complete, ready to transport system. For this particular project, which was designed to separate both ethyl alcohol and isopropyl alcohol from water, three columns of relatively small diameter were positioned with associated components and instrumentation on a prefabricated structure. Although insulation normally would not be supplied in order to preclude the possibility of damage during shipment, major elements of this project were insulated at the customer's request and were trucked 600 miles to the job site without problems. Final size of the package was 65′ × 12′ × 8′.

The packaged approach proved to be beneficial in a number of ways. Factory fabrication and installation of piping was far more economical than field finishing, while erection time for the system was reduced considerably with significant savings in local labor costs.

An alternative approach for larger systems is the fabrication of equipment modules. For one large batch-distillation unit, the column itself can be supplied as one module, the batch tank and reboiler as the second, and the heat exchangers, decanter, pumps, and auxiliary items as the third. This modular construction before shipment was successful despite the fact that the batch tank was more than 8 ft in diameter.

For systems involving columns in excess of 6-ft diameter, prepackaging generally is not feasible. Components of such systems normally have to be installed and piped at the job site.

Figure 8. Packaged distillation system during fabrication.

BATCH DISTILLATION

With process plants becoming increasingly larger, there has been a tendency for the chemical and petroleum industries to focus attention on the use of continuous distillation because this approach becomes progressively more economical as the scale of operation grows. As a result, batch distillation has become a somewhat neglected unit operation. There are, however, many areas where batch distillation can be used more economically than continuous. Batch sizes are available from about 300 to more than 5000 gal. A 300 gal unit is shown in Figure 9.

Advantages

The main advantage of batch distillation is its flexibility. A single unit can, by changing reflux ratios and boil-up rates, be used for many different systems. It also is possible to separate more than two components in the same column, whereas with continuous distillation, at least $N - 1$ separate columns are required to separate N components.

A batch distillation process is very simple to control, and there is no need to balance the feed and draw-off, as is the case with the continuous distillation approach.

Batch distillation is particularly useful when applied to feeds containing residues that have a tendency to foul surfaces as these residues remain in the still and cannot contaminate the rectification column internals. For aqueous systems-handling materials that leave very heavy fouling, it is possible to heat with direct steam injection into the still and thus alleviate problems of buildup on the heat transfer surfaces.

COLUMN DESIGN

Before designing a batch column, it obviously is desirable to have as much detailed information on the system as is possible. Data on vapor liquid equilibria (VLE), vapor and liquid densities, liquid viscosity, and the boiling temperatures of the components are essential if a column is to be properly designed. Failure to have accurate VLE data means that it is necessary to run a small-scale experiment in order to characterize the system. In addition, the customer must identify the product feed composition, the required composition of the residue and distillate, the batch size, and the batch time. Having acquired this data, the vendor finally can commence design work.

To illustrate the design principles, it is useful to study the case of binary mixtures. A typical system is shown in Figure 10. Once the VLE data have been established, it becomes a straightforward task to calculate the number of theoretical stages and reflux ratios.

There are two main techniques for operating a batch column. One is to work with constant reflux ratio during the complete run. The effect of this method is charted in Figure 11. As the composition of the more volatile component (MVC) in the still, x_w, decreases, the fraction of MVC in the top product decreases. For example, to obtain a set composition of 90% in the total amount of top product collected, it will always be necessary to collect initially at a higher composition of about 95% to compensate for a composition below specification at the end of the run. The advantage of constant reflux is that control and operation are very simple.

The second method is to increase the reflux ratio during the run in order to maintain a steady top-product com-

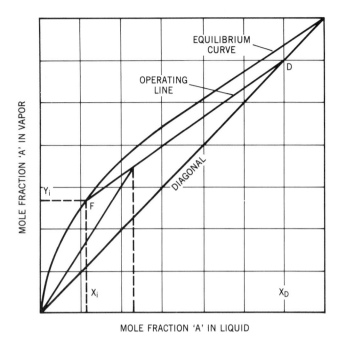

Figure 10. A typical system illustrating the design principles.

The reflux ratio R is given by

$$R = \frac{L/V}{1 - (L/V)}$$

With the VLE data and the top and bottom compositions, it then is possible to calculate graphically by the McCabe and Thiele procedure the minimum reflux ratio, minimum number of theoretical stages, and other such

Figure 9. Batch distillation/solvent recovery silhoutte mask.

position. This is shown in Figure 12, where the increase in the slope of the operating line is obtained by increasing the reflux ratio. The gradient of the operating line (L/V) is obtained from the enrichment equation given below.

$$Y_n = \frac{L}{V} x_{n+1} + \frac{D}{V_n} x_d$$

This equation assumes constant molal overflow

Y_n is the mole fraction of the MVC in the vapor leaving the nth stage

x_{n+1} is the mole fraction of the MVC in the liquid arriving at the nth stage

x_d is the mole fraction of the MVC in the top product

D is the molal flow rate of top product

V is the molal vapor flow rate in the column

L is the molal liquid flow rate in the column

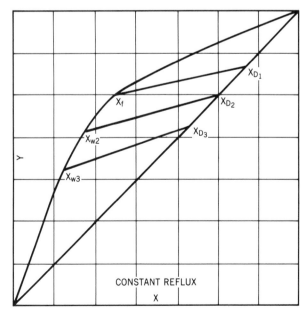

Figure 11. A chart illustrates the effect of the method of working with constant reflux ratio during the complete run. One of two main techniques for operating a batch column.

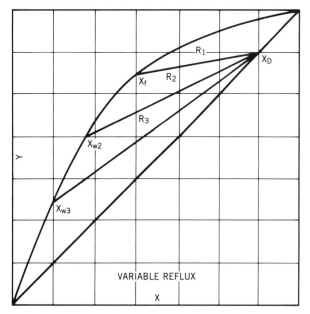

Figure 12. The second method is to increase the reflux ratio during the run in order to maintain a steady top product composition.

parameters. In batch distillation in particular, these procedures are tedious and time-consuming because, of course, the composition of the liquid in the still changes with time and it is necessary to repeat the calculations many times. Obviously, the procedures become even more time-consuming with multicomponent systems.

For this type of design work, computer programs can be used to enable the engineer to produce an efficient design very quickly. One program is extremely flexible and operates in a different series of modes, as is charted in Table 1. Naturally, the VLE data have to be specified for all modes.

The program further assists in the determination of the number of theoretical stages. Other modes also are available to provide different permutations of operation.

To help determine the sizing of the column diameter, various programs can be used to incorporate different proprietary methods, including those of Fractionation Research Inc. (FRI) of California.

Column Internals

The choice of internals for the column depends mainly on the product being processed and the size of the column to be used. To meet virtually every parameter, packed columns as well as sieve, bubble, and ballast trays are available. As a general rule, sieve trays are not used frequently in batch columns because the turndown ratio of most trays of this type is only about 1.5 to 1. This reduces one of the main advantages of batch columns, mainly flexibility. Usually, small batch columns are packed because the efficiency of trays of less than 2-ft diameter often decreases rapidly. Ballast trays, although expensive, are often used for larger columns because they are efficient and have turndown ratios of up to 9 to 1.

Control

The control of batch columns is very simple, and, therefore, required instrumentation usually is quite inexpensive. Constant reflux operation can be handled by a ratio controller or by using a timed deflection of condensate. For operation with variable reflux, the control of the reflux ratio must be tied to some property of the top product that undergoes a sufficiently large change in value for change in composition. Manual control of this type of reflux operation is not feasible.

Numerous batch distillation systems have been supplied to customers who require the separation of many different components. To simplify the operation of these

Table 1. Computer Program Modes

Mode	Specified	Program Calculation
1	Top-product and still composition	Minimum number of theoretical stages Minimum reflux ratio
2	Top-product still composition, reflux ratio	Number of theoretical stages
3	Top-product and still composition Number of theoretical stages	Reflux ratio
4	Top-product composition required Initial and final still composition Still charge Batch time Number of theoretical stages	Reflux ratio at start and end of batch Boiling rate Product quality
5	Average top-product composition Initial and final still composition Still charge Batch time Number of theoretical stages	Reflux ratio required constant throughout batch Boil-up rate Product quality

systems, there are proprietary computers specifically designed for process control. For example, use of certain microprocessors enables the operator to switch from one type of separation to another without having to make many manual adjustments on the control panel. In addition, the computer can be programmed to automatically take care of all necessary tails cuts as well as to clean the system during interims between the various separations.

Batch Systems/Continuous Distillation

In a number of duties, it is essential that the base product contain only very small amounts of one component. For example, in the recovery of solvents from water, it is mandated that the water may retain only trace quantities of solvents before being pumped to waste. Because one disadvantage of a batch system is that there is no stripping section, it is necessary to boil the batch pot for relatively long periods of time in order to reduce the residual solvent to trace quantities.

To resolve this problem, a number of batch distillation systems with the capability of being operated in the continuous mode have been supplied. By incorporating one, two, or three feed points on the column, the feed can be pumped directly to the column instead of to the batch tank. Furthermore, a separate small holdup/reboiler is furnished in certain cases so that the batch tank and its associated reboiler need not be used for the continuous operation.

BIBLIOGRAPHY

This article was adapted from and is used with permission of A. Cooper, *APV Distillation Handbook,* 3rd ed., Lake Mills, Wisc., 1987. Copyrighted APV Crepaco, Inc.

APV Crepaco, Inc.
Lake Mills, Wisconsin

DRYERS: TECHNOLOGY AND ENGINEERING

Throughout the food processing industries, there are many and varied requirements for thermal drying. Some involve the removal of moisture or volatiles from various food ingredients or products that differ in both chemical and physical characteristics. Others involve the drying of solutions or liquid suspensions and different approaches to the problem. To assist manufacturers in arriving at a reasonably accurate first assessment of the type, size, and cost of element for a particular duty, this article describes the most widely used types of both batch and continuous dryers in the food industries and gives an indication of approximate sizes and capital costs for typical installations.

Three basic methods of heat transfer are used in industrial dryers in varying degrees of prominence and combinations, specifically, convection, conduction, and radiation.

In the chemical processing industry, the majority of dryers employ forced convection and continuous operation. With the exception of the indirectly heated rotary dryer and the film drum dryer, units in which heat is transferred by conduction are suitable only for batch use. This limitation effectively restricts them to applications involving somewhat modest production runs.

Radiant, or "infrared," heating is rarely used in drying materials such as fine chemicals or pigments. Its main application is in such operations as the drying of surface coatings on large plane surfaces since for efficient utilization, it generally is true that the material being irradiated must have a sight of the heat source or emitter. There is, however, in all the dryers considered here a radiant component in the heat-transfer mechanism.

Direct heating is used extensively in industrial drying equipment where much higher thermal efficiencies are exhibited than with indirectly heated dryers. This is because there are no heat exchanger losses and the maximum heat release from the fuel is available for the process. However, this method is not always acceptable, especially where product contamination cannot be tolerated. In such cases, indirect heating must be used.

With forced-convection equipment, indirect heating employs a condensing vapor such as steam in an extended surface tubular heat exchanger or in a steam jacket where conduction is the method of heat transfer. Alternative systems that employ proprietary heat-transfer fluids also can be used. These enjoy the advantage of obtaining elevated temperatures without the need for high-pressure operation as may be required with conventional steam heating. This may be reflected in the design and manufacturing cost of the dryer. Furthermore, in addition to the methods listed above, oil- or gas-fired indirect heat exchangers also can be used.

In general, dryers are either suitable for batch or continuous operation. A number of the more common types are listed in Table 1, where an application rating based on practical considerations is given. In the following review, some of the factors likely to influence selection of the various types are discussed for particular applications.

BATCH DRYERS

It will be apparent that batch operated equipment usually is related to small production runs or to operations requiring great flexibility. As a result, the batch-type forced-convection unit certainly finds the widest possible application of any dryer used today.

The majority of designs employ recirculatory air systems incorporating large-volume, low-pressure fans that, with the use of properly insulated enclosures, usually provide thermal efficiencies in the region of 50–60%. However, in special applications of this type of dryer that call for total air rejection, this figure is somewhat lower and is largely related to the volume and temperature of the exhaust air. Capital investment is relatively low, as are in-

Table 1. Product Classification and Dryer Types as an Aid to Selection

	Evaporation Rate, lb/ft²·h Mean Rate = E_{av}	Fluids, Liquid Suspension	Pastes, Dewatered Cake	Powders	Granules, Pellets, Extrudates	Operation
Forced convection (cross-airflow)	0.15–0.25 $E_{av} = 0.2$	Poor	Fair	Fair	Good	Batch
Forced convection (throughflow)	1.0–2.0 $E_{av} = 1.5$	—	—	—	Good	Batch
Agitated pan (sub-atmospheric)	1.0–5.0 $E_{av} = 3.0$	Fair	Fair	Fair	Poor	Batch
Agitated pan (atmospheric)	1.0–5.0 $E_{av} = 3.0$	Fair	Fair	Fair	Poor	Batch
Double-cone tumbler (sub atmospheric)	1.0–3.0 $E_{av} = 2.0$	—	Poor	Fair	Poor	Batch
Fluidized bed (throughflow)	2–50 $E_{av} = 26$	—	—	Good	Good	Continuous
Conveyor band (throughflow)	2.0–10.0 $E_{av} = 6.0$	—	Fair	—	Good	Continuous
Rotary (indirect)	1.0–3.0[a] $E_{av} = 2.0$	—	Poor	Good	Fair	Continuous
Rotary (direct)	2.0–6.0[a] $E_{av} = 4.0$	—	Fair	Fair	Good	Continuous
Film drum (atmospheric)	3.0–6.0 $E_{av} = 4.5$	Good	Fair	—	—	Continuous
Pulumatic or flash	50–250 $E_{av} = 150$	—	Fair	Good	Fair	Continuous
Spray	7.0–33.0 $E_{av} = 20.0$	Good	—	—	—	Continuous

[a] *Note:* Evaporation rates for rotary dryers are expressed in lb/ft³·h.

stallation costs. Furthermore, by using the fan systems, both power requirements and operating costs also are minimal. Against these advantages, labor costs can be high.

In such a plant, the drying cycles are extended, with 24–45 hours being quite common in certain cases. This is a direct result of the low evaporative rate, which normally is in the region of 0.15–0.25 lb/ft²·h.

Following the recent trend and interest shown in preforming feedstock and in particular with regard to the design of extruding and tray-filling equipment for dewatered cakes, it is now possible to obtain the maximum

benefit of enhanced evaporative rates by using through-air circulation dryers when handling preformed materials. Figure 1 shows how a high-performance dryer can product 1950 pounds of dried material in a 24-h period at a terminal figure of 0.5% moisture when handling a preformed filter cake having an initial moisture content of 58%. The very great improvement in performance can readily be seen from the curve, in which it is clear that the corresponding number of conventional two-truck recirculatory units would be between seven and eight for the same duty. The advantage is more apparent when it is seen that respective floor areas occupied are 55 ft² for the

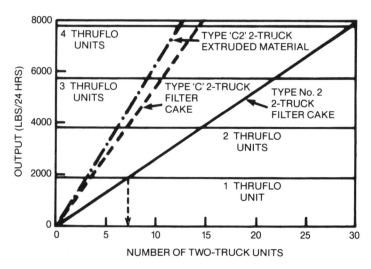

Figure 1. Comparative performance curves for Thruflo and conventional units.

Figure 2. Thruflo dryer.

Thruflo dryer pictured in Figure 2 and 245 ft² in the case of conventional units using transverse airflow.

Reference to the drying curves for the processing of materials in solid or filter cake form or, in fact, in the case of wet powders, clearly indicates that the ultimate rate-governing factor is the rate of diffusion of moisture from the wet mass. This becomes increasingly so during the falling rate period of drying. This situation, however, can be improved by preforming the product in order to increase the effective surface area presented to heat and mass transfer. The logical extension of this technique is to total dispersion drying (flash or pneumatic dryers, fluid beds, etc) where discrete particles can be brought into contact with the hot gas. This produces rapid heat transfer with correspondingly short drying times.

Batch-type fluidized-bed dryers have, therefore, superseded forced-convection units in many cases, notably in the drying of pharmaceuticals. These machines generally are available in a range of standard sizes with batch capacities of 50–200 lb, although much larger units are made for special applications.

When considering this type of dryer, it is important to ensure that the feed material can be fluidized in both initial and final conditions. It also should be remembered that standard fan arrangements are not equally suitable for a variety of materials of different densities. Therefore, it is necessary to determine accurately the minimum fluidizing velocity for each product.

If the feedstock is at an acceptable level of moisture content for fluidization, this type of dryer provides many advantages over a batch-type unit. Simplified loading and unloading results in lower labor costs, high thermal efficiencies are common, and the drying time is reduced to minutes as opposed to hours in conventional units. Current developments of this type of equipment now include techniques for the simultaneous evaporation of water and the granulation of solids. This makes the units ideal for use in the pharmaceutical field.

The various batch dryers referred to operate by means of forced convection, the transfer of thermal energy being

designed to increase the vapor pressure of the absorbed moisture while the circulated air scavenges the overlying vapor. Good conditions thus are maintained for continued effective drying. Alternatively and where the material is thermosensitive—implying low temperatures with consequently low evaporative rates—some improvement can be effected by the use of subatmospheric dryers, ie, by reducing the vapor pressure. Several different configurations are in use and all fall into the category of conduction-type dryers. The most usual type of heating is by steam, although hot water or one of the proprietary heat-transfer fluids can be used.

Two particular types are the double-cone dryer (Fig. 3) with capacities of ≤400 ft³ and the agitated-pan dryer not normally larger than 8 ft diameter, where average evaporative rates per unit wetted area usually are in the region of 4 lb/ft²·h. These units are comparatively simple to operate and, when adequately insulated, are thermally quite efficient, although drying times can be extended. They are especially suitable for applications involving solvent recovery and will handle powders and granules moderately well. They do, however, suffer from the disadvantage with some materials that the tumbling action in double-cone dryers and the action of the agitator in agitated-pan machines can produce a degree of attrition in the dried product that may prove unacceptable.

Similarly, quite large rotary vacuum dryers are used for pigment pastes and other such materials, especially where organic solvents present in the feedstock have to be recovered. These units normally are jacketed and equipped with an internal agitator that constantly lifts and turns the material. Heat transfer here is entirely by conduction from the wall of the dryer and from the agitator. Owing to the nature of their construction, initial cost is high relative to capacity. Installation costs also are considerable. In general, these dryers find only limited application.

Figure 3. Double-cone vacuum dryer.

CONTINUOUS DRYERS

For the drying of liquids of liquid suspensions, two types of dryers can be used: film drum dryers for duties in the region of 600 lb/h for a larger dryer of about 4 ft diameter × 10 ft face length or large spray dryers (as in Fig. 4) with drying rates of approximately 22,000 lb/h. Where tonnage production is required, the drum dryer is at a disadvantage. However, the thermal efficiency of the drum dryer is high in the region of 1.3–1.5 lb steam/lb of water evaporated and for small to medium production runs, it does have many applications.

Drum dryers usually are steam-heated, although work has been done involving the development of units for direct gas or oil heating. Completely packaged and capable of independent operation, these dryers can be divided into two broad classifications, ie, single-drum and double-drum.

Double-drum machines normally employ a "nip" feed device with the space between the drums capable of being adjusted to provide a means of controlling the film thickness. Alternatively, and in the case of the single-drum types, a variety of feeding methods can be used to apply material to the drum. The most usual is the simple "dip" feed. With this arrangement, good liquor circulation in the trough is desirable in order to avoid increasing the concentration of the feed by evaporation. Again, for special applications, single-drum dryers use top roller feed. While the number of rolls is related to the particular application and the material being handled, in general this method of feeding is used for pasty materials such as starches. Where the feed is very mobile, rotating devices such as spray feeds are used.

Figure 4. Conical section of a large spray dryer.

It will be seen from the Figure 5 drawings that there are a number of different feeding arrangements for drum dryers, all of which have a particular use. In practice, these variants are necessary owing to the differing characteristics of the materials to be dried and to the fact that no universally satisfactory feeding device has yet been developed. This again illustrates the need for testing, not only in support of theoretical calculations for the determination of the best dryer size but also to establish where a satisfactory film can be formed.

It must be emphasized that the method of feeding the product to the dryer is of paramount importance to selection or design. There are, of course, certain materials that are temperature-sensitive to such a degree that their handling would preclude the use of an atmospheric drum dryer. In such cases, special subatmospheric equipment may provide the answer, although the capital cost in relation to output generally would restrict its use to premium grade products.

As an alternative, the spray dryer offers an excellent solution to many drying problems. Many materials that would suffer from thermal degradation if dried by other methods often can be handled by spray drying owing to the rapid flash evaporation and its accompanying cooling effect. The continuous method of operation also lends itself to large outputs and with the correct application of control equipment, to low labor cost as well.

SPRAY DRYERS

Fundamentally, the spray-drying process is a simple one. However, the design of an efficient spray-drying plant requires considerable expertise along with access to large-scale test facilities, particularly where particle size and bulk density requirements in the dried product are critical. The sizing of spray dryers on a purely thermal basis is a comparatively simple matter since the evaporation is entirely a function of the Δt across the dryer. Tests on pilot-scale equipment are not sufficient in the face of such imponderables as possible wall buildup, bulk density, and particle size predictions. Atomization of the feed is of prime importance to efficient drying and three basic feed devices are used extensively:

1. Single-fluid nozzle or pressure type
2. Two-fluid nozzle or pneumatic type
3. Centrifugal (spinning disk)

The single-fluid nozzle produces a narrow spray of fine particles. While a multiplicity of nozzles of this type are used in tonnage plants to obtain the desired feedrate, because of the high pressures employed (up to 7000 psig), excessive wear can result, particularly with abrasive products, As an alternative, the two-fluid nozzle with external mixing is used for a variety of abrasive materials. This system generally is limited to small-capacity installations. Normally, the feed is pumped at about 25 psig merely to induce mobility while the secondary fluid is introduced at 50–100 psig, thus producing the required atomization.

Centrifugal atomization achieves dispersion by centrif-

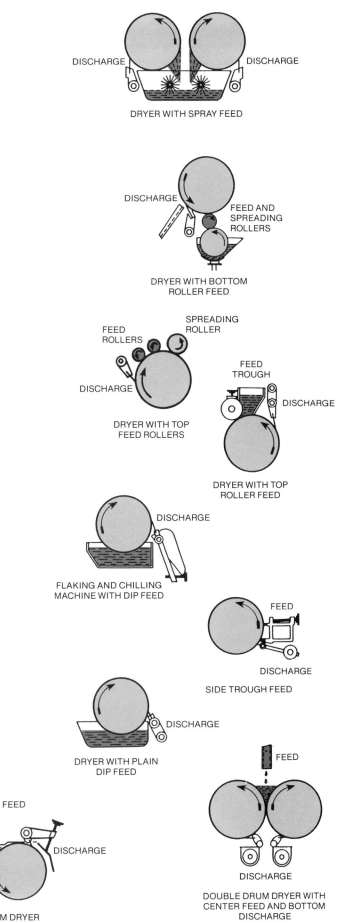

DISCHARGE   DISCHARGE

DRYER WITH SPRAY FEED

DISCHARGE   FEED AND
SPREADING
ROLLERS

DRYER WITH BOTTOM
ROLLER FEED

SPREADING
ROLLER
FEED
ROLLERS

DISCHARGE   FEED
TROUGH

DISCHARGE

DRYER WITH TOP
FEED ROLLERS

DRYER WITH TOP
ROLLER FEED

DISCHARGE

FLAKING AND CHILLING
MACHINE WITH DIP FEED

FEED

DISCHARGE

SIDE TROUGH FEED

DISCHARGE

DRYER WITH PLAIN
DIP FEED

FEED

FEED

DISCHARGE   DISCHARGE

DISCHARGE

DOUBLE DRUM DRYER
WITH CENTER FEED

DOUBLE DRUM DRYER WITH
CENTER FEED AND BOTTOM
DISCHARGE

Figure 5. Feeding arrangements for drum dryers.

ugal force; the feed liquor is pumped to a spinning disk. This system is suitable for and generally used on large productions. When stacked or multiple disks are employed, feed rates of 40,000–60,000 lb/h are not uncommon.

Many spray dryer configurations are in current use along with a variety of airflow patterns. The nature of the chamber geometry selected is strictly related to the system of atomization used. An example of this is the tower configuration designed to accommodate the inverted jet of the two-fluid nozzle whereas the cylinder and cone of the more usual configuration are designed for the spray pattern produced by a disk-type atomizer (Fig. 6).

The product collection systems incorporated in spray-drying installations are many and varied and can constitute a substantial proportion of the total capital investment. In some cases, this can be as high as 20–25% of the installed plant cost. It also must be remembered that to be suitable for spray drying, the feed must be in a pumpable condition. Therefore, consideration must be given to the upstream process, ie, whether there is any need to re-slurry in order to make the feed suitable for spray drying.

It generally is accepted that mechanical dewatering is less costly than thermal methods and while the spray dryer exhibits quite high thermal efficiencies, it often is at a disadvantage relative to other drying systems because of the greater absolute weight of water to be evaporated owing to the nature of the feed. It is, however, interesting to consider this point further. A classic case for comparison is provided by a thixotropic material that may be handled either in a spray dryer using a mechanical disperser doing work on a filtered cake to make it amenable to spray drying or alternatively, drying the same feedstock on a continuous-band dryer. In the latter instance, the cake is fed to an extruder and suitably preformed prior to being deposited on the conveyor band.

The operating costs presented in Table 2 are based on requirements for thermal and electrical energy only. No consideration is given to labor costs for either type of plant since these are likely to be approximately the same. Probably the most obvious figure emerging from this comparison is the 20% price differential in favor of the continuous-band dryer. Furthermore, while energy costs favor spray drying, they are not significantly lower.

It is, of course, impossible to generalize since the economic viability of a drying process ultimately depends on the cost per pound of the dried product and, as mentioned previously, the spray dryer usually has a greater amount of water to remove by thermal methods than other types. In the particular case illustrated, the spray dryer would have an approximate diameter of 21 ft for the evaporation of 6000 lb/h. If, however, the feed solids were reduced to 30% by dilution, the hourly evaporation rate would increase to 14,000 lb and the chamber diameter would be about 30 ft with corresponding increases in thermal input and air volume. The system would, as a result, also require larger fans and product collection systems. The overall thermal efficiency would remain substantially constant at 76% with reducing feed solids. Installed plant costs, however, increase proportionally with increasing

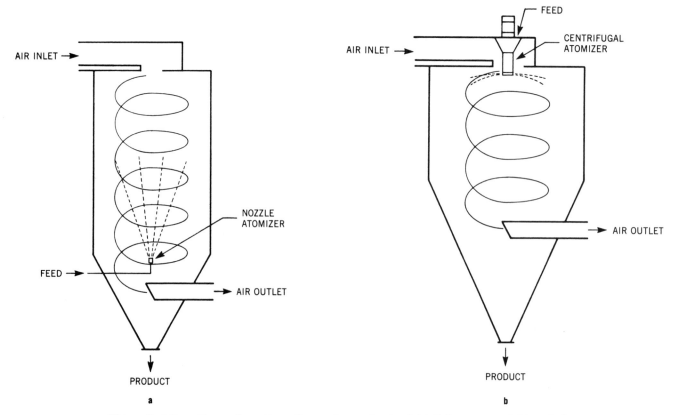

Figure 6. Alternative configurations of spray dryers showing (**a**) tall form type and (**b**) conical.

Table 2. Comparison between Direct-Fired Band Dryer and Spray-Dryer Costs in the Processing of Titanium Dioxide

	Spray Dryer	Band Dryer
Initial moisture content, W/W	50%	50%
Final moisture content, W/W	0.5%	0.5%
Feed, lb/h	12,000	12,000
Evaporation lb/h	5970	5970
Output, lb/h	6030	6030
Inlet air temperature, °F	1150	393
Exhaust air temperature, °F	250	300
Direct gas firing	Gas CV = 1000 Btu/ft^2	Gas CV = 1000 Btu/ft^2
Hourly operating cost gas and electricity	$39.00	$42.26
Thermal efficiency	76.0%	69.5%
Installed horsepower	149	154
Floor area occupied, ft^2	650	2100
Installed cost	$465,000	$330,000
Energy cost per pound of dried product	$0.065	$0.070

dryer size necessary for the higher evaporation involved in producing a dried output equivalent to that shown in Table 2.

Spray drying does have many advantages, particularly with regard to the final product form. This is especially so where pressing grade materials are required, ie, in the production of ceramics and dust-free products such as dyestuffs. It is certain that with the introduction of new geometries and techniques there will be further development into areas such as foods and in the production of powders that may be easily reconstituted.

ROTARY DRYERS

Another type of dryer that is very much in evidence in the chemical and process industries is the continuous rotary dryer. This machine generally is associated with tonnage production and as a result of its ability to handle products having a considerable size variation, can be used to dry a wide range of materials. The principal sources of thermal energy are oil, gas, and coal. While typical inlet temperatures for direct-fired dryers using these fuels is in the order of 1200°F, in certain instances they may be as high as 1500–1600°F depending largely on the nature of the product handled. Where feed materials are thermosensitive, steam heating from an indirect heat exchanger also is used extensively. These dryers are available in a variety of designs but in general, can be divided into two main types: those arranged for direct heating and those designed for indirect heating. As seen from Figure 7, certain variants do exist. For example, the direct–indirect dryer uses both systems simultaneously.

Where direct heating is used, the product of combustion are in intimate contact with the material to be dried while in the case of the indirect system, the hot gases are arranged to circulate around the dryer shell. Heat transfer then is by conduction and radiation through the shell.

With the indirect–direct system, hot gases first pass down a central tube coaxial with the dryer shell and return through the annular space between the tube and the shell. The material being cascaded in this annulus picks up heat from the gases and also by conduction from direct contact with the central tube. This design is thermally very efficient. Again, while there are a number of proprietary designs employing different systems of airflow, in the main these dryers are of two types: parallel and countercurrent flow. With parallel flow, only high-moisture-content material comes into contact with the hot gases, and, as a result, higher evaporative rates can be achieved than when using countercurrent flow.

In addition, many thermosensitive materials can be dried successfully by this method. Such an arrangement lends itself to the handling of pasty materials since the rapid flashing off of moisture and consequent surface drying limits the possibility of wall buildup or agglomeration within the dryer. On the other hand, countercurrent operation normally is used where a low terminal moisture content is required. In this arrangement, the high-temperature gases are brought into contact with the product immediately prior to discharge where the final traces of moisture in the product must be driven off.

In both these types, however, gas velocities can be sufficiently high to produce product entrainment. They therefore would be unsuitable for low-density or fine-particle materials such as carbon black. In such cases, the indirect-fired function-type dryer is more suitable since the dryer shell usually is enclosed in a brick housing or outer steel jacket into which the hot gases are introduced. As heat transfer is entirely by conduction, conventional flighting and cascading of the material is not used. Rather, the inside of the shell is fitted with small lifters designed to gently turn the product while at the same time maintaining maximum contact with the heated shell.

Another type of indirectly heated dryer that is particularly useful for fine particles or heat-sensitive materials is the steam-tube unit. This dryer can be of the fixed-tube variety equipped with conventional lifting flights designed to cascade the product through a nest of square section tubes, or alternatively, a central rotating tube nest can be used. Figure 8 shows a fixed-tube rotary dryer, which normally has an electrical vibrator fitted to the tube nest in order to eliminate the possibility of bridging

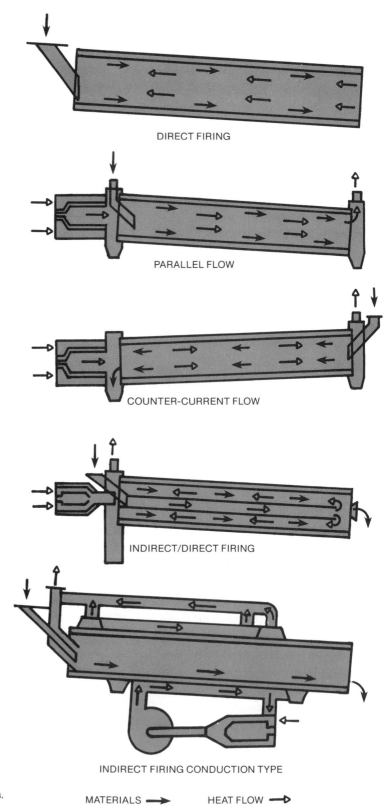

Figure 7. Typical rotary dryer arrangements.

DIRECT FIRING

PARALLEL FLOW

COUNTER-CURRENT FLOW

INDIRECT/DIRECT FIRING

INDIRECT FIRING CONDUCTION TYPE

MATERIALS ➡   HEAT FLOW ⇨

of the product with consequent loss of heat-transfer surface. Since the heat exchanger is positioned within the insulated shell in this type of dryer, the air rejection rate is extremely low and thermal efficiencies are high. In general, this design is suitable only for free-flowing materials.

A considerable amount of work has been done on the development of various types of lifting flights, all designed to produce a continuous curtain of material over the cross section of the dryer shell. Other special configurations involve cruciform arrangements to produce a labyrinth path. The object is to give longer residence times where

Figure 8. Fixed-tube rotary dryer.

this is necessary. When the diffusional characteristics of the material or other process considerations call for extended residence times, these machines no doubt will continue to find application.

PNEUMATIC DRYERS

Where total dispersion of the product in a heated gas stream can be achieved with a significant increase in evaporative rates, pneumatic or continuous fluid-bed dryers are preferred. The capital cost of these alternatives generally is lower and maintenance is limited to such components as circulating fans and rotary valves. When considering these two types of dryers, it is convenient to examine them together since both share similar characteristics. Both employ forced convection with dispersion of the feedstock, and as a result of the intimate contact between the drying medium and the wet solids, both exhibit

much higher drying rates than do any of the other dryers mentioned previously.

In a fluidized-bed dryer, the degree of dispersion and agitation of the wet solids is limited, whereas in a pneumatic dryer, the degree of dispersion is total and the material is completely entrained in the gas stream. This often is turned to advantage as the drying medium is used as a vehicle for the partially dried product. Other operations such as product classification also can be carried out where required. A further feature of fluid-bed and flash dryers is that the method of operation allows many temperature-sensitive materials to be dried without thermal degradation due to the rapid absorption of the latent heat of vaporization. This generally permits high-rate drying, whereas in other types of dryers lower temperatures would be necessary and correspondingly larger and more costly equipment would be required.

A very good degree of temperature control can be achieved in fluid-bed dryers and the residence time of the material can be varied either by the adjustment of the discharge weir or by the use of multistage units. Similarly, the residence time in the flash dryer can be adjusted by the use of variable cross-sectional area and, therefore, variable velocity. In addition, multiple-effect columns can be incorporated to give an extended path length or continuous recirculatory systems employing both air and product recycle can be used as illustrated in Figure 9.

Generally speaking, the residence time in fluidized-bed dryers is measured in minutes and in the pneumatic dryer in seconds. Both dryers feature high thermal efficiencies, particularly where the moisture content of the wet feed is sufficiently high to produce a significant drop between inlet and outlet temperatures. While the condition of the feed in the pneumatic dryer is somewhat less critical than that in the fluid-bed dryer because it is completely entrained, it still is necessary to use backmixing techniques

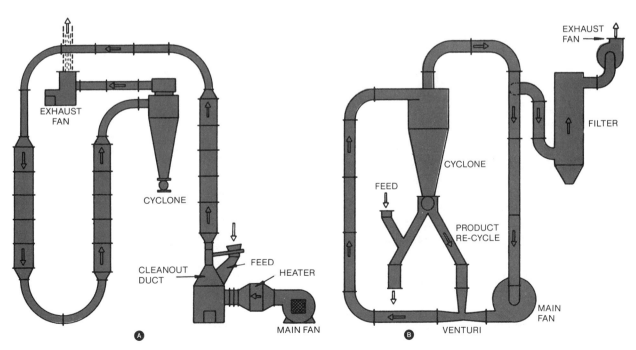

Figure 9. Multipass (A) and air recycle (B) arrangements in flash dryers.

on occasion in order to produce a suitable feed. A variety of feeding devices are used with these machines.

In fluid-bed dryers, special attention must be paid to the nature of the proposed feed since one condition can militate against another. To some extent, this is reflected in the range of variation in the figures given in Table 1 for evaporative rates. If a large or heavy particle is to be dried, the fluidizing velocities required may be considerable and involve high power usage. In such circumstances, if the moisture content is low and the surface/mass ratio also is low, the thermal efficiency and evaporation would be low. This would make selection of a fluid-bed dryer completely unrealistic and probably would suggest a conventional rotary dryer for the application.

Another case is that in which the minimum fluidizing velocity is so low that a dryer of very large surface dimensions is necessary to obtain the required thermal input. This also occurs in problems of fluid-bed cooling and usually is overcome by the introduction or removal of thermal energy by additional heating or cooling media through an extended-surface heat exchanger immersed in the bed.

With both types of dispersion dryer many configurations are available. While the power requirements of each usually is well in excess of other dryers because of the use of high-efficiency product recovery systems, the smaller size of the fluid-bed dryer compared with conventional rotaries and the fact that the flash dryer can be arranged to fit in limited floor space makes them very attractive.

BAND DRYERS

When selecting a dryer, it always is necessary to consider the final product form. When the degree of product attrition common to pneumatic and fluidized-bed dryer operation is unacceptable, continuous-band or "apron" dryers can provide an effective solution. These are widely used where moderately high rates of throughput on a continuous basis are called for. The most commonly used continuous-band dryer is the single-pass machine employing through-air circulation. Alternatively, and where there is limited floor space or a possible need for long residence time, multi-pass units are used with the conveyors mounted one above the other. In similar circumstances, another special type of multi-deck dryer can be used that employs a system of tilting trays so that the product is supported on both the normal working and the inside of the return run of the conveyor band. This arrangement considerably increases the residence time within the dryer and is particularly useful where the product has poor diffusional characteristics.

The method of airflow employed in these dryers is either vertically downward through the material and the supporting band or alternatively upward. Sometimes a combination of the two may be dictated by the nature of the wet feed. It occasionally happens that extruded materials have a tendency to coalesce when deposited on the band, in which case one or more sections at the wet end of the dryer may be arranged for upward airflow to reduce the effect. Wherever possible, through-air circulation is used as opposed to transverse airflow. This results in greatly increased evaporative rates as may be seen from Table 1.

An illustration of the relatively high performance of a band dryer operating on this system as compared with a unit having transverse airflow can be cited in a case involving the processing of a 70% moisture content filter cake. When this material is dried in a conventional unit, the cycle time is in the region of 28 h. This is reduced to 55 min in the through-circulation band dryer largely as a result of using an extruder–preformer designed to produce a dimensionally stable bed of sufficient porosity to permit air circulation through the feed.

In view of this, transverse airflow usually is used only where the type of conveyor necessary to support the product does not allow through-flow or where the product form is not suitable for this method of airflow. The most usual method of heating is by steam through heat exchangers mounted in the side plenums or above the band, although direct oil and gas firing sometimes is used. In such cases, the products of combustion normally are introduced to a hot well or duct at an elevated temperature from where they are drawn off and mixed with circulating air in each zone or section of the dryer.

Another alternative with direct firing is to use a series of small individual burners positioned so that each serves one or more zones of the dryer. Typical single-pass dryers of modular construction are illustrated in Figures 10, 11, and 12.

With this type and size of dryer, the average product throughput is about 5600 lb/h and involves an evaporation of 1600 lb/h moisture. It is not unusual, however, to find equipment with evaporative capacities of 3000 lb/h. Such outputs involve quite a large band area with correspondingly large floor area requirements. Various types of feeding arrangements are available to spread or distribute the wet product over the width of the band. Here again, the nature of the feed is an important prerequisite for efficient drying. Steam-heated, finned drums have been used as a means of producing a partially dried, preformed feed. While the amount of predrying achieved is reflected in increased output for a given dryer size or, alternatively, enables a smaller dryer to be used, these items are usually much more costly than many of the mechanical extruders that are available.

Generally, these extruders operate with rubber covered rollers moving over a perforated die plate with feed in the form of pressed cakes or more usually, as the discharge from a rotary vacuum filter. Others of the pressure type employ a gear pump arrangement, with extrusion taking place through a series of individual nozzles, while some use screw feeds that usually are set up to oscillate in order to obtain effective coverage of the band. Alternative designs include rotating cam blades or conventional bar-type granulators, although the latter often produce a high proportion of fines because of the pronounced shearing effect. This makes the product rather unsuitable because of the entrainment problems that can occur.

Each of the types available is designed to produce continuous–discontinuous extrudates or granules, with the grid perforations spaced to meet product characteristics. In selecting the proper type of extruder, it is essential to carry out tests on semiscale equipment as no other valid assessment of suitability can be made.

As a further illustration of the desirability of using a preforming technique, tests on a designated material ex-

Figure 10. Continuous conveyor band dryer arrangement for direct gas firing.

hibited a mean evaporative rate of 1.9 lb/ft² · h when processed in filter-cake form without preforming. When extruded, however, the same material being dried under identical conditions gave a mean evaporative rate of 3.8 lb/ft² · h. This indicates, of course, that the effective band area required when working on extruded material would be only 50% of that required in the initial test. Unfortunately, the capital cost is not halved as might be expected since the feed and delivery ends of the machine housing the drive and terminals remain the same and form an increased proportion of the cost of the smaller dryer. While the cost of the extruder also must be taken into account in the comparison, cost reduction still would be about 15%. Of course, other advantages result from the installation of the smaller dryer. These include reduced radiation and convection losses and a savings of approximately 40% in the floor area occupied.

This type of plant does not involve high installation

Figure 11. Multistage band dryer.

Figure 12. Band dryer and extruder with dye stuffs.

costs, and both maintenance and operating labor requirements are minimal. Since they generally are built on a zonal principle, with each zone having an integral heater and fan, a good measure of process control can be achieved. Furthermore, they provide a high degree of flexibility because of the provision of variable speed control on the conveyor.

SELECTION

The application ratings given in Table 1 in which are listed the approximate mean evaporative rates for products under a generic classification are based on APV experience over a number of years in the design and selection of drying equipment. It should be appreciated, however, that drying rates vary considerably in view of the variety of materials and their widely differing chemical and physical characteristics. Furthermore, drying conditions such as temperature and the moisture range over which the material is to be dried have a very definite effect on the actual evaporative rate. It is important, therefore, when using the figures quoted that an attempt is made to assess carefully the nature of the product to be handled and the conditions to which it may be subjected in order to achieve greater accuracy. With these factors in mind, it is hoped that the foregoing observations on drying techniques along with the appropriate tables and curves will provide a basis for making an assessment of the type, size, and cost of drying equipment.

In making a preliminary assessment for dryer selection, there are a number of further points to consider:

1. What is the nature of the upstream process? Is it feasible to modify the physical properties of the feed, eg, mechanical dewatering to reduce the evaporative load?

2. Does the quantity to be handled per unit time suggest batch or continuous operation?

3. From a knowledge of the product, select the type(s) of dryer that it appears would handle both the wet feedstock and the dried product satisfactorily and relate this to the equipment having the highest application rating in Table 1.

4. From a knowledge of the required evaporative duty, ie, the total mass of water to be evaporated per unit time and by the application of the approximate E_{av} figure given in Table 1, estimate the size of the dryer.

5. Having established the size of the dryer on an area or volumetric basis, refer to the appropriate curve in Figures 13 and 14 and establish an order of cost for the particular type of unit.

Although a great deal of fundamental work has been carried out into the mechanics of drying, which provides a basis for recommendations, it is most desirable for pilot plant testing to be done. This is necessary not only to support theoretical calculations but also to establish whether a particular dryer will handle the product satisfactorily. In the final analysis, it is essential to discuss the drying application with the equipment manufacturer who has the necessary test facilities to examine the alternatives objectively and has the correlated data and experience from field trials to make the best recommendation.

EFFICIENT ENERGY UTILIZATION IN DRYING

It generally is necessary to employ thermal methods of drying in order to reach what is termed a "commercially dry" condition. As a result, drying forms an important

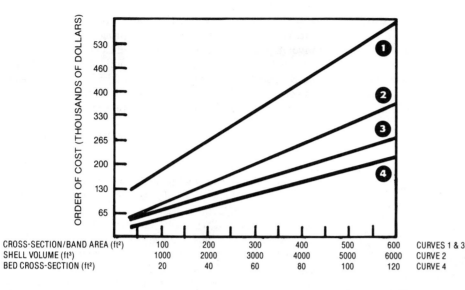

Figure 13. Approximate capital costs of (1) spray dryers, (2) rotary dryers (direct-oil-fired, mild steel, excluding product collection system), (3) continuous-band dryers steam-heated with roller extruder), (4) continuous-fluid-bed dryers (direct-oil-fired with primary cyclone).

part of most food and chemical processes and accounts for a significant proportion of total fuel consumption.

The rapid escalation of fuel costs over the past 10 yr, together with the prevailing uncertainty of future availability, cost, and possible supply limitations, highlights the continued need to actively engage in the practice of energy conservation. Some of the factors affecting dryer efficiency and certain techniques designed to reduce the cost of the drying operation are discussed in the next few sections.

In making an appraisal of factors affecting dryer efficiency, it is useful to draw up a checklist of those items that have a significant bearing on both operation and economy. It also is appropriate prior to examining various options to emphasize that in the final analysis, the primary concern is the "cost per unit weight" of the dried product. This single fact must largely govern any approach to dryer selection and operation. Additionally, there is an increasing necessity to consider the unit operation of drying in concert with other upstream processes such as mechanical dewatering and preforming techniques in a "total energy" evaluation.

In considering which factors have a bearing on dryer efficiency and what can be done to maximize that efficiency, the following aims should be kept in mind:

1. Maximum temperature drop across the dryer system indicating high energy utilization. This implies maximum inlet and minimum outlet temperature.

2. Employ maximum permissible air recirculation, ie, reduce to an absolute minimum the quantity of dryer exhaust, having due regard for humidity levels and possible condensation problems.

3. Examine the possibility of countercurrent drying, ie, two-stage operation with exhaust gases from a final dryer being passed to a predryer, or alternatively, preheating of incoming air through the use of a heat exchanger located in the exhaust gases.

4. Utilize "direct" heating wherever possible in order to obtain maximum heat release from the fuel and eliminate heat-exchanger loss.

5. Reduce radiation and convection losses by means of efficient thermal insulation.

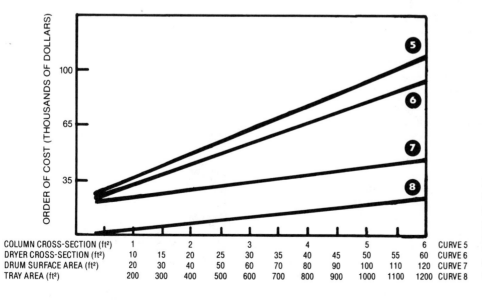

Figure 14. Approximate capital costs of (5) pneumatic dryers (direct-fired, stainless steel, including primary cyclone), (6) double-cone dryers (stainless steel, excluding vacuum equipment), (7) drum dryers (cast-iron drums, dip feed), (8) tray dryers (steam/electric-heated).

While the above clearly are basic requirements, there are a number of other areas where heat losses occur in practice, including sensible heat of solids. Furthermore, other opportunities exist for improving the overall efficiency of the process. These are related to dryer types, methods of operation, or possibly the use of a combination of drying approaches to obtain optimum conditions.

Types of Dryers

Considering the requirements for high inlet temperature, the "flash" or pneumatic dryer offers great potential for economic drying. This stems from the simultaneous flash cooling effect that results from the rapid absorption of the latent heat of vaporization and allows the use of high inlet temperatures without thermal degradation of the product. This type of dryer also exhibits extremely high evaporative rate characteristics, but the short gas–solids contact time can in certain cases make it impossible to achieve a very low terminal moisture. However, a pneumatic dryer working in conjunction with a rotary or a continuous fluidized-bed dryer provides sufficient residence time for diffusion of moisture to take place. Such an arrangement combines the most desirable features of two dryer types and provides a compact plant and conceivably, the optimum solution.

As a further example, it is common practice in the process industries to carry out pretreatment of filter-cake materials using extruders or preformers prior to drying. The primary object is to increase the surface area of the product in order to produce enhanced rates of evaporation and smaller and more efficient drying plants. It is interesting to examine the improvements in energy utilization in a conveyor-band dryer resulting from a reduction in overall size of the dryer simply because of a change in the physical form of the feed material.

Consider a typical case. As a result of preforming a filter cake, the evaporative rate per unit area increases by a factor of 2, eg, 1.9 to 3.8 lb/ft$^2 \cdot$ h. This permits the effective dryer size to be reduced to half that required for the nonpreformed material and, for a plant handling 1 ton/h of a particular product at a solids content of 60%, the radiation and convection losses from the smaller dryer enclosure show an overall reduction of some 140,000 Btu/h. Although it is necessary to introduce another processing item into the line to carry out the pretreatment, it can be shown that the reduction in the number of dryer sections and savings in horsepower more than offset the power required for the extruder. This differential is approximately 15 kW and the overall saving in total energy is approximately 10% ie, 466 kW compared with 518 kW. Furthermore, the fact that the dryer has appreciably smaller overall dimensions provides an added bonus in better utilization of factory floor space.

Drying Techniques

If, as stated, the prime concern is the cost per pound of dried product, then major savings can be achieved by reducing the amount of water in the feedstock to a minimum prior to applying thermal methods of drying. Again, since it generally is accepted that the mechanical removal of water is less costly than thermal drying, it follows that considerable economies can be made when there is a sub-

stantial amount of water that can be readily removed by filtration or centrifuging. This approach, however, may involve changing the drying technique. For example, whereas a liquid suspension or mobile slurry would require a spray dryer for satisfactory handling of the feed, the drying of a filter cake calls for a different type of dryer and certainly presents a totally different materials handling situation.

In approaching a problem by two alternative methods, the overall savings in energy usage may be considerable as illustrated in Figure 15, which is a plot of feed moisture content versus dryer heat load. As may be seen, the difference in thermal energy used in (A) drying from a moisture content of 35% down to 0.1%, or alternatively (B) drying the same material from 14% down to 0.1% is 24.3 × 10^6 Btu. Route A involves the use of a spray dryer in which the absolute weight of water in the feedstock is considerably greater than in route B, which involves the use of a thermoventuri dryer. Both types, incidentally, come under the classification of dispersion dryers and generally operate over similar temperature ranges. As a result, their efficiencies are substantially the same.

It will be apparent from Table 3 that the spray dryer is at a considerable disadvantage because of the requirements for a pumpable feed. This highlights the need to consider the upstream processes and, where a particular "route" to produce a dry product requires a slurried feed, to consider whether a better alternative would be to dewater mechanically and use a different type of dryer.

Comparing these two alternative methods for drying a mineral concentrate, the spray dryer uses a single-step atomization of a pumpable slurry having an initial moisture content of 35% and employs thermal drying techniques alone down to a final figure of 0.1%. The alternative method commences with the same feedstock at 35% moisture content but employs a rotary vacuum filter to mechanically dewater to 14%. From that point, a pneu-

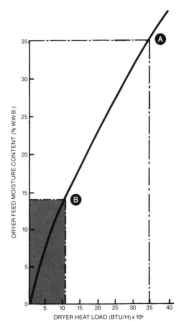

Figure 15. Thermal energy required for drying versus feed moisture content (duty as in Table 1).

Table 3. Comparison of Operating Conditions and Energy Utilization of Pneumatic and Spray Dryers Processing Concentrates

Parameter	Pneumatic Dryer	Spray Dryer
Feed rate	88,200 lb/h	88,200 lb/h
Filtrate rate	21,500 lb/h	—
Evaporative rate	9200 lb/h	30,700 lb/h
Production rate	57,500 lb/h	57,500 lb/h
Initial moisture content	14%	35%
Final moisture content	0.1%	0.1%
Air inlet temperature, °F	752	752
Air outlet temperature, °F	230	230
Total thermal input	21.43×10^6 Btu/h	66.2×10^6 Btu/h
Basic air volume (BAV) at NTP	28,100 NCFM	86,700 NCFM
Fuel consumption	1080 lb/h	3290 lb/h
Total dryer horsepower	187 kW	530 kW
Total filter horsepower	295 kW	—
Total system horsepower	482 kW	530 kW
Total thermal input expressed as kilowatts	6280 kW	19,400 kW
Total energy input to system	6762 kW	19,930 kW

Note: An assumed volumetric flow of 3 ft³/min · ft² of filter area with approximately 1 hp/10 ft³ · min has been used as being typical of the flow rates and energy requirements for the filtration or mechanical dewatering equipment.

matic dryer handles the cake as the second of two stages and thermally dries the product to the same final moisture figure.

While the difference in thermal requirements for the alternative routes already has been noted, it also is necessary to take into account the energy used for mechanical dewatering in accordance with assumptions outlined in the footnote to Table 3. From reference to this chart, it will be evident that the two-stage system requires approximately one-third of the energy needed by the single-stage operation but that the savings are not limited to operating costs. There also are significant differences in the basic air volumes required for the two dryers, which means that ancillaries such as product collection–gas cleaning equipment will be smaller and less costly in the case of the pneumatic dryer. The same applies to combustion equipment, fans, and similar items. Actually, the chart shows that capital cost savings are in the nature of 50% in favor of the filter and pneumatic drying system.

This illustration amply demonstrates the need for a more detailed consideration of drying techniques than perhaps has been the case in the past. It also points to the desirability of examining a problem on a "total energy" basis rather than taking the drying operation in isolation. Such a full evaluation approach often will prove it advantageous to change technology, ie, to use a different type of dryer than the one that possibly has evolved on the basis of custom and practice.

Operating Economies

Looking at dryer operations in terms of improving efficiency, it is interesting to see the savings that accrue if, for example, a pneumatic dryer is used on a closed-circuit basis, ie, with the recycle of hot gases instead of total rejection. Briefly, the "self-inertizing" pneumatic dryer consists of a closed loop as shown in Figure 16 with the duct system sealed to eliminate the ingress of ambient air. This means that the hot gases are recycled with only a relatively small quantity rejected at the exhaust and a

correspondingly small amount of fresh air admitted at the burner. In practice, therefore, oxygen levels of the order of 5% by volume are maintained. This method of operation raises a number of interesting possibilities. It permits the use of elevated temperatures and provides a capability to dry products that under normal conditions would oxidize. The result is an increase in thermal efficiency. Furthermore, the amount of exhaust gas is only a fraction of the quantity exhausted by conventional pneumatic dryers. This is clearly important wherever there may be a gaseous effluent problem inherent in the drying operation.

A comparison of a self-inertizing versus a conventional pneumatic dryer is detailed in Table 4 with the many advantages clearly apparent. Drying with closed-circuit operation is tending toward superheated vapor drying, and in practice 40–50% of the gases in circulation are water vapor. This suggests that the specific heat of the gas will be approximately 0.34 Btu/lb · °F compared with about 0.24 Btu/lb · °F in a conventional total rejection dryer. Since the mass of gas for a given thermal capacity is appreciably less than in a conventional dryer and, as previously mentioned, oxygen levels are low, much higher operating temperatures may be used. This permits a re-

Table 4. Comparison of Self-Inertizing Thermoventuri Dryer Versus Total Rejection Thermoventuri Dryer

Parameter	Closed Circuit	Total Rejection
Evaporation, lb/h	1,250	1,250
IMC, % wwb	85	85
FMC, % wwb	10	10
Feed, lb/h	1,500	1,500
Air inlet temperature, °F	932	662
Air outlet temperature, °F	302	248
Exhaust dry airflow, lb/h	1,368	16,400
Exhaust losses, Btu/h	79,700	744,000
Heat losses, Btu/h	120,000	150,000
Total heat input, Btu/h	1,680,400	2,375,700
Efficiency, %	87.9	62.4

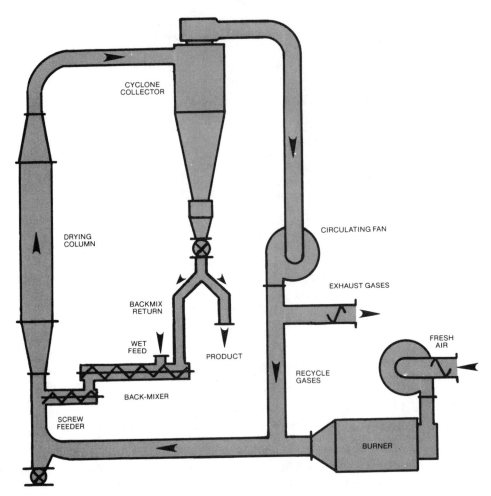

Figure 16. Self-inertizing pneumatic dryer with backmix facility.

duction in the size of the closed-circuit dryer. For the case given in Table 4 and comparing the two dryers on the basis of their cross-sectional areas, the total rejection dryer would have rather more than double the area of the closed-circuit system. There are obvious limitations to the use of such a technique, but the advantages illustrated are very apparent, especially the savings in thermal energy alone, viz, 695,000 Btu/h representing about 29.2% of the total requirements of the conventional drying system.

Effect of Changing Feed Rates

Since it has a significant bearing on efficiency, another factor of major importance to consider when designing dryers of this type is the possible effect of reducing feed rate. What variation in quantity of feed is likely to occur as a result of operational changes in the plant upstream of the dryer and as a result, what turndown ratio is required of the dryer? With spray dryers and rotary dryers, the mass airflow can be varied facilitating modulation of the dryer when operating at reduced throughputs.

This, however, is not the case with pneumatic and true fluidized-bed dryers since the gases perform a dual function of providing the thermal input for drying and acting as a vehicle for transporting the material. Since the mass flow has to remain constant, the only means open for modulating these dryers is to reduce the inlet temperature. This clearly has an adverse effect on thermal efficiency. It

therefore is of paramount importance to establish realistic production requirements. This will avoid the inclusion of excessive scale-up factors or oversizing of drying equipment and thereby maximize operating efficiency.

Figure 17 illustrates the effects on thermal efficiency of either increasing the evaporative capacity by increased inlet temperatures or alternatively, reducing the inlet temperature with the exhaust temperature remaining constant at the level necessary to produce an acceptable dry product. While the figure refers to the total rejection case of the previous illustration where design throughput corresponds to an efficiency of 62.4%, the curve shows that if the unit is used at only 60% of design, dryer efficiency falls to 50%. The converse, of course, also is true.

In this brief presentation, an attempt has been made to highlight some of the factors affecting efficiency in drying operations and promote an awareness of where savings can be made by applying new techniques. While conditions differ from one drying process to another, it is clear that economies can and should be made.

SPRAY DRYING: NEW DEVELOPMENTS, NEW ECONOMICS

In the field of spray drying, the past 10 years have witnessed many new developments initiated mainly to meet three demands from food and dairy processors—better en-

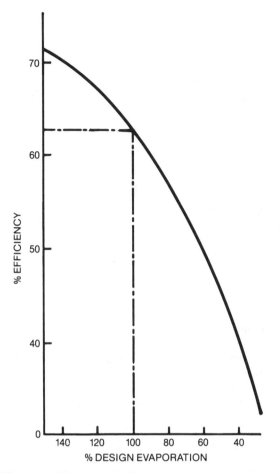

Figure 17. Variation of efficiency with dryer turndown.

ergy efficiency, improved functional properties of the finished powder, and production of new, specialized dried products. This has been accompanied by progress made in the automation of drying systems and in stricter requirements for the reduction of environmental impacts resulting from the operation of spray dryers.

While demands for energy efficient drying may have become somewhat relaxed recently with the stabilizing of energy costs, the fact remains that spray dryers and evaporators are among the most energy-intensive processing systems and the future of energy costs is still uncertain.

The extra functionality that may be required of the powders is to a large extent determined by the drying conditions, including pre- and posttreatment of the concentrate and final powder. Desirable characteristics are dust-free and good reconstitution properties that involve carefully controlled drying conditions, often with added agglomeration and lecithination steps.

The interest in the production of high-value specialty dried products instead of straight commodity products has been accentuated by the government's reduced support to dairy producers, whereby there will be less milk solids available. Processors naturally are looking into ways to better use these available supplies by producing products with improved functional, nutritional, and handling properties and, coincidentally, products with higher profit margins.

This is particularly noticeable in the utilization of whey and whey product derivatives. A larger percentage of the total whey produced today is handled and processed so that it will be suitable as food ingredients for human consumption. It also is finding new applications by being refined into whey protein concentrate and lactose through crystallization, hydrolyzation, and separation by ultra-filtration.

To handle the increase in plant size and complexity and to meet demands for closer adherence to exact product specifications, dryer instrumentation has been improved and operation gradually has evolved from manual to fully automated. Today, it is normal for milk drying plants to be controlled by such microprocessor-based systems as the APV ACCOS, with automation covering the increasingly complex startup, shutdown, and CIP procedures as well as sophisticated automatic adjustment of the dryer operating parameters to variations in ambient-air conditions.

Finally, more stringent demands governing environmental aspects of plant operations have resulted in that virtually all spray dryers today have to be equipped with exhaust air pollution control devices such as wet scrubbers or bag filters.

Principle and Approaches

Spray drying basically is accomplished in a specially designed chamber by atomizing feed liquid into a hot-air stream. Evaporation of water from the droplets takes place almost instantaneously, resulting in dried particles being carried with the spent drying air. The powder is subsequently separated from the air in cyclone separators or bag filters and collected for packaging. The powder particles being protected during the drying phase by the evaporative cooling effect results in very low heat exposure to the product, and the process is thus suitable also for heat-sensitive materials.

A number of different spray dryer configurations are available, distinguished by the type of feed atomization device and the drying chamber design.

Atomization Techniques

Proper atomization is essential for satisfactory drying and for producing a powder of prime quality. To meet varying parameters, APV offers a selection of atomizing systems: centrifugal, pneumatic nozzles, or pressure nozzles.

Centrifugal Atomization. With centrifugal or spinning-disk atomization (Fig. 18), liquid feed is accelerated to a velocity in excess of 800 ft/s to produce fine droplets that mix with the drying air. Particle size is controlled mainly be liquid properties and wheel speed. There are no vibrations, little noise, and small risk of clogging. Furthermore, the system allows maximum flexibility in feed rate, provides capacities in excess of 200 tons/h and operates with low power consumption.

Steam Injection. To produce a product with significantly increased bulk density and fewer fines, the APV steam injection technique (Fig. 19) has been refined to a point that allows its use with centrifugal atomizers in large drying operations.

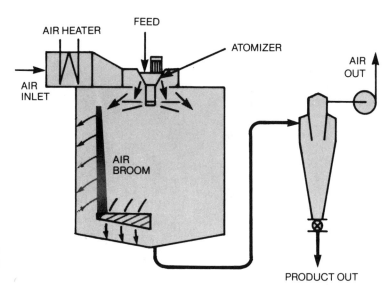

Figure 18. Flat-bottomed drying chamber incorporates rotating powder discharger and can be equipped with an air broom.

During centrifugal atomization, air around the rotating wheel may be entrapped within the atomized droplets.

When heat is transferred from the drying air into the feed droplet, water evaporates and diffuses out through the surface, at the same time creating a hard shell around the particle and some hollow spaces inside. If the droplet already contains some air bubbles, this incorporated air will expand to fill out the created vacuoles and the particle will not shrink much during the drying process, resulting in porous particles as shown in Figure 20 microphoto of spray-dried autolyzed yeast.

In the steam injection process the air atmosphere around the spinning disk is replaced with a steam atmosphere, thus reducing the amount of air incorporated in the droplets. This means that there is no air that will expand within the particle and the created vacuoles will collapse, thus shrinking the particle and resulting in the type of dense, void-free particles illustrated in the Figure 21 microphoto. Control of the amount of steam injected permits a precise adjustment in powder bulk density. Furthermore, reduction of air-exposed surfaces often reduces product oxidation and prolongs powder shelf life.

Pressure Nozzles. With the pressure nozzle system (Fig. 22), liquid feed is atomized when forced under high pressure through a narrow orifice. This approach offers great versatility in the selection of the spray angle, direc-

tion of the spray, and positioning of the atomizer within the chamber. It also allows cocurrent, mixed-current, or countercurrent drying, with the production of powders having particularly narrow particle size distribution and/or coarse characteristics. Since the particle size is dependent on the feed rate, dryers with pressure nozzles are somewhat limited relative to changing product characteristics and operating rates.

Pneumatic (Two-Fluid) Nozzles. This type of nozzle uses compressed air to accomplish the atomization of the feed product. The advantage with this type of atomizer is that it allows a greater flexibility in feed capacity than what can be obtained with pressure nozzles. However, particle size distribution is not as good as what can be obtained with the centrifugal or high-pressure nozzle-type atomizers, and because of the large quantities of compressed air required, operating costs tend to be quite high.

Single-Stage Drying

Defined as a process in which the product is dried to its final moisture content within the spray-drying chamber alone, the single-stage dryer design is well known throughout the industry, although some difference in airflow patterns and chamber design exist.

As illustrated, the drying air is drawn through filters, heated to the drying temperature, typically by means of a

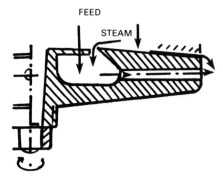

Figure 19. Steam is added around and into the atomizing disk to minimize air in the atomized liquid droplet.

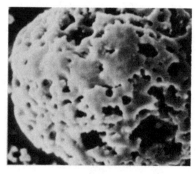

Figure 20. Spray-dried autolyzed yeast (2000× magnification). Particle is hollow and filled with crevices.

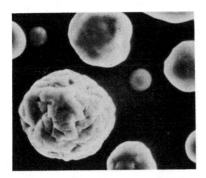

Figure 21. Spray-dried autolyzed yeast with steam injection (2000× magnification). Particle is solid and essentially void-free.

direct-combustion natural-gas heater, and enters the spray drying chamber through an air distributor located at the top of the chamber. The feed liquid enters the chamber through the atomizer, which disperses the liquid into a well-defined mist of very fine droplets. The drying air and droplets are very intimately mixed, causing a rapid evaporation of water. As this happens, the temperature of the air drops, as the heat is transferred to the droplets and used to supply the necessary heat of evaporation of the liquid. Each droplet thus is transformed into a powder particle. During this operation, the temperature of the particles will not increase much owing to the cooling effect resulting from water evaporation.

The rate of drying will be very high at first but then declines as the moisture content of the particles decreases. When the powder reaches the bottom of the drying chamber, it has attained its final moisture and normally is picked up in a pneumatic cooling system to be cooled down to a suitable bagging or storage temperature. Powder that is carried with the air is separated in cyclone collectors. As the powder is cooled in the pneumatic system, it also is

subjected to attrition. This results in a powder that is fine, dusting, and has relatively poor redissolving properties.

Single-stage dryers are made in different configurations. The basic ones are described below.

Box Dryers

A very common type of dryer in the past, the drying air enters from the side of a boxlike drying chamber. Atomization is from a large number of high-pressure nozzles also mounted in the side of the chamber. Dried powder will collect on the flat floor of the unit, from which it is removed by moving scrapers. The exhaust air is filtered through filter bags that also are mounted in the chamber. In some instances, cyclones will be substituted for the filter bags.

Tall-Form Dryers

These are very tall towers in which the airflow is parallel to the chamber walls. The atomization is by high-pressure nozzles. This type of dryer is used mainly for the production of straightforward, commodity-type dried powders.

Wide-Body Dryers

While the height requirement is less than for the tall-form type, the drying chamber is considerably wider. The atomizing device can be either centrifugal or nozzle type. The product is sprayed as an umbrella-shaped cloud, and the airflow follows a spiral path inside the chamber. As shown in Figs. 18 and 22, this type of dryer is designed either with a flat or conical bottom.

Flat-Bottom Dryer. The flat-bottom dryer is not a new development, but the design has gained renewed interest in recent years.

The chamber has a nearly flat bottom from which the powder is removed continuously by means of a rotating

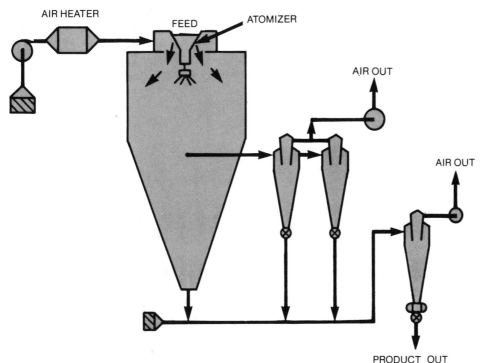

Figure 22. Typical spray dryer with conical chamber; arranged with high-pressure atomizer and air broom.

pneumatic powder discharger. The dryer has the obvious advantage of considerably reduced installed height requirements compared to cone-bottom dryers, and it is also easy for operators to enter the chamber for cleaning and inspection. Another advantage is that the rotating powder discharger provides a positive powder removal from the chamber and a well-defined powder residence time. This has been shown to be important in the processing of heat-sensitive products such as enzymes and flavor compounds. Depending on the product involved, a pneumatic powder cooling system also may be installed.

Typically, this type of dryer is used for egg products, blood albumin, tanning agents, ice cream powder and toppings. The flat- and bottom chamber, incidentally, may be provided with an air broom, which is indicated by the colored section in Figure 18. By blowing tempered air onto the chamber walls while rotating, this device blows away loose powder deposits and cools the chamber walls to keep the temperature below the sticking point of certain products. Some items with which the air bottom technique has been successful are for fruit and vegetable pulp and juices, meat extracts, and blood.

Cone-Bottom Dryer

Figure 22, meanwhile, shows a conical-bottom chamber arrangement with a side air outlet, high-pressure nozzle atomization, and pneumatic product transport beneath the chamber. This single-stage dryer is very well suited to making relatively large particles of dairy products or proteins for cattle feed. If the product will not withstand pneumatic transport, it may be taken out unharmed directly from the chamber bottom.

Multiple-Stage Dryers

The best way to reduce energy usage in spray drying is, of course, to try and reduce the specific energy consumption of the process. With advances in atomizer and air distributor designs it has been possible with many products to operate with higher dryer inlet temperature and lower outlet temperature. While this procedure substantially cuts energy needs and does not harm most heat-sensitive products, care must be taken and proper balance struck. The nature of the product usually defines the upper limit, ie, the denaturation of milk protein or discoloring of other products. A higher inlet temperature requires close control of the airflow in the spray-drying chamber, and particularly around the atomizer. Furthermore, it must be noted that a lower outlet temperature will increase the humidity of the powder.

Generally speaking, the drying process can be divided into two phases: (1) the constant-rate drying period when drying proceeds quickly and when surface moisture and moisture within the particles that can move by capillary action are extracted and (2) the falling-rate period when diffusion of water to the particle surface becomes the determining factor. Since the rate of diffusion decreases with the moisture content, the time required to remove the last few percent of moisture in the case of single-stage drying takes up the major part of the residence time within the dryer. The residence time of the powder thus is essentially the same as that of the air and is limited to

between 15 and 30 s. As the rate of water removal is decreasing toward the end of the drying process, the outlet air temperature has to be kept fairly high in order to provide enough driving force to finish the drying process within the available air residence time in the chamber.

In multiple-stage drying, the residence time is increased by separating the powder from the main drying air and subjecting it to further drying under conditions where the powder residence time can be varied independently of the airflow. Technically, this is done either by suspension in a fluidized bed or by retention on a moving belt. Since a longer residence time can be allowed during the falling-rate period of the drying, it is possible and desirable to reduce the drying air outlet temperature. Enough time to complete the drying process can be made available under more lenient operating conditions.

The introduction of this concept has led to higher thermal efficiencies. Fifteen years ago, the typical inlet and outlet temperatures of a milk spray dryer were 360–205°F. Today, the inlet temperature is often above 430°F, with outlet temperatures down to about 185°F.

Two-Stage Drying. In a typical two stage drying process as shown in Figure 23, powder at approximately 7% moisture is discharged from the primary drying chamber to an APV fluid bed for final drying and cooling. The fact that the powder leaves the spray dryer zone at a relatively high moisture content means that either the outlet air temperature can be lowered or the inlet air temperature increased. Compared to single-stage drying, this will result in better thermal efficiency and higher capacity from the same size drying chamber. As the product is protected by its surrounding moisture in the spray-drying phase, there normally are no adverse effects on product quality resulting from the higher inlet air temperature.

The outlet air from the chamber leaves through a side air outlet duct, and the powder is discharged at the bottom of the chamber into the fluid bed. This prolongs the drying time from about 22 s in a typical single-stage dryer to

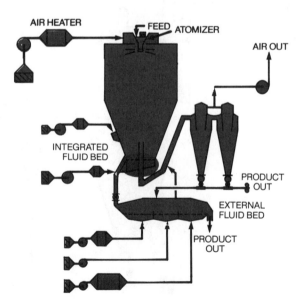

Figure 23. Two-stage drying with external fluid bed.

more than 10 min, thus allowing for low-temperature use in the fluid bed.

In the development of this type of drying system, an initial difficulty was to provide a means to reliably fluidize the semimoist powders. This stems from the fact that milk powder products, and especially whey-based ones, show thermoplastic behavior. This makes them difficult to fluidize when warm and having high moisture content, a problem that was largely overcome by the development of a new type of vibrating fluid bed.

The vibrating fluid bed has a well-defined powder flow and typically is equipped with different air supply sections, each allowing a different temperature level for a optimum temperature profile. The last section normally is where the product is cooled to bagging and storage temperature.

While the specific energy consumption in the fluid-bed process may be relatively high, the evaporation is minor compared with the spray-drying process and the total energy use therefore is 15–20% lower.

Figure 24 shows the specific energy costs as a function of the water evaporated in the fluid bed or the residual moisture in the powder from the spray drying to the fluid-bed process. The curves are only shown qualitatively be-cause the absolute values depend on energy prices, inlet temperatures, required residual moisture content, and other such parameters. The total cost curve shows a minimum that defines the quantity of water to be evaporated in the fluid bed to minimize the energy costs.

The introduction of two-stage drying also created the potential for a general improvement in powder quality, especially as far as dissolving properties are concerned. The slower and more gentle the drying process produces more solid particles of improved density and solubility and opens up the possibility of producing agglomerated powders in a straight-through process. If desired, the fluid bed can be designed for rewet instantizing or powder agglomeration and for the addition of a surface-active agent such as lecithin.

Three-Stage Drying

This advanced drying concept basically is an extension of two-stage drying in which the second drying stage is integrated into the spray-drying chamber with final drying conducted in an external third stage (Fig. 25). The design allows a higher moisture content from the spray drying zone than is possible in a two-stage unit and results in an

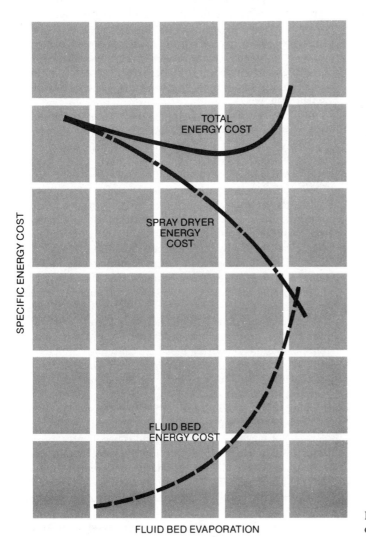

Figure 24. Specific energy cost as a function of the water evaporated in fluid bed.

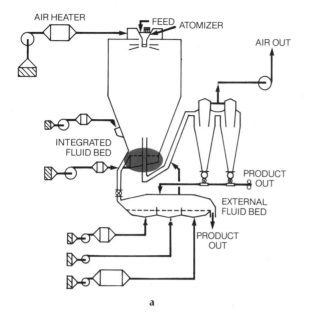

a

b

Figure 25. (a, b) Three-stage dryer combines conical drying chamber, integrated stationary fluid bed with directional air flow and external vibrating fluid bed. Provides optimum conditions for producing nondusty, hygroscopic, and high-fat-content products.

even lower outlet air temperature. An added advantage is that it improves the drying conditions for several difficult products. This is accomplished by spray drying the powder to high moisture content but at the same time avoiding any contact with metal surfaces by handling the powder directly on a fluidized powder layer in the integrated fluid bed. Moist powder particles thus are surrounded by already dry particles, and any tendency to stick to the chamber walls will be reduced.

In this drying concept, it is not necessary to vibrate the chamber fluid bed, although often the powders entering from the spray drying zone are difficult to fluidize because of their high moisture content and thermoplastic and hygroscopic characteristics. The fluidization characteristics are improved as the wet powder from the spray-drying zone is mixed and coated with dryer powder in the integrated fluid bed. The fluidization is further assisted by the use of a special type of perforated plate with directional

airflow. The stationary fluid bed operates with high fluidizing velocities and high bed depths, both optimized to the product being processed.

The air outlet of the integrated fluid-bed system is located in the middle of the fluid bed at the bottom of the chamber. This forms an annulus around the air outlet duct and creates an aerodynamically clean design, completely eliminating the mechanical obstructions found in older two-point discharge designs. The dryer is equipped with tangential air inlets, so called Wall Sweeps, for conditioned air. These are important to the operation of the dryer in two ways: they will cool the wall and remove powder that may have a tendency to accumulate on the wall, and they also serve to stabilize the airflow within the chamber. The dryer thus operates with a very steady and well-defined airflow pattern that minimizes the amount of wall buildup. Atomization can be either by pressure nozzles or by a centrifugal-type atomizer.

The three-stage system is exceptionally suitable for the production on nondusty, hygroscopic, and high-fat-content products. The dryer can produce either high-bulk-density powder by returning the fines to the external fluid bed, or powder with improved wettability by straight-through agglomeration with the fines reintroduced into an atomizing zone. The concept also allows the addition of liquid into the internal fluid bed, thus opening the path to the production of very sophisticated agglomerated and instantized products, ie, excellent straight-through lecithinated powders. Whey powder from such a system shows improved quality because of the higher moisture level present when the powder enters the integrated fluid bed, providing good conditions for lactose after crystallization within the powder.

A three-stage dryer also offers a high production capacity in a small equipment volume. The specific energy requirement is about 10% less than for a two-stage dryer.

Spray-Bed Dryer

While the concept of a three-stage drying system with an integrated fluid evolved from traditional dryer technology, it served to spur the development of a modified technique referred to by APV as the "Spray Bed Dryer." This machine is characterized by use of an integrated fluid bed at the bottom of the drying zone but with drying air both entering and exiting at the top of the chamber. Atomization can be with either nozzles or a centrifugal atomizer.

The chamber fluid bed is vigorously agitated by a high fluidization velocity. Particles from the spray-drying zone enter the fluid bed with a moisture content as high as 10–15% depending on the type of product and are dried in the bed to about 5%. Final drying and cooling take place in an external fluid bed.

The structure of the powder produced in the Spray Bed differs considerably from the conventional. It is coarse, consists of large agglomerates, and consequently has low bulk density but exhibits excellent flowability. The dryer is very suited to the processing of complicated products having high contents of fat, sugar, and protein. Two of the many possible variations of this type drying are shown in Figures 26 and 27.

Spray-drying systems divided into two or more stages

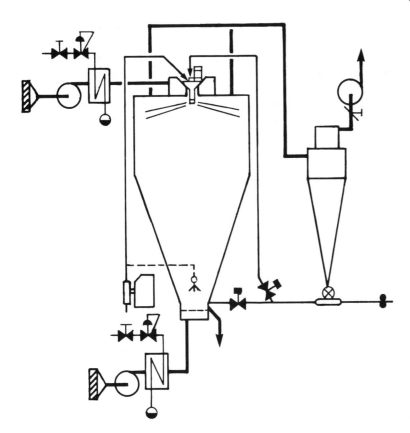

Figure 26. One variation of APV Crepaco Spray Bed Dryer arrangements (see Fig. 27 for other variation). Basic dryer with centrifugal atomizer and integrated fluid bed, which is agitated by high fluidation velocity.

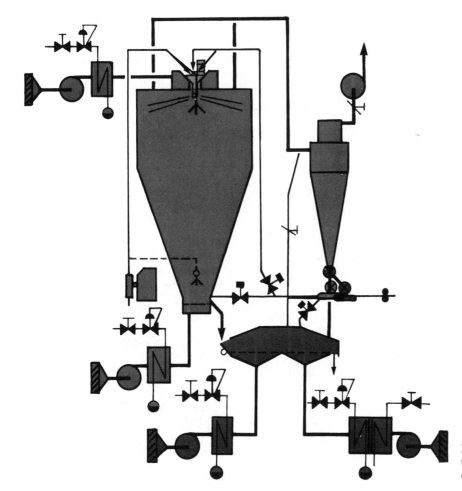

Figure 27. Spray bed dryer with added external fluid bed for final powder drying and cooling.

undoubtedly will be a characteristic of almost all future dryer installations. The advantages resulting from this technology will provide dairy and food processors with the necessary flexibility and energy efficiency required to meet today's uncertain market and whatever changes will be called for in the future.

Pollution-Control Devices

Virtually, all plants have a system to ensure clean exhaust air while collecting powder at an efficiency of about 99.5%, and may be supplied with secondary pollution-control collection equipment if necessary. In most new spray dryers this is accomplished using bag filter collectors after the primary collection device. In the case of such products as acid whey where the use of bag collectors is not practical, high-efficiency scrubbers are used. Some multipurpose plants have both plus an elaborate system of ducts and dampers to switch between the devices as different products are dried.

An added benefit from the use of sanitary bag collectors is an improvement in the yield of sale product. Although this increase typically is not more than 0.5%, it still can be of substantial value over a year of operation.

Heat-Recovery Equipment

Although it is possible to reduce the direct heating energy consumption by the use of multiple-stage drying, optimum thermal conditions generally require that heat recuperators be used as well. A few approaches are shown in Figure 28.

Despite the many recuperators available for recovering heat from drying air, only a few are suited for the spray-drying process. This is due to the dust-laden air that is involved in the process and that tends to contaminate heat

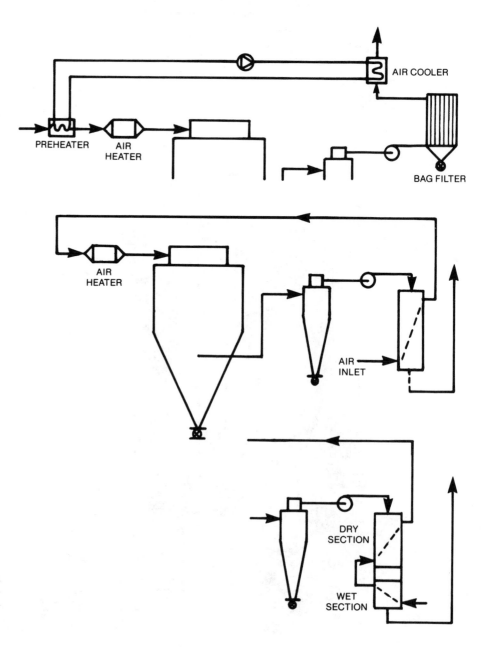

Figure 28. Spray-drying plant with (top) liquid coupled recuperator, (middle) single-stage recuperator, and (bottom) two-stage recuperator.

exchange surfaces. To be effective, therefore, recuperators for spray-dryer use should have the following properties:

- Modular system
- High thermal efficiency
- Low pressure drop
- Automatic cleaning
- Large temperature range
- Stainless steel
- Large capacity
- Low price

Two such recuperators are the air-to-air tubular and the air-to-liquid plate designs.

Air-to-Air Heat Recuperators. For optimum flexibility, this heat exchanger is of modular design, each module consisting of 804 tubes welded to an end plate. Available in various standard lengths, the modules can be used for counterflow, cross-flow, or two-stage counterflow as sketched in Figure 29. While the first two designs are based on well-known principles, the two-stage counterflow (patent pending) system is an APV development. This recuperator consists of two sections, a long dry element and a shorter wet section. During operation, hot waste air is passed downward through the dry module tubes and cooled to just above the dew point. It then goes to the wet

section where latent heat is recovered. The inner surfaces of these shorter tubes are kept wet be recirculating water. The system, therefore, not only recovers free and latent heat from the waste air but also scrubs the air and reduces the powder emission by more than 80%. Effectiveness and economy are such that the system replaces a conventional recuperator–scrubber combination.

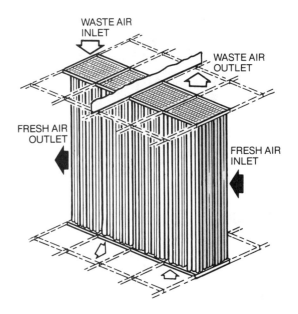

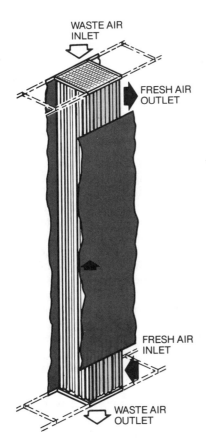

Figure 29. Air-to-air tubular heat recuperators.

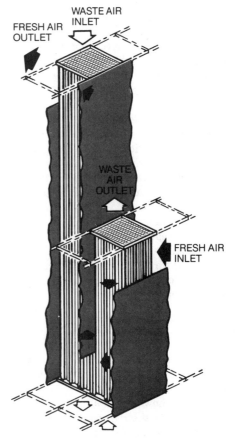

Figure 29. (*continued*)

WATER OUTLET

WATER INLET

Figure 30. Thermoplate for air-to-liquid plate recuperator.

Air-to-Liquid Recuperator. In cases where an air-to-air recuperator is impractical or uneconomical because of space limitations or the length of the distance between the outlet and inlet air to be preheated, a liquid heat-recovery system is available. This plate recuperator as shown in Figure 30 uses various standard plates as the base with waste air being cooled in counterflow by circulating a water/glycol mixture. At high temperature, circulating thermal oils are used. Heat recovered by this means generally is recycled to preheat the drying air but can be used for other purposes such as heating water, clean-in-place (CIP) liquids, or buildings.

Other Heat-Recovery Methods

While many spray-drying plants are equipped with bag filters to minimize emissions to the environment and recover valuable powder, the filtration system can be coupled with a finned-tube recuperator for heat-recovery purposes. This type of recuperator is particularly effective when the air is not dust-loaded. It is compact, very flexible and normally, very inexpensive.

Some degree of heat recovery can also be accomplished by using available low-temperature waste streams to preheat the drying air through a finned-tube heat exchanger. Examples of such sources of waste heat include evaporator condensate and scrubber liquids.

SPRAY DRYERS: FLUID-BED AGGLOMERATION

Most powderlike products produced by spray drying or grinding are dusting, exhibit poor flow characteristics, and are difficult to rehydrate.

It is well known, however, that agglomeration in most instances will improve the redispersion characteristics of a powder. Added benefits of the agglomerated powder is that it exhibits improved flowability and is nondusting.

All of these are characteristics for which the demand has increased in recent years.

Depending on the application or the area of industry where the process is being used, the process sometimes is also referred to as granulation or instantizing.

Instant Powders

Powders with particle size less than about 100 μm typically tend to form lumps when mixed with water and require strong mechanical stirring in order to become homogeneously dispersed or dissolved in the liquid. What is happening is that as water aided by capillary forces penetrates into the narrow spaces between the particles, the powder will start to dissolve. As it does, it will form a thick, gel-like mass that resists further penetration of water. Thus, lumps will be formed that contain dry powder in the middle and, if enough air is locked into them, will float on the surface of the liquid and resist further dispersion.

In order to produce a more readily dispersible product, the specific surface of the powder has to be reduced and the liquid needs to penetrate more evenly around the particles. In an agglomerated powder with its open structure, the large passages between the individual powder particles will assist in quickly displacing the air and allow liquid to penetrate before an impenetrable gel layer is formed. The powder thus can disperse into the bulk of the liquid, after which the final dissolution can take place.

Although there always is some degree of overlap between them, the reconstitution of an agglomerated product can be considered as consisting of the following steps:

1. Granular particles are wetted as they touch the water surface.
2. Water penetrates into the pores of the granule structure.

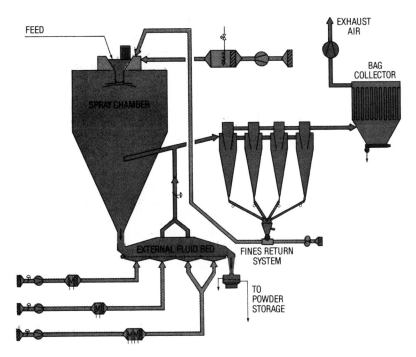

Figure 31. Straight-through agglomeration.

3. The wetted particles sink into the water.
4. The granules disintegrate into their original smallest particles, which disperse in the water.
5. The small dispersed particles dissolve in the water.

It is important to realize that it is the total time required for all these steps that should be the criterion in evaluating a product's instant properties. It is not unusual to see products characterized only on their wettability. This neglects the importance of the dispersion and dissolution steps, the time for which may vary considerably with different agglomeration methods.

For powders that are produced by spray drying, there are a number of ways in which the agglomeration can be accomplished in the spray dryer itself. This often is referred to as the "straight-through" process and is illustrated in Figure 31. Note that fines powder from the cyclone is conveyed up to the atomizer, where it is introduced into the wet zone surrounding the spray cloud. Cluster formation will occur between the semimoist freshly produced particles and the recycled fines. The agglomerated product then is removed from the bottom of the drying chamber, cooled, and packaged. This method produces a degree of agglomeration that is sufficient for many applications.

An alternative approach to agglomeration is referred to as the "rewet method." This is characterized by processing an already existing fine dry powder into an agglomerate using fluidized-bed technology.

The Agglomeration Mechanism

Two particles can he made to agglomerate if they are brought into contact and at least one of them has a sticky surface. This condition can be obtained by one or a combination of the following means:

1. Droplet humidification whereby the surface of the particles is uniformly wetted by the application of a finely dispersed liquid.
2. Steam humidification whereby saturated steam injected into the powder causes condensation on the particles.
3. Heating—for the thermoplastic materials.
4. Addition of binder media, ie, a solution that can serve as an adhesive between the particles.

The steam condensation method usually cannot provide enough wetting without adversely heating the material and is used less frequently on newer systems.

After having been brought into a sticky state, the particles are contacted under such conditions that a suitable, stable agglomerate structure can be formed. The success of this formation will depend on such physical properties as product solubility and surface tension as well as on the conditions that can be generated in the process equipment.

For most products, possible combinations of moisture and temperature can be established as shown in Figure 32. Usually, the window for operation is further narrowed down by the specifications for product characteristics. Once the agglomerate structure is created, the added moisture is dried off and the powder cooled below its thermoplastic point.

Agglomeration Equipment

While slightly different in equipment design and operation, most commercially available agglomeration processes fundamentally are the same. Each relies on the formation of agglomerates by the mechanism already described. This is followed by final drying, cooling, and size classification to eliminate the particle agglomerates that

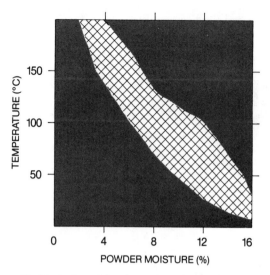

Figure 32. Typical combinations of conditions for agglomeration.

equipment as the process depends on the creation of conditions where the material becomes sticky.

Typically, equipment designs are very complicated, probably reflecting the fact that agglomeration actually is a complicated process. Despite the complexity of the process, however, it is possible to carry out agglomeration by means of comparatively simple equipment that involves the use of a fluidized bed for the rewetting and particle contact phase. This approach provides the following advantages:

1. There is sufficient agitation in the bed to obtain a satisfactory distribution of the binder liquid on the particle surfaces and to prevent lump formation.

2. Agglomerate characteristics can be influenced by varying operating parameters such as the fluidizing velocity, rewet binder rate, and temperature levels.

3. The system can accept some degree of variation of the feed rate of powder and liquid as the product level in the fluid bed always is constant, controlled by an overflow weir. Thus, the rewetting section will not be emptied of powder. Even during a complete interruption of powder flow, the fluidized material will remain in the rewet section as a stabilizing factor in the process.

4. By using fluid bed drying and cooling of the formed agglomerates, it is possible to combine the entire agglomeration process into one continuously operating unit.

5. Startup, shutdown, and operation of the fluid bed agglomerator are greatly simplified owing the stabilizing effect of the powder volume in the rewet zone.

Proper implementation of a fluid-bed agglomeration system requires a detailed knowledge of the fluidization technology itself. Fluidization velocities, bed heights, airflow patterns, residence time distribution, and the mechanical design of vibrating equipment must be known.

are either too small or too large. Generally, designs involve a re-wet chamber followed by a belt or a fluid bed for moisture removal. Such a system is shown in Figure 33.

It is obvious that this system is quite sensitive to even very minor variations in powder or liquid rates. A very brief reduction in powder feed rate will result in overwetting of the material with consequent deposit formation in the chamber. Conversely, a temporary reduction in liquid rate will result in sufficiently wetted powder and, therefore, weak agglomerates. Many designs rely on the product impacting the walls of the agglomeration chamber to build up agglomerate strength. Other designs include equipment for breaking large lumps into suitably sized agglomerates before the final drying. Obviously, deposit formation always will be a concern in agglomeration

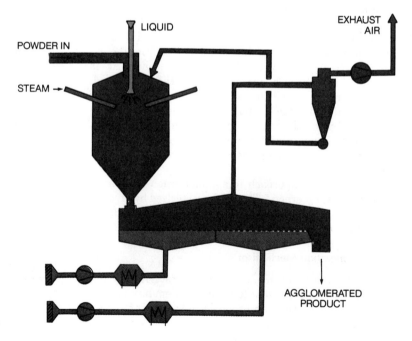

Figure 33. Typical agglomeration system.

Features of Fluid-Bed Agglomeration

Figure 34 shows a typical agglomerator system where the process is implemented through the use of a vibrated, continuous fluid bed.

The powder is fed into the agglomerator by a volumetric screw feeder. As a result of the previously mentioned stabilizing effect of the material already in the fluid bed, the reproducibility of a volumetric feeder is satisfactory and there is no need for a complicated feed system such as a loss-in-weight or similar type.

The fluid bed unit is constructed with several processing zones, each with a separate air supply system. The first section is the rewet and agglomeration section where agglomerates are formed. Here, the powder is fluidized with heated air in order to utilize its thermoplastic characteristics.

The binder liquid almost always is water or a water-based solution, whereas steam, as already explained, rarely is used. The binder liquid is sprayed over the fluidized layer using two fluid nozzles driven by compressed air. For large systems, numerous nozzles are used. Powder deposits are minimized by accurate selection of spray nozzle angles and nozzle position patterns. Powder movement is enhanced by the vibration of the fluid-bed unit itself and by the use of a special perforated air distribution plate with directional air slots. A proper detailed design is vital in order to have a trouble-free operation.

From the agglomeration zone, the powder automatically will flow into the drying area where the added moisture is removed by fluidization with heated air. In some instances, more than one drying section is required, and in such cases, these sections are operated at successively lower drying temperatures in order to reduce thermal exposure of the more heat-sensitive dry powder.

The final zone is for cooling where either ambient or cooled air is used to cool the agglomerates to a suitable packaging temperature.

During processing, air velocities are adjusted so that fine, unagglomerated powder is blown off the fluidized layer. The exhaust air is passed through a cyclone separator for removal and return of entrained powder to the inlet of the agglomerator. When there are high demands for a narrow particle size distribution, the agglomerated powder is passed through a sifter, where the desired fraction is removed while over- and undersize material is recycled into the process.

As with all rewet agglomeration equipment, the operation must be performed within certain operating parameters. Overwetting will lead to poor product quality, while underwetted powder will produce fragile agglomerates and an excessive amount of fines. However, fluid-bed agglomeration does offer a great degree of flexibility in controlling the final result of the process. The characteristics of the formed agglomerate can be influenced by operating conditions such as binder liquid rate, fluidizing velocity, and temperature. Typically, the fluid-bed rewet method will produce agglomerated products with superior redispersion characteristics.

As indicated by this partial list, this method has been used successfully with a number of products.

Dairy products

Infant formula

Calf milk replacer

Flavor compounds

Corn syrup solids

Sweeteners

Detergents

Enzymes

Fruit extracts

Malto dextrines

Herbicides

Egg albumin

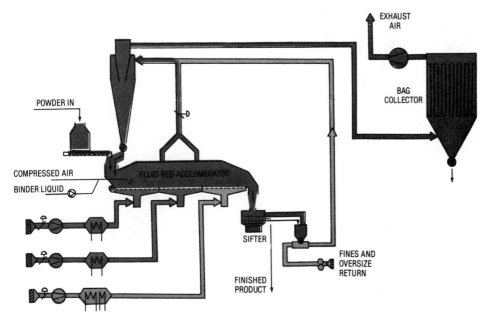

Figure 34. Fluid bed rewet agglomeration system.

Table 5. Frequently Used Binders

For Food Products	For Chemical Products
Malto dextrines	Lignosulfonates
Gum arabic	Poly(vinyl alcohol) (PVA)
Starch	Any of the food
Gelatin	product binders
Molasses	
Sugar	

In most cases, the agglomeration can be accomplished using only water as a rewet medium. This applies to most dairy products and to malto dextrine-based flavor formulations. For some products, increased agglomerate size and strength has been obtained by using a solution of the material itself as the binder liquid. In the case of relatively water-insoluble materials, a separate binder material has been used, but it must be one that does not compromise the integrity of the final product. The addition of the binder material may have a beneficial effect on the end product at times. This is seen, for example, in flavor compounds when a pure solution of malto dextrine or gum arabic may further encapsulate the volatile flavor essences and create better shelf life. In other instances, the added binder can become part of the final formulation as is the case with some detergents.

For some materials, the addition of a binder compounds is an unavoidable inconvenience. At such times, the selected binder must be as neutral as possible and must be added in small quantities so that the main product is not unnecessarily diluted. An example of this is herbicide formulations, which often have a very well defined level of active ingredients.

For products containing fat, the normal process often is combined with a step by which the agglomerates are coated with a thin layer of surface-active material, usually lecithin. This is done by mounting an extra set of spray nozzles near the end of the drying section where the surfactant is applied.

Variations of the fluid-bed rewet technology have been developed whereby the system serves as a mixer for several dry and wet products. An example of this may be seen in the APV fluid mix process for the continuous production of detergent formulations from metered inputs of the dry and wet ingredients. This provides an agglomerated end product that is produced with minimum energy input when compared to traditional approaches.

Spray-Bed Dryer Agglomeration

While the fluid-bed rewet agglomeration method produces an excellent product that is, in most respects, superior to that made directly by the straight-through process, a new generation of spray dryers has evolved that combines fluid-bed agglomeration with spray drying. These are referred to as "spray-bed" dryers. The concept was developed from spray dryers having a fluid bed integrated into the spray chamber itself and is depicted in Figure 35. What distinguishes the spray-bed dryer is that it has the drying air both entering and leaving at the top of the chamber. Atomization can be with nozzles or by a centrifugal atomizer.

During operation, the chamber fluid bed is vigorously agitated by a high fluidization velocity, and as the parti-

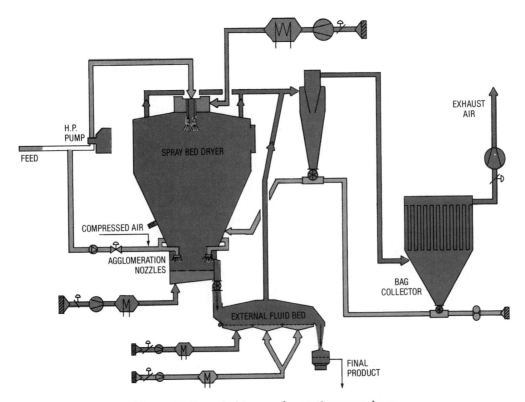

Figure 35. Spray bed-type agglomerating spray dryer.

Table 6. Reconstitutability and Physical Structure of Different Types of Skim Milk Powder

	Ordinary Spray Dried Powder	Integrated Fluid-Bed Agglomeration	Rewet Agglomerated Powder
Wettability, s	>1000	<20	<10
Dispersibility: %	60–80	92–98	92–98
Insolubility index	<0.10	<0.10	<0.20
Average particle size, μm	<100	>250	>400
Density, lb/ft^3	40–43	28–34	28–31

cles from the spray-drying zone, they enter the fluid bed with a very high moisture content and agglomerate with the powder in the bed. Fines carried upward in the dryer by the high fluidizing velocity have to pass through the spray cloud, thus forming agglomerates at this point as well. Material from the integrated fluid bed is taken to an external fluid bed for final drying and cooling.

The spray-bed dryer is a highly specialized unit that can produce only agglomerated powder. It is best suited for small to medium-sized plants since the efficiency of the agglomeration process unfortunately decreases as the plant increases in size. This is because the spray zone becomes too far removed from the fluid-bed zone as the size of the dryer increases.

Agglomerates from the spray-bed dryer exhibit excellent characteristics. They are very compact and show high agglomerate strength and good flowability.

Considerations and Conclusions

While the agglomeration process improves the redispersion, flowability, and nondustiness of most fine powders, it invariably decreases the bulk density. The comparison in Table 6 clearly shows that agglomeration improves the powder wettability and dispersibility. Individual powder particles with a mean diameter of less than 100 μm are converted into agglomerates ranging in size from 250 to 400 μm, with the rewet method being able to produce the coarser agglomerate. The powder bulk density will decrease from about 43 lb/ft^3 to approximately 28 lb/ft^3. Use of the rewet method will expose the product to one additional processing step that, in this case, will somewhat affect the proteins and result in a slightly poorer solubility.

Since fluid bed agglomeration can be operated as an independent process, it can be used with already existing power-producing equipment. It offers great flexibility and ease of operation, and provides a convenient way to add functionality, nondustiness, and value to a number of products.

SPIN FLASH DRYERS

Background Information

While mechanical dewatering of a feed slurry is significantly less expensive than thermal drying, this process results in a paste or filter cake that cannot be spray dried and can be difficult to handle in other types of dryers. The spin flash dryer is one option available for continuous

powder production from pastes and filter cakes without the need for grinding.

Powders generally are produced by some form of drying operation. There are several generic types of dryers, but all must involve the evaporation of water, which can take anywhere from 1000 to 2500 Btu/lb depending on dryer type. The most common of these dryers probably is the spray dryer because of its ability to produce a uniform powder at relatively low temperatures. However, by its definition, a spray dryer requires a fluid feed material to allow its atomization device to be employed. Generally, there is a maximum viscosity limitation in the range of 1000–1500 SSU (Saybolt Seconds Universal viscosity unit) (see Fig. 36).

Figure 37 illustrates the amount of water that must be evaporated to produce 1 lb of bone dry powder from a range of different feed solids. It can clearly be seen that even a 5% increase in total solids will reduce the water evaporation and, hence, the dryer operating costs by about 20%. If this water removal can be done mechanically by, for example, filtration or centrifuging, the cost will be infinitely lower than that required to heat and evaporate the same water. The direct energy cost can be calculated

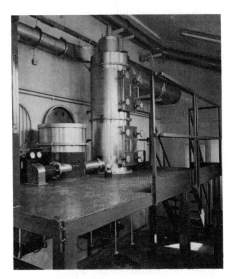

Figure 36. Typical APV Anhydro spin flash dryer with general dryer characteristics. Dryer characteristics: drying method—direct gas contact; flow—cocurrent; food material—dilatent fluids, cohesive paste, filter cake, moist granules; drying medium—air, inert gas, low-humidity waste gas; inlet temperature—up to 1800°F; capacity—up to 10 tons per hour of final product; product residence time—5–500 s.

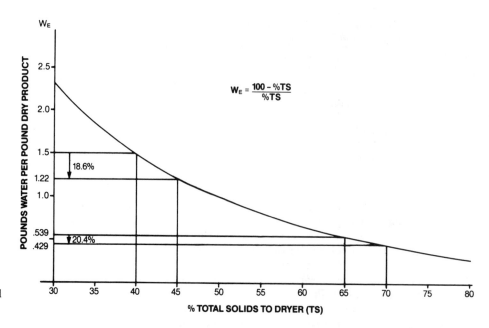

Figure 37. Water evaporation/total solids ratio.

as equivalent to 3–8 Btu/lb, compared to an average 1500 Btu/lb for evaporation. This increase in solids, however, inevitably will result in an increase in viscosity that may exceed the limitations of a spray dryer.

Options available for drying these higher-viscosity feed materials are listed in Table 7. The spin flash dryer is among the newest of the dryer options and has the capability of drying most materials ranging from a dilatent fluid to a cohesive paste.

Description

The spin flash dryer was developed and introduced in 1970 in response to a demand by the chemical industry to produce a uniform powder on a continuous basis from high-viscosity fluids, cohesive pastes, and sludges.

The spin flash dryer can be described as an agitated fluid bed. As shown in Figure 38, the unit consists primarily of a drying chamber (9), which is a vertical cylinder with an inverted conical bottom, an annular air inlet (7), and an axially mounted rotor (8). The drying air enters the air heater (4), is typically heated by a direct-fired gas burner (5), and enters the hot-air inlet plenum (6) tangentially. This tangential inlet, together with the action of the rotor, causes a turbulent whirling gas flow in the drying chamber.

Table 7. Dryer Options for High-Viscosity Materials

Direct Suspension Dryers
  Pneumatic or flash dryers
  Spin flash dryers
  Fluid-bed dryers
Direct Nonsuspension Dryers
  Tray dryers
  Tunnel dryers
  Belt dryers
  Rotary dryers
Indirect Dryers
  Screw conveyor dryers
  Vacuum pan dryers
  Steam tube rotary dryers

The wet feed material, typically filter cake, is dropped into the feed vat (1) where the low-speed agitator (2) breaks up the cake to a uniform consistency and gently presses it down into the feed screw (3). Both agitator and feed screw are provided with variable speed drives.

In the case of a dilatent fluid feed, the agitated vat and screw would be replaced with a progressive cavity pump and several liquid injection ports at the same elevation as the feed screw.

As the feed material is extruded off the end of the screw into the drying chamber, it becomes coated in dried powder. The powder-coated lumps then fall into the fluid bed and are kept in motion by the rotor. As they dry, the friable surface material is abraded by a combination of attrition in the bed and the mechanical action of the rotor. Thus, a balanced fluidized bed is formed that contains all intermediate phases between raw material and finished product.

The dryer and lighter particles become airborne in the drying airstream and rise up the walls of the drying chamber, passing the end of the feed screw and providing, in effect, a continuous backmixing action within the heart of the dryer. At the top of the chamber, they must pass through the classification orifice, which can be sized to prevent the larger particles from passing on to the bag collector. These larger lumps tend to fall back into the fluid bed to continue drying.

Air exiting from the bag collector (10) passes through the exhaust fan (12) and is clean enough for use in a heat-recovery system. Dried powder is discharged continuously from the bottom of the bag collector through the discharge valve (11).

Two important features make the spin flash dryer suitable for products that tend to be heat-sensitive: (1) the dry powder is carried away as soon as it becomes light enough and therefore is not reintroduced into the hot-air zone, and (2) the fluid bed consists mainly of moist powder, which constantly sweeps the bottom and lower walls of the drying chamber and keeps them at a temperature lower than the dryer air outlet temperature. In addition to this self-cooling capacity, the lower edge of the drying cham-

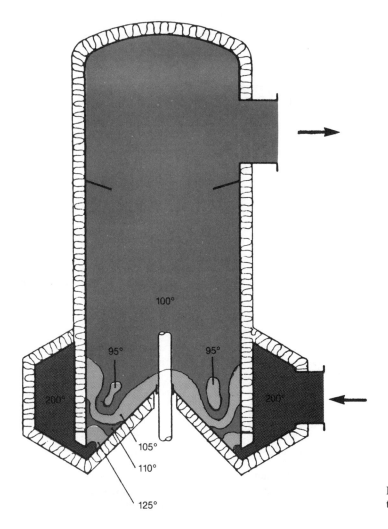

Figure 38. Fluid bed provides rapid air temperature reduction.

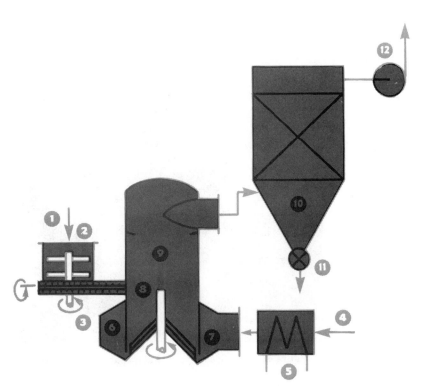

Figure 39. Standard spin flash dryer configuration.

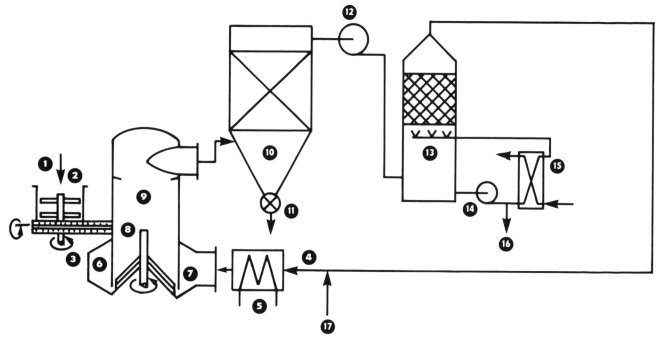

Figure 40. Spin flash dryer in closed-cycle arrangement.

ber directly above the hot-air inlet can be provided with an auxiliary cooling ring.

Figure 39 illustrates the very rapid reduction in air temperature that occurs as a result of the high heat-transfer rate obtained in the fluid bed.

Operating Parameters

Inlet temperature of the drying air introduced into the chamber is dependent on the particular characteristics of the product being dried but generally would be similar to that used on a spray dryer for the same product.

Outlet temperature is selected by test work to provide the desired powder moisture and is controlled by the speed of the feed screw. Since the spin flash dryer produces a finer particle size than does a spray dryer, it has been found that a slightly lower outlet temperature may be used to obtain the same powder moisture. This provides an increase in thermal efficiency.

Air velocity through the cross section of the drying chamber is an important design factor and is determined in part by the final particle size that is required. A lower velocity will tend to reduce the final dried particle size carried out of the chamber. The major factor, however, is the stability of the very complex bed that must neither settle back into the air distributor nor blow out of the top of the chamber. Once the maximum velocity has been determined by test work for a given product, the diameter of a drying chamber can be selected to provide the desired water evaporation rate.

Capacity can be adjusted to suit the output from the preceding process equipment, which may be difficult to control and slow in response time. This is achieved by a cascade control from a feed vat level sensor to the inlet temperature controller set-point. The feed vat sometimes can be oversized to accept the batch discharge from a pre-

ceding filter press while allowing the dryer to operate continuously.

Closed-Cycle Drying

Once a decision has been made to increase feed solids prior to drying, the small size and lower airflow requirements of the spin flash dryer make it practical to design the system as a closed-cycle dryer with nitrogen as the drying medium. This type of system can be used to dry a solvent-based powder, allowing complete recovery of the solvent.

The simplified Figure 40 schematic shows a possible configuration of a closed-cycle spin flash dryer. The operating process as described earlier, is extended, with ex-

Figure 41. Spin flash dryer being prepared for shipment.

Table 8. Products Dried on Spin Flash Equipment

	Unit	Sludge of Various Metal Hydroxides	Dolomite	Yellow Iron Oxide	Alumina	Aluminum Silicate	Calcium Carbonate with Binder	Calcium Carbonate (Pure)	Ferrite	Tartrazine (Azo Dye)	Food Yellow (Color Index 15985)	Calcium Stearate	Polyvanadate	Nickel Catalyst on Diatomaceous Earth	Barium Sulfate	Titanium Dioxide
Inlet air temperature	F	482	590	617	482	842	536	968	932	410	437	212	914	662	1112	1292
Outlet air temperature	F	185	293	212	194	212	194	338	257	203	203	126	320	311	275	257
Temperature of the feed	F	60	60	55	65	60	55	32	55	50	70	70	55			
Total solids in feed	%	70	34	35	29	20	58	67	62	50	28	43	35	70	60	60
Residual moisture	%	4.5	0.4	0.6	12.5	5.5	0.3	0	1.0	5.0	9	0.32	2.7	33	78	65
Mean particle size	μm	40	15	5	70	20	50	5	15	10	10	16	20	3	0.1	0.5
Bulk density	gr/cm³	0.8	0.45	0.3	0.4	0.2	0.45	0.8	0.9	0.7	0.3	0.14	0.76	0.32	1.6	0.6

Table 9. Size and Cost Comparison: Spray Dryer vs. Spin Flash Dryer[a]

Space Requirements	Spray Dryer Type III-K No. 70/71	Spin Flash Type III No. 59
Space Requirements		
Heater type (gas)	Direct-fired	Direct-fired
Chamber diameter, ft	14′0″	2′7½″
Building floor area, ft²	650	325
Building height, ft	46′0″	16′6″
Building volume, ft³	24,700	5,300
Capacity		
Powder, lb/h	880	880
Performance Data		
Feed solids, %	30	45
Feed rate, lb/h	2,995	1,951
Water evaporation, lb/h	2,115	1,071
Powder moisture, %	0.4	0.4
Gas consumption, SCF/h[b] (1000 Btu/SCF)	4,416	2,192
Power consumption (kWh)	40	30
Investment Cost, U.S. $		
Building	75,000	20,000
Dryer equipment	174,000	147,000
Baghouse	22,000	18,000
Installation	35,000	22,000
	$306,000	$207,000

Assumptions

Gas cost, $/M Btu $ 4.32
Electricity cost, $/kWh $ 0.07
Salaries per hour $15.00
Manpower required = 1 operator

Variable Operating Costs per Hour		
Wages	$15.00	$15.00
Gas	19.08	9.47
Electricity	2.80	2.10
Total	$36.88	$26.57
Cost per lb of powder	$0.042	$ 0.03

[a] It can be seen that the investment in a spin flash dryer plant is about 32% lower than a spray dryer for the same capacity and that spin flash operating costs are approximately 29% lower. These figures, however, do not include the capital investment in filtration equipment.
[b] Standard cubic feet.

haust gas from the baghouse being scrubbed and cooled in a condenser (13) using cooled solvent from an external plate heat exchanger (15) as the scrubbing medium. Recovered solvent is bled off at (16), downstream from the scrubber recirculation pump (14) at a controlled rate based on scrubber liquid level.

The drying chamber would be maintained at a pressure slightly higher than ambient using a pressurized nitrogen purge (17). The heater (5) would use either a steam coil or a thermal fluid system with an external heater.

A less expensive alternative to the closed-cycle approach is a "Lo-Ox" (low-oxygen) system where a low excess air burner is used in a direct-fired heater. The products of combustion are recirculated through the condenser, and the surplus gas is vented to atmosphere. The oxygen level in such a system can be controlled to less than 5%. (See also Figure 41.)

Spin-Flash-Dried Products

While Table 8 charts some typical products being dried on spin flash equipment, it should be noted that in some cases the inlet temperatures are limited by the heat source used and not by the spin flash process.

Other products test dried on a laboratory scale spin flash dryer include the following:

Product	Feed Solids, %	Powder Moisture, %
Iron oxide	70	2
Effluent treatment sludge	22	40
Lignin	49	15
Gum	40	13
Chitosan gel	6	40
Crab meat paste	30	9
Dutched cocoa cake	70	2

Cost Benefits

In comparison to a spray dryer, the spin flash dryer has a much shorter residence time and consequently is considerably smaller and requires less building space. Its ability to dry to even higher solids than a spray dryer results in operating cost savings. Table 9 shows a detailed size and cost comparison based on actual test drying of yellow iron oxide.

Conclusion

Despite its obvious size and cost advantages, there are many instances when a spin flash dryer cannot replace a spray dryer. Typically, such cases occur when a free-flowing spherical particle of a particular size range is required or when agglomeration is needed.

There are, however, many situations in both the food and chemical industries where the particular capabilities of the spin flash dryer to produce powders from paste warrant careful consideration of its use.

PARTIAL LIST OF PRODUCTS PROCESSED ON VARIOUS TYPES OF APV DRYERS

Spray Dryer

Albumen
Alcoholic extracts
Alfalfa
Alginates
Amino acids

Baby food
Bananas
Beef extract
Beer wort
Bile extract
Bouillon
Brain
Butter
Buttermilk

Calcium carbonate
Calcium salts
Carbon, active
Carboxymethylcellulose
Carrageenan
Casein hydrolysate
Caseinates
Catalysts
Cellulose acetate
Cellulose hydrate
Cheese
Chelates
Chlorophyll
Chocolate milk
Citrates
Colors
Cornstarch
Corn steep liquor
Corn syrup
Cream

Dextranes
Dextrin maltose
Dextrose
Diastase
Diatomaceous earth
Dietetic products
Distillers waste

Egg white
Egg yolk
Emulsifiers
Enzymes

Fat-enriched milk
Fatty alcohol sulfonates
Fish meat, hydrolyzed
Fish pulp
Fish solubles
Flavors
Fruit juice
Fruit pulp fullers earth

Garlic
Gelatin
Glands
Glauber salt
Glazes
Glucoheptonate
Glucose
Gluten
Glyceryl monostearate
Gum arabic

Hormones
Hydrolysates
Hypochlorites

Ice cream mix

Lactates
Lactose
Licorice
Lignin
Liver extract

Magnesium carbonate
Magnesium oxide
Magnesium salts
Magnesium sulfate
Malted milk
Malt extract
Mango
Meat extract
Milk-cocoa
Molasses with filler

Nitrates

Olive paste

Pancreas
Papain
Peanut milk

Pectin
Penicillin
Pepsin
Peptones
Phosphates
Pigments
Potassium sulfate
Potatoes
Potato waste liquor
Proteins

Quaternary salts

Rennet

Saponin
Seaweed extract
Senna
Sequestering agents
Silica-alumina gels
Silicates
Skim milk
Sodium adipate
Sodium phenate
Sorbate
Sorbose
Soybean milk
Soybean protein

Soy flour
Starch products
Stearic acid
Stearyl-tartrate
Sulfonates

Tannin extract
Tapioca
Tea extract
Thiamin
Thymus
Tomato

Vegetable extracts
Vegetable proteins

Wetting agents
Whey
Whey nonhygroscopic
Whole egg
Whole milk

Yeast
Yeast autolysate
Yeast hydrolysate

Zinc ammonium chloride
Zinc chromate
Zinc stearate

Manure
Meat extract
Meat pieces
Milk crumb
Milk (skimmed)
Molasses

Oat flour
Oats
Onion pulp

Palm-oil residue
Pectin
Phenylglycine casein
Potato (waste)
Press liquors
Protein (hydrolyzed)

Rice starch

Seaweed extract
Semolina

Single-cell protein
Sorghum and maize
Starch
Stearate
Stearic acid
Steep liquor
Stick liquor
Sugar solution

Tapioca starch

Vegetable extract
Vegetable protein

Wax
Whale solubles
Wheat starch
Whey

Yeast cream
Yeast extract

Thermoventuri Dryer

Alginates
Alginic acid

Bread crumbs

Cocoa residue
Cornstarch
Coumarin

Dark grains
Demerara sugar
Dextrose

Fishmeal
Flour

Glucose starch
Groundnut residue

Lactose

Maize starch

Organic acids

Pectin
Potato

Spent grains
Sphagnum moss
Starches
Stearates
Sulfur waste

Tartaric acid

Wheat germ
Wheat gluten
Wheat starch

Yeast

Film Drum Dryers

Agar
Apple pectin
Avocado pulp

Baby food
Beef, comminuted
Beef extract
Beef protein (hydrolyzed)
Beer concentrate
Blood
Buttermilk

Calcium acetate
Calcium carbonate
Calcium propionate
Caramel
Cattle food
Cellulose
Cereal (baby)
Chicken offal
Cider lees
Corn syrup

Dextrine

Effluent
Essence

Fat
Fatty alcohol sulfates
Fermentation waste
Ferric ammonium syrup
Ferrous sulfate
Fish food
Fish stick liquor
Flavors
Flour

Gelatin
Glauber's salts

Herbicide
Hydrolyzed protein

Maize gluten
Mango flake

Conveyor-Band Dryer

Almonds
Apples
Apricots
Assorted biscuits

Beans
Bran
Bread crumbs
Breakfast cereals

Cabbage
Carrots
Celery
Chocolate milk crumb
Coconuts
Corn grit

Dog biscuits
Dog food

Filter board	Pastilles
Food concentrates	Peanuts
French fries	Peas
	Pectin
Garlic	Peppers
Gelatin	Pet food
Grapes	Plums
	Potatoes
Iced biscuits	Puffed cereals
	Rice
Jellies	
	Sausage rusk
Meat	Seaweed
Mints	Soya protein
Molding powders	Sugar dough
	Sugar- and/or Honey-coated cereals
Nougat	
Nuts	Turnip
Onions	Wheat flakes

BIBLIOGRAPHY

1. D. Noden, "Industrial Dryers Selection, Sizing, and Costs," *CPI Digest* **8**(1), 6–23, (1985).
2. D. Noden, "Efficient Energy Utilization in Drying," *dfi News* **7**(4), 8–14, (1985).
3. B. Bjarekull, "Spray Drying New Developments, New Economics," *dfi News* **8**(2), 6–13 (1985).
4. B. Bjarekull, "Fluid Bed Agglomerization for the Production of Free-Flowing Dustless Powders," *dfi News* **8**(2), 6–13 (1985).
5. S. G. Gibson, "Spin Flash Dryers for Continuous Powder Production from Pastes and Filter Cakes," *APV Dryer Handbook*, Oct. 1985, pp. 36–42.

Copyrighted APV Crepaco, Inc. Used with permission.

APV CREPACO, INC.
Lake Mills, Wisconsin

DRY MILK

HISTORY

The development of the dry milk industry apparently stems from the days of Marco Polo in the thirteenth century. It is reported that Marco Polo encountered sun-dried milk on his journeys through Mongolia and that from this beginning dry milk products evolved.

Through early pioneering scientists, such as Appert and Borden, the basic methods were developed for the emergence of processes for drying milk products. Ekenberg and Merrill have been acknowledged as the developers of the first commercial roller- and spray-process drying systems, respectively, in the United States.

Since the initial development of commercial drying systems, significant technological advances have been made resulting in such widely recognized and used dry milk products as nonfat dry milk, dry whole milk, and dry buttermilk. These may be manufactured by roller or spray dryers (now mostly by the latter) or by more unique processes such as foam or freeze drying. The same products may also be processed to make them readily soluble; these products are referred to as instantized.

PROCESSING

The steps in a typical dry milk processing operation are the following:

- Receipt of fresh, high-quality milk from modern dairy farms, delivered in refrigerated stainless-steel bulk tankers.
- Separation of the milk (if nonfat dry milk is to be manufactured) to remove milk fat, which commonly is churned into butter. If dry whole milk is to be manufactured, the separation step is omitted but may be replaced by clarification.
- Pasteurization by a continuous high-temperature–short-time (HTST) process whereby every particle of milk is subjected to a heat treatment of at least 71.7°C for 15 s.
- Holding the pasteurized milk at an elevated temperature for an extended period of time (76.7°C for 25–30 min). This step is used only in the manufacture of high-heat nonfat dry milk, which commonly is used as an ingredient in bakery or meat products.
- Condensing the milk by removing water in an evaporator or vacuum pan until a milk solids content of 40% is reached.
- Delivery of the condensed product to the dryer.

Commercial U.S. drying processes are of two types: roller (drum) and spray. The former is currently used to a limited extent, primarily for dry whole milk. In this process, two large rollers, usually internally steam heated and located adjacent and parallel to each other, revolve in opposite directions contacting a reservoir of either pasteurized fluid or condensed milk. During rotation, the fluid milk product dries on the hot roller surface. After approximately three-quarters of a revolution, a carefully positioned, sharp, stationary knife detaches the milk product, then in the form of a thin dry sheet. The dry milk is next conveyed by an auger to a hammer mill where it undergoes physical treatment to convert it into uniformly fine particles, which then are packaged, usually in 50-lb polyethylene-lined multiwall bags.

Two basic configurations of spray dryers are presently in use: horizontal (box) and vertical (tower). In both, pasteurized fluid milk, which has been condensed to a total solids of 40% or greater, is fed under pressure to a spray nozzle, or to an atomizer, where the dispersed liquid then comes into contact with a current of filtered, heated air. The droplets of condensed milk are dried almost immediately and fall to the bottom of the fully enclosed stainless-steel drying chamber. The dry milk product is continuously removed from the drying chamber, transported through a cooling and collecting system, and finally conveyed into a hopper for packaging, usually in 50-lb bags or in tote bins.

THE PRODUCT

The primary dehydrated dairy products manufactured domestically are nonfat dry milk, dry whole milk, and dry buttermilk.

Nonfat Dry Milk. Nonfat dry milk is the product resulting from the removal of fat and water from milk. It contains lactose, milk proteins, and milk minerals in the same relative proportions as the fresh milk from which it is made. Nonfat dry milk contains not over 5% by weight of moisture. The fat content is not over 1.5% by weight unless otherwise indicated.

Dry Whole Milk. Dry whole milk is the product resulting from the removal of water from milk and contains not less than 26% milk fat and not more than 4% moisture. Dry whole milk with milk fat contents of 26 and 28.5% are commonly produced.

Dry Buttermilk. Dry buttermilk is the product resulting from the removal of water from liquid buttermilk derived from the manufacture of butter. It contains not less than 4.5% milk fat and not more than 5% moisture.

All of these products may be processed to become instantized, which substantially improves their dispersion and reliquefaction properties. Single-pass and agglomerating processes are used to make instant dry milk products; the primary method is agglomeration.

Table 1 reflects the annual domestic production of nonfat dry milk, dry whole milk, and dry buttermilk during the period 1960–1988. Standards for dry milk products were initially developed by the industry in 1929. Since that time additional product standards have been developed, and revised by both industry and government agencies. These standards are based on various general and specific product characteristics, which serve as a basis for determining the overall quality of the dry milk products. Table 2 shows the approximate composition and food

Table 1. Yearly U.S. Production of Dry Milks[a]

Year	Nonfat Dry Milk	Dry Whole Milk	Dry Buttermilk
1960	1,818.6	98.0	86.4
1961	2,019.8	81.7	89.0
1962	2,230.3	86.1	86.4
1963	2,106.1	91.0	87.5
1964	2,177.2	87.6	92.0
1965	1,988.5	88.6	87.4
1966	1,579.8	94.4	76.2
1967	1,678.7	74.3	72.6
1968	1,594.4	79.8	70.4
1969	1,452.3	70.2	66.5
1970	1,444.4	68.9	59.5
1971	1,417.6	72.2	51.7
1972	1,223.5	75.2	49.5
1973	916.6	78.0	43.2
1974	1,019.0	67.7	45.3
1975	1,001.5	63.1	42.8
1976	926.2	78.1	46.3
1977	1,106.6	69.4	53.2
1978	920.4	74.6	47.6
1979	908.7	85.3	44.7
1980	1,160.7	82.7	43.9
1981	1,314.3	92.7	43.8
1982	1,400.5	102.2	38.6
1983	1,499.9	111.2	46.5
1984	1,160.7	119.6	43.5
1985	1,390.0	118.9	51.5
1986	1,284.1	122.4	65.7
1987	1,056.8	145.9	55.6
1988	978.5	172.3	58.7

[a] In millions of pounds.

Table 2. Approximate Composition and Food Value of Dry Milks

Constituents	Nonfat Dry Milk	Dry Whole Milk	Dry Buttermilk
Protein ($N \times 6.38$), %	36.0	26.0	34.0
Lactose, %	51.0	38.0	48.0
Fat, %	0.7	26.75	5.0
Moisture, %	3.0	2.25	3.0
Minerals, ash, %	8.2	6.0	7.9
Calcium, %	1.31	0.97	1.3
Phosphorus, %	1.02	0.75	1.0
Vitamin A, IU/lb	165.0	4,950.0	2,300.0
Riboflavin, mg/lb	9.2	6.7	14.0
Thiamin, mg/lb	1.6	1.2	1.2
Niacin, mg/lb	4.2	3.1	4.5
Niacin equivalent[a], mg/lb	42.2	30.6	40.6
Pantothenic acid, mg/lb	15.0	13.0	14.0
Pyridoxine, mg/lb	2.0	1.5	2.0
Biotin, mg/lb	0.2	0.185	0.2
Choline, mg/lb	500.0	400.0	500.0
Energy, cal/lb	1,630.0	2,260.0	1,700.0

[a] Includes contribution of tryptophan.

value of the three most commonly manufactured dry milk products.

UTILIZATION

An industrywide survey of end uses for dry milk products is conducted annually by the American Dairy Products Institute. Data compiled in the survey are published under the title *Census of Dry Milk Distribution and Production Trends*. Included in the census data are end use markets for nonfat dry milk, dry whole milk, and dry buttermilk. For the past several years the largest markets for nonfat dry milk (NDM) have been dairy, home use (instant nonfat dry milk), prepared dry mixes, and bakery. These market areas accounted for 567 million lb of NDM in 1988, approximately 85% of the domestic sales. The primary market for dry whole milk is confectionery. In 1988 this market area accounted for 134.5 million lb of dry whole milk, nearly 92% of domestic sales for this product. Dry buttermilk is used in three principal markets: dairy, bakery, and prepared dry mixes. In 1988 these mar-

kets used 66 million lb of dry buttermilk, representing 81.5% of domestic sales.

BIBLIOGRAPHY

General References

American Dairy Products Institute, *Standards for Grades of Dry Milks Including Methods of Analysis,* Bulletin 916, rev., American Dairy Products Institute, Chicago, Ill., 1990.

American Dairy Products Institute, *Census of 1988 Dry Milk Distribution and Production Trends,* Bulletin 1000, American Dairy Products Institute, Chicago, Ill., 1989.

C. E. Beardslee, *Dry Milks—The Story of an Industry,* American Dry Milk Institute, Chicago, 1948.

C. W. Hall and T. I. Hedrick, *Drying of Milk and Milk Products,* 2nd ed., AVI Publishing Co., Inc., Westport, Conn., 1971.

WARREN S. CLARK, JR.
American Dairy Products
Institute
Chicago, Illinois